Topics in Ecological and Environmental Microbiology

Topics in Ecological and Environmental Microbiology

Edited by

Thomas M. Schmidt and Moselio Schaechter

AMSTERDAM • BOSTON • HEIDELBERG • LONDON • NEW YORK • OXFORD
PARIS • SAN DIEGO • SAN FRANCISCO • SINGAPORE • SYDNEY • TOKYO
Academic Press is an imprint of Elsevier

Academic Press is an imprint of Elsevier
225 Wyman Street, Waltham, MA 02451, USA
525 B Street, Suite 1900, San Diego, California 92101–4495, USA
Radarweg 29, PO Box 211, 1000 AE Amsterdam, The Netherlands
The Boulevard, Langford Lane, Kidlington, Oxford, OX51GB, UK

Library of Congress Cataloging-in-Publication Data
Topics in ecological and environmental microbiology / edited by Thomas M. Schmidt & Moselio Schaechter.
p. ; cm.
Abridgement of: Encyclopedia of microbiology. 3rd ed. / editor-in-chief, Moselio Schaechter. c2009.
Includes bibliographical references and index.
ISBN 978-0-12-383878-0 (alk. paper)
I. Schmidt, Thomas M. (Thomas Mitchell), 1956- II. Schaechter, Moselio. III. Encyclopedia of microbiology.
[DNLM: 1. Environmental Microbiology. QW 55]
LC classification not assigned
579'.1718–dc23 2011032791

British Library Cataloguing-in-Publication Data
A catalogue record for this book is available from the British Library.

ISBN: 978-0-12-383878-0

For information on all Academic Press publications
visit our Web site at www. elsevierdirect.com

Printed and bound in USA
11 12 13 9 8 7 6 5 4 3 2 1

Contents

8. Environmental Viral Pool

9. Aeromicrobiology

10. Endophytic Microbes

Part II
Metabolism and Behavior of Diverse Microbes

11. Methanogenesis

12. Methanotrophy

13. Microbial Photosynthesis

14. Regulation of Carbon Assimilation

15. Microbial Resistance to Heavy Metals

Part V
Biotechnological Topics

38. Biodeterioration - Including Cultural Heritage

39. Biosensors

40. Gnotiobiotic and Axenic Animals

41. Microbial Corrosion

42. Petroleum Microbiology

Contributors

G.L. Andersen, Lawrence Berkeley National Laboratory, Berkeley, CA, USA

P. Assmy, Alfred Wegener Institute for Polar and Marine Research, Bremerhaven, Germany

M.D. Baker, Princeton University, Princeton, NJ, USA

D.A. Bazylinski, University of Nevada at Las Vegas, Las Vegas, NV, USA

W. Berelson, University of Southern California, Los Angeles, CA, USA

P.S. Berger, US Environmental Protection Agency (Retired) Cincinnati, OH, USA

A.R. Bielefeldt, University of Colorado – Boulder, Boulder, CO, USA

J. Bruckner, California Institute of Technology, Pasadena, CA, USA

P. Cabello, Universidad de Còrdoba, Còrdoba, Spain

R.W. Castenholz, University of Oregon, Eugene, OR, USA

F. Castillo, Universidad de Còrdoba, Còrdoba, Spain

L. Cegelski, Washington University, School of Medicine, St. Louis, MO, USA

T.E.A. Chalew, Johns Hopkins Bloomberg School of Public Health, Baltimore, MD, USA

R.M. Clark, US Environmental Protection Agency (Retired) Cincinnati, OH, USA

S. Collins, McMaster University Medical Centre, Hamilton, ON, Canada

C. Conley, National Aeronautics and Space Administration, Washington, DC, USA

J.W. Costerton, University of Southern California, Los Angeles, CA, USA

Frank B. Dazzo, Department of Microbiology and Molecular Genetics, Michigan State University, East Lansing, MI, USA

J.W. Deming, University of Washington, Seattle, WA, USA

Paul V. Dunlap, University of Michigan, Ann Arbor, MI, USA

A.S. Frisch, CIRA/Colorado State University, Ft. Collins, CO, USA

G.M. Gadd, University of Dundee, Dundee, Scotland, UK

Stephan Gantner, Department of Microbiology and Molecular Genetics, Michigan State University, East Lansing, MI, USA

M.B. Geuking, McMaster University Medical Centre, Hamilton, ON, Canada

J.H. Golbeck, The Pennsylvania State University, University Park, PA, USA

T.K. Graczyk, Johns Hopkins Bloomberg School of Public Health, Baltimore, MD, USA

G.D. Griffin, Fitzpatrick Institute for Photonics, Duke University, Durham, NC, USA

J.-D. Gu, The University of Hong Kong, Hong Kong, PR China

J.F. Holden, University of Massachusetts, Amherst, MA, USA

S.J. Hultgren, Washington University, School of Medicine, St. Louis, MO, USA

B. Jagannathan, The Pennsylvania State University, University Park, PA, USA

C.A. Jerez, University of Chile and ICDB Millennium Institute, Santiago, Chile

P.J. Johnsen, University of Tromsø, Tromsø, Norway

D.B. Johnson, Bangor University, Bangor, UK

D.M. Karl, University of Hawaii, Honolulu, HI, USA

C.A. Kellogg, US Geological Survey, St. Petersburg, FL, USA

J. Kirundi, McMaster University Medical Centre, Hamilton, ON, Canada

A. Konopka, Pacific Northwest National Laboratory, Richland, WA, USA

A.K. Korgaonkar, The University of Texas at Austin, Austin, TX, USA

L.G. Leff, Kent State University, Kent, OH, USA

P. Lens, Wageningen University, Wageningen, The Netherlands

R. Letelier, Oregon State University, Corvallis, OR, USA

E. Levetin, University of Tulsa, Tulsa, OK, USA

B. Lighthart, US EPA, Monmouth, OR, USA

U. Lins, Universidade Federal do Rio de Janeiro, Rio de Janeiro, Brazil

F.E. Lucy, Institute of Technology, Sligo, Ireland Environmental Services Ireland, Co. Leitrim, Ireland

K.R.M. Mackey, Stanford University, Stanford, CA, USA

A.J. Macpherson, McMaster University Medical Centre, Hamilton, ON, Canada

Y. Mashinski, Johns Hopkins Bloomberg School of Public Health, Baltimore, MD, USA

R. Massana, Institut de Ciències del Mar, Barcelona, Catalonia, Spain

K.D. McCoy, McMaster University Medical Centre, Hamilton, ON, Canada

L.A. Miller, GlaxoSmithKline Collegeville, PA, USA

C. Moreno-Viviàn, Universidad de Còrdoba, Còrdoba, Spain

J.C. Murrell, University of Warwick, Coventry, UK

K.H. Nealson, University of Southern California, Los Angeles, CA, USA

K.M. Nielsen, University of Tromsø, Tromsø, Norway

J.A. Nienow, Valdosta State University, Valdosta, GA, USA

S. Osman, California Institute of Technology, Pasadena, CA, USA

R.J. Parkes, Cardiff University, Cardiff, UK

D. Paterno, US Army ECBC, Aberdeen, MD, USA

A. Paytan, University of California Santa Cruz, Santa Cruz, CA, USA

L.T. Phung, University of Illinois, Chicago, IL, USA

J. Plumbridge, Institut de Biologie Physico-Chimique (UPR9073-CNRS), Paris, France

J.S. Poindexter, Barnard College, Columbia University, NY, USA

J.A. Poupard, Pharma Institute of Philadelphia, Inc., Philadelphia, PA, USA

M.M. Ramsey, The University of Texas at Austin, Austin, TX, USA

G. Ranalli, University of Molise, Campobasso, Italy

J.L. Ray, University of Tromsø, Tromsø, Norway

D.J. Reasoner, US Environmental Protection Agency (Retired) Cincinnati, OH, USA

C. Rensing, University of Arizona, Tucson, AZ, USA

E.W. Rice, US Environmental Protection Agency Cincinnati, OH, USA

M.D. Roldàn, Universidad de Còrdoba, Còrdoba, Spain

B.P. Rosen, Wayne State University, School of Medicine, Detroit, MI, USA

R.-A. Sandaa, University of Bergen, Bergen, Norway

J.W. Santo Domingo, US Environmental Protection Agency Cincinnati, OH, USA

H. Sass, Cardiff University, Cardiff, UK

B.F. Sherr, Oregon State University, Corvallis, OR, USA

E.B. Sherr, Oregon State University, Corvallis, OR, USA

S. Silver, University of Illinois, Chicago, IL, USA

A. Singh, University of Waterloo, Waterloo, ON, Canada

V. Smetacek, Alfred Wegener Institute for Polar and Marine Research, Bremerhaven, Germany

C.L. Smith, Washington University, School of Medicine, St. Louis, MO, USA

T.J. Smith, Sheffield Hallam University, Sheffield, UK

C. Sorlini, University of Milan, Milan, Italy

K.R. Sowers, University of Maryland Biotechnology Institute, Baltimore, MD, USA

J.B. Stock, Princeton University, Princeton, NJ, USA

D.N. Stratis-Cullum, US Army Research Laboratory, Adelphi, MD, USA

M. Tadych, Rutgers University, New Brunswick, NJ, USA

Andreas Teske, University of North Carolina at Chapel Hill, Dept. of Marine Sciences, Chapel Hill, NC, USA

J.D. Van Hamme, Thompson Rivers University, Kamloops, BC, Canada

K. Venkateswaran, California Institute of Technology, Pasadena, CA, USA

G. Voordouw, University of Calgary, Calgary, AB, Canada

O.P. Ward, University of Waterloo, Waterloo, ON, Canada

J.F. White, Rutgers University, New Brunswick, NJ, USA

M. Whiteley, The University of Texas at Austin, Austin, TX, USA

A.A. Yayanos, University of California, San Diego, La Jolla, CA, USA

E. Zanardini, University of Insubria, Como, Italy

Preface

The central importance that microbes occupy in nature has been a steady concern of microbiologists since the inception of microbiology. Many of those who founded this science, including Pasteur, and later Winogradsky and Beijerinck, worked extensively to elucidate the roles that bacteria play in the environment. Although this has been a continuous subject for inquiry, it was overshadowed by the intense research done on all aspects on microbial life in the laboratory, mainly with pure cultures. It is in the past few decades that we have seen a burst of interest and activity in ecological and environmental microbiology. In fact, it is now recognized that this is the only way in which we can try to understand the "planet of the microbes."

This surge in interest in all matters ecological has been aided by the development of large assortment of techniques, some old, many novel. These have ranged from the simple use of tangential filters for gathering viral or cellular samples to sequencing a genome starting with a single microbial cell. Hardly any sort of environment seems to have escaped attention of microbial ecologists, from the driest and coldest deserts of Antarctica to the steam emerging from fumaroles. Name any esoteric-sounding habitat, and microbial ecologists have probably already sampled it and tried to determine the complexity of its microbial communities.

Given the vibrancy of the field, a book that includes authoritative treatments of many aspects of ecological and environmental microbiology seems timely. It pleases us to present a compendium of chapters on these topics derived from the *Encyclopedia of Microbiology*, third edition. A few chapters from other sources have also been included; all were written by investigators with high standing in their fields. Where appropriate, these contributions have been updated with current references and sections on recent developments. The chapters are presented in an order we believe is convenient to the readers.

Thomas M. Schmidt and Moselio Schaechter
In Natura, Veritas

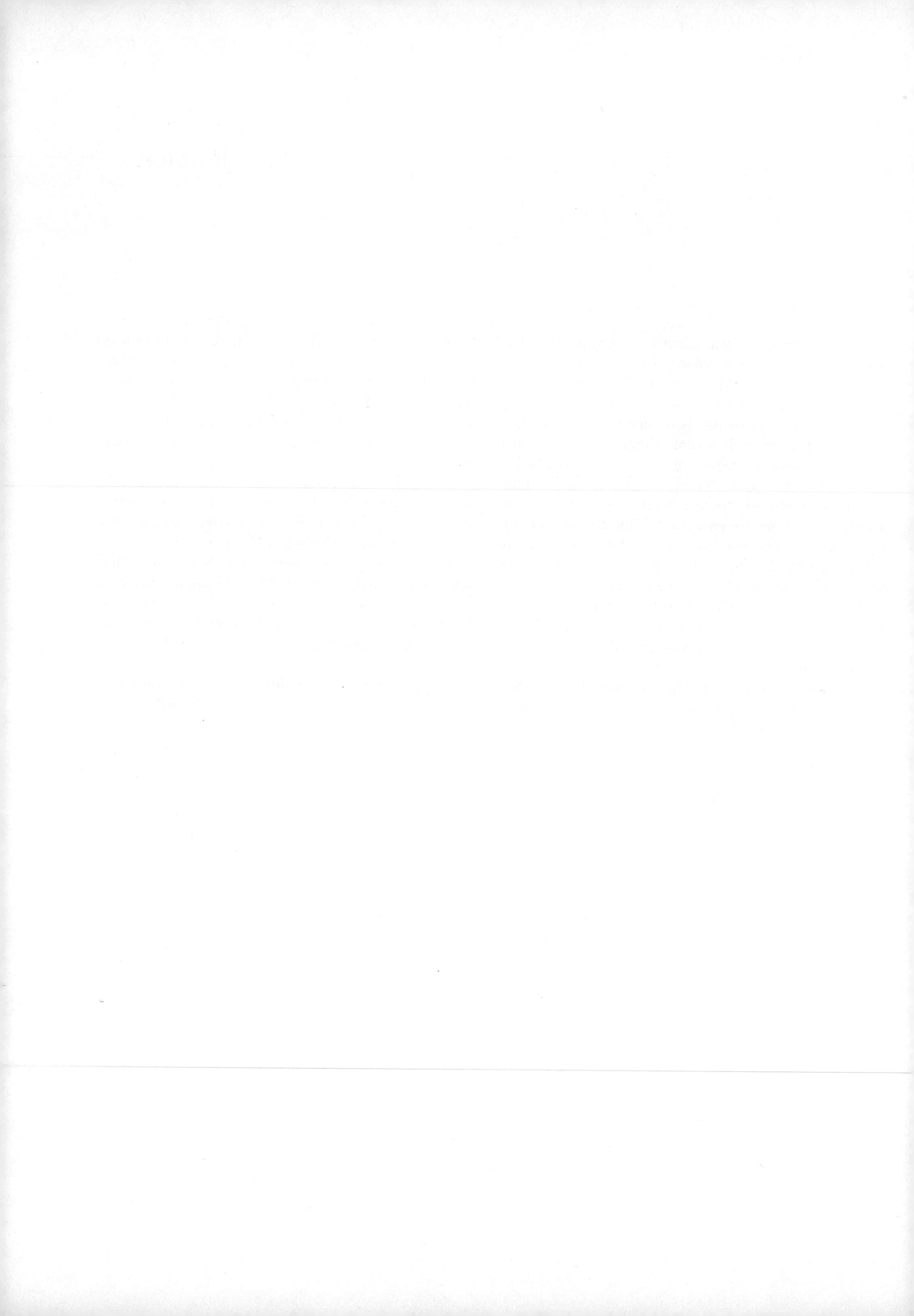

Part I

Microbial Ecology

Chapter 1

Principles of Microbial Ecology

A. Konopka
Pacific Northwest National Laboratory, Richland, WA, USA

Chapter Outline

ABBREVIATIONS

DOC Dissolved organic carbon
HGT Horizontal gene transfer
HPK Histidine protein kinase
PCR Polymerase chain reaction
RTFs Resistance transfer factors
SIMS Secondary ion mass spectrometry

DEFINING STATEMENT

This article provides an overview of the field of microbial ecology by delineating its principles, historical development, areas of study, and experimental approaches.

WHAT IS MICROBIAL ECOLOGY?

The term ecology was devised by Ernst Haeckel from the Greek *oekologie* meaning "the study of the household of nature" in 1866. Hence, microbial ecology concerns the interactions of microbes with both their abiotic and biotic environments. Microorganisms are ubiquitous – they grow not only in soil, freshwater and marine habitats, and on or within plants or animals, but also in so-called extreme environments with chemical and physical conditions that exclude most if not all other forms of life. Thus, the field is extraordinarily broad; a number of other articles in this encyclopedia discuss specific topics within this broad field. This article represents a synthesis of developments that may have arisen from the study of different environments or different organisms, and puts them into a historical context of how microbial ecology has developed and (more importantly) where it may be heading in the future.

The definition of microbial ecology is very important, because it leads to some important points regarding the discipline. Microorganisms (by definition) are too small to be seen with the naked eye; bacterial, fungal, and protistan cells typically have diameters between 1 and 100 μm, with bacterial sizes heavily skewed to the smaller end of this distribution and protists at the larger end. Consequently, the "environment" that these organisms are sensing must also be at a micron scale (the microenvironment). This scale is far too small for humans to sense directly. We can analyze larger, macroscopic samples and infer what occurs at the microscale, although there is a great risk that the properties we measure in the large sample merely represent the average of a rich heterogeneity at the microscale. Because of this, technological innovations are important to microbial ecology when they allow authentic analysis of the *in situ* environment. The definition of microbial ecology also

Topics in Ecological and Environmental Microbiology.
Published by Elsevier Inc.

indicates that the intellectual interests and approaches of the microbial ecologist are quite distinct from those of the laboratory-oriented microbiologist who focuses on analysis of pure cultures of microbes. Microbes in nature interact not only with their abiotic environment but also within biological communities of other microbes and perhaps plants and animals. These complex systems may exhibit behaviors that are not predictable from study of the individual organisms in isolation.

That being said, most microbial ecologists have been trained as laboratory-oriented microbiologists rather than as ecologists; as a result, their understanding of ecological theory is often rudimentary. This can lead to conceptual problems, either because problems in microbial ecology do not become well integrated into ecological theory or because there is no clear recognition where the ecology of microbes has elements that are unique from those of plants or animals.

Historical Development of Microbial Ecology

The realization that there was a microbial world depended upon technological innovation – van Leeuwenhoek improved techniques in lens grinding that led to his visualization of "animalacules" in the late 1600s. However, the discipline of microbiology did not really develop until the late nineteenth and early twentieth centuries when Pasteur and Koch took on practical problems that arose from the interaction of microbes with their environments (such as the germ theory of disease). Their studies were not ecological, but they did develop technologies for the fundamental manipulation of microbes that are still used today. Also, in the early twentieth century, Beijirnck and Windogradsky became interested in the activities of microbes in natural habitats. They took into consideration the chemical conditions in the natural environment and devised selective conditions (enrichment cultures) by which they could isolate organisms that carried out specific biogeochemical activities. However, their focus was not on the habitat *per se* but on the physiology of the microorganism, and their studies did not focus on ecological interactions.

Modern microbial ecology had its origins in the 1950s, in the study of the rumen ecosystem by Robert Hungate and his colleagues. Hungate is well known for his development of techniques for the cultivation of strict anaerobes. In addition, he had an interest in quantitative analysis of microbial ecosystems. This followed from his studies with C. B. van Niel on cellulose hydrolysis in gut microflora first in beetles and termites and subsequently in the bovine rumen. He pioneered studies in which the analysis did not end at the isolation and characterization of pure cultures, but necessarily included the enumeration of organisms in the habitat and quantification of their activities. This quantitative analysis distinguished major from minor microbial species and catabolic pathways in the rumen ecosystem.

As in other areas of science, progress in microbial ecology has been intimately tied to the development of new technologies. The availability of radioisotopes beginning in the 1950s was a major impetus for the measurement of biogeochemical process rates in nature. A major problem is that samples must usually be removed from the habitat and incubated under artificial conditions. As incubation time lengthens, the risk of bottle effects that produce results different from *in situ* rates increases. Thus, there is a premium on sensitive detection methods in which incubation time is minimized. The use of $^{14}CO_2$ allowed analysis of rates of primary production by phototrophs and chemoautotrophs, and both ^{14}C- and ^{3}H-isotopic forms of a broad variety of organic compounds have been applied to analyze fundamental processes such as nutrient uptake and assimilation, mineralization to CO_2, and microbial community growth rates. $^{32}PO_4$ and ^{35}S isotopes have been applied to measure biogeochemical process rates in the phosphorus and sulfur cycles. Unfortunately, no practically useful N radioisotope is available, although advances in the sensitivity of stable isotope analyses may make ^{15}N analyses at short incubation times feasible. Another significant breakthrough in analysis of microbial process rates came in the 1980s, when microelectrodes sensitive to certain chemical species were developed. These electrodes have a spatial resolution of 50–100 μm, and have produced great insights into the spatial and temporal dynamics of microbial processes in habitats such as microbial mats.

Quantitative analyses of biogeochemical process rates blossomed in the period after radioisotopes became widely available, but these studies were incomplete because the specific microbial catalysts were not identified. The techniques of pure culture microbiology were broadly recognized to be ineffective in cultivating numerically abundant organisms from nature. However, over the past 20 years, a set of cultivation-independent techniques have evolved to determine the relative abundance of microbes in a habitat. Carl Woese demonstrated that the sequence of the small subunit ribosomal RNA molecule could be used as a molecular chronometer to analyze the phylogenetic relationships among all forms of life. Norm Pace took this seminal idea and applied it to the analysis of "who's there" in natural environments. That is, by isolation of nucleic acids directly from a natural environment, amplification of small subunit rRNA gene sequences by polymerase chain reaction (PCR), DNA sequencing of the amplicons, and comparison of the retrieved sequences to the ever-growing database of sequences from pure cultures and environmental DNA, it was unnecessary to cultivate the organism to determine their presence. This has become a major activity in the field of microbial ecology over the past 10 years, and has provided tremendous insights into the phylogenetic diversity present within microbial habitats. However, the determination of what microbes are present in a habitat does not directly resolve the problem of specifically linking specific microbes to their quantitative biogeochemical role in habitats. This remains a difficult technical problem in the discipline.

The Scope of Microbial Ecology

Microbes are present in almost all aquatic and terrestrial environments, and associated with all plant and animal species. Although each of these habitats may have unique aspects that select for particular strategies among microbes, in this general discussion of microbial ecology, the focus will be on elements that are common to different physical habitats.

Microbes' impact at different spatial scales

The biosphere can be construed as a series of interlinked systems that operate at different spatial scales. Microbes physically operate at a micrometer scale, but both impact and are impacted by processes that operate at larger scales. At the largest scales (global and landscape) are aquatic and terrestrial biomes. Biomes are large geographical areas defined by macroecologists on the basis of their distinctive flora and fauna, which presumably reflect selection and adaptation to climate and geography. For example, terrestrial biomes include temperate deciduous forests, grasslands, rainforests, deserts, taiga, and tundra. Marine aquatic biomes would include polar, temperate, and tropical oceans, and freshwater biomes comprise flowing water (rivers and streams), lakes, and ponds, or wetlands. Whereas biogeography (the study of the distribution of organisms over space and time) is well established for plants and animals, the concept has been less intensively studied for microbes until very recently. There are studies in which there is a correlation between geographic distance and genetic distance for specific bacterial groups. In addition, an increase in the number of taxa detected with increases in sampling area has been found for microbes, as has been found many times for plants and animals. However, the underlying mechanisms responsible for the species-area relationship may be quite different in microbes, due to unique mechanisms for phylogenetically broad genetic transfer and (in some cases) high rates of recombination.

At the global and landscape scale, biotic and abiotic planetary processes can impact each other across biomes. For example, oceanic photosynthesis may be limited by the amount of iron; iron is a cofactor for electron transport proteins essential for conversion of solar energy to a biologically useful form of energy. Arid conditions in deserts produce iron-rich dust particles that can be transported thousands of kilometers through the atmosphere and deposited in the ocean. As another example, the global dispersal of animal viruses such as West Nile virus and influenza virus H5N1 can be mediated by migratory birds.

At the ecosystem scale, there are rich, complementary interactions between physical, chemical, and biological components. For example, photosynthetic sulfur bacteria may occur in freshwater lakes; they require light energy, sulfide as an electron donor that will be used in large part to reduce CO_2 to the oxidation level of cell organic matter, and anaerobic conditions. This combination of factors only occurs in lakes where morphometry, nutrient inputs, and biology are constrained within certain limits (Figure 1).

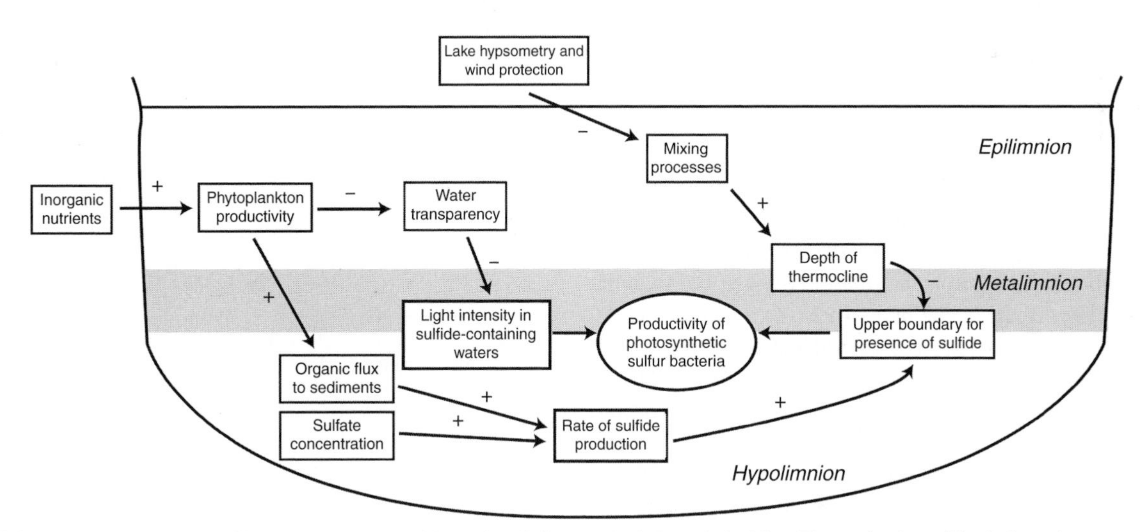

FIGURE 1 System interactions affect the occurrence and intensity of photosynthetic bacteria in lakes. These microbes will only be active at strata where light, anaerobic conditions, and reduced S cooccur. The diagram depicts important physical, chemical, and biological factors, and whether an increase in intensity of one factor produces an increase (+) or decrease (–) in a related factor. For example, increases in inorganic nutrient concentrations stimulate phytoplankton growth. This has a negative affect on water transparency at depth (necessary for photosynthetic bacteria) but also results in a larger flux of organic matter to sediments. This stimulates sulfide production, which is also required for photosynthetic bacterial production. The morphometry of the lake basin and its degree of wind protection drive the intensity of lake mixing, which affects the depth from the sediment to which anaerobiosis can extend. *Reproduced from Parkin, T.B., & Brock, T.D. (1980). Photosynthetic bacterial production in lakes – The effects of light intensity.* Limnology and Oceanography, 25, *711–718.*

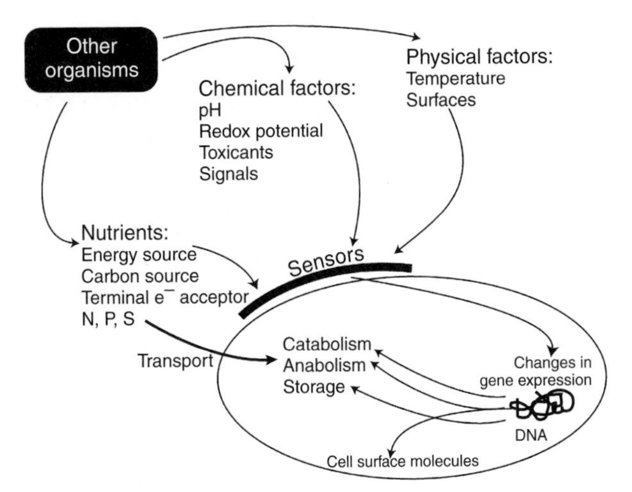

FIGURE 2 The microenvironment. The volume within a few micrometers from the microbial cell surface has the greatest impact upon it. External resources are transported across the cell membrane via membrane-bound permeases. Microbial cells contain sensors at their cell surfaces that can detect chemical signals in the environment. This can lead to changes in gene expression and alterations in microbial cell activity. *Reproduced from Konopka, A. (2006). Microbial ecology – Searching for principles.* Microbe, 1, *175–179.*

However, to understand the mechanistic basis for interactions of a microbe with its environment, one must consider the conditions within a few microns of the cell surface, the microenvironment (Figure 2). The cell's surface interacts with the external physical environment – this includes physical factors such as temperature or light but may also involve interactions with the surface of minerals or microbial or metazoan cells. A broad suite of chemicals are present – some are required nutrient resources, but others may be toxic or represent signals arising from other cells. The chemicals may arrive at the cell surface by diffusion from the bulk phase liquid, or via intimate contact in dense biofilms. For nutrient resources, there are membrane-bound permeases that transport molecules into the cell. As single-celled organisms, microbes have evolved a series of sensors on cell surfaces to detect and respond to chemical cues. One such system results in a molecular memory of past chemical concentrations and controls the rotation of bacterial flagella in a way that effects chemotaxis. In N_2-fixing root nodule symbioses, chemical cues are passed between *Rhizobium* species and specific plant legumes; flavonoids excreted from plant roots are the key signals in initiation of nodule formation in the nitrogen-fixing symbiosis. They bind to NodD proteins in the rhizobial cytoplasmic membrane and these sensors also activate transcription of nodulation genes critical for the initiation of root nodule formation. Some of these *nod* genes specify enzymes that synthesize lipochitooligosaccharides, molecules that are sensed by root cell sensors and lead to morphological changes in them. Environmental sensors may also be important in animal pathogenesis. The low pH of the human stomach is a defense mechanism against bacterial infection of the gastrointestinal tract. In the human pathogen *Vibrio cholerae*, however, transcription of genes required for colonization and pathogenesis is induced by exposure to acid and higher temperatures as would occur after a host ingested the organism from the environment.

A phylogenetically widespread regulatory system that can detect environmental signals and produce changes in bacterial gene expression are two-component regulatory systems, in which one component is a histidine protein kinase (HPK). Approximately 5000 have been identified by bioinformatic analysis of more than 200 bacterial genomes. The proportion of the genome that encoded HPKs increased as genome size increased, and many of the organisms with very high proportions of HPK exist in anaerobic sedimentary habitats in which there are steep chemical gradients over short scales (such as δ- and ε-proteobacteria) or have relatively complex life cycles (the cyanobacterium *Nostoc* and aquatic chemoheterotroph *Caulobacter crescentus*). On the other hand, bacteria that are intracellular parasites of eukaryotes might be expected to encounter a more constant external environment and they were found to have few HPK genes.

Categorizing microbial ecosystems

A focus on the microscopic scale suggests a set of microbial ecosystems (Table 1), based upon the relevant physical and chemical characteristics at that scale. The quintessential

TABLE 1 Microbial-scale ecosystems

Ecosystem Type	Examples	Characteristics
Planktonic	Open ocean, lakes	"Oligotroph lifestyle" high affinity for uptake of multiple nutrients
Surface-associated saturated water	Freshwater and ocean sediments, subsurface sediments, microbial mats, biofilms	"Gradient lifestyle" hydrodynamic processes and fluid flow determine nutrient fluxes Biomass density affects gradient steepness
Surface-associated unsaturated water	Surface and vadose zone soils	Water availability as limiting factor for activity and dispersal Patchy nutrient distribution Dormancy
Macroorganism associated	Gastrointestinal tract, rhizosphere, epiphytes	Coevolution, specific molecular interactions with surface-associated molecules

Reproduced from Konopka, A. (2006). Microbial ecology – Searching for principles. Microbe, 1, *175–179.*

characteristic of aquatic habitats is the planktonic (free-floating) lifestyle. Growth of planktonic chemoheterotrophic bacteria is generally limited by labile, dissolved organic carbon (DOC) present as sugars, amino acids, and organic acids that arise from larger organisms via excretion, lysis, or hydrolysis of macromolecules. The total DOC concentration in ocean waters may approach 1 mg l^{-1}, but this represents a pool of several hundred different organic molecules, each present at very low (nmol l^{-1} to μmol l^{-1}) concentrations. The concentrations are low because the bacteria have evolved high-affinity transport systems, and competitive forces select for microbes that can exploit multiple substrates at low concentrations.

Other microbes must deal with strong concentration gradients of chemical substances, because they are relatively fixed in a solid matrix (such as an aquatic sediment, subsurface aquifer, microscopic biofilm, or macroscopic microbial mat). Strong spatial gradients form in these systems because the chemical sources and sinks differ in location. If microbes are sinks or sources for chemical species, their numbers and activity will affect the steepness of the gradients. These gradient systems may be stable in time or show significant temporal fluctuations. For example, microbial mats contain photoautotrophic cyanobacteria in the top few millimeters of the mat and a variety of chemoheterotrophic bacteria are located deeper. Experiments with microelectrodes have shown that there are large and rapid changes in chemical species such as O_2, H_2S, and H^+ just after the sun rises and after it sets (Figure 3).

Terrestrial soil habitats that are not saturated with water may be likened to vast deserts in which there is an occasional convergence of water, nutrients, and microbial cells on a particle that results in a burst of microbial activity. Soils may seem to be rich with microbes (10^9 bacterial cells and thousands of different bacterial genomes per gram), but these numbers occupy <0.01% of the soil volume. The spatial location of microbial activities in soil can be conceptualized as a series of three-dimensional maps that overlie each other within a matrix of pores. These maps describe the spatial location of (1) different microbes (which have distinct biochemical functionalities), (2) chemical resources (organic and inorganic), and (3) water and gases. Each is necessary but none by themselves is sufficient – activity will proceed only where appropriate levels of all necessary factors intersect. The result is a complex three-dimensional landscape in which the physical structure, determined by not only the mineral matrix but also by plant roots, microbial cells, and extrapolymeric substances, sets the conditions for the activity of microbes and the fate of organic and inorganic C, N, and other elements.

In the case of plants or animals as microbial ecosystems, there is an additional dimension to the chemical gradients and physical structure of the ecosystems described above. That is the capacity for very specific molecular interactions with biomolecules on cells of the plant or animal, and the capacity for (co)evolution by both the microbe and its "environment."

Thinking like a microbe

Microbial activity at the microscopic system level produces effects at the macroscopic and landscape level. These effects have relevance for scientific disciplines such as macroecology, biogeochemistry, and geology. However, there are principles unique to the ecology of microbes, and they only emerge from "thinking like a microbe"; that is, by considering interactions with the environment at the microscopic

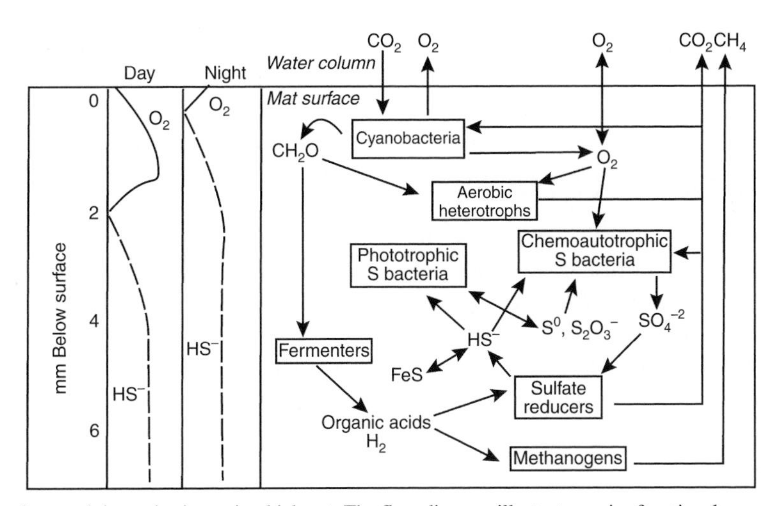

FIGURE 3 Microbial interactions and dynamics in a microbial mat. The flow diagram illustrates major functional groups of microbes and transformations in the biogeochemical C and S cycles. The depth profiles on the left present idealized gradients of O_2 and H_2S during the day (when photosynthetic oxygen evolution occurs) and at night (when respiratory processes lead to anoxic conditions through much of the mat). *Reproduced from DesMarais, D. J. (2003). Biogeochemistry of hypersaline microbial mats illustrates the dynamics of modern microbial ecosystems and the early evolution of the biosphere.* The Biological Bulletin, 204, *160–167.*

scale. It is important to keep in mind some characteristics that distinguish microbes from macroorganisms:

- Small size – Optical amplification (microscopy) is an invaluable tool to assess the distribution of microbes in natural habitats.
- Fast rates of growth and metabolism – Per unit biomass, microbes grow and metabolize much more rapidly than macroorganisms. Generation times of 1–2 days in natural environments are not unusual.
- Exponential growth of populations – The growth of individual microbial cells is autocatalytic; a cell grows to form two cells, each of which can catalyze further growth. When coupled with the relatively fast growth rates of microbes, favorable environmental conditions can produce explosive increases in microbial populations over short period of times.
- Rapid dispersal can occur over long distances – Due to their small size, microbes can be carried for long distances by air or water turbulence. In addition, animals may transport bacteria via mechanisms such as fecal transmission.
- Genetic flexibility – Although Bacteria, Archaea, and many fungi lack sexual recombination mechanisms, many microbes possess genetic elements that mediate vectorial gene transfer from donors to recipients. Analysis of microbial genomes shows that gene transfer is not limited to within a species, but may occur across broad phylogenetic boundaries. In the cases of resistance to newly introduced antibiotics and metabolism of novel xenobiotic organic compounds, evolution has occurred over a period of decades rather than millennia.
- Adaptability to environmental extremes – Most microbial species tolerate and grow within what we consider the normal range of environmental factors such as temperature pH and chemical concentrations. However, species that are optimally adapted to these environmental extremes have evolved.

PRINCIPLES OF MICROBIAL ECOLOGY

A statement of general principles is difficult, because the attention of microbial ecologists is diffused over many different habitats. Thus, there is a tendency to focus on the idiosyncratic rather than the general. In this section, an attempt is made to synthesize phenomena and search for fundamental principles among the unique features of diverse ecosystems.

Overarching Principles

Principle 1. The primary role of microbes in the biosphere is as catalysts of biogeochemical cycles. That is, they mediate kinetically inhibited but thermodynamically favorable reactions. Some microbes may also have strong effects on the ecology of specific micro- or macroorganisms, via their roles as pathogens or symbionts.

Solar energy is the external energy source that drives the biosynthesis of organic matter from inorganic C, N, P, S, and trace elements by photosynthetic primary producers. These inorganic nutrient sources are present in finite amounts; microbes are often largely responsible for catabolism of organic compounds by which the mineral forms used by primary producers are regenerated.

In terrestrial ecosystems, plants are responsible for the great bulk of primary production. Plant organic matter becomes available for microbial decomposition in forms that vary in molecular structure, susceptibility to degradation, and temporal availability. These include root exudates, the fine meshwork of root tissue in soil, and the seasonal fall of plant litter onto the soil surface. Each source may contain complex mixtures of organic substrates that range from low-molecular weight sugars, organic acids, and amino acids to macromolecules. Macromolecule catabolism involves initial hydrolysis to monomers by microbial extracellular hydrolases and subsequent intracellular metabolism. Not only is organic carbon oxidized back to CO_2, but catabolism converts organic N, P, and S to NH_4^+, PO_4^{-3}, and SO_4^{-2}. In this process, energy is released by breaking C–C covalent bonds; some of the released energy is conserved by microbes and used to drive their growth.

The cycling of organic C in soil intersects with biogeochemical reactions in which inorganic chemical species serve either as terminal electron acceptors or as electron donors. These reactions are particularly important with respect to N, because these transformations affect the bioavailability of inorganic N for plants. For example, NH_4 is a reduced form of N that is used as an energy source by nitrifying bacteria, and NO_3^- is the end product of nitrification. However, NO_3^- is more mobile than NH_4^+ in soil solution, and hence more susceptible to loss. Furthermore, if the soil becomes anoxic, NO_3^- can substitute for O_2 as a terminal electron acceptor for chemoheterotrophic microbial metabolism. The oxidized end products, N_2O and N_2, are gaseous and thus represent a loss of N for plants.

In addition to their biogeochemical function in nature, the specific interaction that microbes have with plants or animals can have major consequences, not only on the affected organisms but also on the ecosystem. Frank pathogens of plants and animals can generate selection pressures to evolve resistance determinants upon their hosts. For example, humans with the sickle cell trait (one allele of an altered form of hemoglobin along with one normal allele) are more likely to survive disease caused by the malaria protozoan parasite than individuals homozygous for normal hemoglobin. The incidence of sickle cell trait in humans is highly correlated with areas in which malaria is endemic.

Microbes can also positively affect macroorganisms via mutualisms. A well-studied example relates to another aspect of the nitrogen biogeochemical cycle. Bacteria such as *Rhizobium* and *Frankia* can form specific associations within nodules on plant roots. These associations make photosynthetically produced organic C available to the microbes. When in association with plants, these microbes express nitrogenase, an enzyme that fixes N_2 into NH_4^+; microbial metabolism is regulated such that >90% of fixed N is translocated to the plant.

Principle 2. Although generally unseen to the naked eye, microbes comprise nearly half of all biomass on Earth. All habitats suitable for plants and animals also harbor microbial populations. In addition, some microbes are adapted to grow under physical and chemical conditions that are too extreme for plant and animal growth.

Determining the amount of microbial biomass on a global scale is difficult to do with precision, but estimates suggest that the amount of organic carbon in Bacteria and Archaea (350–550 Pg) is approximately equal to that found in land plants. When one factors in the amount of eukaryotic microbes (e.g., fungi in terrestrial habitats), more than half of the biomass on Earth is likely to be microbial. The largest stocks of microbial biomass are not in habitats near the surface of the earth, but rather in the oceanic and terrestrial subsurface (Table 2). Although population densities (cells per cm^3) in these habitats are 1–2 orders of magnitude smaller than in the near-surface aquatic or soil environments, the volume of these habitats is very large. Thus, the deep subsurface appears to be a major reservoir for microbial life, about which we have very limited information. On the other hand, near-surface terrestrial and aquatic habitats make larger contributions to overall microbial growth and activity. There are few measurements of growth rate in the deep subsurface, but existing measures suggest a generation time for microbes measured in many years; this may reflect occasional periods or locations of microbial growth with most of the microbes surviving in a metabolically quiescent state.

TABLE 2 Estimate of bacterial and archaeal cells on earth

Habitats	Cell numbers ($\times 10^{28}$)
Salt and freshwater surface	12
Oceanic subsurface	355
Surface soils	26
Terrestrial subsurface	25–250
Total	415–640

Reproduced from Whitman, W. B., Coleman, D. C., & Wiebe, W. J. (1998). Prokaryotes: The unseen majority. Proceedings of the National Academy of Sciences of the United States America, *95, 6578–6583.*

Microbes associated with plants or animals can exert strong ecological effects via parasitism or symbiosis. However, in terms of global biomass, these organisms are minor players; they account for less than 1/100,000th of microbial biomass on Earth. Bacteria that function in digestion and nutrition of animal herbivores (ruminants and termites) comprise the largest proportion, due to the large biomass of these animals on Earth.

Humans characterize habitats in which the physical or chemical conditions preclude the existence of plants or animals as "extreme." However, some microbes have evolved adaptations in molecular structure such that they are optimally adapted to growth in a particular extreme environment such as very acidic or basic pH, very cold or hot temperatures, or high salt concentrations. These adaptations usually mean that the organisms cannot grow under the conditions we humans consider normal. In the case of thermophiles and psychrophiles, evolution in cellular macromolecules has occurred. For example, cellular enzymes in thermophiles retain structure and activity at higher temperatures; this may be a consequence of more extensive secondary structure and reduced content of certain labile amino acids. A significant problem at low temperature is fluidity of membranes, and the lipids in psychrophiles have increased content of fatty acids that will remain fluid at lower temperatures.

Principle 3. The effects of microbial catalysis (the rates of chemical and energy flow through an ecosystem) depend upon the population size of microbes (the number of catalysts) and their physiological activity (as determined by extant physical and chemical conditions). In stable ecosystems, small numbers of microbes can still produce significant effects over geological time spans.

Microbial population levels and cell-specific activities can range over many orders of magnitude in nature. In aquatic ecosystems such as lakes and coastal marine areas, excessive nutrient enrichment leads to undesirable ecological effects (eutrophication). The amount of P_i loaded into an aquatic ecosystem will often determine the amount of photoautotrophic biomass (Figure 4) in the water column, and systems with higher levels of photoautotrophic biomass exhibit increased levels of primary productivity. However, physiological studies of phototrophs have shown that the cell-specific rate of photosynthesis will vary significantly as a function of irradiance, and that even at saturating irradiances the photosynthetic rate may vary 5- to 20-fold, depending upon the nature and stringency of nutrient limitation.

The effects of temperature upon activity may be even more striking, when one realizes that much of the biosphere (the deeper waters of oceans and lakes, and many subsurface terrestrial systems) has ambient temperatures of <10 °C. As microbes are exposed to temperatures near their minimum, not only do the catalytic rate of enzymes decline, but membrane lipids stiffen, thereby decreasing the effectiveness of

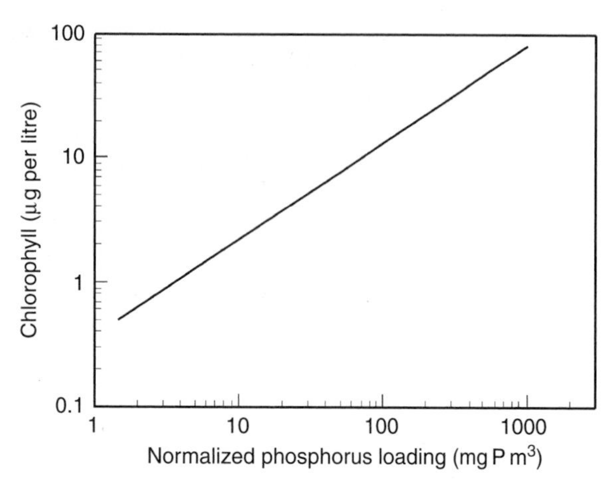

FIGURE 4 The Vollenweider model of the relationship between the amount of P that enters a lake (P loading in mg P per m^3 of lake volume) and the standing crop of phytoplankton biomass (measured as chlorophyll *a*).

membrane-bound transport proteins. As a result, substrate affinity declines and cells are less effective in concentrating substrates from dilute environments. The physiological consequences are that limiting resources can have even more acute affects on activity in cold habitats.

Some processes may occur very slowly, yet have significant impacts on biogeochemical cycles. For example, the anaerobic oxidation of methane in deep-sea sediments is biogeochemically important, given the vast expanse of seabed floor. However, the process occurs at rates of nmol cm^{-3} year^{-1}, and enrichment cultures of the consortium of methanogen and sulfate-reducing bacteria have a doubling time of 7 months.

Principle 4. The results of microbial catalysis can have profound effects on the physical and chemical characteristics of the macroenvironment. However, microbial activity is determined by the physical and chemical characteristics of the microenvironment.

There are many habitats in which spatial heterogeneities at the submillimeter scale produce a richer diversity of habitats and niches than is apparent from measurement of bulk average concentrations of chemicals. In the open ocean, there is a spatial mosaic of nanometer-sized organic colloids as well as sheets of organic compounds that may be up to 100 µm in size. Aggregation of organic matter and organisms into millimeter-sized particles create a habitat for a rich and diverse community of phytoplankton, protists, and bacteria. In well-aerated soil, local deposition of labile organic matter will create a site of intense activity by aerobic bacteria in which the rate of oxygen consumption exceeds its diffusion and an anoxic microsite in a soil particle can arise.

Strong concentration gradients in environments that result from physico-chemical processes, high densities of active microbes, or (most commonly) the interaction between abiotic and biotic forces also produce strong effects at the microscale. The gastrointestinal tract of insects is a fermentation chamber with a high concentration of microbes; the fermentation products (organic acids) are absorbed through the wall of the gut to fuel the nutrition of the animal. However, O_2 can also diffuse from the trachea across the gut wall, so that much of the gut volume is microoxic rather than fully anaerobic. Furthermore, H_2 is a fermentation product that arises in the anoxic lumen of the gut and is not used by the insect. It is an energy source for aerobic chemoautotrophic bacteria, which have a niche where overlapping gradients of O_2 and H_2 occur in the lumen. In dense microbial mat systems, diurnal changes in light irradiance drive changes in the rate of photosynthetic oxygen evolution that result in rapid temporal changes over millimeter distances in the profiles of O_2, H_2S, and pH in the mats.

Principle 5. The microbial world exhibits much greater metabolic versatility than is found among macroorganisms. This versatility is exhibited in several different ways:

- Essentially all reduced chemicals found on Earth (both organic and inorganic) can be used as an energy source by some microbe if its oxidation can be coupled to reduction of a terminal electron acceptor in a thermodynamically favorable reaction.

Many biogeochemical cycles operate because some microbe can extract biologically useful energy from the reduced form of an element (e.g., H_2S) and transfer the electrons to a molecule (such as O_2) that has a greater tendency to become reduced (the terminal electron acceptor). The oxidized product (SO_4^{2-}) can then be reduced to H_2S when it serves as a terminal electron acceptor for a different physiological group of microbes.

For organic compounds, all naturally occurring organic compounds and most synthetic organic chemicals can be used as energy sources by microbes. It is easier to list the recalcitrant exceptions – lignin (under anaerobic conditions), highly halogenated organic compounds (under aerobic conditions), branched alkanes, and insoluble synthetic polymers (e.g., polyethylene).

- Individual organisms may be capable of metabolizing a broad range of different substrates.

From an ecological perspective, much of the organic matter that becomes available to chemoheterotrophic bacteria arises via cell lysis from the death of other organisms (plants, animals, or other microbes). The enzymatic hydrolysis of cell macromolecules will produce a rich mixture of various sugars, amino acids, and organic acids. Thus, it is not surprising that individual microbes may possess transporters and enzymes that provide the potential for utilization of many components of these mixtures.

The study of biodegradative enzymes in bacteria has provided insights into some mechanisms whereby a broad substrate utilization range may arise. In some cases, a key enzyme with a broad substrate range has evolved. A halidohydrolase from an *Arthrobacter* sp. is active against more than 50 different halogenated organic compounds that differ in their halogen substituent and range from aromatic to saturated or unsaturated alkanes. The evolution of biodegradative metabolic pathways has also made it facile for an organism to utilize multiple organic substrates. There is a large diversity of substituted aromatic compounds that can be aerobically catabolized by microbes. Catabolic pathways entail the action of enzymes that funnel different substrates to one of a few common intermediates (usually dihydroxy benzene derivatives), which are then further catabolized by a common set of enzymes. Thus, the growth substrate range can be augmented by acquisition of enzymes to process specific substrates, once it possesses a core aromatic catabolism pathway.

- Complex microbial communities may be metabolically redundant and contain a number of different populations that can metabolize the same substrate. This metabolic diversity increases resiliency to environmental perturbations, as extreme conditions are less likely to kill all species performing a particular ecological function.

A critical issue in all areas of ecology is to understand the relationship between biodiversity and the functioning and stability of ecosystems. Imagine an ecosystem that is devoid of microbes, and that different species are introduced one by one. Initially, each new species is likely to add a new function to the ecosystem, because it can utilize substrates that the previous species could not. At some point, however, new species are likely to be redundant – an existing species can already carry out that function. The ecosystem possesses its maximum microbial functionality. However, higher biodiversity (with its associated functional redundancy) is valuable if the ecosystem is exposed to changes in environmental conditions such as a change in pH or addition of a toxic metal. If functional redundancy were low, then all species that can carry out a particular function might be eliminated. At higher functional redundancy the probability that at least one species is tolerant of the new environmental condition is higher and the ecosystem then retains this function and is perceived as functionally stable.

For example, microbes in soils contaminated with both lead and chromium were examined for the distribution of sensitivity to these metals. The physical structure of soil as well as the chemical properties of the metals (i.e., their low aqueous solubility) meant that there was a very high degree of spatial heterogeneity in the distribution of microbes and bioavailable concentration of metals within the soil matrix. The sensitivity of the microbial community was determined by exposing cell suspensions to a graded series of metal concentrations, and measuring activity via incorporation of ^{3}H-leucine (Figure 5). The broad concentration range over which reduction in activity occurred suggests that there were subpopulations with different tolerance levels to Pb or Cr.

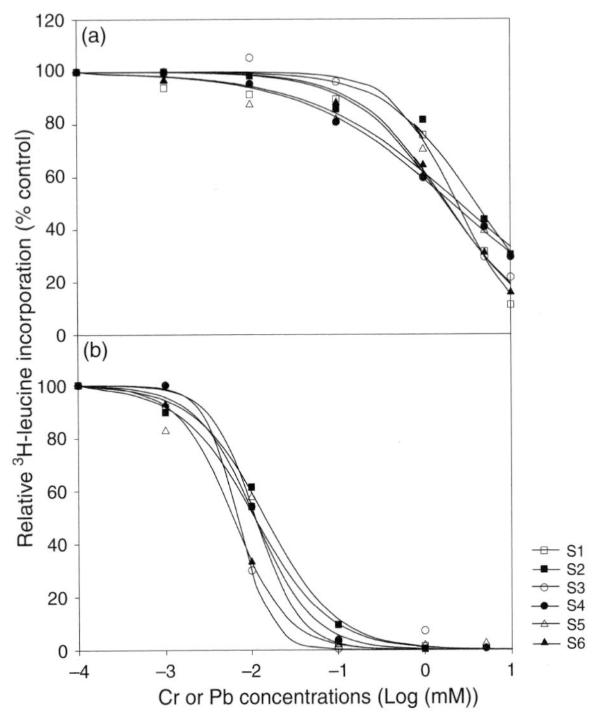

FIGURE 5 Response of microbial populations extracted from heavy-metal-contaminated soils to a series of (a) CrO_4^{2-} and (b) Pb^{2+} concentrations. The responses were measured by the incorporation of ^{3}H-leucine into cell material in 2h incubations. The samples (S1–S6) were soils with different levels of Pb and Cr contamination. *Reproduced from Shi, W., Bischoff, M., Turco, R., and Konopka, A. (2002). Association of microbial community composition and activity with lead, chromium, and hydrocarbon contamination.* Applied and Environmental Microbiology, *68, 3859–3866.*

With pure laboratory cultures, a reduction in activity from 80% to 20% of the maximum occurred over a change in metal concentration of about twofold, whereas in these natural communities, this level of activity reduction occurred over a much broader concentration range. Subsequent work on these soils showed that when carbon sources (such as glucose) that presumably could be utilized by many different microbes in the community were added, community metabolism persisted at high levels of metal input, whereas when xylene (in which the functional redundancy of the community is expected to be much lower) was added, community activity was abolished at much lower metal concentrations.

- Microbial populations are capable of rapid evolutionary responses to environmental perturbations; this is due to mechanisms that generate relatively high mutation frequencies and a capacity for horizontal gene transfer (HGT) across large phylogenetic boundaries.

Evolution via differential reproduction under environmental selection pressures was the central premise of Darwin's work. In the case of plant and animal species, genetic variation (the raw material upon which selection acts) is generated within one species via the shuffling of genes during sexual reproduction. In microbes, however, there are mechanisms of HGT that can cross broad phylogenetic boundaries and generate completely novel gene combinations in much shorter time frames.

Humans have performed two great natural experiments that illustrate this point – the large-scale introductions of antibiotics and xenobiotic organic compounds. Antibiotic-resistant forms of a pathogen have often arisen within 10–30 years after introduction of a new antibiotic. The acquisition of multiple antibiotic resistance by clinical isolates is conferred by resistance transfer factors (RTFs), broad host range plasmids that contain a set of transposons, each of which has a gene for resistance to a different antibiotic. From where did these plasmids arise? A set of clinical isolates of Enterobacteriaceae that had been archived in the preantibiotic era was found to contain plasmids, and their core genes for replication, maintenance, and conjugal transfer were similar to those found in modern RTFs. However, they did not contain transposons with antibiotic resistance determinants. The implication of these findings is that the use of antibiotics in clinical practice and agriculture results in a strong selection pressure for cells with RTFs in which the appropriate transposon has integrated.

Why can microbial evolution proceed more quickly than in the case of macroorganisms? Microbial generation times can be comparatively short, and each replication of genetic material will result in a small number of mutations. When one considers the total population size of microbial cells in habitats (millions to billions per gram), there are ample opportunities for genetic variation to arise.

The "evolution" of specific taxa is particularly enhanced by bacterial mechanisms of gene transfer (the uptake of naked DNA by a cell, or transfer of DNA into a recipient bacterium by direct contact with a donor bacterium or infection with a bacterial virus) that may permit horizontal transfer of genetic material across broad phylogenetic barriers. Gene transfer frequencies are directly proportional to cell density – crowded habitats such as animal gastrointestinal tracts, soil microcolonies, the plant rhizosphere, biofilms, and marine snow probably represent important foci. In some instances, the acquisition of a single gene (e.g., an antibiotic resistance gene or exotoxin genes within prophage of *Corynebacterium diphtheriae* or *Clostrdium botulinum*) can substantively alter the organism's ecological properties. In other instances, there are functional modules of multiple genes that confer novel properties. Some examples include sets of genes for organic catabolism or pathogenicity islands that contain virulence genes.

Principles in Population Ecology

Principle 6. The types and numbers of different microbes present in a habitat are a function of the diversity and amounts of nutrient resources available; in ecology, this is termed bottom-up control. However, the abundance of a microbe will also be modulated by its ability to withstand forces that eliminate it from the habitat. These loss forces may be biotic (due to the activity of predators and parasites and termed top-down control), physical (flushing or scouring action), or physiological (nutrient starvation).

The effect of nutrient resources upon microbial activity and growth will be discussed below in the context of physiological ecology. Although much of the historical effort in microbial ecology has been directed to the study of bottom-up control, there have been significant studies since the 1980s on the effects that biotic loss factors may have on structuring microbial communities. The recognition of the role played by eukaryotic protists in biogeochemical cycling in the ocean was important in this regard. Bacteria have been found to utilize up to 50% of the carbon fixed by photosynthesis in the ocean. The "microbial loop" (Figure 6) provides a means for these resources to be reintroduced into the traditional food chain (via zooplankton to fish). Bacterial numbers are controlled by nanoflagellates and to a lesser degree small ciliated protists, which in turn are preyed upon by microzooplankton. As these predators are of similar size to phytoplankton, they can be consumed by zooplankton and the resources reenter the planktonic food chain. Detailed investigation of other habitats has shown that bacteriovorous protists are important in a range of ecosystems, including soil.

There is a vigorous debate between those who contend that bottom-up mechanisms control microbial biomass levels and those who emphasize top-down control. The truth likely lies in the middle, and may differ depending upon season and ecosystem productivity. Both empirical and theoretical evidence suggest that bacterial numbers may be more tightly controlled by protistan predation in nutrient-poor (oligotrophic) ecosystems, whereas in nutrient-rich habitats competition for nutrients has primacy. Intuition might lead one to the opposite inference, but eutrophic habitats will have higher population levels of higher-level

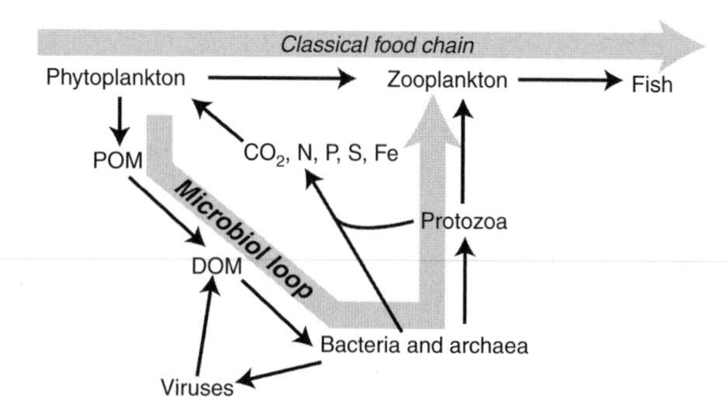

FIGURE 6 The microbial loop in aquatic food chains. Filter feeding by protists upon bacteria can result in the return of primary production back to the classical food chain, as the protists are preyed upon by zooplankton. POM, particulate organic matter; DOM, dissolved organic matter.

predators (zooplankton and fish) that keep tighter control on the abundance of bacteriovorous protists. As a result, predation pressure on bacteria is reduced and higher bacterial numbers result in more stringent nutrient limitation.

Protistan grazing not only can control bacterial numbers, it can also shape bacterial community composition. Protists employ either filter or interception feeding. As a result, bacterial cells with diameters of 1–3 μm are most susceptible to predation. As a result, aquatic protists are very effective predators on rapidly growing single-celled bacteria. Thus, there will be strong selective pressure for antipredator strategies in bacteria that result in smaller or larger particles. Strategies that yield larger particles include morphological changes such as filament formation or production of exopolymers to produce cell aggregates. Ultramicrobacteria (those with a cell diameter of 0.3 μm or less) can be isolated from natural marine samples; these organisms are sized below the optimal feeding range of protists. Chemical defenses have also been found – the purple pigment of *Chromobacterium violaceum* has been found to induce apoptosis in flagellates when a bacterial cell is ingested.

Over the past 15 years, aquatic ecologists have come to recognize that viruses are also significant causes of microbial mortality. There are approximately 10^8 virus particles per milliliter in productive, coastal marine waters. There is significant genetic diversity among these viral particles, and within that diversity are viruses that can infect algae, bacteria, or protists. It has been technically challenging to accurately measure the proportion of microbes that are lysed by viruses; current estimates range from 10% to 20% of the bacterial population per day and the impact of viruses as a loss factor for bacterial production may equal that of bacteriovorous protists in marine systems.

Principle 7. Microbial populations and communities often exhibit much larger dynamics in biomass, composition, and activity than do plant and animal populations.

These complex, nonequilibrium dynamics are thought to be primarily driven by the temporal frequency of changes in important environmental factors (and hence, changes in selective environmental forces). An obvious example is a tidal salt marsh, but there are periodicities that exist at shorter and longer frequencies to which microbes can respond. Temporal changes can occur over very different spatial scales, from kilometers in coastal marine upwellings to millimeters in laminated microbial mats. In addition to these environmentally driven changes, there are also internal biological cycles in ecosystems that alter selective forces. Predator–prey cycles are perhaps the best known, but the identification of cross-species molecular signaling in the laboratory suggests that population dynamics may have a larger biological component than we have appreciated.

Microbes respond more dynamically to these changes than macroorganisms as a consequence of their biological properties: a rapid pace of physiological adaptation to changing environmental conditions, a broad variety of positive and negative interactions between different microbes, and the capacity for rapid genetic change. An important consequence of principles 6 and 7 is that the observed temporal changes in microbial populations (whether termed "community dynamics," "successions," or "oscillations") are not solely determined by environmental or biological factors, but rather by an intricate interaction among these factors.

Principles in Physiological Ecology

Microbial physiological ecology is a field in which the molecular details of gene regulation can be evaluated in the context of ecological effects under the selective pressures experienced by microbes in nature.

Principle 8. The amount of one essential limiting resource is most likely to determine the amount of microbial biomass in an ecosystem.

A nutrient resource that is taken up may be (1) conserved in functional or structural molecules (i.e., become biomass), (2) utilized and excreted for bioenergetic purposes (energy source and terminal electron acceptor), or (3) function both as a component of biomass and in bioenergetics. All organisms must acquire an energy source (light of the appropriate wavelengths for phototrophs and a reduced chemical for chemotrophs) and the appropriate molecular form of atoms that comprise biomass. The most important atoms are C, N, P, S, and several metals (Fe is often the most quantitatively important) that serve as enzyme cofactors. Chemotrophic bacteria also require a terminal electron acceptor, a molecule to which electrons are ultimately donated after oxidation of the energy source. Phototrophs require an electron donor that is used to generate reducing power via photosynthesis.

The molecules that comprise this elementary set of requirements can be called "complementary substrates" in that all are necessary, but none by themselves are sufficient for growth. However, within each requirement, there may be substitutable forms. All microbes use NH_4^+ as an N source, a number can use NO_3^-, and some fix N_2. There are hundreds of different organic substrates in the environment; if a specific microbe has uptake systems and enzymes to catabolize them, these represent substitutable substrates.

The limiting nutrient is not necessarily the one present in the lowest absolute concentration in the habitat. Rather, it is the resource in scarcest supply relative to its requirement for biomass synthesis (Liebig's law of the minimum). It is important to recognize that cellular composition is not immutable. Despite the general biochemical similarity of all organisms, some taxa may have unique requirements. For example, diatoms require silica for their extracellular frustules whereas other groups of phytoplankton do not have this requirement. In addition, physiological conditions may substantially affect the elemental stoichiometry of biomass. Iron is required in a variety of electron transfer proteins such as cytochromes and nonheme iron proteins; it usually comprises 0.2% or less of a bacterium's dry weight. However, in N-limited environments where N_2-fixing bacteria would have a selective advantage, the synthesis of Fe-rich nitrogenase and ferredoxin would increase cellular iron content by at least fivefold.

Principle 9. Growth of microbes on the limiting nutrient (or a set of substitutable nutrients) will depress its concentration to a value that limits microbial growth rate. The nutrient flux rate in an ecosystem may become physically limited by hydrodynamics or mass transport.

The "limiting nutrient" in a habitat has two distinguishable effects on microbial growth. It limits the total amount of biomass. If microbes are inoculated into a sterile environment, they will grow at their maximum rate and at some point the total demand of biomass for nutrient uptake exceeds its supply rate, and the concentration of limiting nutrient decreases. At some low concentration (in the nmol l^{-1} to low μmol l^{-1} range), transport rate into the cell is determined by this low extracellular substrate concentration and a reduced intracellular flux produces a reduced growth rate.

Principle 10. Competition among microbes for low concentrations of limiting substrate and temporal variability in nutrient availability is a strong selective force in natural habitats. A variety of physiological and morphological adaptations can have adaptive value under specific conditions.

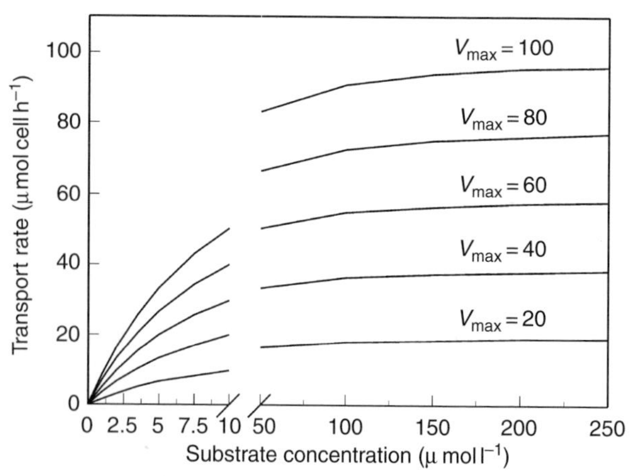

FIGURE 7 The effect of increasing the number of membrane-bound permeases per cell, at both low, transport-limiting external substrate concentrations and saturating concentrations. Calculations were made employing the Michaelis–Menten equation at K_s of 10 μmol l^{-1}, and different V_{max} values. V_{max} is proportional to the number of permeases per cell.

1. The number, diversity, and affinity of transporters in the cell membrane: Nutrient concentrations in the low μmol l^{-1} range are likely to constrain the growth rate. The organisms that can maximize nutrient flux into the cytoplasm will have the highest growth rate and therefore a competitive advantage. In the case of chemoheterotrophic bacteria in the open ocean, organic substrates are likely to be the limiting nutrient. The dissolved organic C level in seawater totals ~1 mg C l^{-1}, but this represents nmol l^{-1} concentrations of a wide variety of specific substrates derived from the lysis of phytoplankton cells and the hydrolysis of their macromolecules. One physiological strategy would be to express genes for high-affinity transporters, because at these low external substrate concentrations, the specific affinity (the initial slope of the uptake rate vs. substrate concentration function) provides a competitive advantage. Furthermore, derepression of gene expression to increase the number of transporters inserted in the membrane will enhance substrate flux via the law of mass action. The maximum rate of transport (V_{max}) normalized to a per cell basis is a function of the number of transporters per cell. Increases in V_{max} not only provide higher transport rates at saturating substrate concentrations, but also increase flux rates at rate-limiting concentrations (Figure 7). When there are approximately 1500 transporters of one specific type in the membrane, they compete with each other for substrate molecules, and further investment in additional transporters of this type is not productive. However, only 2% of the membrane surface is covered with transporters; synthesizing transporters with many different specificities will maximize the total flux of limiting organic C into the cell, if other organic substrates are available. Thus, organic C-limited chemoheterotrophs are predicted to express the capacity for the transport and catabolism of multiple organic substrates. They need only generate a fraction of their total growth rate from the transport flux of each individual organic compound.
2. Regulation of the capacity for substrate uptake versus substrate assimilation: The capacity for nutrient uptake is maximized by increasing the number, diversity, and specific affinity of transporters. However, in a competitive environment this will push ambient substrate concentrations to values where the growth rate is far below maximal. The optimal use of resources under these conditions will regulate decreased amounts of the macromolecular machinery for biosynthesis and assimilation. The most obvious example of this is the direct relationship between growth rate and cellular ribosome content. Ribosomes are energetically expensive to produce; a cellular content that exceeds the required rate of protein synthesis is energetically inefficient.
3. Cell size: When bacteria are grown in a chemostat at a variety of growth rates, slow-growing cells have a smaller size and volume than fast-growing populations. The direct physical reason for this may be the differences in ribosome content mentioned above. A 10-fold increase in cellular ribosome content can only be accommodated

by increasing the volume of a cell. Direct microscopic examination of bacteria in aquatic environments indicates that many cells are quite small, with diameters of 0.5 μm. Size-selective predation by protists may play a role here, but the low growth rates typically found in planktonic aquatic ecosystems (doubling times are often measured in the range of 16–50h) also lead to small cells. The physical nature of small cells has another beneficial effect for nutrient-limited habitats – the surface-to-volume ratio increases as cell diameter decreases. Membrane-bound surface area is the location of substrate transporters, whereas the cytoplasmic volume contains soluble biosynthetic machinery. Thus, decreases in cell size found when nutrient-limited growth is slow have the consequence of increasing the ratio of transporters-to-assimilation machinery.

4. Storage polymers: The physiological responses to nutrient limitation in nature include derepression of nutrient transport systems and downregulation of biosynthetic pathways. However, a microbe may be exposed to pulses of higher nutrient concentration either on a regular diurnal cycle or at irregular intervals (e.g., by encountering particulate organic matter in an aquatic ecosystem). At this physiological state, transport of the limiting nutrient occurs at a much faster rate than can be assimilated into proteins and nucleic acids. The solution to this mismatch is intracellular storage as a polymer that can be assimilated later when external substrate concentration is again limiting. Storage polymers of C and energy include glycogen, starch, and polyhydroxyalkanoates. P is stored as polyphosphate by a number of microbes. N is the macronutrient for which there is no storage polymer, with the exception of cyanophycin produced in many cyanobacteria and a few heterotrophic bacteria. The polymer consists of an aspartic acid backbone to which arginine residues are linked. The synthesis and degradation of glycogen by the cyanobacterium *Oscillatoria aghardii* during a diurnal light–dark cycle illustrates the interplay between resource acquisition, storage, and cell growth. During the light period, fixed CO_2 is used for immediate growth (synthesis of proteins and nucleic acids) and stored as glycogen. Of particular note is that growth rate (measured as rate of protein synthesis) during the light period is lower than in cells cultured in continuous light. However, during the dark period the intracellularly stored glycogen is used as both a carbon and an energy source – the rate of protein synthesis in the dark is very similar to that measured in the light. That is, the microbes grow chemoheterotrophically in the dark on stored glycogen. This physiological strategy produces a higher daily growth rate than if the cyanobacteria had a growth rate of μ_{max} in the light and 0 in the dark.

STRATEGIC APPROACHES TO THE STUDY OF MICROBIAL ECOLOGY

An examination of articles published in microbial ecology journals will illustrate not only the breadth of habitats that are studied by microbial ecologists, but also a variety of technical approaches. Some of the major categories are described below.

Descriptive Studies of the Distribution and Activity of Microbes in a Natural Habitat

"Descriptive studies" has become a pejorative term in science, but this type of analysis is a necessary first step when initiating a study on a new habitat. In addition, field studies have an overhead not found in well-controlled laboratory studies – the collection of routine data on physical, chemical, and biological characteristics of the system. It is essential to monitor the natural variation that lies behind any hypothesis-driven field experiments. These routine data can be mined using statistical techniques to uncover relationships between environmental factors and biological activities. These relationships suggest hypotheses that can be experimentally tested.

Analysis of *In Situ* Rates of Growth or Biogeochemical Activity

These analyses can rarely be accomplished under true *in situ* conditions, as they entail addition of chemical tracers or measurement of changes in biomass. Thus, natural samples are collected and confined in containers, but incubated under the physical conditions found *in situ*. The removal of a sample from the open conditions of the natural system to a confined bottle will lead to changes with time. Therefore, it is advantageous to keep the incubation time as short as possible. The use of radiotracers (i.e., the addition of a radioactive molecule in an amount that is small enough to have a negligible impact on *in situ* chemical concentration) has been invaluable in rate measurements of biogeochemical processes. Bacterial growth rates have been estimated by measuring the rate at which radioisotopic forms of biosynthetic precursors (such as thymidine or leucine) are incorporated into cellular macromolecules. These calculations require several assumptions, and give an "average" growth rate for a heterogeneous population of microbes. These results generally suggest that the doubling times of aquatic bacterioplankton are in the range of 1–2 days, much longer than what is achievable with pure cultures in the laboratory. *In situ* growth rate is a fundamental parameter of interest in ecological studies, but it remains a difficult one to accurately and conveniently measure for microbial populations in nature.

Experimental Manipulation of Natural Samples under Simulated Conditions (Microcosms)

The intent of microcosm experiments is generally not to accurately measure *in situ* activity rates, but rather to perform experimental manipulations of natural microbial communities to test hypotheses regarding the effects of environmental perturbations upon microbial community activity or composition. Incubation times are likely to be much longer than when measuring *in situ* activity rates; therefore, caution in interpretation is warranted as confinement of the sample can introduce major effects.

An example study examined the population structure and activity of a microbial community in soils that were cocontaminated with lead, chromium, and aromatic hydrocarbons. It was difficult to separate the effects of different contaminants by direct analysis of spatial differences in community composition and environmental contaminants. Therefore, microcosm studies were set up in which soil samples from both contaminated and uncontaminated sites were challenged with different terminal electron acceptors and a graded series of heavy metal concentrations, to analyze the effects upon microbial activity (the mineralization of organic carbon) and community composition. The experiments illustrated that if one wanted to detect changes in community composition, it was important to add an energy source (organic substrate) that could drive significant changes in composition. Of course, this energy source will itself represent a strong selective force in structuring the microbial community. However, by comparison of sample treatments, it was possible to demonstrate that whereas resource factors such as the organic energy source and terminal electron acceptor were a major force in structuring species richness, the effects were modulated by toxic factors such as heavy metals. In addition, the factorial design of the experiments (in which a series of combinations of energy sources, terminal electron acceptors, and heavy metal concentrations were tested) showed that microbial activities carried out by a diverse group of functionally redundant community members were more resistant to high toxic metal inputs than activities for which a more restricted diversity of microbes was likely to be present.

Model Laboratory Systems

A distinction between microcosm experiments and model laboratory systems is that the former are initiated with many of the physical and chemical characteristics as well as the taxon richness of the original habitat whereas the latter is often a highly simplified version of nature. In many cases, it may be the study of a pure culture in the laboratory. This can be controversial in microbial ecology, because the laboratory system will lack a large number of features found in nature. However, ecophysiological studies in the laboratory provide the means to rigorously ascertain the physiological responses made by a microbe to environmental changes. These can be approached with a degree of experimental control and replication that is not possible in the habitat. If the study remains housed entirely in the laboratory, it probably ceases to be an ecological one. However, there are advantages to integration of field and laboratory studies, and recursion between real-world complexity and the experimental control that can be applied in the laboratory.

Ecological studies on planktonic cyanobacteria have provided an example of the value of this type of recursion. In eutrophic lakes, cyanobacteria proliferate to high cell densities (blooms) in which >90% of the biomass in the sample consists of one morphological type of cyanobacterium. Under these unique conditions, one can measure physiological properties in whole samples (photosynthetic rate, allocation of fixed C to different macromolecules, rates of macronutrient uptake) and reasonably ascribe these properties to the dominant biomass component. From these measurements on field populations, hypotheses arise regarding the regulatory responses made by the organism to limiting environmental factors. These hypotheses are testable in the laboratory by experiments with cyanobacterial isolates cultivated in chemostats under control of growth rate and stringency of nutrient limitation. As deeper insights are gained into the details of ecophysiological responses from laboratory experiments, it is important to revisit the field and ascertain that those details actually operate under field conditions. This iteration between laboratory and field is straightforward when the natural population has a high relative abundance of the organism of interest. A remaining challenge for microbial ecology is to develop technologies that allow application of this iterative approach to community members present at low relative abundance. This requires the capacity to rapidly detect specific taxa and make physiological measurements on single cells.

FUTURE OF MICROBIAL ECOLOGY

Microbiology at the Microscale: Single-Cell Microbiology

It is a central principle that the direct interaction of microbes with their environment occurs at a spatial scale of micrometers. Thus, analysis of the authentic ecology of microbes is enhanced by technologies that increase the spatial resolution of experimental measurements. This applies not only to analysis of the physical and chemical environment around a cell, but also to measuring the physiological and genetic properties of a set of individual microbial cells. Single-cell microbiology can provide the basis to understand not only physiological responses at the microscale, but also the distribution of properties among a population of cells within a natural habitat.

Over the past decade the capacity to apply advanced spectroscopic and mass spectrometric techniques at small spatial scales has risen dramatically. Raman and FTIR microspectroscopy have been applied to the analysis of microbial cells, and nuclear magnetic resonance can be used to image biofilms and also provide information on chemical gradients within them. New fluorescence probes and optical imaging systems provide the capacity to interrogate physiological state, gene expression, and phylogenetic identity of individual cells in a natural sample.

Radioisotopes have been a powerful technology in advancing microbial ecology; however, they also have practical limitations for performing *in situ* experiments and handling waste. Stable isotope methodologies have the potential to revolutionize the field of microbial ecology. Sophisticated improvements in nuclear magnetic resonance, mass spectrometry, and Raman spectroscopy are approaching the point that allow the kinetics of metabolic function in microbial communities to be determined with great sensitivity. In contrast to radioisotope-based techniques, stable isotopes provide a means to specifically isolate cellular macromolecules that have incorporated the label and therefore have a unique signature. There remains great scope for technical refinements and advancements in the use of stable isotopes for metabolic and molecular analyses of microbial communities; these future methodological innovations will provide the means to mechanistically understand *in situ* function of microbial communities in a quantitative way. With technology developments regarding sensitivity and spatial resolution, this has the potential to provide a direct link between measurements of community function (metabolism) and expression of functional genes (via labeling and analysis of either mRNA or protein). An example of this has been the application of nanosecondary ion mass spectrometry (SIMS) in microbial ecology, which makes possible the determination of chemical and stable isotope composition at the submicron level.

Within the next 10–15 years, single-cell genomics may also be deployed in microbial ecology. Imagine that 300 individual microbes from a habitat are physically separated, and each are subjected to single-molecule genome sequencing. One has in hand a community metagenome in which environmental sequences are inherently placed in their genomic context, and information on the relative abundance of different genomes (and genetic microdiversity within a specific clade) is directly available.

Theory in Microbial Ecology

Technology is not enough. New technologies will generate a blizzard of data at an ever-increasing rate, but understanding can only come from a synthesis of that information. Data-rich technologies carry the risk that they drive science to idiosyncratic pathways that do not yield deep understanding of fundamental principles. The antidote is to ground studies to sound theory, a synthesis of how we imagine microbial populations and communities operate. In comparison to the ecology of plants and animals, microbial systems are much more experimentally tractable, which provides the opportunity to test theoretical syntheses that are either unique to microbial ecology or derived from the rich body of macro-organismal ecological theory.

There are several means to articulate a theoretical synthesis; a simulation model is one of them. This approach has the virtue of forcing a qualitative, descriptive discipline to become a quantitative, predictive one. The challenges are great, particularly at the community level, because functional redundancy will make precise, unambiguous prediction of community composition uncertain. However, robust model predictions may hold at the cellular, population, and ecosystem levels and lead to deeper insights into principles of microbial ecology than presented here. The greatest value of a simulation modeling approach is an iterative interaction with experimentation. The test of a simple model against ecological reality illuminates its weaknesses, and aspects in which greater depth of detail is necessary. Sensitivity analyses of a refined model suggest the most insightful experiments to carry out in the field, which lead to another round of model refinement.

Innovations in Cultivation

The conventional wisdom in microbial ecology is that less than 1% of bacteria in nature can be cultivated. Data on the large discrepancy between direct microscopic counts and viable counts have been available for many decades, but in 1985 this point was highlighted as "the Great Plate Count Anomaly." The development of cultivation-independent analyses of nucleic acids has been a great boon to understanding "who's there" in habitats. However, the future entails understanding the functional potential of individual microbial ecotypes (metabolic versatility) as well as functional redundancy in microbial communities, and molecular approaches have fundamental weaknesses. In particular, metagenomic approaches can uncover novel gene sequences or identify familiar ones, but in the end only allow the formulation of hypotheses that still must be subjected to biochemical analysis for rigorous proof. In contrast, propagation of an isolate under controlled conditions permits application of a broad array of tools to investigate its specific ecophysiology and genetics. It is important to state that the conventional wisdom regarding culturability is no longer true. A variety of techniques that employ substrate concentrations at *in situ* levels or the addition of substances such as signaling molecules or resuscitation factors have been successful in cultivating 10% (and sometimes significantly more) of the direct microscopic count. A concentrated effort in this field could yield quantitative recovery of the overwhelming majority of numerically dominant microbes from common terrestrial and aquatic habitats. In these environments, bacterial clades that have many cultured representatives, such as the Proteobacteria, Actinobacteria, and Firmicutes, are often in high relative abundance.

RECENT DEVELOPMENTS

Technology Advances in Molecular Ecology

The advent of "next generation" DNA sequencing has led to massive increases in the capacity to acquire many thousands to millions of nucleic acid sequence reads at an ever decreasing cost. This trajectory is likely to continue over the next few years, as the subsequent generation of nucleic acid sequencing technology evolves. This has had several important consequences for microbial ecology. (1) The problem of "undersampling" can be addressed. In habitats with thousands of distinct microbial taxa, it is feasible to obtain adequate phylogenetically-related information (16 S rRNA sequences) to detect all but the rarest taxa. If one is interested in the "rare" biosphere, many hundreds of thousands of sequences can be analyzed from a single sample. (2) Adequate replication of ecological experiments can be achieved. For analysis of differences in community composition between habitats (beta diversity) or across environmental gradients, 1000 sequences per sample can provide appropriate statistical rigor. In mid-2010, a single sequencing run could process >400 such samples, at a materials cost of < $50 per sample. The important consequence is that molecular microbial ecologists are no longer bound by costs that have limited them to idiosyncratic, phenomenological studies; they now have the capability to fulfill molecular ecology's promise to provide synoptic and mechanistic understanding. To fulfill this promise will also require innovations in information science – computer algorithms to more rapidly analyze millions of nucleic acid sequences and visualization tools by which microbial ecologists can extract meaning from them, as well as the application of statistical tools to assess the relationship of microbial community structure and/or function to ecosystem processes.

These advances make analyses of "microbial community ecology" (principles 6 and 7) ever more tractable. The ecologist Simon Levin has said that "the most important challenge for ecologists remains to understand the linkages between what is going on at the level of the physiology and behavior of individual organisms and emergent properties such as the productivity and resiliency of ecosystems." From the perspective of microbial ecology, there are several important lines of research that lead in this direction. (1) The flow of material and energy through metabolic networks distributed among functionally redundant microbes within an ecosystem could lead to a predictive understanding of how microbes provide ecosystem services, and how provision was susceptible to environmental perturbations (such as climate change). However, this is a very complex undertaking – fertile soil contains hundreds of different organic molecules and 50,000 microbial taxa. (2) A microbe's "environment" consists not only of local abiotic factors but also other microbes. Complex interactions with other organisms will often play a critical role. These interactions may include competition for resources, positive metabolic interactions (e.g., syntrophy and cross-feeding), production of low-molecular weight signaling or allelopathic chemicals, HGT and coevolution. (3) Microbial communities are complex systems and such systems have properties that emerge from the interactions among components rather than as properties of the individual components themselves. For microbial communities, these properties include taxonomic diversity, functional diversity, and functional redundancy (i.e., multiple taxa may fill in to provide a specific ecosystem function), as well as ecosystem stability and resilience. Given the propensity for HGT across microbial taxa, the collective genetic potential of a microbial community may also be considered an emergent property.

The Human Microbiome

The advances in sequencing technology have recently been applied to microbial communities in or on humans (the human microbiome). Many sites on the body contain microbes; in particular, the number of microbial cells in the gastrointestinal tract may exceed the number of somatic cells in the human body by a factor of 10. These molecular ecology surveys have illustrated the rich diversity of microbes in the human microbiome, but immediately beg the same questions listed above for microbial community ecology. Emerging data indicates that the physiological state and health of animals (including humans) is dependent upon the gastrointestinal tract microbiota. For example, metabolite concentrations in urine, blood, and organs are altered in animals that lack their gastrointestinal flora. Thus, perhaps we should take an ecological view of "super-organism" that comprises not just the animal, but also the associated microflora. There are suggestions that microbial communities, especially the gut microbiota, may play a role in human diseases such as inflammatory bowel disease, diabetes, and obesity. In addition, microbes in the gastrointestinal tract can metabolize a broad variety of chemicals, including pharmaceutical drugs. Perhaps differences in drug efficacy among humans is not only due to their genetics but also to differences in resident gut microbiota. An ecological analysis of the functional properties of the human microbiome will have significant impact upon our understanding of human health and disease over the next decade.

FURTHER READING

Azam, F., & Malfatti, F. (2007). Microbial structuring of marine ecosystems. *Nature Reviews Microbiology*, *5*, 782–791.

Battin, T. J., Sloan, W. T., Kjelleberg, S. *et al.* (2007). Microbial landscapes: New paths to biofilm research. *Nature Reviews Microbiology*, *5*, 76–81.

Brune, A., Frenzel, P., & Cypionka, H. (2000). Life at the oxic-anoxic interface: Microbial activities and adaptations. *FEMS Microbiology Reviews*, *24*, 691–710.

Button, D. K. (1998). Nutrient uptake by microorganisms according to kinetic parameters from theory as related to cytoarchitecture. *Microbiology and Molecular Biology Reviews*, *62*, 636–645.

Fenchel, T., King, G. M., & Blackburn, H. (1998). *Bacterial Biogeochemistry: The Ecophysiology of Mineral Cycling*. San Diego: Academic Press.

Konopka, A. (2009). What is microbial community ecology? *ISME Journal*, *3*, 1223–1230.

Madsen, E. L. (2005). Identifying microorganisms responsible for ecologically significant biogeochemical processes. *Nature Reviews Microbiology*, *3*, 439–446.

O'Donnell, A. G., Young, I. M., Rushton, S. P., Shirley, M. D., & Crawford, J. W. (2007). Visualization, modelling and prediction in soil microbiology. *Nature Reviews Microbiology*, *5*, 689–699.

Pernthaler, J., & Amann, R. (2005). Fate of heterotrophic microbes in pelagic habitats: Focus on populations. *Microbiology and Molecular Biology Reviews*, *69*, 440–461.

Prosser, J. I., Bohannan, B. J. M., Curtis, T. P. *et al.* (2007). Essay – The role of ecological theory in microbial ecology. *Nature Reviews Microbiology*, *5*, 384–392.

Schmidt, T. M. (2006). The maturing of microbial ecology. *International Microbiology*, *9*, 217–223.

Sorensen, S. J., Bailey, M., Hansen, L. H., Kroer, N., & Wuertz, S. (2005). Studying plasmid horizontal transfer *in situ*: A critical review. *Nature Reviews Microbiology*, *3*, 700–710.

Sterner, R. W., & Elser, J. J. (2002). *Ecological Stoichiometry: The Biology of Elements from Molecules to the Biosphere*. Princeton, New Jersey: Princeton University Press.

Suttle, C. A. (2005). Viruses in the sea. *Nature*, *437*, 356–361.

Whitman, W. B., Coleman, D. C., & Wiebe, W. J. (1998). Prokaryotes: The unseen majority. *Proceedings of the National Academy of Sciences of the United States America*, *95*, 6578–6583.

Chapter 2

Microbial Food Webs

E.B. Sherr and B.F. Sherr
Oregon State University, Corvallis, OR, USA

Chapter Outline

ABBREVIATIONS

CFB Cytophaga–Flavobacteria–Bacteriodes
DMS Dimethyl sulfide
GGE Gross growth efficiency
NPZ Nutrient–phytoplankton–zooplankton
TEP Transparent exopolymer particles

DEFINING STATEMENT

In natural ecosystems, microbial food webs consist of predator–prey interactions of unicellular prokaryotes and eukaryotes. In this article, we focus on the structure, and ecological and biogeochemical importance of microbial food webs in aquatic ecosystems, particularly in the oceans.

INTRODUCTION

In his book *The Ecological Theater and the Evolutionary Play*, limnologist G. Evelyn Hutchinson proposed that the environment provides the stage for the drama of evolution of species. If so, microbes are the stage hands, ceaselessly building and remolding the set. For a long time in Earth's history, microbes were the only actors as well, carrying out in microscale the script of production, predation, and dissolution; in other words, microbial food webs. Microbial food webs are similar in some ways to the familiar lynx and hare predator–prey interactions of nature shows. In other ways, they are not. A notable difference is that microbial food webs have an enormously important role in decomposing the plant carbon and all of the other components feces, dead bodies, and so forth, of the macroscopic world.

Most microbes are single prokaryotic or eukaryotic cells, although some form either filamentous chains or colonies of single cells, and fungi producing multicellular fruiting bodies. Microbes form multispecies communities, and thus food webs, throughout the biosphere, including some habitats where multicellular life cannot exist. In natural systems, a large proportion of prokaryotic, or bacterial cells present in an environment may be relatively inactive, or dormant, but able to start growing when conditions are favorable. Unicellular eukaryotes, or protists, may also form resting stages. Ecologists refer to the total complement of microbes, both active and inactive, in a habitat as the microbial assemblage. The microbial community refers to the subset of microbes that are actively growing and metabolizing at any one time.

Microbial food webs are organized into trophic levels, or compartments, depending on their function. Primary producers make up the first level, or bottom, of a food web. Decomposing organisms, heterotrophic bacteria, and fungi grow on nonliving organic matter. Phagotrophic protists can feed on single or multiple compartments of a food web, depending on the consumers' size and feeding capability. The combined activities of microbial communities result in large-scale cycling of bioactive elements – carbon, nitrogen, phosphorus, sulfur, and trace metals – in ecosystems.

Topics in Ecological and Environmental Microbiology.

UNDERSTANDING MICROBIAL FOOD WEBS

The concept of systems ecology was crucial to understanding the role of microbes in ecosystems. The text *Fundamentals of Ecology* by Eugene and Howard Odum, first published in 1953, established the systems approach, which focused on ecosystem function rather than on specific populations, and followed the flows of elements such as carbon or nitrogen, or of energy, either solar energy or energy from the respiration of organic compounds, through food web compartments. Initial formulations were linear food chains, from primary producers, which captured solar energy and produced organic matter from carbon dioxide and other inorganic compounds, to primary and secondary consumers; for example, grass to antelopes to lions. Decomposing organisms, heterotrophic bacteria and fungi, were known to be important components of ecosystems, but did not comfortably fit into such a food chain. Early studies of the roles of heterotrophic microbes in ecosystems focused on the rates of respiration of organic carbon to carbon dioxide and on regeneration of bioactive elements from organic compounds into inorganic compounds such as ammonium and phosphate, which algae and higher plants could use for growth.

The notion of trophic interactions between different groups of microbes was developed during research on marine ecosystems during the 1970s and 1980s. Microbial ecologists found that microbial plankton were responsible for the bulk of respiration in the sea. They also highlighted the role of small flagellated protists as consumers of bacteria in seawater. Bacterivorous flagellates kept bacterial stocks in check and at the same time regenerated much of the nitrogen and phosphorus accumulated in bacterial cells. Subsequent work showed that the bacterivorous flagellates were in turn grazed by larger-sized protists, establishing a microbial food chain that resulted in virtually all of the organic carbon used by bacteria being recycled back to carbon dioxide and inorganic nutrient compounds. Similar microbial food chains, from bacteria to flagellates to larger protists, were found in freshwater ecosystems and in benthic habitats in marine and freshwater environments.

Further research on microbes in aquatic ecosystems showed that this microbial food chain, termed the microbial loop, was too simplistic. Both small flagellates and larger protists also fed on autotrophic cells, including photosynthetic bacteria, and on algae of all sizes. Mixotrophic phytoflagellates could ingest bacteria. Viruses infected and lysed both bacterial and algal cells, causing a short circuit of the microbial loop. A more sophisticated view of a complex microbial food web that included autotrophic, mixotrophic, and heterotrophic microbes and formed the basic food resource for metazoans such as copepods and larvae of pelagic and benthic animals emerged. Microbial food webs in terrestrial systems are simpler in that primary production is carried out by large multicellular plants, and microbes are limited to either decomposition of plant material or feeding on decomposing microbes. In this article, we will focus on the microbial food webs of aquatic ecosystems, particularly of the ocean.

COMPONENTS AND PATHWAYS

Aquatic microbial food webs consist of producer and consumer compartments. Examples of microbial producers and consumers in aquatic food webs visualized via epifluorescence microscopy are shown in Figure 1. Epifluorescence microscopy is a method used by microbial ecologists to inspect cells that are either autofluorescent by virtue of their pigments, such as chlorophyll or phycobilins or made fluorescent by staining with dyes that bind to organic compounds in microbial cells and fluoresce at selective wavelengths. In Figure 1, autofluorescent cells fluoresce red, while heterotrophic prokaryotes and phagotrophic protists fluoresce blue due to added DAPI stain.

Microbial species also include a large size range (Figure 2). Most planktonic bacteria are about half a micron in size. The largest-sized phytoplankton and protists are 100–200 μm in length. Cell size is important in microbial food webs since most, although not all, phagotrophic protists feed on organisms smaller than themselves. The smallest microbial cells are less than 2 μm in diameter. This size category, which includes most aquatic prokaryotes, heterotrophic, and autotrophic Bacteria and Archaea, and the smallest eukaryotic cells, is termed picoplankton. Nanoplankton, cells 2–20 μm in size, includes most species of flagellates, autotrophic, heterotrophic, and mixotrophic, along with some smaller-sized nonflagellated green algae and diatoms and the smallest species of dinoflagellates and ciliates. Microplankton, cells and chains of cells 20–200 μm long, covers the larger-sized phytoplankton, mainly single cells and chains of diatoms and larger species of photosynthetic dinoflagellates, and the larger-sized phagotrophic protists, ciliates, and heterotrophic dinoflagellates. Phagotrophic protists in the plankton greater than about 20 μm are termed microzooplankton and are major consumers, or grazers of phytoplankton in marine and freshwater systems. Viruses that occur in all aquatic systems and are less than 0.2 μm, or 200 nm, are categorized as femptoplankton.

At the bottom level of microbial food webs are the primary producers, usually photosynthetic bacteria and algae. However, in hypoxic habitats in the water column and the benthos, primary producers may be chemosynthetic bacteria that are able to gain energy by the oxidation of reduced chemicals such as sulfide, reduced iron, or methane. Such bacteria form the base of food webs at hot springs on land and at hydrothermal vents and methane seeps on the seafloor.

Trophic level in aquatic food webs is not easily segregated according to the size of the microorganism, since

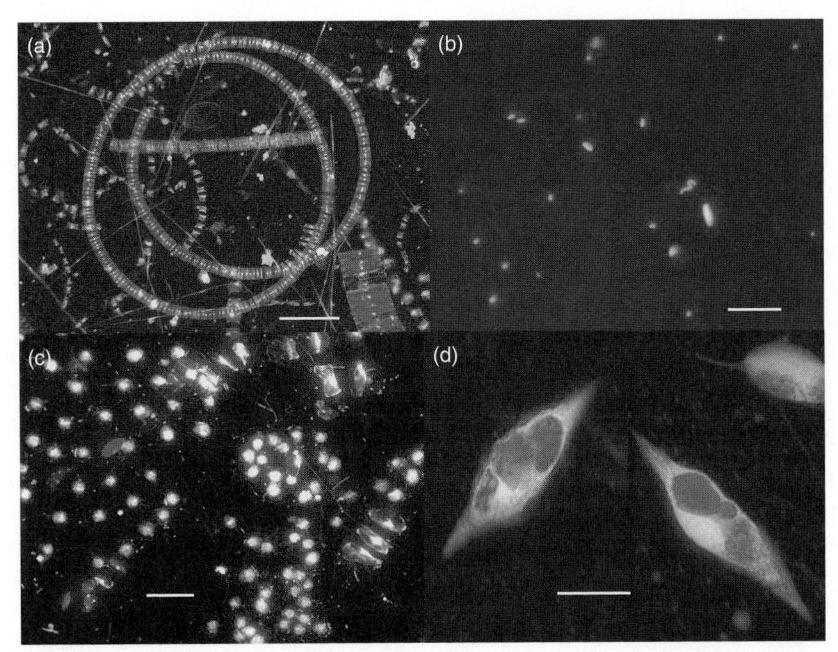

FIGURE 1 Examples of autotrophic and heterotrophic microbes in microbial food webs, visualized by epifluorescence microscopy with DAPI staining and sized using an image analysis system consisting of a Cooke Sensicam QE CCD camera with Image Pro Plus software mated to an Olympus BX61 Microscope with a universal fluorescence filter set. Red color indicates autofluorescence of photosynthetic pigments, blue color indicates DAPI staining of nonfluorescent cells and cytoplasm; the brightest blue staining occurs in the nucleus of the cells. (a) A mixed species bloom of diatoms, major algal producers in aquatic ecosystems, in the western Arctic Ocean, scale bar = 50 μm. (b) Planktonic prokaryotes stained with DAPI in water collected in a Georgia salt marsh estuary, scale bar = 2 μm. (c) Bacterivorous nanoflagellates, probably choanoflagellates, in a decaying diatom bloom in the western Arctic Ocean, scale bar = 10 μm. (d) Two heterotrophic gyrodinium-type dinoflagellates with red-fluorescent food vacuoles full of picoplankton-sized autotrophic cyanobacteria and picoeukaryotes from the western Pacific Ocean off the coast of Oregon, USA, scale bar = 20 μm.

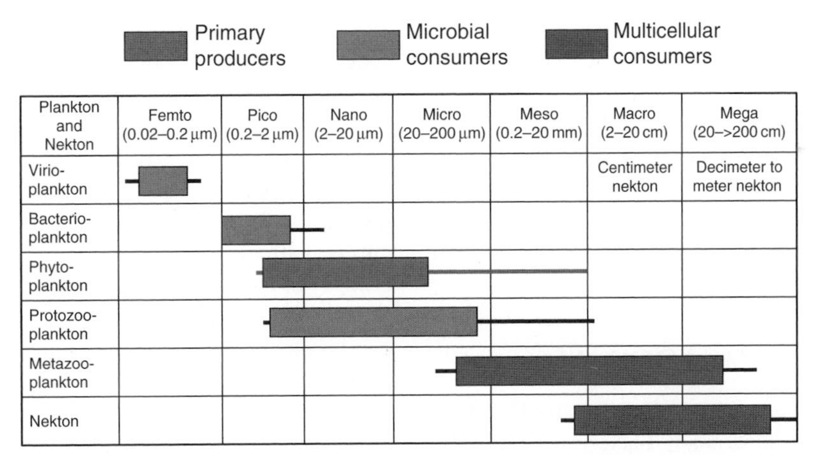

FIGURE 2 Distribution of different taxonomic-trophic compartments of plankton in a spectrum of size fractions, with a comparison of size ranges of zooplankton and nekton. Solid rectangles denote size of most organisms in each size group, bars denote approximate minimum/maximum size range of group. Blue bars, heterotrophic microbes; green bar, autotrophic microbes (phytoplankton); purple bars, animals. Figure is updated from a figure published by John Sieburth and colleagues in an article in Limnology and Oceanography in 1978. (For interpretation of the references to color in this figure legend, the reader is referred to the Web version of this chapter.)

individual species of both heterotrophic and autotrophic microbes occur across the entire range of microbial size categories. In the open ocean and in large lakes, less than 2-μm-sized coccoid-shaped cyanobacteria can be important primary producers. In coastal and shallow water systems, massive blooms of microplankton-sized algae, either diatoms or dinoflagellates, can occur. To make matters more complicated, some species of eukaryotic microbes in all size ranges are mixotrophic. Species of autotrophic flagellates, or phytoflagellates, in both picoplankton and nanoplankton size ranges are capable of ingesting heterotrophic bacteria and even small phototrophic cells. It is likely that most photosynthetic dinoflagellates are also phagotrophic, preying on both autotrophic and heterotrophic protists. Some ciliates temporarily hold chloroplasts from their algal prey just below the cell membrane and use sugars produced by the chloroplasts for supplemental nutrition. One species of marine ciliate, *Myrionecta rubra*, has taken this form of mixotrophy to the extreme. Its captured chloroplasts have retained DNA from the original cryptophyte algal prey and are capable of division along with the host ciliate. *Myrionecta*, unlike other ciliates, is thus primarily autotrophic and can even form blooms in the ocean, although the ciliate is still capable of phagotrophy. There are other types of symbiotic mixotrophy in which one type of microbe lives in or on, and contributes to the metabolism of, another microbe. Examples include nitrogen-fixing cyanobacteria that live in diatom and heterotrophic dinoflagellate host cells, helping them survive in nitrogen-poor environments, and chemosynthetic sulfur-oxidizing Bacteria that live on the cell membranes of benthic ciliates, which essentially farm the chemosynthetic bacteria as their main food resource.

Although trophic interactions between microbes in aquatic food webs are more complicated than trophic interactions in macroscopic food webs, the cast of characters in microbial food webs is less so. Most water column, or pelagic food webs in both the ocean and lakes have a consistent assortment of major groups of microbes with similar trophic roles. Taxonomic groups of microbes appear to be much more uniformly distributed in the ocean, and in lakes, compared to the distribution of species of, for example, copepods and fish, in aquatic systems. Of course, this apparent uniformity may only be a result of the lack of knowledge of genetic differences among strains of distinct microbial species in different habitats. The general taxonomic groups of microbes in aquatic food webs, and their functions, are listed below.

MICROBES IN AQUATIC FOOD WEBS

Heterotrophic prokaryotes: Most of these are less than 1-μm-sized species in the domain Bacteria and live by assimilating dissolved organic compounds from water or by degrading nonliving detrital organic matter. Species in the domain Archaea are also present everywhere in the sea and in freshwater habitats. There are four groups of Archaea in the marine pelagic environment; the most abundant of these are Marine Group 1 Archaea in the Crenarchaea, the same subdomain as sulfur-oxidizing Archaea living in hot springs or in hydrothermal vents. The other marine Archaea (Marine Groups 2, 3, and 4) are in the Euryarchaeota and include methanogenic and halophytic prokaryotes. Very little is known about the modes of metabolism of most of the marine Archaea. Some Marine Group I Archaea have been found to assimilate amino acids. However, archaeal cells are most abundant in the sea at depths of 200–4000 m, where there is little organic carbon. It has been established that the cold-temperature Crenarchaeota present in the ocean can gain energy for growth by oxidizing ammonium. A member of the Marine Group I Archaea isolated from a marine aquarium tank has been shown to grow chemoautotrophically in culture by oxidizing ammonium and assimilating carbon dioxide.

Microscopic and flow cytometric methods used to enumerate heterotrophic prokaryotes in aquatic systems are based on fluorescent staining of cells, which does not distinguish between species of Bacteria and Archaea. For this reason, the general term bacteria, which has long been used by microbial ecologists to mean heterotrophic prokaryotes, is assumed to include cells in both domains. Bacterioplankton refers to heterotrophic prokaryotic cells suspended in seawater or freshwater.

Most strains of aquatic bacteria are in the bacterial phylum Proteobacteria. In the sea, marine bacteria are typically strains of alpha-Proteobacteria, which includes the most abundant open ocean ribotype, the SAR-ll clade, and of gamma-Proteobacteria, which includes fast-growing opportunistic strains of the genera *Pseudomonas* and *Vibrio*. In freshwater habitats and in some coastal and estuarine environments, strains of beta-Proteobacteria are abundant. In eutrophic and benthic habitats, heterotrophic Bacteria grow on surfaces, forming colonies and biofilms that are an important food resource for small invertebrates. Bacteria adapted to attach to and grow on surfaces are phylogenetically diverse, and commonly include ribotypes in the *Cytophaga–Flavobacteria–Bacteriodes* (CFB) group as well as in other groups of Bacteria.

Autotrophic prokaryotes: Coccoid cyanobacteria, photosynthetic Bacteria less than 2 μm in diameter; are ubiquitous in marine and freshwater systems. There are two major groups of these picocyanobacteria: orange-fluorescing *Synechococcus* spp., which have chlorophyll *a* and phycobiliprotein accessory pigments, and red-fluorescing *Prochlorococcus* spp., which have modified chlorophyll pigments, divinyl chlorophyll *a* and divinyl chlorophyll *b* as the main accessory pigment. *Prochlorococcus* spp. are smaller in size than *Synechococcus* spp., and are typically abundant in open ocean habitats, while *Synechoccocus* spp. are most abundant in nearshore to outer continental shelf waters. *Synechococcus* spp. are also abundant in the water column of lakes, while *Prochlorococcus* spp. are predominantly marine. Filamentous cyanobacteria are common in polluted freshwaters and hot springs. In the ocean, the filamentous cyanobacteria *Trichodesmium* spp. form blooms in subtropical regions and are globally important nitrogen fixers. Nonoxygen-producing, bacteriochlorophyll-containing Bacteria, which require a source of reduced compounds to grow, can be significant primary producers in lakes and marine systems that have subsurface anoxic water masses rich in sulfide. In addition, many strains of heterotrophic bacteria living in oxic aquatic habitats have been found to contain either bacteriochlorophyll or bacteriorhodopsin pigments, which may be used to generate extra ATP by using energy harvested from light.

In benthic habitats, chemosynthetic Bacteria can support both microbial and macroscopic food webs. These autotrophs include free-living single-celled and filamentous strains of sulfur-oxidizing Bacteria, such as species of *Beggiatoa* and *Thiospirillum*, and symbiotic sulfur-oxidizing Bacteria living in or on both single-celled and multicellular eukaryotes. Examples are the sulfur-oxidizing Bacteria that grow on the benthic ciliate *Zoothamnium niveum*, which lives at the oxic–anoxic interface in sandy marine sediments and provides its bacterial crop with both sulfide from below and oxygen from above, and the symbiotic sulfur-oxidizing Bacteria that grow in special organs of gutless hydrothermal vent tube worms, providing the worms with all of their nutrition. At methane seeps on the seafloor, methane-oxidizing Bacteria grow in the sediments and in the gill tissues of seep mussels, providing most of the primary production for these ecosystems.

Autotrophic eukaryotes: Also termed algae, single-celled photosynthetic eukaryotes are the most significant primary producers both in the sea and in lakes. Algal cells are diverse both in size and in taxonomic diversity. The smallest algal cells are 0.8 µm-diameter marine *Ostreococcus* spp. and 1–2 µm diameter *Micromonas* spp, both abundant in the open ocean. There is a great diversity of algal species in the nanoplankton size range. Most of these are golden brown-pigmented, flagellated chrysophytes and prymnesiophytes, orange-pigmented cryptophytes, and green-pigmented prasinophytes, although nonflagellated chlorophytes and diatoms also occur in this size range. Algae larger than 20 µm are less abundant than smaller-sized phytoplankton, but at times form dense blooms in coastal waters or in lakes. Bloom-forming algae greater than 20 µm are typically diatoms or autotrophic dinoflagellates. Many flagellated algae, including chloroplast-bearing nanoflagellates and dinoflagellates, can ingest other microbial cells. Some species of algae capable of ingesting bacteria cannot grow in the absence of prey.

Heterotrophic eukaryotes: These microbes were known as protozoa, and researchers still use that term often. However, many species of heterotrophic eukaryotes are close kin to photosynthetic species. The word Protozoa, which means first animal life in Greek, is thus not an appropriate label for these microbes, and we prefer to use the term heterotrophic protist. Heterotrophic protists, which do not have chloroplasts, are as ubiquitous and as diverse as autotrophic protists, the algae. There are some protist lineages: the bodonids, the choanoflagellates, and the kinetoplastids, that do not have any chloroplast-containing species, and are strictly phagotrophic. The choanoflagellates are of particular interest to molecular geneticists as they are the group of single-celled protists most closely related to multicellular animals. Phagotrophic protists have the potential to be major predators in microbial food webs because they are in the same general size range as their microbial prey, bacteria, algae, and other heterotrophic protists (Figure 2), and because protist growth rates are on the same temporal scale, hours to days, as those of their prey. The high rate of metabolism of these small, unicellular predators also facilitates carbon and energy flux through ecosystems.

The smallest heterotrophic protists are 1- to 2-µm-sized flagellated species, which occur in several protist groups, including the chrysophytes and bodonids. Heterotrophic protists 2–20 µm in size, mainly nanoflagellates (e.g., Figure 1c), are very diverse taxonomically and are major consumers of picoplankton and smaller-sized nanoplankton cells in aquatic systems. Some species of ciliates and heterotrophic dinoflagellates are also less than 20 µm. Phagotrophic protists larger than 20 µm are predominately ciliates and nonchloroplast-containing dinoflagellates, Examples of these protists are shown in Figure 3. This size class of phagotrophic protist, termed microzooplankton, is abundant in the sea and in lakes, and these protists are major consumers of phytoplankton and of heterotrophic nanoflagellates. Planktonic ciliates are mainly spherical or conical spirotrichs, cells with cilia grouped around an oral end. One subgroup of spirotrichous ciliates, the tintinnids, build species-specific houses, or loricae, in which they live, serving as a protective shelter (Figure 3b). Heterotrophic dinoflagellates may have rigid cellulosic plating, or armor, which gives them a distinctive shape (e.g., the cell in Figure 3d) and makes it difficult, though not impossible, for them to ingest prey cells directly. Many armored dinoflagellates instead feed externally by extruding a hollow tube into, or a pseudopodial veil around, their algal prey. The dinoflagellate injects digestive enzymes into the prey cells and sucks the digested prey cytoplasm back into itself by these feeding structures. Most species of heterotrophic dinoflagellate are nonarmored, and have an elastic cell membrane that allows them to ingest algal prey up to a size equal to, or even greater than, the dinoflagellate, often greatly distending the dinoflagellate cell in the process (Figure 3e and 3f).

In benthic habitats, where the fluid environment interacts with surfaces such as grains of sand, clay particles, or organic detritus, the protist community is dominated by surface-feeding hymenostome ciliates similar to *Paramecium* spp., whose main food is bacteria. Some of these are very long and thin, adapted to move through narrow spaces between sediment particles, and some have stiff ventral cilia that allow them to scrape bacterial biofilms off particle surfaces. Amoebae and amoeboid flagellates are also common in benthic environments and on detrital particles in the

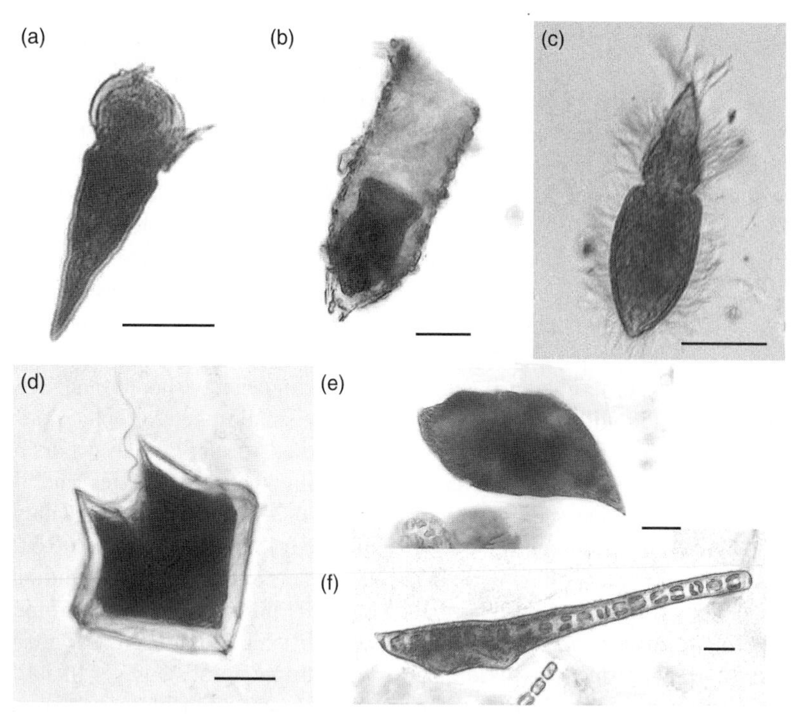

FIGURE 3 Examples of aquatic phagotrophic protists in the 20–200 μm, or microzooplankton, size class. Cells were preserved and stained with iodine-based acid Lugol solution, which gives them a brown color, visualized via light microscopy, and sized and photographed using an image analysis system consisting of a Cooke Sensicam QE CCD camera with Image Pro Plus software mated to an Olympus BX61 Microscope. (a) *Strombidium* sp. ciliate, consumer of nanoplankton-sized microbial prey from the western Arctic Ocean; (b) *Tintinnopsis* sp. tintinnid ciliate in an aggregated lorica, consumer of nanoplankton-sized prey from the western Arctic Ocean; (c) Haptorid ciliate, species unknown, a predatory ciliate that feeds on cells as large as itself from the western Arctic Ocean; (d) *Protoperidinium* sp. armored heterotrophic dinoflagellate that feeds on large cells including diatoms using an extracellular pseudopodial feeding veil from the Oregon upwelling system of the western Pacific Ocean; note flagellum trailing behind the cell; (e) *Gyrodinium* sp. nonarmored heterotrophic dinoflagellate associated with a diatom bloom in the western Arctic Ocean; (f) *Gyrodinium* sp. nonarmored heterotrophic dinoflagellate distended from an ingested chain of the diatom *Thalassiosira* sp., associated with a diatom bloom in the Oregon upwelling system. All scale bars = 20 μm.

water column. Phagotrophic protists are important components of microbial food webs in extreme habitats such as sea ice, solar salterns with salinities as high as 150 parts per thousand, and the deep ocean.

MARINE PELAGIC HABITATS

The oceans, which cover more than two-thirds of the earth's surface, provide large and variable habitats for microbial food webs. In the euphotic zone, the upper lighted depths of the sea, approximately from the surface to 50–200 m depending on the region and season, photosynthetic microbes, both prokaryotic and eukaryotic, provide a continuous source of organic carbon and prey cells for heterotrophic bacteria and protists. At subeuphotic depths, where there is insufficient sunlight for photosynthesis, heterotrophic microbes live on organic carbon sinking down from the euphotic zone. At these depths, the microbial food web regenerates most of the sinking organic carbon back to carbon dioxide and inorganic nutrients.

The largest oceanic habitats are the slowly rotating gyres at the center of each of the great ocean basins. These subtropical open ocean regions have warm surface waters and low amounts of inorganic nutrients for phytoplankton growth. They are oligotrophic or nutrient limited. Phytoplankton cells are small sized, dominated by cyanobacteria and small algal species, and the total biomass of primary producers is low. Microbial food webs are complex, with small heterotrophic protists and mixotrophic flagellated algae ingesting heterotrophic bacteria and cyanobacteria, slightly larger protists feeding on nanoalgae and on bacterivorous flagellates, and the largest protists feeding on algal cells and on heterotrophic protists. Most of the primary production is respired in this multicompartment microbial food web, with little remaining for macroscopic marine life.

At the edge of continents, the oceans cover continental shelves of varying depths and widths. Continental shelf waters are dynamic, and occur in subtropical, temperate, and polar regions. Most shelf systems are mesotrophic, with higher biomass of phytoplankton and of other microbes compared to the oligotrophic gyres. Upwelling or injection of nutrient-rich subsurface water frequently occurs at the edge of the shelf, resulting in eutrophic conditions with high phytoplankton biomass and production. Some regions, such as the narrow continental shelves of the Pacific Northwest of the United States, Northwest Africa, and Peru and Chile in South America, have seasonal or persistent upwelling of nutrient-rich seawater extending to the coast, which results in massive phytoplankton blooms, usually of species of diatoms. Phytoplankton blooms also occur each spring in temperate and polar regions due to increase in day length from winter to summer. Much of the organic matter produced in these mass phytoplankton blooms sinks down to subsurface depths or to the sediment. Some of the bloom production is utilized by heterotrophic bacteria and by microzooplankton-sized protists capable of preying on large algal cells and chains of cells (e.g., the dinoflagellate in Figure 3f).

At the margins of the oceans are shallow nearshore and estuarine habitats influenced by inputs of freshwater from rivers or land runoff and by interactions between the water column and the sediments. Depending on their geographic location and local conditions, nearshore and estuarine systems may be oligotrophic, mesotrophic, or eutrophic. Agriculture and other human activity has resulted in a large increase in the amount of plant nutrients such as nitrate and phosphate carried by rivers and by land runoff to nearshore marine systems. As a result of enhanced phytoplankton growth due to these nutrients, microbial food webs in some nearshore regions respire enough organic matter to deplete the water of oxygen, resulting in hypoxic or anoxic dead zones. A classic example is the large and growing zone of low oxygen water off the Mississippi River in the Gulf of Mexico. Hypoxic and anoxic marine habitats also occur naturally beneath persistent upwelling regions and in some enclosed fjords and basins, for example, the Black Sea.

Most of the pelagic habitat of the ocean is subsurface, below 200 m. Only a small fraction of primary production in the euphotic zone sinks out to depths below 200 m. Microbes that live below that depth are much less abundant than in the euphotic zone and must adapt to highly stressful conditions of low food, cold temperatures, and high pressure. Both Bacteria and Archaea, as well as heterotrophic flagellates, occur at depth in the sea. Cold-temperature Crenarchaeota are relatively more abundant compared to Bacteria in this habitat. Currently, not much is known about how deep ocean bacteria survive, or about deep ocean microbial food webs.

BENTHIC HABITATS

Microbial food webs in aquatic sediments are shaped by the characteristics of benthic habitats. In sediments, microscale structure results in large changes in environmental conditions, for example, redox potential, oxygen and nutrient concentrations, over millimeter to centimeter spatial scales. Particles, both inorganic and organic, are a dominant component of the benthic environment; thus, interactions with particles, for example, attachment to, and grazing on, surfaces, are of major importance for microbes living in and on sediments. Benthic ecosystems are mainly heterotrophic; the microbial food web is based on input of organic matter that settles to the bottom as a result of primary production carried out in the water column. Exceptions are hydrothermal vents and methane seeps, where chemosynthetic sulfur-oxidizing Bacteria or methanotrophic Bacteria form the basis of local food webs. Where the input of organic matter to sediments is high relative to the supply of oxygen, suboxic/anoxic habitats dominate in the benthos. Depth zones in the sediment reflect sequential utilization of compounds as electron acceptors with depth: oxygen, nitrate, sulfate, and carbon dioxide. Food webs in anoxic environments are shorter compared to food webs in oxic environments due to lower growth efficiencies of anaerobic microbes.

The fine-scale habitat differentiation in benthic habitats, in terms of water chemistry, oxidizing/reducing conditions, and sediment texture, yields a high diversity of potential niches for microbes. Since particle surfaces predominate in the benthos, microbial biofilms are prevalent on both organic detrital particles and inorganic grains of sand or clay. Microbial exopolymers, high molecular weight polysaccharide or mucopolysaccharide secretions, are copiously made by benthic bacteria and microalgae such as diatoms. Exopolymers create a microenvironment around a microbial cell, buffering it from rapid environmental changes in pH, salinity, dessication, or nutrient regimes.

The depth to which oxygen is present in sediments depends both on sediment composition: loose sandy sediments allow oxygen to penetrate to a greater depth compared to compact clay sediments, and on the amount of organic matter reaching the sediments: the more organic matter, the greater the rate of oxygen utilization. In coastal waters, usually only the sediment–water interface and the upper few millimeters or centimeters is near oxygen saturation. Oxygen is supplied mainly by diffusion from the overlying water; however, in shallow sediments, some oxygen may be provided by microalgal photosynthesis, and in all sediments, bioturbation by invertebrates results in local oxidizing zones in the top few centimeters, or deeper, in the sediment.

In marine unperturbed sediments there is a standard sequence of redox zones, compounds used as electron acceptors, and associated metabolic processes of microbes with depth (Figure 4). In the upper layer of the sediment in contact with overlying waters, respiration of oxygen by prokaryotes and protists occurs. Where overlying water is rich in nitrate, anaerobic nitrate respiration to nitrite by prokaryotes and some protists, or denitrification to nitrogen gas by denitrifying Bacteria, dominates when oxygen concentration is depleted. Deeper in the sediment, both oxygen and nitrate are exhausted, but the interstitial seawater is still rich in sulfate. In this zone, sulfate-respiring Bacteria grow on hydrogen and fatty acids produced by anaerobic fermenting

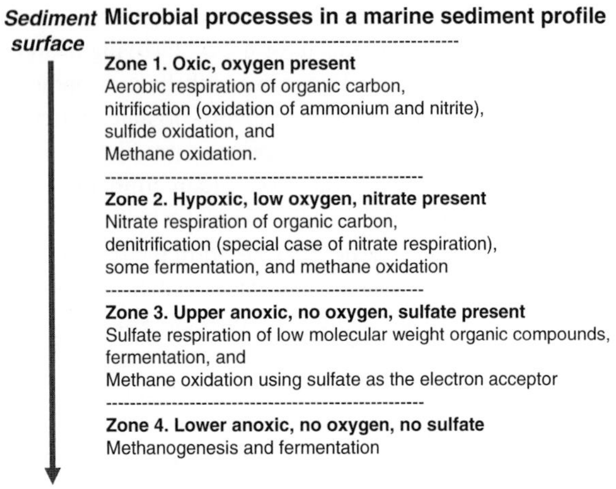

FIGURE 4 Sequence of redox zones and associated microbial processes with depth in an idealized sediment profile. Moving down from the sediment surface the sequence is as follows: (Zone 1) oxic zone: high oxygen concentration at the sediment surface; aerobic respiration, sulfide oxidation, nitrification, and methane oxidation; (Zone 2) hypoxic zone: low oxygen and measurable nitrate concentration; anaerobic nitrate-based respiration, denitrification, some fermentation, and methane oxidation; (Zone 3) upper anoxic zone: no oxygen but sulfate present, sulfate respiration, sulfide formation, fermentation, and methane oxidation using sulfate as the electron acceptor; (Zone 4) lower anoxic zone: no oxygen or sulfate, methanogenesis using carbon dioxide as the electron acceptor and fermentation.

microbes. The end product of sulfate respiration is sulfide, which builds up in anoxic marine habitats, producing a characteristic rotten egg smell. Still deeper, sulfate is depleted and microbial metabolism is mainly based on methanogenesis by Archaea and fermentation by Bacteria and protists.

In the anoxic zones of both marine sediments and oxygen-depleted water masses, sulfate-respiring Bacteria and methanogenic Archaea compete for the metabolites of fermenting bacteria: hydrogen and low molecular weight organic compounds, particularly acetate. Sulfate-reducing Bacteria are better competitors and can grow at lower hydrogen concentrations than can methanogens. Thus, in marine anoxic habitats, sulfate reducers outcompete methanogens in zones where there are significant concentrations of sulfate. Because sulfate respirers and methanogens can utilize only low molecular weight organic substrates, fermenting microbes are primarily responsible for degradation of particulate detritus and high molecular weight organic compounds in anaerobic sediments.

Two major chemoautotrophic processes occur in marine sediments and water columns where there is an interface between oxic and anoxic habitats. Nitrification occurs when oxygen and ammonium are present together. The two phylogenetically distinct groups of nitrifying Bacteria produce energy for carbon fixation by the respective oxidation of ammonium to nitrite and nitrite to nitrate. When oxygen and sulfide are present together, sulfur-oxidizing Bacteria produce energy for carbon fixation by oxidation of reduced sulfur compounds. This is the major source of fixed carbon at hydrothermal vents.

The microbial food web of most marine sediments and anoxic water masses consists mainly of detrital organic matter consumed by heterotrophic prokaryotes, which in turn are consumed by heterotrophic protists. However, in coastal and intertidal marine sediments where there is sufficient light, benthic algae are a component of food webs. Ciliates are abundant in benthic habitats and consume both bacteria and algae; the roles of heterotrophic flagellates and amoebae in sediments are less well known.

Anaerobic metabolism is inherently less energetically efficient than is aerobic metabolism. Fermenting organisms typically grow with a gross growth efficiency (GGE) of substrate use of 10%, while aerobic microbes can transform 40% of assimilated organic carbon into biomass. Any process that serves to enhance the low growth efficiencies of anaerobic organisms would give a competitive edge to such organisms. Anaerobic ciliates are characteristic of both marine and freshwater anoxic habitats. These protists generate energy by fermentation of organic compounds obtained by ingesting other microbes, primarily bacteria. In these ciliates, the fermentative processes resulting in oxidation of pyruvate and production of hydrogen occur in unique organelles, hydrogenosomes. Hydrogenosomes appear to be modified mitochondria that have lost the electron transport system. Many species of anaerobic ciliates are full of endosymbiotic prokaryotes. When excited by blue light, the cells fluoresce blue-green, a characteristic of methanogenic Archaea. The observation of methane generation in these protists, along with molecular genetic analysis of the endosymbionts, has confirmed that they are in fact methanogens. The cytoplasm of the fermenting ciliate is a microhabitat with high abundance of hydrogen, acetate, and carbon dioxide – waste products of the host and substrates for methanogens. This is of particular significance in marine anoxic habitats, where high concentrations of seawater sulfate foster the growth of anaerobic sulfate-respiring Bacteria, which outcompete nonsymbiotic methanogens for available hydrogen and fatty acids. In turn, the endosymbiotic methanogens make the metabolism of the ciliate more efficient by decreasing fermentation waste products in the cell and by serving as a food resource for the ciliate. This unique microbial collaboration is a classic case of syntropy, literally feeding together, in which two organisms grow in a mutually beneficial, intimate association.

ROLE OF MICROBIAL FOOD WEBS IN BIOGEOCHEMICAL CYCLING

Microbes can be viewed as the chemical engineers of the biosphere. Biogeochemical cycles of carbon, nitrogen, and sulfur cannot occur without specific metabolic capabilities of various groups of microorganisms. For many of the cycles of bioactive elements, interactions of both prokaryotic and eukaryotic species in microbial food webs are required for completion of the pathways of the elements. Conversion of organic carbon, organic nitrogen, and organic phosphorus into inorganic compounds, namely, carbon dioxide, ammonium, and phosphate, is facilitated by consumption of prokaryotic and eukaryotic prey cells by phagotrophic protists.

The dominant degradation pathway of organic matter produced by autotrophic microbes, that is, primary production, in aquatic ecosytems is assimilation and respiration by heterotrophic prokaryotes, both Bacteria and Archaea. Prokaryotes utilize organic matter in many forms and steps in aquatic food webs: in the water column as dissolved organic matter released by growing algae, as nonliving particulate organic matter, or organic detritus, produced during decaying phytoplankton blooms and as waste products of protist and metazoan consumers, and in sediments from the settling of organic particles.

Part of the primary production assimilated by prokaryotes is regenerated via cellular catabolism back to inorganic compounds, carbon dioxide, ammonium, and phosphate, and part is converted into cell biomass. The proportion of organic carbon assimilated by microbes that is used in anabolic processes to produce more cell biomass is termed the GGE (Figure 5). For prokaryotic microbes, the GGE is simply the amount of cell biomass produced as a fraction of the total amount of organic carbon assimilated. The rest of the assimilated carbon is respired to carbon dioxide. This growth efficiency is sometimes termed bacterial growth efficiency (BGE). Although the theoretical maximum BGE is 67%, in nature the community BGE is much lower, and generally ranges between 10% and 40%. A major problem with ascertaining the actual BGE of the community of growing cells in the natural environment is that a variable portion of cells in the prokaryotic assemblage are dead or dormant, which makes it difficult to scale assimilation and growth rates measured for the total assemblage to just the active community.

The relative fractions of assimilated organic nitrogen and phosphate that are released as inorganic compounds by prokaryotic metabolism depend on the elemental ratios of carbon, nitrogen, and phosphate in the organic matter on which the microbial cells are growing. Elemental C:N:P ratios of phytoplankton are variable and depend on factors such as the availability of inorganic nitrogen and phosphorus in the environment, species composition, and growth state of the phytoplankton. The classic C:N:P atom ratio of phytoplankton in the sea is 106 C:16 N:1 P, and the ratio between carbon and nitrogen is 6–7:1. This is known as the Redfield ratio, after the oceanographer Alfred C. Redfield who proposed it as an explanation of why the general elemental ratio between nitrate and phosphate in the sea was about 16:1. However, bacterial cells have a higher requirement for both nitrogen and phosphorus, and a C:N ratio of 4–5:1. Prokaryotic cells also have a higher biomass-specific concentration of iron and other trace metals compared to eukaryotic cells. Thus, prokaryotic cells tend to sequester nitrogen and phosphorus and metal ions, which are only released back to the environment by predation or by viral lysis.

Ingestion and digestion of other microbial cells by phagotrophic protists is of vital importance in complete regeneration of the elements fixed by phytoplankton into organic compounds back to inorganic compounds that

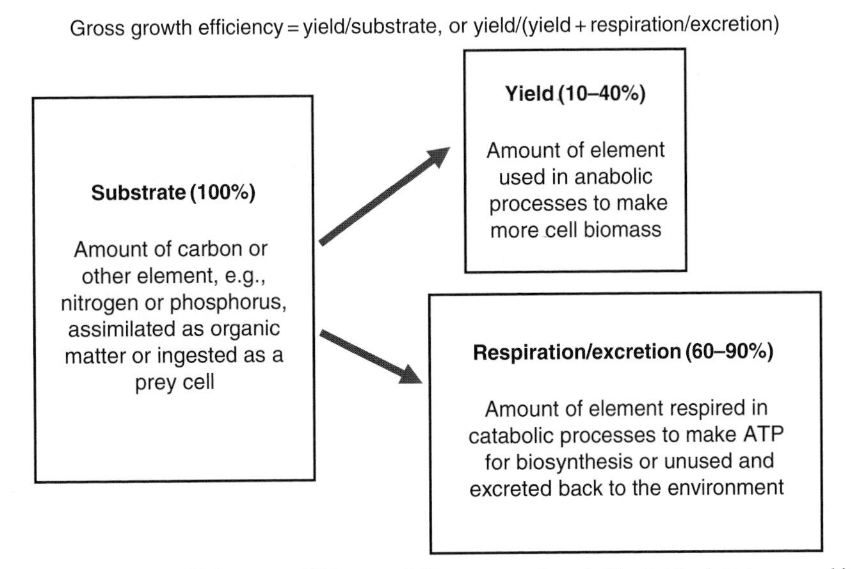

FIGURE 5 Diagram showing gross growth efficiency as cell biomass yield as a proportion of either total substrate or prey biomass assimilated or as a portion of the sum of yield plus amount of ingested food either respired or excreted.

can be reutilized by autotrophs for further primary production. In both marine and freshwater systems, phagotrophic protists, both flagellates and ciliates, are major consumers of prokaryotic cells. Phagotrophic protists are also major consumers of algal cells, even the large-sized phytoplankton characteristic of mass blooms. The carbon-based growth efficiency of phagotrophic protists is about 40% of ingested prey biomass. Protists release undigested components of ingested prey as dissolved and particulate organic matter, as well as metabolic waste products as dissolved inorganic compounds such as ammonium and phosphate. Thus, protistan grazing provides organic and inorganic substrates for further growth of their prey, both heterotrophic bacteria and autotrophic cells. Protists have much higher biomass-specific rates of nutrient excretion than do larger-sized zooplankton, and they regenerate nutrient elements bound up both in bacteria and in phytoplankton. Thus, protist consumption of microbial cells is a major process in regeneration of nitrogen and phosphorus compounds in aquatic systems.

The capacity of many species of autotrophic flagellates to phagocytize gives these phytoplankters an advantage in the acquisition of nutrients in a chemically dilute environment. Mixotrophic algae ingest bacteria and eukaryotic prey to gain both organic substrates and inorganic nutrients. In oceanic systems in which iron is a limiting micronutrient, consumption of iron-rich bacterial cells is an adaptive strategy for phagotrophic algae. Bacterivorous flagellates may also experience iron limitation, and thus ingestion of prokaryotic prey with high iron concentrations can be important to heterotrophic as well as to autotrophic protists.

FOOD RESOURCE FOR METAZOANS

Microbial production forms the base of aquatic food webs. A variable, and at times large, part of the production consumed by aquatic animals is direct consumption of algae. Prokaryotic biomass is a food resource for some animals, for example, rotifers and cladoceran zooplankton such as *Daphnia* spp. in brackish coastal systems and in lakes, and deposit feeding worms in benthic habitats. However, phagotrophic protists play a significant role in channeling microbial, both prokaryotic and algal, production at the base of the food web to higher trophic levels. In addition, phagotrophic protists consume other heterotrophic protists. Species of heterotrophic dinoflagellates and ciliates have been shown in culture to readily ingest heterotrophic flagellates as well as phytoplankton prey. This trophic link in aquatic microbial food webs, although in theory quite important, has, to date, received surprisingly little attention.

There has been a debate about the quantitative significance of trophic transfers involving protists; but there is no doubt that heterotrophic protists represent food for a variety of other consumers. The largest body of studies on this subject deals with ciliates and heterotrophic dinoflagellates as food for mesozooplankton. In regions of the ocean where most phytoplankton are less than 5 μm, which is too small for most multicellular zooplankton to capture, protists may be a primary source of food for copepods and other zooplankters. For example, in an oligotrophic atoll lagoon in the tropical Pacific Ocean, phytoplankton biomass was dominated by coccoid cyanobacteria and algal cells less than 3 μm. Grazing rate assays showed that the major pathway of carbon flow in this food web was from phytoplankton to phagotrophic protists, which formed the main food resource for copepods in the lagoon. Even in mesotrophic systems characterized by diatom blooms, phagotrophic protists can serve as an important trophic link between phytoplankton and mesozooplankton. Heterotrophic dinoflagellates, which are rich in fatty acids and sterols, represent a high-quality food for copepods and enhance their rate of reproduction. Phagotrophic protists in the plankton can also serve as a significant food resource for filter-feeding benthos such as oysters.

MODELING MICROBIAL FOOD WEBS

Conceptual, or box, models, such as the ones shown in Figures 6 and 7, and simulation models, which put the flows between the boxes of conceptual models on a mathematical basis, are a standard approach to understanding how the various components of ecosystems function interactively. The first quantitative models of pelagic food webs, dating from the 1940s, were simple nutrient–phytoplankton–zooplankton (NPZ) simulations based on transfers of

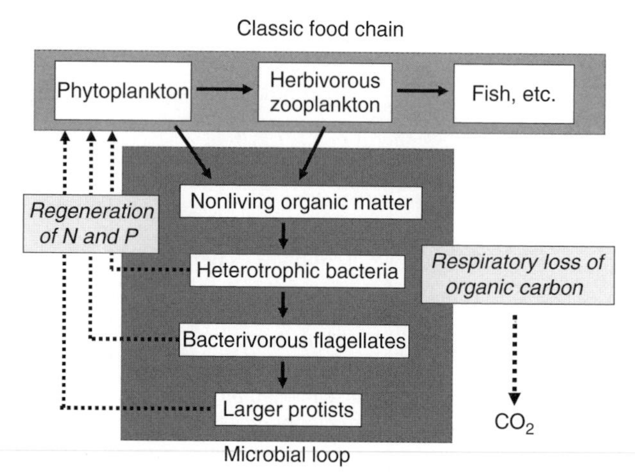

FIGURE 6 Initial conceptualization of the place of microbes in aquatic food webs based on the box model diagram of the microbial loop concept, redrawn from Figure 1 of Ducklow (1983). In this conceptualization, a linear microbial food chain of heterotrophic bacteria to heterotrophic protists to larger protists is added on to the classic phytoplankton to zooplankton to higher consumers food chain. The role of the microbial loop is viewed in this concept as mainly a sink for primary production and a major pathway of regeneration of inorganic nutrients.

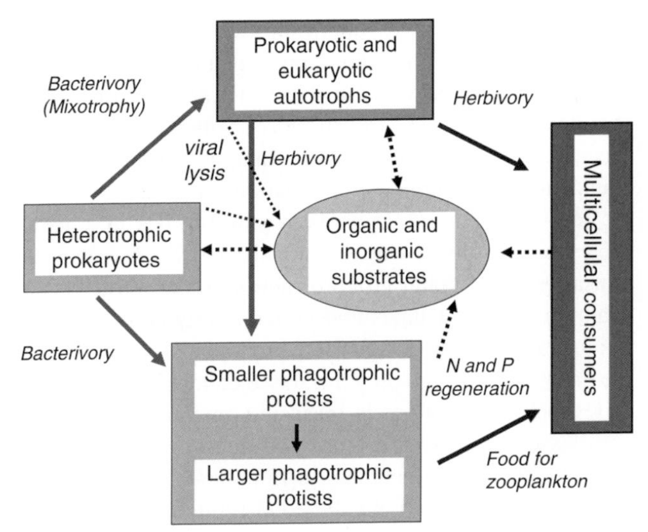

FIGURE 7 Current conceptual model of the major compartments and roles of the microbial food web in aquatic ecosystems. In this conceptualization, the microbial loop is embedded in a larger microbial food web that includes additional pathways of mixotrophy by phytoflagellates, consumption of phytoplankton by phagotrophic protists, viral lysis of both bacteria and phytoplankton, and phagotrophic protists as an important food resource, along with phytoplankton, for multicellular zooplankton.

nitrogen between an inorganic nutrient (nitrate plus ammonium) compartment and phytoplankton and zooplankton compartments. Phytoplankton production was dependent on nutrient availability, and zooplankton consumption of phytoplankton and regeneration of phytoplankton nitrogen back into inorganic nitrogen as ammonium depended on phytoplankton production. The microbial food web, including both heterotrophic prokaryotes and protists, was either ignored or put into an extra organic detritus compartment in the model to account for nitrogen regeneration from decaying phytoplankton cells or zooplankton fecal pellets.

After subsequent research findings proved that heterotrophic microbes played significant and central roles in aquatic food webs, microbes began to be formally included in model diagrams and simulations. The first formulation was to add a decomposing microbial food chain composed of detrital organic matter, heterotrophic prokaryotes, bacterivorous flagellates, and larger protists that fed on the flagellates to the standard, or classic, food chain of phytoplankton to zooplankton to larger consumers (Figure 6). This microbial food chain was termed the microbial loop, and its role in the overall food web was to respire a large fraction, 50% or more, of overall phytoplankton production into carbon dioxide and to regenerate nitrogen and phosphorus nutrients for further primary production.

Further work showed that the heterotrophic microbial food chain was embedded in a much larger, more complex microbial food web in which protists of all sizes fed on autotrophic prey and in which viral lysis of prokaryotes and algal cells at times acted to short-circuit the microbial loop (Figure 7). Recent simulation models of pelagic food webs have explicitly included microzooplankton as consumers of bacteria and phytoplankton, and as food for mesozooplankton. The proportion of phytoplankton carbon that flows through a multistep microbial food web versus a shorter phytoplankton–mesozooplankton food chain has implications for the capacity of marine ecosystems to sequester organic carbon or to efficiently produce fish biomass. Two theoretical scenarios have been proposed in which pelagic systems characterized by an active microbial food web will export less organic carbon compared to systems in which activity of heterotrophic bacteria and protists is relatively low. The first scenario is termed a microphagous food web, dominated by phagotrophic protists consuming prokaryotic and small algal cells, and the second scenario a macrophagous food web in which large-sized algae such as diatoms are consumed by copepods. To a large extent the factor that determines the degree to which pelagic food webs are microphagous versus macrophagous is the proportion of plankton biomass that consists of heterotrophic bacteria and phytoplankton cells less than about 5 μm.

In an empirical test of the theory, it was found that in the St. Laurence River estuary in Canada export flux from the water column was the same during both the spring diatom bloom with a small microbial food web and the postbloom with a larger, more dynamic microbial food web. However, the nature of the sinking material was different. During the spring bloom the vertical flux was primarily in the form of organic aggregates consisting primarily of sedimenting phytoplankton cells; during the postbloom period the major flux was in the form of fecal pellets from omnivorous copepods feeding on heterotrophic protists. Obviously, food web structure alone is not always a good predictor of the quality or quantity of sinking organic carbon. One must also include trophic flux studies coupled with hydrodynamic measurements across time and space. It is also important that studies on microbial food webs include data on all the major components of a pelagic ecosystem, not just the microbes. Since the ocean is the largest reservoir of inorganic carbon that freely exchanges with the atmosphere, understanding the influence of microbial food webs on the ability of the ocean to store atmospheric carbon dioxide is an important research theme for biological oceanographers.

In modeling microbial food webs, whether microbial stocks and rates of biomass production are controlled by bottom-up or by top-down processes is of critical importance. Bottom-up processes are characterized by availability of requirements for growth, for example, inorganic nutrients and light for phytoplankton and quantity and quality of organic substrates for heterotrophic prokaryotes. Top-down processes are mortality processes, mainly due to predation but also including viral lysis, as well as cell death due to unfavorable environmental conditions. An enduring question for aquatic microbial ecologists is why bacterial abundances range over only about 1.5 orders of magnitude in the euphotic zone of the ocean, from several hundred thousand cells per milliliter to one or two million cells per milliliter; while phytoplankton biomass, as measured by concentration of the main photosynthetic pigment chlorophyll *a*, varies over 3 orders of magnitude, from 0.05 in oligotrophic open ocean gyres to 30–50 μg chl *a* l^{-1} in coastal phytoplankton blooms. This results in bacterioplankton biomass being equal to or even greater than phytoplankton biomass in oligotrophic open ocean systems where most primary producers are prokaryotes or small-sized algae, while in coastal marine systems, bacterial biomass is often much less than phytoplankton biomass. One reason for the disparity may be that at the lower end of their abundance range, heterotrophic prokaryote cells enter into a metabolic condition termed starvation survival and persist without dying or disappearing, while at the upper end during phytoplankton blooms, bacteria may either be inhibited from growing to high biomass or have enhanced rates of mortality, that is top-down controls, from grazing and viral lysis when bacterial abundances exceed a threshold abundance of about a million cells per milliliter.

A model that can predict whether marine bacterial communities were controlled mainly by bottom-up or by top-down processes has been proposed. Researchers assumed that in natural environments, when bacterial abundances were close to the carrying capacity of the local environment, bacterial growth rates would be high and variable, and mainly limited by the availability of organic substrate, or bottom-up-controlled. However, when bacterial abundances were far from the carrying capacity of the local environment, growth rates would tend to be lower, and bacterial growth was likely to be limited by mortality processes, or top-down-controlled. Thus, if there is a negative relationship between bacterial growth rate and abundance, it should indicate that the bacterial abundances are near the carrying capacity of the local environment, and that the bacterial community is limited by substrate availability. However, if bacterial abundance and growth rate are not related, it suggests that bacteria are top-down-controlled, with a small range of possible growth rates. This idea was tested by comparing environmental data on bacterioplankton abundance and assemblage growth rates collected from various open ocean habitats. The empirical data sets in fact showed a negative relationship between bacterial growth rate and bacterial abundance in eutrophic regions, indicating bottom-up or resource control, but no relationship between bacterioplankton abundance and growth rate in oligotrophic regions, suggesting top-down or mortality control. Experiments were also carried out in the oligotrophic systems in which water was screened through 0.8-μm-pore-sized filters to remove heterotrophic protists, the main source of mortality. In most cases the bacterial growth rate in screened water increased compared to growth in the presence of phagotrophic protists, confirming top-down regulation.

The main processes controlling the abundances and growth rates of phagotrophic protists in aquatic ecosystems are less well understood. Comprehensive data sets on protist abundance and *in situ* growth rates are lacking. Mortality processes are likely to be important in determining the abundances and biomass of populations of protists.

Top-down control of heterotrophic protists in aquatic food webs results in trophic cascades in which enhanced mortality of a trophic group of heterotrophic protists, for example, nanoflagellates that preferentially consume bacteria and less than 5-μm-sized phytoplankton, increases the growth rate of the prey. Factors that set up trophic cascades include variations in intrinsic growth rates of different classes of grazing protists and dependence of protist growth rate on prey abundance. In natural systems, phagotrophic protists are typically food-limited, and they are poised to rapidly increase their growth rate when they encounter higher prey abundance. Another factor is how the available prey in a microbial food web is partitioned among various groups of grazing protists, particularly by differences in selection of prey by size, as different populations of protist grazers in a system tend to have different prey size preferences. Finally, copepods and other zooplankton can exert a strong top-down control on protists larger than about 10 μm.

Trophic cascade effects have been studied by manipulation experiments in which marine or freshwater samples are treated to exclude predators, and effects of the manipulation on growth of various groups of microbes in a food web are followed over time. As an example, an experiment that demonstrated trophic cascades in a microbial food web was carried out in the northern Baltic Sea. To remove various groups of phagotrophic protists, separate volumes of seawater were filtered to yield four size fractions: a less than 0.8 μm fraction that only included heterotrophic bacteria and autotrophic picoplankton, a less than 5 μm fraction that included heterotrophic bacteria, small-sized phytoplankton, and small-sized heterotrophic flagellates, a less than 10 μm fraction that also included 5–10 μm phytoplankton and flagellates, and a less than 90 μm fraction that included most components of the microbial food web but excluded metazoan grazers. The fractionated water samples were incubated *in situ* in dialysis bags with a molecular weight cutoff of 12–14 kDa, which allowed dissolved nutrients and organic substrates to pass in and out of the incubation bags. The development of the plankton community in the various size-fractioned water samples was followed over 8 days. The results showed that both picoplankton and nanoflagellates were top-down-controlled. Removal of all protists via the 0.8 μm filtration or inclusion of larger-sized protists in the less than 90 μm fraction increased the net growth rates of heterotrophic bacteria and picoplankton, while removal of larger protists by 5 μm or 10 μm filtration led to enhanced growth rates of bacterivorous nanoflagellates and decreased growth rates of picoplankon. The experiment suggested omnivory among phagotrophic protists, which led to trophic cascades in the food web.

CHEMICAL INTERACTIONS BETWEEN MICROBES

Microbial cells, including prokaryotes, algae, and heterotrophic protists, have patterns of behavioral response to environmental cues that affect food web interactions. Many of these cues are specific chemicals dissolved in the aqueous medium or on surfaces of other cells or nonliving particles. Motile bacterial cells can sense gradients of chemical compounds in their environment and respond by moving up or down the gradient. A large portion of prokaryotic strains cultured from seawater are motile, and it is postulated that this motility allows bacterial cells to move toward microsites of high substrate concentration, for example, phytoplankton cells or rich organic particles. The process of chemoreception is based on bacteria having specific recognition sites on their cell membranes for a particular substance. *Escherichia coli*, for instance, has at least 20 chemoreceptor sites that allow the bacterium to respond positively, that is, be attracted to, or negatively, that is, be repelled by, various chemicals. Generally, aerobic bacteria move toward higher concentration of small molecular weight organic substrates such as amino acids and sugars and toward higher oxygen concentration. Aerobic bacteria move away from metabolic poisons, for example, sulfide and heavy metal cations.

Bacteria can sense changes in chemical concentration over time as they swim through concentration gradients. Motile bacteria respond to changes in their environment by changing their swimming behavior or direction. For bacteria having two or more flagella, the normal swimming behavior is straight-line swimming runs with the flagella operating synchronously and occasional tumbles with the flagella flailing in opposite directions, which allows the cell to orient to a new straight-line direction. An increase in concentration of an attractant results in a shift to straight-line swimming alone, while a decrease in concentration of an attractant results in dramatic increase in tumbling, leading to frequent changes of direction to facilitate the search for a more favorable concentration of substrate. Conversely, increasing the concentration of a repellent chemical results in increased tumbling, and decreasing the repellent concentration leads to straight-line swimming. A bacterium that has only a single flagellum can change in rotational direction of the flagellum, so in this case the cell only moves in forward or reverse direction, and does not tumble.

An example of bacterial behavior based on chemosensory response can be observed in the aggregation of bacteria under a coverslip: strains of bacteria that prefer a high oxygen tension will congregate at the edges of the slip, while microaerobic bacteria will remain in the middle. Motile marine bacteria have been shown to aggregate around organic-rich particles and to track the path of a phytoflagellate, presumably by chemosensory response to organic compounds emanating from the particle or algal cell. This ability allows motile bacteria to aggregate around detrital particles or algal cells, forming locally elevated prey concentrations for heterotrophic protists.

Chemosensory response is thus also important in the feeding behavior of phagotrophic protists. Marine bacterivorous flagellates exhibit chemoattraction to amino acids and to bacterial cells, as has been demonstrated. Marine ciliates whose preferred prey is algae exhibit strong positive chemosensory response to some species of algae, and neutral or negative response to other prey species. Species of heterotrophic dinoflagellates have also been found to move toward algal cells and algal cell lysate, and to preferentially prey on some strains of algae. A modeling exercise designed to estimate the advantage conferred on a phagotrophic protist that is able to chemically detect prey cells showed that this capacity conserves energy used to search for food, and would confer the greatest advantage when prey are scarce.

Bacterial and algal cells may deter protist predation directly by producing chemicals that are either toxic or unpalatable to phagotrophic protists. The pigment violacein, a bacterial secondary metabolite, inhibited predation by nanoflagellates on bacteria containing the pigment. Ingestion of just one to three violacein-containing bacterial cells was found to cause rapid death of the flagellate cells. The heterotrophic dinoflagellate *Oxyrrhis marina* was shown to have a much lower feeding rate on one particular strain of the coccolithophorid algae *Emiliania huxleyi* compared to a second strain of *E. huxleyi*. In the presence of the dinoflagellate, the strain of *E. huxleyi* that the dinoflagellate avoided produced a concentration of dimethyl sulfide (DMS) and acrylic acid that was an order of magnitude higher than in the second strain of *E. huxleyi*. DMS was found to inhibit predation of algal prey by several other species of phagotrophic protists, and appears to be a chemical deterrent to protist feeding. The deterrent mechanism is not known.

The biochemical mechanism for chemosensory response in both prokaryotic and eukaryotic microbes is based on chemoreceptor sites associated with the cell membrane. The chemical compound being sensed binds to a specific chemoreceptor site, triggering signal transduction, the release of secondary messenger compounds within the cell, which results in change in motility, for example, direction and/or speed, or in ingestion of a prey cell by a protist. Such cellular processes are obviously important for survival and growth of aquatic bacteria in terms of locating microsites of higher substrate concentration and for aquatic protists in determining prey location, selection, capture success, and rates of ingestion. Understanding the extent to which cell surface chemoreceptor binding and consequent signal transduction pathways operate in bacterial and protist chemosensory behavior is vital to a predictive understanding of the structure and function of marine food webs.

SPATIAL STRUCTURE OF MICROBIAL FOOD WEBS

Initially, the environment encountered by aquatic microorganisms was perceived as being largely fluid, relatively homogeneous, and governed by diffusive processes; and it was assumed that microbes suspended in water were more or less uniformly distributed. Research on how chemosensory behavior of motile microbes can result in microbial aggregations, combined with observations of patchy distribution of phytoplankton, detrital organic matter, and high molecular weight colloidal dissolved organic substances, has led to the understanding that microbial food webs in fact occur in a nonuniform structured habitat. Phytoplankton and bacteria release polymeric material directly, other polymeric material is produced during feeding and egestion by protists and zooplankton. Biopolymer gels form when polymer chains are hydrated and cross-linked or aggregated, resulting in a three-dimensional network. These polymeric microgels then aggregate to form gel-like sheets, strings, and webs that provide surfaces, form barriers against diffusion, and furnish refuges against predators. Detrital particles and microbial cells are interspersed in these gel webs. The interactions of bacteria with the organic matter continuum from dissolved organic compounds to large particles, and the behavioral response of microbes to the patchy distribution of these particles, create microscale features, hotspots of microbial activity, and food web interactions, with distinctive natures and intensities of biogeochemical transformations.

The larger organic particles formed in this way are termed marine snow because these aggregated particles, which are usually greater than 0.5 mm in diameter, strongly resemble snowflakes when seen in the water. The basic glue holding marine snow particles together is thought to consist of fibrillar, long-chain polysaccharide polymers termed transparent exopolymer particles (TEP). Marine snow particles are initially colonized by heterotrophic prokaryotes that produce extracellular enzymes that hydrolyze the organic matrix of the particle into low molecular weight dissolved organic compounds, for example, monomers or small polymers of sugars and amino acids. The colonizing bacteria produce more of these organic substrates than can be assimilated; thus, a plume of organic material is released from the particle that attracts motile bacteria. Higher abundances of bacteria in turn attract bacterivorous flagellates, which may then attract larger protists and zooplankton. The organic particle and its plume may thus become the focus of a complex food web. Marine snow particles can sink at speeds greater than 100 m in a day, allowing them to travel from the surface to the subsurface ocean within a matter of days. As they sink, the organic material in the particles is continually degraded by microbes. Any organic material that reaches the sea floor is either consumed by benthic organisms or incorporated into sediments. Sinking of marine snow particles is important to the ocean's capacity to sequester atmospheric carbon dioxide.

Biofilms, which, as mentioned previously, form on surfaces, are similar to suspended organic particles in that they are essentially an organic polymer gel with living microorganisms embedded in it. Biofilms form on most abiotic surfaces in aquatic systems and are sites of enhanced microbial activity. The polysaccharides produced by surface-growing microbes act to cement sediment particles together and represent a food resource for benthic animals. The microbial species colonizing the biofilm and the organic composition of the matrix largely determine the physical properties of the biofilm. Once attached to a surface, bacteria begin to produce extracellular polysaccharides. The amount of biopolymer produced can exceed the mass of the bacterial cell by a factor of 100 or more. The gel is mostly water, but has properties that influence the transport of materials at the surface of the biofilm. Changes in transport rates can create unique niches within the biofilm for the proliferation of a variety of microbial species. Bacterial extracellular polysaccharides also tend to absorb cations and organic molecules from the overlying water. The reduced diffusion rates of substrates within a biofilm serve to create localized concentration gradients and the possibility of well-defined spatial relationships between individual bacterial cells. These gradients may result in suboxic and even anoxic regions in the interior of these structures, contributing to the coexistence and metabolic interaction of both aerobic and anaerobic microbes.

RECENT DEVELOPMENTS BIOGEOGRAPHY OF FORM AND FUNCTION IN MICROBIAL FOOD WEBS

Major recent discoveries about patterns and processes in microbial food webs stem from sophisticated molecular genetic and cell biology approaches to understanding the biogeography of microbial community composition and metabolic function. Metagenomics, analysis of DNA sequences of microbial genomes present in an environmental sample, has revealed astoundingly high species or strain diversity in communities of prokaryotic and eukaryotic microbes. Genetic fingerprinting methods have demonstrated that communities of marine bacteria have richer diversity in warm tropical waters compared to cold polar waters, a pattern similar to that found for phytoplankton and zooplankton. On a finer scale, these genetic methods have shown that at a coastal ocean station community structure of planktonic bacteria can shift dramatically as environmental conditions vary, in patch sizes of several kilometers and over time scales of months. Metabolic capabilities also vary. New data has confirmed that the Crenarchaea which dominate marine microbial communities at depths greater than 200 m can be autotrophic nitrifyers, gaining energy for carbon fixation from ammonium oxidation. Analysis of the complete genome of the archaeal nitrifyer *Nitrosopumilus maritimus*, which grows autotrophically in culture, has found structurally different enzymes and pathways for ammonium oxidation and carbon fixation compared to those of nitrifying Bacteria. Metagenomic studies of marine bacterial communities have identified abundant genes coding for proteorhodopsin, a light-harvesting pigment that can serve as an energy-generating proton pump.

Proteomics, the study of proteins expressed by individual species and communities of microbes, links genetic potential to in situ biochemical functions. A detailed analysis was made of the proteins expressed in bacterial cell membranes in surface waters of the South Atlantic Ocean. The dominant proteins were TonB-dependent transporters, which require a trans-membrane proton gradient to move substrates across the cell membrane. Diverse rhodopsins were also detected. A close association of these transporter molecules with rhodopsin proton pumps in bacterial membranes suggests that marine bacteria may use solar power to facilitate transport of substrates for growth into the cell, an effective strategy for existence in nutrient-poor habitats.

Recent studies have also identified high diversity among eukaryotic microbes. As found for prokaryotes, protistan communities have different phylogenetic compositions in the ocean surface compared to those at subsurface depths. A common group of marine algae, the haptophytes, has an extreme degree of biodiversity; their evolutionary success may be due, in part, to mixotrophic ingestion of bacteria. Picobiliphytes have been identified as a major new taxonomic class of marine phytoplankton. Diverse ribotypes related to the Syndiniales, parasitic dinoflagellates, are ubiquitous in the sea, but whether these open ocean strains are indeed parasitic is unknown. Microscopic fungi such as yeasts and chytrids are found in both freshwater and marine aquatic ecosystems, and may be significant as decomposers of organic matter as well as parasites of algae and zooplankton. Understanding how the rich biodiversity of prokaryotic and eukaryotic organisms is partitioned among aquatic habitats and how this diversity affects biogeochemical processes and food web interactions continues to be a central theme for microbial ecologists.

FURTHER READING

Azam, F., Fenchel, T., Field, J. G., Meyer-Reil, R. A., & Thingstad, F. (1983). The ecological role of water-column microbes in the sea. *Marine Ecology Progress Series*, *10*, 257–263.

Caron, D. A. (2009). Past president's address: Protistan biogeography: why all the fuss? *Journal of Eukaryotic Microbiology*, *56*, 105–112.

Countway, P. D., Gast, R. J., Dennett, M. R., Savai, P., Rose, J. M., & Caron, D. A. (2007). Distinct protistan assemblages characterize the euphotic zone and deep sea (2500 m) of the western North Atlantic (Sargasso Sea and Gulf Stream). *Environmental Microbiology*, *9*, 1219–1232.

Ducklow, H. W. (1983). Production and fate of bacteria in the oceans. *BioScience*, *33*, 494–501.

Fenchel, T., & Finlay, B. J. (1995). *Ecology and evolution in anoxic worlds*. New York: Oxford University Press.

Gasol, J., Pedros-Alio, C., & Vaque, D. (2002). Regulation of bacterial assemblages in oligotrophic plankton systems: Results from experimental and empirical approaches. *Antonie van Leeuwenhoek*, *81*, 435–452.

Guillou, L., Viprey, M., Chambouvet, A., Welsh, R. M., Kirkham, A. R., Massana, R. *et al.* (2008). Widespread occurrence and genetic diversity of marine parasitoids belonging to Syndiniales (Alveolata). *Environmental Microbiology*, *10*, 3349–3365.

Fuhrman, J. A., & Steele, J. A. (2008). Community structure of marine bacterioplankton: Patterns, networks, and relationships to function. *Aquatic Microbial Ecology*, *53*, 69–81.

Fuhrman, J. A., Steele, J. A., Hewson, I., Schwalbach, M. S., Brown, M., Green, J. L. *et al.* (2008). A latitudinal diversity gradient in planktonic marine bacteria. *Proceedings of the National Academy of Sciences*, *105*, 7774–7778.

Jobard, M., Rasconi, S., & Sime-Ngando, T. (2010). Diversity and functions of microscopic fungi: A missing component in pelagic food webs. *Aquatic Sciences*, DOI 10.1007/s00027-010-0133-z.

Kirchman, L. D. (Ed.) *Microbial ecology of the oceans* (2000). New York: Wiley-Liss.

Legendre, L., & Le Fevre, J. (1995). Microbial food webs and the export of biogenic carbon in oceans. *Aquatic Microbial Ecology*, 9, 69–77.

Liu, H., Probert, I., Uitz, J., Claustre, H., Aris-Brosou, S., Frada M. *et al.* (2009). Extreme diversity in noncalcifying haptophytes explains a major pigment paradox in open oceans. *Proceedings of the National Academy of Science*, *106*, 12803–12808.

Matz, C., Deines, P., Boenigk, J. *et al.* (2004). Impact of violacein-producing bacteria on survival and feeding of bacterivorous nanoflagellates. *Applied and Environmental Microbiology*, *70*, 1593–1599.

Morris, R. M., Nunn, B. L., Frazar, C., Goodlett, D. R., Ting, Y. S., & Rocap, G. (2010). Comparative metaproteomics reveals ocean-scale shifts in microbial nutrient utilization and energy transduction. *The ISME Journal*, *4*, 673–685.

Not, F., Valentin, K., Romari, K., Lovejoy, C., Massana, R., Töbe, K. *et al.* (2007). Picobiliphytes, a new marine picoplanktonic algal group with unknown affinities to other eukaryotes. *Science*, *315*, 252–254.

Pohnert, G., Steinke, M., & Tollrian, R. (2007). Chemical cues, defence metabolites, and the shaping of pelagic interspecific interactions. *Trends in Ecology and Evolution*, *22*, 198–204.

Pomeroy, L. R. (1974). The ocean's food web, a changing paradigm. *BioScience*, *24*, 499–504.

Rivkin, R. B., Legendre, L., & Deibel, D. *et al.* (1996). Vertical flux of biogenic carbon in the ocean: Is there food web control?. *Science*, *272*, 1163–1166.

Sakka, A., Legendre, L., Gosselin, M., & Delesalle, B. (2000). Structure of the oligotrophic planktonic food web under low grazing of heterotrophic bacteria: Takapoto Atoll, French Polynesia. *Marine Ecology Progress Series*, *197*, 1–17.

Samuelsson, K., & Andersson, A. (2003). Predation limitation in the pelagic microbial food web in an oligotrophic aquatic system. *Aquatic Microbial Ecology*, *30*, 239–250.

Sherr, E. B., & Sherr, B. F. (2008). Understanding roles of microbes in marine pelagic food webs: A brief history. D. Kirchman (Ed.) *Advances in microbial ecology of the oceans* 27–44. Hoboken N.J.Wiley-Blackwell.

Sieburth, J. Mc. N., Smetacek, V., & Lenz, J. (1978). Pelagic ecosystem structure: Heterotrophic compartments of the plankton and their relationship to plankton size fractions. *Limnology and Oceanography*, *23*, 1256–1263.

Walker, C. B., de la Torre, J. R., Klotz, M. G., Urakawa, H., Pinel, N., Arp, D. J *et al.* (2010). *Nitrosopumilus maritimus* genome reveals unique mechanisms for nitrification and autotrophy in globally distributed marine crenarchaea. *Proceedings of the National Academy of Sciences*, DOI/10.1073/pnas.0913533107

Wolfe, G. V. (2000). The chemical defense ecology of marine unicellular plankton: Constraints, mechanisms, and impacts. *Biological Bulletin*, *198*, 225–244.

Chapter 3

Biofilms

J.W. Costerton
University of Southern California, Los Angeles, CA, USA

Chapter outline

ABBREVIATIONS

ENT Ear nose and throat
FISH Fluorescent *in situ* hybridization
SRB Sulfate-reducing bacteria

DEFINING STATEMENT

Biofilms are surface-associated bacterial communities that predominate in natural and pathogenic ecosystems. The matrix-enclosed bacterial cells in these communities assume a phenotype that differs profoundly from that of their planktonic counterparts, and this mode of growth protects them from so many antibacterial factors that they constitute protected enclaves in hostile environments and chronic infections.

BIOFILM STRUCTURE AND FUNCTION

Direct observations of a very large number of natural and pathogenic ecosystems have shown that the vast majority of bacteria in these systems live in matrix-enclosed communities attached to surfaces. The term "biofilm" was coined to describe these sessile communities, and the definition of this term has been refined to be "a community of bacterial cells enclosed in a matrix, at least partially of its own production, which functions as a physiologically integrated community." Further examinations, using the methods of modern microbial ecology, have shown that most of the cells in biofilms are alive, and that cells of different species are often spatially arranged in patterns that facilitate metabolic cooperation. The discovery that sessile cells in biofilms express genes that are profoundly different from those expressed by their planktonic (floating) counterparts led us to describe the biofilm phenotype, and the realization that these patterns of gene expression differ by as much as 70% offers at least a partial explanation of the enormous survival value of biofilm formation. A biofilm is not simply a slime-enclosed mass of planktonic cells. A biofilm is a multicellular community in which each sessile cell expresses its genes in different ways, and this biofilm phenotype is profoundly different from that of planktonic cells in a single-species culture.

We have proposed that biofilms have predominated in all natural ecosystems, from the earliest of times, because these sessile communities provide protection from biological predators and antibacterial chemicals, and because biofilm formation anchors these communities in favorable locations. The life of the planktonic cells shed from biofilms, in the primitive earth, would be "nasty and short" (Oscar Wilde), except in rare instances in which they discovered virginal surfaces in halcyon environments, but the biofilm from which they set forth would remain protected and inviolate. This distinction is particularly true in pathogenic ecosystems in which planktonic cells are killed by antibodies produced in response to modern vaccines and by antibiotics, while any biofilm that can become established in the body persists for years in myriad chronic infections. Cells within biofilms in chronic infections persist for years, in spite of the focused attacks of many functional host defense systems, and biofilms in natural ecosystems withstand heavy grazing pressure from protozoa and from specialized predators such as snails. We are just beginning to discover the mechanisms of this protection, which may involve compounds that paralyze phagocytes and poison grazing protozoa, but the simple fact remains that biofilms persist and predominate in virtually all ecosystems. The "life cycle" of biofilms is diagrammed in Figure 1.

The complexity of the architecture of mature microcolonies, like that in the right foreground in Figure 1, suggest that the process of biofilm formation must be under the control of some kind of signaling system. Many researchers have now discovered which of the several cell–cell signaling systems control biofilm formation in many bacterial species; one "master system" controlling cell detachment has also been discovered by David Davies. This major discovery reveals that the sessile bacteria in biofilms are sentient of the presence of neighboring organisms, and of environmental conditions, so the groundwork is laid for the concept of biofilms as multicellular communities whose component cells can communicate with each other. This logical process could lead to a situation in which a stimulus applied to one part of a biofilm could trigger reactions that would ramify throughout the community, and produce reactions in locations far distant from the stimulated area. Preliminary indications of this ability to sense changes, and to react on a community basis, have been seen in bacterial biofilms.

If we look at the structural complexity of biofilms, and their behavioral characteristics, we become motivated to discover the mechanisms that enable this complexity and these reactions. When we see that sessile cells in energy-deprived areas of biofilms can receive energy from more favored areas, we discover electrically conductive nanowires that can be used for energy sharing and may even be used for electrical communication. When we note that the frequency of horizontal gene transfer in biofilms exceeds that in planktonic cell suspensions, we find large numbers of pili that may facilitate conjugation, and may also position each cell in a predetermined location in the community. When we note that some *de facto* biofilms (e.g., myxobacterial swarms) move through soil containing stationary biofilms of thousands of species, without losing contact with each other, we note that these cells produce large numbers of signal-containing vesicles that may be "addressed" exclusively to cells of the same species. The notion of biofilms as sophisticated and integrated communities is new, and we are just beginning to find the first few mechanisms that enable this sophistication and this integration. So we are well advised to marvel at the biofilm's accomplishments, and to ferret out the mechanisms that enable each marvelous attribute, until the whole complex-integrated apparatus is revealed.

The revelation that biofilms predominate, in virtually all natural and pathogenic ecosystems, must drive immediate changes in the ways that microbiologists study bacteria. The reductionist approach of studying a single "type" strain, growing as planktonic cells in a defined liquid medium, may reveal the arcane secrets of protein synthesis but it has nothing at all to do with bacterial processes in nature or disease. When we select a type strain we make an arbitrary choice, from a spectrum of genomes

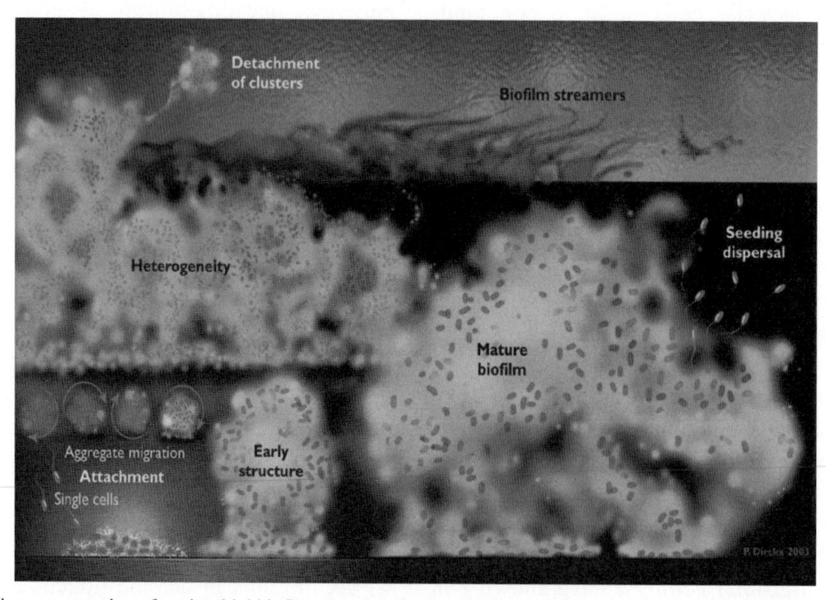

FIGURE 1 Diagrammatic representation of a microbial biofilm showing the development of these communities, from the attachment of planktonic cells to the development of complex mature microcolonies with open water channels and hollow areas from which planktonic cells have dispersed. The metabolic integration of biofilms is implied, in the middle distance, by the juxtaposition of clusters of physiologically different organisms in arrangements that would facilitate interactions. In the far distance, the diagram suggests a structure somewhat like a kelp bed, in which individual microcolonies are anchored on a surface, but are free to respond elastically to shear forces operative in the environment.

that comprise the supergenome of the species concerned, and we exclude as many as 1500 genes from consideration. When we study gene expression using a chip made from a type strain, we are blinded from any information concerning the thousands of genes in the supergenome that are not in the type strain. When we study planktonic cells, we study a phenotype that may vary from the biofilm phenotypes that grow in natural and pathogenic systems by as much as 70%, in terms of the genes that are expressed. We can use mutants to study metabolic processes in planktonic cells, and add more exquisite details to the well-worn cycles that are memorized in Microbiology 101, but we cannot use cultures to study mutations that affect the fitness of an individual cell to function as a member of an integrated community.

Modern microbial ecology has long since abandoned the general practice of extrapolating from cultures to ecosystems, and the other subdivisions of microbiology must soon follow. Ecologists analyze ecosystems of interest by harvesting bacterial DNA and sequencing the 16 S rRNA gene to determine which species are present. Fluorescent *in situ* hybridization (FISH) probes are then constructed, so that individual cells of the species concerned can be identified in microscopic examinations of the real community growing in the ecosystem. These probes also reveal the distribution of cells of each species, in spatial relationships to those of other species, and to the surface on which the biofilm has formed. The whole community can be studied, intact and in its real surrounding, and the effects of various stresses on the biofilm can be assessed using parameters such as carbon fixation or the output of specific products (e.g., organic acids). The somewhat draconian parameter of cell death can be assessed, *in situ* in real ecosystems, using the live/dead BacLite probe and the confocal microscope. The parallel systems of microscopy-based and nucleic acid-based ecological methods have recently been joined, by the PALM "capture" microscope and the MDA amplification system, which allows us to visualize a group of bacteria in a real biofilm and then to excise those cells, extract their DNA, and sequence their genome.

Because biofilms predominate in virtually all natural and pathogenic ecosystems, serious students of bacteria must examine the biofilm phenotype of their minute subjects. In the rare instances in which a biofilm is formed by single species, as in certain device-related infections, a single-species biofilm can be grown on an inert surface in the laboratory and valid extrapolations from the culture to the infection may be made. The mixed species biofilms that predominate in most ecosystems are much more difficult to study in the laboratory, even though some simulations may be useful, and direct observations will provide the best data. As we undertake new large-scale projects, such as the NIH roadmap project on the Human Microbiome, we will turn to direct examination of nucleic acids for population analysis and direct confocal and electron microscopy for community mapping, and microbiology will have moved on to a new phase.

BIOFILMS IN NATURAL ECOSYSTEMS

Perhaps because of the historical tendency of microbiologists to "sample" natural ecosystems, and to head straight for the laboratory with these samples, the bulk of the microbiology of natural ecosystems has involved "grab" samples of the bulk water phase. When we have profiled natural ecosystems including streams, lakes, and near-shore marine and subsurface environments, we have observed that >99.9% of the bacteria grow in biofilms adherent to surfaces, and only a few stray planktonic cells inhabit the bulk water phase. For this reason, complex organic compounds placed in contact with samples of the bulk water of such systems as the Athabasca River, in the region of the tar sands, show only very slow rates of bacterial degradation. When the complex organics (e.g., bitumen) are placed in contact with the enormous biofilm populations, which have developed in response to the continuous availability of these energy-rich substrates, their degradation is very rapid and complete. The validity of these analyses is attested by the fact that the tons of bitumen washed into the Athabasca River by erosion are completely biodegraded by the time the river reaches Lake Chippewa (48 miles downstream), and by the fact that local oil spills are resolved very rapidly by the biofilm system of the river. If we take a rational view of the microbiology of natural ecosystems, based on direct observations of the location and activity of all of the bacteria, we can assess the real potential of each system for the processing of organic molecules, including pollutants. Our analyses of the Peace/Athabasa/MacKenzie river system of Northern Canada suggest that that this system could be designated as an "oil-adapted" corridor through which oil from that region, and oil from the North Slope deposits in Alaska, could be shipped with complete ecological impunity. The omission of biofilm populations from ecological evaluations, including impact statements, is bad science and very bad public policy.

The addition of the biofilm component has solved many ecological mysteries that have led classically inclined microbiologists to throw up their hands, and to declare that the ways of bacteria are simply "wondrous strange." The puzzle concerning the bovine rumen arose because a fastidiously anaerobic bacterial population, dependent on ammonia, grew and functioned happily in an animal organ that was continuously perfused with oxygenated blood containing large amounts of urea. Then we discovered a special biofilm population on the rumen wall, which developed in the first few days of life, and both scavenged oxygen at the tissue surface and changed urea to ammonia, while deriving energy from the proteolytic degradation of shed epithelial cells.

Rumen ecologists also grappled with the problem of laboratory model systems in which biofilms of primary cellulose degrading bacteria broke down cellulose at rates 100-fold slower than the rates seen in the functioning rumen, until they discovered that elusive treponema cells could enter the biofilms for short excursions and remove the butyrate that was slowing the digestive process. Other classical microbiologists have been troubled by the fact that PCR analyses of fruit and produce have indicated the presence of potential pathogens (e.g., *Listeria* and *Escherichia coli* 0157), while cultures of the same materials have yielded negative results. The revelation that biofilm cells do not grow when spread on the surfaces of agar plates stimulated direct examinations of tissue surfaces using FISH probes and confocal microscopy, and the mystery was solved when well-developed biofilms of these organisms were found. We submit that the analysis of any natural ecosystem is incomplete and likely to become stalled by anomalous data, if we examine only planktonic populations, but that the invocation of the biofilm concept has the potential to solve many of these mysteries.

BIOFILMS IN INDUSTRIAL ECOSYSTEMS

For several decades, the control of the very serious problem of corrosion of metals by bacteria was predicated on a planktonic model, in spite of the obvious fact that the attack on stationary metals by mobile bacterial cells seems ludicrous at a basic conceptual level. Several microbial villains were designated, amongst whom the sulfate-reducing organisms (SRB) were prominent, and surveillance of the world's pipelines was initiated using an elaborate whole bottle system to detect SRBs using a lactate medium with iron filings. Sick pipelines were detected and treated with biocides to kill the deadly SRBs, but continuing metal loss caused catastrophic pipe failures, and the planktonic bacterial counts always returned when the biocide treatment was over. Biocide salesmen declared that the SRB in the most affected pipelines had become "resistant" to the biocide in question, and offered new and better (and more expensive) biocides to save the day and keep the oil flowing. In the meantime, engineers with scant knowledge of microbiology observed that regular "pigging" of pipelines with mechanical scrapers that removed tons of "slime" were effective in controlling corrosion, especially when coupled with the use of biocides. David White had a medical degree, and a background in physical chemistry, and he led the team that solved the corrosion puzzle in the 1980s, by figuring out that bacterial biofilms cause corrosion by building structured communities on the metal surface that actually constitute classic "corrosion cells." The biofilms are a living energy-driven cathode, the metal is the matching anode, and a well-organized bacterial biofilm can drill a neat hole through 5/8 inch steel pipe in a couple of months. So the biofilm concept has solved another microbiological mystery and we now detect SRBs in biofilms (not bulk water samples), we pig all "piggable" lines with relentless regularity, and we follow the pig with biocides to kill the bacteria we have just blasted off of the pipe wall with mechanical and shear forces. The solution to this problem is poignant, in terms of communication between scientists in the same area, because the external surfaces of the thousands of miles of metal pipe we bury in the ground are always protected from microbial corrosion by the application of "cathodic protection." For at least four decades, oil companies paid dearly for courses in cathodic protection, which prevents external pipe corrosion by overriding the electrical potentials of biofilm-driven corrosion cells, while sponsoring corrosion classes for the same employees, in which they were taught how to prevent internal corrosion by the use of biocides! Again, we need to know where the bacteria are, in industrial systems ranging from pipelines to cooling towers, and useful answers simply come to light when we know their locations and their mode of growth.

BIOFILMS IN MEDICAL SYSTEMS

The paradigm that has developed in medical microbiology is especially unfortunate, in view of the phenomenal success of the pioneers of this field in the virtual eradication of acute epidemic disease. The paradigm depends on culture methods for the detection and recovery of the bacteria that cause a particular disease, the cultivation of these bacteria in single-species cultures in defined media, and the extrapolation from culture data to define the etiology of the disease and to suggest therapeutic strategies. The success of this paradigm, in Koch's heyday and in the first part of the twentieth century, may have blinded us to its scientific faults and to the fact that it has not been successful in the definition or the treatment of the burgeoning number of chronic bacterial diseases that currently beset us. In terms of the biofilm concept, the failure of the traditional paradigm to accommodate the fact that 80% of the bacteria in modern chronic infections live in biofilms, and fail to produce colonies when spread on agar surfaces, is a pivotal deficit. But it is equally disturbing to note that any given strain of bacteria that is recovered from an infection is only one of many strains that may constitute the super genome of the species, and that any chosen "type" strain may lack hundreds or even thousands of genes that are present in other strains in the same infection. Tragically, because of the general conservation of "housekeeping" genes, the genes that are missing from any type strain may control important pathogenic processes, and even this depleted complement of genes may be further depleted by genetic drift as the strain is carried in serial culture. Coupled with the fact that the biofilm phenotype differs so profoundly from the planktonic phenotype of the same strain, any attempts to extrapolate from a single strain grown as planktonic cells in a defined medium to a disease caused by multiple strains growing as biofilms in infected tissues seems futile, and possibly fraudulent.

The persistence of the traditional planktonic paradigm has puzzled clinicians, many of whom see their patients suffering from infections that produce obvious symptoms and demonstrable tissue damage, while lab tests yield negative cultures. The problem of overt culture-negative infections has bedeviled clinicians in orthopedic surgery, ear nose and throat (ENT), and urology, and modern detection methods have uniformly detected bacteria where cultures have failed. Clinicians from these different specialties do not often read each other's literature, so many are unaware that the problem of culture-negative chronic infections has been solved and that large numbers of biofilm bacteria have been found in each case. At the Mayo Clinic, Robin Patel has pioneered the use of PRC-based nucleic acid technologies to detect the presence of bacteria in infections of orthopedic devices, and many new initiatives will gradually replace cultures with DNA-based methods. Roger Lasken, of the Craig Venter Institute, will soon sequence the entire genomes of a large number of single bacterial cells, isolated from a cystic fibrosis lung by micromanipulation, and we will finally know how many different strains of how many different species are present in a well-defined biofilm infection. On a practical note, it should be noted that many clinicians have ignored negative culture data, because their patients are obviously infected, and have developed empirical therapeutic strategies based on the physical removal of biofilms (where possible) and high-dose antibiotic therapy. As in natural and industrial ecosystems, horse sense has triumphed, and the modern DNA-based methods and direct observations that have brought microbial ecology forward have provided the rationale for this sensible approach.

One area in which the traditional paradigm still lingers, with very invidious effect, is in the treatment of multispecies infections. If a bacterium has been regularly cultured from a type of chronic bacterial infection, that organism is enshrined as the causative agent of that particular infection, and all strategies from prevention to therapy are predicated on that assumption. In fact, when a predisposing condition such as a burn offers a hospitable environment for members of a very large spectrum of bacterial species, many species may colonize the tissue concerned, because many organisms are present and nothing predisposes to favor colonization by any particular organism. Culture media have been devised and refined to favor the growth of certain infamous pathogens, and we can always culture *Staphylococcus aureus*, *Pseudomonas aeruginosa*, and *Enterobacter faecalis* from wounds, the cystic fibrosis lung, and failed root canals, even though we may not be able to culture many of the organisms that coinfect with them. This preoccupation with "pathogens of note" harkens back to the era of the "one pathogen = one disease" theorem, and it is a virtual ecological impossibility that specific pathogens occur exclusively in lesions that are open to colonization from the environment, which is an exuberant microbiological zoo. Furthermore, there was some justification in this thesis if the pathogen possessed specific aggressive toxins and other pathogenic factors, but the common link between biofilm pathogens that cause chronic infections is that their pathology is mediated by the inflammatory response to their continued presence in the affected tissues. There is, therefore, absolutely no valid reason to suspect that a multispecies infection, like that of a wound, is caused by a single bacterial species because only one species is recovered in cultures. DNA-based analysis has revealed the presence of dozens of bacterial species in wounds that have yielded positive cultures for only *S. aureus*, and our ongoing work with FISH probes indicates that several bacterial species form biofilms in the wound bed. A detailed examination of the bacterial biofilms in the wound bed, and an analysis of the extent to which each species causes cytokine production and leukocyte mobilization, will tell us which bacterial species are involved and how they contribute to the sustained infection.

POSTLUDE

Direct observations indicate, unequivocally, that bacteria in all ecosystems live predominantly in matrix-enclosed biofilms attached to surfaces. These sessile biofilms have many attributes that have not been previously associated with prokaryotic organisms, and it is now clear that they function as metabolically integrated communities whose sophistication and internal communications rival those of multicellular eukaryotes. As we approach conceptual problems, in all subdivisions of microbiology, the simple addition of the biofilm concept to the traditional culture-based microbiological paradigm provides us with an intellectual basis for understanding these puzzles by locating and enumerating the bacteria, of all species, that are present and active in the ecosystem.

SUMMARY

Direct observations of bacteria growing in natural and pathogenic ecosystems have shown that these organisms grow predominantly in matrix-enclosed biofilms. These sessile bacteria assume a distinct phenotype that differs from that of their planktonic counterparts, renders them resistant to antimicrobial agents, and makes them incapable of producing colonies when dispersed on agar plates. In mature biofilms, very effective metabolic interaction is achieved by the juxtaposition of cooperative species, and equally effective communication is achieved by means of electrically conductive nanowires, and by cell–cell signals that spread by simple diffusion or by transport in specialized vesicles.

Biofilms recycle organic matter with remarkable efficiency, they cooperate with many plants (e.g., legumes) and animals (e.g., ruminants), and they may exert protective effects on many tissues that they colonize to the exclusion of pathogenic species. Biofilms also mediate the attack of bacteria on metals (e.g., pipelines), by concentrating ions and electrical fields at a specific location of the affected surface, and cause damaging inflammations of tissues by allowing bacteria to persist for decades in human tissues. If we are to enhance these beneficial activities, and to thwart these destructive tendencies of biofilm bacteria, we must apply the biofilm paradigm throughout modern microbiology.

FURTHER READING

Costerton, J. W. (2007). *The biofilm primer*. Heidelberg: Springer, 200 pp.

Costerton, J. W., Stewart, P. S., & Greenberg, E. P. (1999). Bacterial biofilms. A common cause of persistent infections. *Science*, *284*, 1318–1322.

Fuqua, W. C., & Greenberg, E. P. (2002). Listening in on bacteria: Acyl-homoserine lactone signaling. *Nature Reviews Molecular Cell Biology*, *3*, 685–695.

Fux, C. A., Shirtliff, M., Stoodley, P., & Costerton, J. W. (2005). Can laboratory reference strains mirror 'real-world' pathogenesis? *Trends in Microbiology*, *13*, 58–63.

Fux, C. A., Stoodley, P., Hall-Stoodley, L., & Costerton, J. W. (2003). Bacterial biofilms; a diagnostic and therapeutic challenge. *Expert Review of Anti-infective Therapy*, *1*, 667–683.

Hall-Stoodley, L., Costerton, J. W., & Stoodley, P. (2004). Bacterial biofilms: From the natural environment to infectious diseases. *Nature Reviews Microbiology*, *2*, 95–108.

O'Toole, G. A., Kaplan, H. B., & Kolter, R. (2000). Biofilm formation as microbial development. *Annual Review of Microbiology*, *54*, 49–79.

Pratt, L. A., & Kolter, R. (1999). Genetic analysis of biofilm formation. *Current Opinion in Microbiology*, *2*, 598–603.

Stewart, P. S., & Costerton, J. W. (2001). Antibiotic resistance of bacteria in biofilms. *Lancet*, *358*, 135–138.

Stoodley, P., Sauer, K., Davies, D. G., & Costerton, J. W. (2002). Biofilms as complex differentiated communities. *Annual Review of Microbiology*, *56*, 187–209.

Chapter 4

Microbial Mats

R.W. Castenholz
University of Oregon, Eugene, OR, USA

Chapter Outline

ABBREVIATIONS

DBL Diffusion boundary layer
EPS Extracellular polysaccharides
ESSA Exportadora de Sal
ITS Internal transcribed spacers
MAAs Mycosporine-like amino acid derivatives
NCBI National Center for Biotechnology Information
PCR Polymerase chain reaction
PS Photosystem
TEM Transmission electron microscope

DEFINING STATEMENT

Microbial mats are top accreting, cohesive microbial communities that are often laminated and are found growing at the sediment–water interface in shallow fresh or saline waters or on intertidal flats. They may be ephemeral (seasonal) and only a few millimeters thick or perennial (i.e., remaining for several years) and acquiring a thickness of several millimeters to more than a meter. A few become hardened with calcium carbonate, but most at the present time remain organic and relatively soft and gelatinous. They may be regarded as ecosystems in which an energy input (usually solar radiation) and various nutrient salts result in a cascading tier of biochemical transformations. Biofilms are very thin cohesive microbial communities that develop on many surfaces and in some cases may develop into "mats." Although microbial mats are often referred to as modern or living stromatolites, the latter term technically refers only to "fossilized" or lithified structures. Proterozoic stromatolites (those $<2 \times 10^9$ bp) are of certain biogenic origin. These originated as organo-sedimentary microbial mats that are produced by sediment trapping, binding, and followed by cementation and/or precipitation of crystalline carbonate minerals, the latter often associated with the growth and metabolic activity of microorganisms. These structures also show laminations that are of the same scale as those in living microbial mats (i.e., in the order of millimeters). Modern microbial mats seldom lithify except in certain tropical regions where calcium carbonate precipitates, often with the aid of cyanobacteria.

What are normally referred to as stromatolites were formed by microbial mats that developed in great abundance in shallow seas of uncertain salinity along continental margins throughout the Proterozoic (~2.5–0.6 $\times 10^9$ years bp)

Topics in Ecological and Environmental Microbiology.

and apparently also to some extent during earlier times of the Archaeon (perhaps as early as 3.5×10^9 years before the present, bp).

INTRODUCTION

Even laymen are now beginning to understand the general meaning of a "microbial mat," although this knowledge is still largely superficial even with specialists who work in the field of microbial ecosystems. Much is known of the upper, light-exposed, few millimeters, those microlayers composed mainly of photosynthetic bacteria, both O_2-producing (oxygenic) cyanobacteria and algae as well as non-O_2-producing (anoxygenic) photosynthetic bacteria, such as purple sulfur bacteria, green sulfur bacteria, and bacteria of the *Chloroflexus* phylum (the Chloroflexi). However, the mainly anoxic (O_2-free) zone that lies usually only a few millimeters or centimeters below the zone of visible light and IR penetration is still very much a "black box," although much new information is accumulating about the activity of sulfide-producing bacteria and methane-producing archaea in this oxygen-free zone.

The living community or ecosystem known as a microbial mat is usually a "gelatinous" or even leathery sheet or carpet that often holds together as a unit that can be lifted or excised intact from the substrate, whether the substrate be sand, mud, soil, or rock. In a few cases, these mats may become calcified by biologically enhanced precipitation, although this scenario was more common in the distant past. These mats grow (accrete) from the top downward with the uppermost microorganisms providing the new organic material through the photosynthetic incorporation of CO_2. Microbial mats, although they may be perennial formations that persist for years or decades or more in very stable environments, must by definition begin as biofilms. Biofilms can be defined as films of microbial cells, often definitively layered, within an extracellular matrix of polymeric substances (e.g., polysaccharides), the same or similar compounds that bind the various microorganisms of thicker structures termed microbial mats. Biofilms and microbial mats have much in common. They may sequester nutrients or at least make use of nutrient turnover that takes place within the mat, rather than being released. Both may be protected from herbivory or from the input of toxic substances, but on the other hand, the microorganisms are tightly packed and competition for nutrients may be intense. If genetic exchange is possible among some of the organisms, the crowded condition of the mat could facilitate that phenomenon. It is apparent that growth rates of organisms in general within a mat or biofilm may be very low when compared to an independent, free-living existence, except perhaps for the topmost cell layers that are frequently or constantly bathed in moving water containing inorganic nutrients.

It is difficult to identify the microorganisms in a mat complex by using conventional light microscopy or by transmission electron microscopy (TEM). Consequently, cultures of the many types seen microscopically can be made, but many are recalcitrant to common isolation and culturing methods. It should be realized that the present taxonomy used is often incorrect phylogenetically. Several of the groupings presently used are artificial and not related to genetic relatedness. However, the nearest relatives based on the Polymerase Chain Reaction (PCR) and the comparative sequences of the 16S rDNA gene in prokaryotes or the 18S rDNA gene in eukaryotes or on other genes for either group can be used to identify the organisms. Since the database is incomplete, no near relative may pop up, and the species or genus must rest in limbo, although a name and description based on the visual morphotype can be applied temporarily. This works with many types of cyanobacteria and algae that have structural complexity that may be visualized microscopically, but not with the vast majority of small, unicellular bacteria and eukarya that are unicellular rods or spheres. In addition to PCR and the comparative sequences of various genes, biomarkers (namely specific fatty acids, lipids, hopanoids, carotenoids), and a few other cellular components may be used to distinguish naturally related groups of microorganisms. Cyanobacteria generally form the top layer of modern microbial mats. This community of photosynthetic cells use primary CO_2 from the bathing water to build the organic cellular compounds that eventually, through death and decomposition or by leakage or excretion, support the large numbers of nonphotosynthetic microorganisms in the mat underlayer.

Some cyanobacteria in many planktonic habitats (particularly in freshwater lakes) are well-known producers of inhibitory compounds such as a great variety of potent hepatotoxins and neurotoxins that presumably aid in the dominance of the producers by hampering the feeding activity of potential herbivores. However, there is little evidence that cyanobacteria of microbial mats produce these compounds, although it is probable.

HABITAT DISTRIBUTION

Although microbial mats apparently occurred globally in shallow ancient seas of the Precambrian (the "age of stromatolites" earlier than 0.6×10^9 years ago), a time before the advent of efficient, grazing invertebrates, they are now almost entirely relegated to extreme environments where these grazers, such as gastropods, do not exist or are rare.

Mats of Hypersaline Waters

Salinities above about 65‰ total salts (normal seawater being approximately 30–35‰) exclude almost all invertebrate animals, except for brine shrimp and relatives as well as a few copepods and nematodes (meiofauna). However, protists (protozoa) may often occur at these and higher salinities. Hypersaline environments occur in various cul-de-sacs of the oceans (e.g., Shark Bay, Australia, sabhkas (salinas) of the Red Sea (Gulf of Aqaba), lagoons near Guerrero Negro, Mexico) and also in human-manufactured salinas (or salterns) where salinity of seawater is purposely raised through water retention and evaporation. Many show laminations in the same scale as in lithified stromatolites (Figure 1). This results in downstream shallow ponds of increasing salinity, resulting eventually in a saturated fluid and NaCl precipitation (sea salt). Although most hypersaline mat-inhabiting waters are derived from seawater, some inland saline lakes with quite different chemistries also develop microbial mats. The top layer of most hypersaline mats is usually composed mainly of cyanobacteria (formerly called blue-green algae), although diatoms may also occupy a predominant role. Both of these microorganisms photosynthesize in the same manner as green plants and thus evolve O_2 as a "waste product." Hypersalinity indicates salinity that is several times higher than that of seawater. Consequently, special adaptations to compensate for osmotic stress have evolved. Most cyanobacteria use internal concentrations of sucrose, trehalose, glucosylglycerol, or glycine betaine (trimethyl glycine), the last being the most halotolerant species. Similarly, glycine betaine is the most common osmoprotectant in the anoxygenic purple bacteria that inhabit the undermat in most saline habitats. These osmoprotectants are compounds not used for metabolic processes but serve as soluble compounds that simply match the concentration of salts in the environment, thus preventing plasmolysis or cell bursting.

FIGURE 1 An excised section of a laminated microbial mat from a hypersaline pond near Guerrero Negro, Baja California Sur. A green cyanobacterial layer is seen just below the topmost layer. Note that 8 mm intervals are seen on the 1-ml syringe. The orange coloration is due to the carotenoids of cyanobacteria (and sometimes partially from Chloroflexus-like bacteria). However, the orange color of Haloarchaea ("halobacteria") does not contribute to these mats, since they are, at most, a minor component of microbial mats. (For interpretation of the references to color in this figure legend, the reader is referred to the Web version of this chapter.)

Since marine-derived hypersaline waters contain relatively high concentrations of sulfate (SO_4^{2-}), sulfate-reducing bacteria in the darkened undermat reduce sulfate (or elemental S) to sulfide (HS^- or H_2S). The photosynthetic purple and green sulfur bacteria thrive in the low-intensity IR radiation immediately underneath the cyanobacteria or diatoms (that absorb almost all of the light of wavelengths less than 700 nm), using the sulfide as the reductant in photosynthesis instead of H_2O. These bacteria do not produce O_2 during photosynthesis, but their special types of bacteriochlorophyll (rather than the chlorophylls of the cyanobacteria and diatoms) are able to use low levels of near IR radiation that is invisible to the human eye for photosynthesis.

Although O_2 from the surface water, and particularly from O_2 produced by cyanobacteria and/or algae from the surface layers, may be supersaturated in the surface layers of mat in the daytime and diffuse downward into the dark undermat, and as soon as light fails or decreases sufficiently, the O_2 disappears often within a few minutes because of respiration of the phototrophs and numerous other microorganisms (Figure 2).

Many microbial mats, particularly of saline or hypersaline waters, are in habitats that are nitrogen-limited. This condition enriches cyanobacteria that are capable of fixing dinitrogen (N_2), the sole property of some, but not all, prokaryotes. However, most of the cyanobacteria that inhabit microbial mats require darkness and anoxia in order to carry out this process, since the activity of the enzyme complex, nitrogenase, does not tolerate O_2. Thus, most of the conversion of N_2 to ammonia (NH_3) and eventually to amino acids and other nitrogenous cellular compounds in the cell takes place at night. Although cyanobacteria with specialized cells called heterocysts fix N_2 in the daytime, these species are relatively uncommon in submerged microbial mats in saline environments.

Mats of Geothermal Springs

Another habitat of microbial mat development is found in geothermal springs, particularly alkaline types (pH 7–9). Functionally, these mats are in many ways similar to the mats of hypersaline waters. There is usually a cyanobacterial top layer (1 to a few millimeters thick) underlain by purple or green sulfur bacteria or by other anaerobic photosynthetic bacteria such as members of the Chloroflexi. Such hot springs are found on numerous islands and all continents except Antarctica, although hot steam vents occur on two of the active volcanoes there. Although less common, acid hot springs may also develop mats. Hypersaline habitats have salts that are concentrated from seawater. Hot spring waters, however, differ greatly in their chemistry, and are usually less than 2–3‰. Saline hot springs on land are rare, but are common at various depths in the oceans

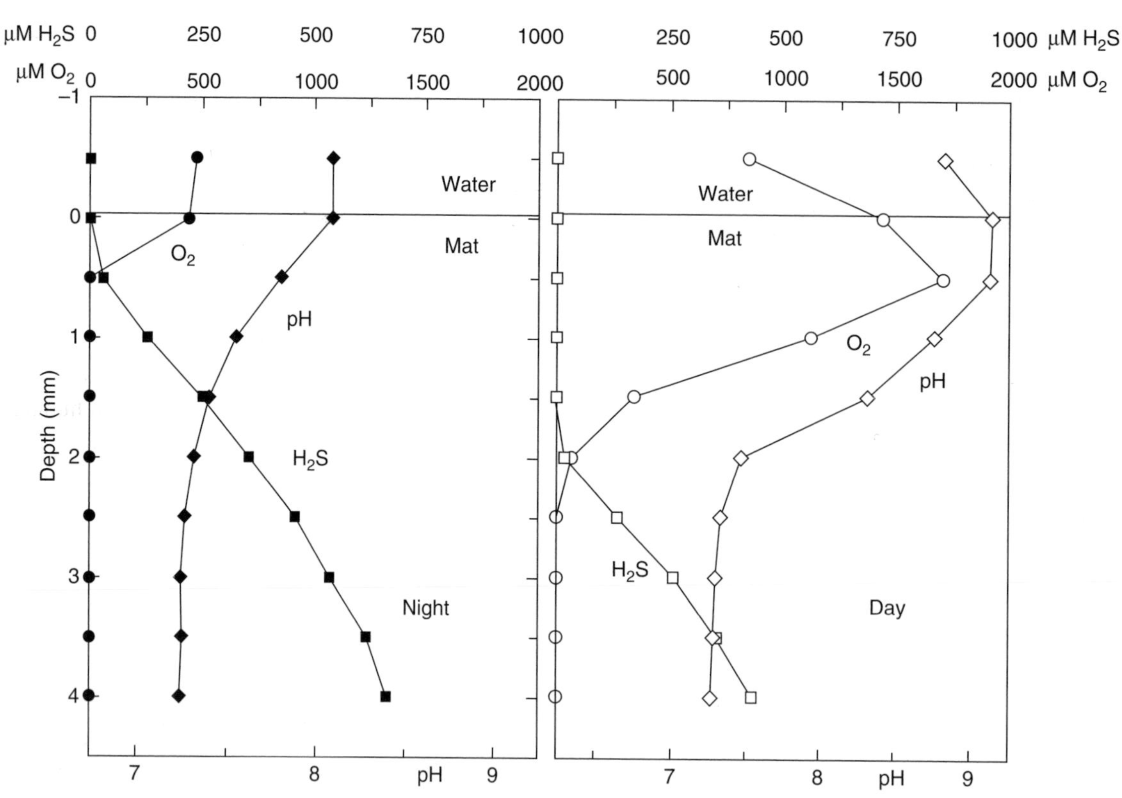

FIGURE 2 O_2, sulfide, and pH shown as profiles with depth (mm scale) in a microbial mat from Solar Lake, Egypt: Nighttime (left panel), daytime (right panel). *Reproduced with permission from Stal, L. (2000). Cyanobacterial mats and stromatolites. In B. A. Whitton, & M. Potts (Eds.).* Ecology of cyanobacteria: Their diversity in time and space *(pp. 61–120). Dordrecht: Kluwer Academy.*

(hydrothermal vents), usually below the limit of light penetration. The most-studied terrestrial springs are alkaline in which sodium, chloride, and silicates are the major ions with various concentrations of minor ions, including possibly limiting nutrients such as phosphorus or combined nitrogen. Many hot spring mats are laminated in a manner similar to that of hypersaline mats (Figure 3). The water of hot springs is commonly flowing, and therefore, the microbial components of the top layer need to be bound in some manner. Usually, this is owing to the extracellular exudates of several members of the community.

Hot spring mats of neutral to alkaline pH waters

Hot springs in North America that issue at temperatures well above 73–74 °C, with pH values of 7 or above, usually show a gray or nongreen zone where nonphotosynthetic prokaryotes occur. As temperatures cool to about 73–74 °C, a visible green to yellow-green biofilm of *Synechococcus* occurs (Figure 4). As temperatures decrease to about 68 °C in North America, the biofilm thickens because of the development of an undermat of *Chloroflexus* or other related Chloroflexi (Figure 4). At temperatures above about

FIGURE 3 Core of laminated mat from Octopus Spring, Lower Geyser Basin, Yellowstone National Park. A green layer the cyanobacterium, *Synechococcus*, is seen at the top. The orange color is imparted by Chloroflexus-like bacteria and partially degraded cyanobacteria. The core was taken from the mat at 55 °C, pH 8. *Reproduced with permission from Ward, D. M., Weller, R., Shiea, J., & Castenholz, R. W. (1989). Hot spring microbial mats: Anoxygenic and oxygenic mats of possible evolutionary significance. In: Y. Cohen, & E. Rosenberg (Eds.).* Microbial mats: Ecological significance of benthic microbial communities *(pp. 3–15). Washington, DC: American Society for Microbiology.* (For interpretation of the references to color in this figure legend, the reader is referred to the Web version of this chapter.)

FIGURE 4 View of Imperial Geyser pool, Midway Geyser Basin, Yellowstone National Park. The light yellow-green zone marks the beginning of the high-temperature biofilm mat of the cyanobacterium, *Synechococcus*, at about 73 °C. The deeper orange-brown region begins at about 68–69 °C, and includes an undermat of *Chloroflexus*-like anoxygenic bacteria. (For interpretation of the references to color in this figure legend, the reader is referred to the Web version of this chapter.)

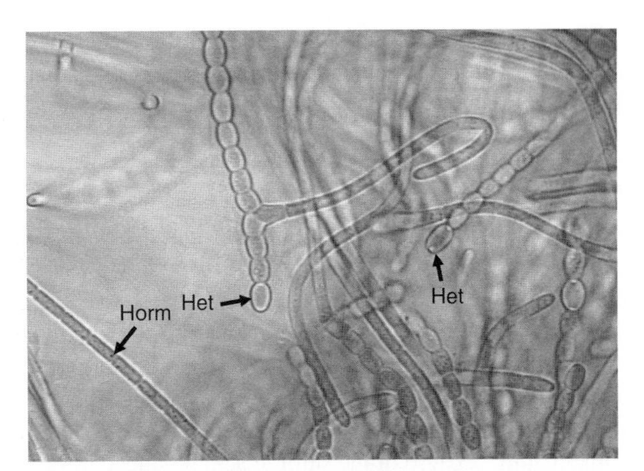

FIGURE 5 Photomicrograph of *Mastigocladus* (=*Fischerella*) showing branching, heterocysts (het) and a hormogonium (horm) that is a motile gliding stage. The width of the hormogonium is ~3.5 μm.

45 °C, the top layer of photosynthetic microorganisms is composed of cyanobacteria, since very few eukaryotic algae can tolerate temperatures above 45 °C. This layer may be only 1–2 mm thick, since the attenuation of visible light (i.e., wavelengths of ~400–700 nm) by the cyanobacteria is almost complete when the depth of the cyanobacterial zone is exceeded. This would be equivalent to 1 to >100 m in a water column of a lake or the sea. Cyanobacteria may be of the unicellular type (e.g., *Synechococcus*). Various species or varieties of this genus may form biofilms or mats up to a temperature of ~73 °C in the hot springs of western North America and eastern Asia (the Pacific Rim). However, high-temperature forms of *Synechococcus* are apparently missing from the rest of the globe. For example, *Synechococcus* species and other cyanobacteria of hot springs in Europe and New Zealand grow only up to temperatures of about 60–62 °C. Thermophilic *Synechococcus* species are missing altogether in the numerous alkaline hot springs of Iceland. In all alkaline hot springs of the globe that issue at temperatures of 60–62 °C or below, numerous genera and species of cyanobacteria occur and many of these are filamentous. Some species appear to be restricted to one geographic area (e.g., New Zealand), while another filamentous type (generally referred to as *Mastigocladus laminosus*) is present up to 57–58 °C in neutral to alkaline hot springs almost everywhere, including Iceland (Figure 5). This cyanobacterium is thought to spread easily, since it is quite capable of surviving desiccation, low temperature, and freezing. A few alkaline hot springs contain sulfide (i.e., HS^- or H_2S) in the source water of the spring and this compound excludes many species of cyanobacteria when the concentration in the bathing waters is high (i.e., >15–20 μM). Some cyanobacteria of alkaline hot springs, namely *Mastigocladus*, produce heterocysts when the waters are low in combined nitrogen (Figure 5). These specialized cells generate internal anoxia and thus are able to fix N_2 in the daytime. Other cyanobacteria of hot springs, which do not produce heterocysts, are known to shift to N_2 fixation as darkness settles, as in many saline or hypersaline mats.

Below the cyanobacterial top layer, in the virtual absence of visible light, there is often a reddish-to-orange layer that consists primarily of phototrophic bacteria that lack the ability to produce oxygen, as in hypersaline mats. These are very prevalent in hot springs and are composed of one or more members of the Chloroflexi phylum, such as *Chloroflexus* and *Roseiflexus*. Their primary method of metabolism and growth is to use the IR radiation wasted by the cyanobacteria for their energy source, but they usually use organic compounds for cell building and growth, compounds that are ultimately derived from the primary producing cyanobacteria.

Below about 45 °C in alkaline hot springs, eukaryotic algae may also occur. However, mat building is not common, because below this temperature, grazing animals such as ephydrid fly larvae (and adults) and small crustaceans such as ostracods and amphipods tend to decimate the various mat-producing microbiota.

Most of the mats of alkaline hot springs described above are soft, but in some areas the deposition of silica (SiO_2) is rapid and the mats become brittle and hardened (e.g., Upper Geyser Basin, Yellowstone National Park). There are also several locations globally where the spring waters are rich in calcium and bicarbonate rather than sodium, chloride, and silicates (e.g., Mammoth Hot Springs, Yellowstone). The pH at the source is usually about 6.5 and this rises to 7.0 or over when the charge of CO_2 is lost through effervescence. These springs deposit large quantities of $CaCO_3$ (with some magnesium) often in the form of travertine terraces. The microbial mats are usually thinner, and sometimes become embedded within the travertine. Although precipitation of the travertine would and does occur without the cyanobacterial/microbial mat, the microorganisms influence the morphology or shape of the deposits to some degree.

Mats of acid hot springs

Although common only in regions of current or recent volcanic activity, pH levels below about 4.0 and down to ~0.0 exclude all photosynthetic bacteria, including the cyanobacteria. However, there is an abundant group of unicellular eukaryotic algae (species of three or more genera of the red algal order Cyanidiales) that forms a dominant feature in warm to hot acidic streams and pools (Figure 6). These springs are acidic mainly because of the sulfuric acid produced through oxidation of sulfide. These streams are often a brilliant blue-green color (or more yellow-green under the high light of summer) due to the sheer numbers and great biomass of this one and only type of photosynthetic microorganism that lives in acid waters and in a temperature range of about 40–56 °C. In some cases, particularly in Yellowstone National Park, stable microbial mats may accrete to thicknesses of over 0.5 cm. In these mats, however, the most prominent heterotrophic utilizer of the organic compounds produced by the photosynthesizer may be one or more types of fungi.

Microbial Mats of Antarctic and Arctic Ponds and Lakes

It was a surprising discovery that freshwater to saline melt ponds in the Antarctic and high Arctic also developed microbial mats dominated by cyanobacteria (Figure 7). In fact, without specific identification of the organisms, they mimic, visually, the mats of many hypersaline and hot spring habitats in temperate climates that Cyanobacteria had previously been known to dominate primarily in warmer water. However, the cyanobacteria known from these cold waters that are ice-free for only 2–3 months of the year are very slow growing both in culture and apparently *in situ* at the low temperatures experienced in these waters (4–10 °C). It is clear that these slowly accreting and often perennial mats develop because of the absence of efficient grazers, and the inability of many potential eukaryotic algal competitors or herbivores to tolerate the winter conditions when most of these ponds freeze throughout the water column, not simply on the surface as in temperate climates. In many of these melt ponds that are slightly saline, a small amount of liquid bottom water remains after freeze-up, and this consists of a highly concentrated brine. Thus, the microbes located there must be halotolerant in addition to tolerating extreme cold as well as darkness for many months. In ponds at 78° S (e.g., McMurdo Ice Shelf) the only herbivores of note are rotifers, nematodes, and unicellular protists.

FIGURE 6 View of Lemonade Creek, Yellowstone National Park at ~45 °C and pH 2. The green color of the mat is imparted by a member of the Cyanidiales, namely Galdieria sp. (For interpretation of the references to color in this figure legend, the reader is referred to the Web version of this chapter.)

FIGURE 7 View of a melt pond (Casten Pond) on the ablation moraine of the Ross Ice Shelf, Antarctica, a few hundred meters from Bratina Island. The submerged mats are seen in orange-brown color near the shores and particularly in the narrows between the two sections of the pond. (For interpretation of the references to color in this figure legend, the reader is referred to the Web version of this chapter.)

Terrestrial Microbial Mats and Crusts

Surprisingly, extensive microbial mats or crusts, with many of the same attributes and microbial components of submerged mats, occur in hot and cold deserts of the world (Figure 8). For example, extensive mats occur on the Colorado Plateau, and on the Antarctic Peninsula. They are slow developers and therefore do not withstand the disturbance of humans or livestock. In undisturbed areas, they may form a ground cover between and around desert shrubs. However, if disturbance occurs, recovery may take many years or never, even if the cause of disturbance is removed.

Ephemeral Mats

Many relatively thin microbial mats develop for only a short time, that is, seasonally. For example, major oil spills in the Persian Gulf have formed thickened tar-like

FIGURE 8 Desert microbial crust, dominated by scytonemin-containing cyanobacteria, near Moab, Utah. The light-colored patches are areas from which the crust has been removed. *Reproduced with permission from Whitton, B. A., & Potts, M. (Eds.) (2000).* Ecology of cyanobacteria: Their diversity in time and space. *Dordrecht: Kluwer Academy.*

coverings in the intertidal and have developed mats that are similar to mats in other saline habitats, with cyanobacteria forming the dominant top layer. The degradation of petroleum products takes place as a result of the activities of the entire community, not simply of the cyanobacteria. Other ephemeral mats, although thin, are well known, particularly in tropical sandy habitats, although they might best be regarded as ephemeral cyanobacterial biofilms.

However, many mats that persist for many months do develop over gently sloping, extensive intertidal flats (e.g., sand flats of the North Sea). Some mats are able to persist, even for years (not very ephemeral), because of the multiple conditions that exclude grazers. These conditions include evaporative drying with increasing salinity during periods of neap tides or the persistence of high sulfide in some marine intertidal marshes that tend to exclude gastropods.

Mats on Tropical Corals

Black band disease is a disease of certain shallow subtidal corals that actually consists of a creeping microbial mat dominated by a dark phycoerythrin-rich oscillatorian cyanobacterium and other rather typical microbial components, including sulfate-reducing bacteria. The sulfide produced may be involved in the death of the coral underlying the mat.

COMPREHENSIVE INVESTIGATIONS

Rigorous studies of microbial mats have concentrated on mats located in relatively few areas. Only two are discussed here. Hypersaline mats have been extensively studied in the lagoons and salinas in the area of Guerrero Negro, Baja California Sur, Solar Lake, Sinai, Egypt, and the Camarque region of southern France. Although much is now known about the microorganisms of these and other mats, only recently has the common presence of phages (bacterial viruses) been documented. Most of the detailed studies of neutral to alkaline thermophilic mats have been in Yellowstone National Park, Wyoming, USA (e.g., Octopus and Mushroom Springs, and the travertine terrace-building springs of the Mammoth area). Acid spring microbial mats have also been extensively studied in Yellowstone Park, particularly in the Norris Geyser Basin. Microbial mats of Antarctica have been studied extensively in the ponds in the Bratina Island area of the Ross Ice Shelf, and in the region of the Antarctic Peninsula. Terrestrial mats and crusts of the Colorado Plateau have been studied intensively, as have the mats of the inselbergs (hammock-like limestone outcrops) of Venezuela and the "Guayanas."

Microbial mats are not static entities. Within various vertical zones or strata, various biochemical transformations are taking place in a diel manner or seasonally, and many cases of mutualism and dependence exist among the various individual types of microorganisms and also among guilds of organisms that more or less serve the same function. There is also movement, particularly vertical movement, that may be initiated by light intensity, the changing light regime, and the chemical gradients that exist or develop. Cyanobacteria and other photosynthetic microorganisms on the mat surfaces produce organic compounds that may be excreted, leaked, or released with cell lysis. These compounds (sugars, amino acids, glycollate, various organic polymers, etc.) may then be metabolized by fermentative bacteria in the anoxic undermat that largely excludes O_2 at least during nighttime. The production of H_2, CO_2, and acetate by some of the fermenters promotes the growth of methane-producing archaea, although in marine or some hot spring mats the presence of high levels of sulfate (SO_4^{2-}) favor the activity of sulfate-reducing bacteria that use most of the same organic compounds as methanogens or hydrogen gas, with the resultant production of H_2S.

The microenvironment within microbial mats has been accessed by various types of microelectrodes that can penetrate mechanically into mats at intervals of a few microns. The most common types of electrodes used today are those that individually measure O_2, pH, CO_2, sulfide, and several other ions. Thus, 24h measurements of these conditions allow the researcher to know when conditions are anoxic or oxic (Figure 2). Microprobes that measure light intensity at different depths within a mat have also been developed and used. For example, a complete spectrum of penetrating light can be completed within seconds using a recording spectroradiometer with a microprobe. This is not a trivial type of data collection, since the changes in various conditions and activities within a few microns depth in a mat may be equivalent to what is happening over many meters deep in the water column of a lake or the sea.

CASE STUDIES OF SELECTED MICROBIAL MAT ECOSYSTEMS

Microbial Mats of Guererro Negro, Mexico

The seawater evaporation ponds of Exportadora de Sal (ESSA) have been studied extensively by biologists and geochemists from many parts of the globe; more often than not, they were under the guidance of Dr. David J. Des Marais of NASA. They are adjacent to the Laguna Ojo de Liebre at 28° N on the west side of Baja California. In summer the water temperatures generally range from 22 to 29 °C, whereas winter temperatures are between 14 and 22 °C. In ponds above about 65‰ total salinity, the invertebrate population is restricted to meiofauna that are not efficient or abundant enough to prevent the accretion of microbial mats dominated by cyanobacteria and diatoms on the surface. From about 65 to ~100‰, the cyanobacteria are predominantly species of *Microcoleus*, *Oscillatoria*, and *Spirulina*. At higher salinities (>100 to about 200‰), unicellular types tend to predominate at the surface (e.g., *Aphanothece*, *Cyanothece*). In some ponds or lagoons where NaCl is at saturation and may be precipitating (> ~ 265‰), the green alga, *Dunaliella salina* (which accumulates much β-carotene), occurs along with the ever abundant orange-red, nonphotosynthetic haloarchaea (usually referred to as halobacteria), forming dense planktonic populations. However, these haloarchaea occur in abundance only in the highest salinity crystallizing ponds or lagoons where no microbial mats occur. These archaea contain the carotenoid (bacterioruberin) imparting the bright orange-red coloration. However, if O_2 deficits occur, these organisms also synthesize bacteriorhodopsins and/or halorhodopsins in special membranes (purple membranes) that create a light-driven proton pump that produces ATP as a supplement to their normal chemoheterotrophic metabolism. However, these haloarchaea are planktonic (containing gas vesicles) and are an insignificant component of microbial mats. However, even above 125‰, embedded in the precipitated $CaSO_4$ (gypsum), filamentous oscillatorian-type cyanobacteria are abundant. At the surface of the soft mats, there is a diffusion boundary layer (DBL) of a few millimeters through which flow is restricted, although diffusion and exchange of nutrients takes place. The undermat at these ponds is highly anoxic a few to several millimeters below the surface in the daytime, depending on how translucent the mat is with respect to light penetration and how deep the O_2 diffuses. However, at night anoxia essentially persists at the surface (see Figure 2). The production of sulfide (H_2S, HS^-, S^{2-}) by anaerobic sulfate-reducing bacteria is extensive wherever anoxia occurs, and the sulfide diffuses to the surface during the night. In this general region, members of the Chloroflexi also occur. These are phototrophic filamentous bacteria that use dim light, especially in the IR region, and one at least ("*Chlorothrix*") uses sulfide as the photosynthetic reductant. Vertically migrating filamentous sulfide-oxidizing, but nonphototrophic, bacteria (e.g., *Beggiatoa*) tend to follow the narrow, moving boundary of O_2 and sulfide on a diel (24 h) basis, since most of these sulfide oxidizers require both sulfide and O_2 for respiration. In the saline ponds of Guerrero Negro, much soluble organic matter is released from the mats, and the waters that flow slowly from pond to pond of increasing salinity foam easily with wind and wave action.

Depending on light intensity, several of the filamentous cyanobacteria make a diel movement downward within the mat (in bright light and UV) and upward in low light or darkness (see "UV stress").

It should be pointed out that these microbial mats develop and accrete to nearly a meter in thickness in some ponds with only about 1–2 m of water above them. They do not precipitate calcium carbonate and thus remain soft throughout their existence. Accumulation of mat, however, is possible only because efficient grazers, such as gastropods, do not inhabit these saline waters, and higher aquatic plants and macroalgae do not compete because of their absence. However, in much earlier times (during the Precambrian and up to about 0.57×10^9 years bp), no efficient grazing animals were present or abundant enough to require microbial mats to "take refuge" in hypersaline waters. Thus, it appears from the fairly extensive Precambrian record of stromatolites that mats such as these developed abundantly in shallow seas almost everywhere, whether saline or not.

It appears that the actual growth rate of the mat (i.e., its accretion rate) is quite slow and that most of the time this compact mass of organisms in a gel-like matrix is simply maintaining itself and recycling its essential nutrients (e.g., N and P), with some loss of inorganic (i.e., CO_2, HCO_3^-) and organic carbon to the water above.

Along the widespread intertidal flats of the natural shoreline of Laguna Ojo de Liebre an almost complete coverage by a thin leathery and persistent mat of the cyanobacterium *Lyngbya* sp. occurs, a mat that is covered by tidal waters of about 35–50‰ salinity only during the twice monthly occurrences of spring tides (Figure 9). Bordering the upper edge of this mat is a corrugated mat of *Calothrix* sp., a cyanobacterium that is covered by water even less often than that of *Lyngbya*. However, when covered, it fixes N_2 in the daytime by use of its heterocysts, whereas *Lyngbya* fixes N_2 at night during anoxic conditions, since heterocysts are not produced by this cyanobacterium.

FIGURE 9 Intertidal flats at Laguna Ojo de Liebre, Baja California Sur, Mexico. The dark area is covered by the scytonemin-containing cyanobacterium, *Lyngbya* sp. This area is covered by seawater of 35–50‰ salinity during days of spring tides (i.e., every 2 weeks).

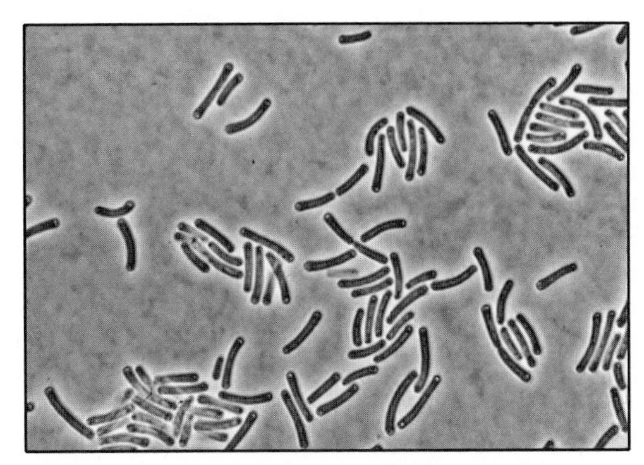

FIGURE 10 Photomicrograph of thermophilic *Synechococcus* sp. isolated in culture from an alkaline hot spring in Yellowstone National Park. A few of the cells may be seen dividing by fission across the midsection of the cell. The width of a typical cell is ~1.2–1.5 μm.

Microbial Mats of Octopus and Mushroom Springs, Yellowstone National Park

Alkaline hot springs in the Lower Geyser Basin of Yellowstone National Park have been studied extensively by associates, students, and visitors of the lab of Dr. David M. Ward at Montana State University, as well as by other investigators from many parts of the globe. They are structurally quite similar to the submerged hypersaline mats of Guerrero Negro, although the species present are quite different, but with cyanobacteria again forming the top accreting layer (Figure 3). The description of hot spring microbial mats in the Section "Hot Spring Mats of Neutral to Alkaline pH Waters" is based primarily on the situation in Octopus and nearby Mushroom Springs. However, more details are added here. Instead of basing species identification solely on morphology, pigments, and physiology of culture isolates, a large effort using molecular/genetic methods has been made. Most of the work has focused on the cyanobacterial genus *Synechococcus* (Figure 10). Not only are there a few species indicated by the sequence comparisons of the 16S rDNA gene, but slight differences in these sequences and greater differences in the sequences of the internal transcribed spacers (ITS) separating 16S and 23S rRNA genes have indicated that there are many more variants within a cluster of seemingly identical cyanobacteria, each of which may be classified as a separate ecotype, meaning that certain small sequence differences may be correlated with differences related to dissimilarity in environment, such as temperature, light intensity, and chemistry optima and tolerances. One of the major contributions from the study of these springs has been the collection of environmental and genetic data over 24 h periods, showing that gene expression for certain physiological traits is not static. For example, when healthy high-temperature *Synechococcus* biofilms encounter normal high solar radiation in summer during late morning, inhibition of photosynthesis occurs (less so when UV radiation is filtered out), and recovery does not occur until the following morning, suggesting that repair and rehabilitation occurs mainly during the dark. Recent work from the Ward lab has shown that previously unknown N_2 fixation genes of *Synechococcus* are turned on during the evening after anoxic conditions develop, that N_2 fixation occurs then, and that genes for fermentation in *Synechococcus* are also switched on during the night period, while genes involved in photosynthesis and aerobic respiration are turned off as would be expected in a dark anoxic environment. Presumably, fermentation provides enough ATP required for N_2 fixation, a process that requires much ATP. Previously, without the intensive 24 h molecular and physiological monitoring, the fact that *Synechococcus* was capable of N_2 fixation was unknown. Additional work with these hot spring mats has shown that differences in ecotypes occurred with relatively small changes in the gradients of temperature and vertical light intensity within the mat, but all within the same species of *Synechococcus*. This type of work has just begun to reveal the complexity of mat interrelationships and functions.

In addition to the photoautotrophic *Synechococcus*, anoxygenic members of the Chloroflexi that make up a major portion of the undermat have been classified by molecular means in these springs. These include the phototrophic members of the genus *Roseiflexus*, *Chloroflexus*, and undescribed members of this phylum. Although sulfate-reducing bacteria operate in these springs, their role is minor compared to the methanogenic archaea that use H_2, CO_2, and/or acetate to produce methane (CH_4) as a waste product.

STRESSES EXPERIENCED BY MICROBIAL MATS

It is likely that most of the organisms within microbial mats are under some form of stress most of the time. Even if the mats occur in flowing systems only the uppermost cell layers have the benefit of continuously renewed nutrients, and at the same time these cells may be under the day-time stress of too much light and UV radiation. Most of the multitude of aerobic and anaerobic microorganisms within the crowded mat matrix may be nutrient starved, reductant starved, light starved, or high light inhibited depending on the circumstances. The combination of various factors normally results in a very slow growth rate or simply represents maintenance of life in the stationary phase with death of some cells being replaced by newly divided cells, and this over long periods. It has been shown, however, that an area recently denuded of mat (naturally by waves or hail storms or experimentally) will be colonized rapidly with a new biofilm and a relatively fast growth until the thickening materializes into a mat.

High Light and UV Stress and Responses by Microorganisms

Although most of the microorganisms inhabiting the uppermost layers of microbial mats are phototrophs (i.e., requiring light for energy), many biologists do not realize that too much light can be extremely detrimental. Since cyanobacteria generally constitute the uppermost layer of organisms in microbial mats, they have evolved various strategies to cope with the damaging effect of high solar radiation, which includes the most detrimental portion of this radiation, the UV component. Photons (expressed as quanta of energy) are absorbed by light-harvesting pigment molecules and then transferred from molecule to molecule to a reaction center where photochemistry occurs. However, the reaction center can become "saturated" and the excess quanta can excite O_2, for example, and produce the destructive singlet state ($^1O_2$). Carotenoid pigments in excess can quench such reactions and dissipate the energy as heat. Carotenoids are also important in preventing reactions that would ordinarily result in the production of the very destructive OH radical. The phenomenon of light intensity-dependent pigment regulation in cyanobacteria and algae has been known for many decades. In general, the exposure to high light results in the active regulated degradation of pigments, especially of phycocyanin (the blue pigment of cyanobacteria), but also of chlorophyll. In some phototrophs, however (e.g., anoxygenic purple bacteria), the response to high light may be simply the cessation of pigment synthesis (e.g., of bacteriochlorophyll) and the consequent dilution of the pigment with each new cell division. In each case, the result is a lower pigment matrix for absorbing photons. As mentioned above, absorbance of photons in excess of what can be used photosynthetically usually results in the production of detrimental species of oxygen, causing photoinhibition or death. However, when light intensity is low, photons may become scarce, and this results in the synthesis of a greater content of light-harvesting pigments.

Although the shorter wavelengths of UVR (UVC) ~190–280 nm no longer impact the surface of the earth, the ozone/O_2 shield in the early Precambrian are missing, and a damaging intensity of this germicidal wavelength would have reached the earth. At present, a small percentage of UVB (280–320 nm) does reach the earth's surface and directly affects the DNA, resulting in dimeric photoproducts between adjacent pyrimidines and other mutagenic damage, some of which can be repaired by photoreactivation or excision repair if the cells are metabolically active. Many other physiological processes and compounds are affected negatively by UVR. UVA (320–400 nm) has less direct effect on DNA, but is primarily associated with the production of reactive oxygen species that cause lipid peroxidation, chlorophyll bleaching, phycobilin degradation, and inhibition of growth and differentiation through a variety of targets. In metabolically active cyanobacteria, the process of repair and resynthesis of damaged compounds usually takes place at a rate that compensates for the damage done during periods of high solar radiation.

However, several species of cyanobacteria that occupy exposed surfaces, such as microbial mats in shallow water, produce UV-shielding compounds that function even when the cyanobacterium is inactive metabolically, as during suboptimal temperature exposure, conditions of nutrient starvation, and during periods of desiccation or freezing. Filamentous, colonial, or unicellular cyanobacteria that produce an extracellular sheath or extracellular slime (both often simply referred to as EPS, i.e., extracellular polysaccharides) synthesize, with exposure to UVA radiation in particular, a yellowish-brown pigment called scytonemin that accumulates in the sheath and absorbs up to 97% of UVA radiation, thus preventing exposure within the cells. The content of scytonemin may be over 5% of the total cellular dry weight. It is also an effective absorber in the more dangerous UVB region. Scytonemin is a dimeric indole alkaloid with a molecular weight of 544. In many higher plants anthocyanins and flavones provide protection against UV radiation. Most of this information has come from the work of Dr. Ferran Garcia-Pichel in the lab of Richard W. Castenholz.

A family of mycosporine-like amino acid derivatives (MAAs) represent another type of passive protective compound in cyanobacteria. These are a series of water-soluble low-molecular-weight condensation derivatives of a cyclohexonone ring and amino acid (or imino alcohol) residues. Some MAAs absorb maximally at wavelengths as low as 310 nm; others absorb maximally at longer wavelengths (e.g., 320–360 nm – UVA by definition). These compounds are widespread throughout the cyanobacteria, especially in hypersaline waters. These compounds also occur in eukaryotic algae, but in both groups of organisms they usually reside within the cytoplasm, and are thus less effective as screens. However, at least one species of the cyanobacterium, *Nostoc*, that often forms mats or is a part of terrestrial crusts accumulates MAA–oligosaccharide complexes in the extracellular sheath along with scytonemin, thus forming a twofold UV shield. Either MAAs or scytonemin or both may have evolved early in the Precambrian when the atmospheric penetrance of the UV solar flux was much greater than at present, a time when even the shorter wavelength UVC reached the earth's surface.

The vertical movements of motile cyanobacteria constitute another strategy to avoid the detrimental effects of high-light and UV radiation. Much of this work has come from the lab of Richard W. Castenholz. Although a few types of unicellular cyanobacteria are known to respond to changing light intensities by a slow upward or downward movement or by vertical or horizontal alignment, the most striking movements are by filamentous cyanobacteria in which the trichomes are able to move by gliding motility (i.e., a sliding movement in contact with a solid or semisolid substance; the mechanism is still uncertain) (Figure 11). In the bright light of midday, the migrating cyanobacteria are situated a millimeter or more below the mat surface that may consist of a gravelly surface or of nonmotile cyanobacteria using another strategy. By late afternoon with declining light, the motile cyanobacteria gradually move to the surface where they remain throughout the night, only to descend again with the approach of high light the next morning. It has been shown in the hypersaline mats of the ponds at Guererro Negro, Baja California Sur, that the response cue is primarily to UV radiation. The cyanobacteria that migrate in this manner, when artificially subjected to the high light and UV of midday, are drastically inhibited photosynthetically, probably enough to cause death. The advantage of this type of escape strategy is that these filaments keep their high content of light-harvesting pigments that enables them to take advantage of the low light of morning and afternoon, and of overcast days. The migration system can easily be manipulated by covering portions of a midday mat with filters (e.g., neutral density). This results in the upward migration of the motile cyanobacteria within about 45–60 min. This response is much faster than the pigment regulation method of nonmotile species in preventing the damage caused by high radiation (see below). An almost identical response may be seen in some hot spring mats where a rapidly gliding oscillatorian cyanobacterium exhibits a similar diel upward and downward pattern.

Other types of motile bacteria of microbial mats may also respond to the daily regime of light intensity changes. *Beggiatoa*, a nonpigmented but light-sensitive sulfide-oxidizing chemoautotroph may often be seen at or near the surface of mats during darkness where they take advantage of the free sulfide and low O_2 in the dark surface waters. In a few cases investigated, purple sulfur bacteria exhibit a somewhat similar response. These flagellated photosynthetic bacteria (e.g., *Chromatium*, *Thermochromatium*) occupy a protected position just below the surface in the day, and perform anoxygenic photosynthesis in which sulfide is used as the electron donor (reductant) rather than H_2O. However, much of the sulfide is only partially oxidized, and consequently elemental sulfur is stored in the cells. At night, however, these cells may swarm to the surface of water and respire the internal sulfur to sulfate for energy with O_2 as the oxidant, thus acting as chemoautotrophs. Many of the migratory activities described here are known in hypersaline, marine, hot spring, and even terrestrial mats or crusts when the last named are moist with rain.

FIGURE 11 Photomicrographs of *Spirulina* (helical trichome) and *Oscillatoria* (nontwisted trichome) from a hypersaline pond near Guerrero Negro, Mexico. These are the most prominent cyanobacteria that have been shown to perform diel vertical migrations. The width of the Oscillatoria trichome is ~3.5 μm.

Desiccation and Osmotic Stress

Liquid water is a requirement for life on earth. Thus, the ability of some organisms to survive without water for extended periods of time is one of the many adaptations to life in extreme environments. Terrestrial mats have different types of cells from all three domains of life that are able to withstand moderate to severe dehydration. When faced with desiccation, most microbial eukaryotes and many types of bacteria make a resting cell, such as a spore. However, there are some microbes in which the vegetative (photosynthetic) cells are able to withstand essentially complete dehydration for prolonged periods of time. Cyanobacteria are best known for this ability. Many occupy the top layer of terrestrial microbial mats. Desiccation-tolerant cyanobacteria can be found in environments ranging from hot or cold desert crusts to hypersaline intertidal mats.

The effects of desiccation at the cellular level have been studied in some detail, especially by Malcolm Potts of Virginia Polytechnic University. The main consequence is the deformation of large molecules, such as proteins, nucleic acids, and lipids and their loss of function. These processes can occur during the loss of water and during rehydration. In addition, while in the desiccated state, cells are more susceptible to other stresses, particularly UV radiation, which may cause significant damage to DNA. The presence of water also helps maintain the fluidity of the cytoplasmic membrane. Thus, loss of water can lead to both hardening of the membrane and loss of structural integrity. Nevertheless, many prokaryotes, such as cyanobacteria, are able to recover from this stress.

Temperature Stress

The stress of high temperature becomes obvious in the hot spring mats of North America and portions of the Pacific Rim where species composition changes, with fewer and fewer species present as the temperature rises to about 70–73 °C, above which typical microbial mats are unknown. In other areas, such as New Zealand, Iceland, Italy, the upper limit of mat development is at about 60–63 °C, simply because of the absence of a high-temperature form of the cyanobacterium, *Synechococcus*. As with hypersalinity, temperatures above about 45 °C provide refuge for mat development in the absence of most grazers.

Low temperature, on the other hand, is a less obvious stress for microbial mat formation. Freshwater and saline ponds in the Antarctic, for example, at 78° S, microbial mats, dominated primarily by cyanobacteria, accrete slowly during the short summer thaw. Although most of the cyanobacteria cultured from such ponds grow extremely slowly at their native temperatures of 4–10 °C, grazing competitors are absent except for a few meiofauna. However, low temperature alone may not be responsible for the absence of herbivores, but the fact that most of these ponds freeze solidly (top to bottom) during the long winter, and if not, concentrate the remaining water at the bottom as a hypersaline brine (see Section "Microbial Mats of Antarctic and Arctic Ponds and Lakes"). All of these conditions, plus possible problems in dissemination from other latitudes, tend to eliminate most eukaryotic microorganisms, allowing the cyanobacteria to develop mats that structurally are very similar to those of hypersaline or hot spring mats in more temperate regions.

The Stress of Low pH, Metal, Metalloid, and Sulfide Toxicity

Many habitats that are of low pH (i.e., <4) due primarily to the abiotic and biotic oxidation of sulfide relegate mat formation to specialized microorganisms. With respect to the photosynthetic top accreting layer, only unicellular members of the red algal order Cyanidiales occur when temperatures are between 40 and 56 °C. At lower temperatures few mats occur, except for a few green algae-dominated types, and none of these resemble typical microbial mats (see Section "Mats of Acid Hot Springs"). Another common property of acidic environments is the high concentration of heavy metals such as iron, manganese, lead, cadmium, copper, mercury, as well as aluminum and the metalloid arsenic. Many of the Cyanidiales are tolerant of many of these combinations and, at least, form thin mats. For still unknown reasons phototrophic bacteria are mainly restricted to habitats with pH values in the neutral and alkaline zones, although a few species tolerate pH values below 5 (and possibly 4.5). This does not apply to nonphotosynthetic bacteria or archaea, many of which thrive at low pH, as do several species of fungi. A high sulfide concentration in the sources of some alkaline hot springs is also a factor in modifying the species composition of cyanobacteria, particularly in hot springs. Nevertheless, microbial mats with sulfide-tolerant cyanobacteria and various anoxygenic phototrophic bacteria do develop under these conditions.

The Stress of Low Nutrient Availability

Depending on location, many mineral nutrients may be in short supply in various microbial mats. Phosphate can be limited in waters overlying mats. However, cyanobacteria have an extraordinary ability to take up and concentrate PO_4^{3-} within their cells in the form of polyphosphates, even though the environmental concentration may be very low. In many microbial mats, nitrogen appears to be the most limiting nutrient, and this condition usually results in the selection of cyanobacterial species that have the ability to fix N_2 internally to ammonia, followed by incorporation into N-containing cellular compounds. Unlike combined forms of N (e.g., nitrate, ammonium, organic N), N_2 is essentially available at all times. Only some species of cyanobacteria and other bacteria or archaea have this ability. Cyanobacteria with this ability are common in many microbial mats, giving some indication that nitrogen in the combined form is limited. However, little is known about nutrient stress in microbial mats.

CONCLUDING REMARKS

The portrayal of various types of microbial mats has been rather superficial and has not included the numerous detailed studies of the physiological and biochemical processes that are occurring either constantly or intermittently, depending on light–dark cycles and seasons. Since almost all microbial mats depend on the light regime that fuels the photosynthetic top layers on which mat accretion depends, there is need for more studies on the effect of low or high light intensities and the negative effects of UV radiation. Microbial mats are not static ecosystems, but are dynamic systems that depend on physical aspects of the environment and the changes and huge differences in the chemistry of the bathing waters. The future of microbial mat studies will certainly involve the detailed dissection of the temporal (24h), seasonal, and functional features of the major microbial components. The microbial components will be identified in much greater detail by molecular (genetic) techniques and related to the *in situ* abundance and activity of each. The major and minor components will also need to be cultured to provide physiological and biochemical confirmation of the results of other methods, even though many of the microorganisms present are quite resistant to current culture methods.

FURTHER READING

Castenholz, R. W. (1994). Microbial mat research: The recent past and new perspectives. In L. J. Stal, & P. Caumette (Eds.) *Microbial mats: Structure, development and environmental significance* (pp. 3–18). Heidelberg, New York: Springer-Verlag.

Castenholz, R. W., & Garcia-Pichel, F. (2000). Cyanobacterial responses to UV-radiation. In B. A. Whitton, & M. Potts (Eds.) *Ecology of cyanobacteria: Their diversity in time and space* (pp. 591–611). Dordrecht: Kluwer Academy.

Cohen, Y., Castenholz, R. W., & Halvorson, H. O. (Eds.) (1984). *Microbial mats: Stromatolites*. New York: Alan R. Liss, Inc.

Cohen, Y., & Rosenberg, E. (Eds.) (1989). *Microbial mats: Ecological significance of benthic microbial communities* (494 pp). Washington, DC: American Society of Microbiology.

Des Marais, D. J. (1995). The biochemistry of hypersaline microbial mats. *Advances in Microbial Ecology*, *14*, 251–274.

Oren, A. (2000). Salts and brines. In B. A. Whitton, & M. Potts (Eds.) *Ecology of cyanobacteria: Their diversity in time and space* (pp. 281–306). Dordrecht: Kluwer Academy.

Potts, M. (1994). Desiccation tolerance of prokaryotes. *Microbiological Review*, *58*, 755–805.

Schopf, J. W., & Klein, C. (Eds.) (1992). *The Proterozoic Biosphere: A Multidisciplinary Study* (p. 1348). Cambridge/New York: Cambridge University Press. [especially pp. 247–342].

Stal, L. (2000). Cyanobacterial mats and stromatolites. In B. A. Whitton, & M. Potts (Eds.) *Ecology of cyanobacteria: Their diversity in time and space* (pp. 61–120). Dordrecht: Kluwer Academy.

Stal, L. J., & Caumette, P. (Eds.) (1994). *Microbial mats: Structure, development and environmental significance*. Heidelberg, New York: Springer-Verlag.

Vincent, W. F. (2000). Cyanobacterial dominance in the polar regions. In B. A. Whitton, & M. Potts (Eds.) *Ecology of cyanobacteria: Their diversity in time and space* (pp. 321–340). Dordrecht: Kluwer Academy.

Ward, D. M., & Castenholz, R. W. (2000). Cyanobacteria in geothermal habitats. In B. A. Whitton, & M. Potts (Eds.) *Ecology of cyanobacteria: Their diversity in time and space* (pp. 37–59). Dordrecht: Kluwer Academy.

Whitton, B. A., & Potts, M. (Eds.) (2000). *Ecology of cyanobacteria: Their diversity in time and space*. Dordrecht: Kluwer Academy.

Chapter 5

Horizontal Gene Transfer: Uptake of Extracellular DNA by Bacteria

K.M. Nielsen, J.L. Ray and P.J. Johnsen
University of Tromsø, Tromsø, Norway

Chapter Outline

ABBREVIATIONS

CDS Coding sequence
GMM Genetically modified microbes
GMO Genetically modified organism
HFIR Homology-facilitated illegitimate recombination
HGT Horizontal gene transfer
MEPS Minimum efficient processing segment
MMR Methyl-directed mismatch repair
RM Restriction modification
SSBP Single-strand binding proteins
USS Uptake signal sequences

DEFINING STATEMENT

Here, we present experimental models that examine the uptake of chromosomal DNA in bacteria and discuss some of the advantages and limitations of the models used. The relationship between DNA uptake frequencies versus selection is examined, and the implications for the open release of genetically engineered microbes is discussed.

PROCESS OF NATURAL UPTAKE OF EXTRACELLULAR DNA IN BACTERIA

Extracellular DNA released from a donor organism can be "horizontally" acquired by bacteria (recipients) through the process of natural transformation. Natural transformation occurring with chromosomal DNA fragments can be divided into several steps:

1. development of competence in the recipient bacterium simultaneously with exposure to transforming DNA;
2. DNA uptake/translocation into the bacterial cytoplasm;
3. heteroduplex formation of transforming DNA with similar DNA sequences in the recipient chromosome;
4. homologous recombination-mediated integration of the transforming DNA strand into the recipient bacterium's chromosome;
5. expression of the acquired trait(s) leading to an altered recipient phenotype; and
6. stable inheritance and maintenance of the acquired trait(s) at the individual recipient cell level and at the bacterial population scale level.

Topics in Ecological and Environmental Microbiology.

Although the last two steps are not strictly part of the transformation process, they are usually an integral part of most natural transformation studies in bacteria. Extracellular DNA is only accessible to naturally transformable bacteria when they are in a competent state. Competence is a genetically encoded physiological state in which bacteria express protein complexes that facilitate DNA uptake into their cytoplasm. In bacteria such as *Streptococcus pneumoniae* and *Bacillus subtilis*, competence is under tight regulation by a cell-to-cell signaling pathway. For other bacteria, such as *Neisseria gonorrhoea*, competence is constitutive and DNA can be taken up during all phases of growth. Competence development is usually linked to particular growth conditions or perturbation of those conditions. For instance, *Acinetobacter baylyi*, a soil and water bacterium, achieves maximum competence for DNA uptake during early exponential phase, with competence peaking at approximately midexponential phase. In *B. subtilis*, random peak levels of the ComK protein trigger transition to a competent stage. Bacteria that can express natural genetic competence are found in diverse habitats, and it is suspected that discovery of more naturally competent species and strains will continue with further improvement of culture- and nonculture-based methods of detection of DNA uptake. The reasons for the existence of DNA uptake/integration mechanisms in bacteria are still debated; incoming DNA fragments may be used for nutrition after further degradation, as a template for recombinational repair of DNA, or for the generation of genetic variation of which some may accelerate adaptation. Natural transformation may serve different functions in different bacterial species, as evidenced by expression of competence at dissimilar stages in their life cycles and to varying extents.

The model organisms *B. subtilis*, *S. pneumoniae*, *A. baylyi*, *N. gonorrhoea*, and *Haemophilus influenzae* have provided most of the information about the molecular aspects of the natural transformation process. The DNA translocation apparatus is a putative pore-forming multimeric protein complex that spans the inner membrane/periplasm/outer membrane in Gram-negative bacteria, and the inner membrane/cell wall in Gram-positive bacteria. Extracellular DNA first binds to the translocation apparatus on the cell surface by a poorly understood mechanism. While *N. gonorrhoea* and *H. influenzae* require specific uptake signal sequences (USS) for successful DNA binding and translocation, the majority of bacteria will bind extracellular DNA nondiscriminately with respect to sequence. Both linear and circular DNA are taken up during natural transformation. Structural proteins associated with extracellular DNA do not hinder the DNA-binding or uptake process, as bacterial cell lysates also efficiently transform competent bacteria. In most cases, although the DNA is initially double-stranded, it enters the bacterial cytoplasm in single-stranded DNA (ssDNA) form, providing evidence for an endonuclease within, or associated with, the membrane-bound translocation apparatus that degrades one strand during DNA uptake. The uptake of DNA by competent cells occurs at rates of up to 100 bp s^{-1}. Studies on the size of chromosomal DNA substrates in bacteria have demonstrated that the initial DNA fragment size is not correlated to uptake efficiency but to integration efficiency. Longer DNA fragments (5 kb or more) are more likely to yield detectable transformants in *in vitro* assays than shorter fragments. In some cases, part of the translocated DNA can be degraded in the cytoplasm (e.g., approximately 500 bp at each end of the linearized strand) prior to chromosomal integration.

Few details are available on the molecular interactions of ssDNA once it enters the cytoplasm of studied bacterial species. There is some evidence that single-strand binding proteins (SSBP) or other proteins bind to the DNA fragment and provide protection from rapid nucleolytic degradation. Although the majority of DNA molecules that enter the bacterial cytoplasm will be degraded, some of the ssDNA may subsequently pair with the bacterial chromosome by homology-guided base-pairing at the cognate locus in the recipient. The resulting duplex molecule acts as a substrate for resolution by the RecA protein. For heterogamic transformation (i.e., recombination between chromosomal DNA from divergent species), the formation and stability of the heteroduplex molecule formed by donor strand base-pairing is determined by the degree of sequence similarity between the donor and recipient DNA. If the incoming chromosomal DNA is less than 30% divergent from the recipient genome, integration by homologous recombination may occur into the host genome. Natural transformation with chromosomal DNA in bacteria seems to depend strongly on DNA sequence similarity. Lack of DNA similarity and, hence, stable heteroduplex formation is, for example, the most likely hindrance to interspecific recombination between *Bacillus* spp., *Streptococcus* spp., and *Acinetobacter* spp.

The most common type of recombination in bacteria is thought to occur when sequence similarity between the donor and recipient DNA is uniformly present over the entire heteroduplex region. In cases where the donor and recipient sequences are identical along their entire lengths, recombination with the donor DNA strand will not result in a detectable genetic or phenotypic change. In cases where some minor sequence dissimilarity between donor and recipient result exists, successful transformation events yield the acquisition of donor polymorphisms in transformants (i.e., allelic replacement). This process is called substitutive recombination, as the resulting DNA sequence in a transformant originates by substitution of the recipient sequence with (parts of) the donor sequence. Such recombination events do not introduce major insertions or deletions and typically do not disrupt the overall coding sequence (CDS) and reading frame if a protein-encoding gene is present in the recombined region.

A second type of recombination process occurring in natural transformation is called additive integration. Such a recombination process occurs when a defined DNA sequence present only in the donor is flanked on both sides by sequences common to both donor and recipient bacteria. Heteroduplex formation occurring at flanking regions with high DNA similarity results in recombination of those sequences, as well as integration of the intervening foreign sequence, into the recipient chromosome. Additive integration may also occur in the presence of only one-sided homology/DNA similarity between donor and recipient. Such homology-facilitated illegitimate recombination (HFIR) allows additive integration of an incongruous DNA sequence via a recombinational anchor of homologous/similar sequence at one end of the invading DNA strand and a random microhomology (3–12 nucleotides) at the other end. HFIR events often result in nonspecific deletion of a recipient DNA sequence located between the anchor and the downstream microhomology. HFIR has been demonstrated as a mechanism of foreign DNA integration by natural transformation in *A. baylyi*, *S. pneumoniae*, and *Pseudomonas stutzeri*. Where DNA integration is additive, the activity of the DNA maintenance machinery, combined with mutations and selection of the host chromosome, may over time result in the nucleotide composition of the integrated fragment resembling that of the recipient. Also, a gradual elimination of nonpositively selected DNA may occur by processes not fully understood.

A third type of recombination, illegitimate recombination, could hypothetically occur also in the absence of DNA sequence similarity (e.g., >70% DNA similarity) between the donor and the recipient bacteria. However, in contrast to many eukaryotic systems, illegitimate recombination of nonmobile chromosomal DNA in competent bacteria has been rarely reported. Random insertion of foreign DNA in bacterial genomes most often results in a reduction in the relative fitness of the transformant. It can, therefore, be hypothesized that such events are extremely rare in bacterial populations and the forces modulating recombination frequencies in bacteria are complex (Figure 1).

Natural transformation with DNA sequences encoding their own mobility and stabilization functions (e.g., mobile DNA such as plasmids) may occur independent of recombination with the bacterial chromosome as these can recircularize and replicate in the bacterial cytoplasm. The stable uptake of plasmid DNA in competent bacteria usually occurs at lower frequencies than the integration of chromosomal DNA fragments due to reassembly constraints on the plasmid fragments in the cytoplasm.

The resulting recombinant locus in the transformant bacterium may contain genetic alterations in existing noncoding (e.g., regulatory) or protein-coding sequence, or larger alterations such as changed gene composition, gene order (synteny), and so on. The various combinations of these factors may result in favorable, unfavorable, or near-neutral fitness changes to the transformed bacterium that will ultimately determine the impact and the survival of the transformant in a dynamic bacterial population over time.

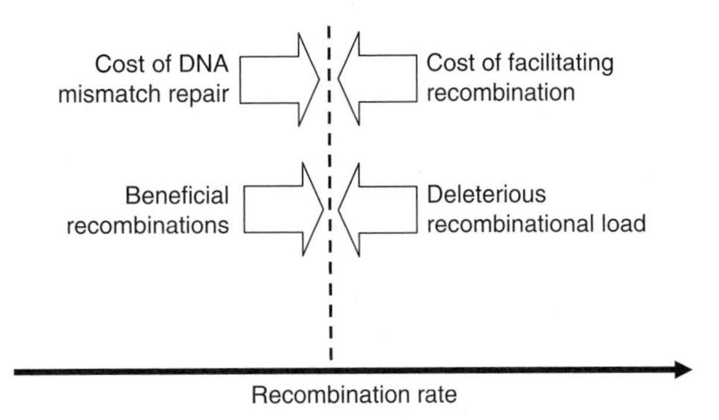

FIGURE 1 Recombination frequencies may vary according to the metabolic (fitness) costs and benefits for the transformant. Recombination events can increase transformant fitness via rare acquisition of novel beneficial traits. Recombination rates are kept low by the high deleterious recombinational load associated with accumulation of deleterious sequences and the costs associated with production and maintenance of DNA translocation/recombination machinery. Recombination rates may be raised by the metabolic costs associated with frequent removal of unsuccessful recombination intermediates by DNA mismatch repair.

UPTAKE OF EXTRACELLULAR DNA BY BACTERIA PRESENT IN VARIOUS ENVIRONMENTS

Extracellular DNA molecules are released into the environment from decomposing cells, disrupted cells, or viral particles, or via excretion from living cells. Release of intact DNA from decomposing cells depends on the activity and location of intracellular nucleases and reactive chemicals. DNA released from dead cells will be usually associated with the constituents of the cell cytoplasm, membrane, and DNA-binding proteins.

The first indication of the capacity of bacteria to take up extracellular DNA was published in 1928 by Griffith, when avirulent nonencapsulated Streptococci in mice were "transformed" into the virulent, capsulated form after exposure to a capsulation "factor" from heat-killed virulent Streptococci. The experiments of Avery, MacLeod, and McCarty in 1944 identified deoxyribonucleic acid (DNA) as the transforming factor in Griffith's experiments. Natural transformation studies of *B. subtilis* and *H. influenzae* followed during the 1960s. Researchers then expanded studies to investigate interspecific genetic exchange in *Streptococcus* spp., between *H. influenzae* and *Haemophilus parainfluenzae*, and between different species of *Bacillus*. The ability to take up extracellular DNA by natural transformation has since been detected in a range of divergent subdivisions of bacteria, including representatives of the Gram-positive, cyanobacteria, *Thermus*, *Deinococcus*, Green-sulfur bacteria, and numerous Gram-negatives. Recent studies have shown that some archaea and the single-celled eukaryote *Saccharomyces cerevisiae* (Baker's yeast) can develop competence for natural transformation as well. The phylogenetically widespread ability to acquire genetic information by natural transformation suggests that it may be functionally important in the environment. The limited number of transformable species identified may be explained by an overall lack of competence in the test populations, inability to obtain sufficient testable population sizes and time scales in the laboratory, or inability to recreate competence-inducing conditions as they occur under natural conditions.

Uptake of DNA by Soil Bacteria

A number of divergent bacterial species present in soil, including *A. baylyi*, *B. subtilis*, *P. stutzeri*, and *Thermoactinomyces vulgaris* are known to be naturally transformable. So far, all the published studies with soil-derived bacteria have been conducted in laboratory microcosms with sterile or nonsterile soil samples. Often, the soils have been amended with clay minerals or nutrients prior to or during the transformation experiments. Natural transformation of bacteria in the field remains to be shown, possibly due to the experimental challenges in both identifying and quantifying low-frequency events occurring in bacterial communities reaching 10^9 bacterial cells per gram of soil. Some of the first transformation studies on soil reported that the commonly occurring Gram-positive bacterium *B. subtilis* takes up chromosomal DNA when grown in autoclaved potting soil. Many subsequent studies have investigated natural transformation of *B. subtilis* and *Bacillus amyloliquefaciens* cells in soil with chromosomal and plasmid DNA. Often, the addition of clay minerals (1–10%, w/w) such as montmorillonite has been shown to increase DNA stability and transformation frequencies. Natural transformation of *P. stutzeri* cells has been shown in sterile soil slurry with chromosomal DNA, and in a nonsterile soil microcosm with plasmids, chromosomal DNA, and intact bacterial cells. While most studies in soil microcosms have relied on the addition of purified DNA, natural transformation of the Gram-negative *A. baylyi* has also been shown after the addition of isogenic cell lysates or lysates from *Pseudomonas fluorescens* or *Burkholderia cepacia*. The detection of DNA uptake was based on recombinational repair of a partially deleted neomycin phosphotransferase gene (*npt*II) in the recipient with a functional *npt*II gene transferred from the donor bacteria. While most studies of natural transformation in soil have relied on the addition of competent cells to the microcosms, some studies have also shown that competence for natural transformation can develop *in situ*.

Several investigations have explored the possibility that plant transgenes, such as antibiotic resistance markers, can be released from decaying genetically modified plants and be exposed to competent soil- and plant-associated bacteria. Short fragments of plant DNA have been shown to remain stable for up to several years in agricultural soil. Soil microcosms have been used to investigate the possible transfer of the kanamycin and hygromycin resistance genes (*npt*II, *hpt*) from tissue homogenates of transgenic tobacco (*Nicotiana tabacum*) plants into indigenous soil bacteria. However, no bacterial transformants could be recovered. This investigation hypothesized that any integration of the plant marker genes into the genome of exposed soil bacteria would take place after illegitimate recombination. The only study that has shown uptake of DNA fragments isolated from transgenic plants into a soil-residing bacterium utilized a genetically engineered strain of *A. baylyi* with inserted sequence similarity to the plant marker gene *npt*II. Using this system, the uptake of the plant marker gene in the bacterium was shown in a sterile soil microcosm at frequencies of 1×10^{-7} transformants per plant-harbored copy of the *npt*II gene. The amount of DNA added was several orders of magnitude higher than the concentrations expected to be released from plants under their natural growth cycle. There is no evidence to date for stable incorporation and inheritance of plant DNA in bacteria in the absence of supplied DNA sequence similarity.

Soil is a spatially and structurally heterogeneous environment in which numerous microhabitats exist. The current empirical knowledge of natural transformation in soil has been exclusively collected in soil microcosms. Most often, the studies have relied on external addition of high concentrations of DNA, bacterial cells, nutrients, or clay minerals to either sterile soil (all studies prior to 1997) or nonsterile soil samples. Moreover, the transformation frequencies recorded in soil represent so far a broad distribution across a heterogeneous milieu. The detection of rare natural transformation events under more realistic conditions depends on advances in methodology to allow more sensitive identification and quantification in the locations in soil that are conducive to natural transformation.

Uptake of DNA by Plant-Associated Bacteria

Both transduction and conjugation have been shown to occur on plant surfaces. However, few studies have examined horizontal gene transfer (HGT) by natural transformation on plant surfaces or plant tissues. A growing interest in gene transfer mechanisms active in these environments has been stimulated by the presence of engineered genes (transgenes) in genetically modified plants. Candidates for exposure to plant transgenes are bacteria that interact closely with plant cells, for example, epiphytes, endophytes, rhizosphere bacteria, and plant pathogens. Several studies have examined the potential of DNA uptake in plant-associated bacteria. Uptake of plant marker genes has been examined in the plant pathogen *Agrobacterium tumefaciens* in tobacco crown-galls, in transgenic potato tubers infected with a pathogenic *Erwinia chrysanthemi* strain, and *Ralstonia solanacearum*-infected tomato plants. A range of studies has been performed with the soil bacterium *A. baylyi*. So far, all attempts to detect horizontal transfer of plant transgenes into *naturally* occurring plant pathogenic bacteria have been unsuccessful. The general 1000-fold larger size of a plant genome, as compared with a bacterial genome, effectively dilutes the concentration of the selectable gene examined for transfer. Transgenes inserted into plant organelles may have higher copy numbers and higher sequence similarity to bacterial chromosomes than transgenes inserted into the plant chromosomes. Some recent studies have shown that bacteria can access plant DNA during colonization of plants, or after mechanical disruption of plant tissues, if regions of high DNA similarity exist between the plant and bacterial DNA. Thus, the main mechanistic barrier to the uptake and integration of plant transgenes in bacteria seems to be the limited ability of plant transgenes to act as a substrate for heteroduplex formation and homologous recombination.

Uptake of DNA by Bacteria in Water and Sediment

Marine environments can contain significant concentrations of dissolved DNA. Several studies have examined the ability of both introduced and native bacteria to take up DNA by natural transformation in water and sediment microcosms. The studies can be divided into those that introduce defined bacterial inoculants (recipients) or purified selectable DNA (e.g., containing an antibiotic resistance marker gene) into marine microcosm environments to detect DNA uptake events, and those that expose selected members of the indigenous population to DNA containing selectable markers *in vitro*. In studies utilizing introduced bacterial recipients, natural transformation has been shown in the *Vibrio* spp. strain WJT-1C (later reclassified as a pseudomonad) with plasmid DNA in small-scale marine water and sediment microcosms sampled from the Eastern Gulf of Mexico; in marine sediments inoculated with the *P. stutzeri* and added chromosomal DNA; in *Acinetobacter calcoaceticus* strain BD413 (recently renamed *A. baylyi*) with chromosomal DNA in groundwater samples from a drinking water well, sterile aquifer material, and in sterile groundwater microcosms; and in *B. subtilis* with chromosomal and plasmid DNA bound to mineral aquifer material.

Escherichia coli K-12 strains have been found to develop competence and take up naked DNA (pUC18) in water samples taken from calcareous regions. The high Ca^{2+} concentrations found in the mineral springs produced up to 20,000 transformants of *E. coli* per 10^8 strain JM109 cells. In the first experiments to describe natural transformation in open systems, an introduced *A. baylyi* strain was capable of being transformed to prototrophy by bacterial cell lysates or live donor cells in different river systems. The recipient *A. baylyi* bacteria and transforming DNA were immobilized on filters secured to stones on the riverbed. Transformation frequencies of 10^{-6}–10^{-3} were reported.

In native bacterial populations, natural transformation of marine bacterial populations has been recorded after exposure to plasmid DNA or DNA in cell lysates of rifampicin-resistant *Vibrio* strains. Three out of 30 marine bacterial isolates were found to be competent for uptake of the plasmid DNA in *in vitro* assays. These included isolates of the genera *Vibrio* and *Pseudomonas*. Moreover, 15 out of 105 sensitive isolates obtained from Tampa Bay (FL, USA) were found to acquire rifampicin resistance in *in vitro* assays. In another set of experiments, plasmid DNA exposure of 14 different whole bacterial communities sampled from a variety of marine environments, such as sediment, surface, and deep water, and from various organisms revealed that bacteria present in 5 out of the 14 communities examined could take up DNA. This study estimated that between 0.00005 and 1 transformant occurred per liter of water per day, suggesting that natural transformation may be an important mechanism for plasmid transfer among marine bacterial communities. In general, experimental studies suggest that DNA present in freshwater, in marine water, and on sediment surfaces is available for natural transformation of competent bacteria of the genera *Acinetobacter, Bacillus, Pseudomonas*, and *Vibrio*, albeit at variable frequencies that depend on environmental conditions and the type of DNA present. The introduction of defined recipient bacteria into an environmental sample results in competition between the introduced recipient population and indigenous communities for nutrients, habitats, and transforming DNA. This presents a challenge in designing experiments that accurately reflect the environmental conditions, species compositions, and concentrations encountered by recipient bacterial strains when present in natural habitats.

Uptake of DNA by Bacteria Present in the Digestive System

The animal digestive system is hypothesized to be an environmental hot spot for bacterial gene transfer due to the high concentrations of nutrients, actively growing bacteria, and surface. However, due to methodological constraints and other undetermined factors, little information is available on DNA uptake processes active in the digestive system of animals. The potential for competence has been identified in few bacteria isolated from this system. Most studies have focused on members of the Gram-positive genus *Streptococcus* that are ubiquitous in the oral cavity and rumen of many higher animals including cattle and sheep. Several microcosms and *in vitro* systems have been applied to detect DNA uptake in *Streptococcus* spp.

Sampling the teeth of 70 human individuals, it was found that at least 20 out of 129 isolates identified as *Streptococcus mutans*, which forms biofilms in dental plaques, could develop competence for acquisition of chromosomally borne streptomycin resistance *in vitro*. In a more recent study, high natural transformation frequencies were reported, reaching up to three transformants per 100 exposed recipients, using *S. mutans* cells grown in biofilms on polystyrene microtiter plates and saturating concentrations of DNA. The transformation frequencies of the surface-bound cells were 10- to 600-fold higher than those observed for planktonic *S. mutans* cells. Natural transformation was also observed when *S. mutans* cells were exposed to biofilms of a heat-killed isogenic strain harboring an erythromycin resistance gene. Natural transformation of *Streptococcus gordonii*, also found in the human oral cavity, has been reported by plasmid or chromosomal DNA in saliva samples at frequencies up to 7×10^{-4} transformants per recipient cell. DNA released from bacteria or food sources within the mouth can, therefore, be potentially taken up by naturally transformable bacteria. The natural transformation of a ruminal bacterium was first reported in 1999 with *Streptococcus bovis* cells at a frequency of 1×10^{-5} transformants per microgram plasmid in a liquid culture medium.

Currently, only a few bacterial species localized to the digestive system of higher animals have been found to express competence *in vitro* and none have been found *in situ*. Few studies have examined or reported the potential development of competence by bacteria colonizing invertebrates. The scarcity of available data may be due to insufficient incentive and methods to investigate limited bacterial transformability in such habitats. Although several members of the genus *Streptococcus* have been found to develop competence *in situ*, and other bacteria inhabiting the animal digestive tracts like *Helicobacter pylori* and *Campylobacter* spp. can develop competence *in vitro*, the biological significance of horizontal acquisition of extracellular DNA in the digestive system remains unclear.

Uptake of DNA by Bacteria Present in Food

Food provides an excellent growth substrate for many types of microorganisms. Some evidence exists for the uptake of extracellular DNA by bacteria residing in food. Natural transformation of *B. subtilis*, a common contaminant in milk, ranges from 10^{-4} to 10^{-3} transformants per recipient, with the highest frequency of 3×10^{-3} obtained in chocolate milk after an incubation time of 12h at 37°C. Plasmid transfer by natural transformation of *E. coli* in various foods, including milk, soy drink, tomato juice, carrot juice, vegetable juice, supernatants of canned cabbage, soy beans, shrimps, and various mixtures of canned vegetables, has also been shown to occur. Natural transformation occurred in all growth substrates, although at variable frequencies. Typically, fewer than 10^{-8} transformants per recipient cell were observed, a frequency three or four orders of magnitude lower than those of transformation experiments conducted under optimized laboratory conditions. No correlation was found between the content of divalent ions such as Ca^{2+}, which are considered prerequisites for artificial transformation of *E. coli*, in the foods and transformation frequencies. The above studies exemplify that food sources may provide the conditions required for natural transformation to occur and that the bacterial contaminants of food can experience competence-inducing growth conditions.

SOME LIMITATIONS OF THE DNA UPTAKE MODEL SYSTEMS USED

The data generated from transformation assays used to determine the availability and uptake of DNA in bacterial recipients are linked to the specific laboratory conditions employed. The informative value of such data on the DNA uptake processes occurring under natural conditions must be assessed case by case. Some technical constraints often embedded in the studies include the practice of adding bacterial recipients of a single strain, and often also DNA and nutrients, to a laboratory-maintained microcosm under semiartificial conditions. With few exceptions, studies of DNA uptake by natural transformation have been conducted *in vitro*, or in microcosms intended to represent soil, plants, or the gastrointestinal tract. The ability to establish representative bacterial habitat locations and population sizes after introducing externally cultured bacterial recipients into structurally organized microcosm models containing heterogeneously distributed indigenous bacteria is questionable. Few, if any, studies have examined DNA uptake processes in indigenous microbial communities with the DNA naturally present and spontaneously released on site. Moreover, the concerted action of DNA uptake and selection in determining the genetic compositions of bacterial populations undergoing adaptation to environmental changes is rarely studied in combination.

Several explanations for the dearth of such studies can be found. Most importantly, the range of gene transfer frequencies that can be practically measured in the laboratory is around 1 transformants per 10^8–10^9 exposed cells. The lower relative cell densities of a given species that are usually present in natural bacterial communities effectively limit the gene transfer frequencies that can be measured. Another limitation is that only a fraction of the indigenous bacteria from any given environment can be cultivated for subsequent selection, making DNA uptake difficult to verify in uncultivable bacteria. Moreover, the use of selectable marker genes as a tool for DNA uptake identification is sometimes problematic due to frequently encountered background resistance to most antibiotics and a limited ability of rare transformants to adequately express the marker gene. Only a minor portion of bacteria from any given complex environment is, therefore, susceptible to positive selection by antibiotics and will take part in any screen for competence development. Because most detection methods of DNA transfer in bacteria are based on the uptake of resistance genes, the enumeration procedures usually performed on selective media disturb the sites and conditions that induced the gene transfer. Methods are now being developed for the *in situ* detection of discrete bacterial cells that allow potential DNA uptake events to be identified and quantified *in situ*. For instance, fluorescent protein (e.g., green fluorescent protein) markers have been used to monitor horizontal acquisition of single genes *in situ*.

Most studies on natural transformation have been performed with monocultures with high population densities of the bacterium grown under laboratory conditions. Transforming bacterial species may, however, have specific requirements for competence development and resource utilization in complex environments. Hence, conditions conducive for gene transfer by natural transformation vary and the conditions established in the laboratory may not be those promoting competence development under natural conditions. It, therefore, follows that species currently thought to be nontransformable may show competence under yet unmonitored conditions.

TABLE 1 Some examples of interspecies/heterogamic recombination in bacteria

Transformation-Mediated Recombination in Chromosomally Localized Loci	Sequence Divergence of Recombining DNA Molecules (%)
Bacillus licheniformis and *B. mojavensis*	17
Streptococcus intermedius and *S. pneumoniae*	18
Acinetobacter sp. strain 01B0 and *Acinetobacter* sp. strain ADP1	20
Rhodococcus erythropolis and *Acinetobacter* sp.	24
Other recombination systems in nonchromosomally localized loci	
In vitro recombination between M13 and fd phage	3
Conjugal gene transfer between *Escherichia coli* and *Salmonella typhimurium*	17
S. pneumoniae and various cloned Streptococcal fragments located on plasmids	18
E. coli and *S. typhimurium* genes located on compatible multicopy plasmids	25
λ phage–plasmid cointegration in *E. coli*	35[a]

[a]*263/405 bp.*

FACTORS AFFECTING THE STABLE UPTAKE OF DNA IN SINGLE BACTERIAL CELLS

Molecular barriers define the sexual isolation or the degree to which two bacteria are prevented from exchanging genetic material by recombination. Sexual isolation may be expressed in terms of observed recombination frequencies between two bacteria, or more specifically as the ratio of homogamic recombination within species to heterogamic recombination between species. Molecular barriers act upon chromosomal DNA uptake from its first interaction with the outside of the cell through its integration in a heritable state.

Studies of natural transformation with various degrees of divergent DNA (Table 1) show that low DNA sequence similarity is a strong barrier to the integration of chromosomal DNA in bacteria. Indeed, studies have consistently demonstrated an absolute requirement for sequence similarity for detectable transformation of bacteria with chromosomal DNA. For some bacteria, the minimum length of 100% DNA similarity necessary to facilitate efficient resolution of heteroduplex molecules has been experimentally determined and is referred to as the minimum efficient processing segment (MEPS). The minimum amount of base-pairing required for a single region to initate heteroduplex formation with linear DNA substrates is 153 bp in *S. pneumoniae* and 183 bp in *A. baylyi*. Fewer details are available for the exact minimum requirements for double-sided DNA similarity for heteroduplex formation to occur. Such quantification is also not straightforward as it is likely to depend on the species and strain, and the specific nucleotide composition (GC%, etc.) of the recombining regions. For circular DNA intermediates (present in double-stranded form in the cytoplasm), DNA similarities of as little as 25 bp can mediate a single-crossover event and integration into the bacterial chromosome.

Base pairing of the single-stranded donor DNA with the complementary recipient strand within the heteroduplex molecule results in the formation of a double-stranded region that are both recognized and acted upon by host restriction modification (RM). With the exception of *H. influenzae*, transforming DNA becomes single-stranded during the translocation process, which should theoretically render it insensitive to immediate digestion by restriction enzymes. Nevertheless, studies indicate that restriction systems reduce heterogamic transformation six-fold in *B. subtilis* and 100-fold in *P. stutzeri*.

The methyl-directed mismatch repair (MMR) systems of bacteria contribute to sexual isolation by binding to mismatches in heteroduplex regions formed during DNA-strand invasion, resulting in abortion of heterospecific recombination events. While the contribution of MMR to sexual isolation in some bacteria (e.g., *E. coli*) is large, it appears to be less important in naturally transformable bacteria such as *B. subtilis*, *S. pneumoniae*, *P. stutzeri*, and *A. baylyi*. In bacteria, defective MMR can lead to increased rates of heterogamic recombination in addition to increased rates of mutation, as cells are unable to identify/repair mismatches in heteroduplex DNA formed during heterologous strand invasion. For instance, inactivation of the *mutS* gene, encoding the protein that recognizes and binds to mismatches, results in less than 20-fold increase in heterogamic transformation and recombination with DNA substrates of 20% sequence divergence in *A. baylyi*, threefold in *B. subtilis* (at 14.5% sequence divergence), and sixfold in *P. stutzeri* (at 14.6% sequence divergence). These results are in contrast to plasmid–phage recombination systems in *E. coli* or *Salmonella enterica* subspecies *Typhimurium*, where *mutS* deletions increase heterogamic recombination rates by 1000- to 10,000-fold. Reasons for this dichotomy in MMR inhibition of heterogamic recombination are unknown, but it may be due to the molecular machinery of the individual bacterium or due to the nature of DNA interacting with the recipient chromosome (single-stranded in *B. subtilis*, *P. stutzeri*, and *A. baylyi* vs. double-stranded during phage or plasmid integration in *E. coli* and *S. enterica*).

CONSIDERATIONS OF THE LONG-TERM PERSISTENCE OF HORIZONTALLY ACQUIRED DNA IN BACTERIAL POPULATIONS

The observed frequencies of DNA uptake, and prevalence and diversity of transformants in a bacterial population are determined by (1) the molecular barriers to natural transformation (see above); (2) the limited access to relevant DNA as determined by the nucleotide diversity present and accessible (physical and geographic isolation); and (3) the reduced fitness of bacterial transformants, after uptake of most random DNA fragments, that leads to their removal from the bacterial population by purifying selection.

Experimental transformation studies have most often focused on quantifying the frequencies at which particular gene segments may be taken up by bacteria at one or a few chromosomal loci. The DNA uptake rates quantified are often described in the scientific literature as "low" to "high." It is important to note that such grading only reflects the range of frequencies that can be practically measured in limited sample sizes in the laboratory. The grading is not, as often erroneously implied, linked to the subsequent biological impact of the acquired DNA in the bacterial population. It may be that frequencies below those that can be observed in a 24–48-h DNA uptake experiment in the laboratory are significant in the long run. Most successful HGT events identified through comparative DNA analysis of bacterial genomes are estimated to have occurred over a timescale of millions of years. Therefore, experimentally measured DNA uptake frequencies collected over minute timescales in the laboratory may not necessarily provide relevant information to understand the occurrence and biological impact of infrequent DNA uptake events taking place in larger bacterial populations over several years.

In addition to considerations of the relevant timescale of DNA uptake events, the timescale of transformant population expansion may also be considerably longer than what can be practically measured in the laboratory. The long-term fate of any DNA uptake event/transformant is determined by selection and genetic drift and not the DNA uptake frequencies themselves. This is because it is the population trajectories of the descendants of the primary transformants that will determine the long-term survival and impact of the transformant. For instance, single DNA uptake events may take place within the time span of a few bacterial generations. Nevertheless, it may subsequently take many thousands of generations and, hence, many years before the trait will become widespread in the overall bacterial population after clonal division and directional selection of the transformant. The latter process can only be approximated through modeling (e.g., probabilistic or deterministic) of bacterial populations, if the strength of the selection of the transformant is known. The timescale needed to understand DNA uptake processes as they occur naturally depends on the objective of the study, the anticipated directional selection of the transformant bacteria, and their population structure, size, and generation time.

PREDICTORS OF DNA UPTAKE IN BACTERIA AND IMPLICATIONS FOR THE RELEASE OF GENETICALLY MODIFIED MICROBES (GMM)

Genomic information and laboratory evidence support both the occurrence of genetic recombination between bacteria of variable genetic relatedness and the existence of mechanistic and selective barriers that maintain divergence between bacterial lineages. A challenge in microbiology is to reconcile these two observations to assess the recombinogenic potential and impact thereof between different bacterial entities. The ability to accurately predict the potential for two separately evolving bacterial lineages to undergo genetic exchange remains, however, in its "adolescence." This is illustrated by our inability to predict the impact of future HGT events on bacterial genome evolution, although data on DNA uptake frequencies can be extracted from many experimental model systems. Knowledge of the factors governing horizontal transfer and selection of transformed bacteria is, nevertheless, crucial to the evaluation of the potential for unintended spread of recombinant DNA from GMM. GMM can offer benefits in, for example, improving food production and utilization and in disease prevention. A potential hazard arising from the indiscriminate use of GMM is adverse alteration of indigenous bacterial populations after horizontal transfer of recombinant DNA (recDNA). For instance, the potential of unintended transfer of antibiotic resistance determinants, used as selective markers in the construction of GMMs, to other members of the microbial community has been extensively evaluated.

Most countries and many international organizations have developed legislation and recommendations on the types of assessments necessary prior to the commercial use of GMMs. For instance, the Codex Alimentarius Commission of the United Nations have developed principles for risk analysis and guidelines for safety assessments of foods derived from modern biotechnology, including recDNA microorganisms. In most assessments of genetically modified organisms (GMOs), the starting point is the familiarity and the history of safe use/behavior/consumption of the parent organism. The concept of "substantial equivalence" is used to structure the assessment relative to the conventional counterpart and to focus the assessment on the determination of similarities and differences. Thus, risk assessment is based on the introduced biological changes in the modified organism and does not usually address the biological safety of the parent microorganism itself. In the context of unintended horizontal transfer of recDNA, some general safety recommendations for the construction of GMMs have been made.

1. The genetic modification performed should be limited to the intended trait, and the final GMM or product thereof should not contain unnecessary DNA sequences, for example, antibiotic marker genes, DNA sequences that confer or stimulate genetic mobility of the recDNA, or DNA sequences that can confer or affect pathogenic properties.
2. A chromosomal location of the recDNA is desired because extrachromosomal elements such as plasmids are often self-transferable or mobilizable between cells and species. Plasmids also harbor replication functions that ensure their stability in an extracellular state over bacterial generation.
3. The use of recDNA (genes) that mediate a selective advantage to unintended bacterial recipients should be avoided to prevent dissemination of recDNA in bacterial communities.

The general recommendations made above are precaution-based and seek to minimize both the likelihood of uptake of recDNA in bacterial populations, and the potential positive selection of unintended bacterial transformants carrying recDNA, if rare DNA uptake events occurred. To reduce the likelihood of occurrence of unintended HGT, the recDNA may be inserted into the chromosome in the absence of sequences conferring mobility. Empirical and theoretical data provide a basis for the identification of parameters of importance for limiting the transfer potential of chromosomal DNA between bacteria. Nucleotide differences in DNA sequence between bacterial strains and species are probably the most important barrier to recombination of chromosomal DNA in bacteria. The lack of DNA sequence similarity between the donor and the recipient DNA preclude the ability of the transforming (rec)DNA fragment to form a heteroduplex with the recipient chromosome. In general, experimental studies in transformable bacteria have shown that DNA substrates with more than approximately 30% sequence divergence will not be successfully integrated in the recipient genome. Knowledge of the sequence divergence between species can be a useful predictor of their overall recombination potential. Different approaches are available to estimate the overall DNA sequence similarity between bacteria, including DNA–DNA hybridization, whole genome comparisons, or comparisons of housekeeping genes, including 16 S rRNA. While predictors of overall genomic sequence relatedness can be useful to understand the potential for chromosomal recombination between species, knowledge of local sequence divergence and conservation in the area spanning the recDNA will more accurately reflect its recombination potential into unintended recipients.

The technical approaches to limit the potential of unintended recDNA transfer from GMM are required because a precise understanding of bacterial fitness of unintended recipients of recDNA is lacking. As discussed above, it is not the transfer frequencies of chromosomal DNA that will cause a biological impact, but directional positive selection of transformants carrying recDNA. Integrated DNA fragments (including recDNA constructs) are likely to cause phenotypic changes that negatively affect the transformant fitness relative to the larger bacterial population and communities. Transformants carrying recDNA that are not competitive in growth and cell division will not remain in the larger bacterial population for long. Identifying conditions that may promote positive selection of transformants is, therefore, essential to understand the fate of recDNA in any bacterial population. The use of recDNA (genes) that is likely to mediate a selective advantage to unintended bacterial recipients should be avoided to prevent unchecked dissemination of recDNA in bacterial communities. Unfortunately, available methodology does not readily facilitate the collection of empirical data on the selection coefficients of GMMs and unintended bacterial recipients of recDNA in agricultural settings or the gastrointestinal tract. Expert evaluation and inference from a history of safe use and expected behavior (of both the recDNA donor[s] and the recDNA recipient bacterium), therefore, remain an essential component in the case-by-case assessment of GMMs.

FURTHER READING

Bushman, F. (2002). *Lateral gene transfer*. Cold Spring Harbor, New York: CSHL Press.

Codex Alimentarius Commission. (2004). *Foods derived from biotechnology*. Rome, Italy: Joint FAO/WHO Food Standards programme FAO.

Cohan, F. M. (2004). Concepts of bacterial biodiversity for the age of genomics. In C. M. Fraser, T. Read, & K. E. Nelson (Eds.) *Microbial genomes* (pp. 175–194). Totowa, NJ: Humana Press, Inc.

De Vries, J., & Wackernagel, W. (2002). Integration of foreign DNA during natural transformation of *Acinetobacter* sp. by homology-facilitated illegitimate recombination. *Proceedings of the National Academy of Science of the United States of America*, *99*, 2094–2099.

Dubnau, D. (1999). DNA uptake in bacteria. *Annual Review of Microbiology*, *53*, 217–244.

Feil, E. J., Holmes, E. C., & Bessen, D. E. *et al.* (2001). Recombination within natural populations of pathogenic bacteria: Short-term empirical estimates and long-term phylogenetic consequences. *Proceedings of the National Academy of Science of the United States of America*, *98*, 182–187.

Lawrence, J. G., & Hendrickson, H. (2003). Lateral gene transfer: When will adolescence end? *Molecular Microbiology*, *50*, 739–749.

Lorenz, M. G., & Wackernagel, W. (1994). Bacterial gene transfer by natural genetic transformation in the environment. *Microbiological Reviews*, *58*, 563–602.

Majewski, J. (2001). Sexual isolation in bacteria. *FEMS Microbiology Letters*, *199*, 161–169.

Mullaney, P. (Ed.). (2005). *The dynamic bacterial genome*. Cambridge, MA: Cambridge University Press.

Nielsen, K. M., & Townsend, J. P. (2004). Monitoring and modeling horizontal gene transfer. *Nature Biotechnology*, *22*, 1110–1114.

Pettersen, A. K., Bøhn, T., Primicerio, R., Shorten, P. R., Soboleva, T. K., & Nielsen, K. M. (2005). Modeling suggests frequency estimates are not informative for predicting the long-term effect of horizontal gene transfer in bacteria. *Environmental Biosafety Research*, *4*, 223–233.

Ray, J., & Nielsen, K. (2005). Experimental methods for assaying natural transformation and inferring horizontal gene transfer. In E. Zimmer, & E. Roalson (Eds.), *Molecular evolution: Producing the biochemical data* (pp. 491–520). San Diego, CA: Elsevier Academic Press. Part B.

Syvanen, M., & Kado, C. I. (Eds.). (2002). *Horizontal gene transfer* (2nd ed.). San Diego, CA: Academic Press.

Thomas, C. M., & Nielsen, K. M. (2005). Mechanisms of, and barriers to, horizontal gene transfer between bacteria. *Nature Reviews Microbiology*, *3*, 711–721.

Zawadzki, P., Roberts, M. S., & Cohan, F. M. (1995). The log-linear relationship between sexual isolation and sequence divergence in *Bacillus* transformation is robust. *Genetics*, *140*, 917–932.

Chapter 6

Picoeukaryotes

R. Massana
Institut de Ciències del Mar, Barcelona, Catalonia, Spain

Chapter Outline

ABBREVIATIONS

ARISA Automated ribosomal interspacer analysis
BAC Bacterial artificial chromosomes
ciPCR Culture-independent PCR
DGGE Denaturing gradient gel electrophoresis
DOM Dissolved organic matter
FISH Fluorescent *in situ* hybridization
HNF Heterotrophic nanoflagellates
HP Heterotrophic picoeukaryotes
HPLC High-performance liquid chromatography
MALV Marine alveolates
MAST Marine stramenopiles
PNF Phototrophic nanoflagellates
PP Phototrophic picoeukaryotes
TEM Transmission electron microscopy
T-RFLP Terminal-restriction fragment length polymorphism

DEFINING STATEMENT

Microorganisms play fundamental roles in marine ecosystems, sustaining food webs and driving biogeochemical cycles. Unicellular eukaryotes smaller than 2–3 μm (picoeukaryotes) are recognized as important members of microbial assemblages in terms of both biomass and activity. They show a large functional and phylogenetic diversity and include many poorly or entirely uncharacterized taxa.

INTRODUCTION

What are Marine Picoeukaryotes?

Marine picoeukaryotes are a heterogeneous assemblage of very small eukaryotic organisms. Although some examples of cultured picoeukaryotes exist, most have only been detected in the last 5–6 years using culture-independent sampling techniques and therefore have not been characterized in any detail. Nonetheless, we now know that picoeukaryotes are ubiquitous throughout the marine environment, occupying a wide variety of habitats. This includes distinct layers of the water column, each with its own dominant biogeochemical regime, marine sediments, and unique ecosystems such as hydrothermal vents.

Salinity is the main parameter that differentiates marine and freshwater habitats, and these habitats are for the most part populated by very different species. This article will focus on planktonic marine picoeukaryotes, those that live suspended in seawater, and especially those living in the upper water column where photosynthesis occurs. Since most primary production in marine systems is due to photosynthesis by planktonic microorganisms (cells smaller than 200 μm), the microbial component plays a key role in marine food webs. Among these organisms, picoeukaryotes, many of which have only recently been discovered, are qualitatively and quantitatively important. Marine primary production accounts for roughly half of the Earth's production, indicating that the oceans and their microorganisms are crucial in sequestering inorganic carbon from the atmosphere and potentially in mitigating global change.

Picoeukaryotes have a typical eukaryotic cell structure in a miniaturized state. This includes the presence of a nucleus, an endomembrane system (endoplasmic reticulum, Golgi body, and vesicles), mitochondria, and in the case of photosynthetic picoeukaryotes, a chloroplast. Nonetheless, these cells are extremely small, the diameter ranging from 0.8 μm in the case of *Ostreococcus tauri*, the smallest known eukaryote, to an upper range of 2–3 μm. Due to this small size, picoeukaryotes are largely indistinguishable by light microscopy, the usual method for studying eukaryotic microbial diversity. Thus, very few picoeukaryotes have been isolated and characterized.

In 1978 a scheme for classification of marine organisms according to size was delineated largely based on sieving technology. Microorganisms were operationally split into three categories: picoplankton (0.2–2 μm in cell diameter), nanoplankton (2–20 μm), and microplankton (20–200 μm). Initially, the picoplankton was thought to be almost exclusively made up of prokaryotes and the nanoplankton mostly of small single-celled eukaryotes. However, the existence and abundance of protists within the picoplankton size class was soon recognized. Today, the term picoeukaryotes is often used a bit loosely to include protists with a size up to 3 μm. Direct inspections of marine protist assemblages indicate that the 2-μm limit often falls in the middle of the size spectra and that a more coherent group is delimited using a 3-μm upper boundary.

Picoeukaryotes thus defined (protists smaller than 3 μm) are abundant in the marine plankton. They include diverse phototrophic and heterotrophic cells, and they play crucial roles as primary producers, bacterial grazers, and parasites. In recent years their diversity, abundance, and widespread distribution has begun to be recognized and they are attracting more attention.

Method-Driven History of Marine Picoeukaryotes

Microorganisms are invisible to the unaided human eye, so the history of their study is inevitably linked to the development of new methods for their observation and characterization (Figure 1). The existence of very small cells in the marine plankton was known from the beginning of the twentieth century to phytoplanktologists, who inspected concentrated samples of seawater by light microscopy. The first cultured picoeukaryote, the pigmented flagellate *Micromonas pusilla* (formerly, *Chromulina pusilla*), was described in 1952. Cultured picoeukaryotes provided the basis for many early microscopic and physiological studies, leading to the description of new species. The easy cultivability of *M. pusilla* allowed the initial estimations of its abundance by the serial dilution method. This showed that it is widely distributed in the marine environment and can reach abundances as high as 10^4 cells ml^{-1}. Despite this particular culturing success with *M. pusilla*, microbial ecologists soon became aware that the dominant microorganisms were often not easily cultured, so direct inspections of natural samples and culture-independent approaches were still much needed.

Electron microscopy, which had been used on cultured material since the 1950s, started to be applied to inspect natural protist assemblages during the 1970s. Transmission electron microscopy (TEM), which allows the inspection of intact specimens, revealed conspicuous features of nanoplanktonic protists. However, these studies rarely targeted picoeukaryotes. The latter were first detected by TEM in thin sections of centrifuged natural samples in 1982. *M. pusilla* and an unknown prasinophyte (later identified as *Bathycoccus prasinos*) were seen in many coastal and oceanic samples, sometimes being relatively abundant. These observations also provided the first clue on mortality of picoeukaryotes, since cells often appeared infected with large viruses. Soon after this study, it was shown that a large fraction of marine primary production in offshore regions is due to picophytoplankton (cyanobacteria and picoeukaryotes).

The first accurate counts of marine picoplankton were obtained by epifluorescence microscopy in the early 1980s. This was based on the fluorescence (natural or stain induced) emitted by cells retained quantitatively on the surface of flat polycarbonate filters. Bacterial numbers revealed by this approach were orders of magnitude higher than expected from previous cultivation-dependent techniques. Protists of different sizes were also evident at abundances in the thousands of cells per milliliter. In addition, phototrophic

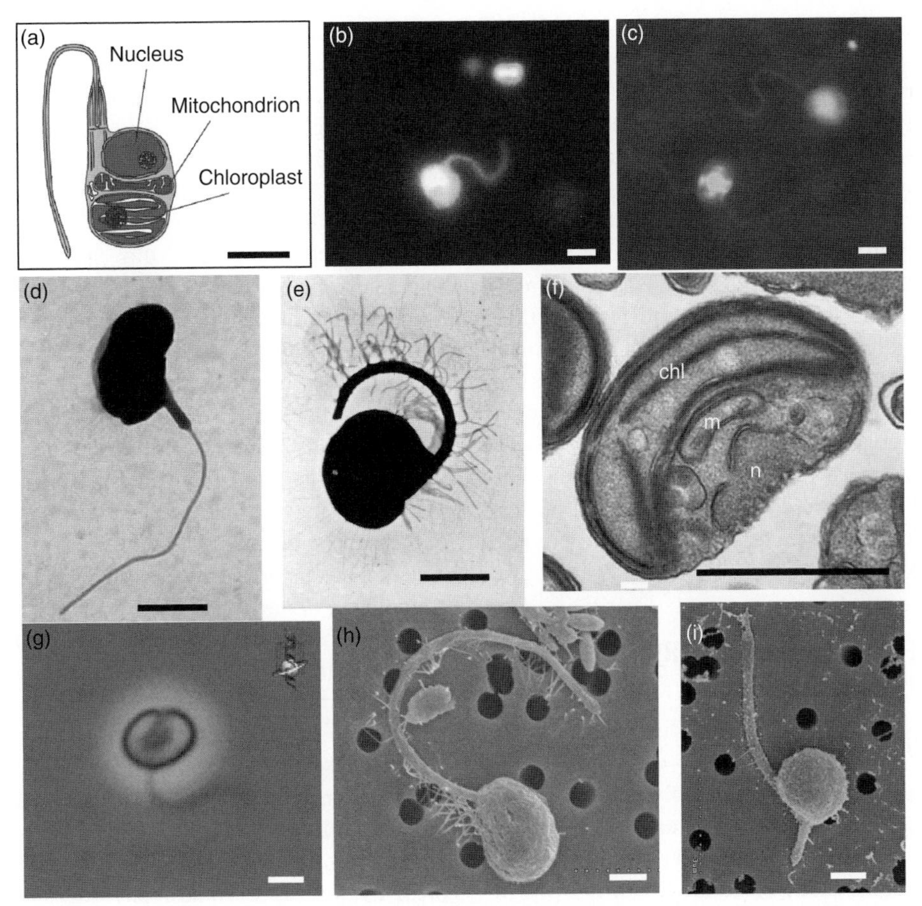

FIGURE 1 Examples of marine picoeukaryotes. (a) Drawing of *Micromonas pusilla*. Reproduced from Slapeta, J., Lopez-Garcia, P., & Moreira, D. (2006). Global dispersal and ancient cryptic species: The smallest marine eukaryotes. *Molecular Biology and Evolution*, *23*, 23–29. (b) Unidentified flagellate seen by epifluorescence under UV radiation after DAPI staining. (c) Unidentified phototrophic (left) and heterotrophic (right) flagellates seen by epifluorescence under blue light. Photo courtesy by Dolors Vaqué. (d) Stained whole mount of *M. pusilla*. Guillou, L., Eikrem, W., Chrétiennot-Dinet, M.-J., *et al.* (2004). Diversity of picoplanktonic prasinophytes assessed by direct nuclear SSU rDNA sequencing of environmental samples and novel isolates retrieved from oceanic and coastal marine ecosystems. *Protist*, *155*, 193–214. (e) Stained whole mount of *Symbiomonas scintillans*. Reproduced from Guillou, L., Chrétiennot-Dinet, M.-J., Moon-van der Staay, S. Y., Boulben, S., & Vaulot, D. (1999) *Symbimonas scintillans* gen. and sp. nov. and *Picophagus flagellatus* gen et sp. nov. (Heterokonta): Two new heterotrophic flagellates with picoplanktonic size. *Protist*, 150, 383–398. (f) Thin section through *Ostreococcus tauri* with chloroplast (chl), mitochondrion (m), and nucleus (n). Guillou, L., Eikrem, W., Chrétiennot-Dinet, M.-J., *et al.* (2004) Diversity of picoplanktonic prasinophytes assessed by direct nuclear SSU rDNA sequencing of environmental samples and novel isolates retrieved from oceanic and coastal marine ecosystems. *Protist*, *155*, 193–214. (g) *Pelagomonas calceolata* by phase contrast. Reproduced from http://starcentral.mbl.edu/microscopeportal.php. (h and i) Unidentified flagellates by scanning electron microscopy (SEM). Scale bar = 1 μm.

or heterotrophic protists, which play fundamentally different ecological roles, could be counted separately based on the presence or absence of chlorophyll autofluorescence. Epifluorescence microscopy still remains the method of choice (albeit time-consuming) for counting heterotrophic picoeukaryotes (HP). Phototrophic picoeukaryotes (PP), on the other hand, can be more easily counted by flow cytometry, a technique imported from biomedicine based on the laser detection of single cells flowing through a small aperture. Flow cytometry was first applied to marine ecology in the 1980s, and was instrumental in the discovery of the most abundant phototroph on Earth, the cyanobacterium *Prochlorococcus*. This tool has been extensively applied to describe the global distribution of marine PP.

The twenty-first century started with a fair knowledge of the global abundance and distribution of marine picoeukaryotes in the sea and a sizeable collection of characterized cultures. However, there was a remarkable lack of knowledge about which species dominate natural assemblages, and whether or not the cultured strains represented relevant ecological models. This was because the techniques used for identification (culturing, electron microscopy) were not quantitative, and conversely the techniques used for quantification (epifluorescence microscopy, flow cytometry) did not permit species identification.

This changed radically with the advent of molecular tools, particularly culture-independent sampling of ribosomal RNA sequences (phylotypes) from selected environments (see Section "Molecular Tools to Study Picoeukaryote Ecology"). These provided exciting results for marine bacteria and archaea during the 1990s, revealing whole new phyla and possibly even kingdoms of organisms. The first studies on the *in situ* diversity of marine picoeukaryotes by cloning and sequencing environmental 18S rRNA genes were published in 2001. Similar to studies of marine prokaryotes, these revealed new major groups of eukaryotes. Later, specific phylogenetic groups were targeted by fluorescent *in situ* hybridization (FISH), which allows for directly observing and quantifying natural cells in the environment. The FISH approach was particularly successful for organismal lineages seen in clone libraries but with no known close relative in culture. The use of molecular tools has revealed an unexpected diversity of protists, including the presence of novel groups, suggesting that marine picoeukaryotes represent a large reservoir of unexplored biodiversity.

BIOLOGY OF CULTURED MARINE PICOEUKARYOTES

Cultured Strains

Marine picoeukaryotes have been isolated using standard methods, such as mineral media and light for phototrophic cells and rice (or yeast extract) enrichment media for heterotrophic cells, often inoculated with 2–3 μm filtered seawater. To describe a new species, isolates must be characterized by a set of complementary techniques. Optical microscopy reveals the cell shape and motility pattern. Electron microscopy uncovers the cell ultrastructure: architecture of mitochondria, chloroplast, and flagellar apparatus, number, size, and ornamentation of flagella, and presence of external structures. Molecular markers, mostly 18S rDNA but also other genes, allow for phylogenetic placement of new isolates. Biochemical markers, such as pigment analysis by high-performance liquid chromatography (HPLC), storage products, or fatty acid profiles also provide specific information. A subset of these techniques may be used to assign new isolates to a given species, but molecular markers are used most commonly because they are faster and easier to gather. Once formally described, strains are deposited in culture collections, such as the Provasoli–Guillard National Center for Culture of Marine Phytoplankton in the United States (http://ccmp.bigelow.org/) or the Roscoff Culture Collection in France (http://www.sb-roscoff.fr/).

Since the first culture of *M. pusilla* was isolated in 1952, there has been some success in isolating additional picoeukaryotes (Table 1). Most are phototrophic and belong to the green algae and the stramenopiles. Cultured green algal picoeukaryotes belong mostly to the order Mamiellales in the Prasinophyceae (*Micromonas*, *Ostreococcus*, and *Bathycoccus*), although Pedinophyceae and Trebouxiophyceae also contain very small representatives. Stramenopile picoeukaryote cultures all belong to novel algal classes, such as Pelagophyceae (described in 1993), Bolidophyceae (in 1999), and Pinguiophyceae (in 2002). Apart from these two groups, the only example of a cultured picophytoeukaryote is the cryptophyte *Hillea marina*, although this species is still not well characterized.

There are few HP in culture, and all belong to the chrysomonads or the bicosoecids. This scarcity probably reflects the difficulty of isolating and maintaining heterotrophic protists, which usually requires culturing with other organisms, typically bacteria in the case of phagotrophic picoeukaryotes. In addition, there are parasitic protists that release very small heterotrophic cells as free-living dispersal zoospores and can only be maintained in culture with their specific host. For instance, some strains of the alveolates *Amoebophrya* and *Parvilucifera*, parasites of marine dinoflagellates, have zoospores as small as 3 μm.

In addition, some larger cultured species have a minimal cell dimension ≤3 μm. These are not strictly picoeukaryotes but would pass through the 3-μm pore size prefilter used to select picoeukaryotes in environmental surveys. Phototrophic species of this size category belong to all classes with picoeukaryotes shown in Table 1 (except Bolidophyceae that only contains picoeukaryotes) as well as Prymnesiophyceae and some additional stramenopiles (e.g., Bacillariophyceae, Dictyochopyceae, and Eustigmatophyceae). Some of these, such as the Bacillariophyceae (diatoms) and the Prymnesiophyceae, include very important marine phytoplankters. Additional heterotrophic species span the breadth of the eukaryotic tree including cercomonads (supergroup Rhizaria), kinetoplastids and jakobids (supergroup Excavata), choanoflagellates (supergroup Opisthokonts), and apusomonads (unclear affiliation).

Some picoeukaryotic species have been named based on direct observations of natural samples, enrichments, or temporary cultures. Examples are the heterotrophic flagellate *Pseudobodo minimum* (a bicosoecid 2.0 μm in size) and the green alga *Chlorella minima* (1.5–3 μm). Morphological descriptions are sometimes accompanied by ultrastructural characters obtained by electron microscopy, such as for the cercozoan *Phagomyxa odontellae* (a diatom parasite with zoospores of 3–4 × 2–3 μm) or the Parmales. The latter is an intriguing algal group, commonly observed in the sea, which includes the picoplankton species *Tetraparma pelagica* (2.2 × 2.8 μm). Although tentatively classified within the class Chrysophyceae, the phylogenetic position of Parmales is still unknown, and it is likely that molecular analyses, once they are possible, will reveal that they deserve a new class rank, as has occurred with other algal stramenopile lineages.

TABLE 1 Examples of marine picoeukaryotes in culture, including the cell size (minimal and maximal length), trophic mode, and where the culture is deposited

Taxonomic Group	Species	Size (μm)	Trophic	Culture Collection
Archaeplastida				
Pedinophyceae	*Marsupiomonas pelliculata*	3.0–3.0	P	PCC
	Resultor micron	1.5–2.5	P	SCCAP
Prasinophyceae	*Bathycoccus prasinos*	1.5–2.5	P	RCC, CCMP
	Dolichomastix lepidota	2.5–2.5	P	
	Micromonas pusilla	1.0–3.0	P	RCC, CCMP
	Ostreococcus tauri	0.8–1.1	P	RCC, CCMP
	Picocystis salinarum	2.0–3.0	P	CCMP
Trebouxiophyceae	*Chlorella* sp.	2.0–3.0	P	RCC
	Picochlorum eukaryotum	3.0–3.0	P	RCC, CCMP
Stramenopiles				
Bicosoecida	*Caecitellus pseudoparvulus*	2.0–3.0	H	RCC
	Cafeteria roenbergensis	3.0–3.0	H	RCC
	Symbiomonas scintillans	1.2–1.5	H	RCC, CCMP
Bolidophyceae	*Bolidomonas pacifica*	1.0–1.7	P	RCC, CCMP
Chrysophyceae	*Picophagus flagellatus*	1.4–2.5	H	RCC, CCMP
Pelagophyceae	*Aureococcus anophagefferens*	1.5–2.0	P	RCC, CCMP
	Pelagomonas calceolata	2.0–3.0	P	RCC, CCMP
Pinguiophyceae	*Pinguiochrysis pyriformis*	1.0–3.0	P	MBIC
CCTH				
Cryptophyceae	*Hillea marina*	2.0–2.5	P	

P, phototrophic; H, heterotrophic; PCC, Plymouth Culture Collection; SCCAP, Scandinavian Culture Centre for Algae and Protozoa; RCC, Roscoff Culture Collection; CCMP, Provasoli–Guillard National Center for Culture of Marine Phytoplankton; MBIC, Marine Biotechnology Institute Culture Collection.

Cellular Organization

Picoeukaryotes are miniaturized unicellular organisms that nonetheless retain all typical eukaryotic subcellular structures. They mostly divide asexually, a common feature in many protists, and a complete life cycle with the plausible presence of a sexual phase is totally unknown. The algal class with most cultured picoeukaryote species is the Prasinophyceae. Picoprasinophyte cells have a single chloroplast (often with a starch granule) with typical prasinophyte pigments (chlorophyll *b* and prasinoxanthin), one mitochondrion, and one Golgi body. The three most common cultured genera illustrate the variability within the class. *B. prasinos* lacks flagella and is covered by spider web-like organic scales. *M. pusilla* is naked, has a single flagellum with a short wide base and long thin distal end and a characteristic swimming behavior. *O. tauri* is coccoid, nonmotile, and naked. For each of these, strains with indistinguishable ultrastructural features have been isolated from distant geographic sites. For *B. prasinos* these strains appear to be genetically similar, but there is a clear genetic structure among *M. pusilla* (at least five clades) and *O. tauri* (at least four clades) strains. It has been proposed that these clades can be viewed as ecotypes with specific adaptations to their environments (see later). The remaining picoeukaryotic green algal species show similar minimal cell structure but have different flagellar architecture, pigment signatures, and swimming behavior.

The second group with a significant number of cultured picoeukaryote strains is the stramenopiles. This is a vast and extremely diverse clade of autotrophic and heterotrophic taxa, most of which have two unequal flagella, the longer being covered by tripartite hairs that reverse the thrust from the flagellum. Photosynthetic stramenopiles have chlorophyll *c* as the main accessory pigment and the chloroplast is surrounded by an endoplasmid reticulum continuous with the outer nuclear membrane. Picostramenopiles have a simplified cell structure, with a single mitochondrion, one chloroplast, and one Golgi body. *Pelagomonas calceolata* is covered by a thin organic theca and has only one flagellum with two rows of bipartite hairs, and even lacks the basal body of the second flagellum. Its main carotenoid is 19′-butanoxylofucoxanthin. *Bolidomonas pacifica* is naked and has two unequal flagella, the longer with tubular hairs similar to those of *P. calceolata*. It can swim vigorously, up to 1.5 mm per second, and contains fucoxanthin as the main carotenoid, like diatoms. *Pinguiochrysis pyriformis* is coccoid, naked, lacks flagella, and produces large amounts of polyunsaturated omega-3 fatty acids. HP within the stramenopiles also have a simplified cell structure, with a nucleus, a single Golgi body, 1–2 mitochondria, and no chloroplast. *Symbiomonas scintillans* is naked and has a single flagellum (and only one basal body) with two rows of tripartite tubular hairs. It contains several endosymbiotic bacteria located close to the nucleus. *Picophagus flagellatus* is naked, has two unequal flagella, the longer with two rows of tripartite hairs, and swims energetically.

Physiological Parameters

Cultured picoeukaryotes provide necessary material not only for ultrastructural, biochemical, and molecular studies, but also for defining physiological properties. Some physiological parameters deal with how fast unicellular organisms use environmental resources, such as inorganic nutrients and light for phototrophs and prey for phagotrophs. For instance, the relationship between ingestion rate and prey availability in a phagotrophic protist (functional response) depends on prey concentration: at low levels, the ingestion rate increases linearly with prey concentration; at medium levels, the rate still increases but not linearly; and at high levels, the rate reaches a maximum. This relationship can be described by several hyperbolic models. The most commonly used is analogous to the Michaelis–Menten equation for enzyme kinetics, which is based on the maximal ingestion rate (*U*m) and the prey concentration allowing half *U*m (*K*m). Both parameters are characteristic of a given species and have interesting ecological implications: *U*m gives the upper limit of a species grazing capacity and *K*m roughly indicates the prey concentration at which the species is adapted to live. The uptake of inorganic nutrients by a phototroph can be described by the same equation. The hyperbolic relationship between light and photosynthetic activity, on the other hand, uses a different model to incorporate the inhibitory effect of high irradiances.

The differential use of resources translates into different growth rates. In the case of phagotrophic protists, a similar relationship can be found between growth rate and prey availability, known as numerical response, again modeled by the Michaelis–Menten equation. A crucial parameter for phagotrophs is growth efficiency, the fraction of the food ingested that is converted to biomass. This relates ingestion rates and growth rates and has strong implications for respiration and nutrient remineralization. For phototrophic protists, growth rate is often the activity parameter measured to follow their relationship with inorganic nutrients and light. The plasticity with which protists change their reproductive rates according to the available resources is remarkable.

This simplistic view of resource–activity relationships becomes more complex when taking into account the properties of given resources, providing an additional set of more realistic species-specific physiological parameters. In photosynthetic protists, the quality of light and the chemical state of inorganic nutrients can be important. Accessory pigments can tune the cell to a given region of the light spectrum, whereas membrane transporters, genetically codified, determine the nutrient state that can be used. Heterotrophic protists can choose prey depending on size, phylogenetic composition, surface properties, and motility behavior. Finally, there are other species-specific responses related to environment parameters, such as temperature. Altogether, physiological parameters for a given species may explain its competitive advantage and success in the environment. Conceptual models illustrating how variations in these parameters may induce similar species (or ecotypes of the same species) to occupy different ecological niches is shown in Figure 2.

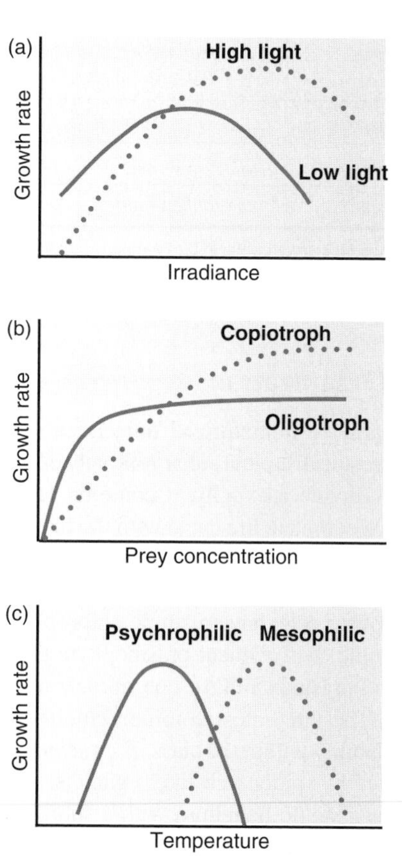

FIGURE 2 Conceptual models illustrating how physiological responses induce ecological adaptation. (a) The low-light adapted phototrophic ecotype grows better at low irradiances, and the high-light adapted ecotype at high irradiances. (b) The oligotroph phagotropic ecotype grows better at low prey concentration and the copiotroph at high levels. (c) The psychrophilic ecotype grows better at cold temperatures and the mesophilic ecotype at warmer temperatures.

O. tauri represents a good example of ecotype differentiation. Twelve *O. tauri* strains are ultrastructurally indistinguishable but form distinct genetic clades using rDNA sequences. One clade is formed by strains isolated from the bottom of the photic zone (~100 m deep), which grow well at low irradiances but are inhibited at irradiances typical of surface waters. The clade formed by strains isolated from surface waters, on the other hand, represents a high-light-adapted ecotype only growing at surface irradiances. The low-light ecotypes possess additional photosynthetic pigments absent from the high-light ecotypes. These different ecotypes, which together are able to exploit a wide range of light levels, might explain the success of *O. tauri* throughout the photic zone. Ecotype differentiation has also been observed in *M. pusilla*, where strains of one clade seem to be adapted to live in polar waters. It is plausible that ecotype diversity is a widespread phenomenon in the microbial world.

The Implications of Being Small

Cell size is the single trait that most influences the physiological and ecological properties of a given organism. Smaller cells, by virtue of their higher surface-to-volume ratio as compared to larger cells, are generally more efficient in resource acquisition and therefore may have higher specific metabolic rates. Very crudely, physiological rates are inversely proportional to body length, the so-called allometric relationship. Thus, picoeukaryotes would be the protists with the highest growth rates and better adapted to oligotrophic conditions. Picoeukaryotes live in an environment with low Reynolds numbers, where their motility is dominated by viscous forces and inertial forces are negligible. This implies that all movements have to be active and that cells do not sink passively. Finally, from an ecological perspective, cell size is the best indicator of the level an organism occupies in the trophic food web. Although there are many exceptions, phagotrophs eat organisms smaller than themselves with a general predator-to-prey ratio of 10:1 (length). So, picoeukaryotes would always be near the base of food webs, and their biomass would arrive at macroscopic trophic levels only after several trophic transfers.

PICOEUKARYOTES IN THE MARINE ENVIRONMENT

Bulk Abundance and Distribution

Very small protists are found in essentially all seawater samples inspected by epifluorescence microscopy. Since these protists were considered nanoplanktonic and many have flagella, they are routinely referred to as phototrophic nanoflagellates (PNF) if they contain chlorophyll, and heterotrophic nanoflagellates (HNF) if they are colorless. However, few cells are larger than 5 μm and many are ≤2 μm. A recent effort in protist counting and sizing by epifluorescence in contrasting marine systems reveals that 84% of phototrophic protists (between 75% and 91% in different systems), and 76% of heterotrophic protists (between 64% and 84%) are 3 μm or smaller. So, most PNF and HNF cells would qualify as picoeukaryotes in all marine habitats studied. It should be noted that, due to the absence of flagella in some picophytoeukaryotes, the terms PP and HP are probably more appropriate when referring to the epifluorescence counts of small protists.

There is an extensive database on epifluorescence counts of marine PP and HP. These operationally defined groups are ubiquitous throughout the photic zone at concentrations of thousands of cells per milliliter. PP are generally the most abundant picoeukaryotes and show a large variability, with cell counts typically increasing with the trophic state of the sample. Conversely, HP are several times less abundant than PP but vary only moderately across systems, generally less than one order of magnitude, and are often correlated with bacterial abundance. In coastal systems, typical ranges for PP are 1.1–8.5 × 10^3 cells ml^{-1}, with episodic peaks well above 10^4 cells ml^{-1}, whereas HP concentrations typically vary between 0.6 and 3.1 × 10^3 cells ml^{-1}. Cells are less abundant in offshore, more oligotrophic systems: typical PP concentration ranges from 1.0 to 3.3 × 10^3 cells ml – ^{1} and HP ranges from 0.6 to 1.5 × 10^3 cells ml^{-1}. For example, in offshore Indian Ocean samples, average cell counts were 1.7 × 10^3 cells ml^{-1} for PP and 0.45 × 10^3 cells ml – ^{1} for HP.

Understanding of the patterns of PP distribution has been considerably expanded by the semiautomatic counts provided by flow cytometry. Inspection of marine samples by this technique reveals an assemblage of photosynthetic picoeukaryotes, generally at thousands of cells per milliliter, yielding counts consistent with those of PP counted manually by epifluorescence. Flow cytometry is routinely used in oceanographic cruises and monitoring programs and allows direct comparisons of picophytoplankton groups. Results indicate that picoeukaryotes generally covary with *Synechococcus* and dominate in coastal waters, whereas these protists are less abundant offshore where *Prochlorococcus* dominates. The ubiquity, abundance, and constancy of picoeukaryotes suggest they must be important players in the photic zone and that their growth and mortality rates are relatively tightly coupled.

Photosynthesis does not occur below the photic zone, in the mesopelagic (200–1000 m deep) and bathypelagic (1000–3000 m) regions; so this extensive biome is largely devoid of photosynthetic protists. Nevertheless, bacterial production can occur in these deep waters based on sedimenting organic matter from the photic zone and on chemolithoautotrophic bacteria gathering their energy from reduced inorganic compounds. However, both bacterial abundance and production is still 1–2 orders of magnitude lower here than at the surface, and the numbers of heterotrophic picoeukaryotes as seen by epifluorescence microscopy are also lower, ranging between 10 and 100 cells ml^{-1}.

Ecological Role of PP

It has long been known that microorganisms are responsible for most of the marine primary production. However, prior to 1983 we were unaware of the importance of picoplankton in this crucial process. Picophytoplankton generally dominate primary production in offshore, oligotrophic systems, and can also be important in coastal systems on a seasonal basis. For instance, picoplankton averaged 71% of photosynthetic biomass and 56% of primary production during four Atlantic Ocean cruises (50° N–50° S) crossing coastal, upwelling, and central oceanic regions. Picocyanobacteria, specifically *Synechococcus* in nutrient-rich and *Prochlorococcus* in oligotrophic regions, can reach abundances of up to 10^5 cells ml – ^{1}, so they were initially thought more important in supporting food webs than picoeukaryotes. However, albeit less abundant, picoeukaryotes are larger (their biovolume can be 100 times that of picocyanobacteria), so they can contribute significantly to biomass and primary production. For example, a study in the central North Atlantic indicated that while only 10% of the surface picoplankton were eukaryotes, these contributed 61% of the biomass and 68% of the primary production.

The size spectrum of marine primary producers strongly influences the food web complexity and has crucial implications for transfer efficiency to higher trophic levels. Large cells such as diatoms and large dinoflagellates can be readily consumed by copepods and directly sustain fish larvae. However, primary production based on very small cells has to pass through several trophic steps (flagellates, ciliates) and a significant fraction of this organic matter is respired at these lower levels. Thus, even though picoeukaryotes appear to be optimal food for small predators, systems largely reliant on their production are less efficient in sustaining higher trophic levels. Other mechanisms, such as viral lysis, also potentially diminish trophic efficiency by causing direct remineralization of picoeukaryotic biomass. However, the role of viruses as mortality agents in picoeukaryotes (and other algae) remains largely unexplored.

The magnitude of the ocean's biological pump (sequestration of atmospheric inorganic carbon to the deep ocean by biological export) is also considered to be heavily influenced by the cell size and species composition of phytoplankton. That is, larger cells are expected to contribute more to inorganic carbon sequestration. This is because larger cells, often with mineralized structures, can sediment directly (alone or in aggregates) or be readily consumed by copepods that excrete fecal pellets that sink out of the surface mixed layer. However, recent data suggest that picoplankton may also contribute to carbon export from the surface at a level proportional to its net primary production.

Ecological Role of HP

HP are mostly phagotrophs and are central in the microbial loop, a concept that has driven a fundamental revision of models of marine food webs. The microbial loop is based on the premise that there is a substantial supply of dissolved organic matter (DOM) created by phytoplankton exudation or inefficient zooplankton feeding that supports active bacterial production. These active bacteria are grazed upon by small protists that in turn are food for larger grazers. So, incorporated into the classical food web (phytoplankton, copepods), there is a microbial loop (DOM, small grazers, copepods) that potentially transfers energy from dissolved pools to higher trophic levels. The reality is closer to a web than a loop, since small grazers also consume picophytoplankton, often the dominant producers. Regardless, the main grazers of bacteria and picophytoplankton are picoeukaryotes, as seen by microscopic inspections of protists with ingested bacteria and by size fractionation experiments showing that most bacterial grazing occur in the fraction below 3 μm.

Grazing by HP represents an important mortality factor for bacterial assemblages over large oceanic scales. Other mortality agents like viruses may contribute in coastal systems but seem less important in oligotrophic regions. Mixotrophic protists, cells capable of both phagotrophy and photosynthesis, can contribute to half of bacterial mortality in coastal systems, but their relevance in offshore systems is not well defined. Bacterial grazing, by either heterotrophic or mixotrophic protists, is the first of a multistep food chain, during which bacterial production is mostly respired and inorganic nutrients bound to bacterial biomass (often enriched in P and N) are released. Thus, the main ecological roles of phagotrophic picoeukaryotes are controlling bacterial (and other picoplankton) abundances and remineralizing inorganic nutrients in the photic zone, allowing sustained primary production in oligotrophic systems.

HP may have other nutritional modes besides phagotrophy. Strictly osmotrophic protists are probably unable to compete with heterotrophic bacteria, which are likely more efficient in using the diluted and refractory marine DOM. Nevertheless, the existence of phagotrophic protists that can supplement their diet with DOM is plausible and ecologically relevant. Parasitism has long been known to occur in the sea, and there are many descriptions of parasites infecting protists (e.g., diatoms, dinoflagellates) and metazoans (e.g., copepods, crabs). These parasites always have a free-living dispersal form (zoospore), colorless and often very small, which would be considered as HP. Parasitism is fundamentally different from phagotrophy because the parasite is smaller than the host, has a limited host range, and does not always kill its host. Given the high diversity of microbial assemblages and the dilute environment in which they live, parasitism was not considered a major process in the marine plankton. However, recent environmental molecular surveys (see Section "*In Situ* Phylogenetic Diversity") reveal many sequences of putatively parasitic protists retrieved from the picoplankton, even at the more oligotrophic stations studied. The ecological relevance of parasitism in coastal and offshore microbial assemblages is an open field for future research.

MOLECULAR TOOLS TO STUDY PICOEUKARYOTE ECOLOGY

Elusive View of *In Situ* Diversity by Nonmolecular Tools

Despite the ecological importance of picoeukaryotes as primary producers, bacterial grazers, and parasites, it is remarkable how little we knew about *in situ* diversity before the application of molecular tools. This is because picoeukaryotes cannot be identified by the techniques that provide accurate cell counts, such as epifluorescence microscopy or flow cytometry. Specific ultrastructural features can often be revealed by electron microscopy, but this cannot be applied routinely to all cells of an assemblage, although sometimes it has served to identify dominant species, such as *Micromonas* and *Bathycoccus*.

Culturing is an excellent approach to obtain biological models (see Section "Biology of Cultured Marine Picoeukaryotes"). However, many cells do not grow easily in the lab, so culturing provides a severely biased view of microbial diversity. In a few cases, there are dominant populations in the sea that are also easily cultured, so they can be counted by diluting and culturing, yielding a most probable number. An excellent example is the prasinophyte *M. pusilla*, which is widely distributed in the sea in numbers ranging from 10^2 to 10^5 cells ml^{-1}, being more abundant in coastal areas but also found in the oligotrophic open sea.

Pigment analysis by HPLC yields information on the composition of marine phytoplankton, since each algal group has a specific pigment signature, and has often been used to compare samples during oceanographic cruises. When applied to samples prefiltered through 3 μm, it can provide a general view of picophytoplankton composition. Since the pigment profile derives from a complex assemblage, algorithms have been developed to infer the contribution of different algal classes to total chorophyll *a* (the only pigment common to all phytoplankters). However, translating pigment profiles to diversity depends on the pigment ratios generated in cultured species, which may not totally represent natural populations. Even, in the best scenarios, HPLC pigment analysis only provides identification to an algal class, not to lower ranks such as genus or species.

Cloning and Sequencing Environmental Genes

Molecular tools were introduced in marine microbial ecology relatively recently. The most widely used gene for these studies is the SSU rDNA (16S rDNA in prokaryotes, 18S rDNA in eukaryotes), which codes for the small subunit ribosomal RNA. This gene has the distinct advantages that codifies for the same function in all organisms, it has both highly conserved and variable regions, and its product (rRNA) is present in high abundance in all living cells. SSU rDNA has been used to delimit the three domains of life and to classify organisms within a given class, genus, and species. The basis of molecular protist ecology is identifying cells *in situ* by directly retrieving their rDNA. This is achieved by extracting DNA from marine picoplankton assemblages, amplifying the 18S rRNA genes by PCR, cloning and sequencing the PCR products, and comparing the sequences with SSU rDNA databases.

The first culture-independent PCR (ciPCR) studies on marine picoeukaryotes appeared in 2001. These showed that picoeukaryotes are extremely diverse, so the indistinguishable cells seen by epifluorescence hide phylogenetically different organisms. In fact, environmental picoeukaryote sequences are scattered throughout the eukaryotic tree of life (Figure 3). Furthermore, while some of these sequences match well-known species, others form phylogenetic clades (sets of related sequences) that represent novel and unexplored diversity. Studies from widely separated sites often show similar phylogenetic clades, suggesting that few biogeographic barriers exist in the marine realm and similar protists thrive when conditions are similar. Besides 18S rDNA, other genes have been used for similar purposes, such as the chloroplast genes 16S rDNA, *rbcL*, and *psbA*. These target only the phytoplankton, and complement and expand the results obtained with 18S rDNA.

The cloning and sequencing approach is critical to unveil novel microbial diversity but is limited in its ability to reveal the true community composition. First, samples are obtained by sequential filtration (through 3 and 0.2 μm pore size filters), and small cells can break during the process and be lost from the picoplanktonic sample. Organisms larger than 3 μm, on the other hand, can break into small fragments or squeeze through pores and be collected as picoplankton. This surely explains the finding of metazoan and ciliate sequences in clone libraries of picoeukaryotes. Second, some species can be more resistant to DNA extraction than others, depending on their cell structure and outer layers. Third, the required PCR step may bias the original gene abundance, both by primer mismatches

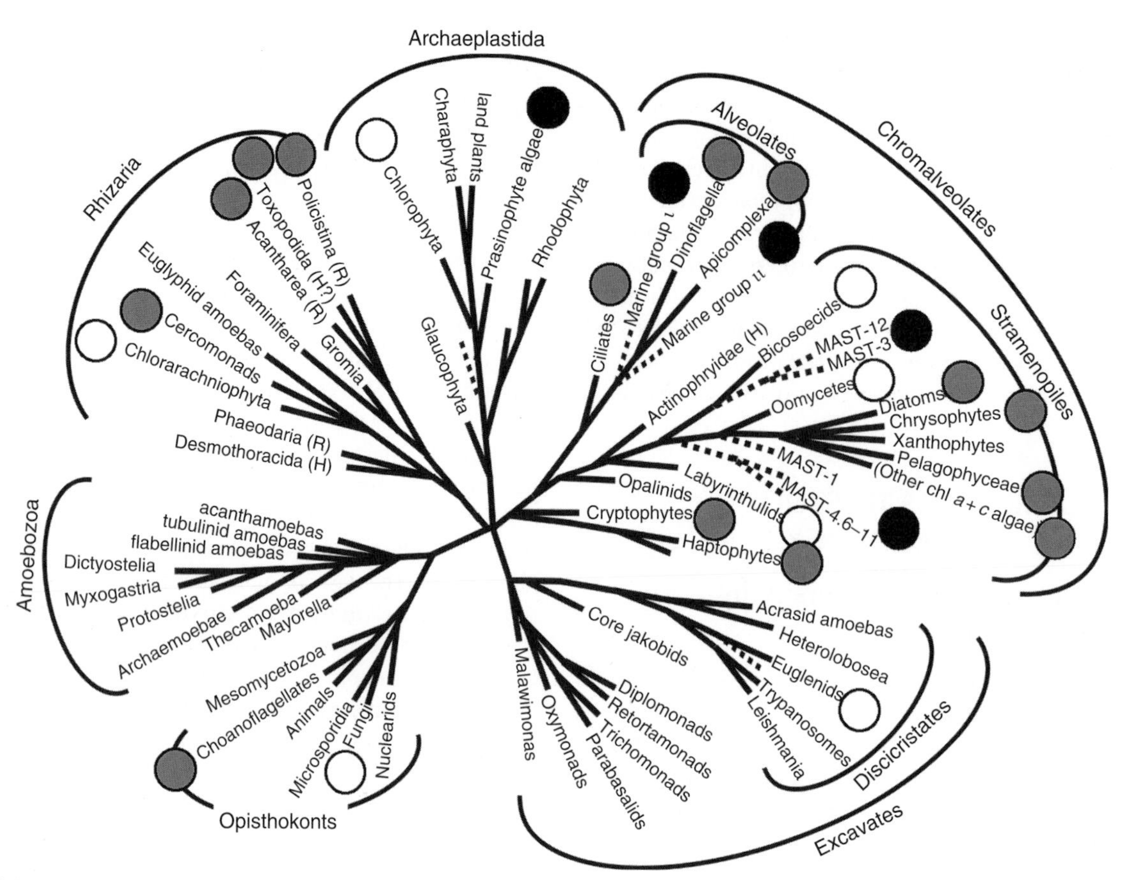

FIGURE 3 Representation of eukaryotic phylogeny displaying all lineages in a few supergroups (modified from Baldauf, S. L. (2003) The deep roots of eukaryotes. *Science*, *300*, 1703–1706). The supergroup CCTH (cryptomonads, centrohelids, telonemids, and haptophytes) has been proposed recently after phylogenomic analysis (Burki *et al.* (2009) Large-scale phylogenomic analyses reveal that two enigmatic protist lineages, Telonemia and Centroheliozoa, are related to photosynthetic chromalveolates. *Genome Biol. Evol. 231*, 213–18). Groups found in picoplankton 18S rDNA libraries are marked with a dot.

and by preferential amplification in some groups. Finally, the rDNA copy number varies widely between eukaryotic species, from one to several thousands, and this obviously influences clonal representation. Thus, 18S rDNA clone libraries are fundamental and informative but do not completely describe *in situ* protist diversity.

Beyond Clone Libraries: FISH and Fingerprinting Techniques

After environmental sequencing has identified dominant phylogenetic groups, their abundance and distribution in the sea can be assessed by FISH, a technique that elegantly complements the rRNA approach. FISH also reveals critical features of novel phylogenetic clades such as cell size, thus confirming whether they are picoeukaryotes. During the FISH process, fixed microbial cells immobilized on a filter are hybridized with a taxon-specific 18S rDNA probe labeled with a fluorochrome. So, target cells fluoresce under epifluorescence microscopy. A crucial step in FISH is probe design, the selection of a short DNA sequence (around 20 nucleotides) specific for a given group, be it a species, genus, class, or domain. For probes targeting a novel clade, no culture is available in which to optimize hybridization conditions, so instead a natural sample where these sequences have been detected can be used. Once the FISH protocol is optimized, it is relatively easy to apply, albeit time-consuming. FISH has been successfully used for three groups of marine picoeukaryotes: prasinophytes, stramenopiles, and picobiliphytes. It has provided data on their cell size and abundance, thus giving unambiguous estimates of their contribution to the bulk picoeukaryote assemblage. Presently, the FISH approach is mostly limited by the probes available, so major efforts in probe design and optimization are required.

Fingerprinting techniques are popular in microbial ecology because they allow a fast comparison of the community composition among samples. They provide a characteristic fingerprint for each sample, typically a banding pattern where each band represents a specific taxa. The most commonly used techniques are denaturing gradient gel electrophoresis (DGGE), terminal-restriction fragment length polymorphism (T-RFLP), and automated ribosomal interspacer analysis (ARISA). All start with a PCR amplification of rDNA genes from the assemblage in question using general primers. Individual sequences in this mixture are then separated by electrophoresis based on their melting domains (DGGE), the location of their first restriction site (T-RFLP), or the size of their ITS region (ARISA). Fingerprinting techniques are used to reveal spatio-temporal patterns of picoeukaryotic assemblages in the sea. Picoeukaryotic assemblages appear to have a relatively wide distribution, with large oceanic areas with similar composition, but they also show significant changes with depth, along onshore–offshore gradients and across frontal regions. Fingerprinting techniques are also very useful to follow microbial dynamics during manipulative experiments.

IN SITU PHYLOGENETIC DIVERSITY

Overview of 18S rDNA Libraries

The phylogenetic diversity contained within marine picoeukaryotes is astonishing, as only one eukaryotic supergroup is not represented in 18S rDNA clone libraries (Figure 3). Some of the retrieved sequences are very similar to cultured picoeukaryotes (Figure 4a). Others clearly affiliate within a given group but are distantly related to its cultured representatives, so likely represent new species or genera (Figure 4b). Other sequences show a profoundly deep distance from all cultured organisms and can only be unambiguously affiliated to a supergroup, such as the marine alveolates (MALV) and the marine stramenopiles or MAST (Figure 4c). Finally, in a few cases the retrieved sequences form a novel high-rank taxonomic entity that cannot be assigned to any given supergroup, for example, the picobiliphytes.

A summary of the clonal representation in 18S rDNA libraries published so far from surface samples is shown in Table 2. These libraries derive from Atlantic, Pacific, Indian, and Southern oceans and Mediterranean and North seas (35 libraries and 2175 clones). The better-represented phylogenetic groups are alveolates (42.5% of clones), stramenopiles (22.8%), and prasinophytes (15.0%). Some groups are more abundant in coastal libraries (ciliates, MALV, cryptophyta, prasinophyta, and cercozoans), whereas others are more abundant in offshore libraries (MAST, pelagophytes, prymnesiophytes, and radiolarians). The different levels of phylogenetic novelty shown in Figure 4 are discussed below.

Relatively Well-Known Groups

Prasinophytes show the best correspondence between molecular and culturing approaches. They are well represented in both approaches, and environmental sequences are often identical to that of cultured cells, particularly from *M. pusilla*, *O. tauri*, and *B. prasinos*. Using FISH probing, *M. pusilla* has been found to be very abundant in coastal systems (up to 10^5 cells ml^{-1}), supporting previous views from culturing, electron microscopy, and pigment analysis. It appears that a substantial fraction of picoeukaryotes in coastal systems are prasinophytes (50–90%), and these account for a lower percentage of cells (>20%) in offshore samples. Other picophytoplankton groups with cultured species are poorly represented in clone libraries: only 1% of the clones affiliate with pelagophytes and 0.5% with bolidophytes have been detected, while the pinguiophytes have never been detected. The pelagophyte sequences are nearly identical to *P. calceolata* (Figure 4a). This species appears to have a wide distribution, but its true abundance remains uncertain, since the attempted FISH probing still gives unclear results. Nevertheless, HPLC pigment analysis suggests a significant contribution of pelagophytes in open ocean systems.

Prymnesiophytes are important marine algae that seem to contribute significantly to the picoplankton according to HPLC analysis. Also, many cultured prymnesiophytes have

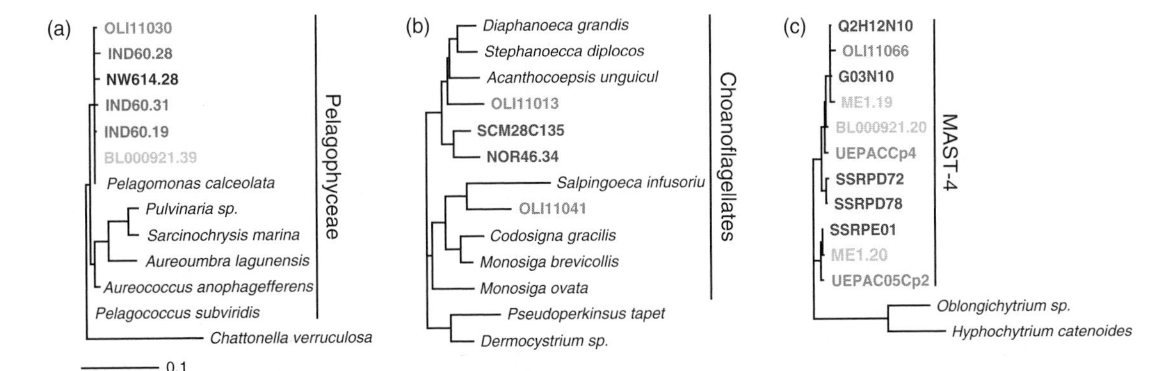

FIGURE 4 Pelagophytes (a), choanoflagellates (b), and MAST-4 (c) phylogenetic trees with 18S rDNA sequences from the marine picoplankton and closest cultured representatives. Environmental sequences are colored depending on their origin: Atlantic (red), Pacific (green), Indian (gray), Arctic (blue), and Mediterranean (yellow). (For interpretation of the references to color in this figure legend, the reader is referred to the Web version of this chapter.).

TABLE 2 Clonal representation of phylogenetic groups in 18S rDNA libraries prepared from surface picoplankton from coastal (23 libraries and 1349 clones from Mediterranean, North Sea, English Channel, and Pacific Coast) and offshore (12 libraries and 826 clones from the Mediterranean Sea, Antarctica, and Indian, Pacific, and Atlantic oceans) systems

		% Total	% Coast	% Offshore	Tendency
Alveolates	Ciliates	4.9	7.6	2.7	Coastal
	Dinoflagellates	5.3	6.1	4.7	
	MALV	32.4	35.8	29.3	Coastal
Stramenopiles	Chrysomonads	2.7	2.2	3.1	
	Diatoms	2.5	0.6	4.0	
	Dictyochales	1.1	1.5	0.9	
	MASTs	13.4	9.0	17.0	Open
	Pelagophytes	1.0	0.1	1.8	Open
Archaeplastida	Prasinophytes	15.0	19.3	11.6	Coastal
Rhizaria	Cercozoans	2.6	4.8	0.9	Coastal
	Radiolarians	5.6	0.3	9.8	Open
CCTH	Cryptophytes	2.4	3.6	1.4	Coastal
	Prymnesiophytes	4.5	1.2	7.2	Open
	Picobiliphytes	0.9	1.2	0.7	
Opisthokonts	Choanoflagellates	1.0	0.9	1.0	
Remaining groups		4.7	5.8	3.9	

the smallest cell dimension ≤3 μm. Nevertheless, their contribution to clone libraries is moderate (4.5%), and the environmental sequences often form novel clades. FISH probing shows that prymnesiophytes are rather abundant in coastal and offshore samples (more than their clonal abundance would suggest), reaching up to 30% of the picoeukaryotic cells. Other algal groups detected in clone libraries are the dinoflagellates (5.3%), important marine protists with phototrophic and heterotrophic species. Since the smallest dinoflagellates observed are around 5 μm, their molecular signal has been interpreted as filtration artifacts, although the presence of picodinoflagellates cannot be excluded. Similar concerns apply to cryptophytes (2.4% of clones) and diatoms (2.5% of clones), although the latter have very small centric and slim pennate species that would pass the 3-μm prefilter. A few sequences are close to the dictyochophyte *Florenciella parvula*. Finally, some sequences affiliate distantly to the chlorarachniophytes, but it cannot be determined if they represent phototrophic protists.

Libraries show a large diversity of putatively heterotrophic protists at low clonal abundance. Some of these groups contain cultured heterotrophic pico- and nanoflagellates: apusomonads, bicosoecids, cercozoans, chrysomonads, choanoflagellates, katablepharids, and *Telonema*. Other groups detected in marine surveys are known to contain osmotrophic or parasitic forms, such as apicomplexa, fungi, labyrinthulids, oomycetes, and pirsonids. A significant number of sequences relate to ciliates (4.9%), important components of the nano- and microzooplankton but at least 10 μm in size, so their presence is most likely a filtration artifact. The chrysomonads also show a moderate clonal representation (2.7%) and contain sequences close to well-known heterotrophic flagellates, such as *Paraphysomonas*, as well as novel clades. The trophic mode of these novel chrysomonads is uncertain, but recent data obtained in plastid rDNA libraries suggest an unexpected importance of chrysomonads within the picophytoeukaryotes. Finally, a substantial fraction of clones affiliate with the radiolarians. This is surprising, since the radiolarian species known so far are rather large (typically between 20 and 100 μm) and most possess mineralized skeletons. They are virtually absent from coastal systems, and reach a significant clonal abundance in the surface of the open sea (9.8%). Moreover, the fraction of radiolarian sequences increases with depth in the water column, and can reach 20–30% of clones at the bottom of the photic zone and below. Marine radiolarian sequences are diverse and form several clades related to acanthareans and polycystinea. The existence of these diverse radiolarian sequences in the picoplankton is a current enigma.

Marine Alveolates and Marine Stramenopiles

Environmental surveys have revealed novel clades that affiliate to a given eukaryotic supergroup but without a clear affiliation to any of its members. Among these, the MALV and MAST clades are particularly interesting because they appear in virtually all marine surveys at high clonal abundance. MALV are divided into five main groups, of which MALV-I (five clades) and MALV-II (16 clades) are the most widely represented. The first described MALV-II species was *Amoebophrya*, a dinoflagellate parasite, and additional parasite sequences have been recently published within both groups. So, it now appears that the whole MALV assemblage might consist of parasites of marine organisms. The specific interaction with different hosts could explain their high level of genetic diversity. This opens new avenues for exploration of the role of parasitism as a trophic mechanism in the open sea.

MASTs form more than ten clades at the basal part of the stramenopile tree, where all protists are heterotrophic, free-living flagellates, parasites, osmotrophs, or commensals. This suggests these MASTs are heterotrophs, which has been confirmed by FISH for several clades. On average, MAST sequences account for 13.4% of the picoeukaryotic clones, and most of this signal is explained by clades MAST-1, -3, -4, and –7. These are colorless protists, with a size from 2 to 8 μm, able to grow in the dark and to ingest bacteria (Figure 5). MASTs are widely distributed and account for a significant fraction of heterotrophic flagellates. For example, the very small MAST-4 protist (2–3 μm) is found in all samples except the polar ones, averages 130 cells ml^{-1}, and accounts for 9% of heterotrophic flagellates.

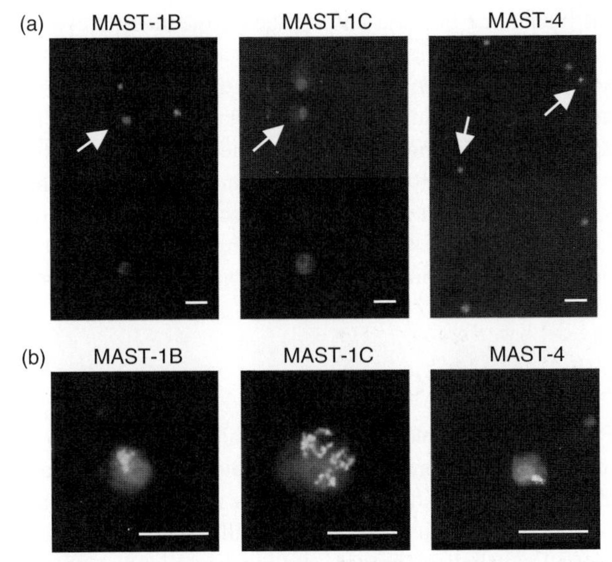

FIGURE 5 Epifluorescence images of MAST-1B, -1C, and –4 cells. (a) Same microscopic field observed by UV (upper panels: DAPI-stained blue cells) and green light (lower panels: fluorescent *in situ* hybridization (FISH)-stained orange cells). MAST cells in the upper panels are indicated with a white arrow. (b) Combination of three images in several MAST cells: DAPI staining (blue nucleus), FISH (orange cytoplasm), and ingested FLBs (yellow spots). Scale bar = 5 μm. (For interpretation of the references to color in this figure legend, the reader is referred to the Web version of this chapter.)

This shows that still-uncultured groups can be dominant in marine picoeukaryote assemblages and that model cultured organisms might have limited use for understanding how marine ecosystems work.

Novel High-Rank Phylogenetic Groups

Early molecular surveys claimed the discovery of novel groups at the highest taxonomic rank, in an effort to highlight the potential of this approach. However, it was soon recognized that some of these novel groups were artifacts due to the presence of undetected chimeras, misplacement of fast-evolving lineages, and incomplete representation of cultured strains. Nevertheless, it is clear that some novel high-rank groups are real and form robust and deep phylogenetic clades that cannot be assigned to any known eukaryotic supergroup. These are generally found at low clonal abundance, so probably are not very important ecologically. Instead, their interest resides in their evolutionary novelty. Perhaps the best example is the picobiliphytes, a novel phytoplanktonic class that probably have phycobilin-containing plastids and can be locally abundant. However, most of these novel potentially high-rank groups await formal characterization.

Biogeography

There is considerable debate about the extent of diversity and biogeographical distribution of microorganisms. It has been argued that given the huge population sizes and the potential for distant dispersal, most microbial species must be cosmopolitan and the total number of species must be relatively low. This is what is observed using the morphospecies concept to identify protists. However, this concept cannot account for cryptic species, the level of which may not be insignificant among protists, given that strains from the same morphospecies can belong to genetically, reproductively, and ecologically isolated groups. Thus, the use of molecular tools to compare assemblages from distant systems provides a more systematic and objective assessment of the extent of protist biogeography and diversity.

Clone libraries from widely separated picoeukaryote assemblages yield similar phylogenetic groups, suggesting that their overall community structure is comparable on a worldwide scale. Further, there appears to be no particular geographic separation as identical 18S rDNA sequences have been retrieved from distant oceans (Figure 4). Altogether, we find little sign of geographic barriers for picoeukaryotes in the marine environment, and most groups seem to be roughly globally distributed. This supports the cosmopolitan view mentioned above for the marine habitat, although finer resolution markers than 18S rDNA might be needed to detect locally adapted populations.

Global distribution has been used as an argument for low protist diversity as it should prevent speciation by geographic isolation or allopatry, the speciation mechanism considered most important. However, most groups detected in environmental surveys show a large 18S rDNA sequence variability, thus strongly disagreeing with the view of low protist diversity. The significance of this large intragroup rDNA diversity is presently not understood but can be relevant, since genetic distances as small as 0.01 (equivalent to 99% similarity) in the 18S rDNA might imply millions of years of evolutionary divergence. It is also not clear how this large phylogenetic diversity of protists translates into functional diversity in the marine environment or how this diversity is generated and maintained. One possibility is to view the seemingly homogeneous pelagic habitat as a continuum of environmental niches. Thus, there is a large genetic diversity of marine protists, which may have a global distribution, but the ecological implications of this large diversity remain to be investigated.

THE GENOMIC ERA

Genome Projects on Cultured Picoeukaryotes

Currently there are 129 complete published genome projects of eukaryotes. These include a few free-living marine protists such as the diatoms *Thalassiosira pseudonana* and *Phaeodactylum tricornutum*, the prasinophytes *Ostreococcus tauri, O. lucimarinus*, and *Micromonas pusilla*, the green algae *Chlamydomonas reinhardtii*, the choanoflagellate *Monosiga brevicollis*, and the ciliates *Paramecium tetraurelia* and *Tetrahymena thermophila*. In addition, there are 1326 ongoing eukaryotic genome or total mRNA (EST) sequencing projects, including relevant marine strains affiliating to the dinoflagellates, chrysophytes, pelagophytes and haptophytes.

The first marine picoeukaryote genome published was *O. tauri*. This prasinophyte has a 12.56-Mb haploid nuclear genome organized in 20 chromosomes. The genome is highly compacted, mainly due to the reduction in size of intergenic regions and the low copy numbers of most genes. The 8166 identified genes include all basic cell functions, such as photosynthesis, central metabolism, and cell–environment interaction. Genes encoding enzymes for C_4-photosynthesis have been identified, which may help the cells adapt to the limiting CO_2 concentrations of phytoplankton blooms. There are genes for transport and assimilation of nitrate, ammonium, and urea; the larger number of ammonium transporters as compared to nitrate transporters indicates that *O. tauri* could be a strong competitor for ammonium, which is uncommon in eukaryotic algae. *O. tauri* also has two chromosomes that differ structurally, being biased toward a lower GC content and a larger number of transposable elements. The first seems to have an alien origin, while the second could represent a sexual chromosome, preventing recombination with similar but not identical strains. Genes related to meiosis have been identified and are apparently functional, suggesting that this protist that usually reproduces asexually may also possess a sexual phase never observed. Thus, the genome of *O. tauri* follows prediction of compaction that might be driven by its specific lifestyle and ecology.

Environmental Genomics or Metagenomics

A striking advance in microbial ecology has been the gathering of gene content of natural communities, an approach that is neither culture-dependent nor hypothesis-driven. There are two main strategies depending on the size of the DNA being cloned. Large DNA fragments (40–200 kb) are cloned in bacterial artificial chromosomes (BAC) or fosmid vectors, and clones with these large inserts are screened before being completely sequenced. This may provide complete operon information, and in the case of prokaryotes may also link phylogenetic and functional markers. Alternatively, small DNA fragments (3–5 kb) may be cloned by routine methods and randomly sequenced at high throughput, which is known as shotgun sequencing. This provides a huge amount of genetic data, but the short sequences obtained are difficult to assemble due to the large diversity of natural assemblages.

Metagenomic approaches have been particularly useful for marine bacteria and archaea, to identify previously unknown metabolic pathways, such as novel uses of light and reduced compounds to generate energy, and putatively novel enzymes, antibiotics, and signaling molecules. This approach is still not routinely applied to picoeukaryotes, which are prefiltered from the samples. However, these will be included in more recent projects such as the Global Ocean Sampling Expedition, which has provided so far 6.3 billion base pairs of sequencing information from 41 distant marine sites.

CONCLUDING REMARKS

Picoeukaryotes are very small organisms (≤3 μm in the maximal dimension) that are morphologically and structurally simple but with all components required for an independent life. It is fascinating to find all organelles necessary for cell metabolism, growth, and reproduction assembled in such a small package. The few model cultures available so far provide useful tools for ecophysiological and genomic studies. In the marine environment, picoeukaryotes are ubiquitous, account for a significant share of planktonic biomass, and play key ecological roles as primary producers, bacterial predators, and parasites of marine life. The true diversity of marine picoeukaryotes has been recently unveiled by the use of molecular tools, which have revealed a very high diversity within this assemblage and the presence of many novel eukaryotes that are uncharacterized and uncultured. This increase in diversity occurs at almost all possible phylogenetic scales: high-rank novel groups, novel clades within supergroups, and putatively new orders, families, genera, and species within known lineages. The challenge is to retrieve in culture these novel organisms and to determine their ecological roles. The implication of this large and novel eukaryotic diversity for biodiversity surveys and ecosystem functioning opens new avenues for future research.

RECENT DEVELOPMENTS

The study of marine picoeukaryotes is a field in expansion that is attracting new scientists, mostly inspired by the possibilities provided by molecular tools to investigate novel ecological, evolutionary and phylogenetic aspects of these minute cells. Recent results have confirmed previous observations regarding their large genetic and functional diversity and the presence of novel groups. In addition, the most innovative contributions have derived from recent technical advances like NGS (Next Generation Sequencing) technologies that offer an unprecedented sequencing capacity, and flow cytometry routines that allow cell sorting before molecular or physiological analysis.

Studies of a few genomes of cultured marine free-living protists have been published during 2008 and 2009, including three additional prasinophytes (*Ostreococcus lucimarinus* and two *Micromonas* strains), one diatom (*Phaeodactylum tricornutum*) and one choanoflagellate (*Monosiga brevicollis*). Each genome has been analyzed under a different evolutionary story, covering gene organization, speciation patterns, or the origin of multicellularity. These studies are expected to reveal the gene basis of the ecological success of these minute cells. At present, genomic projects are limited by the availability of relevant ecological models into culture, since it is well known that a large fraction of the *in situ* diversity is not represented in culture. This occurs at all phylogenetic scales, but it is more critical for the high-rank taxa such as the uncultured MAST or the picobiliphytes. The culturing gap is difficult to bridge, since it requires dedicated laboratory work and the design of innovative culturing strategies. The recent culture of *Triparma* sp. (cf. *Triparma laevis)* exemplifies how original attempts can promote the growth of novel diversity. The parmales is an algal group formed by silica-covered cells that were described by electron microscopy. *Triparma* sp. has been brought into culture by using fluorescent precursors that target the deposition of silica during cell growth. Based on its 18S rDNA sequence, it affiliates with the bolidophytes, the sister group of diatoms, and not to the chrysophytes as was proposed. Additional ecologically relevant cultures will surely solve other scientific conundrums.

Flow cytometry has become a standard technique in oceanography for counting viruses, bacteria, cyanobacteria and phototrophic picoeukaryotes. A recent study explains how to count heterotrophic flagellates (HF) as well. The protocol is based on DNA-staining of microbial cells and separating HF from heterotrophic bacteria by their larger cell size and higher DNA content. This seems to work better in oligotrophic conditions, where there are very few large bacteria than can interfere with HF counts. The full potential of flow cytometry is attained when the counting routine is accompanied with cell sorting capabilities. Cell populations with a particular flow cytometric signature have been sorted to identify their contribution to a particular process or to study their diversity by classical molecular tools. Thus, sorted pigmented protists are contributing to half of oceanic bacterivory, highlighting the importance of mixotrophy in aquatic habitats. Molecular analyses of sorted pigmented picoeukaryotes have expanded their diversity, with novel lineages of prasinophytes, chrysophytes, and haptophytes being identified. Finally, the capacity to isolate individual cells by flow cytometry is opening new possibilities for single cell analyses of microorganisms. These single cells can be used as inoculum to start pure cultures, or as a template for whole genomic amplification, which can be then used in genomic projects. This single cell approach has been successfully applied to marine bacteria but not to picoeukaryotes as yet.

Thirty eight papers have been published so far studying the diversity of marine protists by clone libraries of environmental rDNA genes, eleven of these during 2008 and 2009. These papers confirm and expand previous results and provide a better coverage of the distribution of specific lineages. In addition, useful complementary data has been obtained with slight modifications of the general approach, such as the analysis of flow cytometry sorted populations (as mentioned above), the use of group-specific primers (typically expanding the diversity of the group), or the targeting of other genes than rDNA. Comparing the diversity obtained with a standard rDNA library and a library constructed after environmental ribosomes has revealed that some groups with high clonal abundance in the rDNA analysis (MALV, radiolarians) are little represented in the rRNA approach. So, cells from these two groups could be less active, could have a disproportionately high rDNA copy number or could be more represented in detrital DNA. The *in situ* diversity of marine protists has also been approached using the high sequencing capability provided by NGS technologies like 454 pyrosequencing of short rDNA amplicons. Several studies have already been published and more are in the pipeline. It is expected that NGS will allow the assessment of the true extent of phylogenetic diversity and provide exhaustive datasets for comparative purposes. These analyses indicate a large contribution of rare taxa, although at present it is difficult to fully discriminate new diversity from sequencing errors. A more ambitious use of the NGS technology is to obtain a full inventory of all genes in natural assemblages. Thus, the GOS metagenomic dataset has been explored for the presence of several anchor eukaryotic genes, and the diversity displayed after the detected 18S rDNA genes is strikingly similar to that observed in standard clone libraries. This culture and PCR-independent approach can potentially link diversity markers with putative functions, especially if large pieces of DNA are cloned before sequencing. Metagenomic studies on marine microorganisms, including picoeukaryotes, are expected to burst in the near future.

The natural complement of sequencing surveys is the FISH technique that labels cells belonging to a given rDNA clade, which can be formed by environmental sequences only. When combined with functional experiments and natural observations, FISH allows linking novel lineages to specific ecological roles. Like culturing, this technique is time-consuming and can only target a single taxa at a time, and at present there is plenty of room for more probe design to address critical questions. Combining FISH with bacterivory experiments has revealed important functional differences between several uncultured MAST lineages, which seem to have different grazing rates and consume different bacteria. Also, it has been recently demonstrated that several MALV lineages are parasites of specific dinoflagellates that are controlling the host abundances and dynamics in a coastal marine system.

In summary, the study of marine picoeukaryotes is a very active discipline that is benefiting by methodological improvements that allow opening new windows to unexplored fields. Challenges for the future include culturing the most abundant species, FISH probing to study abundances and biogeochemical implications of the novel groups, and obtaining a full inventory of *in situ* phylogenetic diversity to study biogeographical patterns and understand the evolutionary and ecological factors driving the large diversity observed.

FURTHER READING

Amaral-Zettler, L. A., McCliment, E. A., Ducklow, H. W., & Huse, S. M. (2009). A method for studying protistan diversity using massively parallel sequencing of V9 hypervariable regions of small-subunit ribosomal RNA genes. *PLoS ONE*, *4*, e6372.

Azam, F., Fenchel, T., Field, J. G., Gray, J. S., Meyer-Reil, L. A., & Thingstad, F. (1983). The ecological role of water-column microbes in the sea. *Marine Ecology Progress Series*, *10*, 257–263.

Baldauf, S. L. (2003). The deep roots of eukaryotes. *Science*, *300*, 1703–1706.

Bowler, C., Allen, A. E., & Badger, J. H. (2008). The *Phaeodactylum* genome reveals the evolutionary history of diatom genomes. *Nature*, *456*, 239–244.

Burki, F., Inagaki, Y., Bråte, J. *et al.* (2009). Large-scale phylogenomic analyses reveal that two enigmatic protist lineages, Telonemia and Centroheliozoa, are related to photosynthetic chromalveolates. *Genome Biology and Evolution*, *231*, 213–218.

Chambouvet, A., Morin, P., Marie, D., & Guillou, L. (2008). Control of toxic marine dinoflagellate blooms by serial parasitic killers. *Science*, *322*, 1254–1257.

Derelle, E., Ferraz, C., Rombauts, S. *et al.* (2006). Genome analysis of the smallest free-living eukaryote *Ostreococcus tauri* unveils many unique features. *Proceedings of the National Academy of Sciences of the United States of America*, *103*, 11647–11652.

Hughes Martiny, J. B., Bohannan, B. J. M., & Brown, J. H. *et al.* (2006). Microbial biogeography: Putting microorganisms on the map. *Nature Reviews Microbiology*, *4*, 102–112.

Ichinomiya, M., Yoshikawa, S., Kamiya, M., Ohki, K., Takaichi, S., & Kuwata, A. (2010). Isolation and characterization of Parmales (Heterokonta/Heterokontophyta/Sramenopiles) from the Oyashio region, Western North Pacific. *J. Phycology*, 29 NOV DOI: 10.1111/j.1529-8817.2010.00926.x.

Johnson, P. W., & Sieburth, J. M. (1982). *In situ* morphology and occurrence of eucaryotic phototrophs of bacterial size in the picoplankton of estuarine and oceanic waters. *Journal of Phycology*, *18*, 318–327.

Jürgens, K., Massana, R. (2008). Protistan grazing on marine bacterioplankton. In D. L. Kirchman (Ed.). *Microbial ecology of the oceans*. (2nd ed.) (pp. 383–441). Hoboken, NJ: John Wiley & Sons, Inc.

King, N., Westbrook, M. J., Young, S. L. *et al.* (2008). The genome of the choanoflagellate *Monosiga brevicollis* and the origin of metazoans. *Nature*, *451*, 783–788.

Li, W. K. W. (1994). Primary production of prochlorophytes, cyanobacteria, and eucaryotic ultraphytoplankton: Measurements from flow cytometric sorting. *Limnology and Oceanography*, *39*, 169–175.

Massana, R., Unrein, F., Rodríguez-Martínez, R., Forn, I., Lefort, T., Pinhassi, J. *et al.* (2009). Grazing rates and functional diversity of uncultured heterotrophic flagellates. *The ISME Journal*, *3*, 588–596.

Moon-van der Staay, S. Y., De Wachter, R., & Vaulot, D. (2001). Oceanic 18S rDNA sequences from picoplankton reveal unsuspected eukaryotic diversity. *Nature*, *409*, 607–610.

Not, F., Valentin, K., Romari, K. *et al.* (2007). Picobiliphytes: A marine picoplanktonic algal group with unknown affinities to other eukaryotes. *Science*, *315*, 252–254.

Not, F., del Campo, J., Balagué, V., de Vargas, C., & Massana, R. (2009). New insights into the diversity of marine picoeukaryotes. *PLoS ONE*, *4*, e7143.

Piganeau, G., Desdevises, Y., Derelle, E., & Moreau, H. (2008). Picoeukaryotic sequences in the Sargasso Sea metagenome. *Genome Biology*, *9*, R5.

Richardson, T. L., & Jackson, G. A. (2007). Small phytoplankton and carbon export from the surface ocean. *Science*, *315*, 838–840.

Rodríguez, F., Derelle, E., Guillou, L., Le Gall, F., Vaulot, D., & Moreau, H. (2005). Ecotype diversity in the marine picoeukaryote *Ostreococcus* (Chlorophyta, Prasinophyceae). *Environmental Microbiology*, *7*, 853–859.

Shi, X. L., Marie, D., Jardillier, L., Scanlan, D. J., & Vaulot, D. (2009). Groups without cultured representatives dominate eukaryotic picophytoplankton in the oligotrophic South East Pacific ocean. *PLoS ONE*, *4*, e7657.

Vaulot, D., Eikrem, W., Viprey, M., & Moreau, H. (2008). The diversity of small eukaryotic phytoplankton ((3 (m) in marine ecosystems. *FEMS Microbiology Reviews*, *32*, 795–820.

Viprey, M., Guillou, L., Ferréol, M., & Vaulot, D. (2008). Wide genetic diversity of picoplanktonic green algae (Chloroplastida) in the

Mediterranean Sea uncovered by a phylum-biased PCR approach. *Environmental Microbiology*, *10*, 1804–1822.

Worden, A. Z., & Not, F. (2008). Ecology and diversity of picoeukaryotes. In D. L. Kirchman (Ed.) *Microbial ecology of the oceans.* (2nd ed.) (pp. 159–205). Hoboken, NJ: John Wiley & Sons, Inc.

Worden, A. Z., Lee, J. H., Mock, T. *et al.* (2009). Green evolution and dynamic adaptations revealed by genomes of the marine picoeukaryotes *Micromonas*. *Science*, *324*, 268–272.

Woyke, T., Xie, G., Copeland, A. *et al.* (2009). Assembling the marine metagenome, one cell at a time. *PLoS ONE*, *4*, e5299.

Zubkov, M. V., & Tarran, G. A. (2008). High bacterivory by the smallest phytoplankton in the North Atlantic Ocean. *Nature*, *455*, 224–227.

RELEVANT WEBSITES

http://www.icm.csic.es/bio/projects/icmicrobis/bbmo/. – Blanes Bay Microbial Observatory (BBMO)

http://keydnatools.com/. – KeyDNATools

http://www.arb-silva.de/. – SILVA

http://www.genomesonline.org/. – Genome Projects Home (GOLD).

http://www.tolweb.org/tree/. – Tree of Life Web Project.

http://www.sb-roscoff.fr/. – Station Biologique de Roscoff.

http://ccmp.bigelow.org. – The Provasoli–Guillard National Center for Culture of Marine Phytoplankton, CCMP.

Chapter 7

Algal Blooms

P. Assmy and V. Smetacek
Alfred Wegener Institute for Polar and Marine Research, Bremerhaven, Germany

Chapter Outline

ABBREVIATIONS

DMS Dimethyl sulfide
DMSP Dimethylsulfoniopropionate
HABs Harmful algal blooms
HNLC High-nutrient, low-chlorophyll
H_2S Hydrogen sulfide
N_2O Nitrous oxide
PCD Programmed cell death
PSP Paralytic shellfish poisoning
SML Surface mixed layer

DEFINING STATEMENT

Dense aggregations of phytoplankton cells of one or more species are loosely referred to as algal blooms. They play a central role in the ecology and biogeochemistry of all water bodies from ponds to the ocean, but a mechanistic understanding of the factors leading to the rise and fall of blooms is still lacking.

INTRODUCTION

"Algal bloom" is a term of convenience applied to an outbreak of phytoplankton cells well above the average for a given region or water body. Blooms show up as peaks in the annual cycle of phytoplankton biomass and chlorophyll concentrations. They are ephemeral phenomena that arise when growth rates of one or more species of the phytoplankton assemblage exceed their mortality rates. As a result, cells of these species accumulate in the water column until their growth is checked by resource depletion, generally a nutrient such as phosphorus, reactive nitrogen, or iron. The magnitude of the bloom peak is determined by the nutrient concentrations prior to outbreak of the bloom. Since all phytoplankton contain chlorophyll *a*, which is easily assessed by a range of methods, the size of a bloom is conveniently expressed in units of chlorophyll. In productive water masses with high-nutrient availability, such as eutrophic lakes and coastal seas, chlorophyll concentrations at bloom peaks can reach >20 mg Chl m^{-3}, but over much of

Topics in Ecological and Environmental Microbiology.

the ocean tenfold lower values are also commonly referred to as blooms because their concentrations again are five- to tenfold higher than the concentrations prevailing there over much of the year.

Phytoplankton blooms develop in the surface mixed layer (SML), so the total biomass or standing stock of a bloom is expressed as the concentration (cell numbers or Biomass m^{-3}) multiplied by the depth of the mixed layer (cell numbers or Biomass m^{-2}). The latter varies from less than a few meters in lakes to tens of meters in the ocean, so standing stocks of blooms in both regions are broadly similar because the higher concentrations of algae generally found in shallower mixed layers are compensated by the greater depth of the surface layer where concentrations are lower. It is the standing stock of the bloom that determines how much carbon is fixed per unit area and hence how much food is available to the larger organisms of higher trophic levels and to the microbes and animals (benthos) of the underlying sediments. It follows that the highest standing stocks (largest blooms) develop in SMLs of intermediate depths.

Algal blooms are short-lived phenomena superimposed on the ubiquitous microbial food web, which is the characteristic pelagic ecosystem prevailing in the surface layer of most of the oceans and large lakes (Figure 1). The system maintains itself by recycling regenerated nutrients (ammonia, phosphate, and iron) between phytoplankton, comprising $<2\,\mu m$ solitary cyanobacteria and small ($<10\,\mu m$) autotrophic flagellates, and heterotrophs, comprising bacteria, their nanoflagellate grazers, and larger protozoa in particular ciliates and dinoflagellates that feed mainly on the eukaryotic components of the food web. They in turn are grazed by copepods and other suspension-feeding zooplankton. Algal biomass is generally $<20\,mg\,Chl\,m^{-2}$, that is, concentrations of $<2\,mg\,Chl\,m^{-3}$ and $<0.5\,mg\,Chl\,m^{-3}$ in coastal and open-ocean regimes, or eutrophic and oligotrophic lakes, respectively. Heterotrophic biomass is generally several fold that of the autotrophs. The opposite situation prevails during blooms: additional or new nutrients, generally in the form of nitrate, are introduced by admixture of nutrient-rich water, which disrupts the quasi-steady state of the microbial food web. The microbial community also profits from the new nutrients, but the bulk of the biomass increase is due to larger phytoplankton inaccessible to smaller grazers. These are the bloom-forming species.

In the following section, we first deal with the so-called bottom-up factors that influence bloom formation – the physical and chemical properties of the environment. After an overview of the major genera and species contributing to blooms we deal with the top-down factors – pathogens, parasitoids, and predators – that kill phytoplankton cells. Since the term parasitoid is not very commonly used, we define it as an organism that kills the host as a result of parasitism. The various types of algal blooms – recurrent and sporadic – are discussed followed by brief accounts of harmful algal blooms (HABs) and artificial blooms induced

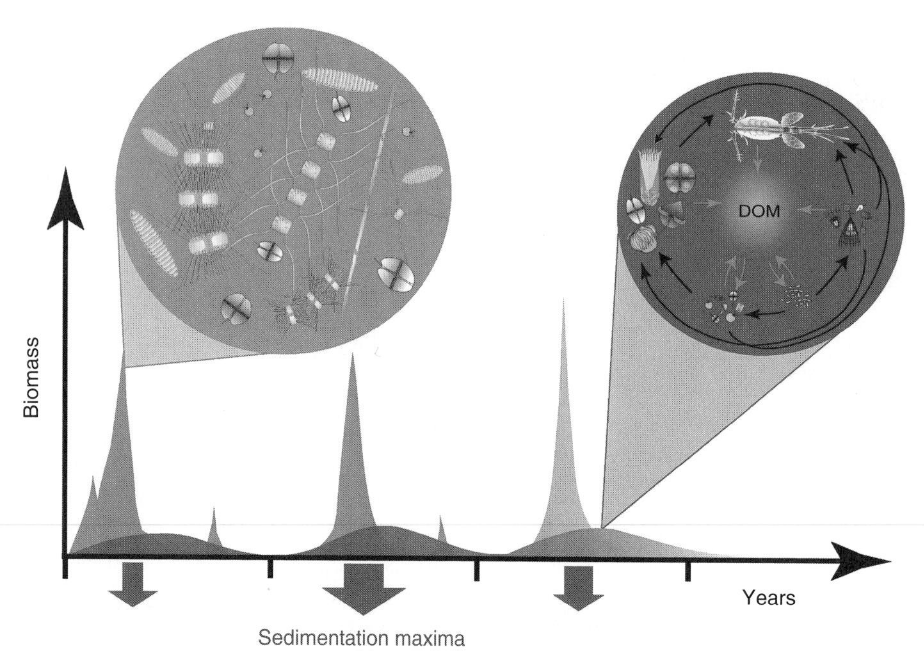

FIGURE 1 A schematic representation of the relationship between seasonality of the microbial network (regenerating system) and superimposed blooms. The latter are followed by sedimentation pulses. DOM, dissolved organic matter. *Modified by Christine Klaas from Smetacek, V., Scharek, R., & Nöthig, E. M. (1990). Seasonal and regional variation in the pelagial and its relationship to the life history cycle of krill. In K. R. Kerry, & G. Hempel (Eds.).* Antarctic ecosystems: Ecological change and conservation, *1st ed. (pp. 103–114). Berlin, Heidelberg: Springer. (© Christine Klaas).*

by iron fertilization in the open ocean. We conclude with some avenues that need to be explored to further our mechanistic understanding of the rise and fall of algal blooms.

Physical Environment of Blooms

The depth of the SML is generally determined by wind mixing, which in turn is a function of wind force and fetch, which is why the SML is shallowest in small, protected lakes and deepest in the open ocean. The density difference between surface and deeper layers constrains the depth of mixing because the greater the gradient the more energy required to mix across it. Generally, temperature determines density, so the SML is at its shallowest during the summer and in low latitudes. The steep density gradient between the warm SML and the underlying colder water is known as the thermocline. Exceptions are found in estuaries, brackish seas, and at the sea-ice edge during melting where the vertical salinity gradient can have a greater effect than temperature in constraining depth of the SML. Here the boundary between the fresher SML and the saltier underlying water is known as the halocline. Pycnocline is the collective term for a density gradient whether caused by temperature, salinity or both.

The depth and rate of vertical mixing of the SML is of critical importance in bloom formation because it determines the light climate experienced by the phytoplankton suspended in it. Light intensity decreases exponentially with depth because light is absorbed by water and converted into heat. Suspended particles and some classes of dissolved organic matter such as humic acids (gelbstoff) also absorb light or scatter it, contributing to attenuation of light intensity with depth. In the clear blue water of the open ocean, living phytoplankton are the major light-absorbing particles because detritus plays a minor role and most protozoa and zooplankton are transparent to avoid visual predators. At the other extreme, in the green waters of turbid, shallow lakes and tidally mixed shallow seas, suspended particles (detritus and lithogenic particles) scatter and absorb a far greater percentage (>90%) of incoming light than the phytoplankton. So, with increasing depth and turbidity of the SML, the light climate experienced by the phytoplankton population as a whole declines.

Net growth of an algal cell occurs only when carbon fixed by photosynthesis exceeds remineralization by respiration. The light intensity at which the two opposing processes balance each other is termed as the compensation light intensity and is in the order of 0.1% of incoming radiation at the surface. Its location in the water column sets the lower boundary of the euphotic zone within which net growth occurs. The depth of the euphotic zone can range from decimeters to meters at low solar angles (winter of high latitudes) and in highly turbid water, to >50 m in the open ocean. The relationship between the depths of the euphotic zone and the SML determines the rate of bloom growth. If the SML is much deeper, then loss processes outweigh gains by photosynthesis, and phytoplankton biomass cannot increase. In the reverse case, given the presence of adequate nutrients, growth can be sufficiently rapid to attain bloom proportions.

However, the relationship between SML and euphotic zone depths on bloom development is rendered complex by the biology of plankton organisms. Thus, phytoplankton can adjust their photosynthetic machinery to work efficiently at low light levels (shade adaptation) of a deepwater column, but the cells suffer damage at high light intensities prevailing closer to the surface if transported there by vertical mixing. A day or two of full sunshine in otherwise cloudy weather can have the same negative effect on shade-adapted cells. Clearly, the species with the highest growth rates will have achieved the optimal trade-off with respect to the inherently fluctuating physical environment. On the other hand, loss rates due to pathogens and grazers also vary widely and are generally of much greater importance than respiratory losses of the phytoplankton.

Satellite observations of chlorophyll distribution indicate that the occurrence of algal blooms is related to nutrient rather than light fields. Light limitation of growth rate is restricted to the few winter months at high latitudes. It should be mentioned that the daily photon flux in midsummer at the poles is equivalent to that in the tropics because low solar angle is compensated by the much greater day length. Temperature also has a secondary effect on algal growth rates because of cold adaptation. Unlike land plants that are exposed to widely fluctuating temperatures in the transition phase from dormant to growth seasons, aquatic plants can adapt to a narrow temperature range. Thus, maximum division rates close to the freezing point can be as high as one per day, which is more than half that attained in the tropics. Indeed among the highest rates of primary production measured anywhere ($<10\,g\,C\,m^{-2}$) were those recorded in dense blooms developing in the nutrient-rich, shallow SML of the Bering Sea inflow to the Arctic Ocean.

Chemical Environment of Blooms

A prerequisite for the growth of blooms is the availability of essential elements of which the macronutrients phosphorus and reactive nitrogen and the micronutrient iron generally limit build up of blooms. Phytoplankton grow by taking up dissolved nutrients and incorporating them into biomass. With the exception of lipids, organic matter is denser than water, so particularization eventually results in loss of nutrients from the SML due to sinking particles. Nutrients are supplied by the mixing of deep, nutrient-rich water to the SML. In high and mid-latitudes, this is invariably due to cooling and breakdown of thermal stratification by convective mixing during autumn and winter. In the tropics, upwelling of deeper water due to large-scale lateral advection of the surface layer is the major source of nutrients although runoff from land is also locally important. The passage of storms also mixes nutrients into the SML.

Bloom biomass is limited by the concentration of the nutrient in shortest supply. There is now a general consensus that it is phosphate in lakes and some oligotrophic seas (e.g., the Mediterranean), nitrate in iron-rich coastal waters, and iron in the open ocean with the exception of most of the North Atlantic. The concentrations of phosphate and iron are limited by their geochemical properties. Thus, maximum winter concentrations of ~1 µmol PO_4 l^{-1} seem to be the rule in coastal regions where surface water is in interaction with oxygenated sediments. The corresponding iron concentrations are <1 nmol Fe l^{-1}, which are well in excess of the P:Fe ratio required by phytoplankton. In contrast, dissolved molecular nitrogen is abundant and can be fixed to reactive forms by cyanobacteria, which results in maintenance of N:P ratios commensurate with the demands of phytoplankton. Because of the extreme insolubility of ferric hydroxide (rust), iron is selectively lost from the SML and becomes the limiting nutrient in land-remote regions. The subarctic North Pacific Ocean and the entire Southern Ocean harbor perennially high concentrations of nitrate and phosphate because of iron limitation of phytoplankton growth. The same is true of the equatorial Pacific. Phytoplankton growth is not continuously iron-limited in the high latitude and equatorial Atlantic presumably due to iron input by dust emanating from the adjacent arid regions.

As unicellular algae are too small to experience shear stress caused by the kinetic energy of water they do not require investment in supporting structures such as carbohydrates to maintain shape and position as do all larger metaphytes. They also do not store reserves such as starch or lipids, which, in sessile plants, enable spurt growth at the beginning of the growth season as a means to overgrow competitors for space and light. As a result, the C:N:P ratio of aquatic unicellular organisms resembles that of living cytoplasm and varies <20% around the global average of 106:16:1. This is known as the Redfield ratio after the scientist who first discovered its relative constancy not only in suspended matter but also in dissolved nutrients in the deep ocean. It should be pointed out that variation in the Redfield ratio within species as a result of adaptation to the growth environment can be as great as between the algal classes.

MAJOR CONTRIBUTORS TO ALGAL BLOOMS

Of the several thousand phytoplankton species described so far <5% have been reported to contribute significantly to algal blooms in lakes and oceans, which implies that the ability to form blooms is not an indication of evolutionary success. Many species are widely distributed and regularly contribute to blooms throughout their range, implying that adaptation to a specific set of environmental conditions, such as nutrient ratios or light conditions, is not a prerequisite. Bloom-forming species differ widely from one another in shape, size, and behavior, and – because they belong to different phylogenetic groups – also differ in their pigment composition and biochemistry. Not surprisingly, their blooms also have very different impacts on the ecosystem and biogeochemistry. Their degree of recurrence also varies. Some species bloom every year in the same water body and are an integral part of the seasonal cycle. Others reach bloom proportions only in some years but always in the same season. Yet others form blooms seemingly haphazardly. It is thus not possible to formulate generalizations about bloom-forming species other than that their size (cell diameter, chain length, and colony diameter) tends to be larger than 10 µm. Since smaller cells tend to have higher growth rates, the ability to grow fast *per se* cannot be regarded as a prerequisite for bloom formation. Apparently, the ability to withstand attack by pathogens and herbivores by means of larger size and other defenses is hence of greater importance in enabling accumulation of cells. Here, we refer to species differentiated under the microscope, that is, on the basis of their morphology (Figure 2). Genetic studies are revealing "cryptic" species within morphologically similar cell walls. In the following section we deal with the bloom-forming species according to taxonomic groupings with some brief notes on their ecology and impact where appropriate.

Cyanobacteria

This group, formerly known as blue-green algae, are prokaryotes and the progenitors of the chloroplasts of all eukaryotes. They possess essentially the same photosynthetic machinery as the algae but, unlike all algae, some species of cyanobacteria are able to fix molecular nitrogen and channel it into the ecosystem. Another unique property of the larger species is their ability to secrete gas bubbles within the cell, which renders them positively buoyant and results in formation of conspicuous scum on the surface. Since nitrogen fixing is an energy-intensive process, it is hypothesized that the gas bubbles enable them to stay at the surface.

Cyanobacteria tend to be rare when fixed nitrogen in the form of nitrate or recycled ammonia is available, but increase their numbers when nitrogen becomes limiting for algal growth but other nutrients, such as phosphate and iron, are still available. This situation invariably arises in the summer months when the N:P ratio, normally at 16:1, reaches its annual minimum. In oxygenated waters, characteristic of most of the ocean as well as oligotrophic deep lakes, this happens because P is recycled more rapidly than N which is selectively lost via sinking particles. In shallow eutrophic lakes with suboxic deep water, nitrate is oxidized to N_2 by

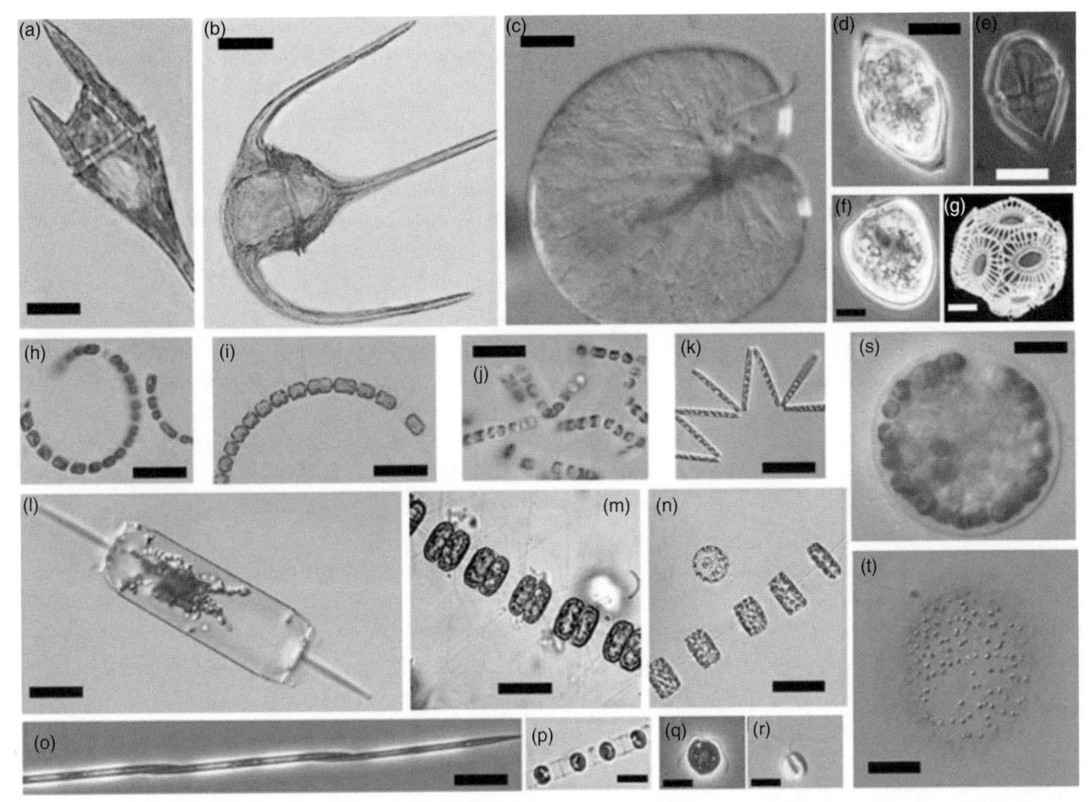

FIGURE 2 Pictures of prominent bloom-forming species. (a) *Ceratium furca*, (b) *Ceratium tripos*, (c) *Noctiluca scintillans*, (d) *Heterocapsa triquetra*, (e) *Scrippsiella trochoidea*, (f) *Prorocentrum minimum*, (g) *Emiliania huxleyi*, (h) *Chaetoceros debilis*, (i) *Chaetoceros curvisetus*, (j) *Chaetoceros socialis*, (k) *Thalassionema nitzschioides*, (l) *Ditylum brightwelli*, (m) *Thalassiosira nordenskioeldii*, (n) *Thalassiosira rotula*, (o) *Pseudo-nitzschia lineola*, (p) *Skeletonema costatum*, (q) *Chrysochromulina polylepis*, (r) *Phaeocystis antarctica* solitary cell, (s) *P. antarctica* compact small colony, and (t) *P. antarctica* large colony. Scale bars = 2 μm (g), 5 μm (f), 10 μm (d, e, m, and p–r), 15 μm (a), 20 μm (l, n, o, and s), 30 μm (b), 50 μm (h, j, and k), 60 μm (i), 100 μm (t), 200 μm (c). Light micrographs (o), (r), (s), and (t) have been kindly provided by Marina Montresor. Scanning electron micrograph (g) was taken by Philipp Assmy. All other light micrographs were taken from the open access repository for plankton-related information PLANKTON*NET (URL: http://planktonnet.awi.de.) of which (a), (b), (h–k), (m), and (p) were kindly provided by Mona Hoppenrath; (d) and (f) by Regina Hansen; (l) and (n) by Tanya Morozova; (c) by Susanna Knotz; (q) by http://www.algaebase.org/index.lasso; and (e) by Alexandra Kraberg.

denitrifying bacteria, which further lowers the N:P ratio. The lowest N:P ratios (<10:1) are found in water bodies overlying anoxic sediments that mobilize not only P but also Fe. It follows that the highest concentrations of cyanobacteria occur in highly eutrophic lakes and ponds.

The most common bloom-forming genera not only in freshwater but also brackish seas such as the Baltic, are dealt with next. The ubiquitous genus *Microcystis* builds spherical colonies, but all the other genera occur either as solitary cells (such as the bacteria-sized *Synechococcus*) or as chains (*Nodularia*, *Aphanizomenon*, and *Anabaena*). The latter generally form aggregates in the shape of bundles or globular clumps of chains. These species have specialized cells known as heterocysts in which nitrogen is fixed. Cyanobacteria are generally well defended and some species, particularly *Microcystis*, *Anabaena*, and *Nodularia*, sometimes secrete potent toxins. Blooms of these species can pose a serious problem in reservoirs and lakes. The only genus that regularly forms blooms in the ocean is *Trichodesmium*, which grows in sheathed chains (trichomes) that form large aggregates that often color the surface of many warm ocean regions a reddish-brown. Toxins are apparently not produced, but this genus is not relished by grazers.

Diatoms

Most algal blooms in oceans and lakes are dominated by a broad range of species belonging to this group (Figure 2h–p). Diatoms have an obligate need for silicic acid because the cell is encased by silica shells comprising two half-boxes that fit tightly into one another. The shell surfaces are dotted with minute pores that only allow dissolved substances to enter into the cells. The fact that, in contrast to other algal groups, only few instances of viral infection of diatoms have been reported so far suggests that diatoms are well defended against pathogens possibly due to the silica shells. Since silica shell formation is an energetically cheap process, diatoms tend to have high growth rates. They can also regulate buoyancy despite the ballast effects of the shell. Shape and size of bloom-forming diatoms vary widely from long-needle-shaped cells up to few millimeters long to small cylindrical cells in long chains. Some species have long silica spines or chitin threads that protrude outward from the cells. Bloom-forming species are scattered over all diatom lineages (raphid and araphid pennates and various families of centrics) and many genera.

The most prominent recurrent feature of the seasonal plankton cycle in temperate and boreal systems is the spring bloom, which is generally dominated by comparatively few species of unrelated genera of diatoms. Typical examples of recurrent bloomers are species of the cosmopolitan diatom genera *Skeletonema*, *Thalassiosira*, and *Chaetoceros* (Figure 2h–j, m, n, and p). *Skeletonema* blooms occur throughout the world with the exception of the polar oceans. They are prominent not only in the spring blooms of the brackish Baltic and the open Atlantic, but also in upwelling blooms of the tropics. Formerly only one species, *Skeletonema costatum* (Figure 2p), was recognized but recent detailed studies on the morphology and genetics of this species have so far revealed eight distinct entities, which differ in their geographic distribution and seasonal occurrence. A wide variety of species within the genus *Thalassiosira* form blooms worldwide. A few examples include *Thalassiosira antarctica* that can form dense blooms in layers underlying sea ice in the Antarctic; *Thalassiosira weissflogii*, *Thalassiosira nordenskioeldii*, and *Thalassiosira rotula* (Figure 2m and n) that contribute to blooms in temperate and boreal areas; and *Thalassiosira partheneia* that builds large (>1 cm diameter) colonies and is restricted to the upwelling region of northwestern Africa.

The bloom-forming *Chaetoceros* species belong to the subgenus *Hyalochaete*, many of which are bipolar. Three prominent species are *Chaetoceros socialis*, *Chaetoceros debilis*, and *Chaetoceros curvisetus* (Figure 2h–j), all of which are cosmopolitan, with the former two species showing a more polar-to-cold temperate distribution and the latter species a warm temperate distribution. *C. socialis* grows in spherical colonies that bear superficial resemblance to those of the haptophyte *Phaeocystis*. The species not only dominates in open-ocean, meltwater-associated blooms in the Antarctic and Arctic, but also in blooms of upwelling areas. Most bloom-forming *Chaetoceros* species are capable of converting vegetative cells into thick-walled resting spores that overwinter on the sediment surface in shallow environments, inside the sea ice in the seasonal sea-ice zone, and in the pycnocline in the open ocean. The widespread and consistently high accumulation rates of *Chaetoceros* resting spores in the Southern Ocean throughout the last glacial and their restricted occurrence during the Holocene provide compelling evidence for substantially higher productivity and organic carbon export to the deep ocean during the last glacial and highlight the potential of resting spores as biological proxies for paleoceanography.

The cosmopolitan needle-shaped genus *Pseudo-nitzschia* comprises many species that are difficult to differentiate under the light microscope; some of the species can contribute substantially to both open-ocean and coastal blooms. Detailed studies on *Pseudo-nitzschia delicatissima* and *Pseudo-nitzschia pseudodelicatissima*, two morphologically distinct entities, have shown reproductive isolation among sympatric cryptic species in the Gulf of Naples. The factors responsible for these subtle differences are as yet unknown but likely represent either reproductive isolation or adaptations at the physiological but not the morphological level such as defenses against specific pathogens or predators.

The needle-shaped, thin-shelled centric genus *Rhizosolenia* commonly forms blooms later in the year. A dense bloom in which the cells were aggregated into mats was reported from the equatorial Pacific where the mats apparently carried out vertical migration between the surface and the nutrient-rich deeper layer. Surprisingly, this effective mechanism of nutrient acquisition, although looked for, has not been found since. Another widespread species that is common in the summer months is *Ditylum brightwelli* (Figure 2l), which has, however, once formed an almost monospecific spring bloom in the southern North Sea.

Haptophytes (Prymnesiophytes)

This group comprises small flagellates equipped with a prehensile organ, the haptonema, which is apparently used to feed on small particles such as bacteria (Figures 2g and q–t). A common characteristic of this group is the copious production of dimethylsulfoniopropionate (DMSP), which is split by an enzyme into volatile dimethyl sulfide (DMS) and the noxious acrylic acid. Several functions of the widespread molecule DMSP have been suggested such as a compatible solute in osmoregulation, defense against predators, and as an antioxidant. DMS released from haptophyte blooms is a major global source of sulfur to the atmosphere where it is oxidized to hygroscopic sulfate, which acts as a cloud condensation nucleus. Large amounts of sulfur in the atmosphere result in smaller droplets, that is, whiter clouds that scatter sunlight back into space, thus contributing to cooling. The global significance of haptophyte blooms in cooling the planet is under debate.

The calcareous coccolithophores belong to this group and are second in importance to the diatoms in their contribution to extensive oceanic blooms. These normally arise in the aftermath of the spring diatom bloom and, because of the high reflectance of their calcareous plates, are visible from space. The species *Emiliania huxleyi* (Figure 2g) is the most prominent member of this group and forms blooms in both coastal and open-ocean regions. Coccolithophores exhibit highly complex life cycles in which haploid, diploid, and polyploid stages, some bearing different types of coccoliths, alternate with one another. The adaptive significance of this life cycle is not understood. Blooms of *E. huxleyi* are regularly terminated by the spread of viral infection within the population. Interestingly, despite its numerical dominance in the surface layer, *E. huxleyi* contributes only a minor fraction of the total calcareous material accumulating in the deep North Atlantic. Less abundant but more heavily calcified coccolithophores like *Calcidiscus leptoporus* contribute the bulk of accumulated calcareous sediments.

Another group of haptophytes involved in extensive blooms in both temperate and polar waters are species of the genus *Phaeocystis*. At least six species have been identified worldwide based on morphological and genetic characteristics but only three species – *Phaeocystis pouchetii* (Arctic), *Phaeocystis globosa* (temperate and tropical waters), and *Phaeocystis antarctica* (Antarctic, Figures 2r–t) – dominate blooms. These are also the only species thus far observed to form colonial stages comprising many hundreds of cells. The others occur only as the solitary nanoflagellate stage. The flagellate stage suffers much higher mortality by both viral infection and protozoan grazing than the colonial one, indicating that colony formation is a defense strategy. The fact that colonies in their early stages are often found attached to the spines of diatoms (e.g., *Chaetoceros*, *Corethron*) also suggests that small colonies are more vulnerable and find protection from protozoan grazing on these large diatoms (Figure 3). In the North Sea, *P. globosa* tends to attain dominance in the later spring, in the aftermath of the diatom bloom following Si exhaustion. Extensive blooms dominated by *P. antarctica* are a regular feature of the Ross Sea but, for unknown reasons, are more sporadic elsewhere around Antarctica. On rare occasions dense blooms of the solitary flagellate genera *Chrysochromulina* and *Prymnesium* occur in some regions. Some species are highly toxic and result in mass mortality of other organisms.

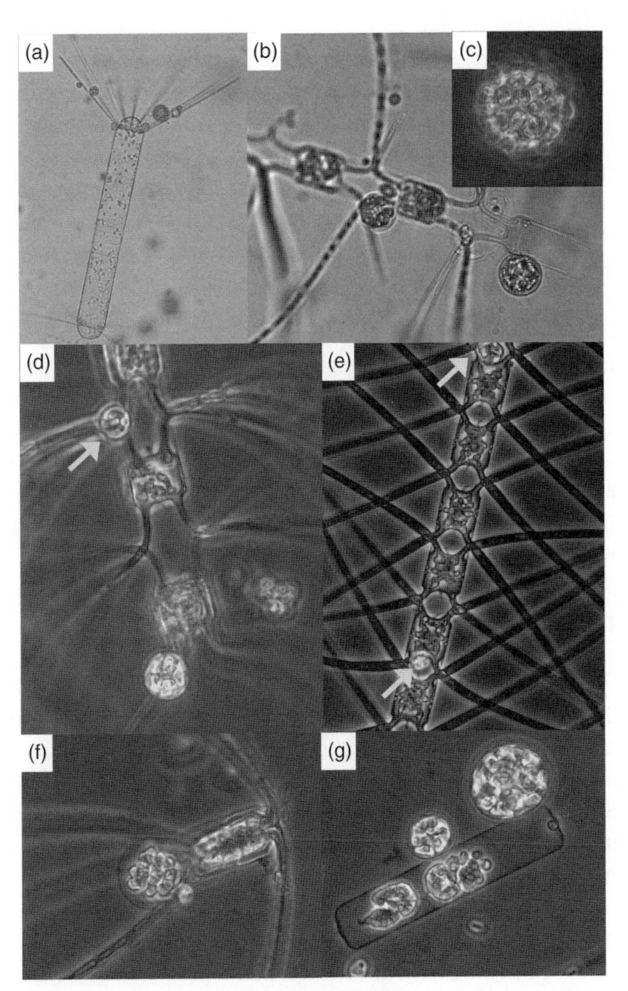

FIGURE 3 Colonies of *Phaeocystis antarctica* attached to different diatom species (from field material): (a) spines of *Corethron pennatum*; (b) setae and cell wall of *Chaetoceros dichaeta*; (c) close-up of colony; (d) *Chaetoceros dichaeta*; (e) *Chaetoceros atlanticus*; (f) *Chaetoceros peruvianus*; (g) *Guinardia cylindrus*. Light micrographs by Philipp Assmy.

Dinoflagellates

This is an ancient, heterogeneous group comprising autotrophic, heterotrophic, and mixotrophic modes of nutrition across a wide range of shapes and sizes (Figure 2a–f). Interestingly, the morphology of photosynthetic and predatory forms does not differ, signifying that their shapes and sizes have not evolved to maximize resource acquisition. Like the coccolithophorids, small-celled dinoflagellates (~10 μm) encased in cellulose armor (e.g., *Prorocentrum*, *Heterocapsa*, *Scrippsiella*) (Figure 2d–f) commonly form blooms in the aftermath of the spring diatom bloom in coastal and shelf areas. Larger, often unarmored forms such as *Gymnodinium* form sporadic blooms not only in the summer months but also in upwelling regions. Spectacular blooms of the heterotrophic bioluminescent *Noctiluca scintillans* (Figure 2c), colored by pigments from their algal food, cause discoloration of large stretches of surface waters mainly in sheltered bays or estuaries. Because of their motility and comparatively large size, some dinoflagellates have been observed to migrate vertically to take up nutrients below a shallow SML. Many species also form dense layers at density discontinuities, which reach the surface along hydrographic fronts where they are highly conspicuous. The standing stocks of such blooms are low, despite their dense concentrations because they are restricted to thin layers.

The largest dinoflagellate blooms in both lakes and the ocean are formed by species of the large, armored genus *Ceratium* during the autumn. For unknown reasons, grazing pressure on this genus is low, suggesting that survival rather than fast growth rates is the major reason for bloom build up over the late summer. Although there are many species, only a few cosmopolitan ones (e.g., *Ceratium tripos*, *Ceratium furca*, *Ceratium fusus*) (Figure 2a and b) form blooms throughout their range.

Other Groups

The autotrophic, cosmopolitan ciliate *Mesodinium rubrum* (also known as *Myrionecta rubra*) is an active swimmer that, like dinoflagellates, carries out vertical migration and sometimes forms dense reddish layers at the surface. Its cells are packed with cryptophyte chloroplasts that still carry the cryptophyte nucleus. Blooms of this species are common and have been reported from habitats as diverse as tropical upwelling regions, the inner Baltic and the Arctic shelf. Large blooms of this ciliate tend to be sporadic but patchy blooms occur regularly in the North and Baltic seas in early summer. Other groups reported to contribute significantly to algal blooms are the euglenophytes, cryptophytes, and raphidophytes; their blooms are only of local importance. In lakes, various species of chlorophytes and chrysophytes form blooms under eutrophic conditions.

PATHOGENS AND GRAZERS OF ALGAL BLOOMS

The important role of viruses in the sea has only come to light over the past two decades. It is now known that bacterial populations are susceptible to viral infection and that their density is partly regulated by this factor. Less is known about viral infection of algae but coccolithophorid blooms in mesocosms have been reported to have been terminated by mass infection. So far, only few viruses have been found to infect diatoms, and the pathogens of other groups have barely been studied. This also applies to pathogenic bacteria that are known to occur but whose influence is unknown.

A wide range of grazers is present in the plankton: puncturing and ingesting protists that prey on individual cells and chains, suspension-feeding copepods that select and ingest particles individually, and filter-feeding larger metazooplankton such as euphausiids (krill) that collect particles *en masse* and ingest them indiscriminately. Complexity is introduced by the tendency of copepods to selectively feed on protistan grazers of phytoplankton, in particular ciliates and heterotrophic dinoflagellates, which results in a trophic cascade favoring accumulation of algal cells.

Protistan grazers have growth rates equivalent to their phytoplankton prey and many have evolved techniques to cope with prey sizes as large as or even larger than themselves. Thus, small flagellates belonging to various protistan lineages can attach themselves to large diatoms and insert feeding tubes through pores in the frustule or between the girdle bands. The cell multiplies while feeding on the diatom plasma, releasing large numbers of flagellated cells that can infect other diatom cells. These specialized grazers are known as parasitoids, because of their small size relative to their prey. They are generally species-specific in prey selection and have the potential to decimate prey populations during blooms. However, this has rarely been observed presumably because they are kept in check by grazing of larger protists, in particular ciliates that are specialized grazers of small flagellates, although many ciliate species can ingest cells as large as themselves.

Dinoflagellates have evolved the broadest range of feeding techniques: some ingest prey cells up to their own size, others pierce the cell wall and suck out the contents with a feeding appendage known as the peduncle. Several genera of armored dinoflagellates are able to extrude a feeding veil known as the pallium, which can envelope an entire chain of diatoms many times the size of the predator and digest it outside the cell wall (Figure 4). The pallium is retracted after digestion is completed. The entire process of prey capture and pallium retraction lasts for 10–20 min and the pallium-feeding dinoflagellates can divide faster than their algal prey. Again, decimation of algal blooms by dinoflagellates is rarely reported although they have the potential to do so. In all likelihood, the ubiquitous small copepods selectively feed on large heterotrophic protists, which release grazing pressure on the algae, particularly diatoms.

In contrast to unicellular herbivores, metazooplankton, in particular copepods and euphausiids have complex life cycles involving a range of larval stages that take weeks or months to complete, depending on temperature in opportunistic species, or innate life cycle traits in species with pronounced seasonality. The latter generally have larval stages that overwinter at depth and ascend to the surface in spring. It is widely believed that blooms arise because copepod populations are unable to match their appearance at the surface with the timing of bloom initiation and subsequently gear their population size to the growing bloom. However,

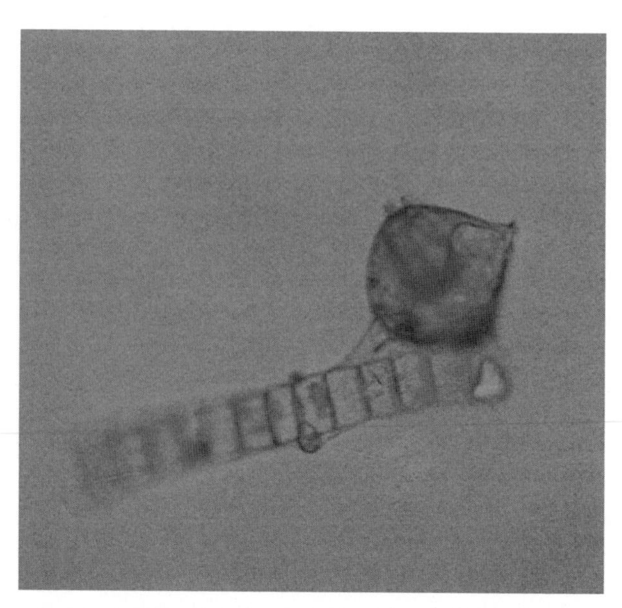

FIGURE 4 A pallium-feeding dinoflagellate of the genus *Protoperidinium* capturing a pennate chain-forming diatom. Light micrograph by Christine Klaas.

this neither explains why other grazers, in particular unicellular herbivores, are unable to suppress blooms, nor does it explain why copepods, if they are indeed so inefficient at utilizing potential food sources, nevertheless manage to maintain the bulk of zooplankton biomass in oceans and oligotrophic lakes.

It has long been appreciated that, unlike terrestrial vegetation, the bulk of phytoplankton cells are grazed within the growth environment. So, given the diversity of phytoplankton grazers and their potential to overgraze their food, it is indeed surprising that blooms develop at all. However, it should be remembered that only a few of the species present contribute significantly to algal blooms. The other species either do not respond with maximized growth rates to the advent of favorable conditions or are grazed faster than the ones making the bloom. The conclusion is that understanding bloom dynamics requires basic knowledge of the ecophysiology and species-specific regulatory mechanisms inherent in the bloom formers as well as a deeper understanding of the complex interactions and potential trophic cascades within pelagic food webs. Given the great variability inherent to complex biological systems, it is indeed surprising that there is any degree of recurrence in the annual cycles of species composition and biomass in lakes and the sea.

RECURRENT AND UNUSUAL ALGAL BLOOMS

Algal blooms can be broadly differentiated into seasonally recurrent and unusual stages in the annual cycles of plankton of a given water body or region. The former tend to be initiated by seasonal change in the physical regime. At higher latitudes, this is due to warming and shallowing of the SML by thermal stratification in the spring, and its breakdown by convective mixing during the autumn. The former process improves the light supply of the cells trapped in the SML and the latter reintroduces nutrients lost by sinking from it. Spring and autumn blooms, respectively, are the result of these physical processes (Figure 1). Blooms are also a regular feature of the upwelling regions of lower latitudes. Blooms that are not recurrent features of the annual cycle or that only occur in some regions are dealt with separately. In particular, long-term data series on the annual cycles of phytoplankton gathered from coastal sites all around the world have provided valuable information on many of the recurrent and unusual aspects of bloom phenomena outlined in the following paragraphs.

Spring Blooms

Algal blooms, almost always dominated by diatoms, are regular events in the annual cycles of all lakes and coastal oceans that arise when deep mixing during the winter is stopped by the advent of surface stratification during spring (Figure 1). The timing of the spring bloom is influenced as much by the hydrography of the region as its latitude (light intensity). Thus, spring blooms along the melting ice edge off Greenland occur well before the bloom on the Norwegian shelf 1000 km (~10° latitude) further south. This fact was first noticed in the 1930s and attributed to earlier improvement of the light climate by shallow stratification due to meltwater in the Arctic as compared to the slower establishment of thermal stratification in the stormy northern North Sea. Similarly, blooms start about a month earlier in protected bays and fjords than in the adjacent, open shelf. Turbidity of the water also plays an important role in spring bloom initiation, which occurs much later in the shallow, highly turbid water overlying the tidal mud flats of the German Bight in the southern North Sea than in the clearer water of the tide-free Baltic Sea on the other side of the Danish Peninsula.

The dynamics and composition of the spring bloom varies widely from region to region and also from year to year in the same region, depending on weather conditions, phytoplankton seeding stocks, and prevailing grazing pressure. Thus, when bloom initiation coincides with prolonged calm, sunny weather the bloom peak is reached within 2–3 weeks and is often dominated by only one or two diatom species with a high initial seeding stock. At the other extreme, bouts of stormy weather alternating with calmer periods at intervals of a few days can prolong the life time of the bloom by many weeks. In such cases different species dominate biomass across a succession of smaller peaks until nutrient exhaustion is reached and the recycling community established.

The fate of bloom biomass varies depending on its dynamics. In the first case, algal biomass is much higher than that of the heterotrophs and the bloom is terminated by mass sinking of ungrazed cells within a few days resulting in a nutrient impoverished SML. Although the termination phase of blooms is less well studied than the longer growth phase, it is likely that morphology and behavior of the species involved influence the depth and extent to which bloom biomass reaches the sea floor. Thus, diatom species that form resting spores tend to sink out quantitatively and the long chains of the spiny genus *Chaetoceros* clump into aggregates, aided by sticky mucilage that entrains other cells in the large, rapidly sinking flocs. Such boom-and-bust-type blooms transfer a large amount of organic matter to the benthos and, when they overlie deeper water, play a significant role in sequestering carbon in the deepwater column and sediments.

Blooms prolonged by fluctuating favorable and unfavorable weather conditions should, theoretically, enable grazer populations to keep pace with the slower accumulation rate of biomass. As a result, more organic matter should be channeled to the zooplankton and higher pelagic trophic levels, leaving a smaller percentage of bloom biomass for export to deeper water and the sediments. However, food web relationships in the plankton during prolonged blooms are not well documented and are likely to be more complex than the theoretical case portrayed above.

Consensus is emerging that diatoms are ubiquitous in spring blooms not only because they have high growth rates but also because they are better defended than algae from other groups and hence do not represent an ideal food source. This is at odds with the long-held view of diatoms representing the pastures of the sea. The first and most obvious line of defense constitutes the silica shell that provides mechanical protection against a range of crustacean predators, wards off protistan grazers through long and spiny cell appendages and restricts access to the encased cell for piercing predators, parasites, and pathogens. Although copepods have evolved silica-edged mandibles to crack diatom shells, laboratory experiments indicate and field observations suggest that they prefer protozoa over diatom food. Why this is so is still under debate. One line of explanation suggests that diatoms are nutritionally less valuable than protozoa; another explanation is that diatoms produce substances noxious to copepods, either directly or by reducing their reproductive efficiency. Thus, it has been shown that some diatom species contain enzymes that cleave structural lipids into toxic aldehydes when the shells are crushed. Such wound-induced herbivore deterrents are commonplace in terrestrial plants but systematic investigation of this aspect of planktonic food webs has commenced only recently. Diatom aldehydes are teratogens, that is, the feeding adults are not affected but egg viability is reduced and the larvae develop malformations. Clearly, such a mechanism will select for individuals that eat food other than just diatoms. However, only some diatoms possess the enzymes that form aldehydes, and some copepod species are more susceptible to its effect than others. So this is by no means a universal defense but merely one of the weapons in an arsenal of unknown diversity. Again, such species-specific variation in defense mechanisms among herbivores and plants is well known from terrestrial, benthic, and limnic pelagic ecosystems, but this evolutionary arms race has been neglected in the marine plankton.

Perhaps not surprisingly, the genera that tend to dominate spring blooms – *Skeletonema* and *Thalassiosira* – have species that produce large amounts of aldehydes; however, they do not always dominate bloom biomass. The other prominent genus *Chaetoceros* has been shown to harbor lipids toxic to copepods. These three morphologically dissimilar genera dominate spring blooms worldwide, from the coasts to the open ocean. This is not to say that one of them will invariably be present at every spring bloom but that they are more likely to contribute significantly to spring bloom biomass everywhere than other genera. Their contribution varies widely both spatially and temporally, possibly with implications for the fate of bloom biomass.

The dominant spring bloomers in lakes also tend to be diatoms whereby the centric genera *Melosira* and *Cyclotella* and the pennate species *Asterionella formosa* often dominate. Indeed the latter species has always dominated the spring bloom of Lake Windermere over the past 60 years of continuous observation. In other lakes, its presence fluctuates annually as a result of infection by a parasitoid. Summing up, there is still a lot we do not understand about the dynamics and fate of spring blooms, particularly at the species level.

A pronounced spring bloom, generally dominated by diatoms, was long considered an invariable feature of high and midlatitude seasonal cycles but recent analyses of long-term data indicate otherwise. Thus, a winter–spring bloom of *Skeletonema* used to occur every year over decades in Narragansett Bay, Rhode Island, USA, but is absent since about a decade. Blooms now occur later in the year and have different characteristics to the spring bloom. A similar situation has been reported from San Francisco Bay and could be related to the spread of an invasive filter-feeding bivalve that prevented the bloom from developing. The situation changed when a basin-scale shift in the climate regime led to more input of shelf water to the Bay, which introduced fish predators that checked benthic filter feeders from grazing down the phytoplankton. These examples indicate the complex interactions between bottom-up and top-down factors in shaping the annual cycles of pelagic ecosystems. Long-term data are normally recorded adjacent to research institutes and it is likely that similar variation occurs in deeper, offshore water.

Autumn Blooms

Autumn blooms arise when enhanced vertical mixing due to stormy weather and cooling deepens the summer thermocline, thereby introducing new nutrients to the SML (Figure 1). The biomass and species composition of autumn blooms vary regionally more than those of the spring and they do not occur in all high- and midlatitude oceans. However, much less effort has been invested in their study as compared to spring blooms. In many regions, the autumn bloom is dominated by diatoms that, in the Kiel Bight, can reach their peak as late as mid-November. The species involved are similar to the spring bloom but genetic studies have revealed that the spring *Skeletonema* species is different from the autumn one. Although the characteristic spring genera are present in autumn diatom blooms, the species assemblage tends to be more diverse and includes dinoflagellates and larger ciliates. The latter is probably due to the low number of copepods. Where studied, the bulk of bloom biomass sinks out of the water column.

In regions where dinoflagellates dominate late summer plankton, an early autumn bloom comprising species belonging to this assemblage develops in the initial stages of thermocline erosion. Most prominent are species of the armored genus *Ceratium*, in particular the cosmopolitan species *C. tripos*, *C. fusus*, and *C. furca*. These species are apparently unpalatable although their defence mechanisms, other than their cellulose armor plating, are not known. They have comparatively low growth rates but because their populations accumulate over the summer months, a few divisions over several weeks in the nutrient-rich, autumn water column suffices to build up large biomasses that, in some regions, can rival those of the spring bloom. Growth of their bloom is generally accompanied by declining zooplankton populations, which can be interpreted as a sign of success in the evolutionary arms race.

Ceratium blooms regularly occur in many, but not all, coastal environments from the tropics to the Arctic but they are absent around Antarctica. They are apparently terminated by mass cell death and disintegration in the water column followed by sinking of phytodetritus, but the sinking rates and the depth to which they sink are not known. Recurrent large *Ceratium* blooms regularly lead to anoxic events in the Laholm Bay of southern Sweden. A massive *C. tripos* bloom, which for unknown reasons survived through the winter and sank the following spring, caused widespread anoxia and collapse of the commercial scallop fishery in New York Bight in 1976. This was a one-time event. The same can be said of a massive *C. furca* bloom that caused extensive anoxia in the late summer of the North Sea in 1981, but has not occurred since.

Blooms in Upwelling Regions of Low Latitudes

Upwelling of nutrient-rich, cold, deep water occurs seasonally or for longer periods along the western continental margins. The intensity of upwelling varies regionally and over periodic cycles of several years of which the El Nino oscillation is the best known. Dense blooms of diatoms develop in regions with intense upwelling but, because of the high sun angle and rapid warming of the newly replenished SML, algal growth rates are much higher and the bloom peak is reached faster than in spring blooms. Nevertheless, as mentioned in the section on diatoms, much the same diatom genera typical of the midlatitude spring blooms – *Skeletonema*, *Thalassiosira*, *Chaetoceros* – are prominent contributors to biomass, although *Phaeocystis* blooms have only been reported from the Arabian Sea. Most diatom species of these genera are different but a few of the prominent species are the same. Thus, genetic analyses indicate that the same species *S. costatum* that makes dense blooms off the Indian coast during monsoon upwelling also contributes to North Sea blooms. The *Phaeocystis* species – *P. globosa* – is also the same. Given the vast differences in light climate and nutrient regimes (Si:N:P concentrations and ratios) it is surprising that the same cosmopolitan species can outgrow locally evolved competitors (species present only in one or the other region), casting doubt on the role of resource competition (bottom-up factors) in selecting species dominance in blooms. There is considerable regional and interannual variability in bloom magnitude and composition depending on the intensity of upwelling. Thus, the depth from which the upwelling water emanates can favor either diatom or dinoflagellate blooms. Extensive blooms of *M. rubrum* have also been reported off Peru.

The structure and response of the food web is generally similar to that of spring blooms and again, much the same genera of protozoa and copepods are represented. However, in some upwelling regions, the local sardine-like fishes have developed fine-meshed gill rakers to graze directly on the diatoms, thus short-circuiting the food chain. Upwelling regions harbor large stocks of pelagic fishes indicating that a significant proportion of primary production is retained in the surface pelagic system. However, in regions where upwelling is prolonged over many months, the sinking flux to deeper layers and the underlying sea floor can be substantial. Thus, suboxic conditions can develop in deeper layers over extensive regions. These zones play a crucial role in oceanic nutrient cycles because denitrifying bacteria reduce nitrate to N_2, thus lowering the N:P ratio. Nitrous oxide (N_2O), which is 300 times more potent as a greenhouse gas than CO_2, is also formed. In the underlying anoxic sediments methane and hydrogen sulfide are produced, which reach such high concentrations off the Namibian coast that methane bubbles, together with H_2S gas, reach the surface and cause mass death of the local benthos. H_2S is oxidized to elemental sulfur in surface water and the resulting reflective particles can be seen in satellite images. Since global warming steepens the temperature gradient between land and sea, stronger and more intense upwelling winds are expected in the future. This will result in greater input of methane and N_2O to the atmosphere further exacerbating global warming.

Miscellaneous Algal Blooms

Blooms that do not fall under the above categories are grouped here. Thus, blooms caused by vagaries of the weather acting on hydrography are generally singular and local events. For instance, the passage of storms over stratified, nutrient-poor waters of summer months causes deep mixing and upward transport of nutrients, often resulting in a bloom. Generally, the standing stock of these blooms is lower than seasonal ones, but where motile forms that concentrate their populations along hydrographic gradients (nutricline) are involved, algal densities in thin layers can be very high. Tilting of the pycnocline along hydrographic fronts introduces this layer to the surface resulting in conspicuous stretches of discolored waters, generally referred to as red tides. The spatial extent of the discolored patches or streaks ranges from meters in protected bays to many kilometers in the open sea, where they appear as meandering structures in satellite images.

In many regions characterized by surface fronts between water masses with different properties, such as at the entrance of the English Channel, frontal blooms are regular features. However, because they tend to be dominated by a single species, their composition from year to year can vary more than in the case of seasonal blooms. Most often dinoflagellates dominate frontal blooms but in estuaries cryptophyte and haptophyte blooms have also been reported. Because of their local nature, these blooms do not have much impact on their environment and tend to go unnoticed. However, when the species involved in the bloom are toxic or cause harm to the environment, they are grouped under HABs. Not surprisingly, these blooms are among the best-studied phenomena in pelagic ecosystems and are dealt with separately next.

HARMFUL ALGAL BLOOMS

The term HAB is in common usage to denote proliferations of algal species that have a detrimental effect on human use of lakes and the sea, from drinking water reservoirs and recreation to aquaculture. The spread of artificial fertilizers in agriculture in the second half of the last century led to increased nutrient input to lakes and estuaries often resulting in algal blooms. Because both tourism and aquaculture have intensified and spread over the past decades, the incidence of HABs in coastal waters has accordingly increased. Thus, they can be regarded as the aquatic equivalent of terrestrial weeds, which by definition are plants that interfere with human usage of the region. There are no "weeds" in pristine ecosystems. It follows that the number of HABs increases with the intensity and diversification of usage. Furthermore, they can be controlled, or at least kept in check, by adopting practices that minimize their proliferation, analogous to the situation in agriculture or gardening. Thus, algal blooms caused by nutrient runoff from fertilized fields, referred to as eutrophication and particularly prominent in lakes, estuaries, and coastal regions, are being brought under control in Europe and North America by improving agricultural practices and installing sewage treatment plants. Another successful measure has been aeration of deep water in shallow, eutrophied lakes. The proliferation of cyanobacteria during the summer months has been checked as a result.

Since HABs are grouped according to their effect on the environment, their dynamics and species composition vary widely. Here, it is impossible to provide an exhaustive review of HABs, so we shall restrict ourselves to an overview of the major types illustrated with some examples. In many regions they are a regular feature of the annual cycle, in others they develop only in some years. The species composition is crucial in determining the degree of harm caused by the bloom. The harm is inflicted either by toxins in the algal species or by massive buildup of biomass resulting in oxygen depletion in deeper layers.

The occurrence of toxic algal blooms was first brought to the attention of the European public by the notorious bloom of a little-known, small haptophyte flagellate, *Chrysochromulina polylepis*, in May 1988. In addition to killing off most marine life in contact with the bloom, this species caused extensive fish kills in aquaculture farms along the Swedish and Norwegian coasts. It built up biomass as a dense layer in the pycnocline of the Belt Sea, which moved north with outflowing Baltic water and reached the surface along the Swedish and Norwegian coasts. The bloom of this toxic species was associated with increasing eutrophication at the time but the absence of a similar bloom in subsequent years despite similar weather conditions to those of May 1988 cast doubt on the explanation of a single cause-and-effect relationship. The same is true of other singular blooms that caused widespread anoxia such as in New York Bight in 1976 and in the North Sea in 1981.

The dinoflagellates have the greatest number of harmful species some of which produce potent toxins. Many of the species are responsible for fish kills as well as human poisoning via shellfish. Saxitoxin was the first algal toxin to be described; it blocks the sodium channels of neurons causing muscular paralysis and, at high dosage, death. The condition is termed paralytic shellfish poisoning (PSP). Some related genera of dinoflagellates, particularly various species of *Alexandrium*, cause PSP. The function of the toxins is not known. Not only mussel farms but also wild mussel beds are now closely monitored for toxins, but in the past human deaths due to PSP occurred regularly. A number of other toxic molecules have since been identified. Thus, toxins produced by the dinoflagellate *Karenia brevis* escape to the atmosphere and cause respiratory problems in humans. Blooms of this species have been reported for the Florida coast of the Gulf of Mexico where they also cause massive fish kills.

Diarrhetic shellfish poisoning is caused by toxins present in some species of the common dinoflagellate genera *Prorocentrum* and *Dinophysis*. Although not fatal, the toxins can cause severe discomfort to the digestive system. The *Dinophysis* toxin ocadaic acid is potent enough to cause closure of clam and mussel maricultures at cell concentrations >1000 cells l^{-1}. *Pseudo-nitzschia* is the only diatom genus known to produce a potent toxin, in this case domoic acid, which bioaccumulates in shellfish and has caused human deaths by amnesic shellfish poisoning. Seals and birds are also affected, particularly along the California coast.

Blooms that cause anoxia or unsightly accumulations along beaches are also grouped under HABs. Thus, the two exceptionally dense and extensive blooms of *Ceratium* species dealt with under autumn blooms fall under the former category. Blooms of the small-celled *Aureococcus anophagefferens* appeared suddenly along the northeast coast of the USA in the 1980s and spread southward in subsequent years reaching the Gulf of Mexico. The blooms produced unsightly brownish colored water along the coast known as brown tides. The blooms have since subsided. Copious amounts of mucus produced by the large, centric diatom *Coscinodiscus wailesii* that invaded the North Sea from the North Pacific during the late 1970s caused damage to the fisheries by clogging and tearing fishing nets. However, although this species is now established in the North Sea, blooms comparable to those of the 1970s have since not been reported. The slime blooms of the Adriatic appear in some years and are also a nuisance for tourism and fisheries. Since phytoplankton densities in the slime accumulations are low, they cannot be attributed to blooms. Where mucous accumulations settle out on the sea floor, the underlying benthos is asphyxiated.

IRON-FERTILIZED BLOOMS

Large expanses of the open ocean have perennially high macronutrient concentrations (nitrate and phosphate) but low phytoplankton biomass: the phytoplankton of these high-nutrient, low-chlorophyll (HNLC) ocean regions – the subarctic and equatorial Pacific and the entire Southern Ocean – have been shown by *in situ* iron fertilization experiments to be limited by iron availability. All but one of the ten experiments carried out in the HNLC regions induced diatom blooms. The smaller phytoplankton of the microbial food web were also stimulated by iron addition but their biomass failed to increase, indicating that the increase in growth rates was compensated by a corresponding increase in mortality rates presumably due to pathogens and herbivores. Only the mortality rates of the diatoms were sufficiently low to enable accumulation of biomass. The maximum biomass concentration of 23 mg Chl m^{-3} was reached in an experiment (SEEDS I) carried out in a 10 m SML, whereas the highest standing stock of 280 mg Chl m^{-2} developed in a 100 m deep SML in the Southern Ocean.

The species composition of the blooming diatoms varied widely and depended on the seeding stock present at the time of fertilization. In some experiments many different species accumulated biomass in about equal proportions, others were dominated by one or a few species with exceptionally high accumulation rates. In the experiment carried out in HNLC waters off Japan, which yielded the highest chlorophyll concentrations (SEEDS I) a single species – the cosmopolitan neritic species *Chaetoceros debilis* – contributed >90% to total biomass at the time of nutrient exhaustion reached within 10 days. A subsequent experiment carried out under the same conditions (SEEDS II) failed to induce a bloom although it did elicit a strong physiological response of the phytoplankton assemblage. The grazer community was apparently not exceptionally large, but *C. debilis* was absent. Southern Ocean experiments also yielded different species compositions: in one case the cosmopolitan pennate species *Pseudo-nitzschia lineola* contributed 25% to diatom biomass with a rising tendency when the experiment had to be ended after 3 weeks. The same ubiquitous species was also present in other experiments but did not accumulate cells faster than the other species. The puzzling results demonstrate the need to carry out more *in situ* experiments because, unlike laboratory and mesocosm experiments, the complexity of factors enabling bloom formation can only be unraveled under natural physical and biological conditions, in particular an intact mortality environment comprising the natural gamut of pathogens and zooplankton.

The iron fertilization experiments were carried out to test the iron hypothesis, proposed by John Martin, that increased iron transport via dust to the Southern Ocean during the dry glacials led to phytoplankton blooms that sank out of the surface layer thereby transporting carbon from the atmosphere to the deep sea. Martin hypothesized that this carbon sink contributed substantially to the lowering of atmospheric CO_2 levels during glacial as compared to interglacial periods. The four experiments carried out so far in the Southern Ocean have supported the first condition of Martin's hypothesis, that iron addition elicits massive diatom blooms. However, the fate of bloom biomass remains under debate because most experiments were not designed to track particles sinking from the bloom. If Martin's hypothesis is correct and artificial iron fertilization elicits the same response as an outfall of dust, then simulating the glacial ocean by large-scale ocean fertilization should help in sequestering some of the anthropogenic CO_2 now accumulating at an alarming rate. The maximum amount that could be sequestered annually is less than one-third of the accumulation rate in the atmosphere but it is too large to be ignored in the upcoming struggle to mitigate the adverse effects of global warming. The widespread belief that iron fertilization "will not work" is premature.

RECENT DEVELOPMENTS

Marine Genomics

The advent of high-throughput sequencing techniques has enabled biological oceanographers to study complete genomes of model species from both prokaryotic and eukaryotic algal lineages. Whole genome sequences are already available for representatives of most algal lineages being constantly amended by new genomes. The complete genomes of two diatom species, the centric diatom *Thalassiosira pseudonana* and the pennate diatom *Phaeodactylum tricornutum*, have shed new light on the evolutionary origins of diatoms and their ecological success. It was originally proposed that diatoms originated through a secondary endosymbiosis in which a red alga became incorporated by a heterotrophic eukaryote and took on the role of the diatom's plastid. Although red algal genes have been found in both the *T. pseudonana* and the *P. tricornutum* genomes providing evidence for a red algal endosymbiont ,recent findings indicate a predominance of green algal genes suggesting that the green algal endosymbiont preceded the red alga. These findings cast doubt on the assumption that the red lineages succeeded in the ocean whereas the green lineages dominated on land and indicates that diatoms were derived from a serial secondary endosymbiosis and not a single event. Diatoms not only inherited genes from their different endosymbionts and the exosymbiont but also a large number of bacterial genes through horizontal gene transfer. This unique combination of genes has permitted novel metabolic pathways in diatoms, for example carbon concentrating mechanisms (CCMs), urea cycle, genes encoding the iron storage protein ferritin, and contributed to their predominance in the ocean. Whole genomes of individual species are now being supplemented by metagenomic and metatranscriptomic approaches that aim at circumscribing the genomic variation of natural communities in case of the former and the gene expression profile of environmental mRNA in case of the latter. The combination of genomic approaches with conventional oceanography will improve our understanding of the factors determining species dominance and succession in phytoplankton blooms and thus, help to understand how different algal classes will respond to future climate change.

Abandoning Sverdrup's Critical Depth Hypothesis

For half a century Sverdrup's critical depth hypothesis has served biological oceanographers as a model to explain the vernal phytoplankton bloom in temperate and polar latitudes. It posits that blooms can only occur when the SML shoals beyond a critical depth defined by the point where phytoplankton growth exceeds losses through respiration, sinking, grazing, viral infection, parasitism, and horizontal and vertical dilution. A multi-year satellite record from the North Atlantic has now challenged the traditional view that blooms are caused by enhanced growth rates in response to improved light climate, rising temperatures, and increased stratification. The satellite record shows that bloom initiation coincides with maximum mixed layer depth in winter despite the fact that phytoplankton instantaneous growth rates (μ) are minimal in winter. This apparent discrepancy can now be resolved in the Dilution-Recoupling Hypothesis coined by M.J. Behrenfeld. During deep winter mixing, phytoplankton cells and their grazers are diluted thus lowering encounter rates between predator and prey and decoupling net population growth rates (r) from grazing pressure. This illustrates the major flaw in Sverdrup's model, namely that loss rates are constant over time. Recoupling of phytoplankton growth (μ) and loss rates during spring stratification due to entrainment of grazers within the shallow mixed layer compensates for the higher μ attained under more favorable light conditions in spring. Therefore, peak net population growth rates (r) are as likely to occur in midwinter as in spring and along the seasonal cycle r is generally inversely related to μ. These new findings highlight the tight coupling of phytoplankton growth (μ) and loss rates and the superior role of grazing mortality over improvement of the light environment for phytoplankton bloom dynamics.

The Role of Grazing in Suppressing Non-diatom Blooms and Recycling Iron in Pelagic Ecosystems

The most recent Indo-German iron fertilization experiment LOHAFEX (loha is the hindi word for iron) was conducted in severely silicon-depleted waters of the productive southwest Atlantic sector of the Antarctic Circumpolar Current (ACC). Diatom increase was thus limited by the lack of silicon and the other large algal species, which form extensive blooms in coastal waters, were heavily grazed by the large population of zooplankton, chiefly copepods. The bloom was hence composed of naked, motile algae, known as flagellates, smaller ($<5\,\mu m$) than the size range accessible to copepod grazing. Nevertheless, these flagellates exhibited only a muted biomass increase in response to iron addition because they were efficiently kept in check by their protistan grazers, chiefly ciliates, that were themselves channelled to higher trophic levels (copepods) illustrating the efficiency of trophic cascading effects in flagellate dominated phytoplankton communities. The results from LOHAFEX have thus shown that only diatoms are able to escape the grazer gauntlet exerted by the larger zooplankton stocks on an areal basis in oceanic as compared to coastal waters and that despite high growth rates biomass build-up of non-diatom phytoplankton is top-down controlled. This result is supported by observations from previous iron fertilization experiments performed in silicon-replete waters that were dominated by diatoms and provides a mechanistic explanation for the observation that massive phytoplankton blooms in the open ocean are almost exclusively due to diatoms provided that silicon is not limiting. Another striking aspect of the LOHAFEX experiment were the low loss rates of carbon by particles sinking from the surface layer indicating that iron fertilization under these conditions does not lead to significantly more carbon being sequestered in the ocean. The low carbon loss rates despite high copepod grazing pressure and concomitant fecal pellet production rates illustrate the high recycling efficiency of this grazer dominated system and its potential to retain essential nutrients within the productive surface layer. Indeed, elevated iron concentrations were measured in copepod fecal pellets during LOHAFEX providing a potential source of regenerated iron for continued algal growth. These findings are being supported by measurements of iron content in baleen whale faeces that is approximately ten million times that of Antarctic sea water, suggesting that defecation of baleen whales is another way of natural iron fertilization apart from continental sources and that pre-whaling populations of whales must have recycled more iron in surface waters thereby stimulating primary productivity through this positive feedback loop. The prominent role of higher trophic levels, in particular the megafauna, for ecosystem functioning was long known for terrestrial ecosystems and is only now emerging for the marine realm. Unfortunately, industrial whaling and fishing has nearly depleted all of the marine apex predators rendering the assessment of their ecosystem wide impact close to impossible. Allowing whale and fish populations to recover will likely help to restore the ecosystem services they provide and contribute to an overall improvement of ocean productivity.

FUTURE RESEARCH AVENUES

The phylogenetic diversity of bloom-forming species has interesting implications for our understanding of the evolutionary ecology of marine plankton. Indeed, the fact that the vast majority of phytoplankton species do not contribute significantly to blooms indicates that the density of individuals required for a species population to survive and evolve, that is, to maintain fitness in the marine plankton, is two orders of magnitude below that achieved by bloom-forming species. Clearly, the >5% of the species reported to form blooms have not outcompeted the >90% background species over evolutionary time scales, implying that the life cycle of bloom-forming species is one among many viable life cycle strategies in the phytoplankton. The phenology of bloom-forming species, that is, their mode of gearing to seasonality and hydrography of their environment together with their specific defense strategies, needs to be addressed if we are to make headway in understanding the driving forces shaping algal blooms.

Given their importance for food webs and biogeochemical cycles, the fate of blooms requires more dedicated study. Thus, mortality is not just caused by an external agent, that is, a pathogen or grazer. New results show that autocatalytic cell death or apoptosis, analogous to programmed cell death (PCD) in multicellular organisms, does occur in unicellular algae and has already been reported from cyanobacteria, green algae, coccolithophores, dinoflagellates, and diatoms. PCD seems to play a crucial role in the coordinated collapse of phytoplankton blooms but the mechanisms are still not fully understood. The PCD pathway involves the expression and biochemical coordination of a specialized cellular machinery that includes receptors, adaptors, signal kinases, proteases, and nuclear factors. Within this cellular machinery a specific class of intracellular cysteinyl aspartate-specific proteases (i.e., caspases) is of particular interest. These caspases play a ubiquitous role in both initiation and execution of PCD through the cleavage of various essential proteins in response to proapoptotic signals. This cellular self-destruction process in phytoplankton is triggered by specific environmental stresses that range from cell age, nutrient limitation, high light and/or UV exposure, and oxidative stress.

The collapse of the annual bloom of *Peridinium gatunense* in Lake Kinneret, Israel, is, for example, triggered by CO_2 limitation followed by oxidative stress that initiates a PCD-like cascade. Interestingly, a signaling molecule, a protease, excreted by senescing *P. gatunense* cells triggers synchronized cell death of the whole population by sensitizing younger cells to oxidative stress. Other studies have shown PCD in the cyanobacterium *Trichodesmium* spp. and the diatom *Thalassiosira pseudonana* in response to nutrient starvation (e.g., phosphorous and iron limitation). Furthermore, it has been shown that lytic viral infection and autocatalytic PCD in the coccolithophore *E. huxleyi* work in tandem. Common morphological observation of cells displaying signs of PCD include DNA fragmentation, degradation of cell organelles, and cytoplasmatic shrinking, while the plasma membrane remains intact. Depriving viruses of their host populations and parasitoids and predators of their food could explain the evolutionary fitness logic of PCD in the pelagic realm. These examples illustrate some of the avenues that need to be explored to further our mechanistic understanding of algal blooms.

ACKNOWLEDGMENT

Philipp Assmy was supported by the Bremen International Graduate School for Marine Sciences (GLOMAR) that is funded by the German Research Foundation (DFG) within the frame of the Excellence Initiative by the German federal and state governments to promote science and research at German universities.

FURTHER READING

Behrenfeld, M. J. (2010). Abandoning Sverdrup's critical depth hypothesis on phytoplankton blooms. *Ecology*, *91*, 977–989.

Bowler, C., Vardi, A., Allen, A. E. (2010). Oceanographic and biogeochemical insights from diatom genomes. *Annual Review of Marine Science*, *2*, 333–365.

Boyd, P. W., Jickells, T., Law, C. S. *et al.* (2007). Mesoscale iron enrichment experiments 1993–2005: Synthesis and future directions. *Science*, *315*, 612–617.

Cembella, A. D. (2003). Chemical ecology of eukaryotic microalgae in marine ecosystems. *Phycologia*, *42*, 420–447.

Cloern, J. E. (1996). Phytoplankton bloom dynamics in coastal ecosystems: A review with some general lessons from sustained investigations of San Francisco Bay, California. *Reviews of Geophysics*, *34*, 127–168.

Falkowski, P. G., Katz, M. E., Knoll, A. H. *et al.* (2004). The evolution of modern eukaryotic phytoplankton. *Science*, *305*, 354–360.

Franklin, D. J., Brussaard, C. P. D., Berges, J. A. (2006). What is the role and nature of programmed cell death in phytoplankton ecology? *European Journal of Phycology*, *41*, 1–14.

Hamm, C., Smetacek, V. (2007). Armor: Why, When, and How? In P. Falkowski, A. Knoll (Eds.) *The evolution of aquatic photoautotrophs* (pp. 311–332). Amsterdam: Elsevier.

Margalef, R. (1978). Life-forms of phytoplankton as survival alternatives in an unstable environment. *Oceanologica Acta*, *1*, 493–509.

Martin, J. H. (1990). Glacial-interglacial CO_2 change: The iron hypothesis. *Paleoceanography*, *5*, 1–13.

Nicol, S., Bowie, A., Jarman, S. *et al.* (2010). Southern Ocean iron fertilization by baleen whales and Antarctic krill. *Fish and Fisheries*, *11*, 203–209.

Raven, J. A., Waite, A. M. (2004). The evolution of silicification in diatoms: Inescapable sinking and sinking as escape? *New Phytologist*, *162*, 45–61.

Richardson, K. (1997). Harmful or exceptional phytoplankton blooms in the marine ecosystem. In J. H. S. Blaxter, A. J. Southward (Eds.) *Advances in marine biology* (pp. 301–385) Vol. 31. San Diego, London, New York, Boston, Sydney, Tokyo, Toronto: Academic Press, Inc.

Smayda, T. J. (2004). What is a bloom? A commentary. *Limnology and Oceanography*, *42*, 1132–1136.

Smetacek, V., Cleorn, J. E. (2008). On phytoplankton trends. *Science*, *319*, 1346–1348.

Smetacek, V., Assmy, P., Henjes, J. (2004). The role of grazing in structuring Southern Ocean pelagic ecosystems and biogeochemical cycles. *Antarctic Science*, *16*, 541–558.

Smetacek, V., Scharek, R., Nöthig, E. M. (1990). Seasonal and regional variation in the pelagial and its relationship to the life history cycle of krill. In Kerry, K. R., Hempel, G. (Eds.) *Antarctic ecosystems: Ecological change and conservation* (pp. 103–114) (1st ed). Springer: Berlin, Heidelberg.

Tillmann, U. (2004). Interactions between planktonic microalgae and protozoan grazers. *Journal of Phycology*, *51*, 156–168.

RELEVANT WEBSITES

http://www.algaebase.org/index.lasso.– AlgaeBASE
http://planktonnet.awi.de.–Plankton Net.

Chapter 8

Environmental Viral Pool

R.-A. Sandaa
University of Bergen, Bergen, Norway

Chapter Outline

ABBREVIATIONS

DGGE Denaturing gradient gel electrophoresis
DMS Dimethylsulfide
DMSP Dimethylsulfoniopropionate
DOC Dissolved organic carbon
dsDNA Double-stranded DNA
dsRNA Double-stranded RNA
EFM Epifluorescent microscopy
EhV *Emiliania huxleyi* virus
FCM Flow cytometry
LGV Large genome viruses
MCP Major capsid protein
MGV Middle genome viruses
MPN Most probable number
PCR Polymerase chain reaction
PFGE Pulsed-field gel electrophoresis
PFU Plaque-forming unit
SGV Small genome viruses
ssDNA Single-stranded DNA
ssRNA Single-stranded RNA
TEM Transmission electron microscopy
VBR Virus-to-bacterium ratio
VLPs Virus-like particles

DEFINING STATEMENT

Viruses are the most numerous biological entities in the environment. They may also be the most diverse. Viruses have been found everywhere there is cellular life and are regarded as important players in various ecological processes, such as nutrient cycling, gene transfer, and biodiversity.

INTRODUCTION

Viruses are the most numerous and probably most diverse biological entities in the environment. There are an estimated 10^{31} viruses on Earth, and most of them infect bacteria. Although the presence of viruses in the sea has been known since the 1950s, this knowledge was not initially appreciated because the vast pool of aquatic microbes had not yet been discovered. At that time, environmental microbiology consisted mainly of culturing microbes from different environments, and it was believed that most environments contained relatively few microbes. Thus, it was thought that these microbes had only minor importance in ecological processes. In 1989, Bergh and colleagues showed that viruses are the most abundant biological entities in the ocean, with their numbers often exceeding the

Topics in Ecological and Environmental Microbiology.

host (microbe) abundance by several orders of magnitude. Almost at the same time, as new molecular techniques facilitated the study of the total microbial communities without the need to culture the organisms, it was discovered that the number and diversity of microbes in the environment is also enormous. These findings resulted in a major shift in thinking regarding the function of microbes in the marine environment. As viruses are infectious particles that are dependent on a host for replication, it was speculated that viruses may also be important players in various ecological processes such as nutrient cycling, gene transfer, and generation of biodiversity.

VIRUSES IN THE ENVIRONMENT

Viruses have been found in every environment in which there is cellular life, from polar ice caps to hot springs. As viruses depend on their hosts for replication, the relative abundance of specific virus types roughly parallels that of the organisms they infect. Viral hosts span the three domains of life, Eukarya, Archaea, and Bacteria; however, most of the hosts that have been identified in the environment are bacteria. Consequently, bacteriophage are by far the most abundant viruses in the environment. The global population of environmental phage has been estimated to be of the order of $\sim 10^{31}$ particles. If these phage were laid end to end, they would span approximately 10 million light years, or the distance between the Earth and the Sun 10^{13} times. Numbers like these point to an important role for these biological entities in the environment.

Aquatic Viruses

Microscopic observation indicates the presence of $\sim 3 \times 10^7$ virus-like particles (VLPs) in 1 ml of marine or fresh water. This gives a virus-to-bacterium ratio (VBR) in aquatic environments of between 10 and 15. This ratio is important because it demonstrates that the probability (likelihood) of a virus meeting a prospective host in order to reproduce itself before it disintegrates is high. This is critical because the half-life of a viral particle in the marine environment is approximately 2–4 days. This means that a virus population must be sufficiently large to ensure that at least one of its members meets up with an appropriate host during that time period.

Current estimates indicate that viruses constitute ~90% of the nucleic acid-containing particles in the ocean, while prokaryotes (Bacteria and Archaea) and protists (unicellular heterotrophic and autotrophic eukaryotes) only constitute 10% and 1%, respectively. However, because of their small sizes, viruses constitute only ~5% of the total microbial biomass in the ocean, while the prokaryotes constitute ~90%. If we convert the biomass of viruses into an equivalent amount of carbon, the total amount of viruses in the sea would contain more carbon than 75 million blue whales. This makes viruses the second largest component of biomass in the ocean after prokaryotes. Most information on viral ecology in aquatic systems is from the marine environment; however, those studies available from freshwater and estuarine environments have shown similar numbers and dynamics to those seen in marine environments.

Viruses that infect prokaryotes: Bacteriophage

On the basis of studies of cultured marine bacteriophage, it is assumed that most marine phage are double-stranded DNA (dsDNA) viruses. A high proportion of these viruses, at least in the surface ocean, belong to three phage groups, the *Podoviridae*, *Siphoviridae*, and *Myoviridae* (Figure 1). These are phage that infect members of the phyla Proteobacteria, Firmicutes, and Cyanobacteria and so far there is no evidence that they infect Archaea. The *Podoviridae* are phage with a very short tail and an icosahedral head with a diameter of ~60 nm. The *Siphoviridae* have long flexible noncontractile tails, and also have an icosahedral head of diameter similar to podoviruses. Viruses classified as *Myoviridae* have a contractile tail; the head is isometric to prolate in shape

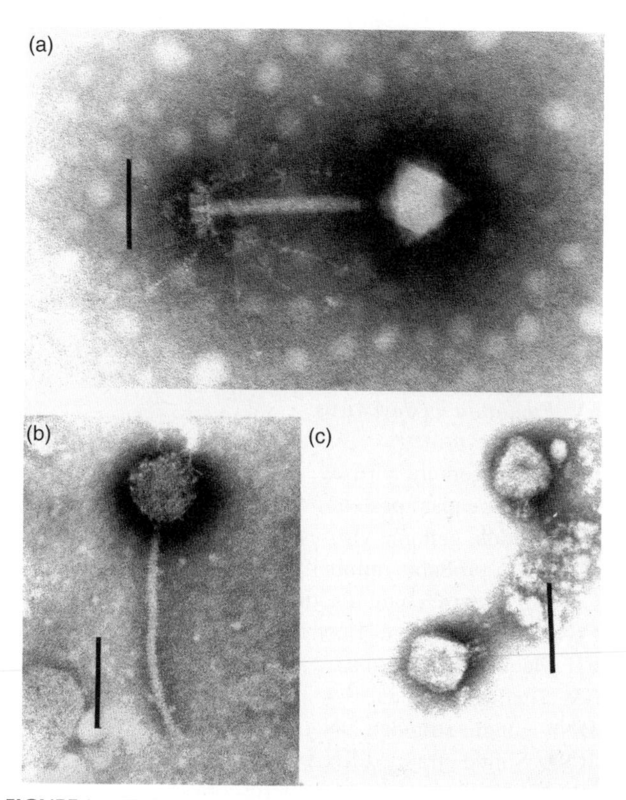

FIGURE 1 TEM of representatives of the most common phage families in the surface ocean: *Myoviridae*, represented by the cyanophage S-PM2 (a), *Siphoviridae*, represented by the bacteriophage VP6 (b), and *Podoviridae*, represented by the bacteriophage H100/1 (c). Scale bar = 100 nm. Photo: Professor Hans-Wolfgang Achermann, Laval University, Canada.

and has a diameter of 50–110 nm. All three types are non-enveloped viruses with linear dsDNA genomes. The size of the genomes seems to be roughly distinct: podoviruses and siphoviruses tend to have smaller genomes (>100 kb) than myoviruses (100–300 kb).

Because lifestyles differ among phage with different tail morphologies, this morphology may provide clues to host range and viral replication. Myoviruses are typically lytic and often have a broader host range than other tailed phage, even infecting different genera of bacteria. In contrast, the host range of podoviruses is generally narrower, while siphoviruses tend to have intermediate host ranges. The podoviruses are also typically lytic, while some siphoviruses have the capability to integrate into the host genome (temperate phage). Analyses using transmission electron microscopy (TEM) have shown that the majority of the phage in the ocean have short noncontractile tails, as described for viruses assigned to the *Podoviridae*. Nevertheless, most of the viruses that have been cultured so far have contractile tails (e.g., viruses belonging to the *Myoviridae*) or long flexible tails (e.g., viruses belonging to the *Siphoviridae*) (Table 1).

TABLE 1 The diversity of viruses isolated from the aquatic environment

Virus	Family	Morphology	Capsid Diameter	Nucleic Acid	Genome Size (kb)	Hosts
Prokaryotic hosts						
T7-like podoviridae	*Podoviridae*	Nonenveloped, icosahedral head, short noncontractile tail	~60	Linear dsDNA	38–43	Proteobacteria and Firmicutes
P-SSP7	*Podoviridae*	Nonenveloped, icosahedral head, short noncontractile tail		Linear dsDNA	45	Cyanobacteria
T4-like myoviridae	*Myoviridae*	Nonenveloped, isometric toprolate contractile tail	50–110	Linear dsDNA	164–255	Proteobacteria and Firmicutes
P-SSM2	*Myoviridae*	Nonenveloped, isometric to prolatecontractile tail		Linear dsDNA	252	Cyanobacteria
ΦHSIC	*Siphoviridae*	Nonenveloped, icosahedral head, long noncontractile tail		Linear dsDNA	39	Proteobacteria and Firmicutes
P-SS1	*Siphoviridae*	Nonenveloped, icosahedral head, long noncontractile tail	nd	Linear dsDNA	nd	Cyanobacteria
Eukaryotic hosts						
CeV	*Phycodnaviridae*	Icosahedral head, no tail	160	dsDNA	510	Microalgae
PpV	*Phycodnaviridae*	Icosahedral head, no tail	130–160	dsDNA	460	Microalgae
MpV	*Phycodnaviridae*	Icosahedral head, no tail	115–135	dsDNA	77–110	Microalgae
MpV	*Reoviridae*	Icosahedral head, no tail	65–80	dsRNA	11 segments, total 25.5	Microalgae
PoV	*Phycodnaviridae*	Icosahedral head, no tail	180–220	dsDNA	560	Microalgae
Chlorella virus	*Phycodnaviridae*	Icosahedral head, no tail	190	dsDNA	330–380	Macroalgae
EhV	*Phycodnaviridae*	Icosahedral head, no tail	160–180	dsDNA	425	Microalgae
HaV	Unclassified	Icosahedral head, no tail	25	ssRNA	9.1	Microalgae
Mimivirus	*Mimiviridae*	Icosahedral head, no tail	750	dsDNA	1200	Amoeba
SssRNAV	Unclassified	Icosahedral head, no tail	25	ssRNA	10.2	Fungoid protist

Further information of some of these viral isolates can be found in Breitbart et al. (2004), Brussaard (2004), Filee et al. (2005), Paul et al. (2005), Raoult et al. (2004), and Sullivan et al. (2005).

Cyanophage were one of the first viral groups to be isolated from the marine environment, probably due to the high abundance of their hosts, the cyanobacteria. These hosts are generally species of *Synechococcus* or *Prochlorococcus*, which are the main prokaryotic component of the picophytoplankton in the photic zone of the world's oceans. Together, these bacteria contribute up to ~90% of the primary production in the oligotrophic regions of the oceans. Cyanophage are, like their hosts, ubiquitous in marine environments. They are found in concentrations of up to 10^6 particles per milliliter in coastal waters during the summer and are considered to be a significant factor in determining the dynamics of cyanobacterial populations. Most of the isolated cyanophage belong to the *Myoviridae*; however, a few isolates have been characterized as phage belonging to the *Podoviridae* and *Siphoviridae*.

Viruses that infect eukaryotes

Very large viruses, often with capsid diameters of ~100–200 nm, can be found in surface waters. Some of these viruses probably infect single-celled photosynthetic eukaryotes (microalgae). An even larger virus (400 nm) has been isolated growing in an amoeba in the water of a cooling tower. This virus, called the mimivirus, is an icosahedral virus with a dsDNA genome of 1200 kb and is so far the largest virus that has been isolated.

Most algal viruses are host-specific, infecting one species or even a single host strain. They have been shown to be a major cause in the decline of phytoplankton blooms, such as the massive blooms of the coccolithophorid, *Emiliania huxleyi*. Furthermore, it has been suggested that some algal viruses may prevent blooms by keeping the host community in check. Viruses that infect blooming phytoplankton may occur at densities more than 10^6 VLPs/ml. They are widely distributed and have been isolated from different, sometimes widely separated geographic locations. Most of the virus–algal systems that have been cultured to date are classified as belonging to the family *Phycodnaviridae*, a family that infects a wide variety of algae including members of the Chlorophyta, Haptophyta, and Heterokontophyta (Stramenopiles). These viruses constitute a genetically diverse, but morphologically homogenous group of icosahedral viruses with large dsDNA genomes of approximately 100–560 kb. Only a few genomes of algal viruses have been completely sequenced, and there are few available genes with unique properties that can be used to study the diversity of these viruses in the marine environment. Genes that have been used for such studies are the gene for the major capsid protein (MCP) and the DNA polymerase gene (Table 2). Recent results have shown that diversity within the algal viruses is higher thanpreviously believed, including viruses with single-stranded DNA (ssDNA) and both single-stranded RNA (ssRNA) and double-stranded RNA (dsRNA) genomes. This means that these viruses belong to families other than the *Phycodnaviridae*. Thus, we are probably just looking at the tip of the iceberg when considering the diversity of algal viruses.

Diversity

The results of investigations of uncultivated marine viral communities using metagenome sequencing have shown that marine viruses are extremely diverse and largely uncharacterized. Although marine viral communities have been assumed to consist almost entirely of dsDNA viruses, recent studies have demonstrated that ssDNA viruses and RNA viruses are also important groups in the marine viral community. The results of initial studies using shotgun libraries to clone and sequence environmental viral DNA from near-shore surface seawater and sediment samples have shown that most of the sequences (70%) have no similarity

TABLE 2 Conserved genes detected in viruses that can be used for diversity studies

Taxonomic Group	Gene (partial)	Product	Coding Properties	Host
Cyanophage	psbA	Part of photosynthetic gene D1	Catalytic	*Synechococcus* and *Prochlorococcus*
Cyanophage	psbD	Part of photosynthetic gene DII	Catalytic	*Synechococcus* and *Prochlorococcus*
Myoviridae	g20	Structural gene in head assembly	Structural	*Synechococcus* and G-bacteria
Myoviridae	T4-DNA polymerase	Part of DNA polymerase	Catalytic	G-bacteria
T7-like *Podoviridae*	T7-DNA polymerase	Part of DNA polymerase	Catalytic	G-bacteria
Phycodnaviridae	MCP	Part of major capsid protein	Structural	Microalgae
Phycodnaviridae	DNA polymerase	Part of DNA polymerase	Catalytic	Microalgae

For more information about these genes in viral diversity studies, see Breitbart et al. (2004), Filee et al. (2005), Hambly et al. (2001), Millard et al. (2004), and Short and Suttle (2002).

to sequences found in the genome database. Mathematical modeling based on overlapping sequences predicts the presence of approximately 5000 viral genotypes/200 l of seawater. A recent study using high-throughput pyrosequencing to examine the diversity of marine viruses also demonstrated that most of the viral sequences were dissimilar to those in the current genome databases. Similar results were found in a metagenome study of an estuarine environment, where 61% of the metagenome sequences had not been previously observed in currently cultivated or well-studied organisms.

Pulsed-field gel electrophoresis (PFGE) can be used to size fractionate whole viral genomes. This approach to the study of total viral diversity has shown substantial temporal and spatial changes in genome size composition of viral communities, suggesting a dynamic link between the viral and hosts communities. Numerous PFGE studies have shown that the viral assemblage in the marine environment has a genome size range from approximately 15 to 630 kb. The most dominant populations of marine viruses are those whose genome sizes range between 20 and 80 kb (small genome viruses, SGV). Less abundant are viruses whose genome sizes range between 80 and 280 kb (middle genome viruses, MGV), and viruses whose genome sizes range between 280 and 500 kb (large genome viruses, LGV) constitute the least abundant group. In general, the distribution of environmental genome sizes corresponds with the genomic size range of viruses in culture (Table 1) and the abundance of the respective host community in the marine environment. Bacteria are the most abundant host group in the marine environment, and current knowledge of the genomic size range of marine bacteriophage (39–243 kb) correlates with the two most abundant genome size groups found in the marine environment (SGV and MGV). Of these bacteriophage, *Podoviridae* are thought to be the most numerous group and the information that is available on the genome sizes of these phage (39–60 kb) correlates with the most abundant size group in the ocean (SGV). Viruses that infect algae often have genome sizes of between 280 and 500 kb, and the abundance of the host is often 10 times lower than the number of prokaryotes. The low abundance of the hosts also correlates with the least abundant viral genomic size class (LGV) in the marine environment.

Several conserved viral genes that are representative of specific subsets of viral communities have been found (Table 2). Studies targeting these genes have uncovered an enormous genetic diversity in viral communities. The first study to apply this approach targeted a fragment of the structural gene g20 (Table 2) of T4-like cyanophage that belong to the *Myoviridae* family and infect marine cyanobacteria of the genus *Synechococcus*. The study examined the variation in genetic composition that occurred with depth along a transect bisecting the North and South Atlantic Ocean and in the coastal waters of British Colombia, off the west coast of Canada. The results from these studies revealed pronounced differences in the genotype of viral communities that were separated by only a few meters. This shows that there is a tremendous spatial heterogeneity in phage communities.

Despite the high diversity found in the viral communities, there are groups of viruses, for example T7-like podophage, T4-like myophage, and some groups of algal viruses, which are geographically widely distributed in the marine environment. The widespread distribution of these viral groups, which show high sequence similarity, is a strong evidence for a shared evolutionary origin of these marine viruses. This is further supported by results of several surveys showing that 80% of phage genotypes are distributed in samples from hugely different marine environments (the Arctic, Gulf of Mexico, British Columbia coast, and Bermuda). However, the relative abundance of the different phage genotypes in these different marine environments is significantly different. This implies that viruses from an individual marine sample are extremely diverse, but that viruses pooled from many samples collected from a geographic region are no more diverse than the viruses from a single sample. Together, these observations support the ideas laid out in a model called the "Bank model," which suggests that there is a globally distributed pool of viruses present at relatively low abundances. Individual viruses from this pool become abundant in a given sample depending on their ability to infect the most abundant microbes.

Viruses in Sediments

Benthic viral abundance is typically 10–100 times higher than in the water column, with numbers in the order of 10^8–10^9 VLPs/cm^3. In addition, the total diversity of the viral particles is higher in sediments than in the overlying water column, with 1 million different viral genotypes in 1 kg of marine sediment. Correlations between viral abundances and suspended solids have been shown, suggesting that viruses are adsorbed to suspended material in the water column that may settle out and contribute to the benthic viral population. The relatively few studies of viruses in sediments show that the abundance and production of benthic VLPs correlates with the trophic status of the environment. Along with a decreasing diagenetic activity, bacterial and VLP abundances generally decrease with increasing sediment depth. VLP abundance has also been shown to be strongly correlated with the microbial sulfate reduction rate in estuarine sediments. These observations suggest that viruses in the sediments represent a dynamic component of marine sediments with implications for microbial activity and ecology.

Viruses in Soil

In contrast to our growing knowledge of the role of viruses in marine microbial communities, the roles of viruses in the soil and their impact on ecological processes are poorly understood. By direct counting using TEM, the number of VLPs in the soil has been determined to be ~1.5×10^7/g. The number of VLPs is lower than the bacterial count (~10^9/g), which indicates a much lower VBR in soil compared with other environments. For example, in sediments or aquatic systems, the number of viruses is often 10–15 times higher than the bacterial count. However, the number of viruses may be underestimated in these samples, due to biases in the counting techniques, or the soil community may differ from other communities with respect to the number of viruses in relation to bacterial abundance.

Most of the viral studies performed on soil samples are based on cultured bacterial hosts, such as *Serratia* and *Pseudomonas*, two genera that are often found in soil communities. These studies have shown major differences in the abundance of specific bacteriophage between different parts of the soil profile. The difference in abundance of these specific phage populations is undoubtedly a reflection of both the physiological role and ecological niche of the two host bacteria. Studies of soil viral communities using metagenome sequencing have shown that soil viruses are taxonomically diverse and are distinct from the communities of viruses found in other environments. Most abundant are sequences resembling bacteriophage that infect typical soil bacteria such as *Actinoplanes*, *Mycobacterium*, *Myxococcus*, and *Streptomycetes*, but a phage infecting the halophilic archaeon *Haloarcula* has also been detected. Many of the phage are unclassified, but those that are classified are phage of the types *Mycoviridae*, λ-like siphophage, and corndog-like siphophage. Comparing different soils, such as those found in desert, prairie, and rainforest environments, results in little overlap of the data, and this finding suggests that viruses in soil are globally as well as locally diverse.

Viruses in Extreme Environments

Extremophile organisms are remarkable in their ability to thrive in harsh physical or geochemical conditions including extremes of pH, salinity, desiccation, hydrostatic pressure, radiation, and anaerobiosis. Most known extremophiles belong to the domain Archaea, but extremophiles are also present in numerous and diverse genetic lineages of the domains Bacteria and to a lesser extent in Eukarya. All viruses isolated so far from extreme environments are dsDNA viruses with genome sizes in the range of 14–80kb. They can reach an abundance of 10^9 VLPs/ml, which is higher than that found in many other natural environments. Some of these viruses resemble morphotypes found in other less extreme environments, while others are unique. About 60 viruses that infect members of one or the other of the two major archaeal divisions Crenarchaeota and Euryarchaeota have been isolated. The viruses infecting extreme thermophilic crenarchaea are the most morphologically diverse group, while those infecting extreme halophilic euryarchaea are morphologically less diverse.

Extreme halophilic viruses

In environments where the salinity is greater than 25%, the viral abundance can be as high as 10^9 VLPs/ml. At this level of salinity, the majority of the host organisms are halophilic Archaea, which are in the Euryarchaeota. Viruses infecting the methanogenic and extreme halophilic euryarchaea have linear dsDNA genomes and belong to at least three morphological groups: head-and-tail, spherical, and spindle-shaped. Those with head-and-tail morphology are reminiscent of bacteriophage belonging to the three families *Myoviridae*, *Siphoviridae*, and *Podoviridae*. However, sequence information from these viruses has shown that they have little sequence similarity with known viruses from nonextreme environments. There are strong genetic relationships between the different haloviruses, despite the fact that their hosts belong to a wide range of different species. Most of the isolated haloviruses have genomes in the size range ~14–80kb. The genomic size distribution of uncultured haloviruses is more diverse than that of the isolated haloviruses. Studies of the total viral community in solar salterns have shown that these viruses exhibit genome sizes varying from 10 to 533kb. Not surprisingly, the viral diversity in these salterns resembles the host diversity. The diversity in genome size is highest in the salterns with the lowest salinity (4–22%), and the genome size of these viral populations range from 32 to –340kb. This is undoubtedly because diversity of the host community in these ponds is also broader, with representatives from all three kingdoms, Archaea, Bacteria, and Eukarya. When the salinity increases to above 22%, the viral, as well as host, diversity declines. In these salterns, virus-like genomes in the size range from 10 to 189kb have been detected.

The viral and prokaryotic abundance in alkaline lakes is among the highest reported in any aquatic environment, with the abundance of VLPs and prokaryotes being 10^{11}–10^{12} and 10^{10}–10^{11}/ml, respectively. Only one lytic phage has been isolated from this environment, and it infects a bacterium that is closely related to *Idiomarina baltica*. Studies of the total viral community using PFGE analysis have revealed the existence of dsDNA viral populations with genomes between 14 and 400kb, with the majority having genomes between 30 and 60kb. These viral communities differ at different depths of the lake, suggesting a strong stratification of viral distribution between oxic and anoxic waters. These differences probably reflect differences in the host community structure.

Viruses in deep subsurface sediments

The deep subsurface is one of the least understood habitats on Earth, even though the huge microbial biomass therein probably plays an important role in global biochemical cycles. Recent studies have shown that there is a large number of viruses in these sediments. The viral abundance in deep subsurface sediments appears to follow the bacterial numbers very closely, with an average of up to 10^9 VLPs/g of dry sediment. No information is available about the diversity of these viruses and the interactions between viral and prokaryote communities.

Viruses in extreme thermal environments

Viruses from terrestrial and oceanic hydrothermal environments represent an enormous reservoir of a truly remarkable morphological and genomic viral diversity. The vast majority of the hyperthermophilic viruses isolated from acidic and neutral hot springs (>80 °C) infect a broad spectrum of the extremely thermophilic Archaea, the Crenarchaeaota. These viruses infect representatives of the genera *Sulfolobus*, *Thermoproteus*, *Acidianus*, and *Pyrobaculum*. Typically, these viruses show no clear similarity in their morphology or at the genomic level to either bacterial or euryarchaeal viruses. These crenarchaeote viruses have been classified into seven new families, of which four are approved by the International Committee of Taxonomy of Viruses. These families are lemon-shaped *Fuselloviridae*, filamentous *Lipothrixvidae*, stiff rod-shaped *Rudiviridae*, droplet-shaped *Guttaviridae*, spherical *Globuloviridae*, two-tailed spindle-shaped *Bicaudaviridae*, and bottle-shaped *Ampullaviridae*. The rod-shaped virions of the *Rudiviridae* and *Lipothrixvidae* show some similarity with the tobamoviruses and closteroviruses of vascular plants, respectively, while those of the *Globuloviridae* resemble viruses of the *Paramyxoviridae*, which infect vertebrates. All these hyperthermophilic viruses contain dsDNA genomes that are either linear or circular and their genomic sizes range from 15 to 75 kb.

The deep-sea vent areas are among the most extreme habitats on Earth, characterized by high pressure, wide temperature ranges (10–400 °C), and the presence of a variable composition of fluids and gases, which are often incompatible with life for most eukaryotes, and indeed most prokaryotes as well. Viruses in these systems can reach abundances of between 10^5 and 10^7 VLPs/ml. The morphology of these viruses is diverse, with morphotypes resembling typical archaeal viruses, including lemon-shaped viruses and pleomorphic types such as spoon-shaped and spindle particles with bipolar expansions, but also some morphotypes that are known to infect bacteria. These morphotypes are filamentous and rod-shaped viruses. The only virus that has been isolated from this environment is named PAV1, and it is lemon-shaped measuring 120 × 80 nm, with a short tail terminated by fibers. This virus infects the hyperthermophilic euryarchaeote *Pyrococcus abyssi*. The genome consists of 18 kb of circular dsDNA. Genomic comparison with other viral sequences has resulted in no significant similarity.

Viruses in deserts

Deserts are extremely dry, and are exposed to extremes of UV light irradiation and temperature variations. Nonetheless, both eukaryotic and prokaryotic microbes have adapted to these extreme conditions. VLPs have been found in surface sands from different dessert locations in North Africa. These particles resemble morphotypes representing the three major bacteriophage families: *Myoviridae*, *Siphoviridae*, and *Podoviridae*. They also have dsDNA genomes in the same size range as these viral families, from 45 to 270 kb.

Viruses in polar environments

Microbial life has also been detected in extremely cold environments such as high-latitude glaciers, polar permafrost, and the Dry Valleys of Antarctica, as well as in sea ice. In Arctic and Antarctic sea ice, microbes belonging to the prokaryotes and microeukaryotes, especially microalgae, have been detected together with VLPs. The viral abundance in sea ice is 10–100 times higher (10^6–10^8/ml) than in the surrounding water. This correlates with the bacterial abundance, which is also higher in the sea ice than in the adjacent water column. Viral proliferation appears to be enhanced in the sea ice relative to the open water, and the VBRs are among the highest reported in natural environments. In contrast, the number of VLPs in Arctic glaciers is ~10–100 times lower than the average for marine and freshwater ecosystems in temperate regions. The VBR in this environment is on average >10, and there is a strong positive correlation between viral and bacterial abundance. Three Arctic viral isolates have been cultured and they all infect psychrophilic bacteria whose closest relatives are *Shewanella frigidimarina*, *Flavobacterium hibernum*, and *Colwellia psycherythrae*. All three phage are lytic, have dsDNA genomes, and are morphologically similar to the families *Siphoviridae* and *Myoviridae*. They are adapted to a cold environment, and phage development has a lower maximum temperature than the maximum growth temperature of the host bacterium.

ECOLOGICAL IMPORTANCE

Viral ecology is the study of the interaction of viruses with other organisms and the environment.

Viruses are agents of mortality in both prokaryotes and eukaryotes, and thus they play important roles in the ecology of these organisms. As mentioned earlier, the global population of environmental phage contains ~10^{31} particles. This viral pool turns over every few days, from which it can be calculated that every second somewhere on Earth 10^{23} microbes are infected by a phage. When the host organisms are lysed, nutrients are released to the surrounding environment; in this way viruses are important players in the cycling of carbon and nutrients. Viruses also directly affect the abundance and diversity of host cell communities and contribute to microbial gene exchange, which is important for the evolution of the host community. To date, virtually all information on the interaction between environmental viruses and their hosts has been gathered from aquatic systems, while little information is available on the importance of viruses in soil or in extreme environments. Consequently, most of the following discussion of viral ecology will focus on aquatic viruses.

Viral Effect on Nutrient Recycling

As the main drivers of oceanic biogeochemical cycles, microorganisms play a significant role in regulating the ecosphere. Marine viruses also play an important role in the microbial food web as catalysts that accelerate the transformation of nutrients from the particulate to the dissolved state. From this the nutrients can then be incorporated by microbial communities. The marine microbial food web consists of a consortium of heterotrophic and autotrophic prokaryotes, as well as their predators (viruses and protozoa). The structure of the food web determines the transfer of energy and nutrients (nitrogen and phosphate) to higher trophic levels and greatly influences global carbon and nutrient cycles (Figure 2). Viruses are a crucial and ubiquitous component of this microbial food web and play their part by lysing their host organisms. It is estimated that phage kill between 4% and 50% of the bacteria produced every day and that 2–10% of phytoplankton primary production is channeled through "the viral shunt" in the microbial food web. Cell lysis implies that organic material is lost from the grazing food chain and becomes available to bacteria, which thrive on dissolved organic material and nutrients. The net effect of this is to increase nutrient recycling and respiratory loss of organic carbon in the lower part of the food chain. In other words, viruses short-circuit the flow of carbon and nutrients from phytoplankton and bacteria to higher trophic levels. This means that viral activity has a direct effect on the carbon budget of the ocean, and hence on the global climate.

Viruses can also have an indirect effect on the climate by influencing the formation of dimethylsulfoniopropionate (DMSP), which is the precursor of dimethylsulfide (DMS). Viral infections in algae have been found to cause increased production of DMSP. In the ocean DMSP is degraded to DMS by the activity of bacteria and phytoplankton. A portion of this DMS diffuses from the ocean to the atmosphere,

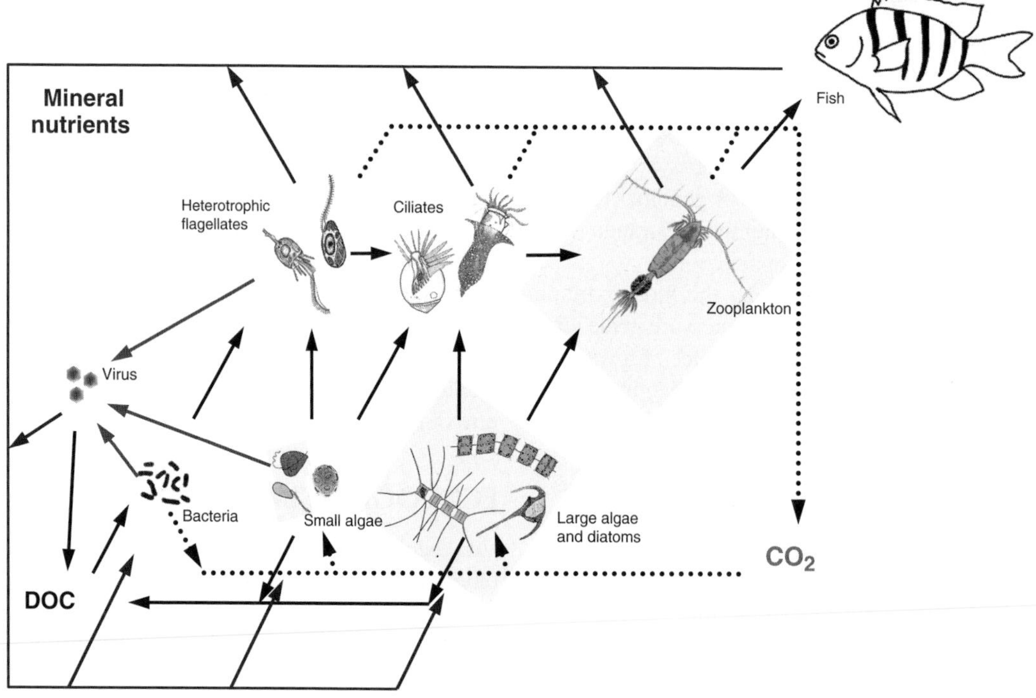

FIGURE 2 Microbial food web and virus-mediated carbon flow. The red arrows represent virus-mediated pathways. The black arrows represent transport of dissolved organic carbon (DOC) in the microbial food web. The black dotted arrows show the contribution to the CO_2 cycle and the blue lines the transport of mineral nutrients in the microbial food web. DOC is the largest biogenic pool of carbon in the ocean. The dynamics of DOC may have an indirect impact on global carbon cycling, contributing to the control of atmospheric CO_2. DOC is only accessible to microbes, primarily heterotrophic bacteria. When viruses lyse the host, carbon is transformed from the particulate form (POC) into the dissolved form (DOC). Figure: Professor Gunnar Bratbak, Department of Biology, University of Bergen. (For interpretation of the references to color in this figure legend, the reader is referred to the Web version of this chapter.)

where it is involved in cloud formation and also causes acid rain. Clouds affect the radiation balance of the Earth and thereby strongly influence its temperature and climate.

Virus–Host Interaction and its Effect on Microbial Diversity

Viruses can influence the genetic diversity of their hosts in various ways. Simple models have been developed that assume that prokaryotic diversity is controlled by a combination of predation, viral lysis, and substrate limitation. If substrate (nitrogen or phosphate) is limiting (bottom-up control), it can control which species are present because different species are specialized to use different substrates. This leads to a system where all species are limited by different substrates, and organismal diversity reflects the diversity of substrates. The host community can also be regulated by specific viral lysis and/or grazing by protozoa (top-down control). The so-called "killing the winner" model explains the role of viruses in the maintenance of high biodiversity in marine systems, where coexisting hosts compete for the same resource. Because viruses control the most abundant or fast-growing host populations, less competitive or slower growing populations can coexist with the dominant fast-growing hosts (Figure 3). Population reduction (lysis) of the most dominant host populations creates open niches, which enable new hosts to become abundant. This provides an opportunity for less competitive host populations to exist in the presence of more competitive hosts, thus enabling maintenance of high diversity in the marine environment.

Empirical evidence concerning the intensity or the outcome of the regulation of viral activity on the host community structure is still scarce, and contradictory results have

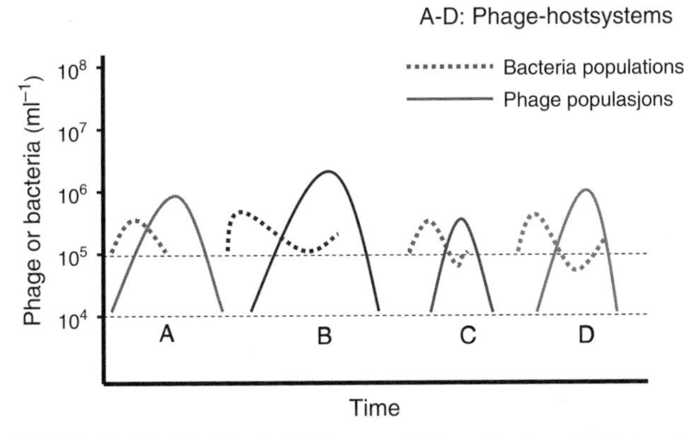

FIGURE 3 Model of Lotka–Volterra oscillations in virus–host interactions. For each phage–host system, a selective factor stimulates growth of a specific host. An epidemic phage infection begins at a critical host cell density and the abundance of a specific phage increases. Phage lysis causes the abundance of host cells to decline to baseline levels and thus prevents dominance of a single host species. At the end of the epidemic, the number of infective phage declines to a baseline level at a decay rate specific to each phage.

been reported. The picture is quite complicated because prokaryotic hosts can be susceptible to multiple phage infections and because sensitivity toward phage infection is strain-specific and exists as a continuum from highly sensitive to highly resistant. Nevertheless, the results of some studies clearly demonstrate that there is a link between viral and host biodiversity and that coexistence of nutrient-competing bacterial hosts might indeed be controlled by viral lysis. The results of other studies, however, indicate limited effects of viral activity on the bacterial community structure; these studies suggest that viral infections affect only a few abundant phylotypes, while other phylotypes may be resistant to infection. Alternatively, it has been suggested that the most dominant hosts are dominant because they are less susceptible to viral lysis. This means that rare marine bacterial groups might be the most susceptible and therefore the losers in the competition over resources. If, however, there is a trade-off between competitive ability among the hosts in terms of nutrient uptake and defensive ability (resistance) against viral infection, fast-growing competition specialists (r-strategists) would presumably be the least abundant, while their associated viral populations will have the highest abundance. Furthermore, the hosts that are defense specialists (K-strategists), being more or less immune to viral attack, will grow more slowly but will be more abundant and have an associated viral population of low abundance.

A complementary view of the dichotomy between simple and complex viruses regards simplicity and complexity as two fundamentally different strategies of virus–host interaction. Simple viruses rely almost entirely on the host for their replication and expression, with a minimal set of essential functions encoded in the viral genome, and no or few genes dedicated to coping with the host defense system. These viruses usually have small genomes, and the reproductive strategies of these viruses are optimized to outrun the host by rapid replication. For example, viruses belonging to the *Podoviridae* family may be regarded as r-strategists as the viruses are abundant and virulent, and have a rapid replication rate, a high burst size, and small genomes. In contrast, complex viruses with large genomes have evolved considerable autonomy of their replication and expression systems, in addition to having accrued elaborate gene ensembles to overcome host defenses. The mimivirus and some temperate phage, which exhibit a lysogenic lifestyle, may be examples of the latter. These viruses may be regarded as K-strategists, because they have low burst size, large genomes, and a slow decay time, which is the time it takes for the virus to be degraded in the environment.

Viral Effect on Gene Transfer and Evolution of the Host

Viruses affect the microbial community through the introduction of new genetic traits via horizontal gene transfer. The most well-known cases of horizontal gene transfer by phage are in the context of pathogenesis, where phage can carry toxin genes and other virulence factors. In an environmental context, horizontal gene transfer is an essential factor in evolution. Phage are major conduits of genetic exchange, and transduce an estimated 10^{25}–10^{28} bp of DNA per year in the world's oceans.

Gene transfer can occur during the viral replication cycle when some of the host genes are by "mistake" packed inside the viral particle. During infection, these genes may be transferred to a new host. Viruses have several different life or replication cycles. The two most dominant are the lytic and the lysogenic (latent) cycles. Lytic (or virulent) viruses infect a cell, replicate, and are released following lysis of the host cell. Temperate viruses infect the host, but the viral DNA stays within the host cell, often physically integrated into the host genome. The viral genome then replicates along with the host genome, until external factors, such as DNA damage, induce the lytic cycle. For prokaryotes in the ocean, it has been suggested that in nutrient-rich waters, which are characterized by a high abundance of hosts, lytic phage dominate. It has also been speculated that viruses might use the lysogenic cycle as a survival strategy in harsh or low host abundance environments. This might be particularly important for the survival of viruses in extreme environments.

Lysogeny has been suggested to be a beneficial lifestyle for viruses in Arctic environments. There are arguments both for and against such a theory, because all the phage that have been isolated from Arctic freshwater and glaciers so far are lytic. In the Arctic marine environment, however, a high percentage of marine bacterial genomes harbor integrated viral genomes (prophage). This finding demonstrates that viruses with lysogenic lifecycles are common in the Arctic marine environment. In fact, metagenome studies of the Arctic viral community have shown that all the five most abundant recognizable phage sequences represented prophage. This is in stark contrast to sequences obtained from temperate regions, which are dominated by genes from lytic phage.

In other environments, such as salt lakes and hot springs, temperate phage have been isolated together with a few lytic viruses. However, the majority of these viruses exhibit a virus–host relationship that is different from those observed in less extreme environments. In this relationship, the infected hosts are not lysed, but virus particles are continuously produced. This is consistent with an equilibrium state being established between viral replication and host cell multiplication. Moreover, the viruses persist in host cells in a stable state and are not lost during continuous growth of the infected cultures. It has been suggested that such a survival strategy may help the virus population to avoid direct, and possibly prolonged, exposure to the harsh conditions of the host habitat. Nevertheless, large numbers of viruses have been reported in some hot springs, at densities of 10^6/ml. Hot-spring viruses have been shown to be resistant to shifts in temperature: 75% of such viruses were still intact after incubation on ice, showing that they are able to withstand extreme environmental conditions.

The dynamic nature of the oceanic gene pool has resulted in genetic diversification of both donor and recipient genomes and has also probably guided their functional diversification. As an example, genes involved in photosynthesis have recently been detected in cyanophage. These genes (*psbA* and *psbD*) code for two proteins, D1 and D2, which form the reaction center of photosystem II (PSII). The D1 protein is common to all oxygenic phototrophs and has a high turnover rate as a result of photodamage. Both the *psbA* and *psbD* genes have been reported in cyanophage infecting *Synechococcus* and *Prochlorococcus*, and have been identified in BAC clones (metagenomes) and polymerase chain reaction (PCR) amplicons from environmental samples. These viral photosynthetic genes may be expressed during infection of the host, which suggests that they probably have a functional role during the infection cycle. The genetic transfer of photosynthetic genes between cyanobacteria and phage might thus have significant implications for the evolution of both hosts and phage (coevolution).

Host genes retained in a particular phage could reflect key selective forces in the host environment. Indeed, phosphate sensing and acquisition genes have been found in phage that infect organisms in low phosphate environments. Likewise, the most abundant viral-encoded enzymes found through extensive sequencing of marine viral communities appear to be involved in scavenging host nucleotides (e.g., riboreductases) and supporting host metabolism throughout the infection cycle (e.g., carboxylases and transferases). Thus, it might be speculated that phage genes are important players in different metabolic processes in their hosts, increasing the fitness of both the host and the phage, and that the functionality of the phage may be coupled to the physiological state of the host.

METHODS FOR STUDYING ENVIRONMENTAL VIRUSES

Documentation of both community diversity (richness of species) and composition (evenness of species distribution) are fundamental elements of ecological research. Consequently, information about abundance, diversity, and community structure is of crucial importance to understanding the ecological roles of viruses in the environment. For this purpose, a number of both qualitative and quantitative methods are available for describing their diversity and dynamics in the analysis of viral communities. Several methods are also available for determining the abundance of these particles in environmental samples.

Enumeration and Measurement of the Concentration of Viruses

There are essentially two different ways to enumerate viruses: the indirect viable count and the direct total count. Viable count techniques are based on the lysis of a cultivable host, while direct counts involve counting the viruses in the environmental sample without the need for a cultivable host. The methods available for performing viable counts are the plaque-forming unit (PFU) and most probable number (MPN) techniques. Direct counts may be performed using TEM, epifluorescent microscopy (EFM), or flow cytometry (FCM).

PFU and MPN techniques are both used to quantify the number of particles released from a lysed host. This means that the host must be cultivable, which is not the case for most of the microbes in the environment. The PFU method is used to determine the number of viruses that cause lysis of bacteria, cyanobacteria, or algae that can grow on a solid medium. Initially, the host has to be grown in a liquid medium, and then it is mixed with a sample that contains the virus. The mixture of virus and host is then combined with molten agar (soft agar) and poured on to a plate in which the agar content is high and contains a medium that the host can utilize. After incubation, the plaques, which are areas in which cells have been killed (or transformed), can be counted and the number of infective virus particles in the original suspension estimated.

The MPN technique is used to quantify the number of infectious viral units for hosts unable to grow on solid media. Serial dilutions of the virus are made in replicates. The last dilution in which the host is lysed (no growth) is considered to contain one infectious viral particle. The number of infectious particles can then be calculated based on the number of replicates in each dilution where lysis has occurred. The PFU and MPN methods can also be used to capture new environmental viral isolates and to purify viral isolates in culture in order to obtain a clonal virus.

Direct counts

Direct counts of viruses from the environment provide the total number of VLPs without the need for culturing, and often give estimates that are 100–1000 times higher than PFU counts. For TEM, viruses are harvested directly on to electron microscopy grids by centrifugation or are concentrated by ultrafiltration (see below) and then transferred to grids. Thereafter, the viruses are stained with an electron-dense material, for example, uracil acetate (Figure 4). This method can also be used to describe the VLPs according to their morphology, a trait used in viral taxonomy. In this way, TEM can be used to give information that enables the tentative taxonomic affiliation of the viruses to be established. In addition, TEM can be used to visualize infected cells and thus to estimate the burst size of individual cells. For EFM, the viruses are concentrated on membrane filters and then stained with a fluorescent dye that binds specifically to the nucleic acids of the viral particles (Figure 5).

FCM is the newest method that has been used to enumerate aquatic viruses (Figure 6). It is a technique for

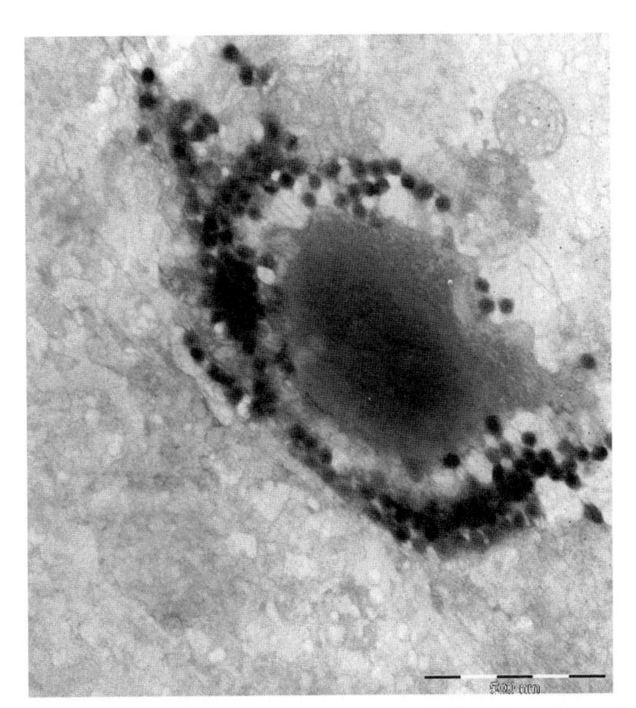

FIGURE 4 Transmission electron micrograph of a lysed bacterium and phage particles in a seawater sample. Scale bar = 500 nm. Photo: Senior Scientist, Mikal Heldal, Department of Biology, University of Bergen.

FIGURE 5 Epifluorescence micrograph of prokaryotes and viruses in a seawater sample stained with a fluorescent dye, SYBR Green I. This dye specifically stains doubled-stranded DNA (dsDNA). The smallest dots are viruses and larger ones are prokaryotes (bacteria or archaea). The prokaryotes are approximately 0.5 μm in diameter. Photo: Prof. Gunnar Bratbak, Department of Biology, University of Bergen.

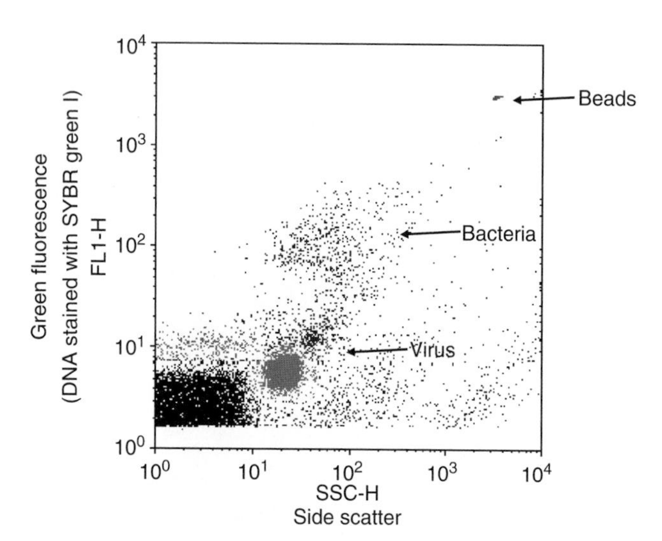

FIGURE 6 Biparametric flow cytometry plot showing characteristic viruses (pink, blue, and green dots) in a seawater sample stained with a fluorescent dye, SYBR Green I. The particles are discriminated using side scatter versus green fluorescence signals. The green-colored dots represent *Emiliania huxleyi* virus (EhV) infecting the microalga *Emiliania huxleyi*. The beads are 1 μm in diameter. Figure: Dr. Aud Larsen, Department of Biology, University of Bergen. (For interpretation of the references to color in this figure legend, the reader is referred to the Web version of this chapter.)

counting, examining, and sorting microscopic particles that are suspended in a stream of fluid. FCM allows populations to be analyzed based on the physical and/or chemical characteristics of single cells. This accurate high-throughput method also allows the quantification of subpopulations of fluorescent-stained viruses that differ in their characteristics of fluorescence and light scattering. The major advantages of FCM are its ability to analyze a large number of cells rapidly and the provision of data suitable for statistical analysis.

Concentration

For most viral analyses, it is necessary to concentrate viruses from the environmental sample prior to performing the specific analysis. Concentration is necessary not only for counting and isolation of virus particles, but also for methods used to study the viral community structure. The most commonly used methods for concentrating viral particles from environmental samples are ultrafiltration and ultracentrifugation. During ultracentrifugation, the viral particles are pelleted by high-speed centrifugation, whereas ultrafiltration is based on filters with very small pore size that retain the viral particles in the sample. The latter method involves a prefiltration step in which either 0.2 or 0.45 μm pore size filters are used to remove microbes from the sample. Thereafter, the viruses are concentrated in the sample by using filters with either a 30-kDa or a 100-kDa cutoff. The viruses in a sample of 200 l of seawater can be concentrated into a volume of ~50 ml, and thus be concentrated by a factor of 4000, using the ultrafiltration system.

Viral Diversity

Culture-dependent methods

Culture-based studies for estimating environmental viral diversity are based on hosts that are able to grow either in cultures or on solid media. Thus, the information available on diversity is based only on a small fraction of the total viral community. Nevertheless, such information is of crucial importance in increasing our knowledge of the ecology of environmental viruses. Studies of virus–host interactions under cultivation conditions may provide valuable information about factors that influence such relationships. In addition, sequencing of whole viral genomes is important to provide information that might give us new insight into the ecological functions of these viral particles. One method that is used for such a purpose is the plaque assay, which makes it possible to obtain and purify the viral isolates infecting one specific host. Another method for isolating viruses is to add concentrated viruses from an environmental sample to different potential hosts and monitor these for cell lysis. To ensure that the lysis has occurred due to virus propagation, the host cells are investigated using TEM to confirm the presence of VLPs inside the host cells.

Culture-independent methods

The introduction of culture-independent molecular techniques combined with the unraveling of phylogenetic information harbored in genes that are common to all microbes, such as 16 S rDNA and 18 S rDNA, has enabled the study of the total diversity and dynamics of environmental samples. The results of these studies have shown that most microbes in the environment belong to as yet uncultured groups. The fact that most hosts have not yet been cultured has severely limited studies of viral diversity. In addition, there is no single universal gene, analogous to 16 S rDNA and 18 S rDNA in prokaryotes and eukaryotes, which is present in all (or even most) viruses. This makes it difficult to study the total viral diversity and dynamics of natural phage communities. However, whole genome comparisons have shown that conserved structural and functional genes, which are shared among all members within certain viral taxonomic groups, do exist and can be used for studies of viral diversity (Table 2) Thus, it is possible to study the diversity within certain taxonomic viral groups using PCR amplification and sequencing on viral concentrates. The dynamics of different viral groups can also be studied by the use of denaturing gradient gel electrophoresis (DGGE) and subsequent sequence analysis. The DGGE technique is a molecular fingerprinting method that is used to separate PCR-generated DNA products on the basis of sequence differences leading to different melting behavior of the DNA.

Studies that are based on conserved viral genes are not able to investigate the total viral community. To access such information, it is necessary either to produce and sequence metagenome clone libraries of the total environmental viral community, or to sequence viral DNA isolated from the environment directly. Both these approaches are based on concentration of the environmental viruses by ultrafiltration. For metagenome cloning, the DNA is cut into short fragments either by enzymes or through random shearing of the DNA. These fragments are then cloned into vectors, such that each vector contains DNA from one viral particle. This sorting by cloning is not necessary with direct sequencing (now by pyrosequencing), in which individual DNA molecules are sorted on beads. A new high-throughput technique, 454 pyrosequencing, generates shorter fragments than the conventional techniques; however, this limitation is compensated for by the very large number of sequences that can be generated quickly and at low cost. Both techniques rely on extensive sequencing, and advances in bioinformatics, refinement of DNA amplification, and increased computational power have greatly aided the analysis of DNA sequences recovered from environmental samples.

Another method that is available for study of the viral community structure without the need to employ sequencing and analysis of conserved genes is PFGE. This method exploits one characteristic of viruses that varies over a wide range and can be easily determined, namely, the size of the genome. The PFGE technique provides separation over the full size range of intact viral genomes. Environmental dsDNA viral genomes are reported to range from a few to several hundred thousand kilobases in size. The variation in genome size for RNA and ssDNA viruses is not as dramatic and these genome types are smaller, rarely exceeding 20 kb of total nucleic acid. Thus, genome size is a phenotypic characteristic of sufficient variability and universality to characterize the most dominant communities of dsDNA viruses. The method was first used in this way to analyze the bacteriophage community in the rumen of sheep, and it was demonstrated that one PFGE band consisted of DNA from one single phage genotype. This approach has been used recently in several studies that have explored the dynamics of communities of dsDNA viruses in the marine environment. The method allows investigation of the nonculturable fraction of the viral community and, when applied together with PCR and specific primers, it is possible to characterize dominant unculturable viral populations.

CONCLUSIONS

Viruses are found in every environment in which there is microbial life. They occur in high numbers and exhibit extraordinary diversity. As a result, viruses influence the composition of their host communities and are a major force behind biochemical cycles. They also affect the evolution of both host and viral assemblages through viral-mediated gene transfer. However, our understanding of these processes is far from complete, and the story of environmental viruses is still emerging as methodology improves. The area of viral research is quickly advancing and new discoveries seem to be forcing a paradigm shift in the thinking on viral ecology.

FURTHER READING

Breitbart, M., Miyake, J. H., & Rohwer, F. (2004). Global distribution of nearly identical phage-encoded DNA sequences. *FEMS Microbiology Letters*, *236*, 249–256.

Breitbart, M., & Rohwer, F. (2005). Here a virus, there a virus, everywhere the same virus? *Trends in Microbiology*, *13*, 278–284.

Brussaard, C. P. D. (2004). Viral control of phytoplankton populations – A review. *The Journal of Eukaryotic Microbiology*, *51*, 125–138.

Edwards, R. A., & Rohwer, F. (2005). Viral metagenomics. *Nature Reviews Microbiology*, *3*, 504–510.

Filee, J., Tetart, F., Suttle, C. A., & Krisch, H. M. (2005). Marine T4-type bacteriophages, a ubiquitous component of the dark matter of the biosphere. *Proceedings of the National Academy of Sciences*, *102*, 12471–12476.

Hambly, E., Tetart, F., Desplats, C., Wilson, W. H., Krisch, H. M., & Mann, N. H. (2001). A conserved genetic module that encodes the major virion components in both the coliphage T4 and the marine cyanophage S-PM2. *Proceedings of the National Academy of Sciences*, *98*, 11411–11416.

Millard, A., Clokie, M. R., Shub, D. A., & Mann, N. H. (2004). Genetic organization of the psbAD region in phages infecting marine Synechococcus strains. *Proceedings of the National Academy of Sciences*, *101*, 11007–11012.

Paul, J. H., Williamson, S. J., Long, A., Authement, R. N., John, D., Segall, A. M. *et al.* (2005). Complete genome sequence of phi HSIC, a pseudotemperate marine phage of *Listonella pelagia*. *Applied and Environmental Microbiology*, *71*, 3311–3320.

Raoult, D., Audic, S., Robert, C., Abergel, C., Renesto, P., Ogata, H. *et al.* (2004). The 1.2-megabase genome sequence of Mimivirus. *Science*, *306*, 1344–1350.

Romancer, M. L., Gaillard, M., Geslin, C., & Prieur, D. (2007). Viruses in extreme environments. *Reviews in Environmental Science and Biotechnology*, *6*, 17–31.

Short, S. M., & Suttle, C. A. (2002). Sequence analysis of marine virus communities reveals that groups of related algal viruses are widely distributed in nature. *Applied and Environmental Microbiology*, *68*, 1290–1296.

Sullivan, M. B., Coleman, M., Weigele, P., Rohwer, F., & Chisholm, S. W. (2005). Three Prochlorococcus cyanophage genomes: Signature features and ecological interpretations. *PLoS Biology*, *3*, e144.

Suttle, C. A. (2005). Viruses in the sea. *Nature*, *437*, 356–361.

Suttle, C. A. (2007). Marine viruses – Major players in the global ecosystem. *Nature Reviews Microbiology*, *5*, 801–812.

Weinbauer, M. G. (2004). Ecology of prokaryotic viruses. *FEMS Microbiology Reviews*, *28*, 127–181.

Weinbauer, M. G., & Rassoulzadegan, F. (2004). Are viruses driving microbial diversification and diversity? *Environmental Microbiology*, *6*, 1–11.

Chapter 9

Aeromicrobiology

G.L. Andersen[1], A.S. Frisch[2], C.A. Kellogg[3], E. Levetin[4], B. Lighthart[5] and D. Paterno[6]

[1]*Lawrence Berkeley National Laboratory, Berkeley, CA, USA*

[2]*CIRA/Colorado State University, Ft. Collins, CO, USA*

[3]*US Geological Survey, St. Petersburg, FL, USA*

[4]*University of Tulsa, Tulsa, OK, USA*

[5]*US EPA, Monmouth, OR, USA*

[6]*US Army ECBC, Aberdeen, MD, USA*

Chapter Outline

ABBREVIATIONS

CA Cluster analysis
CAB Culturable atmospheric bacteria
MAP Microbial air pollution
OAF Open air factor
PCA Principal components analysis
PCR Polymerase chain reaction
TAB Total atmospheric bacterial

DEFINING STATEMENT

This article is a brief survey of information concerning the major populations of airborne viruses, bacteria, and fungi, given in terms of their spatial and temporal abundance, flux, simulation models, and methods of analysis. Collection devices are also listed.

INTRODUCTION

By some estimates as much as two billion metric tons of dust are lifted into the Earth's atmosphere every year.... There are typically about one million bacteria per gram of soil, but let's be conservative and suppose there are only 10,000 bacteria per gram of airborne sediment. Assuming a modest one billion metric tons of sediment in the atmosphere, these numbers translate into a quintillion (10^{18}) sediment-borne bacteria moving around the planet each year – enough to form a microbial bridge between Earth and Jupiter. (Griffin, et al. (2002) American Scientist 90: 230–237)

Viable (i.e., live or 0.1–10% of the total) and nonviable (i.e., dead) microorganisms enter the atmosphere as single or multiple, naked or rafted (e.g., on plant or soil debris and sea spray) cells from many sources that are planktonic while airborne, and finally are deposited on surfaces by either impaction or gravitational settling. Huge natural long-distance transport sources such as dust storms and many more local natural sources contribute to the so-called background microbial population while human activities supply what might be called microbial air pollution (MAP), all contributing to the ambient atmospheric population. Finally, these populations are modulated spatially and temporally by ambient meteorological conditions during their transient transport in the atmosphere. A simplified diagram of the atmospheric microbial sources and dynamics is shown in Figure 1.

Topics in Ecological and Environmental Microbiology.
Published by Elsevier Inc.

Thus, changes in atmospheric microbial density vary as a function of the magnitude of their sources and duration of their input/output (gain/loss) rates as modulated by atmospheric upwind turbulence. Spatial variation in the population densities is generally from nil at high altitude to relatively denser populations near sea level with periodically very high densities (Figure 2).

More specifically, atmospheric bacterial and fungal population density variations have been observed to occur as functions of air mass differences, altitude with respect to the inversion layer, population density reduction upon traversing from land to sea, and frontal activity. On a smaller scale, wind gusts may be significant contributors to microbial liberation from plant and soil surfaces. These variations may occur quickly with changes that may be over several orders of magnitude, for example, 1600% change of culturable atmospheric bacteria (CAB) within a 2-min sampling period.

Airborne survival of these relatively naked cells is critical to their subsequent propagation. Most microbial life forms do not survive well under atmospheric conditions; however some do, for example, many fungi and bacterial spores. Some of the most well-known deleterious conditions are solar UV radiation, relative humidity, temperature, sometimes oxygen and air pollutants such as ozone, carbon monoxide, sulfur dioxide, nitrogen oxides, and particularly open air factor (OAF). Some dissolved substances are also known to protect airborne cells, for example, trehalose.

At the time of this writing, the study of airborne microbes is in a transient state, going from almost all the information concerned with relatively well-known CAB and fungal fraction, to the relatively little known but more inclusive total atmospheric bacterial (TAB), viral, and fungal populations. The latter is the result of increased use of molecular biological techniques of identification and quantification.

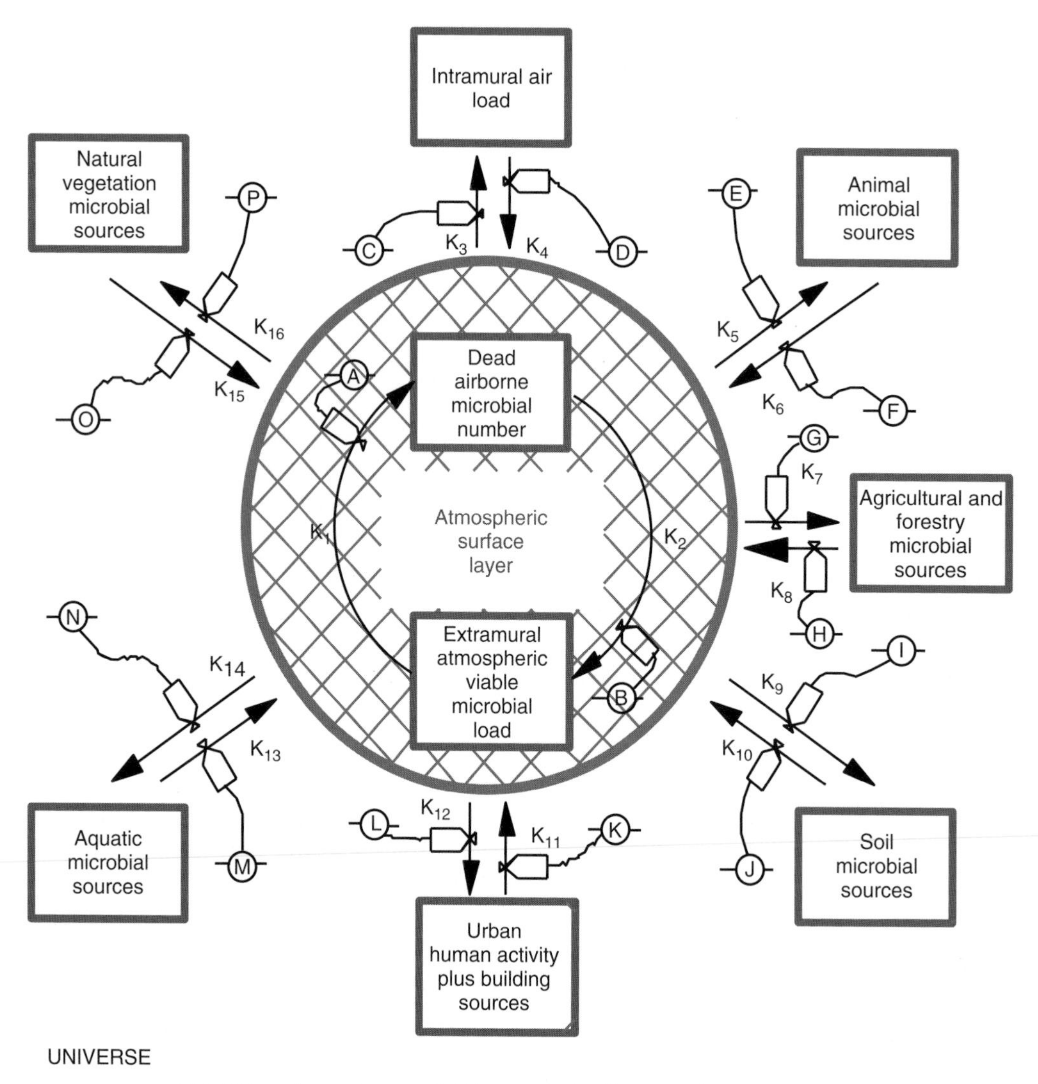

FIGURE 1 Compartment model diagram of microbial aerosol sources and sinks.

FIGURE 2 Giant dust storm above northwestern Africa blows out into the Atlantic over the Canary Islands on 26 February 2000. Such dust storms appear to carry infectious microbes and toxic chemicals, and are now believed to pose a significant hazard to ecosystems in the Caribbean and other parts of the Americas. This image was made by the Seaviewing Wide Field-of-view Sensor (SeaWiFS) on the OrbView-2 satellite, which has observed similar storms every year since its launch in August 1997. Courtesy of the SeaWiFS Project, NASA/Goddard Space Flight Center and ORBIMAGE.

TABLE 1 List of some CAB concentration values found in the outdoor atmosphere

Location	Concentration (CFU m^{-3})
Ocean shore bacteria	113
Forest bacteria	1190
Maximum	384
Minimum	715
City street bacteria	
Grass seed field bacteria	
Maximum	703
Minimum	126
Combine harvester (100 m downwind)	
Bacteria	18,520
Fungi	33,740
Sewage treatment plant	
(Edge of aerator unit)	30,000
(Sewage sludge composting)	55,000

MICROBIAL SOURCES AND SPATIAL AND TEMPORAL DISTRIBUTIONS

Bacteria

Temporal quantity and quality

The sources of natural airborne bacteria are anywhere they can grow or be deposited, that is, virtually anywhere and as shown in Figure 1. Most bacteria probably are liberated from natural sources such as vegetation and soil surfaces by atmospheric turbulence and marine and freshwater spray sources. Bacteria may be transported great distances as indicated in Figure 2. Rainsplash also contributes to the atmospheric populations. Anthropogenic activities contribute to fewer, but significant, usually local populations.

As one might imagine, the concentration of microorganisms in the atmosphere varies as a function of location (Table 1), time of year, and time of day. Generally, the annual CAB concentration distribution in the atmosphere has been found to be maximal in the dry summer/autumn conditions and minimal in the wet or snow-covered winter conditions as shown Figure 3. The diurnal CAB concentration appears to be maximal in the early afternoon and minimal just before dawn as shown in Figure 4.

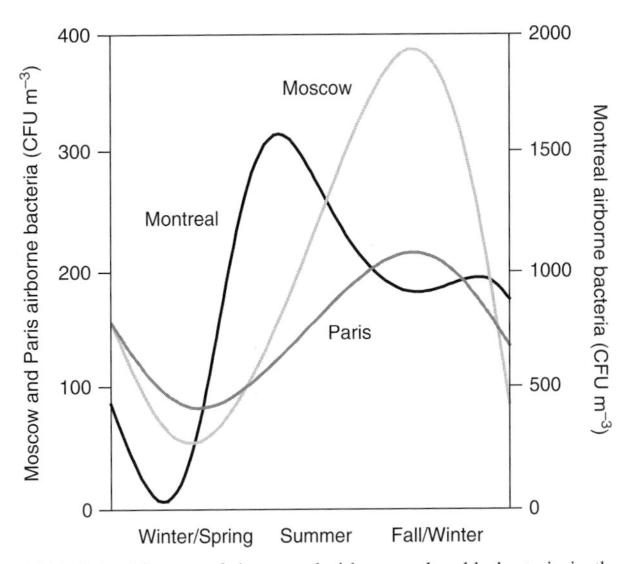

FIGURE 3 Diagram of the annual airborne culturable bacteria in the atmosphere of Paris, Moscow, and Montreal. *Reproduced from Lighthart, B. (2000). Mini-review of the concentration variation found in the alfresco atmospheric bacterial populations.* Aerobiology, 16, *7–16.*

Culture-independent methods for analyzing atmospheric bacterial community structure have found very high levels of biodiversity, suggesting multiple sources of input bacteria. Classifying bacterial species by multiple, unique probes targeting the 16S rRNA gene, molecular biologists have used a high-density Phylochip microarray to monitor

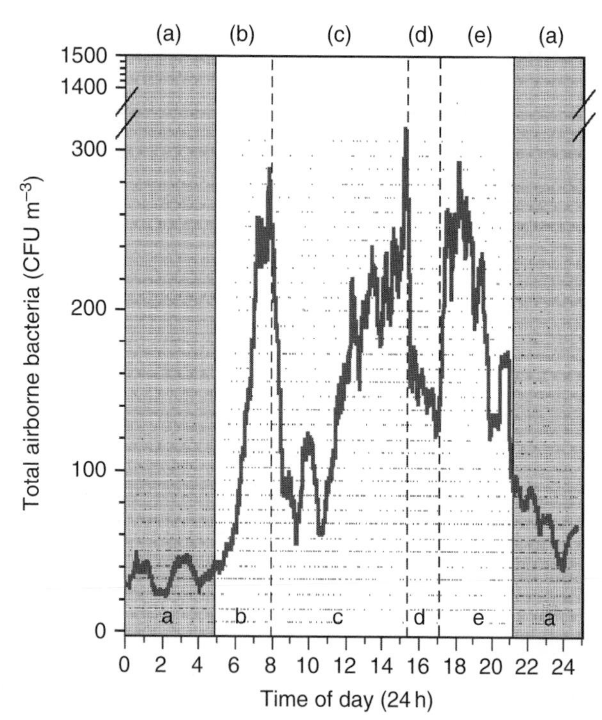

FIGURE 4 One hundred point running average of culturable atmospheric bacteria (CAB) for approximately six thousand 2 min sample intervals on three typical summer days in the mid-Willamette River Valley, Oregon: (a) nighttime clean marine replacement air, or settling out, (b) sunrise convectively driven peak of near-ground air, (c) midday flux accumulation trapped in the inversion layer, (d) afternoon intrusion of clean sea breeze air below the daily inversion, and (e) replacement by microbially polluted air below the inversion on subsidence of the sea breeze. The solar peak and inversion accumulation are speculated to be a general global phenomenon while the sea breeze is a local phenomenon. *Reproduced from Lighthart, B., & Shaffer, B. T. (1995) Airborne bacteria in the atmospheric surface layer: Temporal grass seed field.* Applied and Environmental Microbiology, *61(4), 1492–1496.*

bacterial populations in Austin and San Antonio, Texas, over 17 weeks. Representatives from 26 of the 58 known bacterial phyla were identified, representing over 1800 diverse bacterial types and matching species richness estimates from corresponding 16S rRNA gene clone library sequences. Interestingly, the location of the sample (taken from either Austin or San Antonio) had very little effect on species composition. Meteorological conditions were found to have the greatest impact on atmospheric bacteria. Over 90% of the variability in the microarray data could be explained by four meteorological parameters: (1) average weekly temperature, (2) maximum PM 2.5 μm particulate count, (3) average weekly wind speed, and (4) average sea-level air pressure. Bacterial groups within the Actinomycetes and Firmicutes phyla were most responsive to weather conditions, perhaps due to their spore dispersal strategies (Table 2).

Integrating the annual and diurnal atmospheric bacterial cycles, a general temporal bacterial concentration pattern in the atmosphere might be as shown in Figure 5. Further, extending this idea to a rotating Earth, one could imagine the maximum TAB is found in reference to the solar zenith as a standing wave in the rotating coordinate system of the moving Earth. Conditions of weather, topography, source strength, and human activities (i.e., MAP) modulate the cyclic patterns of TAB loadings (Figure 6).

Flux

It is important to evaluate the flux (bacteria/m^2-time) of the bacteria into and out of the atmosphere in order to understand their population dynamics. The net upward and downward CAB flux has been measured at few locations (Table 3).

At one location, the diurnal upward, downward, and consequent net CAB flux in the mid-Willamette Valley, Oregon, in the summer was observed to have an upward positive net flux between approximately 07.30 and

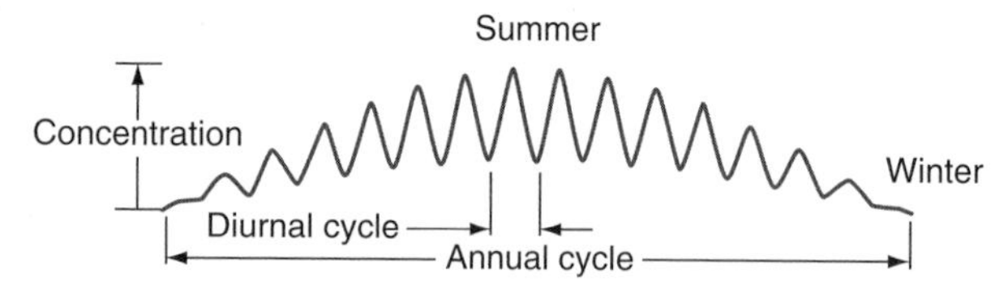

FIGURE 5 Diagrammatic representation of a theoretical alfresco airborne bacteria concentration in the Earth's atmospheric surface layer at a temporal zone location. *Reproduced from Lighthart, B. (1999). An hypothesis describing the general temporal and spatial distribution of alfresco bacteria in the earth's atmospheric surface layer.* Atmospheric Environment, *33(4), 611–615.*

TABLE 2 Bacterial species highly responsive to weather conditions

Phylum	Species	Weather Parameter	Correlation
Actinobacteria	*Saccharothrix tangerinus*	Weekly temperature (average)	Positive
Firmicutes	*Thermoactinomyces intermedius*	Weekly temperature (range of minimums)	Positive
Bacteroidetes	*Bifissio spartinae*	Weekly PM 2.5 (maximum)	Negative
α-Proteobacteria	*Sphingomonas adhaesiva*	Weekly sea-level pressure (minimum)	Positive
β-Proteobacteria	*Polaromonas vacuolata*	Weekly temperature (average)	Positive
γ-Proteobacteria	*Pseudomonas oleovorans*	Weekly wind speed (maximum)	Negative

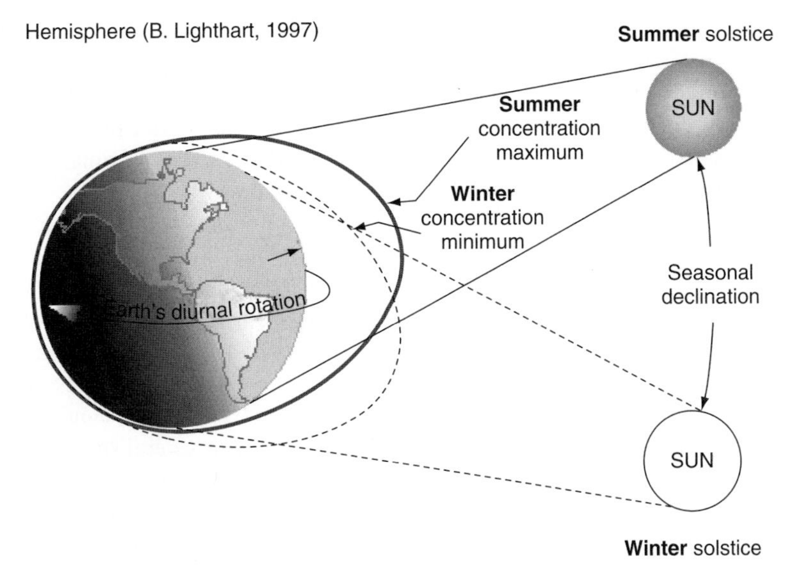

FIGURE 6 Diagrammatic representation of the theoretical alfresco airborne bacterial concentration in atmospheric surface layer at the summer and winter solstices in the northern hemisphere. *Reproduced from Lighthart, B. (1999). An hypothesis describing the general temporal and spatial distribution of alfresco bacteria in the earth's atmospheric surface layer.* Atmospheric Environment, *33(4), 611–615.*

TABLE 3 Tabulation of some CAB flux measurements

Location	Net flux (CFU m^{-2} s^{-1})	Reference
Over winter wheat canopy	57	Lindemann et al. (1982). *Applied Environmental Microbiology, 44,* 1059.
Over bean canopy	499	
Over bare soil	124	
Over alfalfa canopy	543	
Over bean canopy		
Upward in 1980	0.798	Lindemann and Upper (1985). *Applied Environmental Microbiology, 50,* 1229.
Downward in 1980	0.682	
Upward in 1982	0.63	
Downward in 1982	0.59	
Dry soil	43 ± 9.2	
Wet soil	154 ± 57	
High desert (HNR)	4.67	Lighthart and Shaffer (1994). *Atmospheric Environment,* 28(7), 1267.

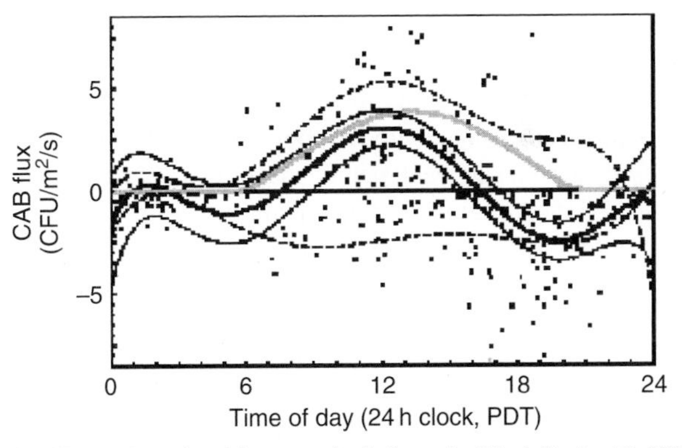

FIGURE 7 Net culturable atmospheric bacteria (black line) with 95% confidence limits (dotted lines), ascending and descending fluxes (upper and lower dashed lines), and solar radiation (gray line) and data points. (Flux values with zero wind velocities were not used in the analyses.)

16.00 h, during the solar heating-generated convection hours, and the rest of the time a net downward CAB flux (Figure 7). Integration of the net CAB flux curve indicated a slight overall CAB input into the atmosphere at the observation site.

Fungi

Distribution and methods

Fungal spores and hyphal fragments are ubiquitous components of the atmosphere and can occur in high concentrations unless the ground is covered with snow or ice. Fungi reproduced by spores, which are produced by either sexual or asexual methods, and the majority of fungal spores are adapted for airborne dispersal. Spores may originate from fungal saprobes, pathogens, or symbionts. Fungi growing on living plants and on plant debris in the soil are important contributors to the air spora. Spore levels can be especially high during harvesting, under certain meteorological conditions, as well as in contaminated indoor environments.

Fungal spores can impact human health as triggers of allergic reactions or as the cause of infectious disease. Although many fungal spores are allergenic, only a limited number of species are considered human pathogens. Immune-compromised individuals are at the greatest risk with regard to fungal infections. The impact of fungi on plant health is also of major significance. Approximately 10,000 species of fungi are recognized as plant pathogens; in fact, the majority of plant diseases are caused by fungi.

Dispersal mechanisms

Spores are released into the atmosphere through both passive and active mechanisms. Passive methods generally rely on either wind or rain to disperse spores into the atmosphere. In temperate regions of the world, the most abundant airborne spores are dispersed by wind. The atmospheric concentration of many spores depends upon the ease with which the spores are detached from the parent hyphae or spore-bearing structures. Many wind-dispersed spores are formed on erect structures that elevate the spores above the substrate. Warm dry conditions promote wind dispersal, and the entrained spores are typically referred to as the dry air spora. Components of the dry air spora include conidia of *Cladosporium*, *Alternaria*, *Drechslera*, *Curvularia*, *Pithomyces*, *Epicoccum*, *Botrytis*, smut teliospores, and rust uredospores. Some of these fungi such as *Cladosporium*, *Alternaria*, and *Epicoccum* are commonly found on leaf surfaces as saprobes or weak pathogens; others such as the smut fungi and rust fungi are obligate plant pathogens.

Spores are also passively discharged by rain through several different mechanisms. Members of the dry air spora may be dispersed by a raindrop striking a substrate such as a leaf with fungi growing on the surface. The shaking caused by the raindrops thrusts the spores into the air. Release of basidiospores by puffballs occurs by a similar puffing mechanism when a raindrop strikes the peridium or wall of a mature fruiting body. Splash dispersal occurs with many fungi. Splash-borne spores are typically surrounded by a mucilage layer, which protects the spores from desiccation but also prevents their removal by wind. Raindrops dissolve the mucilage and leave a spore suspension free for subsequent dispersal by additional raindrops.

Active dispersal mechanisms are also common in fungi and the ballistics of some species is quite impressive. In many ascomycetes, the explosive release of ascospores from the ascus is tied to atmospheric moisture. Hygroscopic compounds within the ascus absorb available moisture, causing the ascus to swell and develop high osmotic pressure. The pressure induces the ascus to burst, explosively shooting the spores into the atmosphere. The need for atmospheric moisture generally restricts dispersal to periods during and after rainfall or at times of high humidity. The active release of basidiospores from the basidia of mushrooms and bracket fungi also depends on atmospheric moisture but not rainfall. The spores are actively propelled from the basidium and then fall from the fruiting body and are picked up by the air currents. Peak atmospheric basidiospore levels occur when humidities are high. Because of the need for atmospheric moisture, both the ascospores and the basidiospores are considered members of the moist or wet air spora. Certain rainsplash fungi are also components of the moist air spora.

Once airborne, by either passive or active mechanisms, the spores are transported vertically and horizontally by wind in the atmospheric boundary layer. They can be carried upward by thermals and have been recovered several kilometers above the Earth's surface. Under normal turbulence, the majority of spores are deposited close to the source; however, long-distance transport has been well documented with several spore types (Table 4). The movement or *Puccinia graminis* (i.e., wheat rust) spores from the southern United States and Mexico into the wheatbelt in the northern United States and Canada is one of the most well-known examples of long-distance transport. The long-distance dispersal of *Phakopsora pachyrhizi* (i.e., soybean rust) spores is responsible for the recent introduction of this pathogen into North America from South America.

Distribution of airborne fungal spores

Airborne fungal spore concentrations show a great deal of variability, with the major causes of the variability due to diurnal rhythms of spore discharge, seasonal effects, spatial effects, and weather-related effects. The diurnal variability generally coincides with the time of day when condi-

TABLE 4 Plant disease spread by long-distance dispersal of airborne fungal spores[a]

Plant disease	Global or Continental Spread
Sugarcane rust	Cameroon in West Africa to Dominican Republic in Latin America – 1978
Coffee leaf rust	Angola in West Africa to Bahia in Brazil, South America – 1970
Wheat yellow (stripe) rust	Eastern Australia to New Zealand – 1980
	China – periodic spread
Potato late blight	Europe – 1845, 1977
Tobacco blue mold	North America – periodic spread
Stem rust of wheat	North America – periodic spread
	Europe – periodic spread
	South Africa to Australia – 1969
Brown rust of wheat	India – periodic spread

[a]*Adapted from Brown JKM and Hovmøller MS (2002) Aerial dispersal of pathogens on the global and continental scales and its impact on plant disease.* Science, *297, 537–541.*

tions are most suitable for dispersal. During dry weather, the members of the dry air spora typically show an early to mid-afternoon peak in spore discharge and an early morning low. The peak typically occurs when wind speeds are greatest and humidities are low. In temperate areas the dry air spora dominate the atmosphere, with *Cladosporium* commonly the most abundant spore type identified (Tables 5–7). Basidiospores typically show peak concentrations during late night and early morning hours when humidities are high, and have low concentrations during the afternoon (Figure 8). Although the release of many ascospores is limited to periods of rainfall, high humidity provides sufficient moisture for other species. As a result, some ascospores also show a late night and early morning peak similar to basidiospores.

TABLE 6 Dominant culturable fungi in the Tulsa, Oklahoma, atmosphere

Fungal Taxa	% of Culturable Fungi[a]
Cladosporium	71.60
Nonsporulating colonies	11.65
Alternaria	7.47
Yeasts	2.91
Penicillium	2.89
Aspergillus	1.40
All other taxa	2.08

[a]*Mean percent of samples collected weekly for 8 months.*

TABLE 5 Dominant airborne spores in the Tulsa, Oklahoma, atmosphere registered by a Burkard spore trap

Fungal Taxa	% of Total Spores[a]
Cladosporium conidia	57.57
Ascospores	16.96
Basidiospores	10.69
Alternaria condia	3.17
Smut teliospores	2.46
Penicillium/Aspergillus conidia	3.82
All other taxa	5.33

[a]*Mean percent from samples collected for 5 years.*

TABLE 7 Median concentration of the most common culturable airborne fungi recovered from outdoor environments in the United States[a]

Taxa Identified	Median Concentration CFU m^{-3}	95% CI CFU m^{-3}
Cladosporium	200	18–1849
Nonsporulating colonies	100	12–901
Penicillium	50	12–377
Aspergillus	20	12–170

[a]*Adapted from Shelton, B. G., Kirkland, K. H., Flanders, W. D., & Morris, G. W. (2002). Profiles of airborne fungi in buildings and outdoor environments in the United States.* Applied and Environmental Microbiology, 68, *1743–1753.*

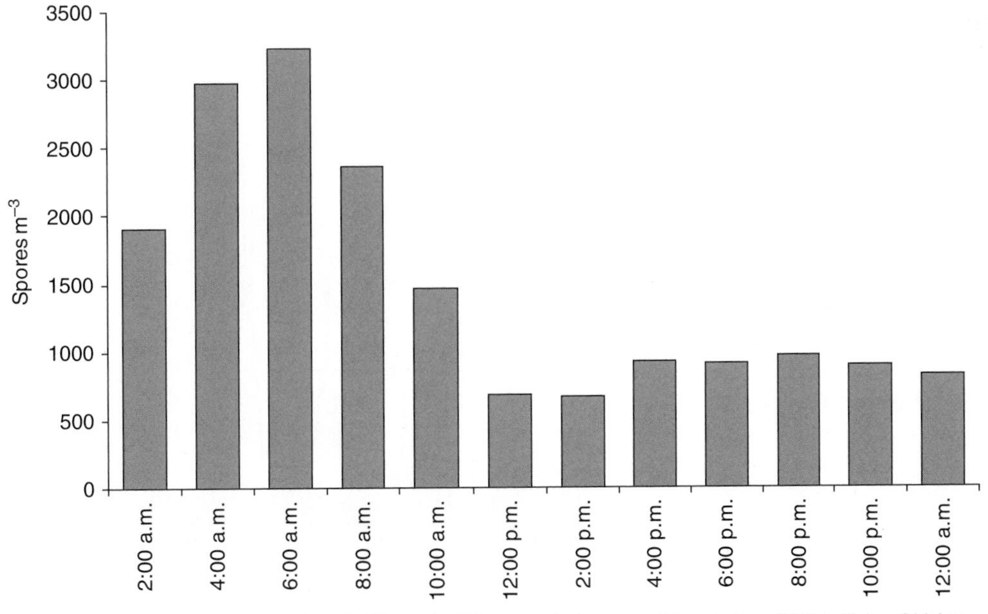

FIGURE 8 Mean hourly concentration of airborne basidiospores during several days in June 2002 in Tulsa, Oklahoma.

Although spores can be present year round, in temperate areas spore levels tend to increase in spring showing significant correlation with increasing temperatures. Total spore levels typically are at their highest concentrations in late summer and early fall (Figure 9). In tropical and subtropical climates, airborne spore levels may show little fluctuation during the year, or may follow local rainfall patterns. Some fungi, especially some mushrooms and bracket fungi, produce fruiting bodies during a specific season each year. In many temperate areas, this is often in the fall; as a result, these regions show a fall basidiospore peak. Some fungal plant pathogens show peak levels that coincide with the flowering or harvesting of the host plant. This is especially pronounced among the smut fungi, which have only a single reproductive phase each year. Although smuts can be found in the atmosphere from spring through fall, different smut species are prevalent in different seasons.

Although many components of the air spora are similar worldwide, at any one time or place, the atmosphere may be dominated by nearby sources of spores. During crop harvesting, very high levels of airborne spores may be released into the atmosphere; many of these originate from leaf surface fungi that are dislodged by the mechanized equipment. Near forests and wooded areas, basidiospores may be the dominant spore type, whereas ascospore concentrations may be the most abundant spore type near standing or running water.

Viruses

Sources

Unlike bacteria, fungi, or pollen, viruses are not typically considered a natural part of the aeromicrobiota. Instead, most are viewed as being briefly injected into the air from infected individuals (people or animals) or contaminated materials. Viruses can be aerosolized by a variety of methods, including but not limited to sneezing, coughing, speaking, vomiting, flushing toilets, shaking contaminated materials, and spray irrigation with treated sewage or 'gray water'. However, there is also the environmental aerosolization of bacteriophage and other viruses that occurs in conjunction with aerosolized soils (e.g., African dust) and marine aerosols in the surf zone.

Most of the studies conducted on aerosolization of viruses have been done in a controlled setting by creating an artificial spray of a particular virus and then determining whether it caused an infection in exposed test subjects. Any virus that can survive aerosolization and the stresses of the atmosphere potentially may be transmitted by air; however, it is often difficult to determine if a virus was directly acquired from the air, or whether it came from contaminated surfaces exposed to the aerosol. A good example is the common cold, originally thought to be spread primarily by air, which we now know is usually acquired by direct contact with a virus-contaminated surface, followed by self-inoculation.

Quality and quantity

The absence of a standard methodology or protocol for producing and sampling aerosolized viruses makes it difficult if not impossible to compare information generated by the previous studies. Because viruses are more difficult to culture than bacteria (requiring live hosts or tissue cultures of live cells), there are few if any studies that have attempted to quantify viruses in the alfresco atmosphere, other than in proximity to a large potential source (e.g., a sewage treatment plant). Additionally, it has been estimated that due to difficulties associated with sampling aerosols for viruses,

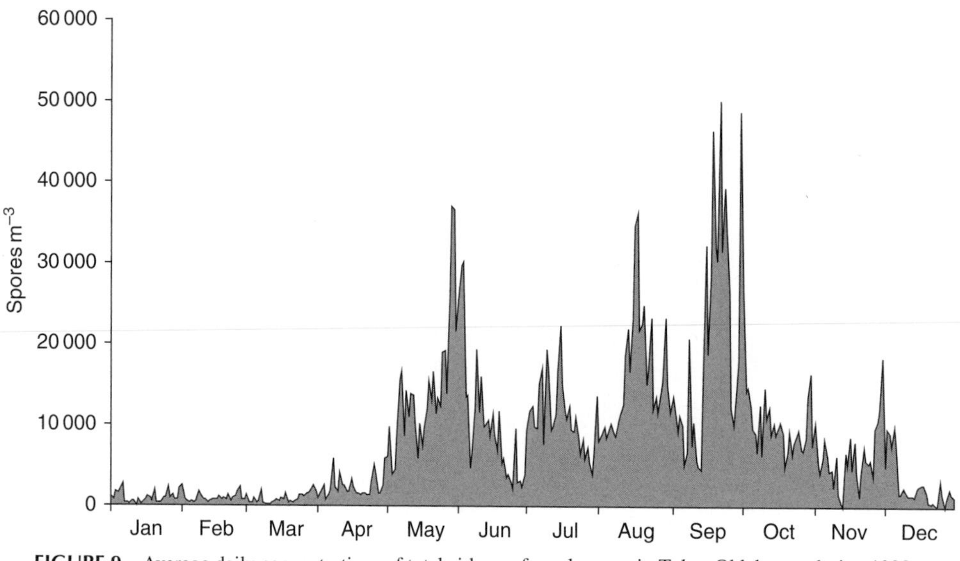

FIGURE 9 Average daily concentrations of total airborne fungal spores in Tulsa, Oklahoma, during 1998.

any recovery data may be several orders of magnitude below the actual number of viruses present in the sample. Modern molecular methods such as polymerase chain reaction (PCR) make it easier to detect viruses in aerosols by eliminating the need for cultivation in a host; however, these methods cannot differentiate between infectious and inactivated particles and are often presence/absence tests rather than strictly quantitative.

Once aerosolized, viruses may become inactivated, meaning either a loss of infectivity or physical degradation of the particle. Inactivation occurs by interactions between the virus particles and UV radiation, oxygen, pollutants (e.g., sulfur dioxide), and may be accelerated by temperature and relative humidity. Laboratory experiments have shown that hydrophilic viruses, such as bacteriophage and poliovirus, survive longer at higher relative humidity (>75%), as compared to lipid-enveloped viruses such as influenza that survive longer at lower relative humidity (>30%). The relationship between survival and relative humidity is hypothesized to be due to differences in denaturation between proteins and lipoproteins on the viral capsids.

Temporal and spatial

The suspension time and distance traveled in the atmosphere by a given virus are highly variable and are affected by airflow and turbulence. Viral particles are <1 μm in size but are typically aerosolized in droplet nuclei that are <5 μm. Airborne particles in this size range may remain suspended for days to weeks before being deposited by downward-moving air.

Many airborne virus transmission studies have focused on indoor environments, such as hospitals, laboratories, schools, and animal husbandry facilities. The capacity for transmission is much greater in these environments, due to issues such as recirculated air, confined space, and proximity to a viral source. It was also believed that human and animal viruses were unlikely to spread in an outdoor environment because of the many inactivation factors and the huge potential for dilution. However, studies in the past two decades have proven the long-distance transmission of animal viruses (e.g., foot-and-mouth disease virus and pseudorabies virus) that have caused outbreaks of infection hundreds of kilometers downwind. Human enteric viruses have been detected up to 100 m downwind from irrigation sprinklers.

Less is known about the intercontinental transport of viruses. Viruses have been detected in association with fog, clouds, smoke, and dust in the atmosphere. Large-scale dust storms, such as the Asian dust from China's Gobi and Taklamakan deserts, travel across the Pacific to impact the United States and Europe, and African dust from the Sahara/Sahel region travels across the Atlantic to the Caribbean and both North and South America.

ANALYSES

Due to the large number of meteorological variables that must be measured over extended periods of time, if a meaningful understanding of the dynamics of airborne microorganisms is to be understood, a statistical approach for the analysis is recommended. A similar argument applies to the large number of bases that must be analyzed to provide a meaningful taxonomic delineation for bacteria isolated from the atmosphere.

Statistics

A useful approach to understanding the dynamics of bacterial transport in the atmosphere has been to identify the prevailing local meteorological conditions and then to associate the microorganisms attendant with these conditions. Principal components analysis (PCA) provides a means of identifying those measured meteorological variables that constitute unique sets of prevailing meteorological conditions at a site (Figure 10), while cluster analysis (CA) allows independent association of these microbes' quantity or quality contained in the PCA-identified atmospheric environmental conditions (Table 8).

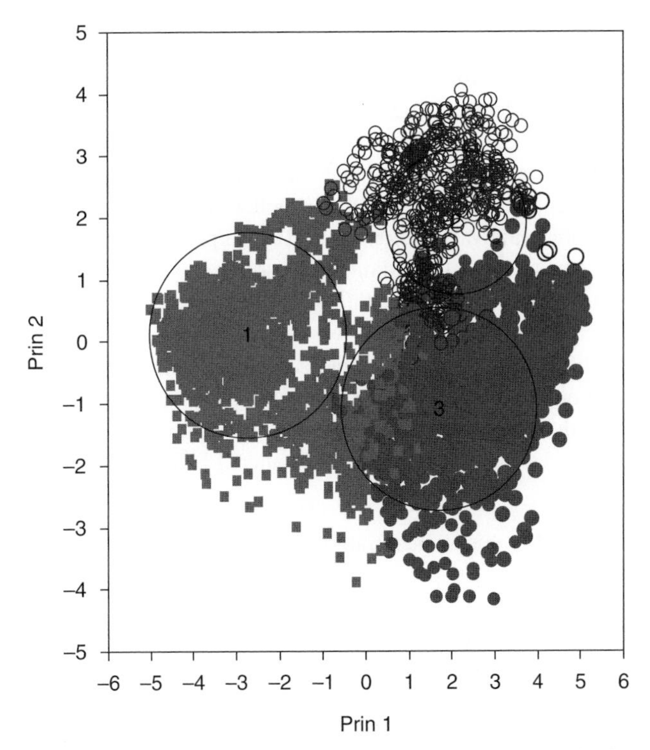

FIGURE 10 Biplot showing the points in three distinct clusters by the principal components analysis of the first two principal components of the 13 meteorological parameters. Circles are drawn around the cluster centers with the area proportional to the number of points in the cluster. (The non-overlapping 3D clusters overlap in this 2D projection.)

TABLE 8 Tabulation of the culturable atmospheric bacteria (CAB) simultaneously found associated with three unique clusters of a PCA of meteorological parameters found in the atmospheric surface layer in the summer time at a mid-Willamette Valley, Oregon, site

			Bacteria (CFU m^{-3})		
Cluster	Statistics	Time of Day (PDT)	Total	Pigmented	Unpigmented
1	Mean	8.77	81	30.6	50.4
	Std Dev	7.396	71.18	38.08	37.48
	N	2074	226	226	226
	HSD[a]	C	B	B	B
2	Mean	13.29	165.7	90.8	75
	Std Dev	2.927	165.2	92.45	84.5
	N	1452	177	177	177
	HSD	B	A	A	A
3	Mean	18.19	45.7	19.1	26.7
	Std Dev	1.889	26.92	14.28	16.73
	N	624	70	70	70
	HSD	A	C	C	C

[a]*HSD, honestly significant difference for comparisons between clusters.*

Survival of Airborne Microbes: A Model

Microorganisms enter the atmosphere from various sources as a result of turbulent and convective processes such as wind gusts and solar heating. For subsequent consequences, it is critical that viable microbes survive after being transported in the atmosphere. Their survival may be estimated by a mathematical simulation model. One such model, termed the airborne microbial survival function (eqn [1]), incorporates varying sensitivities to several environmental variables in their life cycle phases, for example, logarithmic growth and death phases. This function estimates survival over time of multigrowth-phased, multispecies populations of microorganisms in a multivariable environment (Figure 11). This function is appropriate for use in a dynamic model that simulates downwind atmospheric transport, for example, HYSPLIT4 and VLSTRACK.

$$\text{Microbial concentration}_t = \sum_{j=1}^{j=n}\sum_{i=1}^{i=m} b_i^j \mathrm{e}^{-k_j^i(\theta,\mathrm{rh},\ldots,\mathrm{uv})t} \quad (1)$$

where t is the time from release, b_j^i is the concentration of microorganisms for the ith phase of the jth species at $t = 0$, and k_j^i (e.g., θ, rh, uv, θrh, θuv, rhuv) are the death rate constants, and interaction terms, for that phase and species for the ambient environmental, for example, variable temperature (θ), relative humidity (rh), and ultraviolet light (uv). These k_i^j's must be determined experimentally for the different conditions for each microorganism of interest.

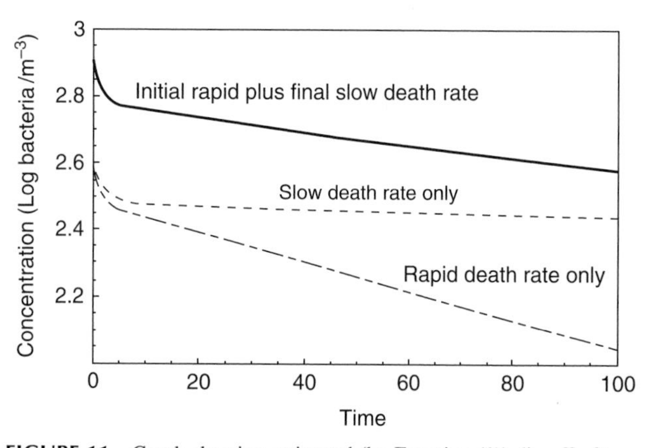

FIGURE 11 Graph showing estimated (by Equation (1)) die-off of two species, one with comparatively slow and one with rapid decay rates for two growth-phase microbial populations injected into the atmosphere. One population is more sensitive than the other to the atmospheric conditions.

METEOROLOGY

In order to estimate the number of viable microbes downwind from some source, several properties of the atmosphere and how the microbes survive as a function of the different meteorological conditions must be known. In addition, how the microbes get from the surface where they are residing and into the air must be understood.

The mechanism for getting microbes from the ground, vegetation, or water sources into the atmosphere depends on various atmospheric conditions. Increases in surface winds can dislodge particles from the ground (such as dust storms), and one can speculate that changes in the electrical properties in the atmosphere can reduce the electrical attraction of microbes from plants and soil. Once the microorganisms are airborne, where they go and their concentration will depend on wind speed, direction, height of the atmospheric boundary layer, and the intensity of atmospheric turbulent motions. In order to estimate the concentration of bacteria downwind from a source for example, we would need to know the following:

1. The flux of bacteria from the source and the size of the source.
2. The wind speed and direction downwind from the source.
3. The height of the atmospheric boundary layer, which is a function of the solar heating of the surface during the day; the nighttime boundary layer is usually much more shallow.
4. The temperature, relative humidity, and UV light intensity along the trajectory.

This information loaded into the airborne microbial survival equation (e.g., Equation (1)) can be used to estimate the viable microbes downwind from the source.

ATMOSPHERIC MICROBIAL SAMPLER METHODS AND SAMPLER SOURCES

Sampling Methods

There are a range of instruments and methods of analysis for studying airborne fungi. The most widely utilized instruments are impaction samplers, and microscopy and culturing are the most common methods of analysis at the present time. Impaction samplers separate particles from the air stream by using the inertia of the particles; this results in spore deposition onto an adhesive-coated solid surface or an agar surface as the air stream bends to bypass the surface. A wide variety of impaction samplers are available, including slit impactors for total spores, rotating arm impactors for total spores, and sieve impactors for culturable fungi. No single sampler or method of analysis is appropriate for all applications; the research objective and the type of data to be collected should determine the methods to be used.

In 1952, Hirst designed a volumetric spore trap that deposited particles from the atmosphere onto a coated microscope slide that moved past a 2-mm slit orifice. This instrument made it possible to accurately estimate the total spore and pollen concentrations of the atmosphere for any hour of the day. Several impaction samplers that are widely used today, including the Burkard spore trap and the Lanzoni sampler, are based on the Hirst design. These samplers generally operate 24 h per day for either 7 days or 1 day depending on the model. A number of portable spore traps that have an orifice based on the Hirst trap are also available; these samplers are used for short-term grab samples of 5–15 min duration. These have broad applications for indoor air sampling. Samples collected with these instruments are analyzed by microscopy for spore identification, generally to the genus level. Spores of some taxa cannot be accurately identified to the genus level by morphology and are grouped with similar spores. For example, *Penicillium* and *Aspergillus* spores cannot be differentiated by microscopy and are grouped into the *Penicillium/Aspergillus* category. Also the sexual spores of many ascomycetes and basidiomycetes are frequently grouped into broad categories of ascospores and basidiospores.

Andersen samplers are sieve impactors that are commonly employed for culture-based sampling; these have multiple holes that allow the impaction of spores onto the agar surface of one or more Petri dishes. Available models include six- and two-stage samplers, which separate particles by size, and the single stage, which has 400 holes for collecting spores with no size separation. Other manufacturers produce similar single-stage samplers including the BioCassette sampler, which is a disposable sieve impactor with an agar plate. These samplers are grab samplers that are generally operated for 1–5 min sampling periods. After exposure, the Petri dishes are incubated at room temperature for 3–7 days for mesophilic fungi. Higher incubation temperatures and shorter incubation times are used for thermophilic species. Colonies resulting from spore impaction are identified by spore development, spore morphology, and colony characteristics.

Spore trap samplers and culture-based samplers give different pictures of the air spora. Spore traps capture all spores, viable and nonviable; however, culture-based sampling is only useful for fungal spores that can germinate and grow on the culture medium used and the incubation temperatures employed. Culturing favors anamorphic fungi (formerly referred to as the deuteromycetes or imperfect fungi) and zygomycetes; anamorphic fungi are mainly the asexual stages of ascomycetes. These include many common airborne taxa, such as *Cladosporium*, *Alternaria*, *Epicoccum*, *Curvularia*, *Penicillium*, and *Aspergillus*. Although *Cladosporium* may be the most abundant genus identified by both types of sampling, spore trap samplers will commonly show ascospores, basidiospores, and smut spores among the most abundant taxa. Air samples collected in Tulsa, Oklahoma, show the differences in sampling methods. Data collected continuously with a Burkard spore trap for 5 years showed that six taxa made up close to 95% of the spore load (Table 5). More than 15 other taxa were routinely identified and counted, but individually these were normally >1% of the yearly total. *Cladosporium* conidia, ascospores, and basidiospores were consistently the top three taxa; however, the ranking of next three spore types varied during the years. Data in Table 7 were obtained from air samples collected once a week for 8 months with an Andersen sampler for culturable fungi. All colonies were identified to genus or species level. *Cladosporium* and nonsporulating colonies were consistently the top two colony types identified. Both sampling methods identified *Cladosporium*, *Alternaria*, *Penicillium*, and *Aspergillus* as some of the most abundant airborne fungi; however, approximately 30% of the total spore load (ascospores, basidiospores, and smut spores) were not identified from the Andersen samples. In a study involving over 2400 outdoor air samples for culturable fungi collected from all regions of the United States and during all seasons, similar data were obtained. *Cladosporium*, *Penicillium*, *Aspergillus*, and nonsporulating colonies were the most abundant fungi identified (Table 7).

Filter samplers collect particles from the air stream by trapping them in a fibrous or porous substrate. These instruments range from small personal-cassette samplers worn by individuals to determine personal exposure levels to large, high-volume devices that process thousands of liters of air per hour. Personal samplers typically hold filters that are 25–47 mm in diameter. Various types of filter material are available depending on the type of analysis desired. Filters can be cleared for direct microscopy or washed for culturing, biochemical, immunochemical, or molecular analysis of the trapped spores. Molecular analysis using PCR is becoming more widely used and may soon become a standard method of analysis for airborne fungi.

Microbial Sampler Sources

Commercially available qualitative and quantitative collecting instruments along with some of their design characteristics are listed in Tables 9 and 10.

RECENT DEVELOPMENTS

Prior to the publication of the Encyclopedia of Microbiology (EoM), 3rd Edition, measurements of airborne viruses, bacteria, and fungi were almost exclusively the culturable microorganisms (e.g., less than 10% of the airborne bacteria). These measurements included: temporal quantity, both diurnal and annual, at numerous geographical locations; altitude; survival protection or injury as functions of relative humidity, atmospheric gases both natural (e.g., O_2) and pollutant (e.g., CO, SO_2, NO_x); weather conditions; particle size; solar radiation, particularly UV; flux from ground and vegetation sources; and survival and distributional simulation models. Further, some species when airborne are thought to be rain droplet nuclei. Also, Microbiological Pest Control Agents (MPCA) may be delivered as an aerosol that drifts away from the target site and becomes one in the set of Microbiological Air Pollutants (MAP).

Moreover in the past few decades, atmospheric microbial research has evolved from the limitations of the living culture methods to molecular genetic techniques where the entire collected population's, live and dead, DNA is evaluated (i.e., metagenomics).

Viruses

The study of airborne viruses continues to be complicated by the limitations of both sampling and detection methods currently available. Attempts are being made to minimize the loss of viruses during collection as well as increasing the sensitivity of their detection in aerosols, but no method has emerged as the "gold standard." Greater attention is being paid to the role of climate, as changes in rainfall effect the dispersal of airborne viruses.

Bacteria

Recently, progress in atmospheric bacteriology has occurred in several areas:

1. Development of a global weather and climate simulation model EMAC (or ECHAM5/MESSy1.5-Atmospheric Chemistry; see Burrows, 2008, for further description) was modified to include biological particle transport (e.g., bacteria). In addition to meteorological variables, 10 globally distributed ecosystem types, each with its measured (or a few with approximations) mean bacterial concentrations, were input to the model. Simulated global concentrations of bacteria emitted into the near-surface atmosphere by the modified model are shown in Figure 1. Model output indicates that residence time alone is insufficient to explain seasonal variation in particulate culturable bacteria and indicates that seasonal variation in culturability and/or emission strength are the main drivers for the variation. Finally, simulation results indicate that the estimated total annual global emission of bacteria containing particles to the atmosphere to be 7.6×10^{23} to 3.5×10^{24} a^{-1} (Burrows *et al.*, 2009a,b,c) (Figure 12).
2. Recent results indicate that microorganisms in cloud water may act either as sinks or sources of organic carbon, and may have to be considered as actors in cloud chemistry. For example, 60 culturable bacterial species isolated from cloud water were found to metabolize lactate, succinate, acetate, formate, formaldehyde, Species that metabolize the latter three have the highest capacity for biodegradation found in cloud water. Additionally, ^{13}C labeled formaldehyde could be transformed into formate and/or methanol by cloud water bacteria (Amato et al., 2007).

TABLE 9 Widely used, commercially available samplers for collecting bioaerosols

Sampler[a]	Principle of Operation	Sampling Rate (l min^{-1})	Manufacturer/ Supplier[b]	Commercial NAME	Application[c]		Calculated cutpoint d_{50}	Calculated cutpoint Other
Slit Agar Impactors								
1. Rotating slit or slit-to-agar impactors (vp, sc)	Impaction onto agar in 10- or 15-cm plates on rotating surfaces	28	BC	Mattson-Garvin Air Sampler	C		0.53	0.5
		175, 350, 525, 700	CAS	Airborne Bacteria Sampler	C			
			NBS	Slit-to-Agar Air Samplers:				
		15–55		STA-203	C			
		15–30		STA-204	C			
Multiple-hole-impactors								
2. 12-hole impactors (vp, sc)	Impaction onto agar in 10-cm plates		VAI	Sterilizable Microbiological Atrium (SMA):				
		28–71		SMA Micro Sampler	C			
		28, 142		SMA MicroPortable Viable Air Sampler	C			
3. 100-hole impactor (sc)	Impaction into agar in 10-cm plates	10	BMC/SBI	Portable Air Sampler for Agar Plates	C			
		20			C		4.18	
		(28)			C		2.56	
4. 219- or 220-hole impactors (sc)	Impaction onto agar in 5.5-cm contact plates		PBI/BSI	Surface-Air-Sampler (SAS):				
		90		Super 90	C		1.97	
							1.94	2.24
		90		Super 90 CR	C			
		100		HiVAC Impact	C			
		180		High Volume Sampler	C		1.35	1.39
								1.45
	9-cm plates	100		HiVAC Petri	C			
	5.5-cm contact plates	100	PAR/SBI	MicroBio Air Samplers: MB1, MB2	C		1.8	
5. 400-hole impactor (vp)	Impaction onto agar in 10-cm plates	28	AND	Anderson Single-Stage Viable Particle	C		0.57	
				Sampler; N6 Single-Stage Viable Impactor			0.58	
								0.65
6. Two-stage, 200-hole impactor (vp)	Impaction onto agar in 10-cm plates	28	AND	Anderson Two-Stage Viable	C	Stg 0		8.0
				Sampler/Cascade Impactor			6.28	
						Stg 1		0.95
							0.83	

(Continued)

TABLE 9 Widely used, commercially available samplers for collecting bioaerosols—Cont'd

Sample[a]	Principle of Operation	Sampling Rate (l min^{-1})	Manufacturer/ Supplier[b]	Commercial Name	Application[c]	Calculated cutpoint		
							d_{50}	Other
7. Six-stage, 400-hole impactor	Impaction onto agar in 10-cm plates	28	AND	Anderson Six-Stage Viable Sample/ Cascade Impactor	C	Stg 1	6.24	7.0
						Stg 2	4.21	4.7
						Stg 3	2.86	3.3
						Stg 4	1.84	2.1
						Stg 5	0.94	1.1
						Stg 6	0.58	0.65
					C	Stg 1	6.61	
						Stg 6	0.57	
8. Eight-stage, personal impactor (vp)	Impaction onto 3.4-cm substrates	2	AND	Personal Cascade Impactor,	O	Stg 1		21.3
				Series 290 Marple Personal Cascade Impactor		Stg 2		14.8
						Stg 3		9.8
						Stg 4		6.0
						Stg 5		3.5
						Stg 6		1.55
						Stg 7		0.93
						Stg 8		0.52
Filters								
9. Cassette filters (vp)	Filtration, generally 25, 37, or 45 mm filters in disposable cassettes or reusable filter holders	1–5	COR, MIL, PGS, SKC	Aerosol Monitor, Air Monitor, Air Monitor Cassette, Particulate Monitor	H, M, O			
Centrifugal Samplers								
10. Centrifugal agar impactors (sc)	Impaction onto agar in plastic strips		BDC	Reuter Centrifugal Samplers (RCS):				
		40		Standard RCS	C		7.5	4
		50		RCS Plus	C		6	0.82
11. Wetted cyclone samplers (vp, sc)	Tangential impingement into thin liquid layer	50–55	HG	AEA Technology PLC Aerojet	C, M, O			
		167		Cyclones	C, M, O			0.8
		500			C, M, O			1.5
		167	PAR/SBI	MicroBio MB3 Portable Cyclone	C, M, O			
		700–1000	LRI	Aerojet-General Liquid-Scrubber	C, M, O			
12. Dry cyclone sampler (sc)	Reverse flow cyclone	≤20	BMC	Cyclone Sampler	H, M, O			1.2

13. Three-jet, tangential impactor (sc)	Tangential impaction onto glass or filter surface or impingement into liquid	12.5	SKC	BioSampler	C, M, O			
Liquid Impingers								
14. All-glass impingers (vp)	Impingement into liquid	12.5	AGI, HG/MIL	All-Glass Impingers (AGI): AGI-4,	C, M, O		0.30	
				AGI-30			0.31	
15. Three-stage impingers (vp)	Impingement into liquid	20	BMC	Multiple-Stage Liquid Impinger	C, M, O	Stg 1		≥10
						Stg 2		4–10
						Stg 3		4
		10, 55	HG	Three-Stage Impinger	C, M, O	Stg 1		≥7
						Stg 2		≥3
						Stg 3		≥1
		20			C, M, O	Stg 1		10
						Stg 2		4
						Stg 3		≥4
(see also Sampler 12)								
Pollen, Spore, and Particle	Impactors							
16. High-volume liquid impinger	Impingement into liquid	500	DYC	XMX-CV, XMX-2L-MIL Aerosol Collection Systems	CMO			
17. 1–7-day tape/slide impactors (sc)	Impaction onto rotating drums with tape strip or glass slide	10	BMC	1–7-Day Recording Volumetric Spore Trap, 2 × 14 mm slit	M		5.2	
		10	LAN	Pollen and Particle Sampler	M			
18. Moving slide impactors (sc)	Impaction onto moving glass slides	15	ALL	Allergenco Air Sampler (MK-3)	M		2.0	
		10	BMC	Continuous Recording Air Sampler	M			
		10	LAN	Volumetric Pollen/Particle Sampler	M			
19. Stationary slide impactor (sc)	Impaction onto stationary glass slides	10	BMC/SBI	Personal Volumetric Air Sampler	M		2.52	
20. Cassette slide impactor (vp)	Impaction onto stationary glass slides	15	ZAA/SKC	Air-O-Cell Sampling Cassette	M		2.3	
21. Rotating rod impactors (sc)	Impaction onto rotating rods	48	STI	Rotorod	H, M			

[a]Letters in parentheses: vp, requires a vacuum pump and flow control device, which the sampler manufacturer/supplier may provide; sc, self-contained with built-in air mover.
[b]See Table 10.
[c]C, culture of sensitive and hardy microorganisms (e.g., vegetative bacterial and fungal cells and spores); H, culture of hardy microorganisms only (e.g., spore-forming bacteria and fungi); M, microscopic examination of collected particles; O, other assays (e.g., immunoassays, bioassays, chemical assays, or molecular detection methods).

TABLE 10 Manufacturers and suppliers of bioaerosol samplers

AGI Ace Glass Incorporated P.O. Box 688 1430 Northwest Boulevard Vineland, NJ 08360–0688 609-692-3333 800-223-4524 fax: 800-543-6752 web: www.aceglass.com.	COR Corning Incorporated 45 Nagog Park Acton, MA 01720–3413 978-635-2200 800-492-1110 fax: 978-635-2476 web: www.scienceproducts.corning.com.	PBI International PBI Via Novara, 89 20153 Millan Italy 39 2-40-090-010 fax: 39 2-40-353695 web: www.wheatonsci.compbiexp@interbusiness.it.
ALL Allergen LLC dba Allergenco/Blewstone Press P.O. Box 8571 Wainwright Station San Antonio, TX 78206–0571 210-822-4116 fax: 210-822-4116 *51, 210-805-8518 web: www.txdirect.net/corp/allergenallergen@txdirect.net.	DYC Dycor Technologies Ltd1851 94 Street, Edmonton, AB T6N 1E6Canada780-486-0091800-663-9267fax: 780-486-3535web: www.dycor.com.	PGS Pall Gelman Sciences 600 S. Wagner Road Ann Arbor, MI 48103–9019 734-665-0651 800-521-1520 fax: 734-913-6114 web: www.pall.com/gelman/.
AND Andersen Instruments 500 Technology Court Smyma, GA 30082–5211 770-319-9999 800-241-6898 fax: 770-319-0336 web: www.thermo.com.	HG Hampshire Glassware 77–79 Dukes Road, Hampshire Southampton SO14 0ST England 44(0)1703-553755fax: 44(0)1703-553020	SBI Spiral Biotech, Incorporated 7830 Old Georgetwon Road Bethesda, MD 20814–2432 301-657-1620 fax: 301-652-5036 infosbi@spiralbiotech.com web: www.spiralbiotech.com.
BC Barramundi Corporation P.O. Drawer 4259 Homosassa Springs, FL 34447–4259 352-628-0200 fax: 352-628-0203 barra@citrus.infi.net web: www.mattson-garvin.com.	LAN Lanzoni, S.R.L. Via Michelino 93/B 40126 Bologna Italy39(0)51-504810, (0)51-501334 fax: 39(0)51-6331892 web: www.lanzoni.it.	SKC Incorporated 863 Valley View Road Eighty Four, PA 15330–1301 724-941-9701 800-752-8472 fax: 724-941-1369 800-752-8476 web: www.skicinc.com.
BDC Biotest Diagnostics Corporation 66 Ford Road, Suite 131 Denville, NJ 07834–1300 973-625-1300 800-522-0090 fax: 973-625-9454 web: www.biotest.com.	LRI Life's Resources, Incorporated 114 E. Main Street P.O. Box 260 Addison, MI 49220–0260 517-547-7494 fax: 517-547-5444 web: lifes-resources .com	STI Sampling Technologies Incorporated 10801 Wayzata Boulevard Suite 340 Minnetonka, MN 55305–1533 612-544-1588 800-264-1338 fax: 612-544-1977 or 800-880-8040 web: www.rotorod.comemail:rotorod@rotorod.com.
BMC Burkard Manufacturing Company, Limited Woodcock Hill Industrial Estate Rickmansworth, Hertfordshire WD3 1PJ England 44(0)1923-773134 fax: 44(0)1923-774790 web: www.burkard.co.uk.	MIL Millipore Corporation 80 Ashby Road Bedford, MA 01730–2271 617-533-2125 fax: 781-533-8891 web: www.millipore.com.	VAI Veltek Associates, Incorporated Environmental Control Monitoring Division 1039 West Bridge Street Phoenixville, PA 19460–4218 610-983-4949 fax: 610-983-9494 web: www.sterile.com/~vai.
BSI Bioscience International 11607 Magruder Lane Rockville, MD 20852–4365 301-230-0072 fax: 301-230-1418 web: www.biosci-intl.com.	NBS New Brunswick Scientific Company, Incorporated P.O. Box 4005 44 Yalmadge Road Edison, NJ 08818–4005 800-631-5417 732-287-1200 fax: 732-287-4222 bioinfo@nbsc.com web: www.nbsc.com.	ZAA Zefon Analytical Associates 2860 23rd Avenue North St. Petersburg, FL 33713–4211 813-327-5449 800-282-0073 fax: 813-323-6965 web: www.zefon.com.
CAS Casella Limited Regent House, Wolseley Road Kempston, Bedford MK42-7JYEngland 44(0)1234-841441 44(0)1234-841468 fax: 44(0)1234-841490 web: www.casella.co.uk.	PAR F.W. Parrett Limited 65 Riefield Road London SE9 2RA England 44(0)1181-8504226 [UK 0181–8593254] fax: 44(0)1181-8504226	

3. Recently, Georgakopoulos et al. (2009) have reviewed existing techniques for detection, quantification, physical and chemical analysis along with aerosol sampling techniques. They discuss some emerging spectroscopic techniques and they delineate some questions and capabilities necessary to understand the chemical and physical characteristics of bioaerosols.
4. Discussions in the literature have appeared for future directions of research into the ambient bacteriological effects on the meteorology of their atmospheric transport medium, for example, atmospheric chemistry, cloud formation, precipitation, and radiative forcing (see Deguillaume et al., 2008; Morris et al., 2008).

Another recent study in northwestern Colorado by Bowers et al. (2009), used SSU rRNA gene analysis for a culture-independent approach to measure the airborne microbial communities at a high-elevation site. The bacterial taxa were dominated by the Betaproteobacteria and the Gammaproteobacteria including *Burkholderiales* and *Moraxellaceae* (*Psychrobacter* spp.). Interestingly, although very few sequences from known ice-nucleating (IN) bacterial species were recovered, the abundance of atmospheric IN particles was found to correlate with increasing relative humidity.

Fungi

Since the publication of the 3rd edition of the Encyclopedia of Microbiology, metagenomic analysis has been used to study the diversity of airborne fungi from air samples in several locations. These studies have found that fungi from the Divisions Ascomycota and Basidomycota were the dominant sequences identified. Some studies identified specific genera and species through molecular methods while others stopped at division or class level. Those identifying specific taxa found that *Cladosporium*, *Alternaria*, and *Penicillium* were frequently identified sequences, paralleling information from culture-based and spore trap studies. In one study, the majority of the sequences identified were from the Basidiomycota, particularly the order Agaricales. These studies using DNA-based methods also detected significant numbers of unknown fungal sequences suggesting that the atmosphere is still an unexplored reservoir of biological diversity. In a similar manner, there have also been recent improvements in molecular diagnostics, particularly quantitative PCR for identifying human and plant pathogens from air samples.

In the high elevation Colorado site mentioned above, members of the classes *Dothidiomycetes* and *Eurotiomycetes* dominated the fungal component of the air samples, which were also the most common members of the sediment in the nearby receding glacier.

The possible influence of climate change on airborne fungi has been the focus of recent investigations. Climate change may have both direct and indirect effects on fungi that result in greater concentrations of airborne spores. Increasing temperatures can directly cause more growth and reproduction of soil and leaf surface saprobes, the two major sources of airborne spores. Indirect effects relate to the enriching effects of elevated CO_2 concentrations on plant growth. Studies have found increased fungal biomass and increased fungal spores in soil and leaf liter of plants grown with elevated CO_2. Increasing temperatures have resulted in longer growing sea-

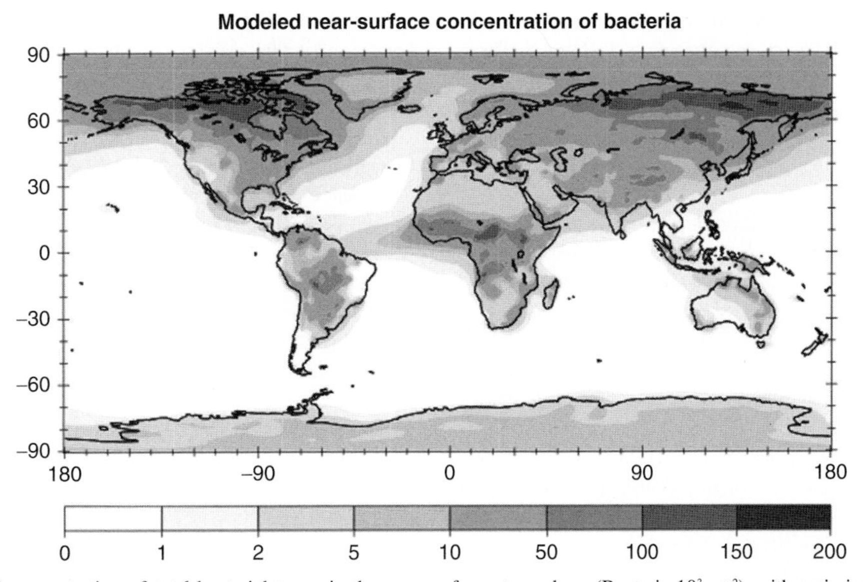

FIGURE 12 Modeled concentration of total bacterial tracer in the near-surface atmosphere (Bacteria 10^3 m^{-2}), with emissions given by the median of the ensemble of positive constrained emissions estimated *(adapted from Burrows, S. M., Butler, T., Jockel, P., Tust, H., Kerkweg, A. et al. (2009c). Bacteria in the global atmosphere – Part-2 Modeling of emissions and transport between different ecosystems.* Atmospheric Chemistry and Physics, 9, *9281–9297.).*

sons for many plant species. This also produces increased plant biomass and, therefore, more fungal saprobes. In addition, a longer growing season may result in more disease cycles and consequently more airborne spores for polycyclic plant pathogens. Finally, newly emerging diseases of plants, animals, and humans may also be a consequence of climate change. Recently, an outbreak of the human and animal pathogen *Cryptococcus gattii* was reported from the Pacific Northwest of both Canada and the United States. This fungus is originally from tropical and subtropical biomes where it is associated with *Eucalyptus* trees. Although the spread of this disease is not directly tied to climate change, it illustrates the potential spread of a novel fungal pathogen.

ACKNOWLEDGMENTS

B.L. would like to thank Susannah Burrows comments concerning his synopsis of her research in this addendum.

FURTHER READING

Amato, P., Demeer, F., Melaouhi, A., Fontanella, S., Martin-Biesse, A. S., Sancelme, M. *et al.* (2007). A fate for organic acids, formaldehyde and methanol in cloud water: their biotransformation by micro-organisms. *Atmospheric Chemistry and Physics, 7*, 4159–4169.

Babich, H., & Stotsky, G. (1974). *Air pollution and microbial ecology. Critical reviews in environmental control Vol. 4*(3) (pp. 353–420). Cleveland OH: CRC Press, Inc.

Bowers, R. M., Kayberm, C. L., Wiedinmyer, C., Hamady, M., Hallar, A. G., Fall, R. *et al.* (2009). Characterization of airborne microbial communities at a high-elevation site and their potential to act as atmospheric ice nuclei. *Applied and Environmental Microbiology, 75*, 5121–5130.

Brown, J. K. M., & Hovmøller, M. S. (2002). Aerial dispersal of pathogens on the global and continental scales and its impact on plant disease. *Science, 297*, 537–541.

Burrows, S. M. Modeling global transport of bacterial aerosol in EMAC. Diplomarbit, Institut fur Physik der Atmosphare Fachbereich Physik, Mathematik und Informatik, Johan-Gutenberg-universitat, Mainz; (2008).

Burrows, S. M., Elbert, W., Lawrence, M. G., & Podvhl, U. (2009a). Bacteria in the global atmosphere – Part-1. Review and synthesis of literature data for different ecosystems. *Atmospheric Chemistry and Physics, 9*, 9263–9280.

Burrows, S. M., Elbert, W., Lawrence, M. G., & Poschl, U. (2009b). Online Supplement to *"Bacteria in the global atmosphere – Part-1 Modeling of emissions and transport between different ecosystems"*. Correspondence to: S.M. Burrows (Susannah.burrows@mpic.de).

Burrows, S. M., Butler, T., Jockel, P., Tust, H., Kerkweg, A. *et al.* (2009c). Bacteria in the global atmosphere – Part-2 Modeling of emissions and transport between different ecosystems. *Atmospheric Chemistry and Physics, 9*, 9281–9297.

Byrnes, E. J. I. I. I., Li, W., Lewit, Y., Ma, H., & Voelz, K. (2010). Emergence and pathogenicity of highly virulent Cryptococcus gattii genotypes in txhe Northwest United States. *PLoS Pathogens, 6*(4), e1000850. doi: 10.1371/journal.ppat.1000850.

Cox, C. S. (1987). *The aerobiological pathway of microorganisms*. New York: John Wiley and sons. p. 293.

Cox, C. S., & Wathes, C. M. (Eds.), (1995). *Bioaerosol handbook* (p. 323). Boca Raton: CRC Press, Inc.

Deguillaume, L., Leriche, M., Amato, P. A., Delori, A. M., Poschl, U., & Chaumerliac, N. *et al.* (2008). Microbiology and atmospheric processes: Chemical interaction of primary biological aerosols. *Biogeosciences Discuss 5*. 841–870. www.biogeosciences-discuss.net/5/841/2008/.

Dimmick, R. L., & Akers, A. B. (Eds.) (1969). *An introduction to experimental aerobiology* (p. 494). New York: Wiley-Interscience.

Edmonds, R. L. (Ed.) (1979). *Aerobiology: The ecological systems approach* (p. 386). Stroudsburg, PA: Dowden, Hutchinson and Ross, Inc..

Fierer, N., Liu, Z., Rodríguez-Hernández, M., Knight, R., Henn, M., & Hernandez, M. T. (2008). Short-term temporal variability in airborne bacterial and fungal populations. *Applied and environmental microbiology, 74*, 200–207.

Fröhlich-Nowoisky, J., Pickersgilla, D. A., Després, V. R., & Pöschla, U. (2009). High diversity of fungi in air particulate matter. *PNAS, 106*, 12814–12819.

Georgakopoulos, D. G., Despres, V., Frohlich-Nowoisky, J., Psenner, R., Ariya, P. A., Posfai, M. *et al.* (2009). Microbiology and atmospheric processes: biological, physical and chemical characterization of aerosol particles. *Biogeoscience*, 6721–6737.

Gregory, P. H. (1973). *The microbiology of the atmosphere* (2nd ed.) (p. 377). New York: John Wiley and Sons.

Griffin, D. W., Kellogg, C. A., Garrison, V. H., & Shinn, E. A. (2002). An intercontinental river of dust, microorganisms and toxic chemicals flows through the Earth's atmosphere. *American Scientist, 90*(3), 228.

Grinshpun, S. A., Buttner, M. P., & Willeke, K. (2007). Sampling for airborne microorganisms. In C. J. Hurst (Ed.), *Manual of environmental microbiology* (3rd ed.) (pp. 939–951). Washington, DC: ASM Press.

Hers, J. F. Ph. (Ed.) (1973). *Airborne transmission and airborne infection* (p. 610). Utrecht, The Netherlands: Oosthoek Publishing Co.

Hurst, C. J., Crawford, R. L., Garland, J. L., Lipson, D. A., Mills, A. L., & Stetzenbach, L. D. (Eds.) (2007). *Manual of environmental microbiology* (3rd ed.). Washington, DC: ASM Press.

Lacey, M., & West, J. (2006). *The air spora*. Dordrecht, The Netherlands: Springer.

Levetin, E. (2007). Aerobiology of agricultural pathogens. In C. J. Hurst (Ed.) *Manual of environmental microbiology* (3rd ed.) (pp. 1031–1047). Washington, DC: ASM Press.

Lighthart, B. (2000). Mini-review of the concentration variations found in the alfresco atmospheric bacterial populations. *Aerobiology, 16*, 7–16.

Lighthart, B., & Mohr, A. J. (1994). *Atmospheric microbial aerosols* (p. 397). New York: Chapman and Hall.

Lipson, D. A., Wilson, R. F., & Oechel, W. C. (2005). Effects of elevated atmospheric CO_2 on soil microbial biomass, activity, and diversity in a chaparral ecosystem. *Applied and Environmental Microbiology, 71*, 8573–8580.

Macher, J. (1999). *Bioaerosols assessment and control* (pp. 11–14–11-6). Cincinnati, Ohio: ACGIH.

Morris, C. E., Sands, D. C., Bardin, M., Jaenicke, R., Vogel, B., Leyronas, C. *et al.* (2008). Microbiology and atmospheric processes: an upcoming era of research on bio-meteorology. *Biogeosciences Discuss, 5*, 191–212. 2008. www.biogeosciences-discuss.net/5/191/2008/

Shelton, B. G., Kirkland, K. H., Flanders, W. D., & Morris, G. W. (2002). Profiles of airborne fungi in buildings and outdoor environments in the United States. *Applied and Environmental Microbiology, 68*, 1743–1753.

Spendlove, J. C. (1974). Industrial, agricultural, and municipal microbial aerosol problems. *Developments in Industrial Microbiology, 15*, 20–27.

Ziska, L. H., Epstein, P. R., & Schlesinger, W. H. (2008). Rising CO_2, climate change, and public health: exploring the links to plant biology. *Environmental Health Perspectives, 117*, 155–158.

Chapter 10

Endophytic Microbes

M. Tadych and J.F. White

Rutgers University, New Brunswick, NJ, USA

Chapter Outline

ABBREVIATIONS

VOC Volatile organic compound

DEFINING STATEMENT

Endophytic associations between microbial and autotrophic organisms in nature are ubiquitous. Because of the complexity of the endophytic associations with their hosts we are only starting to understand these interactions. In this article, we compare the biology and ecology of different groups of endophytes associated with hosts. The functional role of endophytic microbes is also discussed. We also evaluate the potential of endophytic microorganisms as sources of bioactive metabolites with potential use in agriculture and medicine.

INTRODUCTION

Microorganisms are found virtually in every biotic and abiotic niche on earth. This includes extremophiles living in deserts, rocks, thermal springs, freshwater, marine, and arctic environments, and associated with terrestrial and aquatic animals. A wide diversity of microorganisms may be isolated from most terrestrial and marine plants. Microorganisms are present on the surface of and within tissues of most parts of plants, especially the leaves. When microbial organisms colonize a plant and the plant tissue is healthy, the relationship between the microorganism and its host plant may range from latent pathogenesis to mutalistic symbiosis. Therefore, the microbial organisms may be latent pathogens, epiphytes (epibionts), or endophytes (endobionts).

Endophytes are generally any organisms that under normal circumstances are contained within tissues of living plants (usually autotrophs) without causing noticeable symptoms of disease, and the host tissues remain intact and functional. However, the same organisms may also be described as saprobic or pathogenic at another time; for example, when the host is stressed, some endophytes may become pathogenic. The delicate balance between host and endophytic organisms seems to be controlled in part by chemical factors, for example, herbicidal natural products produced by the fungus versus antifungal metabolites biosynthesized by the host plant.

The term 'endophyte' is derived from the Greek *endon* (within) and *phyte* (plant), and was introduced by Heinrich Anton de Bary in 1866 and applied to any organism found within a plant. Traditionally, the term had been broadly applied to fungi in plant tissue, including the mycorrhizal fungi in plant roots. However, some authors do not include mycorrhizal fungi in this group. The term endophyte has also been adapted to other microorganisms, such as endophytic bacteria. The associations of endophytic organisms with their host plants are varied and complex and we are only starting to understand these interactions. Endophytic microbial organisms often contribute to the normal health and development of their hosts in exchange for a relatively privileged niche. Microbial endophytes have been isolated from tissues of algae, mosses and hepatics, ferns and fern allies, grasses, other herbaceous plants, and trees growing in tropical, temperate, boreal, and arctic environments.

There are vast regions of the world where fungus–plant associations are unknown, and endophytes represent a large reservoir of undiscovered genetic diversity. Species composition of endophyte assemblages and infection frequencies vary according to (1) host species, (2) growth stage of the host, (3) tissue type, (4) the age of host organs and/or tissues, (5) position in the canopy and associated vegetation, and (6) site characteristics, such as elevation, exposure, and latitude, and (7) anthropogenic factors. Usually, one to a few species dominates the endophyte community, while the majority of the species are rare. Distribution of rare and incidental species is influenced more by site than by host, and the number of rare and incidental species isolated is proportional to the intensity of sampling.

Many plant fungal endophytes are transmitted horizontally by production of conidia or other spores on plants that may spread to adjacent uninfected plants, and are not directly inherited through germplasm from the previous generation of the host. Therefore, with each host generation, environmental fungi must compete to colonize the empty niches within new host individuals. Alternatively, some endophytes, including *Epichloë* species and all of their asexual relatives (*Neotyphodium* spp.), grow intercellularly throughout aboveground parts of plants and are transmitted vertically via seeds. These endophytic fungi are spread by hyphae growing into the developing seeds or vegetative parts of infected maternal host plants.

Epibiosis is another type of association of two organisms, the epibiont and the basibiont. The majority of organisms that exist on the surfaces of plants are referred to as epibionts or epiphytes. In general usage the terms epiphyte and epibiont are used interchangeably. The epibiotic organism may be sustained entirely by nutrients and water received nonparasitically from within the host (basibiont) on which it resides. However, the interaction of epibiont with basibiont is usually unknown. In some epibionts water and nutrients are taken up entirely from suspended soils and other aerial sources such as dead host tissues, airborne dust, mist, and rain. Any negative effect on the host, if it occurs, is indirect.

The assemblage of epiphyllic flora on the surface of healthy plant leaves is usually composed of bacteria, yeasts, and filamentous fungi that belong to different systematic categories. The density and the number of epiphyllic microorganisms can rapidly change and it strongly depends on environmental conditions, host species, habitat of the host, and the age and the surface structure of the organ on which the epiphyllic organism resides.

While some fungi are adapted to the plant surface, the community also includes propagules of airborne species, fungi that would not otherwise be considered epibionts. The surface of the plant, especially the leaf is an extreme environment. Though plants may leak nutrients through the cuticle, the surface is dry, waxy, and affected by UV radiation. Epibionts have many obvious characteristics that enable continuation through the stressful conditions of the leaf surface. For example, some epibionts can digest lipidic substrates, and thus may utilize the waxy layer covering the leaf. Epibionts are likely to be melanized, and thus able to resist UV radiation. Epibiotic yeasts are able to multiply during the short periods of appropriate environmental conditions. Although epibiotic organisms differ markedly from endophytes, it has been shown that some endophytic fungi are also able to produce epiphytic stages with fungal reproductive structures on the surface of the host plant.

FUNGAL ENDOPHYTES

Many groups of fungi exist as plant endophytes. The fungal endophytes have commonly been divided into two major groups. The first group comprises a smaller number of specialized fungal species (clavicipitaceous fungi) that colonize some monocotyledonous hosts, and the second group that includes a large number of fungal species (nonclavicipitaceous fungi) with a broad range of host plants. Every plant species examined to date maintains endophytic associations with fungi and this association is a ubiquitous and cryptic phenomenon in nature. Although most of the endophytic fungi belong to the phylum Ascomycota and its anamorphs, Glomeromycota and some Basidiomycota and Zygomycota are also known. The host range, ecology, and geographical distribution of most groups of endophytic fungi are still poorly known. A recent study by Betsy Arnold (University of Arizona) and François Lutzoni (Duke University) indicated that endophyte infection of host plants increases along latitudinal gradient from the Arctic to the Tropics, with less than 1% to more than 99% of tissue segments ($2\,mm^2$) with endophytes, respectively. This study also showed that foliar endophyte assemblages, at both an individual host species and the community level, increases in diversity with decreasing latitude.

Diversity and composition of endophytic fungal communities vary considerably and their detection depends on biotic, abiotic and experimental factors. Endophyte community of almost all plant species examined to date have been assessed mostly by culture-based approaches and have been identified using morphological characteristics with support of molecular analysis for identification of some endophytes that remain sterile in culture. Abundance and diversity of unculturable endophytes is still mostly unknown, limiting our understanding of endophyte infection frequencies, taxonomic composition, and diversity. Many researchers suggest that culture-based methods alone underestimate diversity and misrepresent the taxonomic composition of endophyte communities. Therefore, recently cultivation-independent approaches, such as DNA cloning, denaturing gradient gel electrophoresis (DGGE), or terminal restriction fragment length polymorphism (TRFLP or sometimes T-RFLP), have gained popularity as a means to assess the diversity and composition of uncovering endophytes with obligate host association, that is, fungal species that do not grow on standard media or fungi that grow slowly and are lost during the culturing process in competitive interactions. Regardless some limitations and needs for further improvement, these approaches are gaining popularity along with the culture-based methods as means to assess the entire diversity and composition of endophytic fungal communities present within host plant.

Fungal endophytes of grasses (also known as clavicipitaceous endophytes or e-endophytes) are widely studied; with most work concentrating on fungi of the family Clavicipitaceae (Hypocreales; Ascomycota). The following genera of this family have been identified as grass endophytes or epiphytes: *Atkinsonella*, *Balansia*, *Balansiopsis*, *Echinodothis*, *Epichloë*, *Myriogenospora*, and *Parepichloë*. Clavicipitaceous endophytes, also known as e-endophytes, are widespread in grasses; approximately over 10% of all grass species are estimated to harbor clavicipitaceous fungal endophytes. The genus *Epichloë*, with anamorphs in *Neotyphodium*, is the most studied group of grass endophytes. These endophytes are found growing systemically in the aboveground tissues of some temperate grass species of the cool-season grasses. Based on the relative costs and benefits to the hosts, these grass–fungal endophyte symbioses have been showed to range from pathogenic to mutualistic. Endophytic species from the genus *Epichloë* are often considered to be pathogenic and may cause partial or complete sterilization of hosts due to the production of a fungal stroma on the flowering culms (choke disease) of the host. However, the degree of effect on host reproduction varies with species. In mutualistic associations, that is, in some *Epichloë* spp. and *Neotyphodium* spp., the endophyte grows systematically within its host (Figure 1), including the developing seeds (Figure 2), and is entirely dependent on the survival and growth of the grass host plant for its own growth.

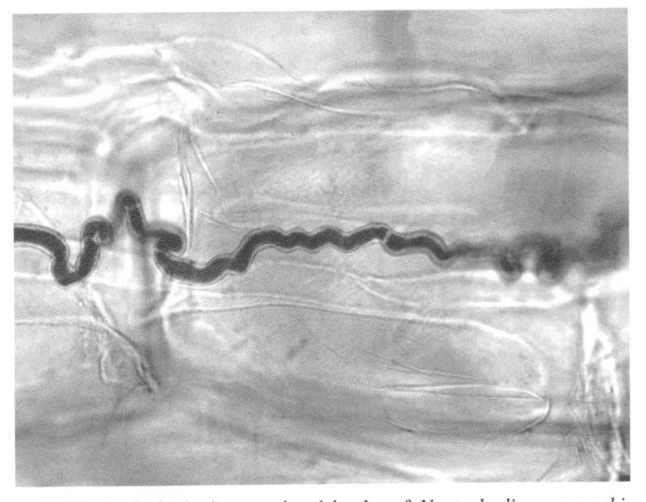

FIGURE 1 Endophytic convoluted hypha of *Neotyphodium coenophialum* in a culm-scraping preparation from *Festuca arundinacea* (× 1600).

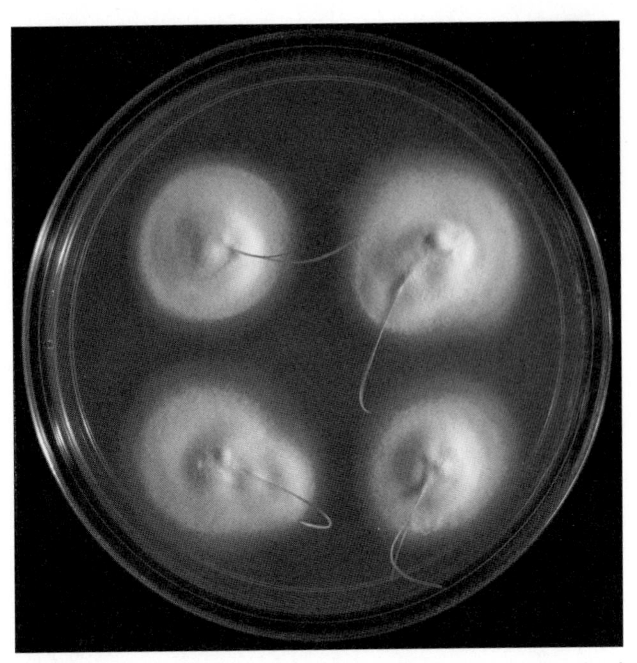

FIGURE 2 Four-week-old culture of *Neotyphodium* sp. from seeds of *Poa ampla* grown on potato dextrose agar at room temperature (× 0.7).

It is believed that fungal endophytes of grasses have two modes of reproduction. The sexual species of genera *Epichloë* and *Balansia* can be horizontally transmitted through development of ascospores. Because *Epichloë* species are obligately outcrossing ascomycetes, development of the sexual spores is dependent upon transfer of spermatia of one mating type to an unfertilized stroma of the opposite mating type occurring on different individuals of the host plants. Transfer of spermatia of *Epichloë typhina* (Pers.) Tul. & Tul. is accomplished by flies of genus *Botanophila* (Anthomyiidae; Diptera), which visit stromata for feeding and oviposition. Immediately after cross-fertilization of the fungus, perithecia begin to develop on the stroma. During flowering of the host plant, the ascospores produced within the perithecia of infected individuals in the population are forcibly ejected. The ascospores, possibly dispersed by air currents, may land on another healthy grass plant and may initiate infection. In contrast, many *Epichloë* species and all of their asexual mutualistic *Neotyphodium* relatives are typically transmitted vertically from maternal plants to their offspring. For most of their life cycle the endophytes inhabit, asymptomatically and systemically, the apoplasts of the aboveground organs of infected host plants, including the embryos of viable seeds, and can be disseminated vertically to successive generations of the host plant. Infected seeds and vegetative tillers of infected host plants are the only known modes of propagation of these endophytes. Additionally, several *Epichloë* spp., such as *Epichloë festucae* Leuchtm., Schardl & Siegel, are represented by species that have a remarkable mixed-transmission strategy, that is, horizontal and vertical transmission modes. In these cases, some tillers produce stromata while other tillers on the same plant are asymptomatic and produce normal, vigorous, endophyte-infected seeds.

During the last decade some *Neotyphodium* endophytes have been shown to emerge from plants and produce a network of mycelium, conidiophores, and conidia on the surfaces of plants (Figure 3). This epiphyllous mycelium is most evident on leaf blades of certain species of the Pooideae, including the bent grasses, fescues, forest hedgehog grass, ryegrasses, wild barley, and some blue grasses. Currently, nothing is known regarding the ecology or implications of production of the epiphyllous conidial stages in the life cycles of endophytes. The epiphyllous conidia are viable and spread via water currents, rain splash, or drip splash to adjacent plants. It is likely that epiphyllous conidia are responsible for some of the horizontal transport and parasexual recombination 'hybridization' that may occur in the *Epichloë*/*Neotyphodium* endophytes.

In addition to the clavicipitaceous endophytes a number of other seed-transmitted grass–fungal endophytes have been described. Certain cool-season grasses harbor nonclavicipitaceous fungal endophytes that are less frequent than clavicipitaceous. These fungi belong to two different groups, that is, p-endophytes and a-endophytes. The p-endophytes include *Gliocladium*- and *Phialophora*-like endophytes ('p' for penicillate disposition of conidiophores, common to *Gliocladium*- and *Phialophora*-like fungal species) that can be isolated from culms and seeds of numerous festucoid grasses, for example, from perennial ryegrass (*Lolium perenne* L.), tall fescue (*Schedonorus arundinaceus* (Schreb.) Dumort.), meadow fescue (*Schedonorus pratensis* (Huds.) P. Beauv.), Arizona fescue (*Festuca arizonica* Vasey), and giant fescue (*Schedonorus giganteus* (L.) Holub). These endophytes are represented by species of the order Eurotiales (Ascomycota). The a-endophytes are a group of grass endophytes found in Italian ryegrass (*Lolium multiforum* Lam.), other annual species of the *Lolium* genus, and *Festuca paniculata* (L.) Schinz & Thell. The a-endophytes belong to parasitic species of *Acremonium* similar to *Acremonium chilense* Morgan-Jones, White & Piontelli (Hypocreales; Ascomycota), an endophyte of orchard grass (*Dactylis glomerata* L.). At present, the ecological and physiological importance of p- and a-endophytes of grasses are not determined or understood.

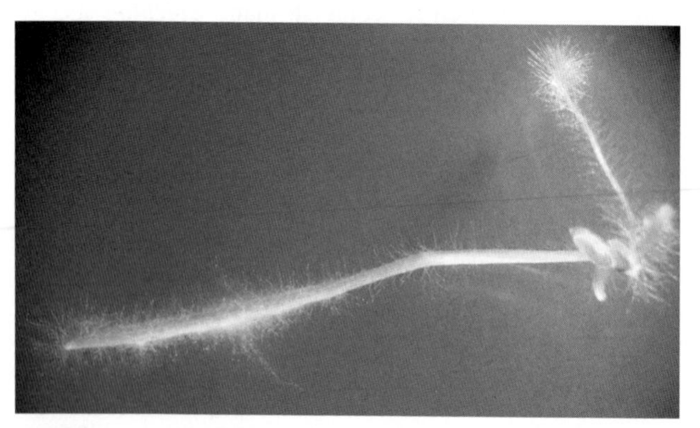

FIGURE 3 Epiphyllous growth of *Neotyphodium* sp. on a 2-week-old *Poa ampla* seedling (×3).

The warm-season grass, *Trichachne insularis* (L.) Nees, was found to harbor a nonclavicipitaceous, seed-transmitted endophyte identified as *Pseudocercosporella trichachnicola* White & Morgan-Jones, an anamorphic species of genus *Mycosphaerella* (Mycosphaerellaceae; Ascomycota). Another seed-transmitted fungus *Fusarium verticillioides* (Sacc.) Nirenberg (Nectriaceae; Ascomycota) has been isolated from maize (*Zea mays* L.), sorghum (*Sorghum* spp.), and other plants in the Poaceae family as a symptomless endophyte. The horizontally transmitted entomopathogenic fungus *Beauveria bassiana* (Bals.-Criv.) Vuill. (Clavicipitaceae; Ascomycota) is also known to form an endophytic association with maize.

Nonsystemic grass endophytes present another group of nonclavicipitaceous fungal endophytes. Because most studies of grass endophytes have focused on systemic endophytes, at present little is known about nonsystemic endophytic fungi that inhabit grass species. However, some studies indicate that in addition to systemic infections, endophytes that form localized infections in grasses may be also diverse and widespread. Generally, dominant nonsystemic grass endophytes are represented by the genera *Alternaria*, *Cladosporium*, *Epicoccum*, *Fusarium*, *Phoma*, and pathogens typical of grass hosts. Because of their mode of dissemination, diversity and dispersion of this group of grass endophytes is probably more variable than that of systemic grass endophytes and depends on availability and viability of spores of the fungus.

Nonclavicipitaceous endophytes associated with healthy organs of nongrass plants are still poorly known. However, fungal surveys conducted during the last 30 years have reported endophytic fungi from over 100 plant families and demonstrated that tissues of the vast majority of plants are colonized by endophytic microfungi. In general, fungal endophytes within the host may inhabit many different tissues of roots, stems, branches, twigs, bark, leaves, petioles, flowers, fruits, and seeds, including xylem of all available plant organs. Endophytic nonclavicipitaceous microfungi in tissues of plant hosts are usually highly diverse and occur as numerous localized infections that increase in number, density, and species diversity with organ age. It may be because the fungal endophytes associated with nongrass plants generally appear to be transmitted horizontally (i.e., from plant to plant in populations). Most fungal surveys of angiosperms show that nonclavicipitaceous fungal endophytes represent polymorphic assemblages mostly from the Ascomycota or Basisiomycota. Most belong to the Ascomycota with Pezizomycotina (Dothideomycetes,

Eurotiomycetes, Leotiomycetes, Pezizomycetes, and Sordariomycetes) that are especially well represented, although some Saccharomycotina are also known. Within the Basidiomycota members of the Agaricomycotina, Pucciniomycotina, and Ustilagionmycotina are known endophytes, although they are reported less frequently than ascomycetous endophytes. Particularly, fungi from the genera *Acremonium*, *Alternaria*, *Chaetomium*, *Cladosporium*, *Cryptocline*, *Cryptosporiopsis*, *Curvularia*, *Fusarium*, *Glomerella*, *Leptostroma*, *Phoma*, *Phomopsis*, *Phyllosticta*, *Physalospora*, and *Trichoderma* are well represented in endophyte assemblages. The profile of fungal endophyte assemblages in specific organs can be completely different from those in other plant organs or tissues. Generally, the main tissues that have been analyzed for the presence of endophytes in woody perennials are the leaves, twigs, branches, and roots; a few studies have looked for endophytes in meristems, flowers, or fruits of woody perennials. However, the profile of fungal endophyte assemblages in flowers and fruits could be radically different than that in other types of plant structures because these organs are young and rapid in development.

Other mostly unknown group of fungal endophytes is represented by marine fungi. Marine fungi are not taxonomically well defined. In general, the marine environment includes obligate marine fungal species, which are considered to be fungi that grow and sporulate exclusively in a marine habitat. Marine fungi that do not germinate in the natural marine habitat are not included in these groups. It also includes facultative marine fungi also called marine-derived fungi, that is, those that can be isolated from both terrestrial and marine environments, and are adapted to and isolated from various marine habitats, like near-shore or estuarine environments. However, not well-documented fungal endophytes harbor by many marine organisms may also protect their host from herbivory, pathogenic and fouling organisms. The marine environment may offer a variety of unexplored epi- and endosymbiotic microorganisms. Due to the complicated nature of marine symbiotic associations and experimental limitations, the ecology and chemistry of these interactions has not been well studied. Only in a few cases a specific microbe–host relationship, for example, symbiosis or permanent association, has been proven. Most of the literature reports describe the sporadic occurrence of variable microorganisms for certain hosts. Mostly bacteria but also fungi are to be found in various marine organisms, including sponges, algae, mangrove plants, and marine animals as epi- and endobionts. Even though metabolite-producing marine-derived fungi are being isolated from sponges, there is still no evidence for the presence of fungal mycelia growing in sponges. One of the better-documented relationships between marine-derived fungi and other marine organisms is the fungal–algal symbiotic association 'mycophycobiosis.' It was estimated that one-third of all known marine fungi are associated with marine algae and these fungi reside inside the algal tissues. Studies of marine-derived fungi indicate the enormous diversity of the fungal community in the world's oceans and their biochemical uniqueness.

SOME COMMON ENDOPHYTE–GRASS ASSOCIATIONS

Tall Fescue–*Neotyphodium coenophialum* Association

Tall fescue (*Schedonorus arundinaceus* (Schreb.) Dumort. = *Festuca arundinacea* Schreb. = *Lolium arundinaceum* (Schreb.) Darbysh.) is one of the best-known examples of a grass with an endophyte that causes toxicity. The grass was brought to the United States from Europe in the late 1800s. It was officially discovered in Kentucky in 1931, tested at the University of Kentucky, and released in 1943 as 'Kentucky 31.' From the mid-1940s it became popular with farmers, spreading quickly throughout the midwestern and southern United States. Today it accounts for well over 16 million hectares of pasture and forage land in the United States. The problem of livestock neurotoxicosis ('fescue toxicosis' also known as 'fescue foot' or 'fescue lameness') became a major concern in the United States. Several studies show that consumption of endophyte-infected tall fescue decreases the feed intake of cattle and therefore lowers animal weight gains. Affected cattle also produce less milk, have higher internal body temperatures and respiration rates, develop a rough hair coat and demonstrate an unthrifty appearance, salivate excessively, have poor reproductive performance, and maintain reduced serum prolactin levels. By the end of the 1970s, the association of this toxicity with the endophyte had been discovered. The endophyte was originally identified as a strain of *E. typhina*, but was later described as *N. coenophialum* (Morgan-Jones & Gams) Glenn, Bacon & Hanlin. *N. coenophialum* produces several alkaloids, particularly, the alkaloid ergovaline that is structurally related to ergotamine, a major factor in ergot poisoning due to ingestion of *Claviceps purpurea* 'ergots' contaminating rye flour.

Perennial Ryegrass–*Neotyphodium lolii* Association

Perennial ryegrass (*L. perenne* L.) is a valuable forage and soil stabilization plant. In New Zealand the neurotoxic disease 'ryegrass staggers' (also known as 'perennial ryegrass staggers') of sheep and cattle has long been reported and attributed to the consumption of perennial ryegrass. In 1898 for the first time the presence of a fungal endophyte in the seeds of *L. perenne* was observed, and 40 years later the endophytic fungus from *L. perenne* was isolated and grown in agar culture. The association between the *L. perenne* endophyte and 'ryegrass staggers' was finally established in 1981. *N. lolii* (Latch, Christensen & Samuels) Glenn, Bacon & Hanlin that infects perennial and hybrid ryegrasses was shown to synthesize several alkaloids. Three of these are known to be particularly important to pasture management, specifically (1) lolitrem B, a tremorgenic molecule responsible for livestock 'staggers,' (2) ergovaline, an ergopeptine that has vasoconstrictive effects and causes heat stress in grazing animals, also responsible for 'fescue toxicosis,' and (3) peramine, a tripeptide that deters some insects, particularly the Argentine stem weevil, from feeding on ryegrass but is not toxic to mammals. Perennial ryegrass staggers is a very serious problem for grazing livestock (sheep, cattle, horses, deer) in New Zealand; however, mortality rates are generally low. Ryegrass staggers occur sporadically in North and South America, Europe, and Australia.

Darnel Ryegrass–*Neotyphodium occultans* Association

Darnel ryegrass = darnel (*Lolium temulentum* L.) is another common endophyte-infected grass species. This species has a long recorded history as a plant poisonous to humans and animals. Darnel was known as a weed and as a poisonous plant in earlier times. The earliest written references to darnel indicating it to be a noxious and toxic weed that caused problems for humans and animals may be found in the Gospel of Matthew of the New Testament and in authors such as Plautus, Virgil, Ovid, Dioscorides, and Shakespeare. Darnel seed from 4000-year-old archeological materials in ancient Egypt contained endophyte mycelium. It was generally understood that human beings would be poisoned by ingestion of flour or baked products containing seeds of darnel as a contaminant. Contamination was also undesirable because of the strong taste of darnel seeds, which, for example, resulted in inferior bread. Darnel seeds were sometimes added to beer as a flavoring. By the end of the 1800s it was discovered that a fungus infects seeds of darnel. Currently we know that darnel seeds contain an endophytic fungus, and this symbiotic fungus is known as *N. occultans* Moon, Scott & Christensen.

The Sleepygrass–*Neotyphodium chisosum* Association

Sleepygrass (*Achnatherum robustum* (Vasey) Barkworth =*Stipa robusta* (Vasey) Scribn.) is a perennial grass forming stout, erected clumps in dry plains, hills, and open woods. The grass is native to North America, and is abundant in southwestern United States. It was found to be toxic and narcotic to grazing animals, that is, to horses, cattle, and sheep. Animals, after consumption of relatively small quantities of the grass, go to sleep for 2–3 days, and then gradually recover. Ergot alkaloids, that is, ergonovine, ergonovinine, lysergic, and isolysergic acid amides, have been identified as the sleep-inducing agents in sleepygrass. These alkaloids are produced by *N. chisosum* (White & Morgan-Jones) Glenn, Bacon & Hanlin, the endophyte isolated from this plant and related species of the genus. However, recent studies suggest that the level of ergot alkaloids in native sleepygrass populations may be highly variable within and among populations despite the level of endophyte infection. Animal toxicity is localized in particular areas where particular strains of the endophyte may predominate.

The 'Drunken Horse Grass'–*Neotyphodium gansuense* Association

Drunken horse grass (*Achnatherum inebrians* (Hance) Keng = *Stipa inebrians* Hance) is a perennial bunchgrass, distributed on alpine and subalpine grasslands of northwestern China and Mongolia. Horses that have grazed *A. inebrians* develop a stagger; the animals walk as if drunk, with some being unable to stand after falling. The symptoms persist for 6–24 h, and usually they appear to be completely recovered after 3 days. Severely affected animals die within 24 h of consumption of the grass. Toxicity of *A. inebrians* was reported by Marco Polo and the Russian explorer Nikolai Mikhaylovich Przhevalsky. In the last decade the presence of *N. gansuense* Li & Nan endophyte in this grass species was confirmed. In the study of alkaloid content it was found that infected *A. inebrians* contains two major toxic alkaloids: ergonovine and lysergic acid amide.

The 'Dronkgras'–*Neotyphodium melicicola* Association

Staggers grass = dronkgras (*Melica decumbens* (L.) Weber) is a tufted perennial, very coarse grass with usually rolled and very rough leaves. *M. decumbens* is endemic to Africa and has a limited distribution in South Africa, being found only in the arid areas of central South Africa. It grows amongst rocks, under trees, and shrubs on hill- and mountainsides, occasionally in areas along roadsides. The name 'staggers grass' comes from the fact that this grass has narcotic effects on cattle, horses, donkeys, and sheep. The tremorgenic neurotoxins produced by the endophyte *N. melicicola* Moon & Schardl were found to be responsible for the narcotic effect of *M. decumbens* on grazing livestock. Usually, consumption of grass is not lethal and the animal recovers. This is probably because the leaves of the grass are very coarse, and animals do not graze it frequently, except when the grass is very young.

Forest Hedgehog Grass–*Neotyphodium* Association

Forest hedgehog grass (*Echinopogon ovatus* (G. Forst.) P. Beauv.) is a tufted perennial, mesophytic grass often found in moist forested areas (in wet sclerophyllic woodlands and by creeks). The grass is endemic to Australia, New Guinea, New Zealand, and Tasmania. Young plants of *E. ovatus* cause 'staggers,' sometimes known as 'wobbles' in stock. The grass was found to harbor endophytes *Neotyphodium aotearoae* Moon & Schardl and *Neotyphodium australiense* Moon & Schardl. The presence of indole-diterpenoid and loline alkaloids in endophyte-infected *E. ovatus* was confirmed.

'Hueću' Grasses–*Neotyphodium tembladerae* Association

Poa huecu Parodi and several other grasses of South America (e.g., *Bromus auleticus* Trin. ex Nees, *Festuca argentina* (Speg.) Parodi, *Festuca hieronymi* Hack., *Poa magellanica* Phil. ex Speg., *Melica stuckertii* Hack., and some other *Poa* spp.) are colonized by the endophytic fungus *N. tembladerae* Cabral & White. The association of the endophyte with *F. hieronymi* and *P. huecu* is probably responsible for toxicosis syndromes, called 'tembladera' or 'hueću' in grazing animals. 'Tembladera' is from the Spanish word that means 'tremble,' and word 'hueću' is from the indigenous Araucanian language of the tribes that lived in the region and means 'intoxicator.' Hueću toxicosis results from consumption of *P. huecu* and is frequently lethal to animals. Studies on *N. tembladerae* suggest that toxicity of the infected grasses to mammals is associated with the ergot alkaloids, lolitrems, and some glycoproteins, produced by the fungus.

ECOLOGICAL IMPACTS OF ENDOPHYTES

Abiotic and biotic stresses due to noninfectious and pathogenic plant diseases, pests, and unfavorable growing conditions are major causes for plant productivity losses. Productivity of cultivated plants relies heavily on high chemical inputs. Recently, natural and biological control of diseases and pests affecting cultivated plants has gained much attention as a way of reducing the use of chemical products in agriculture. Biological control offers an alternative or supplement to chemical pesticides in plant protection. This approach incorporates one or more organisms to maintain another pathogenic organism below a level at which it is no longer an economical problem. Use of endophytic organisms could enhance plant growth and productivity on a worldwide basis.

Some groups of endophytic microorganisms have been described as 'defensive mutualists' that protect host plants against biotic and abiotic stresses. In return, the endophytic symbionts acquire nutrients from their host plants. Endophytic arbuscular mycorrhizal fungi, which live in mutualistic symbiosis with at least 80% of plants, effectively protect host plant from many root diseases, reduce infection by nematodes, increase stress tolerance like drought resistance, tolerance to heavy metals and salinity, and increase phosphorous uptake in phosphorous-deficient soils. Endophytes of upper parts of grasses and other plants also benefit their hosts. The benefits frequently reported include systemic resistance against pathogens, reduced herbivory, increased drought resistance, improved tolerance to heavy metals, and generally enhanced growth and an increase in the plant's fitness to environmental extremes. The fungal endophytes may also influence the plant pathogen assemblages, and reduce their diversity and abundance. For example, in *Schedonorus arundinaceus*, endophyte infection reduced seedling blight caused by *Waitea circinata* Warcup & Talbot and crown rust caused by *Puccinia coronata* Corda relative to endophyte-free plants, but colonization by *N. coenophialum* did not affect rust caused by *Puccinia graminis* Pers. Also, Alternaria leaf spot caused by *Alternaria triticina* Prasada & Prabhu was significantly more common on endophyte-free *Panicum agrostoides* Spreng. than on plants with endophyte. Endophyte-infected grasses may also experience a lower incidence of disease when insects, vectors of plant pathogens, are deterred by endophytes. Although the ecological roles played by endophytic fungi are more diverse and varied, these benefits arise in part from the production of fungal metabolites (usually alkaloids) by the endophyte or endophyte–plant complex. However, in some cases the mechanisms of defense are not fully understood and require further study.

In contrast, in horizontally transmitted endophytes associated with phytosynthetic tissues such as leaves of nongrass plants, the benefits are less clear. Generally, these endophytes are believed to also function as defensive mutualists of host plants; however, in most cases their ecological roles have not been assessed experimentally. Recently, several studies have provided evidence for important roles of class 3 endophytes in enhancing plant defenses against biotic stresses, such as protection from herbivores, pathogenic fungi, nematodes, and neighboring plants' competition, and abiotic stresses. Some studies have shown that the endophyte *Trichoderma ovalisporum* Samuels & Schroers that colonized stems and fruits of *Theobroma gileri* Cuatrec. has the ability to parasitize and antagonize the necrotrophic mycelium of pathogenic *Moniliophthora perniciosa* (Stahel) Aime & Phillips-Mora, and *Moniliophthora roreri* (Cif.) Evans, Stalpers, Samson & Benny. Another study showed that seedlings of *Theobroma cacao* L. inoculated with seven endophytic fungal species and exposed to *Phytophthora* spp., species that are pathogens of the plant, were more resistant than endophyte-free plants. Leaves of the control endophyte-free plants died in greater numbers and suffered much more pathogen damage than endophyte-infected plants. Also, the relative benefit of endophyte inoculation was higher in older than in younger leaves.

Habitat-adopted symbioses of plants with endophytic fungi from class 2 contribute to and may also be responsible for the adaptation of host plants to environmental stresses. For example, the heat-tolerant perennial panicgrass (*Dichanthelium lanuginosum* (Elliott) Gould) is symbiotic with the fungus *Curvularia protuberata* R.R. Nelson & Hodges. The endophytic fungus was isolated from the grass collected from geothermally heated soils of Yellowstone National Park (Wyoming, Montana, and Idaho, USA) and Lassen Volcanic National Park (California, USA). Geothermal soils of Yellowstone and Lassen Volcanic National Parks may reach temperatures as high as 57 °C, and on an annual basis the grass is exposed to high temperatures as well as periods of drought conditions. Laboratory and field studies conducted by Rusty Rodriguez (US Geological Survey/University of Washington), Regina Redman (US Geological Survey/University of Washington), and Joan Henson (Montana State University) have shown that the endophytic fungus confers thermotolerance to the host plant and that this fungal–plant association is responsible for survival of both species in geothermal soils. When these organisms were grown asymbiotically under controlled conditions, the maximum growth temperatures of *D. lanuginosum* and *C. protuberata* were 40 and 38 °C, respectively; while with the endophyte plants grew at considerably higher temperatures. A similar effect was observed with *Fusarium culmorum* (W.G. Sm.) Sacc., which colonizes a costal dunegrass (*Leymus mollis* (Trin.) Pilg.). The plant did not survive and the development of the fungus was retarded when both partners were grown separately and they were exposed to levels of salinity that they experienced in their native habitat. Another example is the fungus *Piriformospora indica* Verma, Varma, Rexer, Kost & Franken originally isolated from a spore of mycorrhizal fungus *Glomus mosseae* (T.H. Nicolson & Gerd.) Gerd. & Trappe. Some studies showed that *P. indica* is able to endophytically colonize roots of various crop plants. It has been shown that isolates of *P. indica* in roots of barley plants enhanced development of the host plant and increased grain yield. It was speculated that this endophytic fungus elevated antioxidative status of the infested roots and therefore protected roots from root pathogens like *F. culmorum* and *Cochliobolus sativus* (Ito & Kurib.) Drechsler ex Dastur. Also, the systemic plant response, as a result of induced resistance by this endophyte, causes a reduction of powdery mildew (*Blumeria graminis* (DC.) Speer) infection of barley leaves. The fungus also protects the barley plants from salt stress.

Production of Nontoxic Endophytes

It is well documented that endophytic fungi have been implicated in toxicity of some poisonous plant species. Class 1 endophytes of grasses are a particularly important group of fungi that have been found to naturally produce a range of mycotoxins, alkaloids, and physiologically active chemical compounds. Some of these substances cause problems for livestock and are recognized as, for example, the causative agents of economically important livestock toxicoses, such as 'fescue toxicosis' and 'ryegrass staggers.' It has been found that some endophytes of ryegrasses and some other related grass species have reduced toxicity to grazing livestock while at the same time they enhance tolerance to pests and/or abiotic stresses. It has also been discovered that some toxic and beneficial alkaloids have separate biosynthetic pathways, allowing the selection or development of endophyte strains that show low animal toxicity but still possess anti-insect qualities. Axenic cultures of selected grass–fungal endophytes may be produced and inoculated into seedlings of grasses. The discovery and commercialization of low-toxicity *Neotyphodium* endophytes along with the technology for their reinoculation resulted in the development of elite grass cultivars that are persistent, productive, and enhance animal performance, for example, Greenstone tetraploid hybrid ryegrass with Endosafe endophyte, and tall fescue cultivars with MaxQ endophyte.

ENDOPHYTES AS SOURCES OF BIOACTIVE METABOLITES

Microorganisms demonstrate many unique characteristics. Among these characteristics is the production of a vast range of biologically active substances and secondary metabolites. Microorganisms are one of the richest resources of biologically active substances and secondary metabolites with novel structures and potential activities. Secondary metabolites, also known as idiolites, have been defined as low molecular weight and naturally produced substances that often possess chemical structures quite different from primary metabolites, such as amino acids, organic acids, and sugars from which they are produced. Their functions in the producing organisms are not obvious. They are usually hypothesized to function as chemical defenses for the hosts, but may be potential viable weapons against other organisms. In addition, secondary metabolites may be agents of symbiosis and agents of metal transport. Finally, these metabolites may act as sex hormones, plant growth stimulants, and as effectors of differentiation. For many reasons these compounds may have tremendous economic importance to humankind and an extraordinary impact on the quality of human life. Some fungal secondary metabolites are beneficial (antibiotics such as cephalosporin and penicillin) while others are harmful (carcinogens such as aflatoxin and ochratoxin). The secondary metabolites produced by microbes are generally large and complex chemicals that are not readily synthesized using recombinatorial synthetic approaches. However, a number of drugs derived from fungal metabolites have been developed as their modified analogs. Antibiotics, antifungal, anticancer, immunosuppressive agents, and hypocholesterolemic agents that are derived from fungal compounds have been used for over 50 years.

During the last two decades the fungi living internally in living tissues of plants have been targeted as valuable sources of new bioactive compounds and they have become a mainstay of natural product screening programs. Growth of endophytic fungi within hosts without causing apparent disease symptoms and metabolic interaction of endophytes with hosts may favor the synthesis of biologically active secondary metabolites. According to a study performed by Barbara Schulz (Technical University of Braunschweig) and her colleagues, the proportion of novel structures produced by endophytic isolates (51%) was considerably higher than that produced by soil isolates (38%). Among the fungal genera, endophytic fungi from *Acremonium*, *Chaetomium*, *Colletotrichum*, *Fusarium*, *Pestalotiopsis*, *Phoma*, and *Phomopsis* are known to produce bioactive compounds of medical importance. In addition, endophytic fungi growing in axenic culture can also produce biologically active compounds, including several alkaloids, antibiotics, and plant growth-promoting substances. However, the amount and kind of compounds that are produced by a fungus will be affected by factors like temperature, degree of aeration, and the composition of the medium used for culturing. Some endophytic fungi have been identified as sources of anticancer, antidiabetic, and immunosuppressive compounds. For example, anticancer molecule paclitaxel (Taxol, Figure 4), originally discovered in the bark of the Pacific yew tree (*Taxus brevifolia* Nutt.), has also been found to be produced by the endophytic fungus *Taxomyces andreanae* Strobel, Stierle, Stierle & Hess. Another endophytic fungus *Muscodor albus* Worapong, Strobel & Hess has an ability to produce a mixture of volatile organic compounds (VOCs) that were fatal to a wide variety of human- and plant-pathogenic fungi and bacteria.

Endophytic fungi are also more often recognized as a group of organisms capable of providing a source of novel bioactive compounds and secondary metabolites for biological control in agriculture as well as for biotechnology and industrial applications. It was shown that production of herbicidally active substances by endophytic fungi is two and three times higher than phytopathogenic fungi and soil fungi, respectively.

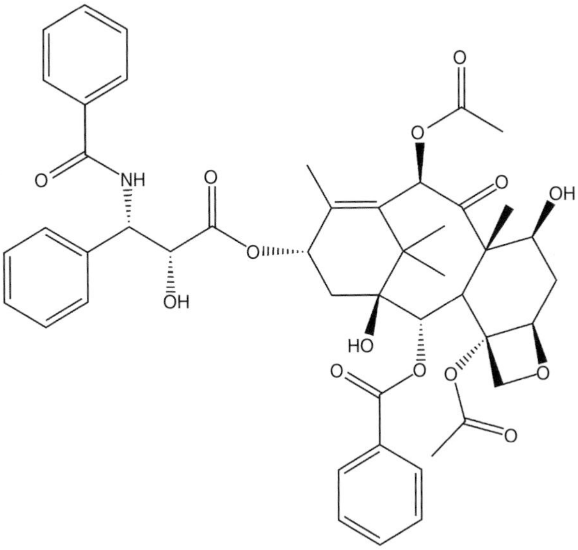

FIGURE 4 Paclitaxel (Taxol).

Endophytic fungi play an increasingly important role in the integrated pest management programs in agriculture. Many of the secondary metabolites produced by *Epichloë*/*Neotyphodium* grass endophytes significantly increase deterrence of vertebrate herbivores, insects, and nematode pests. It has been demonstrated that the secondary metabolites produced by the fine fescue endophyte *E. festucae* are inhibitory to other fungi. It has also been shown that endophyte-infected grasses contain a range of biologically active compounds that either are derived from endophyte or are produced as a result of the association. Four main classes of defensive compounds have been associated with endophyte presence in grasses and the following effects have been described.

Ergot Alkaloids

Ergot alkaloids cause toxicoses in grazing mammals with a range of symptoms from stupor and appetite suppression through reproductive problems and dry gangrene to death. The ergot alkaloids (Figure 5) are produced by a few representatives of filamentous fungi, members of the Trichocomaceae family including the *Aspergillus* and *Penicillium* genera, and Clavicipitaceae family, with the above-mentioned grass endophytes of the genera *Epichloë* with its *Neotyphodium* anamorphs and the genera *Claviceps* and *Balansia*.

Indole-Diterpene Metabolites

Indole-diterpene metabolites produce a number of biological effects, including anti-insect activity (feeding deterrence, modulation of insect receptors functions, toxicity) and mammalian tremorgenic activity (staggers).

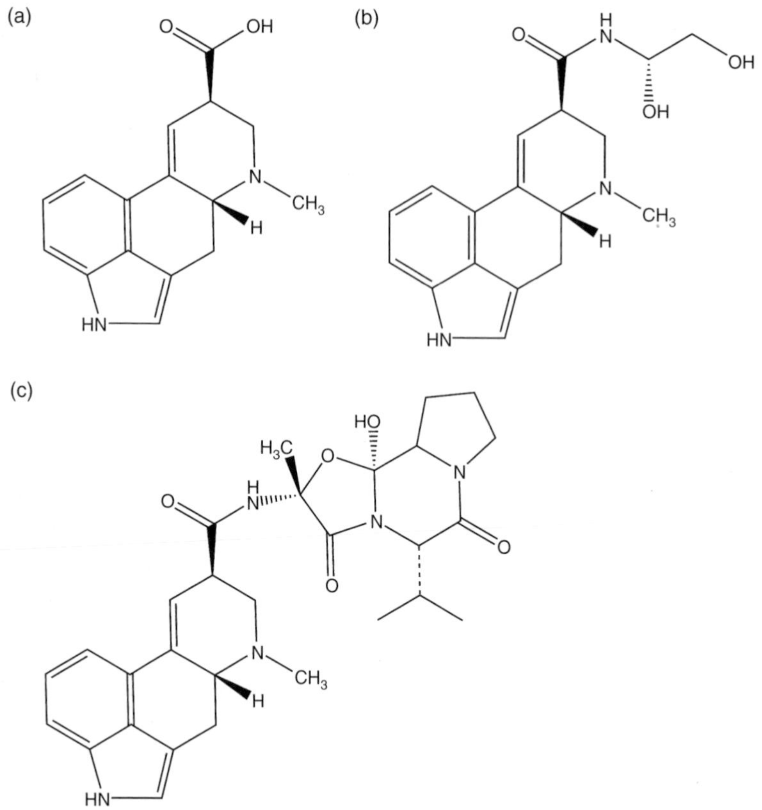

FIGURE 5 Ergot alkaloids: (a) lysergic acid, (b) ergonovine, and (c) ergovaline.

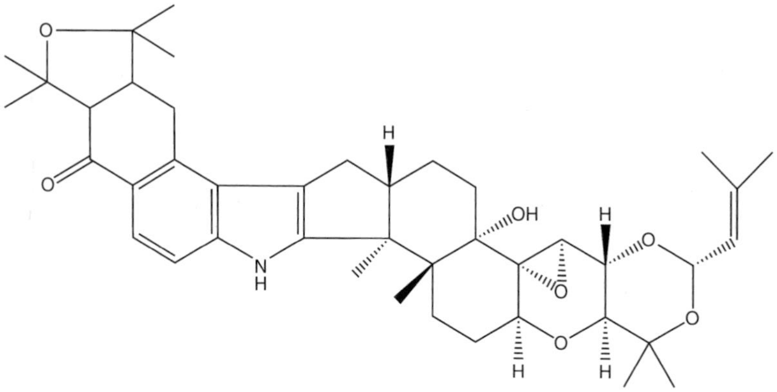

FIGURE 6 Indole-diterpene: lolitrem B.

Indole-diterpenoides (Figure 6) have been reported from *Epichloë* and *Neotyphodium* as well as from some fungi of the genera *Claviceps*, *Aspergillus*, and *Penicillium*.

Peramine

Peramine is an unusual pyrrolopyrazine, and is produced by most *Epichloë/Neotyphodium* endophyte species symbiotic with grasses (Figure 7). It is a metabolite that has specific biological activity as an insect feeding deterrent. Peramine is deterrent to invertebrate herbivores, especially the Argentine stem weevil, and is also active in preventing feeding activity of aphids, but is not toxic to mammalian herbivores.

Lolines

Lolines are classified as pyrrolizidines, a class that also includes plant alkaloids known for their insecticidal activity (Figure 8). These substances are insecticidal alkaloids with insect-deterrent activities, possessing little or no activity against large mammals. Lolines are neurotoxic to a broad range of insects, and when produced by endophytes in plants they have been shown to defend the plants from

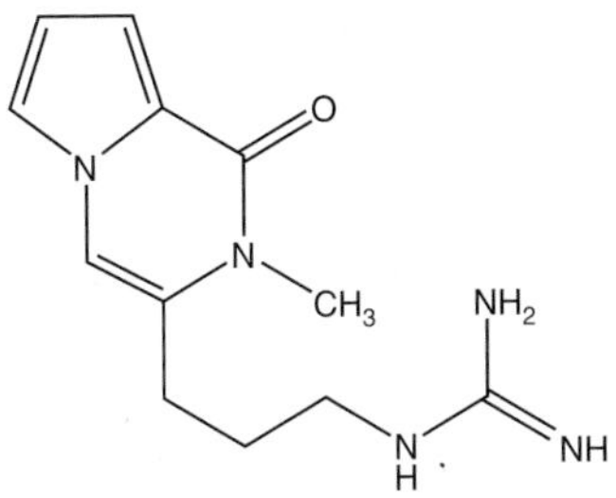

FIGURE 7 Peramine.

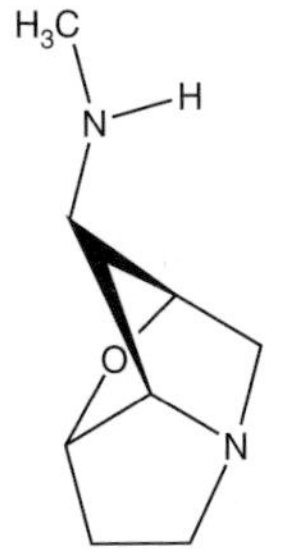

FIGURE 8 Loline.

aphids. Lolines are only known from endophyte-infected grasses, and plants of the genus *Adenocarpus* (Fabaceae) and *Argyreia mollis* (Burm.f.) Choisy (Convolvulaceae). It is possible that undiscovered fungal symbionts might be responsible for loline production in *Adenocarpus* and *Argyreia* species. In fact, clavicipitaceous epibiotic fungus has recently been discovered on leaf surfaces of *Ipomoea asarifolia* (Desr.) Roem. & Schult. and related plants (Convolvulaceae).

Many, but not all, *Epichloë/Neotyphodium* species produce up to three classes of these alkaloids. In addition to the above, several simple indole alkaloids have been isolated from cultures of endophytic fungi; these include the simple auxins, for example, 3-indoleacetic acid (IAA, Figure 9) and the indole glycerols, for example, 3-indolybutanetriol.

Reports of the secondary metabolites from true marine endophytes are relatively rare. This may be due to the relative difficulty in collecting and culturing of marine endosymbionts and their usually slow growth in laboratory conditions rather than lack of ability to produce secondary metabolites. In recent years, an increasing number of natural products from marine-derived fungal endophytes have been reported. Described below are some examples of secondary metabolites from marine-derived endophytic fungi. Isolated from the green alga *Enteromorpha* sp., the endophytic fungus *Wardomyces anomalus* Brooks & Hansford (Microascaceae; Ascomycota), collected around Fehmarn island in the Baltic Sea, showed (1) antimicrobial effects of the crude extract toward *Microbotryum violaceum* (Pers.) Deml & Oberw. and *Aspergillus repens* (Corda) Sacc. and (2) inhibition of HIV-1 reverse transcriptase (HIV-1-RT). Investigation of the extract yielded several xanthone derivatives. Xanthones are a unique class of biologically active compounds possessing numerous bioactive capabilities, such as antimicrobial, antitubercular, antitumor, antiviral, and antioxidant properties. Xanthone derivatives occur in a number of higher plant families and fungi. Some fungal species are well known as sources of xanthone derivatives, for example, *Penicillium raistrickii* G. Sm., *Phomopsis* sp., *Actinoplanes* sp., *Ascodesmis sphaerospora* Obrist, and *Humicola* sp. The cultivation of the marine fungus *Apiospora montagnei* Sacc. isolated from inner tissue of the North Sea alga *Polysiphonia violacea* Grev. led to the isolation of several new secondary metabolites, where some of them exhibited significant cytotoxicity against human cancer cell lines. From the green alga *Ulva* sp., the endophytic and obligate marine fungus *Stagonosporopsis salicorniae* (Magnus) Died. was isolated. This fungus was found to produce, among others, the unusual tetramic acid-containing metabolites ascosalipyrrolidinones A and B. Ascosalipyrrolidinone A has antiplasmodial activity toward (1) strains K1 and NF 54 of *Plasmodium falciparum* Welch (causing malaria in humans) and (2) general antimicrobial activity. Penostatins, new cytotoxic agents toward leukemic cell lines, were reported from a *Penicillium* sp., inhabiting the marine environment and originally isolated from the marine alga *Enteromorpha intestinalis* (L.) Nees. Antimicroalgal substances, halymecins, were isolated from *Fusarium* and *Acremonium* spp. isolated from a marine alga *Halymenia dilatata* Zanardini.

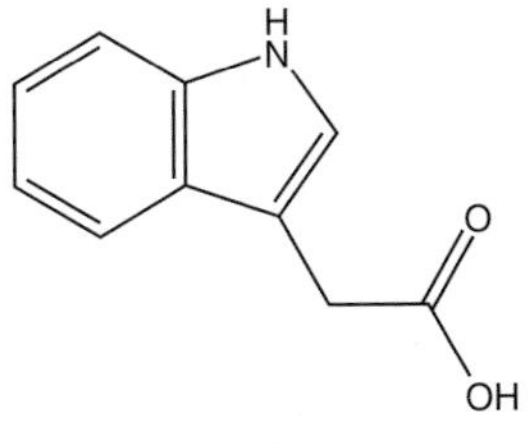

FIGURE 9 Auxin – 3-indoleacetic acid (IAA).

RECENT DEVELOPMENTS

Although in the last few years we noticed significant activity in the study of fungal endophytes associated with host plants, our understanding of their complex and diverse interactions with hosts as well as their ecological functions is still limited. As some researchers have pointed out, if we want to better understand host community dynamics and ecosystem function, we need to better understand the importance of endophytes in host biology. In general, our approaches to study microbe–host interactions are focused on one individual association but individual hosts usually comprise communities of many microorganisms, including viruses, bacteria, microalgae, and fungi that vary in their symbiotic associations with hosts. These fungal–plant associations may be even more complex.

In a last decade it has been documented that phylogenetically diverse endohyphal bacteria are associated with living hyphae of several different groups of fungi, including hyphae of some soilborne pathogenic or mycorrhizal fungi in Mucoromycotina, Glomeromycota, Basidiomycota, and Ascomycota, and have recently been also observed in hyphae of foliar fungal endophytes, members of four classes (Dothideomycetes, Eurotiomycetes, Pezizomycetes, and Sordariomycetes) of filamentous Ascomycota. These bacteria can alter morphology and physiology of the fungal host, and thus fungal interactions with host plants may depend on that association or may be modified in diverse ways. For example, the vertically transmitted bacterium *Candidatus* Glomeribacter gigasporarum colonizes spores and hyphae of the arbuscular mycorrhizal fungi *Gigaspora margarita* Becker & Hall, *Scutellospora castanea* Walker *and Scutellospora persica* (Koske & Walker) Walker & Sanders (Diversisporales, Glomeromycota). Although, *Candidatus* Glomeribacter gigasporarum is not essential to the survival or reproduction of the fungal host, removal of the bacterial partner from the fungal spores suppresses fungal growth and development, altering the morphology of the fungal cell wall, vacuoles, and lipid bodies. In another example, it was reported that a soilborne plant pathogen, *Rhizopus microsporus* Tiegh. (Mucoromycotina) harbors endosymbiotic bacteria *Burkholderia endofungorum* and *B. rhizoxinica*, producers of antimitotic polyketide metabolites, rhizonin and rhizoxin. These toxins are responsible for the pathogenicity of the fungus, are important in securing nutrients for the fungus and protecting the nutrient sources from competitors. In addition to toxin synthesis, the endosymbionts control the ability of the fungal host to reproduce asexually. Fungal mycelia cured of endosymbionts are unable to form asexual spores, which are critical for dispersal of the fungal host. Experimental assessments of such effects will provide key evidence to understanding the degree to which bacterial associates influence the nature of endophytic symbioses. Therefore, complex and multidisciplinary approaches need to be applied to study endophytic fungal–host interactions. Currently, researchers are trying to develop and apply approaches that will take into consideration the host organism as an individual system of various interactions, which is part of the habitat in a specific landscape, and multiple hosts within a single habitat or across different landscapes.

Further Classification of Fungal Endophytes

As previously described, fungal endophytes have frequently been divided into two major groups based on differences in taxonomy, host range, colonization and transmission patterns, tissue specificity, and ecological functions. The first group comprises a smaller number of specialized fungal species (clavicipitaceous endophytes also known as e-endophytes or C-endophytes) that typically colonize some monocotyledonous hosts, and the second group that includes a large number of fungal species (nonclavicipitaceous endophytes or NC-endophytes) with a broad range of host plants. However, complexity and varied phylogenetic origins of the nonclavicipitaceous endopytes, their ecological significance in nature, and continuously growing knowledge about the endophytic associations enabled Rusty Rodriguez (US Geological Survey/University of Washington) and colleagues recently to differentiate endophytic fungi into four separate functional classes. Widespread clavicipitaceous fungal endophytes of grasses of the family Clavicipitaceae (Hypocreales; Ascomycota) are referred as class 1 endophytes. As noted previously, these endophytes are growing systemically in the aboveground tissues and are horizontally and/or vertically transmitted. The second group, the nonclavicipitaceous endophytes, has been divided into three separate functional classes (Class 2, Class 3, and Class 4 endophytes). Class 2 endophytes are represented mostly by the Ascomycota with a minority of Basidiomycota. The class 2 endophytes usually form extensive infections within plants and are able to colonize above- and belowground tissues of a host, but they have low abundance in rhizosphere. They are able to transfer horizontally as well as vertically via seeds, seeds coats, and/or rhizomes. Class 2 endophytes typically have high infection frequencies in plant growing in high-stress habitats. Class 3 endophytes as functional group primarily or exclusively occur in aboveground tissues; they form highly localized limited infections, are horizontally transmitted, and are extremely diverse in host plants. The majority of class 3 endophytes are found in the Ascomycota, but some also belong to the Basidiomycota. Class 4 endophytes also known as dark septate endophytes (DSE) are distinguished from the other two classes of nonclavicipitaceous endophytes based on the presence of darkly melanized septa. Class 4 endophytes are primarily conidial or sterile ascomycetous fungi, and are transmitted horizontally. DSE are restricted to plant roots where they form melanized structures such as inter- and intracellular hyphae and microsclerotia. Class 4 endophytes are associated with mycorrhizal plants, but are also associated with nonmycorrhizal plants. It is assumed that these endophytes have little host or habitat specificity. DSE appear to be ubiquitous and abundant across diverse ecosystems worldwide, and are found to be associated with over 600 plant species. It is almost a hundred years after their first observation and still little is known about the role of DSE in their hosts.

New Endophytic Associations

Clearly, many plants contain, as yet, uncharacterized species and strains of endophytes, and there are vast regions of the world where fungal-plant associations are unknown and where many new fungi await discovery. For example, some endophytic fungi, known as 'endolichenic' fungi were reported to occur within asymptomatic lichens. During the last 3 years another seven new *Neotyphodium* taxa of grass endophytes had been found and described. That includes new species for some *Achnatherum–Neotyphodium* associations previously characterized in this chapter, that is, Sleepygrass and the 'Drunken Horse Grass' associations. It was found that the North American *Achnatherum robustum* species is associated with a new *Neotyphodium* species, *N. funkii* Craven & Schardl, instead of *N. chisosum* as originally described. For the Asian grass *Achnatherum inebrians*, another *Neotyphiodium* endophyte was described as *N. gansuense* var. *inebrians* Moon & Schardl, which is phylogeneticaly closely related to *N. gansuense* previously reported from this host species.

Novel Mechanism of Hyphal Growth

It is generally believed that fungal vegetative hyphae growth is exclusively apical by extension at the hyphal tip. However, Michael Christensen (AgResearch – Grasslands Research Center) and colleagues presented evidence suggesting that vegetative hyphae of *Epichloë* and *Neotyphodium* species infect elongating grass leaves via a unique and novel mechanism of growth, intercalary hyphal division and extension. According to their observations hyphae within the grass shoot apical meristem invade leaf primordia as they form on the shoot apical meristem. At the beginning, the fungus grows amongst dividing plant cells to form a heavily branched mycelium in leaf primordia. Next, the hyphae attach to adjacent plant cells of an emerging leaf, and when the plant cells divide and then enlarge to form a leaf, hyphal compartments are stretched, which causes intercalary extension of the filament accompanied by cellular division along the length of the filament. These findings suggest that intercalary extension of the hypha in the leaf expansion zone along with increased number of compartments formed during their growth, enable a fungus to grow and keep pace with extended leaf tissues. As a result, the growth of endophytic hyphae is correlated with the host's life cycle; it means that hyphal compartments in grass leaves are of similar age to neighboring cells of the host. Beyond the leaf expansion zone hyphal expansion ceases, suggesting that intercalary growth of the hypha of the endophyte may be activated by stretching. Although endophytic fungi in old, matured leaf tissues no longer are growing, nevertheless they remain metabolically active.

Consequently, these observations might suggest that both modes of growth of vegetative hyphae, that is, apical growth and intercalary hyphal extension, exist in *Epichloë* and *Neotyphodium* species. It may be possible that apical growth accounts for hyphal growth at the beginning of colonization when hyphae are intensively branching among dividing plant cells of leaf primordia in the meristematic zone. In contrast, the intercalary hyphal extension and division is responsible for hyphal expansion later during further development of leaf.

Defensive Mutualism and Mechanisms of Stress Tolerance in Hosts

As previously described in this chapter some grasses from geothermal, hypersaline, and other extreme habitats depend on endophytic fungi to tolerate the high-stress conditions. It has been shown that geothermal endophytes confer heat tolerance but not salt tolerance; coastal endophytes confer salt tolerance but not heat tolerance. Fungal endophytes from agricultural crops conferred disease resistance but not heat or salt tolerance. It was also experimentally shown that the agricultural, coastal and geothermal plant endophytes when inoculated into other plant species conferred the tolerance that was exhibited in the original host. This habitat-specific phenomenon was defined as habitat-adapted symbiosis and it was hypothesized that it is responsible for the establishment of plant communities in high-stress habitats. It is clear that these endophytes provide a mechanism for plant adaptation to habitat stresses but the mechanism responsible for stress tolerance remains unclear. It was recently proposed that the beneficial effects of endophytes on host plants stress tolerance may be the result of the production of reactive oxygen species (ROS) by endophytes. Other investigations have demonstrated that some endophytic fungi, clavicipitaceous endophytes of grasses, produce and secrete ROS to limit host colonization and maintain mutualisms, while other endophytic fungi (class 2 endophytes) reduce ROS production to possibly mitigate the impact of abiotic stresses. In response, the tissues of symbiotic plants produce antioxidants, which increase resistance of plants to oxidative stresses produced by plant pathogens, droughts, heavy metals, and other oxidative stressors. However, this hypothesis that antioxidants are responsible for enhanced stress tolerance in endophyte-infected plants requires further experimental evaluations.

Bioactive Compounds

Recent reviews further demonstrate the importance of endophytic and epiphytic fungi as potential sources of bioactive secondary metabolites as new lead structures for medicine as well as for plant protection. Especially, marine-derived fungi regardless of their origin have shown promising potential and developed into an important source of new and structurally unprecedented metabolites, and more than 330 new metabolites were described between 2002 and 2006. This indicates a growing interest in marine and marine-derived fungi as sources of new bioactive compounds and shows that further progress in sampling, isolation, identification as well as fermentation and extraction was obtained.

New Characterization Methods in the Study of Endophytes

Recently, study of the ecology and evolution of the epiphytic and endophytic interactions between fungi and plants or marine organisms has undergone rapid expansion from using traditional methods to molecular systematic and ecological genomics and further to develop new DNA- and RNA-based tools and associated bioinformatics approaches. In addition, some biochemical, for example, stable isotope profiling (SIP) and metabolic incorporation of nucleotide analogs such as bromodeoxyuridine (BrdU) technologies are now being applied. Regardless some limitations and needs for further improvement, these techniques would not only help to assess the entire diversity and composition of endophytic fungal communities but also would allow focusing at a co-evolutionary context of epiphyte and endophyte associations at different scales. These new tools may allow us to better understand processes that shape epiphytic and endophytic fungal-plant interactions at local as well as global scales.

CONCLUSION

Endophytic microbes are relatively common in all families of plants from polar to tropical regions. The endosymbionts have in several cases proven to play adaptive and/or defensive roles in the ecology of host plants. Relatively little is known about the relationships between fungi and their hosts especially with respect to chemical ecology. Nonetheless, some observations suggest the importance of the host as well as the ecosystem in influencing the general metabolism of endophytic microbes. Although endophytes are still a poorly investigated group of organisms, with the exception of the clavicipitaceous endophytes of grasses, many studies have proven that they are relatively rich and promising sources of bioactive and chemically novel compounds with a wide variety of potential uses in medicine, agriculture, and industry.

ACKNOWLEDGMENT

This work was partly supported by Fogarty International Center, NIH under U01 TW006674 for the International Cooperative Biodiversity Groups and Specialty Crop Research Initiative, USDA.

FURTHER READING

An, Z. (Ed.). (2005). *Handbook of industrial mycology*. New York: Marcel Dekker.

Arnold, A. E., & Lutzoni, F. (2007). Diversity and host range of foliar fungal endophytes: Are tropical leaves biodiversity hotspots? *Ecology*, *88*, 541–549.

Bacon, C. W., & White, J. F., Jr. (Eds.). (2000). *Microbial endophytes*. New York: Marcel Dekker.

Cheplick, G. P., & Faeth, S. H. (2009). *Ecology and evolution of the grass–endophyte symbiosis*. New York: Oxford University Press.

Clay, K., & Schardl, C. (2002). Evolutionary origins and ecological consequences of endophyte symbiosis with grasses. *The American Naturalist*, *160*, S99–S127.

Dighton, J., White, J. F., Jr., & Oudemans, P. (Eds.). (2005). *The fungal community: Its organization and role in the ecosystem*. New York: Taylor & Francis.

Gloer, J. B. (1997). Applications of fungal ecology in the search for new bioactive natural products. In D. T. Wicklow, & B. E. Soderstrom (Eds.), *Environmental and microbial relationships. The mycota* (Vol. IV, pp. 249–268). New York: Springer-Verlag.

König, G. M., Kehraus, S., Seibert, S. F., Abdel-Lateff, A., & Müller, D. (2006). Natural products from marine organisms and their associated microbes. *Chembiochemistry*, *7*, 229–238.

Kuldau, G., & Bacon, C. (2008). Clavicipitaceous endophytes: Their ability to enhance resistance of grasses to multiple stresses. *Biological Control*, *46*, 57–71.

Mueller, G. M., Bills, G. F., & Foster, M. S. (Eds.). (2004). *Biodiversity of fungi: Inventory and monitoring methods*. New York: Elsevier Academic Press.

Rodriguez, R. J., White, J. F., Jr., Arnold, A. E., & Redman, R. S. (2009). Fungal endophytes: Diversity and functional roles. *New Phytologist*, *182*, 314–330.

Schardl, C. L., Leuchtmann, A., & Spiering, M. J. (2004). Symbioses of grasses with seedborne fungal endophytes. *Annual Review of Plant Biology*, *55*, 315–340.

Schardl, C. L., Scott, B., Florea, S., & Zhang, D. (2009). Epichloë endophytes: Clavicipitaceous symbionts of grasses. In H. B. Deising (Ed.), *Plant Relationships. The Mycota* (2nd ed., Vol. V, Chapter 15, pp. 275–306). K. Esser (Ed.). *The Mycota-A Comprehensive Treatise on Fungi as Experimental Systems for Basic and Applied Research*. Berlin: Springer-Verlag.

Schulz, B., Boyle, C., Draeger, S., Römmert, A.-K., & Krohn, K. (2002). Endophytic fungi: A source of novel biologically active secondary metabolites. *Mycological Research*, *106*, 996–1004.

Smith, S. E., & Read, D. J. (2008). *Mycorrhizal symbiosis* (3rd ed.). Boston: Academic Press.

Strobel, G. A. (2003). Endophytes as sources of bioactive products. *Microbes and Infection*, *5*, 535–544.

Tadych, M., Bergen, M., Dugan, F. M., & White, J. F., Jr. (2007). The potential role of water in spread of conidia of the *Neotyphodium* endophyte of *Poa ampla*. *Mycological Research*, *111*, 466–472.

White, J. F., Jr., & Torres M. S. (Eds.). (2009). *Defensive Mutualism in Microbial Symbiosis*. Boca Raton: CRC Press – Taylor and Francis Group.

White, J. F., Jr., Bacon, C. W., Hywel-Jones, N. L. & Spatafora, J. W. (Eds.). (2003). *Clavicipitalean Fungi: Evolutionary Biology, Chemistry, Biocontrol, and Cultural Impacts*. New York: Marcel Dekker.

Wilson, D. (1995). Endophyte – The evolution of a term, and clarification of its use and definition. *Oikos*, *73*, 274–276.

Zhang, H. W., Song, Y. C., & Tan, R. X. (2006). Biology and chemistry of endophytes. *Natural Product Reports*, *23*, 753–771.

Part II

Metabolism and Behavior of Diverse Microbes

Chapter 11

Methanogenesis

K.R. Sowers
University of Maryland Biotechnology Institute, Baltimore, MD, USA

Chapter Outline

ABBREVIATIONS

%G + C Percentage guanosine + cytosine
5′-UTR 5′ Untranslated region
BRE B recognition element
CoA Coenzyme A
CRISPRs Clustered regularly interspaced short palindromic repeats
H_4MPT Tetramethanopterin
H_4SPT Tetrahydrosarcinapterin
H_4THF Tetrahydrofolate
MCM Minichromosome maintenance
MFR Methanofuran
MT1 Methanol:5-hydroxybenzinidazolyl
MT2 Methylcobamide:CoM methyltransferase
NAD Nicotinamide adenine dinucleotide
RNAP II RNA polymerase II
TCA Tricarboxylic acid
TFB Transcription factor B
TFE Transcription factor E
tRNA Transfer RNA

DEFINING STATEMENT

Biological methanogenesis is catalyzed exclusively by prokaryotic single-cell microorganisms, classified as methanogenic *Archaea*. Although these microorganisms require highly reduced, anaerobic conditions for growth, methanogenesis is ubiquitous in the environment and has a significant role in the global carbon cycle. Biochemical analyses combined with recent advances in genetic systems have revealed many of the unique metabolic pathways in the methanogens.

HISTORICAL OVERVIEW

The generation of combustible gas, presumably CH_4, has been reported by Pliny as early as during the Roman Empire. Legendary manifestations of methanogenesis include the will-o'-the-wisp, hypothesized to have resulted from the spontaneous combustion of marsh gas, and fire-breathing dragons, conjectured to have resulted from the accidental ignition of gas from CH_4-belching

Topics in Ecological and Environmental Microbiology.

ruminants. The close association between decaying plant material and the generation of "combustible air" was first described by the Italian physicist Alessandro Volta in 1776, when he reported that gas released after disturbing marsh and lake sediments produced a blue flame when ignited by a candle. Bechamp, a student of Pasteur, was the first to establish that methanogenesis was a microbial process, which was corroborated by others throughout the remainder of the nineteenth and early twentieth centuries. Because of the methanogens' requirement for strict anaerobic conditions, the first isolates were not reported until the 1940s. The approach used for isolation, the shake culture, involved adding microorganisms to molten agar growth medium containing a chemical reductant, such as pyrogallol carbonate, to prevent O_2 from diffusing into the agar. However, this approach was not suitable for isolating and maintaining methanogens in pure culture for long periods of time, as the medium was not sufficiently anaerobic. It was not until 1950 that a simple, effective technique was developed that provided the rigorous conditions required for routine isolation and culturing of methanogenic *Archaea* (Figure 1). The technique, referred to as the "Hungate technique," employs gassing cannula, O_2-free gases, and cysteine-sulfide reducing buffers to prepare a highly reduced, O_2-free medium. Boiling initially deoxygenates medium, and a cannula connected to an anaerobic gas line, such as N_2 or CO_2, is inserted into the vessel to displace air as the medium cools. The medium is dispensed into culture tubes or serum vials while purging with anaerobic gas and then the medium is sealed with a rubber stopper or septum. The vessel containing reduced medium and anaerobic gas effectively becomes an anaerobic chamber for culture growth. The development of the anaerobic glove box has further simplified culturing of methanogens by providing a means for colony isolation in Petri plates containing anaerobic medium solidified with agar (Figure 2). Inoculated plates are then transferred to an anaerobe jar that is purged with anaerobic gas and hydrogen sulfide to create conditions necessary for growth.

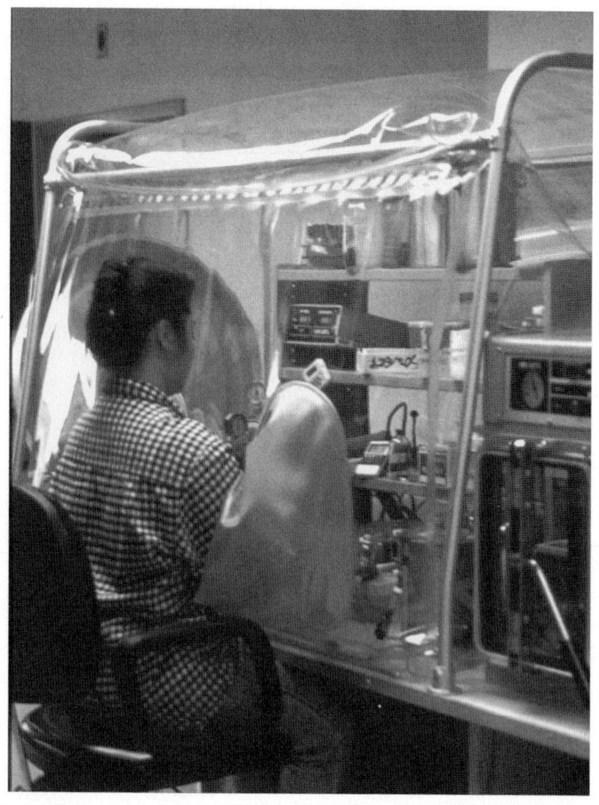

FIGURE 2 Anaerobic glove box used for growing methanogens. Materials are introduced into the glove box through an air lock located on the side.

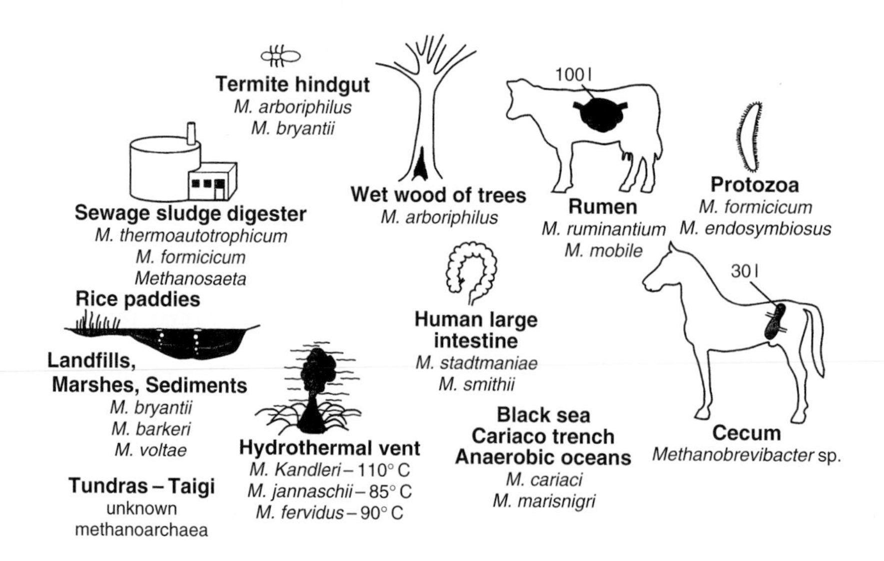

FIGURE 1 Habitats of the methanogenic Archaea. *Reproduced from Wolfe, R. S. (1996). 1776–1996: Alessandro Volta's combustible air.* ASM News *62: 529–534.*

DIVERSITY AND PHYLOGENY

The methanogens are members of the *Archaea*, one of three domains of life proposed by C. Woese on the basis of 16S rRNA sequence, which also include the *Bacteria* and *Eukarya* (Figure 3). *Archaea* have morphological features that resemble the *Bacteria*; they are unicellular microorganisms that lack a nuclear membrane and intracellular compartmentalization. In contrast, several molecular features of the *Archaea* have similarity to the *Eukarya*; these features include histone-like DNA proteins, a large multicomponent RNA polymerase, and eukarya-like transcription initiation. Despite the similarities to the other domains, *Archaea* also have unique characteristics that distinguish them from the *Bacteria* and *Eukarya*. These distinguishing features include membranes composed of isoprenoids ether linked to glycerol or carbohydrates, cell walls that lack peptidoglycan, synthesis of unique enzymes and enzyme cofactor molecules. An additional unifying characteristic among the *Archaea* is their requirement for extreme growth conditions, such as high temperatures, extreme salinity, and, in the case of the methanogens, highly reduced, anoxic environments.

Although the methanogens are a phylogenetically coherent group and have a limited substrate range, they are morphologically and physiologically diverse. They include psychrophilic species from Antarctica that grow at 1.7 °C to extremely thermophilic species from deep submarine vents that grow at 110 °C; acidophiles from marine vents that grow at pH 5.0 to alkaliphiles from alkaline lake sediments that grow at pH 10.3; species from freshwater lake sediments that grow at saline concentrations below 0.1 $mol\,l^{-1}$ to extreme halophiles from solar salterns that grow at nearly saturated NaCl concentrations; autotrophs that use only CO_2 for cell carbon and methylotrophs that utilize methylated compounds. Despite the range and diversity of growth habitats where methanogens are found, methanogens have one common attribute: they all generate CH_4 during growth.

There are currently over 120 described species of methanogens in six orders within the archaeal kingdom *Euryarchaeota* (Figure 3). Characteristics of methanogenic *Archaea* are described in Table 1.

The order *Methanococcales* includes marine autotrophs that grow exclusively by CO_2 reduction with H_2 or in some cases use formate for growth and methanogenesis. Morphologically, these species form irregularly shaped cocci. Instead of a rigid cell wall, typical of most *Bacteria*, these species form an S-layer, composed of an array of protein subunits, and are subject to osmotic lysis at NaCl concentrations below seawater. This order includes several mesophilic *Methanococcus* spp., moderately thermophilic *Methanothermococcus* spp., and the extremely thermophilic genera *Methanocaldococcus* and *Methanotorris*.

The order *Methanobacteriales* is predominantly composed of rod-shaped cells that grow by CO_2 reduction with H_2. One exception is the genus *Methanosphaera*, which grows as cocci and uses H_2 to reduce the methyl group of methanol instead of CO_2; other species can also reduce methyl groups with hydrogen, oxidize alcohols to reduce CO_2 or use formate for growth and methanogenesis. Cells have a rigid cell wall approximately 15–20 nm thick and, when stained for thin-section electron microscopy, resemble the electron-dense monolayer cell wall of Gram-positive bacteria. These archaeal cell walls are composed of pseudomurein, which is chemically distinguishable from bacterial murein by the substitution of *N*-talosaminuronic

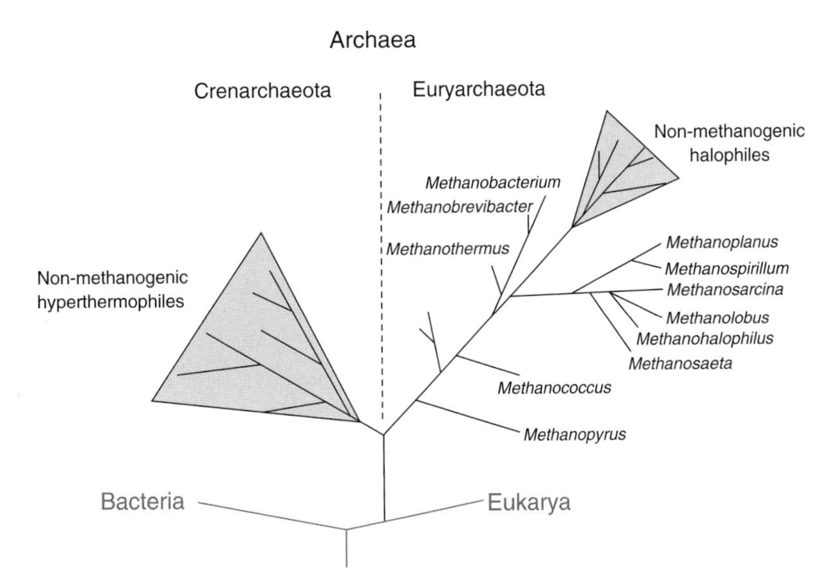

FIGURE 3 Phylogenetic tree based on 16S rRNA sequence showing selected genera of methanogens within one of the two major phyla of *Archaea*, the *Euryarchaeota*.

TABLE 1 Description of methanogenic Archaea

Taxonomic Epithet	Morphology	Substrates[a]	Optimum Growth Conditions		Isolation Source
			pH	Temp (°C)	
Order *Methanobacteriales*					
Family *Methanobacteriaceae*					
Genus *Methanobacterium*					
aarhusense	Rod	H	7.5–8.0	45	Marine sediment
alcaliphilum	Rod	H	8.4	37	Alkaline lake sediment
beijingense	Rod	H, F	7.2–7.7	37	Beer waste digestor
bryantii	Rod	H, 2P, 2B	6.9–7.2	37–39	Sewage digestor
congolense	Rod	H	7.2	37–42	Cassava peel digestor
espanolae	Rod	H	5.6–6.2	35	Kraft mill sludge
formicicum	Rod	H, F, 2P, 2B	6.6–7.8	37–45	Sewage digestor
ivanovii	Rod	H	7.0–7.4	45	Sewage digestor
oryzae	Rod	H, F	7.0	40	Rice field
palustre	Rod	H, F, 2P, CP	7.0	37	Peat bog
subterraneum	Rod	H, F	7.8–8.8	20–40	Deep granitic groundwater
thermoaggregans	Rod	H	7.0–7.5	65	Cattle pasture
uliginosum	Rod	H	6.0–8.5[b]	40	Marsh sediment
veterum	Rod	H, H/ME, H/MA	5.5	28	Ancient permafrost
Genus *Methanobrevibacter*					
acididurans	Coccobacillus	H	6.0	35	Alcohol distillery waste
arboriphilicus	Coccobacillus	H, F	7.8–8.0	30–37	Cotton wood tree
curvatus	Curved rod	H	7.1–7.2	30	Termite hindgut
cuticularis	Rod	H, F	7.7	37	Termite hindgut
filiformis	Filamentus rod	H	7.0–7.2	30	Termite hindgut
gottschalkii	Coccobacillus	H	7	37	Horse feces
millerae	Coccobacillus	H, F			Bovine rumen
olleyae	Coccobacillus	H, F			Ovine rumen
oralis	Coccobacillus	H	6.9–7.4	36–38	Human subgingival plaque
ruminantium	Coccobacillus	H, F	6.3–6.8	37–39	Bovine rumen
smithii	Coccobacillus	H, F	6.9–7.4	37–39	Sewage digestor
thaueri	Coccobacillus	H	7	37	Bovine feces
woesii	Coccobacillus	H, F	7	37	Goose feces
wolinii	Coccobacillus	H	7	37	Sheep feces
Genus *Methanosphaera*					
cuniculi	Coccus	H/ME	6.8	35–40	Rabbit rectum
stadtmanae	Coccus	H/ME	6.5–6.9	36–40	Human feces

TABLE 1 Description of methanogenic Archaea—cont'd

			Optimum Growth Conditions		
Taxonomic Epithet	Morphology	Substrates[a]	pH	Temp (°C)	Isolation Source
Genus *Methanothermobacter*					
defluvii	Rod	H, F	6.5–7.0	60–65	Methacrylic waste digestor
marburgensis	Rod	H	6.8–7.4	65	Sewage digestor
thermoautotrophicus	Rod	H, CO	7.2–7.6	65–70	Sewage digestor
thermoflexus	Curved rod	H, F	7.9–8.2	55	Methacrylic waste digestor
thermophilus	Rod	H	8.0–8.2	62	Digestor methane tank
wolfeii	Rod	H, F	7.0–7.7	55–65	Sewage/river sediment
Family *Methanothermaceae*					
Genus *Methanothermus*					
fervidus	Rod	H	6.5	83	Solfataric hot spring
sociabilis	Rod	H	6.5	88	Solfataric mud
Order *Methanococcales*					
Family *Methanocaldococcaceae*					
Genus *Methanocaldococcus*					
fervens	Irreg. coccus	H	6.5	85	Marine hydrothermal vent
indicus	Irreg. coccus	H	6.6	85	Marine hydrothermal vent
infernus	Irreg. coccus	H	6.5	85	Marine hydrothermal vent
jannaschii	Irreg. coccus	H	6.0	85	Marine hydrothermal vent
vulcanius	Irreg. coccus	H	6.5	80	Marine hydrothermal vent
Genus *Methanotorris*					
formicicus	Irreg. coccus	H, F	6.7	75	Marine hydrothermal vent
igneus	Irreg. coccus	H	5.7	88	Marine hydrothermal vent
Family *Methanococcaceae*					
Genus *Methanococcus*					
aeolicus	Irreg. coccus	H, F	6.6	46	Marine sediment
maripaludis	Irreg. coccus	H, F	6.8–7.2	35–39	Marine marsh sediment
vannielii	Irreg. coccus	H, F	8.0	36–40	Marine sediment
voltae	Irreg. coccus	H, F	6.7–7.4	32–40	Estuarine sediment
Genus *Methanothermococcus*					
okinawensis	Irreg. coccus	H, F	6.0–7.0	60–65	Marine hydrothermal vent
thermolithotrophicus	Irreg. coccus	H, F	6.5–7.5	65	Thermal coastal sediment
Order *Methanomicrobiales*					
Family *Methanocorpusculaceae*					
Genus *Methanocorpusculum*					
aggregans	Irreg. coccus	H, F	6.4–7.2	35–37	Sewage digestor
bavaricum	Irreg. coccus	H, F, 2P, 2B, CP	7.0	37	Sugar plant wastewater

(Continued)

TABLE 1 Description of methanogenic Archaea—cont'd

			Optimum Growth Conditions		
Taxonomic Epithet	Morphology	Substrates[a]	pH	Temp (°C)	Isolation Source
labreanum	Irreg. coccus	H, F	7.0	37	Tar pit lake
parvum	Irreg. coccus	H, F, 2P, 2B	6.8–7.5	37	Whey digestor
sinense	Irreg. coccus	H, F	7.0	30	Distillery wastewater
Family *Methanomicrobiaceae*					
Genus *Methanoculleus*					
bourgensis	Irreg. coccus	H, F	7.4	37	Tannery waste digestor
chikugoensis	Irreg. coccus	H, F, 2P, 2B, CP	6.7–7.2	25–30	Rice field
marisnigri	Irreg. coccus	H, F, 2P, 2B	6.2–6.6	20–25	Marine sediment
palmolei	Irreg. coccus	H, F, 2P, 2B, CP	6.9–7.5	40	Palm oil wastewater reactor
receptaculi	Irreg. coccus	H, F	7.5–7.8	50–55	Oil field
submarinus	Irreg. coccus	H, F	4.8–7.7	45	Deep marine sediment
thermophilicus	Irreg. coccus	H, F	7.0	55	Thermal marine sediment
Genus *Methanofollis*					
aquaemaris	Irreg. coccus	H, F	6.5	37	Mariculture fish pond
ethanolicus	Irreg. coccus	H, F, E, 1P, 1B	7.0	37	Lotus field
formosanus	Irreg. coccus	H, F	6.6–7.0	37	Mariculture fish pond
liminatans	Irreg. coccus	H, F, 2P, 2B	7.0	40	Industrial wastewater
tationis	Irreg. coccus	H, F	7.0	37–40	Solfataric hot pool
Genus *Methanogenium*					
cariaci	Irreg. coccus	H, F	6.8–7.3	20–25	Marine sediment
frigidum	Irreg. coccus	H, F	6.5–7.9	15	Antarctic lake
frittonii	Irreg. coccus	H, F	7.0–7.5	57	Lake sediment
organophilum	Irreg. coccus	H, F, E, 1P, 2P, 2B	6.4–7.3	30–35	Marine sediment
Genus *Methanolacinia*					
paynteri	Irreg. rod	H, F, 2P, 2B, CP	6.6–7.2	40	Marine sediment
Genus *Methanolinea*					
tarda	Rod	H, F	7.0	50	Sewage sludge
Genus *Methanomicrobium*					
mobile	Curved rod	H, F	6.1–6.9	40	Bovine rumen
Genus *Methanoplanus*					
endosymbiosus	Irreg. disk	H, F	6.6–7.1	32	Marine ciliate
limicola	Plate	H, F	7.0	40	Drilling swamp
petrolearius	Plate	H, F, 1P	7.0	37	Offshore oil field
Family *Methanospirillaceae*					
Genus *Methanospirillum*					
hungatei	Sheathed spiral	H, F	6.6–7.4	30–37	Sewage sludge

TABLE 1 Description of methanogenic Archaea—cont'd

Taxonomic Epithet	Morphology	Substrates[a]	Optimum Growth Conditions pH	Temp (°C)	Isolation Source
Genus *Methanocalculus*[c]					
chungsingensis	Irreg. cocci	H, F	7.2	37	Mariculture fish pond
halotolerans	Irreg. coccus	H, F	6.5–7.5	35	Marine waste leachate site
pumilus	Irreg. coccus	H, F	7.6	38	Oil field
taiwanensis	Irreg. coccus	H, F	6.7	37	Estuarine sediment
Genus *Methanosphaerula*[c]					
palustris	Irreg. coccus	H, F	5.5	28–30	Minerotrophic fen
Order *Methanocellales*					
Family *Methanocellaceae*					
Genus *Methanocella*					
paludicola	Rod	H, F	7.0	35–37	Rice paddy
Order *Methanosarcinales*					
Family *Methanosarcinaceae*					
Genus *Methanosarcina*					
acetivorans	Irreg. coccus, pseudosarcina	AC, ME, MA, DMS, MMP, CO	6.5–7.5	35–40	Marine sediment
baltica	Irreg. coccus, pseudosarcina	AC, ME, MA	6.5–7.5	25	Brackish sediment
barkeri	Irreg. coccus, pseudosarcina	H, AC, ME, MA, CO	6.5–7.5	30–40	Sewage digestor
lacustris	Irreg. coccus	H, ME, MA	7.0	25	Lake sediment
mazei	Irreg. coccus, pseudosarcina	AC, ME, MA	6.5–7.2	30–40	Sewage digestor
semesiae	Irreg. coccus	ME, MA, DMS, MT	6.5–7.5	30–35	Mangrove sediment
siciliae	Irreg. coccus	AC, ME, MA, DMS, MMP	6.5–6.8	40	Marine sediment
thermophila	Irreg. coccus, pseudosarcina	AC, ME, MA	6.0	45–55	Sewage digestor
vacuolata	Irreg. coccus, pseudosarcina	H, AC, ME, MA	7.5	40	Methane tank sludge
Genus *Methanococcoides*					
alaskense	Irreg. coccus	MA	6.3–7.5	23.5	Cold marine sediment
burtonii	Irreg. coccus	ME, MA	7.7	23.4	Antarctic saline lake
methylutens	Irreg. coccus	ME, MA	7.0	30–35	Marine sediment
Genus *Methanohalobium*					
evestigatum	Irreg. coccus	ME, MA	7.4	50	Salt lagoon sediment
Genus *Methanohalophilus*					
halophilus	Irreg. coccus	ME, MA	7.4	26–36	Marine cyanobacterial mat

(Continued)

TABLE 1 Description of methanogenic Archaea—cont'd

			Optimum Growth Conditions		
Taxonomic Epithet	Morphology	Substrates[a]	pH	Temp (°C)	Isolation Source
mahii	Irreg. coccus	ME, MA	7.4	35–37	Saline lake sediment
portucalensis	Irreg. coccus	ME, MA	6.5–7.5	40	Solar salt pond
Genus *Methanolobus*					
bombayensis	Irreg. coccus	ME, MA, DMS	7.2	37	Marine sediment
taylorii	Irreg. coccus	ME, MA, DMS	8.0	37	Estuarine sediment
oregonensis	Irreg. coccus	ME, MA, DMS	8.6	35	Saline alkaline aquifer
psychrophilus	Irreg. coccus	ME, MA, DMS	7.0–7.2	18	Cold wetland soil
profundi	Irreg. coccus	ME, MA	6.5	30	Gas field
tindarius	Irreg. coccus	ME, MA	6.5	37	Lake sediment
vulcani	Irreg. coccus	ME, MA	7.2	37	Submarine fumarole
zinderi	Irreg. coccus	ME, MA	7.0–8.0	40–50	Coal seam
Genus *Methanosalsum*					
zhilinae	Irreg. coccus	ME, MA, DMS	9.2	45	Alkaline lake sediment
Genus *Methanimicrococcus*					
blatticola	Irreg. coccus	H/ME, H/MA	7.2–7.7	39	Cockroach hindgut
Genus *Methanomethylovorans*					
hollandica	Irreg. cuccus	ME, MA, DMS, MT	6.5–7.0	34–37	Freshwater sediment
thermophila	Irreg. coccus	ME, MA	6.5	50	Paper mill wastewater digestor
Family *Methanosaetaceae*					
Genus *Methanosaeta*					
concilii	Sheathed rod	AC	7.1–7.5	35–40	Pear waste digestor
harundinacea	Sheathed rod	AC	7.2–7.6	34–37	Beer waste digestor
thermacetophila	Sheathed rod	AC	7.4–7.8	35–40	Thermophilic sludge digestor
thermophila	Sheathed rod	AC	6.5–6.7	55–60	Thermophilic sludge digestor
Family *Methermicoccaceae*					
Genus *Methermicoccus*					
shengliensis	Coccus	ME, MA	6.0–6.5	65	Oil field
Order *Methanopyrales*					
Family *Methanopyraceae*					
Genus *Methanopyrus*					
kandleri	Sheathed rod	H	6.5	98	Geothermal marine sediment

Type strain descriptions.
[a]*H, hydrogen/carbon dioxide; F, formate; AC, acetate; ME, methanol; MA, methylamines; MT, methanethiol; CO, carbon monoxide; H/ME, methanol reduction with hydrogen; H/MA, methylamine reduction with hydrogen; E, ethanol; 1P, 1-propanol; 2P, 2-propanol; 1B, 1-butanol, 2B, 2-butanol; CP, cyclopentanol; DMS, dimethylsulfide; MMP, methylmercaptopropionate.*
[b]*Only one range reported.*
[c]*Family epithet currently uncertain.*

acid for *N*-acetylmuramic acid and substitution of β(1,3) for β(1,4) linkage in the glycan strands, and substitution of D-amino acids for L-amino acids in the peptide cross-linkage. *Methanothermus* spp. also have an additional cell wall layer composed of glycoprotein S-layer that surrounds the pseudomurein. *Methanobrevibacter* spp. are all mesophilic; *Methanobacterium* includes mesophiles and moderate thermophiles with optimal growth temperatures as high as 65 °C; *Methanothermobacter* spp. are exclusively moderate thermophiles, and *Methanothermus* spp. are extreme thermophiles, with maximum growth temperatures as high as 97 °C.

The order *Methanomicrobiales* contains genera that are diverse in morphology and physiology. Most species grow as cocci and rods. In addition, *Methanoplanus* forms flat plate-like cells with characteristically angular ends. Another species, *Methanospirillum hungatei*, forms a helical spiral. The cell walls in this order are composed of a protein S-layer and are sensitive to osmotic shock or detergents. In addition to the S-layer, *M. hungatei* also has an external sheath that is composed of concentric rings stacked together. Species range from nonhalophilic to slightly halophilic. Temperature tolerances include the only currently described psychrophile, *Methanogenium frigidum*, mesophiles, and moderate thermophiles. Most species grow by CO_2 reduction with H_2, but some species also use formate or secondary alcohols as electron donors for CO_2 reduction.

The order *Methanosarcinales* includes the most catabolically diverse species of methanogens. Whereas most methanogenic species grow by obligate CO_2 reduction with H_2, species within this order can also grow by methyl reduction with H_2, aceticlastic fermentation of acetate, or methylotrophic catabolism of methanol, methylated amines, and dimethylsulfide. *Methanosarcina acetivorans* was recently reported to grow also with CO by conversion to acetate and formate, rather than methane, as the major metabolic end products. Whereas some species of *Methanosarcina* can grow by all five catabolic pathways, *Methanosaeta* species are obligate acetotrophs and all other genera are obligate methylotrophs. All species have a protein S-layer cell wall and most species grow as irregularly shaped cocci. However, several species of *Methanosarcina* also synthesize a heteropolysaccharide matrix external to the S-layer. This external layer can be up to 200 nm thick and is primarily composed of a nonsulfonated polymer of *N*-acetylgalactosamine and D-glucuronic or D-galacturonic acids. The matrix is called methanochondroitin because of its chemical similarity to a mammalian connective tissue component, known as chondroitin. At freshwater NaCl concentrations, *Methanosarcina* spp. that synthesize methanochondroitin grow in multicellular aggregates rather than as single cells, but when grown at marine salt concentrations or with high concentrations of divalent cations, such as Mg^{2+}, they no longer synthesize methanochondroitin and grow as single cells. *Methanosaeta* species have an external sheath that appears similar in structure to that described for *M. hungatei*.

Two orders presently include only 1 described species, The recently proposed order *Methanocellales* includes the species *Methanocella paludicola,* which is a hydrogen- and formate-utilizing rod. This is the first isolate from a previously described clonal lineage (Rice Cluster I) from rice paddies. The order *Methanopyrales* is the most deeply branching methanogenic archaeon and presently includes only one species, *Methanopyrus kandleri*. This hyperthermophilic species is an obligate hydrogenotroph and grows as a rod with a pseudomurein cell wall surrounded by a protein S-layer, similar to that described for *Methanothermus*.

HABITATS

Interspecies H_2 Transfer

As described above, methanogens utilize a limited number of simple substrates. In most habitats, they depend on other anaerobes to convert complex organic matter into substrates, which they can catabolize. Therefore, unlike aerobic habitats, where a single microorganism can catalyze the mineralization of a polymer by oxidation to H_2O and CO_2, degradation in anaerobic habitats requires consortia of interacting microorganisms to convert polymers to CH_4 and CO_2. These interactions are dynamic with the methanogens affecting the pathway of electron flow, and, consequently, carbon flow, by a process called interspecies H_2 transfer. In this association, the H_2-utilizing methanogens maintain a low H_2 partial pressure that allows certain reactions that do not occur under standard conditions to be thermodynamically favorable. One physiological group of microorganisms affected by this process is the H_2-producing acetogens. The reactions carried out by these microorganisms for growth are not thermodynamically favorable (i.e., $+\Delta G^{0\bullet}$; under physiological growth conditions), as the H_2 they generate accumulates and growth is subsequently inhibited. However, in association with H_2-consuming microorganisms, such as the methanogens, the H_2 partial pressure is maintained at levels low enough to make the reaction thermodynamically favorable (i.e., $-\Delta G^{0\bullet}$; under physiological growth conditions). Because of their dependence on the H_2-consuming microorganisms, the H_2-producing acetogens are often referred to as obligate syntrophs.

Another physiological group of microorganisms affected by interspecies H_2 transfer is the fermentative anaerobes that synthesize a hydrogenase. In many of these microorganisms, substrate oxidation is linked to the electron carrier nicotinamide adenine dinucleotide (NAD), which has higher redox potential (−320 mv) than H_2 (−414 mv) under standard conditions. However, if the H_2 partial pressure is maintained at a low level (<10 Pa) by H_2-utilizing microorganisms, then

H_2 production from NAD becomes thermodynamically favorable. This enables the fermentative microorganisms to reoxidize NADH by reducing protons to form H_2 rather than reducing pyruvate to form dicarboxylic acids and alcohols. This synergistic process enables the fermentor to conserve ATP by synthesizing more acetate and less further reduced products. In addition, the net products, acetate and H_2, serve as substrates for growth by the methanogen. The result is that carbon and electron flow is directed toward more efficient degradation by the consortium to the gases CH_4 and CO_2. CO_2 reenters the carbon cycle directly and CH_4 is either oxidized by methylotrophs to CO_2 as it diffuses into the aerobic zone or as it enters the atmosphere.

Synergistic interspecies substrate transfer has also been reported to occur with formate and acetate. Removal of formate by formate-utilizing methanogens and the removal of acetate by acetotrophic methanogens have been shown to create a thermodynamic shift that favors butyrate and propionate degradation in syntrophic cocultures of acetogenic *Bacteria* and methanogenic *Archaea*.

Freshwater Sediments

As organic matter accumulates on lake and river basins, the sediment becomes anaerobic, because oxygen is depleted by the activities of aerobic microorganisms. In eutrophic environments with high organic loading, the anaerobic region occurs immediately below the sediment surface and even into the water column if the activities of the aerobic microorganisms exceed the rate of oxygen diffusion. Once oxygen and alternative electron acceptors such as NO^{3-}, Fe^{3+}, Mn^{4+}, and SO_4^{2-} are depleted, methanogenesis becomes the ultimate degradative process. The flow of carbon in anaerobic freshwater sediments is shown in Figure 4. Fermentative bacteria that synthesize hydrolytic enzymes, such as cellulases, proteases, amylases, and lipases, catalyze degradation of complex polymers to soluble monomers. The fermentative bacteria then ferment the soluble products to H_2, CO_2, simple alcohols, and fatty acids, including significant generation of acetate, due to interspecies H_2 transfer. H_2-producing acetogenic bacteria then catalyze the oxidation of alcohols and fatty acids to H_2, CO_2, and acetate. The third consortium group of microorganisms, the methanogenic *Archaea*, utilizes the simple substrates H_2, CO_2, and acetate generated by the fermentative and acetogenic bacteria, to generate CH_4. The net result of this consortium is that carbon and electrons are directed toward the synthesis of CH_4 and CO_2, which then reenter the global carbon cycle. CH_4 is oxidized by aerobic methanotrophic bacteria as it diffuses through the aerobic regions of the water column and reenters the atmosphere as CO_2. However, some CH_4 escapes from shallow water bodies, such as rice fields, through the vascular systems of aquatic plants, and enters the atmosphere as a greenhouse gas.

Anaerobic Bioreactors

Methanogenic bioreactors are used for the conversion of organic wastes to CH_4 and CO_2. These anaerobic digestors are found in nearly all sewage treatment plants where wastes in the form of sewage sludge, polymeric particulate material generated by settling of raw sewage, are converted to CH_4 and CO_2 by a consortium of microorganisms similar to that previously described in freshwater sediments. In contrast to most sediment environments, microbial metabolism is higher in a bioreactor because of greater rates of organic loading and higher temperatures (35–40 °C) generated by heating the reactor, often by combusting the CH_4 generated

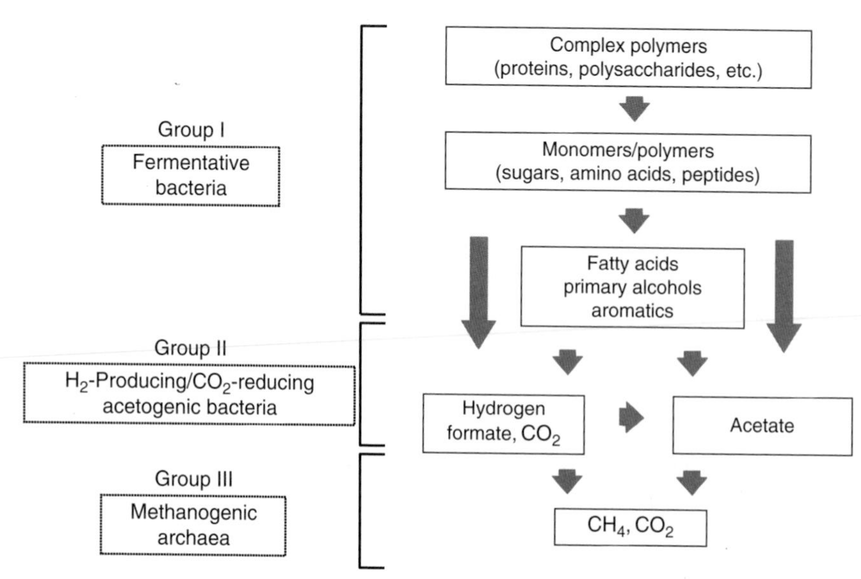

FIGURE 4 Carbon flow in an anaerobic microbial consortium from freshwater sediments.

from the degradative process. As a result, the rate-limiting factor in a bioreactor is the slow growth rates of the acid-consuming H_2-producing acetogens and the acetate-utilizing methanogens, which require retention times of 14 days or more, to compensate for their slow metabolic activities. Another critical factor is the requirement for H_2-utilizing methanogens to maintain H_2-partial pressures below 10 Pa. Perturbations in the process that result in inhibition and subsequent "washout" of either of these metabolic groups will result in a drop in pH, resulting in acid accumulation and subsequent inhibition of the entire reactor process. Perturbations can include sudden overloading with a readily fermented organic substrate that result in rapid accumulation of H_2 and fatty acids or introduction of a toxic compound that disrupts the microbial balance.

Anaerobic bioreactors have also been tested as a low-cost method for treatment of other types of particulate organic wastes, including animal manure and crop wastes. Nonparticulate industrial wastes, including many food processing by-products, and organic solvents, such as chlorinated aliphatic and aromatic compounds, are often too dilute to be economically treated by standard bioreactor configurations, which would require long retention times. In order to decrease the hydraulic retention time of the waste without washing out the biomass, high-rate anaerobic bioreactor configurations have been developed. Examples of these reactors include fixed-film reactors, in which biofilms of microbial consortia are retained in the vessel on solid supports, such as plastic or ceramic matrixes, glass beads, or sand grains. Other designs exist, such as the anaerobic upflow sludge blanket process, in which biomass is immobilized by the aggregation of microbial consortia into distinct granules (Figure 5). The granules usually consist of three discrete layers of microorganisms that include acetate-utilizing methanogens in the interior, H_2-producing acetogens and H_2-utilizing methanogens in the middle layer, and fermentative microorganisms in the outermost layer. Settler screens separate the granules from the treated water and gas is collected at the top of the reactor. In contrast to particulate waste reactors, in which microbial biomass is generated at the same rate of hydraulic washout, requiring retention times as long as several days, growth of biomass in these high-rate reactors is uncoupled from retention time of the waste and can have retention times as short as 1 h. Hydraulic retention times have also been reduced further in anaerobic fermentors that operate at temperatures of up to 60 °C, since the metabolic rates of thermophilic microorganisms, including acetate-utilizing methanogens, are greater than those of mesophiles.

Marine Habitats

In marine habitats where substrates are often limited, sulfate-reducing bacteria outcompete methanogens as the terminal members of the anaerobic consortium (Figure 6). Their predominance is a result of the availability of the electron acceptor sulfate in seawater (~30 mmol l^{-1}). Their lower *K*s for substrate utilization enable the sulfate-reducing bacteria

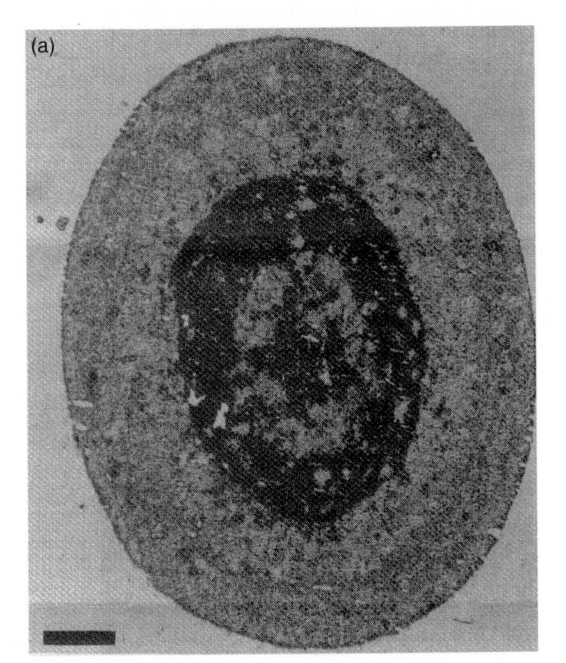

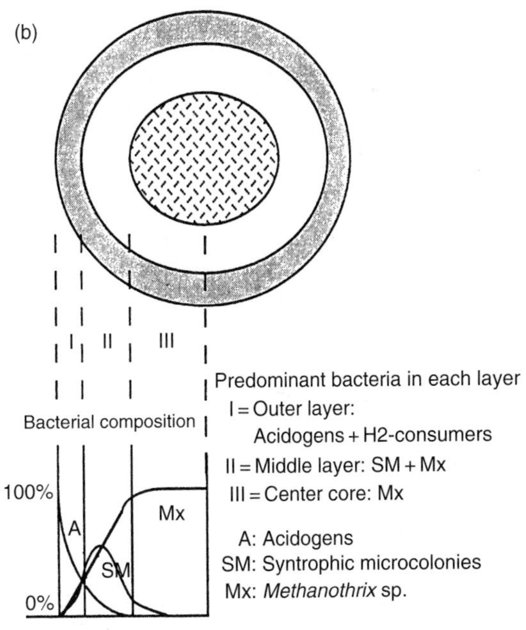

FIGURE 5 An anaerobic granule from a brewery wastewater digestor. (a) A thin-section photomicrograph of a granule under a bright-field microscope. Scale bar = 0.25 mm. (b) The bacterial composition of a granule. *(Reproduced with permission from Fang, H. H. P., Chui, H. K., & Li, Y. Y.(1994). Microbial structure and activity of UASB granules treating different wastewaters.* Water Science and Technology, *30, 87–96. Pergamon Press; and Fang, H. H. P., Chui, H. K., & Li, Y. Y. (1995). Microstructural analysis of UASB granules treating brewery wastewater,* Water Science and Technology. *31, 129–135 Pergamon Press).*

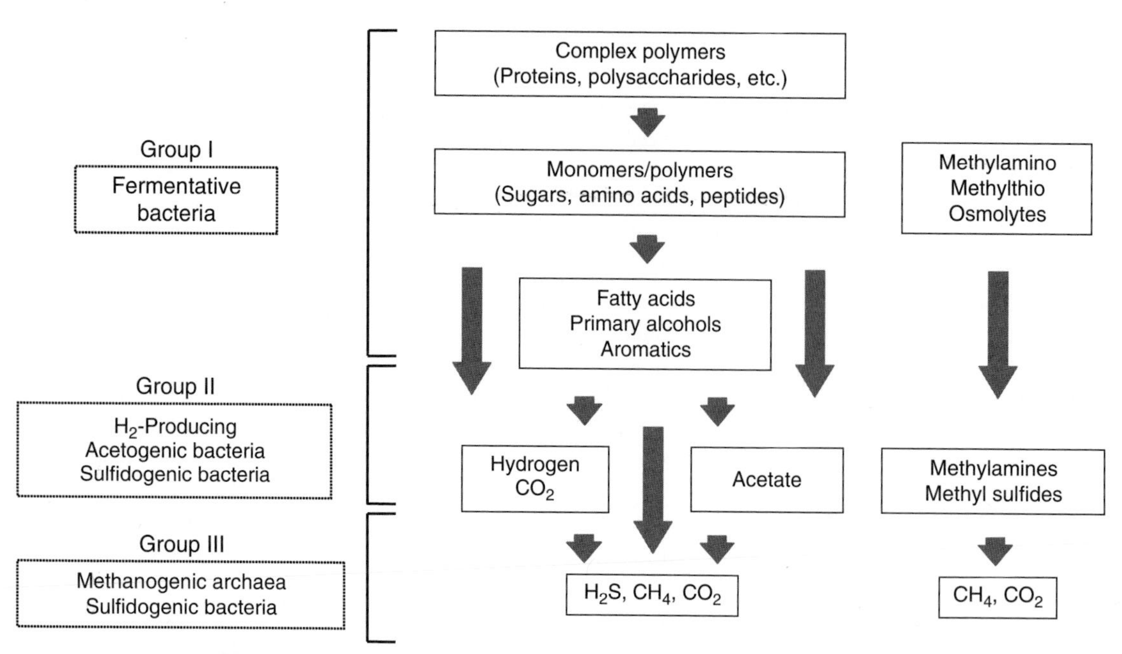

FIGURE 6 Carbon flow in an anaerobic microbial consortium from marine sediments.

to use low concentrations of H_2 and acetate at rates that are greater than those of methanogens. However, methanogenesis is predominant in marine environments where SO_4^{2-} has been depleted. These environments include the lower depths of sediments and elevated coastal marshes, where the rate of SO_4^{2-} reduction is greater than the rate of diffusion from seawater, and also sediments that receive large amounts of organic matter, such as eutrophic coastal regions and submarine trenches. In these regions, the three-member methanogenic consortium is similar to that in freshwater environments, but it is composed of halophilic and halotolerant species. Although the acetotrophic methanogens *Methanosarcina* and *Methanosaeta* have been isolated from marine methanogenic enrichments, isotope studies performed in sediment suggest that most of the acetate is oxidized by an H_2-producing syntroph rather than by fermentation to CH_4. Methanogens also generate CH_4 from methylated amines and thiols, which are readily available in the marine environment as metabolic osmolytes. Since methylated amines are not used by SO_4^{2-}-reducing bacteria, this class of compound is "noncompetitive" and are used by methanogens in habitats that contain high SO_4^{2-} concentrations.

CH_4 generated in sediments is often consumed as it diffuses through the SO_4^{2-}-reducing region of the sediments, before reaching the aerobic regions. Although the microbes that catalyze this anaerobic oxidation have not been isolated, analysis by fluorescence *in situ* hybridization with species-specific probes indicates that the methane is oxidized by a consortium of two microorganisms: a methanogen that oxidizes the methane by reverse methanogenesis coupled with an unknown intermediate to sulfate reduction by a sulfate-reducing bacterium. The metagenome and mRNA expression analysis of the microbial consortia mediating the anaerobic oxidation of methane with sulfate (ANME) are consistent with acetate and formate as putative electron shuttles. Expression of putative secreted multiheme c-type cytochromes suggests direct electron transfer as an additional possible mode to shuttle electrons from ANME-1 to the bacterial sulfate-reducing partner, but a methanogenic isolate and confirmation of the actual mechanism for this anaerobic process remain elusive. Although much of the CH_4 generated in sediments is consumed in the SO_4^{2-}-reducing regions, the water column in the open ocean is supersaturated with CH_4, compared with the atmospheric concentration. This may result from a combination of unoxidized CH_4 that escapes from sediments, the activity of microbial communities in planktonic aggregates known as marine snow, and methanogenic activity in the gastrointestinal tracts and fecal material of marine animals. Biologically generated CH_4 in some organic-rich buried sediments can accumulate as gas deposits. Since natural gas deposits generated abiotically are used as indicators of petroleum, these biologically generated CH_4 deposits can act as false indicators of petroleum deposits during oil exploration. Under high hydrostatic pressures generated in deep-ocean sediments, biogenic CH_4 can also accumulate as solidified CH_4 hydrates.

Ruminant Animals

Ruminant animals include both domestic (e.g., cows, sheep, and camels) and wild (e.g., deer, bison, and giraffes) animals. These animals have a large chamber, called the rumen, before the stomach, in which polymers, such as cellulose,

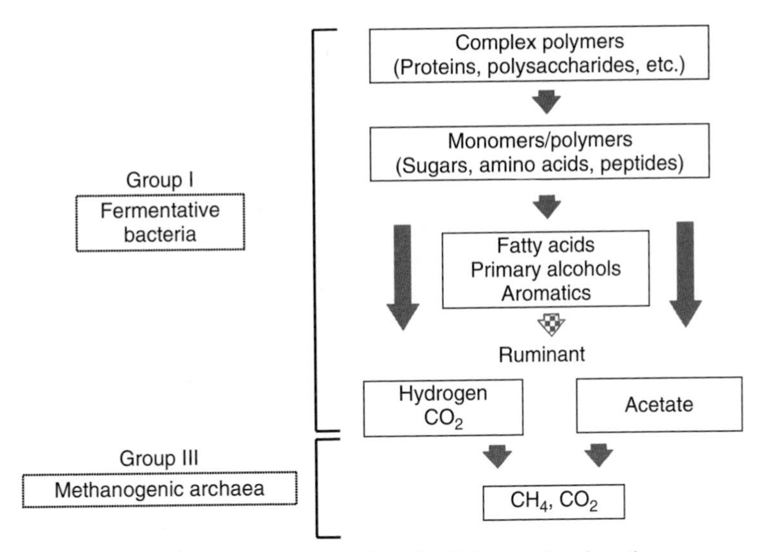

FIGURE 7 Carbon flow in an anaerobic microbial consortium from the rumen.

are fermented by bacteria to short-chain fatty acids, H_2, CO_2, and CH_4. The rumen is similar to an anaerobic bioreactor, except that the short retention time created by swallowing saliva is less than the generation time of H_2-producing fatty acid oxidizers and acetate-utilizing methanogens (Figure 7). Instead, acetate, propionate, and butyrate are not degraded by the consortium, but are absorbed into the bloodstream of the host animal as carbon and energy sources. The volatile gases (e.g., CH_4, CO_2) are removed from the animal by belching. Acidification of the system by the acids is prevented by bicarbonate in the saliva of the animal. Carbon diverted to CH_4 and belched into the atmosphere represents a loss of energy to the animal and a significant global source of greenhouse gas emissions. Ruminant nutritionists have been attempting to increase feed efficiency by adding methanogenic inhibitors, such as monensin, to feed, thereby, diverting carbon flow to metabolites that can be utilized by the animal. One approach currently being tested is to target specific methanogen proteins with antibodies and harness the immune system of ruminants to increase feed efficiency and mitigate greenhouse gas emissions from agriculture.

Xylophagous Termites

All known termites harbor a dense microbial community of anaerobic bacteria, and, in the case of lower termites, they also contain cellulolytic protozoa that catalyze the digestion of lignocellulose from wood. As in the rumen, these microorganisms have a synergistic relationship with the termites by converting polymers to short-chain fatty acids that are used as carbon and energy sources for their host. The carbon flow in the hindgut of soil-feeding and fungus-cultivating termites is similar to that in rumen, but in wood- and grass-eating termites, most H_2–CO_2 is converted to acetate instead of CH_4. Generally, methanogenesis outcompetes H_2–CO_2 acetogenesis and the factors that cause the predominance of acetogenesis in some termites in not known. Three species of H_2-utilizing *Methanobrevibacter* have been isolated from the hindgut of the subterranean termite that exhibit catalase activity and are particularly tolerant to oxygen exposure. Oxygen tolerance may be important for recolonization by bacterial consortia after expulsion of the hindgut contents during molting. Reinoculation is achieved by transfer of hindgut contents from other colony members, which are exposed to air during the process.

Protozoan Endosymbionts

Methanogens are present as endosymbionts in many free-living marine and freshwater anaerobic protozoa, where they are often closely associated with hydrogenosomes, organelles that produce H_2, CO_2, and acetate from the fermentation of polymeric substrates. The products of the hydrogenosomes are substrates for methanogenesis. It is conceivable that the methanogens have a synergistic role by lowering the H_2 partial pressure to create a favorable thermodynamic shift in the protozoan's fermentation reaction. Also, evidence suggests that excretion of undefined organic compounds by the methanogen provides an advantage to the protist host. Endosymbionts are also found in flagellates and ciliates that occur in the hindgut of insects, such as termites, cockroaches, and tropical millipedes. Although rumen ciliates do not harbor endosymbionic methanogens, many have ectosymbionic methanogens that may have an analogous function.

Colonization in Humans

The human colon serves as a form of hindgut where undigested polymers and sloughed off intestinal epithelium and mucin are dewatered, fermented by bacterial consortia, treated with bile acids, and held until defecation. Fatty acids generated by fermentation are absorbed into the bloodstream and can provide approximately 10% of human nutritional needs. Methanogenesis occurs in 30–40% of the human population, with the remaining population producing H_2 and CO_2 instead. Acetogenesis and SO_4^{2-} reduction from H_2–CO_2 also occur in the human colon, but studies with colonic bacterial communities suggest that these activities are only prevalent in the absence of methanogenic activity. The level of CH_4 produced by an individual corresponds to the population levels of *Methanobrevibacter*, but factors controlling the occurrence of methanogens in the human population are not known. Diet does not appear to have a significant role in determining whether an individual harbors an active methanogenic population, but hereditary and individual physiological factors might have a role. For example, methanogens might be absent from individuals who excrete higher levels of bile acids, which are inhibitory to methanogens. Methanogenic *Archaea* have also been isolated from the human vagina and oral cavity.

Epidemiological studies have revealed a relationship between the severity of periodontal disease and the relative abundance of *Methanobrevibacter*-like phylotypes. It has been suggested that the methanogens have a syntrophic relationship with periodontitus-causing fermentative bacteria that promotes their colonization. There have also been studies that show an increase in methanogenesis associated with colon cancer patients, but other reports contradict these conclusions. Although annotation of methanogen genomes reveals properties characteristic of many pathogenic bacteria including putative RelBE toxin/antitoxin genes and evidence of lateral gene transfer with bacteria, methanogens have not been shown unequivocally to cause or promote human disease.

Other Habitats

Habitats that have a source of organic carbon and high water content can become anaerobic as a result of respiratory depletion of oxygen, and, subsequently, support methanogenic communities. Examples include soils waterlogged by heavy rainfall, marshes, rice paddies, rotting heartwood of trees, and landfills. Most of the CH_4 generated in these habitats is released into the atmosphere. However, many landfills are now vented to prevent a buildup of potentially combustible CH_4 underground or in nearby dwellings, and some communities harvest the vented biogas for heat and energy.

Methanogenic habitats are also found in geo-hydrothermal outsources, such as terrestrial hot springs and deep-sea hydrothermal vents. Methanogens from these environments use geothermally generated H_2–CO_2 for methanogenesis and most are hyperthermophilic, requiring growth temperatures as high as 110 °C. In terrestrial sites, methanogens are usually associated with microbial mats composed of photosynthetic and heterotrophic consortia. However, in deep-sea hydrothermal vents, where there is no light available for photosynthetic production of organic carbon, the methanogens and other autotrophic bacteria serve as primary producers of cell carbon for a complex community of heterotrophic microbes and animals that accumulate in the vicinity of a hydrothermal vent.

Methanogenesis occurs in high-saline environments, such as Great Salt Lake, Utah, Mono Lake, California, and solar salt ponds. Methanogens from these environments generate CH_4 from methylated amines and dimethylsulfide, which are synthesized by animals and plants as osmolytes. Methanogenesis has been detected in subsurface aquifers over 400 m deep, where it has been proposed that H_2 generated by an abiotic reaction between iron-rich minerals in basalt and ground water is used as a substrate for methanogenesis. Methanogenesis has also been detected in deep subsurface sandstone, where it is hypothesized that methanogenic consortia use organic compounds that diffuse from adjacent organic-rich shale layers..

PHYSIOLOGY AND BIOCHEMISTRY

Catabolic Pathways

Methanogens generate CH_4 during growth by four catabolic pathways: autotrophic CO_2 reduction with H_2, formate, or secondary alcohols; methyl reduction with H_2; fermentation of acetate or pyruvate; and methylotrophic dismutation of methanol, methylated amines, or methylthiols. In addition, some species grow on CO and M. acetivorans grows non-methanogenically on CO with acetate and formate as the end products. The reactions for methanogenesis are shown in Table 2. Six coenzymes serve as the principle carbon carriers in the methanogenic pathway (Figure 8). Methanofuran (MFR) is an analogue of molybdopterins, which occur in enzymes that catalyze similar reactions in the *Bacteria* and *Eukarya*. Tetramethanopterin (H_4MPT) is an analogue of tetrahydrofolate (H_4THF), which is also a one-carbon carrier in bacterial and eukaryal systems. Although H_4MPT was initially found to be unique to methanogens, it has since been detected in other *Archaea* and, more recently, enzymes catalyzing the methyl transfer for MFR and H_4MPT have been found to coexist with the H_4THF pathway in the CH_4-utilizing methanotrophic *Bacteria*. The methyl carrier coenzyme M (CoM-SH), 7-mercaptoheptanoylthreonine phosphate (HS-HTP), and cofactor F_{430} are currently unique to the methanogens.

Methanogenesis from H_2, formate, or secondary alcohols involves the sequential reduction of CO_2 with electrons generated by substrate oxidation (Figure 9a). The methanogenic sequence is initiated by a two-electron reduction of CO_2 and MFR by formyl-MFR dehydrogenase (a) to form

TABLE 2 Reactions and free energy yields from methanogen substrates

Substrate	Reaction	$\Delta G^{0\prime}$ (kJ mol^{-1} CH_4)
Hydrogen/carbon dioxide	$4H_2 + HCO_3^- \rightarrow CH_4 + 3H_2O$	−135
Formate	$4HCOO^- + 4H^+ \rightarrow CH_4 + 3CO_2 + 2H_2O$	−145
Ethanol[a]	$2CH_3CH_2OH + HCO_3^- \rightarrow 2CH_3COO^- + H^+ + CH_4 + H_2O$	−116
Hydrogen/methanol	$CH_3OH + H_2 \rightarrow CH_4 + H_2O$	−113
Methanol	$4CH_3OH \rightarrow 3CH_4 + HCO_3^- + H_2O + H^+$	−105
Trimethylamine[b]	$4CH_3NH^+ + 9H_2O \rightarrow 9CH_3 + HCO_3^- + 4NH_4 + 3H^+$	−76
Dimethylsulfide[c]	$2(CH_3)_2S + 3H_2O \rightarrow 3CH_4 + HCO_3^- + 2H_2S + H^+$	−49
Acetate	$CH_3COO^- + H_2O \rightarrow CH_4 + HCO_3^-$	−31
Pyruvate	$4CH_3COCOOH + 2H_2O \rightarrow 5CH_4 + 7CO_2$	−31
Carbon monoxide	$4CO + 5H_2O \rightarrow CH_4 + HCO_3^- + 3H^+$	−196
Carbon monoxide[d]	$4CO + 4H_2O \rightarrow CH_3COO^- + HCO_3^- + 3H^+$	−165
Carbon monoxide[d]	$CO + H_2O \rightarrow HCOO^- + H^+$	−16

[a]Other short-chain alcohols are utilized.
[b]Other methylated amines are utilized.
[c]Other methylated sulfides are utilized.
[d]Nonmethanogenic reaction.

formyl-MFR (Figure 10). The formyl group is then transferred to H_4MPT by formyl-MFR:H_4MPT formyltransferase (b) yielding formyl-H_4MPT. A homologue of H_4MPT, tetrahydrosarcinapterin (H_4SPT), found in *Methanosarcina* spp., differs by an additional glutamyl moiety in the substituted R group. Formyl-H_4MPT cyclization to methenyl-H_4MPT is catalyzed by N5,N10-methenyl-H_4MPT cyclohydrolase (c). N5,N10-methylene-H_4MPT dehydrogenase (d) and N5,N10-methylene-H_4MPT reductase (e) catalyze the sequential reduction of methenyl-H_4MPT by the electron carrier coenzyme F_{420} to methylene-H_4MPT and methyl-H_4MPT. The methyl group is then transferred to CoM-SH by N5-methyl-H_4MPT:CoM-SH methyltransferase (f) forming methyl-S-CoM. Methyl-CoM reductase (g) catalyzes the terminal reduction of methyl-S-CoM by two electrons from HS-HTP to CH_4. CoM-SS-HTP is the product of the terminal reaction, which is subsequently reduced by heterodisulfide reductase to regenerate the reduced forms CoM-SH and HTP-SH.

The acetotrophic pathway for acetate catabolism proceeds by initial "activation" of acetate by formation of acetyl coenzyme A (CoA). *Methanosarcina* spp. synthesize acetyl-CoA by sequential activities of phosphotransacetylase and acetyl kinase (Figure 9b). In contrast, activation of acetate to acetyl-CoA by *Methanosaeta* spp. is catalyzed in a single step by acetyl-CoA synthase. In both genera, cleavage of the CC and CS bonds of acetyl-CoA is then catalyzed by the acetyl-CoA decarbonylase/synthase complex, yielding enzyme-bound methyl and carbonyl groups. The complex contains CO:acceptor oxidoreductase, Co-β-methylcobamide:tetrahydropterine methyltransferase, and acetyl-CoA synthase activities. The five subunit complex consists of a two-polypeptide CO-oxidizing nickel/iron–sulfur component, a two-polypeptide corrinoid/iron–sulfur component, and a single polypeptide of unknown function. The nickel/iron–sulfur component catalyzes the cleavage of acetyl-CoA, the oxidation of the bound carbonyl to CO_2, and methyl transfer to the corrinoid/iron–sulfur component. The methyl group is sequentially transferred to H_4SPT by a currently unknown process and to HS-CoM by H_4MST:CoM-SH methyltransferases. Methyl-S-CoM is reductively demethylated to CH_4 by methylreductase as previously described. The enzyme-bound carbonyl group is oxidized to CO_2. *Methanosarcina* spp. and autotrophic methanogens growing on H_2/CO_2 synthesize acetate, presumably by reversing the direction of the acetyl-CoA decarbonylase/synthase complex in a reaction analogous to acetyl-CoA synthase in acetate-utilizing *Clostridia*.

Methylotrophic catabolism of methanol and methylated amines requires three polypeptides for the initial reaction. Methanol is catabolized by transfer of its methyl group to a corrinoid-binding protein, which is methylated by a substrate-specific methyltransferase, methanol:5-hydroxybenzinidazolyl (MT1). The methyl group is then transferred from the corrinoid protein to coenzyme HS-CoM by methylcobamide:CoM methyltransferase (MT2) (Figure 9c). Trimethylamine, dimethylamine, and

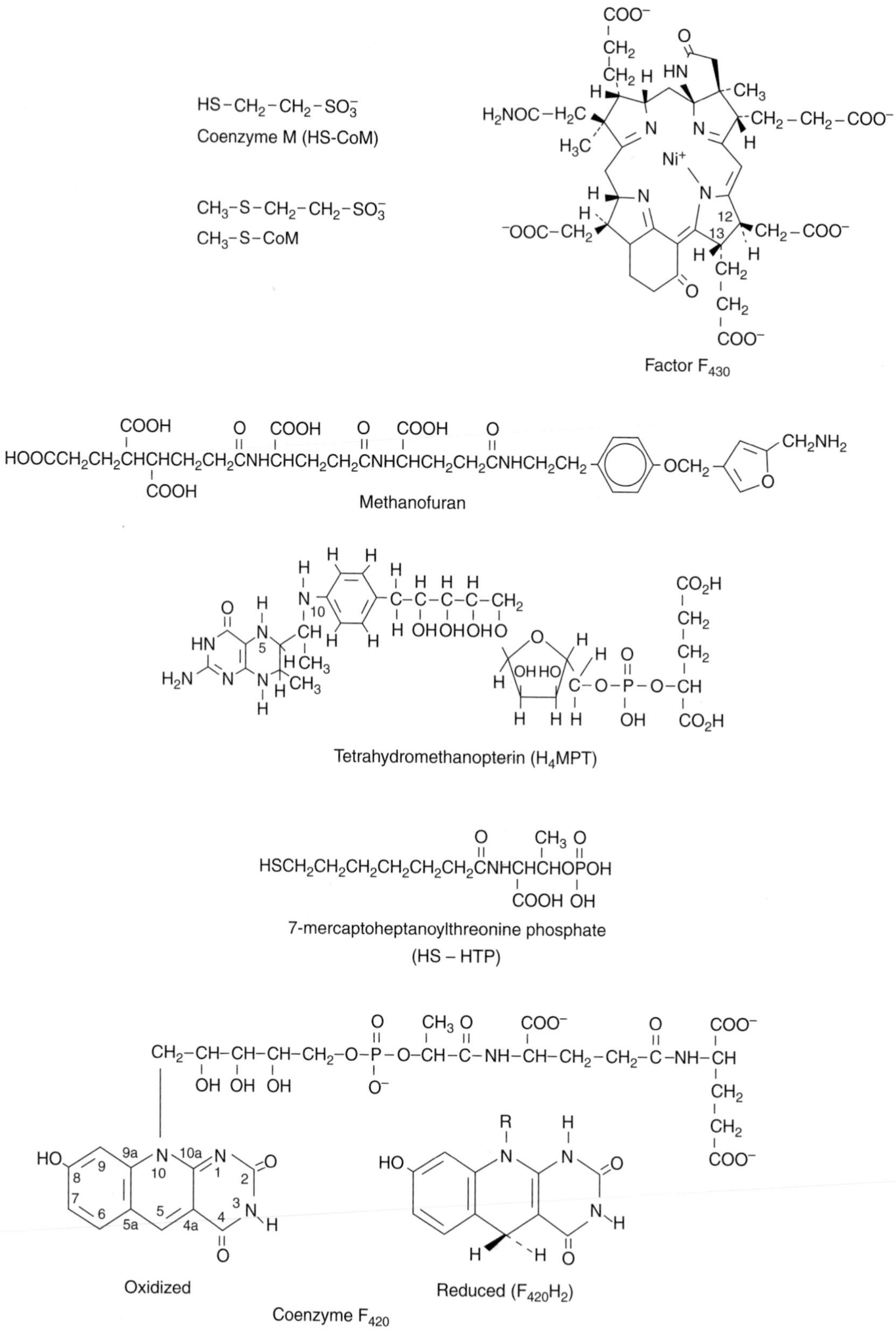

$HS-CH_2-CH_2-SO_3^-$
Coenzyme M (HS-CoM)

$CH_3-S-CH_2-CH_2-SO_3^-$
$CH_3-S-CoM$

Factor F_{430}

Methanofuran

Tetrahydromethanopterin (H_4MPT)

$HSCH_2CH_2CH_2CH_2CH_2CH_2CNHCHCHOPOH$

7-mercaptoheptanoylthreonine phosphate
(HS – HTP)

Oxidized

Reduced ($F_{420}H_2$)

Coenzyme F_{420}

FIGURE 8 Structures of coenzymes that participate in the methanogenic pathway. *(Reproduced with permission from Rouviere, P. E., & Wolfe, R. S., (1988). Novel Biochemistry of Methanogenesis.* The Journal of Biological Chemistry, 263, *7913–7916. American Society for Biochemistry and Molecular Biology, Inc.)*

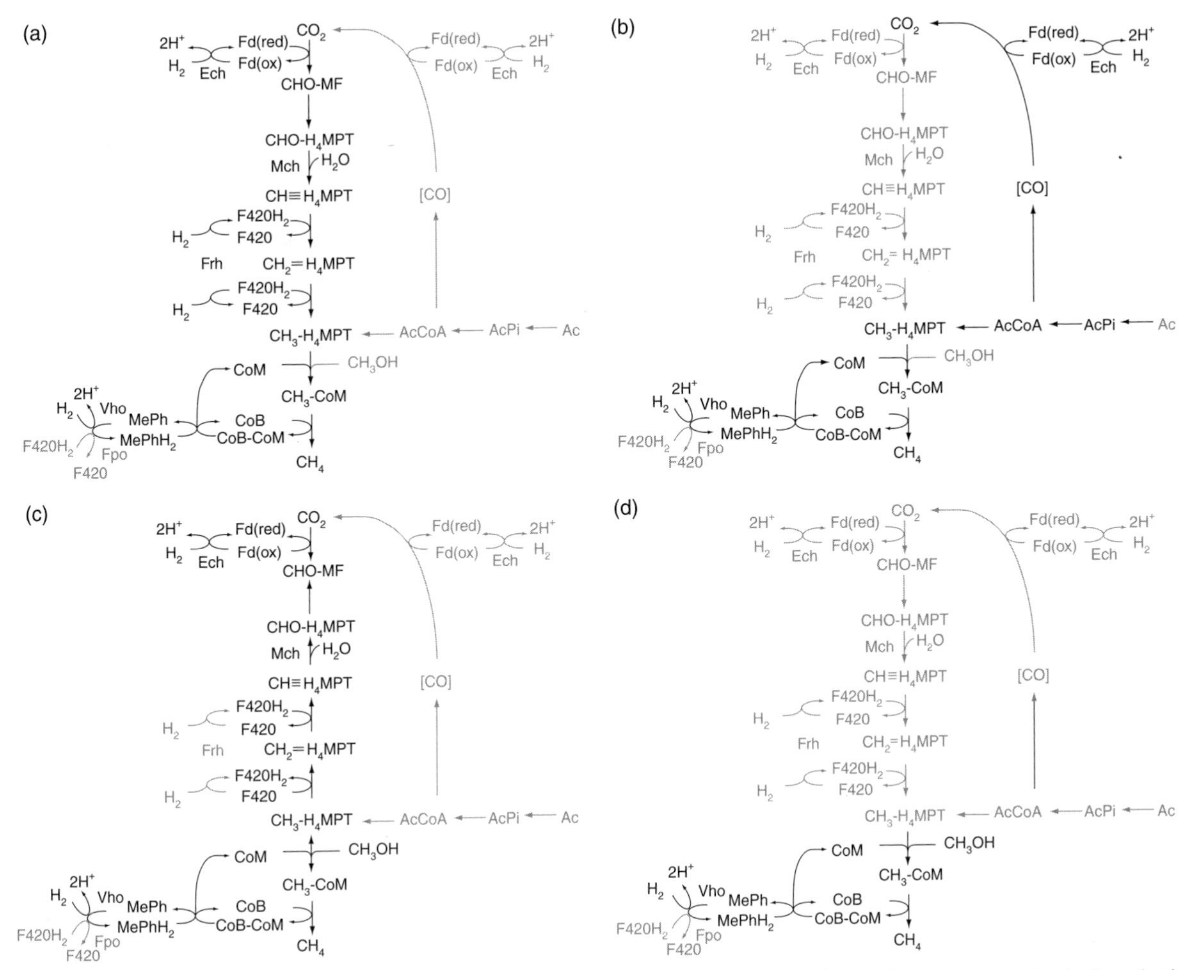

FIGURE 9 Four overlapping methanogenic pathways found in *Methanosarcina barkeri*. All four pathways share a common step in the reduction of methyl-CoM to methane; however, they differ in the pathway used to form methyl-CoM, and in the source of the electrons used for its reduction to methane. (a) Many methanogens reduce CO_2 to methane using electrons derived from the oxidation of H_2 via the hydrogenotrophic pathway. (b) In the aceticlastic pathway, acetate is split into a methyl group and an enzyme-bound carbonyl moiety. The latter is oxidized to CO_2 to provide the electrons required for reduction of the methyl group to methane. (c) In the methylotrophic pathway, C–1 compounds such as methanol or methylamines are disproportionated to CO_2 and methane; one molecule of the C–1 compound is oxidized to provide electrons for reduction of three additional molecules to methane. (d) In the methyl reduction pathway, C–1 compounds can be reduced using electrons derived from hydrogen oxidation. Ech, ferredoxin-dependent hydrogenase; Frh, F_{420}-dependent hydrogenase; Vho, methanophenazine-dependent hydrogenase; Fpo, F_{420} dehydrogenase; CHO–MF, formyl-methanofuran; CHO–H_4 MPT, formyl-tetrahydromethanopterin; CH≡H_4MPT, methenyl-tetrahydromethanopterin; CH_2=H_4MPT, methylene-tetrahydromethanopterin; CH_3–H_4MPT, methyl-tetrahydromethanopterin; CH_3–CoM, methyl-coenzyme M; CoM, coenzyme M; CoB, coenzyme B; CoM–CoB, mixed disulfide of CoM and CoB; Meph/MephH$_2$, oxidized and reduced methanophenazine; F_{420}/$F_{420}H_2$, oxidized and reduced factor 420; Fd(ox)/Fd(red), oxidized and reduced ferredoxin; Ac, acetate; Ac-Pi, acetyl-phosphate; Ac-CoA, acetyl-coenzyme A. *(Modified from Guss, A. M., Mukhopadhyay, B., Zhang, J. K., & Metcalf, W. W. (2005). Genetic analysis of* mch *mutants in two* Methanosarcina *species demonstrates multiple roles for the methanopterin-dependent C-1 oxidation/reduction pathway and differences in H_2 metabolism between closely related species.* Molecular Microbiology, *55, 1671–1680.Blackwell Publishing Ltd.)*

monomethylamine each require a distinct corrinoid-binding protein, which is methylated by a substrate-specific methyltransferase. The methyl group is then transferred from the corrinoid protein to coenzyme HS-CoM by a common MT2 homologue. In contrast, catabolism of the methylthiols, dimethylsulfide, and methylmercaptopropionate is catalyzed by only two polypeptides: a corrinoid-binding protein tightly bound to a methylcobamide:CoM methyltransferase homologue of MT2. Methyl-S-CoM generated from methanol, methylated amines, and methylthiols is reduced to CH_4 in the methanogenic pathway, as has been described. A portion of the methyl groups generated from methylotrophic catabolism is oxidized in reverse sequence in a pathway identical to the CO_2 reduction pathway, after what appears to be a direct transfer of the methyl groups to H_4MPT. However, the mechanism of this transfer is not yet known. This oxidative sequence generates electrons for the reduction of CoM-S-S-HTP in the methyl-S-CoM reductase system. Methylotrophic substrates can also be converted to methane by reduction of their methyl groups via hydrogen oxidation instead of partial oxidation of their methyl groups (Figure 9d).

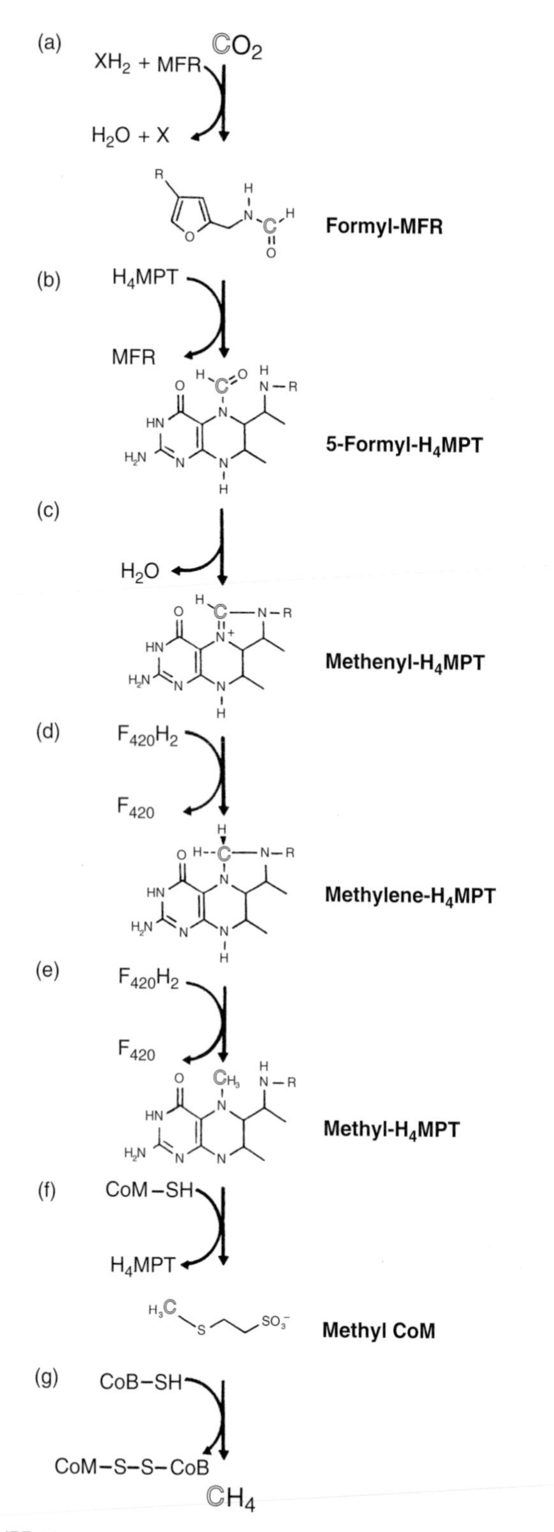

FIGURE 10 Methanogenic pathway in which CO_2 is sequentially reduced as coenzyme-bound intermediates to form CH_4. Details are described in the text. *(Reproduced with permission from Weiss, D. S., & Thauer, R. K. (1993). Methanogenesis and the unity of biochemistry.* Cell *72: 819–822. Cell Press.)*

Bioenergetics

The methanogenic *Archaea* derive their metabolic energy from autotrophic CO_2 reduction with H_2, formate, or secondary alcohols, cleavage of acetate, or methylotrophic dismutation of methanol, methylated amines, or methylthiols (Table 2). Currently, there is little evidence to convincingly support substrate-level phosphorylation. Evidence that redox reactions in the catabolic pathways are catalyzed, in part, by membrane-bound enzyme systems and are dependent upon electrochemical sodium ion or proton gradients indicates that electron transport phosphorylation is responsible for ATP synthesis. Both gradients generate ATP for metabolic energy via membrane ATP synthases. Several reactions in the methanogenic pathway are sufficiently exergonic to be coupled with energy conversion, including reduction of methyl-CoM (−29 kJ mol^{-1}) and methyl transfer from H_4MPT/H_4SPT to HS-CoM (−85 kJ mol^{-1}).

In addition, the oxidation of formyl-MF (−16 kJ mol^{-1}) during methyl oxidation in the methylotrophic pathway and CO oxidation (−20 kJ mol^{-1}) in the acetotrophic pathway are also exergonic.

During growth on H_2–CO_2 or H_2-methanol, the H_2:heterodisulfide oxidoreductase system catalyzes the H_2-dependent reduction of CoM-S-S-HTP to HS-CoM (Figure 9). Electron translocation across the membrane generates a proton gradient for ATP synthesis via an A_1A_0 ATPase. During methylotrophic growth of methanol or methylamines, heterodisulfide dehydrogenase is linked to F_{420} dehydrogenase instead of hydrogenase for CoM-S-S-HTP reduction and generation of a proton gradient. Reduced F_{420} is generated by methylene-H_4MPT dehydrogenase and methylene-H_4MPT reductase. A third type of heterodisulfide oxidoreductase system in the acetotrophic pathway is linked to acetyl-CoA decarbonylase via ferridoxin.

In addition to proton gradients, methanogens also use sodium ion gradients to drive endogonic reactions and generate ATP. During growth on H_2–CO_2 or acetate, vectoral Na^+ translocation is coupled to methyl-H_4MPT:CoM-SH methyltransferase. The Na^+ gradient generates ATP via an F_1F_0 ATPase. During growth on methanol and methylamines, a sodium ion gradient is formed by a Na^+/H^+ antiporter, and the methyl-H_4MPT:CoM-SH methyltransferase sodium pump is used in reverse to drive the endergonic methyl transfer from methyl-CoM to methyl-H_4MPT, for subsequent oxidation of the methyl groups.

Formyl-MFR dehydrogenase is involved in the bioenergetics of autotrophic growth on H_2–CO_2 or methylotrophic growth on methanol. During hydrogenotrophic growth, the exergonic CO_2 reduction to formyl-MF is driven by a hydrogenase-generated H^+ or Na^+ gradient. The reaction is reversed during CO_2 generation by methylotrophs, resulting in a net H^+ or Na^+ gradient.

Biosynthetic Pathways

Most methanogens assimilate carbon by CO_2 fixation or acetate uptake, and current evidence indicates that assimilation occurs via acetyl-CoA synthesis in a pathway analogous to the Ljungdahl–Wood pathway in acetotrophic clostridia. This conclusion is supported by (1) labeling studies; (2) the presence of acetyl-CoA synthase in autotrophic methanogens; and (3) the absence or partial absence of enzymes required for other routes of CO_2 fixation, including the Calvin cycle, reverse tricarboxylic acid (TCA) cycle, the serine pathway, and ribulose monophosphate cycle of the methylotrophs and the hydroxypropionate pathway of *Chloroflexus*. Acetotrophic methanogens likely utilize acetyl-CoA directly synthesized by acetate kinase and phosphotransacetylase in the catabolic acetate pathway. In the autotrophic methanogens, current evidence suggests that methyl-H_4MPT, formed by CO_2 fixation in the H_4MPT reductive pathway, likely donates a methyl group to a corrinoid protein containing an iron–sulfur center where it is subsequently transferred to HS-CoA by acetyl-CoA synthase. This pathway is a reversal of the acetotrophic pathway for acetate catabolism. During growth on methanol or methylated amines, methyl-H_4MPT, formed by direct methyl transfer to H_4MPT, is also the likely precursor for acetyl-CoA synthesis in methylotrophs.

Acetyl-CoA is reductively carboxylated by pyruvate oxidoreductase and enters the predominant biosynthetic pathways as pyruvate via an incomplete TCA. Both reductive and oxidative partial TCA cycles are detected in methanogens and their distribution among species is based on the bioenergetics of α-ketoglutarate synthesis. The oxidative incomplete TCA pathway requires a second acetyl-CoA and is restricted to acetotrophic methanogens, which readily synthesize acetyl-CoA from acetate via acetate kinase and phosphotransacetylase. In contrast, the reductive incomplete TCA pathway requires two additional reductive reactions and a reductive carboxylation to α-ketoglutarate synthesis. This pathway is commonly found in autotrophic methanogens, which have a steady reducing potential from H_2. Studies based on labeling and identification of specific enzymatic activities in selected species, thus far, indicate that biosynthesis of the aromatic aspartate, glutamate, histidine, pyruvate, and serine families of amino acids occurs by pathways similar to those described in the *Bacteria* and *Eukarya*. Genetic studies indicate that aromatic amino acids in some species are synthesized by the *de novo* via chorismate and by activation of aryl acids via the CoA thioesters indoleacetyl-CoA, phenylacetyl-CoA, and *p*-hydroxyphenylacetyl-CoA, which are then reductively carboxylated by indolepyruvate oxidoreductase to indolepyruvate, phenylpyruvate, or *p*-hydroxyphenylpyruvate, respectively. Biosynthesis of hexoses, required as precursors for cell walls and reserve polysaccharides, such as glycogen and trehalose, proceeds via gluconeogenesis from phosphoenolpyruvate. Pentoses are formed from fructose 1,6-phosphate and glyceride 3-phosphate via transketolase and transaldolase in *Methanobacterium* and *Methanococcus*. In contrast, *Methanospirillum* forms pentoses by oxidation of glucose-6-phosphate. Labeling studies in *Methanospirillum* and *Methanobacterium* indicate that pyrimidine and purines are synthesized by the expected pathways and they also have a purine salvage pathway. The polar core lipids, which include diphytanyl glycerol ethers and diphytanyl diglycerol tetraethers, are synthesized via the mevalonate pathway for isoprenoids.

MOLECULAR GENETICS

Genome Structure

The methanogens have a single circular chromosome that ranges from 1.6×10^6 to 5.7×10^6 bp in size. The percentage guanosine + cytosine (%GC) ranges from 23% to 62% and high %GC content is not always associated with high growth temperatures. Transposable elements are detected in all methanogen genomes. A three-way comparison between *M. acetivorans*, *Methanosarcina barkeri*, and *Methanosarcina mazei* reveals that genomes are well conserved with respect to the region proximal to the origin of replication, but the second half of the genomes most distant from the origin are disordered and marked by increased transposase frequency and a decrease in gene density and conserved colocalization (synteny). The apparent genome plasticity might contribute to these species' ability to adapt to a broad range of environments as a result of genome elongation and enrichment for favorable phenotypes. Clustered regularly interspaced short palindromic repeats (CRISPRs), which can cause large-scale rearrangements, and all four known CRISPR-associated, or *cas*, genes have been detected in methanogen genomes, but they do not appear to be associated with large-scale DNA rearrangements.

Extrachromosomal DNA has been detected in methanogens in the form of large extrachromosomal elements, plasmids, viruses, and virus-like particles. At this time, two large extrachromosomal elements have been detected in *Methanocaldococcus jannaschii* (58.4 kb) and *M. barkeri* (36 kb) and 12 smaller plasmids, ranging in size from 4.4 to 20 kb, have been detected in four genera of the methanogenic *Archaea*. The large extrachromosomal element in *M. jannaschii* encodes a putative minichromosome maintenance (MCM) helicase and the element in *M. barkeri* encodes a homologue of the DNA replication initiation protein, Cdc6, in a region of highly repetitive sequence, which suggests that replication of the large extrachromosomal elements is synchronous with cell division. In contrast, homologues of site-specific recombinase and plasmid DNA replication initiator protein RepA detected in *M. acetivorans* suggests that this plasmid replicates by a rolling circle mechanism associated with many bacterial plasmids and phage. Other open reading frames annotated in methanogen plasmids include a type II DNA restriction–modification system, glycosylase transferase, and recombinases. However, genes involved in central metabolism have not yet been detected and the functions of the most methanogen plasmids are cryptic at this time. Lytic viruses have been detected in three strains of *Methanothermobacter* and one strain of *Methanobrevibacter smithii*. These viruses show a varying range of host specificities. A temperate virus-like particle has been isolated that can integrate into the chromosome of *Methanococcus voltae*.

In contrast to the similarities of DNA structure in methanogenic *Archaea* and *Bacteria*, DNA-modifying proteins in the methanogens share features common to both the *Bacteria* and *Eukarya* (Table 3). All cells must package their genomic DNA within the limited space of the cells. Although the *Bacteria* do not appear to have a conserved mechanism for DNA packaging, all *Eukarya* and some *Archaea* (one crenarchaeon and all Euryarchaeota) chromosomal DNAs are compacted by histones into defined structures called nucleosomes, which are further assembled to form chromatin. Methanogen histones form a structural motif called a histone fold, but lack N-and C-terminal sequences associated with higher order nucleosome polymerization and undergo modification in eukaryotes. Unlike the four histones in eukaryotes, methanogens typically encode 1–6 histones. Interestingly, *Methanosarcina* spp., which have the largest genomes, only encode one histone and it is not clear how this affects dimer polymerization. Archaeal histone-like proteins increase the melting temperature of linear DNA *in vitro* by as much as 25 °C and likely protect DNA from heat denaturation *in vivo*. Reverse gyrase, also detected in *Methanothermus fervidus*, may contribute to heat resistance of DNA by creating stable positive supercoils in inter-histone regions. Histone-like proteins isolated from mesophilic *Methanosarcina* species cause concentration-dependent inhibition or stimulation of gene transcriptions *in vitro*, which suggests that they might also have a role in gene regulation. Direct evidence that *M. jannaschii* histones does not have posttranslational modifications indicates that gene regulation by histone modification is unlikely, but this does not rule out the possibility that different sequence affinities might regulate genes by selective proximity of nucleosomes to regulatory regions of DNA.

DNA Replication, Repair, Modification, and Metabolism

Unlike replication in the *Bacteria*, which starts at a single origin, *in silico* analysis indicates that there are up to three origins in some species of methanogens. Multiple origins have been experimentally determined in the *Crenarchaea*, which suggests that some species of methanogenic *Archaea* may also have multiple origins. At least one of the origins lies immediately upstream of genes encoding homologues of the eukaryal DNA replication initiation proteins Orc and Cdc6. The proximity of the replication factors to the origin suggests that the methanogenic *Archaea* use a bacterial-like mode of replication, but with eukarya-like machinery. DNA replication proteins in methanogens are similar to those of other members of the euryarchaeal kingdom and most similar to eukaryal enzymes. Studies on the replication

TABLE 3 Molecular features of three phylogeneic domains

Feature	Eukarya	Bacteria	Archaea
Genome	Multiple linear	Single circular[a]	Single circular[a]
Chromatin	Histone mediated	No conserved mechanism	Histone mediated
Extrachromosomal DNA	Plasmids, viruses, IS units	Plasmids, viruses, IS units	Plasmids, viruses, IS units
DNA polymerase	Families A[b], B, and X	Families A, B, and C	Families B and X
RNA polymerase	Three classes, complex	One class $\beta\beta'\alpha_2\sigma$	One class, complex[c]
Gene structure	Single gene	Multiple gene operon	Multiple gene operon
Transcription promoter	TATA box	−35/−10 sequence	TATA box
Transcription terminator	AAUAAA sequence	Intrinsic, σ-dependent	Intrinsic, oligo T
Ribosomal RNA	28S-5.8S/18S	23S-5S/16S	23S-5S/16S
Translation initiation	5′ cap	Ribosomal-binding site	Ribosomal-binding site
Initiator transfer RNA	Methionine	Formylmethionine	Methionine

[a]Some species have multiple copies of the same genome.
[b]Mitochondrial.
[c]AB′B″C based on homology to eukaryal RNA polymerase II.

machinery of a number of methanogenic *Archaea* show that most have homologues of the eukaryotic initiation proteins Orc/Cdc6, which are thought to function in origin recognition and helicase loading. All methanogenic *Archaea* also contain at least one homologue of the replicative helicase, MCM. The replicative machinery also includes the proteins needed for the elongation phase including primase, DNA polymerases (B-and D-type), the PCNA processivity factor and its loader RFC, the GINS complex, thought to function in coordination of leading and lagging strand replication, and ligase, Fen1 and RNase H needed for Okazaki fragment maturation.

DNA repair mechanisms have also been identified in methanogens. Annotation of genome sequences reveals putative genes for alkyltransferase, base-excision repair, photoreactivation, nucleotide excision repair, *trans*-lesion synthesis, mismatch repair, and homologous recombination among species of the methanogenic *Archaea*. A photoreactivation system in *Methanothermobacter thermoautotrophicus* is mediated by class II photolyase with homology to metazoan photolyases. Homologues of putative DNA include eukaryal repair proteins RAD2, RAD25, and RAD51; and bacterial DNA repair proteins uvrABC, RadA, mutL, and mutS. Although our understanding of DNA repair in the *Bacteria* and *Eukarya* is well advanced, there has been comparatively little functional analyses of these processes in the methanogenic *Archaea* at the present time.

Gene Structure and Transcription

The organization of methanogenic genes in tightly linked clusters is similar to the operon configuration found in the *Bacteria*. As in the *Bacteria*, the archaeal operons are transcribed from an upstream promoter into polycistronic RNAs such that individual mRNA molecules encode multiple genes and transcription and translation are coupled. However, archaeal genes are transcribed by a multicomponent RNA polymerase that is structurally homologous to eukaryal RNA polymerase and recognizes a promoter with high sequence homology to the consensus TATA motif found in eukaryal promoters and B recognition element (BRE) immediately upstream. Unlike the variable upstream distance of eukaryal promoters, the archaeal promoter element is located at a consistent distance, approximately 20–30 bp upstream from the transcription initiation site, a range similar to the conserved −35 bp region observed in the *Bacteria*. Site-directed deletion studies conducted with methanogenic *Archaea* by *in vitro* techniques indicate that efficient transcription and start-site selection is dependent upon a TATA promoter. This arrangement closely resembles the core structure of promoters recognized by RNA polymerase II (RNAP II) in the *Eukarya*. Purified archaeal RNAPs from *M. thermoautotrophicus* and *M. voltae* fail to initiate site-specific transcription without the addition of a protein that binds to the TATA promoter (TBP) and transcription factor B (TFB). Yeast and human TATA-binding proteins can substitute for TBP in a *Methanococcus*-derived archaeal cell-free transcription system, indicating that they are functionally homologous. An additional gene that has sequence similarity to a eukarya-like transcription factor E (TFE) has also been identified in the methanogenic *Archaea*. Site-specific transcription initiation in the *Archaea* is analogous to that in the *Eukarya*. The mechanism involves binding of the TBP to the archaeal TATA box, stabilization of the complex by association of TFB with TBP and interaction with the BRE, and recruitment of the RNAP TFE facilitates binding of TFB to RNAP and further stabilizes the initiation complex by closing the RNAP clamp on the DNA strand.

Mechanisms of transcriptional regulation in highly regulated genes have been recently reported. Regulatory protein Ptr2 in *M. jannaschii*, an archaeal homologue of the bacterial leucine-responsive regulatory protein Lrp, is a transcriptional activator for expression of ferridoxin A and rubredoxin 2. However, unlike bacterial activators that interact with RNAP, Ptr2 activates transcription by recruiting transcription factor. Negative regulators have also been described in the methanogens. TrpY is an allosteric effector that both autoregulates itself and regulates transcription of the tryptophan biosynthesis operon. TrpY forms a dimer that inhibits TrpY expression by blocking RNAP recruitment. When tryptophan binds to TrpY a conformational change increases its affinity for an adjacent, divergent operator for the tryptophane encoding operon (*trp*), thereby competing with TBP binding to the promoter. NrpR regulates nitrogen fixation by repressing the operon encoding nitrogenase (*nif*) during growth with alternative nitrogen sources ammonia or alanine. In the absence of ammonia or alanine, an indicator of intracellular nitrogen level, 2-oxyglutarate, binds to NrpR and reduces its affinity for the *nif* operator, thereby derepressing transcription of the *nif* operon. Regulators involving the 5′ untranslated region (5′-UTR) of the mRNA have also been reported in the methanogens. A 113 nucleotide 5′-UTR upstream of an RNA helicase in the Antarctic species *Methanococcoides burtonii* contains bacteria-like regulatory elements that appear to be involved in cold adaptation. Long 5′-UTRs associated with methanol methyltransferase and carbon monoxide dehalogenase in *Methanosarcina* spp. have also been shown to be involved in regulation. Although the mechanisms of regulation have not been determined, five genes designated msrABCDE – identified by deletion analysis – are involved in repression or induction of methanol methyltransferase isozymes. In the case of carbon monoxide dehalogenase, putative secondary structure has been identified in the 5′-UTR that might be involved in differential termination of transcriptional elongation. Genes encoding three distinct TBPs have been detected in the *M. acetivorans* genome, which raises the possibility that gene regulation is mediated by differential TBP–promoter pairing. Only TBP-1 is expressed during growth of *M. acetivorans*, but this does negate the possibility that the different TBPs might be differentially expressed under different growth conditions, such as a response to stress. This possibility is supported by evidence of differential TBP – promoter pairing in the closely related halophilic *Euryarchaeota*. Generally, transcription is terminated following an inverted repeat sequence located downstream of methanogen genes. Transcription termination sites are similar to the ρ-independent terminators in the *Bacteria*, and likely form a stem-loop secondary structure to mediate termination. A second type of transcription terminator, which consists of a single or several tandemly arranged oligo-T sequences, is found in hyperthermophilic methanogens. The occurrence of this terminator in hyperthermophiles suggests that the stem-loop structures characteristic of σ-independent promoters may be unstable at higher growth temperatures.

RNA Structure and Translation

The stable RNAs transcribed by methanogen genes have been investigated in some detail. Methanogen ribosomes resemble bacterial ribosomes. They are composed of two protein subunits of 30S and 50S and three rRNA components of 23S, 16S, and 5S, which, assembled, yield a ribosome of 70S. Archaeal and bacterial ribosomal proteins are functionally homologous and can be interchanged to create an active ribosome *in vitro*. Genes encoding rRNA are arranged in the order 16S-23S-5S and the number of operon copies varies in number from 1 to 4. The organization and order of genes encoding methanogen ribosomal proteins also resembles that found in the *Bacteria*. Methanogen transfer RNAs (tRNAs) contain the sequence 1-methylψ CG substituted for the sequence TψCG, typically found in the arm of bacterial and eukaryal tRNAs. Introns have been detected in genes encoding tRNA. Several methanogenic *Archaea* lack cysteinyl-tRNA synthetase. In these cases, a class II-type *O*-phosphoseryl-tRNA synthetase specifically forms Sep-tRNACys, which is then converted to Cys-tRNACys by Sep-tRNA:CystRNA synthase. Several methanogenic *Archaea* also synthesize two nonstandard amino acids: selenocysteine, found also in *Bacteria* and animals, and pyrrolysine, which is thus far unique to methanogenic *Archaea* with genes encoding dimethylamine or trimethylamine methyltransferases. Both selenocysteine and pyrrolysine are encoded by the RNA stop codons UGA and UAG, respectively. However, unlike selenocysteine that is synthesized on the charged tRNA, pyrrolysine is charged onto a pyrrolysine-specific tRNA.

Some methanogen mRNAs have poly-A^+ tails, but, as in the *Bacteria*, they average only 12 bases in length. Protein-encoding genes employ the same genetic code as *Bacteria* and *Eukarya*, and codon preferences reflect the overall base composition of the genome and the level of gene expression. The codon ATG is frequently used as an initiation codon as well as GTG and TGG. A ribosomal-binding site located upstream of structural genes, when transcribed, is complementary to the 3′ terminal sequence of methanogen 16S rRNA. Compared with *Bacteria* and *Eukarya*, there is little information on the mechanisms of translation based on biochemical experimentation. Computational analyses of methanogen genomes have identified six putative initiation factors, two elongation factors, and one release factor that are most similar to eukaryal homologues. Inteins have been identified by genomic sequencing of *M. jannaschii*, which suggests that protein splicing occurs in methanogens. RNA processing by endonucleolytic cleavage 12–16 nucleotides upstream of the translation start site of 30 protein-coding genes was identified in *Methanocaldococcus jannaschii*. In addition, evidence has been found for phosphorylation of proteins at a tyrosine residue, which is a mechanism of post-translational control in the *Bacteria* and *Eukarya*.

Genomics and Gene

Function analysis

The genomes of over 25 methanogenic strains have been completely sequenced and annotated. In general, similar genes are found for the CO_2-reducing catabolic pathways in all four genomes. Genes encoding multiple methyltransferases and acetyl-CoA decarbonylase/synthases are found only in the genomes of *Methanosarcina* spp., which is consistent with the ability of this species to also grow by methylotrophic and aceticlastic pathways. The large genome size of *Methanosarcina* spp. likely reflects these species' ability to use a greater range of substrates, adapt to a broader range of environments, and form complex multicellular structures compared with the more limited capabilities of the hydrogen-utilizing species. In contrast, the hyperthermophiles are more "minimalist" exhibiting a paucity of genes encoding proteins for signaling and gene regulation, *grp*E–*dna*J–*dna*K heat-shock operon, proteasome–chaperonin, several DNA repair proteins, DNA helicases, nitrogenase subunits, ribonucleotide reductase, and proteases. Despite their sensitivity to oxygen, superoxide dismutase and catalase have been identified in methanogen genomes. Superoxide dismutase and catalase activities have been confirmed and characterized in methanogens, and genes encoding both enzymes are transcriptionally upregulated in *M. barkeri* upon oxidative stress. Interestingly, a putative cytochrome *d* has been identified in *Methanosarcina* genomes, but its role has not been determined. Overall, the majority of archaeal open reading frames with similarity to bacterial sequences include genes for small molecule biosynthesis, intermediary metabolism, transport, nitrogen fixation, and regulatory functions.

Archaeal open reading frames with similarity to eukaryal sequences include genes for DNA metabolism, transcription, and translation. The presence of Cdc6 homologues and histones suggests that DNA replication initiation and chromosome packaging is eukaryal, but detection of cell division gene *ftsZ* suggests that bacteria-type cell division occurs. Additional unique features include an archaeal B-type DNA polymerase with two subunits, putative RNAP A′ subunits that suggest possibility of additional mechanisms for gene selection, and two introns in the same $tRNA^{Pro}$ (CCC) gene that establishes a new precedent.

The availability of genome sequences of methanogens and current activities to sequence others make the development of archaeal gene-transfer systems essential for confirmation of gene function. Two gene-transfer systems have been developed for species of *Methanococcus* and *Methanosarcina*. Both systems utilize hybrid shuttle vectors derived from native archaeal plasmids. Since plasmids occur in low copy number in the methanogens, the DNA to be transferred is ligated into the vector, which is then amplified in *E. coli* with ampicillin as a selectable marker. The modified plasmid is then transferred into the methanogen using transformation methods that employ either cells made competent for DNA uptake by polyethylene glycol treatment (*Methanococcus* sp.) or uptake of DNA embedded in liposomes (*Methanosarcina* spp.). A second selectable marker in the plasmid, such as *pac* (puromycin acetyl transferase) controlled by a highly expressed methanogen promoter *mcr* (methyl CoM reductase), provides selection in the methanogen. Both autonomously replicating plasmids for introducing specific phenotypes and integration plasmids for disrupting genes by homologous recombination have been designed. These systems are highly efficient, yielding 10^7–10^9 transformants per μg DNA. Methods for gene disruption based on homologous recombination and transposon-mediated mutagenesis, gene reporters for expression analysis, and protein overexpression have also been developed. One current limitation is the limited number of selectable and phenotypic markers that are currently available. This has been addressed by the development of a "markerless" system in which the genetic marker is removed by homologous recombination after selection. A transducing phage has been isolated from *M. thermoautotrophicus*, but it is not currently useful for gene transfer because of its limited burst size (~6/cell). Conjugation has not been observed in methanogens. Although gene-transfer systems are somewhat limited at this time, development of new and more sophisticated systems is ongoing.

RECENT DEVELOPMENTS

There have been several recent discoveries in the field of methanogenesis.

Diversity and Phylogeny – Several new species were described. Among these species, *Methanocella paludicola* represents the first isolate from a previously described clonal lineage associated with rice paddies. A sixth order, *Methanocellales*, was proposed for this new species based on comparative sequence analyses of the 16S rRNA and *mcrA* genes.

Habitats – The metagenome and mRNA expression analysis of the microbial consortia mediating the anaerobic oxidation of methane with sulfate (ANME) are consistent with acetate and formate as putative electron shuttles. Expression of putative secreted multiheme c-type cytochromes suggests direct electron transfer as an additional possible mode to shuttle electrons from ANME-1 to the bacterial sulfate-reducing partner, but a methanogenic isolate and confirmation of the actual mechanism for this anaerobic process remain elusive.

Molecular genetics – In the area of nucleic acids processing, binding of the NrpR repressor protein in *Methanococcus maripaludis* was discovered to occur simultaneously at the two overlapping operator sequences of *glnK* on opposite faces of the double helix. In a recent report the first operator is shown to have a primary role in regulation and the second an enhancing role demonstrating for the first time functional overlapping operators in the Archaea. Systematic examination for evidence of RNA processing upstream of protein-coding genes RNAs in *Methanocaldococcus jannaschii* indicates that endonucleolytic cleavage occurs 12–16 nucleotides upstream of the corresponding translation start site in 30 genes. This processing alters the sequence of several genes in the RNA pool and could affect significantly the perceived balance of proteins in the cell.

Genomics and Genes – A number of additional genomes have been sequenced. Among them, the genome of a rumen methanogen *Methanobrevibacter ruminantium* was sequenced recently to identify new possibilities for mitigating methanogenic activity in ruminants. One approach currently being tested is to target specific methanogen proteins with antibodies and harness the immune system of ruminants to increase feed efficiency and mitigate greenhouse gas emissions from agriculture.

Methanosarcina barkeri mutants for the proton-pumping F420H2 dehydrogenase *fpo* display no measurable phenotype on methanol whereas *frh* mutants lacking the cytoplasmic F420-reducing hydrogenase (Frh) are severely affected during growth on methanol. These data suggest that the preferred electron transport chain involves a hydrogen-cycling via cytoplasmic production, diffusion through the membrane and subsequent generation of a proton motive force that can be used by the cell for ATP synthesis. The relatively rapid enzymatic turnover of hydrogenases may allow a competitive advantage by promoting faster growth rates. A second coenzyme B-coenzyme M heterodisulfide, HdrED, was detected in the genome of *Methanosarcina* spp. and shown to be up-regulated in response to growth with methanol. Gene mutation analysis suggests that HdrABC is not essential and that bifurcation might occur between the cytoplasmic HdrABC and the methanosarcinal membrane-bound HdrED described previously.

CONCLUSION

The methanogenic *Archaea* have a pivotal role in the global carbon cycle by complementing aerobic processes that ultimately lead to the oxidation of organic carbon to CO2. However, a steady increase in the levels of atmospheric CH_4 that has coincided with the increase in the human population is a cause for concern, since CH_4 is a greenhouse gas. Methane's contribution to global warming results from its high infrared absorbance and its role in complex chemical reactions in the stratosphere that affect the levels of ozone. Increased waste disposal activities, such as landfills, are a significant source of atmospheric CH_4. Another significant source results from agricultural activities, such as the increased use of domesticated ruminants for production of meat and dairy products and the increased development of rice paddies. Understanding the properties of methanogens and the roles they have in the global carbon cycle will have important implications in addressing the issue of global warming as the human population increases. The ability of geological hydrogen-utilizing CO_2-reducing methanogens to serve as primary producers of reduced carbon compounds in deep subsurface bedrock and submarine vents has significant implications also in astrobiology and the potential for extraterrestrial life.

The application of methanogens in biotechnology has been largely limited to waste management, which is often coupled to limited biogas production. Although the petroleum crisis of the 1970s led to an interest in methanogenic biogas production by fermentation of sources ranging from agricultural products to marine kelp, cost-effective technologies were never fully developed as interest waned with the drop in petroleum prices. Recent rises in petroleum prices and international concern about global warming have renewed interest in alternative energy sources. Methanogenic biogas is an attractive energy source for applications such as heating and electric generation plants as it requires minimal treatment and can be distributed in current infrastructures used for natural gas. Unlike petroleum, biogas is generated from organic waste and CO_2 released from combustion of biogas does not cause a net increase of greenhouse gases into the atmosphere. Similarities in the DNA expression machinery of mesophilic methanogens and *Eukarya* combined with recent advances in methanogen genetic systems provides a potential platform for functional analyses of mammalian gene and proteins. Other potential applications for methanogens include the production of novel pharmaceuticals, corrinoids, and thermo-stable enzymes. To date, methanogens have yielded only a few restriction endonucleases as commercial products. However, significant advances in our understanding of the physiology and biochemistry of methanogens over the past two decades, combined with the recent developments in gene transfer systems and advances in genome sequencing, make the application of methanogens for biotechnology more likely in the near future.

FURTHER READING

Barry, E. R., & Bell, S. D. (2006). DNA replication in the Archaea. *Microbiology and Molecular Biology Reviews*, *70*, 876–887.

Cavicchioli, R. (2007). *Archaea – Molecular and cellular biology*. Washington, DC: ASM Press.

Deppenmeier, U., Muller, V., & Gottschalk, G. (1996). Pathways of energy conservation in methanogenic Archaea. *Archives of Microbiology*, *165*, 149–163.

Ferry, J. G. (1993). Methanogenesis. In C. A. Reddy, A. M. Chakrabarty, A. L. Demain, & J. M. Tiedje (Eds.). *Microbiology series* 536. New York: Chapman and Hall.

Geiduschek, E. P., & Ouhammouch, M. (2005). Archaeal transcription and its regulators. *Molecular Microbiology*, *56*, 1397–1407.

Lange, M., Westermann, P., & Ahring, B. K. (2005). Archaea in protozoa and metazoa. *Applied Microbiology and Biotechnology*, *66*, 465–474.

Rother, M., & Metcalf, W. W. (2005). Genetic technologies for Archaea. *Current Opinion in Microbiology*, *8*, 745–751.

Sowers, K. R., & Ferry, J. G. (2002). Marine methanogenesis. In G. Bitton (Ed.). *The encyclopedia of environmental microbiology* (pp. 1913–1923). New York: John Wiley & Sons, Inc.

Sowers, K. R., & Schreier, H. J. (1995). Methanogens. In F. T. Robb, K. R. Sowers, S. DasSharma, A. R. Place, H. J. Schreier, E. M. Fleischmann (Eds.) *Archaea: A laboratory manual* (p. 540). Cold Spring Harbor: Cold Spring Harbor Laboratory Press.

Thauer, R. (1997). Biodiversity and unity in biochemistry. *Antonie van Leeuwenhoek International Journal of General Microbiology*, *71*, 21–32.

Chapter 12

Methanotrophy

T.J. Smith[1] and J.C. Murrell[2]
[1]Sheffield Hallam University, Sheffield, UK
[2]University of Warwick, Coventry, UK

Chapter Outline

ABBREVIATIONS

FISH Fluorescence *in situ* hybridization
JGI Joint Genome Institute
MDH Methanol dehydrogenase
MMO Methane monooxygenase
pMMO Particulate MMO
RuBisCo Ribulose bis-phosphate carboxylase/oxygenase
RuBP Ribulose bis-phosphate
RuMP Ribulose mono-phosphate
SCP Single-cell protein
sMMO Soluble MMO
TCE Trichloroethylene

DEFINING STATEMENT

Methanotrophic (methane-oxidizing) bacteria are diverse, environmentally ubiquitous microorganisms that are important in controlling global warming. The methane monooxygenase (MMO) enzymes that they produce can oxidize the unreactive methane molecule and, by virtue of their wide substrate ranges, have current and potential applications in bioremediation and synthesis of fine and bulk chemicals.

INTRODUCTION

Methane oxidizing bacteria, or methanotrophs, are a widely distributed group of aerobic microorganisms that use methane as their sole source of carbon and energy. They synthesize all of their cellular carbon-containing molecules from methane, and can also co-oxidize a wide range of hydrocarbons. Thus, they have found applications in production of single-cell protein (SCP) from natural gas and bioremediation of trichloroethylene (TCE)-contaminated groundwater. The majority of methanotrophs are Gram-negative bacteria that are classified as type I or type II methanotrophs according to whether they belong to the *Gammaproteobacteria* and *Alphaproteobacteria*, respectively. Methanotrophs contain the unique enzyme MMO. MMOs oxidize methane to methanol and catalyze a very large number of adventitious oxidation reactions that are the basis of the ability of methanotrophs to carry out processes in biocatalysis and bioremediation.

ECOLOGY AND TAXONOMY OF METHANOTROPHS

Methanotrophs play an important role in the oxidation of methane in the natural environment. They oxidize methane produced geothermally and by the anaerobic metabolism of methanogenic bacteria, thereby reducing the release of methane to the atmosphere from landfill sites, wetlands, and rice paddies. Methanotrophs can also oxidize very low concentrations of methane (~2 ppmv) found in the atmosphere. They are therefore key players in the global methane cycle. Methanotrophs are ubiquitous in nature and have been isolated from many environments including soils, peatlands, rice paddies, sediments, freshwater and marine systems, acidic hot springs, mud pots, alkaline soda lakes, cold environments, and tissues of higher organisms.

Topics in Ecological and Environmental Microbiology.

Pioneering work by Whittenbury and colleagues in the late 1960s provided a taxonomic framework for methanotrophs, based on their cell morphology, intracellular membrane structure, carbon assimilation pathways, and cell wall components. This divided methane oxidizers into type I and type II methanotrophs (*Gammaproteobacteria* and *Alphaproteobacteria*, respectively). Methanotrophs were generally mesophilic, neutrophilic bacteria of the genera *Methylomonas*, *Methylobacter*, *Methylococcus* (type I), and *Methylosinus* and *Methylocystis* (type II). Since then several new genera of methanotrophs have been described, including psychrophiles (*Methylosphaera*), thermophiles (*Methylocaldum*, *Methylothermus*), halophiles (*Methylohalobius*), and moderate acidophiles (*Methylocella* and *Methylocapsa*). Recent discoveries have revealed a previously unknown diversity among methanotrophs. Two filamentous methanotrophs, *Crenothrix polyspora* and *Clonothrix fusca*, have been described, as well as the first fully authenticated facultative methanotroph, *Methylocella silvestris*, which can grow on either methane or multicarbon compounds. Even more remarkable is the recent isolation of three thermoacidophilic methanotrophs, which belong to the bacterial phylum *Verrucomicrobia* and are hence only distantly related to the proteobacterial methanotrophs. These three new isolates, provisionally named *Methylokorus*, *Acidimethylosilex*, and *Methylacidiphilum*, remarkably grow at pH 1.5 and at 65 °C. The characteristics of all current genera of methanotrophs are described in Table 1.

Despite remarkable successes in isolation and characterization of a wide range of methanotrophs from various environments, with the advent of molecular microbial

TABLE 1 Classification of methanotroph genera

Genus name	Phylogeny	MMO type	C_1 Assimilation	ICM type[a]	N_2 Fixation	G + C (mol.%)	Major PLFA[b]	Trophic Niche
Methylobacter	γ Proteobacteria	pMMO	RuMP	type I	No	49–54	16:1	some psychrophilic
Methylosoma	γ Proteobacteria	pMMO	not known	type I	Yes	49.9	16:1	not extreme
Methylomicrobium	γ Proteobacteria	pMMO+/– sMMO	RuMP	type I	No	49–60	16:1	halotolerant; alkaliphilic
Methylomonas	γ Proteobacteria	pMMO+/– sMMO	RuMP	type I	some	51–59	16:1	some psychrophilic
Methylosarcina	γ Proteobacteria	pMMO	RuMP	type I	No	54	16:1	not extreme
Methylosphaera	γ Proteobacteria	pMMO	RuMP	ND[c]	Yes	43–46	16:1	psychrophilic
Methylococcus	γ Proteobacteria	pMMO +sMMO	RuMP/ Serine	type I	Yes	59–66	16:1	thermophilic
Methylocaldum	γ Proteobacteria	pMMO	RuMP/ Serine	type I	No	57	16:1	thermophilic
Methylothermus	γ Proteobacteria	pMMO	RuMP	type I	No	62.5	18:1/16:0	thermophilic
Methylohalobius	γ Proteobacteria	pMMO	RuMP	type I	No	58.7	18:1	halophilic
Methylocystis	α Proteobacteria	pMMO+/– sMMO	Serine	type II	Yes	62–67	18:1	some acidophilic
Methylosinus	α Proteobacteria	pMMO +sMMO	Serine	type II	Yes	63–67	18:1	not extreme
Methylocella	α Proteobacteria	sMMO	Serine	NA[d]	Yes	60–61	18:1	acidophilic
Methylocapsa	α Proteobacteria	pMMO	Serine	type III	Yes	63.1	18:1	acidophilic
Crenothrix	α Proteobacteria	pMMO		type I				not extreme
Clonothrix	α Proteobacteria	pMMO		type I				not extreme
Methylokorus	Verrucomicrobia	pMMO	Serine, RuMP?	type IV	No?	?	18:0	thermoacidophilic
Acidimethylosilex	Verrucomicrobia	pMMO	?	?	?	?	?	thermoacidophilic
Methylacidiphilum	Verrucomicrobia	pMMO	?	?	?	?	?	thermoacidophilic

[a]ICM, intracellular membrane.
[b]PLFA, phospholipid fatty acid.
[c]ND, not determined.
[d]NA, not applicable because ICMs are very limited in this genus.

ecology techniques it is now clear that there may well be many more methanotrophs out in the environment which we do not have in culture in the laboratory. A number of molecular biology methods have recently been developed to study the structure and function of methanotrophs directly in environmental samples without the need for cultivation. The taxonomic framework for extant methanotrophs has provided 16S rRNA gene sequence information for a large number of representative methanotrophs. Isolation of DNA directly from environmental samples, PCR amplification, and subsequent analysis of 16S rRNA genes present in the environment can to some extent provide a snapshot of the totality of methanotrophs that are present. In addition, a number of "functional gene probes" have been developed, based on genes encoding key enzymes of the methane oxidation pathway (Figure 1). These include genes that are unique to methanotrophs, such as *pmoA* (encoding an active site subunit of particulate MMO (pMMO), found in nearly all methanotrophs) and *mmoX* (encoding the active site subunit of the hydroxylase of soluble MMO (sMMO), which is found in some but not all methanotrophs) (see Table 1). Another important target for functional gene probing is *mxaF*, encoding the large subunit of methanol dehydrogenase (MDH), which catalyzes the second step in biological methane oxidation and is also involved in bacterial metabolism of other one-carbon substrates.

The availability of a large database of methanotroph gene sequences allows researchers to isolate these genes directly from the environment by PCR and assess the diversity of methanotrophs present in that environment. The 16S rRNA gene sequence information can subsequently be used to design fluorescence *in situ* hybridization (FISH) experiments where fluorescently labeled nucleic acid probes are used for specific visualization and quantitation of methanotrophs in environmental samples. Quantitative PCR can also be used to directly assess the number of copies of functional genes (and hence abundance of methanotrophs) in environmental samples. More rapid screening of methanotroph gene sequences retrieved directly from the environment is facilitated by the availability of a functional gene microarray, which contains *pmoA*-specific oligonucleotides, targeting all known pMMO genes. This has been used to determine the community structure of methanotrophs in a number of environments including landfill cover soils and wetlands. Methanotrophs have been an excellent model system for the development of molecular ecology techniques. Stable isotope probing using $^{13}CH_4$ has been used to determine the diversity of active methanotrophs in several environments including peatlands, caves, and soda lakes. Such studies have revealed the remarkable diversity of methanotrophs in the environment and it is clear that their activity plays an important role in cycling of methane in many environments, which helps mitigate the effects of this potent global warming gas.

PHYSIOLOGY AND BIOCHEMISTRY OF METHANOTROPHS

MMO, the principal defining enzyme of methanotrophs, exists in two structurally and biochemically distinct forms, pMMO and sMMO. pMMO is a copper-containing enzyme that is associated with unusual intracellular membranes that take the form of vesicular disks in type I methanotrophs and paired peripheral layers in type II organisms. sMMO is a cytoplasmic nonheme iron enzyme complex.

Methanol, the initial oxidation product of methane, is oxidized to formaldehyde by a PQQ-dependent MDH. Formaldehyde is an important branch point in methylotrophic metabolism (Figure 1) and it appears that multiple pathways for metabolism of formaldehyde are common in methanotrophs. The methanotrophs possess two cyclical pathways for fixation of formaldehyde, the serine and the ribulose monophosphate (RuMP) and serine cycles, which are active in type I and type II methanotrophs, respectively. In addition, some methanotrophs can fix carbon dioxide into biomass via the ribulose bis-phosphate (RuBP) cycle and its key enzyme ribulose bis-phosphate carboxylase/oxygenase (RuBisCo).

Both forms of MMO, sMMO, and pMMO, can not only oxidize methane to methanol but also co-oxidize a range of hydrocarbons and chlorinated pollutants, and hence are responsible for much of the biotechnological potential of methanotrophs. In methanotrophs that can express either

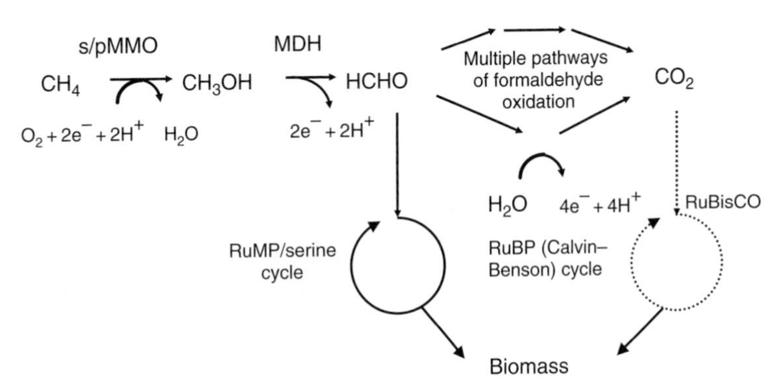

FIGURE 1 Pathways of methane oxidation and carbon assimilation in methanotrophs.

form of MMO, expression is regulated by the availability of copper ions. The copper-containing pMMO is produced at high copper-to-biomass ratios and the nonheme iron-containing sMMO at low copper-to-biomass ratios.

sMMO is a three-component binuclear iron active center monooxygenase that belongs to a large group of bacterial hydrocarbon oxygenases known as the soluble diiron monooxygenases. sMMO, encoded by a six-gene operon *mmoXYBZorfYmmoC*, has three components: (1) a 250-kDa hydroxylase with an $(\alpha\beta\gamma)_2$ structure in which the α-subunits (MmoX) contain the binuclear iron active center where substrate oxygenation occurs; (2) a 39-kDa NAD(P)-H-dependent reductase (MmoC) with flavin adenine dinucleotide and Fe_2S_2 prosthetic groups; (3) a 16-kDa component (MmoB) known as protein B or the coupling/gating protein that contains no prosthetic groups or metal ions. There are X-ray crystal structures for the hydroxylase component (Figure 2) and NMR data on the structures of protein B and the flavin domain of the reductase. The catalytic cycle of sMMO has been extensively studied and excellent progress has been made toward understanding the mechanism of oxygen and hydrocarbon activation at the binuclear iron center. The crucial reaction intermediate compound Q accumulates when the reduced (Fe^{II}–Fe^{II}) hydroxylase is reacted with O_2 in the presence of protein B. In compound Q, which is kinetically competent to oxidize methane and other substrates, the diiron center is most likely in the diferryl (Fe^{IV}–Fe^{IV}) state. The mechanism via which sMMO breaks the unreactive C–H bond of methane continues to be intensely debated. Current evidence, based on the use of radical clock substrates and theoretical studies, suggests a reaction with multiple pathways and the possible involvement of a captive substrate-derived radical species.

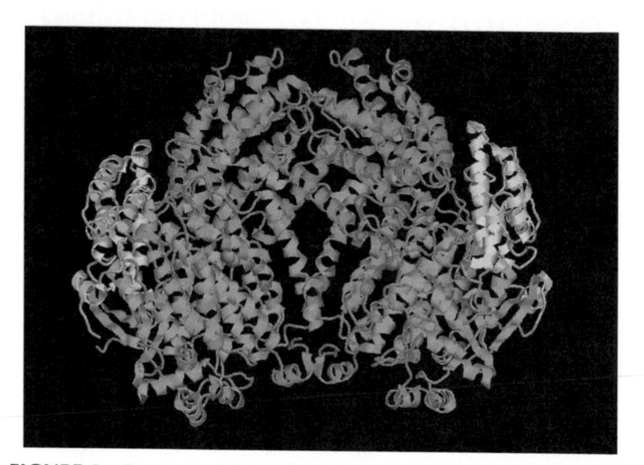

FIGURE 2 Structure of the hydroxylase component of sMMO. The α-, β-, and γ-subunits, encoded by the *mmoX*, *Y*, and *Z* genes, are colored blue, green, and yellow, respectively. The iron atoms of the binuclear iron centers are shown as orange spheres. Figure was constructed using X-ray crystallographic data (PDP accession code 1MMO). (For interpretation of the references to color in this figure legend, the reader is referred to the Web version of this chapter.)

Recent results from Lipscomb and colleagues have established the involvement of quantum mechanical tunneling of hydrogen nuclei in breaking the C–H bond of methane.

The site of substrate binding of sMMO is a hydrophobic cavity deeply buried in the protein. The side chain of residue Leu 110 in the α-subunit of the hydroxylase partly blocks the aperture between the substrate-binding pocket and the innermost of a chain of cavities that communicate between the active center and the outside and may form the route for substrate entry and product exit. Recent site-directed mutagenesis studies have indicated that this "leucine gate" is important in determining how substrates enter and are presented at the active site. The availability of a good, reliable expression system for sMMO paves the way for further studies to elucidate the structural features of this remarkable enzyme, which make it such a versatile biocatalyst.

pMMO is a copper-containing, membrane-associated enzyme that probably catalyzes most of the methane oxidation by methanotrophic bacteria in the environment. The enzyme has proved difficult to purify in active form and therefore, until recently, the biochemistry of pMMO has not been as well developed as that of sMMO. pMMO contains three polypeptides, of about 49, 27, and 22 kDa, encoded by the genes *pmoB, A, and C*, respectively. The X-ray crystal structure of pMMO shows that the enzyme has an $(\alpha\beta\gamma)_3$ stoichiometry (Figure 3). Native pMMO forms a complex with MDH, which may supply electrons to the enzyme. All active preparations of pMMO contain copper, although there has been conflicting evidence about number and roles of copper ions in the active form of the enzyme. Associated with the enzyme is a powerful copper chelator methanobactin, which is almost certainly required to sequester copper ions for activity and/or structural stability of pMMO. Other evidence has shown that pMMO may also contain a diiron centre. A recent study found copper-dependent monooxygenase activity with methane and propene substrates in a recombinant protein comprising the two soluble domains of PmoB component of pMMO expressed in *Escherichia coli*. This suggests that the principal active site of the enzyme (which appears to be a binuclear copper centre) is located in PmoB, although in the native pMMO complex it is the PmoA subunit that is labeled by the suicide substrate acetylene. The presence of MMO activity in this 33-kDa protein lacking the integral membrane portions of PmoB may advance the construction of small-molecule mimics of pMMO for conversion of methane to methanol and other biotransformations.

METHANOTROPHS IN BIOTECHNOLOGY

Initial interest in methanotrophs in the 1970s stemmed from plans to use them to produce SCP from fossil methane. The fall in the price of soya protein in the 1980s stymied these early efforts, but more recently interest in production

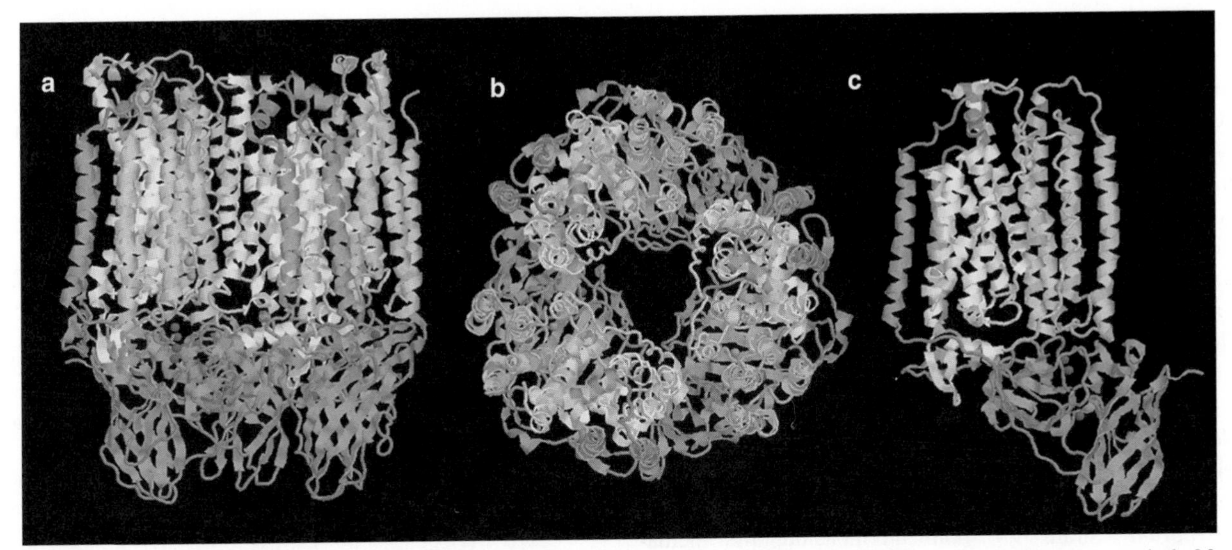

FIGURE 3 Structure of pMMO. The 49-, 27-, and 22-kDa subunits, encoded by *pmoB*, *A*, and *C*, are colored lilac, yellow, and green, respectively. Metal atoms in the protein are shown as spheres, copper red, and zinc orange. (a) $(\alpha\beta\gamma)_3$ enzyme complex with the predominantly α-helical presumed membrane-spanning region uppermost; (b) view looking down on (a) from above; (c) one αβγ protomer showing the mononuclear and dinuclear copper centers associated with each α (PmoB)-subunit and the zinc within the membrane-spanning region. This zinc site, which has been suggested by certain workers to be occupied by one or two iron atoms in the native enzyme, is a candidate for the active center since it contacts the β (PmoA)-subunit that is labeled by the inhibitor acetylene. Figure was constructed using X-ray crystallographic data (PDB accession code 1YEW). (For interpretation of the references to color in this figure legend, the reader is referred to the Web version of this chapter.)

of SCP from methane and methanol has been revived in Norway and Denmark for production of added-value protein products as amino acid-balanced feed for farmed fish and other animals. Methanotrophs have also been investigated for production of the bioplastic polyhydroxybutyrate from methane.

Great advances in genetic methods for methane oxidizers have made possible the use of these organisms as hosts for production of recombinant and heterologous proteins, including β-glucoronidase and genetically engineered monooxygenases. Recently, DuPont reported the engineering of *Methylomonas* sp. to produce the carotenoid astaxanthin, providing proof of concept for metabolic engineering of methanotrophs to synthesize new small-molecule products.

Interest in methanotrophic bacteria as biocatalysts for synthetic chemistry and bioremediation stems almost exclusively from the unique catalytic properties of the two MMO systems, most importantly their ability (1) to oxidize methane to methanol and (2) to co-oxidize a wide range of other substrates. Both systems require an exogenous source of reductant for the monooxygenation reaction, which in whole-cell applications can be supplied from added methanol or formate, via the principal enzymes of methylotrophic metabolism that are also present in the cells (Figure 1). In addition, the presence of oxygen-stable hydrogenase activity in methanotrophs enables hydrogen to be used as the reductant.

sMMO has a remarkably wide substrate range, oxidizing not only methane (the natural substrate) but also co-oxidizing a large number of organic compounds. Alkanes are hydroxylated mostly at the terminal and subterminal positions. Ring hydroxylation of mono-aromatics occurs primarily at the para position. sMMO oxygenates alkenes to epoxides with retention of stereochemistry around the CC double bond. Dalton and colleagues developed a pilot process with sMMO-expressing *Methylococcus capsulatus* cells for production of epoxypropane from propane. Methanol was used as the reductant, and inhibition of sMMO by the epoxide product was overcome by operating the process in a continuous two-stage system. The process gave good productivity and at 45 °C the epoxypropane product was easily recovered from the gas phase. With cells at $30\,g\,l^{-1}$ the epoxypropane production rate was $250\,g\,l^{-1}\,day^{-1}$.

The initial oxygenated products formed from halogenated substrates may decompose rapidly via nonenzymic pathways that result in the loss of halogen substituents. The priority pollutant TCE is a substrate for both forms of MMO and, by a combination of enzyme-catalyzed oxygenation and nonenzymatic steps, pMMO-expressing methanotroph cells can lead to its mineralization to CO_2, water, and chloride. Several studies have used methanotrophs for bioremediation of groundwater and effluents contaminated with TCE and other chlorinated solvents. For example, a TCE-contaminated aquifer at Kimitsu City (Japan) has been periodically biostimulated with methane and inorganic nutrients to promote growth of methanotrophic bacteria to degrade the TCE, where a 10% decrease in TCE concentrations (200 ppb) was observed 40 days after biostimulation with methane. Pilot studies have also used methanotrophs in *ex situ* systems for bioremediation of chlorinated organic solvents such as TCE and *cis*-1,2-dichloroethylene.

Bioremediation considerations include the potential supply of reductant for MMO enzymes, often formate or methanol instead of the natural substrate methane. Manipulation of sMMO enzymes in the future may make them more effective in degrading mono- and diaromatic pollutants (including polychlorinated biphenyls) and even polyaromatic hydrocarbons.

POSTGENOMICS OF METHANOTROPHS

The first publicly available methanotroph genome sequence, that of *Mc. capsulatus* (Bath), was completed in 2004 and the genome sequence of the facultative methanotroph *Methylocella silvestris* BL2 is now available. A draft genome sequence *for "Methylacidiphilum fumariolicum"* strain SolV and a complete genome sequence of *"Methylacidiphilum infernorum"* strain V4 (both thermoacidophilic methanotrophs of the phylum *Verrucomicrobia*) are now available. The genomes of other methanotrophs, including those of *Methylocapsa*, *Methylomonas*, *Methylomicrobium album* and *Methylosinus trichosporium*, are currently being sequenced by the Joint Genome Institute (JGI). Comparative genomics of these methanotrophs with the genome sequence of the facultative methanotroph *Methylocella silvestris* BL2 and with other obligate and facultative methylotrophs may well reveal the molecular basis for the obligate nature of most extant methanotrophs. Proteomic analyses of *Methylococcus capsulatus* (Bath) have been carried out; these and future genomic studies will provide a wealth of information for studying the regulation of methane oxidation and may provide insights into how methanotrophy is regulated under differing environmental conditions. The availability of genome sequences is now facilitating microarray experiments with *Methylococcus capsulatus* and *Methylocella silvestris*, which are yielding insights into the regulation of methane oxidation processes. Future postgenomic studies, using genome sequence information as a blueprint for hypothesis testing, will undoubtedly lead to further advances in our knowledge of the biology of these fascinating bacteria and allow further exploitation in biotechnology.

FURTHER READING

Anthony, C. (1982). *The biochemistry of methylotrophs*. New York: Academic Press.

Borodina, E., Nichol, T., Dumont, M. G., Smith, T. J., & Murrell, J. C. (2007). Mutagenesis of the 'leucine gate' to explore the basis of catalytic versatility in soluble methane monooxygenase. *Applied and Environmental Microbiology*, *73*, 6460–6467.

Dalton, H. (2005). The Leeuwenhoek Lecture 2000. The natural and unnatural history of methane oxidizing bacteria. *Philosophical Transactions of the Royal Society London. Series B, Biological Sciences*, *360*, 1207–1222.

Hakemian, A. S., & Rosenzweig, A. C. (2007). The biochemistry of methane oxidation. *Annual Reviews of Biochemistry*, *76*, 223–241.

Hanson, R. S., & Hanson, T. E. (1996). Methanotrophic bacteria. *Microbiological Reviews*, *60*, 439–471.

McDonald, I. R., Bodrossy, L., Chen, Y., & Murrell, J. C. (2008). Molecular ecology techniques for the study of aerobic methanotrophs. *Applied and Environmental Microbiology*, *74*, 1305–1315.

Semrau, J. D., DiSpirito, A. A., & Yoon, S. (2010). Methanotrophs and copper. *FEMS Microbiology Reviews*, *34*, 496–531.

Smith, T. J., & Dalton, H. (2004). Biocatalysis by methane monooxygenase and its implications for the petroleum industry. In R. Vazquez-Duhalt, & R. Quintero-Ramirez (Eds.). *Petroleum biotechnology: Developments and perspectives studies in surface science and catalysis* (vol. 151). (pp. 177–192). Amsterdam: Elsevier.

Trotsenko, Y. A., & Murrell, J. C. (2008). Metabolic aspects of aerobic obligate methylotrophy. *Advances in Applied Microbiology*, *63*, 183–229.

Zheng, H., & Lipscomb, J. D. (2006). Regulation of methane monooxygenase catalysis based on size exclusion and quantum tunneling. *Biochemistry*, *45*, 1685–1692.

Chapter 13

Microbial Photosynthesis

B. Jagannathan and J.H. Golbeck

The Pennsylvania State University, University Park, PA, USA

Chapter Outline

ABBREVIATIONS

3PG 3-Phosphoglycerate
ATP Adenosine triphosphate
CBB Calvin–Benson–Bassham
Chl Chlorophyll
DHAP Dihydroxyacetone phosphate
EXAFS Extended X-ray absorption fine structure
F6P Fructose-6-phosphate
FBP Fructose-1,6-biphosphate
FNR Ferredoxin-NADP$^+$ oxidoreductase
G1P Glucose-1-phosphate
G3P Glyceraldehyde-3-phosphate
NADPH Nicotinamide adenine dinucleotide phosphate
PS Photosystem
RuBP Ribulose-1,5-biphosphate

DEFINING STATEMENT

The aim of this article is to move beyond the textbook equation of photosynthesis and describe the design principles behind photosynthetic electron transfer. The events that constitute a photosynthetic cycle are described in the exact order they occur, using the cyanobacterial system as a model. Photosynthesis in lesser-known phototrophs is also discussed.

INTRODUCTION

Photosynthesis is the biochemical process carried out by certain bacteria, algae, and higher plants in which light is converted into chemical bond energy. The process is crucial as nearly all life on earth depends on sunlight either directly or indirectly for energy, food, and O_2. The advent of photosynthetic prokaryotes with the ability to consume CO_2 and produce O_2 from H_2O resulted in a hospitable environment on earth for advanced forms of life. Fossil records indicate that the first oxygenic photosynthetic bacteria appeared around 3.5×10^9 years ago. Earlier, organisms survived by anaerobic metabolism, a process that generates only a fraction of the energy produced by aerobic metabolism. It is likely that in the absence of oxygenic photosynthesis, advanced forms of life would not have emerged and only microorganisms would now exist. Today, as the primary means of carbon fixation, oxygenic photosynthesis forms one half of the energy-carbon cycle. Phototrophic organisms reduce CO_2 to carbohydrates, which are oxidized back to CO_2 by heterotrophic (as well as phototrophic) organisms. The energy released during the oxidation reaction is stored in the form of NADH and ATP, which are subsequently used for growth, metabolism, and reproduction. In addition, prehistoric plants and algae were largely responsible for the generation of the vast reserves of fossil fuels that are now being mined for their energy value. They provided a large portion of the initial biomass, which was converted into oil and coal over millions of years through pressure, heat, and microbial action.

The general process of photosynthesis is described by Van Niel's equation:

$$2H_2A + CO_2 \rightarrow 2A + CH_2O + H_2O \qquad (1)$$

where H_2A is the reductant and A is the oxidized product.

Van Niel's equation can be applied to oxygenic photosynthesis as:

$$6CO_2 + 6H_2O + \text{light} \rightarrow C_6H_{12}O_6 + 6O_2 \qquad (2)$$

Although complete, this equation belies the overwhelming complexity of the process. For example, the generation of the light-induced charge-separated state and its subsequent stabilization over time requires a large number of pigments and cofactors arranged in a specific protein environment. The splitting of H_2O into O_2 is extremely difficult to replicate in the laboratory, yet plants and cyanobacteria perform the task repeatedly with seeming ease. The conversion of CO_2 into sugars is another intricate process that requires an extensive set of physical and chemical reactions to occur in a highly coordinated fashion.

In this chapter, we will expand on this simple equation. In addition to describing the general design principles behind the sophisticated biomachinery involved in photosynthesis, we will provide structural and functional details, placing special emphasis on light-induced electron transfer in aerobic and anaerobic organisms.

HISTORICAL PERSPECTIVE

The first experiments on photosynthetic organisms were performed in the 1770s when Joseph Priestley showed that plants were capable of generating a gas that could support combustion. Building on his work, Jan Ingenhousz established that sunlight was required, and Jean Senebier and Nicolas Theodore de Saussure demonstrated the indispensability of CO_2 and H_2O. In 1845, Julius Robert von Meyer postulated that plants convert light into chemical energy during photosynthesis. Early scientists believed that the O_2 was produced from the splitting of CO_2, and it was not until the 1930s that Cornelius van Niel proposed, correctly, that H_2O was the source of O_2. It is interesting that 75 years later the exact biochemical mechanism of H_2O splitting remains to be elucidated.

Photosynthesis research has had its share of Nobel laureates. Melvin Calvin won the chemistry prize in 1961 for identifying most of the intermediates in the conversion of CO_2 into carbohydrates. Peter Mitchell was the sole recipient of the chemistry award in 1978 for his work on the chemi-osmotic theory of proton translocation. Johann Deisenhofer, Robert Huber, and Hartmut Michel won the chemistry prize in 1988 for solving the first crystal structure of a photosynthetic reaction center. Rudolph Marcus's investigation of the factors guiding electron transfer in chemical systems remains the paradigm for theoretical calculations of electron transfer in photosynthetic reaction centers. He was awarded the Nobel Prize in chemistry in 1992. More recently, Paul Boyer and John Walker were awarded the prize in chemistry in 1997 for elucidating the enzymatic mechanism underlying the synthesis of ATP.

Artificial photosynthesis has seen a recent spurt of activity, largely due to an increased awareness of the depletion of fossil fuel reserves and the effect of their combustion products on the earth's climate. The aim is to synthesize inexpensive and long-lasting organic and inorganic molecules that convert light into chemical energy, thereby mimicking the basic process of photosynthesis. This has brought new disciplines such as material science and bioengineering into photosynthesis, making the field truly interdisciplinary.

CLASSIFICATION OF PHOTOSYNTHETIC ORGANISMS

There exist five bacterial phyla with members capable of chlorophyll (Chl)-based phototrophy: Firmicutes, Chloroflexi, Chlorobi, Proteobacteria, and Cyanobacteria. With the recent discovery of *Chloracidobacterium thermophilum*, Acidobacteria have become the sixth known phylum to carry out the process of photosynthesis.

All photosynthetic organisms can be classified as either oxygenic or anoxygenic. Oxygenic phototrophs employ H_2O as the source of electrons and liberate O_2 as the by-product. Anoxygenic phototrophs derive their electrons from organic or inorganic molecules, and hence they do not evolve O_2. Of the five well-established phototrophic bacterial phyla, only the Cyanobacteria are capable of performing oxygenic photosynthesis. In addition, all eukaryotic phototrophs such as higher plants and algae, which evolved later than cyanobacteria, produce O_2 during photosynthesis.

The remaining four phyla include anaerobes such as the purple nonsulfur bacteria, purple sulfur bacteria, green sulfur bacteria, and heliobacteria, which survive only under low concentrations of O_2. The recently discovered Acidobacteria have been reported to live under oxic conditions, although a detailed physiological characterization of this organism remains to be carried out.

We will discuss oxygenic photosynthesis first, using cyanobacteria as the model organism. Cyanobacteria are photosynthetic prokaryotes that are found in every conceivable habitat from oceans to fresh water to soil. These Gram-negative bacteria are responsible for generating the majority of the O_2 in the earth's atmosphere. The most widely used cyanobacterial strains for current experimental research are *Synechocystis* sp. PCC 6803, *Synechococcus* sp. PCC 7002, and *Thermosynechococcus elongatus*.

In cyanobacteria, photosynthesis is associated with a well-organized system of internal membranes in the cytoplasm. These are called thylakoids, from the Greek word *thylakos* meaning sac. These membranes are highly folded, allowing the cell to pack a large amount of surface area into a small space. The interior space enclosed by the thylakoid membrane is termed the lumen and the matrix surrounding the thylakoids is termed the stroma. The thylakoids are home to the integral membrane protein complexes that are involved in the light reactions of photosynthesis.

Eukaryotic organisms such as higher plants and algae conduct photosynthesis in membrane-bound organelles called chloroplasts. They consist of an outer, freely permeable membrane and a selectively permeable inner membrane that encloses the stroma. The sac-like thylakoids immersed in the stroma are similar in organization to the comparable membranes in cyanobacteria. Chloroplast thylakoids, however, tend to form well-defined stacks called grana, which are connected to other stacks by intergrana thylakoids called lamellae. It is widely thought that chloroplasts evolved from an endosymbiotic relationship of a heterotrophic prokaryote with a cyanobacterium.

THE CONSTITUENT PROCESSES OF PHOTOSYNTHESIS

Equation (1) is the end product of a large number of events that occur during a typical photosynthetic cycle. The basic processes that constitute oxygenic photosynthesis are:

- Absorption of light by pigment molecules and transfer of the excitation energy to two reaction centers, Photosystem II (PS II) and Photosystem I (PS I).
- Light-induced transfer of an electron across the photosynthetic membrane and splitting of H_2O into O_2 by PS II.
- Light-induced excitation and transfer of an electron across the photosynthetic membrane, generating reducing equivalents in the form of nicotinamide adenine dinucleotide phosphate (NADPH) by PS I.
- Production of ATP using the proton gradient generated across the membrane from both H_2O splitting and electron transfer through the cytochrome b_6f complex.
- Conversion of CO_2 into carbohydrates using ATP and the reducing power of NADPH.

The division of photosynthetic labor is relatively straightforward. All the light reactions occur within or on the thylakoid membrane. The ATP and NADPH produced by the light reactions are released into the stroma where the dark reactions of CO_2 fixation are carried out. We focus first on the overall design philosophy of the process of converting light to stable chemical energy.

ABSORPTION AND TRANSFER OF LIGHT ENERGY

The Light-Absorbing Chromophores

Photosynthesis in cyanobacteria and plants is driven by light in the visible (380–750 nm) region of the electromagnetic spectrum. Phototrophic organisms such as purple bacteria, green sulfur bacteria, and heliobacteria extend this region to the near-infrared so as to exploit unique ecological niches. All of this makes evolutionary sense as the majority of the sun's energy that reaches the earth's surface lies in this range. Ultraviolet radiation and far-infrared radiation are both limited in amount; also, the former is too energetic and is capable of breaking chemical bonds, while the latter contains insufficient energy to be useful for most photochemical processes.

Primary chromophores

Photosynthetic organisms use a range of chromophores to efficiently capture photons in the visible and near-IR regions. The most abundant chromophore involved in photosynthesis is Chl, a molecule structurally similar to, and produced by, the same metabolic pathway as porphyrin pigments such as heme. The basic structure of the Chl molecule is a chlorin ring coordinated to a central magnesium atom (Figure 1). The addition of a long phytol tail makes Chl insoluble in water. There are four common types of Chl molecules in photosynthetic organisms, named chlorophyll *a*, *b*, *c*, and *d*. Their overall structure is similar, with minor changes in the side-chain groups that result in slightly different absorption spectra (Figure 2). Cyanobacteria employ chlorophyll *a* (Chl *a*), while plants utilize both chlorophyll *a* and *b*. Some species of algae contain chlorophyll *c*, and a few species of cyanobacteria contain chlorophyll *d*. Chlorophylls absorb primarily in the blue and red regions of the visible spectrum and have a high molar extinction coefficient. They have an inherently high fluorescence yield, which guarantees a long-lived excited singlet state, making them the ideal chromophore.

Accessory chromophores

Besides containing Chls, photosynthetic organisms contain accessory pigments that extend the range of absorbed wavelengths. Carotenoids are the main accessory pigment found in cyanobacteria, algae, and higher plants. They belong to the tetraterpenoid family, that is, contain 40 carbon atoms, and absorb light in the 400–500 nm region. Structurally, these compounds are composed of two small six-carbon rings connected by a polyene chain of carbon atoms. They are insoluble in water and are normally attached to proteins that are attached to the membrane. There are over 600 types of carotenoids, which are classified as either carotenes or

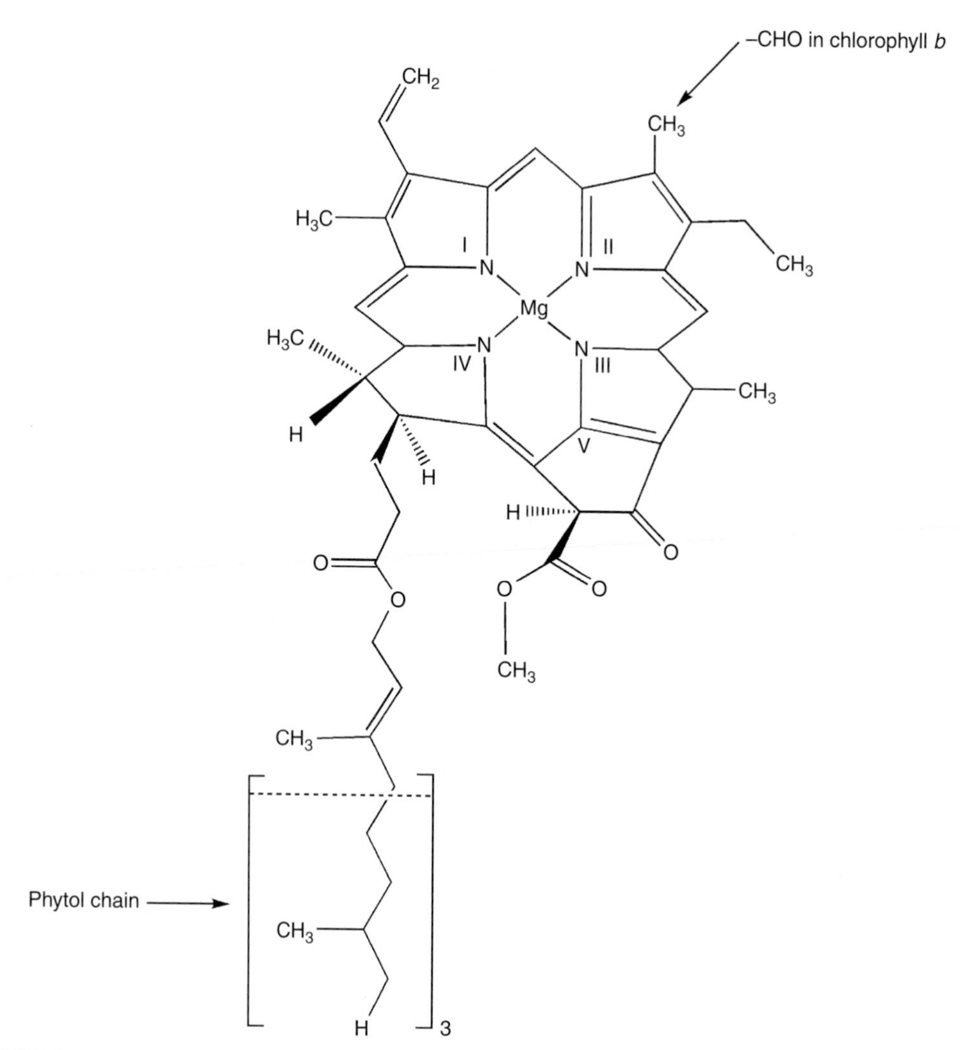

FIGURE 1 Chemical structure of chlorophyll *a*. Chlorophyll *b* has a –CHO group instead of the –CH_3 group in ring II.

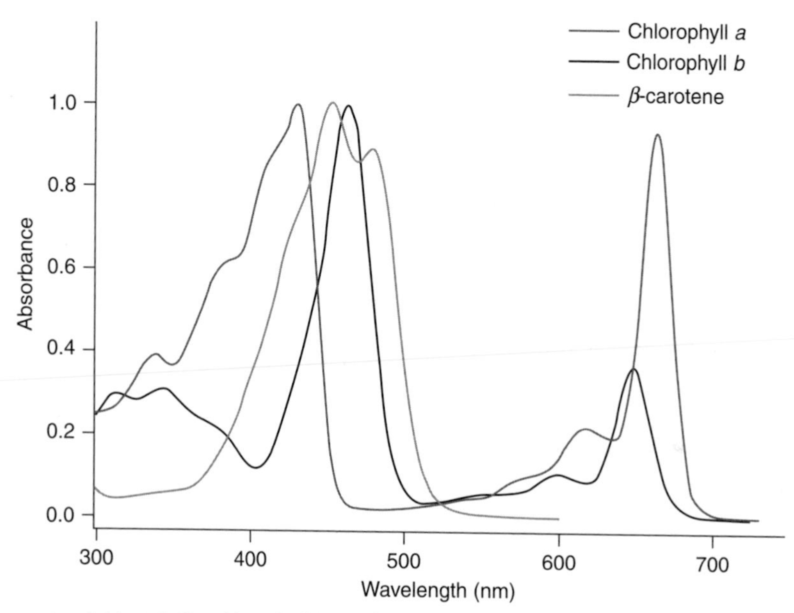

FIGURE 2 The absorption spectra of chlorophyll *a*, chlorophyll *b*, and β-carotene in solution. (For interpretation of the references to color in this figure legend, the reader is referred to the Web version of this chapter.)

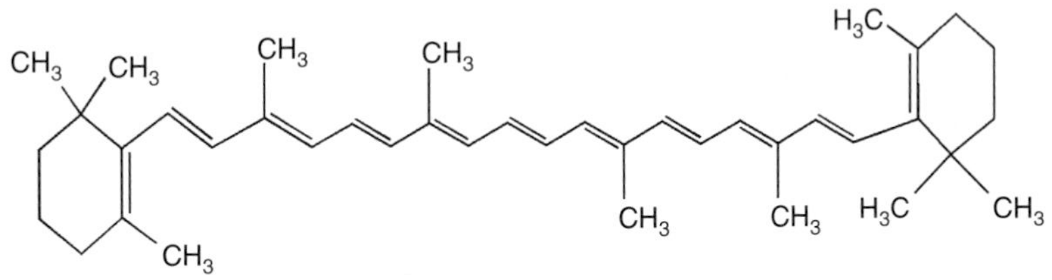

FIGURE 3 Chemical structure of β-carotene.

xanthophylls. Carotenes consist exclusively of carbon and hydrogen, while xanthophylls also contain oxygen. The most abundant carotenoid in cyanobacteria is β-carotene, which is the same pigment that gives carrots its distinctive color (Figure 3). In addition to functioning as an accessory pigment, carotenoids play a vital role in dissipating excess light energy, which would otherwise lead to the generation of superoxide radicals. These radicals are highly reactive to chemical bonds and could be potentially lethal to the cell if left unchecked.

Cyanobacteria and certain types of algae contain additional pigments called phycobilins, which absorb light between 500 and 650 nm. Phycobilins consist of an open chain of four pyrrole rings and are water-soluble. They are attached to proteins termed phycobiliproteins and they pass on the absorbed light energy to nearby antenna Chl molecules.

Plants and cyanobacteria therefore use a combination of Chls and accessory pigments to effectively blanket a large majority of the visible spectrum. Both appear dark green or blue-green because the few photons that are not absorbed lie between the blue and red regions of the spectrum.

The Light-Gathering Structures and Resonance Energy Transfer

The task of the photosynthetic reaction center is to convert the energy stored in the excited singlet state of Chl to a form useful for work. In photosynthesis, work refers to the creation of a charge-separated state consisting of a donor, D^+, and an acceptor, A^-, pair. At one extreme of time, the creation of the singlet excited state occurs within 10^{-15} s of absorbing a photon. At the other extreme, the captured light energy must be utilized within 10^{-8} s, otherwise the energy will be lost as heat or fluorescence as the excited state decays. The generation of the charge-separated state must occur within this window of time.

A network of closely spaced Chl molecules, termed the antenna system, absorbs the photon and the resulting excited state migrates to a neighboring antenna Chl by a process known as resonance energy transfer. This occurs on a timescale of 10^{-12} s and is a nonradiative process. The excited state, known as an exciton, randomly wanders about the antenna system until it chances upon the specialized reaction center Chls associated with PS I and PS II. The energy levels of these specialized Chls are slightly lower than the antenna Chls because they are in a different protein environment. This allows these specialized Chls to trap the exciton and use it to create a charge-separated state. In most photosynthetic reaction centers, this state is generated within 10^{-10} s following photon absorption. Accessory pigments also transmit the absorbed energy to antenna Chls by a similar process of resonance energy transfer.

THE WATER-SPLITTING COMPLEX

We now turn our attention to the source of electrons in oxygenic photosynthesis. The catalytic redox center that carries out H_2O splitting is termed the O_2-evolving complex, and is an integral component of PS II. The water-splitting reaction can be summarized by the following equation:

$$2H_2O \rightarrow O_2 + 4e^- + 4H^+ \quad (3)$$

The 3.0-Å X-ray crystal structure of PS II from *T. elongatus* (PDB ID 2AXT), as well as extended X-ray absorption fine structure (EXAFS) studies on PS II crystals, has led to a structural model of the O_2-evolving complex (Figure 4). This structure is the starting point for discussion on the mechanism of O_2 evolution.

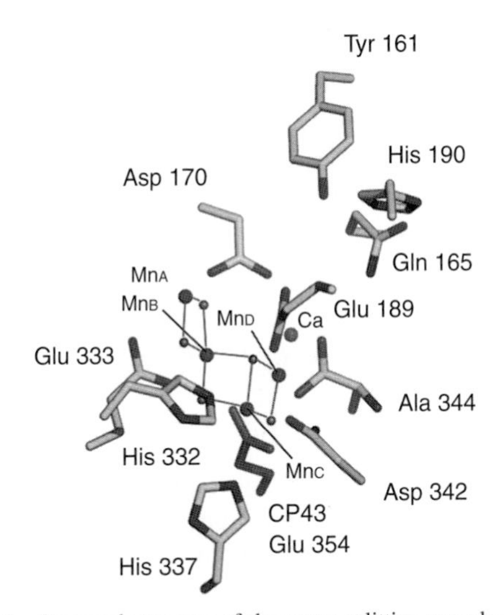

FIGURE 4 Proposed structure of the water-splitting complex of PS II, based on the 3.0-Å resolution X-ray crystal structure and on extended X-ray absorption fine structure (EXAFS) data. The spheres represent manganese (red), calcium (green), and the bridging oxygen ligand atoms (gray). (For interpretation of the references to color in this figure legend, the reader is referred to the Web version of this chapter.)

A cluster of four manganese atoms and a calcium atom is responsible for stripping four electrons from two H_2O molecules, liberating O_2 in the process. The protons are released into the thylakoid lumen, thereby generating a portion of the pH gradient that is used to synthesize ATP.

The structural model of PS II shows the metal atoms arranged in an extended cubane structure, with three manganese and one calcium at the corners, and the fourth manganese located immediately to one side. The metal atoms are connected to each other by mono-μ-oxo, di-μ-oxo, and/or hydroxo bridges, but the amino acids that contribute the ligands are not known with complete certainty.

Since 1970, the paradigm for understanding O_2 evolution has been the S-state cycle proposed by Bessel Kok. This model includes five oxidation states (S-states) for the 4Mn–Ca cluster. The cluster is oxidized in one-electron steps from S_0 (most reduced) to S_4 (most oxidized) by a successively photooxidized reaction center Chl in PS II. The 4Mn–Ca cluster thus accumulates four equivalents of oxidizing power and uses it to split two H_2O molecules. An O_2 molecule is released after the S_4 state, returning the 4Mn–Ca cluster to the S_0 state.

A large number of questions remain unsolved concerning the catalytic mechanism of water splitting. Most importantly, the S_4 state of the cycle is fleeting and has not been observed spectroscopically. This state is critical because it may be the starting point for O–O bond formation. Without knowledge of its chemistry, the precise catalytic mechanism of the water-splitting complex is difficult to formulate. The binding site of the two H_2O molecules represents another uncertainty, although a recent proposal suggests that the calcium coordinates one of the H_2O molecules, with a manganese binding the other.

The electrons from the water-splitting complex are donated to an oxidized tyrosine residue termed Tyr_Z. According to the X-ray crystal structure, the calcium atom is positioned between the 4Mn–Ca cluster and Tyr_Z. A chloride ion is bound near the water-splitting complex in the vicinity of Tyr_Z. It is believed that its role in O_2 evolution is to neutralize accumulated charge.

TRANSMEMBRANE ELECTRON TRANSFER

Stabilization of the Charge-Separated State

The electrons obtained from the splitting of H_2O are ultimately used to reduce CO_2. To achieve this goal, the electrons must be transferred against a highly unfavorable thermodynamic gradient from the lumenal side of the membrane to the stromal side, where the dark process of carbon fixation occurs. The electrons traverse a distance equal to the thickness of the membrane, which is around 40 Å. This distance appears short, but it is significant on the scale of an electron.

Consider a hypothetical donor–acceptor pair, D–A, where D becomes excited to D* and donates an electron to A, thereby creating a charge-separated state D^+–A^-:

$$D-A \rightarrow D^*-A \rightarrow D^+-A^-$$

Due to their close proximity, the D^+–A^- charge-separated state is unstable and short-lived. Adding a second closely spaced acceptor molecule, A_1, can extend the lifetime of the charge-separated state:

$$D-A-A_1 \rightarrow D^*-A-A_1 \rightarrow D^+-A^--A_1 \rightarrow D^+-A-A_1^-$$

The driving force for electron transfer is a drop in Gibbs free energy between A^- and A_1, thus altering the equilibrium constant in favor of A_1. This results in the expenditure of some of the original energy of the photon, but it is a necessary trade-off to extend the lifetime of the charge-separated state.

Adding a second closely spaced donor molecule, D_1, has a similar effect:

$$D_1-D-A \rightarrow D_1-D^*-A \rightarrow D_1-D^+-A^- \rightarrow D_1^+-D-A^-$$

The positive charge will migrate from D^+ to D_1 due to a drop in Gibbs free energy between D_1 and D, thus altering the equilibrium constant in favor of D.

Photosynthetic complexes adopt both strategies and trade off some of the energy of the original photon to extend the lifetime of the otherwise short-lived charge-separated state.

Factors Affecting the Rate of Electron Transfer: Marcus Theory

To further understand this process, we must delve further into the factors that govern the rate of electron transfer in proteins. Marcus theory states that the rate of electron transfer between a donor and an acceptor pair depends on two factors. The first is the Frank Condon factor, which includes the change in the Gibbs free energy, the reorganization energy, and the temperature. Mathematically, it is expressed as:

$$k_{et} \propto \exp\frac{(-\Delta G_0+\lambda)^2}{4\lambda k_B T}, \qquad (4)$$

where k_{et} is the rate of electron transfer, ΔG_0 is the difference in free energy between the product and the reactant, λ is the reorganization energy, k_B is the Boltzmann's constant, and T is the absolute temperature. This is the equation of a parabola; as ΔG increases, the rate first increases, then attains a maximum, and finally decreases. The difference in Gibbs free energy between the product (D^+–A^-) and the reactant (D–A) translates into the thermodynamic driving force for the reaction. The reorganization energy corresponds to the amount of energy required to alter the microenvironment of the reactants before electron transfer so that it resembles the equilibrium microenvironment of the products after electron transfer. This term reflects small changes in bond lengths and reorientation of dipoles around the redox centers to reflect the new pattern of electric fields before and after the electron transfer event. It is difficult to measure experimentally and is usually assumed to be a constant 0.7 eV in proteins.

The second factor that influences the rate of electron transfer is the matrix coupling element. It relates electron transfer rate to the distance between the donor and the acceptor pairs. The following equation describes the relationship:

$$|H_{AB}|^2 \propto \exp(-\beta R) \qquad (5)$$

where $|H_{AB}|$ is the coupling probability between the donor and acceptor wave functions, β is a constant whose value is 1.4 $Å^{-1}$ in proteins, and R is the edge-to-edge distance between the donor and the acceptor molecules.

In proteins, this logarithmic relationship translates to a 10-fold change in the rate of electron transfer for every 1.7 Å change in distance. Accordingly, an electron would take nearly a century to traverse a distance of 45 Å. In bioenergetic membranes, large distances are traversed by introducing several cofactors into the membrane so as to shorten the distance between any two redox pairs. In photosynthetic systems, an electron traverses the width of the thylakoid membrane in less than 1 μs due to the presence of multiple electron transfer cofactors. With detailed knowledge of the factors that affect electron transfer in proteins, we now turn to how nature has incorporated these concepts into the design of a photosynthetic reaction center.

The Photosynthetic Reaction Center

The pigment–protein complex that is responsible for translocating the electron across the photosynthetic membrane is termed the photosynthetic reaction center. It comprises the antenna pigments, the organic and inorganic molecules that function as electron transfer cofactors, and the proteins that provide the scaffold for these components.

Oxygenic phototrophs employ two photosynthetic reaction centers in series for achieving transmembrane electron transfer.

Photosystem II

The net reaction performed by PS II can be summarized as:

The structural model of cyanobacterial PS II (Figure 5) shows that the redox cofactors bound to the D1 and D2 proteins form two branches that are arranged symmetrically along a pseudo-C_2 axis of symmetry (Figure 6).

The electron transfer chain in PS II starts at a special pair of Chl *a* molecules termed P_{680}, named for its peak absorbance in the visible region. The magnesium atoms between the two Chl *a* molecules (P_{D1} and P_{D2}) are separated by a distance of only 7.6 Å. When the exciton reaches P_{680} via the antenna system, it becomes excited to the singlet state and the electron is transferred to the primary acceptor, a pheophytin molecule. A bridging Chl *a* molecule acts as an intermediate between P_{680} and the pheophytin. The bridging Chl *a* molecules on either side of the pseudo-C_2 axis (Chl_{D1} and Chl_{D2}) are located 10.4 and 10.3 Å, respectively, from P_{D1} and P_{D2}. The two pheophytin molecules ($Pheo_{D1}$ and $Pheo_{D2}$) are located at a distance of 10.5 and 10.7 Å, respectively, from Chl_{D1} and Chl_{D2}.

The charge-separated state between P_{680}^+ and $Pheo^-$ is stabilized by the transfer of the electron from $Pheo^-$ to Q_A and then to Q_B, both of which are molecules of plastoquinone. A nonheme iron (Fe^{2+}) located between Q_A and Q_B aids in the electron transfer. The Q_A site is 13 Å distant from $Pheo_{D1}$, while the Q_A and Q_B sites are separated by 18 Å, with the nonheme iron precisely in the middle. Because a quinone molecule (Q_B) acts as the terminal electron acceptor, PS II is classified as a Type II reaction center.

Water, through the action of the O_2-evolving complex, reduces oxidized P_{680} via Tyr_Z, thereby precluding charge recombination between P_{680}^+ and Q_B^- and thus forming the $P_{680}Q_AQ_B^-$ charge-separated state. When a second exciton reaches P_{680}, the process is repeated, creating an unstable $P_{680}Q_A^-Q_B^-$ charge-separated state. Two protons are taken up from the medium, and the stable $P_{680}Q_A(Q_BH_2)$ state is generated. Plastoquinol (Q_BH_2) has a low affinity for the Q_B-binding site. It diffuses within the membrane to the cytochrome b_6f complex, which oxidizes plastoquinol to plastoquinone, and by virtue of the protonmotive Q-cycle translocates up to two protons per electron across the membrane. The regenerated plastoquinone diffuses back to the Q_B-binding site to participate in another round of light-induced electron transfer. The cytochrome b_6f complex passes the electron to a soluble carrier such as plastocyanin (a copper protein) or cytochrome c_6 (a heme protein), which diffuses laterally along the lumenal space, donating its electron eventually to PS I.

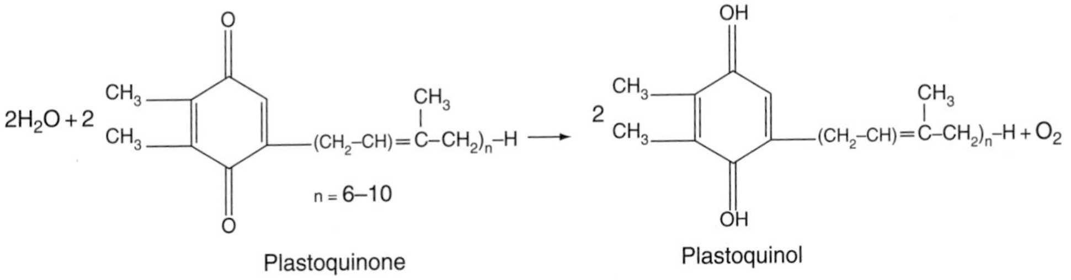

$$2H_2O + 2\ \text{Plastoquinone}\ [CH_3, CH_3, O, O, (CH_2\text{–}CH{=}C(CH_3)\text{–}CH_2)_n\text{–}H;\ n = 6\text{–}10] \longrightarrow 2\ \text{Plastoquinol}\ [CH_3, CH_3, OH, OH, (CH_2\text{–}CH{=}C(CH_3)\text{–}CH_2)_n\text{–}H] + O_2 \qquad [6]$$

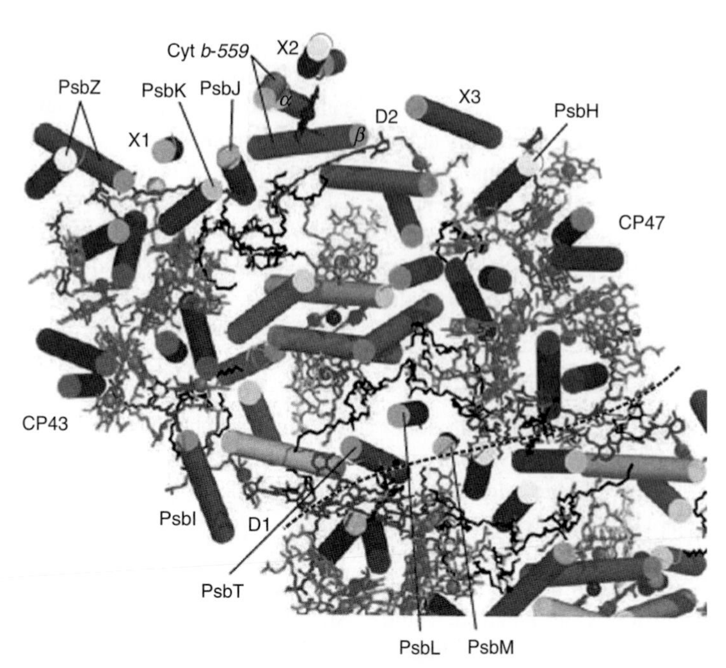

FIGURE 5 Overall view of the PS II monomer. Transmembrane *α*-helices are represented by cylinders. The main subunits are colored as follows: reaction center subunits D1 (yellow) and D2 (orange), antenna subunits CP43 (magenta) and CP47 (red), and the *α*- and *β*-subunits of cytochrome *b559* (green and cyan, respectively). Low molecular mass subunits are colored gray. Cofactors are colored green (Chl *a*), yellow (pheophytin), red (carotenoid), blue (heme), violet (quinone), and black (lipids). (For interpretation of the references to color in this figure legend, the reader is referred to the Web version of this chapter.)

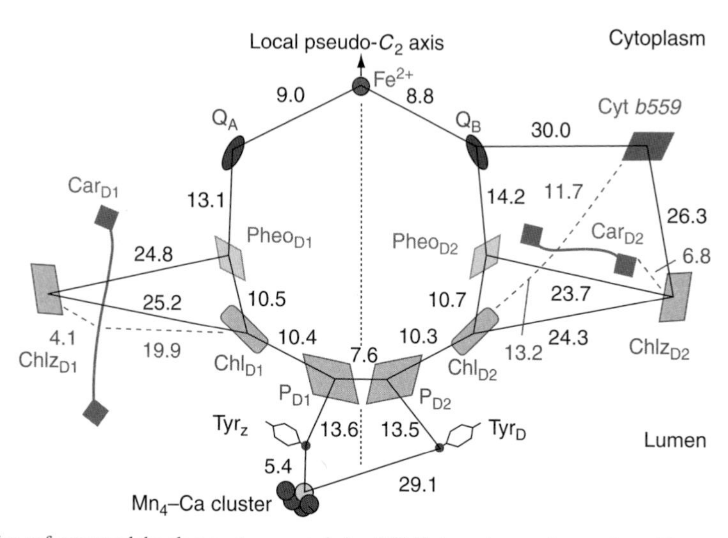

FIGURE 6 View of redox-active cofactors and the electron transport chain of PS II along the membrane plane. The pseudo-C_2 axis is represented by the dotted line with the arrow. Selected distances (in angstrom) are drawn between cofactor centers (black lines) and edges of *π*-systems (red dotted lines). (For interpretation of the references to color in this figure legend, the reader is referred to the Web version of this chapter.)

The 38-kDa D1 protein (also known as PsbA) and the 39-kDa D2 protein (also known as PsbD) contain all the electron transport cofactors of PS II. The 56-kDa CP47 protein (also known as PsbB) and the 50-kDa CP43 protein (also known as PsbC) harbor most of the antenna Chls and carotenoids associated with PS II. The 9-kDa PsbE and the 4.5-kDa PsbF proteins constitute the α- and the β-subunits of cytochrome *b559*, which is present to prevent radical formation under conditions of suboptimal electron flow. PS II contains additional subunits (denoted PsbG–PsbT), many of which have poorly understood roles.

The kinetics of electron transfer between most of the redox cofactors have been determined to a high degree of precision (Figure 7). The electron is transferred from P_{680} to Pheo in 4 ps and then to Q_A in 200 ps. The $Q_A \rightarrow Q_B$ step requires about 100 and 200 μs when Q_B becomes singly and doubly reduced, respectively.

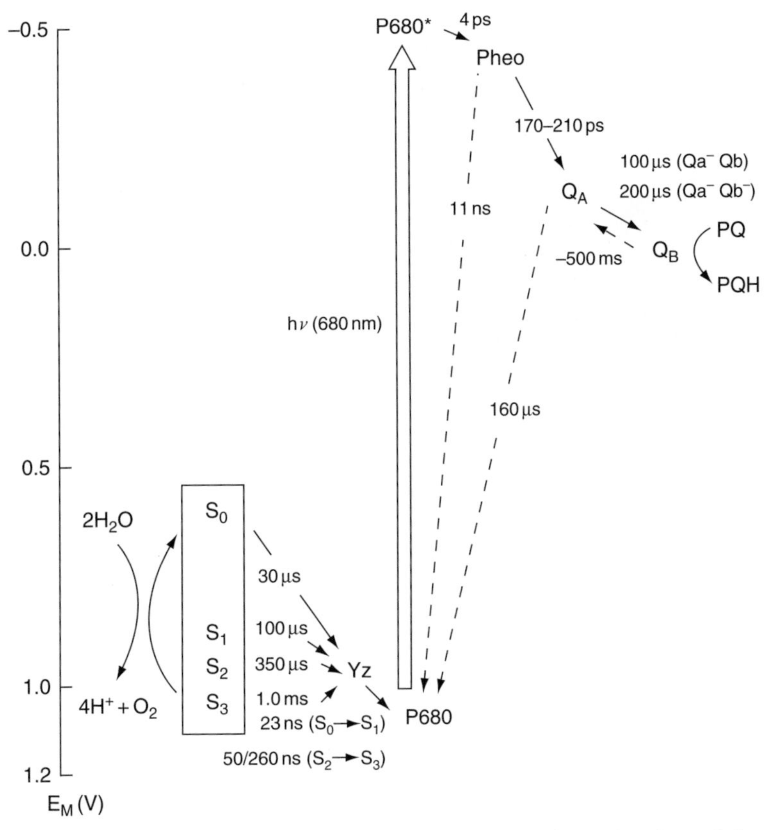

FIGURE 7 The components of the PS II reaction center depicted with redox potential on the *y*-axis and rate of electron transfer on the *x*-axis.

Even though any given reduced acceptor can recombine with the P_{680}^{+} cation, forward electron transfer to the next acceptor is at least 1000 times faster than charge recombination. Thus, the quantum yield (i.e., the number of charge-separated events created divided by the number of photons absorbed) is extremely high.

The bioenergetics of electron transfer further reflect the artful design of the photosynthetic reaction center. All electron transfer steps on the acceptor side of PS II are thermodynamically downhill, that is, there is a loss in Gibbs free energy at every step that "pays" for the successively longer charge time of charge separation.

Photosystem I

The net reaction performed by PS I can be summarized as:

$$2NADP^{+} + 2H^{+} + 4e^{-} \rightarrow 2NADPH \quad (7)$$

PS I employs the same design principles as PS II to generate and stabilize a charge-separated state. The electron transfer chain starts at a special pair of Chl *a* molecules termed P_{700}, named for its peak absorbance in the visible region. When P_{700} becomes excited to the singlet state, its electron is transferred to the primary acceptor, a Chl *a* monomer. An additional molecule of Chl *a* located between P_{700} and A_0 acts as an electron transfer bridge. The charge-separated state is stabilized by rapid transfer of the electron to a molecule of bound phylloquinone (A_1) and then to a series of [4Fe–4S] clusters termed F_X, F_A, and F_B, which function as an electron transfer wire. Because the terminal acceptors are iron–sulfur clusters, PS I is a Type I reaction center.

A model based on the 2.5-Å resolution crystal structure of the trimeric PS I reaction center from the cyanobacterium *Synechoccoccus elongatus* (PDB ID 1JB0) has provided detailed structural information on the arrangement of proteins and cofactors (Figure 8). Each monomer of the PS I heterodimer contains 96 Chl *a* molecules, 22 carotenoids, and 4 lipids. All of the organic cofactors in the electron transport chain are arranged in two pseudo-C_2 symmetric branches along the PsaA–PsaB heterodimer (Figure 9). However, unlike PS II, in which electron transfer occurs only along one branch, both the branches in PS I participate to varying degrees. In higher plants and algae, electron transfer along the PsaB branch is nearly as active as along the PsaA branch, whereas in cyanobacteria electron transfer is more asymmetrical, with only a small minority of electrons transported along the PsaB branch.

However, which branch is most active has little practical consequence, as the two branches converge at F_x, a unique interpolypeptide [4Fe–4S] cluster ligated by two cysteine residues from PsaA and two cysteine residues

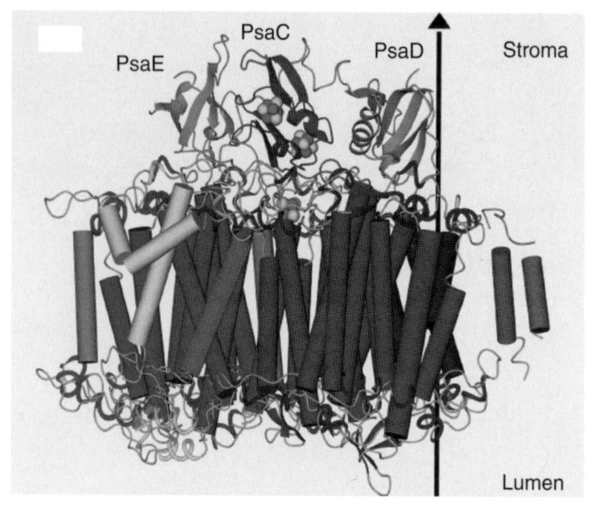

FIGURE 8 Side view of the arrangement of all proteins in one monomer of PS I. The main subunits are colored as follows: PsaA (blue), PsaB (red), PsaC (pink), PsaD (turquoise), PsaE (green), and PsaF (yellow). The iron–sulfur clusters in PsaC are colored yellow. The antenna pigments have been omitted for clarity. The vertical line (right) depicts the crystallographic C_3 axis. (For interpretation of the references to color in this figure legend, the reader is referred to the Web version of this chapter.)

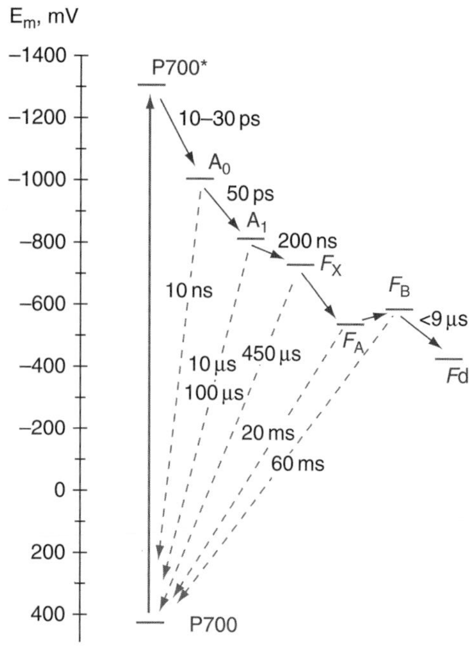

FIGURE 10 The components of the PS I reaction center depicted with redox potential on the *y*-axis and rate of electron transfer on the *x*-axis.

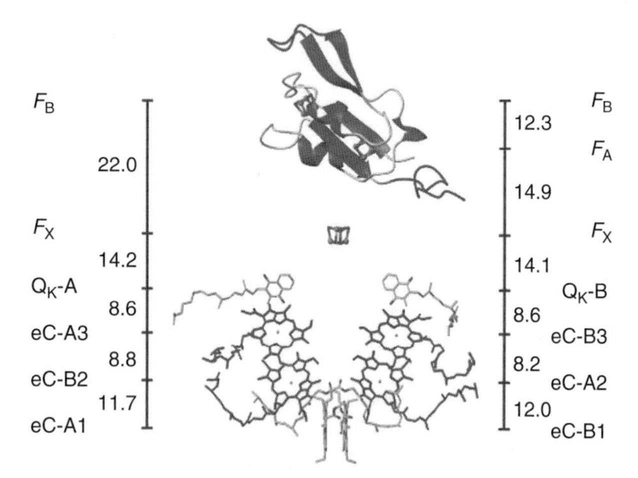

FIGURE 9 The arrangement of cofactors in PS I with center-to-center distances (in angstrom). The Chl *a* molecules, eC-A_1 and eC-B_1, constitute the special pair of chlorophylls P_{700}. The chlorophylls eC-A3 and eC-B3 represent A_0 on the PsaA and PsaB branches, respectively. Q_K-A and Q_K-B represent phylloquinone (A_1) on the PsaA and PsaB branches, respectively. The three [4Fe–4S] clusters are F_X, F_A, and F_B.

from PsaB. F_X is possibly the most reducing iron–sulfur cluster in biology, with a measured midpoint potential of −710 to −730 mV. The terminal [4Fe–4S] clusters, F_A and F_B, are located in the PsaC subunit, which is located on the stromal side of the membrane. To our knowledge, PsaC is the only membrane-associated protein whose three-dimensional structure has been solved in both the bound and unbound forms. PsaD and PsaE are two additional stromal proteins that flank PsaC, stabilizing its association with the heterodimeric core. These stromal proteins are vital for the docking of soluble electron acceptors such as ferredoxin and flavodoxin. Cyanobacterial PS I contains seven additional subunits (PsaF, PsaI, PsaJ, PsaK, PsaL, PsaM, and PsaX) whose roles are not well established. Some function as additional antenna proteins by binding one or two Chl *a* molecules. Others, such as PsaL, along with PsaI and PsaM, contribute to the trimerization of cyanobacterial PS I. It is interesting that the absence of a portion of the C-terminus in the PsaL subunit leads to a completely monomeric PS I in higher plants and algae.

The electron transfer kinetics among the bound cofactors in PS I are roughly comparable to those in PS II (Figure 10). The primary electron donor, P_{700}, becomes oxidized in ~20 ps, and the electron is transferred from A_0 to A_1 in 50 ps. The electron transfer step from A_1 to F_X is biphasic with lifetimes of ~200 and 20 ns. It is likely that the slower kinetic phase represents electron transfer along the PsaA branch of cofactors, while the faster phase represents electron transfer along the PsaB branch. The rates of electron transfer between F_X, F_A, and F_B are not known with certainty; however, soluble ferredoxin is known to be reduced within 500 ns. The reduced ferredoxin interacts with the enzyme ferredoxin-$NADP^+$ oxidoreductase (FNR) to transfer its electron to $NADP^+$. NADPH is used along with the ATP generated from the transmembrane proton gradient to fix CO_2 into hexose sugars.

The extremely low midpoint potentials of A_1 and F_X are intriguing and make PS I an excellent model for studying the influence of the protein environment on redox cofactors in biological systems. Our current knowledge of this topic is largely restricted to theoretical predictions based on influences such as uncompensated charges and polarities of amino acid side groups, the fixed dipole moment of the peptide bond in the polypeptide chain, and solvent accessibility to the cofactors. A change in the protein environment introduced by mutagenesis will usually lead to a change in the redox potential of one or more electron transfer cofactors, which, in turn, results in an altered rate of electron transfer. The change in kinetics can be related to a change in midpoint potential using Marcus theory. As an example, the substitution of a negatively charged aspartate near the A_1 quinone in PS I with a neutral or positively charged residue results in a decrease in the rate of electron transfer to F_X, which implies a lower driving force between the two cofactors. Conversely, removal of a negative charge near the F_X cluster in PS I leads to higher rates of electron transfer from A_1 and hence a greater driving force between the two cofactors. The experimental data agree well with the electrostatic calculations based on the X-ray crystal structure of PS I.

Cyanobacterial PS I is an ideal candidate for these experimental studies, as it is relatively straightforward to determine the rates of electron transfer using time-resolved optical or EPR spectroscopy. The presence of a well-established transformation system facilitates the construction of site-directed variants. The availability of a 2.5-Å resolution X-ray crystal structure of PS I is an added advantage because it allows accurate predictions to be made, which can then be tested by experiments.

Although PS II and PS I use similar design principles in achieving charge separation, they have complementary roles in photosynthesis by functioning at the two extremes of the biological redox scale. PS II has evolved to generate a strong oxidant to split H_2O, whereas PS I has evolved to generate a strong reductant to produce NADPH.

With knowledge of how the electron from H_2O is transported across a thermodynamic gradient to produce NADPH, we now focus on the biochemistry of converting CO_2 into carbohydrate.

CALVIN–BENSON–BASSHAM (CBB) CYCLE

The CBB cycle of carbon fixation can be divided into two stages. The first involves the trapping of CO_2 as a carboxylate and its subsequent reduction to glyceraldehyde-3-phosphate (G3P). The second involves the regeneration of the acceptor molecule for the next cycle of carbon fixation. A schematic view of the CBB cycle is depicted in Figure 11.

Atmospheric CO_2 diffuses into the stroma where it is added to the five-carbon acceptor molecule, ribulose-1,5-biphosphate (RuBP). The enzyme Rubisco, which is the most abundant protein on earth, catalyzes the reaction. The product is a six-carbon intermediate that is cleaved into two molecules of 3-phosphoglycerate (3PG). At this point, CO_2 has already been fixed into a carbohydrate. The remainder of the cycle is dedicated to the formation of hexose sugars and to the regeneration of RuBP.

Each molecule of 3PG is phosphorylated by the ATP-dependent enzyme phosphoglycerate kinase, liberating ADP in the process. The 1,3-biphosphoglycerate so formed is then reduced to G3P by NADPH, with the accompanying loss of

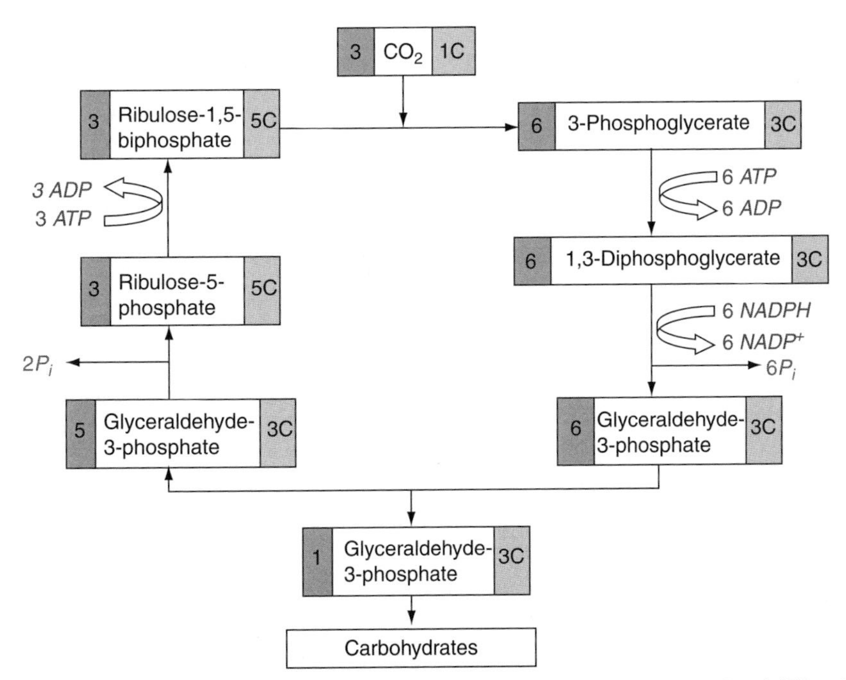

FIGURE 11 Schematic view of the Calvin–Benson–Bassham (CCB) cycle of carbon fixation. The net uptake of ATP and NADPH at each step is indicated.

a phosphate. This reaction is catalyzed by the enzyme G3P dehydrogenase. For each CO_2 molecule that passes through these steps, two molecules of ATP are hydrolyzed and two molecules of NADPH are oxidized. Every new molecule of hexose requires six CO_2 molecules to enter the cycle. That requires the formation of 12 G3P and therefore 12 ATP and 12 NADPH are required.

The conversion of G3P into carbohydrates is a complex multistep process involving several enzymes. G3P can be isomerized to dihydroxyacetone phosphate (DHAP) by triose phosphate isomerase. Thus, the 12 G3P molecules essentially form an interconvertible pool of G3P and DHAP. At this stage, the pathway bifurcates toward two goals: the production of hexoses and the regeneration of RuBP. Six molecules of G3P (four G3P plus two DHAP) are diverted to the regeneration pathway, while the remaining six molecules (three G3P plus three DHAP) are used for carbohydrate synthesis.

Three molecules of G3P combine with three molecules of DHAP via the enzyme fructose biphosphate aldolase to yield three molecules of fructose-1,6-biphosphate (FBP). The FBP is then dephosphorylated to provide three molecules of fructose-6-phosphate (F6P).

Of these, two molecules will be used in the regeneration pathway, leaving one as the net product of the CCB cycle. F6P is isomerized to glucose-6-phosphate (G6P) and finally to glucose-1-phosphate (G1P), which is the precursor for oligosaccharide and polysaccharide formation. G1P can be hydrolyzed to form glucose, or it can be converted to amylose and sucrose via separate pathways.

The input molecules for the regeneration cycle are four G3P, two DHAP, and two F6P molecules. Enzymes termed *trans*-ketolases and aldolases perform molecular rearrangements that are necessary to form five-carbon molecules from six-carbon and three-carbon molecules. The final step in the regeneration of RuBP is a phosphorylation reaction catalyzed by the ATP-dependent enzyme ribulose-5-phosphate kinase. An additional six ATP are required for six rounds of this step. The overall CCB cycle can thus be summarized as:

$$6CO_2 + 18ATP + 12NADPH + 12H_2O \rightarrow C_6H_{12}O_6 + 18ADP + 18P_i + 12NADP^+ + 6H^+. \quad (8)$$

ANOXYGENIC PHOTOSYNTHESIS

As the name suggests, anoxygenic photosynthetic bacteria do not evolve O_2 as a by-product of photosynthesis. These descendants of ancient microbes contain only one type of reaction center and hence the electrons used to reduce CO_2 are taken from highly reduced molecules such as succinate and sulfide. Although most photosynthetic bacteria use the CCB cycle to fix carbon, some are able to fix atmospheric CO_2 by other biochemical pathways. Most anaerobic phototrophs can survive only under very low concentrations of O_2.

Despite these differences, the general principles of energy transduction in anoxygenic photosynthesis are similar to those in oxygenic photosynthesis. The primary chromophore belongs to a family of molecules called bacteriochlorophylls. There are six types of bacteriochlorophylls, denoted bacteriochlorophyll (BChl) *a*, *b*, *c*, *d*, *e*, and *g* (Figure 12). They are similar to Chls, but absorb light in the near-infrared region (Figure 13). As in aerobic photosynthesis, electron transfer is coupled to the generation of a proton gradient that is used to synthesize ATP. The energy required to reduce CO_2 is provided by ATP and NADH, a molecule similar to NADPH but lacking the phosphate.

Purple Bacteria

Purple photosynthetic bacteria are a versatile group of proteobacteria that can be classified further into purple nonsulfur bacteria and purple sulfur bacteria. All purple bacteria use a Type II reaction center to generate a proton gradient for ATP synthesis, that is, there is no formation of NADPH. The reductant for carbon fixation is derived from organic compounds such as succinate and malate (nonsulfur bacteria) or from inorganic sulfide (sulfur bacteria). Light-driven electron transfer in purple bacteria is cyclic and hence no net oxidation or reduction occurs.

Purple nonsulfur bacteria are found in ponds, mud, and sewage. Purple sulfur bacteria are obligate anaerobes and are found in illuminated anoxic zones of lakes where H_2S accumulates and also in geothermal sulfur springs. Both fix carbon via the CBB cycle.

All purple bacteria have a very efficient antenna system consisting of BChl *a*, BChl *b*, and carotenoids. The presence of purple carotenoids such as spirilloxanthin gives these bacteria their distinct color. The first three-dimensional X-ray crystal structures of a photosynthetic reaction center were from purple nonsulfur bacteria (*Rhodopseudomonas viridis* and *Rhodobacter sphaeroides*). The basic composition of their reaction centers is similar to that of PS II. The primary donor is a special pair of BChl *a* molecules, which, after excitation by light, transfer the electron to bacteriopheophytin, the primary electron acceptor. The charge-separated state is stabilized by successive electron transfer to two ubiquinone molecules, Q_A and Q_B. After two cycles of reduction, two protons are taken up from inside the membrane to form the doubly reduced dihydroubiquinol in the Q_B site. Dihydroubiquinol diffuses to the cytochrome bc_1 complex, where it becomes oxidized, regenerating ubiquinone. The cytochrome bc_1 complex employs the protonmotive Q-cycle and translocates up to two protons per electron across the membrane. The energy stored in the resulting electrochemical proton gradient is used to synthesize ATP via the membrane-bound ATP synthase complex. The cytochrome bc_1 complex completes the cycle by transferring the electron back to the primary donor via the soluble carrier protein cytochrome *c*.

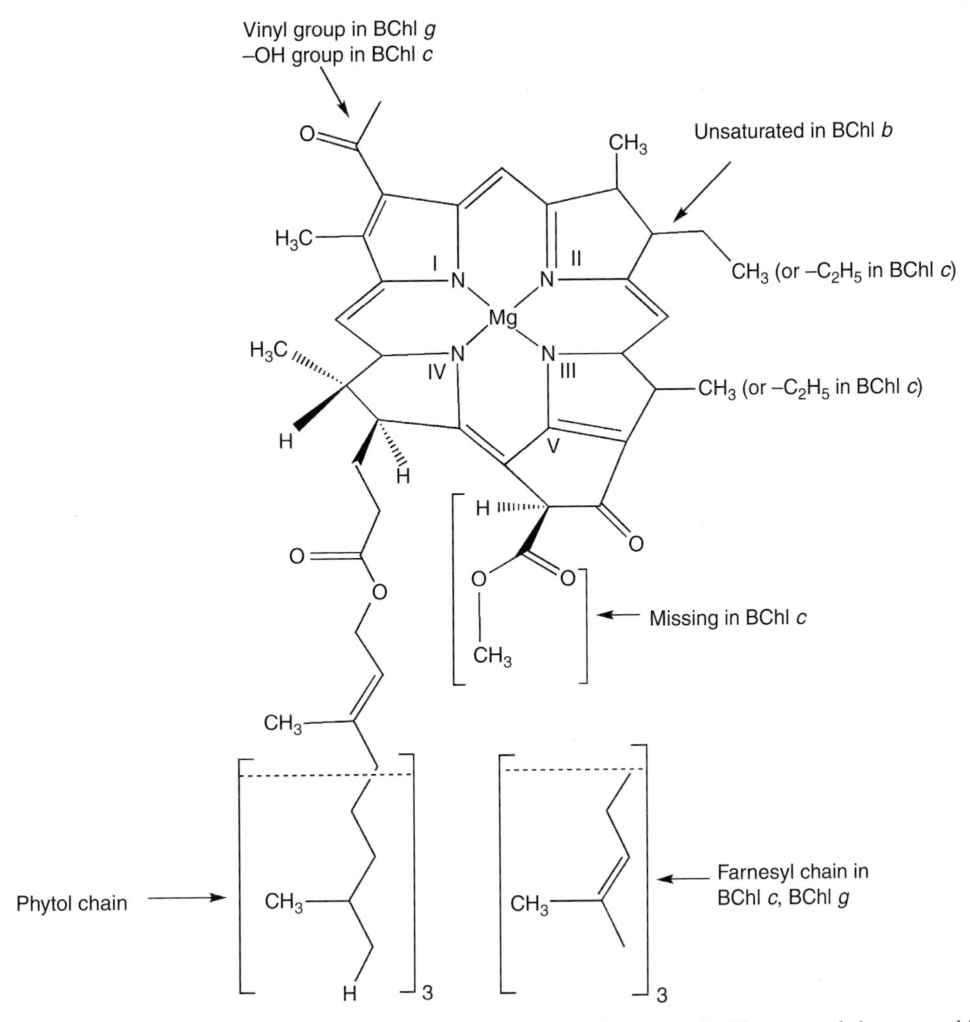

FIGURE 12 Chemical structure of BChl *a*, one of the pigments employed by anoxygenic phototrophs. The structural changes resulting in other forms of bacteriochlorophyll are also indicated.

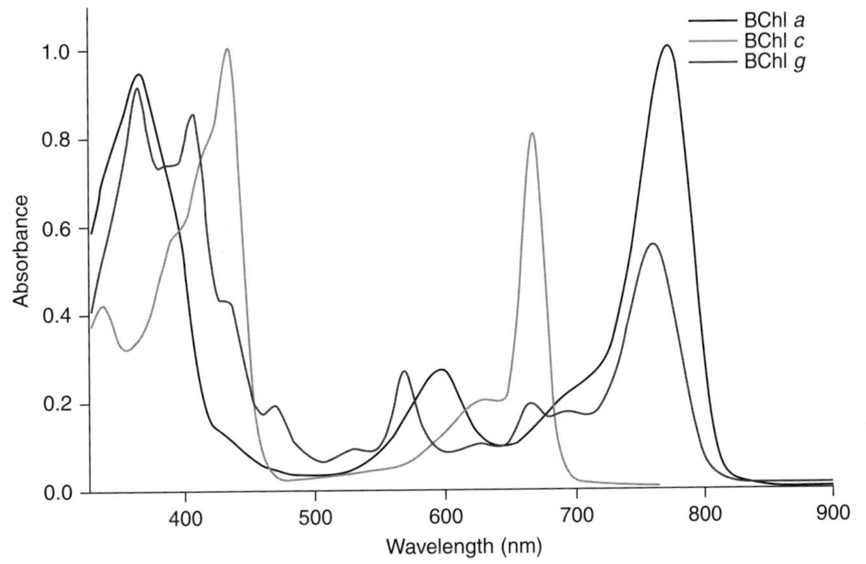

FIGURE 13 Absorption spectrum of BChl *a*, BChl *c*, and BChl *g* in solution. (For interpretation of the references to color in this figure legend, the reader is referred to the Web version of this chapter.)

Green Sulfur Bacteria

Green sulfur bacteria such as *Chlorobium tepidum* and *Chlorobium vibrioforme* belong to the phyla Chlorobi and are strictly anaerobic photoautotrophs. They use reduced sulfur compounds as their electron donors and fix carbon using the reverse TCA cycle. Unlike purple bacteria, light-induced electron transfer is noncyclic in green sulfur bacteria; hence NADPH is generated. These bacteria live in sulfur-rich environments that have characteristically low light intensities. They employ a unique antenna complex termed the chlorosome, which comprises BChl *c*, BChl *d*, and BChl *e*. It is the largest known antenna structure in biology, with each chlorosome containing ~200,000 BChl molecules. The habitats of green sulfur bacteria necessitate such an extensive antenna system, requiring a very large optical cross section to capture the few available photons. The light energy is transferred to a homodimeric Type I reaction center via the BChl *a* containing the Fenna–Matthews–Olsen (FMO) protein. The FMO protein is soluble in water, and was the first Chl-containing protein to have its three-dimensional structure solved. The reaction center core is a homodimer of PscA, and it contains most of the redox cofactors. Electron transfer begins at P_{840}, a special pair of BChl *a* molecules, and proceeds through the primary acceptor, a Chl *a* molecule monomer, and three [4Fe–4S] clusters F_X, F_A, and F_B. It is uncertain whether a quinone functions as an intermediate electron transfer cofactor between A_0 and F_X. F_A and F_B are bound to a protein named PscB, which has an unusually long N-terminal segment of proline, lysine, and arginine residues. A protein named PscD is thought to be involved in the docking of soluble ferredoxin and in the stabilization of the FMO protein. Another protein, PscC, is a tightly bound cytochrome, c_{551}, that donates electrons to P_{840}.

Heliobacteria

Heliobacteria (e.g., *Heliobacterium mobilis* and *Heliobacterium modesticaldum*) are members of the phylum Firmicutes and are the only known Gram-positive photosynthetic organisms. They were discovered 25 years ago in soil on the campus of Indiana University, Bloomington. Heliobacteria are anaerobic photoheterotrophs that fix nitrogen and are commonly found in rice fields. They can grow on selected organic substrates like pyruvate, lactate, and butyrate. Heliobacteria do not contain ribulose-1,5-bisphosphate or ATP-citrate lyase, the two enzymes commonly used in carbon fixation, but rather incorporate carbon via an incomplete reductive carboxylic acid pathway. These bacteria use BChl *g* as their primary pigment and employ a simple homodimeric Type I reaction center to perform noncyclic electron transfer. The components of the electron transfer chain are similar to green sulfur bacteria except that the pigment used as the special pair (P_{798}) is BChl *g*. The reaction center core is a homodimer of PshA, and it contains the primary donor and acceptor Chls and the F_X iron–sulfur cluster. The F_A and F_B iron–sulfur clusters are harbored on a low molecular mass polypeptide termed PshB. Similar to the reaction centers in the phylum Chlorobi, the participation of a quinone as an electron transfer cofactor between A_0 and F_X is still under debate.

Little or no structural information is available on any homodimeric Type I reaction center. Based on analogy with PS I, it is believed that a bifurcating electron transfer chain with two equivalent branches of cofactors exists in these reaction centers, but there is no spectroscopic evidence yet to support this proposal.

Other Photosynthetic Bacteria

Some species of photosynthetic bacteria do not fall under any of the previously discussed categories. The green gliding bacteria (*Chloroflexi*), also known as green filamentous bacteria, can grow photosynthetically under anaerobic conditions or in the dark by respiration under aerobic conditions. Like green sulfur bacteria, they harvest light by using chlorosomes, but like purple bacteria, they employ a Type II reaction center. These poorly studied organisms fix CO_2 via the 3-hydroxypropionate pathway.

The most recent addition to the list of photosynthetic microbes is an acidobacterium, *C. thermophilum*, which reportedly synthesizes BChl *a* and BChl *c* in aerobic environments. This organism was isolated from microbial mats at an alkaline hot spring and is thought to contain chlorosomes and a homodimeric Type I reaction center. Further studies are needed to determine whether the photosynthetic apparatus has new and interesting features, or whether it falls into a typical Type I class.

THE EVOLUTION OF PHOTOSYNTHESIS

The origin of photosynthesis is such an ancient event that the details of how this biological process developed may be irretrievably lost. Nevertheless, an extensive interdisciplinary effort involving researchers in biochemistry, genetics, biophysics, geology, biogeochemistry, and bioinformatics has led to new and important insights into the evolutionary history of photosynthesis.

There is widespread agreement that the first photosynthetic organisms were anoxygenic prokaryotes and that the cyanobacteria, with their much more complicated system of two linked photosystems, were a later development. However, the origin of the most primitive photosynthetic machinery remains shrouded in mystery.

According to one school of thought, a common single polypeptide could have been the ancestor to both Type II and Type I reaction centers. This primordial reaction center would have diverged into a homodimeric reaction center and

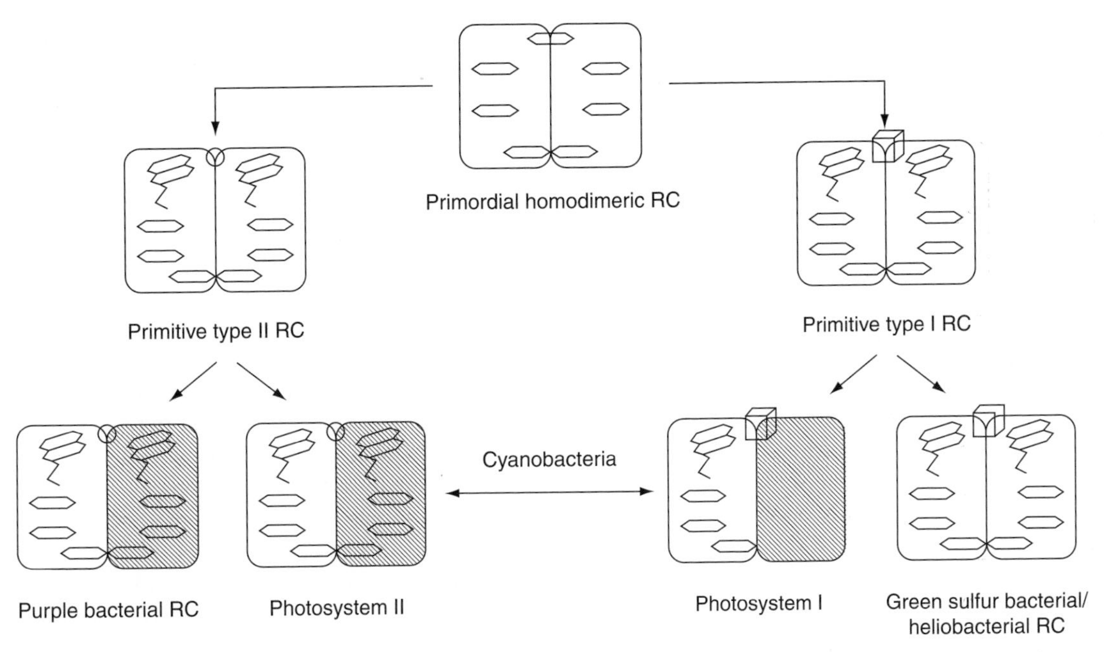

FIGURE 14 Schematic representation of the homodimeric "primordial reaction center" model of photosynthetic evolution.

subsequent development would have led to the present reaction centers found in anoxygenic phototrophs. Alternately, the formation of a homodimeric reaction center might have been the initial photosynthetic event (Figure 14). From a geometrical perspective, a three-dimensional surface is easiest to construct by abutting two two-dimensional surfaces. A highly protected environment at the interface of the two identical proteins would neatly solve the problem of shielding the electron transfer cofactors from water without requiring a highly folded three-dimensional structure such as the interior of a protein. Inherent in both ideas, the formation of a heterodimeric core would have been a result of gene divergence from a homodimeric core. The similar motifs of charge separation and charge stabilization in both Type I and Type II reaction centers are certainly compatible with a common origin. Genetic fusion between two organisms, one having a Type II and the other a Type I reaction center, would have led to the colocation of both in the same organism. This fusion organism likely evolved into the present-day cyanobacteria.

The terminal electron acceptor of the primordial reaction center in either case could have been a quinone or an iron–sulfur cluster. The recruitment of a bacterial ferredoxin later in evolution would account for the presence of F_A and F_B in heterodimeric as well as homodimeric Type I reaction centers. It is also possible that a completely different moiety could have functioned as the primordial terminal acceptor, with subsequent modifications that led to the emergence of Type II and Type I reaction centers.

In Type II reaction centers, the transition from the homodimeric to the heterodimeric state might have occurred to specialize the function of the Q_A and Q_B quinones as one-electron and two-electron gates, respectively. The reason for the transition is less clear in Type I reaction centers, which do not require a division of labor between the two quinones. It is nevertheless interesting that heterodimeric Type I reaction centers are exclusively associated with cyanobacteria, algae and plants, organisms that employ accessory antenna Chl proteins. They might have evolved from a homodimeric to a heterodimeric state to provide specialized binding sites for these additional structures. The two branches of redox cofactors are highly symmetrical in heterodimeric Type I and Type II reaction centers, suggesting that this transition has not resulted in a significant alteration of the protein environment surrounding the electron transport cofactors.

Although analysis of protein similarity is a highly effective method to predict evolutionary events, it does not help in understanding the development of complex pigment molecules such as Chl. The original Granick hypothesis holds that biosynthetic pathways for the formation of Chl recapitulate their evolution. It proposes that the pathway is built forward, with each step fulfilling a function, eventually being replaced by the next step selected for improved utility. The problem with this proposal is that the synthesis of bacteriochlorophyll proceeds through a Chl-like intermediate step. A strict interpretation of the Granick hypothesis would therefore imply that bacteriochlorophyll-containing organisms (anoxygenic) evolved later than Chl-containing organisms (oxygenic). Because the opposite is most likely the case, the reaction centers and Chls may have followed a dissimilar history in evolution.

The origin of the water-splitting complex is also uncertain, although some groups have postulated that it evolved from a manganese-containing catalase. According to this idea, weak electron donors such as H_2O_2 and $Fe(OH)^+$ once provided the electrons to PS II. The incorporation of the water-splitting complex might have been driven by the necessity to replace these rare electron donors with the most abundant electron donor on earth, H_2O.

FUTURE RESEARCH DIRECTIONS

Although extensive research in the last four decades has led to a good understanding of the overall design philosophy of photosynthetic electron transfer, there remain many unanswered questions. The mechanism behind the splitting of H_2O is still unknown, primarily because the S_4 state of the manganese cluster, which is the starting point for O–O bond formation, has not been observed. High-resolution crystal structures are invaluable, but they only provide a static ground-state depiction of the proteins and cofactors. Sophisticated spectroscopic techniques, particularly electron paramagnetic resonance, are required to probe the excited states, providing precise details of the critical steps in oxygenic photosynthesis.

Homodimeric Type I reaction centers remain poorly understood. The presence of two branches of electron transfer cofactors in identical environments begs the question of how charge separation is initiated and how the electron "chooses" which of two equivalent pathways to take. The main hindrance to progress lies in the inability to spectroscopically distinguish electron transfer on either branch.

Photosynthetic reaction centers are undoubtedly the best model system for studying the influence of the protein environment on the redox potentials of organic and inorganic cofactors. This is best exemplified by comparing the quinones in Type II and Type I reaction centers, which are structurally similar but differ in redox potential by hundreds of millivolts. Q_A and Q_B, which are both plastoquinones, have midpoint potentials of −150 and +100 mV, respectively, when bound to PS II. An even more dramatic case is provided by A_1, the phylloquinone bound to PS I. Its redox potential in organic solvent is similar to that of plastoquinone, but when bound to PS I, it has a midpoint of −800 mV. Thus, the redox cofactors are tuned by the protein to provide an appropriate midpoint potential for a given electron transfer step. Detailed knowledge of how the tuning occurs will be exceedingly useful in the designing of artificial photosynthetic systems.

Indeed, the next decade will probably see a major advance in artificial photosynthesis, mainly due to the concern over the need to develop new sources of energy after the depletion of fossil fuels. The basic outline of solar energy conversion is now known from the knowledge of natural photosynthesis, and these principles can be used to design artificial organic and inorganic systems of high efficiency. A solar cell consisting of photosynthetic reaction centers from spinach and from purple bacteria layered on a silver electrode has already been shown to generate an electric current. Efforts to develop synthetic counterparts of the reaction center and the oxygen-evolving complex have had limited success due to the inability to replicate the complex environment of a living cell. However, the coming years will undoubtedly see new developments in this field.

Several research groups are also trying to use photosynthetic organisms to generate H_2 that can be used as an alternate source of stored solar energy. Several promising ideas include the manipulation of cyanobacterial genes to enhance H_2 production in whole cells and the synthesis of a PS I–hydrogenase hybrid complex.

It is somewhat ironic that after depleting the fossil fuels that were produced by photosynthetic organisms in the first place we are again looking to photosynthesis for our long-term energy needs. Photosynthesis truly has come full circle.

ACKNOWLEDGMENTS

Research in this laboratory is funded by grants from the National Science Foundation (MCB-0519743) and the United States Department of Energy (DE-FG-02-98-ER20314).

FURTHER READING

Blankenship, R. E. (1992). Origin and early evolution of photosynthesis. *Photosynthesis Research*, *33*, 91–111.

Blankenship, R. E., Madigan, M. T., & Bauer, C. E. (Eds.). *Anoxygenic photosynthetic bacteria* (1995). Dordrecht (The Netherlands): Kluwer Academic Publishers.

Golbeck, J. H. (2002). Photosynthetic reaction centers: So little time, so much to do. *Biophysics Textbook Online*, http://www.biophysics.org/education/golbeck.pdf.

Golbeck, J. H. (Ed.) *Photosystem I: The light driven plastocyanin: Ferredoxin oxidoreductase* (2007). Dordrecht (The Netherlands): Springer.

Heinnickel, M., & Golbeck, J. H. (2007). Heliobacterial photosynthesis. *Photosynthesis Research*, *92*, 35–53.

Wydrzynski, T. J., & Satoh, K. *Photosystem II: The light-driven water: Plastoquinone oxidoreductase* (2005). Dordrecht (The Netherlands): Springer.

Chapter 14

Regulation of Carbon Assimilation

J. Plumbridge
Institut de Biologie Physico-Chimique (UPR9073-CNRS), Paris, France

Chapter Outline

ABBREVIATIONS

ABC ATP-binding cassette
AC Adenylate cyclase
AR1 Activating Region I
CAP Catabolite activator protein (=CRP)
CCA Carbon catabolite activation
CcpA Catabolite control protein
CCR Carbon catabolite repression
crc Catabolite repression control
cre Catabolite response element
CRP cAMP receptor protein (=CAP)
crr Carbohydrate repression resistant
ED Entner–Douderoff
EI Enzyme I
EMP Embden–Meyerhof–Parnas pathway
PEP Phospho-*enol*-pyruvate
PQQ Pyrroloquinoline-quinone dependent
PRD PTS regulation domain
PTS Phosphotransferase system
RAT Ribonucleotide antitermination targets
RNAP RNA polymerase
UAS Upstream activating site
α-NTD N-terminal domain of the alpha subunit of RNAP
α-CTD C-terminal domain of the alpha subunit of RNAP

DEFINING STATEMENT

Bacteria must adapt to the nutrients available in their environmental niche. One important regulatory mechanism to optimize carbon use is carbon catabolite repression (CCR).

The different strategies adopted by bacteria to optimize their use of available carbon sources are described.

INTRODUCTION

The Need to Regulate

Like all living organisms, bacteria have to adapt to their environment. Their major concern is to find the nutrients that they need to survive. A source of assimilable carbon is a primary requisite and, except for phototrophic (photosynthetic) bacteria or other autotrophs, this means some form of organic carbon compound. The ability of bacteria to use an incredible range of carbon sources is indicative of the equally wide range of habitats where they are found. Their ability to inhabit a particular environment is reflected in the cohort of genes present in their genomes, for taking up and degrading the variety of compounds likely to be available in a particular location. These range from the nutrient-rich environments of lactic acid bacteria to the low-nutrient aqueous environments frequented by *Caulobacter crescentus* and ocean-dwelling bacteria. There is a rough correlation between genome size and metabolic capacity. Lactic acid bacteria, which are endogenous to food-related habitats, have slimmed down their genomes, whereas bacteria adapted to more diverse or desert-like conditions have retained all their biosynthetic capacities plus diverse catabolic pathways for a wide range of potential nutrients.

Most experimental studies have been carried out in the completely artificial environment of the Erlenmeyer flask supplied with high concentrations of a single carbon source or a defined mix of nutrients during exponential growth. Bacteria in nature will rarely find themselves in a richly supplied medium and, at best, pass between times of relatively high levels of nutrients to a stage when all has been used up and the bacteria must either go searching for a new supply or wait until more food comes their way. This has led to the "feast or famine" concept. In the last decade, attempts have been made to mimic a more natural state by using carbon limitation in chemostats.

It has long been known that not all carbon sources are equal and some are certainly more equal than others; that is to say one carbon source is used preferentially before others. In general, this equates to a rapidly metabolizable carbon source being used first. This observation poses another question of why some carbon sources are used more rapidly than others. The general consensus (dogma) is that uptake is the rate-limiting step. Limitation at the level of uptake could be due to the amount of the transporter available, the activity of the transporter for a specific substrate, or the supply of energy to energize the transport system. However, other intracellular steps, such as the amount or activity of an enzyme needed to metabolize a compound or the supply of an intermediate that is required for its metabolism, for example, ATP or NADH, could be limiting. It is frequently observed that overproduction of one key enzyme within a pathway does not increase flux through the pathway because something else is limiting. Metabolic flux analysis has addressed the problem of flux control during glycolysis and for *Escherichia coli* the rate of use of ATP, by the anabolic pathways, appeared to be the rate-limiting step. Artificially increasing ATP hydrolysis increased the flux through glycolysis. However, in a similar study on *Lactococcus lactis*, control by ATP consumption was not observed, implying that in this organism there is no excess glycolytic capacity.

Diauxie

The classic experiment showing the preferential use of one sugar over another is the phenomenon of diauxic growth, as described by Jacques Monod. *E. coli* supplied with a mixture of glucose and lactose uses the glucose completely first, then effectively stops growing (lag time) while the genes for uptake and degradation of lactose are induced before the bacteria resume growth, at a slower rate, using lactose as the carbon source.

The ability of glucose to inhibit the use of other carbon sources in enteric bacteria was called the "glucose effect" and is now known to be an aspect of the general phenomenon of CCR. This is a term that covers a range of physiological phenomena that enable bacteria to make a hierarchical choice between different sources of carbon. CCR exists in all bacteria but the form it takes is very varied between different species. Generally, glucose is the preferred carbon source but this is not a universal rule. For example, the nitrogen-fixing *Azotobacter vinelandii* uses acetate or galactose before glucose and in *Pseudomonas putida* succinate produces stronger CCR than glucose. CCR has been most extensively studied in two model organisms: *E. coli* for Gram-negative bacteria and *Bacillus subtilis* for Gram-positive species. In these two organisms the molecular mechanisms are quite different, and although they are representative of bacterial species closely related to themselves, it is clear that there are alternative, completely different, mechanisms operational in other bacteria. Studies on carbon control in other microorganisms, particularly those with biotechnological or pharmaceutical applications, such as *Streptomyces coelicolor* or *Corynebacterium glutamicum*, are revealing a very diverse set of mechanisms. Even the dogma of hierarchical choice of nutrients has been questioned since *C. glutamicum* prefers to use several carbon sources simultaneously.

GLOBAL VERSUS SPECIFIC REGULATION

The proteins and genes responsible for CCR are the prime examples of global regulators in bacteria. They control a large number of genes in a coordinated manner, albeit with a certain level of flexibility, depending upon the individual genes under regulation. It has been estimated that 50% of the genes in *E. coli* are regulated by one of the seven global regulators and, of these seven, catabolite activator protein (CAP, also called CRP, cAMP receptor protein) is by far the most important, with about 200 direct targets. Expression of catabolite-regulated genes in *E. coli* usually requires transcriptional activation by the cAMP/CAP complex. The magnitude of the activation by cAMP/CAP of different genes varies considerably and depends upon the sequence and location of the CAP-binding site on the target DNA. This global response means that when the preferred carbon source (glucose in the case of *E. coli*) is present, none of the other carbon utilization systems is expressed. However, in the absence of glucose, the bacteria do not want to express their full repertoire of all other carbon utilization operons, which can amount to many hundred genes. So, superimposed on the generalized response to catabolite regulation, there are specialized regulatory mechanisms dedicated to individual carbon sources. In the case of the diauxie experiments of Monod, the presence of glucose prevented the expression of genes for all other carbon utilization systems, but when the glucose had been used up, the presence of lactose meant that only the genes of the *lac* operon were expressed and not those for the use of other carbon sources such as maltose or arabinose.

More recent results, however, have shown that this idea is too simplistic. The bacterial response to changing energy supplies is not just on/off but is more measured. Global transcriptome analysis of RNA from bacteria grown in the presence of different carbon sources, producing a wide range of growth rates, from 0.97 (glucose) to $0.13\,h^{-1}$ (proline), has shown that, as the quality of the carbon source declines and the growth rate decreases, bacteria gradually upregulate a wide range of genes to broaden their metabolic potential and scavenge for alternate carbon sources. Similarly, a transcriptome analysis of cells undergoing diauxie showed that a range of stress-related genes (controlled by σ^S, the stationary phase sigma factor, *rpoS*) and stringent response genes (controlled by ppGpp) are also activated.

Sensing and Signaling

The general rule is that genes required for the use of a particular carbon source are only expressed at a high level when the compound is available to be assimilated. This requires a signal transduction pathway to sense the external presence of the compound and turn on the expression of the required genes. This type of regulation is nearly always transcriptional, implicating transcription factors acting either positively (activators) or negatively (repressors), of which the classic examples in *E. coli* are the lactose and galactose repressors and the maltose and arabinose activators. A sugar outside the cell enters at a low level, generating a signal, which is detected inside the bacteria, indicating the presence of the carbon source outside. The signal is generally a small molecule related to the pathway under control (allolactose, galactose, maltotriose, and arabinose in the four examples listed). The signal starts to induce the genes necessary for the transport of the sugar, thus amplifying the intracellular signal and the level of induction. There are some alternative methods of signaling an available nutrient. Some signaling is due to phosphorylation cascades, as in the case of two-component regulators. Detection of a signal outside the bacteria provokes the autophosphorylation of a histidine sensor kinase, which in turn phosphorylates and activates a second protein, a response regulator, which is generally a positive transcription factor. There are a few examples of two-component systems regulating the use of a carbon supply, for example, the uptake of hexose phosphates in *E. coli* by the UhpT transporter is controlled by the UhpABC-encoded two-component system, but generally this type of regulation is associated with stress responses. In bacteria, the phosphotransferase system (PTS), a major sugar uptake system, is responsible for many signaling phosphorylation reactions, which are described below (see Sections "Catabolite Repression in Gram-Negative Bacteria: *Escherichia coli*" and "Catabolite Repression in Firmicutes: *Bacillus subtilis*"). A slight variation to the rule that regulation of sugar utilization is at the level of transcription initiation is the case of some PTS sugar operons of *B. subtilis*, whose induction is at the level of transcription elongation, involving a transcription antitermination mechanism (see Section "Transcriptional Antitermination of PTS Operons").

These sugar-specific factors all work in collaboration with the global regulators (cAMP/CAP in *E. coli*) so that together there is a combination of positive induction superimposed upon relief of catabolite repression, to allow the expression of the genes at the required time. The details of each regulated system are unique. Common variations are the number of operator sites involved and the need for other DNA-binding proteins producing complex nucleoprotein structures in either the activated or the repressed state. For example, repression of both the *gal* and *lac* operons in *E. coli* requires DNA loop formation between Lac or Gal repressors bound to three (in the case of *lac*) or two (in the case of *gal*) operators. In the case of the *galETK* operon the histone-like protein, HU, is also required for formation of the *gal* repression loop.

Since the pioneering studies of Jacob and Monod on the *lac* operon in the 1950s and 1960s, people have been dissecting in more and more detail the mechanisms of regulation inherent in the utilization of numerous carbon sources especially in *E. coli* and *B. subtilis*, and also in other more exotic bacteria. The twenty-first century has launched the genomic era, which has revolutionized the study of bacterial carbon use. The sequences of hundreds of bacterial chromosomes allow comparative genomics to predict what carbon sources are likely to be used and in some cases identify a potential regulatory mechanism. It should be noted, however, that structural genes for enzymes are much better conserved than either their gene organization or the intergenic regulatory sequences. Regulatory patterns are only conserved in closely related organisms. Adenylate cyclases (ACs) and CRP homologues are found in a wide range of bacteria but the erstwhile paradigm of catabolite repression by cAMP/CAP only exists in the γ-proteobacterial lineage.

In addition to genomic sequencing the other "-omic" studies of transcriptome, proteome, and metabolome have allowed studies of the global response of bacteria to a wide range of nutrients, stimuli, and mutations. Exploiting the data from these studies has led to the concepts of flux analysis and systems biology to understand the flow of metabolites and generalized coordination of cell growth. These studies have revealed that the number of genes in a regulon, that is, the family of genes coregulated by a transcription factor, is often higher than previously anticipated. Moreover, they have confirmed that the genes of the central carbon pathways in bacteria, the housekeeping genes of glycolysis, pentose phosphate and TCA cycle are not just constitutively expressed but are also subject to complex regulation.

CARBON ASSIMILATION PATHWAYS: CENTRAL CARBON METABOLISM

Carbon assimilation can be divided into two levels: peripheral and central. The peripheral steps transform diverse and varied carbon sources into phosphorylated intermediates, which can then be channeled into one of the major routes for the dissimilation of carbon. These central routes are the three interconnecting pathways of glycolysis (EMP = Embden–Meyerhof–Parnas pathway), the Entner–Douderoff (ED) pathway, and the pentose phosphate pathway, which all feed carbon into the TCA (Krebs) cycle (Figure 1 and Table 1). The enzymes of central metabolism are historically some of the best studied. Together, these reactions generate energy (ATP), reducing power (NADH and NADPH), and, directly or indirectly, the precursor molecules for all the biosynthetic reactions. The 13 intermediates that are drawn away from the central pathways of carbon metabolism are glucose-6P (to be used to make sugar nucleotides), fructose-6P (for amino sugars), pentose phosphates (for various vitamins and cofactors, purine and pyrimidine nucleotides, histidine and tryptophan), sedoheptulose phosphate (for heptoses in lipopolysaccharides), erythrose phosphate (for aromatic amino acids and some vitamins and cofactors), dihydroxyacetone phosphate (for nicotinamide coenzymes and phospholipids), 3-phosphoglycerate (for serine family amino acids, tryptophan, and purine nucleotides), phospho*enol*-pyruvate (PEP, for various vitamins and cofactors and aromatic amino acids), pyruvate (pyruvate family amino acids), acetyl-CoA (for fatty acids, peptidoglycan, and leucine), oxaloacetate (aspartate family amino acids and pyrimidines nicotinamide coenzymes), 2-ketoglutarate (glutamate family amino acids, polyamines, and heme derivatives), and succinyl-CoA (for lysine family amino acids and heme derivatives).

Glycolysis, ED, and Pentose Phosphate Pathways

In glycolysis six-carbon sugars are split into three-carbon compounds and then to PEP, pyruvate, and acetyl-CoA, generating ATP and NADH (Figure 1). Glycolysis (EMP pathway) was considered to be the trunk route of sugar metabolism because it is the predominate route in the model organisms *E. coli* and *B. subtilis*. However, the relative importance of the EMP and ED pathways and the oxidative branch of the pentose phosphate pathway varies in different organisms. In the ED route, glucose is converted to gluconate-6P using the enzymes Glc6P dehydogenase (*zwf*) and phosphogluconolactonase (*pgl*) and then converted directly into glyceraldehyde-3P and pyruvate. In *E. coli*, sugar acids such as gluconate and galacturonate, derived from mammalian and plant tissues, are processed via the ED pathway. The ED route is absent in *B. subtilis* but in many other bacteria it is the major route for glucose metabolism. In some bacteria, like the xenobiotic-degrading *Pseudomonades* and the ethanol-producing *Zymomonas mobilis*, there is no *pfk* gene, for phosphofructose kinase, a key enzyme of glycolysis, which means that all glucose is processed through the ED pathway. It has been hypothesized, based on phylogenetic analysis, that the ED route is the original route to degrade glucose. The oxidative part of the pentose phosphate pathway uses the same first two enzymes (*zwf* and *pgl*) as the ED route to form gluconate-6P, which is then decarboxylated to ribulose-5P, generating NADPH. A series of transketolase and transaldolase reactions produces sugar phosphate precursors (for the biosynthesis of aromatic amino acids, vitamins, and nucleotides) and ultimately glyceraldehyde-3P and fructose-6P (Figure 1).

Glycolysis and Gluconeogenesis

During rapid growth on glycolytic sugars such as glucose, *E. coli* or *B. subtilis* generate most of the energy they need by substrate level phosphorylation. A part of the carbon flux is directed into the TCA cycle to generate 2-ketoglutarate and oxaloacetate (anaplerosis) while the excess carbon is excreted as acetate (a process normally called overflow metabolism). When all the glucose is used up, the acetate

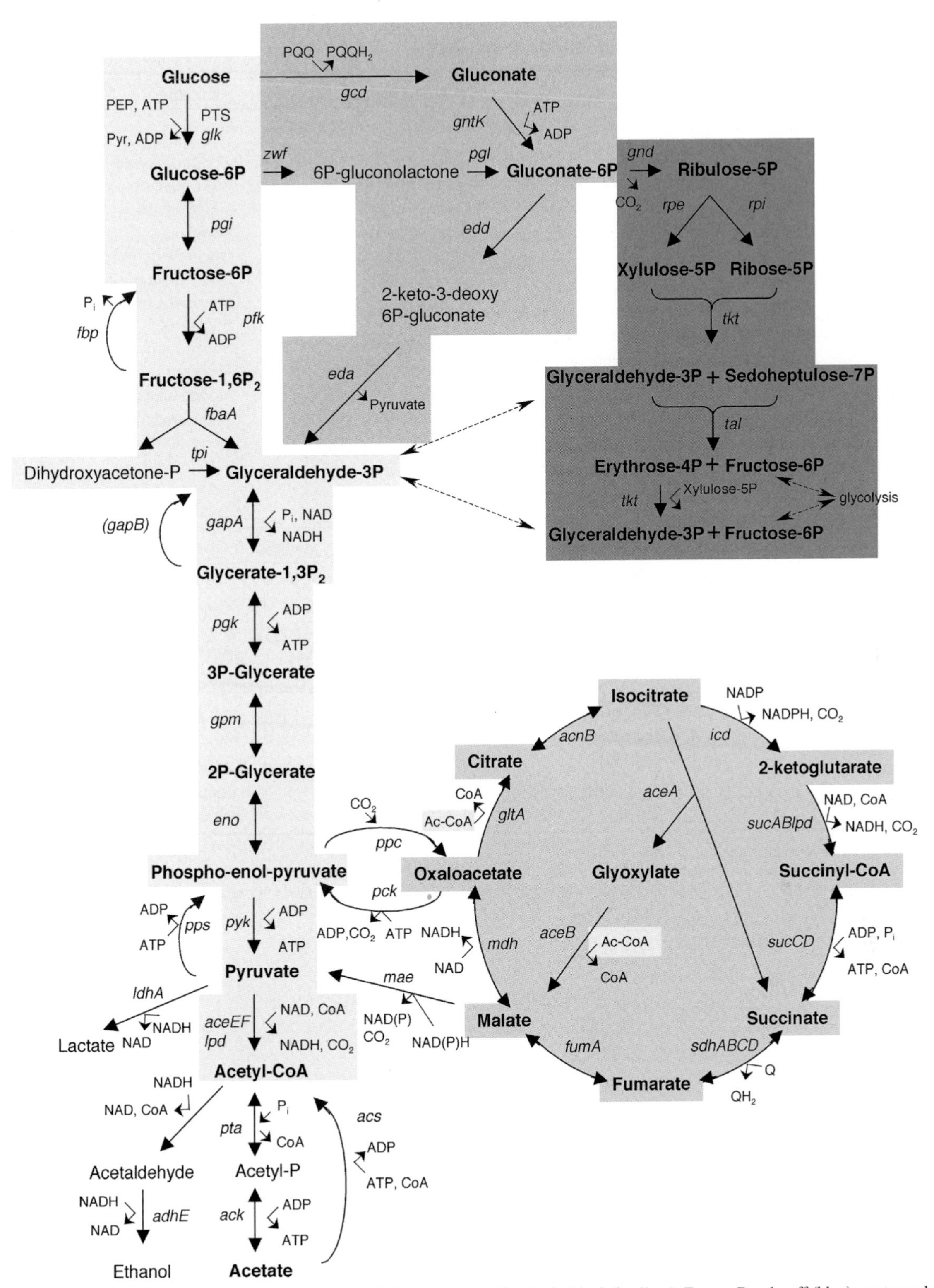

FIGURE 1 Pathways of central carbon metabolism. Routes of glucose use via glycolysis (shaded yellow), Entner–Doudoroff (blue), pentose phosphate pathway (green), and TCA cycle (pink) are shown. Reactions specific to gluconeogenesis are shown with blue arrows. The enzymes encoded by the different genes are listed in Table 1. The pathways shown are those of *E. coli*, certain specific differences existing in *B. subtilis* are indicated in Table 1. The *gapB* gene for gluconeogenic glyceraldehyde-3P dehydrogenase is only present in *B. subtilis*. (For interpretation of the references to color in this figure legend, the reader is referred to the Web version of this chapter.)

TABLE 1 The genes and enzymes of central carbon metabolism

Gene Name		Enzyme
E. coli	*B. subtilis*	
Glycolysis		
glk	*glcK*	Glucokinase
pgi		Phosphoglucose isomerase
pfk	*pfkA*	Phosphofructose kinase
fbaA		Fructose-1,6-bisphosphate aldolase
tpi		Triose phosphate isomerase
gapA		Glyceraldehyde-3P dehydrogenase
pgk		Phosphoglycerate kinase
gpm	*pgm*	Phosphoglycerate mutase
eno		Enolase
pyk		Pyruvate kinase
aceEF lpd	*pdhABCD*	Pyruvate dehydrogenase
Entner–Doudoroff pathway		
gcd	–	Glucose dehydrogenase (PQQ[a])
gntK		Gluconokinase
zwf		Glucose-6P dehydrogenase
pgl (ybhE)	*yqjI*[b]	Phosphogluconolactonase
edd	–	Phosphogluconate dehydratase
eda	– *(kdgA)*[c]	2-Keto-3-deoxygluconate-6P aldolase
Pentose phosphate pathway		
gnd	*gntZ*	Gluconate-6P dehydrogenase
rpe		Ribulose-5P epimerase
rpi	*ywlF?*[d]	Ribulose-5P isomerase
tkt		Transketolase
tal	*ywjH?*[d]	Transaldolase

Gene Name		Enzyme
TCA cycle and glyoxylate shunt		
ppc	–	PEP carboxylase
–	*pyc*	Pyruvate carboxylase
gltA	*citZ (citA)*	Citrate synthase
acnB	*citB*	Aconitase
icd	*citC*	Isocitrate dehydrogenase
sucAB lpd	*odhABpdhD*	2-Ketoglutarate dehydrogenase
sucCD		Succinyl-CoA synthetase (succinate thiokinase)
sdhCDAB	*sdhCAB*	Succinate dehydrogenase
fumA	*citG*	Fumarase
mdh	*citH*	Malate dehydrogenase
maeB sfcA	*ytsJ*[e]	Malic enzyme
aceA	–	Isocitrate lyase
aceB	–	Malate synthase
aceK	–	Isocitrate deyhrogenase Kinase/phosphatase
Gluconeogenesis and growth on acetate		
pps		PEP synthetase
fbp		Fructose bisphosphatase
pck	*pckA*	PEP carboxykinase
acs	*acsA*	Acetyl-CoA synthetase
–	*gapB*	Glyceraldehyde-3P dehydrogenase
Overflow metabolism		
pta		Phosphotransacetylase
ack	*ackA*	Acetate kinase
ldhA		Lactate dehydrogenase
adhE	*adhA*	Alcohol dehydrogenase

The gene names for the enzymes in E. coli are given in the first column. When the name is different in B. subtilis the name is given in the second column. A dash indicates that the enzyme is not found in one or other bacteria. Isozymes exist for many genes, which are active under different conditions, for example, aerobic, anaerobic, oxidative stress, and stationary phase. The name of the gene encoding the predominant enzyme active during aerobic exponential growth is given.

[a]*PQQ, pyrroloquinoline-quinone dependent.*
[b]*Based on 70% identity with pgl (ybhE).*
[c]*Part of the kdgRKAT operon.*
[d]*Indicates putative identification based on homology.*
[e]*Three other paralogous genes encoding malic enzymes also exist in B. subtilis maeA, malS, mleA. YtsJ is required for growth on malate.*

can be reabsorbed and used to generate acetyl-CoA. This has been called the "acetate switch," when the bacteria switches its enzymatic machinery from excreting acetate to using acetate for growth. Acetate and other carbon sources that generate acetyl-CoA feed into the TCA cycle so that for bacteria growing on acetate or dicarboxylic acids, the TCA cycle is used to generate energy. A part of the carbon supply must return backward up the glycolysis pathway, in the process called gluconeogenesis, to produce the sugar phosphates required for the synthesis of sugar nucleotides and other precursors to cell wall and membrane components.

The majority of the reactions of glycolysis are reversible but some are irreversible and specialized enzymes exist to catalyze the reverse gluconeogenic reactions (Figure 1, blue arrows). Phosphofructokinase (*pfk*) is a classic allosteric enzyme, inhibited by PEP and activated by nucleoside diphosphates (ADP), thus signaling the need for more or less energy generation. Fructose bisphosphatase (*fbp*) performs the reverse reaction; it is inhibited by AMP and is active during growth on gluconeogenic carbon sources. Other enzyme pairs performing reverse reactions are pyruvate kinase (*pyk*) generating acetyl-CoA in glycolysis and PEP synthetase (*pps*) synthesizing PEP during gluconeogenesis. In *B. subtilis* the glyceraldehyde-3P dehydrogenase reaction is also catalyzed by two enzymes, encoded by *gapA*, using NAD as the cofactor, for the glycolytic direction and *gapB* (cofactor NADP) for the gluconeogenic step. In *E. coli* the *gapA* enzyme functions in both directions. A gene for a highly homologous enzyme exists in *E. coli* and was initially suspected to be another phosphorylating glyceraldehyde-3P dehydrogenase because it is located in an operon with two other glycolytic enzymes, *pyk* and *fbaA*. However, it was shown to be a nonphosphorylating erythrose-4P dehydrogenase (*epd*) using NADP as reductant, and to be part of the biosynthetic pathway toward chorismate.

In both *E. coli* and *B. subtilis*, control mechanisms exist to direct the carbon flow in the glycolytic or gluconeogenic direction. In *B. subtilis* the genes for the lower steps of glycolysis *gapA*, *pgk*, *tpi*, *pgm*, *eno* are arranged in an operon controlled by the product of the first gene of the operon, *cggR* (central glycolytic genes repressor). Transcriptome analysis suggested that CggR has only one target and is a dedicated regulator of the *gapA* operon. Its inducer is Fru1,6P. The *B. subtilis gapA* operon is also controlled by CcpA, the global catabolite regulator in Gram-positive bacteria (see Section "Catabolite Repression and Regulation by CcpA and the *cre* Element"). Expression of the *B. subtilis* genes specific for gluconeogenesis, *gapB* and *pckA*, is very low in the presence of glucose but is high during growth on succinate. Repression is due to the DNA-binding protein, CcpN (catabolite control protein of gluconeogenic genes), whereas the gene adjacent to *ccpN*, *yqfL* has a positive effect on their expression, but the molecular details are currently unknown.

In *E. coli* the expression of several genes of glycolysis has been shown to increase during growth on glucose, for example, *gapA*, *pgk*, and *fbaA*. The FruR repressor of the fructose PTS, in addition to repressing the *fruBKA* operon 20-fold, more weakly represses the *epd pgk fbaA* operon and other catabolic genes such as *pfk*, *edd*, *eda*, and *ptsH* (part of the PTS, see Section "The Bacterial PTS"). FruR also acts as an activator of the *pps* gene (encoding PEP synthetase), some other genes of gluconeogenesis (*pckA*, *fbp*), and the TCA cycle genes (*aceBA*, *icd*). The binding of FruR is inhibited by fructose phosphates, especially Fru1P but also Fru1,6P, which signals the activity of the glycolytic pathway. FruR thus plays a pleiotropic role in directing the carbon flow in response to metabolic signals and was renamed Cra (catabolite repressor activator) to reflect this function. Allosteric regulation also plays a role in directing the flux by the opposing reactions of phosphofructokinase (*pfk*) and fructose 1,6 bisphosphatase (*fbp*).

The PEP–Pyruvate–Oxaloacetate Node

As in the case of the glycolytic pathway, different sets of enzymes control the direction of the carbon flow between glycolysis and the TCA cycle, which has been called the PEP–pyruvate–oxaloacetate node. PEP carboxylase (*ppc*) catalyzes the anaplerotic formation of oxaloacetate from PEP and carbonate, while PEP carboxykinase (*pck*) carries out the reverse reaction, conversion of oxaloacetate to PEP (Figure 1). Bacteria with *ppc* mutants require a TCA cycle intermediate (succinate) for growth on sugars to replenish the TCA cycle. PEP carboxylase is a tetrameric allosteric enzyme with many effectors including acetyl-CoA, Fru1,6P, GTP, and long-chain fatty acids (positive) and aspartate and malate (negative). *B. subtilis* does not possess a PEP carboxylase but replenishes the TCA cycle from glycolysis by forming oxaloacetate using pyruvate carboxylase (*pycA*). Pyruvate carboxylase is activated by acetyl-CoA. An exchange between TCA cycle and glycolysis can also occur at the level of pyruvate and malate, catalyzed by so-called malic enzymes, of which several isozymes exist: two in *E. coli* and four in *B. subtilis*. They differ in their cofactor requirements, NAD or NADP. In *E. coli* both enzymes are active during growth on dicarboxylic acids. In *B. subtilis* the NADP-linked enzyme (YtsJ) is particularly required for growth on malate. This is most likely related to maintaining the correct NADPH balance. Expression of the operon for the pyruvate dehydrogenase complex (*aceEFlpd*) is repressed by the pyruvate-sensing repressor, PdhR, a GntR family repressor. PdhR also controls some genes of the respiratory electron transfer system.

The TCA Cycle and Glyoxylate Shunt

The TCA cycle completes the oxidation of acetyl-CoA produced by glycolysis and provides intermediates for biosynthesis of aspartate, glutamate, and their derivatives. In fact, the TCA cycle rarely runs as a cycle but mostly as two arms producing succinyl-CoA and 2-ketoglutarate from oxaloacetate. During growth on acetate or fatty acids, it runs as two cycles including the glyoxylate shunt.

The glyoxylate shunt represents a bifurcation of the pool of isocitrate directly toward malate and succinate rather than via 2-ketoglutarate because the latter route involves two irreversible decarboxylations and hence depletion of the TCA intermediates (Figure 1). The direction of carbon flow from isocitrate to glyoxylate or to 2-ketoglutarate is decided by the relative activities of isocitrate dehydrogenase (*icd*) and isocitrate lyase (*aceA*). The activity of isocitrate dehydrogenase is strongly downregulated by reversible phosphorylation carried out by the Icd kinase/phosphatase enzyme (IcdK/P) encoded by *aceK*. This is the third gene of the inducible *aceBAK* operon, where *aceB* and *aceA* encode the two enzymes of the glyoxylate shunt. All three genes are induced by growth on acetate and repressed by growth on glucose, coordinating a simultaneous decrease in Icd activity and an increase in the glyoxylate shunt or *vice versa*. A specialized repressor, IclR, whose inducing signal appears to be PEP rather than acetate, controls the operon.

The genes of the TCA cycle are controlled by several global regulatory circuits. They are subject to CCR, repressed during growth on glucose, and their expression is dependent on the cAMP/CAP complex in *E. coli* (see Section "cAMP/CAP Regulation of Transcription Initiation"). Isozymes for several of the enzymes exist and are subject to different regulation, for example, *fumA* and *acnB* encode the major aerobic fumarase and aconitase activities, whereas *fumC* and *acnA* are expressed under conditions of oxidative or iron stress from an *rpoS*-dependent promoter (*rpoS* encodes σ^S (σ^{38}) the stationary phase/stress-related sigma factor). Expression of several genes for the utilization of acetate, *acs*, *aceBA*, *sucA* and also *tka* and *tal*, is increased at the entry of stationary phase in an *rpoS*-dependent manner. In addition, mutations in *rpoS* also affect TCA cycle activity and acetate use during exponential growth. The ArcA/ArcB redox sensing two-component system is responsible for the repression of the TCA cycle genes in anaerobic and microaerobic conditions, but mutations in *arcA* also increase the flux through the TCA cycle during aerobic growth on glucose. Expression of *sdhCDAB* is repressed by glucose but apparently in a cAMP/CAP-independent manner but which requires PtsG, the principal glucose PTS transporter.

Succinate dehydrogenase (*sdhCDAB*) is the only membrane-bound member of the TCA cycle and produces a reduced quinone to be used by the respiratory chain. It contains a $[4Fe–4S]^{2+}$ cluster. Aconitase (*acnA* and *B*) and fumarase (*fumA* and *B*) are also Fe–S enzymes. Expression of all three enzymes was known to be positively controlled by the Fur protein (iron-specific repressor) such that *fur* mutants do not grow on succinate. The explanation for this unexpected behavior came with the discovery of the RhyB sRNA. Synthesis of the RhyB RNA is under negative control by Fur so that it is only synthesized when intracellular iron is low. The sRNA, by binding to the *sdhCD* mRNA, initiates RNaseE-dependent degradation of the *sdh* mRNA, thus lowering the intracellular content of succinate dehydrogenase (and other nonessential iron proteins) and liberating iron when iron is scarce. In addition, scarcity of iron is thought to destroy the Fe–S clusters in aconitases. Apo-aconitases, from both bacteria and vertebrates, are mRNA-binding proteins and have been shown to repress the translation of their own mRNA, thus constituting a double regulation to conserve precious iron supplies.

In *B. subtilis*, CcpC, a LysR-type regulator, represses the *citZCH* operon (encoding citrate synthase, isocitrate dehydrogenase, and malate dehydrogenase) and *citB* (aconitase) and its own gene. The inducer appears to be citrate. The genes of the glyoxylate shunt are missing in *B. subtilis* and in some related species. These bacteria cannot grow on acetate as the sole carbon source.

Balancing the Redox Potential, NADPH/NADH

A general rule is that catabolic enzymes are coupled to NAD reduction while reactions forming part of anabolic pathways use NADP. Thus, the *B. subtilis* GapA enzyme uses NAD while the gluconeogenic enzyme, GapB, uses NADP. The major routes for generating the anabolic reductant NADPH were thought to be via the pentose phosphate pathway and isocitrate dehydrogenase but in *E. coli* a membrane-bound proton-translocating transhydrogenase (PntAB) accounts for at least one-third of the cellular NADPH generated from NADH during growth on glucose. A second soluble transhydrogenase (UdhA) seems dedicated to the reverse reaction, generating NADH from NADPH under conditions when NADPH is high (growth on acetate or on sugars metabolized via the pentose phosphate pathway). Both the expression and activity of these enzymes must be carefully correlated with the redox potential of the cell to eliminate a futile cycle dissipating NADPH. As the two enzymes have only been detected in enterobacteria, alternative mechanisms must exist in other bacteria to balance their redox potential.

UPTAKE OF CARBON SOURCES FROM THE MEDIUM

Sugar-specific utilization operons generally encode two types of enzymes: those involved in the transport of (usually) hydrophilic carbohydrates across the inner membrane and those encoding the enzymes that transform the more exotic starting materials into one of the compounds of the central carbon degradation pathway. The transporters belong to many protein families and are of many types, including H^+ symporters (e.g., LacY where lactose uptake is associated with proton uptake), ATP-binding cassette (ABC)-type transporters (e.g., MalFGK for maltose uptake), and facilitated diffusion (e.g., GlpF, for glycerol uptake). These systems transport the sugars unchanged, so that inside the cell they must be phosphorylated by specific kinases.

Another important route is the PTS, which is widely distributed in bacteria. In this group translocation mechanism, sugar transport across the inner membrane is associated with the concomitant phosphorylation of the sugar as it enters the cytoplasm. The phosphate is derived from PEP and passes via a protein phosphorylation cascade to the incoming sugar (Figure 2). The proteins of the PTS are intimately involved in CCR in both Gram-positive and Gram-negative bacteria, although the mechanisms of their action are fundamentally different. Therefore, I will first describe the PTS as a sugar transporter and then deal with CCR in Gram-negative and in Gram-positive bacteria.

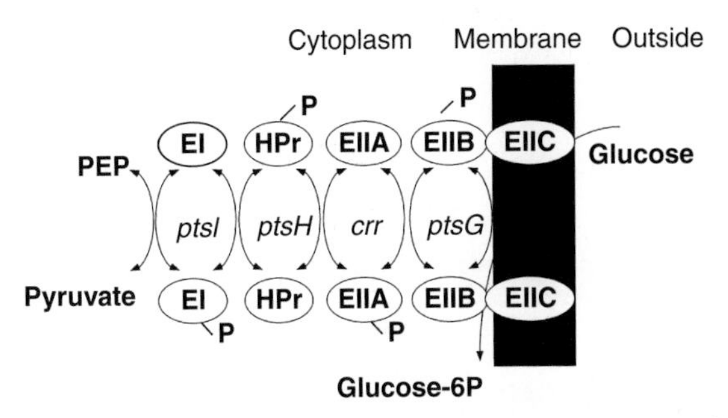

FIGURE 2 The glucose PTS in *Escherichia coli*. The phosphate from PEP is passed via a series of cytoplasmic proteins EI (*ptsI*), HPr(*ptsH*), and $EIIA^{Glc}$ (*crr*) to the glucose-specific transporter $EIICB^{Glc}$ (*ptsG*). Passage of glucose across the membrane via the $EIIC^{Glc}$ domain results in its simultaneous phosphorylation by $EIIB^{Glc}$ to give cytoplasmic Glc6P.

The Bacterial PTS

The basic PTS consists of two cytoplasmic proteins called Enzyme I (EI, gene *ptsI*) and HPr (*ptsH*) common to all sugars, which pass a phosphate derived from PEP to the sugar-specific transporters, called Enzymes II (EII) (Figure 2). The EIIs are multidomain proteins where the three domains, EIIA, EIIB, and EIIC can be found on one, two, three, or, occasionally, four polypeptide chains. EIIA and EIIB are soluble cytoplasmic proteins whereas EIIC is an integral membrane protein, which serves as the passage for the sugar through the membrane.

EI autophosphorylates on a histidine residue and then passes the phosphate to the small protein HPr. HPr distributes phosphates to the EIIA domain of any of the sugar-specific EII transporters within the cell. From EIIA the phosphate is transferred to EIIB, another soluble domain but which is associated with the cytoplasmic side of the inner membrane. From EIIB the phosphate is transferred to the incoming sugar, as it is transported by the EIIC domain across the cytoplasmic membrane. The phosphate is carried by a histidine residue in EI and HPr but usually by a cysteine, in most EIIB domains. In the resting state, that is, in the absence of a PTS sugar, the PTS proteins are predominately phosphorylated. It should be noted that these phosphates are relatively labile and can be easily passed between different EIIs via the HPr protein.

For the glucose PTS of *E. coli* (Figure 2), $EIIC^{Glc}$ and $EIIB^{Glc}$ are present on the same polypeptide, encoded by the gene *ptsG*. The $EIIA^{Glc}$ protein is a separate polypeptide, encoded by the gene *crr* (for carbohydrate repression resistant) a name that reflects the primordial role of $EIIA^{Glc}$ in CCR in *E. coli* (see next section). In fact, the gene *crr* is part of the *ptsHIcrr* operon, encoding the common components of the PTS, whereas *ptsG* is expressed as a single gene elsewhere on the chromosome. The PTS proteins are modular proteins par excellence. The structural genes can be recognized in bacterial genomes by homology-scoring programs and the EIIs have been divided into several subfamilies but the arrangement of their domains both within the polypeptides and at the genetic level can vary enormously. For example, in *E. coli* the *N*-acetylglucosamine (GlcNAc) transporter is encoded by the *nagE* gene ($EIICBA^{Nag}$) whereas the homologous β-glucosidase transporter, *bglF*, has the domains arranged as $EIIBCA^{Bgl}$. The fructose PTS in *E. coli* does not use HPr but has a specialized fusion protein, FPr, including an HPr domain and an $EIIA^{Fru}$ domain. Other unusual fusion proteins exist between both PTS and non-PTS domains, as in the nontransporting dihydroxyacetone kinase operon.

CATABOLITE REPRESSION IN GRAM-NEGATIVE BACTERIA: *ESCHERICHIA COLI*

The key molecule in CCR in *E. coli* is the $EIIA^{Glc}$ protein encoded by the *crr* gene. At least two regulatory circuits are in action during CCR by glucose. Glucose reduces the concentration of cAMP/CAP and so decreases the transcription of target genes and causes inducer exclusion, by preventing the uptake of other carbon sources, and hence prevents induction of the genes required for their metabolism. Both activities depend upon $EIIA^{Glc}$ but their relative contributions to CCR have been the subject of much debate.

Inducer Exclusion

Mutations in EI or HPr prevent the use of both PTS and non-PTS carbohydrates. Mutations called *crr* were isolated as suppressors, which restored growth on a range of non-PTS sugars. The *crr* gene, in which the suppressors were mapped, encodes the EIIA component of the glucose PTS. As such, it exists in two forms: phosphorylated and nonphosphorylated. These two forms were shown to have complementary roles in CCR. The nonphosphorylated form of EIIA, the form expected to predominate when the bacteria are growing on glucose, was shown to interact with and inhibit the activity of numerous transport systems for non-PTS sugars such as lactose, melibiose, raffinose, maltose, and also glycerol kinase (Figure 3). Inhibition of the basal level of these transporters in the bacteria is responsible for the phenomenon of "inducer exclusion." Inhibiting the activity of the transporters for different carbon sources prevents the uptake of these alternative substrates and induction of the genes required for their transport and metabolism. It has been reported that $EIIA^{Glc}$ interacts strongly with transporters only when their specific substrates are present, thus preventing the wasteful sequestration of $EIIA^{Glc}$ to unused transporters. The interaction between $EIIA^{Glc}$ and many of its targets has been demonstrated *in vitro*; specific binding sites on $EIIA^{Glc}$ and on several of its targets have been identified and a IIA^{Glc}-binding site consensus proposed.

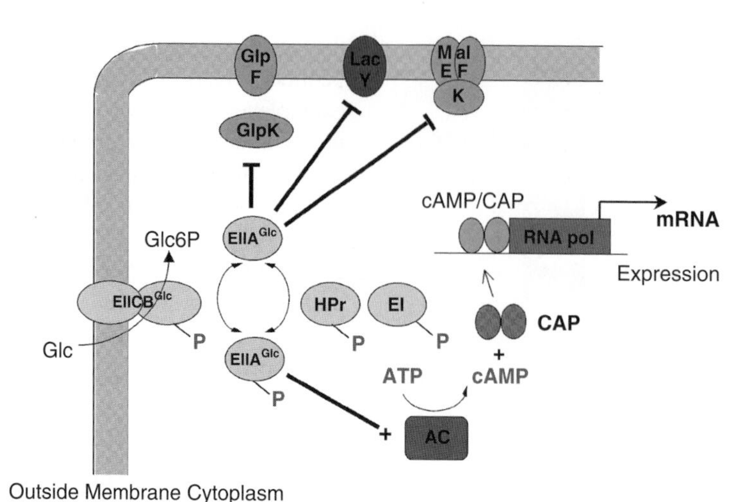

FIGURE 3 Carbon catabolite repression in enteric bacteria: *E. coli*. Transport of glucose by the PTS and EIICB^{Glc} (PtsG) produces predominately dephosphorylated EIIA^{Glc}. EIIA^{Glc} produces inducer exclusion because it inhibits the transport activity of non-PTS transporters like LacY and MalEFK and the glycerol kinase enzyme, GlpK. In the absence of glucose, EIIA^{Glc} is mostly phosphorylated and in this form allosterically activates the enzyme adenylate cyclase (AC), increasing the concentration of cAMP, which forms a complex with CAP. Higher concentrations of cAMP/CAP allow RNA polymerase to bind and transcribe genes for use of alternative carbon sources.

Activation of AC

EII^{Glc} is also largely implicated in changing the functional concentration of the global activator cAMP/CAP. When glucose is used up, EIIA^{Glc} becomes predominately phosphorylated and cAMP concentrations increase many fold. This increase has been shown to be in majority due to an increase in the activity of the enzyme AC (encoded by the gene *cya*). Unphosphorylated EIIA^{Glc} stimulates the enzymatic activity of AC 50-fold (Figure 3). Deletion of the 48 C-terminal amino acids from AC prevented this allosteric activation of cyclase activity, without affecting the intrinsic basal activity. A small transcriptional effect of EIIA^{Glc}-P on *cya* gene expression has been observed (fourfold) but the major regulation is via the activity of the enzyme. Although the *in vivo* evidence was strong, it is only recently that the stimulation of AC activity by purified EIIA^{Glc} has been demonstrated *in vitro*. It required the addition of a crude *E. coli* extract, indicating that some other, so far unidentified factor, is also required.

It has long been known that glucose is not the only sugar that produces catabolite repression in *E. coli*. Growth on other PTS sugars producing fast growth (e.g., *N*-acetylglucosamine or mannitol) provokes dephosphorylation of EIIA^{Glc}, via the reequilibration of phosphates between the different PTS components. In addition, certain non-PTS sugars, including glucose-6P, gluconate, and lactose, can exert repression on bacteria growing on less good carbon sources. Addition of these compounds to bacteria growing on lactate affects cAMP/CAP levels and provokes at least a partial dephosphorylation of EIIA^{Glc}. The level of dephosphorylation of EIIA^{Glc} seemed to correlate with the decrease in the PEP/pyruvate ratio, implying that products from the far end of the glycolysis route are involved in the sensing. It remains to be determined how transport of non-PTS sugars affects the PEP/pyruvate ratio. On the other hand, catabolite repression by glycerol was shown to require just the synthesis of glycerol-3P and not its further metabolism, and catabolite repression by Glc6P can occur in a *pgi* mutant, which cannot further metabolize the Glc6P.

The *crp* and *cya* genes are also subject to multiple levels of regulation. *crp* is autoregulated by cAMP/CAP both negatively (at physiologically relevant cAMP concentrations) and positively (at very high concentrations) whereas *cya* is repressed by cAMP/CAP. How these controls are integrated with the EIIA^{Glc} regulation and whether other mechanisms are also implicated is not known. Mutations in *crp* (called *crp**)-producing proteins that are active in the absence of cAMP have been described but their physiological consequences are not always as predicted. Although under extensive study since the discovery of cAMP in 1960s, the molecular mechanisms governing the physiological control of cAMP/CAP levels inside the cell are still far from clear. There are several unanswered questions: for example Why does Glc6P produce much stronger levels of repression than glucose? Why do bacteria excrete the majority of the cAMP they synthesize.

cAMP/CAP Regulation of Transcription Initiation

Despite our incomplete understanding concerning the physiology of cAMP/CAP, the molecular mechanisms of how cAMP/CAP activates transcription by RNA polymerase (RNAP) have been described in exquisite detail. CAP is undoubtedly the best-understood transcriptional activator. The crystal structures of the free protein and its cocomplex with DNA are known, thus defining its interaction with the DNA operators. CAP operators are found in two distinct locations relative to the promoter. At so-called Class I CAP-dependent sites the center of the CAP operator is positioned at about −61.5 or −72.5 upstream of the +1 transcription start site (Figure 4a). At this position a surface-exposed group of seven amino acids called activating region I (AR1) on the downstream subunit of CAP makes a precise contact with a region in the C-terminal domain of the alpha subunit (α-CTD) of RNAP. α-CTD is attached to the N-terminal domain of the alpha subunit (α-NTD) by a long and flexible linker, allowing α-CTD to adopt various positions at the promoter. At Class I promoters it is probably tucked in between CAP and RNAP and in contact with the DNA and also in contact with the sigma subunit of RNAP. It is presumed that the role of CAP (and transcription factors in general) is to provide an additional point of anchorage to increase the affinity of RNAP for the promoter; this is the recruitment model of transcription activation.

At Class II CAP-dependent promoters the CAP operator is centered at about −41.5 bp from the transcription start site and is nestled adjacent to the RNAP (Figure 4b). At this position at least three types of contact between RNAP and CAP have been detected. ARI is still active but this time it is the upstream subunit of CAP, which is in contact with α-CTD, now located on the DNA on the distal side of CAP. Two other activating regions on CAP have been demonstrated. AR2 on the downstream subunit interacts with amino acids from α-NTD, while AR3 contacts region 4 of σ^{70}. Contacts between CAP and RNAP at these two positions are thought to activate transcription by a postrecruitment mechanism, facilitating isomerization of the closed complex to yield the open RNAP–DNA complex.

The *crp* regulon is one of the most extensive and best characterized. Over 50 targets were identified by classical studies of individually regulated genes. Transcriptome studies of both glucose-grown bacteria and *crp* mutants have expanded this list to at least 200 genes. Glucose affected even more genes, both positively and negatively, and this is partly due to indirect effects. Precise identification of sites on the *E. coli* chromosome that bind CAP has been achieved by chromatin immunoprecipitation and microarrays (ChIP on chip analysis). This identified 68 targets of which only 29 corresponded to previously identified sites. An alternative technique using transcripts synthesized *in vitro* yielded

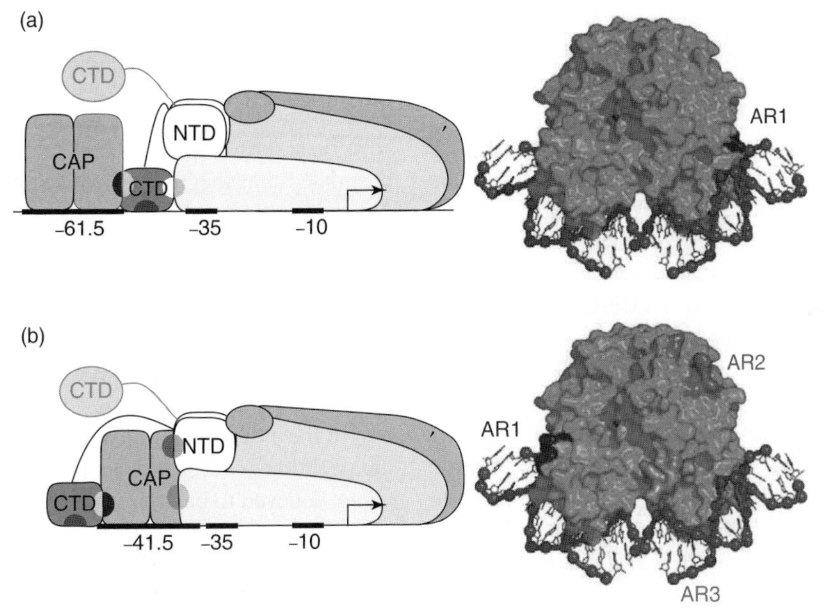

FIGURE 4 Transcription activation by cAMP/CAP. Reproduced from Lawson, C., Swigon, D., Murakami, K., Darst, S., Berman, H., & Ebright, R. H. (2004). Catabolite activator protein: DNA binding and transcription. *Current Opinion in Structural Biology, 14*, 10–20 with permission. (a) Activation at Class I CAP-dependent promoters. Left: Ternary complex of CAP, RNAP, and a Class I CAP-dependent promoter having the DNA site for CAP centered at position −61.5 relative to the transcription start site. Transcription activation involves interaction between activating region I (AR1) on the downstream subunit of CAP (blue) and one α-CTD protomer (yellow). The AR1–α-CTD interaction facilitates the binding of α-CTD to the DNA segment downstream of CAP (red) and Region 4 of sigma (pink). The second α-CTD protomer can interact nonspecifically with upstream DNA. Right: Structure of the CAP–DNA complex showing AR1 on the downstream subunit (blue). (b) Transcription activation at Class II CAP-dependent promoters. Left: Ternary complex of CAP, RNA polymerase, and a Class II-dependent promoter with the CAP site at −41.5. Transcription activation involves three sets of CAP–RNA polymerase (RNAP) interactions: interaction between AR1 of the upstream subunit of CAP (blue) and one α-CTD (yellow), an interaction that facilitates binding of α-CTD to the DNA segment immediately upstream of CAP (red); interaction between AR2 of the downstream subunit of CAP (dark green) and α-NTD (orange); and interaction between AR3 of the downstream subunit of CAP (olive green) and Region 4 of sigma (pink). The second α-CTD promoter can interact nonspecifically with upstream DNA. Right: Structure of the CAP–DNA complex showing AR1 of the upstream subunit (blue), AR2 of the downstream subunit (dark green) and AR3 on the downstream subunit (olive green). (For interpretation of the references to color in this figure legend, the reader is referred to the Web version of this chapter.)

nearly 200 targets of which about 40 were previously known. These experiments have thus confirmed the vast extent of the *crp* regulon and provided valuable information for the study of the previously unidentified target genes.

Other Global Regulators of Carbon Use

As explained above, a specific regulator controls the genes for use of a particular sugar. Two of the repressors of PTS operons in *E. coli* have acquired a more global function. It is perhaps significant that they are the repressors for two of the most abundant sugars: glucose (Mlc) and fructose (FruR, renamed Cra). The role of FruR in directing the flow of carbon in either the glycolytic or the gluconeogenic direction has been described above (see Section "Glycolysis and Gluconeogenesis").

Mlc: The Glucose PTS Repressor

Mlc was initially identified because plasmids expressing the protein slowed down growth on a complex medium containing glucose. It thus prevented an accumulation of acetate and the resulting lowering of the pH, which normally stops growth on glucose; so the bacteria grew for longer, used more of the glucose, and produced larger colonies (hence the mnemonic – making large colonies). Mlc has been shown to repress several genes related to glucose use in *E. coli*. This includes *ptsG*, encoding the major glucose transporter, and *ptsHI*, encoding the central components of the PTS. All genes so far identified as being repressed by Mlc are also controlled by CAP, so that growth on glucose has two opposing effects: it relieves repression by Mlc but at the same time lowers cAMP/CAP concentrations, provoking catabolite repression. In the case of *ptsG* and *ptsH*, expression is greater during growth on glucose than on glycerol but at other genes it is catabolite repression that dominates. It should be noted that even expression of *ptsG* requires a functional cAMP/CAP complex. There is no expression of *ptsG* in a *cya* mutant (i.e., in the complete absence of functional cAMP/CAP) but the *ptsG* promoter can work with the reduced level of cAMP/CAP present in the cell during growth on glucose.

The most remarkable thing about Mlc is the inducing signal, which causes its displacement from its operators. Instead of responding to a small molecule related to glucose, it is sensitive to the activity status of PtsG ($EIICB^{Glc}$). When PtsG is actively transporting glucose it is predominately unphosphorylated; Mlc can bind to the unphosphorylated $EIIB^{Glc}$ domain and is found sequestered to the inner membrane. In this position it is incapable of binding to DNA and repressing its targets. Sites on both Mlc and $EIIB^{Glc}$ that prevent this interaction have been identified by mutagenesis, and a cocrystal structure of Mlc and $EIIB^{Glc}$ is available. The membrane location is an essential part of the inactivation of Mlc; the soluble $EIIB^{Glc}$ domain is not sufficient even though it binds Mlc. The exact role of the membrane is unknown.

The *ptsG* gene is also subject to posttranscriptional regulation. The *ptsG* mRNA is rapidly degraded during conditions of phosphosugar stress (e.g., in a *pgi* mutant growing on glucose). The sugar stress initiates the synthesis of a transcriptional activator (SgrR) that promotes the synthesis of a sRNA, SgrS (sugar-related sRNA). SgrS, in the presence of Hfq, binds to the *ptsG* mRNA, which prevents its translation and initiates *ptsG* mRNA degradation by RNaseE. The logic of this posttranscriptional regulation is to turn down expression of PtsG and prevent the accumulation of toxic sugar phosphates. In fact, the SgrS RNA promotes a second, even more rapid response because, unlike most other sRNA, it also encodes a small polypeptide, which seems to behave as an inhibitor of PtsG transport activity.

The nitrogen PTS

In addition to the sugar-transporting PTS, there is a series of paralogous PTS proteins, equivalents of EI, HPr, and EIIA, not apparently involved in sugar uptake. The three proteins coded by *ptsP* ($E1^{Ntr}$), *ptsO* (NPr), and *ptsN* ($EIIA^{Ntr}$) lack any equivalent of the sugar-specific EII transporters. Because *ptsO* and *ptsN* are usually localized with the *rpoN* gene, encoding σ^{54}, the nitrogen-specific sigma factor, they have been termed the nitrogen or Ntr PTS and it has been postulated that they have a role in coordinating the use of different carbon and nitrogen sources.

The *rpoN* operon in *E. coli* (*rpoN*, *yhbH*, *ptsN*, *yhbJ*, *ptsO*) includes the genes for two unidentified proteins. Further evidence of the involvement of these genes in carbon–nitrogen metabolism comes from the fact that mutations in *yhbJ* accumulate the enzyme GlmS (Glucosamine-6P (GlcN6P) synthase). GlcN6P is the first intermediate in the production of the essential amino sugar-containing components of the peptidoglycan and lipopolysaccharides. The YhbJ protein is believed to be involved in sensing GlcN6P levels and is necessary for the feedback regulation of *glmS*, a control in which two small RNA (GlmZ and GlmY) and mRNA processing by RNaseE have been implicated.

A variety of phenotypes have been ascribed to mutations in one or other of the Ntr PTS genes, ranging from growth defects on certain carbon sources to leucine sensitivity and regulation of TrkA, the K^+ transporter in *E. coli*. The Ntr PTS has been well examined particularly in *P. putida* (see Section "Pseudomonades"). *P. putida* has a total of only five PTS proteins, a specialized fructose PTS *fruBA* (where *fruA* encodes EI:HPr:$EIIA^{Fru}$ and *fruB* $EIIBC^{Fru}$), plus the three proteins of the Ntr PTS, *ptsO*, *ptsN*, and *ptsP*. Both *ptsN* and *ptsO* mutations affect glucose repression of the Pu promoter of the σ^{54}-dependent xylene utilization operon. In *ptsN* mutants Pu was constitutively expressed, while in *ptsO* mutants Pu was permanently repressed. A proteome study showed that *ptsN* affected the expression of a large number of genes, not just those that are glucose-repressed or σ^{54}-dependent, implying it has wide regulatory functions. Growth rates of the *ptsN* mutant on glucose, fructose, glycerol, and succinate were lower than that of the wild type. Moreover, PtsN was shown to undergo growth phase-dependent phosphorylation. The exact molecular function of these pleiotropic PTS regulators remains to be discovered.

CATABOLITE REPRESSION IN FIRMICUTES: *BACILLUS SUBTILIS*

In low-GC content Gram-positive bacteria, as in Gram-negative bacteria, there are several distinct pathways contributing to the preferential use of glucose over other carbon sources. The key molecule in this case is the HPr protein, the small (9 kDa) protein of the central PTS. HPr from Gram-positives, in addition to being phosphorylated by EI of the PTS on His15, can be phosphorylated at a second regulatory site, Ser46, by a specific serine kinase. The same enzyme is also responsible for the dephosphorylation of P-Ser46-HPr and so is called HPr kinase/phosphatase (HprK/P gene *hprK*). HPr can be phosphorylated on His15 and Ser46 at the same time and so four forms of HPr exist in the cell: HPr, P-His15-HPr, P-Ser46-HPr, and the doubly phosphorylated form (Figure 5). The kinase activity is stimulated by Fru1,6P while the phosphatase activity is stimulated by inorganic phosphate (P_i). The phosphatase activity is not hydrolytic, but phosphorolytic, using P_i and producing pyrophosphate. A gene for a pyrophosphatase is encoded in the *hprK* operon. The presence of P-Ser46 on HPr reduces the rate of the PTS phosphorylation on His15 and the P-His15 inhibits the kinase phosphorylation on Ser46. Although the Ser46 residue exists in Gram-negative HPr, the surrounding amino acids are not conserved in enteric bacteria and HprK/P does not phosphorylate *E. coli* HPr.

P-Ser46-HPr and Inducer Exclusion

The relative population of the four forms of HPr depends upon the growth conditions. The kinase activity of HprK/P is high during growth on rapidly metabolized carbon sources, when Fru1,6P concentrations are high. P-Ser46-HPr and the doubly phosphorylated form predominate during rapid growth on glucose with only a small percentage in the single P-His15 form, capable of transferring phosphates to the incoming sugar. Considering the strong inhibitory effect of Ser46-P on PTS-mediated phosphorylation, it would appear that the 5–10% of HPr, which is not phosphorylated on Ser46, must be responsible for and sufficient for the rapid growth on glucose. Indeed, strains with mutations preventing Ser46 phosphorylation grow more slowly on glucose, which could be due to too much glucose uptake. Moreover, in these strains, inducer exclusion was eliminated and simultaneous uptake of glucose and another sugar (mannitol) was observed. On the other hand, mutations that specifically eliminate the phosphatase activity of HprK/P grow very slowly on glucose and all PTS sugars, because all the HPr is now in the P-Ser46 form and is unable to supply phosphates to PTS sugars. Thus, inducer exclusion of other PTS sugars by glucose is thought to be because of competition for P-His15-HPr. Preferential use of non-PTS sugars seems to depend on P-Ser46-HPr regulation of the

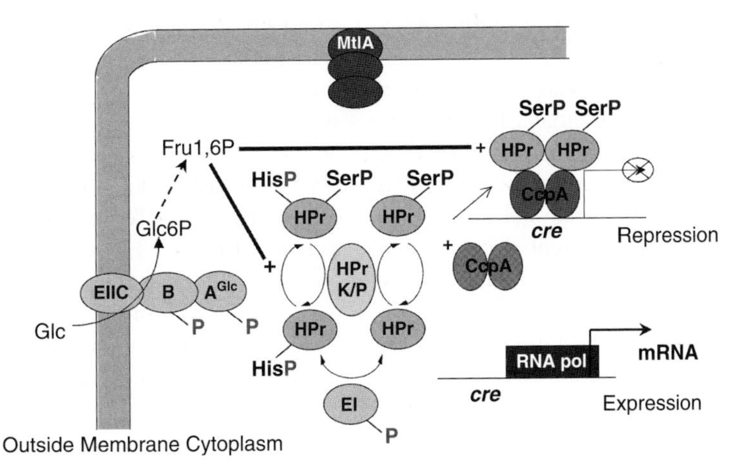

FIGURE 5 CCR in Firmicutes: *Bacillus subtilis*. Transport of glucose by the PTS and EIICBA^{Glc} (PtsG) produces internal Glc6P, which is converted by glycolysis to Fru1,6P. Fru1,6P activates the kinase activity of HprK/P producing P-Ser46-HPr and P-Ser46,P-His15-HPr. P-Ser46-HPr interacts with the transcription factor CcpA and increases its DNA-binding affinity. The P-Ser46-HPr:CcpA complex binds to *cre* sites and inhibits transcription. P-His15-HPr levels are low and the phosphate from His15-HPr is preferentially directed to EIIA^{Glc}. Other PTS transporters such as EIICBA^{Mtl} (MtlA) are inactive due to a lack of phosphates from HPr and the PTS. In the absence of glucose transport when concentrations of Fru1,6P are low, HprK/P dephosphorylates Ser46 of HPr and the P-Ser46-HPr:CcpA complex detaches from *cre* sites. Operons encoding proteins for use of other carbon sources are transcribed in the presence of the appropriate specific signal (not shown).

CcpA repressor, as described in the following sections. The phenomenon of inducer expulsion has also been described in *B. subtilis* and in some other bacteria. After precharging bacteria with a (nonmetabolizable) sugar, the addition of glucose causes the first sugar to be dephosphorylated and secreted into the medium. The molecular mechanism for this phenomenon is not known; experimental evidence suggests it does not require P-Ser46-HPr.

Catabolite Repression and Regulation by CcpA and the *cre* Element

AC and CAP are absent from most Gram-positive bacteria and so an alternative mechanism of controlling expression of genes for different carbon sources was sought to explain catabolite repression in Gram-positive bacteria. A few genes are better expressed in glucose in a process called carbon catabolite activation (CCA) in *B. subtilis*, for example, acetate kinase (*ackA*) and phosphotransacetylase (*pta*), which allow the bacteria to get rid of excess metabolites produced by glycolysis. Cis-acting sites in the vicinity of the transcription start sites of catabolite-repressed genes were found and called *cre* (catabolite response element). The protein that binds to them is a LacI-type repressor called CcpA. Mutation in either the *cre* site or in *ccpA* in many cases reduced or eliminated catabolite repression or catabolite activation (Figure 5). The locations of the *cre* sites vary greatly in different operons from upstream of the promoter (for those subject to activation), to covering the promoter or transcription start site, to near the translational start site or even within the translated region. *Cre* sites near the transcription initiation site inhibit RNAP binding. The mechanism of CcpA inhibition by binding at the sites within the transcribed region has not been investigated but is presumably due to a "road block" mechanism inhibiting the elongating RNAP.

As in the case of CAP, several hundred genes have been identified as true or potential CcpA targets by genomic screens for *cre* sites and transcriptome and proteome analysis of wild-type and mutant strains grown with and without glucose as well as by conventional studies on individual genes. P-Ser46-HPr is an essential cofactor in the repression of CcpA-regulated genes. The *ptsH1* allele encodes HPr with the Ser46Ala mutation, which cannot be phosphorylated by HprK/P. The transcriptomes for the *ccpA* mutant and a strain carrying the *ptsH1* mutation are very similar. The two strains have similar phenotypes because the CcpA protein alone has only low affinity for its DNA targets and requires P-Ser46-HPr as a corepressor (Figure 5). The P-Ser46-HPr concentration is highest during growth on glucose, which is the condition for maximum repression. Interestingly, these studies also revealed that several nitrogen-related genes (e.g., *sigL* encoding σ^{54}) are controlled by CcpA.

The structure of the P-Ser46-HPr:CcpA:*cre* DNA cocrystal shows that P-Ser46-HPr binds to the N-terminal subdomain of the dimerization and effector-binding domain of CcpA. The phosphate is locked into a small pocket on CcpA and comparison with the structure of free (apo) CcpA reveals how the interaction with Ser46-P provokes both local and global structural changes in the subdomains. The so-called hinge helices are formed, which contact the center of the quasi-palindromic *cre* site and kink the DNA and widen the minor groove, analogous to their function in LacI- and PurR-operator structures. This facilitates contact between the *cre* and the helix-turn-helix motif. On the other hand, His15 of HPr is in contact with Arg296 of CcpA so that P-His15 is not compatible with HPr binding to CcpA, thus explaining the discrimination against formation of P-His15-HPr:CcpA. Glc6P and Fru1,6P are reported to stimulate catabolite repression at some loci and a recent structure has shown that they bind in the groove between the two subdomains of the effector/dimerization domain, thus bolstering subdomain movements produced by P-Ser46-HPr binding.

A second protein called Crh (catabolite repression HPr), 45% identical to HPr, is also phosphorylated by HprK/P and can regulate CcpA binding. His15 is absent in Crh, which has no function in PTS sugar transport. *crh* mutations have no effect in a strain with a wild-type HPr but Crh can partially replace HPr function at some operators in *ptsH1* mutants (carrying the HPr Ser46Ala mutation). A much lower cellular level of Crh compared to HPr might account for its limited effect *in vivo*. A homologue of CcpA, called CcpB, also exists; however, its function is still not clearly defined.

PTS Regulation of PRD (PTS Regulation Domain)-Containing Proteins

Although *B. subtilis* does regulate some catabolic genes and operons for use of different carbon sources by classical repressor–operator interactions, which are sensitive to a small metabolite cofactor (e.g., XylR which controls the use of xylose or RbsR, which controls the use of ribose), a large number of transcription regulators carry a sequence element called PRD, for PTS regulation domain, which is subject to both EIIB- and HPr-dependent phosphorylations. These phosphorylations antagonistically affect, positively or negatively, the activity of the regulator. Two major types of proteins carry this domain, which exist in pairs on each protein. One class is transcriptional activators, like the LevR and MtlR proteins, and the other is transcriptional antiterminators.

Transcriptional antitermination of PTS operons

Transcriptional antitermination is a common mechanism of regulating gene expression in *B. subtilis*. A relatively long leader region is transcribed upstream of the structural gene.

This mRNA can fold into alternative structures, including one resembling a classical Rho-independent transcription terminator (GC-rich stem loop followed by a series of U) and another structure destroying the terminator. Various RNA-binding effectors bind to the antiterminator and stabilize it and promote read-through of the terminator and hence expression of the downstream genes. The known types of effectors include uncharged tRNA, in the case of aminoacyl-tRNA synthetase operons, or SAM, in the case of methionine/cysteine biosynthetic genes (in this case SAM binding enhances termination).

In the case of the PTS antiterminators the structured leader RNA region is called a ribonucleotide antitermination targets (RAT). They are found upstream of PTS operons for transport and use of glucose (*ptsGHI*), sucrose (*sacPA* and *sacX*), and aryl β-glucosides (*bglPA*), as well as *bglS* (an endoglucocanase) and *sacB* (a levansucrase). A specific antitermination protein controls each operon: LicT controls *bglPA* and *bglS*, SacY controls *sacB* and *sacX*, SacT controls *sacPA,* and GlcT controls the *ptsGHI* operon. The β-glucosidase operon in *E. coli* (*bglFA*) is also regulated by a similar mechanism using the antiterminator BglG, and in fact was the first discovered. The RAT-binding antitermination protein is usually encoded by the first gene of the operon, separated from the genes for the metabolic functions by a terminator. The RAT-binding antitermination proteins carry an RNA-binding domain (called CAT for co-antiterminator) at their N-terminal, and two PRD domains. Each PRD domain carries two phosphorylatable histidine residues; PRD1 is preferentially phosphorylated by the sugar-specific EIIB and PRD2 by EI and HPr. The relative importance of phosphorylations at the two sites of the two PRDs varies with the different antiterminators but the basic principle is the same. Each antitermination protein is able to undergo two types of antagonistic phosphorylation, one of which enhances the antitermination activity and the other suppresses it (Figure 6a).

In the absence of its particular carbon source, the EIIB domain should be predominately phosphorylated. EIIB-P is able to transfer its phosphate to the PRD1 domain of the corresponding RAT-binding antitermination protein. In this form (PRD1 phosphorylated), the antitermination protein is inactive and cannot bind to the RAT. The loss of this phosphate, retransferred back to the EIIB domain when the corresponding transporter is actively transporting its sugar, activates the antitermination function and allows it to bind to the RAT,

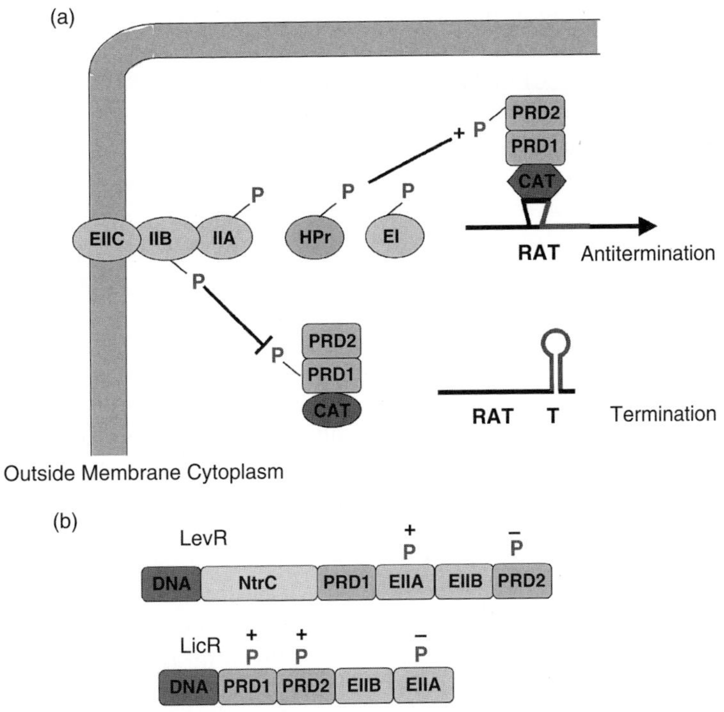

FIGURE 6 PRD-containing proteins. (a) Antitermination by PRD-containing antiterminators. In the absence of its specific sugar substrate, EIIB is predominately in its phosphorylated form. It can transfer a phosphate to the PRD1 domain of the antiterminator, which prevents it binding to the RAT operator within the leader of the mRNA. The terminator stem-loop structure can form so that transcription termination occurs. In the presence of the specific sugar substrate, PRD1 is nonphosphorylated but PRD2 can be phosphorylated by P-His15-HPr, which activates the RNA-binding function of the antiterminator. The N-terminal CAT domain of the activated antiterminator binds the RAT sequence and prevents formation of the terminator, so that transcription continues. (b) Domain organization of PRD-containing transcriptional activators. In LevR the N-terminal DNA-binding domain is followed by an NtrC-like domain and the two PRD domains are interspersed by EIIB and EIIA domains. In LicR the domain order is different. The domains subject to negatively controlling phosphorylation by EIIB and positively controlling phosphorylation by HPr are indicated by + and – signs above the P.

preventing formation of the terminator so that the downstream genes are expressed. For several operons, mutations that knock out EIIB are sufficient to produce constitutive expression of the operon, in the absence of the inducing sugar.

Some antitermination proteins are active in the absence of all phosphorylations, for example, SacY and GlcT. Thus, in *ptsH* or *ptsI* mutants, where all PTS phosphorylation is abolished, *sacB* and *ptsG* are constitutively expressed. Other antiterminators require phosphorylation by HPr on the PRD2 domain. This is the case for SacT and LicT controlling *sacPA* and *bglPH* and also for BglG from *E. coli*. Phosphorylation of PRD2 is responsible for the residual CCR of some antiterminator-regulated PTS operons in a *ccpA* strain. The level of P-His15-HPr during uptake of a rapidly metabolizable carbon source is low, since most HPr is phosphorylated on Ser46 and unavailable for sugar phosphorylation (see Section "P-Ser46-HPr and Inducer Exclusion"). All the P-His-HPr is thus engaged in phosphorylating EIIA and the incoming preferred sugar and PRD2 remains unphosphorylated and in its inactive form.

In both PRDs there are two conserved histidines and both can be phosphorylated; one seems to be the major target for initial phosphorylation in each PRD but the second site plays a greater or lesser role as shown by mutagenesis studies. Both intradomain, interdomain, and interprotein phosphate transfers are possible and complicate the analysis of the role of individual His residues in each antitermination protein. Crystal structures of native LicT and a constitutively active form (where the phosphorylatable histidines are replaced by aspartates, which mimic the phosphorylated state) indicate that phosphorylation produces a massive structural rearrangement of PRD2 relative to PRD1. The aspartate residues are buried in the middle of the dimer and the movement of PRD1 changes the interface between PRD1 and the RNA-binding CAT domain, presumably producing the RNA-binding conformation.

PTS regulation of PRD-containing transcriptional activators

PRD-containing activators are generally large multidomain proteins with an N-terminal DNA-binding domain and two PRD domains interspersed with EIIA and EIIB domains (Figure 6b). There are two classes known. The best characterized is LevR, which regulates the *levDEFGsacC* operon encoding a fructose/mannose PTS and an extracellular levanase. The operon is expressed from a σ^{54} promoter, which correlates with the presence of an NtrC-like domain within LevR. The domain order for LevR is the DNA-binding domain, NtrC-like domain, PRD1, EIIA, EIIB, PRD2. LevR binds to an upstream activating site (UAS) and activates expression of the *lev* operon in response to fructose. As in the case of the PRD antitermination proteins, it is subject to antagonistic phosphorylations. Phosphorylation of PRD2 by the EIIB component of the LevPT (*levE*) inhibits its activity, while phosphorylation by HPr has a positive effect, but unexpectedly occurs on the EIIA domain rather than PRD1.

The second class of activators, for example, LicR or MtlR, have an N-terminal DeoR-type DNA-binding transcription activation domain followed by two PRD domains and then EIIB and EIIA domains. Once again, multiple phosphorylations by HPr and EIIB are responsible for positive and negative regulation of the activators depending upon the presence of the transported sugar.

PTS Regulation of Glycerol Kinase

The use of glycerol is also regulated by PTS-mediated phosphorylation in *B. subtilis*. The phosphorylated target is not a transcription factor but an enzyme, the glycerol kinase, encoded by *glpK*. PEP-dependent phosphorylation of GlpK by HPr on a histidine residue increases the kinase activity about 10-fold, thus enhancing use of glycerol and formation of glycerol-3P. The presence of glucose promotes the dephosphorylation of GlpK (via a reversal of the reaction with HPr) and exerts a form of catabolite repression by reducing the activity of GlpK and hence the concentration of Gly3P. Gly3P is the inducing signal for the GlpP antiterminator protein that controls the *glpFK* operon, encoding the glycerol facilitator transport protein and the kinase, so that the end result is that in the presence of glucose the *glpFK* operon is not induced. However, in the absence of glucose or another preferred PTS carbon source and the presence of glycerol, the activity of GlpK is high, Gly3P concentrations are high, and the *glpFK* operon is derepressed.

CARBON REGULATION AND CATABOLITE REPRESSION IN OTHER BACTERIA

Original mechanisms regulating carbon use and catabolite repression have been uncovered in certain bacteria and a few examples are discussed here, to demonstrate the versatility of regulatory mechanisms in bacteria.

Pseudomonades

Pseudomonades are free-living Gram-negative bacteria of which the best studied are *P. putida* and *P. aeruginosa*. They thrive in a variety of habitats, both terrestrial and aquatic. *P. aeruginosa* is an opportunist pathogen that infects the lungs and is a major agent of nosocomial infections. *Pseudomonades* are metabolically versatile and can degrade a wide range of organic and aromatic compounds especially derived from degrading plant material. Complex aromatic compounds are degraded to simple intermediates that feed into one of four chromosomally encoded central aromatic pathways (homogentisate, catechol, protocatechuate, and phenylacetate), which are responsible for cleaving of the aromatic rings. There is only one PTS sugar – fructose – but it is nonrepressing. Glucose is converted to gluconate and degraded via the ED pathway and produces catabolite repression on some degradative pathways but the strongest repression is exerted by succinate and acetate. A close CAP homologue, Vfr, exists but cAMP levels are constant and although a *vfr* mutation affected the level of about 60 genes, CCR was not affected.

The *crc* (catabolite repression control) gene product is a master regulator of carbon metabolism. CCR is both carbon source- and promoter-specific. Crc controls CCR by succinate and is responsible for the repression of a number of the pathways for degradation of aromatic compounds when bacteria are growing in rich medium, but it has little effect in defined medium. The level of the Crc protein depends upon the carbon source and increases in stationary phase. The Crc protein does not bind DNA but controls gene expression posttranscriptionally and it is an RNA-binding protein. It has been shown to inhibit the translation of transcription factors like AlkS and BenR, which are the direct activators of the alkane and benzoate degradation pathways. Crc control of these catabolic operons is thus indirect, by keeping the levels of the direct activators low. Moreover, by preventing the expression of the first genes in a degradation pathway, Crc prevents the synthesis of the intermediates that act as inducers for subsequent steps in the pathway and so the whole pathway is repressed. Crc (259 aa) has homology with bacterial exonuclease III, an enzyme involved in DNA repair (endonuclease–exonuclease–phosphatase family) but mutations in conserved catalytic residues did not eliminate CCR.

The genes for the degradation of toluene and xylene, encoded by the TOL plasmid, are controlled by XylR, an NtrC-like activator that binds to a UAS and activates a σ^{54}-controlled promoter. Expression of the XylR regulon is controlled by an unusual form of catabolite repression. Glucose (but not succinate) represses expression in a manner requiring synthesis of the ED intermediates gluconate-6P or 2-keto 3-deoxy gluconate-6P. Mutations in the *ptsO* and *ptsN* genes of the Ntr PTS (see Section "The Nitrogen PTS") also affect repression and are likely involved in signaling. Mutations in other genes have been shown to affect CCR under certain conditions, for example, *cyo* encoding a terminal oxidase of the respiratory chain has been implicated in repression of phenol degradation and a novel two-component system, ChbAB, affects the use of a number of carbon compounds and along with NtrCB is involved in maintaining a correct carbon/nitrogen balance. Many questions remain concerning how all these players act together to control carbon use by *Pseudomoanades*. Considering their potential applications in bioremediation (cleaning up contaminated environments) and the industrial transformation of toxic chemicals, there should be rapid progress.

Streptomyces coelicolor

Streptomycetes are aerobic, spore-forming, soil bacteria of the species actinomycetes (high-GC Gram-positives) that are important sources of secondary metabolites including antibiotics. They can use a wide range of carbon sources including insoluble materials found in soil like lignocellulose and chitin and produce a range of biotechnologically useful enzymes, like α-amylases, xylanases, and chitinases. *S. coelicolor*, the best studied, has a PTS system that is apparently dedicated to the use of *N*-acetylglucosamine (GlcNAc). GlcNAc is generated by the action of secreted chitinases on the chitin derived from fungal walls and insect cuticles in the soil. The genes for the cytoplasmic components of the PTS, *ptsH*, *ptsI*, and *crr*, as well as the GlcNAc specific $EIICB^{Nag}$ are induced by GlcNAc and are controlled by a FadR family transcription factor called DasR, which also controls chitobiose uptake. DasR was originally characterized as affecting formation of the aerial mycelium and spores, and high GlcNAc concentrations prevent *S. coelicolor* from entering the developmental pathway leading to sporulating hyphae.

Readily assimilable carbon sources are used preferentially and are used in a hierarchical fashion. Homologues of the *crp* and *cya* genes exist but mutants are affected in germination and not catabolite repression. CCR is absolutely dependent upon the glycolytic enzyme glucose kinase, GlkA, so that *glkA* mutants are no longer glucose repressible. The glucokinase activity *per se* is not sufficient for CCR since heterologous glucokinases can restore growth on glucose without restoring CCR, and *glkA* mutants are defective in CCR by carbon sources like galactose or glycerol, which are not metabolized via glucose kinase. The exact mechanism of CCR by Glk is still elusive. The protein is constitutively expressed but its activity is sugar source dependent, being high during growth on glucose and low in mannitol-grown bacteria. A complex between glucose permease, GlcP, and Glk has been suggested. Glk is a member of the ROK (repressors, ORFs and kinases) family of proteins, as is Mlc, the repressor of the glucose PTS in *E. coli*. There is a fascinating parallel here since activation of Mlc-repressed genes involves the sequestration of Mlc to the membrane via a formation of a complex with the PTS glucose transporter, PtsG (see Section "Mlc: The Glucose PTS Repressor"). Glucose kinase has been implicated in CCR in several related bacteria, for example, *Mycobacterium smegmatis*, and in the less-related *Streptococcus xylosus*, which also has a CcpA-mediated system.

Corynebacterium glutamicum

Corynebacteria are aerobic, rod-shaped, nonsporulating, low-GC Gram-positive soil bacteria. *C. glutamicum* has worldwide importance for the biotechnological production of amino acids, especially glutamate and lysine, both derived from TCA cycle intermediates. In consequence, the central metabolism of *C. glutamicum* has been well investigated in recent years. Original regulators control TCA cycle gene expression, for example, AcnR is a repressor of aconitase. Iron homeostasis is achieved by the combined action of two repressors DtxR and RipA, which control the amount of succinate dehydrogenase and aconitase, both Fe–S proteins, in response to available iron. In *E. coli,* a similar iron control is achieved via FurR and the sRNA RyhB (see Section "The TCA Cycle and Glyoxylate Shunt"). Genes for the use of acetate as carbon source (*ack*, *pta*, *aceB*, *aceA*) are controlled by an opposing activator and repressor pair: RamA and RamB (regulator of acetate metabolism). In addition, the GlxR protein, a CAP homologue, represses the genes of the glyoxylate shunt, *aceBA*. GlxR is sensitive to cAMP, which is reported to be high during exponential growth on glucose but low during growth on acetate.

The activity of the enzyme 2-ketoglutarate (2-oxoglutarate) dehydrogenase (ODHC) is inhibited by binding with the small protein, OdhI. OdhI undergoes phosphorylation by a threonine kinase, PknG, and a second kinase, in response to glutamine availability. The relative levels of phospho-OdhI and PknG determine the flux through ODHC and hence the yield of glutamate from the action of glutamate dehydrogenase on 2-oxoglutarate.

The metabolism of *C. glutamicum* is unusual in that coutilization of several carbon sources occurs, for example, glucose with acetate, lactate, pyruvate, or fructose. On the other hand, the use of ethanol as carbon source is repressed by glucose, even though the *adhA* gene is controlled by the RamA and RamB regulators, responsible for control of the genes for acetate use. Coutilization of glucose and gluconate has been shown to be due to the GntR-specific repressors of the gluconate operon, acting as activators of the *ptsG* enzyme encoding the glucose transporter. The presence of gluconate causes a simultaneous decrease in PtsG levels and glucose transport and an increase in gluconate uptake. There are three PTS transporters, for glucose (*ptsG*), sucrose (*ptsS*), and fructose (*ptsF*). The genes for the central composants *ptsHI* are in an operon with *ptsF* and expression of all five genes are repressed by SugR, a DeoR-type repressor, during growth on gluconeogenic carbon sources. SugR is sensitive to Fru1P and to a lesser extent Fru1,6P.

Lactic Acid Bacteria and Bifidobacteria

Lactic acid bacteria comprise a group of low-GC, acid-tolerant, Gram-positive bacteria, which have been exploited for centuries for the preservation of food and drinks, because the lactic acid they produce lowers the pH sufficiently to prevent the action of other spoilage agents. They carry out either homolactic (e.g., *Lactococcus* and some Lactobacilli) or heterolactic (e.g., *Leuconostoc*) fermentations. Lactose is transported by a PTS and is metabolized via glycolysis to lactate during homofermentation. In heterofermentation some glucose is metabolized via the pentose phosphate pathway (Figure 1) producing CO_2, acetic acid, and/or ethanol in addition to lactic acid. Glucose transported by the PTS is the preferred sugar and produces CCR via CcpA and HPr as described for *B. subtilis* (see Section "Catabolite Repression and Regulation by CcpA and the *cre* Element"). The lowering of the pH through generation of lactic acid affects carbon metabolism and increases glucose consumption mostly through changes in enzyme activities in *L. lactis.*

Bifidobacteria are anaerobic commensals of the intestinal tract and are the most important probiotic organism due to their ability to degrade a wide range of carbon sources including many mono- and disaccharides and complex carbohydrates derived from indigestible plant oligosaccharides of our diet, like pectin, and including the prebiotic xylo-, galacto- and fructooligosaccharides. They are heterofermentive, nonsporulating Gram-positive rods, now classed as Actinomycetes (rather than lactic acid bacteria) and they degrade lactose and other sugars. They lack the reductive half of the Krebs cycle. Hexoses are degraded via an unusual version of the pentose phosphate pathway (the bifido shunt) in which an original enzyme, fructose-6P phosphoketolase (*xfp*), converts Fru6P to erythrose-4P and acetylphosphate. The latter is converted to acetate with the production of a mole of ATP. The erythrose-4P reacts with another molecule of Fru6P to start the pathway of sequential transketolase, transaldolase reactions of the pentose phosphate shunt.

There are a large number of ABC-type transporters for a variety of sugars and oligosaccharides. The gene for a PTS transporter for glucose (*ptsG*) as well as *ptsH* and *ptsI* are found in *B. longum* but are only weakly expressed. A gene for a glucose permease (*glcP*) appears to encode the major glucose uptake system. *ptsG* and *glcP* are transcribed divergently and are separated by a putative RAT-controlled antiterminator. Lactose is used preferentially over glucose and represses expression of *glcP*.

CONCLUSION

Control of carbon use involves numerous regulatory circuits operating at various levels. The enormous range of carbon sources, which bacteria can use, must be specifically converted into simple sugar phosphates, which can be used by one of the ubiquitous pathways of central carbon metabolism. These central pathways consist of about 100 key enzymes that catalyze the formation of the energy, reducing power, precursors, and cofactors necessary for bacterial growth. Modern "-omics" methods, sequence comparisons, transcriptome, proteome, metabolome, and flux analyses have revolutionized the analysis of metabolic pathways in a cellular setting and identified certain original pathways and regulatory strategies. A rather high correlation between metabolic flux analysis data and mRNA levels from transcriptome studies suggests that flux is primarily controlled at the transcriptional level. However, superimposed on this base is metabolic control, due to the large number of allosteric or otherwise regulated enzymes. Metabolic pathways are robust. A block in one pathway causes a local rerouting of the flux. Syntheses of secondary metabolites are important biotechnological applications of microbes, and genetic engineering of carbon use is a major incentive to research in this field. Bacteria are being tailored to enhance production of many commercially important compounds.

A few key metabolites have emerged as important signaling molecules in different bacteria. For example, the glycolytic intermediate Fru1,6P is used to signal high glycolytic flux rates in numerous situations. It is a key molecule to signal CCR in low-GC Gram-positive bacteria since it activates the kinase function of HprK/P and is a corepressor of the complex between CcpA and P-Ser46-HPr to repress catabolite-regulated operon. It is implicated in the switch from use of the PTS and glycolytic sugars to gluconeogenesis since it is an effector for CggR in *B. subtilis*, FruR (Cra) in *E. coli*, and SugR in *C. glutamicum.* It is also an allosteric effector for several enzymes, for example, pyruvate kinase, PEP carboxylase, and glycerol kinase from *E. coli*.

Many novel regulatory small RNA have been discovered in the last few years. These sRNA are mostly implicated in posttranscriptional controls affecting the translation and/or degradation of the corresponding mRNA. A few already have defined roles in carbon metabolism, for example, the SgrS RNA promoting the RNaseE-dependent degradation of *ptsG* mRNA in *E. coli*, in response to a phosphosugar stress (see Section "Mlc: The Glucose PTS Repressor"). An intriguing observation in this context is the presence of the glycolytic enzyme enolase as an inherent component of the *E. coli* degradosome (the RNaseE-based RNA-degrading complex). It suggests a connection between the control of mRNA turnover and the rate of formation of PEP but no precise role is yet known.

The discovery of sRNA-mediated regulation has provided explanations for several long known but unexpected observations, for example, the apparent positive control of *sdhABCD* by Fur in *E. coli* (see Section "The TCA Cycle and Glyoxylate Shunt"). Another example is the discoordinate expression of the *galETK* operon. In the absence of galactose, only *galET* are required for cellular glycosylation reactions. The Spot42 sRNA has been known for decades, but it has only recently been shown that it is an antisense RNA, which hybridizes to the beginning of the *galK* cistron, blocking access of the ribosome to *galK* and thus down-regulating the expression of *galK*. Expression of the Spot42 RNA is repressed by cAMP/CAP; thus high cAMP/CAP concentrations, necessary for use of galactose as carbon source, increase *galK* levels relative to *galET* levels.

Another regulatory system employing sRNA and affecting carbon use is the RNA-binding protein, CsrA, and its two sRNA, CsrB and CsrC. CsrA (carbon storage regulator) represses translation of the *glgC* mRNA and other genes necessary for glycogen biosynthesis. CsrB and CsrC are noncoding RNA carrying many copies of the CsrA RNA-binding site; thus they titrate CsrA. An increased synthesis of CsrB and C in stationary phase results in the sequestration of CsrA by the sRNA and expression of the glycogen biosynthesis genes. Glycogen is thought to play a role in stationary phase survival. CsrA also participates in the switch from glycolysis to gluconeogenesis in stationary phase, repressing or activating several key enzymes. Use of acetate is seriously impaired in *csrA* strains. CsrA homologues exist in numerous bacteria and in some cases have been implicated in the synthesis of secondary metabolites.

Other recent interesting results have demonstrated correlations between the use of different carbon sources and effects on global bacterial functions such as size at cell division, growth in biofilms, or expression of virulence genes. In *B. subtilis* the size at which bacteria initiate cell division is determined by a metabolic sensor detecting the quantity of UDP-Glc. PrfA is a transcription factor of the CAP family. It is the key regulator of virulence in *Listeria monocytogenes.* Activation of virulence genes is inhibited by growth on rapidly metabolizable sugars. However, this is not due to a normal CCR response since it does not depend upon CcpA, the catabolite repressor protein in Gram-positive bacteria, but it does require P-Ser46-HPr, the CcpA corepressor. In conclusion, there are still a lot of carbon source-dependent regulatory mechanisms to unravel in bacteria.

RECENT DEVELOPMENTS

CCR is a topic which has been around for a long while but it can still raise a certain amount of controversy in the literature. Witness a review of the subject by Gorke and Stulke, which provoked some lively discussion in subsequent volumes concerning the relevant contribution of cAMP/CAP to catabolite repression in *E. coli*. CCR is a subject still relevant today. It has important consequences in biotechnological situations. The preferred feeder biomass often is made up of glucose and other simple sugars, which can produce CR and thus limit yields of the desired end products in the process. Thus, there is an active search for efficient, regulatable promoters, which are no longer subject to CR in different biotechnologically relevant bacteria.

It has become more and more evident that Monod's adage, that what is true for *E. coli* is true for elephants, does not apply to catabolite repression in other classes of bacteria or even Gram-negative bacteria. Examples of the different mechanisms responsible for catabolite represssion have already been described. In some cases it appears that they are variations of mechanisms already characterized in other species but with new functions. Crc is responsible for catabolite repression by organic acids, the preferred carbon sources in *Pseudomonas aeruginosa*. Crc is a translational repressor and RNA-binding protein and its activity has been shown to be regulated by a sRNA (*crcZ*) and activated by a two-component system (*cbrAB* encoding a histidine kinase and a response regulator). The *crc, crcZ* pair resembles the well studied *csrA,B,C* system of *E. coli* controlled by the BarA/UvrY two-component regulators. Interestingly both two-component systems are sensitive to small organic acids (succinate for Cbr and acetate and formate for BarA/UvrY), but the respective functions of the regulated proteins in the two bacteria are different. In *E. coli, csrA,B,C* is important for the switch from glycolytic growth to gluconeogenic growth at the entrance to stationary phase. Succinate produces strong catabolite repression in *Pseudomonas aeurginosa* whereas acetate and formate have no such activity in *E. coli*.

Similarly, GlcX, an orthologue of CRP/FNR in *E. coli*, had previously been characterized as a regulator of use of gluconate and the glyoxalate bypass in the Gram-positive bacteria *Corynebacterium glutamicum*. It has now been shown to be a master regulator affecting all aspects of carbon metabolism either directly or by regulation of more specific regulators like SugR, RamA, and RamB. However, unlike CRP in *E. coli*, regulation by GlxR is predominately negative.

Another important theme in recent research is the contribution of carbon metabolism and catabolite repression to pathogenicity. The activity of the major virulence factor in *Listeria monocytogenes* is affected by the transport of different carbon sources and seems to be controlled by the phosphorylation state of one or more PTS components. Another example is the Mga regulator in group A streptococcus, which controls both the expression of the core set of known virulence factors and also various metabolic functions. The latter are probably controlled indirectly. Expression of *mga* is controlled by CcpA and interestingly, the Mga protein contains two PRD domains (see Figure 6b) suggesting that its activity is controlled by the availability of sugars. A similar scenario has been demonstrated for the AtxA transcription factor of *Bacillus anthracis,* which is subject to antagonistic phosphorylations on two separate histidine residues.

In more general terms, many examples are known where the availability or lack of a carbon source affects the developmental program of a bacterium. In some cases, the regulators of carbon utilization genes have become more global regulators. The classical example is that limiting nutrient availability initiates sporulation in Bacilli. In *Streptomyces coelicolor,* the presence of *N*-acetylglucosamine (GlcNAc) blocks development of aerial hyphae and antibiotic production in rich media but enhances both in poor media. Transport of GlcNAc by the PTS is essential for these lifestyle changes and the DasR transcriptional regulator controls both the expression of the genes for the GlcNAc PTS and is also an activator for genes of the *act* and *red* biosynthetic antibiotic clusters. GlcNAc is a pleiotropic signaling molecule, which acting via the PTS determines the developmental path. Formation of biofilms, a major worry in medical contexts because of the resistance of this form of bacterial cultures to antibiotics and standard cleaning methods, is another lifestyle change very dependent upon carbon metabolism.

The numerous examples of how carbon utilization impacts on both the desirable (e.g. in a biotechnological context) and undesirable behavior (e.g. in a medical situation) of bacteria means that understanding how bacteria control of carbon use is still an important experimental subject.

FURTHER READING

Detacher, J. (2008). The mechanisms of carbon catabolite repression in bacteria. *Current Opinion in Microbiology, 11*, 1–7.

Detacher, J., Francke, C., & Postma, P. (2006). How phosphotransferase system-related protein phosphorylation regulates carbohydrate metabolism in bacteria. *Microbiology and Molecular Biology Reviews, 70*, 939–1031.

El-Mansi, M., Cozzone, A., Shiloach, J., & Eikmanns, B. (2006). Control of carbon flux through enzymes of central and intermediary metabolism during growth of *Escherichia coli* on acetate. *Current Opinion in Microbiology*, *9*, 173–179.

Fuhrer, T., Fischer, E., & Sauer, U. (2005). Experimental identification and quantification of glucose metabolism in seven bacterial species. *Journal of Bacteriology*, *187*, 1581–1590.

Gorke, B., & Stulke, J. (2008). Carbon catabolite repression in bacteria: many ways to make the most out of nutrients. *Nature Reviews Microbiology*, *6*, 613–624.

Lawson, C., Swigon, D., Murakami, K., Darst, S., Berman, H., & Ebright, R. H. (2004). Catabolite activator protein: DNA binding and transcription. *Current Opinion in Structural Biology*, *14*, 10–20.

Martinez-Antonio, A., & Collado-Vides, J. (2003). Identifying global regulators in transcriptional regulatory networks in bacteria. *Current Opinion in Microbiology*, *6*, 482–489.

Sauer, U., & Eikmanns, B. (2005). The PEP–pyruvate–oxaloacetate node as the switch point for carbon flux distribution in bacteria. *FEMS Microbiology Reviews*, *29*, 765–794.

Sonenshein, A. (2007). Control of key metabolic intersections in *Bacillus subtilis*. *Nature Reviews Microbiology*, *5*, 917–927.

Wolfe, A. (2005). The acetate switch. *Microbiology and Molecular Biology Reviews*, *69*, 12–50.

Chapter 15

Microbial Resistance to Heavy Metals

S. Silver and L.T. Phung
University of Illinois, Chicago, IL, USA

Chapter Outline

ABBREVIATIONS

CDF Cation diffusion facilitator
RND Resistance, Nodulation, and Division

DEFINING STATEMENT

Toxic heavy metals have occurred on Earth at high levels since before the beginning of life, nearly 4 billion years ago. Microbes exposed to these inorganic toxins early developed resistance mechanisms. Thus, the question of whether toxic metal cation (and oxyanions of some soft metals) resistance systems evolved in microbes in response to human pollution in the last few thousands of years is easily answered: they arose much earlier, almost at the origin of life. The arguments supporting such a broad hypothesis need to be addressed element by element. The primary basis is the widespread occurrence of such resistance systems, from bacterial type to type (and occasionally recognized in Archaea and fungi as well) and with frequencies ranging from a few percent in pristine environments to nearly 100% in heavily polluted environments. The amino acid (and DNA) sequences and structures of the genes and proteins concerned with arsenic resistance indicate an ancient origin, although one cannot conclude whether early bioavailable As occurred in As(III) or As(V).

Genetic and mechanistic studies of toxic metal ion resistance systems have been reviewed frequently, and recent examples are provided in "Further Reading," especially in the special issue of *FEMS Microbiology Reviews* edited by Brown and colleagues. Some toxic elements have been chosen for deeper consideration in this article, as they are better understood and provide models for general mechanisms.

The most frequent mechanism of toxic divalent cation resistance is energy-dependent pumping out from the cytoplasm (Figure 1). The P-type ATPase (CadA for Cd^{2+} and SilP for Ag^{+} are toxic metal efflux examples, but these are found in all cell types – prokaryote and eukaryote) is generally a single large polypeptide embedded in the membrane and has multiple protein domains (a membrane-embedded cation translocation pathway and a three-domain ATPase region; Figure 1). The single-polypeptide membrane potential-driven efflux pump (ArsB for arsenite and ChrA for chromate are well-known examples for toxic oxyanions) moves the substrate out of the cell without ATP energy (Figure 1). Similarly, the three-polypeptide Resistance, Nodulation, and Division (RND) complex (CzcCBA for Cd^{2+}, Zn^{2+}, and Co^{2+} is an example) of Gram-negative bacteria picks up the cation either from the cytoplasm or from the periplasm and moves it through a channel formed by the inner membrane protein (CzcA in Figure 1) and the outer membrane protein (CzcC in Figure 1). The third polypeptide (CzcB) links the inner and outer membrane proteins together and may be involved in energy transduction.

FLAVORS OF ARSENIC AND ITS RESISTANCE

Resistance to both arsenite [As(III) $As(OH)_3$] and arsenate [As(V), AsO_4^{3-}] is widely found among both Gram-negative and Gram-positive bacteria, and even in the known genomes

Topics in Ecological and Environmental Microbiology.

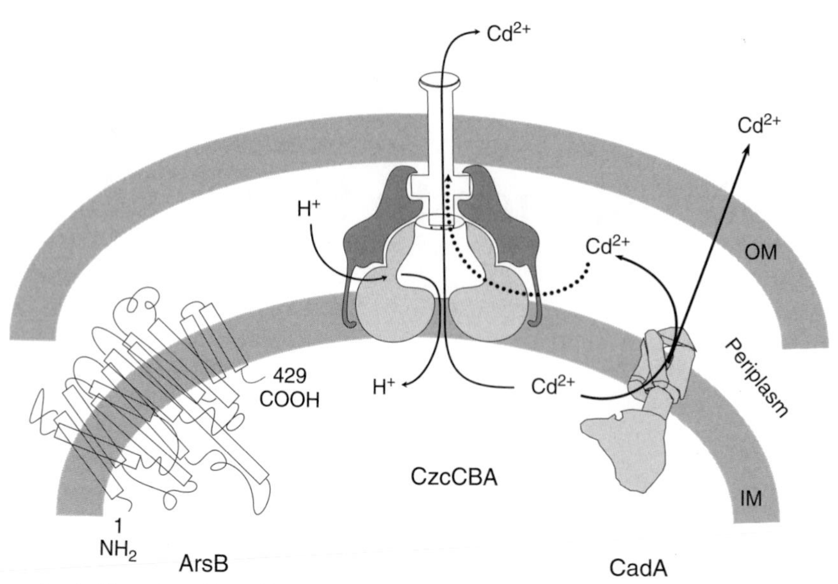

FIGURE 1 Membrane-associated efflux of metal ions. Families of carriers are represented by the single-polypeptide ArsB arsenite efflux protein, the three-component chemiosmotic antiporter system (CzcCBA), and the CadA P-type ATPase. IM, inner membrane; OM, outer membrane.

of all *Escherichia coli* and related clinical bacteria. Usually, it is in the form of an *ars* operon with a minimum of three cotranscribed genes – *arsR* (determining the regulatory, negatively acting, repressor), *arsB* (determining the membrane transport pump), and *arsC* (the determinant of a small intracellular arsenate reductase enzyme). Indeed, the ars operon occurs more widely in newly sequenced bacterial genomes with over 1000 or 2000 genes than do the genes for tryptophan biosynthesis. It has been argued, partially because of the above-mentioned fact, that arsenic resistance is very ancient, probably found in early cells.

Occasionally, two additional genes are found in *ars* operons of Gram-negative bacteria, so the gene order is *arsRDABC*. ArsA is an intracellular ATPase protein that docks onto the ArsB membrane protein, converting its energy coupling from the membrane potential to ATP hydrolysis. The arsenite membrane efflux pump is unique in that it can function either chemisomotically (with ArsB alone) or as an ATPase (with the ArsAB complex). ArsD is thought to function as a polypeptide chaperone carrying the arsenite to the ArsA ATPase protein, functioning with two pairs of adjacent thiol residues, Cys12–Cys13 and Cys112–Cys113, which bind As(III). Sb(III) can bind to ArsD and ArsA as alternatives to As(III).

The ArsB membrane efflux pump is specific for As(III), arsenite (and also for closely related Sb(III), and therefore, As(v), arsenate, resistance requires conversion of As(v) to As(III)). Thus, a small cytoplasmic enzyme, arsenite reductase, ArsC, is found.

The reductase uses oxidized/reduced cysteine thiol cycling to reduce arsenate to arsenite. Three distinct and unrelated families of ArsC arsenate reductase proteins have arisen by convergent evolution, analogous to the wings of birds and insects, which are unrelated but both enable flying. One class of arsenate reductase uses three cysteine thiols within the ArsC protein as an electron source and thioredoxin to regenerate reduced cysteines for the next cycle (Figures 2 and 3). This enzyme is paralogous in structure and in function to a class of cell division enzymes found in microbes and in animals, called low molecular weight protein tyrosine phosphates (Figure 2). The steps involved in the thioredoxin-linked arsenate reductase cycle are shown in Figure 3, as an example of how these enzymes work. Initially, inorganic ionic arsenate forms a covalent As–S bond with a cysteine thiol, from Cys10 in the example shown. Next, arsenate is reduced to arsenite, concomitant with oxidation of two cysteines to cystine (Figure 3, step 2). The third cysteine displaces Cys10 in a cystine regenerating the reduced Cys10 (Figure 3, step 3), and finally, the small soluble cellular protein thioredoxin reduces arsenate reductase fully (Figure 3, step 4). For the other two classes of structurally and evolutionarily unrelated arsenate reductases, the tripeptide glutathione and the small protein glutaredoxin function as thiol electron intermediates, in a manner similar to Cys82 and Cys89 of the thioredoxin-linked arsenate reductase. Although overall they are similar proteins, thioredoxin does not function with glutaredoxin-linked arsenate reductase and glutaredoxin does not function with thioredoxin-linked arsenate reductase.

In more recent years, additional and totally different enzymes involved in arsenic resistance and arsenic redox chemistry have been isolated and their genetic basis studied. These are the periplasmic respiratory arsenite oxidase (genes called aox) and respiratory arsenate reductase (genes called arr). The arsenite oxidase is a resistance mechanism

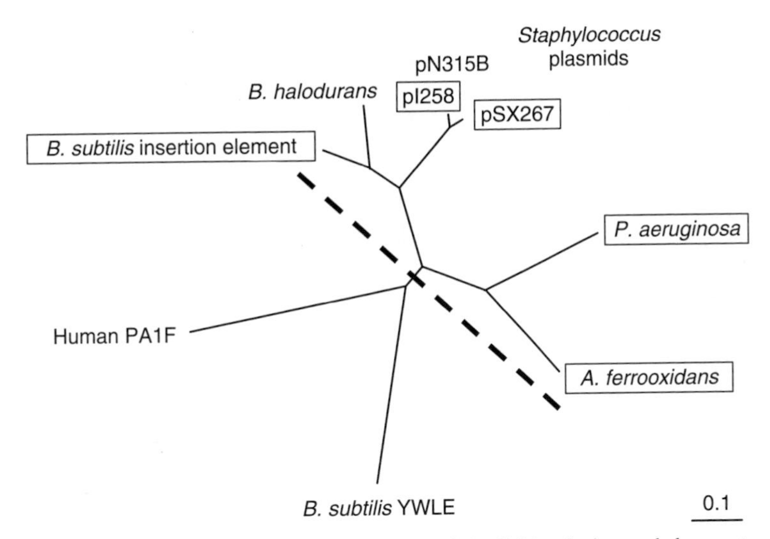

FIGURE 2 Phylogenetic tree showing evolutionary relationships among selected ArsC thioredoxin-coupled arsenate reductases and paralogous low molecular weight protein tyrosine phosphatase.

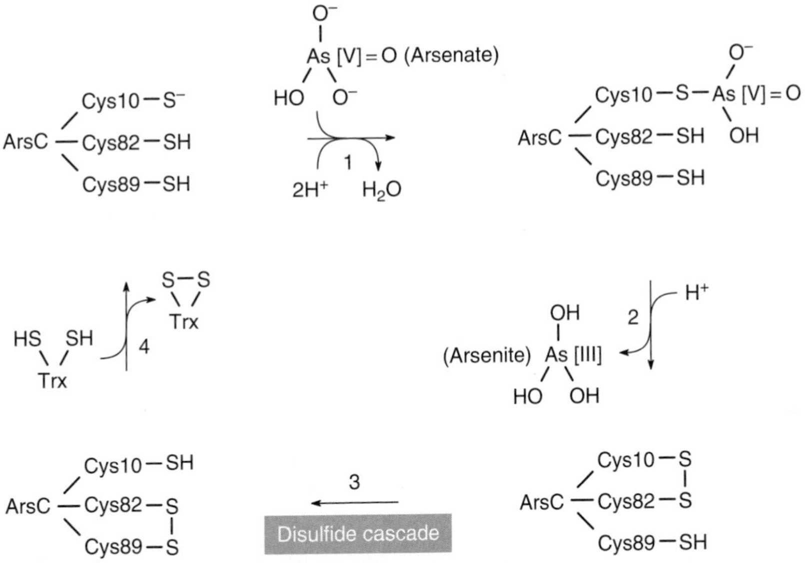

FIGURE 3 The reaction cycle for thioredoxin-linked arsenate reductase with intermediates. *Reproduced from Messens, J., & Silver, S. (2006). Arsenate reduction: Thiol cascade chemistry with convergent evolution.* Journal of Molecular Biology, 362, *1–17.*

converting highly toxic high arsenite levels to relatively less toxic arsenate, while respiratory arsenate reductase functions as a terminal electron acceptor, allowing anaerobic growth in the absence of oxygen. The oxidase and reductase enzymes contain related molybdopterin centers and [Fe–S] cages (Figure 4), and both are coupled to inner membrane respiratory chains, with the oxidase as an initial electron donor and the reductase as the terminal electron acceptor in an anaerobic respiratory process. For the Aox respiratory arsenite oxidase (Figure 4), the oxyanion substrate is thought to enter a shallow conical pit on the enzyme surface, directly contacting the embedded Mo(VI). Concerted two-electron transfer occurs, arsenate is released, and molybdenum reduced to Mo(IV). Next, electrons are transferred to an [3Fe–4S] cage in the large molybdopterin subunit and from the [3Fe–4S] to an [2Fe–2S] cluster in the small subunit (Figure 4). From the small subunit, the electrons are transferred to the inner membrane respiratory chain, and eventually to oxygen, the terminal electron acceptor. For the functionally related Arr respiratory arsenate reductase, the electrons are transferred in the opposite direction, from the respiratory electron transport chain to the enzyme small subunit and finally, from the Mo(IV)-pterin cofactor to the substrate arsenate.

There is an additional aspect of microbial arsenic enzymatic transformations about which less is known.

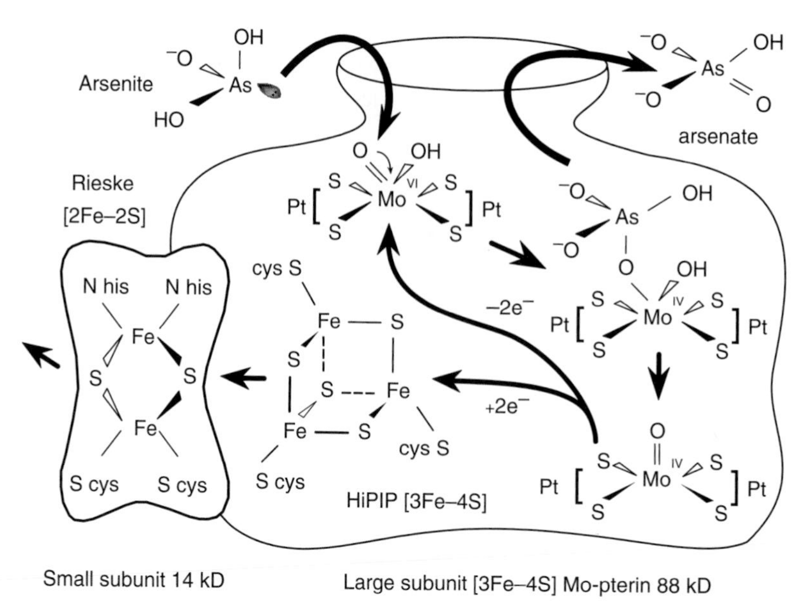

FIGURE 4 The reaction mechanism of arsenite oxidase. *Reproduced from Silver, S., & Phung, L. T. (2005). Genes and enzymes involved in bacterial oxidation and reduction of inorganic arsenic.* Applied and Environmental Microbiology, 71, *599–608.*

Microbes methylate and demethylate arsenic compounds, and inorganic arsenic is incorporated into small organic compounds, such as arsenobetaine, arsenolipids, and arsenosugars. The methylase and demethylation enzymes (and their genes) have recently been identified, so a microbial cycle occurs. Although microbial methylation of inorganic arsenic was recognized over a century ago and attributed to fungi, when poisonous volatile arsenic compounds were released from "moldy wall paper," methylation by prokaryotes and animal tissues has been identified only more recently. Major progress occurred with the isolation and sequencing of the arsenic methylase genes, first from mammals and more recently from prokaryotes. The methyl donor is S-adenosyl methionine and methylation requires arsenite as substrate. At each methylation step, from inorganic arsenite to monomethyl-As, and subsequently to dimethyl-As and trimethyl-As, oxidation to As(V) occurs, and reduction to an As(III) organoarsenical is required before the next methylation step.

MERCURY AND ORGANOMERCURIAL RESISTANCE

Mercury resistance, together with arsenic resistance, is the best understood and most widely found of the toxic metal resistance systems. The same mechanism has been found for mercury resistance widely in all bacterial divisions where it has been sought (and also in some Archaea). This has not been found, however, in any eukaryote. Mercury resistance occurs widely in clinical and industrial isolates, as well as in environmental strains, supporting a wide and ancient occurrence. Mercury resistance genes are frequently found on plasmids and encoded by transposons.

The first mercury-specific gene products for resistance encountered by Hg^{2+} approaching a Gram-negative bacterial cell from the outside are a small monomeric polypeptide in the periplasmic space, MerP, with a single cysteine pair for Hg^{2+} binding, and one of three inner membrane proteins, generally MerT, but sometimes its alternative forms MerC or MerF (all three of which have two sets of Hg^{2+}-binding cysteine pairs that are thought to function as a serial cascade of thiol Hg^{2+}-binding sites, in the absence of redox chemistry). Once at the inner surface of the inner membrane, Hg^{2+} is thought to be transferred in still another cysteine pair to cysteine pair exchange to the N-terminal cysteine pair of the homodimer enzyme mercuric reductase.

Progress has occurred in the understanding of the structures and mechanisms of the two enzymes involved in mercury resistance, mercuric reductase (Figure 5) and organomercurial lyase (Figure 6). Most recognized MerA sequences (with the exception of that from *Streptomyces*) have an N-terminal domain that is homologous in sequence and considered to function as a thiol Hg^{2+}-binding site similar to MerP. Hg^{2+} in the reductase enzyme is next transferred by another cysteine pair to cysteine pair exchange from the N-terminal domain to the C-terminal cysteine pair of the MerA subunit (C557 C558 in Tn*21* MerA numbering; Figure 5, steps 1 and 2), forming an S–Hg–S ring structure. Hg^{2+} bound to mercuric reductase on the C-terminal Cys557 Cys558 of one subunit is then transferred by rapid thiol/thiol exchange to the Cys135 Cys140 thiol pair of the other monomer (Figure 5, steps 3 and 4), where it is reduced by electron transfer from FADH cofactor (Figure 5, step 6) and Hg^0 released. Monoatomic Hg^0 is a membrane soluble gas and rapidly leaves the cell and under aerobic shaking or bubbling conditions is released into the atmosphere.

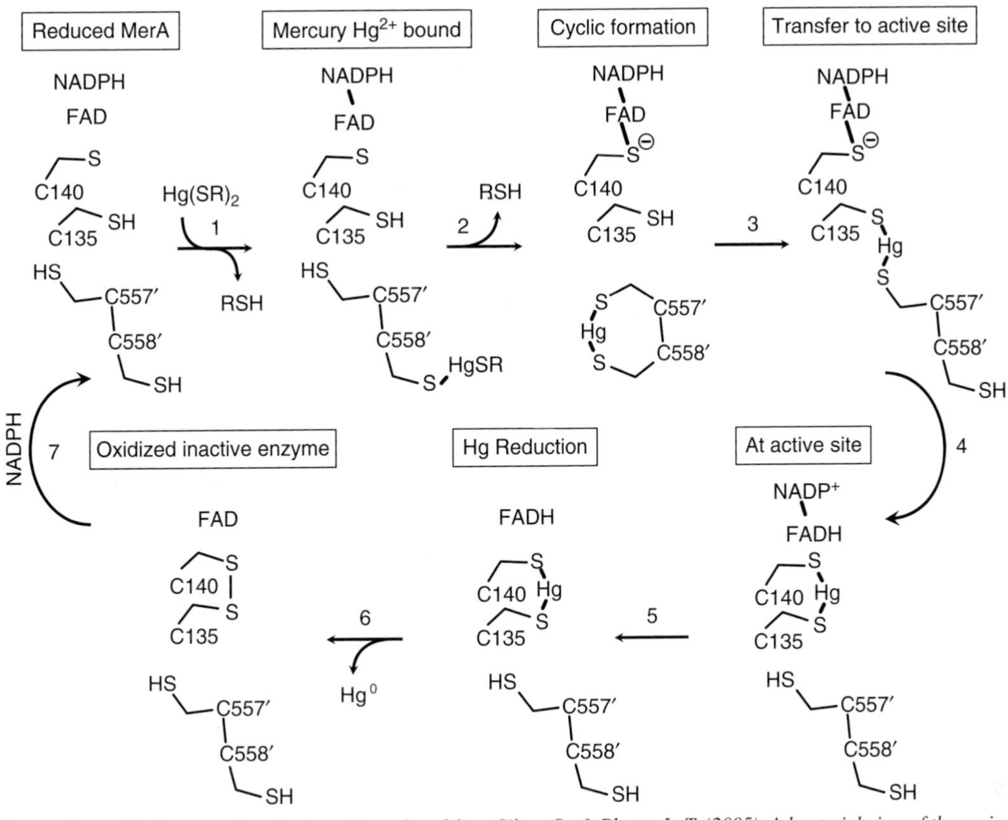

FIGURE 5 The reaction cycle for mercuric reductase. *Reproduced from Silver, S., & Phung. L. T. (2005). A bacterial view of the periodic table: Genes and proteins for toxic inorganic ions.* Journal of Industrial Microbiology and Biotechnology, 32, *587–605.*

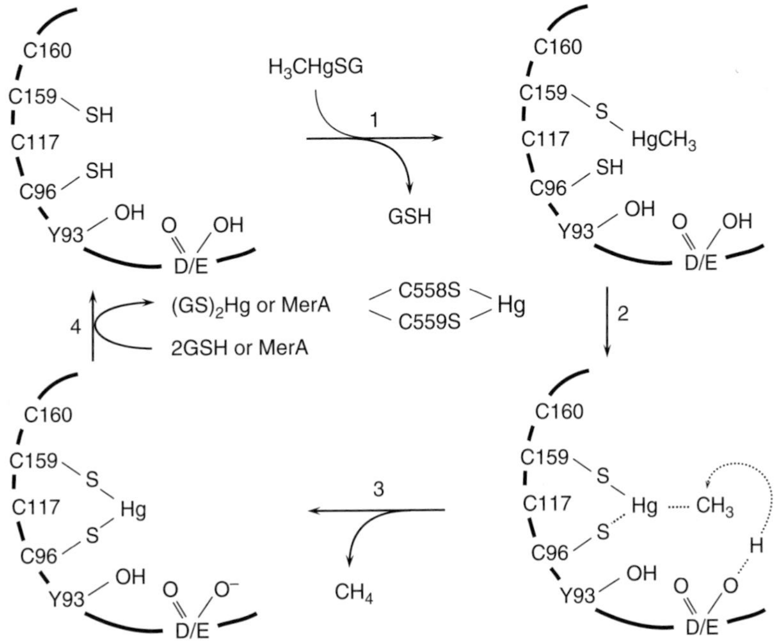

FIGURE 6 The reaction cycle for organomercurial lyase. *Modified from Silver, S., & Phung, L. T. (2005b). A bacterial view of the periodic table: Genes and proteins for toxic inorganic ions.* Journal of Industrial Microbiology and Biotechnology, 32, *587–605 and Lafrance-Vanasse, J., Lefebvre, M., Di Lello, P., Sygusch, J., Omichinski, J. G. (2009). Crystal Structures of the organmercurial lyase MerB in its free and mercury-bound forms: Insights into the mechanism of methylmercury degradation.* Journal of Biological Chemistry, 284, *938–944.*

Organomercurial lyase is a small monomeric enzyme that cleaves the Hg–C covalent bond releasing Hg^{2+} (the substrate of mercuric reductase) and reduced organic compounds, such as methane from methyl mercury (Figure 6) or benzene from phenyl mercury. The MerB primary sequence is unusual, having no paralogous enzymes with related sequences but different substrates. A wide range of MerB sequences have been identified (accessible in GenBank) and a wide range of organomercurial substrates are known, although at this time there is no understanding of how sequence differences determine substrate variation. The enzymatic reaction for organomercurial lyase was shown to be a concerted proton attack on the Hg–C bond by an SE2 reaction mechanism, after the organomercurial is initially bound to a cysteine thiol (Figure 6, step 1), either from a membrane protein, such as MerT, or directly from a small cellular thiol compound, such as glutathione (GSH in Figure 6). It is now clear that the dicarboxylic acid residue, aspartate99 is the source of a proton that adds, for example, to methylmercury, forming methane. Results from recent mutagenesis and structural studies of organomercurial lyase have led to the pathway shown in Figure 6. The five steps are shown in Figure 6, with methylmercury as the organomercurial are of greatest environmental and medical concern. However, experimental work has generally been done with less toxic mercurials. Initially, the organomercurial forms a thiol–mercury covalent bond with the conserved Cys96 (Figure 6, step 1). A second cysteine thiol, from Cys96, possibly forms a bond with the Hg (Figure 6, step 2) and the proton from aspartate (Asp99) attacks the Hg–C bond releasing the methane (Figure 6, step 4). The inorganic Hg^{2+} bound to the organomercurial lyase by the two thiols is released in cell-free assays to added small thiols such as GSH, but more likely in intact cells, Hg^{2+} is directly transferred to the C-terminal vicinal cysteines of mercuric reductase (Figure 6, step 5).

SILVER RESISTANCE

Silver compounds have come into wide use as antimicrobials over the last 40 years, following earlier use, for example, as antimicrobial rinses for the eyes of neonates. Current uses are primarily as burn ointments and on bandages intended for use on burns, trauma wounds, and slow-healing diabetic ulcers. In addition, biocidal silver compounds are used for hygiene in dish and clothes washers, refrigerators, water purifiers, sports clothing, and a wide variety of other uses. It is perhaps not surprising that bacterial resistance to Ag compounds has been repeatedly reported.

Silver ions are highly toxic to all microorganisms, probably due to poisoning of the respiratory electron transport chains and components of DNA replication. The genetic and physiological basis of bacterial Ag^{+} resistance has been analyzed. These efforts are recent enough that one does not know if the same mechanism will be found broadly among bacteria. However, for enteric bacteria, nine genes contribute to Ag^{+} resistance. These are found on large plasmids, and homologues of the central six genes contribute to low-level Ag^{+} resistance governed by the *E. coli* chromosome. The nine silver resistance genes start with the *silE* gene that determines a small periplasmic metal-binding protein that is unrelated to other metal-binding polypeptide domains, except for the PcoE polypeptide of copper resistance. Upstream from *silE*, a gene pair, *silRS*, determining proteins involved in transcriptional gene regulation occurs. The membrane sensor kinase (SilS, with the conserved histidine residue that is predicted to be phosphorylated by ATP) and the regulatory responder protein (SilR, which is *trans*-phosphorylated on an aspartate residue from the SilS histidine) are homologous to other members of the large two-component regulatory family. The remaining six open reading frames in the Ag^{+} resistance system occur in the opposite orientation from *silRSE* with the *silCBA* genes determining a three-polypeptide RND membrane potential-driven cation/proton exchange complex, quite similar to the CzcCBA complex shown in Figure 1. This complex includes the SilA inner membrane proton/cation pump protein, the SilB membrane fusion protein that is predicted to physically contact SilA and SilC, and the SilC outer membrane protein. The product of the last gene of the silver resistance determinant, SilP, is predicted to be a P-type ATPase (similar to CadA shown in Figure 1) and represents an additional subclass with a monovalent cation substrate, differing in cation specificity from other members of the P-type efflux ATPases, most of which have divalent cation substrates. The full silver resistance determinant is unique among resistance systems in encoding both a periplasmic metal-binding protein and chemiosmotic and ATPase efflux pumps, rather than one or the other.

CADMIUM RESISTANCE

Cadmium resistance is found widely in environmental and clinical bacteria, and the mechanism is generally a membrane efflux pump, either a P-type ATPase or a chemiosmotic pump (Figure 1). Occasionally binding by proteins or small anionic metabolites (phosphates or carbonates) has been reported. The best-studied Cd^{2+} efflux pumps are the CadA ATPase of *Staphylococcus aureus* and the CzcCBA complex of *Cupriavidus metallidurans* strain CH34 (the genus and species names of this strain have been changed five times over recent years). There are additional known P-type toxic cation efflux ATPases for Ag^{+}, Cu^{+} (or Cu^{2+}), Zn^{2+}, Ni^{2+}, and Pb^{2+} and three-polypeptide RND (CBA) systems that pump out Cd^{2+}, Co^{2+}, Ni^{2+}, and Zn^{2+}. An additional single-polypeptide cation diffusion facilitator (CDF) chemiosmotic efflux system was first described with the CzcD

Cd^{2+} and Zn^{2+} efflux system of *C. metallidurans*. Additional members of the CDF family include the Zn^{2+} efflux systems ZitB of *E. coli* and ZntA of *S. aureus* and the Fe^{2+} FieF efflux system of *E. coli* and *C. metallidurans*. CDF homologues are found encoded in many bacterial genomes and also in Archaea, yeast, plants, and animals. There are even seven CDF proteins encoded in the human genome. Thus, the CDF family is as widely occurring in different life forms as are P-type ATPases. The RND chemiosmotic proton/divalent cation exchangers are, however, limited to Gram-negative bacteria.

Detailed protein structures of the Cd^{2+} P-type ATPase and CzcCBA chemiosmotic pump have been proposed by analogy to closely related structures that have been solved by X-ray diffraction or NMR analysis (Figure 1). The P-type efflux ATPase has four readily discerned protein domains and five movements between domains and substrate-binding motifs during the reaction cycle, which could be diagrammed in steps similar to those for mercuric reductase as described above. Within the membrane domain, CadA ATPase is thought to have eight predominantly alpha-helical transmembrane stretches. These move relative to one another during the reaction cycle. The cytoplasmic domains are the nucleotide-binding (N), phosphorylation (P), and activator/phosphatase domains. The CadA divalent cation ATPase has a Cys-Pro-Cys motif in alpha helix no. 6, which is conserved in this group of P-type ATPases and is considered to be involved in cation translocation. The N domain includes a shared motif of GDGXNDXP toward the carboxyl-end of the domain sequence, while the P site TGTKD and the dephosphorylation motif TGES are shared by all P-type ATPases. When ATP binds to the N domain, it rotates above the P domain allowing close proximity to the gamma phosphate of ATP with the aspartate. ATP hydrolysis and ADP release result in rotation of the activator/phosphatase domain so that the TGES phosphatase motif is in contact with the phosphoaspartate, which becomes accessible by outward rotation of the N domain, with concomitant movement within the membrane domain responsible for energy-dependent cation efflux.

Understanding of the CzcCBA RND family system is less advanced. There are structures of individual components from homologous systems, but there are no data available supporting polypeptide movements as for the P-type ATPases. It is proposed that the RND systems, such as CzcCBA, pick up cation substrates from either the periplasmic space or the cytoplasm (Figure 1) and release the cation through outer membrane porin proteins. CzcA contains four domains, two membrane-embedded domains each with six membrane-spanning alpha-helical regions (that are thought to form the cation and H^+ pathways) and two periplasmic domains of approximately equal size. The large periplasmic domains of CzcA may form the wall of a large central cavity (Figure 1) in contact at its top with the opening in the outer membrane protein CzcC (Figure 1). CzcC as a monomer may bridge halfway across the periplasmic space and dock with a CzcA trimer across the outer membrane to the cell surface (Figure 1). The pore through CzcC is bordered by alpha helices in the periplasmic space and a beta-barrel structure in the outer membrane region. CzcB is thought to be anchored by its N-terminus to the inner cell membrane and to make contact with the periplasmic domains of both CzcA and CzcC (Figure 1).

CHROMATE RESISTANCE

Two physiological functions are widely associated with bacterial chromate resistance: first, reduction from the highly soluble chromate (Cr(VI), CrO_4^{2-}) oxyanion to the less soluble (Cr(III), $Cr(OH)_3$), and second, efflux of chromate from the cells by the single chemiosmotic ChrA membrane protein. Although both membrane-associated and soluble cytoplasmic chromate reductases have been described, well-defined resistance is invariably associated with the ChrA efflux pump that is frequently genetically contiguous to other required chromate resistance genes.

ChrA is very widely found in GenBank sequences and appears to occur in several subbranches, which as yet have not been assigned to specific functional differences. The most familiar ChrA products determined by Pseudomonas and by *C. metallidurans* strain CH34 have been shown to cross the cytoplasmic (inner) membrane 12 or 13 times.

OTHER BACTERIAL TOXIC METAL RESISTANCE SYSTEMS

Additional bacterial toxic metal resistance systems for which genes have been sequenced and mechanisms proposed include those for Co^{2+}, Cu^{2+} (and Cu^+), lead (Pb^{2+}), Ni^{2+}, tellurite, and Zn^{2+}.

Bacterial lead resistance is often reported from environmental studies. However, it is unclear whether a single mechanism of Pb^{2+} resistance commonly occurs or whether there are different mechanisms in different bacterial types. The DNA sequence of Pb^{2+} resistance for *C. metallidurans* strain CH34 contains six genes, including those for a positively acting (MerR-like) activator PbrR and genes that appear to encode the resistance mechanism. PbrD appears to be an intracellular Pb^{2+}-binding protein with a Cys-rich potential metal-binding motif (Cys-X7-Cys-Cys-X7-Cys-X7-His-X14-Cys). A resistance mechanism including intracellular sequestration may occur. However, the large PbrA P-type ATPase with the conserved CysProCys and other general sequence properties of soft metal cation P-type ATPases including CadA may function in Pb^{2+} efflux. PbrB is predicted to be a small outer membrane lipoprotein, which may be the pathway for removal of Pb^{2+} that was pumped by

PbrA into the periplasmic compartment. In contrast to the Pbr system of *C. metallidurans*, intracellular lead phosphate precipitates have been associated with plasmid-governed Pb^{2+} resistance in Gram-positive bacteria. It also appears that intracellular and extracellular binding of Pb^{2+} may provide additional mechanisms for lead resistance.

FURTHER READING

Barkay, T., Miller, S. M., & Summers, A. O. (2003). Bacterial mercury resistance from atoms to ecosystems. *FEMS Microbiology Reviews*, *27*, 355–384.

Begley, T. P., & Ealick, S. E. (2004). Enzymatic reactions involving novel mechanisms of carbanion stabilization. *Current Opinion in Chemical Biology*, *8*, 508–515.

Benison, G. C., Di Lello, P., Shokes, J. E. *et al.* (2004). A stable mercury-containing complex of the organomercurial lyase MerB: Catalysis, product release, and direct transfer to MerA. *Biochemistry*, *43*, 8333–8345.

Brown, N. L., Morby, A. P., & Robinson, N. G. (Eds.) (2003). *Interactions of bacteria with metals. FEMS Microbiology Reviews* (Vol. 27), pp. 129–447.

Cervantes, C., Campos-Garcia, J., Devars, S. *et al.* (2001). Interactions of chromium with microorganisms and plants. *FEMS Microbiology Reviews*, *25*, 335–347.

Lafrance-Vanasse, J., Lefebvre, M., Di Lello, P., Sygusch, J., & Omichinski, J. G. (2009). Crystal Structures of the organmercurial lyase MerB in its free and mercury-bound forms: Insights into the mechanism of methylmercury degradation. *Journal of Biological Chemistry*, *284*, 938–944.

Messens, J., & Silver, S. (2006). Arsenate reduction: Thiol cascade chemistry with convergent evolution. *Journal of Molecular Biology*, *362*, 1–17.

Muller, D., Medigue, C., Koechler, S. *et al.* (2007). A tale of two oxidation states: Bacterial colonization of arsenic-rich environments. *PLoS Genetics*, *3*(4), 0518–0530. e53.

Schiering, N., Kabsch, W., Moore, M. J., Distefano, M. D., Walsh, C. T., & Pai, E. F. (1991). Structure of the detoxification catalyst mercuric ion reductase from *Bacillus* sp. strain RC607. *Nature*, *352*, 168–172.

Silver, S. (2003). Bacterial silver resistance: Molecular biology and uses and misuses of silver compounds. *FEMS Microbiology Reviews*, *27*, 341–354.

Silver, S., & Phung, L. T. (1996). Bacterial heavy metal resistance: New surprises. *Annual Review of Microbiology*, *50*, 753–789.

Silver, S., & Phung, L. T. (2005). Genes and enzymes involved in bacterial oxidation and reduction of inorganic arsenic. *Applied and Environmental Microbiology*, *71*, 599–608.

Silver, S., & Phung, L. T. (2005). A bacterial view of the periodic table: Genes and proteins for toxic inorganic ions. *Journal of Industrial Microbiology and Biotechnology*, *32*, 587–605.

Silver, S., Phung, L. T., & Silver, G. (2006). Silver as biocides in burn and wound dressings and bacterial resistance to silver compounds. *Journal of Industrial Microbiology and Biotechnology*, *33*, 627–634.

Taylor, D. E. (1999). Bacterial tellurite resistance. *Trends in Microbiology*, *7*, 111–115.

Chapter 16

Microbial Adhesion

L. Cegelski, C.L. Smith and S.J. Hultgren
Washington University, School of Medicine, St. Louis, MO, USA

Chapter Outline

ABBREVIATIONS

CF Cystic fibrosis
DAF Decay-accelerating factor
ECM Extracellular matrix
GbO4 Globotetraosylceramide
IBCs Intracellular bacterial communities
UPEC Uropathogenic *E. coli*
UTI Urinary tract infection
Tafi Thin aggregative fimbriae

DEFINING STATEMENT

Microbial adhesion is crucial to the survival and lifestyle of many microorganisms. Both beneficial and pathogenic relationships forged between microbe and host depend on adhesive events and colonization. This article highlights the highly evolved microbial adhesion mechanisms and discusses the prevalence and implications of adhesion in diverse ecosystems.

INTRODUCTION

From the center of the earth and deep-sea vents to plant roots and the human intestine, microorganisms occupy remarkably diverse niches on our planet. These microbes include bacteria, archaea, fungi, and protista, and are found attached to rocks and soil particles, corals, and ocean sponges. Bacteria, for example, symbiotically colonize plants and humans as well as fish and squid, resulting in mutual benefit to both microbe and host. Pathogenic and unwelcome bacteria can egress from their native niche and adhere to and infect other sites and host tissues, leading to cellular injury and disease. Microbes also adhere to the hulls of ships and to machinery in food-processing factories, resulting in contamination and adverse circumstances. Specific adhesion strategies have evolved in order to facilitate microbial attachment to diverse substrata in both symbiotic and pathogenic associations. Understanding the molecular mechanisms and functional implications of microbial adhesion is crucial for generating complete descriptions of our ecosystems, understanding and predicting ecosystem stability due to globalization and climate change, and attempting to control and prevent the unfortunate and often devastating consequences of infectious diseases. Thus, microbial adhesion is a fundamental component of the field of microbial ecology. This article will focus on the adhesive strategies employed specifically by bacteria, though many parallels can be found in the arsenal of adhesive strategies harbored by the other classes of microbes. We will highlight several exciting and up-to-date scientific discoveries as a platform to illustrate the biological significance and implications of microbial adhesion.

BIOLOGICAL SIGNIFICANCE OF MICROBIAL ADHESION

The propensity for bacteria to associate with surfaces (living or abiotic) in nearly all ecosystems far exceeds the tendency to persist in suspension, living freely in a planktonic state. Attachment to surfaces allows bacteria to persist in advantageous locations where there may be high nutrient concentrations or to provide protection from hostile environments. In numerous instances, bacteria form biofilms – structurally complex and dynamic bacterial communities. The metabolic labor of acquiring nutrients is divided, sometimes according to spatial coordinates in the community, and distribution is promoted through an organized architecture of community members. Protection from harsh environmental conditions is a major benefit of life in a biofilm and the first line of defense is provided by members residing at the edges of the community. Under certain conditions, bacteria disperse from the biofilm, to seek a new environment and potentially readhere and colonize new niches. Adhesion events are crucial to biofilm formation, growth, and development. We set the stage for discussing the mechanisms of microbial adhesion by first illustrating a few examples across a broad landscape in which bacterial adhesion (often followed by biofilm formation) takes place.

Adhesion in the Water

The coral reefs are home to an enormous diversity of marine life, including beautiful fish, mollusks and urchins, and the significantly smaller microorganisms with which they cohabit. Bacteria are, in fact, an integral constituent of the microbiota of healthy corals. They colonize distinct sites in coral tissue including the surface mucous layer and porous components in the coral skeleton where they fix nitrogen, decompose chitin, and provide organic compounds. The molecular mechanisms of adhesion and the sustained interactions between bacteria and their coral hosts are currently not well understood but are key questions being addressed in the emerging field of coral microbiology. Understanding the microbial interactions that promote health versus those that cause disease is important in efforts to preserve and prevent further destruction of coral reefs worldwide.

Adhesion of bacteria in many water environments inevitably has detrimental consequences. Bacterial adhesion and biofilm formation on the hulls of ships creates resistance to water flow, increases drag, and thus decreases the efficiency of movement through the water. Microbial fouling is an economic and environmental burden in this way and in several industrial settings including drinking water pipes and oil pipelines, where the proliferation of sulfide-producing bacteria leads to the deterioration and corrosion of the steel surfaces. Understanding the mechanisms of adhesion is key to developing strategies to control and prevent these adverse and costly consequences.

Adhesion to Plants

Rhizobia are Gram-negative soil bacteria that adhere to and colonize the root cells of leguminous plants, including soybeans and alfalfa. Upon entry into a root hair, rhizobia traverse a distance to the center of the root hair cell and together with proliferating plant cells form a nodule. Here, rhizobia fix nitrogen, converting molecular nitrogen (N_2) from the air into ammonia, nitrates, and other nitrogenous compounds to support plant metabolism. Rhizobia are particularly important to plants in nitrogen-deficient soils. In return, rhizobia receive carbon-rich organic compounds, important for their own energy production, from the plant.

Other beneficial symbionts include *Bacillus thuringiensis*. This bacterium is an important Gram-positive pathogen whose insecticidal properties have gained attention in the development of crops genetically modified to express the bacterium's potent toxin, now referred to as Bt transgenic crops. In the wild, *B. thuringiensis* colonizes the surface of some plants and exists naturally in some caterpillars. The bacterium produces a unique kind of endotoxin, a proteinaceous crystal that is lethal to several pests, including flies, mosquitoes, and beetles, upon ingestion. This symbiosis with plants is dependent on initial host–microbe adhesion events.

The attractive chemical signals and ultimate adhesive interactions of *Agrobacterium tumefaciens* with wounded plants leads to the unfortunate development of tumors on the lower stems and main roots, the hallmark of Crown Gall diseases. Attachment is the first step in the pathogenic cascade and takes place in the soil around the roots – the rhizosphere. In a two-step adhesive process, initial weak binding interactions are followed by the bacterial expression of multiple gene products to synthesize cellulose and anchor the microbe to the host tissue, while enhancing adhesive interactions between bacteria in the microcolony. Adhesive plant proteins called vitronectins are also implicated in the adhesion process. Subsequent DNA transfer and integration of a specific fragment of DNA (the transfer DNA) from the bacterium to a plant cell results in the expression of several oncogenic genes and the formation of tumors.

Adhesion in the Human Host

The normal and healthy human body is composed of approximately 10 times more bacterial cells than human cells. These bacteria comprise our microbiota and are colonized in distinct sites throughout the body, including the skin and mouth, and the small intestine and colon. In the mouth and small intestine, bacterial adhesion is critical to maintenance of microbial populations, where either salivary flow or movement of contents eliminates the nonadherent bacteria. Our microbiota is, in general, beneficial. Bacteria in the gut, for example, attach to undigested by-products and degrade some polysaccharides into carbon and energy sources, for example. Recent results indicate that the balance of bacterial populations in the gut influence caloric intake through complex interbacterial metabolic networks and further study may help to understand and potentially control (decrease or increase) caloric uptake. Indeed, the microbiota is dynamic and shifts in balance that alter the sizes of different bacterial populations can also lead to proliferation of disease-causing opportunistic pathogens. In addition, the inoculation of the human host with bacteria from the environment is a common source of infectious disease, particularly in the hospital setting. The consequences of bacterial adhesion in human infectious diseases are numerous and will be addressed in more detail after the description of specific adherence mechanisms.

MECHANISMS OF MICROBIAL ADHESION

General Physicochemical Factors Affecting Adhesion

Mechanisms of microbial attachment are incredibly diverse and can be generally classified as either general nonspecific interactions or specific molecular-recognition binding events that involve the presentation of specific adhesive proteins on the bacterial cell surface. Of course, multiple mechanisms can act cooperatively to promote adhesion. A successful adhesive event depends on properties of both the bacterium and the substratum. Nonspecific interactions are the primary form of attachment to abiotic surfaces in aquatic and soil environments. Van der Waals interactions are attractive, usually weak, noncovalent forces that can operate at large separation distances (>50 nm) between the bacterium and the surface. At smaller distances (10–20 nm), electrostatic interactions participate and compete as attractive and repulsive forces. The net surface charge of most bacteria is negative due to cell wall and cell membrane components including negatively charged phosphate groups, carboxyls, and other acidic groups, in addition to surface-exposed proteins. Thus, bacteria like to adhere to positively charged surfaces. Typical binding surfaces, however, have a net negative surface charge, creating electrostatic repulsion that must be overcome by other physicochemical factors. The entire binding process is akin to a tug-of-war. The ionic strength of the surrounding medium affects the electrostatic interactions, and the aforementioned repulsion is eliminated, for example, in most aquatic environments due to high ionic strength resulting from high salt concentration. In the range of near contact (0.5–2 nm), hydrophobic interactions are important for bacterial adhesion. Energetically, the association of nonpolar groups on a bacterial surface with hydrophobic surfaces compensates for the unfavorable displacement of water molecules at that surface. When separated by less than 1 nm, stronger interactions including hydrogen bonding and the formation of salt bridges contribute to surface adhesion.

Specific Adhesin–Receptor Mechanisms

On many biotic surfaces, the adhesive forces and interactions described above promote the formation of an initial interface, but require concomitant or subsequent specific adhesive interactions to enable firm adhesion. Adhesin is the term ascribed to the surface-exposed bacterial molecule that mediates specific binding to a receptor or ligand on a target cell. It is not unusual for bacteria to harbor several types of specific adhesive machinery to provide adhesive capacity to multiple receptor molecules or to permit adhesion under changing environmental conditions such as temperature, pH, or nutrient status, where one adhesive strategy may be more effective than another.

Bacteria can produce a diverse array of adhesins with varying specificities for a wide range of host receptor molecules. Adhesion mechanisms can be classified according to the type of adhesin–receptor pair. Many bacterial adhesins function as lectins and the interactions between bacterial lectins and host cell carbohydrates are among the best-characterized attachment processes. Hallmark examples of carbohydrate recognition include *Pseudomonas aeruginosa*, *Haemophilus influenzae*, and *Streptococcus pneumoniae* adhesion in the respiratory tract, *Escherichia coli* adhesion in the urinary tract and intestine, and *Helicobacter pylori* adhesion in the stomach. Other adhesins recognize specific amino acid-recognition motifs in proteins expressed on host cell surfaces. Extracellular matrix (ECM) proteins that are not directly integrated into the host cell also serve as attractive binding platforms for many bacteria, and numerous adhesins bind to these components in order to indirectly hijack the host signaling pathways, often to enable host cell internalization. Another general category of adhesins includes nonproteinaceous molecules such as lipopolysaccharides and teichoic acids, synthesized by Gram-negative and Gram-positive bacteria, respectively.

Most adhesins are incorporated into heteropolymeric extracellular fibers called pili or fimbriae. Bacteria invest enormous cellular resources to assemble fimbriae in order to present adhesins at the right time and the right place to initiate attachment when conditions are favorable and to permit detachment when necessary. Indeed, hundreds of such fibers have been described in Gram-negative organisms, and although they have diverse functions, many appear critical to binding, invasion, and survival of pathogenic microorganisms in the human host. Four distinct assembly mechanisms have emerged as the most well studied and include the chaperone–usher pathway, the general secretion pathway, the extracellular nucleation–precipitation pathway, and the alternate chaperone pathway. Gram-positive pathogens also produce adhesive pili. Unlike their Gram-negative counterparts, Gram-positive pili are formed by covalent polymerization of pilin subunits. A representative set of fimbrial adhesins is provided in Table 1.

Some bacteria present afimbrial adhesins on their surface. These are expressed as monomeric proteins or protein complexes that assemble at the cell surface and recognize host cell surface elements. Adhesins of the Dr family are expressed by *E. coli* strains and mediate recognition of decay-accelerating factor (DAF). DAF is found in the respiratory, urinary, genital, and digestive tracts, and Dr-mediated adhesion is important for binding in the intestine and urinary tract. Adhesive autotransporters represent a class of afimbrial adhesins expressed by a variety of unrelated microorganisms, including species of *Rickettsia*, *Bordetella*, *Neisseria*, *Helicobacter*, and many members of the family Enterobacteriaceae. *H. influenzae*, a causative agent of sinusitis, bronchitis, otitis media, and pneumonia, expresses an adhesive autotransporter termed Hap. Hap mediates binding to laminin, fibronectin, and collagen, all components of the ECM.

The most comprehensive descriptions of bacterial adhesion have emerged from studies of pathogenic bacteria involved in infectious diseases. Examples of these host–microbe interactions as well as some involved in the attachment of bacteria to plants, either as symbionts or as pathogens, are described in more detail below to highlight the remarkable diversity, specificity, and complexity among microbial adhesive strategies.

SELECTED SURVEY OF SPECIFIC ADHESION STRATEGIES

Pilus-Mediated Adhesion to Carbohydrates in the Urinary Tract

Uropathogenic *E. coli* (UPEC) colonize the gut as well as the genitourinary tract and produce numerous important adhesins and adhesive organelles to mediate adhesion in these niches. For example, FimH and PapG adhesins are presented at the tips of type 1 and P pili, respectively. FimH-presenting type 1 pili are required for *E. coli* to cause cystitis, or infection of the bladder, and PapG-presenting P pili are associated with pyelonephritis, infection of the kidney. Type 1 and P pili are composite heteropolymeric structures, with a distal tip fibrillum joined to a thicker rigid helical rod and both are assembled by the chaperone–usher system. More than 100 chaperone–usher systems have been identified through comparative genome analyses and many are well studied and required for the assembly of extracellular adhesive organelles in pathogens including *Salmonella*, *Haemophilus*, *Klebsiella*, and *Yersinia*. In each chaperone–usher system, pilus assembly requires a unique protein pair (a chaperone and a usher) to facilitate the folding, transport, and ordered assembly of pilus subunits at the cell surface. This process begins in the periplasm, after subunit expression and translocation by the general secretory pathway into the periplasm. Periplasmic pilus chaperones

TABLE 1 Representative fimbrial adhesins and disease association

Organism(s)	Adhesin	Assembly Proteins	Associated Fiber	Associated Disease(s)
Escherichia coli	FimH	FimC/FimD	Type 1 pili	Cystitis
	PapG	PapD/PapC	P pili	Cystitis/pyelonephritis
	PrsG	PrsD/PrsC	Prs pili	Cystitis
	SfaS	SfaE/SfaF	S pili	UTI, newborn meningitis
	CooD	CooB/CooC	CS1 pili	Diarrhea
	CsgA	CsgB (nucleator), CsgE/CsgF (assembly), CsgG (secretion)	Curli	Sepsis
Salmonella typhimurium		PefD/PefC	Pef pili	Gastroenteritis
		LpfB/LpfC	Long polar fimbriae	Gastroenteritis
Salmonella enteritidis	AgfA	AgfB (nucleator)	Sef17 (thin aggregative fimbriae)	
Klebsiella pneumoniae	MrkD	MrkB/MrkC	MR/K (type 3) pili	Pneumonia
Bordetella pertussis	FimD	FimB/FimC	Type 2 and 3 pili	Whooping cough
Yersinia enterocolitica		MyfB/MyfC	Myf fimbriae	Enterocolitis
Neisseria gonorrhoea	PilC	General secretion apparatus	Type 4 pili	Gonorrhea
Pseudomonas aeruginosa, *Vibrio cholerae*, *Mycobacterium bovis*	Pilinprotein			Cholera
Haemophilus influenzae		HifB/HifC	Hif pilus	Otitis media, meningitis

consist of two immunoglobulin (Ig)-like domains and bind to folded subunits to keep their interactive surface capped and prevent nonproductive subunit aggregation. Pilin subunits also have an Ig-like fold, but they lack the seventh β strand, thus exposing the hydrophobic core. In a process termed donor strand complementation, the chaperone's G_1 β strand serves as the pilin's seventh strand, catalyzing the folding of the subunit. Chaperone–subunit complexes are targeted to an outer membrane usher to facilitate chaperone uncapping, translocation of subunits across the outer membrane, and pilus assembly. This occurs via a process termed donor strand exchange, in which the G_1 β strand of the chaperone is replaced by an N-terminal extension of the next pilus subunit. Thus, in the mature pilus, each subunit incorporates its neighbor's N-terminal extension as part of its own Ig fold. Subunits have distinct specificity for other interactive subunits, such as the adhesin, and this confers distinct roles in pilus adhesion, initiation, elongation, termination and regulation.

The FimH adhesin, incorporated at the tip of the type 1 pilus, consists of a pilin subunit and the receptor-binding domain (Figure 1). The primary carbohydrate specificity of FimH is mannose. Interestingly, different *E. coli* isolates present FimH variants (specific allelic variations in protein sequence and structure) that exhibit varying specificities for monomannose and trimannose binding. FimH expressed by most commensal isolates of the intestine exhibit a higher specificity for trimannose-presenting glycoprotein receptors, whereas urinary tract isolates encode for a FimH variant with higher affinity for monomannose. In the latter, FimH mediates adhesion to the monomannose-containing glycoprotein uroplakin Ia that is expressed on the surface of superficial facet cells – the epithelial cells that line the lumen bladder (Figure 1).

Presented at the tips of P pili, the PapG adhesin mediates binding to a different carbohydrate receptor, the α-D-galactopyranosyl-(1–4)-β-D-galactopyranoside moiety of glycolipids presented by cells predominantly in the kidney. PapG variants (G-I, G-II, and G-III) exhibit altered specificities for three Galα(1–4)Gal-containing isoreceptors: globotriaosylceramide, globotetraosylceramide (GbO4), and globopentaosylceramide (the Forssman antigen). The demonstrated allelic variation in PapG and FimH binding specificities supports the notion that, through bacterial evolution, pathoadaptive mutations are selected for increasing the fitness of pathogenic organisms in distinct niches in the host.

Adhesion to ECM Components

The ECM contains a diverse array of oligosaccharides, proteoglycans, and proteins and functions to provide structural support and adhesive interactions among cells. Prevalent components include collagen, fibronectin, laminin, and vitronectin, as well as molecules such as heparan sulfate and

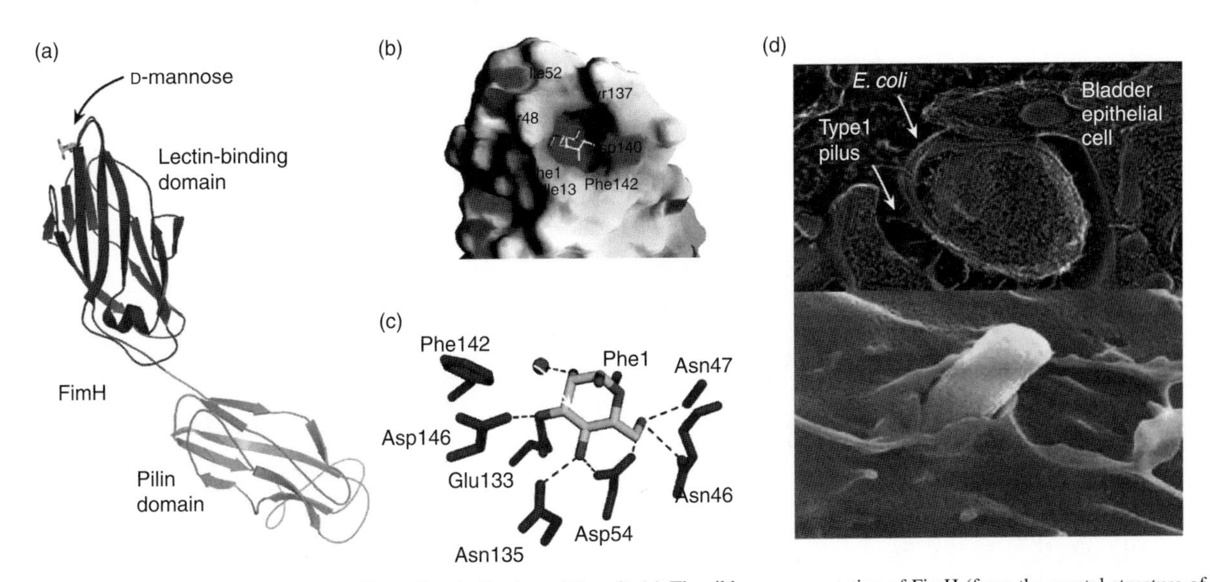

FIGURE 1 The FimH adhesin and type 1 pili-mediated adhesion of *E. coli*. (a) The ribbon representation of FimH (from the crystal structure of the FimCH complex). D-mannose is located at the top of the molecule. (b) Molecular surface representation in which the electrostatic potential surface with positively charged residues is shown in blue, negatively charged residues in red, and neutral and hydrophobic residues in white. Residues defining the hydrophobic ridge around the mannose-binding pocket are labeled. (c) The mannose-binding site with FimH residues. Mannose residues are shown with carbon atoms in yellow, oxygen atoms in red, and nitrogen atoms in blue. (d) Type 1 pili-mediated attachment of uropathogenic *E. coli* (UPEC) to the luminal surface of the bladder epithelium. (Top) High-resolution, freeze-fracture, deep-etch electron micrograph is from Mulvey, M. A., Lopez-Boado, Y. S., Wilson, C. L., *et al.* (1998). Induction and evasion of host defenses by type 1-piliated uropathogenic *E. coli*. *Science*, *282*, 1494–1497. *Reprinted with permission from AAAS. (Bottom) Scanning electron micrograph of a bacterium entering the membrane of bladder epithelial cells is reprinted from Soto, G. E., & Hultgren, S. J. (1999). Bacterial adhesins: Common themes and variations in architecture and assembly.* Journal of Bacteriology, 181, *1059–1071.* (For interpretation of the references to color in this figure legend, the reader is referred to the Web version of this chapter.)

chondroitin sulfate. Fibronectin is present in most tissues and fluids of the body and helps to create a cross-linked network between cells by presenting binding sites for other ECM components, a process that pathogenic organisms exploit to gain a foothold in host tissue. The ability to adhere to ECM components is a primary adhesion mechanism that contributes to the virulence of many pathogenic microorganisms. *Staphylococcus aureus* is a significant cause of nosocomial and often persistent infections. Among other ECM-binding proteins, *S. aureus* expresses the fibronectin-binding proteins FnBP-A and FnBP-B that permit adherence to fibronectin that are bridged to cellular integrins. This crucial binding event leads to host cell cytoskeletal rearrangements and invasion. *Streptococcus pyogenes* is armed with more than 12 fibronectin- and collagen-binding proteins. Like the FnBP-A adhesin in *S. aureus*, the major *S. pyogenes* adhesin, SfbI, and the *Yersinia* adhesin, YadA, bind to fibronectin and bridge the bacteria to integrins, leading to integrin clustering and eventual internalization. Invasin is a *Yersinia* adhesin that bypasses the ECM and binds directly to integrin transmembrane receptors. Other less-ubiquitous ECM components also serve as binding receptors for bacterial adhesins and their sites of expression often relate to the tissue tropism of a particular bacterial pathogen.

Curli-Mediated Multipurpose Adhesion

Curli are a unique class of adhesive extracellular amyloid fibers produced by Gram-negative bacteria, including *E. coli*. The highly homologous fibers produced by *Salmonella* species are called Tafi (thin aggregative fimbriae). The fibers mediate biofilm formation and attachment to host proteins including fibronectin, laminin, and plasminogen, and have been implicated in human sepsis. When expressed together with cellulose, curli and Tafi contribute to a remarkable aggregative phenotype characterized by a patterned assembly of cells radiating from the center when grown on a surface such as agar. Curli are assembled by the nucleation–precipitation pathway, and assembly requires specific molecular machinery encoded by the *csgBA* and *csgDEFG* operons (Figure 2). The major subunit protein (CsgA) and the nucleator protein (CsgB) are secreted to the cell surface in a CsgG-dependent fashion. CsgE and CsgF are assembly factors required for the stabilization and transport of CsgA and CsgB. Transcriptional regulation of the curli operons is complex and responds to many environmental cues including temperature, pH, and osmolarity. The adhesive functionality is attributed to the main fiber subunit, CsgA.

Curli are also implicated in the binding of *E. coli* strains to plant surfaces and are expressed by many strains

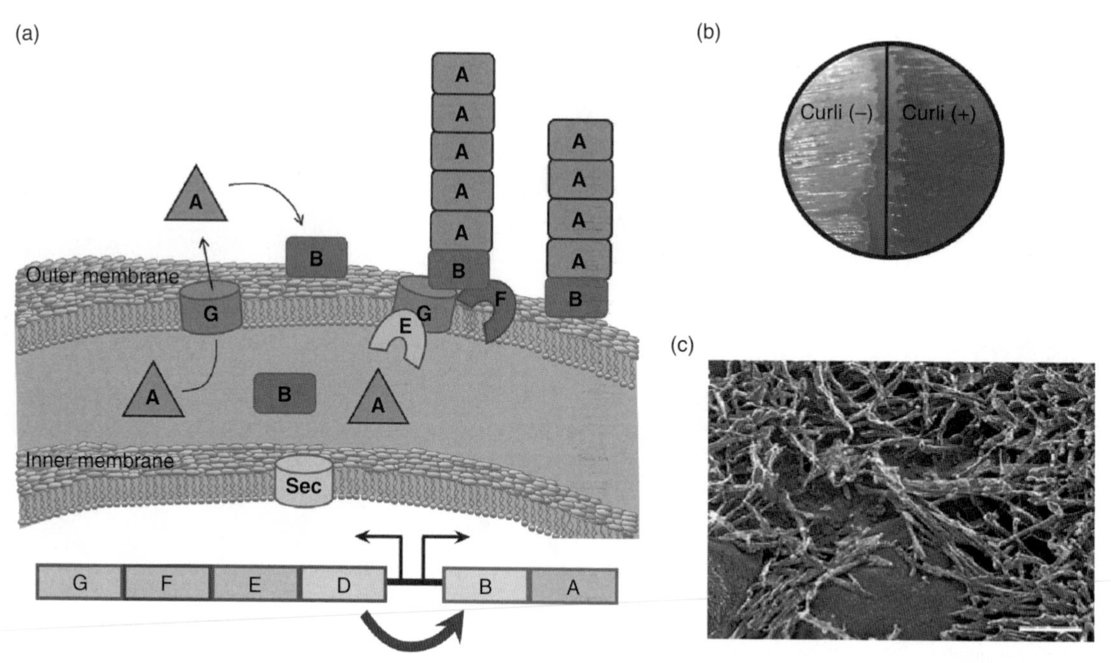

FIGURE 2 Curli biogenesis and biology. (a) Current model of curli biogenesis. Curli assemble through the nucleation–precipitation pathway. Polymerization of the major curli subunit protein, CsgA into β-sheet-rich amyloid fibers depends on the nucleating activity of the minor subunit, CsgB. Proteins CsgE, CsgF, and CsgG are assembly factors required for the stabilization and transport of CsgA and CsgB to the cell surface. (b) Congo red-binding phenotype. Curli are amyloid fibers and bind the hallmark amyloid dyes Congo red and thioflavin T. Curliated *E. coli* grown on Congo red-containing agar medium take up the dye and stain red. Noncurliated cells do not. (c) High-resolution deep-etch electron micrographs of curliated *E. coli*. *From Chapman, M. R., Robinson, L. S., Pinkner, J. S.,* et al. *(2002) Role of* Escherichia coli *curli operons in directing amyloid fiber formation.* Science, *295, 851–855. Reprinted with permission from AAAS.* (For interpretation of the references to color in this figure legend, the reader is referred to the Web version of this chapter.)

associated with food-borne illness, including the prototype strain *E. coli* O157:H7, which has caused several food-borne outbreaks in the United States and around the world. Although the exact nature of binding is still under investigation, curli production is sufficient to permit laboratory strains of *E. coli* to bind plant tissues, such as alfalfa. However, among pathogenic strains such as *E. coli* O157:H7, there appear to be redundant adhesion systems, and under the conditions tested, curli are not required for adhesion. Indeed, external conditions in the environment and in the host may differ as a function of time, and bacteria may depend more on one adhesive system than another in certain circumstances.

The curli bacterial adhesive fiber machinery has gained considerable attention since the discovery of curli as amyloid fibers in 2002. The sticky nature of curli amyloid fibers is like that of amyloid aggregates and plaques associated with eukaryotic amyloid disorders such as Alzheimer's and Parkinson's diseases. Thus, ongoing curli research that aims to elucidate structural features of curli assembly and the functional implications of curli-mediated adhesion may also provide valuable information to the exciting field of amyloid fiber biogenesis and aggregation.

CONSEQUENCES OF MICROBIAL ADHESION IN HUMAN DISEASE

The critical first step in most infectious diseases requires physical contact between a bacterium and host cell. Bacterial adhesins mediate this binding event through the sophisticated adhesion mechanisms described above and allow the pathogen to gain a foothold in the host, initiating complex signaling cascades in both the pathogen and the host. Binding events can lead to extracellular colonization and invasion into underlying host cells. Adhesion is the first step that promotes the cascading sequelae of infectious diseases, particularly important in the pathogenesis of chronic infections including urinary tract infection (UTI), chronic otitis media (middle ear infection), and chronic lung infections.

E. coli and UTI

UPEC engage in an incredibly coordinated and regulated genetic and molecular cascade to assemble type 1 pili, as described above. UTIs are among the most common bacterial infections and nearly 50% of women will be afflicted by at least one UTI in their lifetime, with many experiencing recurrent UTIs. Virtually all clinical UPEC isolates express type 1 pili, enabling them to bind the mannose-containing host receptors, which results in invasion of host bladder epithelial cells. Inside urothelial cells, bacteria form large, densely packed, biofilm-like intracellular bacterial communities (IBCs) of morphologically coccoid bacteria, comprising up to 10^5 bacteria per superficial facet cell. In this intracellular niche, the pathogens are protected from antibodies, the flow of urine, and other host defenses. Yet, this is only the beginning of a sometimes life-long cycle of interactions between pathogen and host. IBC formation is not an end point or dead end for *E. coli*. Upon entry into superficial facet cells, UPEC activate a complex developmental cascade; UPEC eventually detach and disperse, or flux, from the IBC to initiate another round of IBC formation in other urothelial cells (Figure 3). Some fluxing bacteria form filaments, which are resistant to neutrophil phagocytosis. Filamentation facilitates survival of the bacteria and allows them to invade other epithelial cells. Even after acute infection is resolved and the urothelium is seemingly intact, bacteria can remain within the bladder for many days to weeks regardless of standard antibiotic treatments. Thus, the ability of UPEC to adhere to and invade bladder cells appears to facilitate long-term bacterial persistence within the urinary tract.

P. aeruginosa and Cystic Fibrosis (CF)

P. aeruginosa has emerged as an opportunistic pathogen in several clinical settings, causing nosocomial infections such as pneumonia, UTIs, and bacteremia. *P. aeruginosa* adheres

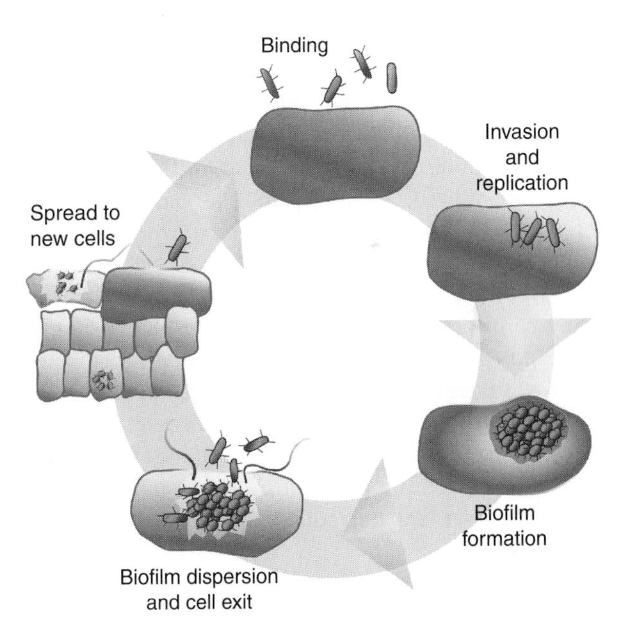

FIGURE 3 Pathogenic cascade of uropathogenic *E. coli* (UPEC). UPEC coordinate highly organized temporal and spatial events to colonize the urinary tract. UPEC bind to and invade the superficial umbrella cells that line the bladder lumen, where they rapidly replicate to form a biofilm-like intracellular bacterial community (IBC). In the IBC, bacteria find a safe haven, are resistant to antibiotics, and subvert clearance by innate host responses. UPEC can persist for months in a quiescent bladder reservoir following acute infection, and challenge current antimicrobial therapies. Quiescent bacteria can reemerge as pathogens from their protected intracellular niche and can be a source of recurrent urinary tract infections (UTIs).

to the respiratory epithelium, leading to chronic lung infections in CF patients, responsible for the eventual pulmonary failure of most CF patients, typically by 37 years of age. Pilus-mediated adherence is important in the adhesion and early stages of epithelial colonization, and additional virulence factors contribute to the subsequent persistence in the lung. Alginate, for example, is a mucoid exopolysaccharide produced by *P. aeruginosa* that forms a matrix of "slime" to surround a forming biofilm and anchors the cells to each other and to their host. Surrounded by alginate, the bacteria are protected from the host defenses and are often resistant to treatment with antibiotics.

TARGETING ADHESION TO INHIBIT BACTERIAL VIRULENCE

The ability to impair bacterial adhesion represents an ideal strategy to combat bacterial pathogenesis because of its importance early in the infectious process. In addition, adhesion is essential to the long-term persistence of bacteria in the pathogenic cascade of several infectious diseases. Moreover, the adhesion process can be targeted without placing life or death pressure on the bacterium, *per se*. Targeting bacterial virulence in this way is an alternative approach to the development of new therapeutics to disarm pathogens in the host that may offer reduced selection pressure for drug-resistant mutations. In addition, virulence-specific therapeutics could avoid the undesirable dramatic alterations of the host microbiota. Indeed, standard antibiotic treatment regimens may lead to the loss of symbiotic benefits and the proliferation of disease-causing opportunistic pathogens.

As emphasized earlier, pathogens are capable of presenting multiple adhesins that can be expressed differentially to permit binding in specific sites and at specific times over the course of a complex infectious cycle. Thus, it may be difficult to develop a universal class of antiadherence drugs. Nevertheless, several specific pathogenic adhesive strategies have emerged as hallmark requirements for virulence in certain infectious diseases, and represent amenable targets for drug discovery and development. Adhesion is sometimes just the first step of many in pathogenic cycles, yet targeting adhesion holds value even after an infection has been established. In biofilm-associated infections, for example, drug development strategies include attempts to induce the dispersal of bacteria from the biofilm and to inhibit the chemical signaling necessary to encourage new biofilm formation. In UTI, the fluxing bacteria are capable of readhering to new host cells, gaining a foothold and potentially invading a new cell to remain undetected until drug pressure subsides and conditions encourage replication and new intracellular biofilm formation. Thus, strategies to prevent microbial adhesion are being considered in combination therapies to both prevent and treat infectious diseases.

Carbohydrate derivatives of host ligands have demonstrated efficacy in blocking the adhesive properties of *E. coli* expressing type 1 and also P pili in biophysical and hemagglutination assays. This approach of using soluble carbohydrates or mimics recognized by the bacterial lectin can be readily extended to other adherent organisms by tailoring the antiadhesive compounds to their receptor specificities.

"Pilicides" are a class of pilus inhibitors that target chaperone function. A new class of pilicides, based on a bicyclic 2-pyridone scaffold, inhibit the assembly of both type 1 and P pili in *E. coli* (Figure 4). The potent molecules inhibit an essential protein–protein interaction between chaperone and usher, required for pilus biogenesis. Chaperone–usher systems are highly conserved among various bacteria including *Salmonella*, *Haemophilus*, *Klebsiella*, and *Yersinia* and it is possible, although not yet demonstrated, that pilicides may exert broad-spectrum activity and be effective against several Gram-negative pathogens.

Compounds have been identified that target the two-component signaling system, AlgR2/AlgR1, that controls the synthesis of alginate by *P. aeruginosa*. Alginate is a key component of the protective exopolysaccharide coat, critical to *P. aeruginosa* adherence, biofilm formation, and CF pathogenesis. The inhibitors of alginate synthesis could be therapeutically employed to render the pathogen more susceptible to host defenses or to standard antibiotics currently in use, and thus, could be effective also in combination therapy. The ability to inhibit microbial adhesion and thus prevent subsequent pathogenic processes holds enormous therapeutic potential and promises to improve the treatment of numerous infectious diseases.

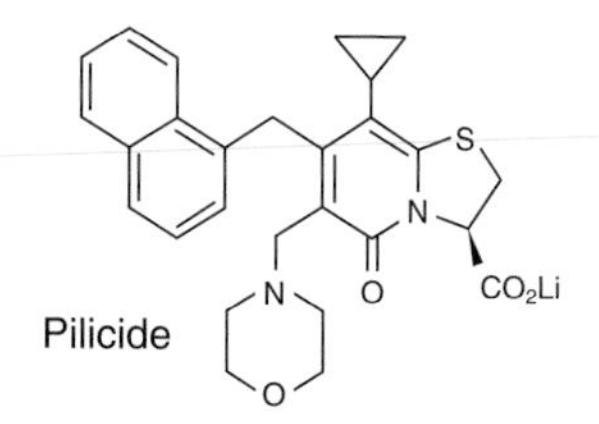

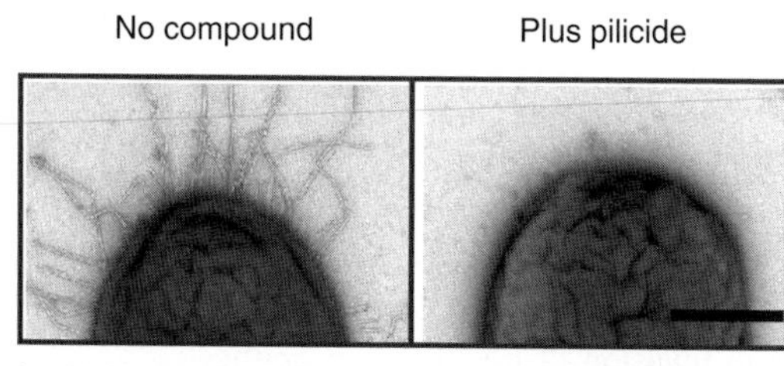

FIGURE 4 Targeting microbial adhesion. Rationally designed "pilicides" inhibit pilus biogenesis by disrupting chaperone–usher protein interactions and reduce piliation levels dramatically. *Electron micrographs reproduced from Pinkner, J. S., Remaut, H., Buelens, F.,* et al. *(2006) Rationally designed small compounds inhibit pilus biogenesis in uropathogenic bacteria.* Proceedings of the National Academy of Sciences of the United States of America, 103, *17897–17902. Copyright (2006) National Academy of Sciences, U.S.A.*

RECENT DEVELOPMENTS

In nearly all ecosystems, the propensity for bacteria to associate with a surface (biotic or abiotic) and with other community members far exceeds the tendency to persist in suspension, living freely in a planktonic state. Adhesion refers to the fundamental event in which a microorganism transitions first from the planktonic state to a surface-associated one. The surface can be anything and ranges from a simple soil particle to another microorganism, an indwelling medical device, or the gut epithelium. Following adhesion, biofilm formation can occur. Bacterial biofilms are complex microbial communities that exhibit reduced sensitivity to conventional antibiotics, host defenses, and external stresses. While there is still much to learn regarding individual attachment strategies conferred by the arsenal of bacterial lectins and surface-exposed adhesins, there exists an enormous body of knowledge including atomic-level structural detail of many of the extracellular adhesive proteins and fibers assembled by bacteria at the cell surface. Bacterial biofilm formation, on the other hand, is a much more complex process and can involve the participation of many adhesive proteins, proteinaceous fibers, and extracellular components including polysaccharides and nucleic acids. Multiple determinants contribute to biofilm development and maintenance, and their requirements in biofilm formation may vary depending on environmental conditions. In addition, factors important in biofilm formation are often functionally redundant and add to the complexity of building biofilm models and developing antibiofilm strategies. Virulent pathogens often harbor multiple adhesive systems that are used in different stages of pathogenesis, supporting the notion that multipronged approaches may be required to target and prevent biofilm formation.

Over the last few years, there have been many discoveries regarding genetic and molecular factors that promote formation and also dispersal of biofilms. Rather than review these, we will take this opportunity to highlight recent developments in the intriguing arena of amyloid-integrated bacterial biofilms and strategies to interfere with their formation. Curli are the most well studied bacterial amyloid fiber and are described more fully in the main article. Compounds that block curli biogenesis, termed "curlicides" were discovered and reported in December 2009. The compounds have a ring-fused 2-pyridone structure, blocked curli assembly in UPEC, prevented biofilm formation, and attenuated virulence in a mouse model of UTI. These discoveries are important from a pure microbial ecology and pathogenesis standpoint, but also offer the provocative possibility that compounds that block the dedicated amyloid biogenesis machinery in prokaryotes may be ideal candidates for blocking amyloid nucleation and protein mis-assembly in eukaryotic disorders. Indeed, lessons learned in our genetically tractable prokaryotic co-habitants hold many opportunities to understand and influence eukaryotic processes.

The Gram-positive organism *B. subtilis* is a well-established model for bacterial cell division, growth, metabolism, and biofilm formation. The discovery of amyloid production by *B. subtilis* was reported by the Kolter and Losick laboratories in February 2010. Their studies also demonstrated that the amyloid fibers, comprised of the protein TasA, contribute to the structural integrity of bacterial biofilms akin to how curli contribute to *E. coli* biofilms, both on agar and at an air-liquid interface. The *B. subtilis* team subsequently reported that D-amino acids self-produced by *B. subtilis* could trigger biofilm dispersal. Exogenously added D-amino acids also could effect the dispersal. This finding not only provides a possible therapeutic avenue for disrupting and preventing biofilm formation, but also adds to the growing appreciation for the role of D-amino acids in biology.

Since the discovery of curli in 1989 and the identification of curli as amyloid in 2002, we now appreciate that functional amyloids are prevalent among microorganisms, and that they are also an integral part of normal mammalian cellular physiology. Assembly of functional amyloids in bacteria is regulated in order to direct polymerization at the right time and place, and to prevent toxicity. The continued dissection of amyloidogenesis in these systems and our ability to interfere with gene-directed amyloid assembly will enhance our perspective on functional amyloid folding pathways and will lead to a broader understanding of amyloid function in microbes.

ACKNOWLEDGMENT

L. Cegelski is the recipient of a Burroughs Wellcome Fund Career Award at the Scientific Interface.

S.J. Hultgren acknowledges funding from the National Institutes of Health (Scor P50 DK64540/ORWH, R01AI029549, R01AI048689, and R01DK51406).

FURTHER READING

Barnhart, M. M., & Chapman, M. R. (2006). Curli biogenesis and function. *Annual Review of Microbiology*, *60*, 131–147.

Blanke, S. R. (2009). Expanding functionality within the looking-glass universe. *Science*, *325*, 1505–1506.

Cegelski, L., Marshall, G. R., Eldridge, G. R., & Hultgren, S. J. (2008). The biology and future prospects of anti-virulence therapies. *Nature Reviews Microbiology*, *6*, 17–27.

Cegelski, L., *et al.* (2009). Small-molecule inhibitors target *Escherichia coli* amyloid biogenesis and biofilm formation. *Nature Chemical Biology*, *5*, 913–919.

Chapman, M. R., *et al.* (2002). Role of *Escherichia coli* curli operons in directing amyloid fiber formation. *Science*, *295*, 851–855.

Kolodkin-Gal, I., *et al.* (2010). D-amino acids trigger biofilm disassembly. *Science*, *328*, 627–629.

Ofek, I., Doyle, R. J., & Hasty, D. L. (2003). *Bacterial adhesion to animal cells and tissues*. Washington, DC: ASM Press.

Ofek, I., Sharon, N. S., & Abraham, S. N. (2006). Bacterial adhesion. *Prokaryotes*, *2*, 16–31.

Olsen, A., Jonsson, A., & Normark, S. (1989). Fibronectin binding mediated by a novel class of surface organelles on *Escherichia coli*. *Nature*, *338*, 652–655.

Pizarro-Cerda, J., & Cossart, P. (2006). Bacterial adhesion and entry into host cells. *Cell*, *124*, 715–727.

Romero, D., Aguilar, C., Losick, R., & Kolter, R. (2010). Amyloid fibers provide structural integrity to *Bacillus subtilis* biofilms. *Proceedings of the National Academy of Sciences of the United States of America*, *107*, 2230–2234.

Rosenberg, E., Koren, O., Reshef, L., Efrony, R., & Zilber-Rosenberg, I. (2007). The role of microorganisms in coral health, disease and evolution. *Nature Reviews Microbiology*, *5*, 355–362.

Wright, K. J., & Hultgren, S. J. (2006). Sticky fibers and uropathogenesis: Bacterial adhesins in the urinary tract. *Future Microbiology*, *1*, 75–87.

Chapter 17

Bacterial Bioluminescence

Paul V. Dunlap
University of Michigan, Ann Arbor, MI, USA

Chapter Outline

ABBREVIATIONS

AFLP Amplified fragment length polymorphism
AHLs Acyl-homoserine lactones
cAMP Cyclic AMP
CEA Ciliated epithelial appendage
CRP cAMP receptor protein

DEFINING STATEMENT

Bioluminescence is the enzymatic production of light by a living organism. Many different kinds of organisms are bioluminescent. Luminous microbes include some fungi, certain unicellular eukaryotes, and several kinds of bacteria. This article summarizes information on bioluminescence in bacteria, from the perspectives of biochemistry, systematics, ecology and symbiosis, genetics, and evolution.

INTRODUCTION

Bioluminescence, which is the enzyme-catalyzed emission of light, is an attribute of many different kinds of organisms. The yellow-green flashes of light made by fireflies at dusk in summer are one of the more commonly observed forms of bioluminescence. Various other terrestrial and many marine organisms are luminescent, including cnidarians, mollusks, annelids, arthropods, echinoderms, and fish. Certain eukaryotic microorganisms, the protist *Gonyaulax*, and certain fungi, for example, also emit light, as do several kinds of bacteria. A common theme in bioluminescence is the use of oxygen as a substrate for the light-emitting enzyme, referred to generically as luciferase. However, in most of these organisms, the substrates that luciferase uses other than oxygen are completely different, and the luciferases themselves exhibit no homology in their nucleotide sequences. The biochemical diversity of extant bioluminescence systems and their lack of DNA sequence homology indicate that bioluminescence has evolved independently many times during the history of life on Earth. This article focuses on the smallest of luminous organisms with presumably the longest evolutionary history, namely, the bioluminescent bacteria. Additional information on bioluminescence in eukaryotic microorganisms can be found in references listed in Further Reading.

Luminous bacteria (Table 1) are those bacteria whose genomes naturally contain genes for bacterial luciferase and for the enzymes that produce a long-chain aldehyde substrate used by luciferase in light emission. Bacterial luminescence and many of the luminous bacteria themselves have been known for some time. During the 1700s and 1800s, various animal products, such as meats, fish, and eggs, the decaying bodies of marine and terrestrial animals, and even human wounds and corpses, were seen to emit light. Those observations were preceded by many years by the demonstration of Robert Boyle in 1668 that the "uncertain shining of Fishes," the light coming from decaying fish, required air, long before the existence of bacteria was known. Light-emitting bacteria were first isolated from nature in the 1880s by the early microbiologists Bernard Fischer and Martin Beijerinck, and they have been subjects of biochemical, physiological, ecological, and, more recently, genetic studies since that time. Although luminous bacteria are in most ways not different physiologically and genetically from other bacteria, studies of these bacteria from the perspective of their light-emitting capability have led to substantial progress in understanding quorum sensing in bacteria, now a major research theme in microbiology, and in understanding the ecology and genetics of mutualistic bacterial symbioses with animals. More

TABLE 1 Species and habitats of luminous bacteria[a]

Species	Habitats[b]
Aliivibrio fischeri	Coastal seawater, light organs of monocentrid fish and sepiolid squids
A. logei	Coastal seawater, gut of arctic mussel
A. salmonicida	Diseased fish, tissue lesions of Atlantic salmon
A. sifiae	Coastal seawater
A. "thorii"	Coastal seawater, light organs of sepiolid squids
A. wodanis	Fish skin and skin ulcers, light organs of sepiolid squids
Photobacterium angustum	Seawater
P. aquimaris	Seawater
P. damselae	Lesions of fish
P. kishitanii	Skin and intestines of marine fishes, light organs of deep-sea fishes
P. leiognathi	Coastal seawater, light organs of leiognathid fishes and loliginid squids
P. mandapamensis	Coastal seawater, light organs of apogonid and leiognathid fishes
P. phosphoreum	Skin and intestines of marine fishes, coastal and open ocean seawater
Photorhabdus asymbiotica	Human skin lesions
P. luminescens	Insect larvae infected with heterorhabditid nematodes
P. temperata	Insect larvae infected with heterorhabditid nematodes
Shewanella hanedai	Seawater and sediment
S. woodyi	Seawater and ink sac of squid
Vibrio azureus	Seawater
V. "beijiernickii"	Coastal seawater
V. chagasii	Intestine of larval fish, seawater
V. cholerae	Coastal seawater, brackish and estuarine waters
V. harveyi	Coastal seawater and sediments
V. orientalis	Coastal seawater, surfaces of shrimp
V. splendidus	Coastal seawater
V. vulnificus	Coastal seawater, oysters

[a] *All are members of Gammaproteobacteria, phylum Proteobacteria, domain* Bacteria. *Species epithets in quotations marks are species not yet formally described.*
[b] *Listed are typical habitats from which luminous strains of the species have been isolated.*

recently, studies of these bacteria using modern phylogenetic approaches, while opening up their evolutionary history, have made contributions to understanding the nature of bacterial species and to bacterial biogeography.

BIOCHEMISTRY OF BACTERIAL LIGHT PRODUCTION

Light emission in bacteria (Figure 1) is catalyzed by bacterial luciferase, which is a heterodimeric protein of approximately 80kDa, composed of an α-subunit (40kDa) and a β-subunit (37kDa). Bacterial luciferase mediates the oxidation of reduced flavin mononucleotide ($FMNH_2$) and a long-chain aliphatic aldehyde (RCHO) by O_2 to produce blue-green light, according to the following reaction:

$$FMNH_2 + O_2 + RCHO \xrightarrow{\text{luciferase}} FMN + H_2O + RCOOH + \text{light}(\sim 490\text{ nm})$$

In the luminescence reaction, binding of $FMNH_2$ by the enzyme is followed by interaction with O_2 and then binding of aldehyde, forming a highly stable enzyme/substrate complex that slowly decays with the oxidation of the $FMNH_2$ and aldehyde substrates and the emission of light. Quantum yield for the reaction has been estimated at 0.1–1.0 photon. The reaction is highly specific for $FMNH_2$, and the aldehyde substrate *in vivo* is likely to be tetradecanal. Synthesis of the long-chain aldehyde and its recycling from the long-chain fatty acid are catalyzed by a fatty acid reductase complex composed of three polypeptides, namely, a nicotinamide adenine dinucleotide phosphate (NADPH) dependent acyl protein reductase (54kDa), an acyl transferase (33kDa), and an adenosine triphosphate (ATP) dependent synthetase (42kDa). The activity of this complex is essential for the production of light in the absence of exogenously added aldehyde.

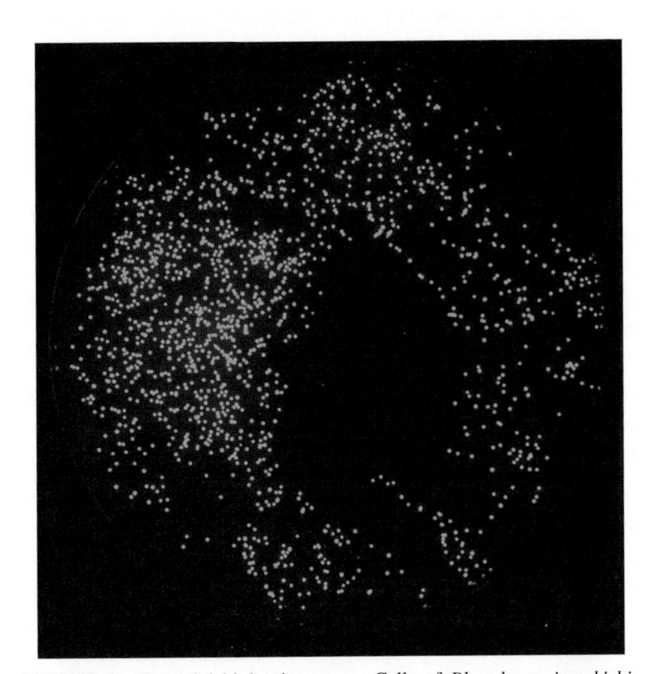

FIGURE 1 Bacterial bioluminescence. Cells of *Photobacterium kishitanii*, a newly described species of luminous bacteria, have formed colonies on this plate of nutrient seawater agar. The bacteria were taken from the ventral light organ of the deep-sea fish *Chlorophthalmus albatrossis* (Chlorophthalmidae). The plate was photographed in the dark by the light the bacteria produce.

The genes for bacterial light production are present as a coordinately expressed set of genes, *luxCDABEG*, which is the *lux* operon (Figure 7). The *luxA* and *luxB* genes encode the α- and β-subunits of luciferase; *luxC*, *luxD*, and *luxE* encode the polypeptides of the fatty acid reductase complex; and *luxG* encodes a flavin reductase. The *luxCDABE* genes are present and have the same gene order in all luminous bacteria examined to date, a defining characteristic of these organisms. More information on the biochemistry of bacterial light production can be found in the references listed in Further Reading.

SPECIES AND SYSTEMATICS OF LUMINOUS BACTERIA

At present, 19 species of luminous bacteria have been identified (Table 1). This list includes some species in which only certain strains are luminous. It should be noted that many more kinds of luminous bacteria remain to be discovered and characterized. The basis for this statement is the recent identifications of new species of luminous bacteria (i.e., *Photobacterium kishitanii*) and of luminous strains of species not previously known to be luminous (i.e., *Aliivibrio salmonicida*), as well as descriptions of new species in progress at this time and not listed in Table 1.

Luminous bacteria are all Gram-negative, non-sporeforming, motile chemoorganotrophs. Most, that is, those in genera *Vibrio*, *Aliivibrio*, *Photobacterium*, and *Photorhabdus*, are facultatively aerobic, able to use oxygen in respiration, and also able to use sugars by fermentation for energy generation when oxygen and other suitable terminal electron acceptors are not available. In contrast, *Shewanella* species use only the respiratory mode of energy generation. Luminous *Vibrio*, *Aliivibrio*, *Photobacterium*, and *Shewanella* species are found in the marine environment, whereas *Photorhabdus* species are terrestrial. Luminous strains of *V. cholerae* have been isolated from coastal seawater as well as from brackish and freshwater environments. Some species of luminous bacteria form highly specific, mutually beneficial bioluminescent symbioses with marine fishes and squids (discussed below).

Luminous bacteria are members of three families, *Vibrionaceae*, *Shewanellaceae*, and *Enterobacteriaceae*, all of which belong to *Gammaproteobacteria*, a class in phylum Proteobacteria of domain *Bacteria*. A phylogeny of luminous bacteria is shown in Figure 2. This phylogeny is based on sequences of the 16S rRNA gene, which encodes the RNA component of the small subunit of the bacterial

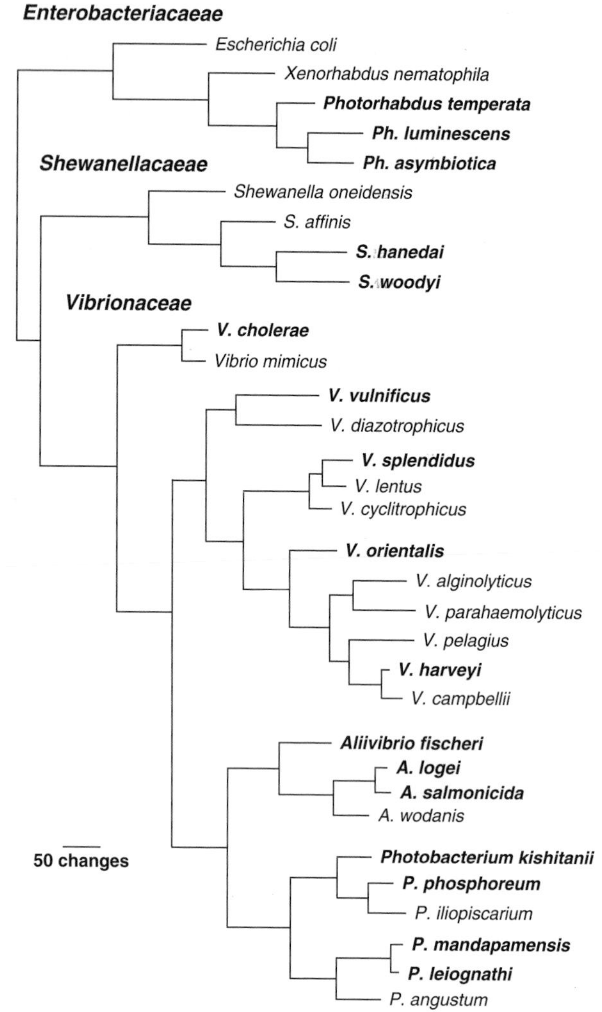

FIGURE 2 Phylogeny of luminous bacteria. This analysis (provided by Dr. Jennifer Ast, University of Michigan) is based on sequences of the 16S rRNA and *gyrB* genes. Luminous species (in boldface) are found in three families, *Vibrionaceae*, *Shewanellaceae*, and *Enterobacteriaceae*. These families contain many more nonluminous species than shown here.

ribosome, and *gyrB*, a housekeeping gene that encodes DNA gyrase subunit B and that, like the 16S rRNA gene, is useful for evolutionary analysis of bacteria. The figure reveals that most species of luminous bacteria are members of the *Vibrionaceae* genera *Vibrio*, *Photobacterium*, and *Aliivibrio*. Species previously known as members of the *Vibrio fischeri* species group, that is, *V. fischeri*, *V. logei*, *V. salmonicida*, and *V. wodanis*, were for many years known to differ from members of *Vibrio* and *Photobacterium*. These bacteria have been reclassified in accord with those differences as members of a new genus *Aliivibrio*. Four species of this genus are currently recognized, namely, *A. fischeri*, *A. logei*, *A. salmonicida*, and *A. wodanis*. The phylogenetic tree includes most of the luminous bacteria but only a few representative nonluminous species in *Vibrionaceae*, *Shewanellaceae*, and *Enterobacteriaceae*; luminous bacteria make up only a small number of the many species in these genera. Within *Vibrio*, *Aliivibrio*, *Photobacterium*, and *Shewanella*, there are many species of nonluminous bacteria; in contrast, genus *Photorhabdus* as presently defined contains only luminous species. Even species characterized as luminous, however, such as *Photorhabdus luminescens*, can contain strains that do not produce light, and some species, such as *V. cholerae* and *V. vulnificus*, contain relatively few strains that are luminous. The phylogenetic relationships among luminous bacteria and the presence of luminous and nonluminous species in several groups provide insights into the evolution of the bacterial luminescence system, a topic discussed in a later section.

Our knowledge of what species of bacteria are luminous is based to a large extent on the production of high levels of light by many commonly encountered luminous species. This criterion, however, does not recover all luminous bacteria. The problem is that some luminous bacteria produce a high level of light under natural conditions but little or no light when grown on laboratory media; "cryptically luminous" bacteria therefore can be missed in screening environmental samples for light-emitting bacteria. Examples include luminous bacteria infecting crustaceans and strains of *A. fischeri* symbiotic with the Hawaiian sepiolid squid *Euprymna scolopes*. Other examples are *V. cholerae*, many strains of which carry *lux* genes but produce little or no light, and *A. salmonicida*, which requires addition of aldehyde to stimulate light production. Enzyme assay and antibody methods previously were used to indicate the presence of luciferase in several nonluminous *Vibrio* spp., and *lux* gene-containing bacteria not producing light in culture have been identified with *luxA*-based DNA probes from various seawater samples. A further complication is that luminescence often is not phenotypically stable; strains that are luminous on primary isolation often become dim or dark in laboratory culture and therefore may be discarded as nonluminous contaminants. It is therefore reasonable to assume that more luminous bacteria exist than are listed here (Table 1). Supporting this view, descriptions of additional new luminous bacteria and of bacterial species newly recognized to contain luminous strains are under way at this time.

Methods for taxonomic identification of luminous bacteria have changed significantly over the past few years, with more emphasis now being placed on DNA sequence-based phylogenies over previously standard descriptive phenotypic and genotypic methods. Analysis of *lux* gene sequences together with sequences of housekeeping genes, such as the 16S rRNA gene, *gyrB*, *pyrH*, *recA*, *rpoA*, and *rpoD*, has proven helpful both for defining evolutionary relationships among luminous bacteria and for the identification of new species. For example, bacteria previously grouped as members of *Photobacterium phosphoreum* based on phenotypic and genotypic traits have been resolved by a multigene phylogenetic approach as three distinct species (Figure 3). The more definitive species-level

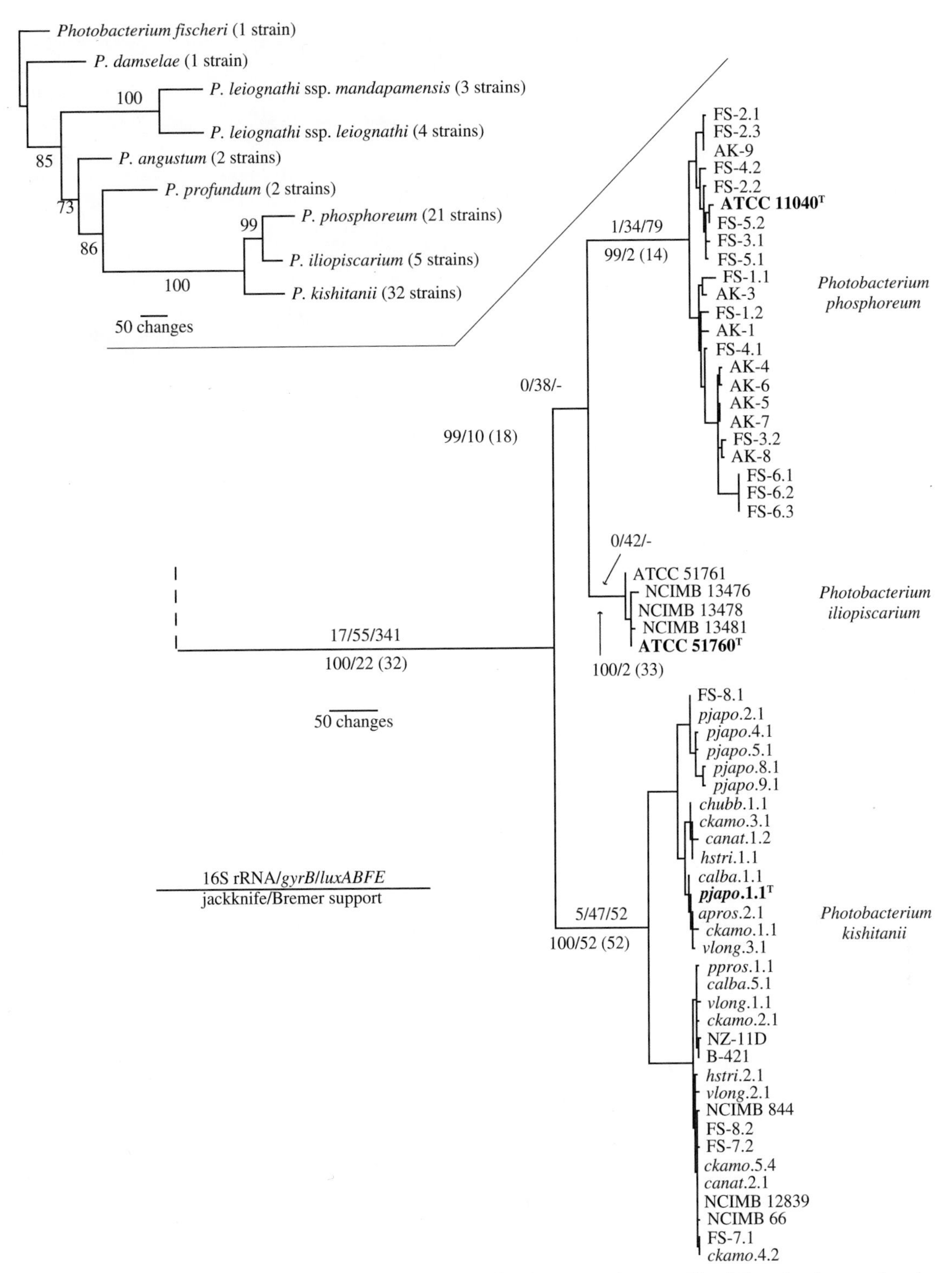

FIGURE 3 Phylogenetic resolution of the members of the *Photobacterium phosphoreum* species group. The phylogenies shown are based on combined analysis of the 16S rRNA, *gyrB*, and *luxABFE* genes. The numbers above the branches represent the steps each locus (16S rRNA gene, *gyrB*, and *luxABFE*) contributed to branch length. The 16S rRNA gene contributed several steps (17 steps) to the separation of the *P. phosphoreum* group from the rest of *Photobacterium*, but few steps (0–1 step) to the branches separating *P. phosphoreum* and *P. iliopiscarium*; the sequences of *P. kishitanii* strains differed slightly (five steps) from those of *P. phosphoreum* and *P. iliopiscarium*. In contrast, the much greater sequence divergence of the *gyrB* and *luxABFE* genes permits resolution of *P. phosphoreum*, *P. iliopiscarium*, and *P. kishitanii* as separate species. Numbers below the branches represent jackknife resampling values and Bremer support values. Species type strains are in boldface. *Reproduced from Ast, J. C., & Dunlap, P. V. (2005) Phylogenetic resolution and habitat specificity of the* Photobacterium phosphoreum *species group.* Environmental Microbiology, *7, 1641–1654.*

resolution provided by molecular phylogenetic analysis is proving to be helpful also in opening up the species-specific ecologies of luminous bacteria.

HABITATS AND ECOLOGY OF LUMINOUS BACTERIA

Marine

Luminous bacteria are globally distributed in the marine environment (Table 1) and can be isolated from seawater, sediment, and suspended particulates. They also colonize the skin of marine animals as saprophytes, their intestinal tracts as commensal enteric symbionts, and their tissues and body fluids as parasites, where they presumably use the organic compounds available in these habitats as sources of energy and carbon. Certain of the luminous bacteria form highly specific bioluminescent symbioses with marine fish and squids. They also colonize marine algae, but agar-digesting luminous bacteria are rare.

In seawater and marine sediments and on the skin of marine animals, luminous bacteria are a consistent but usually small fraction of the bacteria present. Also in the intestinal tracts of marine fishes and other animals, luminous bacteria usually make up a small but consistent percentage of the bacteria present than can grow under aerobic laboratory conditions, although in some cases luminous colonies have been observed to form 50% or more of the colonies arising on plates of seawater-based complete medium inoculated with intestinal contents. Luminous bacteria attain exceptionally high numbers, up to 10^9–10^{11} cells per milliliter, in light organs of fishes and squids, where they exist in mutualistic symbioses, and as incidental parasites reproducing in the hemocoel of marine crustaceans and other marine animals. Luminous bacteria grow readily, for example, on fish muscle tissue (Figure 4). Reproduction in each of these habitats releases cells into seawater, providing inocula for colonization of the other habitats. In colonizing free-living, saprophytic, commensal, and parasitic habitats, luminous bacteria coexist and compete for nutrients with many other kinds of bacteria. The exception to their existence as members of a diverse community of bacteria is bioluminescent symbiosis; light organs of fishes and squids are highly specific to certain species of luminous bacteria and do not harbor nonluminous species.

The distributions of individual species of luminous bacteria and their numbers in a given habitat correlate with certain environmental factors, as is seen for nonluminous bacteria. Primary among these factors are temperature and depth in the marine environment, nutrient limitation, and sensitivity to solar-irradiation-induced photoxidation in surface waters. Temperature, also being an important environmental factor, can influence whether luminous bacteria are detected from environmental samples. For example,

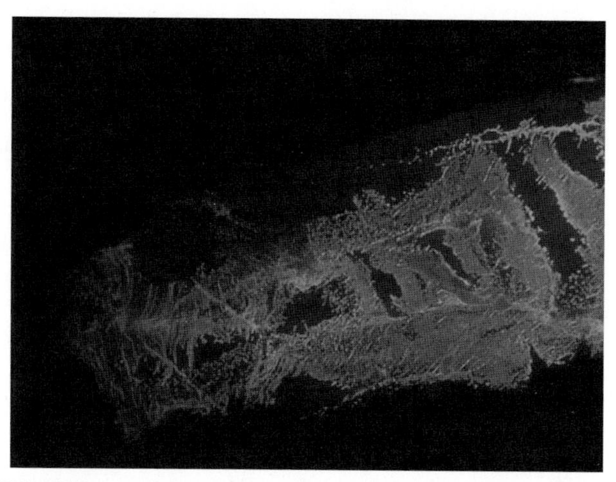

FIGURE 4 Saprophytic growth of luminous bacteria. Luminous bacteria have colonized this slice of fish meat, which was photographed in the dark by the light the bacteria produce. Cells of luminous bacteria injected into the flank muscle tissue of live fish can rapidly reproduce and cause septicemia and death of the fish, indicating the potential of luminous bacteria for pathogenesis.

Shewanella hanedai and some strains of *A. logei*, which are psychrotrophic, grow and produce light at low temperature (e.g., 15 °C), and grow but do not produce light at room temperature (24 °C). Therefore, incubation of platings of environmental samples at the lower temperatures may reveal the presence of other psychrotropic luminous species. Temperature relationships of luminous bacteria appear to be species-specific. For example, *S. woodyi*, characterized from squid ink and seawater in the Alboran Sea near Gibraltar, and *A. fischeri*, species that are closely related to *S. hanedai* and *V. logei*, respectively, grow and produce light at room temperature. Some bacteria appear to be eurythermal, growing and producing light from low to relatively high temperatures.

Freshwater

V. cholerae apparently is the only luminous bacterium to have been isolated from freshwater and brackish estuarine habitats. A luminous non-O1 strain of this species, sometimes called *Vibrio albensis*, was isolated from the Elbe River in Germany in 1893, and non-O1 strains of *V. cholerae* sometimes have been isolated from diseased, glowing crustaceans from freshwater lakes.

Terrestrial

Luminous bacteria are found also in the terrestrial environment, where they infect a variety of terrestrial insects, causing the infected animal to glow. These bacteria presumably are members of *P. luminescens* and *Photorhabdus temperata*, which occur also as the mutualistic symbionts of nematodes, which are common in soil, or *Photorhabdus asymbiotica*, which has been isolated from human clinical infections initiated by spider and insect bites (Table 1).

P. luminescens and *P. temperata* occur as the mutualistic symbionts of entomopathogenic nematodes of the family Heterorhaditidae. They are carried in the intestine of the infective juvenile stage of the nematode, contribute to a lethal infection of insect larvae, and permit completion of the nematode life cycle. When the nematode enters the insect, via the digestive tract or other openings, and penetrates the insect's hemocele, the bacteria are released into the hemolymph, the constituents of which they utilize for growth. The bacteria elaborate a variety of extracellular enzymes that presumably breakdown macromolecules of the hemolymph. Proliferation of the bacteria leads to death of the insect, the carcass of which can become luminous. Through consumption of the bacteria or products of bacterial degradation of the hemolymph, the nematodes develop and sexually reproduce. Completion of the nematode life cycle involves reassociation with the bacteria and the emergence from the insect cadaver of the nonfeeding infective juveniles, the intestines of which are colonized by the bacteria. Cells of *P. luminescens* presumably are present in soil, but association with the nematode may be necessary for their survival and dissemination. Luminescence of the infected insect larva might function to attract nocturnally active animals to feed on the glowing carcass, thereby increasing the opportunities for the bacterium and the nematode to be disseminated. However, luminescence is not required for successful symbiosis with the nematode; not all strains of *P. luminescens* produce luminescence. Furthermore, bacteria in the genus *Xenorhabdus*, which are symbiotic with entomopathogenic nematodes in the family Steinernematidae, are ecologically very similar to *Photorhabdus*, except that they do not produce light. The similarities between the life styles and activities of *Photorhabdus* and *Xenorhabdus* might be a case of ecological convergence.

Bioluminescent Symbiosis

A different kind of symbiotic association is seen with members of several families of marine fishes and squids, which form mutually beneficial associations with luminous bacteria (Figure 5). These associations, of which there are many, are called bioluminescent symbioses. The bacteria are housed extracellularly in the body of the animal in a complex of tissues called a light organ. In fishes, the light organs are outpocketings of the gut tract (most fishes), or are positioned below the eyes (anomalopids), in the lower jaw (monocentrids), or at the tips of tissue extensions (ceratioids). In squids, they are found as bilobed structures within the mantle cavity, associated with the ink sac. Accessory tissues associated with the light organ, that is, lens, reflector, and light-absorbing barriers, direct and focus the light the bacteria produce. The animal uses the bacterial light in various luminescence displays that are associated with sex-specific signaling, predator avoidance, locating or attracting prey, and schooling. In turn, the bacteria, which are present at a high cell density (Figure 6), use nutrients obtained from the host to reproduce and are disseminated from the animal's light organ into the environment.

Five species of marine luminous bacteria have been identified as light-organ symbionts, *A. fischeri* (previously *V. fischeri*), *A. logei* (previously *V. logei*), *Photobacterium leiognathi*, *P. kishitanii*, and *P. mandapamensis*. The families of squids and fishes whose light organs these bacteria colonize are listed in Table 2. *P. phosphoreum* has not been found as a light-organ symbiont, whereas *P. kishitanii*, a newly identified species closely related to *P. phosphoreum*, is found in the light organs of a wide variety of deep, cold-dwelling fishes. Luminous bacterial symbionts of two groups of fishes, the flashlight fish (family Anomalopidae) and deep sea anglerfish (e.g., families Melanocetidae and Ceratiidae in suborder Ceratioidei), have not yet been brought into laboratory culture and therefore remain to be identified; these as-yet uncultured bacteria probably represent new species. In the cases studied, the newly hatched aposymbiotic animals acquire their symbiotic bacteria from the environment with each new host generation.

Based on symbiont acquisition from the environment, the association between the sepiolid squid *E. scolopes* and the *A. fischeri* has emerged as an experimental system with which to examine various aspects of the association. The nascent, rudimentary light organ lobes in aposymbiotic hatchling juvenile squids bear a pair of ciliated epithelial appendages (CEAs) and contain three simple sac-like epithelial tubules embedded in the undifferentiated accessory tissues. The outer portions of these tubules are ciliated and directly connect to the mantle cavity via a lateral pore. Colonization of the epithelial tubules, which is facilitated by ciliary beating of the CEAs, occurs through these pores, which later coalesce, with the formation of a ciliated duct for each light organ lobe. Notably, colonization triggers regression of the CEAs within a few days. Other morphological changes upon colonization include alterations in the epithelial cells of the distal portions of the light organ tubules, which develop a dense microvillous brush border. Nonetheless, the developmental program giving rise to the light organ and accessory tissues runs independently of the presence of the bacteria; these tissues develop normally in aposymbiotic animals, and the light organ can remain receptive to colonization in aposymbiotic animals from hatching to adulthood. At the level of bacterial genes and functions necessary for symbiosis, motility of *A. fischeri*, via polar flagella, is necessary for colonization of the squid light organ. Various other bacterial genes may be involved in the ability of this bacterium to establish bioluminescent symbiosis with *E. scolopes*.

FIGURE 5 Luminous animals. Different fish and squid hosts of light-organ symbiotic luminous bacteria are shown. Counterclockwise from the upper left, the animals are *Monocentris japonicus* (Monocentridae), host of *Aliivibrio fischeri*; *Chlorophthalmus albatrossis* (Chlorophthalmidae), host of *Photobacterium kishitanii*; *Aulotrachichthys prosthemius* (Trachichthyidae) (photo provided by Atsushi Fukui, Tokai University), host of *P. kishitanii*; *Leiognathus splendens* (Leiognathidae), host of *Photobacterium leiognathi*; *Acropoma japonicus* (Acropomatidae) (photo provided by Atsushi Fukui, Tokai University), host of *Photobacterium mandapamensis*; and, *Euprymna scolopes* (Sepiolidae) (dissected to reveal bilobed light organ in this ventral view), host of *A. fischeri*.

Bioluminescent symbiosis is a special class of symbiosis, different in fundamental ways from other kinds of symbiotic associations. In most bacterial mutualisms with animals and plants, the host is dependent nutritionally on its symbiotic bacteria, for bacterial fixation of carbon or nitrogen, the activity of bacterial extracellular degradative enzymes such as cellulases, or bacterial provision of vitamins or other essential nutrients to the host. The absence of symbiotic bacteria consequently can have a profound effect on the growth, development, and survival of the host. In the bioluminescent symbiosis of *A. fischeri* and the sepiolid squid *E. scolopes*, however, animals cultured aposymbiotically from hatching

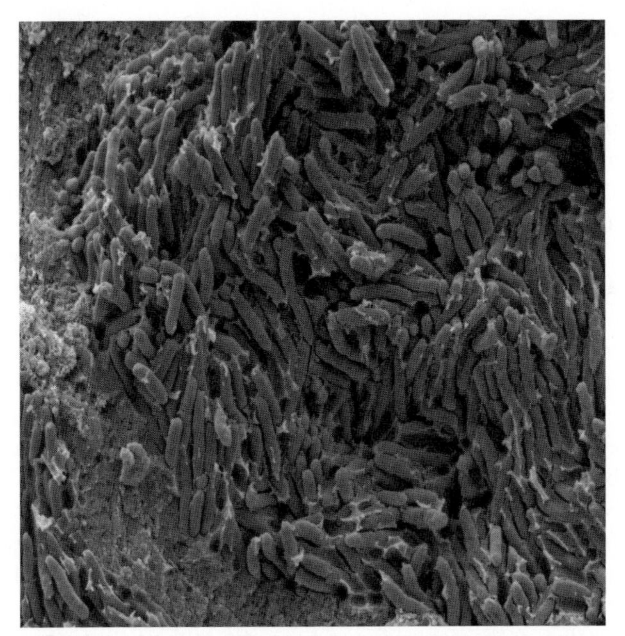

FIGURE 6 Symbiotic light-organ bacteria. This scanning electron micrograph of a section of the light organ of the fish *Siphamia versicolor* (Apogonidae) (micrograph prepared by Sasha Meshinchi, Microscopy and Imaging Laboratory, University of Michigan) shows the exceptionally high density of bacteria that is typical of bioluminescent symbioses.

to adulthood grow, develop, and survive in the laboratory as well as animals colonized by *A. fischeri*. This observation, which indicates that, at least under laboratory conditions, the animal is not dependent on the bacterium for completion of its life cycle, suggests that *A. fischeri* makes no major nutritional contribution to the animal. The metabolic dependency of *E. scolopes* on *A. fischeri* probably is limited to light production, and selection that maintains the association presumably is ecological, not nutritional, at the level of the squid's ability to use light to avoid predators.

Bioluminescent symbioses differ in this and other ways also from endosymbiotic associations, which are mutually obligate relationships in which the symbiotic bacteria are housed intracellularly and are transferred maternally. Symbiotic luminous bacteria are housed extracellularly, and in most cases they are known not to be obligately dependent on the host for their reproduction. Unlike obligate intracellular bacteria, the symbiotic luminous bacteria colonize a variety of other marine habitats, including intestinal tracts, skin, and body fluids of marine animals, sediment, and seawater, where they coexist and compete with many other kinds of bacteria as members of commensal, saprophytic, pathogenic, and free-living bacterial communities. A second major difference with endosymbiotic associations

TABLE 2 Squid and fish families harboring light-organ symbiotic luminous bacteria

Bacterial Species							
Host Family[a]		*Aliivibrio fischeri*	*Aliivibrio logei*	*Photobacterium kishitanii*	*Photobacterium leiognathi*	*Photobacterium mandapamensis*	*Not identified*[b]
Squids							
	Loliginidae				+		
	Sepiolidae	+	+				
Fishes							
	Opisthoproctidae			+			
	Chlorophthalmidae			+			
	Macrouridae	+		+			
	Moridae	+		+		+	
	Melanocetidae						+
	Ceratiidae						+
	Anomalopidae						+
	Trachichthyidae			+			
	Monocentridae	+					
	Acropomatidae			+	+	+	
	Apogonidae					+	
	Leiognathidae				+	+	

[a] *For a listing of the bacteria colonizing individual species of squids and fishes, see Dunlap, P. V., Ast, J. C., Kimura, S., Fukui, A., Yoshino, T., & Endo, H. (2007). Phylogenetic analysis of host-symbiont specificity and codivergence in bioluminescent symbioses. Cladistics 23, 507–523.*
[b] *Not yet in laboratory culture.*

is that symbiotic luminous bacteria are acquired from the environment with each new generation of the host instead of being transferred vertically through the maternal inheritance mechanisms seen for obligate bacterial endosymbionts of terrestrial and marine invertebrates.

Bioluminescent symbioses also lack the strict specificity expected for partners in a mutually obligatory, endosymbiotic association. Two or three different species of bacteria colonize the light organs of members of a single fish or squid family (Table 2), and even light organs of individual fish and squids can contain two bacterial species, a state called bacterial cosymbiosis. Furthermore, the bacteria resident within individual light organs often represent several genetically distinct strain types. The ability of host animals to accept genetically distinct strain types and even different species of luminous bacteria as symbionts suggests that a strict genetically based selection by a host of its specific symbiont probably is not operative in these associations. An alternative explanation for the patterns of symbiont–host affiliation observed in nature is that the species of luminous bacteria most abundant and active where aposymbiotic hatchling animals first encounter bacteria determines which species of luminous bacteria are most likely to initiate the symbiosis. Temperature, for example, influences the presence and relative numbers of the different species of luminous bacteria in the marine environment. Thus, lower temperatures, found in deeper waters, favor the incidence of the more psychrotrophic *A. logei* and *P. kishitanii*. Animals whose eggs hatch out in these waters would be more likely to acquire these luminous species as light-organ symbiont. Conversely, warmer temperatures, found in temperate and tropical coastal waters, favor the incidence of the more mesophilic *A. fischeri*, *P. leiognathi*, and *P. mandapamensis*; animals whose eggs hatch out in these waters would be more likely to encounter and therefore take up these bacteria as light-organ symbionts. The general depth and temperature distributions of bacterially luminous animals in the marine environment and the lack of strict specificity between hosts and symbionts are consistent with this environmental congruence hypothesis. Nonetheless, the apparently complete absence of nonluminous bacteria from light organs indicates some form of selection, possibly for a bacterial activity associated with luminescence.

Another major difference between bioluminescent symbiosis and endosymbiosis is that luminous bacteria and their host animals show no evidence of cospeciation. Endosymbiosis is generally assumed to involve coevolutionary interactions, that is, reciprocal genetic changes in host and symbiont that result from the obligate and mutual dependence of each partner on the other. One manifestation of coevolution is a pattern of codivergence (i.e., cospeciation) in which the evolutionary divergence of the symbiont follows and therefore reflects that of the host. Detailed molecular phylogenies of bacterially luminous fishes and squids, however, are very different from and do not resemble the phylogenies of their symbiotic light-organ bacteria. This lack of host–symbiont phylogenetic congruence demonstrates that the evolutionary divergence of symbiotic luminous bacteria has occurred independently of the evolutionary divergence of their host animals.

Bioluminescent symbioses therefore appear to represent a paradigm of symbiosis that differs fundamentally from associations involving obligate, intracellular, maternally transferred symbionts. While fishes and squids are dependent ecologically on luminous bacteria, the bacteria are not obligately dependent on their bioluminescent hosts. The evolutionary (genetic) adaptations for bioluminescent symbiosis, that is, presence of light organs that can be colonized by luminous bacteria, accessory tissues for controlling, diffusing, and shaping the emission of light, and behaviors associated with light emission, all are borne by the animal. No genetic adaptations have been identified in the bacteria that are necessary for and specific to their existence in light organs compared to the other habitats they colonize. Luminous bacteria therefore seem to be opportunistic colonizers, able to persist in animal light organs as well as in a variety of other habitats to which they are adapted.

REGULATION OF BACTERIAL LUMINESCENCE

Quorum Sensing

Luciferase synthesis and luminescence are regulated in many luminous bacteria. This regulation has been studied in detail in *A. fischeri* and *V. harveyi*. At low population density, these bacteria produce very little luciferase and light, whereas at high population density, luciferase levels are induced 100- to 1000-fold and light levels increase by 10^3- to 10^6-fold. This population density-dependent induction of luciferase synthesis and luminescence is controlled in part by the production and accumulation in the cell's local environment of small secondary metabolite signal molecules, called autoinducers (acyl-homoserine lactones (AHLs) and other low molecular weight compounds), which function via regulatory proteins to activate or derepress transcription of the *lux* operon. Originally called autoinduction, this gene regulatory mechanism is now referred to as quorum sensing to reflect its relationship with population density. As a mechanism by which a bacterium can detect its local population density, quorum sensing might function as a diffusion sensor, mediating whether cells produce extracellular enzymes and other factors for obtaining nutrients, or as a sensor of host association.

Quorum-sensing control of luminescence in *A. fischeri* and *V. harveyi* involves complex regulatory circuits. In *A. fischeri*, quorum-sensing regulatory genes, *luxR* and

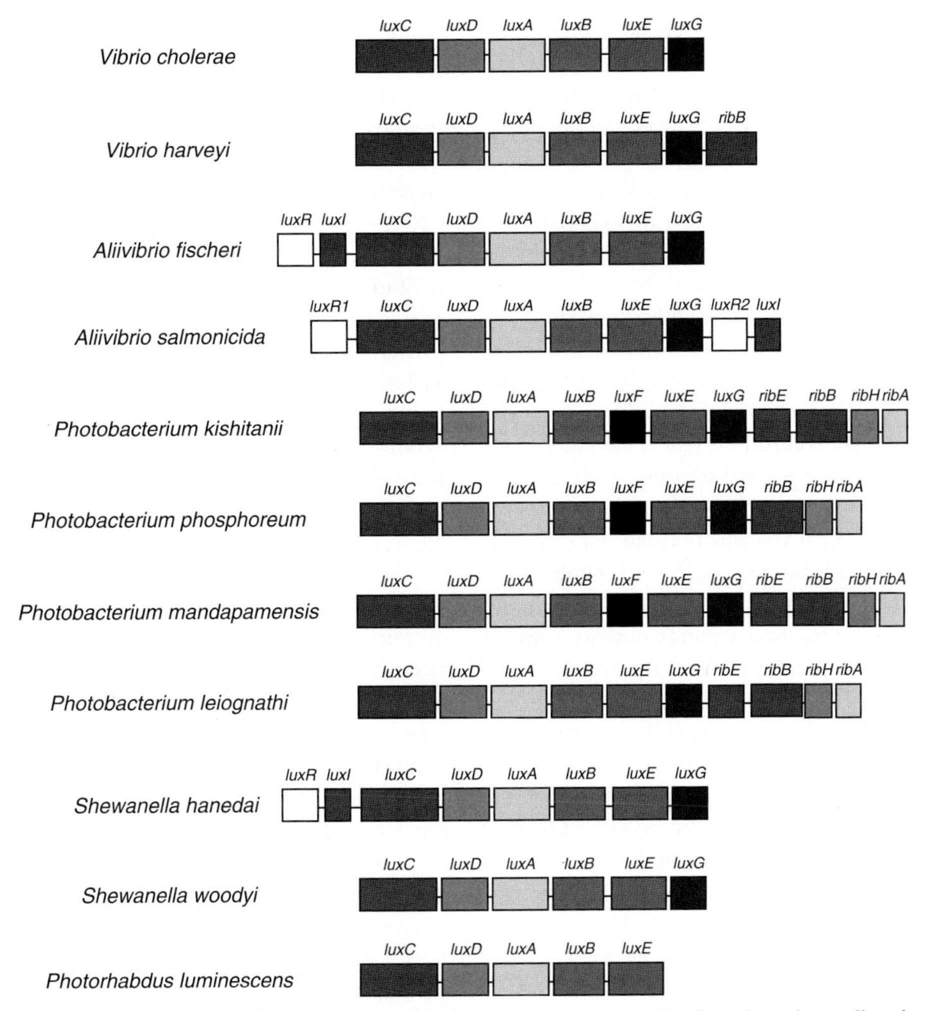

FIGURE 7 The *lux* genes of luminous bacteria. Contiguous genes of the luminescence operons of luminous bacteria are aligned to highlight commonalities and differences (Dr. Henryk Urbanczyk and Dr. Jennifer Ast, University of Michigan, assisted in the preparation of this figure). Note that the *lux* operon in *Photobacterium*, referred to as the *lux-rib* operon, contains the *ribEBHA* genes, which are involved in synthesis of riboflavin. Many strains of *Photobacterium leiognathi* carry multiple, phylogenetically distinct *lux-rib* operons.

luxI, are associated with the *lux* operon (Figure 7). The *luxR* gene, which is upstream of the *lux* operon and divergently transcribed from it, encodes a transcriptional activator protein LuxR, which binds autoinducer *N*-3-oxo-hexanoyl-homoserine lactone (3-oxo-C6-HSL), forming a complex that binds at a site in the *lux* operon promoter and facilitates association of RNA polymerase with the *lux* operon promoter, thereby activating transcription of the genes for light production. The other regulatory gene *luxI* is the first gene of the *lux* operon in this species, *luxICDABEG*; it encodes an AHL synthase necessary for synthesis of 3-oxo-C6-HSL. Between *luxR* and *luxI* is a regulatory region containing the *luxR* and *lux* operon promoters. According to a simple model for luminescence induction in *A. fischeri*, cells produce a low level of 3-oxo-C6-HSL, which, as a membrane-permeant molecule, diffuses out of the cells and away into the environment. In seawater, for example, where the number of *A. fischeri* cells is low, 3-oxo-C6-HSL would not accumulate, and transcription of the *lux* operon would remain uninduced. However, under conditions where cells can attain a high population density, such as in gut tracts of fishes, light organs of fish or squid, or in the laboratory in batch culture, the local concentration of 3-oxo-C6-HSL can build up, both outside and inside cells. Once 3-oxo-C6-HSL reaches a critical concentration, it then interacts inside the cell with LuxR protein, leading to activation of *lux* operon transcription. Because *luxI* is a gene of the *lux* operon, increased transcription leads to increased synthesis of 3-oxo-C6-HSL in a positive feedback manner. The result is a rapid and strong induction of luciferase synthesis and luminescence.

Many other regulatory components contribute to quorum sensing in *A. fischeri*. These include GroEL, which is necessary for production of active LuxR; 3′:5′-cyclic AMP (cAMP) and cAMP receptor protein (CRP), which activate transcription of *luxR* and thereby potentiate the cell's response to 3-oxo-C6-HSL; a second autoinducer, octanoyl-HSL, which is dependent on the *ainS* gene for its synthesis and which interacts with LuxR apparently to delay *lux* operon induction; and several proteins homologous to components of the phosphorelay signal transduction system that controls luminescence in *V. harveyi* (see below). In *A. fischeri*, LuxR and 3-oxo-C6-HSL also control several genes unrelated to luminescence, forming a quorum-sensing regulon in this species.

In *V. harveyi*, the quorum-sensing regulatory mechanism differs substantially from that of *A. fischeri*. A transcriptional activator, called LuxR (which is not homologous to *A. fischeri* LuxR), is necessary for *lux* operon transcription, but the regulatory genes controlling that transcription are not contiguous with the *lux* operon, for example. Expression of the *lux* operon in *V. harveyi* is regulated by a quorum-sensing phosphorelay signal transduction mechanism. The mechanism involves two separate two-component phosphorelay paths, each involving a transmembrane sensor/kinase, LuxN and LuxQ, responsive to a separate quorum-sensing signal. The *luxLM* genes are necessary for synthesis of the *N*-3-hydroxy-butanoyl-HSL (3-OH-C4-HSL) signal. In the absence of 3-OH-C4-HSL, LuxN operates as a kinase, phosphorylating LuxU, a signal integrator, which in turn passes the phosphate on to LuxO, which in phosphorylated form is a repressor of the *lux* operon. In the presence of 3-OH-C4-HSL, the activity of LuxN is shifted from kinase to phosphatase, which draws phosphate from LuxU and thereby from LuxO, which then no longer represses *lux* operon expression. A similar activity is carried out by a second signal, AI-2, identified as furanone borate diester, which requires LuxS for its production. AI-2 operates via LuxP, a putative periplasmic protein, to mediate the kinase/phosphatase activity of LuxQ, which in turn, like LuxN, feeds phosphate to or draws it from LuxO. Previously thought to directly repress *lux* operon expression, LuxO may operate indirectly, by controlling a negative regulator of luminescence. Expression of *luxO* itself is subject to repression by LuxT. In a manner possibly analogous to LuxR in *V. fischeri*, LuxR in *V. harveyi* is autoregulatory and responsive to 3-OH-C4-HSL.

Despite the major differences in the quorum-sensing systems of *A. fischeri* and *V. harveyi*, there are several commonalities. These include homologs in *A. fischeri* of the *V. harveyi luxR* gene (*litR*), *luxO*, *luxU*, and *luxM* (*ainS*), among others. These homologies indicate that a phosphorelay system is likely to be a part of the quorum-sensing system of *A. fischeri*.

Physiological Control

Induction of luciferase synthesis and luminescence are also influenced by physiological factors. Oxygen, amino acids, glucose, iron, and osmolarity have distinct effects, depending on the species studied. Those factors that stimulate growth rate of the bacterium, such as readily metabolized carbohydrates, tend to decrease light production and luciferase synthesis. They do so presumably by causing oxygen and reducing power ($FMNH_2$) to be directed away from luciferase and by lowering cellular levels of cAMP and CRP, which are needed to activate *lux* gene expression. In *A. fischeri*, the *lux* regulatory region between *luxR* and *luxICDABEG* contains a cAMP-CRP binding site, and in *V. harveyi*, a cAMP-CRP binding site is upstream of *luxCDABEGH*. Conversely, factors that restrict growth rate, such as limitation for iron and either high or low salt concentrations, depending on the species, tend to stimulate the synthesis and activity of luciferase. Much remains to be learned about how these factors operate and the relationship between growth physiology of the cell and regulatory elements controlling *lux* gene expression.

Over the past 25 years, there has been a rapid accumulation of information on how cells regulate luminescence by quorum sensing. Based on these studies, quorum-sensing systems biochemically and genetically homologous to that in luminous bacteria have also been identified in a wide variety of nonluminous bacteria in which quorum sensing controls many cellular activities other than light production, particularly the production of extracellular enzymes and other extracellular factors thought to be useful for bacteria at high population density and in host association. More details on quorum sensing in luminous and nonluminous bacteria are provided in Quorum-Sensing in Bacteria.

FUNCTIONS OF LUMINESCENCE IN BACTERIA

The production of light consumes a substantial amount of energy, through the synthesis of Lux proteins and their enzymatic activity. This energetic cost, which may account for the fact that luminescence is regulated, suggests that activity of the luminescence system plays an important role in the physiology of luminous bacteria. Most attention to what that role might be has focused on oxygen. One possibility is that the light-emitting reaction arose evolutionarily as a detoxification mechanism, removing oxygen and thereby allowing otherwise anaerobic organisms to survive. A related possibility is that luciferase, as an oxidase, might function like a secondary respiratory chain that is active when oxygen or iron levels are too low for the cytoplasmic membrane-associated electron transport system to operate. This activity would allow cells expressing luciferase to reoxidize reduced coenzyme even when oxygen levels are low. Reoxidation of reduced coenzyme would permit cells of luminous bacteria in low oxygen habitats, such as in animal gut tracts, to continue to transport and metabolize growth substrates, gaining energy through substrate-level phosphorylation. Another possibility is that light production could facilitate dissemination of luminous bacteria. The feeding of animals on luminous particles (decaying tissues, fecal pellets, and moribund animals infected by luminous bacteria) to which they are attracted would bring the bacteria into the animal's nutrient-rich gut tract for reproduction and dispersal. Other hypotheses for the function of luminescence in bacteria have been put forward, and future studies may provide support for one or more of these hypotheses. It is by no means clear yet what actual benefits, physiological or ecological, accrue to luminous bacteria that lead them to retain and express an energetically expensive enzyme system.

EVOLUTION OF THE BACTERIAL LUMINESCENCE SYSTEM

The natural presence of genes necessary for producing light defines the luminous bacteria. The necessary genes *luxA* and *luxB* encoding the luciferase subunits *luxC*, *luxD*, and *luxE* for the fatty acid reductase subunits as well as *luxG* encoding a flavin reductase are consistently found together as a cotranscribed unit *luxCDABEG*. The reason for this conservation of *lux* genes as a unit is not known, but it might relate to efficient light production; the contiguous presence of these genes as an operon might help promote the coordinated production of luciferase and substrates for luciferase, long-chain aldehyde, and reduced flavin mononucleotide ($FMNH_2$). The conservation of these genes as a unit in nearly all luminous bacteria examined (Figure 7) suggests that the *lux* operon arose in the distant past evolutionarily. Supporting this view, phylogenetic analysis demonstrates that the individual *lux* genes of different bacterial species are homologous, as was suggested by the high levels of amino acid sequence identities of the inferred Lux proteins. This homology implies that the bacterial *luxCDABEG* genes arose one time in the evolutionary past. The use by luciferase of oxygen as a substrate implies that this enzymatic activity originated after oxygenic photosynthesis by ancestors of modern-day cyanobacteria began to increase the level of O_2 on Earth, approximately 2.4 billion years ago, during the Great Oxidation Event. A marine origin for bacterial luminescence seems likely because most species of luminous bacteria are marine (Table 1).

Evolutionary Origin of Bacterial Luciferase

The origins of the individual genes of the *lux* operon remain obscure. Luciferase, however, might have arisen from a primitive flavoprotein, a flavin-dependent, aldehyde-oxidizing protoluciferase that incidentally produced a small amount of light. If the level of light produced by this protoluciferase was sufficient to be detected by phototactic multicellular animals, they might have been attracted to luminous particles containing these early bioluminescent bacteria and may have fed on them, introducing the bacteria into the animal's nutrient-rich digestive tract. This interaction might thereby have enhanced the reproduction of these bacteria and led to selection for more intense light output. The evolutionary steps leading to protoluciferase may also have involved selection for oxygen detoxification activity that permitted early protobioluminescent anaerobic bacteria to survive an increasingly aerobic environment before the evolution of cytochrome-dependent respiration. Bacterial protoluciferase, either a monomer or a homodimer, might have been encoded by a single gene, an early form of *luxA*. Individual α- and β-subunits of luciferase, however, are unable to produce high levels of light *in vitro* or *in vivo*. Alternatively, a light-emitting bacterial protoluciferase might have arisen following gene duplication of the primitive *luxA* gene that is thought to have given rise to *luxB* (see below).

Most luminous bacteria are members of *Vibrionaceae*, which suggests that the bacterial luminescence system arose in the ancestor of this family. Recent identifications of new luminous species and of luminous strains of species not previously known to be luminous support this view. Assuming vertical inheritance of the *lux* genes from that ancestor, through its descendents to modern-day bacteria, gene phylogenies for luminous bacteria (Figure 2) imply a complex history of *lux* gene duplication, gene recruitment, and gene loss within this family. Phylogenetic analyses also suggest that the *lux* genes were acquired by the luminous species of *Shewanellaceae* and *Enterobacteriaceae* from a member or members of *Vibrionaceae*.

Gene Duplication

Based on amino acid sequence identities, a tandem duplication of the ancestral *luxA* gene, followed by sequence divergence, is thought to have given rise to *luxB*, leading to formation of the heterodimeric luciferase present in modern-day luminous bacteria. Similarly, a tandem duplication of *luxB* is thought to have given rise to *luxF*, which encodes a nonfluorescent flavoprotein; *luxF* is present in the *lux* operons, *luxCDABFEG*, of three of the four luminous *Photobacterium* species (Figure 7).

Gene Loss

Most species of *Vibrionaceae* lack the *lux* genes and therefore are nonluminous. Also, most strains of some luminous species, such as *V. cholerae*, are nonluminous. The low incidence of luminous species in the family suggests that the *lux* genes have been lost over evolutionary time from many of the lineages that have given rise to extant species. It also seems likely that nonluminous variants of luminous species can arise frequently through loss of the *lux* operon. The scattered incidence of *lux* genes in *Vibrionaceae* presumably relates to different ecologies of the different species.

It is not clear, however, how having and expressing *lux* genes contributes to the life style of luminous bacteria, because there are no obvious ecological differences between luminous and nonluminous species except in the case of those species that are light-organ symbionts.

With respect to the loss of individual *lux* genes, *P. leiognathi* strains lack *luxF*, a gene that is present in the *lux* operons of other luminous *Photobacterium* species (Figure 7). Presumably, therefore, *luxF*, possibly acquired in the lineage leading to *Photobacterium* through duplication and sequence divergence of *luxB*, was lost from the lineage that gave rise to *P. leiognathi*. This loss might reflect the evolutionary divergence of *P. leiognathi* from other luminous species of *Photobacterium*. Also, some strains of *P. mandapamensis* bear nonsense mutations in *luxF*, further evidence that this gene does not play an essential role in the biology of *Photobacterium* species. Mutations in *luxF* in *P. mandapamensis* might set the stage for loss of this gene from the *lux* operon. Similarly, the *lux* operons of *P. phosphoreum* strains lack *ribE*, one of the genes "recruited" to the *lux* operon in *Photobacterium*.

Recruitment of Genes to the *lux* Operon

Linked to the luminescence genes in *Photobacterium* species, and apparently cotranscribed with them, are genes involved in synthesis of riboflavin, forming an operon of 10 or 11 genes, *luxCDAB(F)EG-ribEBHA*, referred to as the *lux-rib* operon (Figure 7). The presence of the *ribEBHA* genes (just *ribBHA* in *P. phosphoreum*) appears to be a case of gene recruitment to the *lux* operon of *Photobacterium*, because these genes, with the exception in *V. harveyi* of *ribB* (referred to originally as *luxH*), are not contiguous with the *lux* operons of other species of luminous bacteria. Again, one can invoke the notion that the presence of genes for synthesis of riboflavin, a major component of FMN, as part of the *lux* operon might facilitate light production by ensuring coordinate synthesis of luciferase and of substrates for the enzyme. In this regard, it is interesting to note that in *A. fischeri*, *ribB*, which is not linked to the *lux* operon, is controlled by the LuxR/AHL-quorum sensing system that controls *lux* operon expression.

A second example of apparent gene recruitment to the *lux* operon is the presence of regulatory genes *luxI* and *luxR*, which control *lux* operon transcription, in *A. fischeri* and *A. salmonicida*. In *A. fischeri*, the *luxI* gene is part of the *lux* operon, whereas *luxR* is upstream and divergently transcribed. In *A. salmonicida*, the arrangement is somewhat different, with two *luxR* genes flanking the *lux* operon and with *luxI* following the downstream *luxR* gene (Figure 7). In other species, the *lux* regulatory genes either have not been identified or, as in the case of a different (not homologous) *luxR* gene in *V. harveyi*, they are not linked to the *lux* operon. The grouping of regulatory genes with the *lux* operon in *Aliivibrio* might ensure a tight regulation of luminescence under quorum sensing control.

Horizontal Transfer of the *lux* Operon

In addition to gene duplication, loss, and recruitment, evidence is accumulating that *lux* genes have been acquired in some bacteria by horizontal gene transfer. Two species of *Shewanella* are luminous and carry *lux* operons similar to that of *A. fischeri*, suggesting acquisition from *A. fischeri* or an ancestor of this bacterium. Support for the notion of horizontal transfer of the *lux* operon from *A. fischeri* to luminous *Shewanella* species is seen in the presence of *luxR* and *luxI* genes in association with the *lux* operon of *S. hanedai* and in the same gene order as in *A. fischeri* (Figure 7). Recent evidence indicates that three *Vibrio* species, namely, *V. chagasii*, *V. damselae*, and *V. vulnificus*, acquired their *lux* gene by horizontal transfer. The situation for *Photorhabdus* is not yet clear, but an early transfer of *lux* genes from a member of *Vibrionaceae* might have occurred. These considerations suggest that the *lux* genes may have arisen within the lineage leading to modern-day members of *Vibrionaceae* and were then lost from several descendents, retained by some, and transferred relatively recently from a member or members of *Vibrionaceae* to *Photorhabdus* and *S. hanedai* and *S. woodyi*.

Natural Merodiploidy of the *lux-rib* Operon

An intriguing wrinkle in the evolutionary dynamics of the *lux* operon is that many strains of *P. leiognathi* isolated from nature carry two intact and apparently functional *luxCDABEG-ribEBHA* operons. This situation represents an unusual case of natural merodiploidy in bacteria, that is, the presence of two or more copies of the same gene or genes in the genome of a bacterium, because of the large number of genes involved and because the second operon did not arise by tandem duplication of the first. The two *lux-rib* operons are distinct in sequence and chromosomal location. One operon is in the ancestral chromosomal location of the *lux-rib* operon in *P. leiognathi* and related bacteria. The other is located elsewhere on the chromosome and is present in many but not all strains of *P. leiognathi*; it is flanked by genes specifying transposases, which suggests it can transfer between strains. Phylogenetic analysis indicates that the two *lux-rib* operons are more closely related to each other than either is to the *lux* and *rib* genes of other bacterial species. This finding rules out interspecies horizontal transfer as the origin of the second *lux-rib* operon in *P. leiognathi*; instead, the second operon apparently arose in the distant past within, and was acquired by transposon-mediated transfer from, a lineage of *P. leiognathi* that either has not yet been sampled or has gone extinct. Merodiploidy of the *lux-rib* operon in *P. leiognathi* also is the first instance of merodiploid strains of a bacterium having a nonrandom geographic distribution; strains bearing a single *lux-rib*

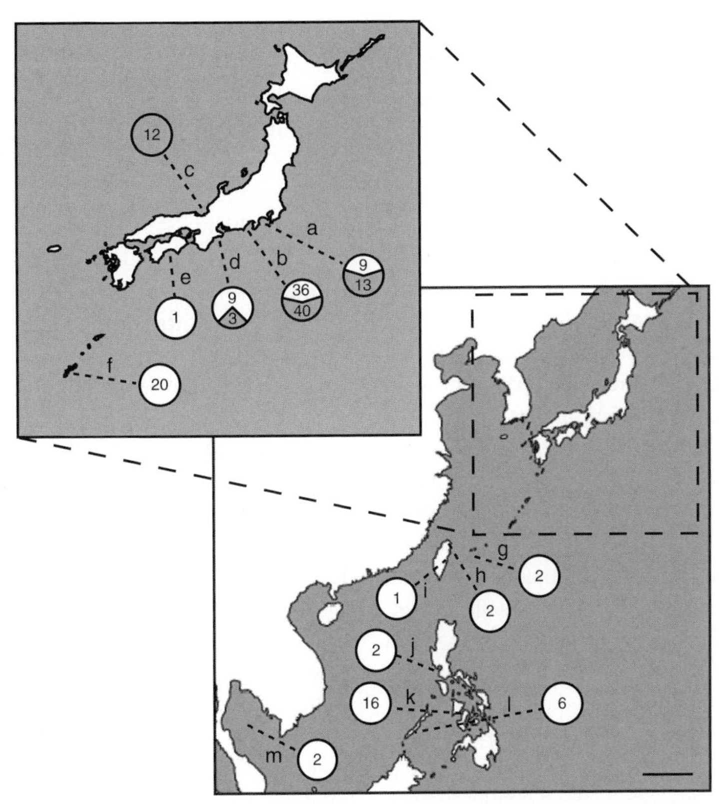

FIGURE 8 Nonrandom geographic distribution of *lux-rib* merodiploid strains of *P. leiognathi*. The numbers next to each location indicate the number of strains identified as bearing a single *lux-rib* operon (white area in circle) or multiple (gray area in circle) *lux-rib* operons. The insert shows an enlarged map of the main islands of Japan, with some landmasses omitted for clarity. The scale bar is approximately 500 km. *Reproduced from Ast, J. C., Urbanczyk, H., & Dunlap, P. V. (2007). Natural merodiploidy of the* lux-rib *operon of* Photobacterium leiognathi *from coastal waters of Honshu, Japan.* Journal of Bacteriology, 189, *6148–6158.*

operon are found over a wide geographic range, whereas *lux-rib* merodiploid strains have been found only in coastal waters of Honshu, Japan (Figure 8). The presence of multiple copies of each of the *lux* and *rib* genes might provide opportunities for sequence divergence and selection that could lead to the evolution of new gene functions in one or the other of the duplicate genes.

ISOLATION, STORAGE, AND IDENTIFICATION OF LUMINOUS BACTERIA

When working with luminous bacteria, and particularly when isolating new strains from nature, the possibility that bacteria could be pathogenic should always be kept in mind and appropriate care to avoid infection should always be used.

Luminous bacteria can be isolated from most marine environments, and two methods, namely, direct plating of seawater and enrichment from marine fish skin, are effective and simple for this purpose. An easily prepared complete medium that is suitable for growing all known luminous bacteria contains natural or artificial seawater, diluted to 70% of full strength, 10 g l^{-1} tryptone or peptone, and 5 g l^{-1} yeast extract, with 1.5 g l^{-1} agar for solid medium. Sugars and sugar alcohols (i.e., glycerol) are unnecessary and can lead to acid production and death of cultures; their use in isolation media should be avoided. For isolations from environments where high numbers of bacteria that form spreading colonies may be present, such as coastal seawater, sediment, and intestinal tracts of marine animals, the use of agar at 4% (40 g l^{-1}) is recommended. This harder, less moist agar limits the ability of bacteria motile on solid surfaces, for example, certain peritrichously flagellated bacteria and bacteria that move by gliding motility, to spread over the plate and cause cross contamination of colonies.

Directly plating of seawater involves simply spreading an appropriate volume, typically 10–100 μl for coastal seawater, of the sample on one or more plates and incubating at room temperature or, preferably, cooler temperatures, such as 15–20 °C. For open ocean seawater and other samples with a lower number of bacteria, larger volumes, for example, 100 ml to 1 l, can be filtered through membrane filters with a pore size of 0.2 or 0.45 μm to collect the bacteria. The filters are then placed, bacteria side up, on plates of the above medium. Once colonies have arisen, usually within 18–24 h at room temperature, the plates can be examined in a dark room. Luminous colonies can then be picked (sterile wooden toothpicks are suitable for this purpose) and streaked for isolation on fresh plates of the same medium. Use of a red light, such as a photographic darkroom light, with variable intensity control can make the picking of luminous colonies easier; by adjusting the red light, colonies of nonluminous bacteria can be made to appear reddish, whereas luminous colonies are blue due to their luminescence. Samples collected from warm waters and incubated at room temperature are more likely to yield *V. harveyi* and related luminous *Vibrio* species, as well as *A. fischeri*, *P. leiognathi*, and *P. mandapamensis*; whereas cold seawater samples plated and incubated at lower temperatures are likely to yield *A. logei*, *P. kishitanii*, *P. phosphoreum*, and *S. hanedai*. It should be noted that some strains of *A. logei* and *S. hanedai* grow well but do not produce light at room temperature; attempts to isolate these and other psychrotrophic bacteria should be carried out at 15 °C.

Enrichment from fish (or squid) can be made using fresh samples and sterile seawater or frozen samples with natural, unsterilized seawater. The tissue, preferably with the skin on, is placed in a tray, skin up, covered halfway with seawater, incubated, and observed daily in the dark for luminous spots, which arise in one to a few days. These spots, colonies of luminous bacteria, can then be picked and streaked for isolation on the medium described above containing 4% agar. From fish and squid, a variety of different species of luminous bacteria can be isolated, especially when different incubation temperatures, such as 4, 15, and 22 °C, are used.

Storage at ultralow temperature, for example, from −75 to −80 °C, in a suitable cryoprotective medium is effective for all luminous bacteria. Cryopreservation of luminous bacteria is recommended to ensure their survival and to avoid the formation of dim and dark variants and the occurrence of other genetic changes. An effective cryoprotective medium for luminous bacteria is filter-sterilized deep-freeze medium (2 × DFM), prepared with 1% w/v yeast extract, 10% dimethylsulfoxide (DMSO), 10% glycerol and 0.2M K_2HPO_4/NaH_2PO_4 (pH 7.0).

Phenotypic and genotypic traits were used in the past for identification of luminous bacteria and descriptions of new species. These methods remain useful, both for practical provisional identifications of new strains and as a complement to the construction of a DNA sequence-based molecular phylogenies for identification. With the advent of rapid, inexpensive DNA sequencing as well as the availability of many sets of primers for various genes whose sequences are useful in bacterial species resolution and an expanded database of sequences for comparisons (e.g., GenBank), a DNA sequence-based approach to identification has become cost effective, rapid, and highly accurate. For luminous bacteria, phylogenetic analysis of *lux* and *rib* genes, together with housekeeping genes such as the 16S rRNA gene, *gyrB*, and *pyrH*, for example, can quickly and accurately place a strain in a species or indicate the possibility that it may be new.

Complete characterization of a new species of luminous bacteria should include a multigene phylogenetic analysis together with examination of biochemical and morphological traits, DNA hybridization analysis, determination of the mol% G + C ratio, fatty acid profile analysis, and comparative genomic analysis such as amplified fragment length polymorphism (AFLP) or repetitive extragenic polymorphic PCR (rep-PCR). The examination of multiple independent isolates of the new entity as well as inclusion in the analysis of the type strains of all closely related species is critically important for accurate and definitive work. Increasingly, multigene phylogenetic analysis is becoming a standard for identification and characterization of luminous bacteria.

CONCLUSIONS

Knowledge of bioluminescent bacteria has increased greatly in recent years through examination of their evolutionary relationships and symbioses and through the identification of new species and strains. Many new species of light-emitting bacteria will likely be identified in future studies as new habitats are examined and as molecular phylogenetic criteria are used to discriminate among closely related species. The ability to distinguish among luminous bacteria using phylogenetic criteria opens up for detailed analysis questions of their habitat specificity, biogeography, and host specificity. Information gained from analysis of the *lux* operons of newly recognized luminous bacteria will likely provide further insight into the evolutionary dynamics of the *lux* operon and its horizontal transfer gene among members of *Vibrionaceae* and to members of other bacterial groups.

RECENT DEVELOPMENTS

Recent studies have contributed substantial new knowledge on the systematics and ecology of bioluminescent bacteria, their symbiotic interactions with squids and fish, and the incidence of horizontal gene transfer. These recent studies are summarized briefly here.

New Species of Luminous Bacteria

Several new species of luminous bacteria, members of *Aliivibrio*, *Photobacterium*, and *Vibrio*, have been identified recently. Currently (as of May, 2010), there are 25 species of marine, brackish water, and terrestrial bacteria that are luminous or have luminous strains (Table 1). Effective species descriptions are increasingly relying on approaches that combine traditional taxonomic criteria – that is, biochemical, morphological, genetic, and ecological traits – with sequence-based phylogenetic analysis using multiple unlinked and functionally independent loci. Additional species will likely be described in the near future, because several luminous strains are known that do not fit into any currently accepted species definition.

Bioluminescent Symbiosis

Strains of luminous bacteria isolated from seawater and light organs of sepiolid squids and identified as *Aliivibrio logei* (previously *Vibrio logei*) based on limited taxonomic criteria have been found by multilocus phylogenetic analysis instead to be members of *A. fischeri*, *A. wodanis*, and a new clade *A. "thorii"* (updated Table 1). At this time, no bona-fide strain of *A. logei* has been isolated from bioluminescent symbiosis. The currently known host animal–bacterial species affiliations in bioluminescent symbiosis are shown in the updated Table 2. There are over 500 species of fish and squid, in 9 orders and 23 families, that form bioluminescent symbiosis with luminous marine bacteria.

Much progress on symbiont–host interactions continues to be made through studies of the bioluminescent symbiosis of *A. fischeri* and the sepiolid squid *Euprymna scolopes*. With respect to the specificity of bacterial colonization of the squid light organ, the gene coding for RcsS, a two-component sensor kinase, is both necessary and sufficient for efficient colonization, through its control over biofilm formation. A fish-symbiotic strain that naturally lacks this gene does not colonize the squid light organ well but gains this capability upon acquisition and expression of the gene (Mandel et al., 2009). Transcriptome analysis of both the squid and the bacterium indicates substantial diurnal fluctuations in gene expression which correlate with the daily expulsion of most the bacteria from the light organ and regrowth of the population from the residual bacterial cells (Wier et al., 2010).

For the squid *E. scolopes*, which undergoes direct development and hatches from eggs as juveniles, light organs are immediately receptive to bacterial colonization by *A. fischeri* upon hatching (Ruby, 1996). Recent studies of fish, however, which undergo indirect development through post-embryonic and larval stages, reveal that development of the light organ begins several days after eggs hatch, and bacterial colonization apparently requires several additional days of larval development (Dunlap et al., 2009). The developmental and colonization delays in fish may provide time for migration of larvae to locations where the symbiotic bacteria are abundant and for development of the vertebrate immune system so that the fish can cope with the bacterial colonization of its nascent light organ.

Horizontal Transfer of the *lux* Operon

In addition to acquisition of luminescence genes by *Shewanella hanedai*, *S. woodyi*, and possibly by members of *Photorhabdus*, luminous strains of two otherwise nonluminous species, namely, *Vibrio chagasii* and *Photobacterium damselae*, have been isolated from nature. These strains apparently obtained their luminescence genes by horizontal transfer from *V. harveyi*. Furthermore, a strain of the luminous species *P. mandapamensis* has been found to have two copies of the *lux-rib* operon, with second *lux-rib* operon apparently obtained by horizontal transfer from *P. leiognathi*. Although these instances of horizontal gene transfer are distinctive, they are the only known instances among several hundred luminous strains examined for horizontal acquisition of the *lux* genes. Therefore, horizontal transfer of the *lux* genes, with stable retention and expression of the transferred genes by the recipient, appears to be an uncommon event.

FURTHER READING

Ast, J. C., & Dunlap, P. V. (2005). Phylogenetic resolution and habitat specificity of the *Photobacterium phosphoreum* species group. *Environmental Microbiology*, *7*, 1641–1654.

Ast, J. C., Urbanczyk, H., & Dunlap, P. V. (2007). Natural merodiploidy of the *lux-rib* operon of *Photobacterium leiognathi* from coastal waters of Honshu, Japan. *Journal of Bacteriology*, *189*, 6148–6158.

Ast, J. C., Urbanczyk, H., & Dunlap, P. V. (2009). Multi-gene analysis reveals previously unrecognized phylogenetic diversity in *Aliivibrio* (*Gammaproteobacteria*: *Vibrionaceae*). *Systematic and Applied Microbiology*, *32*, 379–386.

Baumann, P., & Baumann, L. (1981). The marine gram-negative eubacteria: Genera *Photobacterium, Beneckea, Alteromonas, Pseudomonas*, and *Alcaligenes*. In M. P. Starr, H. Stolp, H. G. Trüper, A. Balows, & H. G. Schlegel (Eds.) *The prokaryotes. A handbook on habitats, isolation, and identification of bacteria* (pp. 1302–1331). Berlin: Springer-Verlag.

Buchner, P. (1965). *Endosymbiosis of animals with plant microorganisms*. New York: Wiley Interscience.

Dunlap, P. V., Ast, J. C., Kimura, S., Fukui, A., Yoshino, T., & Endo, H. (2007). Phylogenetic analysis of host-symbiont specificity and codivergence in bioluminescent symbioses. *Cladistics*, *23*, 507–523.

Dunlap, P. V., & Kita-Tsukamoto, K. (2006). Luminous bacteria. In M. Dworkin, S. Falkow, E. Rosenberg, K. H. Schleifer, & E. Stackebrandt (Eds.) *The prokaryotes. A handbook on the biology of bacteria* (3rd ed.) (pp. 863–892). *Ecophysiology and Biochemistry* (Vol. 2). New York, NY: Springer.

Dunlap, P. V., Kojima, Y., Nakamura, S., & Nakamura, M. (2009). Inception of formation and early morphogenesis of the bacterial light organ of the sea urchin cardinalfish, *Siphamia versicolor* (Perciformes: Apogonidae). *Marine Biology*, *156*, 2011–2020.

Harvey, E. N. (1952). *Bioluminescence*. New York: Academic Press.

Hastings, J. W. (1995). Bioluminescence. In N. Sperelakis (Ed.) *Cell physiology source book* (pp. 665–681). New York: Academic Press.

Hastings, J. W. (2000). Microbial bioluminescence. In J. Lederberg (Ed.) (2nd ed.) (pp. 520–529). *Encyclopedia of Microbiology* (Vol. 1). New York:Academic Press.

Hastings, J. W., & Nealson, K. H. (1981). The symbiotic luminous bacteria. In M. P. Starr, H. Stolp, H. G. Trüper, A. Balows, & H. G. Schlegel (Eds.) *The prokaryotes. A handbook on habitats, isolation, and identification of bacteria* (pp. 1332–1345). Berlin: Springer-Verlag.

Herring, P. J., & Morin, J. G. (1978). Bioluminescence in fishes. In P. J. Herring (Ed.) *Bioluminescence in action* (pp. 273–329). London: Academic Press.

Mandel, M. J., Wollenberg, M. S., Stabb, E. V., Visick, K. L., & Ruby, E. G. (2009). A single regulatory gene is sufficient to alter bacterial host range. *Nature*, *458*, 215–218.

McFall-Ngai, M. J., & Ruby, E. G. (1991). Symbiont recognition and subsequent morphogenesis as early events in animal-bacterial mutualism. *Science*, *254*, 1491–1494.

Nealson, K. H., & Hastings, J. W. (1992). The luminous bacteria. In A. Balows, H. G. Trüper, M. Dworkin, W. Harder, & K. H. Schleifer (Eds.) *The Prokaryotes*. (2nd ed.) (pp. 625–639). Berlin: Springer-Verlag.

Ruby, E. G. (1996). Lessons from a cooperative bacterial-animal association: The *Vibrio fischeri – Euprymna scolopes* light organ symbiosis. *Annual Review of Microbiology*, *50*, 591–624.

Urbanczyk, H., Ast, J. C., Kaeding, A. J., Oliver, J. D., & Dunlap, P. V. (2008). Phylogenetic analysis of the incidence of *lux* gene horizontal transfer in *Vibrionaceae*. *Journal of Bacteriology*, *190*, 3494–3504.

Urbanczyk, H., Ast, J. C., & Dunlap, P. V. (2010). Phylogeny, genomics, and symbiosis of *Photobacterium*. *FEMS Microbiology Reviews 35*, 324–342.

Wassink, E. C. (1978). Luminescence in fungi. In P. J. Herring (Ed.), *Bioluminescence in action* (pp. 171–197). London: Academic Press.

Wier, A. M., Nyholm, S. V., Mandel, M. J., Massengo-Tiassé, R. P., Schaefer, A. L., Koroleva, I. et al. (2010). Transcriptional patterns in both host and bacterium underlie a daily rhythm of anatomical and metabolic change in a beneficial symbiosis. *Proceedings of the National Academy of Science*, *107*, 2259–2264.

Chapter 18

Magnetotaxis

U. Lins[1] and D.A. Bazylinski[2]

[1]*Universidade Federal do Rio de Janeiro, Rio de Janeiro, Brazil*

[2]*University of Nevada at Las Vegas, Las Vegas, NV, USA*

Chapter Outline

ABBREVIATIONS

BCM Biologically-controlled mineralization
BIM Biologically-induced mineralization
MMA Magnetotactic multicellular aggregates
MMO Magnetotactic multicellular organisms
MMPs Many-celled magnetotactic prokaryotes
PCR Polymerase chain reaction

DEFINING STATEMENT

Magnetotactic bacteria represent a diverse group of microorganisms with respect to morphology, physiology, and phylogeny. All synthesize chains of intracellular magnetic crystals enveloped by a membrane comprising the organelle known as the magnetosome. Magnetosomes are responsible for the tactic behavior of these bacteria in magnetic fields, that is, magnetotaxis.

INTRODUCTION

Many bacteria are capable of swimming and as they do so, they modify their random swimming direction to move toward a more favorable environment. This environment, which varies from species to species, is tracked on the basis of sensing changes in a range of stimuli that includes both chemicals like nutrients and toxins and physical parameters like pH, light, temperature, and even the geomagnetic field. The behavioral response to directional stimuli is called taxis. Among the most-studied tactic response are chemotaxis and phototaxis. This article deals with tactic behavior influenced by magnetic fields, that is, magnetotaxis.

The geomagnetic field is considered one of the environmental cues for navigation by many different biological organisms. Different species that have been shown to respond to the earth's magnetic field include homing pigeons, migratory birds, trout, sea turtles, and social insects such as honeybees and ants. Among microbes, the motility of several types of bacteria and protists is affected by magnetic fields. This behavior is called magnetotaxis and the organisms are described as magnetotactic. Magnetotactic microbes represent the best-understood example of magnetically-influenced, biological navigation in the earth's geomagnetic field.

The first description of magnetotactic bacteria was made by Salvatore Bellini and reported in 1963 in a publication of the Microbiology Institute (Instituto di Microbiologia) of the University of Pavia, Italy. His microscopic observations of freshwater sediments revealed large numbers of bacteria that swam in a single, consistent direction. Bellini realized that the bacteria swam toward the North Pole and hence called them magnetosensitive bacteria (batteri magnetosensibili). He presumed that some sort of biomagnetic compass was present within the bacterial cells. His idea proved to be true. Richard P. Blakemore, then a Microbiology graduate student

Topics in Ecological and Environmental Microbiology.

at the University of Massachusetts at Amherst, reported in *Science* magazine in 1975, similar magnetosensitive behavior of bacteria in sediments collected from a freshwater swamp in Woods Hole, Massachusetts. He discovered that these magnetosensitive bacteria contained electron-dense, iron-rich crystalline inclusions. Blakemore also realized that these microorganisms were swimming toward the geomagnetic north in the Earth's magnetic field and named them magnetotactic bacteria and their behavior magnetotaxis. Since then, many multidisciplinary studies have been done on these microorganisms, many of which include recent advances toward understanding the molecular mechanisms underlying the biomineralization and ultrastructural organization of magnetosomes in bacteria. The controlled synthesis and organization of lipid-bilayer membrane-bounded organelles, the magnetosomes, using specific cytoskeleton elements has caused rethinking on some of the traditional boundaries used to distinguish eukaryotic cells from their prokaryotic counterparts, as prokaryotes were thought to not contain lipid-bilayer membrane-bounded organelles or an organized cytoskeleton. Magnetotactic bacteria are increasingly becoming the cell biology model for understanding the molecular mechanisms underlying the basic rules that govern the intracellular ultrastructural organization of prokaryotes.

Despite the efforts of a number of different research groups, only a few representatives of the magnetotactic bacteria have been isolated in axenic culture, and even fewer have been validly named. As a consequence, little is known about their metabolic/physiological characteristics and versatility; many studies on their diverse morphology and phylogeny as well as mineralogy of their magnetosomes have been done using culture-independent methods.

GENERAL FEATURES OF MAGNETOTACTIC BACTERIA

The term "magnetotactic bacteria" has no taxonomic significance and represents a heterogeneous group of bacteria with various cell morphologies, whose swimming direction is affected by magnetic fields. The operational definition of magnetotactic bacteria is that they are a heterogeneous collection of prokaryotes that share the traits of magnetotaxis and the biomineralization of magnetosomes. The magnetosome is defined as an intracellular organelle consisting of a single-magnetic-domain crystal of a magnetic iron mineral enveloped by a lipid-bilayer membrane that contains proteins that are unique to it. Despite their heterogeneity, there are other features shared by these microorganisms.

Magnetotactic bacteria are morphologically diverse, comprising cocci, rods, spirilla, vibrios and multicellular forms. All possess a gram-negative cell wall and are phylogenetically associated with gram-negative groups of prokaryotes (the Proteobacteria and the *Nitrospira* phyla) in the domain Bacteria. However, there is no known reason why magnetotactic bacteria are restricted to the Bacteria although no magnetotactic bacterium has yet been found that belongs to the Archaea.

All known magnetotactic bacteria are motile by means of flagella. Because most studies on uncultivated magnetotactic bacteria relied on the use of a magnetic field for retrieving cells from water and sediment samples, it is possible that nonmotile bacteria that synthesize magnetosomes exist that would not swim toward the magnet and thus would not be collected and observed. By definition, the latter organisms would be magnetic rather than magnetotactic. Flagellation patterns in magnetotactic bacteria vary greatly and include polarly monotrichous, bipolar, or lophotrichous representatives. The number of flagella per cell ranges from one in the marine vibrio, strain MV-1, to over a thousand in the multicellular magnetotactic prokaryotes (MMPs). The velocity of swimming also varies between strains; from about 40 $\mu m\ s^{-1}$ in members of the genus *Magnetospirillum* to over 1000 $\mu m\ s^{-1}$ in the magnetotactic cocci. Although motility is a key factor in magnetotaxis, very little is known about the structural and molecular components of flagella in magnetotactic bacteria. In general, flagella in many magnetotactic bacteria (e.g., the magnetococci) are shorter than those of many other bacteria (e.g., *Escherichia coli*) (Figure 1). Tufts of flagella in the magnetotactic cocci originate from a disk-shaped structure, in a depression or pit. Targeted disruption of the flagellin

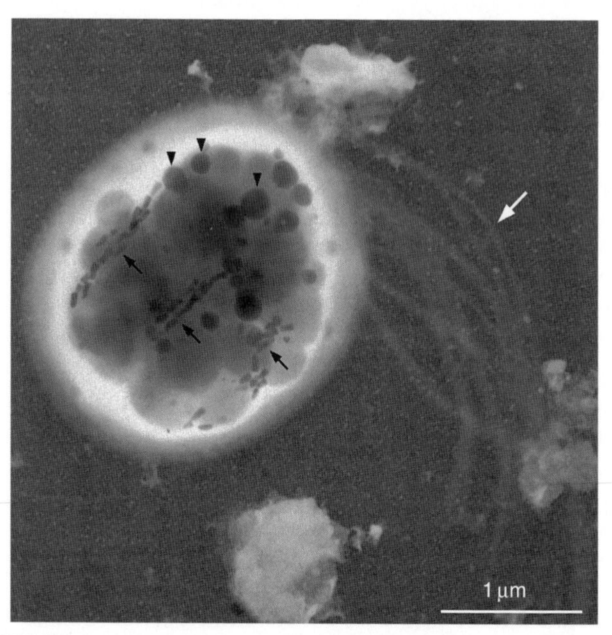

FIGURE 1 Transmission electron micrograph of a freshwater magnetotactic bacterium. Three rows or chains (arrows) of bullet-shaped magnetosomes can be seen aligned parallel to the main axis of the cell. A flagella bundle at one pole of the cell (white arrow) is observed. Several granules, possibly sulfur inclusion, can also be seen (arrowheads).

gene *flaA* was found to eliminate flagella formation and motility in *Magnetospirillum gryphiswaldense*. Large numbers of flagella are distributed on one side of the cells of the MMP; the part of the cell exposed to the environment and thus the entire organism can be considered to be peritrichously flagellated.

Many magnetotactic bacteria swim in helical tracks. The flagella force the cell body of the magnetotactic bacteria to rotate around its axis while the cell is propelled by the flagella. Under the influence of the geomagnetic field, the consequence of the action of the short flagella is a misalignment of the trajectories that enables the cells to change the trajectory direction when they, for example, hit any obstacle in the sediment. The small misalignments during movement might be essential for certain magnetotactic bacteria to escape from physical traps present in sediments where they thrive.

Magnetotactic bacteria are distributed worldwide in aquatic habitats that are neutral or close to neutral in pH. In general, they inhabit many aquatic environments where there is chemical stratification, that is, where vertical chemical (e.g., oxygen) concentration gradients exist. In general, magnetotactic bacteria are obligately microaerophilic and/or anaerobic and generally display a negative tactic response to high concentrations of oxygen but also show a positive aerotactic response to low concentrations of oxygen (they display aerotaxis under these conditions). In natural environments they prefer the oxic–anoxic interface where they are found in their highest numbers. This interface can occur in the water column or in the uppermost millimeters to centimeters of unconsolidated sediments depending on the environment.

In general, most magnetotactic bacteria biomineralize a single type of mineral in their magnetosomes: either the iron oxide magnetite (Fe_3O_4) or iron sulfides, for example, greigite (Fe_3S_4) and its precursors. Magnetite-producers are usually present at the oxic–anoxic interface or slightly above while some magnetotactic bacteria, particularly those that produce the iron sulfide greigite, are generally found below the oxic–anoxic interface in the anaerobic zone. All known freshwater magnetotactic bacteria and some marine, estuarine, and salt marsh morphotypes produce magnetite. To date, greigite-producing magnetotactic bacteria have only been found in brackish to marine environments and none have been reported from freshwater habitats. Several magnetotactic bacteria have been found to produce both mineral crystals sometimes aligned in the same chain but none of these have been isolated and grown in pure culture.

Metabolically, all cultivated strains of magnetotactic bacteria are magnetite producers that have a respiratory metabolism and are unable to ferment substrates, except for *Desulfovibrio magneticus* strain RS-1, which is chemoorganoheterotrophic and ferments pyruvate to acetate, CO_2, and H_2 in the absence of a terminal electron acceptor. Most magnetotactic bacteria are mesophilic with respect to growth temperatures although moderately thermophilic magnetotactic bacteria were recently reported from a series of hot spring pools at the Great Boiling Springs geothermal field, Nevada USA, whose temperatures ranged from 32 to 63 °C. Psychrophilic magnetotactic bacteria have been detected in marine sediment samples (temperature 4 °C) from maritime Antarctica. All cultivated magnetotactic bacteria except *D. magneticus* (has not yet been tested) have been shown to exhibit nitrogenase activity and thus can presumably fix atmospheric N_2. Because N_2 fixation is an oxygen-sensitive process, the microaerophilic magnetotactic bacteria might be further restricted to the oxic–anoxic interface in nitrogen-limited environments.

CULTIVATED MAGNETOTACTIC BACTERIA IN PURE CULTURE

Magnetospirillum species are facultatively anaerobic microaerophiles that show varying tolerances to oxygen. They are chemoorganoheterotrophs that use organic acids (e.g., succinic) as a source of electrons and carbon. Some *Magnetospirillum* species are also chemolithoautotrophic using reduced sulfur compounds as sources of electrons and fix CO_2 through the Calvin–Benson–Bassham cycle for autotrophy. In addition to oxygen, *Magnetospirillum* species can use nitrate as a terminal electron acceptor. This genus is phylogenetically affiliated with the Alphaproteobacteria, a subgroup of the Proteobacteria. Thus far, there are only three validly described species of *Magnetospirillum*: *M. magnetotacticum* strain MS-1, *M. gryphiswaldense* strain MSR-1, and *M. magneticum* strain AMB-1. All strains biomineralize cubo-octahedral crystals of magnetite. Numerous strains of *Magnetospirillum* have been isolated from various aquatic environments. Those include seven strains (MSM strains) isolated from a freshwater pond in Iowa, USA, and strain WH-1, isolated from sediments in North Lake, located in Qingshan, Wuhan, China. Interestingly, several related nonmagnetotactic spirilla with ~95% 16S rDNA sequence similarity have also been isolated. Tractable genetic systems have been developed for *M. gryphiswaldense* and *M. magneticum*.

Desulfovibrio magneticus strain RS-1 is an obligately anaerobic, curved rod-shaped sulfate-reducing bacterium. It can also utilize fumarate as a terminal electron acceptor. This species produces tooth-shaped crystals of magnetite and is phylogenetically associated with the Deltaproteobacteria class.

The cultured, unnamed marine species, including the vibrio strain MV-1, the coccus MC-1, and the spirillum strain MMS-1, are all facultative chemolithoautotrophs and are phylogenetically affiliated with the Alphaproteobacteria class. Strain MV-1 utilizes the Calvin–Benson–Bassham cycle for autotrophy while strain MC-1 utilizes the reverse or reductive tricarboxylic acid cycle. The pathway for autotrophy in strain MMS-1 has not yet been elucidated. All three strains used reduced sulfur compounds (e.g., thiosulfate) as a source of electrons and produce intracellular sulfur globules. In chemically-stratified marine aquatic environments where an oxygen sulfide double inverse gradient exists, the presence of oxygen and sulfide often overlaps at the oxic–anoxic interface. Because these marine species require both compounds for energy, these organisms would also be restricted to the oxic–anoxic interface. Interestingly, many uncultivated cells of magnetite-producing magnetotactic bacteria collected from these environments also contain intracellular sulfur-rich globules, suggesting that they also oxidize reduced sulfur compounds as a source of electrons. Strain MV-1 is a facultatively anaerobic microaerophile while strains MC-1 and MMS-1 appear to be obligate microaerophiles. All three species biomineralize elongated crystals of magnetite.

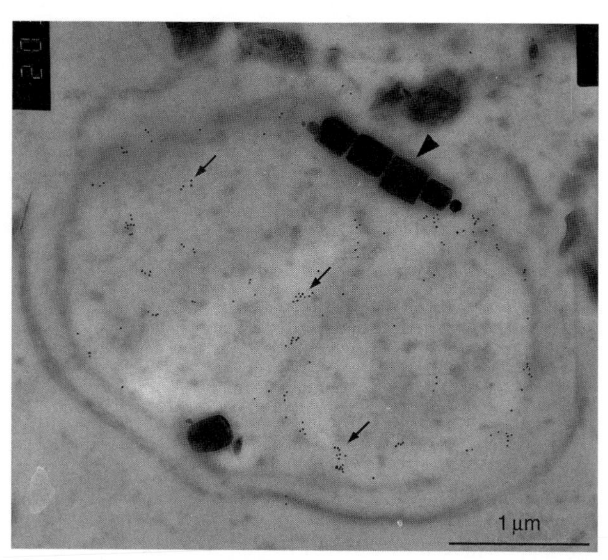

FIGURE 2 Electron micrograph of an ultrathin section of a cell of the magnetic coccus classified as Itaipu-1, which was hybridized with a 16S rDNA-digoxigenin probe and detected by antidigoxigenin antibodies conjugated with 10 nm gold particles (arrows). Magnetosomes can be seen (arrowhead).

UNCULTURED MAGNETOTACTIC BACTERIA

The presence of magnetosomes and the trait of magnetotaxis allow for the magnetic manipulation and easy retrieval of uncultured magnetotactic bacteria from environmental samples with the aid of a magnetic field. Our current knowledge regarding the diversity of magnetotactic bacteria is based more on microorganisms harvested from the environment than on the isolation and characterization of pure cultures. Thanks to modern molecular techniques including the polymerase chain reaction (PCR), it is relatively easy to obtain and analyze the sequence of their 16S rRNA genes and therefore be able to determine phylogenetic relationships of these organisms.

Many studies have been done on uncultured marine and freshwater magnetotactic cocci, the most commonly found morphotype of magnetotactic bacteria in aquatic environments. In addition, these organisms, unlike other magnetotactic morphotypes, persist for long periods in microcosm experiments. All magnetotactic cocci biomineralize elongated crystals of magnetite although the arrangement of the crystals is variable including those with multiple chains and those with a clump (rather than a chain) of magnetosomes at one side of the cell. All known magnetotactic cocci are phylogenetically affiliated with the Alphaproteobacteria. They are not closely related to any other Alphaproteobacterium and form a unique lineage within the group.

The techniques of gold labeling in conjunction with *in situ* hybridization and transmission electron microscopy revealed the presence of at least three different types of uncultured cocci in Itaipu Lagoon, Brazil (Figure 2). In this case, polyribonucleotide probes were labeled with digoxigenin or fluorescein and detected by immunolabeling with antifluorescein and antidigoxigenin antibodies bound to 10 and 15 nm gold particles. The use of this technique allows for the correlation of phylogenetic association with specific magnetosome morphology and organization; that is, to determine which 16S rDNA sequence belonged to which of the three cocci. The cocci were distributed in two clusters, Itaipu-1 and Itaipu-2, both closely related sequences within the lineage of magnetotactic cocci in the Alphaproteobacteria.

Four uncultured magnetotactic bacteria are phylogenetically affiliated with the *Nitrospira* phylum, and, interestingly, all biomineralize bullet-shaped crystals of magnetite in their magnetosomes. Two are large, rod-shaped (5–10 μm long and 1.5 μm wide) bacteria with clusters of multiple rows of magnetosome chains. They are motile by means of a polar tuft of flagella. Each cell contains up to a 1000-magnetosomes and numerous sulfur globules within the cytoplasm. 16S rDNA sequencing showed a sequence similarity of 91% between them. The first organism was retrieved from freshwater lakes in Upper Bavaria, Germany, and tentatively described as *Magnetobacterium bavaricum*, and the second was named *Magnetobacterium bremense* and was collected from a lake near Bremen, Germany. This morphological type does not seem to be restricted to the Northern Hemisphere and has been detected in a South American lagoon, although phylogenetic studies on the latter organism are lacking. The third organism is an unnamed rod-shaped bacteria, strain MHB-1, that also biomineralizes bullet-shaped crystals of magnetite in their magnetosomes. The fourth organism is a moderately thermophilic bacterium (designated strain HMVS-1) found in several springs in the Great Boiling Springs geothermal field in Nevada, USA. Cells of HSMV-1 are small, gram-negative, vibrioid-to-helicoid in morphology and biomineralize a single chain of bullet-shaped magnetite magnetosomes.

Large numbers of a barbell-shaped magnetotactic bacterium were discovered in a chemically-stratified coastal salt pond that displayed unusual magnetotactic behavior (discussed in the "Magneto-aerotaxis" section). This microorganism appears to consist of a chain of two to five cocci and belongs phylogenetically to the Deltaproteobacteria class. Unfortunately, the composition of the magnetosome mineral was not determined.

At this point in time, no greigite-producing magnetotactic bacterium has been isolated in pure culture. However, there have been several studies on these organisms using the culture-independent techniques described above. A relatively large (~5 μm), slow-moving, greigite-producing rod was retrieved from chemically-stratified Salt Pond, Woods Hole, USA and was found to be a member of the Gammaproteobacteria although a more detailed phylogenetic analysis raises doubts to this finding. Recently two cultured magnetite-producing rods were found to unequivocally phylogenetically belong to the Gammaproteobacteria.

Morphologically conspicuous, multicellular magnetotactic bacteria, MMPs, have been reported from a number of different brackish, marine, or hypersaline environments in both the Northern and the Southern Hemispheres, usually in sulfide-rich environments that are very likely to be anaerobic. These very unusual organisms are assemblage of gram-negative cells that generally contain chains of greigite-containing magnetosomes. The "organism" is not a consortium of cells of different species, but an aggregate of similar genetically-identical cells. All those examined are phylogenetically affiliated with the Deltaproteobacteria.

For many years, because of their unique morphology, MMPs were thought to represent a single species. However, phylogenetic analyses of a collection of these organisms from a natural population present in a salt marsh (Falmouth, MA, USA) showed that this particular group of MMPs included those with a number of different 16S rDNA sequences separated by at least 5% divergence, suggesting that they do not represent a single species but a number of distinct species within a single genus. MMPs capable of producing both greigite and magnetite magnetosomes have also been reported (Figure 3). Recently, MMPs that were not magnetotactic and did not contain magnetosomes (called nMMPs) were observed in temperate shallow springs in the Great Boiling Springs geothermal field and Pyramid Lake, both being of low salinity (5–11 ppt) and located in northwestern Nevada, USA. These nMMPs appear to represent a single, new species belonging to the MMP group in the Deltaproteobacteria.

A valid scientific name has not yet been proposed for the MMPs (largely because they have not been isolated and grown in pure culture) and confusing and imprecise

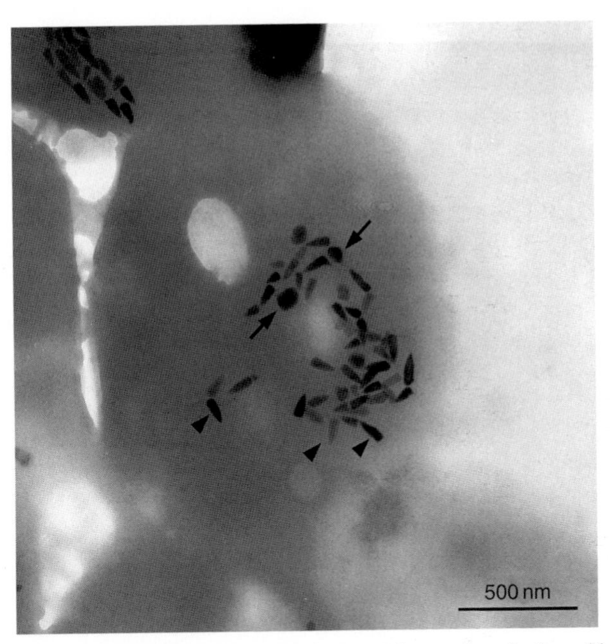

FIGURE 3 Transmission electron microscopy image of a cell of a multicellular magnetotactic bacterium (MMP) with chains of bullet-shaped magnetite (arrowheads) and rounded greigite magnetosomes (arrows).

terminology has been used for them including magnetotactic multicellular aggregates (MMA), MMP, many-celled magnetotactic prokaryotes, and magnetotactic multicellular organisms (MMO). The best studied of these multicellular prokaryotes is a representative found in the Araruama Lagoon near Rio de Janeiro, Brazil. Studies involving its cell architecture, unique life cycle, magnetotactic and magnetic properties, coordinated motility and ecology were enough to warrant naming this organism in the *Candidatus* category; *Candidatus* Magnetoglobus multicellularis (Figure 4). Like other MMPs, *Candidatus* M. multicellularis is a member of the Deltaproteobacteria. The most conspicuous structural feature of these microorganisms is the fact that its cells appear to be precisely organized, polarized, and unable to live individually. Their 10–40 Gram-negative cells are arranged together in a spherical unit (Figure 5). Each cell faces both the external surrounding environment and the internal portion of the aggregate. If a cell leaves the assemblage, for example, when the organism dies, it loses its ability to swim. Most cells also no longer align magnetic field lines.

Candidatus M. multicellularis exhibits a unique method of reproduction. This has been studied in some detail and the following growth stages have been proposed: (1) individual cells grow in size thereby increasing their volume and their number of magnetosomes; (2) the cells then divide doubling the number of cells in the microorganism; (3) individual cell within the microorganism then reorganize; and (4) the microorganism divides into two new identical colonies or units.

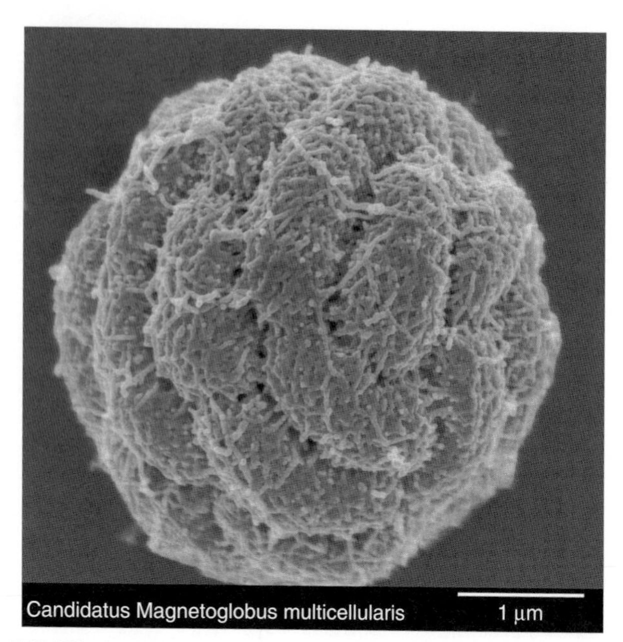

FIGURE 4 Scanning electron microscopy image of *Candidatus* Magnetoglobus multicellularis showing spherical morphology and firmly attached cells.

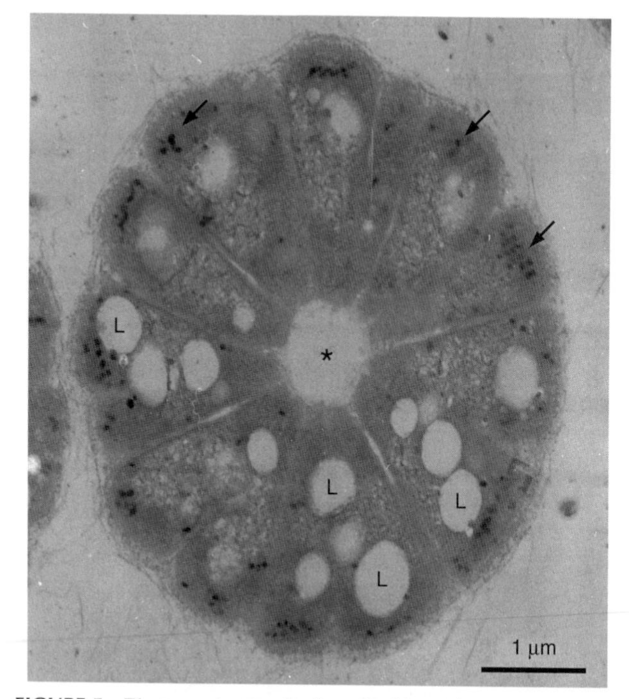

FIGURE 5 Electron micrograph of an ultrathin section of a *Candidatus* Magnetoglobus multicellularis showing the general organization of cells and the internal compartment (asterisk). All cells have a cell wall surface structure of gram-negative bacteria and face both the external environment and the internal compartment. The numerous magnetosomes occur mostly at the periphery of the cells (arrows). Lipid or poly-hydroxy-alkanoate inclusions are marked (L).

A single entire aggregate of *Candidatus* M. multicellularis contains thousands of magnetosomes containing crystals of iron sulfide. The greigite particles are enveloped by a membrane and are generally distributed as chains near the outside periphery of the cells. The chains collectively organize to provide a near optimal magnetic moment. The magnetic polarity (their swimming direction preference – discussed later) is transferred after division to the newly formed colonies.

On the outer surface of the microorganism, cells are covered with numerous flagella that are responsible for its coordinated complex movement. Four different types of motility have been observed in *Candidatus* M. multicellularis: free motion, rotation, walking, and escape. In free motion, the microorganism swims in either straight or helical trajectories along magnetic field lines in a uniform magnetic field. Swimming velocities up to 100 µm s^{-1} have been observed. At the edge of a drop under oxygen in air, a rotation movement is observed. In this movement, the microorganism spins around an axis that passes through its center. In walking movement, individual aggregates of *Candidatus* M. multicellularis swim freely in a complex trajectory when they reach the air–water interface on the top of the drop, maintaining, however, the same sense of organismal rotation. The most peculiar behavior of *Candidatus* M. multicellularis appears to be an "escape" motility, often referred to as the so-called "ping-pong" movement. The movement consists of a backward movement (north-seeking in the Southern Hemisphere or vice versa in the Northern Hemisphere) for tens or even hundreds of micrometers, followed by a forward south-seeking movement. In the backward movement, aggregates decelerate continuously with time, whereas in the forward movement that follows, they display uniform acceleration.

Recently, it was shown that MMPs change their behavior with magnetic field intensity in a way that cannot be explained by the simple compass model used for magnetotaxis in unicellular bacteria. This unexpected behavior led to the hypothesis that MMPs have a sort of magnetoreception mechanism in magnetic fields larger than that of the earth. This magnetoreception mechanism and the unusual "escape" motility might imply transference of information from the magnetic interaction of the magnetosome chains to the flagellar motors. There is, however, no scientific evidence for any type of magnetic "sensor" such as this in magnetotactic bacteria.

MAGNETOSOMES

Many prokaryotic microorganisms facilitate the formation of a large number of various minerals, including magnetosome crystals, through one of two biomineralization processes. These include biologically-induced mineralization (BIM) and biologically-controlled mineralization (BCM). There are several important differences between BIM and

BCM and the most important relate to the control of the synthesis of the biominerals and the functionality of the biomineralized products. Generally, in BIM, the microorganism exerts little control over the biomineral, which has no known recognized function for the bacterium. In BCM, the microorganism exerts strict crystallochemical control over the nucleation and growth of the biomineral, which in some cases has a somewhat defined function. Because of these features, BCM processes are considered to be under specific biochemical and genetic control. Examples of products of BCM in higher organisms are bones, teeth or shells. In the microbial world, most known biominerals are produced by bacteria through BIM. The mineral particles produced by bacteria in BCM are characterized as well-ordered crystals with narrow size distributions, high purity, and specific particle morphologies (Figure 6). Generally, the biomineral particles produce through BIM lack these qualities. The most characterized example of BCM in bacteria is the magnetosome.

Magnetosomes are magnetic iron oxide or sulfide mineral crystals, each enveloped by a lipid-bilayer membrane (the magnetosome membrane). As previously stated, there are two general types of minerals in magnetosomes: magnetite (Fe_3O_4) and greigite (Fe_3S_4). Additional nonmagnetic iron sulfide minerals, including mackinawite (tetragonal FeS) and a cubic form of FeS, have been found in some greigite-producing magnetotactic bacteria and are assumed to represent precursors to greigite. In most magnetotactic bacteria, the magnetosomes are organized in chains within the cell. This has important physical significance and will be discussed later.

The nature of the magnetosome membrane has been studied in detail in several magnetite-producing *Magnetospirillum* species. In this genus, the magnetosome membrane appears to originate as an invagination of the cell membrane and contains a number of proteins unique to it that are thought to be responsible for regulating the size, the chemical purity and the morphology of the magnetosome crystal.

The crystalline phase of the magnetosome has been extensively studied using a number of forms of electron microscopy. Magnetotactic bacterial magnetite typically has cubic 100, octahedral 111, and dodecahedral 110 faces in crystals that have quasi-rectangular projected shapes in the horizontal plane or are bullet-, tooth-, or arrowhead-shaped (Figure 7). Morphologies of greigite crystals are similar

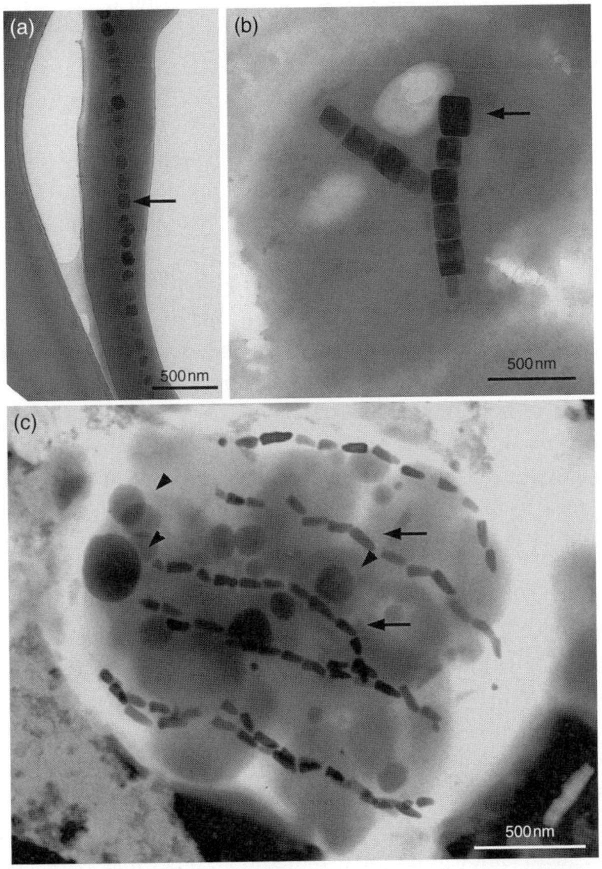

FIGURE 6 Magnetosome morphologies. (a) *Magnetospirillum magnetotacticum* cell with a chain of octahedral magnetosomes (arrow), (b) uncultivated marine coccus with two chains of prismatic magnetosomes (arrows), (c) uncultured freshwater magnetotactic bacterium with rows of bullet-shaped magnetosomes (arrows) and sulfur inclusion (arrowheads).

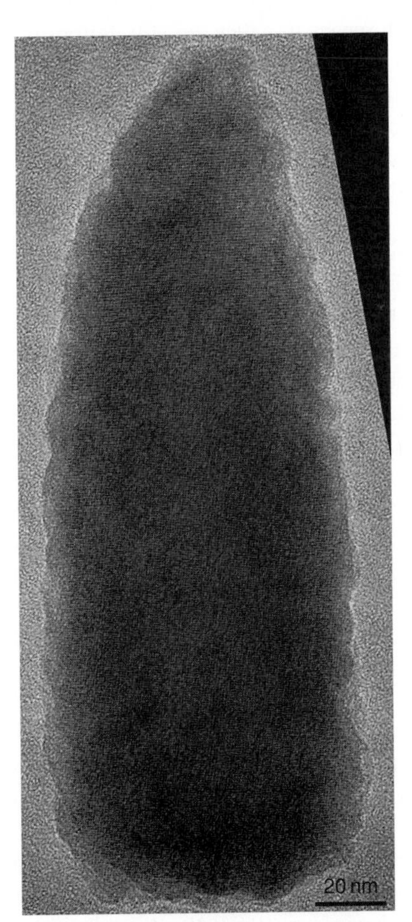

FIGURE 7 High-resolution transmission electron microscopy image of a bullet-shaped magnetite magnetosome.

and include the same forms as magnetite but, in contrast, the crystals often appear "wrinkled" and have an irregular border. The number of magnetosomes varies from a few to thousands per cell.

Regardless of mineral type, magnetosome crystals are generally in the size range of 35–120 nm, which corresponds to the single-domain size range for magnetite or greigite. However, magnetosomes up to 250 nm have been reported. The significance of the size of the mineral crystals is discussed in the next section.

BIOPHYSICS OF MAGNETOTAXIS

As previously stated, the sizes of magnetosome crystals in magnetotactic bacteria are within the permanent single-magnetic-domain size range (~35–120 nm), meaning that they are uniformly magnetized with the maximum magnetic dipole moment per unit volume. Magnetosomes larger than single-magnetic domains are not uniformly magnetized because of the formation of domain walls, resulting in multiple magnetic domains. The overall effect of being larger than the single-magnetic-domain size range is a significant reduction of the magnetic dipole moment. On the other hand, very small particles (<35 nm) are superparamagnetic, which means that they are still uniformly magnetized but their magnetic dipole moments are not constant because of thermally-induced spontaneous reversals. This results in a net magnetic moment of zero over time. Thus, the single-domain size range is the optimum particle size for the maximum magnetic dipole moment in a magnetosome.

Magnetosomes are anchored to the cell membrane in most, if not all, magnetotactic bacteria and are thus fixed in place within the cell. In most magnetotactic bacteria, the magnetosomes are arranged as a chain or multiple chains. Recent experimental evidence with *Magnetospirillum* species shows that supporting structures are necessary (cytoskeletal elements) to prevent the chains from collapsing into a disordered clump of particles. This results in an efficient transfer of torque to the cell body. The magnetosomes attach to protein filaments that arrange the magnetosomes in chains and presumably attach the magnetosome chains to the structural framework of each cell that they occupy.

The magnetosome chains within the cell impart a magnetic moment to the cell that overcomes the thermal forces that tend to randomize the position of the cell in the aqueous environment in which the bacteria live. The magnetosome chain works as a magnetic compass, continuously causing the cell to passively align along local magnetic field lines. This alignment is disturbed mostly by the thermal fluctuations of the aqueous medium. Thus the alignment of magnetotactic bacteria along geomagnetic field lines depends on the ratio the magnetic to thermal energies, or MB/k_BT, where M is the magnetic dipole moment of the cell, B is the geomagnetic field strength, k_B is the Boltzmann's constant, and T is the temperature. The organization in chains spontaneously causes the individual magnetic moments of the single-domain magnetosomes to orient parallel to each other along the chain. When magnetosomes are arranged in a linear chain within the bacterial cell, the total magnetic moment of the cell is the sum of the magnetic moments of the individual crystals and thus is at its maximum value. Electron holography in the electron microscope, magnetic force microscopy and pulsed-magnetic-field remanence measurements on individual cells have demonstrated the permanent magnetic microstructure of the magnetosome chains.

At ambient temperature, 10–20 magnetosomes (50 nm) in a chain would be sufficient for passive orientation of a magnetotactic bacterium along geomagnetic field lines. The efficient orientation in the geomagnetic field results in the migration of the cell along the magnetic field lines as it swims. If the magnetic field is increased (e.g., by exposing the cells to a bar magnet), the time-averaged orientation of the cell along the field is increased and the migration rate of the cell in the magnetic field direction is increased, even though the swimming speed of the cell remains unchanged.

Intriguingly, many magnetotactic bacteria seem to produce more magnetosomes than is necessary for orientation, suggesting that magnetosomes and magnetic mineral biomineralization play other roles than simple magnetic orientation of the cell. Unusually large magnetite-containing magnetosomes (up to 250 nm) were reported in an uncultured magnetotactic coccus, named Itaipu-1, collected from Itaipu Lagoon, near Rio de Janeiro, Brazil. Because of the large size of the magnetic crystals, they would be expected to be outside the single-magnetic-domain range. However, electron holography measurements demonstrated that the crystals are in fact single-magnetic domains while they are organized in chains. Breaking or removal of magnetosomes from the chains results in a complicated magnetic microstructure with domain walls. The functional significance of the unusually large volume of these magnetosome crystals is unknown. It has been speculated that the large crystals in Itaipu-1 occur in order to compensate for the anomalous low magnetic field strength (0.25 gauss) of the region of the south Atlantic, which includes Itaipu Lagoon. However, a calculation of the total magnetic dipole moment for the a typical Itaipu-1 cell results in $MB/k_BT \sim 250$, which is over 10 times the minimum value required for efficient magnetotaxis, suggesting there may be other functions of the magnetosomes in Itaipu-1. In addition, other magnetotactic bacteria collected from the same habitat possess magnetosome magnetite crystals with sizes within the normal range for other magnetite-producing magnetotactic bacteria.

MAGNETOTAXIS

It is important to understand that magnetotaxis is somewhat of a misnomer and does not represent a "true" taxis response as in phototaxis or chemotaxis. Magnetotactic cells do not swim toward or away from a magnetic field source. Instead, magnetotaxis is the combination of the passive orientation of the cell and its active motility by flagella along magnetic field lines. All known magnetotactic bacteria generally swim bidirectionally and some also "loop" while swimming. None are known to exhibit run-and-tumble motility as do some nonmagnetotactic bacteria (e.g., *E. coli*).

Magnetotaxis is considered advantageous to bacteria in vertical chemical concentration gradients (e.g., oxygen) in locating and maintaining an optimum position in the gradient because magnetotactic bacteria use geomagnetic field lines as a fixed reference to swim, unlike other bacteria. Presumably, if a magnetotactic bacterium is displaced from its optimal position, it would use geomagnetic field lines to more efficiently return to its optimal position. Magnetotactic bacteria, after aligning along geomagnetic field lines, would have a one-dimensional search problem in locating their optimal position in a vertical chemical concentration gradient whereas nonmagnetotactic bacteria would have a three-dimensional search problem, presumably expending more time and thus more energy.

The original model of magnetotaxis stated that all magnetotactic bacteria had one of two possible polarities: north-seeking and south-seeking. Polarity is defined as the direction of motility of the cell under oxic conditions. This was based on the observation that in the Northern Hemisphere, magnetotactic bacteria predominantly swim parallel to the magnetic field toward the north, hence the term north-seeking. In the Southern Hemisphere, the opposite was observed and south-seeking bacteria swim antiparallel to the field. The Earth's geomagnetic field lines are inclined everywhere on the planet except at the equator with the magnitude of the inclination (angle of dip) increasing from the equator (0°) to the poles (90°). Thus north-seeking magnetotactic bacteria in the Northern Hemisphere are also downward-seeking. The reverse is true in the Southern Hemisphere. The outcome of this situation is that in both hemispheres the bacteria swim downward toward regions of the sediment with little or no oxygen. In the original model, once a bacterium reached a layer of the sediment with suitable oxygen concentration, it would stop moving and possibly adhere to a sediment particle.

Two important observations were inconsistent with this magnetotaxis model: (1) the observation that north-seeking magnetotactic bacteria in pure culture (strain MC-1) form horizontal bands at certain positions below the meniscus in culture tubes containing an oxygen concentration gradient, instead of swimming persistently to the bottom; and (2) populations of uncultured magnetotactic bacteria were sometimes observed in the water column in chemically-stratified aquatic environments and not always in the sediments. Should the original model be true, all magnetotactic bacteria would swim to the bottom in culture tubes and to the sediment in natural aquatic environments. An expansion of the magnetotaxis model was then proposed based on the observation that magnetotaxis appears to function in conjunction with aerotaxis. This is known as the magneto-aerotaxis model.

MAGNETO-AEROTAXIS

Two different magneto-aerotactic mechanisms have been proposed: axial and polar magneto-aerotaxis. In axial magneto-aerotaxis, cells do not show a polar preference and the magnetic field defines the axis but not the direction of swimming. This mechanism is typical of cells of *Magnetospirillum* species grown in liquid culture. Cells spontaneously reverse their swimming direction in an applied magnetic field. Thus, the distinction between north-seeking and south-seeking does not apply to axial magneto-aerotactic spirilla. Interestingly, however, cells of *Magnetospirillum* species collected from natural environments or those grown in semi-solid oxygen gradient media display polar magnetotaxis.

In polar magneto-aerotaxis, the most common mechanism observed in natural populations of magnetotactic bacteria, cells show a polar preference in their swimming direction. The magnetic field provides both direction and axis for the magnetotactic behavior. This is based on experimental findings of strain MC-1 (*Candidatus* Magnetococcus marinus), a cultivated magnetotactic coccus. In the magneto-aerotaxis model, magnetotactic bacteria can be either in an oxidized or in a reduced state. In the oxidized state, when in the oxic zone of a vertical gradient of oxygen, cells swim downward. They do this by rotating their flagella counterclockwise (Figure 8). However in the reduced state (in the anaerobic zone), the bacteria reverse direction, presumably

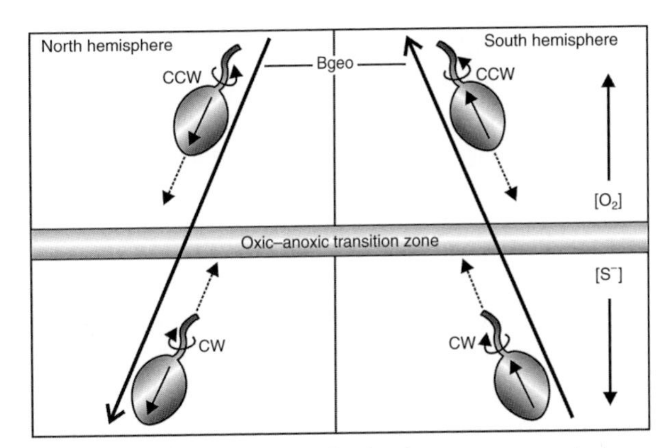

FIGURE 8 Schematic drawing showing how magneto-aerotaxis may help magnetotactic bacteria to find preferred habitats. See text for details.

by reversing the direction of flagellar rotation, without turning around (they would still be aligned along magnetic field lines) and thus move upward, again toward their optimal oxygen concentration. In this model, magnetotactic bacteria still benefit from having a fixed axis for swimming.

The polar magneto-aerotaxis model can be extended to redoxtaxis in habitats in which spatially separated reservoirs of reductants and oxidants coexist. In these cases, some magnetotactic bacteria would swim into anoxic zones of their habitat to accumulate reduced compounds (e.g., reduced sulfur compounds such as hydrogen sulfide) and then swim back to zones in which oxidized compounds are present.

There is evidence that magneto-aerotaxis cannot explain the behavior of all magnetotactic bacteria particularly those in natural environments. For example, a population of mainly south-seeking magnetotactic bacteria (the barbell-shaped bacterium discussed in an earlier section) has recently been discovered in a chemically-stratified environment in the Northern Hemisphere. These microorganisms seem to represent a deviation to the magneto-aerotaxis model for magnetotactic bacteria as they were found in the anoxic zone of the costal pond and were predominantly south-seeking under oxic conditions (south-seeking by definition). Thus, additional models of magnetotaxis are likely needed to explain the behavior of some magnetotactic bacteria.

MAGNETOTACTIC EUKARYOTES

A magnetotactic response was reported in some eukaryotic microbes (protists) collected from coastal areas. Transmission electron microscopy revealed intracellular particles identical to bacterial magnetite-containing magnetosomes. Endosymbiotic magnetotactic bacteria were not observed in the cells. Two possibilities exist for the origin of these magnetosomes. First, the protist itself might biomineralize magnetosomes. This seems to be the case for the Euglenoid alga *Anisonema platysomum* found to be present in water collected from a marine environment in Brazil. This protist produces rows of chains of tooth-shaped single-domain magnetosomes in an elaborate pattern along the axis of the cell. The morphology of the crystals is uniform and the magnetotactic response is consistent with the magnetic polarity of magnetotactic bacteria from the same environment as they show a directional swimming preference in the magnetic field.

Second, bacteria-grazing protozoa may ingest magnetotactic bacteria and their magnetosomes, which may be stable or partially or fully solubilized. They might be incorporated into the cell or present for a short time in food vacuoles, the likely site for solubilization. This seems to be the case for certain biflagellates, dinoflagellates, and ciliates detected in coastal areas. The significance and implications of this, as well as details of protozoal grazing of magnetotactic bacteria, is discussed later. Regardless of the origin of magnetite or greigite crystals in protists, those that contain these crystals contain significant amounts of intracellular iron.

GEOCHEMICAL, GEOPHYSICAL, AND ASTROBIOLOGICAL ASPECTS OF MAGNETOTACTIC BACTERIA

Cultured magnetotactic bacteria are known to facilitate a number of important geochemical reactions in the laboratory and thus it is likely they play important roles in these processes in natural environments. These include sulfide oxidation, sulfate reduction, denitrification, nitrogen fixation, and iron redox reactions (iron cycling) through magnetosome biomineralization and possibly through iron oxidation and reduction not related to magnetosome synthesis. In addition, all cultured marine strains are facultatively autotrophic through either the Calvin–Benson–Bassham or the reverse (or reductive) tricarboxylic acid cycles. The freshwater magnetospirilla either been shown to be capable of autotrophy or show potential for it as most appear to contain an intact ribulose-1,5-bisphosphate carboxylase/oxygenase (RubisCO) gene which encodes for the key enzyme for CO_2 fixation in the Calvin–Benson–Bassham cycle. Thus magnetotactic bacteria have great potential in regulating important geochemical processes and cycles involving several important elements including sulfur, nitrogen, iron, and carbon.

Protozoa are known to be important components of the marine ecosystems as the primary consumers of bacteria and recyclers of certain nutrients, including iron. Magnetotactic protists might play an important role in iron cycling either through direct biomineralization or through grazing on magnetotactic bacteria and solubilizing magnetosome crystals. Iron is a limiting factor in primary production in some oceanic and coastal areas. This element, which is often present in seawater in particulate and colloidal forms, can be digested in the food vacuoles of protozoans during grazing of particulate and colloidal matter. This process might generate more bioavailable iron for other species in the food chain. The chemical stability and thus the "survival" of the magnetosome and/or its mineral crystal depend on its surrounding environment. Magnetite magnetosomes appear to be quite stable in the bacterial cell possibly due to the magnetosome membrane and there is no evidence that magnetotactic bacteria can reuse iron in magnetosomes for another purpose even when cells are starved for iron. Is it possible for protozoans grazing on magnetotactic bacteria to make ingested magnetosome iron more bioavailable through solubilization? One possible mechanism of conversion of biologically unavailable forms of iron into more soluble forms is through the action of acidic food vacuoles on solid or colloidal forms of iron in protozoa. This has been observed to occur with colloidal iron and with greigite magnetosomes from magnetotactic bacteria.

The dissolution of greigite magnetosomes by the ciliate *Euplotes vannus* has been demonstrated in *in vitro* experiments. The ciliates ingested greigite-producing magnetotactic bacteria, the multicellular, colony-forming, magnetotactic prokaryote (MMP) discussed earlier, which became enveloped in food vacuoles known to have an internal acidic pH. Transmission electron microscopy (Figure 9) and X-ray microanalysis of thin-sectioned ciliates showed that within the acidic vacuoles, the greigite crystals dissolved at different rates after the magnetosome membrane was digested. The experiments with *Euplotes* suggest that protozoans that graze on magnetotactic bacteria in natural environments are likely important in iron cycling as they may help to solubilize iron trapped in the magnetosomes, making it bioavailable to other organisms. The impact of magnetite magnetosome digestion by protozoans remains unknown, but, based on the experiments with greigite magnetosomes, there is potential for magnetosome magnetite to also be solubilized by protozoans.

A simple calculation may help estimate the potential of the recycling of magnetosomes produced by magnetotactic bacteria. Each multicellular bacterial unit contains about 30 cells with approximately 100 magnetosomes per cell, which leads to each aggregate containing about 3000 magnetosomes. Using an average magnetosome size of 100 nm, which results in a cubic volume of 10^6 nm^3, and a greigite density of 4.1 g/cm^3, each colony contains an estimated total amount of 1.2×10^{-11} g of greigite that could potentially be solubilized within the ciliates and eventually released to the environment. Marine magnetotactic bacteria are known to reach population densities of 10^4 cells ml^{-1} in some natural environments. Taking into consideration the large amount of iron concentrated by magnetotactic bacteria in intracellular magnetosomes (10^{-13}–10^{-15} g iron per cell) and their estimated population density, in some habitats, nanomolar concentrations of iron may be sequestered by magnetotactic bacteria and then solubilized by grazing protozoans that ingest them, thereby recycling significant amounts of iron.

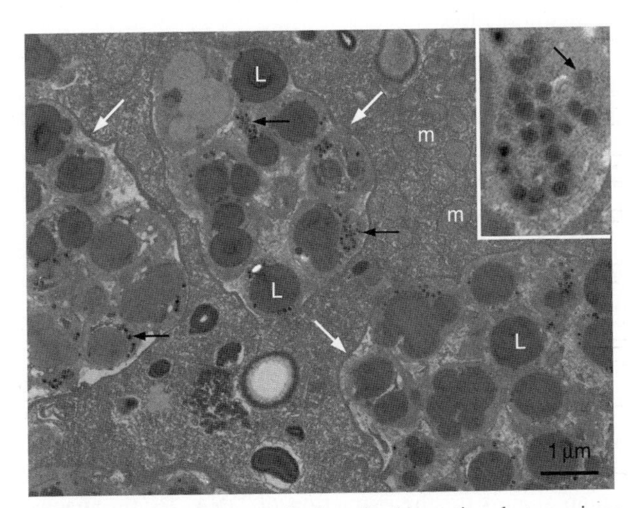

FIGURE 9 Electron micrograph of an ultrathin section electron microscopy of a ciliate protozoan that has ingested a number of magnetotactic multicellular bacteria. The bacterial cells (white arrows) are in vacuoles where they appear to be in the process of being digested. Magnetosome clusters can be seen (dark arrows). The inset shows a high-magnification view of a cluster of magnetosomes in different degrees of dissolution. No magnetosome membrane is seen in heavily digested crystals (arrow); m, mitochondria; L, Lipid inclusions.

Magnetotactic bacteria are considered to be a main source of the magnetization of sediments in a number of freshwater and marine environments. In many of these environments, magnetosome crystals, particularly magnetite, is released from dead, lysing cells and accumulate in sediments and persist relatively unaltered for long periods of time. In some cases, even magnetosome chains persist. Because magnetosomes are single-magnetic-domain particles, they form ideal carriers of paleomagnetic information as they cannot be demagnetized. Nanometer-sized magnetite grains have been recovered from a number of soils and modern and ancient lacustrine and deep sea sediments. In many cases, these particles were identified as biogenic by their shape and size, which were similar to magnetite crystals in magnetotactic bacteria and were therefore referred to as "magnetofossils". Putative fossil magnetotactic bacterial magnetite crystals have been found in stromatolitic and black cherts from the Precambrian, some dating as far back as 2 billion years. If these crystals are actually from magnetotactic bacteria, they represent some of the oldest bacterial fossils on Earth. Many of the crystals, however, found in these rocks were partially degraded, showed indistinct edges, and underwent full or partial oxidation to maghemite. The crystals have been used as supporting evidence that free O_2 had begun to accumulate in the atmosphere before 2000 Ma and that the present level of the earth's magnetic field strength had appeared by 2000 Ma. If magnetosomes are biomineralized by protists or retained by protists after ingestion of magnetotactic bacteria, then these eukaryotes, like the magnetotactic bacteria, might also directly contribute to the magnetization of sediments.

The cubo-octahedral crystal habit is the equidimensional form of magnetite and it is commonly found in inorganically produced magnetites. The elongated, nonequidimensional, pseudo-hexagonal prismatic and tooth-shaped forms of magnetite are very unusual. Both have been found in magnetic separates of sediments and it would seem only the latter elongated forms of magnetite might represent reliable magnetofossils of magnetotactic bacteria and could therefore be used as biomarkers of the past presence of magnetotactic bacteria.

The presence of ultrafine-grained magnetite, pyrrhotite, and greigite in the rims of carbonate inclusions in the Martian meteorite ALH84001 was considered one of the lines of evidence for life on ancient Mars. This meteorite, estimated to be ~4.5 billion years old, contains magnetite crystals about 10–100 nm in size with cuboid, teardrop-shaped and irregular morphologies. About 25% of the magnetite crystals present have an unusual morphology referred to as truncated hexa-octahedral, which is essentially the same morphology of magnetite crystals found in the marine magnetotactic vibrio, strain MV-1.

The elongated magnetite crystals from strain MV-1 have been studied in great detail and show a number of different distinctive properties putatively not found in inorganically produced magnetite crystals. These properties have been considered by some to be suitable criteria for the determination of whether magnetite crystals are biogenic or not. These properties include (1) narrow size range (a non-log-normal size distribution with the mean centered in the single-magnetic-domain size range); (2) restricted width-to-length ratios; (3) high chemical purity; (4) few crystallographic defects (crystals are generally defect-free with the exception of occasional crystal twinning); (5) crystal morphology with unusual truncated hexa-octahedral geometry; and (6) elongation along a specific crystallographic axis of the crystal. The 25% of the magnetite crystals in ALH84001 that resemble those of strain MV-1 seem to fit the above criteria. However, there are several other hypotheses regarding the origin of the magnetite crystals in meteorite ALH84001 that do not involve biology but only geochemistry. Thus, some disagree that these criteria are robust enough to distinguish between biogenically- and abiogenically-produced magnetite crystals and so the debate continues.

BIOTECHNOLOGICAL APPLICATIONS OF MAGNETOTACTIC BACTERIA, MAGNETOSOMES, AND MAGNETOSOME CRYSTALS

The controlled biomineralization of magnetosomes and their single-magnetic-domain particles are of relatively high structural perfection and purity and therefore there is great potential for biotechnological applications. In general, the amount of magnetite and magnetosomes from magnetotactic bacteria is relatively low especially considering the amount needed for specific applications. Thus, in order to produce enough cells, magnetosomes, and magnetite crystals for these applications, cells must be grown in mass culture where the conditions for growth and magnetite synthesis must be optimized. This was a major obstacle to research in this area for years. In the last decade, techniques including those involving continuous culture methods have been devised for the successful mass culture of various *Magnetospirillum* species and strain MV-1.

Cells of magnetotactic bacteria have proven useful in some interesting applications. For example, living north-seeking cells of polar magnetotactic bacteria have been used to determine south magnetic poles in meteorites and rocks containing fine-grained (<1 μm) magnetic minerals. Cells of magnetotactic bacteria have also been used in a number of medical applications including the magnetic separation of cells such as granulocytes and monocytes that phagocytized magnetotactic bacteria. Recently, the possibility of using magnetotactic bacteria in the removal of heavy metals and radionucleotides from wastewater was investigated. The advantage here again is that if the cells can take up significant amounts of heavy metals they can be separated magnetically and the contaminant can be eliminated from an aquatic system. Cells of the sulfate-reducing magnetotactic bacterium, *D. magneticus*, have been used in cadmium recovery using magnetic separation. A recently described application is the trapping of magnetotactic bacteria using a commercial magnetic recording head. This method may be useful in counting magnetotactic bacteria cells in water samples or to detect magnetically-labeled bacteria or magnetosomes.

Magnetosomes contain single-magnetic-domain crystals that have been shown to have interesting and useful magnetic and physical properties. In addition, the organic phospholipid membrane that envelopes the crystals allows for the immobilization of biological molecules such as nucleic acids or nonmagnetosome proteins on their surfaces. Magnetosomes have thus been used in the immobilization of specific enzymes, some of which show higher enzymatic activity when attached to magnetosomes than when attached to commercially synthesized magnetite particles. Antibodies linked to magnetosomes or magnetosome crystals have been developed and have proven useful in various immunoassays involving the detection of allergens, certain cancer cells, and the quantification of immunoglobulins.

Magnetotactic bacterial magnetite crystals have been used to detect single nucleotide polymorphism based on a fluorescence resonance energy transfer technique in which double-stranded labeled DNA is synthesized by PCR and immobilized to the magnetite crystals hybridize to target DNA and a fluorescence signal is detected. Magnetosome membrane proteins on the surface of bacterial magnetite particles have been used as anchor proteins for the assembly of foreign proteins on the surface of magnetite magnetosomes. These so-called protein displays have been designed using the magnetosome membrane proteins MagA, MpsA, Mms16, and Mms13 (MamC). Magnetotactic bacterial magnetite particles have been modified using compounds such as hyperbranched polyamidoamime dendrimers or amino silanes for the extraction of DNA. Biotin attached to a monolayer-modified substrate was detected by streptavidin immobilized to bacterial magnetite particles using a magnetic force microscope showing the potential of magnetosomes in the detection of biomolecular interactions in medical and diagnostic analyses.

PERSPECTIVES

Magnetotactic bacteria and magnetosomes are subjects of an active and interdisciplinary field of research that includes areas such as cell biology, geomicrobiology, palaeomagnetism, microscopy, biomineralization, biophysics and (bio)chemistry. New and exciting findings are constantly being reported on these microorganisms. These new findings should inspire research groups to follow new research directions in this highly multidisciplinary field of research.

FURTHER READING

Abreu, F., Martins, J. L., Silveira, T. S. *et al.* (2007). '*Candidatus* magnetoglobus multicelularis', gen. Nov. Sp. Nov., a multicellular magnetotactic prokaryote from a hypersaline environment. *International Journal of Systematic and Evolutionary Microbiology*, *57*, 1318–1322.

Bazylinski, D. A., & Frankel, R. B. (2004). Magnetosome formation in prokaryotes. *Nature Reviews Microbiology*, *2*, 217–230.

Bazylinski, D. A., Schlezinger, D. R., Howes, B. H., Frankel, R. B., & Epstein, S. S. (2000). Occurrence and distribution of diverse populations of magnetic protists in a chemically stratified coastal salt pond. *Chemical Geology*, *169*, 319–328.

Frankel, R. B., Bazylinski, D. A., Johnson, M. S., & Taylor, B. L. (1997). Magneto-aerotaxis in marine coccoid bacteria. *Biophysical Journal*, *73*, 994–1000.

Jimenez-Lopez, C., Romanek, C. S., & Bazylinski, D. A. (2010). Magnetite as a prokaryotic biomarker: A review. *Journal of Geophysical Research-Biogeosciences*, *115*, G00G03.

Kopp, R. E., & Kirschvink, J. L. (2008). The identification and biogeochemical interpretation of fossil magnetotactic bacteria. *Earth Science Reviews*, *86*, 42–61.

Lins, U., McCartney, M. R., Farina, M., Frankel, R. B., & Buseck, P. R. (2005). Habits of magnetosome crystals in coccoid magnetotactic bacteria. *Applied and Environmental Microbiology*, *71*, 4902–4905.

Martins, J. L., Silveira, T. S., Abreu, F., Silva, K. T., Silva-Neto, I. D., & Lins, U. (2007). Grazing protozoa and magnetosome dissolution in magnetotactic bacteria. *Environmental Microbiology*, *9*, 2775–2781.

Magnetoreception and magnetosomes in bacteria. (2007). *In* D. Schüler, (Ed.). *Microbiology monographs*. Vol. 3. (p. 319). Berlin Heidelberg: Springer.

Simmons, S. L., Bazylinski, D. A., & Edwards, K. J. (2006). South-seeking magnetotactic bacteria in the Northern Hemisphere. *Science*, *311*, 371–374.

Chapter 19

Quorum Sensing

M.M. Ramsey, A.K. Korgaonkar and M. Whiteley
The University of Texas at Austin, Austin, TX, USA

Chapter Outline

ABBREVATIONS

AHL Acylhomoserine Lactone
AIP Autoinducing peptide
CSP Competence-stimulating peptide
CT Cholera toxin
DPD 4,5-Dihydroxy-2,3-pentanedione
EHEC Enterohemorrhagic *E. coli*
EPS Exopolysaccharide
GBAP Gelatinase biosynthesis-activating pheromone
HHQ 4-Hydroxyl-2-heptyl-quinoline
LEE Locus of enterocyte effacement
PQS Pseudomonas Quinolone Signal
SAH *S*-adenosylhomocysteine
SAM *S*-adenosylmethionine
SRH *S*-ribosylhomocysteine
TCP Toxin coregulated pilus

DEFINING STATEMENT

Bacteria utilize numerous mechanisms to monitor and adapt to their external environment. One such mechanism involves the ability to "count" their local population numbers. Once a specific number of cells, or quorum, is reached, bacteria are able to modify their group behavior through a mechanism known as quorum sensing.

INTRODUCTION

Bacterial growth and survival is dependent upon the ability of an organism to sense its environmental conditions and respond to external stimuli. Stimulus can come from a variety of sources including the nutrients available for growth, the presence of secondary metabolites, and the presence of other microorganisms. In the case of

Topics in Ecological and Environmental Microbiology.

nutrients and secondary metabolites, sensor and regulatory proteins exist that affect a change in gene expression as these compounds change concentration. These changes in expression can cause bacteria to synthesize new proteins for catabolism in the case of diauxic growth, can lead to the production of extracellular enzymes to liberate nutrients from the environment in the case of starvation, or can cause changes in motility, chemotaxis, or metabolism that allow the bacteria to avoid or eliminate toxic concentrations of a secondary metabolite. In addition to these basic stimulus–response events, bacteria have evolved a mechanism to "count" the local cell population. This process is known as quorum sensing and is mediated by the bacterium's ability to produce and recognize soluble factors known as quorum signals.

The ability of bacteria to utilize extracellular signals to modify their behavior in a cell density-dependent manner was first described in the early 1970s by Kenneth Nealson, Terry Platt, Woodland Hastings, and Anatol Eberhard. These researchers discovered that the bioluminescent bacterium *Vibrio fischeri* only produced light when bacterial cell numbers were high (Figure 1). They also demonstrated that culture supernatants from high cell density cultures were able to stimulate light production in low cell density cultures. This phenomenon was deemed autoinduction and provided the first clues that bacteria were able to utilize a soluble extracellular signal to monitor population density.

Subsequent work characterized the autoinducer of *V. fischeri* and the genes responsible for its synthesis and detection by the cell. As autoinducer-related genes were studied, it was discovered that many other Gram-negative bacterium contained the genes necessary for autoinducer synthesis and detection. It was demonstrated that autoinducer sensing was dependent upon a critical number of cells in a defined volume. This threshold density of cells was first referred to as a quorum by Clay Fuqua, Stephen Winans, and Peter Greenberg in 1994, and they proposed the term *quorum sensing* to describe this event.

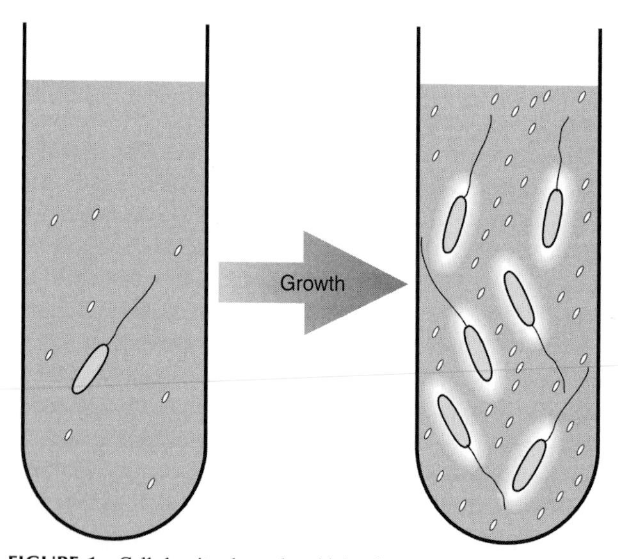

FIGURE 1 Cell density-dependent bioluminescence. As cells multiply, autoinducer concentrations increase. At a critical concentration, they induce synthesis of bioluminescence genes and subsequent light production.

Quorum sensing has rapidly expanded as a field of study, and the discovery of new quorum signaling bacteria as well as new types of quorum signals is increasing at an ever growing rate. Quorum signals in Gram-negative species such as *V. fischeri* were the first to be studied at the genetic and chemical levels; however, a significant amount of work has subsequently been performed with Gram-positive species. It is notable that quorum signal-dependent behavior was observed in Gram-positive bacteria long before the discovery of autoinducers or quorum sensing.

QUORUM SIGNALING IN GRAM-POSITIVE BACTERIA

Although the molecular mechanism was unknown, the physiological effects of quorum signaling were known for a considerably long time in some Gram-positive species. Early work in bacterial genetics determined that some bacterial species were able to participate in horizontal gene transfer through natural competence. Several pioneering studies were done with organisms such as *Streptococcus pneumoniae* well before it was known that the process was quorum signal-dependent. Gram-positive species typically utilize small polypeptide signals as quorum signals. Some of these polypeptides undergo significant modifications to become a soluble extracellular signal (Figure 2).

S. pneumoniae ComC

S. pneumoniae is a common causative agent of pneumonia, otitis media, meningitis, and several other diseases. Genetic transformation studies were carried out as early as 1928 in this organism by Frederick Griffith. Subsequent work by Avery, McCarty, and MacLeod in 1944 demonstrated that genetic transformation in *S. pneumoniae* was due to uptake of extracellular DNA. Competency in *S. pneumoniae* is regulated by several factors but is primarily controlled by the competence-stimulating peptide (CSP) quorum signal encoded by the gene *comC*.

Genetic competence in *S. pneumoniae* occurs during exponential phase and is initiated by CSP as cell density increases. The 17-amino acid CSP is produced by posttranslational modification of the 41-amino acid precursor peptide ComC and is exported from the cell by the ComAB ABC transporter. When extracellular CSP reaches a threshold concentration due to increasing cell number, it is recognized by the sensor histidine kinase ComD. ComD autophosphorylates and subsequently transfers its phosphate group to the response regulator ComE. Phosphorylated ComE activates

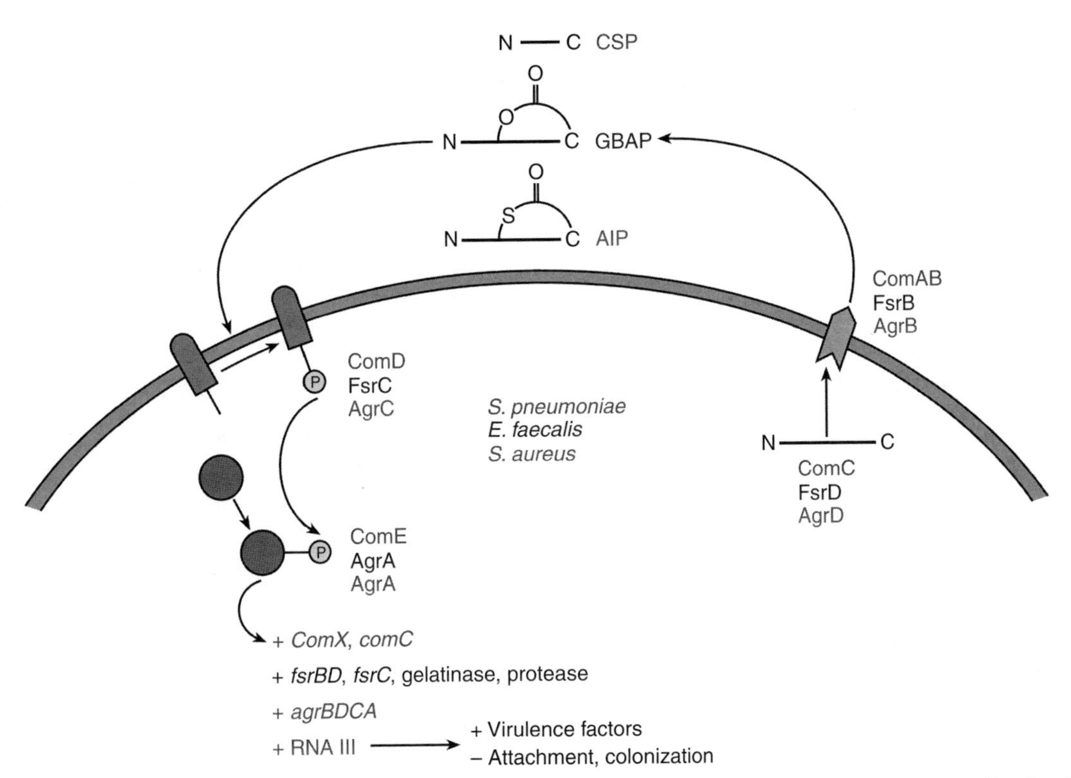

FIGURE 2 Quorum sensing peptides of *Streptococcus pneumoniae*, *Enterococcus faecalis*, and *Staphylococcus aureus*. Polypeptides ComC, FsrD, and AgrD are modified and exported out of the cell by the ComAB, FsrB, or AgrB transporters. Peptide autoinducers bind to the sensor kinases ComD, FsrC, or AgrC on the cell surface and initiate a phosphorelay to a cytoplasmic response regulator ComE or AgrA, which upregulates cell density-dependent gene expression.

transcription of the alternative sigma factor ComX, which goes on to induce expression of multiple genes involved in DNA uptake and recombination (Figure 2).

This peptide-induced, two-component signal transduction system is typical for Gram-positive quorum signaling circuits and was among the first to be characterized. Induction of competency at high cell density allows *S. pneumoniae* to sample and incorporate genetic material from its environmental counterparts. Surprisingly, *S. pneumoniae* competency permits the organism to take up DNA irrespective of its sequence or host origin. Such behavior gives *S. pneumoniae* the ability to adapt to selective pressures by taking up genes from neighboring cells that may have acquired genes or mutations that allow for enhanced fitness in a specific growth environment.

Enterococcus faecalis FsrB/D

E. faecalis is an opportunistic pathogen that causes endocardial, urinary tract, epidermal, and septic infections in clinical environments. *E. faecalis* is resistant to many common antibiotics and has recently acquired resistance to more modern antibiotic types making it a dangerous human pathogen. *E. faecalis* quorum sensing uses a cyclic lactone-modified peptide signal known as the gelatinase biosynthesis-activating pheromone (GBAP). GBAP is an 11-residue, circular polypeptide closed by a lactone moiety not seen in other Gram-positive-produced cyclic peptides. GBAP is not produced by all *E. faecalis* strains, but its production has been observed in all strains that produce the virulence factor gelatinase.

Originally, the *E. faecalis* GBAP signal was thought to be derived from the C-terminal region of the 212-amino acid FsrB protein. Recent studies have shown that a previously unknown gene, *fsrD*, exists in frame with *fsrB*, and GBAP is derived from FsrD by proteolytic activity of the FsrB protein. The FsrB protein was originally thought to autocatalyze by simultaneously forming and modifying GBAP while exporting it out of the cell. This new data show that the N-terminal region of FsrB is necessary for modification of the previously unknown FsrD peptide to produce the mature GBAP signal.

GBAP is produced maximally at the end of exponential phase in *E. faecalis*. When GBAP concentration reaches a density-dependent threshold, it induces autophosphorylation of the sensor kinase FsrC followed by phosphotransfer to the AgrA protein. Phosphorylated AgrA is a response regulator that promotes expression of *fsrBD*, *fsrC*, and the virulence factors gelatinase and serine protease (Figure 2). Recent transcriptome analyses of *fsrBD* mutants have shown indirect regulation of genes involved in biofilm formation, surface protein synthesis, and carbon source uptake, as well as direct regulation of an uncharacterized open reading frame. Interestingly, GBAP production is critical for successful colonization in an animal model. It is hypothesized that density-dependent regulation of gelatinase and serine protease by GBAP at the onset of stationary phase allows *E. faecalis* to liberate nutrient sources in the host as they become limiting.

Staphylococcus aureus AgrD

S. aureus is a human pathogen that causes toxic shock syndrome, food poisoning, and epidermal and endocardial infections. The prevalence of antibiotic-resistant *S. aureus* strains has placed it among the most commonly encountered nosocomial infections. *S. aureus* uses the accessory gene regulation (*agr*) system to regulate numerous genes involved in virulence and colonization in a density-dependent fashion via an autoinducing peptide (AIP) signal.

AIP is a thiolactone containing octapeptide ring derived from the 46-amino acid protein AgrD. To form AIP, AgrD is proteolyzed, modified by addition of a thiolactone moiety and exported from the cell by AgrB. Extracellular AIP is detected by the AgrC sensor histidine kinase, which autophosphorylates when bound to AIP. Phospho-AgrC transfers its phosphate group to the response regulator AgrA. The AgrC-AgrA two-component sensor kinase system upregulates the *agrBDCA* operon, establishing a positive feedback loop at sufficient extracellular AIP concentrations. AgrA also induces transcription of the regulatory RNA, RNAIII. RNAIII is an untranslated RNA that upregulates many *S. aureus* virulence factors including toxins and extracellular proteases while downregulating low-density genes involved in attachment and colonization (Figure 2). Thus, AIP signaling provides *S. aureus* with the ability for transition from colonization and survival at low cell density to pathogenesis and nutrient acquisition at high cell density. This behavior has been hypothesized to allow *S. aureus* to establish a colonization site in the host before virulence factor activity stimulates the host-immune response.

The *S. aureus* quorum sensing and response pathway is very similar to the one found in *E. faecalis*; however, there are notable differences including an increased number of AIP-regulated genes as well as strain specificity of the AIP signal. *S. aureus* AIP has been shown to modulate expression of a larger number of genes, a feat accomplished by induction of RNAIII, and other transcriptional regulators. AIP signaling has also been shown to be strain-specific. There are four known *agrD* specificity groups in *S. aureus* strains. AIP from a single specificity group has been shown to inhibit the AgrC sensor kinase of other groups. Thus, *S. aureus* has acquired a way to compete against other strains of its own species in an infection site when an individual strain reaches a threshold cell density. Therefore, the first strain in a single specificity group to achieve autoinduction of the *agr* pathway will induce its own virulence factor production and block quorum signaling activity in other strains. Quorum sensing systems similar to AIP are present in other *Staphylococcus* species, and these systems exhibit group strain competition as well.

Bacillus subtilis ComX and Phr Peptides

B. subtilis is a soil bacterium that can undergo spore formation or display genetic competence upon entry into stationary phase. Roughly 10% of *B. subtilis* cells entering stationary phase become competent. Competency in this situation is thought to aid *B. subtilis* in DNA repair and contribute to the inheritance of genetic material from *B. subtilis* strains that have successfully grown to high cell density. The decision to become competent or to sporulate is influenced by many factors in a complex regulatory pathway through which quorum signals play a significant part.

The ComX peptide is a 10-amino acid linear peptide derived from a 55-amino acid precursor product of the *comX* gene. The original ComX polypeptide is cleaved, modified, and exported out of the cell by the ComQ protein. Extracellular ComX is sensed by the ComP sensor kinase. Above threshold concentrations of extracellular ComX, ComP phosphorylates the response regulator ComA, which induces the *comS* gene. ComS blocks proteolysis of the transcriptional activator ComK, which upregulates the production of many genes that stimulate competency in the host cell (Figure 3).

A second quorum signaling pathway utilizes multiple proteinaceous pheromone (Phr) signals in a complex regulatory circuit (Figure 3). Phr precursors are proteolyzed and exported out of the cell as linear pentapeptides. Extracellular Phr is taken up into the cell by an ABC oligopeptide transporter. Multiple Phr genes exist and serve to antagonize the activity of several Rap phosphatase enzymes, which regulate many steps along the sporulation and competence pathways. For example, extracellular PhrA enters *B. subtilis* and binds directly to the RapA regulatory phosphatase either preventing it from binding or disassociating it from bound Spo0F-phosphate. RapA bound to the intermediate response regulator Spo0F-phosphate prevents a phosphorelay cascade to the global response regulator Spo0A, thereby preventing sporulation. Thus, increased PhrA levels are one of many factors that allow sporulation to begin by interrupting Spo0F inhibition of Spo0A. In addition to PhrA, there are several other Phr-type signals that are recognized by *B. subtilis* at the end of exponential phase and help direct the organism to specialize in sporulation or competence.

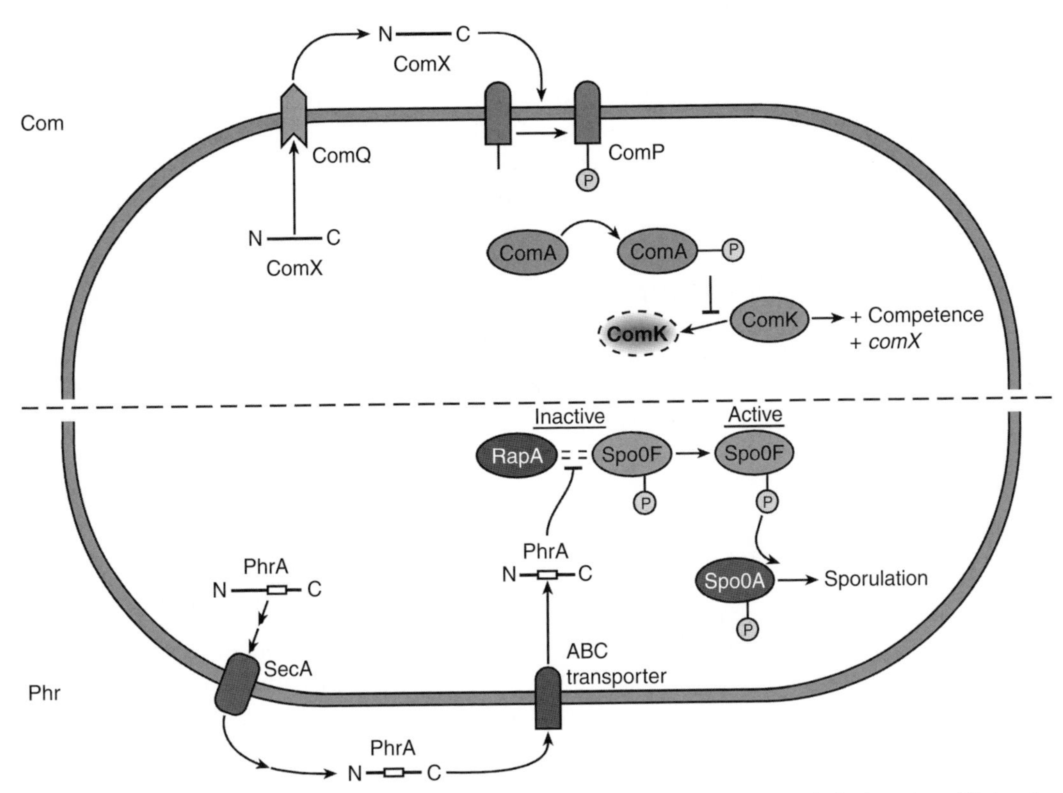

FIGURE 3 Quorum sensing pathways in *Bacillus subtilis*. In the Com system (upper panel), the ComX protein is cleaved, modified, and exported out of the cell by the ComQ transporter. Mature ComX interacts with the ComP sensor kinase that phosphorylates ComA. ComA-phosphate blocks the turnover of the transcriptional regulator ComK, which induces *comX* and competence gene expression. The Phr system (lower panel) utilizes the PhrA protein, which is cleaved and exported out of the cell using the Sec transport system. Mature PhrA returns to the cytoplasm via an oligopeptide ABC transporter and interferes with RapA binding to Spo0F-phosphate. Spo0F-phosphate can then phosphorylate the response regulator Spo0A, which induces sporulation genes.

Lactococcus lactis Nisin

L. lactis is a fermentative bacterium commonly used in the dairy industry to produce buttermilk and cheese. *L. lactis* is classified as a lactic acid bacterium as it produces large amounts of lactate upon fermentation of sugars found in dairy products. *L. lactis* produces the antimicrobial peptide nisin, which was discovered in 1928 due to its ability to inhibit growth of other lactic acid bacteria. In 1988, nisin was approved by the FDA for use as a preservative agent in the food industry.

Nisin is a Class I bacteriocin or lantibiotic characterized by its small size and the inclusion of the amino acids lanthionine and β-methyllanthionine as well as other dehydrated amino acids. Nisin is bactericidal and forms pores in bacterial membranes. This activity is especially effective against other Gram-positive bacteria where cell membrane perforations immediately cause a loss of membrane integrity.

Despite its role as an antimicrobial agent, nisin also serves as a density-dependent signal for *L. lactis*. Nisin is a highly modified 34-amino acid product of the 57-amino acid polypeptide NisA. NisA is modified and exported out of the cell by the NisBCT membrane-associated complex. Upon leaving the cell, modified NisA is cleaved to form nisin by the extracellular NisP protease. Damage of the parent cell membrane by nisin is prevented by the production of several proteins that block nisin activity at the cell surface. Extracellular nisin concentration is sensed by the NisK sensor kinase, which phosphorylates and activates the NisR response regulator. Nisin production is maximal during early stationary phase growth. NisR promotes expression of the *nisA* gene as well as genes that encode for proteins that block nisin activity at the cell surface, allowing for a positive feedback loop to upregulate nisin production as *L. lactis* enters stationary phase. Interestingly, the production of nisin demonstrates the ability of a compound to act as both a quorum signal and an antimicrobial agent. This type of dual function by an extracellular compound provides *L. lactis* with an efficient and powerful quorum signaling system that might provide a notable competitive advantage in polymicrobial environments.

ACYLHOMOSERINE LACTONE (AHL) QUORUM SIGNALS IN GRAM-NEGATIVE BACTERIA

The distribution of Gram-negative and Gram-positive bacterial species throughout the world is highly similar. Both types of organisms inhabit nearly every known niche on the planet and are involved in biogeochemical mineral cycles, biochemical degradation, disease, and the production of a staggering number of extracellular small molecules. While Gram-positive species have largely evolved to utilize polypeptide quorum signals that depend on cell-surface receptors, Gram-negative bacteria often utilize soluble signals known as AHLs. Many AHLs are able to pass through the lipid bilayer and are therefore able to interact with cytoplasmic regulatory proteins; thus, AHLs do not rely on the phosphorelay cascades that Gram-positive quorum sensing pathways commonly use. Both types of signals are effective due to their stability and solubility in environments colonized by the organisms that synthesize them.

V. fischeri LuxI/R

As mentioned previously, quorum signaling was discovered in the luminescent marine bacteria *V. fischeri* and *Vibrio harveyi*. In the early 1970s, researchers observed that supernatants from stationary phase cultures could be added to cells at low density and trigger light production. This suggested the presence of an autoinducing compound made in later phases of *V. fischeri* and *V. harveyi* growth. Further study determined that the factor in spent medium was relatively species-specific and dependent on cell density rather than the nutritional status of the cells. In 1981, the *V. fischeri* autoinducer was purified and determined to be the AHL 3-oxo-hexanoyl-HSL ($3OC_6$–HSL). A variety of AHL compounds have been characterized from other organisms and are collectively referred to as autoinducer-1 (AI-1).

V. fischeri can be found free-swimming and can also participate in a symbiotic relationship with the Hawaiian bobtail squid *Euprymna scolopes*. *V. fischeri* colonizes a cavity on the squid host known as the light organ. *V. fischeri* receives nutrients from the host in the light organ and emits light as its population density and AI-1 concentrations increase. Luminescence from the light organ of *E. scolopes* is thought to help the squid evade predation by masking the organisms' shadow in shallow water. *V. fischeri* has also been observed to enter symbiotic relationships with other marine organisms such as the fish *Monocentris japonicus* where a *V. fischeri*-dependent light organ is used to attract a potential mate.

The *V. fischeri* LuxI protein synthesizes 3-oxo-C6-HSL from *S*-adenosylmethionine (SAM) and acylated acyl carrier protein. Once synthesized, $3OC_6$–HSL freely diffuses across the membranes and out of the cell. At a particular density, a critical concentration of AI-1 is reached that stimulates AI-1 to interact with and activate the response regulator LuxR. Activated LuxR promotes transcription of the *luxR* gene as well as the *luxICDABEG* operon, which serves to produce light and generate more AI-1 (Figure 4). Thus, a positive feedback loop develops at sufficient cell densities, which allows for a substantial increase in light production. AI-1 exists in many Gram-negative species and is thought to function as an intraspecies-specific signal.

Recently, a second AHL quorum signal has been identified in *V. fischeri* and has been shown to function as part of a regulatory network in combination with $3OC_6$–HSL and a third signal, the furanosyl borate diester signal AI-2 (which will be discussed later). This AHL signal is *N*-octanoyl HSL (C_8–HSL) and is produced by AinS (Figure 4). C_8–HSL regulates luminescence at low culture densities by relieving negative regulation of the LuxR protein due to inactivation of the transcriptional regulator LuxO. Inactivation of LuxO allows upregulation of *litR*, whose gene product serves to activate *luxR* transcription. C_8–HSL also interacts with LuxR directly, allowing for initial activation of the *luxICDABEG* operon.

Pseudomonas aeruginosa Las/Rhl AI-1 System

P. aeruginosa is a Gram-negative bacterium commonly isolated from soil environments, which is frequently used as a model organism to study quorum sensing. *P. aeruginosa* is an opportunistic pathogen that can cause an array of infections in immunocompromised individuals, most notably those inflicted with the heritable disease cystic fibrosis. Many of the genes necessary for infection, nutrient acquisition, virulence factor production, and biofilm growth are regulated by the concentrations of two AHL signals produced by LasI and RhlI. Due to its effects on virulence factor production, interference with quorum signaling in *P. aeruginosa* has been proposed as a therapeutic strategy. This novel approach to antimicrobial therapy in a notoriously antibiotic-resistant organism makes further study of quorum sensing in *P. aeruginosa* increasingly important.

The LasI protein synthesizes the signal *N*-(3-oxododecanoyl)-L-HSL ($3OC_{12}$–HSL). $3OC_{12}$–HSL interacts with the LasR response regulator, which upregulates the expression of multiple genes involved in virulence such as the *lasB* protease as well as the *lasI* gene itself, establishing $3OC_{12}$–HSL as an autoinducer. While $3OC_{12}$–HSL is diffusible, it can also partition into the cell membrane and has been shown to be exported by efflux pumps in *P. aeruginosa*.

The LasR–$3OC_{12}$–HSL complex serves to upregulate a second AHL-dependent system by inducing expression of *rhlR* (Figure 5). RhlI synthesizes *N*-butyryl-L-HSL (C_4–HSL), which interacts with the response regulator RhlR to regulate numerous genes including those involved in secondary metabolite production, as well as the *rhlI* gene that establishes a second AHL autoinduction loop. C_4–HSL-dependent regulation overlaps with $3OC_{12}$–HSL-mediated regulation in the control of several genes such as the extracellular protease encoded by the *lasB* gene. It is notable that $3OC_{12}$–HSL inhibits the interaction of C_4–HSL

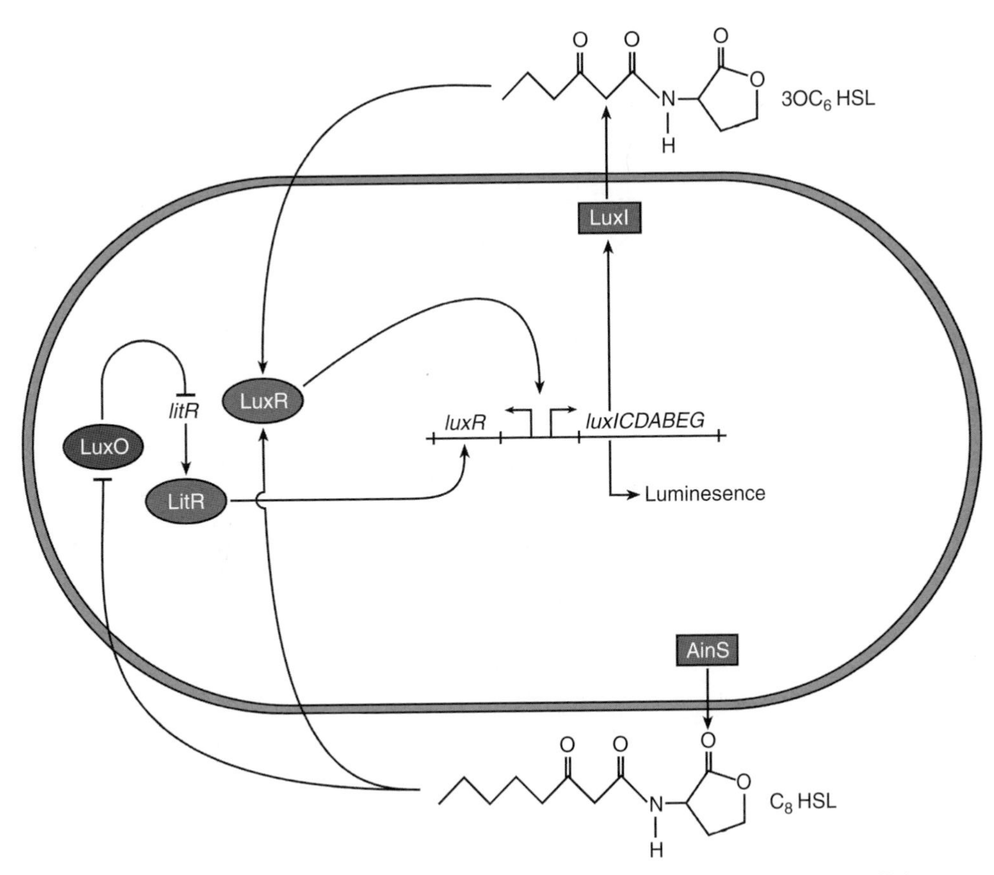

FIGURE 4 Quorum sensing in *Vibrio fischeri*. The LuxI-produced $3OC_6$–HSL diffuses out of the cell until it reaches sufficient concentration to induce the LuxR response regulator. $3OC_6$–HSL–LuxR induces expression of the *luxR* and the *luxICDABEG* genes, which increase synthesis of $3OC_6$–HSL as well as proteins involved in bioluminescence. At low cell density, C_8–HSL, produced by the AinS gene, provides initial activation of LuxR by direct activation as well as by inhibiting LuxO, preventing the *litR* gene product from inducing the *luxR* gene, thereby increasing levels of *luxR* transcript in the cell.

with its response regulator RhlR while inducing expression of the *rhlR* gene. This interaction serves to maintain low levels of C_4–HSL until RhlR concentration increases to levels sufficient to interact with the low levels of C_4–HSL produced in late exponential phase. Thus, the activation of the C_4–HSL–RhlI feedback loop becomes functional only after LasR–$3OC_{12}$–HSL-dependent regulation has occurred.

Both quorum signals in *P. aeruginosa* ultimately serve to regulate the expression of over 300 genes involved in diverse pathways. Such regulation is not achieved by the LuxR-like LasR and RhlR response regulators alone. For example, LasR induces expression of the *vqsR* gene, whose product upregulates genes involved in virulence factor production and quorum signaling. Apart from downstream regulators induced by LasR or RhlR, "orphan" LuxR-type regulators that do not have a cognate LuxI-type produced AHL signal have been discovered. QscR is a LuxR-type response regulator that binds the promoters of several genes in the presence of the LasI-generated signal $3OC_{12}$–HSL. Apart from these interactions, QscR has been shown, at least indirectly, to repress *lasI* expression (and was in fact named for this activity, quorum signaling control repressor). The manner in which QscR represses *lasI* expression is unclear but has been hypothesized to involve heteromultimerization of QscR and LasR. Another hypothesis is that QscR binding to $3OC_{12}$–HSL prevents this AHL from coming in contact with its cognate regulator LasR. Interestingly, QscR has also been shown to be more promiscuous in AHL binding than either LasR or RhlR, suggesting a role in interspecies AHL sensing. Ultimately the HSL-dependent quorum signaling network combined with orphan response regulators contribute a great deal to global gene regulation in *P. aeruginosa*.

Agrobacterium tumefaciens Tra System

A. tumefaciens is a Gram-negative soil bacterium, which is the causative agent of crown gall disease in plants. Some *A. tumefaciens* strains carry a Ti (tumor-inducing) plasmid that contains the LuxI/R-type quorum signaling genes *traI* and *traR*. The Ti plasmid also carries genes that encode for specific plant hormones. After conjugation of the Ti plasmid from *A. tumefaciens* to the host nucleus, plants synthesize these hormones that then stimulate cell proliferation and tumor formation.

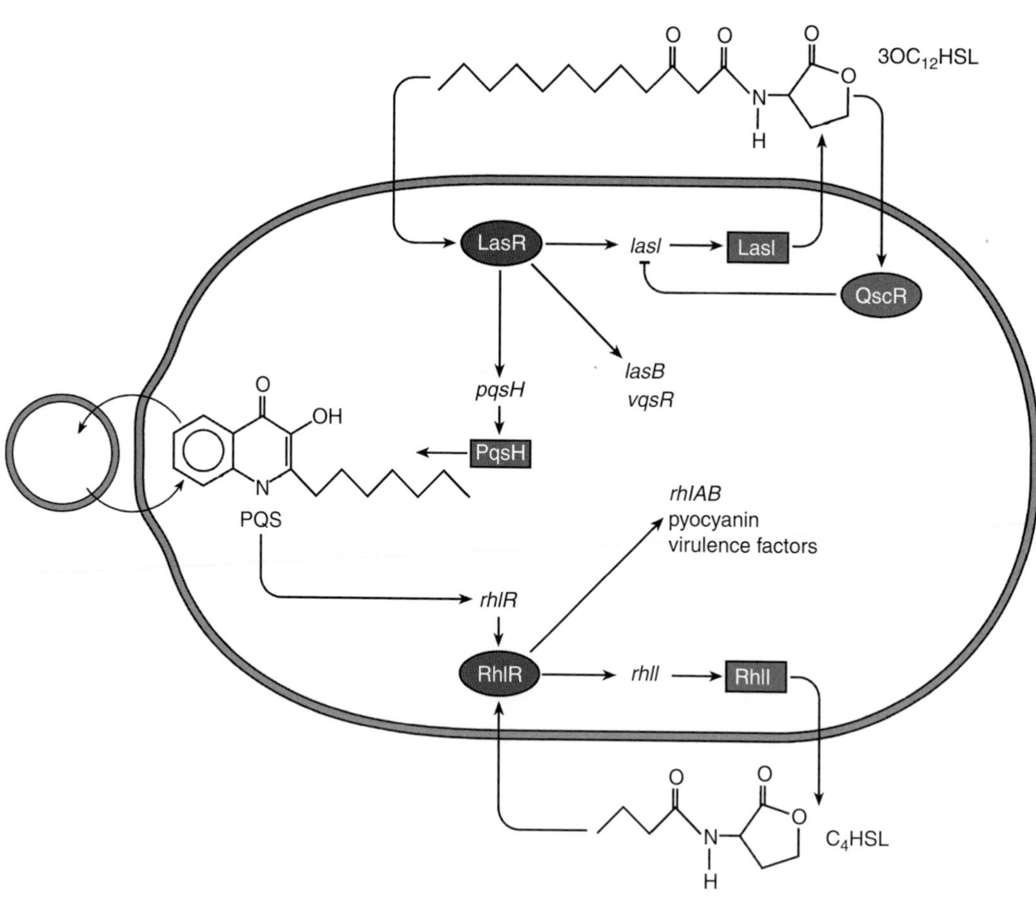

FIGURE 5 Quorum sensing in *Pseudomonas aeruginosa*. $3OC_{12}$–HSL produced by the LasI synthase interacts with the LasR transcriptional regulator, which upregulates *lasI*, *lasB*, *vqsR*, and *pqsH*. $3OC_{12}$–HSL also interacts with the orphan transcriptional regulator QscR to repress *lasI* expression. PqsH produces the quinolone signal PQS that is transported between cells via outer membrane vesicles. PQS induces expression of the *rhlR* gene whose product induces expression of the *rhlI* synthase gene as well as several other virulence factors when induced by the RhlI-produced C_4–HSL signal.

Free-swimming *A. tumefaciens* move throughout the soil by flagellar motility and search for susceptible hosts by chemotaxis toward metabolic intermediates released from wounded plants. Upon contacting the host, *A. tumefaciens* produces cellulose fibrils as well as attachment proteins to anchor it to host tissue. After attachment, *A. tumefaciens vir* (virulence) genes are induced by host-derived compounds that in turn activate synthesis of a secretion apparatus that directs translocation of the Ti plasmid from *A. tumefaciens* into the host. Ti plasmid-encoded hormones then cause cell proliferation and tumor formation. The Ti plasmid also contains genes whose products direct synthesis of opines. Opines are specialized amino acids that *A. tumefaciens* uses as a growth substrate. There are at least two known classes of opine-encoding genes, octapine and nopaline, and *A. tumefaciens* strains or pathovars can be classified by genes that are present on the Ti plasmid. These plasmid-borne genes not only direct host synthesis of opines, but contain opine catabolism genes, allowing *A. tumefaciens* to use opines as a source of carbon and energy.

A. tumefaciens uses the TraI/R quorum sensing system to regulate conjugation of the Ti plasmid between *A. tumefaciens* strains. The *traI/R* quorum signaling genes are similar to the *luxI/R* genes from *V. fischeri*, but there are key differences in regulation. In octapine-utilizing pathovars, the *A. tumefaciens* transcriptional activator OccR is induced by the presence of the opine octapine. OccR positively regulates expression of *traI*, whose product synthesizes the *N*-(3-oxo-octanoyl)-L-HSL ($3OC_8$–HSL) quorum signal. In nopaline-utilizing pathovars, the opine agrocinopine inactivates the *traR* repressor AccR. This interaction also increases expression of *traI* in the cell (Figure 6).

Increased *traI* expression in the cell ultimately leads to an increase in $3OC_8$–HSL. As $3OC_8$–HSL reaches a critical threshold, it induces *traR* expression. When TraR is bound to $3OC_8$–HSL, it induces *traI*, establishing an autofeedback loop. Activated TraR also induces expression of the *trb* operon, which produces the mating pore between two *A. tumefaciens* cells; the *tra* operon, which helps mobilize the Ti plasmid; and the *traM* gene. TraM inactivates TraR, thus providing a means for *A. tumefaciens* to downregulate

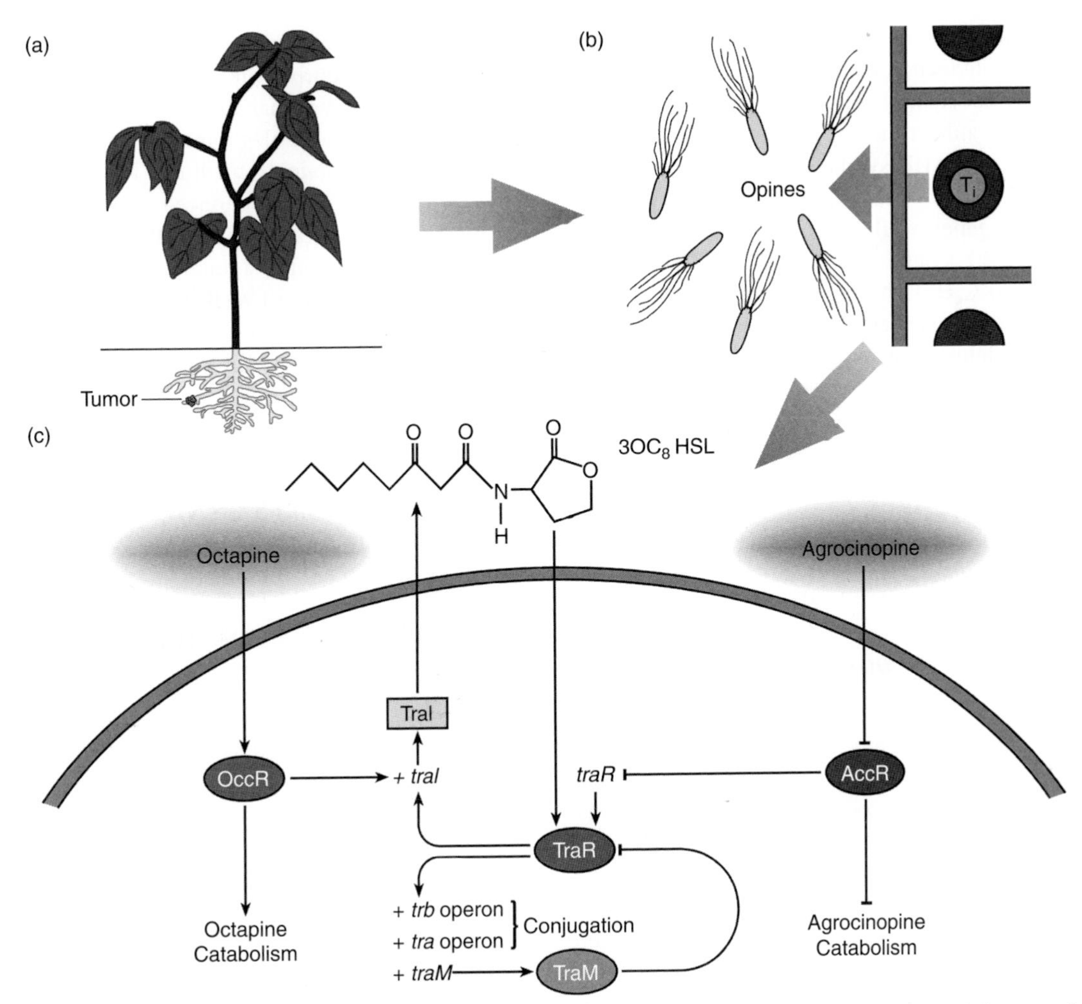

FIGURE 6 Quorum sensing in *Agrobacterium tumefaciens*. (a) A plant infected with *A. tumefaciens* displaying a characteristic tumor. (b) Inside the tumor, proliferating host cells that carry the Ti plasmid produce an opine nutrient for *A. tumefaciens*. (c) Octapine-type opines interact with the OccR transcriptional regulator causing it to upregulate octapine catabolism genes as well as induce *traI* expression. TraI synthesizes the $3OC_8$–HSL signal that interacts with the TraR protein, causing it to further induce *traI* expression and the *trb* and *tra* operons utilized in conjugation. $3OC_8$–HSL–TraR also induces the *traM* gene, whose product inhibits TraR activity.

the quorum signal-dependent synthesis of the conjugation apparatus after it has begun. The *A. tumefaciens* TraI/R system, while similar to the LuxI/R system of *V. fischeri*, has an additional level of regulation dependent on host opine production ensuring that quorum signal-controlled induction of conjugation does not occur prematurely outside of a compatible host. This system serves as an elegant example of interspecies-dependent cues having regulatory input in bacterial signaling pathways.

Pantoea stewartii EsaI/EsaR System

The Gram-negative bacterium *P. stewartii* is the causative agent of Stewart's bacterial wilt and leaf blight in sweet corn and maize. *P. stewartii* is spread by infected populations of the corn beetle, *Chaetocnema pulicaria* that introduces the bacterium into the xylem and intercellular leaf spaces of the plant during feeding. Infection by *P. stewartii* results in wilting due to colonization of the xylem and formation of water-soaked lesions due to bacterial growth within young leaves. Buildup of large amounts of exopolysaccharide (EPS) results in vascular occlusion of plant tissue. EPS secreted by *P. stewartii* is the principal virulence factor and is part of a multistep invasion process.

P. stewartii secretes EPS in a cell density-dependent fashion. Biosynthesis of EPS is encoded by the *cps* gene cluster, while regulation of EPS synthesis is mediated partly by the EsaI/R quorum signaling system. The EsaI/R system also controls the Hrp (hypersensitivity and pathogenicity) regulon. *P. stewartii* requires the Hrp type III secretion system for infection of the intercellular leaf spaces and formation of the characteristic water-soaked lesions.

The *cps* genes are regulated by the Rcs (*r*egulator of *c*apsule *s*ynthesis) two-component signal transduction system that detects environmental signals. The Rcs system is composed of the RcsB cytoplasmic response regulator and the RcsC transmembrane sensory protein. It is thought that for complete induction of the *cps* genes, another protein (RcsA) may be required. Also, in the absence of AHLs, EsaR negatively regulates the *cps* genes.

Interestingly, the AHL synthase gene *esaI* is constitutively expressed and is not autoregulated by *esaR* like many other *luxI/R*-type genes, while *esaR* is autorepressed by the EsaR protein. EsaI synthesizes 3-oxo-C_6HSL and small amounts of 3-oxo-C_8HSL. Mutants in *esaI* do not produce AHLs or EPS and are avirulent. In contrast, *esaR* and *esaI/R* double mutants demonstrate a constitutive hypermucoidy phenotype and are less virulent than the wild type. The hypermucoidy mutants of *P. stewartii* also appear impaired in attachment compared to wild-type *P. stewartii*. This indicates the importance of quorum control of pathogenicity factors in *P. stewartii* as production of EPS at an incorrect location and phase of infection renders the cells avirulent and unable to colonize the host.

Acidithiobacillus ferrooxidans AfeI/R

A. ferrooxidans is a Gram-negative acidophilic chemolithotrophic bacterium that is commonly found in multispecies biofilms on mineral surfaces such as pyrite and elemental sulfur in rock and soil. *A. ferrooxidans* catalyzes the oxidation of iron and sulfur yielding sulfuric acid. It is a causative agent of acid mine drainage, which can lead to groundwater contamination. The ability of *A. ferrooxidans* to dissolve mineral structures can also be used in "biomining" where acid solubilization of rocks and soil can release minerals such as copper and gold that are used in industrial applications. This process, also known as bioleaching, is a slower yet more energy efficient and environmentally containable process as opposed to traditional smelting of ores to release rare metals.

To date, AHL quorum signaling in *A. ferrooxidans* is relatively poorly understood. Recent studies have determined that *A. ferrooxidans* produces nine different medium to long-chain (C8–C16) AHL signals containing an even number of carbons. The AHL signals also contain either oxo or hydroxyl modifications at the third carbon of the molecule. At this time, only one AHL synthesis gene, *afeI*, has been identified and characterized, but another open reading frame has been hypothesized to be an AHL synthase due to its sequence homology to the AHL synthase *hdtS* from *Pseudomonas fluorescens*.

AfeI is similar to LuxI of *V. fischeri* and produces five different AHLs when expressed in an AHL-lacking *Escherichia coli* strain. Immediately downstream of the *afeI* gene in *A. ferrooxidans* is the *luxR* homologue *afeR*. Transcriptional studies show that *afeI* and *afeR* transcripts are present and *afeI* transcript levels increase relative to increases in AHL concentration. Downstream regulatory targets for AfeR in *A. ferrooxidans* have not been fully determined, but it has been shown that *afeI* transcripts are strongly induced during phosphate-limiting conditions. It has also been shown that there are differential increases of specific AHL molecules during growth in sulfate, thiosulfate, or iron-containing medium. This data suggest that the type of AHL signal produced may be influenced by the growth substrate as well as cell density. The fact that *A. ferrooxidans* is often found in multispecies biofilms and produces a variety of AHL types could also suggest a role for *A. ferrooxidans* in modulating density-dependent expression of other species in response to growth substrates *in situ*.

THE STUDY OF AHL-DEPENDENT SIGNAL PATHWAYS DEMONSTRATED THE PRESENCE OF ANOTHER QUORUM SIGNAL AI-2

V. harveyi and the LuxS-Produced Signal AI-2

Previously, we discussed the discovery of quorum signaling in the luminescent marine bacteria *V. fischeri*. Further work with *V. fischeri* identified and characterized the LuxI/R regulatory circuit and the AI-1 HSL signal. However, determining the nature of the density-dependent signal in the closely related luminescent bacterium *V. harveyi* yielded different results. *V. harveyi* luminescence is density-dependent, inducible by spent culture medium, and is dependent upon a 3OHC_4–HSL. Despite these similarities, genes homologous to *luxI* and *luxR* were not found on the *V. harveyi* chromosome.

Mutational analysis determined that *V. harveyi* 3OHC_4–HSL was produced by the AHL synthase proteins LuxL and LuxM. It is not clear which protein is the actual synthase, but the presence of both is necessary for 3OHC_4–HSL production. Neither gene has significant sequence homology to *V. fischeri luxI* but appear to carry out similar reactions to synthesize 3OH–C_4–HSL. It was also found that a *luxN* gene encoded for a sensor kinase protein, which was necessary to sense extracellular AHLs. This mechanism is reminiscent of Gram-positive AIP sensors. LuxN was later shown to autophosphorylate at low 3OHC_4–HSL concentrations and transfer phosphate to the LuxO response regulator via the intermediate LuxU. LuxO-phosphate represses expression of the *luxCDABE* luminescence operon, ensuring that light production is turned off at low AHL concentrations. Knockout mutants in either the *luxLM* synthesis genes or the *luxN* sensor kinase gene did not completely abrogate density-dependent expression of luminescence genes. This suggested the presence of a second quorum signaling system in *V. harveyi*.

The same mutational study that characterized the *luxL*, *luxM*, and *luxN* genes also showed that deletions in the sensor kinase *luxQ* and periplasmic binding protein *luxP* were necessary for density-dependent luminescence independent of the $3OHC_4$–HSL signal. These proteins were shown to be part of a second sensory mechanism in *V. harveyi*. Reporter strains were created using the *luxP* and *luxQ* genes to detect a second signal. Initial biochemical characterizations revealed that a second signal could activate the *luxQP* reporter strains, and this signal was named AI-2. Surprisingly, AI-2 is produced by several other species of bacteria and the *luxS* gene responsible for AI-2 synthesis was identified in *V. harveyi*, *E. coli*, and *Salmonella typhimurium* shortly after.

LuxS synthesizes AI-2 as part of a detoxification reaction in the activated methyl cycle. Methyl transfer reactions are vital in bacterial metabolism and are dependent on the methyl donor SAM. SAM is converted into the toxic *S*-adenosylhomocysteine (SAH) molecule and can be detoxified by two different mechanisms. The first mechanism utilizes the SAH hydrolase gene, which hydrolyzes SAH into the SAM precursors adenosine and homocysteine. Organisms that do not produce AI-2, such as *P. aeruginosa,* utilize this pathway. The second mechanism utilizes the Pfs enzyme, which hydrolyzes SAH into adenosine and *S*-ribosylhomocysteine (SRH). Homocysteine is removed from SRH leaving 4,5-dihydroxy-2,3-pentanedione (DPD) in a reaction catalyzed by the LuxS enzyme; DPD spontaneously cyclizes to form the furanosyl diester AI-2. Recent work has shown that AI-2 in *V. harveyi* incorporates boric acid into its structure, the first known biological use of this molecule. To date, it does not appear that boron-containing AI-2 signals are used outside of the *Vibrio* genus and that AI-2 molecules lacking boron function as signals in other bacteria.

As mentioned above, AI-2 is recognized by the sensor kinase LuxQ via the periplasmic AI-2-binding protein LuxP. When a threshold concentration of AI-2 is reached, LuxQ changes from a kinase to a phosphatase, removing the phosphate from its downstream target LuxU. As LuxU is dephosphorylated, it removes phosphate from the LuxO protein. Unphosphorylated LuxO loses the ability to repress the *luxCDABE* operon, and luminescence genes are then activated by the transcriptional activator LuxR, which has no homology to the *V. fischeri* LuxR regulator. Thus, the *V. harveyi* quorum-signaling pathway utilizes the input of two separate density-dependent signals that act on the central regulator LuxO. This pathway is similar to the *Vibrio cholerae* quorum-signaling pathway described in Figure 7.

It is notable that AI-2 synthesis is directly linked to a key metabolic pathway in organisms that carry the *luxS* gene. Because of this link, it is hypothesized that LuxS production is an indicator of the metabolic status of the organism as well as its population density. This could be similar to the differential AI-1-type signal production seen in *A. ferrooxidans*. It should also be mentioned that only two genomes sequenced to date contain genes for both SAH detoxification mechanisms, and this phenomenon may indicate a point of divergence in bacterial evolution.

AI-2 and *V. cholerae*

V. cholerae is a free-swimming marine bacterium and is the causative agent of cholera in humans. *V. cholerae* is commonly found attached to the surface of zooplankton in the ocean. Cholera outbreaks can be associated with

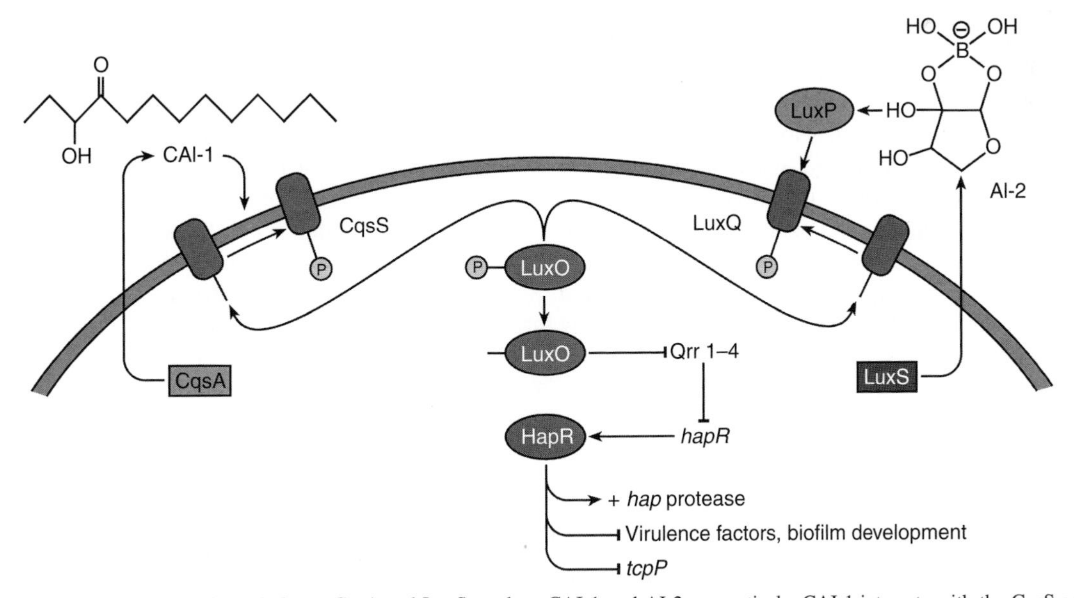

FIGURE 7 Quorum sensing in *Vibrio cholerae*. CqsA and LuxS produce CAI-1 and AI-2, respectively. CAI-1 interacts with the CqsS sensor phosphatase, and AI-2 interacts with the LuxQ sensor phosphatase via the AI-2-binding protein LuxP. Both phosphatases remove phosphate from the LuxO protein. Dephosphorylated LuxO is inactive and no longer induces expression of the small RNAs Qrr1–4, which served to repress *hapR* expression. Translation of HapR induces expression of the *hapR* protease gene and represses expression of genes involved in virulence and biofilm formation.

zooplankton blooms near human populations. During the course of disease, *V. cholerae* is ingested and survives the low pH of the stomach to colonize the host small intestine. During colonization, *V. cholerae* uses motility and mucinase to penetrate the mucus layer of the intestine and gain access to the underlying epithelial cell layer.

Currently, *V. cholerae* strain El Tor is the cause of a global cholera pandemic in underdeveloped nations. *V. cholerae* El Tor contains the HapR regulator that represses expression of virulence factors such as the cholera toxin (CT) and the cholera toxin coregulated pilus (TCP). TCP and CT have been shown to be critical for *V. cholerae* host colonization in animal models. The HapR regulator is acted upon by quorum signals in *V. cholerae* El Tor and links the quorum signaling pathway to virulence factor production. Previous pandemic strains of *V. cholerae* contain mutations within the *hapR* gene and are not considered to be responsive to quorum signaling.

V. cholerae contains a quorum sensing system that is similar to *V. harveyi*. However, one of the autoinducer signals as well as the genes regulated by quorum signaling are different in *V. cholerae*. In *V. cholerae*, the AI-2 signal and the recently characterized signal (*S*)-3-hydroxytridecane-4-one (CAI-1) are recognized by the sensor kinases LuxQ and CqsS, respectively (Figure 7). In low cell density conditions, LuxQ and CqsS act as kinases that phosphorylate LuxU, which then transfers its phosphate to LuxO much like the pathway in *V. harveyi*. Phosphorylated LuxO induces the expression of four small regulatory RNAs (sRNAs) known as Qrr1–4. These sRNAs interact with the sRNA chaperone protein Hfq and induce expression of an uncharacterized regulatory gene (vca0939) as well as destabilize the mRNA for *hapR*, the master quorum signaling regulator. When *hapR* is repressed, virulence genes are upregulated, due in part to removal of HapR repression of the *tcpP* gene, which encodes the signaling protein TcpP that induces expression of the ToxT regulator. ToxT induces expression of the virulence factors TCP and CT, so in conditions of low cell density and autoinducer concentrations, virulence and biofilm development genes are upregulated. TCP allows for cell–cell aggregation of *V. cholerae* as well as attachment to the host. CT activity causes epithelial cells to excrete large amounts of water, ultimately causing the host to rapidly expel its intestinal contents, one of the hallmark symptoms of a *V. cholerae* infection.

At high cell densities, LuxQ and CqsS act as phosphatase in the same fashion as LuxQ and LuxN in *V. harveyi*. LuxQ and CqsS ultimately cause the dephosphorylation of LuxO, which causes expression of the Qrr sRNAs to be decreased. As Qrr sRNA expression decreases, *hapR* is upregulated. HapR blocks expression of virulence factors and genes involved in biofilm development. HapR also induces expression of the *hap* gene, which encodes for a protease that aids *V. cholera* in release from a colonized area. In *V. cholerae*, density-dependent signals provide the cell with a method to change cell behavior from a colonizing, scavenging lifestyle at low density, to a motile, migratory lifestyle at high cell density. Interestingly, this change in behavior occurs at a time when pathogenic effects of *V. cholerae* are inducing the host to expel large amounts of intestinal contents, thus allowing the freshly liberated *V. cholerae* to return to a free-swimming lifestyle. It is also interesting that quorum signal-utilizing pathogens such as *S. aureus* and *P. aeruginosa* upregulate virulence factors at high density to form persistent infections, whereas *V. cholerae* downregulates virulence and rapidly leaves the host once high population density has been achieved.

E. coli, S. typhimurium, and AI-2

The Gram-negative enteric bacteria *E. coli* and *S. typhimurium* both grow in the lower intestine of mammals. While certain *E. coli* strains are pathogenic, many strains are part of the normal flora. *S. typhimurium* is the causative agent of typhoid fever and, while closely related to *E. coli*, is often a more transient inhabitant of the gastrointestinal tract. Both of these bacteria produce AI-2 using the LuxS enzyme, with maximum production occurring at mid-exponential phase growth. Surprisingly, extracellular AI-2 concentrations drop rapidly as either organism enters stationary phase. This suggests a mechanism for AI-2 sequestration or degradation.

AI-2 has been shown to regulate several genes in both *E. coli* and *S. typhimurium*. These genes include the *lsrACDBFGE* operon, the *lsrK* kinase gene, and *lsrR* whose product represses the *lsr* operon. In both organisms, AI-2 is produced during growth and accumulates outside of the cell. When threshold AI-2 concentrations are reached, AI-2 is imported by the LsrABCD ABC transporter. After import into the cell, AI-2 is phosphorylated by the LsrK kinase and catabolized by the LsrE, LsrF, and LsrG proteins. The products of phospho-AI-2 catabolism have not been characterized at the time of this writing. While the import and catabolism of AI-2 by these organisms is regulated in a density-dependent fashion, it is unclear if these organisms utilize AI-2 to direct group activities. The simplest explanation is that AI-2 is produced as a metabolite during SAH detoxification in the activated methyl cycle and is taken as backup to be utilized as a carbon source later in the growth phase.

Despite the possibility that *E. coli* and *S. typhimurium* may utilize AI-2 only as a metabolite, production of AI-2 by these bacteria may affect gene expression of other AI-2 responsive bacteria in polymicrobial environments. In a 2005 *Nature* paper, Karina Xavier and Bonnie Bassler demonstrated that when *V. harveyi* or *V. cholera* were grown in coculture with *E. coli*, they were subject to interference of AI-2-mediated signaling due to the ability of *E. coli* to respond to higher AI-2 concentrations. As *E. coli* cell density increased, it was able to remove AI-2 from the culture and reduce expression of AI-2-dependent genes in either *Vibrio* species. At low cell densities, premature induction of AI-2-regulated genes was observed in *V. harveyi*, as the AI-2 produced by *E. coli* increased the AI-2 concentration that *V. harveyi* would normally encounter at that cell density. It is apparent from these studies that AI-2 signal turnover can have substantial effects on multispecies interaction even if AI-2 does not appear to have a strong role in quorum-signal-dependent behavior in a single species.

luxS-Dependent Interactions in Human Oral Bacteria

The human oral cavity is populated by over 300 different bacterial species. There are several distinct niches within the oral cavity (tooth enamel surface, epithelial cell surface, and the subgingival space), all of which can be colonized by bacteria growing in multispecies biofilms. Many species isolated from the oral cavity contain the *luxS* gene, and in several cases, density-dependent AI-2-mediated behavior has been observed in single and multispecies cultures.

Streptococcus mutans is a common inhabitant of the oral cavity and is the causative agent of dental caries. Merritt and Shi in a 2003 *Infection and Immunity* paper demonstrated that biofilm formation by *S. mutans* was altered in a *luxS* mutant. The *S. mutans luxS* mutant formed a biofilm with denser cell aggregates that was more resistant to detergent and antibiotics. While growth rate and acid production did not appear to be altered in the *luxS* mutant, it was not determined whether extracellular AI-2 concentration or the effects of impeded SAH detoxification were responsible for the mutant phenotype.

Direct AI-2-dependent interaction in oral bacteria was reported in 2003 by McNab and Lamont in the *Journal of Bacteriology*. The researchers demonstrated that *Streptococcus gordonii* and *Porphyromonas gingivalis* formed significantly dense biofilm structures if either organism contained a functional *luxS* gene. Simultaneous *luxS* mutations in both species led to diminished biofilm formation, suggesting that *luxS*-dependent AI-2 formation from one organism can complement the ability of the other to form more fully developed multispecies biofilms.

This phenomenon is similar to another described in a 2004 *Proceedings of the National Academy of Sciences* paper by Egland and Kolenbrander. In that study, *S. gordonii* was shown to upregulate production of amylase only when grown in close proximity to *Veillonella atypica*. This interaction was hypothesized to be dependent on an increase in local concentration of a diffusible signal when the bacteria grow in close proximity to one another. This data suggest that signal production by *S. gordonii* alone is insufficient to upregulate amylase under the conditions tested, and the presence of another signal-producing organism is also necessary. AI-2 is hypothesized to be the diffusible signal in this system, but this has not been unequivocally shown. Both of these cases demonstrate that extracellular signal concentrations are critical to modulate behavior in each organism. This is notable because in the case of *S. mutans* and other literature discussing *luxS*-dependent signaling events, it is often unclear if the phenotypes observed are due to actual AI-2-dependent signaling, or due to metabolic effects caused by an impediment of SAH detoxification.

AI-2-dependent quorum signaling has also been observed in the opportunistic pathogen *Aggregatibacter actinomycetemcomitans*. LuxS activity or exogenous AI-2 addition to cultures have been shown to cause differential expression of iron uptake genes as well as leukotoxin production. AI-2 production by *A. actinomycetemcomitans* has also been shown to complement *luxS*-dependent gene expression in a *P. gingivalis luxS* mutant when cocultured. A 2007 *Infection and Immunity* paper by Shao and Demuth demonstrated that *A. actinomycetemcomitans* is impaired for biofilm production in a *luxS* mutant. Subsequent complementation of *luxS* by *P. aeruginosa sahH* seems to restore the ability to detoxify SAH in the activated methyl cycle, but exogenously added AI-2 was necessary to restore the biofilm phenotype. While *sahH* transcripts were present in the complemented strain, it is not clear if SahH was produced and able to detoxify SAH. Despite the unknown status of SahH activity, this data strongly suggest that AI-2 serves as a signal for community behavior in *A. actinomycetemcomitans*, and not just a metabolite produced in the activated methyl cycle.

AI-2 signaling is responsible for density-dependent regulation of genes in several well-characterized systems. At this time, over 153 separate eubacterial genomes contain a *luxS*-like sequence according to the Kyoto Encyclopedia of Genes and Genomes (KEGG) database. It is evident that in the case of *Vibrio* species and several inhabitants of the mammalian oral cavity, AI-2-dependent signaling plays a critical part of these organisms' lifestyle. However, in many organisms, the presence of a *luxS* gene does not correlate to density-dependent signaling behavior. The "other" function of LuxS as part of the activated methyl cycle is an important component of metabolism, and it cannot be assumed at this time that phenotypic changes in *luxS* knockouts are exclusively due to AI-2 signaling and not disruption of SAH detoxification.

OTHER TYPES OF QUORUM SIGNALS

In addition to AI-1, AI-2, and peptide autoinducer signals, there are many other types of quorum signals utilized by bacteria. We are only beginning to understand the molecular aspects of these systems, and in some cases the structure of the signal is not known.

The AI-3/Epinephrine Quorum Signaling System

E. coli is a common inhabitant of the human lower intestine. Recently, strains have evolved that carry multiple virulence factors giving rise to enterohemorrhagic *E. coli* (EHEC) strains such as *E. coli* O157:H7. EHEC strains are notorious for their ability to cause hemorrhagic colitis and hemolytic-uremic syndrome. As EHEC colonizes the lower intestine, it forms lesions on host epithelial tissue and produces a Shiga-type toxin. A majority of the genes involved in attachment, lesion formation, and virulence exist on the EHEC chromosome in a pathogenicity island known as the locus of enterocyte effacement (LEE).

Expression of *LEE* genes is regulated by factors present in the native *E. coli* chromosome as well as factors encoded on LEE itself. In addition to these, some *LEE* genes appear to be regulated in a cell density-dependent fashion. Preliminary studies of a *luxS* EHEC mutant showed changes in LEE gene expression, indicating that AI-2 may play a role in regulation. However, the addition of exogenous AI-2 to the *luxS* mutant did not complement LEE gene regulation. This data suggested that LEE gene regulation was subject to metabolic changes observed in the *luxS* mutant. It was then shown that LEE gene regulation could be complemented by concentrated extracts from *luxS* mutant supernatants, and that an unidentified compound chemically different from AI-2 was synthesized at a low rate in the *luxS* mutant. LEE gene expression phenotypes were also restored in a *luxS* mutant strain that was complemented with the SAH hydrolase enzyme from *P. aeruginosa*. This experiment determined that homocysteine formation by either the SAH hydrolase enzyme or LuxS restored LEE gene expression. This suggested the presence of an inducer deemed, AI-3, whose biosynthesis utilizes precursors from the activated methyl cycle.

Other studies indicated that LEE gene expression could be influenced by epinephrine and norepinephrine hormones from the host, and it was hypothesized that AI-3 might resemble epinephrine/norepinephrine due to similarities in chemical properties. At this time, the structure of AI-3 is still unclear; however, the addition of exogenous epinephrine/norepinephrine induces similar sets of genes when compared to addition of impure AI-3 extracts. It has also been shown that AI-3 and epinephrine induce *LEE* genes as well as other targets and are dependent on the sensor histidine kinase QseC. EHEC are likely to come in contact with epinephrine/norepinephrine in the lower intestine during colonization as norepinephrine is produced by neurons in the enteric nervous system and epinephrine is secreted as a systemic response to host stress during EHEC disease progression. The discovery of QseC as a receptor for both a host-produced hormone and a bacterially produced quorum signal is astounding and contains many implications for future discoveries in quorum signaling as well as evolution between the bacterium and the host.

Bradyrhizobium japonicum Bradyoxetin Signal

Bradyoxetin is produced by the nitrogen-fixing soil bacterium *B. japonicum*. Genes involved in biosynthesis are unknown at this time, although the biosynthetic pathway is likely similar to that of the siderophore mugenic acid. Similar to siderophores, bradyoxetin production is regulated by Fe^{3+} concentration as well as cell density with bradyoxetin produced maximally at high cell densities or at low Fe^{3+} concentrations. At high cell densities, bradyoxetin appears to upregulate its own synthesis as well as the response regulators *nolA* and *nodD2*, whose products repress the expression of *nod* genes necessary for symbiotic root nodule formation in legumes. This system is yet another instance of quorum signal production being controlled by population density as well as the metabolic state of the organism.

Pseudomonas aeruginosa Pseudomonas Quinolone Signal (PQS)

In addition to the LasI- and RhlI-produced AI-1 AHL signals, *P. aeruginosa* synthesizes a quinolone compound PQS important for quorum signaling gene expression. PQS was discovered in 1999 by Everett Pesci, and its chemical structure is 2-heptyl-3-hydroxy-4-quinolone (Figure 5). PQS is synthesized from anthranilate via the *phnAB* and *pqsABCDE* gene products, which produce the precursor 4-hydroxy-2-heptyl-quinoline (HHQ). HHQ is converted into PQS by the PqsH protein.

PQS functions as part of the AI-1-mediated quorum signaling network in *P. aeruginosa*. Activated LasR (bound to $3OC_{12}$–HSL) can upregulate *pqsH*, ultimately increasing PQS concentration in a $3OC_{12}$–HSL-dependent manner. As PQS concentration increases, it induces *rhlI* expression in a *rhlR*-dependent manner. PQS negative mutants lack the ability to produce many virulence factors that are also absent in *rhlR* mutants. Thus, PQS serves as a link between the LasI/R and RhlI/R quorum signaling systems (Figure 5).

PQS is also interesting due to its hydrophobic nature. The hydrophobicity of PQS suggests that it would be a poorly diffusible signal in aqueous environments. Recently, it has been discovered by our lab that PQS is transported by outer membrane vesicles. This phenomenon allows PQS to be moved by a water-soluble carrier that may also serve to protect the signal from extracellular degradation by other organisms. Another interesting property of PQS is that its production appears to be restricted to *P. aeruginosa* (at least to this point), while its precursor HHQ is synthesized by many other Gram-negative bacteria, notably *Burkholderia*, *Pseudomonas*, and *Alteromonas* species. HHQ has also been shown to act as a signal in *P. aeruginosa* because of its release and uptake by neighboring cells and subsequent conversion into PQS. Thus, the production of 2-alkyl-4-quinolone compounds by these organisms suggests another class of quorum signaling molecules.

Ralstonia solanacearum 3-Hydroxypalmitic Acid Methyl Ester (3-OH PAME)

R. solanacearum is a Gram-negative plant pathogen that causes wilting in many different plant species. *R. solanacearum* infects plants at the root and over several days enters the xylem where it travels to the aerial portions of the plant. *R. solanacearum* colonizes the xylem and expresses several extracellular and intracellular virulence factors. A primary virulence factor in wilting disease is the production of copious amounts of EPS that physically blocks fluid transport through the xylem of the plant, thereby leading to wilting. This disease is very similar to the previously mentioned Stewart's wilt disease caused by *P. stewartii*.

R. solanacearum virulence factor production is ultimately controlled by the concentration of the quorum signal 3-OH PAME (Figure 8). 3-OH PAME is produced by the PhcB protein and released from the cell. When 3-OH PAME reaches a threshold concentration, it interacts with the sensor kinase PhcS, which then transfers a phosphate group to inactive the PhcR protein. PhcR inactivation allows the regulator PhcA to activate. PhcA ultimately controls the expression of many virulence genes in *R. solanacearum* as well as a LuxI/R like signaling system encoded by the *solI/R* genes.

3-OH PAME is significant among quorum signals because it is a volatile compound. Volatility of a quorum signal has positive and negative consequences in that it allows rapid signaling over greater distances than an aqueous signal; however, the signal can also diffuse away from receptive cell populations before a threshold concentration is reached. It is interesting that a volatile signal appears to function sufficiently inside plant tissues, which are highly heterogeneous in terms of fluid and gas density. It is also notable that *R. solanacearum* appears to use multiple quorum signal cascades to regulate virulence gene expression much like *P. aeruginosa*.

EXTRACELLULAR EFFECTORS OF QUORUM SIGNALING SYSTEMS

Quorum signals are produced by many diverse bacterial species in many different environments. The ability to utilize signals to direct efficient and timely changes in gene expression almost certainly provides that species with a strong competitive advantage for nutrient acquisition and defense. Likewise, a competitor or host would gain tremendous benefit if it were able to interfere with a pathogen or competitors' quorum-sensing system. The evolutionary "arms race" between quorum signaling and signal interference by other species could be millions of years old, yet was just recently revealed by researchers.

In 1996, Michael Givskov and Staffan Kjelleberg and colleagues described the production of AHL mimics by the marine seaweed *Delisea pulchra*. *D. pulchra* produces several halogenated furanone molecules that structurally resemble AHL signals. Furanones were shown to inhibit quorum signal-dependent swarming motility of *Serratia*

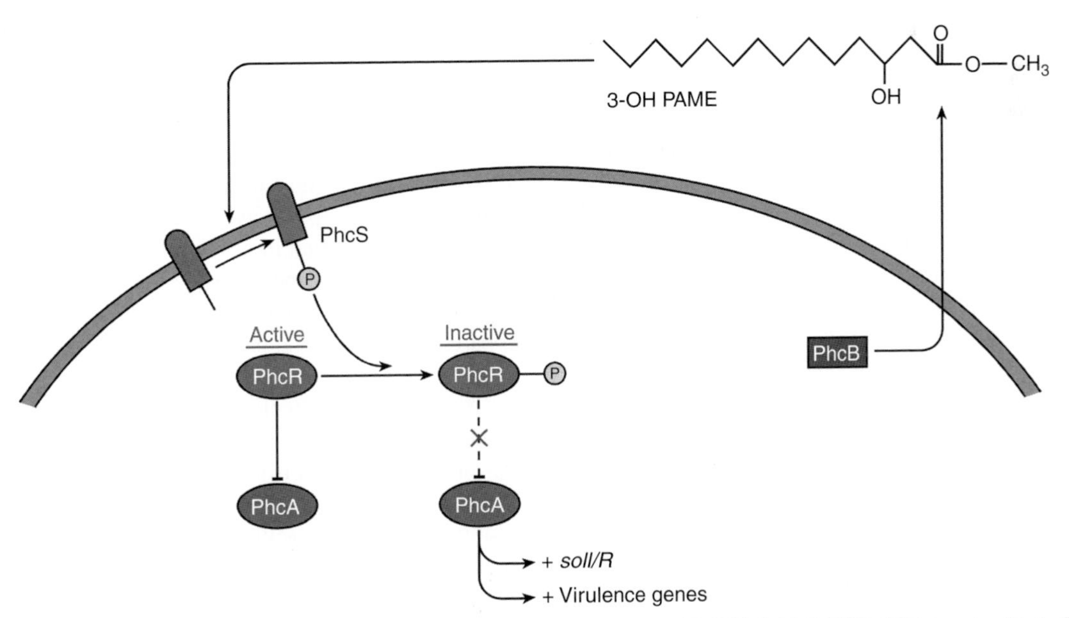

FIGURE 8 Quorum sensing in *Ralstonia solanacearum*. PhcB synthesizes the volatile signal 3-OH PAME. 3-OH PAME interacts with the PhcS sensor kinase to phosphorylate the PhcR regulatory protein. PhcR-phosphate is unable to block activity of the PhcA transcriptional regulator that induces expression of virulence genes as well as the *solI/R*-dependent AHL quorum sensing system.

liquefaciens as well as several marine bacteria. AHL reporter systems in *E. coli* as well as *V. harveyi* were shown to be inhibited by exogenous addition of furanones. Further work on *D. pulchra* furanones showed that they disrupt quorum signaling by binding to LuxR-type proteins in a competitive fashion with the native AHL signal. In the case of *D. pulchra*, furanone-dependent quorum inhibition could protect the plant from colonization by pathogenic bacteria that regulate motility, biofilm, or virulence gene expression through quorum-signaling systems.

A similar phenomenon was observed in the unicellular eukaryotic alga *Chlamydomonas reinhardtii*, which also produces quorum signal mimics. *C. reinhardtii* grows in soil and freshwater environments where it likely encounters many bacterial species that utilize quorum signaling. It was shown that *C. reinhardtii* produced at least 12 uncharacterized compounds that stimulated the LasR and CepR regulators in *E. coli* reporter strains. Interestingly, none of the compounds stimulated LuxR or AhyR regulators in similar reporter strains, suggesting that the putative AHL mimics produced by *C. reinhardtii* are receptor-specific. The chemical properties of the *C. reinhardtii*-produced mimics suggest that they are AHL-like in nature, but subsequent characterization of the compounds revealed that they are not known AHLs. Despite their dissimilar structure, *C. reinhardtii* AHL mimics were shown to increase the production of many AHL-controlled genes to levels observed with the native AHL; however, some proteins were found to be downregulated by the AHL mimics. The full impact of these results is still unclear, but demonstrates that *C. reinhardtii* likely has significant effects on bacterial quorum signaling in the natural environment.

Recent research by several groups has identified eubacterial and eukaryotic enzymes that degrade AHLs. There are many incidences of production of these enzymes in bacteria, and it appears that one function may be to degrade the bacteria's own AHL signal in order to downregulate the quorum signal response after its desired effects are achieved. Another function of these enzymes may be to degrade AHLs from competing organisms as a means of interfering with their quorum signaling systems. The observation that some enzymes produced by AHL-synthesizing bacteria cannot degrade their own AHLs, yet degrade those produced by other organisms, supports the idea of competitive interference between species.

In a 2004 *Proceedings of the National Academy of Sciences* paper, Carlene Chun and E. Peter Greenberg demonstrated that human airway epithelial cells were able to degrade the $3OC_{12}$–HSL from *P. aeruginosa*. As mentioned previously, *P. aeruginosa* can cause respiratory infections in humans. During the course of infection, airway epithelial cells are likely to come in contact with colonizing *P. aeruginosa,* and the ability of the host to interfere with quorum signaling may be of great importance in preventing infection. Surprisingly, this study showed that the degradation of $3OC_{12}$–HSL was contact-dependent and that degradation did not occur in culture supernatants from the airway cells. It was also shown that degradation was specific for $3OC_{12}$–HSL, C_{12}–HSL, and C_6–HSL. No degradation of C_4–HSL or $3OC_6$–HSL was observed. The AHL degradation phenotype varied widely among cell types tested but seemed to occur most frequently in cell lines that were derived from epithelial cells that are most likely to encounter AHL-producing organisms. These data suggest that humans, and potentially other mammals, have evolved mechanisms to specifically defend against AHL-producing bacteria.

QUORUM SIGNAL INTERACTION WITH THE HOST

Eukaryotic hosts utilize both quorum signal-degrading enzymes and mimics to interfere with bacterial quorum signaling. Some organisms produce these substances throughout their life cycle, while others produce them in response to bacterial quorum signals. The ability of eukaryotic organisms to sense and respond to quorum signals is fascinating and further emphasizes the affect that prokaryotes may have on the evolution of eukaryotes and vice versa. We will describe a few representatives from the growing number of cases of eukaryotic responses to bacterially produced quorum signals.

The legume *Medicago truncatula* senses and responds to AHL signals produced by *Sinorhizobium meliloti* and *P. aeruginosa.* Many *M. truncatula* proteins are regulated in a similar manner when exposed to AHLs from either bacterial species; however, some proteins are differentially expressed in response to AHLs from one species compared to the other. This suggests that *M. truncatula* has a conserved response to AHLs but also has the ability to discriminate between AHL signals from these bacteria. AHL signals were found to differentially regulate over 150 proteins in *M. truncatula*, the most notable of which were involved in primary metabolic pathways, stress response, and enzymes used in quorum signal mimic production. At this time, it is unclear how these responses specifically benefit *M. truncatula* or how they affect *S. meliloti* or *P. aeruginosa,* but it is apparent that both organisms stimulate a significant change in the behavior of the plant.

In 2006, Elmus Beale and Kendra Rumbaugh demonstrated that the soil nematode *Caenorhabditis elegans* is able to sense AHL signals from bacteria. *C. elegans* was shown to preferentially move toward AHL-producing organisms as opposed to AHL-lacking species. This AHL-dependent chemotaxis is thought to aid *C. elegans* in locating bacteria as a source of food. Another surprising observation was that *C. elegans* would move toward *P. aeruginosa* despite the fact that *P. aeruginosa* kills *C. elegans*. After exposure to *P. aeruginosa*, surviving *C. elegans* would display avoidant behavior toward AHL-producing organisms including non-harmful species. This work provides an elegant example of a eukaryote behavioral response to bacterial quorum signals. Another study by the same lab described the ability of *P. aeruginosa* AHLs to enter and modulate expression in mammalian cells; thus demonstrating that AHLs are able to cross the host cell membrane and stimulate host gene expression. These changes in gene expression are thought to occur through AHL stimulation of host nuclear hormone receptors. Further work demonstrated that *P. aeruginosa* AHLs can induce inflammation, immunosupression, and apoptosis in immortalized cells as well as primary cell cultures. Many of these events are hallmarks of tissue exposed to *P. aeruginosa* during infection, but it remains to be determined if the responses seen during infection are actually due to *in vivo* AHL signaling or the myriad of other virulence factors *P. aeruginosa* produces.

DIFFUSION SENSING, EFFICIENCY SENSING, AND THE FUTURE OF QUORUM SIGNALING SYSTEMS

In 2002, Rosemary Redfield at the University of British Columbia proposed that quorum sensing may be a property of what she referred to as diffusion sensing. Before her proposal of this concept, quorum signaling had been widely accepted as a cell density-dependent signal that affected gene expression of a population, with the consequence of directing a population to exhibit behavior that affected the group as a whole. The concept of diffusion sensing proposes that individual bacteria sense autoinducer concentrations and subsequent changes in gene expression direct a response that is intended for the individual cell, not the population as a whole. She proposed that evolutionary selection for quorum sensing behavior was selected for by the benefit given to an individual cell that has the ability to use a signal to gather information on properties of the local environment.

While the concept of diffusion sensing is interesting, it must be considered that any receptor-based system should be induced when ligand concentration exceeds that of the receptors' affinity for it. This condition can be achieved either by increased production of signal, decreased mass transfer rate, or decreased local volume around the cell. In quorum signaling, we ascribe that signal-dependent behavior is due to an increase in population and subsequent increase in signal. In many instances of quorum signaling systems, it may indeed be a product of diffusion sensing by the cell. However, observations of autoinducer-producing organisms in some characterized systems seem to indicate behavioral changes that are more beneficial to a group of cells as opposed to an individual. These changes include density-dependent luminescence as well as conjugation, which serve little purpose when local cell numbers are low.

Hense *et al.* in a 2007 *Nature* review attempted to unify the gap between diffusion and quorum sensing and propose the novel concept of efficiency sensing. Efficiency sensing suggests that cells produce signaling compounds to "test" the population density, volume, and mass transfer rate of their environment. If signal concentration increases, then induction of energetically expensive, signal-dependent pathways occurs. This hypothesis takes into account that signal concentration can be affected by local concentration of cell density as well as the mass transfer rates and the volume of the local environment. Hense and colleagues propose that efficiency sensing provides a means for individual as well as community advantages of diffusible signals to overlap and provide evidence through mathematical models that indicate that spatial distribution of cells may be more important than density.

It is possible that diffusion sensing behavior in individuals may be the primary means of positive selection for organisms that have quorum signaling systems of either low complexity or are newer aspects in the organism's evolutionary history. Thus, as bacteria adapt and evolve, diffusion sensing behavior can be recruited into true quorum signaling behavior and exert a community benefit on a population of cells.

WHAT'S NEXT FOR QUORUM SIGNALING?

Research in quorum signaling initially uncovered a few specific cases that demonstrated density-dependent signaling and its impact on a single response in the signaling organism. Phenomena such as luminescence and regulation of competence were largely thought to be the single process in the cell directed by a diffusible signal, and much initial research focused on utilization of a specific signal and its interaction with a single receptor or transcriptional regulator to direct regulation of just a few related genes or operons. The discovery of orphan regulators in *P. aeruginosa* and other species and the convergence of multiple types of signals involved in gene expression in *V. harveyi*, *V. cholerae*, *P. aeruginosa*, *R. solanacearum*, and *B. subtilis* (to name a few) demonstrate that many organisms utilize quorum signals in increasingly complex pathways. Further research into orphan regulators and how bacteria integrate multiple quorum signal pathways may reveal an even greater role for quorum signaling in bacteria than was previously considered.

There are ever increasing observations of interspecies quorum signal interactions between bacteria as well as inter-domain interactions between bacteria and eukaryotic hosts. These studies demonstrate that prokaryotes and eukaryotes may utilize signals not only as a means to detect signal producers, but as a way to defend against them *in situ*. These adaptations shed light on possible therapeutic strategies for infections caused by quorum signal-producing bacteria.

The discovery of quorum signal interference, especially in plants, may be useful in developing new variants of commercial crops that are more resistant to bacterial infections.

A poorly understood aspect of quorum signaling is that some microorganisms utilize quorum signals not only as a strict density-dependent signal but also as a way of monitoring and reporting the nutrient status of the environment. PQS signals in *P. aeruginosa* have been shown to be differentially produced in different environmental concentrations of aromatic amino acids. Also, the types of quorum signals produced by *A. ferrooxidans* vary depending on the substrates available to the organism. These phenomena as well as the production of AI-2 and its dependency on the metabolic status of the cell indicate that quorum signal production may serve functions aside from monitoring cell density.

FURTHER READING

Chun, C. K., Ozer, E. A., Welsh, M. J., Zabner, J., & Greenberg, E. P. (2004). Inactivation of a pseudomonas aeruginosa quorum-sensing signal by human airway epithelia. *Proceedings of the National Academy of Sciences of the United States of America*, *101*, 3587–3590.

Egland, P. G., Palmer, R. J., & Kolenbrander, P. E. (2004). Interspecies communication in *Streptococcus gordonii-Veillonella atypica* biofilms: Signaling in flow conditions requires juxtaposition. *Proceedings of the National Academy of Sciences of the United States of America*, *101*, 16917–16922.

Farah, C., Vera, M., Morin, D., Haras, D., Jerez, C., & Guiliani, N. (2005). Evidence for a functional quorum-sensing type AI-1 system in the extremophilic bacterium *Acidithiobacillus ferrooxidans*. *Applied and Environmental Microbiology*, *71*, 7033–7040.

Gonzalez, J. E., & Keshvan, N. D. (2006). Messing with bacterial quorum sensing. *Microbiology and Molecular Biology Reviews*, *70*, 859–875.

Hense, B. A., Kuttler, C., Muller, J., Rothballer, M., Hartmann, A., & Kreft, J. U. (2007). Does efficiency sensing unify diffusion and quorum sensing. *Nature Reviews Microbiology*, *5*, 230–239.

McNab, R., Ford, S. K., El-Sabaeny, A., Barbieri, B., Cook, G. S., & Lamont, R. J. (2003). LuxS-based signaling in Streptococcus gordonii: Autoinducer 2 controls carbohydrate metabolism and biofilm formation with *Porphyromonas gingivalis*. *Journal of Bacteriology*, *185*, 274–284.

Merrit, J., Qi, F., Goodman, S. D., Anderson, M. H., & Shi, W. (2003). Mutation of luxS affects biofilm formation in *Streptococcus mutans*. *Infection and Immunity*, *71*, 1972–1979.

Miller, M. B., & Bassler, B. L. (2001). Quorum sensing in bacteria. *Annual Review of Microbiology*, *55*, 165–199.

Reading, N. C., & Sperandio, V. (2006). Quorum sensing: The many languages of bacteria. *FEMS Microbiology Letters*, *254*, 1–11.

Rumbaugh, K. P. (2007). Convergence of hormones and autoinducers at the host/pathogen interface. *Analytical and Bioanalytical Chemistry*, *387*, 425–435.

Schell, M. A. (2000). Control of virulence and pathogenicity genes of *Ralstonia solanacearum* by an elaborate sensory network. *Annual Review of Phytopathology*, *38*, 263–292.

Shao, H., Lamont, R. J., & Demuth, D. R. (2007). Autoinducer 2 is required for biofilm growth of *Aggregatibacter (Actinobacillus) actinomycetemcomitans*. *Infection and Immunity*, *75*, 4211–4218.

Sturme, M. H., Kleerebezem, M., Nakayama, J., Akkermans, A. D., Vaugha, E. E., & de Vos, W. M. (2002). Cell to cell communication by autoinducing peptides in Gram-positive bacteria. *Antonie Van Leeuwenhoek*, *81*, 233–243.

Venturi, V. (2007). Regulation of quorum sensing in Pseudomonas. *FEMS Microbiology Reviews*, *30*, 274–291.

Xavier, K. B., & Bassler, B. L. (2005). Interference with AI-2-mediated bacterial cell-cell communication. *Nature*, *437*, 750–753.

Chapter 20

Chemotaxis

J.B. Stock and M.D. Baker
Princeton University, Princeton, NJ, USA

Chapter Outline

ABBREVIATIONS

DPP Dipeptide-binding protein
ETS Electron transport system
GBP Galactose-binding protein
HAMP domain Histidine kinases, adenyl cyclases, methyl-accepting chemotaxis proteins, and phosphatases' domain
MBP Maltose-binding protein
MCP Methyl-accepting chemotaxis protein
RBP Ribose-binding protein

DEFINING STATEMENT

Chemotaxis in microbiology refers to the migration of cells toward attractant chemicals or away from repellents. Virtually, every motile organism exhibits some type of chemotaxis. The chemotaxis responses of eukaryotic microorganisms proceed by mechanisms that are shared by all cells in the eukaryotic kingdom and generally involve regulation of microtubule- and/or microfilament-based cytoskeletal elements. In this chapter, we deal only with bacterial chemotaxis.

All bacteria share a conserved set of six different regulatory proteins that serve to direct cell motion toward favorable environmental conditions. Essentially the same regulatory system operates, irrespective of whether motility involves one or several flagella, or whether it occurs by a mechanism such as gliding motility that does not involve flagella. The same basic system mediates responses to a wide range of different chemicals including nutrients such as amino acids, peptides, and sugars (which are usually attractants) and toxic compounds such as phenol and acid (generally repellents). The same proteins also mediate responses to oxygen (aerotaxis), temperature (thermotaxis), osmotic pressure (osmotaxis), and light (phototaxis). As a bacterium moves, it continuously monitors a spectrum of sensory inputs and uses this information to direct motion toward conditions that are optimal for growth and survival. To accomplish this task, the chemotaxis system has developed molecular correlates of processes such as memory and learning that are widely associated with sensory motor regulation by higher neural systems.

RESPONSE STRATEGY

Biased Random Walk

In a constant environment, motile bacteria generally move in a random walk of straight runs punctuated by brief periods of reversal that serve to randomize the direction of the next run. The chemotaxis system functions by controlling

Topics in Ecological and Environmental Microbiology.

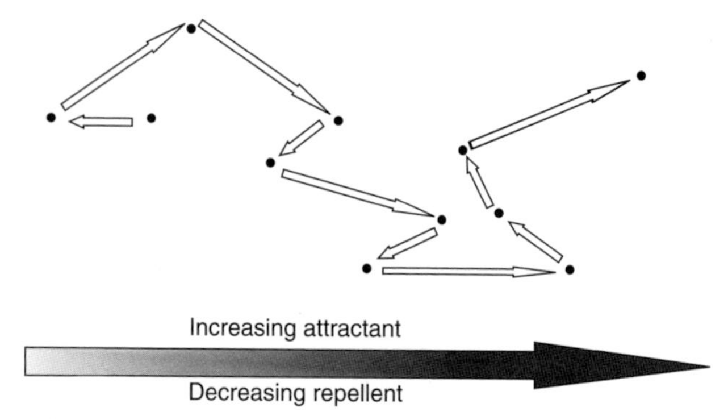

FIGURE 1 Chemotaxis is accomplished by a biased random walk strategy. Bacterial swimming behavior involves a series of runs (indicated by arrows) punctuated by motor reversals that randomize the direction of the subsequent run. Cells migrate toward attractants and away from repellents by increasing their average run lengths in the preferred direction.

the probability of a reversal. If, during a run, the system senses that conditions are improving, then it sends a signal to the motor that suppresses reversals so that the cell tends to keep moving in the preferred direction. If, however, the system senses that conditions are getting worse, then it sends a signal for the motor to change direction. The effect is to bias the random walk so that cells tend to migrate toward attractants and away from repellents (Figure 1). Thus, bacterial chemotaxis is effected by the simple strategy of using environmental cues to modulate the probability of random changes in direction. By using this mechanism, individual cells never have to determine in which direction they want to move. Instead, they determine whether they want to continue on course or change the direction. The biased random walk strategy is essential for bacterial chemotaxis because it provides a mechanism whereby bacteria can direct their motion despite the fact that bacterial cells are far too small to have a sense of direction.

Temporal Sensing and Excitation

Bacterial cells are generally only a few micrometers long. This is too small to possibly measure differences in attractant and repellent concentration over the length of their bodies. In the early 1970s, through the work of Macnab, Koshland, Berg, and others, it was shown that bacteria solve this problem by sensing changes in attractant and repellent concentration in time rather than in space. In other words, chemotaxis depends on a temporal rather than a spatial sensing mechanism. A temporal sensing mechanism is evident when examining a population of randomly moving bacteria that are suddenly transferred from one spatially uniform environment to another that contains a uniform distribution of an attractant or repellent chemical. In this type of experiment, there are no spatial gradients; yet, cells that are exposed to only a temporal change in their environment elicit a chemotactic response. As expected from a temporal sensing mechanism, cells that are suddenly exposed to attractants suppress the tendency to change direction and stay on course (Figure 2a). The opposite effect is seen when cells are suddenly exposed to repellents. All the individuals in the population, despite the fact that they are moving in many different directions, suddenly change the direction of their motion.

Memory and Adaptation

As a cell moves, it constantly compares its current surroundings to those it has experienced previously. If the comparison is favorable, the cell tends to keep going; if not, it tends to change direction. This mechanism implies a memory function whereby the present can be compared with the past to determine whether conditions are getting better or worse as time (and movement in a given direction) proceeds. After a period of time, bacteria that have been transferred to a new environment gradually adapt so that their behavior returns to the same random walk as they exhibited before they were exposed to the attractant or repellent stimulus (Figure 2a). This occurs despite the fact that the attractant or repellent is still present. Thus, bacteria do not respond to absolute concentrations of attractant and repellent chemicals. They respond only to changes.

There is a close relationship between memory and adaptation. If one moves a population of bacteria that has adapted to an environment with an attractant back to an environment lacking the attractant, the cells think they are moving in an unfavorable direction and therefore they all change course as if they had been exposed to a repellent. The opposite happens with cells that are adapted to a repellent. Thus, an increase in attractant concentration is equivalent to a decrease in a repellent concentration and vice versa.

In bacterial chemotaxis, the sense and degree of excitation in response to a new place in time are only determined in relation to previous places, with the memory of the past environment being set by the process of adapting to it. In effect, there must be two core mechanisms at work in chemotaxis – an excitation mechanism that controls the probability of a motor reversal and an adaptation mechanism that modulates the sense and degree of excitation with respect to a preset default value.

GENETICS OF BACTERIAL BEHAVIOR

Chemotaxis Mutants

The most common strategy for isolating mutants that are defective in chemotaxis has involved selecting cells that cannot swarm from a colony inoculated into the center of a dish filled with semisolid nutrient agar. Chemotactic

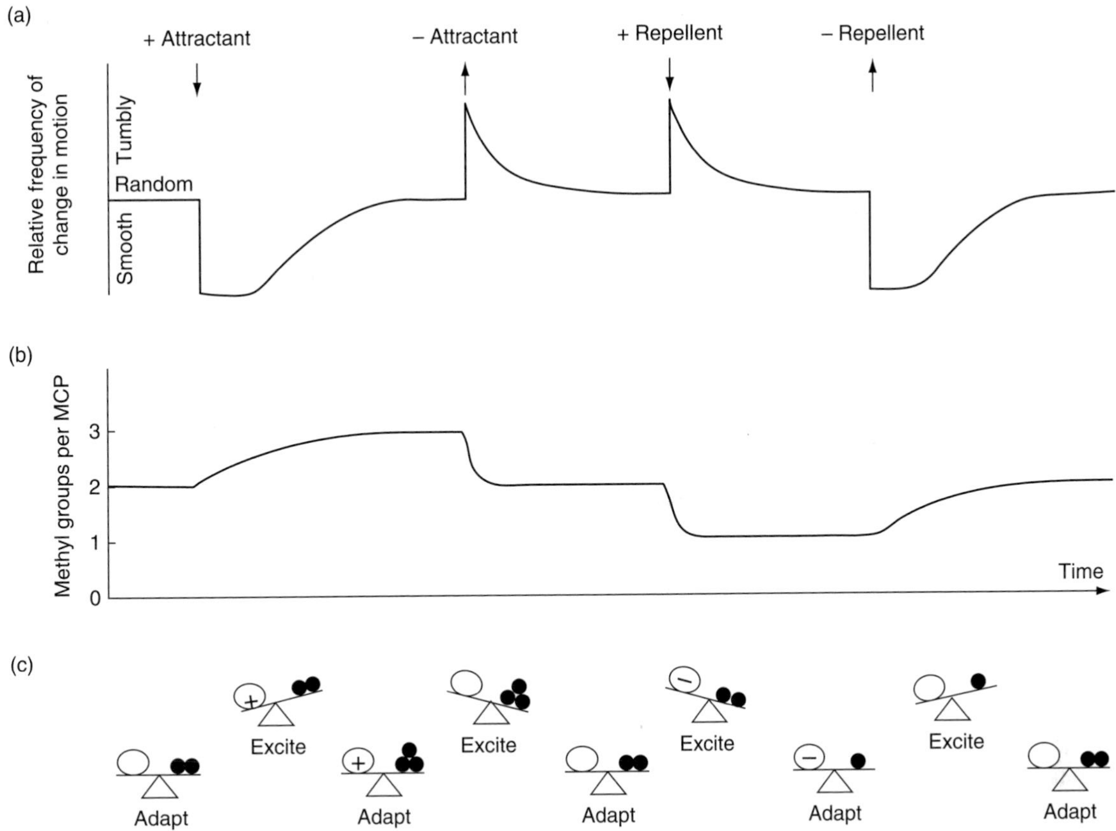

FIGURE 2 Excitation and adaptation in bacterial chemotaxis. (a) Addition of attractant causes cells to continue swimming smoothly without changing direction until they adapt back to their prestimulus random behavior; removal of attractant causes cells to frequently change direction or tumble until adaptation restores random behavior; adding repellent causes the same tumbling response as removing attractant; and removing repellent causes the same swimming response as adding attractant. (b) In the absence of attractants or repellents, a typical MCP has approximately two of four possible methylated glutamates; addition of a saturating attractant stimulus causes addition of approximately one methyl group; removal of attractant causes loss of this group; addition of saturating repellent stimulus causes loss of one more methyl group; and removal of repellent causes addition of a methyl group. (c) MCP signaling depends on the balance between stimulus (○, no stimulus; +, attractant; –, repellent) and the level of methylation (•, one methyl group; ••, two methyl groups, etc.). A sudden change in stimulus causes an imbalance that leads to excitation, and the level of methylation then changes to restore the balance.

cells form a colony at the point where they are initially inoculated and the growing cells consume nutrients in the culture media creating attractant gradients that cause them to swim outward from the center. Mutant cells that are deficient in chemotaxis are left behind at the center. This strategy produces several different classes of mutant strains. By far, the most common are strains that are not motile. In bacteria whose motility depends on flagella, these nonmotile strains can be subdivided into two classes: Fla mutants, which have lost the ability to make flagella; and Mot mutants, which make flagella that are paralyzed. Mutant strains that do not swarm through semisolid agar but are fully motile are categorized into two additional subclasses: Che mutants, which are generally nonchemotactic; and receptor mutants, which are unable to swarm in semisolid agar with one type of chemoattractant but can swarm normally in agar that contains another.

Genetic Analysis of *Escherichia coli* Chemoreceptors

It was through the selection and characterization of chemoreceptor mutants from *E. coli* that Julius Adler and colleagues first demonstrated that chemosensing in bacteria is mediated by specific receptor proteins in the cell envelope rather than by some other mechanism such as nutrient utilization. The first chemosensing system to be identified genetically sensed galactose and ribose through periplasmic binding proteins that had previously been shown to function in the transport of these sugars. The selection of ribose and galactose transport mutants with normal chemosensing abilities established that uptake and metabolism were not required for chemotaxis. Further genetic analysis indicated that ribose and galactose sensing required an additional protein component, a chemoreceptor termed Trg (taxis to ribose and galactose). Early work on the ribose and galactose receptors revealed another important aspect of chemotaxis – the possibility that bacterial cells could exhibit a simple form of learning. The galactose- and ribose-binding proteins (RBPs) are specifically induced by growth in the presence of galactose and ribose, respectively. Thus, whereas a naive cell is unable to respond to either sugar, once it is allowed to grow on ribose or galactose, the corresponding binding protein is induced and the cell has now "learned" to respond to the sugar.

In addition to the sugar chemoreceptors, Adler's group identified two *E. coli* genes, *tar* and *tsr*, that are required for responses to amino acids and several repellents. Tar senses the attractants aspartate, glutamate, and maltose and the repellents cobalt and nickel. The maltose response via Tar, much like the responses to ribose and galactose, requires an inducible sugar-binding protein in addition to the transmembrane Tar protein. Tsr is required for chemotaxis to serine, alanine, and several other amino acids and repellents. In contrast to the genes that encode sugar-binding proteins, expression of *tar, tsr*, and *trg* is controlled within the same regulon as the flagellar genes of *E. coli* so that, as long as a cell is motile, it can sense aspartate and serine.

Che Genes

Analysis of *E. coli* che mutants reveals two major Che complementation groups, designated as *cheA* and *cheB*. Studies by Parkinson and others have established that the CheA locus is composed of two genes (*cheA* and *cheW*), whereas the CheB locus is composed of four genes (*cheR, cheB, cheY*, and *cheZ*). Strains defective in *cheA, cheW, cheY*, or *cheR* exhibit a smooth swimming phenotype, never changing their direction of motion. In contrast, *cheB* and *cheZ* mutants exhibit a constantly changing, tumbly pattern of swimming behavior. Whereas the ability of *cheB* and *cheZ* mutants to change direction is suppressed by the addition of attractants, and *cheR* mutants still reverse in response to repellent stimuli, *cheW, cheA*, and *cheY* were completely unresponsive. From these results CheW, CheA, and CheY proteins were concluded to be essential for excitation, whereas CheR and CheB are involved in adaptation (for a summary of the Che genes and their protein products, see Table 1). CheZ, unlike the five other Che proteins, is not absolutely required for either excitation or adaptation. The fact that mutants defective in excitation were invariably smooth in swimming suggested that the excitation mechanism produced a signal that caused a change in the direction of motion, and that, in the absence of this hypothetical signal, the cell would rarely, if ever, change its direction.

TABLE 1 *Escherichia coli* Che genes

Gene	M_r (kDa)	Function
cheR	32	Methylation of MCPs
cheB	36	Demethylation of MCPs
cheW	18	Coupling CheA to MCPs
cheA	73	Histidine kinase
cheY	14	CheY-P binds to flagellar switch to cause change in swimming direction
cheZ	24	CheY-P phosphatase

ROLE OF PROTEIN METHYLATION IN ADAPTATION

Methionine Requirement for Chemotaxis

One of the most important discoveries from Adler's pioneering work on the *E. coli* chemotaxis system was the serendipitous finding that methionine was required for chemotaxis. In his initial characterization of *E. coli* chemotaxis, Adler employed a capillary assay, which was developed in the late nineteenth century by the great German microbiologist and botanist Pfeffer. In this assay, the tip of a glass capillary tube that contains an attractant chemical is placed into a suspension of bacteria. As the attractant diffuses from the capillary tip, an attractant gradient is established, which the cells follow up into the capillary tube. After about 1 h, the capillary is withdrawn and the bacteria inside are counted to provide a measure of the chemotactic response. Unlike with swarming on semi-solid agar, capillary assays do not require cell growth and are generally performed with cells suspended in a defined buffer solution. Adler observed that a mutant *E. coli* strain that required methionine for growth exhibited chemotactic responses to attractants that were generally depressed. Among all the amino acids, this effect was specific for methionine. Further analysis showed that, as cells became starved for methionine, they lost the ability to adapt causing addition of attractants such as aspartate or serine, resulting in a smooth swimming behavior similar to that observed with *cheR* mutants.

Methyl-Accepting Chemotaxis Proteins (MCPs)

Subsequent studies established that the effect of methionine depletion stemmed from a requirement for the universal methyl donor *S*-adenosylmethionine (AdoMet), which is produced from methionine and adenosine triphosphate (ATP) through the action of AdoMet synthetase:

$$\mathrm{ATP} + \mathrm{methionine} + \mathrm{H_2O} \rightarrow \mathrm{AdoMet} + \mathrm{PP_i} + \mathrm{P_i},$$

AdoMet is required for the methylation of a wide range of different macromolecules, including proteins, DNA, and RNA, as well as numerous different small molecules. It was shown that, in chemotaxis, the requirement for AdoMet is to methylate a set of ~60-kDa membrane proteins that were termed MCPs. These proteins corresponded to the products of genes such as *tar, tsr*, and *trg* that were previously implicated in chemosensing. As is the case with *E. coli* Tar, which directly binds aspartate but senses maltose indirectly via protein–protein interactions with the maltose-binding protein (MBP), each MCP may detect several different stimuli through direct and indirect mechanisms.

Methylation studies, isolation of anti-MCP antibodies, and, most importantly, DNA sequencing have shown that the MCPs are a large and highly conserved family of proteins that are invariably associated with bacterial chemosensing. There are five different MCPs encoded in the *E. coli* genome (Figure 3) including Tar, Tsr, and Trg as well as a sensor for cellular redox potential termed Aer that is responsible for aerotaxis and a receptor that mediates responses to dipeptides termed as Tap. Chemotaxis systems in other species of bacteria have an equivalent or larger number of different MCPs.

Structural and Functional Organization of Sensing and Signaling Domains of Receptor MCPs

The MCPs are the principal sensory receptors of the bacterial chemotaxis system. They have a structural organization, membrane topology, and mode of function that is typical of type I receptors in all cells, including important vertebrate type I receptors such as the insulin, growth hormone, and cytokine receptors. In recent years, the Tar and Tsr proteins have emerged as archetypal MCPs. Tar and Tsr have the typical membrane topology of type I receptors with an N-terminal extracytoplasmic sensing domain connected via a hydrophobic membrane-spanning sequence to an intracellular signaling domain. The sensing and signaling domains can function independently of the membrane and of each other. Most MCPs have this structural organization. Sequence comparison indicates that, as one might expect, the extracytoplasmic sensing domains tend to be highly variable, whereas the cytoplasmic signaling domains, which interact with the Che proteins, are highly conserved. In fact, the tools of genetic engineering have been used to construct several different hybrid receptors with one MCP's sensing domain connected to another's signaling domain. In every case, the hybrids exhibit sensory specificities equivalent to those of the MCP that contributed the N-terminal portion.

The sensory and signaling domain can also be produced as independent soluble protein fragments. This approach has been used with many type I receptors to produce protein fragments that are free of the membrane and can, therefore, be much more easily crystallized for X-ray diffraction studies. Determination of the X-ray crystal structure of the sensing domain of Tar in the presence and absence of aspartate revealed a dimer of two α-helical bundles with aspartate binding at the subunit interface. The sensing domain of Tsr has a homologous structure.

The signaling domain is composed of a highly conserved region that binds the CheW and CheA proteins and thereby connects the receptor to the remainder of the chemotaxis signal transduction system. This region is flanked on both sides by methylated α-helices that together contain four or more potential sites of glutamate methylation and demethylation. Attractants cause increases in the level of methylation and repellents cause decreases. These changes are responsible for

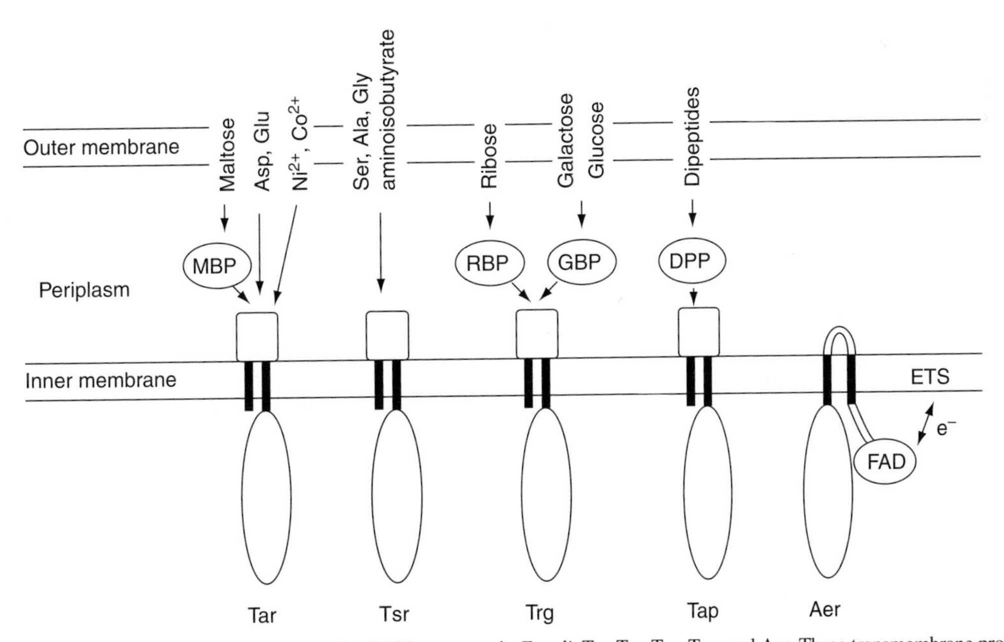

FIGURE 3 *Escherichia coli* chemoreceptors. There are five MCP receptors in *E. coli*; Tar, Tsr, Trg, Tap, and Aer. These transmembrane proteins either bind stimulatory ligands directly or interact with ligand-bound periplasmic binding proteins. Aer differs from the other MCPs in that it has an intracytoplasmic sensing domain with an associated flavin cofactor, FAD. It is thought that Aer senses cellular oxidative potential through redox interaction with the electron transport system (ETS) in the membrane. MBP, maltose-binding protein; RBP, ribose-binding protein; GBP, galactose-binding protein; DPP, dipeptide-binding protein.

adaptation to attractant and repellent stimuli (Figure 2b). The N-terminal methylating helix is attached to the second transmembrane domain by a mostly α-helical HAMP domain (histidine kinases, adenyl cyclases, methyl-accepting chemotaxis proteins, and phosphatases' domain), which is required for transmembrane signaling.

Receptor Methylation Enzymology

The MCP receptors are methyl-esterified at several specific glutamate residues. The methylation reaction is catalyzed by an AdoMet-dependent methyltransferase encoded by the *cheR* gene. Receptor methyl groups are removed through the action of a specific methyl esterase encoded by the *cheB* gene. CheR and CheB are both soluble monomeric proteins. In *E. coli*, CheR is tethered to Tar and Tsr by binding to the C-terminal five amino acids, which are identical in these two MCPs but absent in Trg, Tap, and Aer. Tar and Tsr are present in cells at more than 10-fold higher levels than Trg, Tap, and Aer, and considerable evidence suggests that the latter, so-called minor receptors, function in higher order complexes with the major receptors, Tar and Tsr. CheR binding to Tar and Tsr positions the enzyme to methylate the closely associated minor receptors. Sequence analysis of CheR proteins and their receptor substrates suggests that additional mechanisms of CheR localization may operate in different bacterial species.

The active site of CheB contains a Ser-His-Asp catalytic triad that is characteristic of serine hydrolases. An N-terminal regulatory domain occludes the active site so that the enzyme is relatively inactive. CheB is activated to remove receptor methyl groups by essentially the same mechanism that produces the signal that causes a change in the direction of cell movement in response to repellent stimuli. CheB activation provides a feedback mechanism that contributes to the adaptive phase of the chemotaxis response. Thus, repellent addition or attractant removal produces an excitatory signal to change direction. At the same time, CheB activation leads to a rapid decrease in the level of methylation that causes adaptation. The converse is true with attractant addition or repellant removal. It is as if the receptor–signaling system functions as a balance between the effects of stimulatory ligands and methylation (Figure 2c). The addition of an attractant or repellent offsets the balance to produce a positive or negative excitatory signal, and changes in methylation restore the balance to effect adaptation.

MECHANISM OF SIGNAL TRANSDUCTION IN CHEMOTAXIS

Receptor–CheW–CheA Signaling Complexes

The CheA protein is a histidine kinase that binds ATP and catalyzes the phosphorylation of one of its own histidine residues. The rate of autophosphorylation of the isolated CheA protein is very slow. The physiologically relevant form of CheA involves a complex with CheW and the MCPs. The rate of CheA autophosphorylation in these receptor–signaling complexes can be elevated at least 100-fold or completely inhibited depending on the level of receptor methylation and the binding of stimulatory ligands. Attractants such as serine or aspartate have an inhibitory effect, whereas increased levels of methylation cause dramatic increases in CheA kinase activity. Because of the dimeric nature of the receptor-sensing domain and the fact that CheA is a dimer, it was originally assumed that the receptor–CheW–CheA signaling complex had a 2:2:2 stoichiometry. Considerable evidence indicates a much more complex architecture, with the thousands of receptors in a cell clustering together in a higher order complex with CheW and CheA. Receptors of different types are believed to cooperatively signal within these complexes through their interactions. These clusters of receptors and signaling proteins provide a mechanism for signal integration and amplification. It has been hypothesized that packing interactions within these signaling arrays may function to control kinase activity in response to the binding of stimulatory ligands or changes in the level of receptor methylation.

Motor Regulation and Feedback Control

The level of phosphorylation of the CheY protein controls the probability that a cell will change its direction of motion. CheY is a 14-kDa monomeric enzyme that catalyzes the transfer of a phosphoryl group from the phosphohistidine in CheA to one of its own aspartate residues. CheY phosphorylation induces a conformational change in the protein that allows it to bind to switching proteins at the flagellar motor. Repellent-induced increases in the rate of CheA phosphorylation produce elevated levels of phospho-CheY that binds to the motor to enhance the probability of motor reversal. Phospho-CheY spontaneously dephosphorylates to terminate the response. In *E. coli*, this autophosphatase reaction is dramatically enhanced by the CheZ protein. Thus, addition of attractants inhibits CheA autophosphorylation, and CheZ activity leads to a rapid decrease in the level of phospho-CheY, a reduction in the level of phospho-CheY bound to the motor, and a decrease in the probability that a cell will change direction.

The CheY protein is homologous to the regulatory domain of the CheB protein and, like CheY, the regulatory domain of CheB acts to transfer phosphoryl groups from the phosphohistidine in CheA to its own aspartate residues. Phosphorylation of CheB causes a dramatic increase in demethylation activity, which leads to a decrease in receptor methylation and a concomitant decrease in the rate of CheA autophosphorylation. Thus, the same mechanism that acts to produce a motor response feeds back to cause adaptation. The signal transduction network that mediates *E. coli* chemotaxis is summarized in Figure 4.

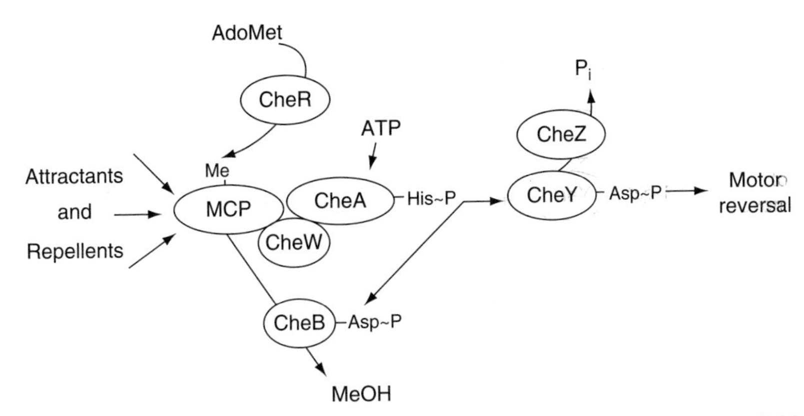

FIGURE 4 Biochemical interactions between the Che proteins that mediate chemotaxis responses in *Escherichia coli*. MeOH, methanol; P_i, inorganic phosphate.

PHYLOGENETIC VARIATIONS

Most of our understanding of the molecular mechanisms that underlie bacterial chemotaxis has come from studies of the system in *E. coli*. Other motile bacteria that have been investigated have MCPs and all the same Che proteins as those in the *E. coli* system except for CheZ, which has only been found in enterics. There appear to be a number of other variations on the *E. coli* scheme, however. Studies in *Bacillus subtilis* indicate that in this species, the system is reversed so that attractants activate CheA, and CheY phosphorylation suppresses the tendency for the cell to change direction. The *B. subtilis* system also appears to have additional components that are not found in *E. coli*.

Many species have several copies of one or more of the chemotaxis genes. In some instances, it is apparent that there are multiple chemotaxis systems functioning in different cell types. A good example of this is provided by *Myxococcus xanthus*, in which at least three different chemosensory systems operate to control two distinct types of motility and cellular differentiation. In contrast, two CheY proteins in *Rhodobacter sphaeroides* seem to supply divergent functions within the same signal transduction network. One CheY interacts with the motor to control swimming behavior, whereas the other CheY functions as a CheA phosphatase to drain phosphoryl groups out of the system.

There is no evidence for any of the chemotaxis components in eukaryotic cells. Although homologous systems are found in archae such as *Halobacterium salinarum* and *Archaeoglobus fulgidus*, the sequences of the component proteins are so closely related to those of *B. subtilis* that one can be fairly certain that they originated by lateral transfer from a *B. subtilis* relative. It, therefore, seems likely that the bacterial chemotaxis system originally evolved in eubacteria.

SOCIAL MOTILITY

Bacterial chemotaxis has generally been viewed as a mechanism by which single cells move through the environment in search of beneficial living conditions. In recent years, it has become more and more evident that bacteria communicate with one another in a process that has come to be known as "quorum sensing." These social interactions are particularly important in virulence and biofilm formation and have also emerged as important determinants of chemotaxis behavior. The sensory specificities of the *E. coli* chemoreceptors can best be understood in terms of metabolic interactions between cells in culture rather than as random nutrient-sensing mechanisms. For instance, the most abundant *E. coli* receptor Tsr mediates responses to L-serine, L-glycine, and L-alanine, all of which are overflow metabolites that are secreted by *E. coli* growing in rich environments. In complex microenvironments, cells are, therefore, attracted to one another so that they aggregate into communities where quorum-sensing mechanisms modulate gene expression changes that influence cellular differentiation and the formation of complex synergistic relationships required for pathogenic virulence and biofilm formation. One of the most exciting areas of future microbial research will involve the elucidation of how chemotaxis mechanisms function to facilitate these types of communal activities.

FURTHER READING

Adler, J. (1975). Chemotaxis in bacteria. *Annual Review of Biochemistry*, *44*, 341–356.

Alexander, R. P., & Zhulin, I. B. (2007). Evolutionary genomics reveals conserved structural determinants of signaling and adaptation in microbial chemoreceptors. *Proceedings of the National Academy of Sciences of the United States of America*, *104*, 2885–2890.

Baker, M. D., Wolanin, P. M., & Stock, J. B. (2006). Systems biology of bacterial chemotaxis. *Current Opinion in Microbiology*, *9*, 187–192.

Berg, H. C., & Brown, D. A. (1972). Chemotaxis in *Escherichia coli* analyzed by 3-dimensional tracking. *Nature*, *239*, 500–504.

Falke, J. J., & Hazelbauer, G. L. (2001). Transmembrane signaling in bacterial chemoreceptors. *Trends in Biochemical Sciences*, *26*, 257–265.

Macnab, R. M., & Koshland, D. E. (1972). The gradient-sensing mechanism in bacterial chemotaxis. *Proceedings of the National Academic Sciences of the United States of America*, *69*, 2509–2512.

Maddock, J. R., & Shapiro, L. (1993). Polar location of the chemoreceptor complex in the *Escherichia coli* cell. *Science*, *259*, 1717–1723.

Park, S., Wolanin, P. M., Yuzbashyan, E. A., Silberzan, P., Stock, J. B., & Austin, R. H. (2003). Motion to form a quorum. *Science*, *301*, 188.

Parkinson, J. S., Ames, P., & Studdert, C. A. (2005). Collaborative signaling by bacterial chemoreceptors. *Current Opinion in Microbiology*, *8*, 116–121.

Sourjik, V., & Berg, H. C. (2004). Functional interactions between receptors in bacterial chemotaxis. *Nature*, *428*, 437–441.

Szurmant, H., & Ordal, G. W. (2004). Diversity in chemotaxis mechanisms among the bacteria and archaea. *Microbiology and Molecular Biology Reviews*, *68*, 301–319.

Wadhams, G. H., & Armitage, J. P. (2004). Making sense of it all: Bacterial chemotaxis. *Nature Reviews Molecular Cell Biology*, *5*, 1024–1037.

Zhulin, I. B. (2001). The superfamily of chemotaxis transducers: From physiology to genomics and back. *Advances in Microbial Physiology*, *45*, 157–198.

Part III

Microbial Habitats

Chapter 21

Acid Environments

D.B. Johnson
Bangor University, Bangor, UK

Chapter Outline

ABBREVIATIONS

AMD Acid mine drainage
DGGE Denaturing gradient gel electrophoresis
FISH Fluorescent *in situ* hybridization
PGM Platinum group metal
RISC Reduced inorganic sulfur compound
SRB Sulfate-reducing bacteria

DEFINING STATEMENT

This article gives an overview of the nature of extremely acidic environments and of the biodiversity of microorganisms found within them. Ways in which acidophiles interact with each other in both positive and negative fashions are described. Finally, the microbial ecology of some of the most widely studied extremely acidic environments on our planet is discussed.

NATURE AND ORIGIN OF EXTREMELY ACIDIC ENVIRONMENTS

Oceanic waters, which constitute the largest biome on planet Earth, are uniformly moderately alkaline (pH 8.2–8.4). In contrast, some lentic and lotic waters, soils, and anthropogenic environments are moderately acidic (pH 3–5) or extremely acidic (pH < 3). In some rare cases, environments that have recorded negative pH values have been documented. Living organisms that are active in extremely acidic environments are now known to be far more diverse than was recognized even a couple of decades ago. As with other extremophiles, acidophiles tend to be specialized life-forms, in that many are unable to grow in neutral pH environments. The majority of acidophiles are prokaryotic microorganisms, and these comprise a large variety of phylogenetically diverse bacteria and archaea, though some single-celled and multicellular eukaryotes are known to grow in highly acidic ponds and streams.

Extremely acidic environments may be formed by processes that are entirely natural, though human activities have became increasingly important in generating such sites. While the scale of human impact has paralleled global industrialization, small-scale anthropogenic generation of acidic, metal-polluted environments probably began in the Bronze Age. Overall, the majority of extremely acidic sites that now exist on planet Earth are associated with one particular human activity – the mining of metals and coal.

Topics in Ecological and Environmental Microbiology.

There are a number of important (in terms of their scale) microbial activities that can generate acidity. Among the most important of these is the formation of organic acids as waste products in either anaerobic (fermentative) or aerobic metabolisms. However, the generation of strong inorganic acids by aerobic microorganisms gives rise to the most acidic environments on our planet. Nitrification (the formation of nitric acid from ammonium) is potentially one of these, though the process is self-limiting in poorly buffered environments, as the majority of nitrifying bacteria are highly sensitive to even mild acidity. In contrast, prokaryotes that oxidize sulfur (in many of the large variety of reduced forms of this element that exist) include species that grow in neutral pH and moderately alkaline environments, as well as those that grow optimally in acidic environments. Indeed, some of the most extremely acidophilic life-forms known are those that obtain energy by oxidizing reduced sulfur to sulfuric acid.

Geothermal Areas

Elemental sulfur may occur in geothermal areas (e.g., around the margins of fumaroles) where it can form by the condensation of sulfur dioxide and hydrogen sulfide, two common volcanic gases (Equation (1)):

$$SO_2 + 2H_2S \rightarrow 2H_2O + 3S^0. \quad (1)$$

Oxidation of sulfur by acidophilic bacteria and archaea generates sulfuric acid (Equation (2)):

$$S^0 + H_2O + 1.5O_2 \rightarrow H_2SO_4. \quad (2)$$

This can result in severe acidification of environments both on the micro- (i.e., microbial habitats) and macroscale. Oxidation of sulfide minerals (see Section "Mine-Impacted Environments") may also contribute to acid genesis in these locations. Whether specific sites develop net acidity depends on how effectively acid generation is counterbalanced by the dissolution of basic minerals such as carbonates. Geothermal, sulfur-rich acidic sites are known as "solfatara" (Figure 1); water temperatures in solfatara fields approach boiling point (~85–100 °C, depending on altitude) but tend to cool rapidly as the water flows from the source of the geothermal spring. These sites may therefore be colonized by a variety of acidophilic microorganisms that have different temperature optima, and are therefore very fertile locations for isolating novel acidophilic microorganisms.

High-temperature environments that host solfatara and acid streams occur in zones of volcanism and in areas where the earth's crust is relatively thin. Examples of terrestrial and shallow marine locations include Yellowstone National Park (USA), Whakarewarewa (New Zealand), Krisuvik (Iceland), the Kamchatka Peninsula (Russia), Sao Michel (Azores), the islands of Volcano and Ischia (Italy), Djibouti (Africa), and some Caribbean islands, such as Montserrat and St. Lucia. Related to these are deep and abyssal submarine hydrothermal systems, such as the Mid-Atlantic Ridge, the East Pacific Rise, the Guaymas Basin, and active seamounts (e.g., around Tahiti). In contrast to many terrestrial sites, submarine hydrothermal systems are generally in the range pH 3–8, and saline, due to the high buffering capacity of seawater.

Mine-Impacted Environments

Many of the most important base metals (such as copper, lead, and zinc) used by humankind are sourced mostly from sulfide minerals. In addition, many precious metals including gold, silver, and PGMs (platinum group metals) are often found in association with sulfidic ores. Mining of metallic ores has, in the past, involved smelting whole rocks, though the advent of concentration techniques (mostly involving froth flotation and separation of target minerals) and

FIGURE 1 (a) Elemental sulfur forming from gases venting the Soufriere Hills volcano on Montserrat, West Indies. (b) An acidic geothermal pool (Frying Pan Hot Spring) in Yellowstone National Park, Wyoming.

nonpyrometallurgical techniques (such as pressure oxidation and biological processing) have had, and continue to have, a major impact on the mining industry. Production of mineral concentrates results in the generation of large quantities of waste minerals that, because of the intensive rock grinding involved, are fine-grained. These waste minerals (referred to as tailings) are usually disposed of in large lagoons, which may, in time, become drained, thereby allowing ingress of oxygen and dissolution of the minerals, a process often paralleled by intensive acidification. Apart from this, waste rocks from mines and the abandoned mines themselves can serve as source points for the generation of metal-rich, acidic wastewaters.

The mining of copper ores, which is carried out in many parts of the world, is a potential source of acid pollution. Copper exists in a variety of sulfide minerals, of which the most important (quantitatively) is the mixed copper–iron sulfide chalcopyrite (generally notated as $CuFeS_2$, though a more accurate mineral formulation is probably $CuFeS_{1.5}$, as both copper and iron occur in their more reduced ionic forms). Other significant copper minerals include single-metal sulfides (chalcocite, Cu_2S; and covellite, CuS) and mixed-metal sulfides bornite (Cu_5FeS_4) and enargite (Cu_3AsS_4). The sulfide moiety in these minerals represents a source of energy that can be utilized by some lithotrophic (literally, rock-eating) prokaryotes, many of which are obligate acidophiles (Figure 2). In addition, the ferrous iron present in chalcopyrite and bornite is a second potential energy source for mineral-oxidizing bacteria and archaea. These microorganisms require both oxygen and water (though little else) to facilitate their attack on the minerals, which is why mine wastes may be safely stored in

FIGURE 2 Partial dissolution of a sulfidic rock by chemolithotrophic bacteria. The rock on the right is freshly exposed and contains grains of pyrite (fool's gold) and other metal sulfides. That on the left has been exposed to attack by chemolithotrophic acidophiles, and the sulfide minerals have been effectively dissolved, leaving a porous remnant composed of inert minerals. Inset: the mineral-oxidizing acidophile *Leptospirillum ferrooxidans* (the scale bar represents 2 μm).

environments that are either totally dry or anoxic. In moist, aerated environments, however, the minerals are prone to oxidative dissolution, resulting in the release and potential solubilization of their component metals.

Acid genesis, however, is relatively limited when copper sulfides are (biologically) dissolved. In contrast, the iron disulfide mineral pyrite (FeS_2; "fool's gold"), which is the most abundant sulfide mineral in the lithosphere and which is invariably associated (often as the dominant mineral) with copper and other metal sulfide ores, generates significant levels of acidity due to its greater sulfur content. Pyrite has served as the model mineral for most studies of microbial attack on sulfide minerals and, although there have been various schemes proposed, that described by Wolfgang Sand of Duisburg–Essen University is generally regarded as the most accurate. In this, the initial attack on the hard, dense mineral is by ferric iron, which is a powerful oxidizing agent in acidic liquors. Ferric iron oxidizes the sulfur moiety of the mineral to thiosulfate, and in so doing is reduced to ferrous iron (Equation (3)):

$$FeS_2 + 6Fe^{3+} + 3H_2O \rightarrow 7Fe^{2+} + S_2O_3^{2-} + 6H^+. \quad (3)$$

The ferrous iron formed is reoxidized to ferric iron by a variety of iron-oxidizing acidophilic bacteria and archaea in an oxygen-consuming reaction (Equation (4)):

$$4Fe^{2+} + O_2 + 4H^+ \rightarrow 4Fe^{3+} + 2H_2O. \quad (4)$$

Thiosulfate is unstable in acidic liquors (particularly when ferric iron is present) and oxidizes to form a variety of other reduced inorganic sulfur compounds (RISCs) such as trithionate ($S_3O_6^{2-}$) and tetrathionate ($S_4O_6^{2-}$), as well as elemental sulfur (S^0). The latter can all serve as substrates for sulfur-oxidizing bacteria and archaea and are oxidized, when oxygen is available, to sulfuric acid, thereby generating the extreme acidity that helps to maintain a suitable pH required by the mineral-oxidizing microorganisms.

A second mechanism has been proposed for minerals, such as sphalerite (ZnS) and galena (PbS), that are soluble in sulfuric acid. In this scenario (the polysulfide mechanism), the metal–sulfur bond is broken by proton attack and hydrogen sulfide (H_2S) is liberated. If ferric iron is also present, concomitant attack by iron and protons results in the proposed formation of H_2S^+, which dimerizes to form free disulfide (H_2S_2), and is further oxidized forming, ultimately, elemental sulfur. In the absence of sulfur-oxidizing prokaryotes, this sulfur accumulates, though in their presence it is oxidized to sulfuric acid.

Dissolution of sulfidic ores not only produces acidity but also generates liquors that contain concentrations of base metals (copper, zinc, manganese, etc.) and aluminum that are far greater than those found in most surface waters. The two reasons for this are (1) the occurrence of these metals in sulfide minerals and others (many aluminosilicates) that spontaneously degrade at low pH, and (2) the far greater solubility of these metals in low pH than in circum-neutral pH solutions. Exceptions to the latter are metals such as molybdenum and vanadium, that occur mostly as oxyanions rather than cations. Metalloids, most significantly arsenic, can also be present at highly elevated concentrations in sulfide ore leach liquors; arsenopyrite (FeAsS) and realgar (As_4S_4) are two other relatively common sulfide minerals. Waters percolating through fissures in worked-out underground mines, as well as those draining stockpiled mine waste rock dumps and mine tailings, become enriched in these soluble metals and metalloids. At their point of discharge from underground mines or tailings ponds, mine waters are frequently devoid of oxygen and appear untainted. However, flowing waters become increasingly aerated, facilitating the oxidation of uncolored ferrous iron (usually the dominant dissolved metal found in mine drainage waters) to highly colored (yellow-red) ferric iron. This is why mine water-impacted waters are, in the main, very obvious sites of water pollution. Depending on pH, the ferric iron formed will either remain in solution or hydrolyze (react with water) to produce a variety of solid-phase minerals (e.g., schwertmannite ($Fe_8O_8(OH)_6SO_4$), ferrihydrite ($5Fe_2O_3 \cdot 9H_2O$), and amorphous ferric hydroxide ($Fe(OH)_3$)):

$$Fe^{3+} + H_2O \rightarrow Fe(OH)_3 + 3H^+. \quad (5)$$

This has two important consequences: First, the reaction generates protons, as illustrated in Equation (5), thereby helping to maintain the acidity of water. Aluminum and manganese also behave similarly, but these metals are generally less abundant than iron in mine waters. Second, the precipitates that form sink to the bottom of the stream, forming a dense coating (known in Europe as "ochre" and in the United States as "yellow boy") that can seriously impact benthic life. In the most extremely acidic mine waters ($pH < 2.5$), ferric iron remains in solution and the resulting red water color is often reflected in the names given to these streams or rivers, most famously, the Rio Tinto in Spain (Figure 3).

Apart from their characteristic low pH and elevated metal contents, the chemistries of mine waters are highly variable,

FIGURE 3 The Rio Tinto, an iron-rich extremely acidic river that flows through southwest Spain (left); the deep red-colored water of the Rio Tinto, due to the presence of elevated concentrations of soluble ferric iron (top right); colonies of rust-colored iron-oxidizing bacteria (*Acidithiobacillus ferrooxidans*) and heterotrophic acidophiles (*Acidiphilium* spp.) isolated from the Rio Tinto (bottom right). (For interpretation of the references to color in this figure legend, the reader is referred to the Web version of this chapter.)

as described in the Section titled "Acid Mine Streams and Lakes." However, concentrations of inorganic nitrogen (generally exclusively ammonium, except where there is input of nitrate from rock blasting at working mines), phosphate, and dissolved organic carbon all tend to be relatively small.

BIODIVERSITY OF EXTREME ACIDOPHILES

Extremely acidophilic organisms are exclusively microbial and include both prokaryotes and eukaryotes. The axiom that as an environmental parameter (in this case, acidity) becomes more extreme biodiversity declines holds true for both groups. Although some angiosperms have been observed to grow in highly acidic lakes, their root systems grow in sediments in which the pH is usually much higher than the water body itself. Many eukaryotic microorganisms that have been observed in extremely low pH environments are acid-tolerant rather than truly acidophilic, and may grow equally well, or better, in circum-neutral pH environments.

Primary Producers in Acidic Environments

The first extremely acidophilic microorganism to be isolated and characterized was the sulfur-oxidizing bacterium *Acidithiobacillus* (*At.*) *thiooxidans* (then referred to as *Thiobacillus thiooxidans*) by Waksman and Joffe in 1921. Some years later, another sulfur-oxidizing bacterium was isolated from water draining a coal mine that had the unique trait (at the time) of also being able to oxidize ferrous iron to ferric. *At. ferrooxidans* has subsequently become the most well-studied of all acidophilic microorganisms. Both of these early isolates are autotrophic chemolithotrophs, that is, they use inorganic electron donors and fixed carbon dioxide. Such a metabolic lifestyle is highly appropriate in extremely acidic environments, which, as noted previously, tend to contain elevated concentrations of potential inorganic energy sources (ferrous iron and reduced sulfur) but often low concentrations of dissolved organic carbon. While more recent studies have led to the isolation of a number of other genera and species of chemolitho-autotrophic acidophiles, a large number of highly biodiverse microorganisms that have very different metabolic lifestyles (e.g., phototrophic microalgae, heterotrophic bacteria and yeasts, and phagotrophic protozoa) have also been shown to be obligate acidophiles (see Section "Acidophilic Eukaryotic Microorganisms").

Primary production (net assimilation of carbon) in extremely acidic environments is carried out by two main groups of microorganisms, the relative importance of which varies from site to site. Chemolitho-autotrophic acidophiles are CO_2 fixers that use either ferrous iron or reduced sulfur (or, in some cases, both) as energy sources. Sulfide (e.g., in minerals), elemental sulfur, and RISCs are far more energetic substrates than ferrous iron (Table 1), though interestingly at least one bacterium (*At. ferrooxidans*) that can use both ferrous iron and sulfur as substrates appears to opt for the former when both are available. Some acidophiles have also been shown to use hydrogen as an electron donor. The significance of this is unknown, though hydrogen may form in these environments from reactions between protons and various minerals in contact with acidic liquors. Other chemolitho-autotrophic metabolisms (e.g., nitrification) have not been observed in extremely acidic environments. Photoautotrophy (the use of solar energy to fuel carbon dioxide fixation) may be the dominant mechanism of primary production in acidic ecosystems, as in most others, though in underground sites (acidic caves and mine caverns) primary production is exclusively mediated by chemolithotrophs. All acidophilic phototrophic microorganisms that have been identified are eukaryotic microalgae. No truly acidophilic phototrophic bacteria (aerobic cyanobacteria or anaerobic purple/green S-bacteria) have been described, though clones of anaerobic photosynthetic green sulfur and purple nonsulfur bacteria have been obtained from an acidic geothermal site in New Zealand (see Section "Geothermal Areas").

TABLE 1 Comparison of free energy changes associated with the oxidation of inorganic substrates used by chemolithotrophic acidophiles

Reaction	Free Energy Change ΔG^o (kJ mole per substrate)
Ferrous iron oxidation	
$4FeSO_4 + O_2 + 2H_2SO_4 \rightarrow 2Fe_2(SO_4)_3 + 2H_2O$	−30 (at pH 2.0)
Hydrogen oxidation	
$H_2 + 0.5O_2 \rightarrow H_2O$	−237
Elemental sulfur oxidation	
$S^0 + 1.5O_2 + H_2O \rightarrow SO_4^{2-} + 2H^+$	−507
Hydrogen sulfide oxidation	
$H_2S + 2O_2 \rightarrow H_2SO_4$	−714
RISC oxidation	
(1) $S_2O_3^{2-} + 2O_2 + H_2O \rightarrow 2SO_4^{2-} + 2H^+$	−739
(2) $S_4O_6^{2-} + 3.5O_2 + 3H_2O \rightarrow 4SO_4^{2-} + 6H^+$	−1225

Source: *Data from Kelly, D. P. (1978). Bioenergetics of chemolithotrophic bacteria. In A. T. Bull, & P. M. Meadows (Eds.)* Companion to Microbiology *(pp. 363–386). London: Longman; and Kelly, D. P. (1999). Thermodynamic aspects of energy conservation by chemolithotrophic bacteria in relation to the sulfur oxidation pathways.* Archives of Microbiology, *171, 219–229.*

Many acidophilic microorganisms that fix carbon dioxide are obligate autotrophs. Some, however, can switch to assimilating organic carbon if and when this becomes available. The metabolic logic for this is obvious, as CO_2 fixation is a highly energy-consuming process (e.g., *At. ferrooxidans* has been estimated to utilize most of the energy it obtains by oxidizing iron on this single process), and using prefixed carbon, assuming that it is readily incorporated and metabolized, avoids this expenditure of energy. Various terms have been used to describe such microorganisms, which include some eukaryotic algae as well as some prokaryotic acidophiles, though the most appropriate (and least ambiguous) is to refer to them as "facultative autotrophs." Whether such acidophiles are net contributors to total primary production depends not only on the presence of metabolizable organic carbon but also (in the case of phototrophs such as *Cyanidium caldarium*) on the availability of solar energy.

Heterotrophic Acidophiles

Unusually for microbial ecology, the first obligately heterotrophic acidophilic bacteria (*Acidiphilium* spp.) were isolated some 70 years after the first chemolithotrophic acidophile – though a heterotrophic acidophilic archaeon (*Thermoplasma* (*Tp.*) *acidophilum*) – was actually described a decade before the first *Acidiphilium* sp. (*Acidiphilium cryptum*). There are now a large number of characterized species of acidophilic bacteria and archaea that are known to use organic compounds as sources of both carbon and energy. Some of these are able to supplement their energy budgets by oxidizing inorganic substrates (ferrous iron or reduced sulfur) when these are also available. In the case of truly mixotrophic acidophiles, such as the iron-oxidizing heterotroph *Ferrimicrobium* (*Fm.*) *acidiphilum*, the inorganic substrate can serve as the sole source of energy and the organic moiety only to meet the carbon requirements of the bacterium.

Bacteria, in particular, are renowned as a collective group of microorganisms for their abilities to degrade a multitude of small and large molecular weight organic compounds, including many synthetic materials. Acidophilic prokaryotes, on the other hand, appear to use a far more restricted range of monomeric organic substrates and few polymeric materials. Simple sugars and alcohols are utilized by many heterotrophic acidophiles, but aliphatic acids (such as acetic acid) tend to be lethal to acidophiles when present in only micromolar concentrations. The reason for this relates to the fact that many small molecular weight organic acids exist as undissociated, lipophilic molecules in low pH liquors. These can freely permeate microbial membranes and accumulate in the circum-neutral pH cell interiors where they dissociate and cause intracellular acidification of the cytoplasm. Di- and tricarboxylic organic acids, such as citric acid, are not so toxic and are actually used as substrates by many heterotrophic acidophiles. Some organic acids, most notably glutamic acid, also serve as appropriate substrates for many acidophiles, though others (e.g., glycine) do not. Complex, nitrogen-rich organic substrates, such as yeast extract and tryptone, are also suitable substrates for isolating and cultivating many heterotrophic acidophiles and supplementing defined organic growth media with, for example, yeast extract often promoting growth of heterotrophic acidophilic bacteria and archaea. One of the few known examples of an acidophile being able to grow on an organic polymer is the archaeon *Acidilobus aceticus*, which grows anaerobically on starch, forming acetate as the main metabolic product.

Aerobic and Anaerobic Acidophiles

The majority of known acidophilic prokaryotes have been classed as obligate aerobes. More detailed examination has revealed that, in a number of cases, they can also grow in the absence of oxygen and are therefore facultative anaerobes. Of the various options that microorganisms use for living in the absence of oxygen, by far the most widespread among acidophiles appears to be ferric iron respiration. This is understandable since iron, as both ferrous and ferric, is usually abundant in extremely acidic environments, particularly those originating from the oxidative dissolution of sulfide minerals. There is also a thermodynamic advantage to be gained from using ferric iron in that the redox potential (E_h value) of the ferrous/ferric couple at low pH is about +770 mV, a value which is not much below that of the oxygen/water couple (+840 mV) and considerably more positive than alternative inorganic electron acceptors such nitrate and sulfate. Most bacteria (and the Euryarchaeote *Ferroplasma* (*Fp.*) *acidiphilum*) that can oxidize ferrous iron in the presence of molecular oxygen can also reduce it when oxygen is absent. Notable exceptions are species of *Leptospirillum*, though this is explained by the fact that these highly specialized bacteria have not been found to use an electron donor other than ferrous iron. Other acidophilic bacteria that also reduce ferric iron to ferrous are obligate heterotrophs, one of which, *Fm. acidiphilum*, is an iron oxidizer, while others (all species of *Acidiphilium*, as well as many *Acidocella* and *Acidobacterium* spp.) are not. This trait is not universal among acidophilic heterotrophic bacteria, however, as illustrated by the fact that *Acidisphaera rubrifaciens* and closely related isolates do not appear to reduce ferric iron. The earlier claim that the sulfur-oxidizing bacterium *At. thiooxidans* can reduce ferric iron was later challenged as probably being an artifact resulting from chemical reduction by RISCs that are produced during sulfur metabolism. The thermotolerant sulfur oxidizer *Acidithiobacillus caldus* also does not appear to reduce ferric iron.

Sulfur respiration (the use of elemental sulfur as electron acceptor) is not uncommon among acidophilic archaea: *Acidianus*, *Stygiolobus*, *Sulfurisphaera* (all thermoacidophilic crenarchaeotes), and *Thermoplasma* (a moderately thermoacidophilic euryarchaeote) can all grow anaerobically by reducing sulfur to hydrogen sulfide. *Acidianus* spp. and *Sulfurisphaera ohwakuensis* are both facultative anaerobes that couple the oxidation of hydrogen to the reduction of sulfur in anoxic environments, while *Stygiolobus azoricus* is an obligately anaerobic thermoacidophile that can do the same. In contrast, both classified species of *Thermoplasma* (*Tp. acidophilum* and *Tp. volcanium*) are facultative anaerobes that couple the oxidation of organic carbon to the reduction of elemental sulfur. No extremely acidophilic sulfur- or sulfate-reducing bacteria (SRB), or sulfate-reducing archaea, have yet been isolated and characterized, though there is evidence that sulfidogens are both present and active in some anaerobic acidic environments. A *Desulfosporosinus*-like isolate (M1), isolated from a geothermal site on Montserrat, West Indies, has been demonstrated to grow in a mixed culture at pH 3.2 and above, but is probably acid-tolerant rather than a true acidophile. Many other apparently acid-tolerant sulfidogenic isolates and putative clones detected in acidic mine waters have also been found to be Gram-positive bacteria.

Clones of methanogenic archaea have also been identified in gene libraries constructed from DNA extracted from some extremely acidic environments, but no extremely acidophilic methanogens are known. Likewise, no acetogenic acidophiles have been isolated, though this may be explained on the basis of the biotoxicity of acetic acid in low pH liquors, as discussed previously. The general toxicity of aliphatic acids may also help to account for the apparent absence of fermentative metabolism among extreme acidophiles, with the exception of the thermophilic archaeon *Acidilobus aceticus*, which can grow by fermenting starch to acetic acid. The scarceness of nitrate in most acidic environments, apart from those in the vicinity of rock blasting, and the fact that acidophiles are more sensitive to nitrate and nitrite than most other bacteria are probably why nitrate respiration is apparently absent in these microorganisms.

Temperature and pH Characteristics of Acidophilic Microorganisms

One of the most widely used methods to categorize acidophilic prokaryotes is based on their temperature characteristics, that is, their optimum temperatures for growth and the range of temperatures within which they are active (Table 2). Three groups of acidophiles have often been recognized in this way: (1) mesophiles, with temperature optima of 20–40 °C; (2) moderate thermophiles, with temperature optima of 40–60 °C; and (3) extreme thermophiles, with temperature optima of 60–80 °C. While some acidophiles (strains of *At. ferrooxidans* and *Acidiphilium*) have been demonstrated to be active at very low (<5 °C) temperatures, all of these have temperature optima well above 20 °C,

TABLE 2 Categorization of validated species and genera of extremely acidophilic prokaryotic microorganisms, based on growth temperature optima

	Carbon Assimilation	Fe^{2+} Oxidation	Fe^{3+} Reduction	S^0 Oxidation	S^0 Reduction
(a) Mesophiles (temperature optima 20–40 °C)					
At. ferrooxidans	OA	+	+	+	+
L. ferrooxidans	OA	+	–	–	–
Fm. acidiphilum	OH	+	+	–	–
At. thiooxidans	OA	–	–	+	–
Thiomonas spp.	FA	+	–	+	–
Acidiphilium spp.	OH	–	+	+	–
A. acidophilum	FA	–	+	+	–
Acidocella spp.	OH	–	+	–	–
Acidobacterium spp.	OH	–	+	–	–
Fp. acidiphilum	OH	+	+	–	–
(b) Moderate thermophiles (temperature optima 40–60 °C)					
L. ferriphilum	OA	+	–	–	–
Sulfobacillus spp.	FA	+	+	+	–

(Continued)

TABLE 2 Categorization of validated species and genera of extremely acidophilic prokaryotic microorganisms, based on growth temperature optima—cont'd

	Carbon Assimilation	Fe^{2+} Oxidation	Fe^{3+} Reduction	S^0 Oxidation	S^0 Reduction
Alicyclobacillus spp.[a]	OH/FA	+/–	+/–	+/–	–
Am. ferrooxidans	FA	+	+	–	–
Fx. thermotolerans	OH	+	+	–	–
Acd. organivorans	OH	–	–	+	–
At. caldus	OA	–	–	+	–
Thermoplasma spp.	OH	–	–	–	+
Picrophilus spp.	OH	–	–	–	–
(c) Extreme thermophiles (temperature optima >60 °C)					
H. acidophilum	OA	–	–	+	–
S. acidocaldarius	OH	–	–	–	–
S. solfataricus	OH	–	–	–	–
S. metallicus	OA	+	–	+	–
S. tokodaii	OH	+	–	+	–
Metallosphaera spp.	FA	–	–	+	–
Sulfurococcus spp.	FA	–	–	+	–
Ac. infernus	OA	–	–	+	+
Ac. ambivalens	OA	–	–	+	+
Ac. brierleyi	FA	+	–	+	+
Sg. azoricus	OA	–	–	–	+
Ss. ohwakuensis	FA	–	–	–	+

Note: *OA, obligate autotroph; FA, facultative autotroph; OH, obligate heterotroph.*
Genera abbreviations: At., Acidithiobacillus; L., Leptospirillum; Fm., Ferrimicrobium; A., Acidiphilium; Sb., Sulfobacillus; Fp., Ferroplasma; Am., Acidimicrobium; Fx., Ferrithrix; Acd., Acidicaldus; H., Hydrogenobaculum; S., Sulfolobus; Ac., Acidianus; Sg., Stygiolobus; Ss., Sulfurisphaera.
[a]Alicyclobacillus *spp. include species that are facultatively autotrophic and obligately heterotrophic, and vary in terms of their dissimilatory transformations of iron and sulfur.*

and are therefore psychrotolerant rather than psychrophilic microorganisms. At the other end of the temperature spectrum, the most thermophilic extreme acidophile known is the facultatively anaerobic sulfur-metabolizing archaeon *Acidianus infernus*, which has a growth temperature optimum of 90 °C and a maximum of about 96 °C. However, relatively few hyperthermophilic acidophiles are known, and the fact that the maximum temperature for growth of an acidophile is about 25 °C lower than that of the most thermophilic life-forms known (neutrophilic *Pyrolobus*-like archaea) is possibly a reflection of the difficulty that living organisms have when challenged by the dual stresses of extreme temperature and acidity. In addition, the pH of high-temperature (>100 °C) abyssal environments around submarine vents is maintained at close to neutral by the strong buffering capacity of seawater, precluding extensive colonization by acidophiles.

As with neutrophilic prokaryotes, extremely thermophilic acidophiles are mostly archaea while mesophiles are predominantly bacteria. The majority of moderate thermoacidophiles are also bacteria, and mostly Gram-positive, while most known Gram-negative acidophilic bacteria grow best at below 40 °C. There are exceptions to this general trend. Indeed, the most thermophilic acidophilic bacteria known – the sulfur-oxidizing autotroph *Hydrogenobaculum acidophilum*, which grows at up to 70 °C, and the heterotroph *Acidicaldus organivorans*, which grows at up to 65 °C – are both Gram-negative.

The ability to tolerate elevated concentrations of protons (strictly speaking, hydronium ions; H_3O^+) is obviously what defines an acidophile. While there is no official cut-off pH value that determines whether an organism is or is not an acidophile, the generally accepted view is that, as a group, these can be divided into extreme acidophiles that have pH optima for growth at pH < 3, and moderate acidophiles that have pH optima of between 3 and 5. As can be anticipated, the most extremely acidic environments have less potential biodiversity than those that are moderately acidic. The number of prokaryotes that are known to grow at pH < 1 is relatively small and includes some Gram-positive bacteria (e.g., *Sulfobacillus* spp.), Gram-negative bacteria (e.g., *Leptospirillum* spp. and *At. thiooxidans*), and archaea (e.g., *Ferroplasma* spp.) that oxidize iron and/or sulfur. The most acidophilic of all currently known life-forms is, however, a heterotrophic archaeon, *Picrophilus*. Two species are known, *Picrophilus oshimae* and *Picrophilus torridus*, both of which have an optimal pH for growth of ~0.7 and grow in synthetic media poised at pH ~ 0. These "hyperacidophiles" are also thermophilic, with optimum temperatures for growth at ~60 °C.

Physiological Versatility in Acidophilic Prokaryotes: Specialized and Generalist Microorganisms

Acidophiles as a group are highly versatile and are able to utilize a wide variety of energy sources (solar and inorganic and organic chemicals) and grow in the presence or complete absence of oxygen at temperatures of between 4 and 96 °C. However, individual species display very different degrees of metabolic versatility. On the one end of this spectrum are members of the genus *Leptospirillum*. Three species are known: *L. ferrooxidans*, *L. ferriphilum*, and *L. ferrodiazotrophum*. All grow as highly motile curved rods and spirilli, and species and strains vary in temperature and pH characteristics. All three species, however, appear to use only one energy source – ferrous iron. Because of the high redox potential of the ferrous/ferric couple (see Section "Aerobic and Anaerobic Acidophiles"), these bacteria, by necessity, have to use molecular oxygen as an electron acceptor, restricting them to being active only in aerobic environments. All three species fix carbon dioxide (but not organic carbon) and two of the three (*L. ferrooxidans* and *L. ferrodiazotrophum*) are also able to fix molecular nitrogen. *Leptospirillum* spp. are, therefore, highly specialized acidophiles. Their metabolic limitations appear, however, to be compensated by their abilities to outcompete other iron-oxidizing bacteria in many natural and anthropogenic environments, such as stirred-tank bioreactors used to bioleach or biooxidize sulfide ores. This is achieved, at least in part, by their greater affinities for ferrous iron and greater tolerance of ferric iron than most other iron oxidizers.

At. ferrooxidans is, in contrast, a more generalist bacterium. Initially it was described as an obligate aerobe that obtains energy by oxidizing ferrous iron, elemental sulfur, sulfide, and RISCs, and fixes CO_2 as its sole source of carbon. The first hint of a more extensive metabolic potential was in a report by Thomas Brock and John Gustafson in 1976 who showed that the bacterium could couple the oxidation of elemental sulfur to the reduction of ferric iron, though it was not confirmed at the time whether this could support growth of the acidophile in the absence of oxygen, though the free energy of the reaction ($\Delta G = -314\,\text{kJ mol}^{-1}$; Equation (6)) suggested that this might be the case.

$$S + 6Fe^{3+} + 4H_2O \rightarrow HSO_4 + 6Fe^{2+} + 7H^+. \qquad (6)$$

Later, Jack Pronk and colleagues at Delft University showed conclusively that *At. ferrooxidans* is, indeed, a facultative anaerobe and can grow anaerobically by ferric iron respiration using not only sulfur as electron donor, but also formic acid (which can also be used as sole energy source under aerobic conditions). The finding that this acidophile can use formic acid, although somewhat unexpected, does not imply that it is capable of heterotrophic as well as autotrophic growth, as C-1 compounds, such as formate and methanol, are also used by other autotrophic prokaryotes. About the same time, it was discovered that some strains of *At. ferrooxidans* (including the type strain) can use hydrogen as an energy source, but that bacteria cultivated on hydrogen are less acidophilic than when grown on sulfide ores. It was shown later that hydrogen oxidation could also be coupled to ferric iron reduction by some *At. ferrooxidans* isolates.

The most generalist of all acidophiles are, however, *Sulfobacillus* spp. These Gram-positive bacteria can grow as chemolithotrophs, heterotrophs, or mixotrophs in aerobic or anaerobic environments. Although there are no reports of *Sulfobacillus* spp. using hydrogen, they can (unlike *At. ferrooxidans*) use a variety of organic compounds (such as glucose and glycerol) as carbon and energy sources, though their capacities for heterotrophic growth are more limited than *Alicyclobacillus* spp. (related acidophilic Firmicutes, some of which can also oxidize ferrous iron and sulfur).

Acidophilic Eukaryotic Microorganisms

Extremely acidophilic organisms are exclusively microbial. While some angiosperms, such as *Juncus bulbosus* and *Eriophorum angustifolium*, can grow in highly acidic (pH < 3) ponds and lakes, their root systems grow in sediments where the pH is usually significantly higher than the water body itself. Many eukaryotic microorganisms that may be found in extremely low pH environments are acid-tolerant rather than truly acidophilic and may grow equally well, or better, in higher pH waters.

All known phototrophic acidophiles are eukaryotic, and both mesophilic and moderately thermophilic species are known. Some photosynthetic acidophiles are also capable of heterotrophic growth in the absence of light, provided that a suitable carbon source is available. Microalgae that can live in highly acidic environments include genera of Chlorophyta, such as *Chlamydomonas acidophila* and *Dunaliella acidophila*; Chrysophyta, such

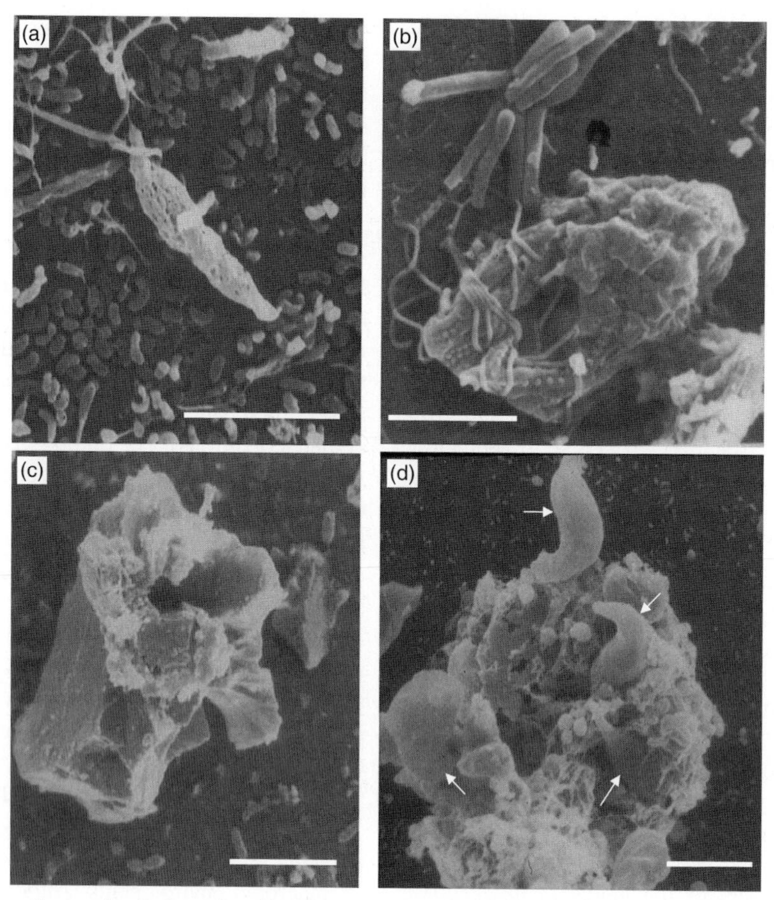

FIGURE 4 Scanning electron micrographs of eukaryotic acidophiles: (a) a *Eutreptia*-like flagellate protozoan, grazing on *Leptospirillum ferrooxidans*; (b) a *Cinetochilum*-like ciliate protozoan, grazing on *Acidithiobacillus ferrooxidans*; (c) a *Vahlkampfia*-like amoeboid protozoan; (d) a bundle of *Euglena mutabilis* (an acidophilic microalga) with individual cells arrowed. The scale bar represents 5 μm in micrographs (a)–(c), and 10 μm in micrograph (d).

as *Ochromonas sp.*; and Euglenophyta, such as *Euglena mutabilis* (Figure 4). Some diatoms, including several *Eunotia* spp., have also been found to colonize extremely acidic waters. A filamentous alga, identified from its morphology as Zygnema and confirmed from biomolecular analysis to be *Zygnema circumcarinatum*, has been found in abundance on surface streamer growths in an extremely acidic (pH ~2.7) metal-rich stream draining a mine adit in southwest Spain. Four species of thermoacidophilic Rhodophyta have been described. Of these, *Galderia* spp. (*G. sulfuraria* and *G. maxima*) can grow as heterotrophs, while *Cyanidioschyzon merolae* and the original strain of *C. caldarium* are strict autotrophs. One *C. caldarium*-like isolate has been reported to grow in synthetic media poised as low as pH 0.2. Chlorella-like microalgae have also been detected in acidic geothermal waters.

Many species of yeasts and fungi can tolerate moderate or even extreme acidity. Truly acidophilic fungi are, however, less common, though these include some remarkable species, such as *Acontium velatum* and *Scytalidium acidophilum*, both of which are copper-tolerant mitosporic fungi that can grow at pH values of below 0.5. Among the most commonly encountered yeasts in metal-rich acidic waters are *Rhodotorula* spp., while some *Cryptococcus* spp. and *Trichosporon dulcitum* are also acidophilic yeasts. Novel acidophilic fungal isolates (proposed name *Acomyces richmondensi*) have been isolated from warm (30–50 °C), extremely acidic (pH 0.8–1.38), and iron/zinc/copper/arsenic-contaminated waters within the Richmond mine at Iron Mountain, California.

Microscopic animal life-forms may also be found in acidic environments. The most biodiverse of these appear to be protozoa (Figure 4). Phagotrophic flagellates (Eutreptia), ciliates (Urotricha, Vorticella, Oxyticha, and Cinetochilum), and amoeba (Vahlkampfia) have all been encountered in acidic mine waters, and some have also been grown in acidic media in the laboratory. Multicellular animal life-forms are relatively uncommon, though rotifers (such as *Cephalodella hoodi* and *C. gibba*) have occasionally been identified in acidic mine waters. The two most acidophilic species of known rotifers appear to be *Elosa woralii* and *Brachionus sericus*, though the latter can also grow at neutral pH *in vitro*. The pioneering crustacean *Chydorus sphaericus* has also been observed in the pelagic community of acid mine lakes in Germany, though it is acid-tolerant rather than acidophilic, with a pH range of 3.2–10.6.

INTERACTIONS BETWEEN ACIDOPHILIC MICROORGANISMS

The study of microbial ecology involves not only understanding the impact of the environment on microorganisms (and vice versa) but also examining how microorganisms interact with each other. Along with increasing awareness of the biodiversity and complexity of life in extremely acidic environments have come fresh insights into the wide range of microbial interactions that occur within them. In some cases, such as grazing by phagotrophic protozoa on acidophilic bacteria, the interaction may be readily observed, though more often it is more clandestine.

Mutualistic Interactions

Mutualistic interactions are where both partners derive some benefit from their association. One way in which this occurs in extremely acidic environments is via redox transformations and transfer of iron and/or sulfur between prokaryotes. As noted in the Section titled "Biodiversity of Extreme Acidophiles," ferrous iron is an energy source that is widely used by acidophilic bacteria and some acidophilic archaea, while ferric iron can act as a highly effective alternative electron acceptor to oxygen in low pH environments. Juxtaposition of aerobic and microaerobic/anaerobic environments can lead to rapid cycling of iron between the two zones. This is aided by the fact that, in contrast to most environments, ferric iron is soluble at pH < 2.5 and is more readily utilized as an electron sink as soluble Fe^{3+} than when present in its various amorphous and crystalline forms. The importance of iron cycling has been illustrated in major acidic environments such as the Rio Tinto, and also demonstrated *in vitro*. Obviously, an extraneous energy source is required for iron cycling to perpetuate. In acidic environments, this may be organic carbon, originating as exudates and lysates from primary producers (phototrophs and chemolithotrophs) that act as electron donors for iron-reducing acidophiles. Cycling of iron may involve more than one species (e.g., the iron oxidizer *At. ferrooxidans* and the iron reducer *Acidiphilium*) or a single species (e.g., *Sulfobacillus*). The situation with sulfur transformations is less clear, due in part to the relative paucity in the knowledge of bacterial sulfate/sulfur reduction in low-temperature acidic environments, and the far greater insolubility of some reduced sulfur compounds (metal sulfides and elemental sulfur) than ferric iron at extremely low pH, which limits their free diffusion. Sulfate produced by aerobic sulfur-oxidizing acidophiles (such as *Acidithiobacillus* and *Thiomonas* spp.) can diffuse into underlying sediments and act as a terminal electron acceptor for any acidophilic/acid-tolerant SRB present. These generate sulfide, which, at low pH, is present almost exclusively as gaseous H_2S. The presence in the sediments of soluble metals, such as copper, that form very insoluble sulfides results in the rapid removal of H_2S, even at very low pH. However, if, as is often the case, the dominant soluble chalcophilic metal present is (ferrous) iron, the lower solubility of the sulfide mineral (FeS) means that it does not form until the pH has risen to ~5. If the sediment pH is <5, at least some of the H_2S can, at least in theory, diffuse into the overlying water and act as an energy source for sulfur-oxidizing acidophiles.

Other examples of mutualistic interaction between acidophilic bacteria involve the metabolism of organic compounds. Autotrophic iron- and sulfur-oxidizing acidophiles fix carbon dioxide to incorporate into cell biomass. Some of this, mostly small molecular weight material, is lost from actively metabolizing cells and has been shown to accumulate in axenic cultures of these prokaryotes grown in the laboratory. Additional organic carbon originates from dead and dying cells as cell lysates. In the environment, much of this organic carbon is metabolized by heterotrophic acidophiles, such as *Acidiphilium* spp., which are adept scavengers. Positive feedback to the autotrophs comes from the fact that many of these are sensitive to small molecular weight organic compounds in general, and aliphatic acids in particular, and the catabolism of these materials by heterotrophic acidophiles therefore reduces or eliminates this potential toxicity hazard. Iron-oxidizing acidophiles vary in their sensitivities to organic materials; *L. ferrooxidans* is, for example, much more sensitive than *At. ferrooxidans*. This is reflected in the far greater mortality rate of the former in spent (substrate-depleted) media, and is the reason why mixed cultures of *L. ferrooxidans* and *Acidiphilium* spp. grown on ferrous iron or pyrite tend to be far more stable than pure cultures of the iron oxidizer. In practical terms, the inclusion of obligately or facultatively heterotrophic acidophiles in mineral-leaching consortia has been shown to improve metal recovery, and commercial-scale stirred tanks used to bioprocess sulfide ores have invariably been found to include organic carbon-degrading acidophiles as well as those that fix CO_2. The same rationale of using heterotrophic acidophiles to remove potentially toxic organic materials has been used to develop solid media for isolating and cultivating iron- and sulfur-oxidizing acidophiles from environmental and industrial samples.

Synergistic Interactions

Microbial interactions that result in the complementary activities of both (or all) participants being more efficient in, for example, degrading a substrate, than the individual species working alone are referred to as synergistic. One such example involves the oxidative dissolution of pyrite by mixed cultures of *L. ferrooxidans* and *At. thiooxidans*.

L. ferrooxidans is an iron oxidizer that is unable to oxidize sulfur, while *At. thiooxidans* has the opposite abilities. Pyrite, being an acid-insoluble sulfide mineral (see Section "Mine-Impacted Environments"), is oxidized by ferric iron produced by ferrous iron-oxidizing *L. ferrooxidans* in an acid-consuming reaction. The RISCs produced as a result of ferric iron attack on pyrite are oxidized to sulfuric acid by *At. thiooxidans*, thereby generating the extremely low pH conditions under which *L. ferrooxidans* thrives and mineral dissolution is accelerated. Pure cultures of *L. ferrooxidans* are, however, also able to accelerate the oxidative dissolution of pyrite, in contrast to the heterotrophic iron oxidizer *Fm. acidiphilum*, which requires a source of organic material, such as yeast extract, as carbon source. When *Fm. acidiphilum* and *At. thiooxidans* are grown in mixed culture, the interaction involves carbon transfer as well as the modulation of acidity. *Fm. acidiphilum* can obtain sufficient carbon to grow from autotrophic partners, such as *At. thiooxidans*. This facilitates ferric iron generation, producing RISCs from the degrading pyrite that serves as the energy source for *At. thiooxidans* and continued fixation of CO_2 and release of organic carbon (Figure 5).

Syntrophic Interactions

In syntrophic relationships, the degradation of a substrate by one species is made thermodynamically possible through the removal of an end product by another species. Neutrophilic SRB have frequently been reported as one of the partners in a syntrophic association. Hydrogen is a common end product of fermentative metabolism and the oxidation of this gas, coupled to sulfate reduction, may result in a change in the overall free energy (ΔG) and allow an otherwise thermodynamically unfeasible reaction to proceed. A syntrophic association involving an acid-tolerant *Desulfosporosinus*-like SRB and an acetic acid-degrading *Acidocella* sp. has been proposed to account for sulfidogenesis in moderately acidic (pH 3.2 and above) media. This mixed culture grows anaerobically using glycerol as sole carbon and energy source, a substrate that the SRB can oxidize but the *Acidocella* cannot. In pure cultures of the *Desulfosporosinus*, acetic acid accumulates in equimolar proportion to the amount of glycerol oxidized, but in the presence of *Acidocella* the appearance of acetic acid is transient and more sulfide is produced. It was postulated that acetic acid is degraded to hydrogen and carbon dioxide, a reaction that is feasible in thermodynamic terms only if at least one of these products is rapidly removed. This role was fulfilled by the *Desulfosporosinus* sp., which was shown to use hydrogen as well as glycerol as an electron donor. The energetic bonus for the SRB (hydrogen), which was only available in the mixed culture, resulted in it generating more hydrogen sulfide than when it was grown in pure culture. Additional support for the hypothesis came from the observation that the mixed culture, but not axenic cultures of the *Desulfosporosinus* isolate, could use acetic acid to fuel sulfidogenesis.

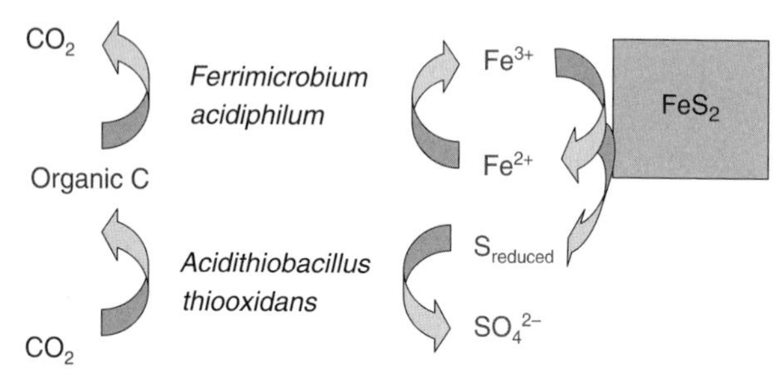

FIGURE 5 Dissolution of pyrite by a mixed culture of *Ferrimicrobium* (*Fm.*) *acidiphilum* and *Acidithiobacillus* (*At.*) *thiooxidans*. Though *Fm. acidiphilum* can oxidize ferrous iron to ferric iron (which is the chemical that directly attacks the pyrite mineral), it requires organic carbon to grow. *At. thiooxidans* can fix CO_2 and releases some of this as organic carbon, but it cannot access its energy source (reduced sulfur) directly from pyrite. Therefore, neither bacterium can grow in organic carbon-free pure cultures, using pyrite as an energy source, but together they form a successful synergistic consortium, via the interactions shown. *Modified from Bacelar-Nicolau, P. & Johnson, D. B. (1999). Leaching of pyrite by acidophilic heterotrophic iron-oxidizing bacteria in pure and mixed cultures.* Applied and Environmental Microbiology, 65, *585–590, with permission from the publisher.*

Predation

The fact that some acidophilic bacteria are predated by protozoa and rotifers has been known for many years. One of the first indicators that some protozoa could grow in highly acidic waters was a report in 1941, when it was noted that the flagellate *Polytomella caeca* could grow over a wide pH range (from 1.7 to 9.2). Early studies of acid mine drainage (AMD) frequently reported the presence of flagellates, ciliates, and amoeba, and the first laboratory study was by Henry Ehrlich at the Rensselaer Polytechnic Institute (in 1963) who found that a *Eutreptia*-like flagellate could grow in enrichment cultures prepared using mine water as an inoculum. A detailed study of another *Eutreptia*-like flagellate was described some 30 years later. This protozoan was found to be obligately acidophilic, with a pH range of 1.8–4.5 for growth. Although it was highly sensitive to some heavy metals (e.g., copper, silver, and molybdenum), it could tolerate very high concentrations of both ferrous and ferric iron. The flagellate was found to graze a wide range of acidophilic bacteria, including *At. ferrooxidans*, *L. ferrooxidans*, and *A. cryptum*. It was noted that the highly motile iron oxidizer *L. ferrooxidans* was less effectively grazed than the less motile acidophile *At. ferrooxidans*, leading to mixed cultures of the two bacteria being dominated by *L. ferrooxidans* when the protozoan was present. Filamentous growth by some acidophiles also appeared to give them some protection from predation by

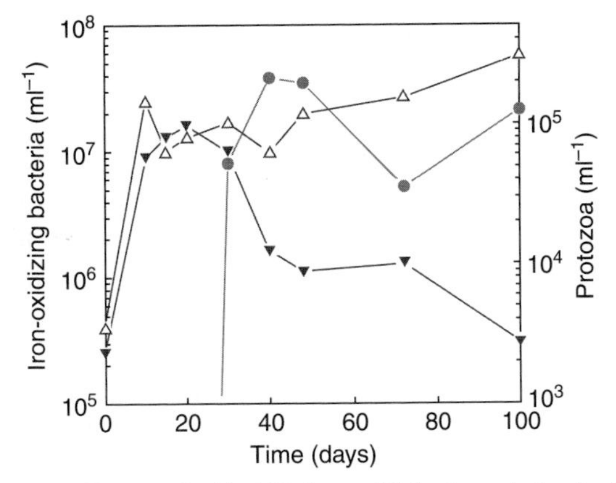

FIGURE 6 Grazing of acidophilic iron-oxidizing bacteria (a mixed culture of *Acidithiobacillus ferrooxidans* and *Leptospirillum ferrooxidans*) by a *Cinetochilum*-like acidophilic ciliate protozoan, showing a classic predator–prey relationship. Numbers of iron-oxidizing bacteria are shown in red: solid symbols show data from a ciliate-containing culture, while hollow symbols show data from a corresponding culture where protozoa were absent. Protozoan numbers are shown in green. *Modified from Johnson, D. B., & Rang, L. (1993). Effects of acidophilic protozoa on populations of metal-mobilizing bacteria during the leaching of pyritic coal.* Journal of General Microbiology, 139, *1417–1423, with permission from the publisher.* (For interpretation of the references to color in this figure legend, the reader is referred to the Web version of this chapter.)

the flagellate. Detailed examination of mixed cultures of acidophilic bacteria and five different acidophilic protozoan isolates (three flagellates, and *Cinetochilum*-like ciliate, and a *Vahlkampfia*-like amoeba) showed that, in each case, population dynamics followed classic predator–prey population dynamics (e.g., Figure 6). Early attempts to cultivate the acidophilic protozoa in media containing pyrite failed, even though large populations of iron-oxidizing and other bacteria were present. This was later shown to be due to the fine size (<61 μm) of the pyrite grains used. When coarser-grain (61–200 μm) pyrite particles were used, all five protozoa were able to grow effectively, suggesting that the phagotrophic protozoa were unable to differentiate between pyrite and bacteria, and that inadvertent ingestion of bacteria-sized pyrite grains resulted in the death of the protozoa. Interestingly, dramatic reductions of numbers of iron-oxidizing bacteria due to protozoan grazing did not necessarily result in decreased rates of pyrite dissolution, possibly because of the overriding influence of mineral-oxidizing bacteria attached to the pyrite, which were not grazed.

Evidence for predation of acidophilic bacteria by other microorganisms has come mostly from microscopic observations. Rotifers, for example, have been seen to feed on acid streamer microbial communities (see Section "Acidophilic Eukaryotic Microorganisms") using their wheel-like cilia to draw a vortex of bacterial cells into their mouths.

Competitive Interactions

As might be anticipated, competition between acidophiles for electron donors and acceptors, inorganic nutrients, and so on, is as important in acidic as in all other environments. One of the most detailed studies of this kind, given the importance of iron-oxidizing bacteria in commercial mineral processing and the genesis of AMD, has been the competition between *At. ferrooxidans* and *Leptospirillum* spp. for their communal substrate (electron donor), ferrous iron. Early assumptions that *At. ferrooxidans* was invariably the dominant iron-oxidizing bacterium in metal-rich, acidic environments have gradually been eroded, with increasing numbers of reports describing *L. ferrooxidans* (though in some cases this is probably *L. ferriphilum*) as the more abundant species. These include mine drainage waters and some (mostly stirred tank) commercial biomining operations. In general, Bacteria classified as *At. ferrooxidans* tend to grow more rapidly than *Leptospirillum* spp., and are better able to exploit acidic environments that contain relatively large concentrations of ferrous iron. In terms often used to differentiate heterotrophic bacteria, *At. ferrooxidans* is a "copiotroph" while *Leptospirillum* spp. are "oligotrophs." This is also the reason why using ferrous iron-rich synthetic media to enrich for iron-oxidizing acidophiles favors *At. ferrooxidans* rather than *L. ferrooxidans*. On the other hand, the greater affinity for ferrous iron and the greater tolerance of ferric iron of *L. ferrooxidans* (and probably other *Leptospirillum* spp.) facilitate their dominance in stirred-tank mineral leachates, where ferric iron concentrations can be many grams per liter and, conversely, in those extremely acidic environments (pH < 2.3) where ferrous iron concentrations are very small. In both situations, redox potentials (which are determined by the relative concentrations, rather than actual concentrations, of ferrous and ferric iron) are commonly above 750 mV, and *Leptospirillum* spp. are known to be far more efficient iron oxidizers than *At. ferrooxidans* under such highly oxidizing conditions. Other important factors that affect competition between these iron-oxidizing autotrophs are temperature and pH. *Leptospirillum* spp. in general (and *L. ferriphilum* in particular) tends to be more thermotolerant than *At. ferrooxidans*, which partly explains their greater importance within the warm interior of the Richmond mine at Iron Mountain (see Section "Acid Mine Streams and Lakes") and in stirred tanks used to bioprocess gold and cobaltiferous ores, which generally operate at around 40 °C. On the other hand, cold-tolerant iron-oxidizing acidophiles have been invariably identified as *At. ferrooxidans*-like. *Leptospirillum* spp. also tend to be more tolerant of extreme acidity (many stains grow at pH 1) than *At. ferrooxidans*, some strains of which do not grow below pH 1.8, though others, including the type strain, can grow at pH 1.5. The higher pH optima for their growth is one of the reasons why *At. ferrooxidans* is often more important in heap-leaching of mineral ores, as engineered mineral heaps are generally not so acidic as stirred tanks.

In one of the few studies to describe competition between two other iron-oxidizing acidophiles, dissolution of pyrite at 45 °C by a mixed culture of the thermotolerant facultatively autotrophic bacterium *Acidimicrobium* (*Am.*) *ferrooxidans* and a thermotolerant strain of the obligate autotroph *L. ferriphilum* was examined. Numbers of the two bacteria (estimated using fluorescent *in situ* hybridization; FISH) remained very similar until the pH of the bioreactor was lowered from 1.5 to 1.2, at which point *L. ferriphilum* emerged as the dominant bacterium. However, when the thermotolerant sulfur oxidizer *At. caldus* was also included in the microbial consortium, *Am. ferrooxidans* was more abundant than *L. ferriphilum* at pH 1.5 and 1.2. The reason for this was probably the additional amount of organic carbon available for the heterotrophically inclined *Am. ferrooxidans* originating from the CO_2-fixing *At. caldus* that used the RISCs produced by ferric iron attack of the pyrite in the bioreactor.

MICROBIAL ECOLOGY OF EXTREMELY ACIDIC ENVIRONMENTS

Geothermal Areas

Geothermal areas occur in discreet zones in various parts of the world, as noted in the Section titled "Nature and Origin of Extremely Acidic Environments." Probably the most well-studied land-based geothermal area is Yellowstone National Park, Wyoming, USA. Much of the early pioneering microbiological work in Yellowstone was carried out by Thomas Brock and coworkers in the 1970s and formed the basis of later research on thermophilic microorganisms (including thermoacidophiles). One of the major breakthroughs around the time was the isolation, by James Brierley, of the first thermophilic acidophile (a *Sulfolobus*-like archaeon, though it was not recognized as such at the time) from an acid hot spring in Yellowstone. Two of the more important solfatara fields in Yellowstone are located in and around the Norris Geyser Basin and at Sylvan Springs, though there are numerous other smaller-scale, often ephemeral sites (such as around the Gibbon River) where acidic ponds and streams may be found. Moderately thermophilic and acidophilic phototrophs (*Galderia*/*Cyanidium*-like rhodophytes) and Gram-positive bacteria (*Sulfobacillus* and *Alicyclobacillus* spp.) have frequently been isolated from these sites. A greater biodiversity was revealed in a study reported in 2003, where strains of what turned out to be novel genera of thermophilic Gram-negative bacteria (*Acidicaldus*) and moderately thermophilic Gram-positive bacteria (*Ferrithrix*) were isolated, as well as novel strains of *Firmicutes* that have not, as yet, been formally classified. Interestingly, many of the moderately thermophilic bacteria isolated from these sites (where the acidity is derived chiefly from the oxidation of sulfur) have been shown to catalyze the oxidation or reduction (and, in some cases, both) of iron. A study of acidic geothermal springs within the Norris Geyser Basin that contained a variety of electron donors that support the growth of chemolithotrophic acidophiles (hydrogen, hydrogen sulfide, arsenic(III), and ferrous iron) revealed complex and changing microbial communities that were determined, at least in part, by changing chemical gradients, which in turn effected major geochemical transformations. It was found that (1) *Hydrogenobaculum* (a hydrogen and sulfur oxidizer) and *Stygiolobus* (a hydrogen-oxidizing anaerobe) were present in the high-temperature (~79 °C) source waters; (2) *Hydrogenobaculum* and *Thiomonas* (a ferrous iron, sulfur, and arsenic(III) oxidizer) were present in zones of rapid As(III) oxidation; and (3) *Metallosphaera* (a sulfur oxidizer), *Acidimicrobium* (an iron oxidizer), and *Thiomonas* were present in areas where As(V)-rich ferric iron oxides were being generated.

In a separate study based in the Ragged Hills area of Yellowstone, the effect of increased geothermal activity on soil microbial diversity across a temperature gradient of 35–65 °C was assessed. The pH of the soil samples analyzed ranged from 3.7 to 5.1. It was found that the DNA profiles of the soil bacteria (estimated using denaturing gradient gel electrophoresis; DGGE) in heated soils were less complex than those that had not undergone geothermal heating. The majority of clones obtained belonged to *Acidobacterium*, cultivated species of which are mostly moderate acidophiles and mesophilic. It was concluded that thermophilic and thermotolerant microbial species are probably widely distributed in soils within Yellowstone, and that localized geothermal activity selects them. The effects of natural hydrocarbon seeps (composed almost entirely of saturated, branched C_{15} to C_{30}, straight-and branched-chain alkanes) on the microflora of acidic (pH 2.8–3.8) sulfate-rich soils in the Rainbow Springs area of Yellowstone were examined in another study. Over 75% of the clones recovered in 16S rRNA gene libraries were related to known species of heterotrophic acidophiles (*Acidiphilum* and *Acidisphaera*), though clones related to *Acidithiobacillus* spp. were also recovered. An alkane-degrading alphaproteobacterium (distantly related to *Acidicaldus* (*Acd.*) *organivorans*, a Yellowstone isolate that has been shown to grow on phenol and other organic substrates) was isolated and partially characterized.

Other geothermal areas where the distribution of acidophilic microorganisms has been studied include New Zealand, the Caribbean island of Montserrat, and northern California. One site that has been studied in New Zealand was an acidic (pH 2.5) stream water on White Island that, in addition to soluble iron and sulfate, contained significant concentrations (2000–4400 mg l^{-1}) of chloride. Among the clones identified from DNA extracted directly from the acid stream were, unusually, those closely related to a green sulfur bacterium (*Chlorobium vibrioforme*); the marine, purple nonsulfur bacterium *Rhodovulum*; and the heterotrophic bacterium *Ralstonia solanacearum*, while those obtained from enrichment cultures also included a bacterium that was closely related to the Yellowstone isolate *Acd. organivorans*. Pure cultures of *Acidiphilium* and *C. caldarium* were also obtained from the site. Results of a large-scale survey of geothermal sites (many of which were also extremely acidic) on Montserrat, carried our shortly prior to the major eruption of the Soufriere Hills volcano in 1996, showed that temperatures of pools and streams in the volcanic southern region of the island ranged from 30 to 99 °C, and pH from 1.0 to 7.4. Most of the acidophilic bacteria that were isolated were similar to known strains, though some *Sulfobacillus*-like isolates had novel traits in being able to grow as mesophiles, or at higher maximum growth temperatures (up to 65 °C) than classified species. Clone libraries constructed from DNA extracted from acidic sites with different temperatures indicated the presence of (1) *Acidiphilium*-like bacteria and *At. caldus* (33 °C site); (2) *At. caldus*, and a putative moderately thermophilic sulfate-reducing bacterium of the *Desulfurella* group (48 °C site); (3) novel *Ferroplasma*-like and *Sulfolobus*-like archaea (78 °C site); and (4) an archaeon distantly related to *Ac. infernus* (98 °C site). Elsewhere, a microbiological survey of high-temperature (82–93.5 °C) acidic (pH 1.2–2.2) hot springs located in the Lassen Volcanic National Park in northern California failed to detect any bacteria, though archaea distantly related to the crenarchaeotes *S. azoricus* and *Sulfolobus solfataricus* (both extremely acidophilic thermophiles), and others more closely related to the moderately acidophilic thermophile *Vulcanisaeta distributa*, were identified in clone libraries.

Acid Mine Streams and Lakes

Waters draining abandoned mines, mine spoils, and tailings deposits are often characterized by low pH and elevated concentrations of soluble metals (particularly iron) and sulfate (Table 3). These are generically referred to as AMD waters (or acid rock drainage in North America). Acidity in such waters derives from the presence of soluble aluminum, manganese, and iron (mineral acidity) as well as hydronium ions. Extremely acidic lakes may develop naturally in volcanic area, for example, Lake Kawah Idjen in Indonesia, which has a pH of ~0.7. Acidic mining lakes, in contrast, are relics of opencast mining, where worked-out voids have not been backfilled and have become progressively filled with rising groundwater or river water. Where the surrounding bedrocks are rich in sulfide minerals (normally chiefly pyrite and marcasite) and contain small amounts of carbonates, the oxidative dissolution of the former can lead to the formation of extremely acidic mine lakes. Acid mine lakes are particularly abundant in central Europe, in parts of Germany, Poland, and the Czech Republic. In past times (up to the end of the twentieth century), the extensive reserves of lignite in these areas were extracted by opencast mining on enormous

TABLE 3 Examples of mine water chemistries (all units are mg l^{-1}, except pH)

	pH	$[Fe_{total}]$	$[Fe^{2+}]$	[Al]	[Cu]	[Zn]	$[SO_4]$
Coal mines							
Bullhouse (UK)	5.9	61	45	1.2	>1	>1	
Ynysarwed (UK)	6.2	160	140	20			460
Oatlands (UK)	5.5	287		0.97	>0.007	0.05	146
Sverdrupbyen (Norway)	2.7	179		27.5	0.168	1.3	1077
Metal mines							
Mynydd Parys (UK)	2.5	650	650	70	60	40	3100
Roeros (Norway)	3.7	6.7		4.3	11	3.76	
Wheal Jane (UK)	3.6	130	130	50	2	130	350
Cwm Rheidol (UK)	2.6–2.7			104–128	1.2–9.35	577–978	250
Sao Domingos (Portugal)	1.7	31,000	10,000				14,850
Iron Mountain (California)	1.5	2670	2470		293	58	14,000

Source: *Data are from Johnson, D. B. (2006). Biohydrometallurgy and the environment: Intimate and important interplay.* Hydrometallurgy, *83, 153–166; and Nordstrom, D. K., Alpers, C. N., Ptacek, C. J., & Blowes, D. W. (2000). Negative pH and extremely acidic minewaters from Iron Mountain, California.* Environmental Science and Technology, *34, 254–258.*

scales, leaving a legacy of a very large number of man-made lakes of varying sizes and chemistries. In the Lusatia district of eastern Germany alone there are an estimated 200 mining lakes of >1 ha that have pH values of >3.

The microbiology of AMD streams has been the subject of a number of reviews in books and journals. Knowledge of how biodiverse these flowing waters can be has expanded considerably since *At. ferrooxidans* was first isolated from an AMD stream draining a bituminous coal mine in the United States in 1947. The most important factors in determining which microbial species are present in AMD appear to be pH, temperature, and concentrations of dissolved metals and other solutes. At the most extreme end of the AMD spectrum, the microbiology of mine waters within the Richmond mine at Iron Mountain, California (which can have negative pH values), has been studied extensively. Within this abandoned mine, pyrite is undergoing oxidative dissolution at a rate that is sufficient to maintain air temperatures of 30 to 46 °C, and produce mine waters containing ~200 g l^{-1} of dissolved metals. A novel iron-oxidizing archaeon, *Ferroplasma acidarmanus*, was found to be dominant in waters within the mine that had the lowest pH and highest ionic strengths, while *L. ferriphilum* and *L. ferrodiazotrophum* were also associated with exposed pyrite faces. *Sulfobacillus* spp. were more important in some of the warmer (~43 °C) waters. *At. ferrooxidans* was rarely found in sites that were in contact with the ore body, though it was found in greater abundance in the cooler, higher pH waters that were peripheral to the ore body. In contrast, a microbiological survey of much cooler and higher pH mine waters at an abandoned subarctic copper mine in Norway showed that an *At. ferrooxidans*-like isolate (closely related to a psychrotolerant strain found subsequently in a mine in Siberia) was the dominant iron oxidizer present. *L. ferrooxidans* was detected only in enrichment cultures using mine water inocula. The Norwegian AMD waters also contained significant numbers of acidophilic heterotrophs related to some species (*Acidiphilium*, *Acidocella*, and *Acidisphaera*) that had previously been observed in acidic environments, and one (a *Frateuria*-like bacterium) that had not.

The importance of *At. ferrooxidans*-like bacteria in cooler (<20 °C) mine waters of pH 2–3 has also been supported at sites in other parts of the world. For example, biomolecular analysis (from clone libraries) of four AMD sites at the Dexing copper mine in the Jiangxi province of China found differences in the distribution of acidophiles with water pH. In the most acidic site (pH 1.5), *Leptospirillum* spp. (*L. ferrooxidans*, *L. ferriphilum*, and *L. ferrodiazotrophum*) were the dominant species in the clone library, while in pH 2.0 AMD *L. ferrodiazotrophum* was the single dominant species detected. In slightly higher pH (2.2) AMD, most clones recovered were related to *At. ferrooxidans*, while in the highest pH waters (3.0) most were related to the heterotrophic moderate acidophile *Acidobacterium*. Where mine waters have pH values of above 3, however, there is increasing evidence that moderately acidophilic iron oxidizers assume a more important role than *At. ferrooxidans*. The dominant iron oxidizer in AMD flowing from an underground coal mine in south Wales was found to be a *Thiomonas*-like bacterium, and similar strains (given the novel species designation *T. arsenivorans*) were isolated from an abandoned tin mine in Cornwall, England, and a disused gold mine (Cheni) in France. Other acidophilic bacteria isolated from the Cornish site included *Acidobacterium*-like and *Frateuria*-like isolates, and an iron oxidizer related to *Halothiobacillus neopolitanus*. Further evidence of the importance of previously uncultured acidophiles in AMD has come from a study of acidic (pH 2.7–3.4) iron- and arsenic-rich water draining mine tailings at Carnoul's in France. The dominant bacteria found in clone libraries were betaproteobacteria, many of which were related to a *Gallionella*-like sequence previously reported in a chalybeate spa in north Wales. The sole *Gallionella* sp. that has been characterized (*Gallionella ferruginea*) is a neutrophilic iron oxidizer that grows best under microaerophilic conditions, and the circumstantial evidence for the existence of an acidophilic (or acid-tolerant) species of *Gallionella* is intriguing. Researchers also found evidence of SRB distantly related to *Desulfobacterium* in AMD at Carnoul's. SRB may also be found in sediments (and microbial mats) underlying AMD, though the pH in such sediments is frequently much higher than the AMD itself.

Microbiological studies of acid mine lakes in Germany have focused on phototrophic eukaryotes as well as acidophilic bacteria and have also examined how dissimilatory microbial reductive processes may be stimulated in order to ameliorate water acidity and immobilize metals. A survey of 14 acidic lakes in Lusatia (ranging in pH from 2.14 to 3.35, and conductivities from 690 to 4460 μS cm^{-1}) found a positive correlation between the relative numbers of the iron-oxidizing heterotroph *Fm. acidiphilum* and concentrations of aluminum. However, it was concluded that indicator groups of bacteria, rather than single species, were better correlated with different lake chemistries. Addition of organic carbon, nitrogen, and phosphorus to enclosed water columns in a pH 2.6 mine lake was shown to induce changes in both water chemistry and microbiology. Treatment of water resulted in increased microbial diversity, and SRB (*Desulfobacter* spp.) were among the microorganisms detected in the amended water columns.

One other important extremely acidic ecosystem that has been studied extensively is the Rio Tinto, a major river, some 92 km in length, located in southwest Spain (Figure 3). The source of the river is the Peña de Hierro (Iron Mountain) in the Iberian Pyrite Belt, and from there it flows though a large and historic area of copper mining (the Riotinto mines), eventually reaching the Atlantic Ocean at

Huelva. Interestingly, even above the Riotinto mines, the river is acidic and enriched with metals, but this is very much accentuated as it flows through the (now abandoned) mining district. The river has a mean pH of about 2.2 and its distinctive red coloration derives from its soluble ferric iron content (~2 g l^{-1}). Primary production in the river is carried out by both photosynthetic and chemoautotrophic acidophiles. A study of the indigenous prokaryotes showed that >80% were bacteria, and that archaea accounted for only a relatively small proportion of cells. A variety of different iron oxidizers (*At. ferrooxidans*, *Leptospirillum* spp., *Fm. acidiphilum*, and *Fp. acidiphilum*), as well as the iron-reducing heterotroph *Acidiphilium*, were identified. A geomicrobiological model involving cyclical oxidation of ferrous iron and reduction of ferric iron has been proposed to account for the remarkable chemical stability of the river ecosystem.

Acid Streamers, Mats, and Slimes

The most obvious and dramatic manifestations of microbial life in extremely acidic environments are macroscopic growths, referred to as "acid streamers," "slimes," and "mats." Streamers may occur in flowing AMD streams inside and outside of abandoned mines; these have distinct filamentous morphologies and each filament may be more than a meter in length. Acid mats are denser in texture and are often found below growths of acid streamers. Acid slimes are thick, macroscopic biofilms that grow on moist surfaces of exposed rock faces. In addition, macroscopic filaments (or pipes) composed of acidophilic microorganisms may attach to, and suspend from, mine roofs and pit props. Where these are small-scale, they have been referred to as "snotites," though larger structures have been described as "microbial stalactites." Because of the macroscopic nature of acid streamer growths, they were among the first life-forms to be reported in extremely acidic environments (in 1938). Most of the early reports described acid streamers as being composed of bacteria embedded in a gelatinous matrix. The first attempts, in the 1970s, to characterize their component microorganisms used classical microbiological (cultivation-based) techniques. In the main, neutrophilic (chiefly spore-forming) *Bacillus* spp. were isolated from the acid streamers examined, leading to the erroneous conclusion that acid streamers were composed of neutrophilic heterotrophic bacteria that maintained circum-neutral pH within the macroscopic growths. In contrast, other researchers noted that streamers found in an iron/sulfur mine in Japan were able to catalyze the oxidation of both ferrous iron and sulfur, though some subsamples of streamers have very limited capacity to oxidize iron. *At. ferrooxidans*-like bacteria were isolated from these growths, leading the researchers to conclude that acid streamers were a mass of *At. ferrooxidans* embedded in a gelatinous matrix. However, it is only after biomolecular tools have been used to analyze acid streamers and related macroscopic growths that their true nature has been elucidated. This approach, coupled with the major advances made in the past two decades in techniques for isolating and cultivating acidophiles in the laboratory, has shown that streamer communities are highly complex and vary from site to site. Superficial similarities in their gross morphologies may mask completely contrasting microbial diversities.

One of the first intensive biomolecular examinations of acid slime (~1 cm thick) and snotite growths was carried out with materials collected from the Richmond mine at Iron Mountain, California. Microscopic examination showed that both growth forms were composed mostly of spirillum-shaped cells embedded in an extracellular polymeric matrix. Phylogenetic analysis based on 16S rRNA genes showed that most of the recovered sequences were novel but related to known iron-oxidizing acidophiles. The single most dominant sequence recovered from slime growths was a novel strain of *Leptospirillum* (subsequently named *L. ferrodiazotrophum*). *L. ferriphilum*-related clone sequences were also identified. Other iron-oxidizing bacteria identified were related to the Gram-positive actinobacteria *Acidimicrobium* and *Ferrimicrobium*, while sequences that affiliated with deltaproteobacteria (which includes anaerobic sulfate and iron reducers) were also detected, suggesting that microzones of low redox potential existed within the macroscopic growths. Archaeal genes were also amplified, and sequences related to *Fp. acidarmanus* were identified. Other archaeal sequences were, however, only distantly related to known archaea.

In contrast to the Richmond mine slimes, acid streamer growths in less extremely acidic and cooler sites in north Wales (one an AMD stream at an abandoned copper mine and the other, water in a chalybeate spa) were both found to be composed predominantly of betaproteobacteria. At the copper mine site, a single novel bacterium was the dominant prokaryote present, while this and a second betaproteobacterium accounted for 90% of bacteria (determined by the quantitative FISH technique) in the spa water streamers. A modified solid medium was developed to isolate the unknown bacterium from the copper mine streamers, and the betaproteobacterium isolate was shown to be the first representative of a novel genus (proposed name, *Ferrovum*) of iron-oxidizing chemoautotrophic acidophiles. The second (spa water) unknown species was identified as being most closely related to *G. ferruginea* (and a clone obtained from the Carnoules mine; see Section "Acid Mine Streams and Lakes") but was not isolated. Although known species of acidophilic bacteria (*At. ferrooxidans*, *Acidiphilium*, *Acidocella*, *Thiomonas*, *Ferrimicrobium*, and *Acidobacterium*) were also isolated from the streamers, these were shown to be present in only relatively small numbers.

Another underground site that has been the focus of intense study is an abandoned pyrite mine (Cae Coch) located in northwest Wales. This mine is home to the most extensive and diverse macroscopic acidophile growths yet reported, with an estimated acid streamer biovolume of $>100\,m^3$ alone, in addition to extensive slime biofilms and mats, microbial stalactites, and snotites. Although the temperature in the mine shows very little seasonal fluctuation (8.5 ± 1°C), other physicochemical factors, including pH, dissolved oxygen, and concentrations of dissolved metals and other solutes, vary from site to site within the mine. This, together with the fact that the underground mine has been largely undisturbed for 90 years, has facilitated the colonization of different niches by different streamer/slime acidophilic communities. Using a combination of biomolecular and cultivation-based techniques, the Cae Coch streamer microflora has been shown to be very different from those of the AMD streams and ponds in which they bathe. All the macroscopic growths were found to be composed of acidophilic bacteria (though protozoa and rotifers were also found in some locations), and the novel iron oxidizer *Ferrovum* was found to be the most abundant single organism present overall. Many of the Bacteria identified were well-known acidophiles (*At. ferrooxidans*, *Leptospirillum*, *Ferrimicrobium*, *Acidiphilium*, and *Acidocella*), while others (including *Frateuria*- and *Ralstonia*-like bacteria) were not. Included in the latter group was a *Sphingomonas* sp. that was detected by biomolecular methods, and also isolated in pure culture where it was shown to be an obligate acidophile, the first such species of *Sphingomonas* to have this trait. The fact that acid streamer communities can be very different from planktonic communities in the same ecosystem was also shown by a study of long (1.5 m) filamentous biofilms found in the Rio Tinto, Spain. Whereas the dominant microorganisms in the Rio Tinto water column are *Acidithiobacillus*, *Leptospirillum*, *Acidiphilium*,

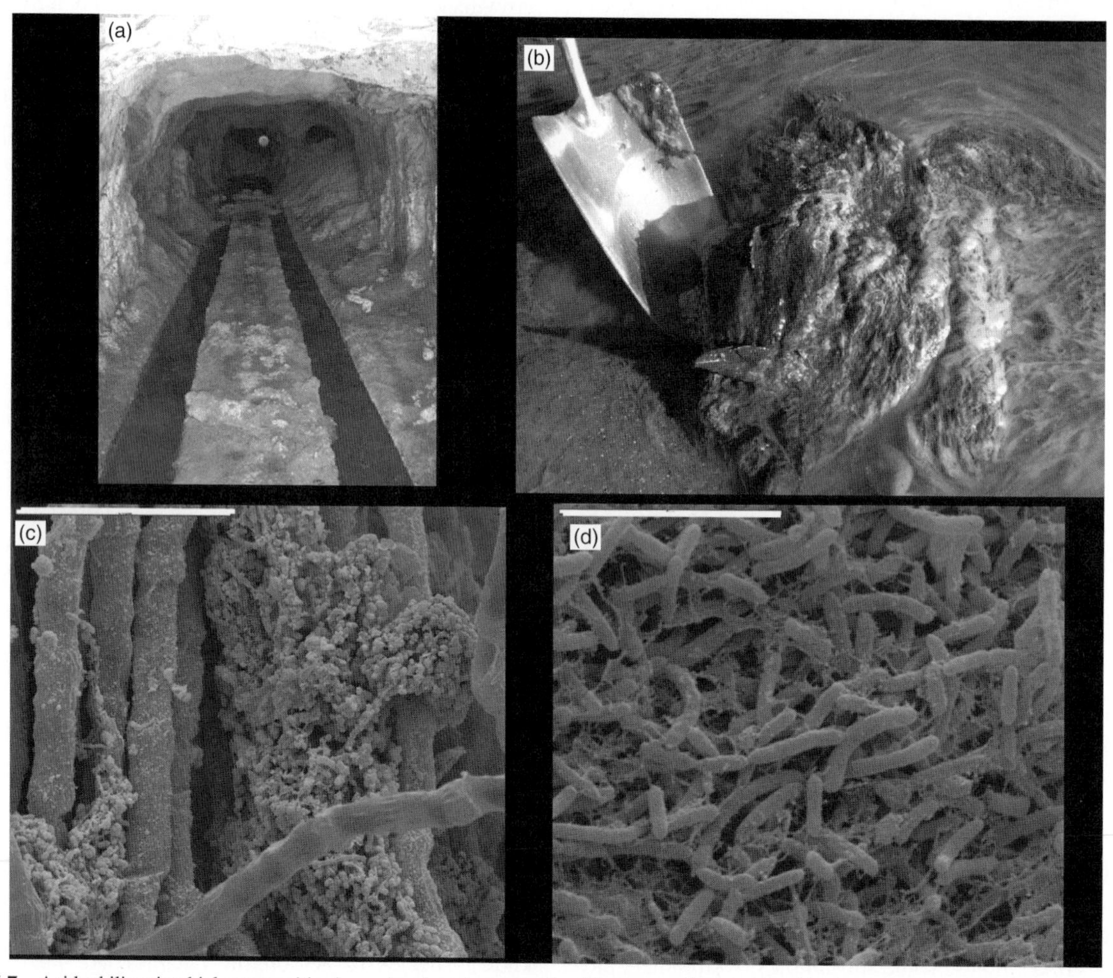

FIGURE 7 Acidophilic microbial communities in an abandoned copper mine (Cantareras, Spain): (a) acid mine drainage (AMD) channels draining the mine adit, showing deposition of copper salts on the adit walls; (b) stratified acid streamer and mat growths in the main drain channel; (c) scanning electron micrograph of the surface streamer layer, showing filaments of microalgae (*Zygnema*) and aggregates of bacteria (the bar scale represents 20 μm); (d) scanning electron micrograph of lower zone streamers, showing rod-shaped bacteria embedded in dehydrated exopolymeric material (the bar scale represents 5 μm).

and archaea (Thermoplasmatales), the streamer-like growths examined were composed of gamma- and alphaproteobacteria; Gram-positive bacteria and betaproteobacteria were also detected in smaller numbers. As with the Cae Coch streamers, *Sphingomonas*- and *Ralstonia*-like bacteria were identified in clone libraries constructed from Rio Tinto streamers.

A more complex streamer/mat community in another mine site in southwest Spain (Cantareras) has been described recently (Figure 7). Both solar and chemical (mostly ferrous iron) energy drives primary production in an adit drainage channel at the site, and consequently acidophilic microalgae and chemoautotrophic iron-oxidizing Bacteria both thrive. The streamer growths that fill the ~100-m-long drain channel show distinct stratification (Figure 7). The surface layer is green due to the presence of *Zygnema*, *Chlamydomonas*, and other phototrophic eukaryotes that both aerate the anoxic mine water and provide organic carbon, which supports the growth of heterotrophic acidophiles. The lower layers are (in sequence) cream-brown, turquoise, and gray-black in color, and are almost exclusively bacterial. In contrast to most other acid streamer communities that have been described, heterotrophic (mostly iron-reducing) acidophiles dominate subsurface layers at Cantareras. This is particularly the case with the bottom mat layer, which is composed almost exclusively of *Acidobacterium*-like bacteria and novel strains of sulfate reducers. Complex biogeochemical cycling of iron (acting as both electron donor and electron acceptor) and sulfur in the Cantareras streamer community helps to sustain the highly diverse population of acidophiles that occurs there (Figure 8).

Acidophilic snotite-like biofilms have also been found in at least one other very different location. The Frasassi complex, located in central Italy, is a large and actively developing sulfidic cave, hosted in limestone rock. Large concentrations (~0.3 m mol l^{-1}) of hydrogen sulfide have been found in groundwater in deep sections of the cave system, and this gas has been shown to support the growth of sulfur-oxidizing acidophiles that grow in biofilms on the cave roof. Although the pH of the cave stream water is 7.0–7.3, droplets of liquid at the tips of the snotites have pH values of between 0 and 1, as a result of microbiological oxidation of sulfide, forming sulfuric acid. A bacterium related to *Halothiobacillus* was isolated from a snotite sample from Frasassi, together with clones related to *Acidithiobacillus* and *Sulfobacillus*. A later

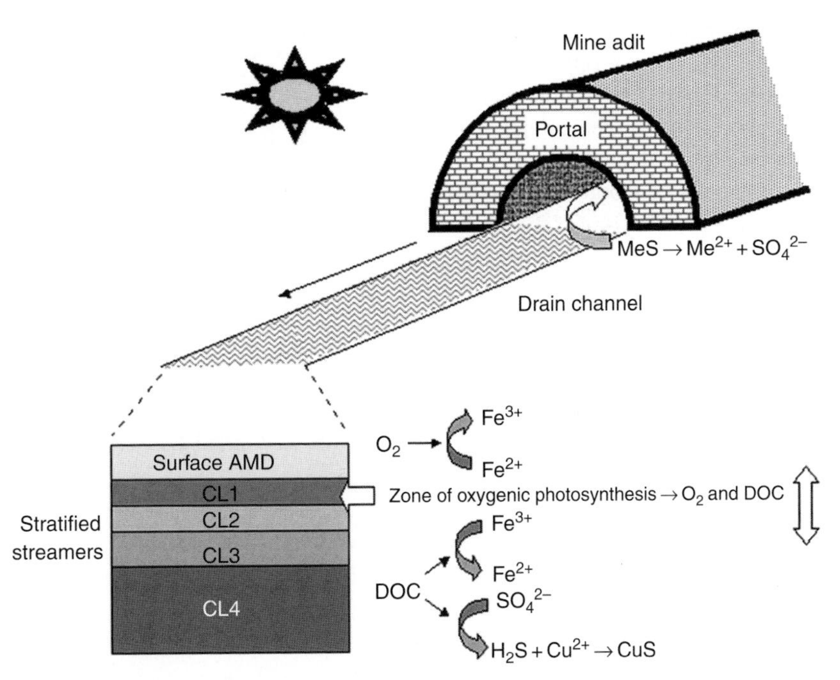

FIGURE 8 Proposed model of the biogeochemical cycling of iron and sulfur at the abandoned Cantareras mine. Dissolution of sulfide minerals in the exposed mine workings gives rise to a highly acidic metal- and sulfate-rich effluent. The anoxic water draining the mine is oxygenated by photosynthetic acidophilic algae in the surface (CL1) layer of the acid streamer growths that develop immediately outside of the adit, which facilitates oxidation of ferrous iron in the surface AMD (catalyzed primarily by *Acidithiobacillus ferrooxidans*). Dissolved organic carbon (DOC) originating from photosynthetic and chemosynthetic primary producers serves as substrates for the (dominantly) heterotrophic bacteria in the deeper zone (CL2–4) streamer layers. Ferric iron is used as terminal electron acceptor in streamer layers CL2 and CL3, while in the thick CL4 layer sulfate is also used, resulting in the deposition of copper sulfide (CuS). The gradual buildup of ferric iron concentrations as the AMD flows through the channel results in the elimination of the microalgae, thereby removing the major primary production system that supports the streamer microbial community. *Reproduced from Rowe, O. F., Śnchez-Espaã, J., Hallberg, K. B., & Johnson, D. B. (2007). Microbial communities and geochemical dynamics in an extremely acidic, metal-rich stream at an abandoned sulfide mine (Huelva, Spain) underpinned by two functional primary production systems.* Environmental Microbiology, *9, 1761–1771, with permission from the publishers.*

study showed that most of the clones obtained (65%) from a snotite sample from Frasassi were related to the mesophilic sulfur-oxidizing acidophile *At. thiooxidans*. The second most phylotype identified (31% of clones) was most closely related to *Am. ferrooxidans*, which is interesting as this is a moderately thermophilic acidophile that can oxidize ferrous iron but not reduced sulfur.

OUTLOOK AND APPLICATIONS

Knowledge of the phylogenetic and physiological diversities of acidophilic microorganisms has expanded greatly in the past 25 years. Data from biomolecular studies of extremely acidic sites, however, suggest that a large number of acidophilic prokaryotes still await isolation and characterization. There is a great deal of interest in acidophiles, not only from the standpoint of understanding how these microorganisms can thrive in conditions that are hostile to most life-forms but also due to their importance in environmental pollution (mine spoils and mine drainage waters) and in biotechnology (their central role in biomining and in removal of metals from contaminated soils). Significant research effort is currently concerned with finding acidophiles that can also tolerate other environmental extremes, such as temperature (above that of currently known thermoacidophiles) and salinity. There will doubtless be new opportunities to exploit existing and novel acidophilic microorganisms in future biotechnologies that will harness their unique abilities to thrive in conditions that are moderately or extremely acidic and mediate transformations of inorganic as well as organic chemicals.

FURTHER READING

Baker, B. J., & Banfield, J. F. (2003). Microbial communities in acid mine drainage. *FEMS Microbiology Ecology*, *44*, 139–152.

Baumann, K. H., Andruleit, H., Bockel, B. *et al.* (2005). The significance of extant coccolithophores as indicators of ocean water masses, surface water temperature, and palaeoproductivity: A review. *Palaontologische Zeitschrift*, *79*, 93–112.

Cros, L., & Fortuno, J. M. (2002). Atlas of Northwestern Mediterranean coccoliths. *Scientia Marina*, *66*(supplement 1), 1–186.

Donati, E. R., & Sand, W. (Eds.) *Microbial processing of metal sulfides* (2007). Dordrecht, The Netherlands: Springer.

Geller, W., Klapper, H., & Salomons, W. (Eds.) *Acidic mining lakes*. (1998). Berlin: Springer-Verlag.

Gross, W. (2000). Ecophysiology of algae living in highly acidic environments. *Hydrobiologia*, *433*, 31–37.

Gross, S., & Robbins, E. I. (2000). Acidophilic and acid-tolerant fungi and yeasts. *Hydrobiologia*, *433*, 91–109.

Hallberg, K. B., & Johnson, D. B. (2001). Biodiversity of acidophilic prokaryotes. *Advances in Applied Microbiology*, *49*, 37–84.

Johnson, D. B. (2007). Physiology and ecology of acidophilic microorganisms. In C. Gerday, & N. Glansdorff (Eds.), *Physiology and Biochemistry of Extremophiles* (pp. 257–270). Washington DC: ASM Press.

Johnson, D. B., & Hallberg, K. B. (2008). Carbon, iron and sulfur metabolism in acidophilic micro-organisms. *Advances in Microbial Physiology*, *54*, 201–255.

Kleijne, A., & Cros, L. (2009). Ten new extant species of the coccolithophore Syracosphaera and a revised classification scheme for the genus. *Micropaleontology*, *55*, 425–462.

Rawlings, D. E. (2002). Heavy metal mining using microbes. *Annual Review of Microbiology*, *56*, 65–91.

Rawlings, D. E., & Johnson, D. B. (2007). *Biomining*. Heidelberg: Springer-Verlag.

Chapter 22

Hot Environments

J.F. Holden
University of Massachusetts, Amherst, MA, USA

Chapter Outline

ABBREVIATIONS

APS Adenosine phosphosulfate
cDPG Cyclic 2,3-diphosphoglycerate
CoA Coenzyme A
DIP Di-*myo*-inositol-1,1′-phosphate
MGD Molybdopterin guanine dinucleotide
PRPP 5-Phospho-D-ribose-1-pyrophosphate
RubisCO Ribulose-1,5-bisphosphate carboxylase/oxygenase

DEFINING STATEMENT

This article examines the microbial ecology of thermophiles and hyperthermophiles by describing factors used for DNA, protein, and cell wall thermostability; CO_2 and acetate assimilation; heterotrophy; and respiration. It also examines the evidence for the upper temperature limit for life based on pure cultures, studies on natural assemblages, and field studies.

INTRODUCTION

Life at high temperatures is classified as being either thermophilic or hyperthermophilic. Thermophiles are those organisms with optimal growth temperatures between 55 and 80 °C, whereas hyperthermophiles grow optimally at and above 80 °C. The lower bound for thermophily is based on the rarity of temperatures above this point in nature because it is very uncommon for eukarya to grow above 55 °C. The hyperthermophile boundary is arbitrary but does largely distinguish those organisms that grow at the highest temperatures. It also has some molecular basis as generally only hyperthermophiles possess the enzyme reverse gyrase, which helps to stabilize double-stranded DNA at high temperatures. The highest known optimal growth temperatures are 105–106 °C where the slight addition of hydrostatic pressure prevents boiling.

High-temperature organisms are of interest for the information they provide on early life history, life in geothermal environments, and biochemistry. It is widely believed that the earliest life on Earth or the last common ancestor of life was either thermophilic or hyperthermophilic. Therefore, the study of modern-day (hyper)thermophiles may provide insight into the biochemical processes that occurred more than 3 billion years ago. Geothermal environments are ubiquitous on land and on the ocean floor and likely extend into vast regions of the Earth's crust. Thermophilic and hyperthermophilic microorganisms likely alter at some level the chemistry and fluid flow within these environments. High-temperature microorganisms are also useful for several industrial applications, including *in vitro* DNA

Topics in Ecological and Environmental Microbiology.

replication (i.e., the polymerase chain reaction, PCR), gas and oil recovery, laundry detergents, sweetener production, and biofuels.

Relating the biochemical characteristics of an organism to its ecology is best called physiological ecology and is the goal of this article. After an introduction to the kinds of thermophiles and hyperthermophiles found in geothermal environments, this article will examine those physiological attributes that permit these organisms to survive at high temperatures and place the information in an ecological context. This will include discussions on molecular stability at high temperatures, the upper temperature limit of life, and metabolic traits and capabilities that are relevant to biogeochemistry and microbial ecology.

DISCOVERY: LIFE AT HIGH TEMPERATURES

The presence of living organisms in geothermal hot springs was noted in *Naturalis Historia* by Pliny the Elder (23–79 AD). The first significant studies began in the nineteenth century, and early researchers recognized their importance on topics such as the structure and organization of life. Early challenges included demonstrating that the organisms were indeed alive in the springs and that the temperature was measured precisely at the organism's location because temperatures can vary significantly over a few centimeters. The first likely report of life above 80 °C came from the observation of high concentrations of microorganisms (or "chlorophylless filamentous Schizomycetes") in an 89 °C hot spring in Yellowstone National Park sampled in 1898.

In 1965, Thomas D. Brock began detailed studies of life in the hot springs of Yellowstone National Park and elsewhere and launched the study of thermophilic microbiology. He found microorganisms were thriving in geothermal pools and streams, in some cases up to the boiling point of the fluid, and cultured the first isolates for study. After 1980, many novel hyperthermophilic genera were cultured and characterized by Karl O. Stetter, Wolfram Zillig, and others from the solfataras of Iceland and shallow marine vents along the coasts of Iceland and Italy. At the same time, microorganisms associated with deep-sea hydrothermal vents were found that utilize the steep temperature and chemical gradients to support growth. More recently, hyperthermophiles have been found in other hot environments, such as petroleum reservoirs, subsurface geothermal pools, and deep mines.

TAXONOMY

Thermo(acido)philes

Although thermophiles are limited to bacteria and archaea, they are found in a wide range of environments, including coal refuse, hot-water tanks, and compost piles. This article will focus only on thermophiles from geothermal environments and are divided into the following groups. Photosynthetic bacteria generally grow up to 70–75 °C and include many cyanobacteria as well as green- and purple-sulfur bacteria such as *Chloroflexus* and *Chromatium*. Spore-forming thermophiles include *Bacillus*, *Clostridium*, and *Moorella* species. Several thermophilic Actinomycetes as well as thermophilic sulfur-oxidizers (e.g., *Thiobacillus*), sulfate reducers (e.g., *Desulfovibrio*), and Gram-negative aerobes (e.g., *Thermus*) have been described. Among archaea, thermophiles are often also acidophiles. *Sulfolobus*, *Acidianus*, and *Metallosphaera* are found in acidic terrestrial hot springs and have pH optima around pH 2. *Aciduliprofundum boonei* is a marine thermoacidophile recently found in deep-sea hydrothermal vents that grows optimally at pH 4.5, which highlights the potential importance of acidophily in marine environments as well. Methanogens are found in the thermophilic temperature range as well as the hyperthermophilic range.

Hyperthermophiles

Hyperthermophily is found primarily within the archaea, which contain 33 of the 37 reported genera (Table 1). The remaining four genera are bacteria. Archaea are further divided into two major phyla: the Crenarchaeota and the Euryarchaeota. Hyperthermophiles are found in both, but most genera are Crenarchaeota. A third archaeal phylum is Nanoarchaeota whose sole representative is a hyperthermophilic epibiont of certain *Ignicoccus* species. A fourth archaeal phylum is Korarchaeota with one cultured representative, a heterotroph that ferments peptides.

In the Crenarchaeota, the Sulfolobaceae contain primarily thermoacidophiles that are found in terrestrial hot springs and grow optimally around pH 2. The Thermoproteaceae are also largely freshwater organisms but grow in more neutral environments. The Desulfurococcaceae and Pyrodictiaceae are marine hyperthermophiles that grow in mildly acidic to neutral pHs. In the Euryarchaeota, the Thermococcaceae are marine heterotrophs, whereas the Archaeoglobaceae are marine heterotrophs and autotrophs. The Methanopyraceae, Methanocaldococcaceae, and Methanothermaceae contain freshwater and marine methanogens. Hyperthermophilic bacteria are found in the Aquificaceae, Thermotogaceae, and Thermodesulfobacteriaceae. The degree of homology between the nucleotide sequences of small-subunit ribosomal RNA from all life suggests that both bacterial and archaeal hyperthermophiles are positioned largely near the root of the phylogenetic tree. This supports the idea that phylogenetically they are the closest extant organisms to early life on the planet.

TABLE 1 Taxonomy of hyperthermophiles at the superkingdom, phylum, family, and genus levels and their growth characteristics

Taxonomic group	Metabolism	Electron Acceptors	T_{opt} (T_{max}) (°C)
Archaea			
Crenarchaeota			
Desulfurococcaceae			
Acidolobus	H	H^+, S^0	85 (92)
Aeropyrum	H	O_2	95 (100)
Caldococcus	H	S^0	92 (96)
Desulfurococcus	H	H^+, S^0	85 (95)
Ignicoccus	C	S^0	90 (98)
Ignisphaera	H		95 (98)
Staphylothermus	H	S^0	92 (98)
Stetteria	C	S^0, $S_2O_3^{2-}$	95 (102)
Sulfophobococcus	H	H^+	85 (95)
Thermodiscus	H	H^+, S^0	90 (98)
Thermosphaera	H	H^+	85 (90)
Pyrodictiaceae			
Hyperthermus	H	S^0	101 (108)
Pyrodictium	C, H	FeO, S^0, $S_2O_3^{2-}$, SO_3^{2-}	105 (110)
Pyrolobus	C	FeO, NO_3^-, O_2, $S_2O_3^{2-}$	106 (113)
Sulfolobaceae			
Acidianus	C	S^0, O_2	88 (95)
Stygiolobus	C	S^0	80 (89)
Sulfolobus	C	O_2	81 (90)
Sulfurisphaera	H	O_2, S^0	84 (92)
Thermoproteaceae			
Caldivirga	H	O_2, S^0, $S_2O_3^{2-}$, SO_4^{2-}	85 (92)
Pyrobaculum	H, FA	Fe^{3+}, FeO, NO_3^-, S^0, $S_2O_3^{2-}$	100 (103)
Thermophilum	H	S^0	88 (95)
Thermoproteus	H, FA	S^0	88 (97)
Vulcanisaeta	H	Fe^{3+}, S^0, $S_2O_3^{2-}$, SO_4^{2-}	90 (99)
Euryarchaeota			
Archaeoglobaceae			
Archaeoglobus	C, FA	SO_4^{2-}	83 (95)
Ferroglobus	C	FeO, NO_3^-, $S_2O_3^{2-}$	85 (95)
Geoglobus	FA	FeO	88 (90)
Methanocaldococcaceae			
Methanocaldococcus	C	CO_2	85 (91)

(Continued)

TABLE 1 Taxonomy of hyperthermophiles at the superkingdom, phylum, family, and genus levels and their growth characteristics—cont'd

Taxonomic group	Metabolism	Electron Acceptors	T_{opt} (T_{max}) (°C)
Methanopyraceae			
Methanopyrus	C	CO_2	98 (110)
Methanothermaceae			
Methanothermus	C	CO_2	88 (97)
Thermococcaceae			
Paleococcus	H	Fe_2S_3, O_2, S^0	83 (88)
Pyrococcus	H	H^+, S^0	100 (105)
Thermococcus	H	FeO, Fe_2S_3, H^+, S^0	87 (93)
Nanoarchaeota			
Nanoarchaeum	H	S^0	90 (98)
Korarchaeota			
Korarchaeum	H	H+	85
Bacteria			
Aquificaceae			
Aquifex	C	NO_3^-, O_2	85 (95)
Thermocrinis	C	O_2	80 (89)
Thermotogaceae			
Thermotoga	H	FeO, H^+, S^0	80 (90)
Thermodesulfobacteriaceae			
Geothermobacterium	C	FeO	90 (100)

Optimum (T_{opt}) and maximum (T_{max}) growth temperatures are the maximum known temperatures for each genus
C, chemolithoautotrophic; H, heterotrophic; FA, facultative autotroph.

MECHANISMS OF THERMOSTABILITY

DNA

DNA denatures *in vitro* from a double strand to two single strands (i.e., melts) at temperatures between 70 and 90 °C. The stability of the DNA secondary structure is a function of the degree of guanosine–cytosine base pairing and salt concentration, with DNA melt temperature increasing with GC content and salt. As a result, thermophiles and especially hyperthermophiles must have *in vivo* mechanisms that prevent this melting. This protection comes in the form of protein–DNA interactions and solutes. Above 90 °C, DNA damage (e.g., depurination) occurs at rates that are 1000-fold higher than at 25 °C. Thus, hyperthermophiles also require a highly efficient DNA repair mechanism.

DNA from the hyperthermophile *Pyrococcus furiosus* was fragmented into pieces that were less than 500 kb when cultures were irradiated with 2.5 kGy of γ-radiation. After irradiation and incubation for 4 h at 95 °C, its genome was fully restored and growth recommenced, thus demonstrating a highly efficient DNA repair mechanism within this organism. DNA thermostability and repair has focused primarily on a protein known as reverse gyrase, which is a type IA topoisomerase with a helicase domain. Reverse gyrase has the unique ability to introduce positive supercoiling into DNA molecules, which counters the effect of DNA melting at higher temperatures and stands in contrast to the negative supercoiling of DNA found in mesophiles. It is the only protein known to be specific to hyperthermophiles in both archaea and bacteria. Reverse gyrase is also recruited to DNA after the induction of DNA damage, showing that it is part of a broader DNA repair mechanism. It recognizes nicked DNA, recruits a protein coat to the site of damage through cooperative binding, and reduces the rate of double-stranded DNA breakage. Thus, it maintains damaged DNA in a conformation that is amenable to repair. However, reverse gyrase is not a prerequisite for hyperthermophilic life. *Thermococcus kodakaraensis* with a disrupted reverse gyrase gene had retarded growth, especially at higher temperatures, but the disruption did not lead to a lethal phenotype. The gene for a RecA-like protein known as RadA and several other putative genes for DNA repair were induced in *P. furiosus* after γ irradiation, suggesting that hyperthermophiles possess a spectrum of DNA repair mechanisms. Some hyperthermophiles possess up to 55 copies of their genomes, which may readily facilitate DNA repair by homologous recombination.

The thermostability of the DNA double helix in hyperthermophilic archaea is also due to DNA interaction with double-stranded DNA-binding proteins. Many Euryarchaeota produce histone-like proteins that show sequence homology to the core fold region of eukaryotic histones but are shorter than eukaryotic histones. These proteins form dimers (or monomeric pseudodimers with twofolds) that must assemble into tetramers to bind DNA. The tetrameric form has been shown to compact DNA both *in vitro* and *in vivo* and adds to the thermostability of the double-stranded DNA. Crenarchaeota lack histone-like proteins but have other DNA-binding proteins that also compact DNA and increase the melting temperature of double-stranded DNA *in vitro*. These archaeal DNA-binding proteins are more likely for DNA packaging and nucleosome formation than for DNA stability.

The secondary structure of nucleic acids is also stabilized by the accumulation of specific organic osmolytes. The melting temperature of DNA from *Methanothermus fervidus* increased with the concentration of cyclic 2,3-diphosphoglycerate (cDPG), a compatible solute that is produced by thermophilic and hyperthermophilic methanogens. Another osmolyte unique to hyperthermophiles, di-*myo*-inositol-1,1′-phosphate (DIP), may also serve as a DNA thermoprotectant as this compound increases in concentration with organism growth temperature (see Section "Protein"). Polycationic polyamines are also observed in hyperthermophiles and increase the melting temperature of DNA. Concentrations of up to 0.4 g% (d.w. cell biomass) of putrescine, spermidine, norspermine, thermospermine, and spermine were detected in various *Sulfolobus* strains. Norspermine and norspermidine occur only in hyperthermophilic archaea, and these organisms typically contain a greater diversity of polyamines than other organisms.

Protein

A remarkable feature of thermophiles and hyperthermophiles is the broad thermostability of proteins across all functional classes and forms. Perhaps the most significant finding is that enzymes from hyperthermophiles, thermophiles, and mesophiles have no pattern of systematic structural differences that provides thermostability. There are small differences within a related group of proteins with varying optimal temperatures, but the changes will be different for a different related group of proteins. Furthermore, only a few amino acid substitutions are necessary to bring about significant changes in protein thermal stability.

There is strong evidence in general terms for an inverse correlation between protein thermal stability and molecular flexibility (i.e., a less flexible protein will be more thermostable). A balance must be struck between stabilizing and destabilizing interactions to meet the conflicting demands of thermostability and catalytic function, respectively. The structural basis of thermostability in hyperthermophilic proteins varies for different proteins with some commonality between certain examples. A comparison of crystal structures of hyperthermophilic, thermophilic, and mesophilic orthologous proteins shows that some of the more common structural changes found in thermophilic proteins are as follows: (1) an increase in the number of ion pairs, (2) an increase in the number of hydrogen bonds between positively charged side chains and neutral oxygens, (3) more extensive α helix secondary structure, (4) a decrease in the number of internal cavities, (5) a decrease in surface-to-volume ratios by shortening surface loops, (6) an increase in the number of hydrophobic residues as the hydrophobic effect increases with temperature, and (7) oligomerization.

In addition to these intrinsic factors, there are numerous extrinsic factors that likewise enhance protein stability at high temperatures. These include a chaperone protein (thermosome) that is unique to hyperthermophilic archaea, is highly abundant at all hyperthermophile growth temperatures, and is the primary protein produced during heat shock. The chaperone from *Sulfolobus solfataricus* prevents the aggregation of denatured target protein, catalyzes the refolding of the protein upon the addition of K^+, and releases the substrate after ATP hydrolysis. Hyperthermophiles also produce nonproteinaceous osmolytes that serve to stabilize proteins at high temperatures. Thermophilic and hyperthermophilic methanogens produce cDPG, which stabilizes thermolabile proteins *in vitro* at superoptimal temperatures. Other hyperthermophiles produce DIP. Like cDPG, DIP stabilized proteins *in vitro* at superoptimal temperatures. The concentration of DIP in *P. furiosus* increases 20-fold during heat shock, further suggesting that DIP functions specifically to stabilize macromolecules at high temperatures. It was also shown that elevated pressure stabilizes numerous hyperthermophilic enzymes, whereas their mesophilic counterparts were unaffected or inhibited by elevated pressure. Elevated pressure is believed to cause increased packing of the molecule, which likewise imparts increased structural rigidity and thermostability to these proteins.

Cell Wall

The maintenance of membrane fluidity is essential for normal cell function, and the mechanisms for maintaining stable and adaptive membranes in hyperthermophilic archaea and bacteria differ significantly from each other. Studies on artificial membranes (i.e., liposomes) demonstrate that the membranes of hyperthermophiles have evolved mechanisms for maintaining a liquid crystal state at high temperatures. In archaea, lipids are composed of isoprenoid alcohol chains that are ether-linked to the glycerol backbone. Ether linkages are also found in low proportions in some thermophilic bacteria and may mark a thermophilic adaptation. The lipid bilayer is also cross-linked at certain points by C_{40} trans-membrane phytanyl chains (i.e., tetraether lipids). Membranes that contain these membrane-spanning lipids are much more thermostable than those formed from phosphodiester lipids. The proportion of tetraether-to-diether lipids in *Methanocaldococcus jannaschii* increase significantly with temperature, supporting the idea that an increased proportion of tetraether lipids in archaeal membranes contributes to cell wall thermostability.

UPPER TEMPERATURE LIMIT OF LIFE

Hyperthermophile Culture Studies

Hyperthermophiles have the highest known growth temperatures for life, and most of what is known about them has come through pure culture studies under well-defined and regulated conditions in the laboratory. From these studies, the highest optimal growth temperature for an organism is 105–106 °C (Table 1). The heterotrophic archaea *Hyperthermus butylicus* and *Pyrodictium abyssi* have maximum growth temperatures of 108 and 110 °C, respectively. They grow on peptides and their growth is stimulated by the addition of H_2, CO_2, and S^0. The obligately chemolithoautotrophic archaea *Pyrodictium occultum* and *Pyrodictium brockii* grow up to 110 °C; require H_2 and S^0, $S_2O_3^{2-}$, or SO_3^{2-} for growth; and are stimulated by the addition of yeast extract (i.e., mixed organic compounds). After prolonged incubation (9 days at 96 °C), very few *Pyrodictium* cells are present singly in suspension. Rather, the cells are connected by ultra-thin fibers that form a network. This may be a mechanism to enhance the thermostability of these cells. The methanogenic archaeon *Methanopyrus kandleri* also grows at temperatures up to 122 °C and requires H_2 and CO_2 to produce CH_4. Other cultured organisms with the highest growth temperatures are the obligately chemolithoautotrophic archaea *Pyrolobus fumarii* and Pyrodictiaceae strain 121, which grow up to 113 and 121 °C, respectively. *P. fumarii* requires H_2 as an electron donor and can use NO_3^-, $S_2O_3^{2-}$, iron oxide hydroxide, and low levels of O_2 as electron acceptors. In contrast, strain 121 can use either H_2 or formate as an electron donor but can only use iron oxide hydroxide as an electron acceptor.

A parameter related to the upper temperature limit of life is pressure. Any life above 100 °C requires pressure >0.1 MPa to maintain a liquid environment. Most marine hyperthermophiles are present in deep-sea hydrothermal vent sites where the *in situ* pressure is 20–45 MPa (pressure increases 0.1 MPa, or 1 times atmospheric pressure, for every 10 m of water depth). Pressure effects on hyperthermophiles are generally favorable for growth at high temperatures. Relative to low pressures (0.1–3 MPa), the maximum growth temperature increases 2–6 °C for *Pyrococcus*, *Thermococcus*, and *Desulfurococcus* species when incubated at *in situ* pressures. For other hyperthermophiles, although their optimum growth temperature does not increase with pressure, their rate of growth does increase significantly at elevated pressure. For *M. jannaschii*, hyperbaric pressure significantly increases its growth rate at 86 °C but does not increase its optimum growth temperature. However, the maximum temperature for CH_4 formation increases from 92 at 0.8 MPa to 98 °C at 25 MPa. Likewise, methanogenesis rates are higher at higher pressures.

Laboratory Studies on Natural Microbial Assemblages

Natural assemblages of microorganisms collected from deep-sea hydrothermal vent sites have also been studied under controlled conditions in the laboratory. Petroleum-rich sediment cores from Guaymas Basin showed maximum sulfate reduction activities at 90 °C with additional activity between 105 and 110 °C. The hyperthermophilic sulfate reducer *Archaeoglobus profundus* was cultured from these same sediment samples, which has a maximum growth temperature of 90 °C and most likely is responsible for the 90 °C sulfate reduction activity peak. The organisms responsible for sulfate reduction between 100 and 110 °C are unknown. A natural assemblage of microorganisms collected from high-temperature black smoker fluids was incubated in solid gel material (GELRITE) that is stable at temperatures up to 120 °C. Colonies of microorganisms formed in the gel at 115 and 120 °C at 7 and 27 MPa. The largest colonies (~0.5 mm diameter) formed at 27 MPa, which again demonstrates the positive influence of pressure on the growth of microorganisms at high temperatures. Growth also may have been enhanced by the presence of a solid matrix to which cells attach themselves.

Field Observations

Analyses for biomolecules and intact cells have been performed on exiting hydrothermal fluids, sulfide ore deposits, and sediment and rock core samples. The primary difficulty has been determining what proportion of the material originated from seawater and cooler sulfide ore material that frequently contaminates samples. An *in situ* incubator containing interior melting point standards was deployed on top of a black smoker at Guaymas Basin (Vent 1). Microbial colonization was observed where a 125-°C sensor had melted but not a neighboring 140 °C sensor, although exposure to 125 °C may have been transient. In a sulfide ore deposit from a deep-sea hydrothermal vent site, the highest concentrations of ether lipids and intact fluorescent cells were found in mineral layers consisting primarily of anhydrite and Zn-Fe sulfides, which suggests that their temperatures were between 100 and 140 °C.

Field evidence for life above 100 °C is circumstantial due to the difficulties of obtaining accurate temperature measurements on the spatial scales necessary, the temporal variations in temperature at a given site, the uncertain origin of the biomass analyzed, and the absence of direct microbial activity measurements. These results are speculative rather than conclusive, and await further detailed analyses for verification and identification of indigenous microorganisms and their metabolic traits.

METABOLISM AND GROWTH

CO_2 and Acetate Assimilation

The assimilation of CO_2 by autotrophs is often associated with the Calvin cycle. However, this pathway is generally absent in thermophilic bacteria and is completely absent in archaea. Alternative CO_2 assimilation pathways are generally used in these organisms and include the coenzyme A (CoA) pathway, the reductive citric acid cycle, the 3-hydroxypropionate cycle, and the 4-hydroxybutyrate cycle. These pathways were discovered in part by studying anaerobic thermophiles. Acetate assimilation can occur using the citramalate cycle, the glyoxylate shunt, or by reversing a portion of the acetyl-CoA pathway. Thermophilic bacteria and archaea use all of these pathways for inorganic carbon assimilation in some capacity. An archaea-specific version of the acetyl-CoA pathway is used by Euryarchaeota with CH_4 production as added steps to the pathway in methanogens (Figure 1). Crenarchaeota use the reductive citric acid cycle, which is lacking in most hyperthermophilic Euryarchaeota, as well as the 3-hydroxypropionate, the 4-hydroxybutyrate, and citramalate cycles. The pathway for CO_2 assimilation in the Pyrodictiaceae is largely unknown and likely represents a novel autotrophic pathway.

Acetyl-CoA pathway and methanogenesis

Moorella thermoacetica and *Moorella thermoautotrophica* (formerly *Clostridium thermoaceticum* and *C. thermoautotrophicum*) are facultative autotrophs and obligately anaerobic spore formers. When grown on glucose, these organisms produce three molecules of acetate per glucose and little CO_2. This led to the suggestion that glucose is first oxidized to two molecules each of acetate and CO_2, followed by the production of a third acetate from the two CO_2 molecules. $^{14}CO_2$ and $^{13}CO_2$ incubation experiments confirmed that both carbons of acetate are labeled during growth. Incubation with ^{13}C-formate led to the labeling of the methyl group in acetate and suggested that formate is an intermediate in the pathway. The isolation of the autotrophic acetogen *Acetobacterium woodii* confirmed CO_2 assimilation can occur by means other than by the Calvin cycle, which is now known as the acetyl-CoA pathway (or Wood–Ljungdahl pathway). In addition to these thermophilic bacteria, an archaeal version of the pathway is used by autotrophic members of the Archaeoglobaceae and by all methanogens growing on H_2 and CO_2.

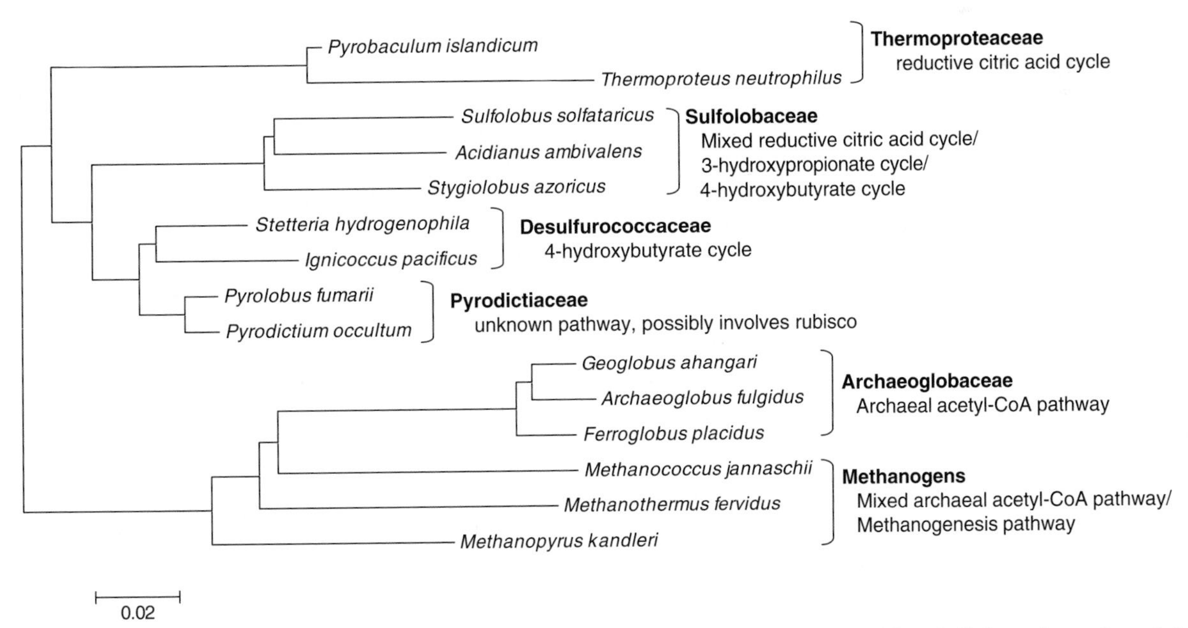

FIGURE 1 Phylogeny of chemolithoautotrophic archaea based on 16S rRNA sequence homologies and the CO_2 assimilation pathways for each family or group of organisms.

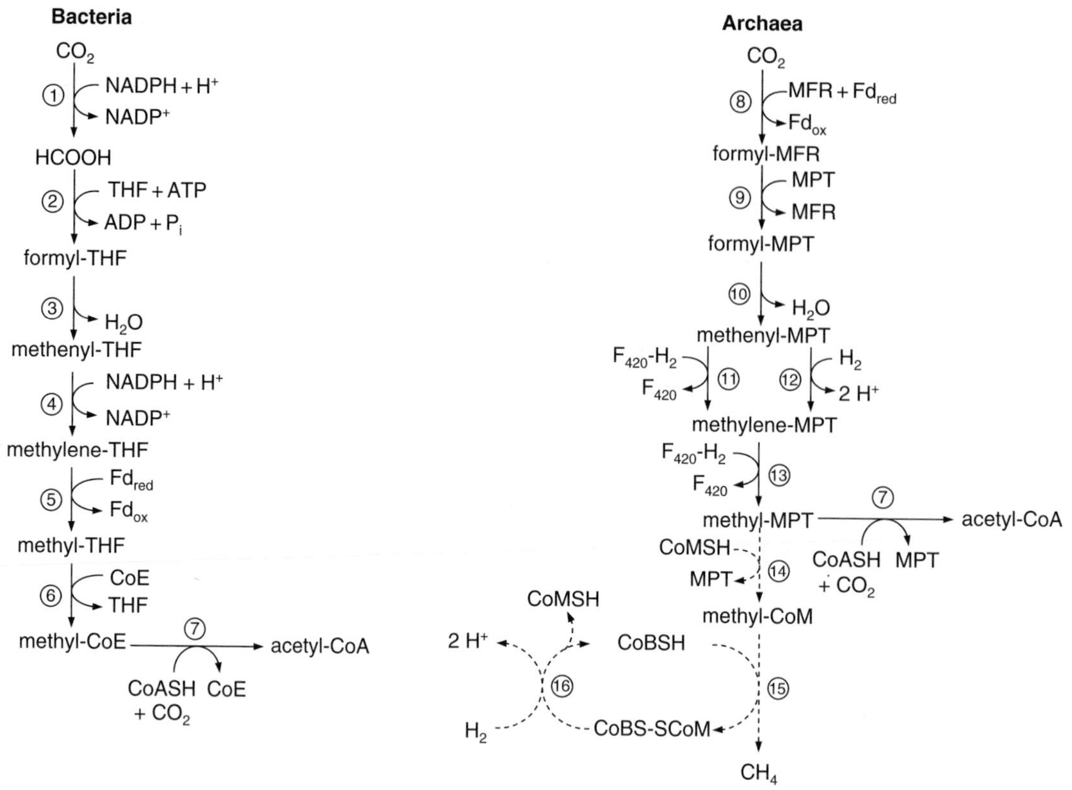

FIGURE 2 The acetyl-CoA pathways in bacteria and archaea. The enzymes are as follows: 1, formate dehydrogenase; 2, formyl-THF synthetase; 3, methenyl-THF cyclohydrolase; 4, methylene-THF dehydrogenase; 5, methylene-THF reductase; 6, methyltransferase; 7, carbon monoxide dehydrogenase; 8, formyl-MF dehydrogenase; 9, formyl-MF:MPT formyltransferase; 10, methenyl-MPT cyclohydrolase; 11, F420-dependent methylene-MPT dehydrogenase; 12, F420-independent methylene-MPT dehydrogenase; 13, methylene-MPT reductase; 14, methyl-MPT:CoMSH methyltransferase; 15, methyl-CoM reductase; and 16, heterodisulfide reductase. THF, tetrahydrofolate; MF, methanofuran; MPT, tetrahydromethanopterin; CoE, coenzyme E; CoM, coenzyme M; CoB, coenzyme B; Fd, electron carrier ferredoxin; and F420, electron carrier coenzyme F420. The dashed lines show the enzyme steps that are unique to methanogenesis.

The acetyl-CoA pathway reduces and condenses two molecules of CO_2 to form one molecule of acetyl-CoA (Figure 2). One CO_2 molecule is reduced in a series of reduction steps to a methyl group ligated to a C_1 carrier. In bacteria, the C_1 carriers are tetrahydrofolate and coenzyme E; in archaea, they are methanofuran and tetrahydromethanopterin. Another difference between the pathways in bacteria and archaea is the use of a soluble 5-deazaflavin called coenzyme F_{420} by archaea that carries two electrons but only one hydrogen. The enzymes in bacteria and archaea that catalyze these steps are conserved functionally but have little-to-no sequence homologies between them. The key enzyme in the pathway is carbon monoxide dehydrogenase/acetyl-CoA synthase, which is a highly conserved protein across superkingdoms. This enzyme reduces the second CO_2 molecule to a carbon monoxyl group, ligates it to the methyl group from the C_1 carrier, and then adds CoA to form acetyl-CoA. Methanogens use three additional enzymes (methyl-MPT:CoM methyltransferase, methyl-CoM reductase, and heterodisulfide reductase) to dispose of electrons during their anaerobic respiration using methyl-MPT as their terminal electron acceptor (Figure 2).

Reductive citric acid cycle

Like *M. thermoacetica* and *M. thermoautotrophica*, it was shown in the photosynthetic green sulfur bacterium *Chlorobium limicola*, the purple-sulfur bacterium *Chromatium vinosum*, and the purple nonsulfur bacterium *Rhodospirillum rubrum* that CO_2 assimilation occurs by a pathway other than the Calvin cycle. The concept of the reductive citric acid cycle (Figure 3) had its origin in the discovery of ferredoxin-dependent reductive carboxylation reactions: pyruvate synthase and 2-oxoglutarate (formerly α-ketoglutarate) synthase. They are driven by the strong reducing potential of reduced ferredoxin and complement the irreversible NAD^+-dependent pyruvate dehydrogenase and 2-oxoglutarate dehydrogenase reactions. Other key findings were enzymes that produce oxaloacetate and acetyl-CoA from citrate and complement the irreversible citrate synthase step in the oxidative citric acid cycle. In *C. limicola* and *Desulfobacter hydrogenophilus*, citrate cleavage is accomplished in a single step by ATP citrate lyase (Equation 1):

$$\text{Citrate} + \text{ATP} + \text{CoA} \rightarrow \text{acetyl-CoA} + \text{oxaloacetate} + \text{ADP} + P_i. \quad (1)$$

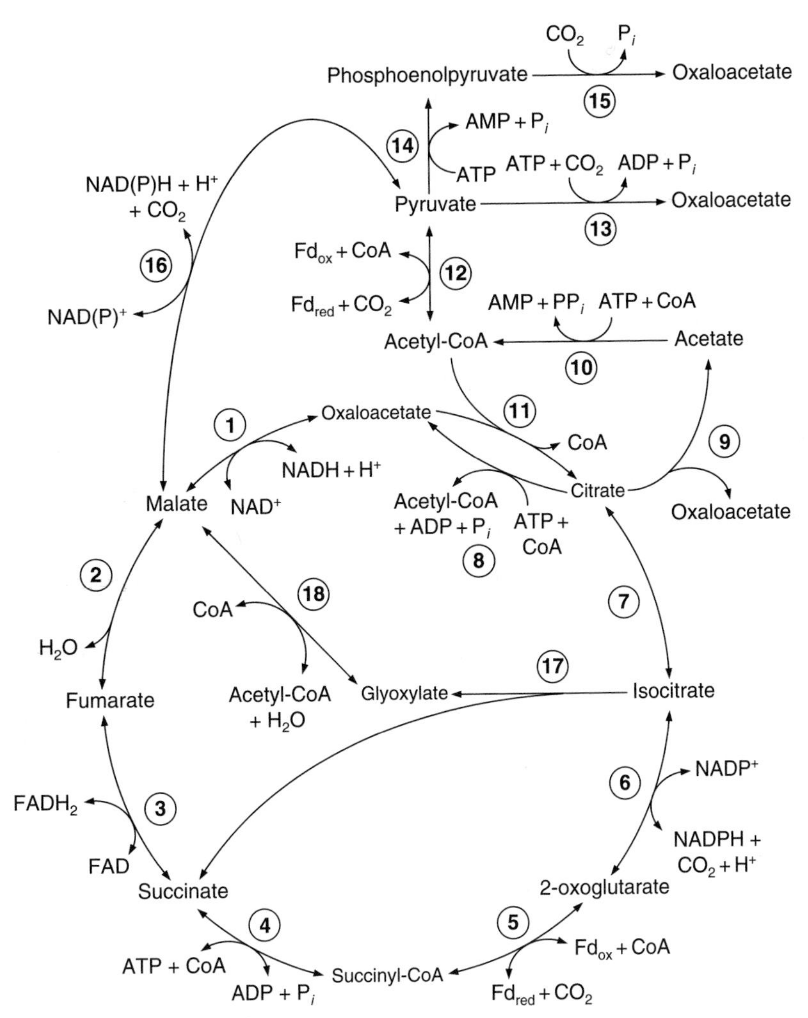

FIGURE 3 The citric acid cycle, the glyoxylate shunt, and related enzymes. The enzymes are as follows: 1, malate dehydrogenase; 2, fumarase; 3, fumarate reductase/succinate dehydrogenase; 4, succinyl-CoA synthetase; 5, 2 oxoglutarate synthase; 6, isocitrate dehydrogenase; 7, aconitase; 8, either ATP citrate lyase (one step) or citryl-CoA synthase and citryl-CoA lyase (two steps); 9, citrate lyase; 10 AMP-forming acetyl-CoA synthetase; 11, citrate synthase; 12, pyruvate synthase; 13, pyruvate carboxykinase; 14, phosphoenolpyruvate synthetase; 15, phosphoenolpyruvate carboxylase; 16, malic enzyme; 17, isocitrate lyase; and 18, malate synthase. Fd, electron carrier ferredoxin. Copyright © American Society for Microbiology, *Journal of Bacteriology*, *188*, 4350–4355, 2006.

This protein is the citrate cleavage enzyme that is most commonly associated with the reductive citric acid cycle. In the thermophilic bacterium *Hydrogenobacter thermophilus* TK-6, citrate cleavage is catalyzed in two steps by citryl-CoA synthetase (Equation 2) and citryl-CoA lyase (Equation 3):

$$\text{Citrate} + \text{ATP} + \text{CoA} \rightarrow \text{citryl-CoA} + \text{ADP} + \text{P}_i, \quad (2)$$

$$\text{Citryl-CoA} \rightarrow \text{oxaloacetate} + \text{acetyl-CoA}. \quad (3)$$

The first evidence for the reductive citric acid cycle in archaea came from the study of thermoacidophile *Acidianus brierleyi* (formerly *Sulfolobus brierleyi*). Pulse labeling of autotrophically grown *A. brierleyi* with $^{14}CO_2$ showed the formation of labeled malate, citrate, aspartate, and glutamate. Initially, all of the citric acid cycle enzymes and pyruvate synthase were measured in cell extracts of *A. brierleyi* grown autotrophically. However, the lack of ATP citrate lyase activity and the presence of 3-hydroxypropionate cycle activities led to the suggestion that the organism uses this latter pathway for CO_2 assimilation. It was since shown that autotrophically grown *A. brierleyi* does possess ATP citrate lyase activity but it requires covalent modification by acetylation for activity. Therefore, the organism appears to use a combination of the reductive citric acid cycle and the 3-hydroxypropionate cycle for CO_2 assimilation.

The hyperthermophilic archaeon *Thermoproteus neutrophilus* was grown autotrophically and pulse labeled with ^{14}C- and ^{13}C-succinate, yielding labeled malate, glutamate, and aspartate. This and the presence of all of the activities of the citric acid cycle enzymes, pyruvate synthase, and ATP citrate lyase suggest that it also uses the reductive citric acid cycle for CO_2 assimilation. The presence of pyruvate synthase, 2-oxoglutarate synthase, and ATP citrate lyase activities in the hyperthermophilic archaeon *Pyrobaculum islandicum* grown autotrophically suggests that this organism likewise uses this pathway. However, for both *T. neutrophilus* and *P. islandicum*, it was suggested that the ATP citrate lyase activities are too low to account for all of the CO_2 assimilated. It was subsequently shown for *P. islandicum* that acetylated citrate lyase (Equation 4) and AMP-forming acetyl-CoA synthase (Equation 5) activities increase significantly in cells grown autotrophically relative to those grown heterotrophically and are higher than the ATP citrate lyase activities measured.

$$\text{Citrate} \rightarrow \text{oxaloacetate} + \text{acetate}, \quad (4)$$

$$\text{Acetate} + \text{CoA} + \text{ATP} \rightarrow \text{acetyl-CoA} + \text{AMP} + \text{PP}_i. \quad (5)$$

Therefore, there appears to be a third mechanism for citrate cleavage in thermophiles that requires covalent modification by acetylation.

All 8 of the citric acid cycle enzymes are present within 20 of the 46 archaeal genome sequences currently available (Figure 4). The complete cycle is found in all organisms within the Thermoproteales, the Sulfolobales, and *Aeropyrum pernix* in the Crenarchaeota and in all Halobacteriales (i.e., extreme halophiles) and Thermoplasmales in the Euryarchaeota. Only a portion of the cycle is found in the Thermococcales, the Archaeoglobales, the Desulfurococcales (except *A. pernix*), and all methanogens. The distributions of the citric acid cycle enzymes in both archaeal phyla and the acetyl-CoA pathway enzymes (only in Euryarchaeota) suggest that the last common archaeal ancestor contained the complete citric acid cycle and that portions of the cycle were lost in the methanogens, Archaeoglobales, and Thermococcales, and currently it only serves to produce intermediates for some biosynthesis reactions in these organisms.

3-Hydroxypropionate cycle

Chloroflexus aurantiacus is a thermophilic green nonsulfur bacterium that is a facultative photoautotroph and an anaerobe. Cultures grown photoautotrophically with H_2 and CO_2 lack ribulose-1,5-bisphosphate carboxylase/oxygenase (RubisCO) and ribulose-5-phosphate kinase (formerly

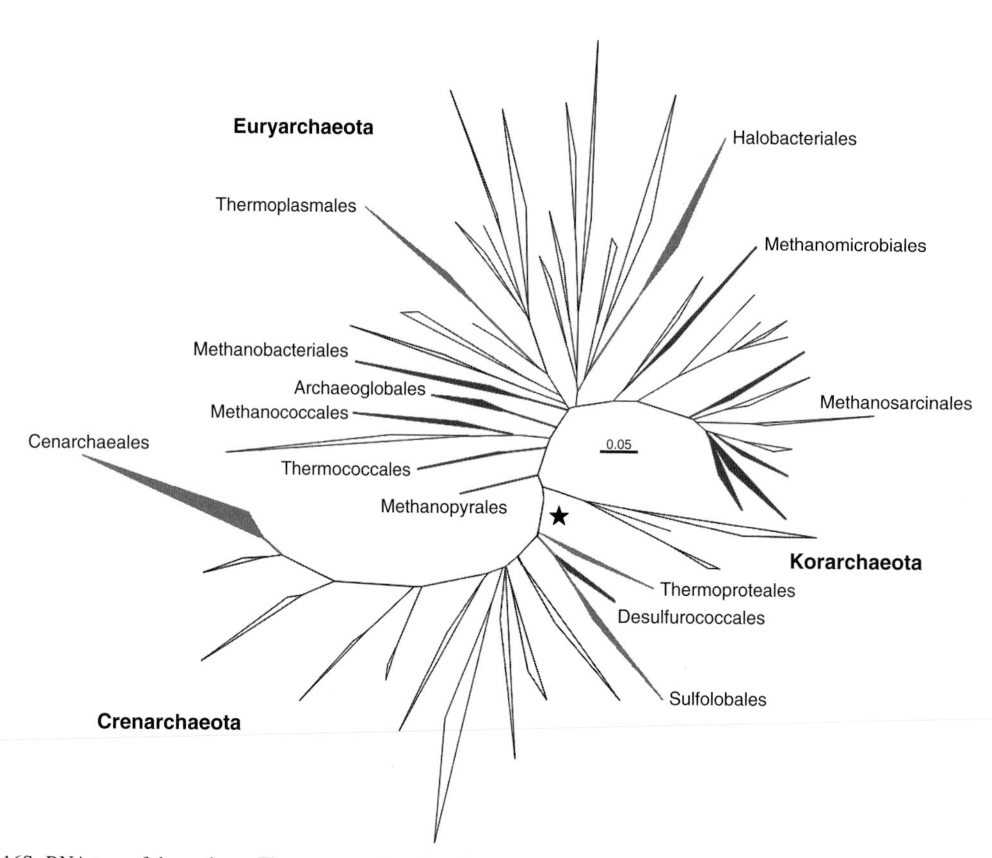

FIGURE 4 A 16S rRNA tree of the archaea. Those taxonomic orders that contain all of the genes that encode for citric acid cycle enzymes are shown in green, whereas those with only a subset of these genes are in red. Uncultivated orders are shown as unfilled groups. The star indicates the location of the root of the tree. The tree was constructed as described previously. The scale bar represents 0.05 changes per nucleotide. *Adapted by permission from Macmillan Publishers Ltd: Schleper, C., Jurgens, G., & Jonuscheit, M. (2005). Genomic studies of uncultivated Archaea.* Nature Reviews, 3, 479–488. *Copyright 2005.* (For interpretation of the references to color in this figure legend, the reader is referred to the Web version of this chapter.)

phosphoribulokinase) activities, which are the key enzymes of the Calvin cycle. They also lack ATP citrate lyase and 2-oxoglutarate synthase activities that are necessary for the reductive citric acid cycle. However, *C. aurantiacus* secretes 3-hydroxypropionate during phototrophic growth, which suggests that it is an intermediate of CO_2 assimilation. This led to the discovery of the 3-hydroxypropionate cycle where two molecules of CO_2 are assimilated to form glyoxylate (Figure 5). The carboxylation reactions are catalyzed at two points in the cycle by a single bifunctional enzyme called acetyl-CoA/propionyl-CoA carboxylase. Many of the enzymes in the cycle overlap with those of the citric acid cycle to form malate, which is then split to form glyoxylate and regenerate acetyl-CoA.

As mentioned previously, the lack of ATP citrate lyase activity in the thermoacidophilic archaeon *A. brierleyi* and the presence of acetyl-CoA carboxylase and propionyl-CoA carboxylase activities led to the suggestion that this organism uses the 3-hydroxypropionate cycle for CO_2 assimilation. Acetyl-CoA carboxylase and propionyl-CoA carboxylase activities were also measured in autotrophically grown cell extracts from *Sulfolobus metallicus* and *Acidianus infernus*. The subsequent discovery of ATP citrate lyase activity in *A. brierleyi* after acetylation and the activities of both the citric acid and 3-hydroxypropionate cycles suggest that this organism uses a combination of these pathways for CO_2 assimilation.

4-Hydroxybutyrate cycle

Ignicoccus species are hyperthermophilic obligately autotrophic archaea that belong to the Desulfurococcaceae. *Ignicoccus pacificus* and *Ignicoccus islandicus* lack ATP citrate lyase, 2-oxoglutarate synthase, carbon monoxide dehydrogenase, and acetyl-CoA/propionyl-CoA carboxylase activities that are indicative of the reductive citric acid cycle, the acetyl-CoA pathway, and the 3-hydroxypropionate cycle, respectively. It was shown that *Ignicoccus hospitalis* assimilates CO_2 in two steps using pyruvate synthase and phosphoenolpyruvate carboxylase (Figure 6), and many of the enzymes are the same as those used in the reductive citric acid cycle. However, because the organism lacks 2-oxoglutarate synthase, it reduces succinyl-CoA in two enzymatic steps to form 4-hydroxybutyrate and eventually

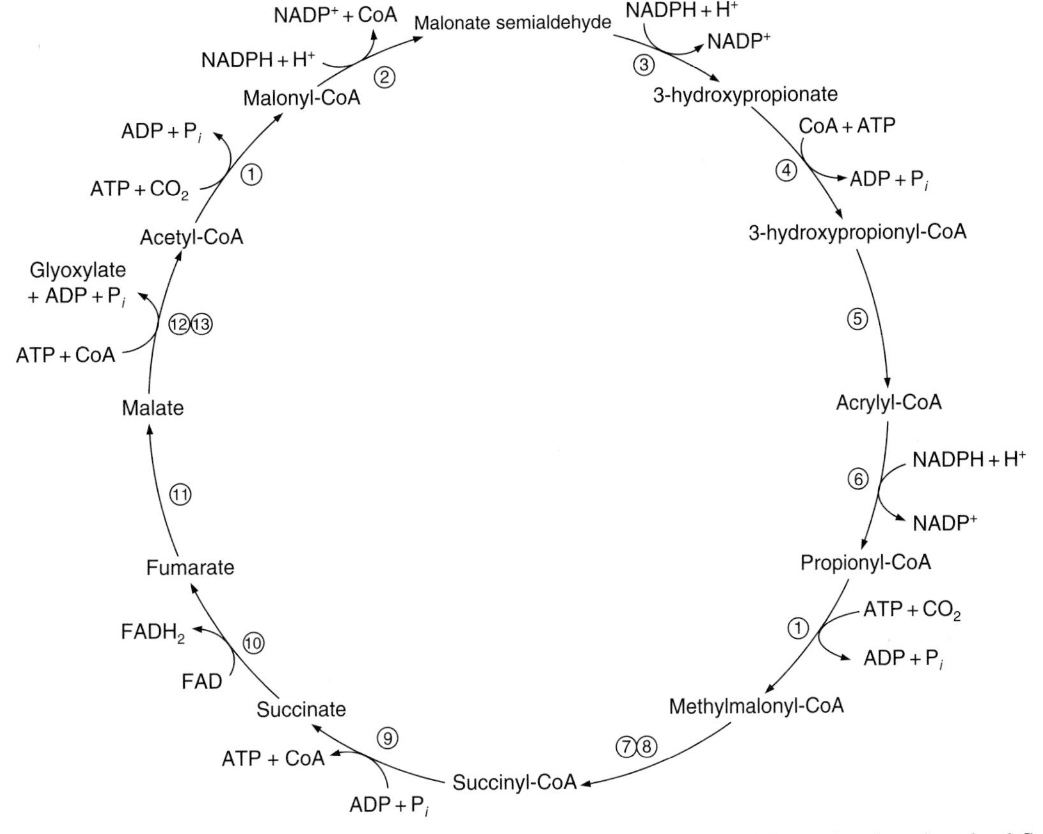

FIGURE 5 The 3-hydroxypropionate cycle. The enzymes are as follows: 1, acetyl-CoA/propionyl-CoA carboxylase; 2, malonyl-CoA reductase; 3, 3-hydroxypropionate dehydrogenase; 4, 3-hydroxypropionyl-CoA hydrolase; 5, acrylyl-CoA hydratase; 6, acrylyl-CoA dehydrogenase; 7, methylmalonyl-CoA epimerase; 8, methylmalonyl-CoA mutase; 9, succinyl-CoA synthetase; 10, succinate dehydrogenase; 11, fumarase; 12, malyl-CoA synthetase; and 13, malyl-CoA lyase.

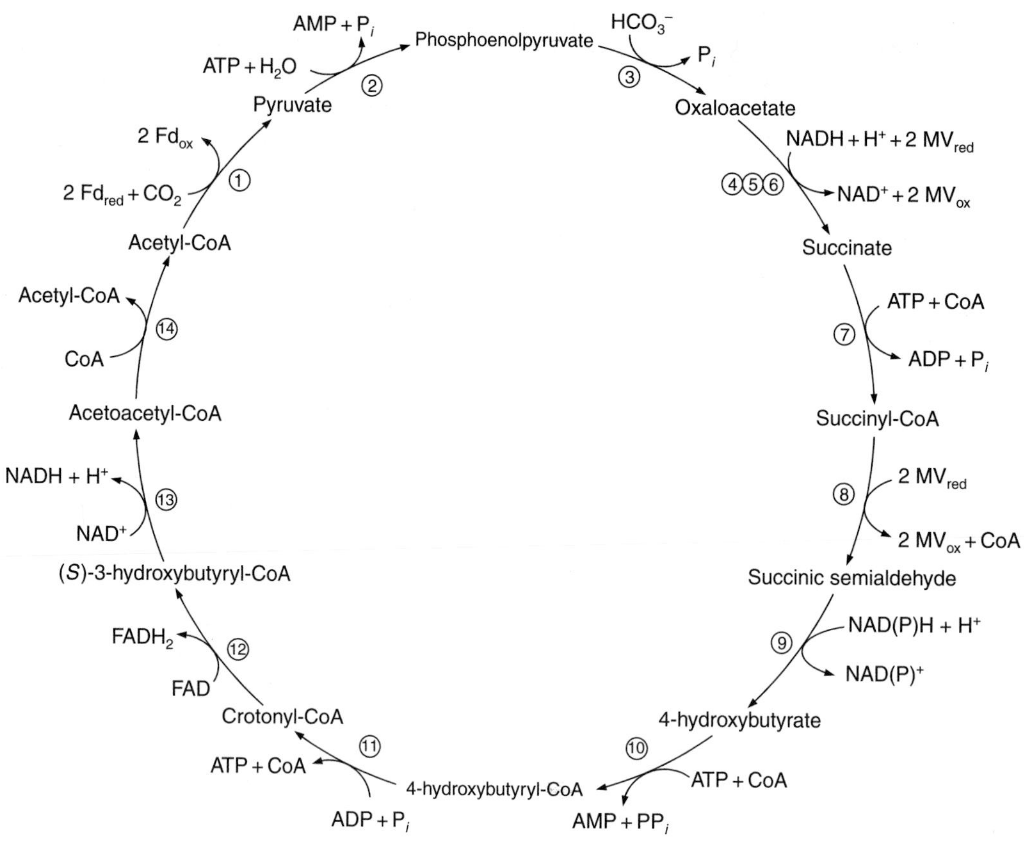

FIGURE 6 The 4-hydroxybutyrate cycle. The enzymes are as follows: 1, pyruvate synthase; 2, phosphoenolpyruvate synthetase; 3, phosphoenolpyruvate carboxylase; 4, malate dehydrogenase; 5, fumarase; 6, fumarate reductase; 7, succinyl-CoA synthetase; 8, succinyl-CoA reductase; 9, succinate semialdehyde reductase; 10, 4-hydroxybuturyl-CoA synthetase; 11, 4-hydroxybutyryl-CoA dehydratase; 12, crotonyl-CoA hydratase; 13, 3-hydroxybutyryl-CoA dehydrogenase; and 14, acetoacetyl-CoA β-ketothiolase.

acetoacetyl-CoA. In the final step, acetoacetyl-CoA is cleaved to form two molecules of acetyl-CoA, one of which brings the cycle back to its starting point, yielding the net formation of one acetyl-CoA.

Interestingly, the thermoacidophilic archaeon *Metallosphaera sedula*, which is a close relative of *Sulfolobus* and *Acidianus* species, uses a mixture of the 3-hydroxypropionate and 4-hydroxybutyrate cycles. CO_2 is assimilated using acetyl-CoA/propionyl-CoA carboxylase, as is found in the 3-hydroxypropionate cycle. However, instead of forming glyoxylate and acetyl-CoA from succinyl-CoA as is found in the 3-hydroxypropionate cycle, *M. sedula* converts succinyl-CoA into two molecules of acetyl-CoA via 4-hydroxybutyrate using the same enzymes found in the 4-hydroxybutyrate cycle. Therefore, across all of the autotrophic Crenarchaeota (with the possible exception of the Pyrodictiaceae), CO_2 assimilation often involves a mixture of the reductive citric acid cycle, the 3-hydroxypropionate cycle, and the 4-hydroxybutyrate cycle, and many of the same enzymes (i.e., malate dehydrogenase, fumarase, succinate dehydrogenase/fumarate reductase, succinyl-CoA synthetase, pyruvate synthase, PEP synthetase, and PEP carboxylase) are used, suggesting that there is an evolutionary relationship between these CO_2 assimilation pathways.

Other possible CO_2 assimilation pathways

Hyperthermophilic archaea belonging to the Pyrodictiaceae may possess novel pathways for CO_2 assimilation. Autotrophically grown *P. fumarii*, *P. abyssi*, and *P. occultum* grown on yeast extract with H_2 and CO_2 all had pyruvate synthase activity but lacked 2-oxoglutarate synthase activity and generally lacked other enzymes of the citric acid cycle needed for the 3-hydroxypropionate and 4-hydroxybutyrate cycles. *P. abyssi* and *P. occultum* also lack carbon monoxide dehydrogenase/acetyl-CoA synthase activity. Therefore, these organisms do not appear to use the acetyl-CoA pathway, the reductive citric acid cycle, the 3-hydroxypropionate cycle, or the 4-hydroxybutyrate cycle for CO_2 assimilation.

In *P. abyssi* and *P. occultum*, there are low levels (5–15 nmol min^{-1} mg^{-1} cell protein) of RubisCO activity, which is the key enzyme of the Calvin cycle. For *P. abyssi*, RubisCO activity increases approximately twofold per 10 °C temperature increase, as expected for most enzymes, and activity requires strictly anoxic and reducing conditions. The product of the reaction is 3-phosphoglycerate. However, ribulose-5-phosphate kinase activity is not measured, suggesting that CO_2 assimilation does not occur via the standard Calvin cycle. Similarly, RubisCO activity is present in several Euryarchaeota including methanogens, the hyperthermophilic heterotrophs in the Thermococcaceae, and *Archaeoglobus fulgidus*. All lack ribulose-5-phosphate kinase activity. Ribulose-1,5-bisphosphate is generated in *M. jannaschii* using 5-phospho-D-ribose-1-pyrophosphate (PRPP) as a substrate, which is an intermediate in nucleotide biosynthesis. A similar pathway is found in *T. kodakaraensis* where adenosine monophosphate is used as the starting material instead of PRPP. First the adenine in AMP is replaced with a phosphate to form ribose-1,5-bisphosphate, then an isomerase converts this to ribulose-1,5-bisphosphate. Although pathways for CO_2 assimilation via RubisCO seem to exist in Euryarchaeota, they do not appear to contribute significantly, if at all, to CO_2 assimilation.

Acetate catabolism

Acetate is potentially an important carbon source in high-temperature environments. It is a common metabolite formed by heterotrophs during the breakdown of organic material and is the primary end product of acetogens. The most common pathway of acetate catabolism in bacteria is via the glyoxylate shunt. First, acetate and CoA are combined to form acetyl-CoA by acetyl-CoA synthase using the energy of ATP and forming AMP + PP_i (Figure 3). Then some of the acetyl-CoA condenses with oxaloacetate to form citrate and enters the citric acid cycle. Isocitrate is formed from citrate. At this point, the fate of the carbon varies. Some isocitrate remains within the citric acid cycle for biosynthesis reactions. The remainder is cleaved into succinate and glyoxylate by isocitrate lyase. The succinate then re-enters the citric acid cycle pool of intermediates. Glyoxylate is then combined with a second molecule of acetyl-CoA to form malate by malate synthase. Malate can either enter the citric acid cycle pool for biosynthesis reactions or be used to form pyruvate using the malic enzyme. The key enzymes of the glyoxylate shunt are isocitrate lyase and malate synthase, and both of these as well as the complete citric acid cycle are found in the thermoacidophilic archaeon *Sulfolobus acidocaldarius*.

Other thermophilic organisms lack isocitrate lyase and pyruvate synthase activities, the two most common means for biosynthesis from acetyl-CoA, when grown on acetate. In these cases, acetate catabolism is accomplished using the citramalate cycle (Figure 7). Acetate (or acetyl-CoA) combines

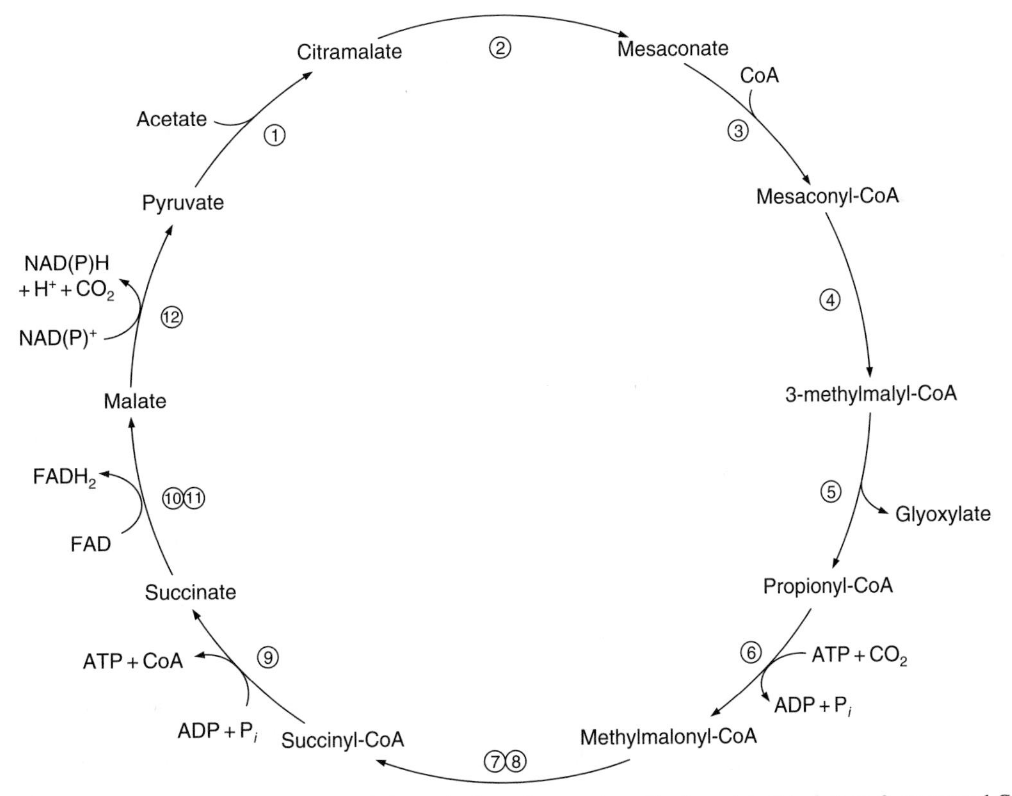

FIGURE 7 The citramalate cycle. The enzymes are as follows: 1, citramalate synthase; 2, citramalate dehydratase; 3, mesaconyl-CoA synthetase; 4, mesaconyl-CoA hydratase; 5, 3-methylmalyl-CoA lyase; 6, propionyl-CoA carboxylase; 7, methylmalonyl-CoA epimerase; 8, methylmalonyl-CoA mutase; 9, succinyl-CoA synthetase; 10, succinate dehydrogenase; 11, fumarase; and 12, malic enzyme.

with pyruvate to form citramalate by citramalate synthase. After a series of enzyme reactions, 3-methylmalonyl-CoA is cleaved to form propionyl-CoA and glyoxylate. The glyoxylate can be used to form malate using acetyl-CoA and malate synthase. Propionyl-CoA is carboxylated and eventually forms intermediates of the citric acid cycle using enzymes found in the 3-hydroxypropionate and citric acid cycles. Pyruvate is recycled from an intermediate in the citric acid cycle. The citramalate cycle was first described in purple bacteria, which grow best on acetate when H_2, CO_2, and low levels of either pyruvate or organic compound are added to the growth medium. Similarly, the hyperthermophilic archaeon *P. islandicum* increases its citramalate synthase and 3-methylmalyl-CoA lyase activities when grown on acetate relative to autotrophic and heterotrophic growth, suggesting that ituses the citramalate cycle in part for acetate metabolism. It grew best on acetate when H_2 and low levels of yeast extract (0.001%) were added to the medium.

The biochemical overlap between the citric acid cycle, the glyoxylate shunt, the 3-hydroxypropionate cycle, and the citramalate cycle highlights the importance of viewing these and other pathways holistically because it is likely that they do not always operate completely independent of the others. Furthermore, these pathways each perform multiple cellular functions for the cell. Some or all of the citric acid cycle is involved in all of the pathways listed above and also functions for energy production and biosynthesis reactions such as amino acid synthesis. Portions of the 3-hydroxypropionate cycle are also used for propanoate metabolism and the steps of the citramalate cycle are also used for leucine biosynthesis. These overlaps may have significant evolutionary implications and their study is important to understand the natural history of catabolic and anabolic pathways.

Heterotrophy

The majority of high-temperature microorganisms are heterotrophs or facultative autotrophs (Table 1). Not surprisingly, the most common organic compounds catabolized at high temperatures are carbohydrates and peptides. However, some thermophiles and hyperthermophiles also oxidize low-molecular-weight organic acids and many thermophiles catabolize hydrocarbons as sources of carbon and electrons.

Carbohydrate metabolism

Carbohydrate metabolism can be divided into four categories: uptake, hydrolysis, glycolysis, and gluconeogenesis. The enzymes for gluconeogenesis are found in most organisms, including autotrophs, whereas those for uptake, hydrolysis, and glycolysis vary. Many of the hyperthermophilic enzymes involved in glycolysis and gluconeogenesis are biochemically and phylogenetically unique to either hyperthermophiles or archaea (Figure 8). For example, glucokinase and phosphofructokinase from *P. furiosus* are ADP-dependent rather than ATP-dependent and do not show any sequence similarity with their ATP-dependent counterparts in mesophilic bacteria. Glyceraldehyde-3-phosphate is oxidized to 3-phosphoglycerate in a single enzyme step with concomitant reduction of ferredoxin rather than NAD^+ and is catalyzed by the unique tungsten-containing protein glyceraldehyde-3-phosphate oxidoreductase. Both $NAD(P)^+$-dependent glyceraldehyde-3-phosphate dehydrogenase and 3-phosphoglycerate kinase are present and are homologous to their counterparts in mesophilic bacteria, but they are used for gluconeogenesis rather than for glycolysis in *P. furiosus*. Furthermore, fructose-1,6-bisphosphate aldolase, fructose-1,6-bisphosphatase, and phosphoglucose isomerase are all unique to either hyperthermophiles or archaea.

Three membrane-bound sugar-binding proteins have been characterized from *P. furiosus* that are specific for maltose/trehalose (MalE), maltose/maltodextrin (MBP), and cellobiose (CbtA). These demonstrate the specificity and coordination of carbohydrate uptake in hyperthermophiles. The MalE binding, permease (MalFG), and ATP-binding transporter (MalK) proteins from *Thermococcus litoralis* and the operon encoding these proteins as well as the regulatory protein (TrmB) for this operon were characterized and it was shown that the sequences, function, and regulation of ATP-binding cassette (ABC)-type transport systems in hyperthermophiles are similar to those found in mesophiles. The *malEFGK* operon in *P. furiosus* is flanked by insertion sequences but is absent in *P. abyssi* and *Pyrococcus horikoshii*, suggesting lateral gene transfer from other organisms. Using proteomics, MalE and CbtA were identified in the membrane cellular fraction of *P. furiosus* grown on a mixture of maltose and peptides, and MalE is one of the most abundant proteins within the membrane.

The genes for MBP and CbtA in *P. furiosus* are likewise part of an ABC-type operon but lack flanking insertion sequences and are found in *P. abyssi* and *P. horikoshii*. Furthermore, the maltodextrin uptake operons in *P. furiosus* and *Pyrobaculum aerophilum* contain the gene encoding for the sugar hydrolase amylopullulanase (*apu*), whereas the cellobiose operon in *P. furiosus* is next to and shares a putative promoter region with the β-mannosidase gene (*bmn*), thus demonstrating a tight coupling between sugar uptake and hydrolysis in these organisms. *P. furiosus* amylopullulanase is an extracellular glycosylase that is active at temperatures up to 140°C. This demonstrates that some proteins are stable well above 110°C and that the biogenic impact of *P. furiosus* in its native environment extends beyond its maximum growth temperature as its extracellular enzymes "forage" for growth substrates.

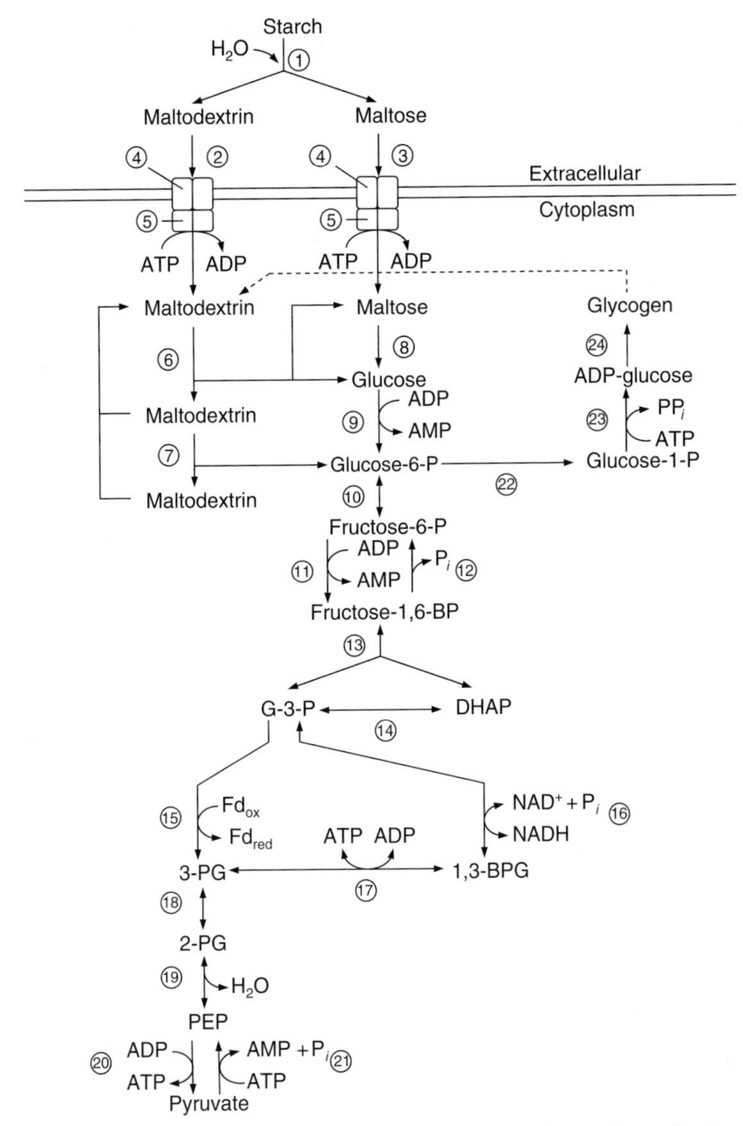

FIGURE 8 Starch hydrolysis, uptake, and glycolysis via the modified Embden–Meyerhof pathway and steps for gluconeogenesis. The enzymes are as follows: 1, amylopullulanase; 2, maltose/maltodextrin-binding protein; 3, maltose/trehalose-binding protein; 4, sugar transport permease; 5, sugar transport ATPase; 6, α-amylase; 7, α-glucan phosphorylase; 8, α-glucosidase; 9, glucokinase; 10, glucose-6-phosphate isomerase; 11, fructose-6-phosphate kinase; 12, fructose-1,6-bisphosphatase; 13, fructose-1,6-bisphosphate aldolase; 14, triosephosphate isomerase; 15, glyceraldehydes-3-phosphate:ferredoxin oxidoreductase; 16, glyceraldehydes-3-phosphate dehydrogenase; 17, 3-phosphoglycerate kinase; 18, 3-phosphoglycerate mutase; 19, enolase; 20, pyruvate kinase; 21, phosphoenolpyruvate synthetase; 22, phosphoglucomutase; 23, ADP-glucose synthase; and 24, glycogen synthase. Fd, the electron carrier ferredoxin.

Peptide metabolism

Peptide metabolism can be divided into three categories that are functionally similar to those found in carbohydrate metabolism: uptake, hydrolysis, and peptidolysis (Figure 9). Unlike the sugar ABC transport system, little is known about the peptide ABC transport system in hyperthermophiles. Using proteomics, a putative membrane dipeptide-binding protein was highly abundant in the membrane fraction of *P. furiosus* cells grown on tryptone and maltose along with the MalE-binding protein. Up to 13 protease activity bands are observed in gelatin-containing zymograms from *P. furiosus* cell extracts, demonstrating the large suite of proteases available with the cells. Four of the 13 predicted transaminases in *P. furiosus* were shown to have varying degrees of specificity for amino acids, although each uses 2-oxoglutarate as the amine group acceptor. The glutamate produced from the transamination reaction is recycled back to 2-oxoglutarate by glutamate dehydrogenase with concomitant reduction of $NADP^+$ in *P. furiosus*. Hyperthermophiles and archaea produce up to four ferredoxin-linked 2-keto acid oxidoreductases that decarboxylate the acid, pass electrons to ferredoxin, and ligate CoA to the remaining compound

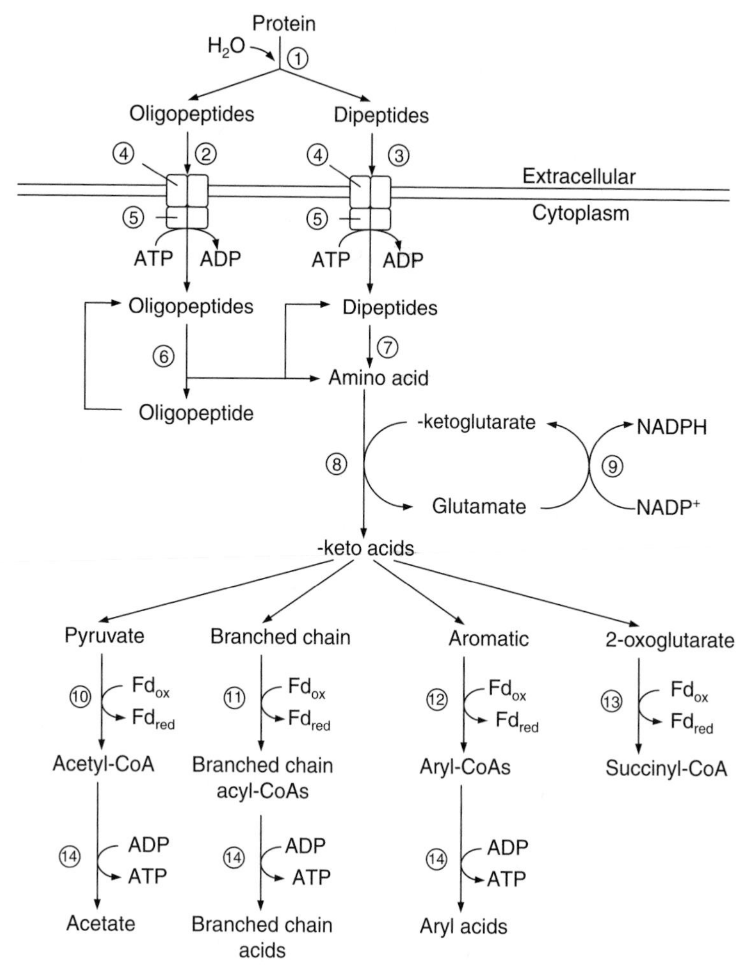

FIGURE 9 Peptide hydrolysis, uptake, and peptidolysis in archaea. The enzymes are as follows: 1, pyrolysin; 2, oligopeptide-binding protein; 3, dipeptide-binding protein; 4, peptide transport permease; 5, peptide transport ATPase; 6, intracellular protease; 7, prolidase; 8, amino acid aminotransferase; 9, glutamate dehydrogenase; 10, pyruvate:ferredoxin oxidoreductase; 11, α-ketoisovalerate:ferredoxin oxidoreductase; 12, indolepyruvate:ferredoxin oxidoreductase; 13, 2-oxoglutarate:ferredoxin oxidoreductase; and 14, acyl-CoA synthetase. Fd, the electron carrier ferredoxin.

(Figure 8). Three of these (IOR, VOR, and OGOR) are unique to archaea. The coenzyme is then cleaved, forming an organic acid with the phosphorylation of ADP to ATP, which is the only substrate-level phosphorylation step within the peptidolysis pathway.

P. furiosus growth on maltose was compared with its growth on peptides using growth kinetics, metabolite analyses, enzyme activities, and DNA microarray analyses. Based on growth rates, *P. furiosus* grows better on peptides than on maltose. As expected, the primary organic acid produced when cultures are grown on maltose is acetate (Figure 7), whereas growth on peptides yields a fairly even mixture of acetate, phenylacetate, (iso)butyrate, and isovalerate (Figure 9). The activities of glutamate dehydrogenase, 2-oxoglutarate oxidoreductase, indolepyruvate oxidoreductase, isovalerate oxidoreductase, formaldehyde oxidoreductase, aldehyde oxidoreductase, acetyl-CoA synthetase I, glyceraldehydes-3-phosphate dehydrogenase, and cytoplasmic hydrogenase were all significantly higher when *P. furiosus* cultures were grown on peptides. Conversely, the activities of glyceraldehydes-3-phosphate oxidoreductase, acetolactate synthase, and α-amylase were higher when cultures were grown on maltose. In both cases, these enzymes appear to follow their proposed physiological functions. Formaldehyde oxidoreductase, aldehyde oxidoreductase, and glyceraldehyde:ferredoxin oxidoreductase are tungsten-containing enzymes and explain in part why hyperthermophiles generally have a tungsten requirement for growth.

Respiration

The consumption of electron donors and the reduction of terminal electron acceptors are among the primary means that microorganisms have of altering the chemistry of their environment. Although several compounds can serve as electron donors for hyperthermophiles, the most common compounds used for this purpose in geothermal environments are H_2, organic compounds, and reduced sulfur compounds. Hydrogen is typically oxidized on the membrane by a hydrogenase where electrons then enter the electron transport pathway. Organic compounds are oxidized as described in Sections "Carbohydrate Metabolism" and "Peptide Metabolism" and result in the production of reduced ferredoxin and NADH.

Respiration is a series of exergonic redox reactions within the cytoplasmic membrane that are coupled with proton translocation across the membrane, which forms an electrochemical gradient (Figure 10). This proton motive force is then used to generate ATP from ADP and phosphate using a membrane-bound ATP synthase. The canonical electron transport chain through the membrane typically begins with the oxidation of NADH by a membrane-bound NADH:quinone oxidoreductase and the direct reduction of a quinone. Often electrons from the quinone are transferred to a cytochrome *c* by a quinol:cytochrome *c* oxidoreductase, typically a bc_1 complex. Electrons from either the quinones or the cytochrome are then passed to a terminal reductase that reduces the terminal electron acceptor. The marvelous aspect of respiration in bacteria and archaea is that the system is modular, and individual components (e.g., terminal reductases) can be exchanged with changes in environmental conditions and electron acceptor availability.

Homologues of NADH:quinone oxidoreductases are found in the genome sequences of most thermoacidophilic and hyperthermophilic archaea. The catalytic (NuoD) and quinone-binding (NuoH) subunits are conserved except for NuoH in methanogens, but all archaea lack the NuoAEFGJK subunits found in bacteria, which includes the NADH binding, flavin, and iron–sulfur cluster containing subunits (NuoEFG). Membrane-soluble archaeal electron carriers have been found in the hyperthermophiles *P. islandicum* and *Pyrobaculum organotrophum*, in the thermoacidophile *S. solfataricus*, and in the mesophilic methanogen *Methanosarcina mazei*. Menaquinones were found in the two *Pyrobaculum* species whereas two novel sulfur-containing quinone-like compounds were observed in *S. solfataricus*. The methanogen uses a 2-hydroxyphenazine derivative called methanophenazine, which could be reduced by a membrane-bound F_{420} dehydrogenase and oxidized by the membrane-bound enzyme heterodisulfide reductase. The use of methanophenazine by methanogens may explain the absence of the quinone-binding subunit in their NADH:quinone oxidoreductase.

The presence of bc_1 complexes in hyperthermophilic and thermoacidophilic archaea is relatively scarce. They are found in some *Pyrobaculum*, *Aeropyrum*, *Sulfolobus*, and *Acidianus* species, all organisms with some capacity for aerobic growth. Homologues of membrane-bound H^+-translocating ATP synthase are found in the genome sequences of all thermoacidophilic and hyperthermophilic archaea. The catalytic subunits (AtpAB) are conserved, but all archaea lack the AtpGHJK subunits found in bacteria.

Reduction of sulfur compounds

The reduction of elemental sulfur is one of the most common traits of thermoacidophiles and hyperthermophiles (Table 1). Elemental sulfur is the terminal electron acceptor for neutrophilic heterotrophs from marine environments (e.g., *Pyrococcus* and *Thermococcus*) and terrestrial environments (e.g., *Thermoproteus* and *Pyrobaculum*), for chemolithoautotrophs from marine environments (e.g., *Pyrodictium*), and for some thermoacidophiles from terrestrial environments (e.g., *Acidianus*). This is not surprising given the abundance of sulfur compounds in their native geothermal environments. Environmental conditions significantly influence the form of sulfur available for respiration. Above pH 5, sulfide anion (HS-) is a nucleophile that reacts with the elemental sulfur ring (S_8) forming polysulfide (S_4^{2-} and S_5^{2-}). Above pH 7 and 75 °C, elemental sulfur disproportionates into thiosulfate and sulfide $\left(S_8 + 6H_2O \rightarrow 2S_2O_3^{2-} + 4HS^- + 8H^+\right)$.

Pyrodictium and *Acidianus* species couple H_2 oxidation with elemental sulfur reduction. They grow at pH 5–8 and pH 1–4, respectively, suggesting that *Pyrodictium* uses polysulfide whereas *Acidianus* uses S8. In *Pyrodictium*, H_2 oxidation is coupled directly to sulfur/polysulfide reduction

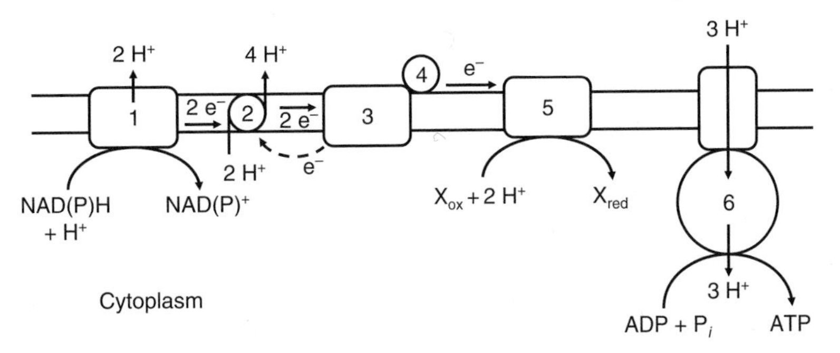

FIGURE 10 Membrane electron transport pathway. The components are as follows: 1, NADH:quinone oxidoreductase; 2, quinone; 3, quinol:cytochrome *c* oxidoreductase (bc_1 complex); 4, cytochrome *c*; 5, generic terminal reductase; and 6, H^+-translocating ATP synthase.

in a membrane-bound multienzyme complex with both hydrogenase and sulfur reductase activities. It contains Fe, Ni, Cu, acid-labile sulfur and hemes b and c but lacks Mo and W. Quinones were required for activity in the *P. brockii* complex but not in the *P. abyssi* complex, suggesting that the complete electron transport chain is contained within the latter complex. Dissimilatory sulfur reductase from *Acidianus ambivalens* is a heterotrimer with a 110-kDa catalytic subunit containing a molybdo-bis-molybdopterin guanine dinucleotide (MGD) cofactor and one Fe-S center, an Fe-S electron transfer subunit, and a membrane anchor. The catalytic subunit contains a twin arginine (Tat) signal peptide sequence, suggesting that it faces the outside of the cell. Mo, but not W, was found in the solubilized membrane. Sulfolobus quinone is used to shuttle electrons from a membrane-bound hydrogenase to the sulfur reductase.

The hyperthermophilic archaeon *P. furiosus* can reduce elemental sulfur when it is separated from the cells by a porous barrier and can use polysulfide as the electron acceptor. *Pyrococcus* and *Thermococcus* differ from *Pyrodictium* and *Acidianus* in that they do not appear to use a membrane-bound sulfur/polysulfide reductase for sulfur respiration, nor do they appear to use quinones or cytochromes as electron carriers. Instead, *P. furiosus* uses a soluble NAD(P)H- and CoA-dependent sulfur reductase whose gene expression increases up to sevenfold when cultures are shifted from growth without elemental sulfur to growth with sulfur. The enzyme is a homodimeric flavoprotein. The mechanism for generating a proton motive force is unknown.

Dissimilatory sulfate, thiosulfate, and sulfite reduction is found in several hyperthermophilic archaea (Table 1). Sulfate is reduced in three steps $\left(SO_4^{2-} \rightarrow APS \rightarrow SO_3^{2-} \rightarrow S^{2-}\right)$ by three enzymes localized in the cytoplasm. The ATP sulfurylase from the hyperthermophilic archaeon *Archaeoglobus fulgidus* is a homodimer and activates sulfate using ATP, yielding adenosine phosphosulfate (APS) and pyrophosphate. This and the ATP sulfurylase from the thermophilic bacterium *Thermus* contain a zinc site that is absent in mesophiles, suggesting that it may be related to thermostability at higher temperatures. Dissimilatory APS reductase from *A. fulgidus* is a heterodimer that contains FAD and two Fe-S clusters. Dissimilatory sulfite reductase from *A. fulgidus* has an $\alpha_2\beta_2$ structure and contains siroheme iron, nonheme iron, and acid-labile sulfide. It uses six electrons to reduce sulfite to sulfide. Our understanding of the source of electrons for these reductases and their relationship with the development of a proton motive force is at a rudimentary level.

Dissimilatory thiosulfate reduction occurs on the membrane producing sulfite and sulfide, and then the sulfite is reduced in the cytoplasm to sulfide as described above. The amount of thiosulfate reductase in the membrane fraction of *P. islandicum* cultures increased dramatically in thiosulfate-grown cultures relative to those grown on elemental sulfur and iron. Like the sulfur/polysulfide reductases described above, the thiosulfate reductase in *P. islandicum* is predicted to be a membrane-bound heterotrimer with MGD and Fe-S cofactors. Dissimilatory sulfite reductase from *P. islandicum* has biochemical properties that are nearly identical to those of *A. fulgidus*.

Reduction of nitrogen compounds

Denitrification is found in a limited number of hyperthermophilic archaea (Table 1). Nitrate is reduced in four steps $\left(NO_3^- \rightarrow NO_2^- \rightarrow NO \rightarrow N_2O \rightarrow N_2\right)$ by four enzymes. In contrast to denitrifying bacteria, all four denitrifying enzymes in the hyperthermophilic archaeon *P. aerophilum* are membrane bound and use menaquinol as an electron donor. Dissimilatory nitrate reductase from *P. aerophilum* is a heterotrimer that consists of a 146-kDa catalytic subunit with an MGD cofactor and one Fe-S center, an electron transfer subunit with four Fe-S centers, and a membrane anchor with biheme b and quinol-oxidizing capability. Like the sulfur reductase in *A. ambivalens*, the catalytic subunit contains a twin arginine (Tat) signal peptide sequence, suggesting that it faces the outside of the cell, which is unlike bacterial nitrate reductases that face the cytoplasm. If so, this would significantly influence the manner in which *P. aerophilum* generates a proton motive force when grown on nitrate. Cultures did not grow without the addition of tungstate; however, concentrations above $0.7\,\mu mol\ l^{-1}$ led to a fourfold decrease in dissimilatory nitrate reductase activity. Therefore, tungsten does not replace molybdenum in this metalloenzyme as it does in other thermophiles but apparently is required by other enzymes in the organism. Variations in tungstate concentrations had no effect on nitrite reductase and NO reductase activities. NO reductase from *P. aerophilum* is homomeric, contains derivatives of heme *b*, and uses menaquinone as an electron donor. Denitrification to N_2O was also measured in *Ferroglobus placidus*.

Oxygen

The majority of thermophiles and especially hyperthermophiles are anaerobes, due in large part to the insolubility of O_2 in water at high temperatures and the lack of fluid contact with O_2. However, there are several organisms that are obligate aerobes, microaerophiles, or facultative anaerobes (Table 1). As expected, these organisms are generally found in geothermal environments such as in hot springs that interface with oxic environments. Aerobic respiration generally requires electrons carried by cytochrome *c* that are passed to O_2 via cytochrome *c* oxidase. Various forms of this enzyme are found in *Aeropyrum* and *Pyrobaculum* whereas quinol oxidases are found in *Sulfolobus* and *Acidianus*. *Pyrobaculum oguniense* has both cytochrome *a* and cytochrome *o* containing heme-copper oxidases.

The bc_1 complex and the cytochrome *o*-containing oxidase are present in the membranes of cells grown aerobically and anaerobically whereas the cytochrome *a*-containing oxidase is only present in aerobically grown cells. The two oxidases have different affinities for O_2 and are specialized for microaerophilic and aerobic growth.

Metal compounds

Two forms of ferric iron are generally used for growth of bacteria and archaea: soluble Fe(III) that is chelated with citrate and insoluble Fe(III) oxide hydroxide (FeO). Several hyperthermophiles grow on FeO whereas only *Pyrobaculum* and *Geoglobus* grow on Fe(III) citrate. Often the end product of FeO reduction is insoluble magnetic iron. *P. islandicum* also can reduce U(VI), Tc(VII), Cr(VI), Co(III), and Mn(IV). Frequent research questions with mesophilic dissimilatory iron-reducing bacteria are whether they are able to reduce FeO without direct mineral contact and whether polyheme *c*-type cytochromes are required. The two most commonly studied iron-reducing bacteria are *Shewanella* and *Geobacter*. Both require polyheme *c*-type cytochromes for iron reduction. *Shewanella* can grow without direct FeO contact by producing an extracellular electron shuttle whereas *Geobacter* requires direct contact unless a soluble mediator is provided. *P. aerophilum* and *Pyrobaculum arsenaticum* can grow without direct FeO contact whereas *P. islandicum* and *Pyrobaculum calidifontis* require direct contact. Genome sequence analyses show that *P. aerophilum*, *P. islandicum*, and *P. arsenaticum* lack polyheme *c*-type cytochromes whereas *P. calidifontis* contains a cytochrome with eight predicted hemes that is highly homologous to those found in *Shewanella* and *Geobacter*. Growth of *P. aerophilum* and *P. islandicum* on Fe(III) citrate and FeO is favored at pHs slightly above neutral and at reduction potentials that are above −220 mV. In contrast, growth of *P. islandicum* on thiosulfate and elemental sulfur is favored at slightly acid pHs and at low reduction potentials (−570 mV). Growth of *P. aerophilum* on nitrate is favored at neutral pH and at reduction potentials above −220 mV.

H_2 production

The anaerobic catabolism of organic compounds often yields low-molecular-weight organic compounds (e.g., acetate) and H_2. Although common, H_2 production (Eo' = −410 mV) by most bacteria is easily inhibited due to their use of NADH (Eo' = −320 mV) as the electron donor for the redox reaction (thermodynamically, the midpoint potential (Eo') of the electron donor should ideally be more negative than that of the electron acceptor). Therefore, this process often requires the presence of a H_2 syntroph such as a methanogen in order to keep H_2 at low partial pressure. In contrast, *Pyrococcus* and *Thermococcus* readily produce H_2 as their primary metabolite when grown in the absence of elemental sulfur, their preferred terminal electron acceptor, without a H_2 syntroph. The electron donor for H_2 production in *P. furiosus* is ferredoxin (*E*m, 95 °C = −471 mV), making the reaction more energetically favorable. The hydrogenase from *P. furiosus* is membrane bound and receives electrons directly from ferredoxin. The reaction is coupled directly with proton translocation across the membrane and the development of a $\Delta\Psi$ and a ΔpH. ATP synthesis on the membrane was likewise shown to be linked to H_2 production. *P. furiosus* also has two cytoplasmic hydrogenases that use NADH as the electron donor, which are upregulated when cultures are grown without sulfur.

RELATIONSHIP BETWEEN ORGANISMS AND THEIR ENVIRONMENT

The high temperatures and geochemistry found in terrestrial and marine geothermal sites are unique. Volcanically derived gases and products from water–rock reactions support chemolithoautotrophic-based microbial communities in what has been termed the deep, hot biosphere. Endolithic microbial communities are pervasive in these environments and likely contribute significantly to subsurface biomass production, which may constitute a significant portion of the total biomass on the planet. The subsurface biosphere is a largely unknown and untapped natural resource. Thermophiles and hyperthermophiles inhabit these environments and serve as model organisms for microbial processes that occur at high *in situ* temperatures. Although known hyperthermophiles may comprise only a small minority of the total microbial population in a geothermal environment, their metabolisms are likely reflections of the kinds of processes occurring within them. Because they are typically not found in nongeothermal background fluids, they can serve as tracers of *in situ* chemical and physical conditions within geothermal environments.

Before one can use these organisms as models of biogeochemical processes in geothermal environments, there are a number of fundamental questions that must be addressed related to the relationship between high-temperature organisms and their environment. For example, what are the physical and chemical constraints on metabolic processes? Are different forms of thermophile and hyperthermophile metabolism spatially and temporally segregated on the basis of fluid chemistry? Clearly, the presence of thermoacidophiles, thermoneutrophiles, and thermoalkaliphiles shows how pH can influence microbial distributions and metabolisms, but can these types of changes be observed on a finer scale even within the same organism? What are the different ways in which organisms assimilate CO_2 or respire a given compound? Are these differences rooted in environmental factors that favor one metabolism over another? Many hyperthermophiles have a requirement for tungsten to meet the needs of certain enzymes found in central metabolic pathways. Are there other unique cofactors used by these organisms? What do these mean with respect to the natural history of these organisms?

In conclusion, extremophiles from hot environments have moved from mere curiosity to a group of organisms that have significant medical and biotechnological applications and are useful for the study of the evolution and biochemistry of metabolic pathways and the biogeochemistry of geothermal environments. Many thermophiles and most hyperthermophiles belong to the Archaea, which is the third superkingdom of life for which there is still much to be learned. Because physiology and ecology go hand in hand, the continued study of high-temperature organisms from these two perspectives should expand our appreciation for these organisms and the function they have in nature.

FURTHER READING

Adams, M. W. W. (1999). The biochemical diversity of life near and above 100 °C in marine environments. *Journal of Applied Microbiology*, *85*, 108S–117S.

Berg, I. A. (2011). Ecological aspects of the distribution of different autotrophic CO_2 fixation pathways. *Applied and Environmental Microbiology*, *77*, 1925–1936.

Brock, T. D. (1978). *Thermophilic microorganisms and life at high temperatures*. New York: Springer-Verlag.

Buchanan, B. B., & Arnon, D. I. (1990). A reverse KREBS cycle in photosynthesis: Consensus at last. *Photosynthesis Research*, *24*, 47–53.

Cabello, P., Roldón, M. D., & Moreno-Vivián, C. (2004). Nitrate reduction and the nitrogen cycle in Archaea. *Microbiology*, *150*, 3527–3546.

Daniel, R. M., van Eckert, R., Holden, J. F., Truter, J., & Cowan, D. A. (2004). The stability of biomolecules and the implications for life at high temperatures. In W. S. D. Wilcock, E. F. DeLong, D. S. Kelley, J. A. Baross, & S. C. Cary (Eds.) *The subseafloor biosphere at mid-ocean ridges* (pp. 25–39). *Geophysical Monograph Series* (vol. 144). Washington, DC: American Geophysical Union Press.

Holden, J. F., & Daniel, R. M. (2004). The upper temperature limit for life based on hyperthermophile culture experiments and field observations. In W. S. D. Wilcock, E. F. DeLong, D. S. Kelley, J. A. Baross, & S. C. Cary (Eds.) *The subseafloor biosphere at mid-ocean ridges* (pp. 13–24). *Geophysical Monograph Series* (Vol. 144). Washington, DC: American Geophysical Union Press.

Kletzin, A. (2007). Metabolism of inorganic sulfur compounds in archaea. In R. A. Garrett, & H. P. Klenk (Eds.) *Archaea: Evolution, physiology, and molecular biology* (pp. 261–274). Malden, MA: Blackwell Publishing.

Petsko, G. A. (2001). Structural basis of thermostability in hyperthermophilic proteins, or "there's more than one way to skin a cat". *Methods in Enzymology*, *334*, 469–478.

Schäfer, G., Engelhard, M., & Müller, V. (1999). Bioenergetics of the Archaea. *Microbiology and Molecular Biology Reviews*, *63*, 570–620.

Schleper, C., Jurgens, G., & Jonuscheit, M. (2005). Genomic studies of uncultivated Archaea. *Nature Reviews*, *3*, 479–488.

Stetter, K. O. (1990). Extremophiles and their adaptation to hot environments. *FEBS Letters*, *452*, 22–25.

Verhees, C. H., Kengen, S. W. M., Tuininga, J. E. *et al.* (2003). The unique features of glycolytic pathways in Archaea. *Biochemistry Journal*, *375*, 231–246.

Woese, C. R., Kandler, O., & Wheelis, M. L. (1990). Towards a natural system of organisms: Proposal for the domains Archaea, Bacteria, and Eucarya. *Proceedings of the National Academy of Sciences of the United States of America*, *87*, 4576–4579.

Wood, H. G., & Ljungdahl, L. G. (1991). Autotrophic character of the acetogenic Bacteria. In J. M. Shively, & L. L. Barton (Eds.) *Variations in autotrophic life* (pp. 201–250). New York: Academic Press.

Chapter 23

Cold Environments

J.W. Deming
University of Washington, Seattle, WA, USA

Chapter Outline

DEFINING STATEMENT

Earth today is a cold planet, with over 80% of its biosphere at temperatures of ≤5 °C and 10–20% of its surface frozen. Widely diverse microbes have evolved specific molecular, cellular, and extracellular adaptations to enable their essential roles in the biogeochemical cycles of the planet.

DISCOVERY OF COLD-ADAPTED MICROBES

Earliest Observations and Terminology

The earliest known report of microbial life in a cold environment dates back to the fourth century BC and the writings of Greek philosopher Aristotle who made observations of what later proved to be photosynthetic Eukarya ("red algae") that turned snow to a reddish color. More than two millennia later in the nineteenth century (1887), the German scientist Forster described the ability of a bioluminescent bacterium, derived from fish preserved in the cold, to reproduce at 0 °C. In the twentieth century (1902), the term "psychrophile" for cold loving was introduced by Schmidt-Nielsen to describe such microorganisms.

Over the next 60 years, the term psychrophile continued to be in use to describe cold-adapted microbes according to the ability to reproduce in the cold, regardless of the upper temperature limit for growth. That limit, with few exceptions, fell above room temperature, thus overlapping with the thermal category of mesophiles. These microbes today are called psychrotolerant (or psychrotrophic), with the descriptor psychrophilic reserved for organisms that fit the more precise definition provided by the American marine microbiologist Morita (1975), based on cardinal growth temperatures: minimal temperature (T_{min}) of 0 °C or lower, optimal temperature (T_{opt}) of 15 °C or lower, and maximal temperature (T_{max}) of about 20 °C (Figure 1). By this definition, which provided much needed clarity at the time and remains in wide use today, the first true psychrophile was discovered by Tsiklinsky as part of a French expedition to Antarctica (1903–1905). Another half-century would pass before the abundance and functional roles of true psychrophiles in cold environments would be appreciated. First, microbiologists had to learn about the sensitivity of cold-adapted microbes to room temperature (even to unchilled pipets) during isolation procedures. Eventually, after many enrichment studies using 4–6 °C (convenient refrigerator temperature) and reports of predominantly psychrotolerant isolates, came the understanding that the temperature of initial enrichment influences the thermal nature of the resulting isolates: enrichments near the freezing point (e.g., at −1 °C for marine samples) are more likely to yield psychrophilic than psychrotolerant bacteria.

Morita expected that future adjustments to his definition of psychrophily, based on new isolates obtained by paying attention to detrimental (and influential) temperatures during isolation procedures, would be in the direction of

Topics in Ecological and Environmental Microbiology.

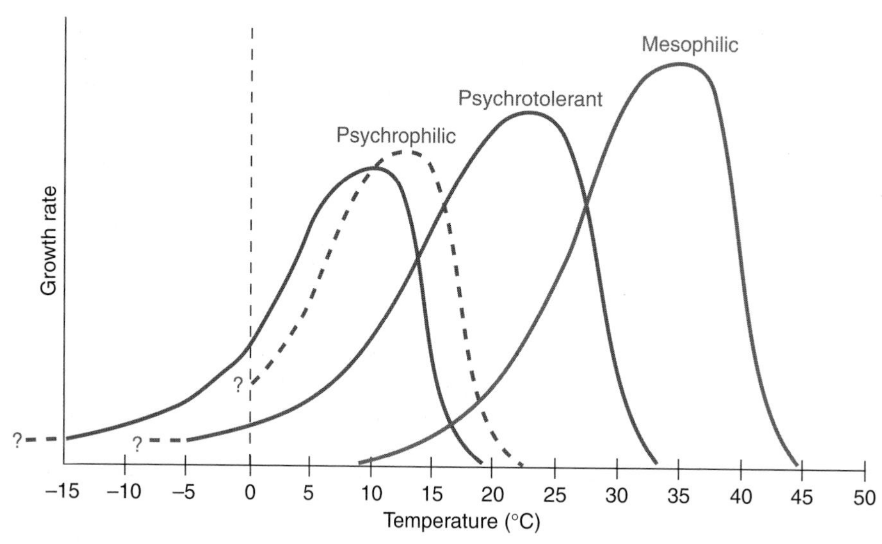

FIGURE 1 Schematic depiction of bacterial growth rate as a function of temperature (at atmospheric pressure) for a psychrophilic, psychrotolerant, and mesophilic microbe. The dotted line depicts the psychrophilic response when grown under elevated hydrotstatic pressure.

lower cardinal temperatures. Microbes with considerably lower cardinal growth temperatures, called extremely psychrophilic, have been reported in the last decade, particularly the current record holders for low-temperature growth, *Psychromonas ingrahamii* (T_{min} = −12 °C, T_{opt} = 5 °C, T_{max} = 10 °C) and *Colwellia psychrerythraea* strain 34 H (T_{min} = −12 °C, T_{opt} = 8 °C, T_{max} = 18 °C) from subzero Arctic sea ice and sediments, respectively. Measurements of microbial metabolic activity at very cold temperatures (down to at least −20 °C) in natural ice formations, where measuring a reproductive rate is methodologically challenging, has led to introduction of the term eutectophile for the most cold-adapted microbes, living in ice near the eutectic temperature with only nanometer-scale films of liquid water available.

Because cardinal growth temperatures cannot fully capture the adaptability of a microorganism to its environment and are unavailable for the vast majority of microbes in nature that evade cultivation, the term psychrophilic (and psychrotolerant) remains most useful for categorizing cultured members of the domains of Bacteria and Archaea. Cold-adapted is used very generally to describe microorganisms, cultured or uncultured, that express a recognizable adaptation to low temperature, while cold-active is often reserved for enzymes and viruses with catalytic or infective activity at low temperature. Other terms based on cardinal growth temperatures are recently in play, particularly stenopsychrophilic and eurypsychrophilic (comparable in operational meaning to psychrophilic and psychrotolerant, respectively), in an attempt to recognize that psychrotolerant microbes are not simply tolerant of the cold (some mesophiles and thermophiles can tolerate the cold) but also adapted to it, in spite of higher T_{max} for growth than psychrophiles.

What constitutes a cold temperature is also subject to perspective. Combining cardinal growth temperatures for psychrophiles and some key environmental temperatures yields the following set of descriptors. Moderately cold is 15–5 °C; that is, from the upper T_{opt} for psychrophilic growth, which is also the average temperature of the surface of the Earth, down to the (upper) temperature of 80% of the biosphere. Cold is 5 to −2 °C, the approximate freezing point of seawater, while very cold is below −2 °C. Extremely cold is below −12 °C, the current lowest T_{min} for microbial growth in culture and the temperature of the Earth's near-million year-old permafrost in polar regions. Cold in the term cold adaptation remains broadly defined by any temperature 15 °C or lower.

Exploration of the Cold Deep Sea

The cold deep sea, as the volumetrically dominant and most persistently cold environment on Earth (over geologic time), has provided an important natural laboratory for studying and advancing understanding of cold adaptation in the microbial realm. Although its temperature is always above the freezing point of seawater and thus not as thermally extreme as most frozen environments, the cold deep sea is considered extreme for other reasons: its elevated hydrostatic pressure, which increases linearly at 10 atm (=1 MPa) per 100 m increase in water depth, and its typically oligotrophic state. The search for psychrophiles in the cold deep sea has often been coupled with the search for barophiles (also known as piezophiles), pressure-adapted microbes that grow more rapidly under elevated hydrostatic pressures than at atmospheric pressure when adequate nutrients in the form of organic substrates are available. The focus of these searches has usually been the heterotrophic bacteria.

Scientific exploration of oceanic life, in general, began with a series of deep-sea expeditions at the end of the nineteenth century (study of the productive surface layers and sea ice would come later). Until the discovery of deep-sea hydrothermal vents toward the end of the twentieth century (in 1977), the deep ocean was understood to be uniformly cold, the temperature not exceeding about 5 °C (except in the deep Mediterranean and Sulu seas where temperatures reach about 15 °C). Yet, early expeditions had no facilities for incubating samples shipboard at such low temperature. The first opportunities to discover cold-adapted microbes from the vast, cold deep sea were thus thwarted by lack of refrigeration. Remarkably, the early French explorer Certes was able to test for pressure-adapted microbes from the deep sea in the 1880 s, even if cold adaptation was beyond reach.

More than a half-century later, American marine microbiologist ZoBell began the study of deep-sea microbes under both *in situ* pressure and temperature, documenting with Morita in 1957 the first psychrophilic barophiles. Although these cultures were lost to future study, similar efforts by several groups beginning from the 1970s eventually yielded sizeable culture collections of psychrophilic barophiles; indeed, all barophiles cultured from the cold deep sea are also psychrophilic. The synergistic effects of temperature and pressure on microbial growth (or other activities) have not been fully explored, but for several psychrophilic bacteria their cardinal growth temperatures can be shifted upward by incubating under higher pressure (Figure 1). When the growth responses of deep-sea bacteria have been examined according to a matrix of three parameters – temperature, pressure, and salinity – salt concentration is also observed to shift cardinal growth temperatures, though the direction of the shift is variable and strain dependent. Microbiological exploration of the cold deep sea has thus raised general awareness that cold adaptation must be understood in relation to other parameters and not exclusively to temperature.

Not least of other parameters is the concentration of available energy sources or organic substrates in the case of the heterotroph. For heterotrophic psychrophilic barophiles the barophilic trait depends upon available substrate concentration: at low substrate concentrations, growth rate is similar under both atmospheric and elevated pressures (barotolerant, within the pressure range for growth), while at high substrate concentrations, growth rate is higher under elevated pressures than at atmospheric pressure (barophilic). When substrates are in adequate supply in the cold deep sea, for example, from the hydrolysis of freshly deposited organic detritus, psychrophilic barophiles will outcompete other cold-adapted microbes that may be present. Genetic work with pressure-adapted psychrophiles indicates that membrane proteins involved in substrate uptake are upregulated under elevated pressures when provided with sufficient substrate.

Somewhat analogous work with marine psychrophiles from shallow cold waters, considering only temperature and substrate concentration, indicates that the lower the temperature, the higher the required substrate concentration to achieve a comparable growth (or oxygen respiration) rate. Low temperature or viscosity-driven reduction of the diffusive flux of various solutes to the cell is insufficient to account for this increase in the required substrate concentration threshold. Resolving the issue is important to understanding why, at least at times, the microbial role in biogeochemical cycles of surface polar waters appears limited until sufficient organic solutes accumulate. Although gene expression work that may help to explain this phenomenon remains to be conducted, the recent whole-genome analyses of several heterotrophic psychrophiles reveals an apparent ability to store substrate reserves intracellularly, a potential means around the problem for the individual cell. Also awaiting a gene-based explanation is the repeated observation that cold temperature favors higher bacterial growth efficiency: a greater fraction of the organic carbon consumed in the cold goes to new biomass (gain to the food chain) than to respiration (loss as carbon dioxide). Warming temperatures in polar waters are thus expected to shift the role of cold-adapted microbes in the cycling of organic carbon to increased remineralization to carbon dioxide.

Exploration of Other Low-Temperature Environments

Many other cold environments have been explored over the past several decades for cold-adapted microbes. Moderately cold to cold environments include a wide range of aquatic and sedimentary environments, both marine and freshwater (rivers, lakes, and subsurface aquifers), terrestrial soil environments of varying degrees of desiccation, and numerous surfaces that support microbial biofilms, from moist rocks in caves and mines to fluid-bathed tissues of vertebrate and invertebrate animals that live in the cold. Unlike the cold deep sea, many of these environments experience fluctuating temperatures (and other parameters) on a seasonal and diurnal basis. Exceptions with stably cold temperatures include animal-associated microbial environments, where the animal host spends its life history in cold water, as well as the many deep subglacial lakes of Antarctica (over 150 have been discovered) with stable temperatures just above the freezing point, given separation from the lower atmosphere and solar radiation by kilometers of glacial ice. With the possible exception of subglacial lakes, which await direct exploration (drilling efforts have only reached ice accreted above the water line), all of these cold environments have in common their successful colonization by cold-adapted microbes, which then play active if not dominant roles in cycling the inorganic and organic materials within them.

Based on classical chemostat studies pitting cold-adapted microbes with different cardinal growth temperatures against each other at constant low temperature, the more psychrophilic organisms are the dominant players.

In contrast are the environments that experience very cold to extremely cold temperatures, including the upper atmosphere where viable microbes have been found associated with microscopic particles and, of course, the major types of ice formations on Earth – freshwater snow and glacial ice (formed from long-term compaction of snow), lake and river ice, polar sea ice, and frozen soils. Except for sea ice (see below) and possibly lake ice (understudied), frozen environments in general can be viewed as preserving an often cosmopolitan suite of microbes, largely inactive in the cold, rather than as actively colonized by cold-adapted microbes with all the attendant successional and adaptive responses. At extremely cold temperatures, the primary limitation to the latter scenario is the absence of sufficient water in the liquid phase. All ice formations on Earth derive from source waters containing impurities of one kind or another that depress the freezing point (especially inorganic salts), so they retain at least some liquid water. Only those that may drop below their eutectic temperature, for example, high-altitude glacial ice on Antarctica, become completely desiccated.

Some of these frozen environments experience seasonal and/or diurnal temperature swings, which intermittently relieve the limitation of insufficient liquid water. They can then support highly productive microbial ecosystems. An example is sea ice, which during spring and summer seasons, with near-continuous sunlight and seawater flushing its base with nutrients, develops algal and microbial communities visible to the naked eye as strongly discolored ice (Figure 2). These communities contribute 25% or more of total primary production in the Arctic with consequent effects on secondary production (the transformation and consumption of this biomass) and higher trophic levels. Sea ice (like lake and river ice), however, is not a stable environment, melting by late summer before reforming in fall. An exception is multiyear sea ice in the Arctic Ocean that until recently could survive 8–10 melting seasons before circulating out of the Arctic into melting Atlantic waters. Climate-driven declines in multiyear (and first-year) sea ice, which had been averaging about 10% per decade since satellite coverage began in the 1970s, recently accelerated beyond all model predictions. This particular frozen environment may soon represent a lost opportunity for the study of cold adaptation.

The upper layers of soil in alpine and polar regions also experience regular and wide fluctuations in temperature

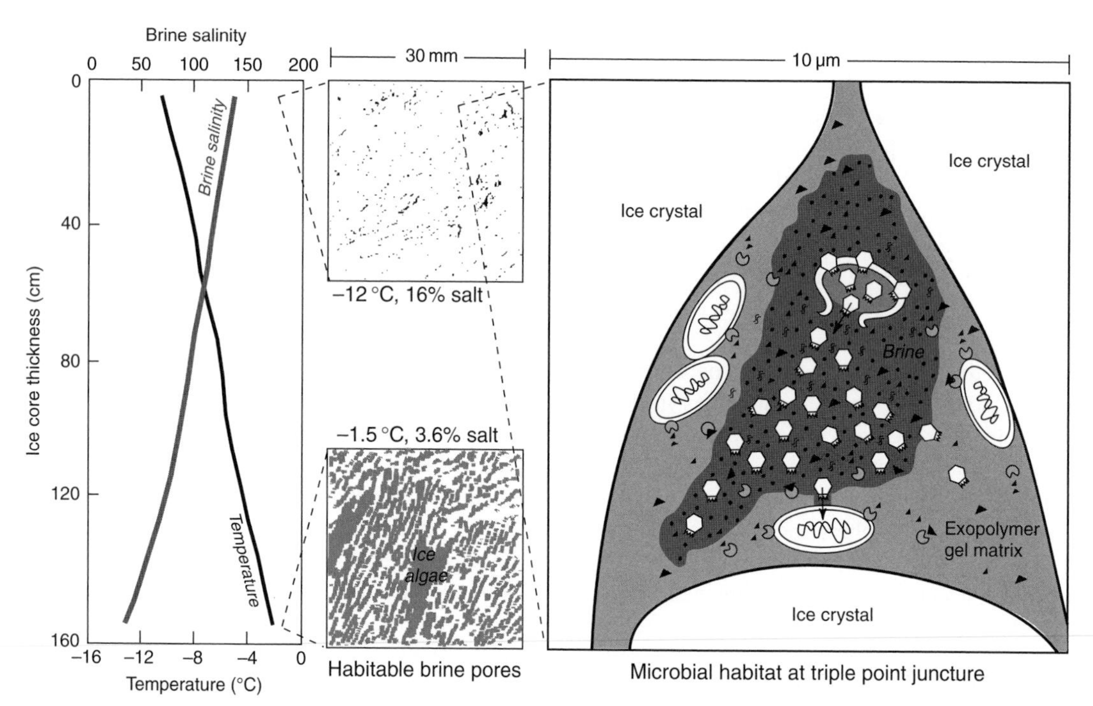

FIGURE 2 Schematic depiction of some characteristics of sea ice. At left are vertical gradients, in temperature and in salinity of liquid inclusions, that develop in winter as temperature of the overlying atmosphere drops but underlying seawater remains near freezing point. Middle panels depict relative size of brine pores in a very cold section of ice versus bottom ice with larger channels flushed by seawater that will support an ice-algal bloom in spring. At right is an enlarged schematic of very cold brine at the juncture of three ice crystals, depicting microbes embedded in a gelatinous matrix of exopolymers, brine, and organic substrates concentrated in the interior, and extracellular enzymes hydrolyzing the substrates. Also shown are proposed cold-active viral enzymes at work, successful infection and viral reproduction, lysis of host and release of free DNA and new viruses, potential agents of horizontal gene transfer.

seasonally, from warm (>20 °C) to extremely cold, as well as climate-driven warming. Even during moderately cold periods, microbial activity increases such that the environment becomes a source of greenhouse gases like carbon dioxide and methane, rather than a sink. In alpine soils, an insulating snow cover promotes substantial microbial activity through the winter. The deeper frozen layers (>50 m) of polar soils removed from atmospheric and solar influences, however, have been permanently frozen (permafrost) in the temperature range of −10 to −12 °C for close to a million years in some Siberian locations. The typical diversity of soil microbes that have been recovered in culture from permafrost, including aerobes, anaerobes, heterotrophs, sulfate reducers, and methanogens, may represent some of the oldest viable forms available to study. Wedged between deep layers of permafrost are cryopegs, recently discovered lenses of very old unfrozen water kept liquid by high salt concentration. Whether these very cold brines are life preserving or actively colonized remains to be determined.

In the upper atmosphere, high-altitude glacial ice and snow that covers Greenland and Antarctica, and Arctic winter sea ice (and its overlying snow), microbes experience extremely cold temperatures, sometimes approaching or reaching the eutectic of −55 °C (for seawater). The thermal gradients inherent to glacial and sea-ice environments (Figure 2) provide natural laboratories to examine the question of the lower temperature limits for microbial growth, activity, and survival. To date, studies of such environments and of artificially produced ices at extreme temperatures suggest that cellular reproduction may be limited to about −20 °C, metabolic activity to about −40 °C, and survival to the lowest temperatures yet to be tested (−196 °C in liquid nitrogen). These general guidelines, however, are subject to change, especially as the field of Astrobiology stimulates increased experimentation under extremely cold conditions.

Early in the study of this varied array of low-temperature environments, a general paradigm emerged: stably cold environments tend to support a greater (culturable) community of psychrophilic microbes, while those with temperatures that fluctuate, especially above the T_{max} of psychrophiles, tend to support a greater community of psychrotolerant (eurypsychrophilic) microbes. The implication was that psychrophily requires a stably cold environment to evolve. Although this paradigm appeals to common sense, much of the early data supporting it relied upon enrichment temperatures that would have favored psychrotolerant microbes. When more stringent enrichment conditions are used (e.g., −1 °C for saline environments), temperature-fluctuating environments that previously yielded mainly psychrotolerant isolates yield a predominance of psychrophiles instead. Psychrophiles have also been observed to reestablish dominance in an environment in a relatively short period of time (days), once fluctuating temperatures have stabilized at a cold temperature. Furthermore, the sea-ice environment, which has always consistently yielded greater numbers of psychrophiles, is an ephemeral one, with inhabitants released during the summer melt period to seawater that then warms under 24 h solar radiation. From initial encasement during ice formation in fall through the winter, spring and summer seasons, sea-ice microbes can experience a temperature fluctuation of more than 40 °C, from −35 in winter to 6 °C (or higher) in melt water. The corresponding fluctuations in other parameters, especially salt concentration (Figure 2), add osmotic and other stresses to the thermal swing. Although psychrophiles have long been considered the more sensitive of cold-adapted microbes, in large part because some express narrow temperature ranges for growth (unlike that shown in Figure 1), environmental robustness is a matter of perspective influenced by considering other parameters. What constitutes an ideal habitat for cold adaptation needs to be reconsidered.

Influence of Astrobiology

The ongoing and planned exploration of other bodies in our solar system for evidence of past or present life opens a new chapter in the history of studying cold-adapted extremophiles. Even the extremely cold environments of Earth are thermally moderate relative to the extraterrestrial sites targeted for exploration. For example, the average surface temperature of the desiccated soils on Mars is about −55 °C, coincidentally the eutectic temperature for Earth's seawater. Where the surface temperature is warmer, not even nanometer-scale films of liquid water are available and radiation intensity (in the absence of a protective atmosphere as on Earth) would be prohibitive. The average temperature of the deeply frozen surface of Europa, the Jovian moon believed to harbor beneath its ice cover a global ocean larger than Earth's, is about −160 °C. Such extremely cold ice formations are not found on Earth, invoking the need to study forms of ice generated in the laboratory. Initial studies of the reactions of a known psychrophile to flash freezing, as might be experienced when Europan seawater rises into cracks of its extremely cold ice cover, suggest the importance of organic (sugar-rich) exopolymers in buffering cells against fatal damage from ice crystals and even enabling the completion of enzymatic reactions begun prior to freezing. Because the presence of exopolymers is also known to alter the microstructure of saline ice in readily detectable ways, similar effects in extraterrestrial ices may constitute a recognizable biosignature.

Although space missions first access only extremely cold surfaces, the subsurface environments of both Mars and Europa, hidden from damaging radiation, hold greater promise for life. They are expected to be more moderate in temperature, favoring liquid water (perhaps in analogy to the brine layers of deeply buried permafrost), and to offer potential energy sources for chemolithotrophic life forms, for example, via the exothermic water–rock reaction known as serpentinization. In Earth's deep sea at a mid-Atlantic site called Lost City, serpentinization is known to yield hydrogen and methane in support of luxurious archaeal biofilms and mats. The study of cold-adapted Archaea and chemolithotrophs in general is in its infancy, relative to the heterotrophic bacteria.

EVOLUTION OF COLD ADAPTATION

Glaciation Periods on Earth

The temperature of the early Earth and its ocean is actively debated, but marine geological evidence points to a very hydrothermally active and thus warm ocean. The first hypothesized period of planetary-scale chilling or glaciation does not occur until about halfway through Earth's history about 2.2 billion years ago during the Proterozoic era. Less than a billion years ago, Earth is believed to have experienced a severe freezing episode resulting in what has been called "Snowball Earth." Since then, the planet has experienced a series of glaciation events, not always global in scale, eventually leading to the glacial/interglacial periods of recent Earth history. Their periodicity is estimated in tens of thousands of years. In between each major ice age, the planet is believed to have been completely free of ice with an average temperature above that permissive of a psychrophilic life style.

Unless the deep sea remained sufficiently cold to provide refuge to a stock of psychrophiles, a difficult hypothesis to test, psychrophilic microbes likely evolved more than once during Earth's history. The implication is that the evolutionary steps between psychrotolerance and psychrophily must be accommodated by the time available between glaciation periods. In addition to vertical gene transfer from an ancestor (inherited beneficial gene mutations), horizontal gene transfer (e.g., mediated by viruses) may have played important roles in achieving these steps. Leading the way to tests of this hypothesis are phylogenetic analyses of extant microbes, cultured and uncultured, comparative genomic evaluations of psychrophiles and other thermal classes of microbes, and experimentation with horizontal gene transfer in the cold.

Phylogeny of Cold Adaptation

The now classic 16S rRNA gene-sequencing approach to deducing relationships among organisms yields a universal tree of life on which known psychrophilic and psychrotolerant microbes can be located. Use of a phylogenetic tree originally designed to highlight hyperthermophilic genera of Bacteria and Archaea (those that grow at 90 °C or higher) for this purpose emphasizes the late arrival of cold adaptation among extant organisms (Figure 3). It also indicates the slightly deeper branching of groups containing only psychrotolerant members, reinforcing the expectation that psychrophiles evolved from psychrotolerant strains.

Most of the major branches within the domain of Bacteria, except those unique to thermophiles, contain psychrophilic members, including all five groups of the *Proteobacteria*,

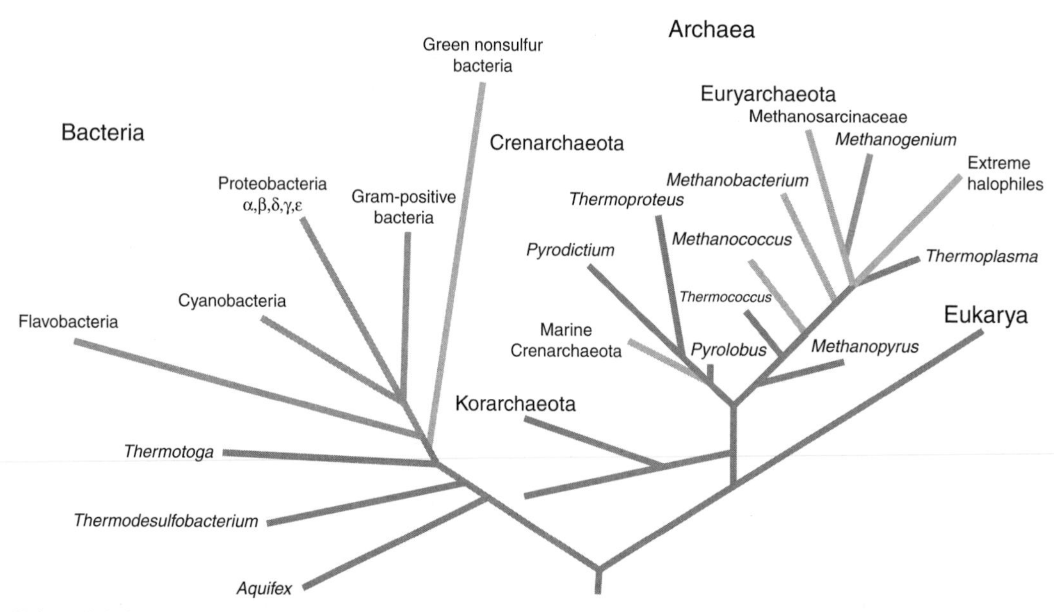

FIGURE 3 Universal phylogenetic tree of life based on 16S rRNA sequences, emphasizing the domains of Bacteria and Archaea. Orange branches indicate hyperthermophiles that grow at ≥90 °C; purple branches, groups that contain known (cultured) psychrotolerant strains; and blue branches, groups that contain known psychrophiles. Note that the (uncultured) marine Crenarchaeota are colored purple because degree of cold adaptation is not known. (For interpretation of the references to color in this figure legend, the reader is referred to the Web version of this chapter.).

aerobes and anaerobes alike, the *Cytophaga-Flavobacteria*, the *Cyanobacteria*, and the Gram-positive bacteria. Some genera of the gamma-Proteobacteria, in particular *Moritella* and *Colwellia*, are comprised mainly or exclusively of psychrophiles. The Green nonsulfur bacteria, however, contain only psychrotolerant isolates so far.

In the domain of Archaea, only a single culture-authenticated psychrophile is known, the methanogen *Methanogenium frigidum*, isolated from an Antarctic lake. Its position within the archaeal domain of the tree also indicates a later evolutionary arrival (Figure 3). The marine Crenarchaeota have a somewhat earlier branching position. Members of this archaeal group often dominate numerically in the cold deep sea and occur in many of the polar waters and sediments that have been examined, but only one of their members has been brought into culture (from an aquarium sample enriched at room temperature). It is not cold-adapted. A well-studied crenarchaeal symbiosis with a sponge host that dwells in cold waters clearly suggests cold adaptation, but whether psychrophilic or psychrotolerant awaits a cultured isolate. Because the marine Crenarchaeota that inhabit the cold deep sea are believed to be involved in the nitrogen cycle, especially the process of nitrification (microbial oxidation of ammonia to nitrite and nitrate) which generates the inorganic nitrogen required by primary producers in surface waters, they are targets of intensive study.

Given that the study of cold adaptation in the microbial realm has historically centered on heterotrophic bacteria, in large part because these organisms are more readily brought into culture than chemolithotrophs or Archaea in general, conclusions from the depicted phylogenetic tree (Figure 3) should be drawn with caution. As more microbes are brought into culture and shown to be cold-adapted, the branching patterns evident today may change. Stable isotope (and other) probing techniques that allow recognition of microbial activity under different temperatures in the absence of cultivation, yet coupled to a sequencing identification, may also bring new information to the tree.

Genetic Mechanisms

Genetic mutation as a means to cold adaptation is evident from studies of the molecular interactions inferring enzymes with catalytic ability in the cold and in comparative analyses of whole-genome sequences from related organisms with different cardinal growth temperatures. In the former case, site-directed mutagenesis and related approaches indicate that, depending on the enzyme and often its size, anywhere from a single amino acid change to numerous amino acid substitutions or chemical alterations can explain the gain (or loss) of cold activity. In the latter case, and despite an oft-cited idea that only a critical subset of an organism's enzymes need be cold-active for it to function as a psychrophile, results of comparative genome studies for both Archaea and Bacteria suggest otherwise. Significant amino acid replacements were observed in over 1000 genome-deduced proteins from the psychrophilic methanogen, *M. frigidum*, relative to proteins from other methanogens covering a range of temperature growth optima, from 15 to 98 °C. When the whole proteome of *C. psychrerythraea* 34H and three-dimensional architectures of its proteins were compared to nearest mesophilic neighbors with available genome sequences, the same general observation was made. In both cases, the interactions and locations of polar and charged amino acids (in particular serine, histidine, glutamine, and threonine) in protein tertiary structures appeared influential in imparting psychrophily.

In other analyses of the whole genomes from the several psychrophiles that have been sequenced so far, specific proteins and other features known from previous culture work to impart cold adaptation (discussed below) have been documented. A new genomic finding, not widely evident from prior culture work, is the prevalence of virally delivered genes in psychrophiles along with the presence of prophage (viruses residing benignly in the bacterial genome and thus implicated in gene transfer). That virally mediated horizontal gene transfer has played an important role in the evolution of psychrophily appears inescapable, although other forms of gene transfer also need to be considered (transformation by direct uptake of free DNA and plasmid exchange by conjugation between cells; Figure 4), particularly since the transfer of plasmid DNA via conjugation between mesophilic and psychrophilic bacteria has been demonstrated.

In the same time frame that genomic analyses of psychrophiles have become available, the study of cold-active virus–host systems in culture has been renewed. While work over a half-century ago had documented infectivity at 0 °C, the environmental and evolutionary implications had only rarely been pursued. Recently, several promising cold-active virus–host systems have been obtained by a number of researchers and the lower (known) temperature limit for infectivity has been pushed to −12 °C (and 16% salt) in simulated sea-ice brines. The goal of demonstrating active horizontal gene transfer under realistic environmental conditions, whether or not mediated by viruses, as means to evolve cold adaptation remains for the future.

MOLECULAR BASIS FOR COLD ADAPTATION

As the temperature of an aqueous solution decreases, for example, from 37 to 0 °C, its viscosity more than doubles, slowing solute diffusion rates, while chemical reaction rates (also influenced by viscosity) decrease exponentially. The cytoplasmic membranes and enzymes of mesophilic microbes tend to rigidify under these conditions and lose

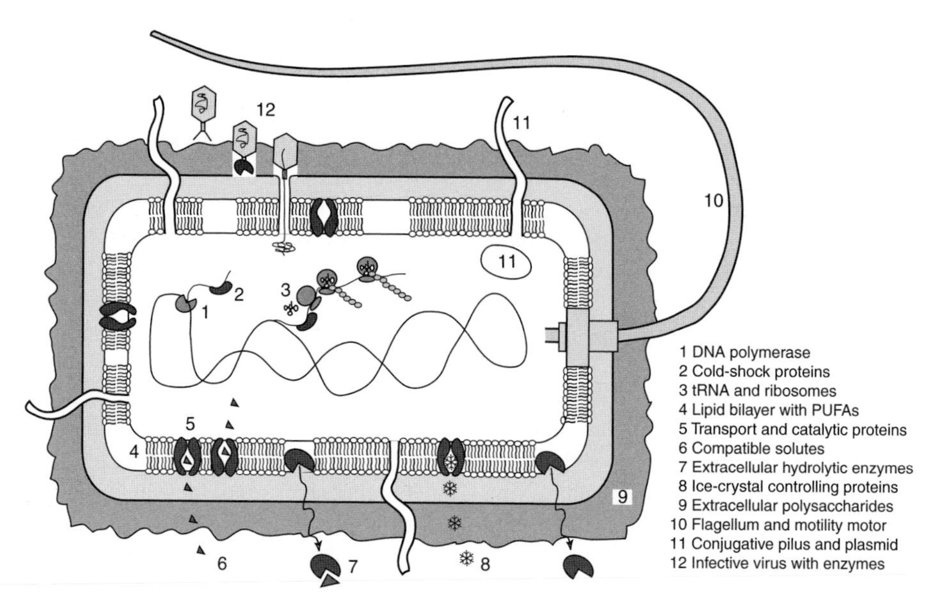

FIGURE 4 Simplified schematic of a bacterial cell depicting some of the known components and processes linked to cold adaptation (see text).

function. Protein folding is impaired and nucleic acids assume secondary or super-coiled structures that interfere with their proscribed activities. A hallmark of microbial cold adaptation at the molecular level is thus to retain sufficient flexibility in its macromolecules such that essential functions can go forward in spite of the challenging effects of low temperatures. When arguably nonessential functions, for example, motility, can also go forward, the microbe becomes even more competitive in the cold. The lower temperature limit for bacterial motility, −10 °C (in a high sugar solution), is held by an extreme psychrophile, *C. psychrerythraea* 34H. Enough direct research on psychrophiles and comparative studies of psychrotolerant and mesophilic bacteria has been accomplished in recent decades to identify key components and aspects of a cell that impart cold adaptation (Figure 4).

Membrane Fluidity

The cellular membranes of cold-adapted microbes must remain flexible enough under rigidifying temperatures to facilitate their essential functions in the transport of nutrients and metabolic byproducts and the exchange of ions and solutes critical to maintaining intracellular integrity. The cold-adapted cell accomplishes this feat by fine-tuning the composition of its membrane lipid bilayer (Figure 4), introducing steric hindrances that change the packing order of the lipids or reduce interactions between them and other membrane components. The genome must encode the ability to produce a flexible membrane in the first place (adaptation), as well as to adjust membrane components on the short-term (acclimation). The list of specific alterations known to enhance flexibility in the cold is extensive, if sometimes strain specific, but typically includes producing a higher content of unsaturated (fewer double bonds between carbon atoms), polyunsaturated, and branched and/or cyclic fatty acids in response to cold. In some cases, shortening the length of the fatty acid can enhance flexibility, while in others, changing the content or size of the lipid head groups helps. Genes for enzymes involved in polyunsaturated fatty acid (PUFA) synthesis are clearly present in psychrophilic genomes, although without tests of their cold-active nature, such findings are not definitive for cold adaptation (some mesophiles also contain them).

Other components of the lipid bilayer include membrane proteins and carotenoid pigments. The manner in which the interaction of membrane proteins with lipids affects both the active and passive permeability of the membrane is important for controlling the exchange of ions and organic solutes (Figure 4). The genome of *C. psychrerythraea* 34H contains numerous gene families involved in the transport of compatible solutes, low-molecular-weight organic compounds (often sugars or amino acids and their derivatives) that accumulate to high intracellular levels under osmotic stress, and are compatible with the metabolism of the cell. They help to maintain cellular volume and turgor pressure and protect intracellular macromolecules in the face of changing salt concentrations exterior to the cell, as occurs in sea-ice brines when the temperature drops (Figure 3). The cold-adapted membrane thus needs to be flexible, especially in very cold environments, to permit both the uptake of energy-yielding substrates and compatible solutes, yet impermeable enough to prevent excessive passive exchange; adjusting the protein components of membranes can help. The sensitivity of some membrane proteins to temperature-driven conformational changes appears to provide a thermal sensor that results in the upregulation of genes involved in subsequent membrane adjustments, making life in a temperature gradient imminently feasible. As temperatures approach T_{max} for growth, cold-adapted microbes must also be able to keep their membranes sufficiently stable (inflexible) to avoid cell leakage and death by lysis. In some cases, adjusting pigment content appears to accomplish this goal.

Cold-Active Enzymes

The exponential drop in chemical reaction rates brought on by decreasing temperature highlights the impressive evolutionary development of all manner of enzymes with high catalytic activity in the cold. Complementing the known membrane adjustments to achieve flexibility in the cold are those expressed by cold-active enzymes, including both essential intracellular enzymes required for nucleic acid and protein synthesis in the cell, membrane permeases for active solute transport and alteration, and extracellular enzymes released to perform hydrolytic functions in the environment (Figure 4). As with achieving a more flexible membrane against the cold, high enzymatic activity at low temperature involves creating greater molecular flexibility than that observed in enzymes active at warm temperatures. It also involves trade-offs; although not a firm rule, for many cold-active enzymes the very traits that impart cold activity also make them unstable at higher temperature. Unlike membrane adjustments, enzyme adaptations over an evolutionary time scale appear more relevant to cold activity than means for short-term acclimation. The strategies for increased flexibility of an enzyme leading to cold activity are numerous but not uniform. In some cases, increased flexibility is linked to a shift in primary structure (amino acid composition) of the entire protein, while in others only direct adjustments to the catalytic site of the three-dimensional macromolecule are involved. For some enzymes, including those released from the cell to perform hydrolytic functions in the environment, adjustments to flexibility are detected in the regions of the protein exposed to the solvent. Keeping an exterior shape firm as temperature drops can translate to keeping a more protected catalytic site flexible.

Considering the interface between the exterior shape of an enzyme and the solvent raise the need to consider enzyme interactions with other components in the environment, independently of the cell. For example, the extracellular polysaccharides released by *C. psychrerythraea* 34H have been observed to stabilize an extracellular protease (that it also produces) against thermal denaturation. The effective work of extracellular enzymes in the cold, for example, in hydrolyzing organic substrates to a size that can be transported into the cell, may be an important trait of heterotrophic psychrophiles. Over half of the enzymes assigned to the degradation of proteins and peptides in the *C. psychrerythraea* genome are predicted to be localized external to the cytoplasm, among the highest percentage in any completed genome (from all thermal classes). The successful infection (and reproduction) of a virus in this same psychrophile at very cold temperatures (−10 to −13 °C) suggests that at least some viruses in cold environments may carry highly cold-active enzymes for penetrating the cell membrane (Figures 3 and 4).

In spite of multiple strategies to achieve enough flexibility for catalytic activity in the cold, intracellularly or extracellularly, some common trends have emerged from enzyme studies regarding specific chemical modifications required to reduce the strength or number of otherwise stabilizing factors for a protein. These include reducing ion pairs, hydrogen bonds and hydrophobic interactions, inter-subunit interactions, cofactor binding, and proline and arginine content. Increasing exposure of apolar residues to the solvent, accessibility to the active site, and the clustering of glycine residues also pertain. Such trends provide means to search both available and future genomes for signs of cold adaptation. They also provide blueprints to identify or engineer proteins for applied uses in the cold, many of which have been identified by the food, detergent, and biotechnology industries, by those seeking means to remediate the contamination of cold environments, and by start-up companies interested in the possible production of cost-effective alternatives to fossil fuels that take advantage of enzymatic hydrolysis in the cold.

Cold-Shock Proteins

For the cold-adapted microbe, all nucleic acids and proteins involved in maintaining (if not synthesizing), transcribing, and translating genetic information intracellularly must be able to function in the cold. In some cases, this feat is thought to be accomplished not by primary alteration of the macromolecule itself, as already described, but by production of specific proteins that bind to them, presumably enabling proper conformation and flexibility, including required periods of destabilization. The production of cold-shock proteins to serve a similar function when mesophiles are subjected to a temperature downshift is well studied, but less is known about related responses of cold-adapted microbes to downshifts in temperature or to continuous life in the cold. Available information on psychrotolerant bacteria indicates that large numbers of cold-shock proteins (related to those in mesophiles) are always present and that production of cold-acclimation proteins is continuous in the cold, as is the expression of housekeeping genes (for basic cellular functions). By contrast, the mesophile *Escherichia coli* carries few cold-shock proteins prior to cold shock, but a temperature downshift immediately results in repression of critical housekeeping genes and induction of cold-shock (but not cold-acclimation) proteins, which is transient.

The continuous production of a variety of binding proteins to maintain proper conformation, flexibility, and function of major macromolecules thus appears to be an important trait of cold adaptation (Figure 4). Although work with live psychrophiles is needed, genomic sequence data support this idea; for example, the genome of *C. psychrerythraea* 34H encodes for multiple common cold-shock proteins. Furthermore, the genetic acquisition of cold-shock proteins may not require the longer term evolutionary process of vertical inheritance but may be facilitated by horizontal gene transfer. Genes for cold-shock proteins known only from the domain of Bacteria have been observed upon genomic sequencing of an uncultured population of marine Crenarchaeota from cold Antarctic waters.

Cryoprotectants and Exopolymers

In very cold environments, the cellular membranes of resident microbes are subject not only to rigidity but also to physical damage from ice-crystal formation during the freezing process. Some cold-adapted microbes are known to produce and release specific proteins that help to control the formation of ice crystals (Figure 4), including ice-nucleating proteins (that provide a template for crystal formation away from the cell) and antifreeze proteins that inhibit ice nucleation by dropping the freezing point or repressing the recrystallization of ice. The rate of freezing experienced by the cell also influences the degree of damage, with faster rates limiting the damage. Natural ice formations on Earth freeze slowly (producing ice crystals) relative to the vitrification process, whereby the liquid phase converts directly to solid without ice-crystal formation. When microbial cultures are vitrified in the laboratory (using liquid nitrogen at −196 °C), their cell membranes remain intact with no morphological sign of damage. When vitrified in the presence of sugars, the likelihood of recovering them in culture after thawing increases.

Small molecular weight sugars (like glycerol) have long been used as cryoprotectants in the deep-freeze (−80 °C) storage of microbes, presumably providing a buffer between cells and ice crystals. Newly discovered, however, is the overproduction of complex extracellular polysaccharides by cold-adapted bacteria, both psychrophilic and psychrotolerant, when subjected to increasingly cold temperatures, especially below the freezing point. Sea ice through its seasonal lifetime is also recently known to harbor high concentrations of sugar-based exopolymers in its liquid brine inclusions (Figure 3). These exopolymers, produced copiously not only by sea-ice algae but also by ice-encased bacteria, are understood to serve as natural cryoprotectants not only against potential ice-crystal damage but also by further depressing the freezing point such that more liquid water remains available within the ice matrix. In this regard, cellular coatings of exopolymers (Figure 4), or exopolymers available in the environment, are believed to provide a hydrated shell that helps to buffer the cell against the osmotic stress of high salt concentrations in winter sea-ice brines (Figure 3). Along with the possible stabilization of extracellular enzymes, this myriad of functions makes exopolymers a cold-adaptive trait worth examining in more detail. The organization of genes for exopolymers on the genome of the psychrophilic bacterium *Psychroflexus torques*, for example, suggests that they may be the result of a series of lateral gene transfer events.

CONCLUSION: A MODEL HABITAT FOR COLD ADAPTATION

Considering that the ocean represents the bulk of Earth's cold biosphere, today and in the past, the annual freezing of its surface waters in polar regions takes on special significance as an important planetary driver of cold adaptation. Astronomical numbers of bacteria (10^5 in a single milliliter of seawater) pass through this frozen gauntlet annually, and over extended periods in geological time. Microbes that experience a winter in sea ice are subjected to the linked stressors of increasingly cold temperature and high brine salinity as shrinking pore space further concentrates all impurities in the source seawater, including microbes (Figure 2). This concentrating factor brings microbes in close proximity to each other, as verified by microscopic observations of DNA-stained cells in unmelted ice, in a liquid environment of abundant low- and high-molecular-weight organic compounds, including complex exopolymers that serve many positive functions for the trapped cells. Model calculations and observed concentrations in melted sea ice indicate that agents of lateral gene transfer (free DNA and viruses) also surround the encased cells (Figure 2). The virus–bacteria contact rate in a winter sea-ice brine may be as much as 600 times higher than in underlying seawater. Active virally mediated gene transfer has not yet been demonstrated, but the sea-ice environment would appear to favor it.

Even if horizontal gene transfer is not operative in sea ice, the vertical inheritance of genes for cold adaptation must be a regular occurrence. The habitat of *P. ingrahamii*, one of the most extremely psychrophilic bacteria on record (shown to reproduce at −12 °C), is sea ice. Isolates of *C. psychrerythraea* (strain 34H also grows at −12 °C) are readily cultured from sea ice. Virtually all of the canonical molecular traits of cold adaptation, along with some new ones, have been documented in the test tube or by genome analyses of such sea-ice psychrophiles. Rather than a stably cold environment, the key to the evolution of cold adaptation may be repeated exposure to the extreme cold and brine of sea ice, selecting for robustness in the face of multiple insults, including future ones like hydrostatic pressure. That salt-heavy water masses form in polar oceans actively growing sea ice and then sink to fill the cold deep ocean over time points to the concept of a cold refuge for cold-adapted bacteria during interglacial times when ice as an evolutionary driver was nonexistent, as we may witness again in this century.

RECENT DEVELOPMENTS

Our planet continues to warm at rates beyond those predicted by climate models. The recent loss of glacial ice on a global scale and sea ice in the Arctic has been dramatic. Even as habitat is lost, however, discoveries in the realm of cold-adapted microbes have continued apace. The increasing problems of pollution in cold regions have led to successful isolations of new microbes effective at degrading xenobiotics in the cold. Recent evidence for water on our Moon, on Mars, and on Enceladus (a moon to Saturn), all extremely cold settings, has enhanced the astrobiological motivation to study the limits of microbial life in the cold compounded by other extremes.

A previously overlooked, extremely cold environment has only recently been explored as a possible microbial habitat: frost flowers, the delicate, centimeter-scale, three-dimensional ice-crystal structures of high brine content that form in large numbers on the surface of new sea ice when atmospheric temperature drops to −8 °C and well below (< −40 °C over the Arctic Ocean in winter). Frost flowers had been attracting much attention since the early 1990s for their physical and chemical characteristics. Their multi-faceted ice-crystal structures reflect more light than a planar ice surface, altering the albedo and heat budget of the polar regions; their inorganic salt content appears linked to large-scale atmospheric ozone and mercury depletion events. Because more new sea ice is forming from open water during the Arctic winter than ever before, the areal extent of frost flowers is increasing in the North; the extensive zone of sea ice that surrounds the entire Antarctic continent already supports vast fields of frost flowers. Now we know that frost flowers also contain an abundance of microbes, concentrated in these structures on the surface of sea ice at greater densities than in the ice or water below and immediately available for dispersal by wind. Because microbial abundance in frost flowers correlates strongly with salt content, the source of the microbes is traced to underlying sea-ice brines wicked upwards into the growing surface structures. Newly applied phylogenetic fingerprinting has revealed that these sea-ice brines contain an array of Bacteria closely related to known (cultured) cold-adapted sea-ice clades and, for the first time (for sea ice of any season), Archaea. Frost flowers have also been determined to contain high concentrations of exopolymers, already known for their cryo- and osmo-protective functions. These discoveries thus provide a more extreme habitat than Arctic winter sea ice to explore from the microbial perspective, even as we are losing ice globally. They open new questions about microbial adaptability to combined extremes of low temperature, high salt and high radiation (in early spring, when temperatures are still cold enough for frost flower formation), long-distance microbial dispersal, biogeography and endemism, and possible microbial roles in large-scale atmospheric phenomena.

In the realm of genomics and proteomics, several benchmarks have recently been established. The genome sequence of the first cold-active virus (infective at 0–15 °C) was reported, revealing the lowest G + C content (30.6%) for a bacteriophage to date. A more extremely cold-active virus (infective at −12 to 8 °C) is in the pipeline, as well as a suite of environmental virome samples from subzero habitats, promising an expanded database for deducing cold-active features of viruses that may be contributing, via lateral gene transfer, to cold adaptation in their hosts. A high-resolution deep sequencing technique (454 pyrosequencing) that detects the rare members (<0.01%) of a microbial community (based on the V6 region of the 16 S rRNA gene) was also applied for the first time in a polar setting. This work revealed that rare microbes, including Archaea, follow cold water mass boundaries in the Arctic Ocean along with the most abundant clades: cold adaptation must also apply to the survival of the rarest. The first statistically robust proteomics analysis of cold adaptation was performed on the psychrotolerant marine bacterium, *Sphingopyxis alaskensis*, quantitatively comparing the suite of proteins it expresses at 10 °C versus 30 °C. This gene expression work revealed cold-specific transcriptional machinery and a remarkable, complete cold-active protein-folding system. It also verified cold-adaptive strategies previously inferred from whole-genome analyses, including storing carbon intracellularly as an energy reserve, shifting from carbohydrate to lipid metabolism for greater energy return, and controlling the cold-quality of membrane, cell wall and exported proteins. Several lines of evidence once again pointed to the importance of producing exopolymers in the cold.

FURTHER READING

Borris, M., Lombardot, T., Glockner, F. O., Becher, D., Albrecht, D., & Schweder, T. (2007). Genome and proteome characterization of the psychrophilic *Flavobacterium* bacteriophage 11b. *Extremophiles, 11*, 95–104.

Bowman, J. P. (2008). Genomic analysis of psychrophilic prokaryotes. In R. Margesin, F. Schinner, J. C. Marx, & C. Gerday (Eds.), *Psychrophiles: From biodiversity to biotechnology* (pp. 265–284). Berlin: Springer-Verlag.

Bowman, J. S., & Deming, J. W. (2010). Elevated bacterial abundance and exopolymers in saline frost flowers and implications for atmospheric chemistry and microbial dispersal. *Journal of Geophysical Research Letters, 37*, L13501 doi: 10.1029/2010GL043020.

Breezee, J., Cady, N., & Staley, J. T. (2006). Subfreezing growth of the sea ice bacterium *Psychromonas ingrahamii*. *Microbial Ecology, 47*, 300–304.

Connelly, T. L., Tilburg, C. M., & Yager, P. L. (2006). Evidence for psychrophiles outnumbering psychrotolerant marine bacteria in the springtime coastal Arctic. *Limnology and Oceanography*, *51*, 1205–1210.

Deming, J. W. (2002). Psychrophiles and polar regions. *Current Opinion in Microbiology*, *3*(5), 301–309.

Deming, J. W. (2007). Extreme high-pressure marine environments. In C. J. Hurst, R. L. Crawford, J. L. Garland, A. L. Mills, & L. D. Stetzenbach (Eds.), *ASM manual of environmental microbiology*. (3rd ed.) (pp. 575–590). Washington, DC: ASM Press.

Deming, J. W. (2010). Sea ice bacteria and viruses. In D. N. Thomas, & G. S. Dieckmann (Eds.), *Sea ice – An introduction to its physics, chemistry, biology and geology*. (2nd ed.) (pp. 247–282). Oxford: Blackwell Science Ltd.

Deming, J. W., & Eicken, H. (2007). Life in ice. In W. T. Sullivan, & J. A. Baross (Eds.), *Planets and life: The emerging science of astrobiology* (pp. 292–312). Cambridge: Cambridge University Press.

Galand, P. E., Casamayor, E. O., Kirchman, D. L., & Lovejoy, C. (2009). Ecology of the rare microbial biosphere of the Arctic Ocean. *Proceedings of the National Academy of Sciences*, *106*(52), 22427–22432.

Gerday, C., & Glansdorff, N. (Eds.), (2007). *Physiology and biochemistry of extremophiles*. Washington, DC: ASM Press.

Helmke, E., & Weyland, H. (2004). Psychrophilic versus psychrotolerant bacteria – Occurrence and significance in polar and temperate marine habitats. *Cellular and Molecular Biology*, *50*, 553–561.

Margesin, R., Schinner, F., Marx, J. C., & Gerday, C. (Eds.), (2008). *Psychrophiles: From biodiversity to biotechnology* Berlin: Springer-Verlag.

Meth, B. A., Nelson, K. E. Deming, J. W. *et al.* (2005). The psychrophilic lifestyle as revealed by the genome sequence of *Colwellia psychrerythraea* 34H through genomic and proteomic analyses. *Proceedings of the National Academy of Sciences of the United States of America*, *102*(31), 10913–10918.

Morita, R. Y. (1975). Psychrophilic bacteria. *Bacteriological Reviews*, *39*, 144–167.

Moyer, C. L., & Morita, R. Y. (2007). *Psychrophiles and psychrotrophs. encyclopedia of life sciences*. New York: John Wiley and Sons. doi:10.1002/9780470015902.a0000402.pub2.

Panikov, N. S., & Sizova, M. V. (2007). Growth kinetics of microorganisms isolated from Alaskan soil and permafrost in solid media frozen down to −35 °C. *FEMS Microbiology Ecology*, *59*, 500–512.

Parrilli, E., Duilio, A., & Tutino, M. L. (2008). Heterologous protein expression in psychrophilic hosts. In R. Margesin, F. Schinner, J. C. Marx, & C. Gerday (Eds.), *Psychrophiles: From biodiversity to biotechnology* (pp. 365–379). Berlin: Springer-Verlag.

Price, B. P., & Sowers, T. (2004). Temperature dependence of metabolic rates for microbial growth, maintenance, and survival. *Proceedings of the National Academy of Sciences of the United States of America*, *101*, 4631–4636.

Priscu, J. C., & Christner, B. C. (2004). Earth's icy biosphere. In A. T. Bull (Ed.), *Microbial biodiversity and bioprospecting* (pp. 130–145). Washington, DC: ASM Press.

Rodriguez, D. F., & Tiedje, J. M. (2008). Coping with our cold planet. *Applied and Environmental Microbiology*, *74*, 1677–1686.

Saunders, N. F., Thomas, T., Curmi, P. M., et al. (2003). Mechanisms of thermal adaptation revealed from the genomes of the Antarctic Archaea *Methanogenium frigidum* and *Methanococcoides burtonii*. *Genome Research*, *13*, 1580–1588.

Ting, L., Williams, T. J., Cowley, M. J., Lauro, F. M., Guilhaus, M., Raftery, M. J., et al. (2010). Cold adaptation in the marine bacterium, *Sphyngopyxis alaskensis*, assessed using quantitative proteomics. *Environmental Microbiology*, 12:2658–2676. doi:10.1111/j.1462–2920.2010.02235.x.

Chapter 24

Dry Environments (including Cryptoendoliths)

J.A. Nienow
Valdosta State University, Valdosta, GA, USA

Chapter Outline

ABBREVIATIONS

ETP Annual potential evapotranspiration
P Annual precipitation

DEFINING STATEMENT

This article summarizes the current state of knowledge concerning the microbial communities inhabiting arid regions, with an emphasis on lithic environments. The constraints put on these communities by their dry environment and their adaptations to survive the desiccated state are also addressed.

INTRODUCTION

It can be argued that dry environments present one of the most extreme set of conditions faced by microorganisms. The cell must be able to withstand the specific biochemical stresses created by the lack of water; also, it may be called upon to withstand a variety of other environmental insults – exposure to high levels of UV radiation, for example, or elevated temperatures during periods when its metabolic activity has been reduced to minimal levels by the same lack of water. On one level, then, it is surprising that any cells possess the required set of adaptations. Yet, when one considers the obvious selective advantages conferred by these adaptations in the 30% of the world not covered by liquid water, perhaps it is not so surprising after all. As will be seen, there are a variety of ways to cope with dry environments and a diversity of microorganisms employing them.

Before proceeding, it may be worthwhile to spend some time discussing what constitutes a dry environment. Clearly, the concept includes some mechanism leading to a reduction in the liquid water content of the cell. In general, the reduction of liquid water can be achieved either by freezing the cell or by the net movement of water out of the cell. Freezing conditions, then, in the broadest sense, constitute a form of dry environment. However, the freezing process involves its own set of biochemical stresses, which are not necessarily related to the lack of liquid water, and, therefore, will not be treated further. Instead, the discussion will revolve around environments leading to the removal of cellular water. Technically, these include any environment with a reduced water activity or a low water potential, including syrups, brines, and even some hypersaline lakes and salinas. While peripheral to the main treatment, the microorganisms occurring in these rather wet "dry" environments provide some insight into some of the mechanisms available to microorganisms living in the truly dry environments – dry soils and all aerial and subaerial environments, especially those in arid regions, the exterior of spacecraft, and dried grains and other foodstuffs – and so will not be explicitly excluded. The discussion will focus, however, on the microorganisms of arid regions, because they experience the widest range of extreme conditions, and, in consequence, display the broadest set of adaptations. Readers interested in a more thorough understanding of hypersaline environments should consult the Further Reading section.

MICROBIAL ASSOCIATIONS IN ARID REGIONS

Traditionally, deserts have been defined as regions with barren landscapes and limited water, with the degree of water limitation usually defined on the basis of annual precipitation (P). It was recognized early on that a definition based strictly on precipitation is incomplete, in that it ignores the effects of temperature and plant cover on the fate of the precipitation. Instead, some researchers have championed a comparison of the actual P with the potential evapotranspiration in the delineation of climatic regions. This approach was followed in the development of some of the first precise global maps of arid regions in the early 1950s. Currently, arid regions *sensu lato* are defined as those parts of the world where the ratio of the mean P to the mean annual potential evapotranspiration (ETP) is less than 1. This definition emphasized the water deficit of the region, which is greater for hotter deserts where the rate of evaporation is greater. Four intergrading zones have been delineated: the hyperarid zone with P/ETP less than 0.03; the arid zone with P/ETP between 0.03 and 0.20; the semiarid zone with P/ETP between 0.20 and 0.50; and the subhumid zone with P/ETP between 0.50 and 0.75. In this system, regions previously classified as deserts on the basis of a mean P of less than 200 mm per year correspond roughly to the hyperarid and arid climate zones. Together, these comprise in excess of 20% of the Earth's terrestrial surface. Semiarid regions, including steppes, savannahs, and similar habitats, cover another 16%. Such a large area encompasses a diversity of habitats and associations. In this article, we touch upon three broad clusters of associations: the soil microflora, biological soil crusts, and rock-inhabiting microorganisms (Figure 1).

The Microflora of Desert Soils

Numerous studies over the past several decades have indicated the presence of a remarkably diverse assemblage of microorganisms in soils from both hot and cold deserts. Included in most assemblages are a number of cosmopolitan species, which gives the overall assemblage an appearance similar to what would be expected from milder climates. This raises questions concerning (1) how harsh the desert soil climate is; (2) what role the cosmopolitan species play in the species assemblage; and (3) whether there exists a microflora indigenous to desert regions. It is possible that a number of the bacteria isolated from desert soils are recent additions, carried in by wind, dust, or human activity, and not truly active members of the ecosystem. This is especially true for the cold McMurdo Dry Valleys region of Antarctica (also known in the literature as the Ross Desert), where the conditions would tend to preserve nonindigenous cells in a nonmetabolic state for extended periods of time. Even with the possible inclusion of a number of nonindigenous forms, in hot deserts there does seem to be a skew in favor of desiccation-tolerant forms. This manifests itself in the proportionally larger numbers of spore-forming Gram-positive bacteria, especially actinomycetes, and in the relative proportion of filamentous fungi. There is a shift toward radiation-resistant forms in arid soils, especially members of the genus *Deinococcus*, with up to nine species present in single soil sample. Such a shift is consistent with the suggestion that radiation resistance in bacteria is actually a side effect of adaptations to water stress. In the interior of the cold McMurdo Dry Valleys region, in contrast, actinomycetes may only comprise about 3% of the cultured soil microflora. The numbers of yeasts and filamentous fungi are also very low. *Deinococcus*, including several newly described species, however, is present. The low numbers of yeasts in McMurdo Dry Valleys soils are somewhat surprising in light of the fact that the dominant yeast in the region, *Cryptococcus vishniacii*, is considered to be both endemic- and desiccation-resistant.

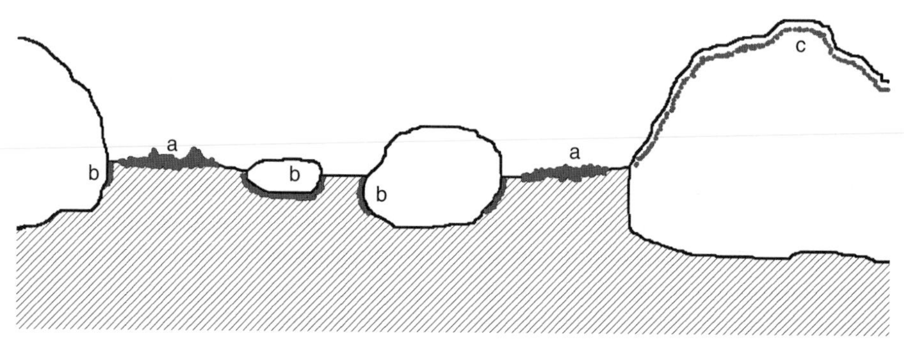

FIGURE 1 Diagrammatic representation of the major microbial associations in dry environments. (a) Biological soil crust. (b) Hypolithic associations. (c) Endolithic association.

Early studies of desert microorganisms relied on standard culture-based techniques. It is now recognized that many of the bacteria in environmental samples cannot be cultured on standard media, or, for that matter, on any media yet devised. Therefore, in recent years a number of molecular techniques have been developed, primarily based on direct or indirect analysis of gene sequences, but including fatty acid analysis and community-level physiological profiling. Application of some of these techniques to desert soils has indicated the presence of a number of organisms not previously seen in culture-based studies, including members of the domain Archaea. Even bacteriophage, either as prophage (viral genetic material incorporated into the bacterial genome) or as virus particles in bacteria in a pseudolysogenic state (infecting the cell, but maintained separately from the genome), but not as free particles, have been detected in the subsurface desert sands. The full significance of these results is yet to be elucidated.

As expected, the numbers of soil bacteria in deserts tend to be lower than in comparable soils from more temperate climates. Reported plate counts from a variety of hot desert soils range from a low of less than 10 to 1.6×10^7 cfu g^{-1}, numbers at least 2 orders of magnitude lower than is typical for agricultural or forest soils. The lowest recorded numbers are from the vicinity of Yungay in the hyperarid core of the Atacama Desert, Chile. This region has generated considerable interest among microbiologists and astrobiologists because it is considered to be one of the oldest and most extreme of the hot deserts, and can be considered a terrestrial analogue of Mars. Some microbiologists, because of the low numbers of bacteria recovered, suggest that the dry conditions in the hyperarid core of the Atacama may be near the boundary for life in desert environments. Others dispute this view on the basis of their own higher counts; they suggest that the boundary for life has yet to be found on Earth. No matter which group is right, it is clear that large-scale shifts in the moisture regime have a significant impact on the total number of desert soil bacteria.

The cold deserts of Antarctica have also been cited as being at or beyond the limits of life; there are several examples of studies where the number of colony-forming units (cfu) in soil from the McMurdo Dry Valleys region of the Antarctica was below detectable limits. However, the apparent sterility of at least some of the samples may have been an artifact of the medium used (standard nutrient agar); true oligotrophs and some nutrient-starved bacteria are unable to grow on this relatively rich medium. Later sampling of similar soils using oligotrophic media containing 0.1% peptone and 0.02% yeast extract enriched with soil extracts from the site yielded significantly higher numbers of bacteria, suggesting that viable cells can be presumed to be present in all soils not containing toxic substances. The absolute number varies widely, depending on the type (e.g., ornithogenic, enriched with bird wastes vs. mineral, with limited organic material), the local climate, and the growth medium used. Away from the coast, the numbers are similar to those encountered in the Atacama Desert, on the order of 10^3–10^5 cfu g^{-1} soil, when procedures similar to those used in the Atacama studies, plate counts on R2A agar, a low-nutrient medium designed to count slow-growing heterotrophic bacteria in stressed environments, are used.

The source of the organic carbon supporting the low numbers of heterotrophic bacteria in the Antarctic Dry Valleys, and other extreme deserts, is not completely clear. In the milder deserts, shrubs and animal mounds provide obvious sources of carbon and other nutrients, creating "islands of fertility" with bacterial populations up to two orders of magnitude larger than is found just a few decimeters away. These sources of nutrients are completely lacking in the more extreme deserts. Therefore, the heterotrophic microbial community must rely on *in situ* production by cyanobacteria and microalgae, importation of carbon from more productive regions, or relic carbon remaining in the soils from periods when the climate was more benign. There is evidence suggesting that, at least for the lower elevations of Taylor Valley, Antarctica, relic organics deposited either as entrained material in glacial tills or as primary production in ancient glacial lakes is the primary source of organic material in the soil. It has been suggested that this legacy material may sustain the microbial soil community in the absence of any other inputs to the system. However, this suggestion may not be tenable. Field measurements of short-term respiration rates, coupled with known meteorological parameters, indicate a carbon release rate of 6.5 g C m^{-2} $year^{-1}$; at this rate the mean residence time for carbon would be 23 years, not the thousands of years expected. This suggests the presence of significant *in situ* production since the carbon signature of the organic material rules out inputs from the surrounding habitats. The legacy carbon may exist in a recalcitrant form unavailable to the community under normal conditions. This two-compartment model of soil carbon, one compartment labile, one compartment recalcitrant, may prove to be of value in the analysis of other desert systems.

Biological Soil Crusts

Biological soil crusts are created by surface-bound assemblages of microorganisms that consolidate the soil into crusts up to a centimeter thick. Much of our current knowledge of this system has been summarized previously; those wishing a thorough review are directed to the Further Reading section. Here we will just touch upon a few highlights (Figure 2).

Biological soil crusts are widespread, occurring on all of the continents and in most biomes, essentially wherever a shortage of water limits the growth of vascular plants, thereby allowing light to reach bare soil, but there is still

FIGURE 2 Biological soil crusts in the Mojave Desert. (a) The desert landscape. Soil crusts appear as darker areas of the soil surface. (b) A closer view of the crust. In this instance, the crust forms a pattern of darker bands. The total length of the scale bar in the foreground is 10 cm.

sufficient water to allow the growth of phototrophic microorganisms. They reach their greatest development in semiarid and arid environments, where they can cover more than 90% of the ground surface. In such cases biological soil crusts may provide a significant fraction of the photosynthetically fixed carbon, serve as sources of fixed nitrogen, reduce soil erosion, and influence local and regional hydrologic cycles. Extensive growths of biological crusts can be mapped using aerial photography and satellite-based imagery.

The distribution, composition, and appearance of biological soil crusts depend primarily on moisture and climatic conditions, secondarily on soil properties. On a broad scale, four distinct types of crusts can be recognized macroscopically. Smooth crusts, as suggested by the name, are recognized by the smoother appearance of the soil surface resulting from activity of cyanobacteria, algae, and fungi. The biological components spend much of the time just below the surface, only appearing when the soil is wet. This type of crust is most prevalent in hyperarid and arid regions with very low precipitation and very high temperatures. Rugose crusts, which give the soil surface a rough appearance, are common in arid regions with a slightly moister climate. They are also dominated by cyanobacteria, green algae, and fungi, but may include varying numbers of lichens and desiccation-resistant mosses. Pinnacled crusts are characterized by the presence of a patchwork of mounds up to 15 cm in height but only a few millimeters in width, caused by a combination of frost heaving during the winter and differential erosion during the rest of the year. Cyanobacteria are the dominant microorganisms, but lichens and mosses may contribute up to 40% of the cover. Pinnacled crusts are found in some of the cooler deserts, where freezing temperatures are possible. Rolling crusts are also associated with frost heaving. However, in this case, a thicker growth of mosses, lichens, and/or cyanobacteria reduces the amount of erosion, leading to a generally smoother appearance than that of the pinnacled crusts. Rolling crusts are characteristic of colder climates with a lower potential evaporation rate. Within these groupings, there is a tendency for sandy soils of low nutritional content to be dominated by cyanobacteria, with the proportion of lichens increasing with the carbonate, gypsum, and silt content of the soil.

From the limited data available, it appears that the abundance of soil crusts decreases as the climate becomes more arid. For example, in Central Asia, biological crusts are described as abundant in the eastern semideserts of Mongolia, present in the central Gobi Desert, and covering about 29% of the land in the relatively drier Gurbantunggut Desert farther west. Within the Gurbantunggut Desert there is a marked decrease in the abundance of crusts in the northern regions than in the southern desert, which corresponds to a decrease in the amount of precipitation. Biological soil crusts in the Sahara Desert may be restricted by climate and surface instability to the marginal regions. However, there has only been a single study of soil crusts from the region. True soil crusts have only been reported from coastal regions in the Atacama Desert, near the port cities of Antofagasta and Caldera, although *Nostoc* crusts are present in southern Peru. In the drier parts of Antarctica, away from the coasts, small patches of cyanobacterial crusts occur sporadically in widely scattered locations.

Common microbial constituents of biological soil crusts are members of the filamentous cyanobacterial genera *Microcoleus*, *Nostoc*, and *Scytonema*. The filaments of *Microcoleus* are formed from numerous trichomes (chains of cells) with a common sheath. The individual trichomes are motile within the sheath and may even, when conditions warrant, leave the sheath. In the latter case, trichomes will group and secrete a new sheath later. The buildup of used and new sheath material contributes to the stability of the soil crust. Members of *Nostoc*, especially *Nostoc commune*, frequently form large masses of coiled trichomes within a communal polysaccharide sheath. Masses of *N. commune* can become large enough to harvest and are apparently used as a supplemental food source for humans in some regions. Heterocysts, specialized nitrogen-fixing cells, are common within colonies. Filaments of *Scytonema* are formed from a single trichome with heterocysts and false branching within a colored sheath. Trichomes reproduce through the formation of small motile filaments called hormogonia. Additional common phototrophic microorganisms include the cyanobacteria *Gloeocapsa* and members of the green algal genera *Chlorococcum*, *Macrochloris*, and *Stichococcus*. It should be kept in mind, however, that the actual diversity of photosynthetic microorganisms in the desert crusts is much greater than indicated here. Molecular analysis of isolates and natural samples from North America suggests that biological soil crusts can almost be considered biodiversity hot spots. Somewhat surprisingly, soil crust fungi and heterotrophic bacteria have so far been virtually ignored.

Rock-Inhabiting Microorganisms

Most deserts contain extensive regions covered by rock outcrops or by bare soils with embedded or loose-lying cobbles, pebbles, and stones. The importance of rock, or lithic, habitats in these regions is suggested by the variety of terms derived from indigenous languages used to describe them. In the Middle East and northern Africa, regions characterized by rock outcrops, large boulders, and limited soil are called hamadas. Regions characterized by smaller stones and pebbles accompanied by sandy or gravelly soil are referred to variously as reg (western Sahara, the Iranian desert), serir (the Libyan and Egyptian Sahara), gobi (Mongolia), or gibber plains (Australia). Desert pavement is used specifically for those parts of the desert covered by embedded stones in a relatively compact arrangement.

Lithic habitats can be divided into three separate habitats corresponding loosely to the extremes of environmental conditions encountered by the microbial inhabitants. The epilithic habitat corresponds to the upper surface of the rock or stone, the part most directly exposed to atmospheric and climatic conditions and, therefore, more directly exposed to environmental extremes. The endolithic habitat corresponds to the interior of the rock or stone. Three distinct subdivisions of this habitat can be identified: the cryptoendolithic habitat, consisting of naturally occurring pore spaces within the rock; the chasmoendolithic habitat, consisting of cracks and fissures in the rock leading to the surface; and the euendolithic habitat, consisting of holes and pits created by the metabolic activity of the microbial community (Figure 3). The hypolithic habitat consists of the lower surface of a stone in association with adherent soil particles. (The term "sublithic" is sometimes used, erroneously, as a synonym for hypolithic.) Recently, the term perilithic has been introduced to distinguish the subsurface sides of large stones from the true hypolithic habitat below the stone. The hypolithic and perilithic habitats are similar, but differ slightly in their moisture and light regimes.

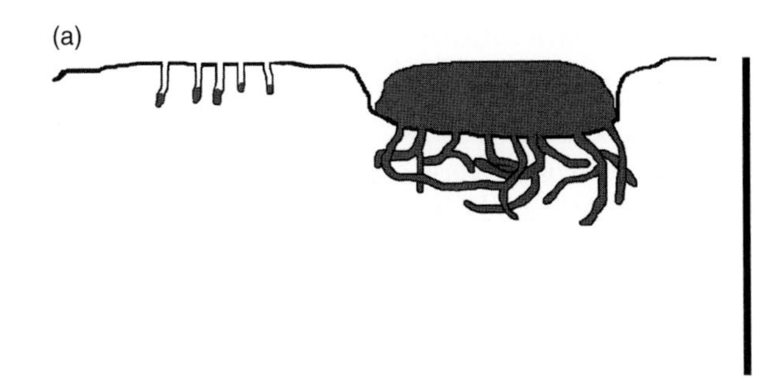

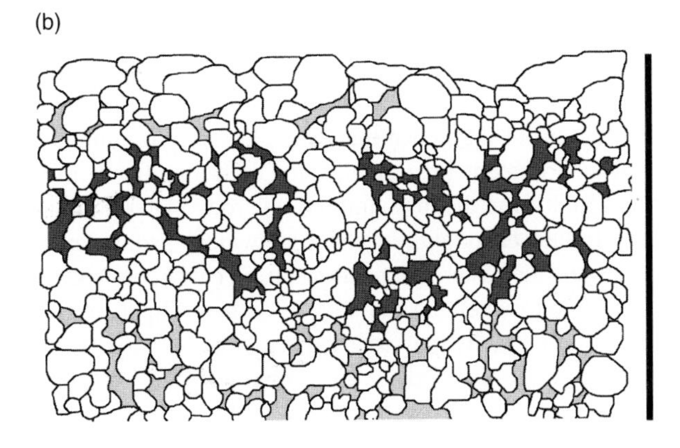

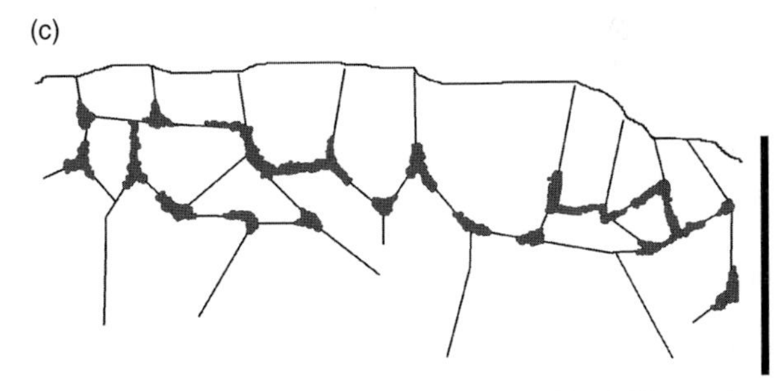

FIGURE 3 Diagrammatic representations of endolithic habitats and associations. (a) Euendoliths, boring into the rock either as an individual (left) or as part of a lichen association (right). (b) A cryptoendolithic association. (c) A chasmoendolithic association inhabiting microfractures in a weathered substratum. The drawings are not to scale. However, scale bars have been included to provide an indication of the relative dimensions of the habitat – in (a) the bar represents 1 mm, in (b) and (c) it represents 1 cm.

Epilithic associations

In relatively moist regions epilithic cyanobacteria and cyanophilous lichens form extensive growths, capable of completely covering the surface of large inselbergs, rock hills arising abruptly from the surrounding plains. In hot arid regions, such growths are essentially absent. Instead, the dominant epilithic microbial community is that associated with rock varnishes. Rock varnishes are thin, dark layered veneers of clay minerals held together by oxides and hydroxides of manganese and iron. They are found in all weathering environments, but are most common in arid lands, where they are referred to as desert varnishes. There is a long-standing debate concerning the origins of rock varnishes with some advocating the predominance of physical–chemical mechanisms and others advocating biological mechanisms. It is also possible that the mechanism may change with changing climatic conditions. Resolving this debate is beyond the scope of the present article. Our interest is in the presence of a ubiquitous and relatively diverse microbial community in what is considered to be a dry environment. Included are actinomycetes, *Bacillus*, *Micrococcus*, *Metallogenium*-like strains, cyanobacteria, and microcolonial fungi. Nearly all of the heterotrophic strains of bacteria isolated are Gram-positive and are capable of precipitating manganese. Populations of bacteria in rock varnishes from the eastern border of the Mojave Desert are on the order of 10^7–10^8 cells per gram of varnish, estimated on the basis of direct microscopic observation and phospholipid fatty acid analysis, somewhat remarkable given their exposed position. Raman spectroscopy of varnishes from Colorado and Nevada indicates the presence of chlorophyll, scytonemin, and β-carotene; the latter two are considered to be bioprotective molecules (see below).

Endolithic associations

Euendolithic associations: Euendolithic microorganisms, which use chemical reactions to bore into the surfaces of rocks, are common in aquatic systems, less so in terrestrial systems. The decrease in abundance in terrestrial systems is caused, presumably, by the need for suitable substrates, generally carbonates although volcanic glasses are possible, and for sufficient external moisture to support the chemical reaction. In fact, moisture is more of a problem in euendolithic systems than in other terrestrial endolithic systems because there is a direct connection between the colony and the external environment. One way around the moisture problem is the formation of a lichen association – the added mass and the structure of the lichen thallus could serve to maintain a moist environment long enough for both significant metabolic activity and the dissolution of the rock material to take place. In fact, euendolithic lichens are widespread, at least in Europe and the eastern Mediterranean, on calcareous rocks in microclimates where air humidity is high and light intensity is low. Their activity may contribute significantly to the deterioration of marble buildings and monuments.

Extensive growths of euendolithic lichens have even been reported from the moister regions of the Negev Desert. Here, the distribution is linked to the local microclimate through the imbibition time, an indicator of how many hours a year a lichen thallus is moist. If the conditions result in more than 450 h of daylight imbibition time epilithic lichens dominate, between 300 and 450 h of imbibition euendolithic lichens dominate, and at lower levels the free-living euendolithic fungus *Lichenothelia* dominate in loose association with cryptoendolithic and chasmoendolithic cyanobacteria and eukaryotic algae. Because the different associations form distinctive patterns on the surface of the rock, which can persist long after the association has disappeared, they can be used as indicators of the climate at the time the community was destroyed by fire or burial. The small pits and other scars in the rock surface may also facilitate the recolonization of the surface by desert microorganisms.

Free-living colonies of euendolithic coccoid cyanobacteria have also been reported from several locations in Israel. If true, these would represent the first such reports from a terrestrial setting. However, it is possible that the colonies were actually lichenized, but with a small proportion of fungi. More recently, a species of the filamentous heterocystous cyanobacterial genus *Matteia* has been isolated from Negev limestones. This species is capable of boring into limestone rocks both in culture and in the Negev, thus demonstrating unequivocally the existence of free-living terrestrial euendoliths.

Outside of the Negev, there are almost no reports of euendolithic associations of any sort from arid regions. The euendolithic lichen *Verrucaria calciseda* is described as widely scattered in temperate and boreal regions rich in limestones, including northern Arizona. The euendolithic lichen *Verrucaria rubrocincta* is present in caliche deposits in the Sonoran Desert.

Chasmoendolithic associations: Like euendolithic systems, chasmolithic systems have a direct connection with the surface, allowing light, moisture, and nutrients to reach the association relatively unimpeded. As a result, a wide variety of rock types, including darker basalts and granites, can sustain chasmolithic growth at some depth. In spite of their widespread occurrence in a variety of climate zones, including desert regions, there have been relatively few detailed investigations of either the habitat or the organisms within. The most extensive studies are from the relatively mild Vestfold Hills and Mawson Rock regions of Antarctica. This region contains a diverse group of microalgae, with members of cyanobacterial genera *Chroococcidiopsis* and *Plectonema* and the green algal genera *Desmococcus* and *Prasiococcus* the most widespread in the habitat. These can be grouped into associations dominated by *Desmococcus*, common in the drier regions, or by *Chroococcidiopsis*, common in the wetter regions of the Mawson Rock site. The domination of *Chroococcidiopsis* in the wetter and not the drier sites was unexpected given the frequency of reports of *Chroococcidiopsis* from arid deserts. A few of the associations displayed a distinct zonation pattern, with either *Prasiococcus* or *Desmococcus* forming an outer green band and *Chroococcidiopsis* forming an inner bluish-green band. At the Vestfold Hills site, extensive development of chasmolithic associations is restricted to the slopes of hills and boulders sheltered from the wind.

Chasmolithic lichen associations and cyanobacterial biofilms from the Granite Harbor region of Antarctica have been shown to develop acidic microenvironments within the association that may be involved in weathering of the substratum – chasmoendolithic associations are often implicated in the weathering of marble and limestone monuments in the semiarid Mediterranean region. Molecular analysis indicated the presence of what may be a relative of the chlorophyll *d*-containing cyanobacterium *Acaryochloris marina* in the association; this is the first record of this group in a terrestrial endolithic environment.

Cryptoendolithic associations: Cryptoendolithic organisms inhabit the interstitial spaces of porous rocks (Figure 4). Because the organisms lie below the surface of the rock, without a direct light path to the surface, only rocks composed of light-colored or translucent grains are colonized by cryptoendolithic phototrophs. Cryptoendolithic communities are easily recognized macroscopically as a pigmented layer running more or less parallel to the surface of the rock at a depth of one to a few millimeters. They can develop in any environment, wherever suitable rocks are exposed to sufficient light. Examples have been reported from weathered granites, gneisses, limestones, and marbles, including some used as building materials, beachrock, gypsum crusts,

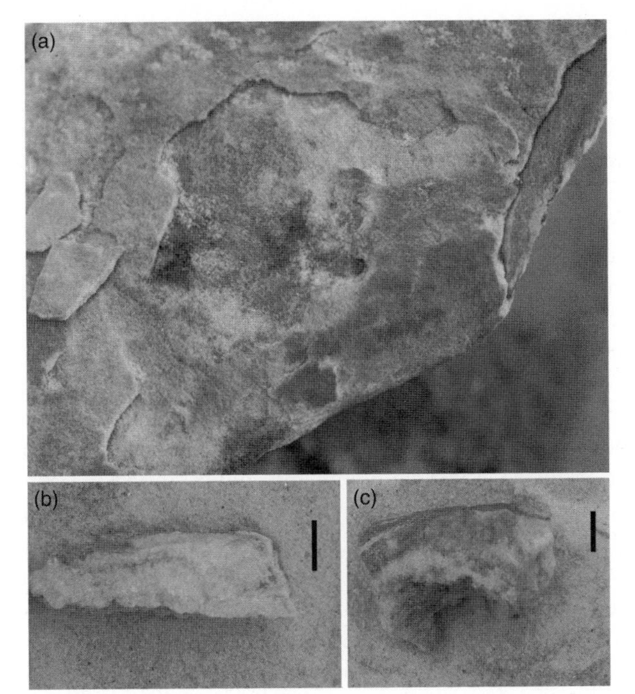

FIGURE 4 Examples of endolithic associations. (a) Cryptoendoliths inhabiting a sandstone in the Dry Valley region of Antarctica. In this instance, the association causes a characteristic weathering pattern visible on the surface of the rock that leads to flaking of the crust surface (seen in the center of the image). The association is visible to the right of the image, where the rock has been broken off. (b) An endolithic association from the Mojave Desert. (c) An endolithic association inhabiting halite in the Atacama Desert. The scale bars in (b) and (c) represent 1 cm.

halites and evaporites, sinters associated with geothermal environments, dark volcanic basalts, and gneisses altered by asteroid or comet impacts. However, the most commonly colonized substrata appear to be porous quartz sandstones.

The most extensively studied cryptoendolithic associations from desert regions are those found in porous sandstones in the McMurdo Dry Valleys regions of Antarctica. One of the most striking features of these associations is the complexity of the system. At least five distinct associations, three of which are widespread, have been described. The most abundant is generally referred to as the lichen-dominated community or the cryptoendolithic lichen community. It can be recognized macroscopically by the presence of parallel black, white, and green bands a few millimeters below the rock surface. The black layer consists of dark-pigmented filamentous fungi and lichenized cells of the green algal genus *Trebouxia* (or, possibly, *Pseudotrebouxia*). The white layer is populated by green algae and hyaline filamentous fungi, possibly a morphological variant of some of the black fungi from the upper layer; it appears white because the iron oxides that normally give the rocks their characteristic brown to orange-red color have been leached from this layer. Included among the dark-pigmented fungi are a number of nonlichenized meristematic black fungi, typically members of *Friedmanniomyces* and *Cryomyces* in the Dothideomycetidae, a taxonomic order known to have many stress-tolerant members. A number of culture-based and molecular studies indicate the presence of a small but diverse heterotrophic bacterial community dominated by actinomycetes, but including members of *Deinococcus*.

The *Hemichloris* community can be recognized as a green band several millimeters below the surface of the rock. Typically, it develops below the white layer of the lichen-dominated community, but can also be found as the sole community inhabiting the lower surface of overhanging sandstone ledges. The community is named after its dominant member, the green alga *Hemichloris antarctica*. Included in the community is a diverse assemblage of cyanobacteria and eukaryotic algae, including a second species of *Hemichloris*, the endemic species *Heterococcus endolithicus*, and members of the green algal genus *Stichococcus* and the cyanobacterial genera *Gloeocapsa* and *Chroococcidiopsis*. Members of *Stichococcus*, *Gloeocapsa*, and *Chroococcidiopsis* are all common constituents of terrestrial systems; *Stichococcus bacillaris* is generally considered to be a cosmopolitan inhabitant of subaerial environments, including lithic environments in Antarctica. The species of *Chroococcidiopsis* in this community is similar to that described in other endolithic and hypolithic communities.

The *Hormathonema–Gloeocapsa* community is a complex cyanobacteria-dominated community only found on Battleship Promontory, where it colonizes white sandstone boulders permanently wetted by snowmelt percolating through the dolerite rubble at the base of the rocks; the drier upper surfaces of large outcrops in the region are colonized by the lichen-dominated community. It can be recognized macroscopically by the presence of two parallel color bands below the rock surface, without a leached zone in between. The upper layer is a complex assemblage of cyanobacteria dominated by members of the genera *Hormathonema* and/or a dark gray species of *Gloeocapsa*. The lower green band is dominated by a species of *Aphanocapsa*, occasionally accompanied by several species of *Gloeocapsa*, *Anabaena*, and *Lyngbya*. The heterotrophic component of the community includes flagellated protozoans, filamentous fungi, and yeasts and up to 39 morphotypes of heterotrophic bacteria. Molecular analysis indicated the presence, among others, of actinomycetes and members of the *Thermus–Deinococcus* group.

The remaining associations are referred to as the *Chroococcidiopsis* community and the red *Gloeocapsa* community. Both are dominated by unicellular cyanobacteria of similar morphology. The *Chroococcidiopsis* community was the first endolithic community discovered in Antarctica. It appears as a single green or brownish band just below the surface crust, and *Chroococcidiopsis* sp. is the sole phototrophic organism. When discovered, it seemed to confirm a pattern just then beginning to be recognized: in the most extreme deserts where environmental conditions preclude eukaryotes and filamentous cyanobacteria, the unicellular *Chroococcidiopsis* becomes the dominant or sole photoautotroph. The relatively limited distribution of the *Chroococcidiopsis*-dominated community in the Antarctic desert suggests that this polar desert may not be the most extreme on Earth, although it remains the closest terrestrial analogue to Mars.

It has been estimated that cryptoendolithic associations colonize 20–30% of the exposed surfaces of Beacon sandstones in the McMurdo Dry Valleys region, equivalent to a total colonized area of between 200 and 300 km^2. Given the abundance of material in an association, 1.4–7.4 g N (Kjeldahl) m^{-2}, or 35–190 g C m^{-2}, this represents a significant pool of organic matter in a desert region and suggests that endolithic organisms play an important role in the ecosystem. How significant is a matter of debate. Short-term laboratory measurements of CO_2 gas exchange under different temperature conditions coupled with multiyear records of environmental conditions with colonized surfaces suggest net and gross primary productivity rates on the order of 600 and 1200 mg C m^{-2} $year^{-1}$, respectively. At this rate of incorporation, a fully developed community could replace itself within 300 years. Multiple lines of evidence suggest that the endolithic communities grow much more slowly than this. Radiocarbon dating of endolithic communities, carbon turnover times, comparisons of the rates of biogenic and abiogenic weathering, and estimates of the number of metabolically active hours per year and the number of divisions per year in the lower levels of the community all indicate that the age of a well-developed community is on the order of 1000–10,000 years. Some of the material, especially low-molecular-weight compounds lost during rewetting, may be leached from the community, ending up, eventually, in the surrounding soils. It is not clear how significant this leaching is. Chemical signatures indicative of the cryptoendolithic community are present in soils from the higher elevations in Taylor Valley, but not in soils from the lower elevations.

Cryptoendolithic communities are widespread outside of Antarctica, although not nearly as intensely investigated. In most cases, they occur in some sort of sandstone substratum. Here, they take on the same macroscopic appearance. There is a colored band beginning just below the surface of the rock, extending inward just a few millimeters at most. The composition of the community depends, at least to a degree, on the aridity of the environment. For example, in the vicinity of the Colorado Plateau in North America, the relative abundance of eukaryotic green algae increases in comparison with cyanobacteria in the milder upper elevations – at the milder sites, members of the green algal orders Chlorococcales and Chlorosarcinales comprise 50–75% of the phototrophic associations, while at the lower, more xeric sites, nearly 100% of the phototrophs are cyanobacteria. Generally speaking, however, most cryptoendolithic associations in hot deserts are dominated by members of the cyanobacterial genus *Chroococcidiopsis*; in many cases *Chroococcidiopsis* appears to be the sole phototrophic taxon present. The reduced importance of eukaryotic algae in hot desert endolithic associations is usually ascribed to temperature, rather than moisture concerns, although this has not been tested in detail; in fact, there are few data concerning the temperature ranges of hot desert forms.

Endolithic associations have also been reported from gypsum rocks in Antarctica, and in the hot Atacama, Jordanian, and Mojave deserts. Preliminary investigations of the Antarctic community indicated the presence of a member of the cyanobacterial genus *Chloroglea* as the principal phototrophic member, with the fungal genus *Verticillium* and the bacterial genus *Sphingomonas* also represented. Overall, the biodiversity appears to be low. The hot desert gypsums contained a much more diverse assemblage with molecular techniques indicating the presence of up to eight phylogenetic groups. The dominant phototrophs were not identified to genus but were morphologically similar to *Chroococcidiopsis*; samples from the relatively mild Mojave also contained a small number of filamentous forms.

A unique endolithic community exists in halite evaporites of nonmarine origin from the hyperarid core of the Atacama Desert. The evaporites form extensive nodular crusts at several sites in the Atacama, including Yungay, Salar de Llamara, and Salar Grande. The community forms a macroscopically visible grayish layer beginning a few millimeters below the surface and extending another 2–5 mm into the rock. The dominant phototrophic organism in the community appears to be a form of *Chroococcidiopsis* with a brown sheath; heterotrophic bacteria are also present, embedded in the sheaths of *Chroococcidiopsis*.

Hypolithic associations

Scientific study of hypolithic associations began with the pioneering work of Vogel in South Africa in the 1950s and Cameron and Blank in the American Southwest in the early 1960s (Figure 5). Since then there have been additional reports from the Namib Desert in southern Africa; the Negev Desert in the Middle East; the southwestern United States; Australia; the Atacama Desert in South America; the arid northwestern region of China; the Dry Valleys region of Antarctica; and the high Arctic. In most cases, colonized stones are composed of quartz, flint, or some other translucent material. These allow small but measurable amounts of photosynthetically active radiation to reach the phototrophic portion of the community, estimated to be on

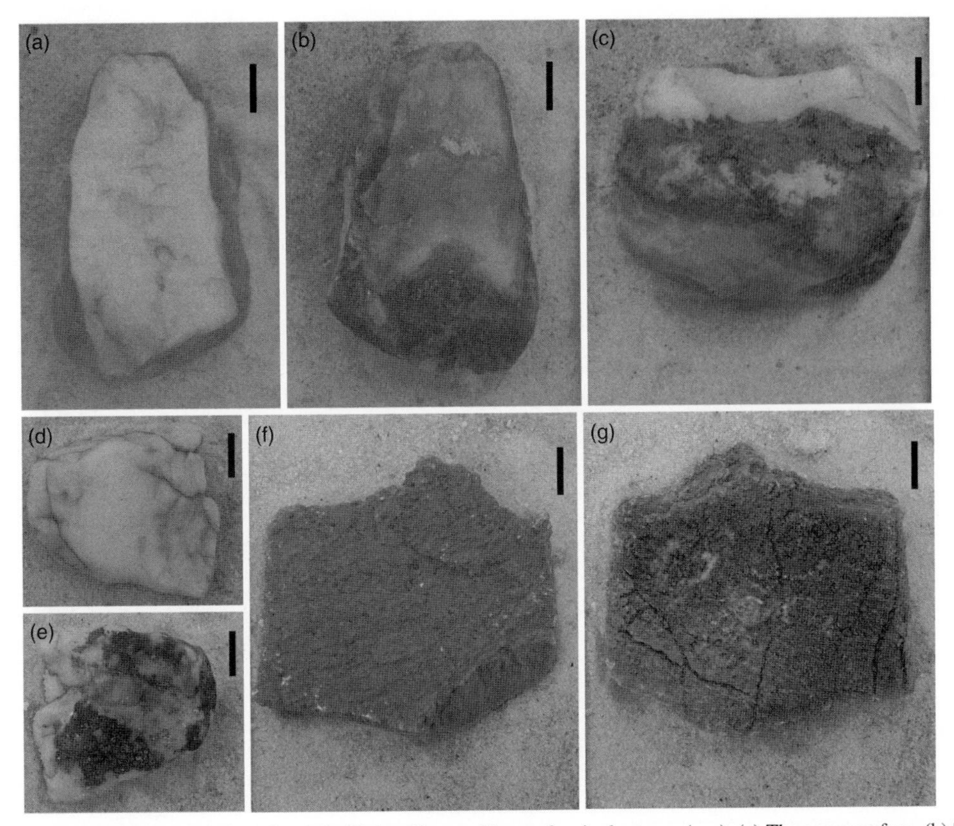

FIGURE 5 Examples of hypolithic associations from the Mojave Desert. Views of a single stone: (a–c). (a) The upper surface. (b) The lower surface. (c) View from one of the sides. Note the limited distribution of the photosynthetic association (greenish band). Two views of a smaller stone from the same region: (d–e). (d) The upper surface. (e) The lower surface. In this case, the photosynthetic association nearly covers the lower surface – the rock was less than 1 cm thick, allowing light to reach the entire lower surface. Two views of a third stone: (f–g). (f) The upper surface. (g) The lower surface, again with an almost complete cover by the hypolithic association. This example is somewhat unusual, because of the dark color of the stone. Upon closer examination, the coloration was found to be superficial; the interior of the stone is white. Scale bar = 1 cm. (For interpretation of the references to color in this figure legend, the reader is referred to the Web version of this chapter.)

the order of 0.01–0.1% of the incident light at the lowest colonized levels; similar numbers have been obtained for cryptoendolithic communities. In one case, colonized dolomites from the Arctic, no light transmission was detected through 1-mm-thick sections of the substratum, suggesting that light reaches the community through the surrounding coarse-grained soil.

Phototrophic hypolithic communities form macroscopically visible green to bluish-green films on the lower surfaces of colonized stones. In most of the instances, the reported community contains a remarkably diverse assemblage of species including a mixture of unicellular and filamentous cyanobacteria, green algae, and, in some cases, diatoms. For example, more than 23 species of hypolithic phototrophs have been recorded from Antarctica, including representatives of eight genera of cyanobacteria, three genera of green algae, two genera of yellow-green algae, and two species of diatoms. In the Namib Desert, representatives of over 50 species of just diatoms have been reported from hypolithic associations.

In more arid regions, however, the diversity decreases sharply. In both the Atacama and the arid regions of China, only cyanobacteria with a *Chroococcidiopsis*-like morphology have been observed, although molecular techniques indicated the possible presence of filamentous *Phormidium*-like cyanobacteria at colder or wetter sites. The percent abundance of hypolithic communities on suitable rocks is also moisture-dependent. In the relatively mild Mojave Desert, nearly 100% of quartz pebbles are colonized by hypoliths. In the nearby hyperarid desert of Baja California, in contrast, only between 26% and 38% of quartz stones were colonized. This shift is in keeping with extensive studies along a moisture gradient in the Atacama Desert. Here, 27.6% of quartz pebbles were colonized at the "moist" Copiapo end of the transect, while at the dry end of the transect, near Yungay, only 3 out of 3723 quartz pebbles were colonized. Thus, the hyperarid core of the Atacama appears to be near the dry limit of metabolic activity for both soil and lithic microorganisms. This makes the presence of what appears to be an active and extensive cryptoendolithic association in evaporites near Yungay even more intriguing (Table 1).

TABLE 1 Characteristics of the more common habitats

Habitats/Associations	Ecological Features
Edaphic habitats	On or in bare soils; surface fully exposed to the environment; exposure decreasing with depth; subject to disturbance (burial and erosion) from wind and flowing water, animal movement
Biological soil crusts	Fully exposed to the environment; insolation, desiccation, and wind erosion may be ameliorated by the structure of the association and the degree of connectivity with the underlying soil
Lithic habitats	
Epilithic	Fully exposed to the environment, organisms experience extremes of insolation, desiccation, long- and short-period temperature variation, wind erosion; generally restricted to relatively moist regions
Endolithic	With varying degrees of protection from environmental conditions: light filtered by the rock substratum; water retained by the physical structure of the substratum; wind erosion reduced by the overlying rock material; short-period temperature variations reduced, long-periods variations similar to those of the epilithic habitat; limited space for growth of the association
Chasmoendolithic	Protection from environmental conditions depending, in part, on the nature of the habitat – macroscopic cracks lead to more exposure to desiccation, but may be more efficient in collecting water and wind-blown nutrients; habitats with microscopic cracks similar to cryptoendolithic habitats
Cryptoendolithic	Generally more protected from environmental conditions than chasmolithic habitats because of the absence of a direct connection to the surface
Euendolithic	Restricted to the upper millimeter of the rock surface, therefore with a high degree of exposure to environmental conditions, especially insolation and desiccation
Hypolithic	Highest degree of protection from environmental conditions: light reduced by transmission through the overlying stone; long- and short-term temperature variations reduced; possible transport of moisture and nutrients to the habitat from the underlying soil
Perilithic	Subset of the hypolithic habitat relying on the condensation of water on the relatively large upper surface of the stone

ADAPTATIONS TO THE ENVIRONMENT

All desert microorganisms are subjected to extended periods in a desiccated, metabolically inactive state. However, in this inactive state, individual cells are subject to a variety of chemical and physical insults, resulting in damage that cannot be repaired until metabolism restarts. If the damage incurred while in the inactive state is too great, the cell is unable to make repairs fast enough, and dies. At the same time, the association as a whole is subject to losses by erosion and predation that cannot be replaced until cellular reproduction restarts. Therefore, any feature tending to prolong the periods of activity or to help the cell survive the periods of inactivity in a desiccated state would be adaptive. A number of such features have been identified. The principal adaptations are summarized below, following a brief review of the nature of the problem.

The Concepts of Water Activity and Water Potential

The net movement of water across a cell boundary is governed by the related concepts of water activity and water potential. Water activity (a_w) is a measure of the concentration of water in a solution, as modified by the type of solutes present in the solution. It is derived from a modification of Raoult's law relating the vapor pressure of a solution (p) to the vapor pressure of the pure solvent (p_0). In the ideal case, both a_w and p/p_0 would equal the mole fraction of water (N_w) defined as $n_w/(n_w + n_s)$, where n_w is the number of moles of water and n_s is the number of moles of all solutes in the system. In reality, unless the solutions are very dilute, the solutes interact with water, necessitating the inclusion of an empirical coefficient (γ_s). The water activity is then defined as

$$a_w = \frac{p}{p_0} = \gamma_s \frac{n_w}{(n_w + n_s)} = \gamma_s N_w.$$

The chemical potential of the water (μ_w) is given by

$$\mu_w = \mathring{\mu}_w + RT\ln(a_w) + V_w P + z_w FE + m_w gh,$$

where $\mathring{\mu}_w$ is the chemical energy of water under standard conditions (J mol^{-1}), R is the universal gas constant (~8.31447 J mol^{-1} K^{-1}), T is the temperature in Kelvin, V_w is the change in volume associated with a change in the amount of water (~$1.8 \times 10^{-5}\,m^3\,mol^{-1}$), P is the hydrostatic pressure ($N\,m^{-2}$), z_w is the charge on a molecule of water, F is the Faraday constant (~$9.6485 \times 10^4\,C\,mol^{-1}$), E is electrical potential ($J\,C^{-1}$), m_w is the molar mass of water (~$1.8 \times 10^{-2}\,kg\,mol^{-1}$), g is the acceleration due to gravity (~$9.8\,m\,s^{-2}$), and h is the height of the cell (m). The electrical term can be ignored because the charge on a molecule of water (z_w) is negligible. The gravitational term can also be ignored because of the small size of the cells. The chemical potential then reduces to

$$\mu_w = \mu_w^\circ + RT\ln(a_w) + V_w P,$$

which can be rewritten as

$$\frac{(\mu_w - \mu_w^\circ)}{V_w} = \frac{(RT\ln(a_w))}{V_w} + P.$$

This equation leads to the definition of the water potential (ψ) of the system as

$$\psi = \frac{(\mu_w - \mu_w^\circ)}{V_w}$$

with units in terms of pressure ($N\,m^{-2}$ or pascals). In this derivation, the water potential is the sum of two separate potentials, one based on the hydrostatic pressure (P), the other based on the chemical activity of water ($RT\ln(a_w))/V_w$. Conceptually, the water potential can be broken into additional terms, as appropriate. A typical formulation gives the total water potential (Ψ) as

$$\Psi = \psi_p + \psi_p + \psi_m + \psi_g$$

where ψ_π is the osmotic potential and refers to part of the water potential associated with dissolved substances in the system, ψ_p is the pressure potential and refers to part of the water potential associated with hydrostatic pressure, ψ_m is the matric potential and refers to part of the water potential associated with the interaction of water molecules with surfaces and interfaces, and ψ_g is the gravitational potential and refers to part of the water potential associated with gravitational forces.

A simple example may be instructive. Suppose a cell is suspended in seawater so that the external matric and gravitational potentials are negligible, and that the external hydrostatic pressure (ψ_p) is 1 atm (1.01325 bars or 1.01325×10^5 Pa) directed inward. Because the activity of seawater at 25 °C is 0.9806, the external osmotic potential (ψ_π) is -2.70×10^6 Pa (−2.70 MPa); the negative sign indicates that the "pressure" is directed outward. The total water potential experienced by the cell ($\Psi_{environment}$) is given by $\psi_\pi + \psi_p = -2.6$ MPa. At equilibrium, the total internal water potential of the cell (Ψ_{cell}), which equals the sum of the internal osmotic, hydrostatic, and matric potentials, would be the same, although none of the individual internal potentials would necessarily match any of the individual external potentials. If that same cell were now plunged into pure water ($a_w = 1.0$), $\Psi_{environment}$ would increase to +0.1 MPa, while Ψ_{cell} would start at −2.6 MPa. Such an energy differential cannot be maintained. Water would begin to flow down the energy gradient into the cell. The additional water would raise the mole fraction and water activity within the cell, thereby bringing ψ_π closer to 0; it is also possible that some of the solutes may be lost. At the same time, the increased volume of water would correspond to an increased hydrostatic pressure directed outward. If the cell membrane and/or cell wall can withstand the increased hydrostatic pressure without rupturing, the cell will eventually come to a new equilibrium, with Ψ_{cell} again equal to $\Psi_{environment}$. Of course, if the cell were moved to more concentrated seawater, with a salinity of 70‰ instead of 35‰ and a water activity of ~0.96, $\Psi_{environment}$ would decrease to −5.82 MPa, and water would flow out of the cell until the new equilibrium is reached.

Seawater would represent an environment with relatively constant, even if reduced, water activity. Of course, seawater, even double-strength seawater, does not constitute a dry environment and a wide range of microorganisms can survive in it without major modification. More relevant environments would be found in hypersaline lakes and salinas, seawater-based evaporites (a_w ~0.75) or in cereals, candy and other confectionaries, and dried fruits (a_w ~0.70). Organisms living in these habitats would be subjected to chronic conditions of low water potential and may even be considered to be under chronic water stress. The constancy of the stress allows for some degree of predictability and biological accommodation. Even so, there are only a limited number of microorganisms capable of growing at water activities below 0.85; only two, the yeast *Zygosaccharomyces rouxii* and the filamentous ascomycete *Xeromyces bisporus*, are capable of growth at water activities below 0.70. Terrestrial microorganisms, especially subaerial and desert forms, face a much more difficult moisture environment. The water activity of the external environment, roughly equivalent to the relative humidity of the air divided by 100, is not constant, but can vary between lows near 0.10 and highs near 1.00, all within a single day. Upshifts in the external water activity due to changes in the relatively humidity are ameliorated somewhat by the slow pace of the diffusion of the water molecules in the air. When rains appear, however, the cells are suddenly plunged into a liquid medium with a much higher water potential. This sudden shift creates an acute stress not experienced by the food- or evaporite-inhabiting microorganisms.

Impacts of Low Water Activity on Unicellular Organisms

The irrefutable feature of exposure to low water activities and low water potentials is the removal of water from the cell. Four distinct states for intracellular water can be identified based on the total water content. In the initial state, the "solution domain," water is free-flowing and metabolic reactions are still possible, albeit at possibly reduced rates. Once a certain water content is reached, ~20% in the desiccation-tolerant yeast *Saccharomyces cerevisiae*, the cell enters what is call a "gel domain," where water molecules no longer form a continuous solvent and metabolic activity ceases. Further reductions in water content lead to two additional states distinguished by how tightly the remaining water is bound.

These decreases in water content are accompanied by changes in the structure of the phospholipid bilayer. As the number of water molecules decreases, the hydration state of the individual molecules forming the bilayer changes, leading to an increase in the temperature at which the membrane changes from the biologically important liquid crystalline state to the gel state. Because of these changes, in the dehydrated state the membrane components are in the gel state at physiological temperatures. As water is added back into the system, and the gel–liquid crystalline transition temperature drops back to physiological levels, there is a transient state when the membrane is part liquid crystalline and part gel. Such two-part systems are highly permeable to small molecules. For non-desiccation-resistant microorganisms, these changes in the cell membrane may be the primary cause for cell death during freeze-drying.

Presumably, changes in the hydration states of individual proteins, similar to those experienced by molecules of the bilayer, will occur as the amount of water decreases. Such changes could impact the three-dimensional structure of the proteins, especially where hydrophobic interactions are involved, causing potentially irreversible denaturation. How much damage is caused by this mechanism is not clear. For the desiccation-tolerant cyanobacteria *N. commune*, proteins involved in lipid biosynthesis and in glycerol, sulfate, and phosphate uptake appear to be undamaged, while phyobiliproteins involved in light harvesting are rapidly degraded.

In contrast with the case for proteins, there is clear evidence that DNA is a major target of desiccation-induced damage. Conformational changes accompanying the loss of hydration shells allow the formation of cross-links between the phosphodiester group of DNA and a protein-bound OH group. These cross-links result in single-strand and double-strand breaks in the DNA, as many as 10 breaks per single-strand genome after 12 min of drying in the desiccation-sensitive strain *Escherichia coli* K-12 AB 1157. Even desiccation-resistant *Bacillus subtilis* spores and *Aspergillus ochraceus* conida display increasing amounts of DNA–protein cross-linking during exposure to desiccation. Depurination of individual nucleotides may also be a significant source of DNA lesions, with the rate for desiccated samples of *N. commune* estimated to be sufficient to depurinate 1% of the genome in 10 years.

The presence of reducing sugars and O_2 can increase the amount of damage sustained by all components. Maillard reactions, a series of browning reactions between reducing sugars and amino acids, may occur, thereby damaging proteins, although the rate would be expected to be low in the desiccated state. There is evidence that reducing sugars can also interact with DNA and membranes. These reactions would also occur in hydrated cells, but in the hydrated case repair mechanisms not available in the desiccated state would be in operation. It is also expected that active species of oxygen (O_2^-, H_2O_2, and OH^-) will form in drying cells under the influence of light. The presence of appreciable amounts of these species would lead to increased rates of single-stranded breaks in DNA, oxidation of proteins, and peroxidation of lipids, all of which cannot be repaired in the inactive state.

Adaptations to Arid Environments

As mentioned previously, a variety of adaptations to arid environments can be identified. In the following discussion, these are grouped according to the feature of the environment that the adaptations appear to address. Note, however, that this is only a loose grouping and that many of these adaptations serve multiple purposes. For example, the presence of a sheath can serve to retard water loss, maintain cell membrane integrity, and protect against photodamage (Table 2).

Adaptations to prolong the period of activity

Higher plants and terrestrial animals remain active in desiccating environments through the formation of complex multicellular structures with essentially watertight surfaces. This option is not open to unicellular forms. However, they can use extracellular structures to accomplish the same end, albeit not as efficiently. For example, nearly all terrestrial cyanobacteria, many eukaryotic algae, and the cryptoendolithic black fungi from Antarctica form extensive mucilaginous sheaths, as much as 98% of which may be water. This water reserve can prolong the period of activity under desiccating conditions from several hours to a few days. In soil crusts and mats, the polysaccharide sheaths may form a physical barrier, which serves to retard evaporation and retain moisture in the lower levels. Soils and the porous rocks and stones of the cryptoendolithic habitat may also serve as water reservoirs for the community, and help to retain water at low water activities either through their physical or through their chemical structure.

TABLE 2 A partial list of the major innate adaptations to the dry environment

Adaptation	Function(s)	Examples
Mucilaginous sheaths (extracellular polysaccharides)	Water storage; cell membrane stabilization	Nearly all terrestrial cyanobacteria, including *Chroococcidiopsis, Gloeocapsa, Microcoleus, Scytonema,* and *Nostoc;* many chlorophytes including *Hemichloris;* black fungi *Friedmanniomyces, Cryomyces*
Production of compatible solutes	Cell membrane stabilization; stabilization of the tertiary structure of proteins and DNA	*Chroococcidiopsis, Nostoc;* the fungi *Zygosaccharomyces, Xeromyces*
Photoprotective pigments in sheaths and walls	Absorption of UV and high-energy visible light	Many cyanobacteria including *Chroococcidiopsis, Gloeocapsa* and *Scytonema* (scytonemin and MAA's); black fungi (melanin)
Carotenoid production	Quenching of free radicals	*Deinococcus,* cyanobacteria such as *Nostoc;* many chlorophyte algae including *Trentepohlia, Haematococcus, Chlorella* spp.
Efficient DNA repair mechanisms	Repair the damage incurred during desiccation or exposure to radiation	*Deinococcus,* possibly the cyanobacterium *Chroococcidiopsis*
Multiple copies of the genome	Provide multiple copies of essential genes, serve as template for DNA repair	Possibly *Chroococcidiopsis, Nostoc*
Spore formation; encystment	Specialized cells able to resist dry conditions	endospore-forming bacteria, members of the Actinobacteria; most fungi
Motility	Avoidance of excessive insolation, possible movement toward higher concentrations of water	*Microcoleus, Oscillatoria,* and some other filamentous cyanobacteria; some flagellated eukaryotic algae

Examples are for illustrative purposes only; the specific adaptations of most microorganisms found in dry environments have not been investigated.

The hypolithic association enjoys a particularly favorable habitat in this regard. The dominant microorganisms, usually a member of *Chroococcidiopsis* or *Gloeocapsa*, form thick sheaths extending some distance from the surface of the stone. In addition, the nonporous stones overlying the association serve as efficient moisture barriers. Therefore, it is not surprising that the moisture content in the soil beneath the association is often much higher than in the surrounding soil. This effect can be enhanced by the thermal properties of the stone. Large stones can generate thermal gradients within the soil that result in the net movement of water vapor through the soil to the hypolithic environment; in effect, the lower surface of the stone acts as a water collection system. However, for the water collection system to be effective, there must be significant quantities of moisture available in the soil. Therefore, it is not too surprising that attempts to measure this effect in the core of the Atacama have been unsuccessful. The upper surfaces of stones are also more efficient dew, fog, and mist collectors than soil surfaces. This may account for the presence of desert varnish communities in semiarid and arid regions. It is also possible for the collected water to pool on the surface and then trickle off the surface to moisten the soil around the edges and just below the stone, thereby supporting the hypolithic association. The effect is enhanced with large, nonporous stones, explaining the preference for colonization of larger quartz stones, especially in the perilithic habitat, in hyperarid deserts.

A number of eukaryotic algae are capable of absorbing sufficient water vapor when the relative humidity is high to maintain some level of metabolic activity. Included in these are a number of forms commonly found in desert environments, including strains of the green alga *Stichococcus*. It is generally believed that this ability is not present in cyanobacteria or other prokaryotic organisms. However, the desert crust cyanobacterium *Microcoleus sociatum* may be able to use water vapor at relative humidities in excess of 95%. The mechanism behind this ability is unknown. In the case of *Microcoleus*, it may result from the structure of the filament – several trichomes within a communal sheath. The compact arrangement could lead to the formation of liquid water within small capillaries. In the case of unicellular forms such as *Stichococcus* or the subaerial taxon *Apatococcus*, it is presumed to be related to the buildup of compatible solutes in the cell (see below), although this is yet to be tested.

Some of the filamentous cyanobacteria present in desert crusts are known to be motile during wet conditions, rising to the surface of the soil and causing a brief "greening" of the desert and then retreating back into the soil to take up residence a few millimeters below the surface. In *Microcoleus*, the movement is apparently caused by a combination of physical exclusion of the living trichomes from the polysaccharide sheath as it swells with water and typical cyanobacterial gliding motility. In other cases, gliding motility alone is sufficient. Much of the movement appears to be directed upward toward the surface of the soil, possibly in a phototactic response, possibly as part of a random process. As the soil dries, at least some members of the association track the moisture gradient back into the soil, in a hydrotactic response.

Adaptations to survive desiccation

Even with the most efficient means of prolonging water availability, desert microorganisms will become desiccated. At present, little is known of the mechanisms by which desert microorganisms survive the drying process. The first step seems to be the formation of some sort of compatible solute. Compatible solutes are low-molecular-weight organic compounds that can serve to lower the osmotic potential and, therefore, the total water potential within the cell while maintaining cellular and molecular integrity. A large number of different types of molecules have been implicated as compatible solutes, including polyols, sugars and sugar derivatives, betaines, some amino acids and amino acid derivatives, and ectoines. Trehalose and sucrose seem to be particularly important in organisms subjected to extreme water stress; thus, it is not surprising that desert *Chroococcidiopsis* sp. and *N. commune* are both known to accumulate trehalose and sucrose under conditions of water stress. Trehalose and sucrose both seem to work by stabilizing cell membranes by replacing the water bound to phosphate groups, and thereby depressing the phase transition temperature of desiccated membranes to physiological temperatures, and by preventing the fusion of membrane vesicles; they may also stabilize protein and DNA structure. The extracellular polysaccharides may work in the same way to stabilize the outer membranes of bacteria or the outer layers of the cell membrane of other organisms. It is interesting, in this respect, to note that cyanobacteria with intact, lamellated sheaths have greater viability than those without.

Adaptations to reduce the damage while in the desiccated state

Phototrophic microorganisms are all subject to the accumulation of photooxidative damage resulting from the reduction of oxygen and other molecules by photosystems only partially active in the desiccated state. In addition, all desert microorganisms are subject to the accumulation of damage caused by UV light. Therefore, the existence of a variety of mechanisms to reduce overall exposure to light is not surprising. Simplest among these is growth in low-light environments. Both hypolithic and cryptoendolithic microorganisms inhabit environments where the maximum photon flux is several orders of magnitude lower than that of the soil or rock surface. Motile crust-forming cyanobacteria, for example, *Microcoleus* and *Oscillatoria*, form layers several millimeters below the surface of the soil during dry conditions; here, again, the photon flux is much reduced. Nonmotile members of desert crusts, for example, *Scytonema*, and cryptoendolithic inhabiting the upper layers of the association usually form some sort of photoprotective pigment. Many desert and subaerial cyanobacteria produce the brown sheath pigment scytonemin and/or water-soluble mycosporine-like amino acids, both of which are known to provide some degree of protection against UV and high-energy visible light. Many lithophytic and lichen-forming fungi produce melanin, mycosporine and mycosporine-like amino acids, and/or carotenoids. Terrestrial eukaryotic algae generally lack colored sheaths and walls, suggesting that they are more exposed to photooxidative damage than cyanobacteria. This may help explain the predominance of cyanobacteria in the more arid deserts.

Adaptations leading to a quick recovery

The final set of adaptations seen in desert organisms are those leading to a quick recovery from damage associated with desiccation, especially damage to the DNA. These have been best studied in *Deinococcus radiodurans*, primarily in regard to damage from ionizing radiation. Several mechanisms have been implicated, including the presence of multiple copies of the genome within individual cells, delays in DNA replication until damage has been repaired, physiologic responses including the induction of a suite of genes following exposure to desiccation and ionizing radiation, protein-based protection of the broken ends of the DNA molecule, RecA-independent double-strand break repair, and novel forms of RecA repair mechanisms. The multiple copies of the genome reduce the risk of inactivating all copies of individual genes at the same time. They also provide reserve of information that can be used to repair damaged sequences. Not all of the stress-induced proteins have been assigned functions, but at least five have been implicated in genome repair. One of these, DdrA, binds to the 3′ end of single-stranded DNA, protecting it from degradation by nucleases *in vitro*. Presumably, within desiccated or irradiated cells it would bind to the broken ends of the DNA, protecting them until repair can be accomplished.

How far the mechanisms used by *Deinococcus* can be applied to other members of the desert community is not clear. Desert strains of *Chroococcidiopsis*, also known to be radiation-resistant, are postulated to rely on multiple genome copies, but this has not been investigated in detail. *N. commune* is thought to rely more on protective mechanisms, high trehalose concentrations, in particular, to reduce the amount of damage occurring during desiccation. Clearly, more research is necessary in this area.

RECENT DEVELOPMENTS

Pattern Formation in Biological Soil Crusts

Recently, attempts have been made to further categorize the types of biological soil crusts encountered. The most comprehensive of these first classifies crusts as epiterranean, endoterranean, or a mixture of the two based on the position of their photosynthetic components relative to the soil surface. This division is reminiscent of that proposed by Friedmann and his colleagues in the 1960s, where epedaphic and endedaphic were used in place of epiterranean and endoterranean. It also gives primacy to the degree of exposure experienced by the association. Next, aspects of organization of the cyanobacterial association are taken into account. Important features include the degree of stratification within the crust and the growth form (horizontal filaments, erect and/or fasciculated filaments, or globose colonies) of the dominant cyanobacterial members of the crust. These latter features are presumed to develop through the interaction of the crust with the surrounding environment. If true, we may be able to make specific inferences concerning environmental conditions based on the type of crust observed.

Relevance to Astrobiology

The search for life on Mars

According to the currently accepted theory, about 3.8 billion years ago conditions on Mars included flowing liquid water and surface temperatures above zero. While these conditions did not last long, it is possible that they were of sufficient duration that life could have arisen independently on Mars at about the same time that it arose on Earth. In addition, much of the surface of Mars, unlike the surface of Earth, remains relatively unchanged since that period. Therefore, Mars may have the best record of the origin of life on Earth-like planets. One focus of astrobiolological research is to develop strategies and techniques for the in situ search for indications of the remnants of these early stages of life. Dry environments on Earth are useful in this context for two reasons. First, as the environment on Mars changed, the surface would have become increasingly drier, approximating the conditions in arid and hyper-arid regions currently in existence on Earth. These regions, therefore, serve as models of the types of ecosystems that may have been present during the final stages of life on Mars. The continued development tools and techniques for the morphological and chemical analysis of the microbial associations found in the arid regions of Earth is clearly central to the development of similar tools and techniques for the search for the remnants of life on Mars.

Interplanetary exchange of microorganisms: Testing panspermia

Allen Hills 84001 (ALH 84001), a meteorite thought to have originated on Mars, gained notoriety when it was claimed that it contained evidence of extra-terrestrial life. While the initial claims remain, at best, controversial, the discussions that ensued led to a renewed interest in panspermia, the theory that organisms can be transferred between planets by natural processes. The current form of the theory, lithopanspermia, suggests that rocks containing microorganisms are ejected from a planetary body, such as Mars, by the impact of a large asteroid or comet. The rocks, with their microbiota relatively intact, eventually reach another planet where the microorganisms take up residence. Because the organisms would have to be present in the interior of the rock before ejection from the planetary body, endolithic communities, especially those from arid regions are considered candidates for testing this theory.

For lithopanspermia to be successful, the microbiota must survive the initial ejection from the planetary body, the subsequent journey through space, and re-entry on a second planetary body. Two of these have been directly tested using endolithic strains of *Chroococcidiopsis;* it has been inferred, based on its desiccation and radiation resistance that, if properly shielded, *Chroococcidiopsis* could survive prolonged exposure to the vacuum of space. In one series of experiments designed to test the survival rate of cells exposed to shock pressures mimicking those of an asteroid impact/ejection event, *Chroococcidiopsis* was able to survive shocks of up to 10 GPa; higher shocks resulted in mechanical damage to the cells. The forces associated with this shock may not be sufficient to eject a stone from a Mars-like planet. However, other microorganisms, including at least one species of lichens and bacterial spores, could withstand much greater shocks, including shocks sufficient for ejection.

The survivability of *Chroococcidiopsis* upon re-entry was tested using the European Space Agency's STONE facility. In the system, pieces of rock are fixed to the surface of the heat shield of a FOTON re-entry vehicle near the ablation point. The system is launched into space, then re-enters the atmosphere at speeds near 8 km s^{-1}, somewhat slower than meteorites. In an early experiment a porous impactite was inoculated with a culture *Chroococcidiopsis* to a depth similar to that of naturally occurring endolithic associations. The inoculum was obliterated by the heat of re-entry; the upper few millimeters of the impactite had melted, forming a fusion crust. In a follow-up experiment, the back of a 2-cm thick piece of rock was inoculated with a culture of *Chroococcidiopsis*. A piece of lichen and bacterial endospores were also included. Again, none of the biological material survived the high-temperatures associated with re-entry, although the remains of the cells and biochemical markers were detected. These results suggest that for lithopanspermia to occur, the microorganism must be protected by several centimeters of rock material.

FURTHER READING

Belnap, J., & Lange, O. L. (Eds.) (2003). *Biological soil crusts: Structure, function, and management*. Berlin: Springer Revised 2nd printing. Ecological Studies 150.

Brown, A. D. (1990). *Microbial water stress physiology*. New York: John Wiley & Sons.

Cockell, C. S. (2008). The interplanetary exchange of photosynthesis. *Origins of Life and the Evolution of Biospheres*, *38*, 87–104.

Crowe, J. H., & Crowe, L. M. (2002). Freeze-drying: Preservation of microorganisms by freeze-drying. In G. Bitton (Ed.) *Encyclopedia of environmental microbiology* (Vol. 3) (pp. 1350–1359). New York: John Wiley & Sons, Inc.

Cox, M. M., & Battista, J. R. (2005). *Deinococcus radiodurans* – The consumate survivor. *Nature Reviews Microbiology*, *3*, 882–892.

Dose, K. (2002). Desiccation by exposure to space vacuum or extremely dry deserts: Effect on microorganisms. In G. Bitton (Ed.) *Encyclopedia of environmental microbiology* (Vol. 2) (pp. 1029–1041). New York: John Wiley & Sons, Inc.

Friedmann, E. I. (Ed.) (1993). *Antarctic microbiology*. New York: Wiley-Liss.

Grant, W. D. (2004). Life at low water activity. *Philosophical Transactions of the Royal Society of London B*, *359*, 1249–1267.

Kieft, T. L. (2002). Hot desert soil microbial communities. In G. Bitton (Ed.) *Encyclopedia of environmental microbiology* (vol. 3) (pp. 1576–1586). New York: John Wiley & Sons, Inc.

Navarro-Gonźlez, R., Rainey, F. A., Molina, P. *et al.* (2003). Mars-like soils in the Atacama Desert, Chile, and the dry limit of microbial life. *Science*, *302*, 1018–1021.

Oren, A., & Gunde-Cimerman, N. (2007). Mycosporines and mycosporine-like amino acids: UV protectants or multipurpose secondary metabolites. *FEMS Microbiology Letters*, *269*, 1–10.

Perry, R. S., & Kolb, V. M. (2004). Biological and organic constituents of desert varnish: Review and new hypotheses. In R. Hoover, & A. Rozanov (Eds.) *Proceedings of the SPIE* (Vol. 5163) (pp. 202–217). Instruments, Methods, and Missions for Astrobiology VII.

Pócs, T. (2008). Cyanobacterial crust types, as strategies for survival in extreme habitats. *Acta Botanica Hungarica*, *51*(1–2), 147–178.

Potts, M. (1994). Desiccation tolerance in prokaryotes. *Microbiological Reviews*, *58*, 755–805.

Selbmann, L., de Hoog, G. S., Mazzaglia, A., Friedmann, E. I., & Onofri, S. (2005). Fungi at the edge of life: Cryptoendolithic black fungi from Antarctic desert. *Studies in Mycology*, *51*, 1–32.

Whitton, B. A., & Potts, M. (Eds.) *The ecology of cyanobacteria: Their diversity in time and space* (2000). Amsterdam: Kluwer.

Wierzchos, J., Ascaso, C., & McKay, C. P. (2006). Endolithic cyanobacteria in halite rocks from the hyperarid core of the Atacama Desert. *Astrobiology*, *6*, 415–422.

Chapter 25

Marine Habitats

D.M. Karl[1], R. Letelier[2]
[1]University of Hawaii, Honolulu, HI, USA
[2]Oregon State University, Corvallis, OR, USA

Chapter Outline

ABBREVIATIONS

ALOHA A Long-term Oligotrophic Habitat Assessment
ATP Adenosine triphosphate
BATS Bermuda Atlantic Time-series Study
CZCS Coastal Zone Color Scanner
DCML Deep Chlorophyll Maximum Layer
DOM Dissolved organic matter
DON Dissolved organic nitrogen
HOT Hawaii Ocean Time-series
HTL Higher trophic level
MODIS Moderate Resolution Imaging Spectroradiometer
NPSG North Pacific Subtropical Gyre
OSP Ocean Station Papa
PAR Photosynthetically Available Radiation
SEATS SouthEast Asia Time-Series
WOCE World Ocean Circulation Experiment

DEFINING STATEMENT

A habitat is the natural abode of an organism. The marine habitat is composed of a diverse spectrum of environments each supporting the proliferation of a diverse assemblage of microorganisms. When habitats vary, for example as a result of seasonal and longer term climate forcing, the diversity and function of the microbial assemblage will also change. The North Pacific Subtropical Gyre (NPSG), one of Earth's largest habitats, is an excellent example of a marine habitat in motion with respect to microbial structure and function.

INTRODUCTION

A habitat is often defined as the natural abode or place of residence. For this reason, the global ocean may be considered as one of the largest and oldest habitats on Earth; it covers 71% of the Earth's surface to a mean depth of 3.8 kilometers and comprises >95% of Earth's probable living space (Figure 1). However, despite the appearance of homogeneity, the ocean is actually a complex mosaic of many different macrohabitats that can be identified, studied, and compared; each macrohabitat has a potentially distinct assemblage of microorganisms. Examples are rocky intertidal, coastal upwelling, deep sea hydrothermal vent, and open ocean habitats.

Some of these habitats are critical to the mass balance of elements in the ocean because they are found at the interface or boundary between terrestrial, or freshwater, and marine systems. For instance, within estuarine habitats, which are located at the freshwater–marine boundary, processes can trap dissolved nutrients and particulate matter, or can export energy in the form of organic matter and nutrients to surrounding coastal habitats. The characteristics of any given estuary will vary depending upon dimensions, hydrology, and geographical location. And, because they represent gradual transition zones, estuaries highlight

Topics in Ecological and Environmental Microbiology.

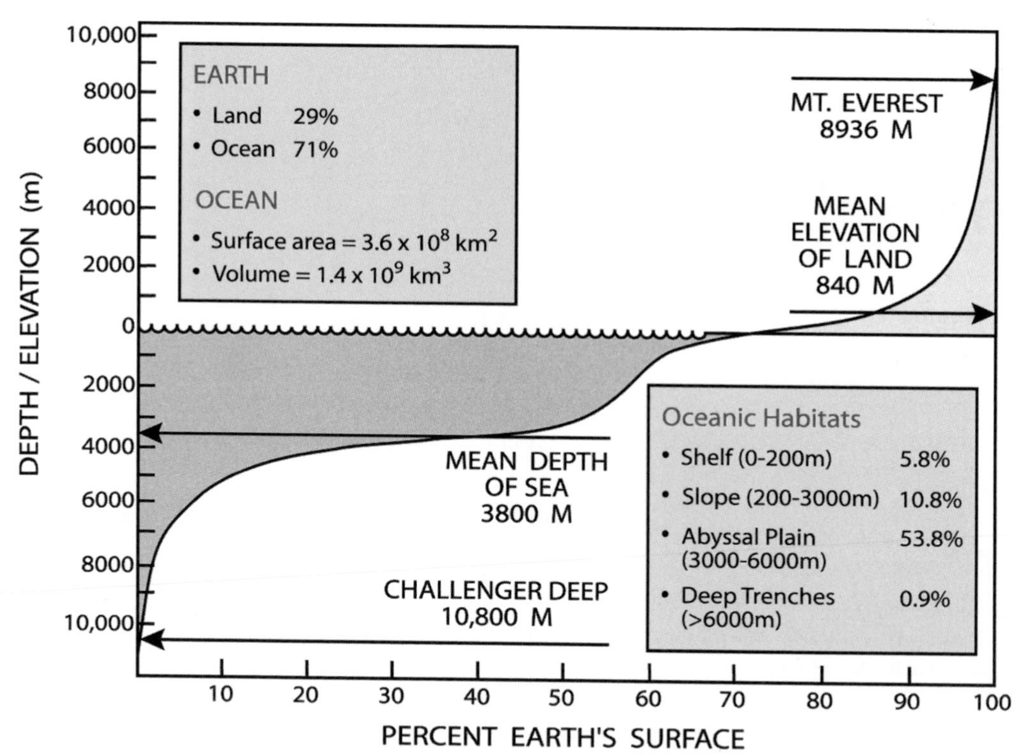

FIGURE 1 The world ocean covers 71% of Earth's surface with the deep blue sea (regions seaward of the continental shelf) accounting for more than 60% of the total. With an average depth of 3800 m, the volume of habitable space on planet Earth is dominated by marine habitats. Because microbes dominate the marine environment, they are directly or indirectly linked to most global processes and are largely responsible for the habitability of our planet. *Reproduced, with permission, from* Microbial Oceanography *(http://www.agi.org)*.

the difficulty of defining the boundaries of marine habitats. The continental–marine boundary also includes many specialized and important microbial habitats including fjords, salt marshes, mangrove stands, coral reefs, kelp forests, and many man-made or human-impacted (e.g., harbors, sewage outfalls, mariculture farms, gas and oil production facilities) zones.

Another approach to define a marine habitat is through the comprehensive list of physical, chemical, and biological parameters experienced directly by the organism during its lifetime; collectively these parameters determine the success or failure of a particular strain, species, or assemblage of microbes. Given this definition, the size, motility, and lifespan of the microbes under consideration define the spatial and temporal scale of their habitat. Hence, it is important to recognize that most microorganisms live in "microhabitats" that are defined on space scales of micrometers to millimeters and time scales of hours to weeks.

Microhabitats are often the sites of elevated microbial biomass and accelerated metabolism. Once colonized by microorganisms, the environmental conditions within a given microhabitat (e.g., pH, redox level, nutrients, and dissolved gas concentrations) can change as a result of the metabolic activities of the microbial assemblage. Consequently, a description of the macroscopic surrounding habitat (scales of meters to kilometers) may not always be a valid representation of the true habitat of existence. This is the reason why otherwise incompatible microbes, for example, obligate aerobic, and anaerobic microorganisms can co-occur in a single environmental sample. Therefore, any given marine habitat – particularly when viewed on macroscale (meter or more) – is likely to be composed of numerous microhabitats that collectively support the growth and proliferation of the microbial assemblage as a whole.

Historically, there has been a concerted effort to define only the physical/chemical properties of a given marine habitat, but more recently the key role of biotic factors in establishing and maintaining microbial community structure and function has been recognized. Many microorganisms in the sea live attached to surfaces including nonliving particulate matter and living organisms. These surfaces help to create and sustain the above-mentioned microhabitats; at the same time they provide local enrichments of organic matter. In addition, bacteria and other microbes can also be found within the digestive tracts of marine macroorganisms, from small crustaceans to large cetaceans; it is likely that every macroorganism in the sea has a unique, species-specific set of microbial partners that are not commonly found elsewhere. And, because many marine macroorganisms (e.g., tuna, squid, and seabirds) have broad geographical distributions and enormous migratory capability, these animal vectors may affect the geographic distribution range of their microbial partners. Additional biotic interactions include virus–host interactions, gene exchange between otherwise unrelated microorganisms, microbe–microbe and microbe–macroorganism symbioses, obligate metabolic partnerships (syntrophy), and coevolution of different groups of microorganisms.

For syntrophic relationships to succeed there needs to be a high probability of juxtaposition of cell types to maximize interspecific interactions. This can be achieved in a homogeneous environment where microbes live freely but in close proximity, or through an ordered spatial structure of the microbial assemblage as a whole. In marine sediments and other "solid" habitats, for example, microbial communities are often established along diffusion gradients with each cell type growing in the most favorable microhabitat within the gradient in order to maximize success. Growth along these stable gradients, where stability is defined as a function of microbial generation time, can lead to the development of microbial (usually bacterial) mats. Analogous features termed microbial lens or plates can occur in highly stratified "fluid" habitats, for example, at the oxic–anoxic boundaries of permanently anoxic basins such as the Black Sea or Cariaco Basin. Metabolic activities of microorganisms in these mats and lens can be very high, thereby producing extremely steep vertical gradients of biomass and other metabolic by-products.

Even though we know that microbes are by far the largest contributors to living matter in the sea and have been responsible for the development of the atmosphere under which terrestrial life evolved, much of the research on the role of the habitat in structuring marine microbial communities and their ecosystem function is fairly new, incomplete, and lacking any formal theoretical description or predictive capability under changing habitat conditions. And, because seascapes are changing, in part, due to the activities of human populations, a comprehensive understanding of sea microbes and their activities under various global environmental change scenarios is a major and urgent intellectual challenge.

NATURE OF MARINE MICROBIAL LIFE

Life on Earth most likely began in the sea; so the marine environment was the original habitat for the growth and proliferation of microorganisms. As the pioneering prokaryotes evolved into more complex life forms, including multicellular macroscopic organisms, and radiated into freshwater habitats and eventually onto land, the imprint of a marine origin remained. Today, virtually all life is intimately dependent upon the availability of water; even desert microbes are aquatic.

Aquatic habitats are built around the unique properties of water. The most important criterion is the fact that water is a polar molecule having positively and negatively charged sides. This characteristic establishes its high dielectric constant and effectiveness as a solvent, setting the stage for the high dissolved ionic (salt) composition of seawater (referred to as salinity). During the HMS *Challenger* expedition of 1872–1876, the "law of constant proportions" was confirmed, namely that the ratios of the major ions in seawater are relatively constant throughout the world ocean. This relative ionic stability has very important implications for the evolution of marine microorganisms. Furthermore, the unique solvation property of water also facilitates nutrient delivery to and waste material export from the cell, thereby sustaining microbial metabolism.

Other water properties including density and gas solubility, which vary with temperature and salinity, can also have major implications for the distribution and abundance of microorganisms. The density of pure water at 4 °C is $1.000\,g\,cm^{-3}$, decreasing slightly to $0.994\,g\,cm^{-3}$ at 35 °C; the average density of seawater is ~1.025 at 25 °C [in marine sciences, the density is often expressed as an anomaly ((density − 1.000) × 1000], to amplify the small differences in density between freshwater and seawater; for example, the density anomaly of average seawater would be ([(1.025 − 1.000) × 1000] or 25.0). This means that river and rain water will float on seawater, as will surface waters warmed by the sun, whereas colder and saltier seawater will sink. As a result, the world's ocean is highly stratified with depth; mass exchange and transport occur mainly within layers of constant density (along isopycnals) or via turbulent mixing of waters with different densities (across isopycnals). Diffusional exchange processes that depend on molecular kinetic energy are, by comparison, very slow. This vertical density stratification, a hallmark of the marine environment, tends to insulate inorganic nutrient-rich deep seawaters from the sunlit surface region where the capture of solar energy by biological systems (photosynthesis) occurs. Consequently, density stratification strongly influences the rate of organic matter production and attendant ecosystem services.

In order for marine organisms to live in the water column they need to remain in suspension. The specific gravity of marine microbes varies with their bulk chemical composition (e.g., protein = 1.5, nucleic acids = 2.0, lipids = $0.9\,g\,cm^{-3}$) which is, in turn, dependent upon nutrient supply, growth rate, and other biotic factors. If the mean cell density is less than the density of seawater, the cells will tend to rise. Conversely, if the bulk cell density is greater than seawater or if microbes are attached to dense particulate materials, cells will settle. Many marine microorganisms adjust their density by the formation (or collapse) of gas vacuoles, alterations in the ionic composition of the cytoplasm, or by adjusting the above-mentioned composition of the cell, for example, by the synthesis of storage components such as carbohydrates that can serve as ballast. The rate at which cells will rise or settle is also related to their size causing small cells to remain suspended in their environment. Furthermore, most microorganisms, even small bacteria, are motile by means of one or more flagella, but movement through a relatively viscous medium at low Reynold's numbers can be difficult (the Reynold's number is a dimensionless metric that determines whether inertial or viscous forces dominate motion of an object in a fluid). Small bacteria and virus particles are also displaced by means of Brownian motion, a process driven by the random movement of water molecules that can act as a counter-force to gravitational settling. However, even though many marine microorganisms in the water column are motile, their directed movements are small relative to the marine currents in which they reside. Hence, they drift with the currents and are commonly known as plankton (from the Greek root *Plankto*, which means "wandering").

In addition to its role in density stratification, solar radiation provides most of the energy required to fuel the biological activity in marine environments. Water has a characteristic solar absorption spectrum that allows electromagnetic radiation between ~350 and 700 nm to penetrate to various depths in the water column depending upon surface solar irradiance, sun angle, and water clarity; the euphotic zone is generally defined as the water column region located above a specific isolume (a constant daily level of irradiance) or above a specific percentage (usually 1 or 0.1%) of the surface irradiance. The water absorption spectrum, with a maximum transmission at 417 nm, creates a sharp gradient, in terms of both the intensity and the quality of light available as a function of depth. For example, while photosynthetic organisms confined to surface waters are exposed to high light intensities within a broad spectrum range within the visible (400–700 nm) and extending into the near-ultraviolet region (350–400 nm), deeper marine habitats experience lower light intensities due to the exponential decrease of light with depth and a shift toward a dominance of blue light. Hence, while organisms in surface waters have had to adapt to protect themselves against excess light and ultraviolet radiation by producing photoprotective pigments, organisms living near the base of the euphotic zone require strategies that increase their capabilities of solar energy capture in the blue region of the spectrum; this is achieved by increasing the number of photosynthetic units per cell and by modifying the spectral absorption characteristics through changes in the photosynthetic pigments associated with them. As a consequence, the light gradient can generate and sustain a highly structured vertical pattern of light-harvesting microorganisms in the marine environment; some microbes are adapted to high and others to lower light fluxes. In addition, the surface light intensity and its propagation through the water column define the region where the photosynthetic rates of the microbial assemblage can exceed its respiration, driving the balance between the uptake (photosynthesis) and remineralization (respiration) of nutrients with depth. It is this balance in microbial activity that ultimately drives the biological sequestration and transport of elements, such as carbon, in the ocean and defines the large-scale distribution and availability of nutrients in the marine environment.

STRUCTURE AND CLASSIFICATION OF MARINE MACROHABITATS

There are numerous criteria that can be used to classify marine habitats. The most widely accepted scheme divides the ocean into two broad categories: pelagic and benthic, depending upon whether the habitat of interest is the overlying water column (fluid) or the seabed (solid), respectively (Figure 1; Table 1). Within each of these main categories, a number of additional subdivisions can be made depending, for example, on increasing water depth from the high tide mark. For the pelagic habitat, major subdivisions include neritic for waters overlying the continental shelves (≤200 m deep) and oceanic, for the vast open sea. The oceanic realm can also be further subdivided (Table 1). Benthic habitats include littoral (intertidal), sublittoral (from the low tide boundary to the edge of the continental shelf), bathyal, abyssal, and hadal. Other classification schemes use the topographic boundaries: continental shelf, continental slope,

TABLE 1 Classification of marine habitats according to Hedgpeth (1957)

Region	Boundaries/Comments
Pelagic Realm (water column)	
Neritic	Waters over the continental shelves (~200 m)
Oceanic	Waters seaward of the continental shelves
-epipelagic	0–200 m; sunlit regions
-mesopelagic	200–1500 m
-bathypelagic	1500–4000 m
-abyssopelagic	>4000 m
Benthic Realm (seabed)	
Supralittoral	Above high-tide mark
Littoral	Between the tides
Sublittoral	Between low tide and the edge of continental shelves (~200 m)
Bathyal	Seaward of continental shelves to ~4000 m
Abyssal	4000–6000 m
Hadal	>6000 m, including trenches

abyssal plain, and deep sea trenches (see Figure 1). Although these and other terms are routinely used, the depth ranges are not always identical, so they should be considered as guidelines rather than rules.

Along the seawater depth gradient, whether in benthic or pelagic habitats, some physical and chemical characteristics systematically change (e.g., decreasing temperature and increasing pressure). In this regard, it is important to emphasize that the most common marine habitat is cold (<12 °C) and exposed to high hydrostatic pressure (>50 bars). Consequently, many marine microbes are cold- and pressure-adapted, even obligately so; indeed, some abyssopelagic bacteria require high pressure (>400 bars) to grow. Other classification schemes based on the availability of sunlight (euphotic (light present) or aphotic (light absent)) or the relative rates of organic matter production (eutrophic (high), mesotrophic (medium), or oligotrophic (low)) have also been used.

Sharp horizontal gradients can also be observed on the surface of the ocean. For example, since the 1960s, oceanographers have used satellite-based remote sensing approaches to map various features of the global ocean, including sea surface temperature, winds, altimetry, and the distributions of photosynthetic microbes as inferred from observations of spectral radiance. The first satellite-based ocean color measurements were obtained using the Coastal Zone Color Scanner (CZCS) aboard the Nimbus-7 satellite that was launched in October 1978; it provided useful data for nearly a decade. The CZCS sensor was eventually replaced with the Sea-viewing Wide Field-of-view Sensor (SeaWiFS), launched in September 1997 followed by Moderate Resolution Imaging Spectroradiometer (MODIS) aboard the Terra and Aqua satellites (1999 and 2002 to present, respectively). And, although these instruments cannot provide information regarding spatial variability below 1 km resolution, they have provided unprecedented observations on the temporal variability or surface ocean macrohabitats (depth-integrated to one optical depth, ~25 m in clear open ocean waters), as well as the mesoscale (10–100 s of km) and large-scale distributions of chlorophyll (chl). Daily synoptic global images can be pieced together to track the dynamics (days to decades) of photosynthetic microbial assemblages in the global ocean and their correlations with other environmental variables in ways that are not possible by any other means (Figure 2). Furthermore, systematic analyses of these ocean color datasets can be used to define spatial habitat structure in oceanic ecosystems, and the partitioning of the global ocean into a suite of ecological provinces or functional habitat units, leading to the novel subdiscipline of marine ecological geography. Unfortunately, there are no satellite-based sensors that can track non-chl-containing marine microbes, although several novel remote detection systems are under development for *in situ* application based on molecular/genetic probes and imaging-in-flow cytometry.

There are distinct water types throughout the ocean that can be easily identified by measuring their temperature and salinity characteristics, which together determine their densities (T–S diagram; Figure 3). Using T–S diagrams as a fingerprinting tool, water types can be traced throughout the world ocean to specific regions of formation. A water mass results from the mixing of two or more water types, and is represented by a line between distinct water types on the T–S diagram (Figure 3). Water masses can also be tracked for great distances throughout the world ocean, and their microbial assemblages can also be sampled and characterized. The global circulation rate, as deduced by nonconservative chemical properties and radioisotopic tracers, has a time scale of hundreds to thousands of years. Consequently, "young" and "old" water masses can be identified based on the time that the water was last in contact with the atmosphere.

Due to unique seafloor topography and interactions with the atmosphere, certain regions "short-circuit" the mean circulation by serving as conduits for a more rapid ventilation of the deep ocean (bringing it into contact with the atmosphere) and the concomitant delivery of nutrient-rich deep water to the surface of the sea. These so-called upwelling regions occupy only ~1% of the surface ocean, but they are important areas of solar energy capture through enhanced photosynthesis and the selection of relatively large algae and short food chains; thereby they support some of the great fisheries of the world (Figure 4). In contrast, the more common condition (90% of the global ocean) is the oligotrophic habitat (low nutrient and low rates of organic matter production) that selects very small primary producers, long and complex microbial-based food webs, and relatively inefficient transfer of carbon and energy to higher trophic levels (HTLs) like fish. These fundamental differences in physics result in marine habitats with diverse structures and dynamics that host dramatically different microbial assemblages, as discussed later in this article.

MARINE MICROBIAL INHABITANTS AND THEIR GROWTH REQUIREMENTS

The marine environment supports the growth of a diverse assemblage of microbes from all three domains of life: *Bacteria*, *Archaea*, and *Eucarya*. The term "microorganism" is a catchall term to describe unicellular and multicellular organisms that are smaller than ~100–150 μm. This grouping includes organisms with broadly distinct evolutionary histories, physiological capabilities, and ecological niches. The only common, shared features are their size and a high surface-to-biovolume ratio. A consequence of being small is a high rate of metabolism and shorter generation times than most larger organisms.

Microorganisms, particularly bacteria and archaea, are found throughout the world ocean including marine

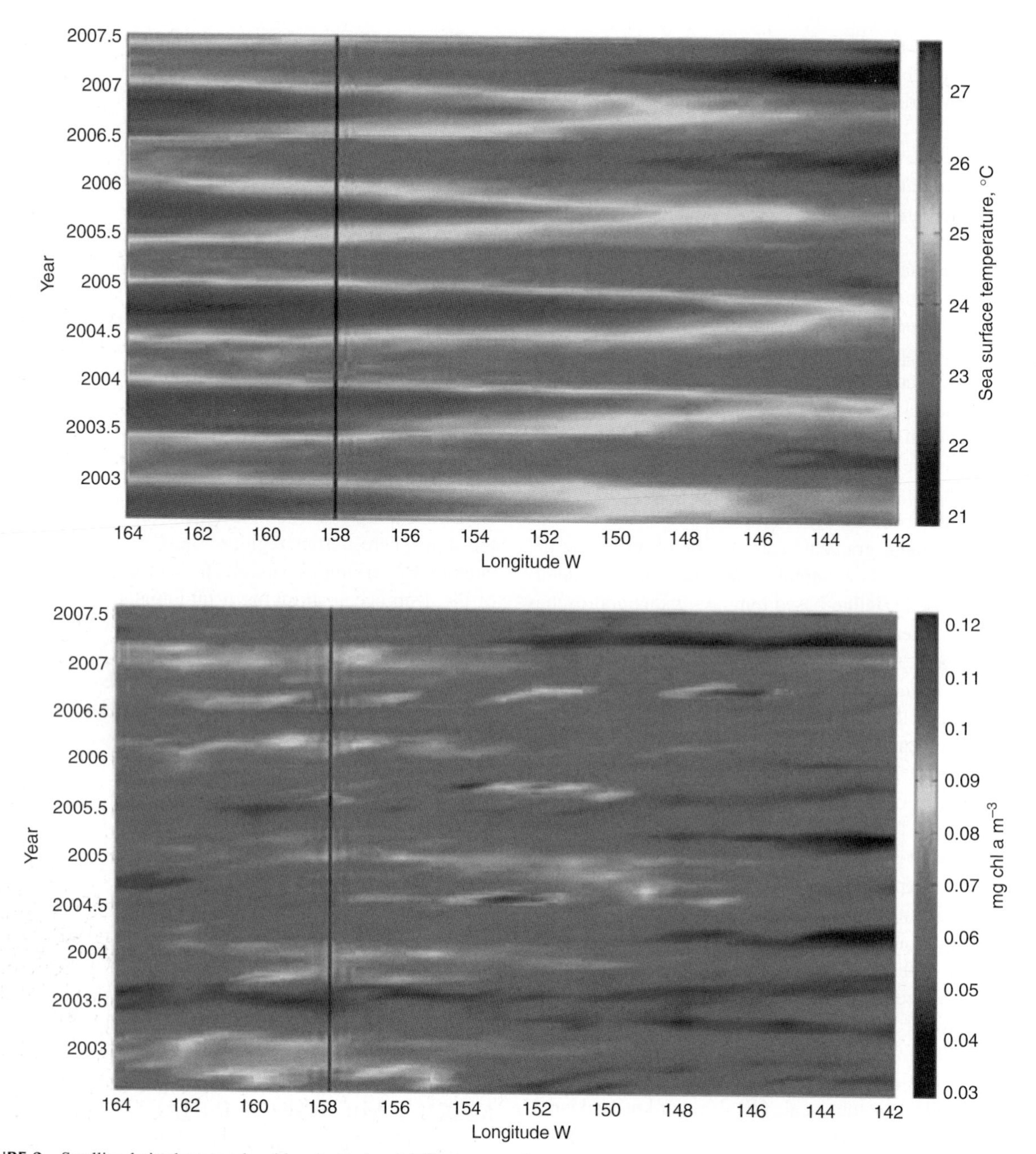

FIGURE 2 Satellite-derived temporal and longitudinal variability in sea surface temperature (Top) and chlorophyll (Bottom) for the region surrounding Station ALOHA (22.75°N, 158°W). The data, available from the NASA Ocean Color Time-Series Online Visualization and Analysis website (http://reason.gsfc.nasa.gov/), have been obtained through NASA's Moderate Resolution Imaging Spectroradiometer (MODIS) sensor on board Aqua between July 2002 and June 2007 and correspond to the latitudinal average between 22.5°N and 23.5°N for the longitude band 142–164°W. The black line marks longitude 158°W where Station ALOHA is located.

sedimentary and subseabed habitats. There is probably no marine habitat that is devoid of microorganisms, with the possible exception of high-temperature (>100 °C) zones. In addition to a physically favorable environment, the metabolism and proliferation of microorganisms also require a renewable supply of energy, electrons for energy generation, carbon (and related bioelements including nitrogen, phosphorus, sulfur), and occasionally organic growth factors such as vitamins. Depending upon how these requirements are met, all living organisms can be classified into one of several metabolic categories (Table 2). For example, photolithoautotrophic microbes use light as an energy source, water as an electron source, and inorganic carbon, mineral nutrients, and trace metals to produce organic matter. At the other end of the metabolic spectrum, chemoorganoheterotrophic microbes use preformed organic matter for energy generation and as a source of electrons and carbon for cell growth. In a laboratory setting, only obligate

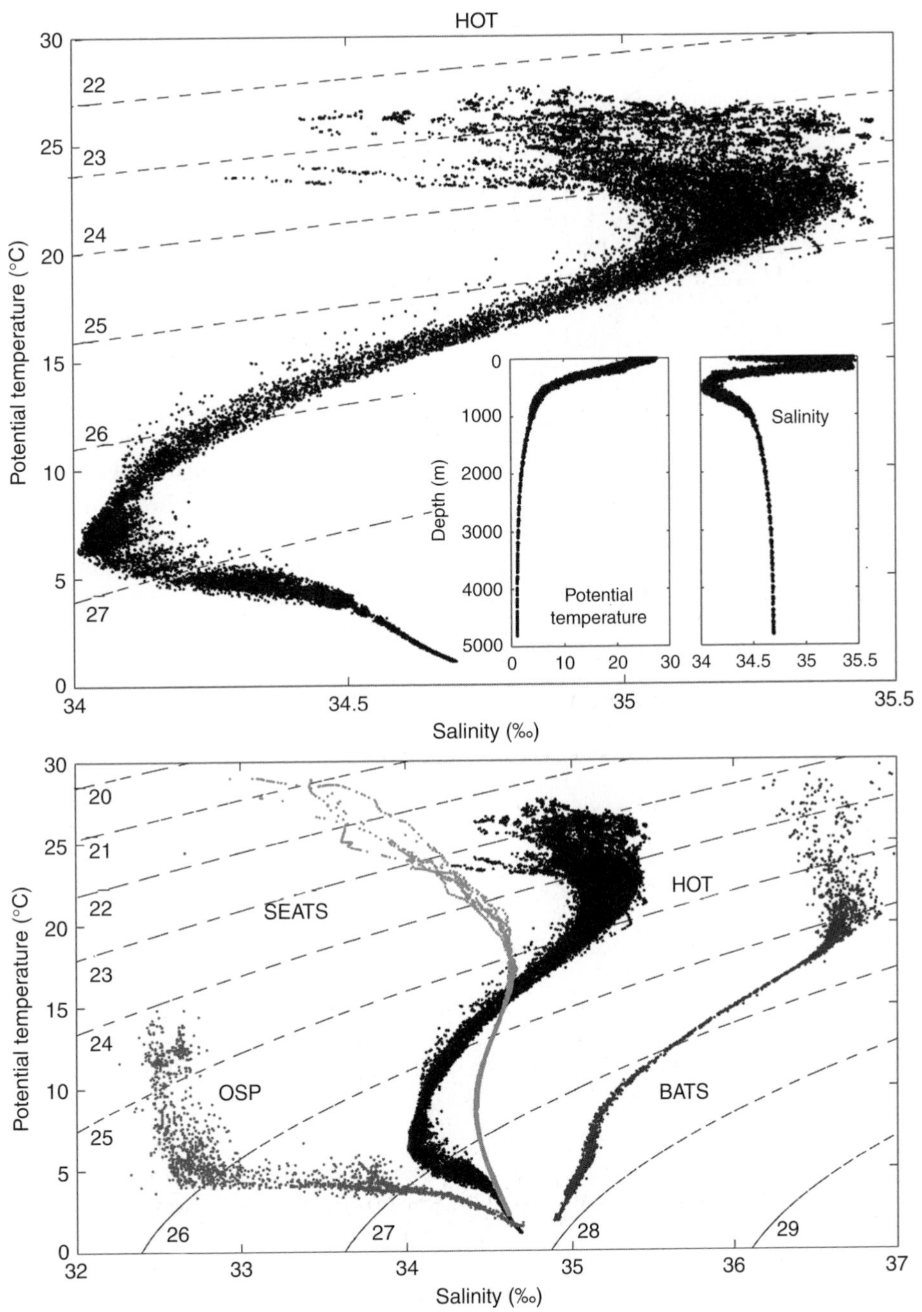

FIGURE 3 Potential temperature versus salinity (T–S) plots are used to identify, trace, and compare distinct water types and water masses in the marine environment. (Top) T–S diagram for the Hawaii Ocean Time-series (HOT) Station ALOHA for the period 1988–2006. The inset shows the depth profiles of potential temperature and salinity. The ALOHA T–S fingerprint shows the presence of numerous water masses at specific depths. The contours show lines of constant density, or isopycnal surfaces, in density anomaly notation (((density in g cm^{-3}) – 1.000) × 1000). In addition to temperature and salinity (density) variations, these distinctive water masses also have distinctive chemical properties and may contain unique assemblages of microorganisms. The large variability of T and S at the top of the graph is a result of seasonal and interannual changes in near-surface water properties. (Bottom) Comparison of T–S fingerprints for a variety of oceanic time-series stations including: Ocean Station Papa (OSP; 50°N, 145°W), SouthEast Asia Time-Series (SEATS; 18°N, 116°E), Hawaii Ocean Time-series (HOT; 22.75°N, 158°W), and Bermuda Atlantic Time-series Study (BATS; 32°N, 64°W). The three North Pacific stations (OSP, SEATS, HOT) have a common deep water mass.

photolithoautotrophs are self-sufficient; all other autotrophs and all heterotrophs rely upon the metabolic activities of other microorganisms. However in nature, even obligate photolithoautotrophs must tie their growth and survival to other, mostly deep-sea, microbes that are vital in sustaining nutrient availability over evolutionary time scales. Most marine microorganisms probably use a variety of metabolic strategies, perhaps simultaneously, to survive in nature. Because needed nutrients in the ocean's surface are often found in dissolved organic molecules, it seems highly

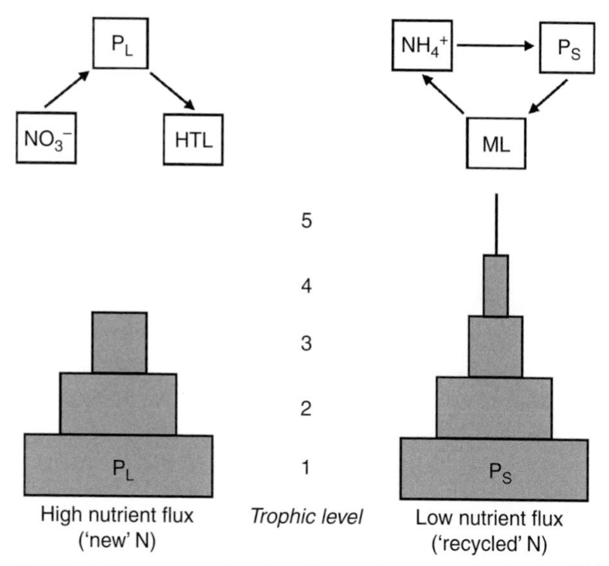

FIGURE 4 Importance of nutrient flux on the size distribution and efficiency of biomass and energy flow in marine habitats. The schematic on the left depicts a habitat where "new" nutrient (as NO_3^-) flux is high (e.g., an upwelling region). This leads to a selection for large phytoplankton cells (P_L) that are efficiently consumed by HTLs including large zooplankton and fish. This results in a short and efficient food chain. In contrast to the upwelling regions, most open ocean habitats have low new nutrient (NO_3^-) fluxes and survive by local remineralization of required nutrients ("recycled"). These conditions select for small phytoplankton cells (P_S) that serve as the food source for long and complex microbial-based food webs (also called microbial loops; ML) that recycle mass and dissipate most of the solar energy that was initially captured. The great marine fisheries of the world are generally found in association with upwelling regions.

TABLE 2 Variations in microbial metabolism based on sources of energy, electrons, and carbon according to Karl (2007)

Source of Energy[a]	Source of Electrons	Source of Carbon
Sunlight	Inorganic	CO_2
photo-	*-litho-* Organic *-organo-*	*-autotroph* Organic *-heterotroph*
Chemical	Inorganic	CO_2
chemo-	*-litho-* Organic *-organo-*	*-autotroph* Organic *-heterotroph*
Radioactive decay	Inorganic	CO_2
radio-	*-litho-* Organic *-organo-*	*-autotroph* Organic *-heterotroph*

[a] *A "mixotroph" is an organism that uses more than one source of energy, electrons, or carbon.*

improbable that sunlit marine habitats would select for obligate photolithoautotrophy as opposed to, for instance, mixotrophic growth.

Across the full metabolic spectrum of possible modes of growth, some microbes are more self-sufficient than others. For example, while most microbes require a supply of chemically "fixed" nitrogen, either in reduced (ammonia or dissolved organic nitrogen (DON)) or in oxidized (nitrate or nitrite) form to survive, a special group of N_2-fixing microbes (diazotrophs) can use the nearly unlimited supply of dissolved N_2 as their sole source of cell N. Additionally, some microbes can manufacture all their required building blocks (e.g., amino acids and nucleic acid bases) and growth cofactors (e.g., vitamins) from simple inorganic precursors, whereas others require that they be supplied from the environment; "auxotrophic" microorganisms are, therefore, ultimately dependent upon the metabolic and biosynthetic activities of other microbes. These "incomplete" microbes, probably the bulk of the total microbial assemblage in seawater, cannot grow unless the obligate growth factors are present in and resupplied to the local habitat. In this regard, most marine habitats provide the laboratory equivalent of a complex or complete medium containing low-molecular-weight compounds (e.g., amino acids, simple sugars, nucleic acid bases, and vitamins), in addition to the mineral nutrients and trace metals. The active salvage and utilization of these biosynthetic precursors, in lieu of *de novo* synthesis, conserves energy, increases growth efficiency, and enhances survival. Over evolutionary time, some unused biosynthetic pathways in particular organisms appear to have been lost from the genome, perhaps, as a competitive strategy for survival in a mostly energy-limited environment. This process has been termed genome streamlining.

Finally, growth and reproduction are often viewed as the most successful stages of existence for any microorganism. However, in many of the low nutrient concentration and low energy flux habitats that dominate the global seascape, the ability to survive for extended periods under conditions of starvation may also be of great selective advantage and ultimately may affect the stability and resilience of microbial ecosystems. The starvation-survival response in marine bacteria leads to fragmentation (i.e., cell division in the absence of net growth) and, ultimately, to the formation of multiple dwarf or miniaturized cells. Other physiological changes, including reduction in endogenous metabolism, decreases in intracellular adenosine triphosphate (ATP) concentrations, and enhanced rates of adhesion are also common consequences. These starved cells can respond rapidly to the addition of organic nutrients. This "feast and famine" cycle has important implications for how we design *in situ* metabolic detection systems and model microbial growth in marine habitats.

DISTRIBUTION, ABUNDANCE, AND BIOGEOGRAPHY OF MARINE MICROBES

The distribution and abundance of microbes is highly variable, but somewhat predictable, across globally distributed marine habitats. For example, phototrophic microbes are restricted to sunlit regions (0–200 m in the open sea) whereas chemotrophic microbes are found throughout the oceanic realm. However, because the abundance and productivity of marine microbes depend on the availability of nutrients and energy, there is often a decreasing gradient in total microbial biomass from the continents to the open ocean, and a decreasing gradient in total microbial biomass from the sunlit surface waters to the abyss. For the pelagic zone, total microbial biomass in near-surface (0–100 m) waters ranges from 30 to 100 mg carbon m^{-3} in neritic waters to 6–20 mg carbon m^{-3} in oceanic waters. For open ocean habitats, this biomass decreases by approximately three orders of magnitude from euphotic zone to abyssal habitats, with values <0.02 mg carbon m^{-3} in the deepest ocean trenches. When scaling these concentrations to the volume of the ocean, the total oceanic microbial biomass, excluding sediments, has been estimated to be $0.6–1.9 \times 10^{15}$ g carbon with approximately half its stock residing below 100 m.

Temperature is an important habitat variable, and may be responsible for structuring microbial assemblages and setting limits on various metabolic processes. However, temperature *per se* does not limit the existence of marine microbes so long as liquid water exists. Accordingly, there are some marine habitats that select for thermophilic microbes ("warm temperature-loving"; e.g., deep-sea hydrothermal vents) and others for psychrophilic microbes ("cold temperature-loving"; e.g., polar latitudes and abyssal regions). Spatial gradients in temperature across open ocean habitats as well as seasonal changes in temperature can also affect the diversity of microbial assemblages in most marine habitats. Finally, for any given microbial species, there is a positive correlation between rates of metabolism and temperature over its permissive range. Generally, for a 10 °C change in temperature there is a two- to threefold increase in metabolic activity, for example, respiration. Photochemical reactions, including photosynthesis, have much smaller temperature coefficients, and it has been hypothesized that low temperature suppression of chemoorganoheterotrophic bacterial activity, relative to photosynthesis, might significantly restrict energy flow through microbial food webs, increasing the efficiency of the transfer of carbon and energy to HTLs via metazoan grazing. This is just one way in which temperature may structure and control microbial processes in the sea.

In the near-surface waters, microbes capture solar energy, which is locally transferred and dissipated as heat, or exported to other surrounding marine habitats in the form of reduced organic or inorganic substrates, including biomass. Apart from very restricted shallow coastal regions where light can penetrate all the way to the seabed for use by benthic micro- and macroalgae, essentially all marine photosynthesis is planktonic (free floating) and microbial. The dynamic range in total marine photosynthesis, from the most productive to the least productive regions of the global ocean, is probably less than two orders of magnitude for a given latitude, and the biomass of chemoorganoheterotrophic bacteria may be even less. There are much steeper gradients in photosynthesis and bacterial/archaeal biomass in the vertical (depth) than in the horizontal (spatial) dimensions. Furthermore, most marine respiration is also driven by microbes, both phototrophs and chemotrophs. For this reason, the mean turnover time of oceanic carbon within biological systems in surface waters is weeks, compared to decades for most terrestrial ecosystems.

Size spectral models and analyses, which relate the relative abundance of organisms as a function of size, have been used to examine the distribution of biomass among various size classes. The emergent patterns from these analyses, particularly between and among different marine habitats, are relevant to issues regarding the environmental controls on microbial community structure and function as well as to the trophic efficiency of marine food webs. In some oceanic habitats, solar energy is captured and completely utilized within microbial-based food webs; in other regions a significant proportion of the energy captured via photosynthesis is passed to large organisms, including fish and humans. An important consideration appears to be the size of the primary producer populations, and this determines the number of trophic transfers that are sustainable in light of the typically inefficient (<10%) transfer of carbon and energy between trophic levels (Figure 4). If, for example, the primary producers are relatively large (>10–20 μm diameter; P_L in Figure 4, left), such as unicellular algae including diatoms, rather than tiny picoplankton (<2 μm; P_S in Figure 4, right), then the grazer/consumer based food chain is shorter, leading to a more efficient transfer of carbon and energy (Figure 4). However, the length of the food chain is not always defined by the difference in size between the primary producer and the top consumer; in some cases, large organisms such as baleen whales have adapted a feeding strategy that relies mostly on very small, planktonic organisms. Nevertheless, the size and structure of marine food webs is determined, in large part, by physical processes such as turbulence, which, in turn, affects the flux of nutrients into the euphotic zone and, therefore, shapes the structure and function of marine ecosystems.

In most sunlit marine habitats there is generally a significant correlation between chl concentrations and the number of bacterial cells, and between net primary production and bacterial production across a broad range of ecosystems. These empirical relationships suggest that phototrophs and chemotrophs grow in response to common factors (e.g., nutrients, temperature), or that phototrophs produce substrates for the growth of chemotrophs, or vice versa.

In addition to living organisms, virus particles – particularly those capable of infecting specific groups of microorganisms – can exert influence on microbial-based processes. For example, through microbial infection and subsequent lysis, viral activity may directly influence the composition of the microbial assemblage. Furthermore, through the release of dissolved organic matter (DOM) into the marine environment during virus-induced cell lysis, an indirect effect on metabolic activity of the chemotrophic assemblage can occur. Viruses can also facilitate genetic exchange between different microbial strains contributing to the metabolic plasticity of certain microorganisms and the redundancy of some metabolic processes in a given environment. It has been reported that virus particle abundances closely track the abundance of bacteria plus archaea, at least in the water column, with virus-to-prokaryote ratios ranging from 5 to 25, and commonly close to 10. This relationship appears to hold even into the deep sea, suggesting a close ecological linkage throughout the entire marine habitat.

From an ecological perspective, understanding and modeling how microbial assemblages emerge as a result of interaction of physics and biology is a primary goal in microbial oceanography. In this context, the study of the distribution of biodiversity over space and time, also known as biogeography, seeks fundamental information on the controls of speciation, extinction, dispersal, and species interactions such as competition. The field of microbial biogeography is just beginning to develop a conceptual framework and analytical tools to examine distribution patterns and to quantify diversity at the ecologically relevant taxonomic scale. For example, recent studies of the marine phototroph *Prochlorococcus* have documented significant intraspecific genomic variability that confers a distinct niche specificity including nutrient and light resource partitioning. What appears at one level to be a cosmopolitan species is actually a group of closely related ecotypes (populations within a species that are adapted to a particular set of habitat conditions); the high- and low-light ecotypes have >97% similarity in their 16S ribosomal RNA gene sequences and share a core of 1350 genes, but vary by more than 30% in their total gene content (and genome size; the high- and low-light adapted ecotypes have genome sizes of 1,657,990 bp and 2,410,873 bp, respectively). An assemblage of related *Vibrio splendidus* (>99% 16S RNA identity) sampled from a temperate coastal marine habitat had at least 1000 distinct coexisting genotypes, and bacterial samples collected from the aphotic zone of the North Atlantic Ocean revealed an extremely diverse "rare biosphere" consisting of thousands of low-abundance populations. The ecological implications of these independent reports of taxonomic diversity are profound; new ecological theory may even be required to build a conceptual framework for our knowledge of marine habitats and their microbial inhabitants.

SUNLIGHT, NUTRIENTS, TURBULENCE, AND THE BIOLOGICAL PUMP

Of all the environmental variables that collectively define the marine habitat, we single out three – namely, sunlight, nutrients, and turbulence – as perhaps the most critical for the survival of sea microbes. Together, these properties control the magnitude and efficiency of the "biological pump," a complex series of trophic processes that result in a spatial separation between energy (sunlight) and mass (essential nutrients) throughout the marine environment. In the sunlit regions of most (but not all) marine habitats, nutrients are efficiently assimilated into organic matter, a portion of which is displaced downward in the water column, mostly through gravitational settling. As particles sink through the stratified water column, a portion of the organic matter is oxidized and the essential nutrients are recycled back into the surrounding water masses. Depending upon the depth of remineralization and replenishment to the surface waters by physical processes, these essential nutrients can be sequestered for relatively long periods (>100 yrs). The vertical nutrient profile, for example of nitrate, shows a relative depletion near the surface and enrichment at depth as a result of the biological pump (Figure 5a and b); regional variations in the depth profiles reflect the combination of changes in the strength and efficiency of the biological pump and the patterns of global ocean circulation (Figures 5a and 6). The highest nutrient concentrations in deep water can be found in the abyss of the North Pacific, the oldest water mass on Earth. The regeneration of inorganic nutrients requires the oxidation of reduced organic matter, so the concentrations of dissolved oxygen decrease with depth and with age of the water mass as a result of the cumulative effect of microbial metabolism (Figures 5a and 6).

Turbulence in marine habitats derives from a variety of processes including wind stress on the ocean's surface, ocean circulation, breaking internal waves, and other large-scale motions that can create instabilities, including eddies, in the mean density structure. Turbulence, or eddy diffusion, differs fundamentally from molecular diffusion in that all properties (e.g., heat, salt, nutrients, and dissolved gases) have the same eddy diffusion coefficient; a typical value for horizontal eddy diffusivity in the ocean is $\sim 500\,m^2\,s^{-1}$, a value that is 10^9 times greater than molecular diffusion. Vertical eddy diffusivity is much lower ($\sim 0.6–1 \times 10^{-4}\,m^2 s^{-1}$) suggesting that the upward flux of nutrients into the euphotic zone is a slower process than movement horizontally in the open ocean. Most near-surface dwelling microbes, particularly phototrophs that are also dependent upon solar energy and are effectively "trapped" in the euphotic zone habitat, depend on turbulence to deliver deep water nutrients to the sunlit habitat.

In addition to the eddy diffusion of nutrients from the mesopelagic zone, wind stress at the surface and other forces can mix the surface ocean from above. If the near-surface

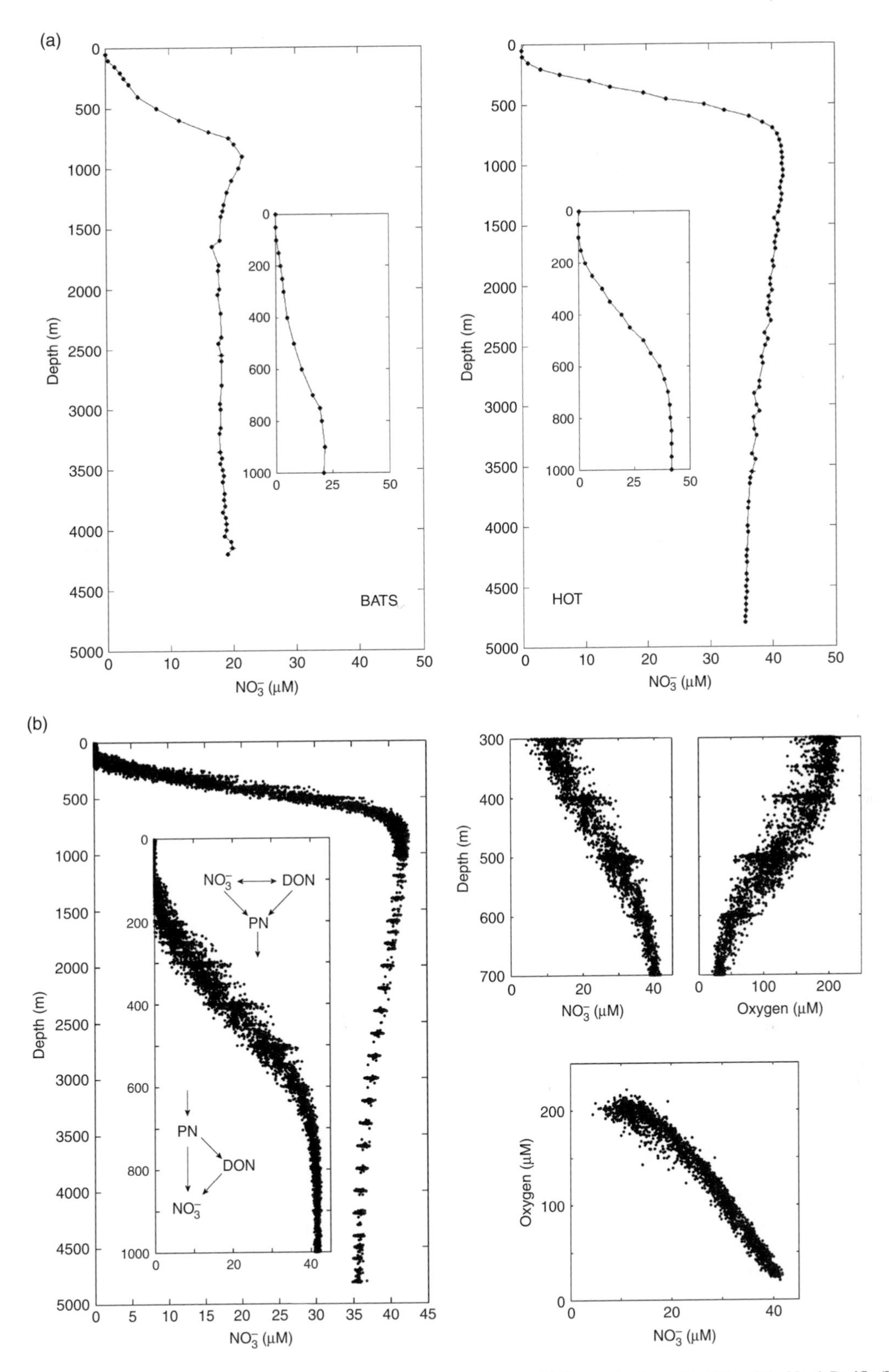

FIGURE 5 (a) Nitrate (NO_3^-) versus depth profiles for the North Atlantic (Bermuda Atlantic Time-series Study; BATS) and the North Pacific (Hawaii Ocean Time-series; HOT) showing significant interocean differences including a steeper nitracline (i.e., a larger change in NO_3^- concentration per meter in the upper mesopelagic zone region) and higher deep water (>4000 m) NO_3^- concentrations for HOT. These differences in NO_3^- inventories and gradients, part of a systematic global pattern (see Figure 6), have significant implications for NO_3^- fluxes into the euphotic zone. Data available at the HOT and BATS program websites (http://hahana.soest.hawaii.edu.; http://www.bios.edu/.). (b) Relationships between the vertical distributions of nitrate (NO_3^-) and dissolved oxygen (O_2) at Station ALOHA in the North Pacific Subtropical Gyre (NPSG). (Left) Graph of NO_3^- (μmol l^{-1}) versus depth (m) showing the characteristic "nutrient-like" distribution of NO_3^- with regions of net NO_3^- uptake and DON cycling and particulate nitrogen (PN) export near the surface, and net NO_3^- remineralization at greater depths. The insert shows these main N-cycle processes, which are most intense in the upper 1000 m of the water column. (Right, top) NO_3^- and O_2 concentration versus depth profiles of the 300–700 m region of the water column at Station ALOHA showing the effects of net remineralization of organic matter. (Right, bottom) A model 2 linear regression analysis of NO_3^- versus O_2 suggests an average consumption of 80 μmol l^{-1} O_2 for each 1 μmol of NO_3^- that is regenerated from particulate and dissolved organic matter. Data available at the HOT program website (http://hahana.soest.hawaii.edu.).

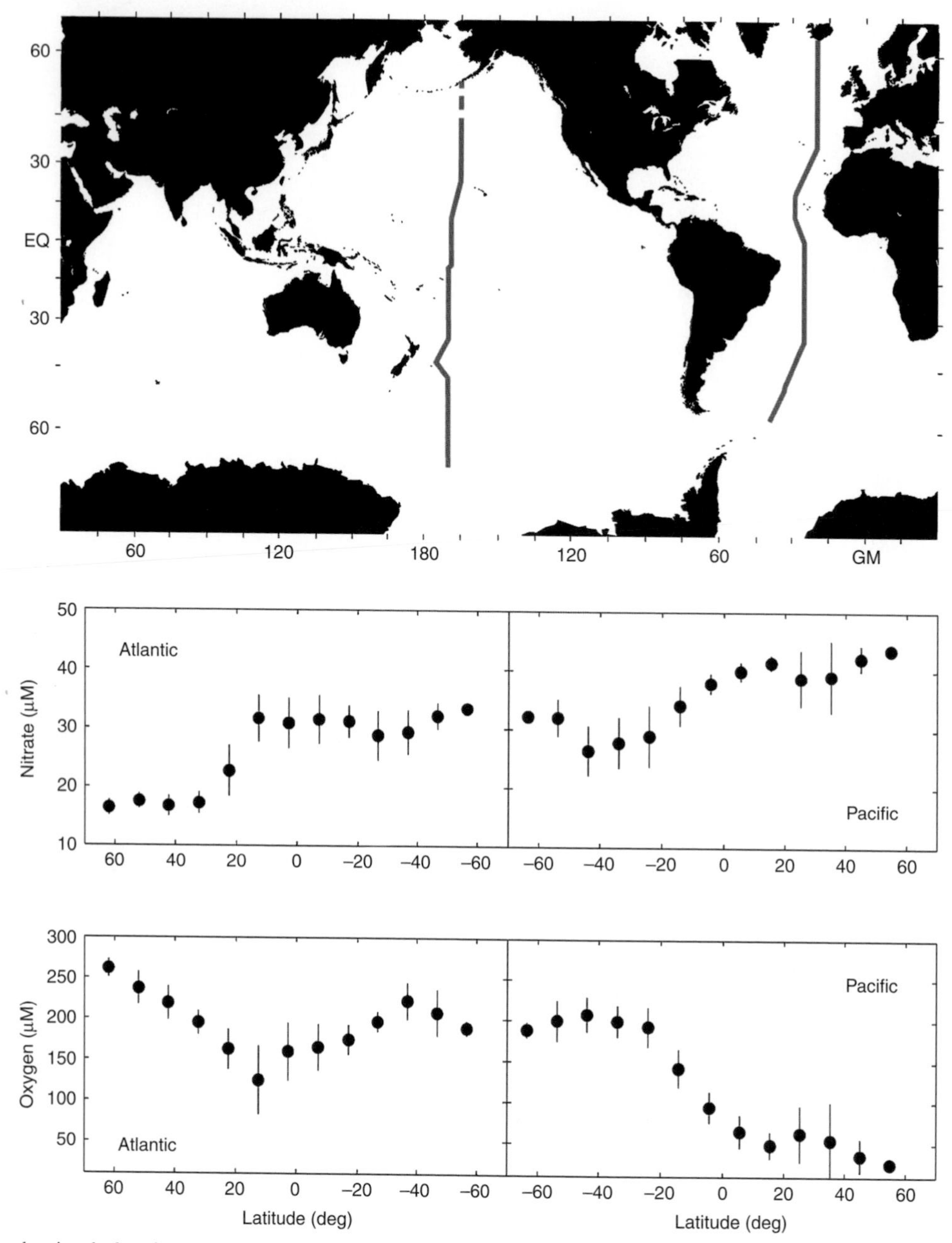

FIGURE 6 Map showing the locations of the World Ocean Circulation Experiment (WOCE) program transects A-16 (Atlantic) and P-15 (Pacific). Data from these cruises were obtained from http://woce.nodc.noaa.gov. and averaged over the depth range of 500–1500 m, then combined into 10° latitude bins and plotted as mean nitrate and dissolved oxygen concentrations (±1 standard deviation). The resultant plot shows a systematic increase in nitrate concentrations at mid-water depths "down" the Atlantic and "up" the Pacific, and an opposite trend for oxygen. These spatial patterns are the result of the time-integrated aerobic decomposition of organic matter along known pathways of deep water circulation.

density stratification is weak or if the mixing forces are strong, or both, then a large portion of the euphotic zone can be homogenized; in selected latitude regions the surface mixing layer can extend to 500 m or more, well below the maximum depth of the euphotic zone. These well-mixed environments usually have sufficient nutrients but insufficient light to sustain photosynthesis because the phototrophs are also mixed to great depths, as in some polar habitats during winter months. Following these seasonal deep-mixing events, the ocean begins to stratify due to the absorption of solar radiation in excess of evaporative heat loss. As the wind forcing from winter storms subsides and the intensity of solar radiation increases, a density gradient develops in the upper water column. Phototrophic microorganisms in the euphotic zone gain a favorable niche with respect to both light energy and nutrient concentrations. Depending upon the presence or absence of grazers, this condition results in an increase in phototrophic microbial biomass, a condition referred to as the spring bloom. A comprehensive formulation of the "vernal blooming of phytoplankton" presented by H. Sverdrup remains a valid representation of this important marine microbial phenomenon.

However, in many portions of the world ocean, particularly in tropical ocean gyres, local forcing due to wind stress is too weak to break down the density stratification, so the nutrient delivery from below the euphotic zone through mixing is not possible. In these oligotrophic regions, the habitat is chronically nutrient-stressed and oftentimes nutrient-limited. Although surface mixed layers can be observed, they rarely penetrate deeper than 100 m. Even within the so-called mixed layer, gradients in chl, nutrients, dissolved gases, and microorganisms can be detected, suggesting that these regions are not always actively mixing. This subtle distinction between a mixing layer, where there is an active vertical transport of physical, chemical, and biological properties, and a mixed layer, which is defined operationally as a layer with weak or no density stratification, has important implications for microbial growth and survival, particularly for phototrophic microorganisms. Consequently, without additional information on mixing dynamics (e.g., a profile of turbulent kinetic energy), the commonly used term mixed layer can be misleading with regard to habitat conditions for microbial growth. The time required to change from a mixing layer to a mixed layer to a density-stratified surface habitat and back again will depend on the habitat of interest.

One approach for distinguishing between a mixing layer and a mixed layer is to measure the near-surface concentrations and temporal dynamics of a short-lived photochemically produced tracer, for example, hydrogen peroxide (H_2O_2). The concentration versus depth profile of H_2O_2 in a mixing layer with a short mixing time scale (≤1 h) would be constant because the concentration of photochemically active DOM and the average solar energy flux would also be relatively constant. On the other hand, the H_2O_2 concentration profile in a nonmixing (or slowly mixing, turnover >1 day) "mixed layer" would approximate the shape to the flux of solar energy decreasing exponentially with depth nearly identical to a density-stratified habitat, assuming that the concentration of photosensitive DOM is in excess. It is also possible to use other photochemical reactions to obtain information on vertical mixing rates.

TIME VARIABILITY OF MARINE HABITATS AND CLIMATE CHANGE

Marine habitats vary in both time and space over more than nine orders of magnitude of scale in each dimension. Compared with terrestrial habitats, most marine ecosystems are out of "direct sight," and, therefore, sparsely observed and grossly undersampled. The discovery and subsequent documentation of the oases of life surrounding hydrothermal vents in the deep sea in 1977 revealed how little we knew about benthic life at that time. Furthermore, because marine life is predominantly microscopic in nature, the temporal and spatial scales affecting microbial processes may be far removed from the scales that our senses are able to perceive. And, due to this physical and sensory remoteness of marine microbial habitats, even today unexpected discoveries about the ocean frontier continue to be made, many of these involving marine microbes.

We have selected the NPSG for a more detailed presentation of relationships between and among habitat structure, microbial community function, and climate. Our choice of the NPSG as an exemplar habitat is based on the existence of the Hawaii Ocean Time-series (HOT) study, a research program that seeks a fundamental understanding of the NPSG habitat. The emergent comprehensive physical, chemical, and biological data sets derived from the HOT benchmark Station ALOHA (A Long-term Oligotrophic Habitat Assessment) is one of the few spanning temporal scales that range from a few hours to almost two decades. More generally, we submit that the sampling and observational components of the HOT program at the deep water Station ALOHA are applicable to other locations that may be representative of key marine habitats.

The NPSG is one of the largest and oldest habitats on our planet; its present boundaries have persisted since the Pliocene nearly 10^7 years before the present. The vertical water column at Station ALOHA can be partitioned into three major microbial habitats: euphotic zone, mesopelagic (twilight) zone, and aphotic zone (Table 3 and Figure 7). The main determinant in this classification scheme is the presence or absence of light. The euphotic zone is the region where most of the solar energy captured by phototrophic

TABLE 3 Conditions for microbial existence in the three major habitats at Station ALOHA in the North Pacific Subtropical Gyre

Habitat	Depth range (m)	Conditions
Euphotic zone (nutrient-limited)	0–200	• high solar energy • high DOM • low inorganic nutrients, trace elements, and organic growth factors
Mesopelagic (twilight) zone (transition)	200–1000	• low solar energy • decrease in reduced organic matter with depth • increase in organic nutrients and trace elements with depth
Abyssal zone (energy-limited)	>1000	• no solar energy • low DOM • high inorganic nutrients and trace elements

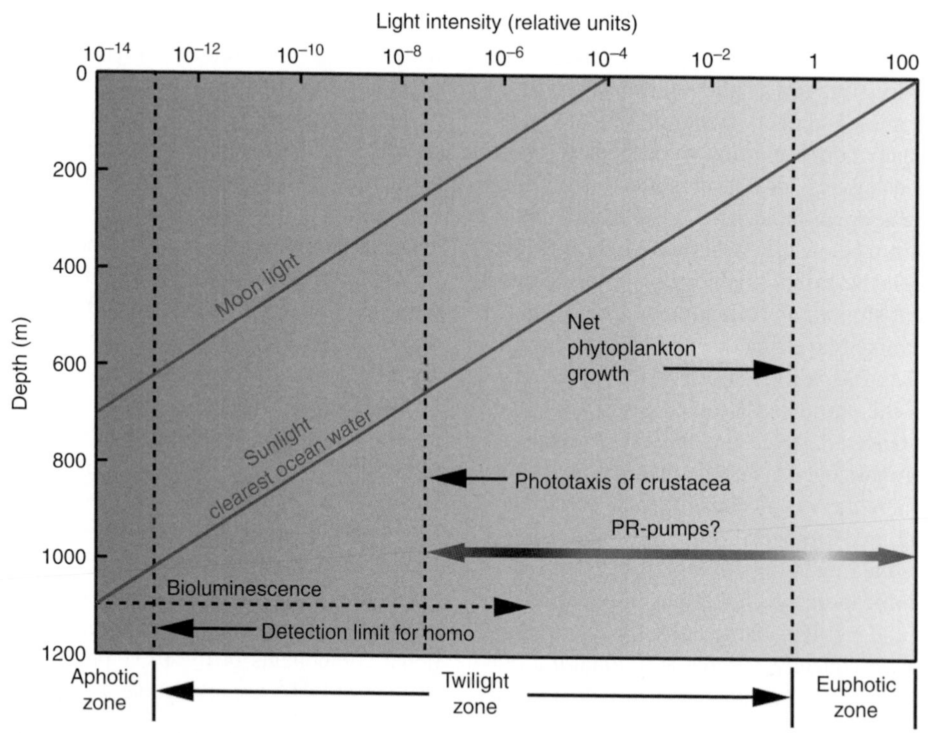

FIGURE 7 Schematic representation of the distribution of light in open ocean marine habitats. The *X*-axis displays light intensity (on a $\log_{10}$ scale in relative units) and the *Y*-axis is water depth. The euphotic zone where net photosynthesis can occur extends to a depth of ~150–200 m but sunlight can be detected by mesozooplankton (crustacean and fish) to depths of 800 m or more. The dark adapted human eye can detect even lower light fluxes. Proteorhodopsin proton pumps that have recently been detected in marine bacteria may also be able to use light but this is not yet confirmed. Moonlight, in contrast, is ~10^{-4} as bright as sunlight, but can also be detected by marine organisms and, perhaps, microbes. Bioluminescence, light production via cellular metabolism that can be found in nearly all marine taxa including microorganisms is found throughout the water column even in the "aphotic" zone.

marine microbes is sufficient to support photosynthetic activity. In the twilight zone (200–1500 m), light is present at very low photon fluxes, below which photosynthesis can occur, but at sufficiently high levels to affect the distributions of mesozooplankton and nekton and, perhaps, microbes as well. At depths greater than ~1500 m, light levels are less than 10^3 quanta cm^{-2} s^{-1}; the aphotic zone is, for all intents and purposes, dark.

Each of these major habitats is characterized by specific physical and chemical gradients, with distinct temporal scales of variability, providing unique challenges to the microorganisms that live there, and resulting in a vertical segregation of taxonomic structure and the ecological function of the resident microbial assemblages. A recent report of microbial community genomics at Station ALOHA, from the ocean's surface to the abyss, has revealed significant changes in metabolic potential, attachment and motility, gene mobility, and host–viral interactions.

The NPSG is characterized by warm (>24 °C) surface waters with relatively high light and relatively low concentrations of inorganic nutrients and low microbial biomass (Figure 8). The euphotic zone has been described as a "two-layer" habitat with an uppermost light-saturated, nutrient-limited layer (0–100 m) which supports high rates of primary productivity and respiration, and a lower (>100 m) light-limited, nutrient-sufficient layer. A region of elevated chl a, termed the Deep Chlorophyll Maximum Layer (DCML), defines the boundary between the two layers (Figure 9). The DCML in the NPSG results from photoadaptation (increase in chl a per cell) rather than enhanced phototrophic biomass; this can also be seen in the near-surface "enrichment" of chl in winter when light fluxes are at their seasonal minimum (Figure 9).

Previously considered to be the oceanic analogue of a terrestrial desert, the NPSG is now recognized as a region of moderate primary productivity (150–200 g carbon m^{-2} year^{-1}), despite chronic nutrient limitation. Furthermore, based on data from the HOT program it appears that the rates of primary production have increased by nearly 50% between the period 1989 and 2006 due in part to enhanced nutrient delivery resulting from climate controls on habitat structure and function.

At Station ALOHA the light-supported inorganic carbon assimilation extends to 175 m, a depth that is equivalent to the 0.05% surface light level (~20 mmol quanta m^{-2} day^{-1}). Most of the light-driven inorganic carbon assimilation (>50%) occurs in the upper 0–50 m of the water column (Figure 8), a region of excess light energy (>6 mol quanta m^{-2} day^{-1}). In addition, chemoorganoheterotrophic microbial activities are also greatest in the upper 0–50 m.

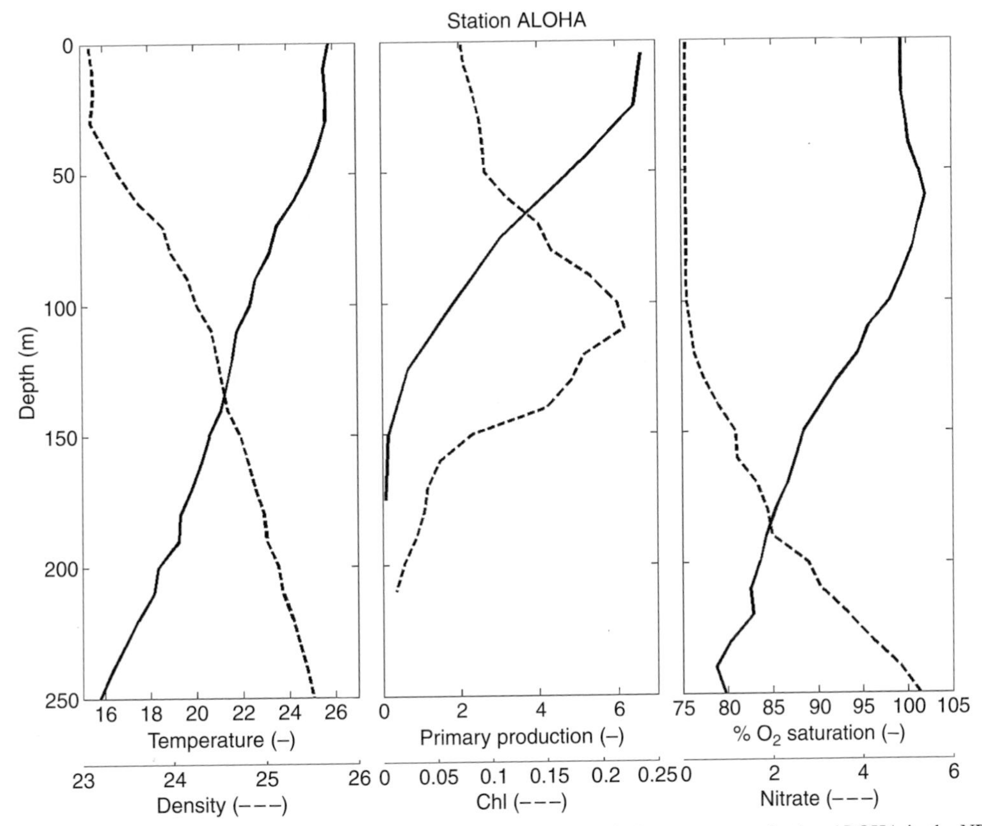

FIGURE 8 Typical patterns of the vertical distributions of selected physical and biological parameters at Station ALOHA in the NPSG. The base of the euphotic zone, defined here as the depth where primary production is equal to zero, is ~175 m. Units are: temperature (°C), density (shown as density anomaly; ((density g cm^{-3})–1.000) × 1000), primary production (mg C m^{-3} day^{-1}), chlorophyll (mg m^{-3}), nitrate (μM), and O_2 (% of air saturation). Compiled from the HOT program database (http://hahana.soest.hawaii.edu.).

However, unlike photolithoautotrophic production (where light is required as the energy source and inorganic carbon is assimilated for growth), the metabolism of chemoorganoheterotrophs is not dependent on light energy so it continues, albeit at a reduced rate, well into the twilight zone and beyond. Recently, it has been observed that "heterotrophic production" at Station ALOHA is enhanced by sunlight, suggesting the presence of microorganisms using light and both inorganic and organic substrate (photolithoheterotrophic) or light and organic substrates (photoorganoheterotrophic) to support their metabolism, or both. Several possible pathways for solar energy capture and carbon flux potentially exist in the euphotic zone at Station ALOHA, and we are just beginning to establish a comprehensive understanding of these processes, their roles and controls, and the diversity of microbes supporting them in the pelagic ecosystem.

As described earlier, physical and chemical depth gradients in the water column affect the vertical distribution of microbial assemblages and their metabolic activities. Furthermore, at a macroscopic scale we can assess how each depth horizon is affected by different temporal patterns of variability, which, in turn, influence the microbial environment. For example, in the upper euphotic zone the variability in solar radiation due to cloud coverage and changes in day length associated with the seasonal solar cycle can affect the rates of photosynthesis. In this habitat, far removed from the upper nutricline (the depth at which nutrient concentrations start to increase), the dynamics of microbial processes will be controlled mainly by the rates of solar energy capture and recycling of nutrients through the food web. Furthermore, upper water column mixing rates also contribute significantly to the variability in the light environment. However, if the variability has a high frequency relative to the cell cycle, then microbes integrate the signal because the energy invested in acclimation may be greater than that gained by maximizing photosynthetic and photoprotective processes along the variability (light) gradient.

Between the base of the mixing layer and the top of the nutricline, the microbial assemblage resides in a well-stratified environment that is nevertheless still influenced by variability in light. Although mixing does not play a significant role in this habitat, unless a deep wind- or density-driven mixing event occurs, the vertical displacements of this stratified layer as the result of a near-inertial period (~31 h at the latitude corresponding to Station ALOHA)

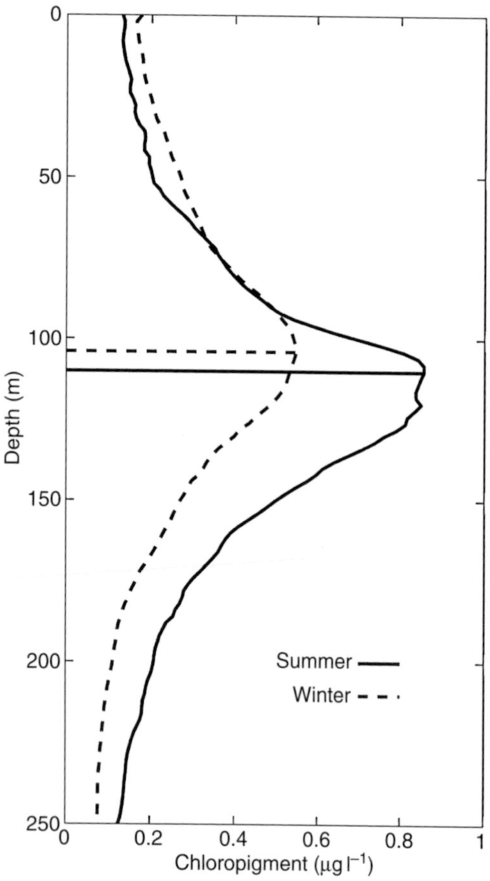

FIGURE 9 Vertical distributions of chloropigments (chlorophyll plus pheophytin) determined from *in vivo* fluorescence measurements and bottle calibrations. These graphs show average distributions at Station ALOHA for summer (June–Aug) versus winter (Dec–Feb) for 1999 showing and documenting changes in both total concentration of chloropigments at the surface and at the depth of the Deep Chlorophyll Maximum Layer (DCML). Both of these seasonal differences are caused primarily by changes in light intensity.

oscillation forces may introduce strong day-to-day variability in the light availability and photosynthetic rates (Figure 10); these vertical motions can affect the short-term balance between photosynthesis and respiration. The variability in solar irradiance described above propagates into the lower euphotic zone, penetrating into the upper nutricline. But the apparent presence of excess nutrients relative to the bioavailable energy that can be derived through photosynthesis in this region indicates that light is the limiting factor supporting microbial activity. For this reason, day-to-day variations, as well as the seasonal cycle of solar irradiance in this layer may trigger successional patterns in the microbial assemblage, and lead to pulses of organic matter export into the deeper regions of the ocean.

At Station ALOHA, as well as in most oceanic regions, the gravitational flux of particles formed in the euphotic zone represents the major source of energy that links surface processes to the deep sea. In addition, these sinking organic particles represent energy- and nutrient-enriched microhabitats that can support the growth of novel microbial assemblages. The remineralization of particles with depth follows an exponential decay pattern indicating that most of the organic matter in these particles is respired in the upper layer below the euphotic zone. If the quality and quantity of organic rain was constant, we would expect to observe stable layers of microbial diversity and activity with depth. However, the long-term records of particle flux to abyssal depths at Station ALOHA suggest that, during certain periods of the year, this flux increases significantly, representing potential inputs of organic matter into these deep layers driven by changes in upper water column microbial processes. Microscopic analysis of these organic matter pulses at Station ALOHA reveal that their composition is dominated by diatoms. These photolithoautotrophic microbes produce an external siliceous skeleton that can act as strong ballast when the cells become senescent.

Several mesoscale physical processes have been observed that can modify the upper water column habitat at Station ALOHA, triggering an increase in the relative abundance of diatoms in surface waters and subsequent cascade of ecological processes. The passage of mesoscale features, such as eddies and Rossby waves, can shift the depth of nutrient-rich water relative to the euphotic zone, leading to a possible influx of nutrients into the well-lit zone that can last from days to weeks. This sustained nutrient entrainment can alter the microbial size spectrum, in favor of rapidly growing, large phytoplankton cells (usually diatoms), resulting in a bloom. In addition, eddies can trap local water masses and transport microbial assemblages for long distances.

A second mechanism triggering changes in the microbial community appears to occur during summer months at Station ALOHA, when the upper water column is warm and strongly stratified. Under these conditions, N_2-fixing cyanobacteria, sometimes living in symbiosis with diatoms, aggregate in surface waters and provide an abundant supply of reduced nitrogen and organic matter to the microbial community. Although it is still not clear what triggers these summer blooms, *in situ* observations suggest that they significantly alter the structure and metabolic activity of the microbial assemblage.

Finally, the mixing layer can periodically penetrate to a depth where it erodes the upper nutricline and delivers nutrients to the surface waters, while mixing surface-dwelling microbes into the upper nutricline. This deepening of the mixing/mixed layer can be driven by sudden events such as the development of a severe storm or the cooling of surface waters by the passage of a cold air mass. And, although each of these three mechanisms can lead to the entrainment of nutrients into the euphotic zone, they generate different microbial responses and interaction. For example, while the first two mechanisms do not involve a change in stratification, the third mixes the water column

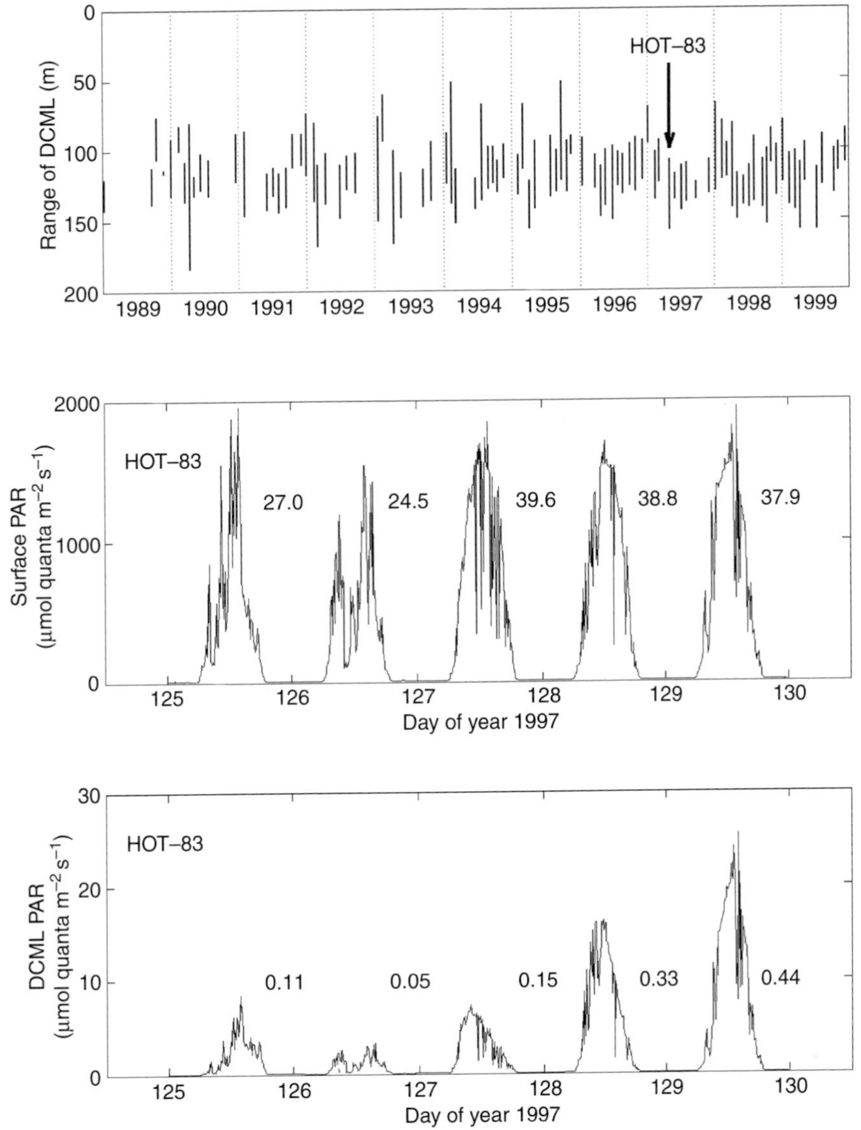

FIGURE 10 Effect of isopycnal vertical displacements in accounting for day-to-day variability of Photosynthetically Available Radiation (PAR) at the DCML at Station ALOHA: (Top) Observed minimum and maximum depth range distribution of the DCML for each HOT cruise based on continuous fluorescence trace profiles obtained from 12 CTD casts deployed over a 36-h sampling period. (Center) Surface PAR measured at the HALE ALOHA mooring location during HOT-83 (5–9 May 1997). (Bottom) Estimated PAR at the DCML based on the vertical displacement of the DCML, surface PAR, and assuming $k_{PAR} = 0.04\,m^{-1}$. Daily integrated PAR values (in mol quanta m^{-2} day^{-1}) are displayed next to each light cycle in (Center) and (Bottom). These day-to-day variations in light caused by inertial period oscillations of the DCML and variations in surface PAR due to clouds are certain to have significant effects on rates of *in situ* photosynthesis. *Reproduced from Karl, D. M., Bidigare, R. R., & Letelier, R. M. (2002). Sustained and aperiodic variability in organic matter production and phototrophic microbial community structure in the North Pacific Subtropical Gyre. In P. J. Williams, B. le, D. R. Thomas, & C. S. Reynolds (Eds.).* Phytoplankton productivity and carbon assimilation in marine and freshwater ecosystems *(pp. 222–264). London: Blackwell Publishers.*

temporarily erasing the physical, chemical, and biological gradients that had existed before the event. Furthermore, while the passage of eddies and Rossby waves introduce nutrients into the base of the euphotic zone, affecting primarily the microbial populations inhabiting the upper nutricline, summer blooms have their strongest effect in the microbial assemblages residing in the upper few meters of the water column. Nevertheless, all these mechanisms appear to generate pulses of particulate organic matter rain that enhance the availability of ephemeral microenvironments, fuel the deeper microbial layers, and carry microbes to the depth.

In addition to mesoscale events and seasonal cycles that seem to support small transient changes in the microbial community structure and function, variability at longer time scales (interannual to decadal) may shift the taxonomic structure of the microbial community. For example, there is evidence suggesting that a significant shift in the dominance of phototrophic taxa may have taken place in the NPSG as a result of changes in ocean circulation and wind forcing during the 1970s. More recently, changes in the stability of the upper water column since the 1997–98 El Niño event may have also triggered long-term changes in the phototrophic community structure.

Ultimately, these long-term habitat changes are the result of processes taking place over a broad range of scales propagating into the habitat experienced by a microbe. In this context, the advent of novel molecular tools such as metagenomic, proteomic, and transcriptomic analyses has provided an unprecedented opportunity to infer the diversity and biogeochemical relevance of microhabitats via the characterization of the genes being expressed in the environment. These new tools may help us better explore how physical and biological processes, by affecting the spatial and temporal distribution of these habitats, shape the microbial diversity and metabolism in the sea. However, understanding how microbial assemblages in different oceanic habitats may evolve over time in response to climate change will require not only a characterization of the microbes' response to physical and chemical changes, but also the development of an understanding of how interactions among microbes contribute to the plasticity and resilience of the microbial ecosystem in the marine environment.

SUMMARY AND PROSPECTUS

All marine habitats support diverse microbial assemblages that interact through a variety of metabolic and ecological processes. The characteristics and dynamics of marine habitats determine the composition, structure, and function of their microbial inhabitants. Many microbial habitats (i.e., microhabitats) are cryptic, ephemeral, and difficult to observe and sample; the spatial and temporal domains of these environments are poorly resolved at present. The changing ocean will lead to different and, probably, novel marine habitats that will select new microbial assemblages. Future ecological research should focus on the relationships among climate, habitat, microbes, and their individual and collective metabolic function. These comprehensive studies demand coordinated, transdisciplinary field programs that fully integrate physical and chemical oceanography with theoretical ecology into the wonderful world of marine microbes.

ACKNOWLEDGMENT

We thank our many colleagues in the HOT and C-MORE programs for stimulating discussions, and the National Science Foundation, the National Aeronautics and Space Adminstration, the Gordon and Betty Moore Foundation, and the Agouron Institute for generous support of our research.

FURTHER READING

Cole, J. J., Findlay, S., & Pace, M. L. (1988). Bacterial production in fresh and saltwater ecosystems: A cross-system overview. *Marine Ecology Progress Series*, *43*, 1–10.

Cullen, J. J., Franks, P. J. S., Karl, D. M., & Longhurst, A. (2002). Physical influences on marine ecosystem dynamics. In A. R. Robinson, J. J. McCarthy, & B. J. Rothschild (Eds.). *The sea* (Vol. 12) (pp. 297–336). New York: John Wiley & Sons, Inc.

DeLong, E. F., Preston, C. M., Mincer, T. *et al.* (2006). Community genomics among stratified microbial assemblages in the ocean's interior. *Science*, *311*, 496–503.

Fenchel, T., King, G. M., & Blackburn, T. H. (1998). *Bacterial biogeochemistry: The ecophysiology of mineral cycling* (2nd ed.). California: Academic Press.

Giovannoni, S. J., Tripp, H. J., Givan, S. *et al.* (2006). Genome streamlining in a cosmopolitan oceanic bacterium. *Science*, *309*, 1242–1245.

Hedgpeth, J. W. (Ed.) (1957). *Treatise on marine ecology and paleoecology*. Colorado: The Geological Society of America, Inc.

Hunter-Cevera, J., Karl, D., & Buckley, M. (Eds.) (2005). *Marine microbial diversity: The key to earth's habitability*. Washington, DC: American Academy of Microbiology.

Johnson, K. S., Willason, S. W., Weisenburg, D. A., Lohrenz, S. E., & Arnone, R. A. (1989). Hydrogen peroxide in the western Mediterranean Sea: A tracer for vertical advection. *Deep-Sea Research*, *36*, 241–254.

Karl, D. M. (1999). A sea of change: Biogeochemical variability in the North Pacific subtropical gyre. *Ecosystems*, *2*, 181–214.

Karl, D. M. (2007). Microbial oceanography: Paradigms, processes and promise. *Nature Reviews Microbiology*, *5*, 759–769.

Karl, D. M., Bidigare, R. R., & Letelier, R. M. (2002). Sustained and aperiodic variability in organic matter production and phototrophic microbial community structure in the North Pacific Subtropical Gyre. In P. J. Williams, B. le, D. R. Thomas, & C. S. Reynolds (Eds.). *Phytoplankton productivity and carbon assimilation in marine and freshwater ecosystems* (pp. 222–264). London: Blackwell Publishers.

Karl, D. M., & Dobbs, F. C. (1998). Molecular approaches to microbial biomass estimation in the sea. In K. E. Cooksey (Ed). *Molecular approaches to the study of the ocean* 29–89. London: Chapman & Hall.

Letelier, R. M., Karl, D. M., Abbott, M. R., & Bidigare, R. R. (2004). Light driven seasonal patterns of chlorophyll and nitrate in the lower euphotic zone of the North Pacific Subtropical Gyre. *Limnology and Oceanography*, *49*, 508–519.

Longhurst, A. (1998). *Ecological geography of the sea*. San Diego: Academic Press.

Mann, K. H., & Lazier, J. R. N. (1996). *Dynamics of marine ecosystems: Biological-physical interactions in the oceans* (2nd ed.). Massachusetts: Blackwell Science.

Martiny, J. B. H., Bohannan, B. J. M., Brown, J. H. *et al.* (2006). Microbial biogeography: Putting microorganisms on the map. *Nature Reviews Microbiology*, *4*, 102–112.

Morita, R. Y. (1985). Starvation and miniaturization of heterotrophs, with special emphasis on maintenance of the starved viable state. In M. Fletcher, & G. D. Floodgate (Eds.), *Bacteria in their natural environments* (pp. 111–130). Orlando: Academic Press.

Platt, T., & Sathyendranath, S. (1999). Spatial structure of pelagic ecosystem processes in the global ocean. *Ecosystems*, *2*, 384–394.

Purcell, E. M. (1977). Life at low Reynolds number. *American Journal of Physics*, *45*, 3–11.

Rocap, G., Larimer, F. W., Lamerdin, J. *et al.* (2003). Genome divergence in two *Prochlorococcus* ecotypes reflects oceanic niche differentiation. *Nature*, *424*, 1042–1047.

Ryther, J. H. (1969). Photosynthesis and fish production in the sea. The production of organic matter and its conversion to higher forms of life vary throughout the world ocean. *Science*, *166*, 72–76.

Sverdrup, H. U. (1953). On conditions for the vernal blooming of phytoplankton. *Journal du Conseil International pour l'Exploration de la Mer*, *18*, 287–295.

Thompson, J. R., Pacocha, S., Pharino, C. *et al.* (2005). Genotypic diversity within a natural coastal bacterioplankton population. *Science*, *307*, 1311–1313.

Yayanos, A. A. (1995). Microbiology to 10,500 meters in the deep sea. *Annual Review of Microbiology*, *49*, 777–805.

RELEVANT WEBSITE

http://www.agi.org. – AGOURON Institute.

http://www.bios.edu/. – BIOS, Bermuda Institute of Oceanic Sciences.

http://hahana.soest.hawaii.edu. – Microbial Oceanography, Hawaii.

http://reason.gsfc.nasa.gov/. – NASA, National Aeronautics and Space Administration.

http://woce.nodc.noaa.gov. – NODC, National Oceanographic Data Center.

Chapter 26

High-Pressure Habitats

A.A. Yayanos
University of California, San Diego, La Jolla, CA, USA

Chapter Outline

ABBREVIATIONS

DSRV Deep Submergence Research Vessel
PTG Pressurized temperature gradient
PUFA Polyunsaturated fatty acid
ROV Remotely operated vehicle

INTRODUCTION

The influence of pressure on biological systems attracted the interest of scientists at Accademia del Cimento in Florence and of Robert Boyle in England during the seventeenth century. Science historian Stephen G. Brush credits Boyle for introducing "a new dimension – pressure – into physics." Euler provided the first mathematical definition of pressure in the eighteenth century. By the end of the nineteenth century, the concept of pressure had been developed into its present-day meaning. Pressure and temperature are today fundamental parameters for physical–chemical theory, for the description of environments, in industrial chemistry and biotechnology, and in both laboratory and ecological investigations of organisms. An indispensable aspect to the analysis of organisms inhabiting high-pressure environments is that temperature and pressure are coordinate variables.

HISTORICAL AND CURRENT BACKGROUND

Seventeenth-century biological research at the Accademia del Cimento and by Boyle quantified and described the effects of decompression from atmospheric pressure. Hot air ballooning began in the late eighteenth century and provided another reason for studies in high-altitude physiology. Human problems similarly aroused interest in elevated pressures. Workers in diving bells, caissons, and deep mine shafts often became ill and experienced physiological difficulties. Bert's classic treatise reviews many of these early studies. The interest in what we now term high-pressure habitats, epitomized by the deep sea, began in earnest in the nineteenth century. Certes, Regnard, and Roger in France did pioneering research in experimental biology. The question of how high pressure influences deep-sea life was, indeed, an impetus for their work. Key references to nineteenth century and to early twentieth century high-pressure research are in the book of Johnson, Eyring, and Pollisar.

Several oceanographic expeditions from the nineteenth century onward established the existence of life, both animal and microbial, in all of the deep sea, which is the largest high-pressure habitat on Earth. Scientific research on HMS *Challenger* during the first round-the-world oceanographic expedition from 1873 to 1876 showed the presence of animals throughout the seas to depths greater than 5000 m. Nearly 80 years later, participants on the Danish *Galathea* Expedition (1950–1952) demonstrated the presence of life in the greatest ocean depths. ZoBell published in 1952 the results of his work done on the Danish research vessel *Galathea*. He found that bacteria from 10,000 m depths in the Philippine Trench grow at pressures as great as 100 MPa whereas bacteria from the upper ocean do not. This work, done with the most probable number method and microscopic examination of cultures incubated at high pressures, was with natural populations of microorganisms. Since 1979, pure cultures of deep-sea bacteria have been isolated by several investigators. Beginning in the late 1970s, expeditions utilizing Deep Submergence Research Vessel (DSRV) *Alvin* led to the discovery of hydrothermal vents at depths approaching 4000 m. At vents, high temperatures and high flow rates of seawater containing copious nutrients result in remarkably localized and productive communities of microorganisms and animals. The seawater effusing from a vent, moreover, may be carrying microorganisms from seafloor depths and locales distant from the vent.

High-pressure habitats in addition to those in the oceans are found within the seafloor and beneath the surface of continents. These subterranean regions, also called subsurface environments, are inhabited by microorganisms. Geochemical analyses showed more than 30 years ago that microbes very likely inhabit the seafloor to depths of hundreds of meters. The continental subsurface has also been studied for many years, as reviewed by Amy and Haldeman (1997). Microbiologists only recently began to sample a wide spectrum of subsurface environments. Although results unusual in terms of pressure have not been found, these are likely to appear as deeper parts of the subsurface environment are explored. Petroleum reservoirs in the seafloor and in subsurface continental locales have yielded thermophilic bacteria. Continental subsurface microbial inhabitants have been studied by scientists in Sweden to understand and quantify microbial processes that could influence plans to bury radioactive waste. Finally, thermophilic bacteria were cultivated from samples collected at a depth of 3700 m in an African gold mine.

The search for life buried in the seafloor has recently accelerated on a number of fronts. First, more scientists are attempting to cultivate microorganisms from cores collected under the auspices of the Deep Sea Drilling Program. Second, geochemists and molecular biologists are looking for patterns in chemical composition along depth profiles (i.e., along the length of a core) that would suggest microbial activity. Third, an examination of the morphology of inorganic surfaces on rocks in cores shows erosion features that provide compelling evidence for biological activity in buried environments. Fourth, the radiolysis of water by ionizing radiation from the decay of Earth's naturally occurring radionuclides has been shown to provide enough hydrogen to fuel microbial metabolism in buried environments. Finally, several clever *in situ* experiments have revealed an abundance of microbial growth in the upper part of the seafloor. In summary, many high-pressure regions of the earth beneath our feet and the seafloor are inhabited.

Definition of Pressure

A stress, σ, is the ratio of the magnitude of a force, ΔF, to the area, ΔA, on which it is acting as $\Delta A \rightarrow 0$. Force acting on an area can be further resolved into components parallel and perpendicular to the area. The parallel components are the shear stresses and the perpendicular ones are the normal stresses. These are summarized by the stress tensor,

$$\sigma_{ij} = \begin{bmatrix} \sigma_{11} & \sigma_{12} & \sigma_{13} \\ \sigma_{21} & \sigma_{22} & \sigma_{23} \\ \sigma_{31} & \sigma_{32} & \sigma_{33} \end{bmatrix}.$$

In fluids at rest and at a uniform temperature, shear stresses are absent or negligible ($\sigma_{ij} = 0$ for $i \neq j$) whereas the three normal stresses are equal to each other ($\sigma_{11} = \sigma_{22} = \sigma_{33}$). The normal stresses are called the hydrostatic pressure or simply the pressure, P, when dealing with fluids of uniform temperature and at rest. At any point in such a fluid, the pressure is the same in all directions. In solids, the pressure may or may not be hydrostatic. For example, some solids can have $\sigma_{11} \neq \sigma_{22} \neq \sigma_{33}$. The stress tensor is a comprehensive expression of stress distribution in a substance. The remainder of this article deals with hydrostatic pressure.

High-Pressure Environments on Earth and Other Planets

The pressure at the surface of the earth arises from the gravitational attraction between Earth and its atmosphere and is $0.101325\ \mathrm{N\,m^{-2}} = 0.101325 \times 10^6\ \mathrm{Pa} = 0.101325\ \mathrm{MPa}$. Since water is considerably denser than air, the pressure increases rapidly with depth in the oceans to reach approximately 110 MPa in its deepest trenches. The pressure in the ocean increases with depth according to the equation

$$dP = g\rho dz,$$

where g is the gravitational "constant," ρ is the density of the seawater, and z is the depth. On Earth at sea level, $g = 9.8\ \mathrm{m\,s^{-2}}$. Although the values of g and ρ vary with latitude, longitude, and depth, these variations in the oceans are small for microbiological considerations so that the pressure in megapascal is approximately given by the depth in meters multiplied by 0.01013. At a depth of 10,000 m, the pressure in Earth's ocean is close to 101.3 MPa.

On Mars, $g = 3.71\,\mathrm{m\,s^{-2}}$. Therefore, if Mars once had an ocean with the same maximum depth as the one on Earth, then the maximum pressure in the Mars Ocean would have been less than 38 MPa. Furthermore, if an ocean on Mars had been as cold as or warmer than the one on Earth, then pressure would not have been a limiting factor for Earth-like life. However, if the Mars Ocean had been substantially colder than that on Earth, then the effects of pressure would have been pronounced even at 38 MPa. Obviously, if life on Mars had been substantially different from that on Earth, then there is no way at present to surmise how pressure influenced it. New evidence suggests that there may be oceans on certain Jovian satellites. Europa, a Jovian satellite slightly smaller than our moon, has a $g \approx 1.23\,\mathrm{m\,s^{-2}}$. Callisto has a $g \approx 1.3\,\mathrm{m\,s^{-2}}$ and is about the size of the planet Mercury. The pressure in any of the oceans on these two moons of Jupiter can be determined by both depth and the local gravitational field. As shown in Table 1, the pressure at the bottom of a possible Europa Ocean that is 30,000 m deep would be approximately 40 MPa.

A list of high-pressure environments is given in Table 1. There is no incontrovertible evidence that the Kola well at a depth of 8000 m or that hydrothermal vents at temperatures much greater than 115 °C are inhabited. Figure 1 assists us in getting an idea of how temperature and pressure influence (1) where life is found; (2) where life is likely absent; and (3) where searches for life may lead to the discovery of new organisms. Figure 1 is a diagram of the *PT*-plane where *P* is the pressure and *T* is the temperature of a locus on Earth. Some of the environments listed in Table 1 are indicated in this figure as lines of constant pressure (isopiests) or constant temperature (isotherms) to show where life is found as well as where it may be found. Only a small portion of all possible *PT*-habitats have been studied. Chief among these are the atmospheric pressure isobar, the cold deep-sea isotherm, the Mediterranean Sea isotherm, hydrothermal vent isopiests up to 40 MPa, and a few subsurface habitats. Potentially interesting habitats that have not been sampled enough are beneath the seafloor at water depths in excess of 5000 m. Figure 1 also serves to underscore the fact that the distribution of organisms is influenced and delimited by both temperature and pressure acting in concert. That is, the temperature range wherein life is found increases with pressure; and, quite possibly, the pressure range of life increases with temperature. There is a line or transition zone of demarcation on the *PT*-plane separating conditions compatible with life from abiotic ones. We know neither where to quite put this line or *PT*-envelope delimiting life processes nor the shape of the envelope. It may, for example, be a continuous concave inward curve. However, this is not yet known. At this time, a theoretical analysis to define a *PT*-envelope for life on Earth has not been made. No one can yet say, for example, that at 60 MPa life will be found at an upper temperature of *X* °C and never 1 °C higher.

TABLE 1 High-pressure environments

High-Pressure Environment	*T* (°C)	*P* (MPa)	Depth (m)
Earth's oceans			
Weddell Basin	–0.5	45.6	4500
Central South Pacific	1.2	50.7	5000
Central North Pacific	1.5	50.7	5000
Peru–Chile Trench	1.9	60.8	6000
Philippine Trench	2.48	101.3	10,000
Tonga Trench	1.8	96.3	9500
Mariana Trench	2.46	110.4	10,915
Celebes Sea	3.26	63.0	6300
Halmahera Basin	7.54	20.7	2043
Sulu Sea	9.84	56.5	5576
Mediterranean Sea	~13.5	50.7	5000
Red Sea	44.6	22.3	2200
Hydrothermal vent	~2–380	25.3	2500
Freshwater bodies			
Lake Baikal	3.5	16	>1600
Lake Vostok	>–5	ca. 35	3800 ice + 500 water
Subsurface of earth's continents			
Kola well in Russia	>155	88.3–205.9	8000
Subsurface of earth's seafloor	0		
Nankai Trough, 30 m into sediments	~5	45.9	4530
6 km water depth, 1 km into seafloor	100	70.9	7000
Planetary environment			
Venus (at its surface)	227	9	0
Europa (conjectured values)	5	40	30,000

Life exists in most of the high-pressure environments listed in Table 1. The world's oceans contain an abundant and diverse group of animals, bacteria, and archaea. The deep Earth is a home for mainly bacteria and archaea. Environments on Mars and Europa could contain microorganisms similar to those on Earth. However, hard evidence must await further exploration. Venus has a surface temperature of 227 °C that experiments and calculations suggest is far too high for Earth-like life processes. However, a final verdict on whether Venus is inhabited must await a determination of its subsurface temperatures.

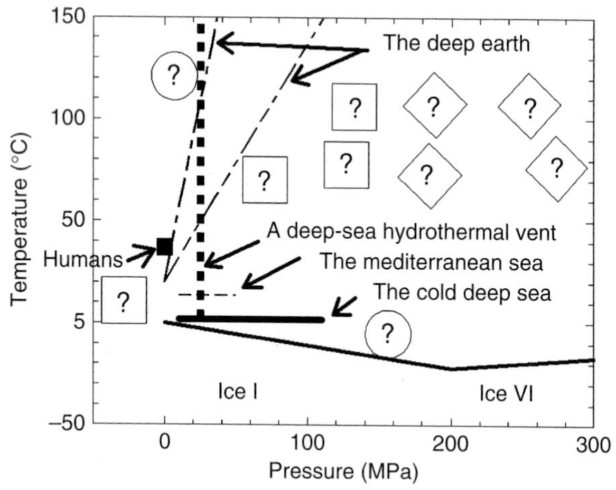

FIGURE 1 Aspects of life in environments of different temperatures and pressures. (1) The heavy black line is an isotherm at 2 °C that represents the cold deep sea inhabited to pressures greater than 110 MPa. The deep sea is populated with animals, archaea, and bacteria and is the most prominent high-pressure habitat on earth. (2) The isotherm at 13.5 °C, drawn as a dashed line, represents the habitats of the Mediterranean Sea and extends to 50 MPa. (3) The vertical dashed line is an isopiest (line of constant pressure) at 25 MPa. This represents the pressure–temperature environment around a hydrothermal vent at 2500 m depth where the maximum temperature observed, in excess of 350 °C, undoubtedly exceeds the tolerance of organisms. The upper temperature limit for life has not been established. Many believe that it will be a temperature less than 150 °C. (4) Two lines show how temperature and pressure might increase along a depth profile into the earth on continents. Similar lines could be drawn beginning, for example, at a pressure of 50 MPa and a temperature of 2 °C to represent a temperature–pressure profile in the seafloor. (5) The two question marks in circles are to indicate that we do not yet know the upper pressure–temperature limits of life. (6) The question marks in squares represent extant temperature–pressure conditions on earth where life could plausibly exist. The three upper squares are with regard to conditions found deep in the seafloor. The square at negative pressures is at temperature–pressure conditions found in the xylem sap of tall trees. (7) The question marks in diamond symbols are positioned at temperature–pressure conditions that possibly do not exist on earth but in which life could exist. (8) The bottom line is approximately along the water–ice phase transition. Price (2007) reviews the possibilities for life and for the origin of life in icy places.

Table 1 and Figure 1 show that Earth's oceans and seafloor have inhabited regions not easily reproduced on other planets and moons of the solar system. That is, almost all of the extraterrestrial conditions that are extreme and habitable in terms of pressure–temperature are represented somewhere on Earth in a marine setting. It is interesting how oceans provide this variety of pressure–temperature habitats. The temperature of the sea surface is warmest at tropical latitudes and coldest at polar regions. The deep sea is a nearly uniformly cold habitat. Seawater at the poles is less than −1.5 °C. The cold surface waters acquire a density less than that of the underlying ocean waters. Enough cold surface water forms and sinks in polar oceans to initiate and sustain deep ocean circulation. Although sinking polar water masses compress in a mostly adiabatic fashion, they become only slightly warmer as they sink. The net result is a deep sea that is preponderantly a cold environment with temperatures close to 2 °C.

Only a few warm deep-sea environments exist and they arise in two distinct ways. One is through the presence of sills, elevated portions of the seafloor that completely separate the deep parts of adjacent basins. The sills block the entry of cold deep water derived from polar regions. Thereby, the abyssal regions of these seas are warm, as shown in Table 1. An example is the Sulu Sea that receives its water from the South China Sea over a sill depth of 400 m, reckoned from the sea surface to the top of the sill. The temperature of the deep sea in the Sulu Sea is 9.8 °C. Another example is the Mediterranean Sea separated from the Atlantic Ocean by the Strait of Gibraltar sill at a depth of 320 m that blocks the entrance of cold deep Atlantic water. Furthermore, the Mediterranean Sea has strong vertical mixing driven by evaporation that forms a dense layer of warm water on the sea surface, which then sinks. Thus, horizontal mixing as well as vertical mixing plays a role in determining the temperature of about 13.5 °C in the deep sea of the Mediterranean Sea. Figure 1 shows an isotherm at 13.5 °C extending to a pressure of 50 MPa found at the greatest depth of the Mediterranean Sea off of the west coast of Peleponesus.

The second cause for warm and hot habitats in the otherwise cold deep ocean is hydrothermal circulation, most notably at midocean ridges. The temperature in geothermal water can be over 370 °C. Mixing with cold ocean water occurs rapidly. A vent environment, called a vent field, is a highly localized region of hundreds of square meters to over a square kilometer. The water depth at most midocean ridges is less than 4000 m. Vents are inhabited to temperatures somewhat above 110 °C. The fluids emanating from vents derive from a sub-seafloor circulation. This implies that these fluids are inhabited for at least part of their journey through the seafloor.

Noteworthy is evidence that the deep sea was not always as cold as it is today. Estimates suggest that the temperature of the deep sea has alternated several times between 15 °C and its present value of 2 °C over the past 800 million years. Thus, the comparative biology of organisms in the cold deep sea, the Sulu Sea, the Mediterranean Sea, and at hydrothermal vents is of value for paleoecology.

Other Parameters of High-Pressure Habitats

In the previous section the case is made that temperature and pressure must be considered as twin parameters of habitats. An important part of the rationale for this is that the values of other essential parameters of habitats and of the intracellular milieu change along with changes in temperature and pressure. Chief among these temperature-dependent and pressure-dependent parameters are those of density, viscosity, dielectric constant, coefficient of thermal expansion, coefficient of compressibility, chemical equilibrium constants, phase transition and stability conditions, and chemical reaction rate constants. The dielectric constant, for example, increases with increasing pressure and decreases with increasing temperature. An understanding of how these essential parameters are affected by both temperature and pressure will ultimately offer two important explanations. One is to provide a basis for the existence of limits of temperature and pressure beyond which life cannot exist. The other is to explain (or, to predict) the temperature and pressure limits for the growth of a given organism.

INSTRUMENTS TO SURVEY AND SAMPLE MICROORGANISMS IN HIGH-PRESSURE HABITATS

There are two distinct methods used to study life in high-pressure habitats. One is the *in situ* approach whereby measurements, observations, and experiments are performed in the deep sea or in other high-pressure habitats. The second is to recover organisms from a high-pressure habitat and conduct research on them in the laboratory. Organisms during the sampling process may suffer decompression, temperature change, contamination, exposure to light, and a change in habitat chemistry. Specially designed sampling instruments minimize or eliminate these changes. Instruments used in marine microbiology include Niskin bottles, pressure-retaining water samplers, pressure-retaining animal traps, pressure-retaining corers, and thermally insulated corers. Sediment microbiologists use cores collected mostly with decompression. Microbiologists sampling the terrestrial subsurface devise methods to minimize or detect contamination. Sometimes they flame the outer surface of a core to inactivate microbial contaminants. They also add to drilling fluids detectable particles whose presence in a core sample serves as a proxy for possible microbial contamination.

DSRV *Alvin* and DSRV *Shinkai* 6500 are particularly effective in collecting samples both without the introduction of additional foreign microorganisms and from precisely identified locales. Scientists Kato, Li, Tamaoka, and Horikoshi of the Japanese Marine Science and Technology Center operated *Kaiko*, a remotely operated vehicle (ROV), using DSRV *Shinkai* 6500 to collect sediments from the deepest part of the Mariana Trench in 1997 without introducing contaminant organisms via the sampling procedure. Axenic sampling of seafloor sediments and subterranean habitats is difficult and likely impossible because these habitats are contaminated constantly with upper ocean organisms as a result of natural processes. It is, however, important to undertake efforts to minimize and assess any additional contamination due to sampling.

If every experiment required samples strictly maintained at habitat conditions during sample collection, then research progress would be slow and expensive. Thus, development of alternate sampling strategies is important. Indeed, most deep-sea microbiology is conducted with decompressed samples collected with Niskin bottles and a variety of coring devices deployed from ships. An inescapable sampling requirement for microbiology of the cold deep sea is the avoidance of sample warming. The justification for this is seen in Figure 2 showing the rapid thermal inactivation of a deep-sea bacterium at temperatures encountered in the upper tropical and temperate ocean through which sampling gear must pass. Also shown in Figure 2 is the inactivation of a bacterium from the deepest part of the ocean while it is held at atmospheric pressure and 0 °C. This shows that decompression per se, although lethal to this bacterium, does not instantly kill it. Death following decompression has so far been seen only in bacteria of the cold deep ocean. Most subsurface microbiology has been done with decompressed core samples from mesophilic and thermophilic habitats and with atmospheric pressure cultivation. Such samples have been adequate to establish enrichment cultures of mesophilic and thermophilic bacteria and archaea. Pressure retention on samples, however, remains an essential aim in work addressing questions on community composition, structure, and function and seeking to identify any variability in the manifestations of pressure adaptation.

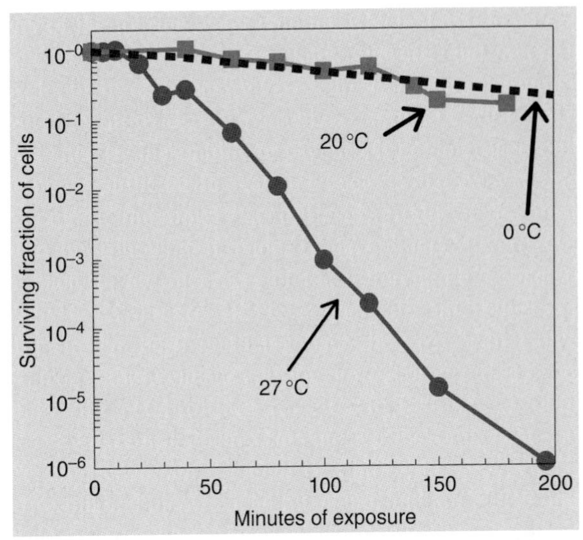

FIGURE 2 Thermal inactivation kinetics for two different bacterial isolates. The data in red are for bacterial isolate CNPT3, from a depth of 5782 m in the Pacific Ocean. One suspension of cells was placed at 27 °C and another at 20 °C. At selected times the cell suspensions were sampled and the pour tube method (Figure 3) was used to determine the fraction of cells that were still able to form colonies. This bacterium looses viability (colony-forming ability) at atmospheric pressure when kept at temperatures greater than 10 °C. Sensitivity to warming has not been determined for very many bacterial strains from the cold deep sea. Nevertheless, high thermal sensitivity is likely one of their traits. One bacterial strain from the Mariana Trench, isolate MT41, looses viability at 0 °C at atmospheric pressure. The dashed line shows the slow rate of death for isolate MT41. This slow death rate may be due to decompression per se. *Reproduced from Yayanos, A. A. (1995). Microbiology to 10500 meters in the deep sea.* Annual Reviews of Microbiology, 49, 777–805. (For interpretation of the references to color in this figure legend, the reader is referred to the Web version of this chapter.)

Research on samples from the Mid-Atlantic Ridge at a depth of 3550m shows that certain hyperthermophilic microbes grew in enrichment cultures only if incubated at high pressure. One of the startling things about this work is that the organism *Thermococcus barophilus* thus isolated also grows at atmospheric pressure, raising the question of why it was not isolated in atmospheric pressure enrichments. *T. barophilus*, moreover, is a hyperthermopiezophile since it exhibits its maximum growth rate at a high pressure when tested over all possible temperatures. This study by Marteinsson and colleagues shows clearly the value of imitating as much as possible the conditions of the sampled environment in enrichment cultures and also the necessity of experimentally determining the pressure–temperature growth responses of organisms under a variety of experimental conditions.

Piezothermophiles survive exposure both to low temperatures and to low pressures whereas piezopsychrophiles do not survive low pressures and temperatures much above 20°C. This means that the dispersal path for the latter is through the cold deep ocean. In contrast, some piezothermophiles survive exposure to cold temperatures for at least 1 year. This gives piezothermophiles many pathways for dispersal from one vent to another. For example, they could be transported by cold deep-ocean currents from one vent field to another. Also, they could emerge from one vent and enter the seabed to become part of the sub-seafloor hydrothermal circulation and then emerge from another vent. Although speculative, these possibilities seem quite feasible.

HIGH-PRESSURE APPARATUS FOR LABORATORY STUDIES OF MICROORGANISMS

Essential tools of microbiologists are enrichment culture technique, plating for isolation of clones and assay of colony-forming ability (viability), replica plating for isolation of mutants, and determination of growth rates in liquid culture. Each of these methods needs to be modified for the study of microorganisms having a strong preference or absolute necessity for high pressure. Yayanos (2001) provides a more thorough treatment of the methods discussed in the remainder of this section.

The key components, commercially available, of a high-pressure apparatus are a pressure gauge, a pump (sometimes called an intensifier), a pressure vessel, valves, and tubing connecting these components. Although pressure vessels can be purchased, the need to fabricate them with features otherwise unavailable often arises. The fluid used to compress a sample in a pressure vessel is usually water. Gases present problems as a hydraulic medium and find infrequent use. For example, they dissolve in biological phases to have effects other than those arising from compression. Also, the need for precautions such as barricades to ensure safety in work with compressed gases makes the work expensive.

One of the technical difficulties in high-pressure microbiology is that cultures must be isolated from the hydraulic fluid. Although liquid cultures have been incubated with success in syringes, the possibility of contaminants entering the culture around the syringe plunger must always be kept in mind, especially in long-term incubations. A great boon to high-pressure microbiological research has been the advent of heat-sealed plastic containers such as bags and polyethylene transfer pipettes. These allow for pressure equilibration, isolation of the culture from microbes and chemicals in the hydraulic fluid, and a degree of control of gas composition in the culture. The latter is achieved through the selection of plastic bag materials with appropriate gas permeability allowing for the exchange of dissolved gases between the culture and the hydraulic fluid.

Clones of bacteria can be obtained at high pressure with pour tubes. Figure 3 shows colonies of a marine bacterium grown at high pressure. Parafilm stretched tightly over the opening of a pour tube sealed it from contamination and served to transmit the pressure to the medium. The pour tube method can also be accomplished with heat-sealed polyethylene transfer pipettes rather than test tubes. The pour tube technique works for high-pressure microbiology

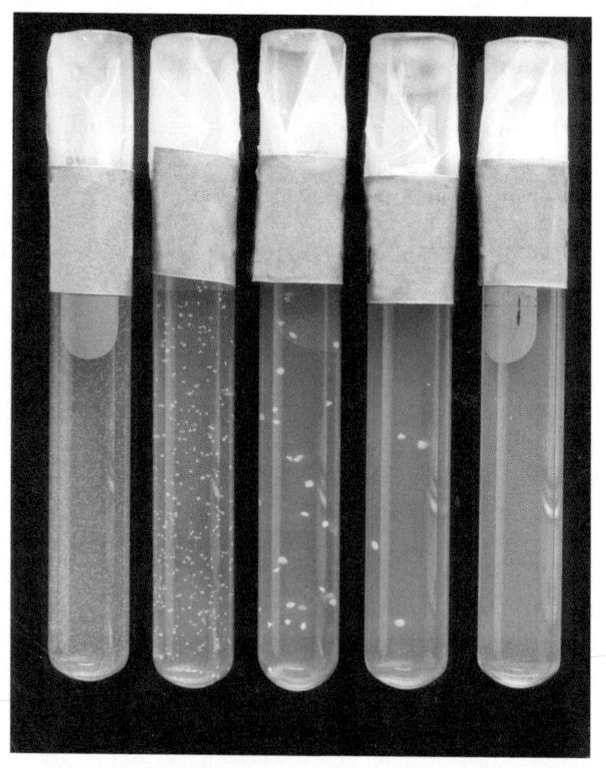

FIGURE 3 Five pour tubes are shown. The one on the right was not inoculated. The other four, beginning with the tube on the left, were inoculated with a serial dilution of a culture of *Micrococcus euryhalis*, covered with parafilm and incubated at high pressure. After several days, visible colonies developed around each immobilized bacterial cell. Note that the colonies are larger at greater dilutions. This pour tube method is ideal for obtaining clones of bacteria in pressure vessels and for assay of cell viability following the exposure of cells to chemical or physical stress.

as petri plates do for atmospheric pressure microbiology with one exception. That is, replica plating cannot be done with pour tubes.

Mesophiles and thermopiles from habitats having a pressure of less than 50 MPa grow in enrichment cultures incubated at atmospheric pressure. For example, methanogens and sulfide oxidizers have been isolated through atmospheric pressure enrichment cultures that were inoculated with samples from deep-sea hydrothermal vent habitats. The enrichment culture method is also known as the ecological method because it is an attempt to grow an organism under the physical and chemical conditions of its natural habitat. Strict application of the ecological method would seem to require incubation of the enrichment cultures at the pressure of the hydrothermal vent habitat. If additional evidence, however, provides little doubt that an organism isolated at atmospheric pressure is an inhabitant of the sampled high-pressure habitat, then the need for conducting enrichments at high pressure seems diminished. Most of the cultures of mesophilic and thermophilic organisms from deep-sea hydrothermal vents and the terrestrial subsurface have so far been found amenable to cultivation and plating at atmospheric pressure. Perhaps, a requirement for strict high-pressure technique will be necessary for the successful isolation of mesophiles and thermophiles from warm and hot habitats having a pressure of 100 MPa or more.

In contrast to mesophiles and thermophiles, the isolation and study of psychrophilic deep-sea bacteria is nearly impossible without the use of high-pressure laboratory methods, especially the pour tube technique modified for high pressures.

An instrument called a pressurized temperature gradient (PTG) allows for the concurrent determination of growth parameters of bacteria incubated at different pressures and temperatures. It also allows for the isolation of clones of bacteria growing at different temperatures and pressures from a single inoculum. Scientists in Germany, Japan, and the United States have published three different designs of a PTG instrument. The fact that a wider use of such instruments has not been adopted suggests that further design improvements are needed.

DISTRIBUTION OF BACTERIA AND ARCHAEA IN HIGH-PRESSURE HABITATS

Bacteria and archaea are ubiquitous in the ocean, including its greatest depths with a pressure of approximately 110 MPa. Since animals are found at all ocean depths, as shown in Figure 4, a highly plausible hypothesis is that animals, bacteria, and archaea could exist at pressures even higher than 110 MPa. It is further likely that microorganisms could live at pressures beyond those limiting animal life. Perhaps an environment having a pressure of 200 MPa would be habitable by bacteria. If such an environment existed and was as cold as or colder than the 2 °C temperature of today's deep sea, then metabolic processes probably would be very slow and confined within narrow ranges of temperature and pressure.

FIGURE 4 Photograph of amphipods swarming to food (dead fish) placed on the seafloor at a depth greater than 10,500 m. The dead fish were tied to one end of an expendable pole and the 35-mm camera in a pressure-resistant housing was mounted at the other end of the pole by means of a timed release mechanism. The pole and other ballast were left on the seafloor when the timed release mechanism was actuated to enable floats to raise the camera to the sea surface. The amphipods are *Hirnodellea gigas*. Pictures such as this one suggest that the animals inhabit this depth. But this does not prove that the animals reproduce at this depth. These animals were not able to survive decompression to even 34 MPa in the few experiments conducted to date.

Bacteria have been found beneath the surface of both continents and the seafloor. Among the principal factors bounding the distribution of life in subsurface environments are restricted space as judged by the size of pores in compacted sediments, the absence of cracks or fissures, and temperatures exceeding approximately 115 °C.

PROPERTIES OF HIGH-PRESSURE INHABITANTS

Before any research had been done on deep-sea bacteria, it was known that physiological and biochemical processes in bacterial and eukaryotic cells were influenced by pressure.

Single-cell organisms from atmospheric pressure habitats typically show morphological and physiological aberrations when placed at 20–50 MPa. Against this background, the finding of pressure adaptation in deep-sea bacteria is expected. Exactly how adaptation is achieved, however, remains an active area of research. Current biochemical and molecular biological research is primarily on pressure adaptation in Gram-negative heterotrophic bacteria isolated from the cold deep ocean and in archaea from both the cold ocean and hydrothermal vents. The most obvious manifestation of adaptation to high pressure was first seen in organisms of the cold deep sea. Only there, for example, have microorganisms that grow exclusively at high pressure been found so far. Such bacteria have been isolated from samples of ocean depths between 6000 and 10,000 m. Nevertheless, pressure adaptation, although less conspicuous, is present in cold deep-sea microorganisms inhabiting depths at least as shallow as 2000 m and in hyperthermophiles from submarine hydrothermal vents. The rest of this article describes some of the manifestations of pressure adaptation.

Rates of Growth

The kinetics of chemical transformations mediated by microbes and of microbial growth are of general interest because the associated rate constants are important parameters in models of food web dynamics and biogeochemical cycles. The question of whether bacteria at a given depth in the sea grow more slowly than do their relatives at a shallower depth is difficult to answer. This is because the growth rate of bacteria is a multivalued function of not only temperature and pressure but also other factors. These include pH of the medium, oxidation–reduction reactions, and nutrient types and levels. *E. coli*, for example, grows 20 times more rapidly in a complex nutrient medium than in a minimal medium with succinate as the sole carbon source. Experiments with a deep-sea bacterial isolate also show slower growth in a minimal medium. Remarkably, the growth rate of this bacterial isolate also exhibited a diminished response to pressure change when grown on glucose or glycerol minimal medium. To summarize, the pressure dependence of the growth rate is dependent on many factors.

Nevertheless, a comparison of growth rates of different heterotrophic bacterial species grown in an identical nutrient-rich medium at their respective habitat temperatures and pressures shows a trend for deeper-living species to have the slowest growth rates. Thus, the fastest-growing deep-sea heterotrophic bacteria are usually from depths of less than 5 km. And, the slowest-growing ones are generally from the greatest ocean depths of about 10 km. Growth rates (at habitat temperatures and pressures) of bacteria isolated from depths of 2–11 km are in the range of 3–35 h, about as ZoBell reported nearly 50 years ago.

PTk-Diagrams

In pure culture, the growth rate of microorganisms can be exponential,

$$dN/dt = kN,$$

where N is the number of cells at time t and k is the exponential growth rate constant. The pressure dependence of the growth rate of an exponentially growing bacterial species is widely used as an index of pressure adaptation. Bacterial growth rates in media having high levels of nutrients provide the following view of cellular adaptation to pressure in the cold deep ocean. In Figure 5, growth rates of five heterotrophic bacterial strains, isolated from six different depths of the Pacific Ocean, are shown as a function of both temperature and pressure. These plots, called *PTk*-diagrams where P is the pressure, T the temperature, and k the specific growth rate constant, summarize a large number of growth rate determinations and reveal relationships otherwise difficult to see. Examples are as follows.

1. The pressure dependence of the growth rate is more pronounced in bacteria from the deepest habitats. Thus, strains MT199 and MT41 from deep habitats do not grow at atmospheric pressure whereas strain PE36 from a relatively shallow habitat of 3.5 km grows at atmospheric pressure.
2. The *PTk*-diagrams show that each bacterial species has a single pressure–temperature condition where its growth rate is a maximum.
3. The pressure where the growth rate has a maximum value is very nearly the pressure of the presumed habitat. This relationship between the pressure of maximum growth rate and the habitat pressure is not observed among heterotrophic bacteria from the warm deep-sea environments of the Sulu and Mediterranean seas. This fact shows very clearly that growth rate as an index of pressure adaptation depends greatly on the temperature of the high-pressure habitat. It is conceivable that physiological parameters other than growth rate will provide useful indices of pressure adaptation for inhabitants of high-temperature, high-pressure environments.
4. A particularly curious characteristic of deep-sea psychrophilic bacteria is that the temperature where the bacteria grow best is greater by 6–10 °C than the constant habitat temperature of 2 °C. This relationship, also evident with bacteria from the Sulu and Mediterranean seas, remains unexplained.

The study of the properties of bacteria from the cold deep ocean and from warm deep seas with the aid of *PTk*-diagrams shows a progression in the properties of bacteria along the temperature and pressure gradients. That is, the growth properties of these microorganisms are in a sense a

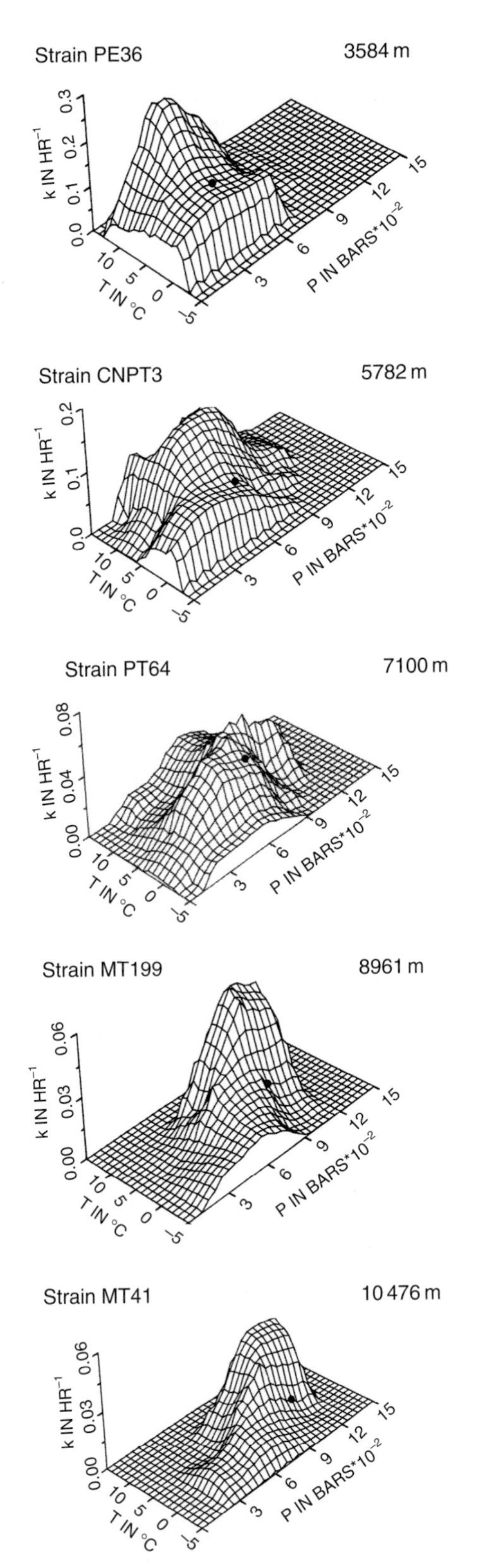

FIGURE 5 *PTk*-diagrams of five deep-sea bacteria. This shows that pressure adaptation becomes more pronounced as the habitat depth of an isolate increases. Each bacterial isolate has a maximum growth rate close to the pressure of its habitat (indicated by the black dot). In this figure the unit of pressure is the bar. 1 bar = 0.1 MPa and 1 atm = 1.01325 bars.

signature for the depth from which they came. Similar conclusions from the study of thermophiles cannot be made as yet. One possible reason for this is that thermophiles grow over large conditions of temperature and pressure. Another compounding effect is that the particular place where a sample was collected to isolate a thermophile may not be the principal habitat of that organism. That is, the process of hydrothermal circulation may be bringing the organism from a habitat of considerably different temperatures and pressures.

Physiological Classification of Bacteria and Archaea with Respect to Temperature and Pressure

Microorganisms are called psychrophiles, mesophiles, thermophiles, or hyperthermophiles based on the temperature range wherein they grow. Although there is likely a continuum of adaptations along the temperature scale, this classification is useful. Even if these microorganisms are from atmospheric pressure habitats, they can grow at high pressures, usually less than 50 MPa. The conditions for maximum rate of growth, however, have always been found, so far, at atmospheric pressure. Since all organisms grow to some extent at high pressure, the older term "barotolerant" to describe an organism that can tolerate high pressures has little significance.

There are two ways in which microorganisms can be classified with respect to their growth as a function of pressure. The traditional classification is to call them "barophiles" if they grow best at a pressure greater than the atmospheric pressure. This binary classification of the pressure-adaptive trait is thus different from the one used for classifying organisms based on their response to temperature. Furthermore, growth studies of bacterial isolates from high-pressure habitats of different temperatures make it apparent that an organism can be "barophilic" at one temperature but not at another. Significantly, there are bacteria showing little or no "barophilic" character at the temperature of their habitat while being distinctly "barophilic" at a greater temperature. This binary categorization of microorganisms with respect to pressure was created before there were any studies with axenic cultures of deep-sea bacteria and of bacteria and archaea from high-pressure habitats having different temperatures.

A newly suggested categorization of a given bacterial isolate is based on its *PTk*-diagram and is analogous to the scheme used to describe temperature adaptation. First, in this new scheme, non-pressure-adapted microorganisms are simply psychrophiles, mesophiles, thermophiles, and hyperthermophiles. They are all "barotolerant" in the old terminology. Second, if the pressure, P_{kmax}, where the maximum growth rate occurs, as determined with a *PTk*-diagram, is

$$0.1\,\text{MPa} < P_{k_{max}} < 50\,\text{MPa}$$

TABLE 2 Categorization of microorganisms

P of k_{max}	$P \approx 0.1$ MPa	0.1 MPa > P > 50 MPa	P > 50 MPa
T of k_{max}			
$T > 15\,°C$	Pychrophiles	Piezopsychrophiles	Hyperpiezopsychrophiles
$15 > T > 45\,°C$	Mesophiles	Piezomesophiles	*Hyperpiezomesophiles*
$45 > T > 80\,°C$	Thermophiles	Piezothermophiles	*Hyperpiezothermophiles*
$80\,°C > T$	Hyperthermophiles	Piezohyperthermophiles	*Hyperpiezohyperthermophiles*

Representative bacteria or archaea in the categories shown in italics have not been found.

then the isolate is a piezophile. Third, if

$$P_{k_{max}} > 50\,\text{MPa}$$

then the isolate is hyperpiezophile. Table 2 shows the different possible types of microorganisms based on where the pressure and temperature of the maximum growth appears on a *PTk*-diagram. Inspection of Table 2 shows that this is a more useful classification than to simply state that an organism is a "barophile." Also evident in Table 1 is that hyperpiezophiles have only been found in the cold deep ocean, that is, among the psychrophiles of a high-pressure habitat. As discussed in the context of Figure 1, the possibly inhabited environments having a pressure of approximately 100 MPa and a temperature in the range of 100 °C may exist only in the seafloor beneath the deepest parts of the ocean. It is there that hyperpiezomesophiles, hyperpiezothermophiles, and hyperpiezohyperthermophiles will be found. Parenthetically, the prefix "piezo" (having the meaning to press) rather than "baro" (having the meaning of weight) is widely adopted in science, in terms such as piezochemistry and piezoelectric, to describe the role of pressure. To summarize, the scheme in Table 1 provides a succinct view of states of adaptation to both temperature and pressure.

Molecular Biology and Biochemistry of Adaptation to High Pressures

The structure of deep-sea bacteria viewed to date with light and electron microscopy is the same as that of bacteria in general. The adaptations making possible their existence at high pressure can be presumed thereby to be at the molecular level. Several macromolecular structures and interactions as well as chemical reactions are altered by pressure change. These include microtubule assembly, ribosome integrity, helix-coil transitions in nucleic acids and proteins, protein conformation, protein–protein interactions, protein–nucleic acid interactions, enzyme activity, transport across membranes, and membrane fluidity. The list of pressure-affected processes is long. Bacterial membrane protein composition is also a function of pressure. The altered composition reflects pressure action on gene expression, on protein synthesis, and on membrane function. The changes observed are further dependent on the physiological state of the cell. For example, protein profiles not only change with pressure but also in a different way when cells grow on different carbon sources. Thus, the adaptation of marine microorganisms to nutritional states alters the manifestation of pressure adaptation. The final view of pressure and temperature adaptation will most likely comprise both an understanding of the pressure and temperature dependencies of many individual processes and a mathematical model of how the cell operates as a dynamical system. The latter is necessary because biological systems are quintessentially greater than the sum of their parts.

The fatty acid composition of the membrane phospholipids of deep-sea heterotrophic bacteria is a function of the growth temperature and pressure. In some bacterial isolates, the trend in fatty acid composition change is consistent with the homeoviscous hypothesis. The crux of this hypothesis is that cells regulate their membrane composition to maintain the membrane in an appropriate physical state (fluidity, in particular). The regulation is done by altering membrane fatty acid composition and phospholipid molecular species. The regulation is activated in response to any physical or chemical factor that causes a change in membrane fluidity. The fact that bacterial membrane lipid composition changes in response to temperature change have long been known. It is now well documented that the composition of the membranes of deep-sea bacteria varies with both the temperature and pressure of growth. Work remains to be done to show that the observed changes are along the lines of the homeoviscous hypothesis. Figure 6 shows the membrane fatty acid composition of a bacterial isolate from the Philippine Trench and its dependence on growth pressure.

Deep-sea bacteria synthesize polyunsaturated fatty acids (PUFAs) for their membrane phospholipids. Figure 7 shows the chemical structure of the principal PUFAs found in many deep-sea bacteria. Although not demonstrated for deep-sea animals, it is generally believed that animals do not synthesize PUFAs and fulfill the requirement for them through their diet. It is conceivable that deep-sea bacteria contribute to this dietary need in the deep sea.

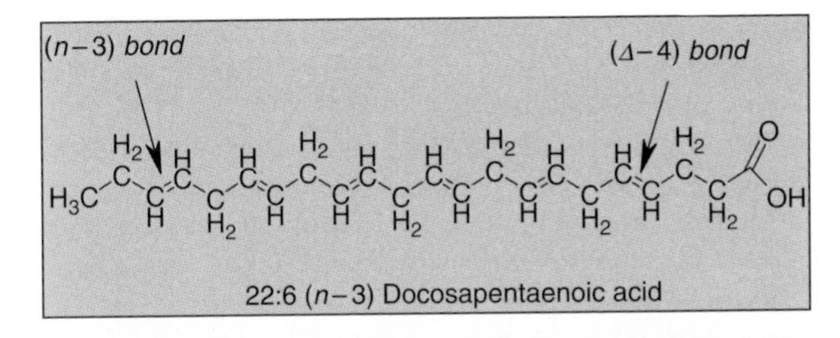

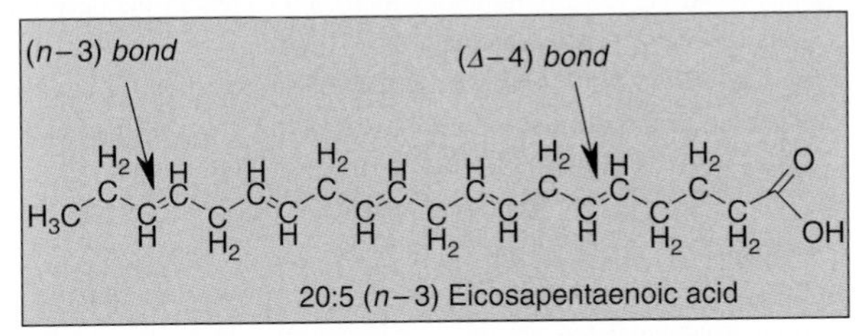

FIGURE 6 The chemical structures of two polyunsaturated fatty acids (PUFAs) found in deep-sea bacteria. Prior to the discovery of these fatty acids in the membrane phospholipids of deep-sea bacteria, it was believed that few if any bacteria could synthesize these lipids. They now appear to be common among heterotrophic deep-sea bacteria.

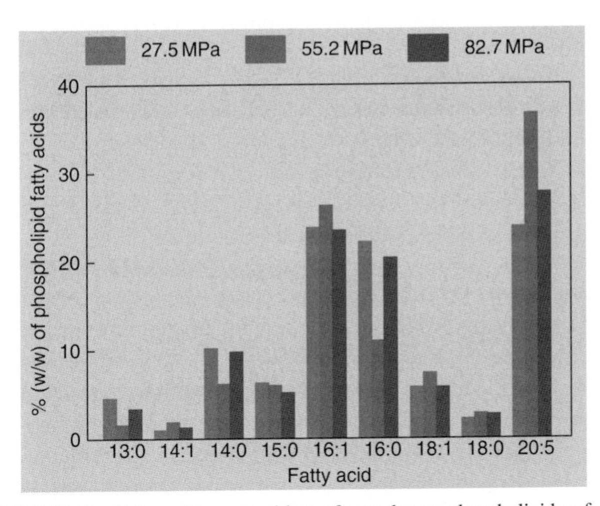

FIGURE 7 Fatty acid composition of membrane phospholipids of a deep-sea bacterium and how it changes with pressure. Plotted from data in DeLong and Yayanos (1985). (For interpretation of the references to color in this figure legend, the reader is referred to the Web version of this chapter.)

The deep sea is also dark. The results in Figure 8 show that one deep-sea bacterial isolate has an extraordinary sensitivity to UV light. Similar results were obtained with four other deep-sea bacterial strains. However, it has not yet been determined if these bacteria have lost the genes to repair UV-damaged DNA. The sensitivity of deep-sea bacteria to ionizing radiation, moreover, is no different than that observed with shallow-water bacteria. This is in accord with the background of natural radioactivity being similar in both shallow water and the deep sea. Since the radiation biology of deep-sea microorganisms is based on scant information, more work is needed before general statements can be made confidently.

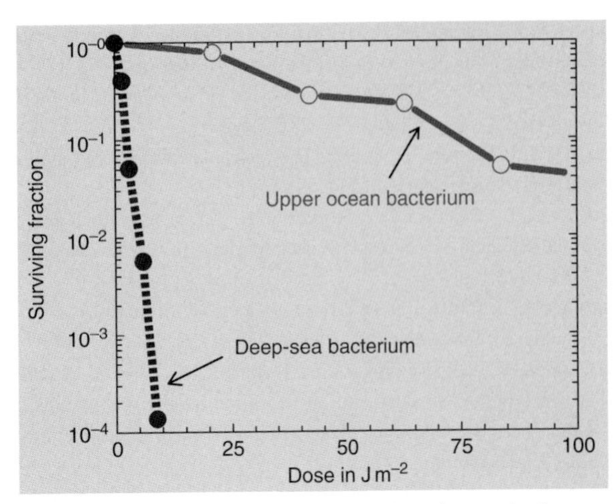

FIGURE 8 The sensitivity of a deep-sea bacterium and of an upper ocean bacterium to UV light. *Reproduced from Yayanos, A. A. (1989). Physiological and biochemical adaptations to low temperatures, high pressures, and radiation in the deep sea. In T. Hattori,* et al. *(Eds.)* Proceedings of the 5th international symposium on microbial ecology *(pp. 38–42). Tokyo: Japan Scientific Societies Press.*

The genome sequences of several deep-sea bacteria have been completed recently. Studies of these genomes will certainly lead to hypotheses ultimately elucidating the nature of adaptation to the diverse high-pressure environments on Earth. As pointed out elsewhere in this article, the understanding of life in all of the extreme habitats on Earth will undoubtedly be of great value in finding life elsewhere in the solar system.

FURTHER READING

Amy, P. S., & Haldeman, D. L. (Eds.) (1997). *The microbiology of the terrestrial deep subsurface* (p. 356). Boca Raton, FL: CRC Lewis Publishers.

Bartlett, D. H. (2002). Pressure effects on in vivo microbial processes. *Biochimica Biophysica Acta, 1595*, 367–381.

Bert, P. (1943). *Barometric pressure. Researches in experimental physiology.* Columbus, OH: College Book. (Translated from the French by MA Hitchcock and FA Hitchcock).

DeFlaun, M. F., Fredrickson, J. K., Dong, H. *et al.* (2007). Isolation and characterization of a *Geobacillus thermoleovorans* strain from an ultra-deep South African gold mine. *Systematic and Applied Microbiology, 30*(2), 152–164.

DeLong, E. F. (1997). Marine microbial diversity: The tip of the iceberg. *Trends in Biotechnology, 15*, 203–207.

DeLong, E. F., & Yayanos, A. A. (1986). Biochemical function and ecological significance of novel bacterial lipids in deep-sea prokaryotes. *Applied and Environmental Microbiology, 51*, 730–737.

DeLong, E. F., Franks, D. G., & Yayanos, A. A. (1997). Evolutionary relationships of cultivated psychrophilic and barophilic deep-sea bacteria. *Applied and Environmental Microbiology, 63*, 2105–2108.

Deming, J. W., & Baross, J. A. (1993). Deep-sea smokers: Windows to a subsurface biosphere. *Geochimica et Comochimica Acta, 57*, 3219–3230.

Furnes, H., & Staudigel, H. (1999). Biological mediation in ocean crust alteration: How deep is the deep biosphere. *Earth and Planetary Science Letters, 166*(3–4), 97–103.

Johnson, F. H., Eyring, H., & Polissar, M. J. (1954). *The kinetic basis of molecular biology*. New York: John Wiley & Sons, Inc, 874 pp.

Karl, D. M. (Ed.) (1995). *The microbiology of deep-sea hydrothermal vents* (p. 299). Boca Raton, FL: CRC Press.

Kato, C., & Bartlett, D. H. (1997). The molecular biology of barophilic bacteria. *Extremophiles, 1*, 111–116.

Kato, C., Li, L., Tamaoka, J., & Horikoshi, K. (1997). Molecular analyses of the sediment of the 11000-m deep Mariana Trench. *Extremophiles, 1*, 117–123.

Lauro, F. M., & Bartlett, D. H. (2008). Prokaryotic lifestyles in deep sea habitats. *Extremophiles, 12*(1), 15.

Marteinsson, V. T., Reysenbach, A. L., Birrien, J. L., & Prieur, D. (1999). A stress protein is induced in the deep-sea barophilic hyperthermophile Thermococcus barophilus when grown under atmospheric pressure. *Extremophiles, 3*(4), 277–282.

Middleton, W. E. K. (1971). *The experimenters: A study of the Accademia del Cimento*. Baltimore: Johns Hopkins Press, 415 pp.

Price, P. B. (2007). Microbial life in glacial ice and implications for a cold origin of life. *FEMS Microbiology Ecology, 59*(2), 217–231.

Sakiyama, T., & Ohwada, K. (1997). Isolation and growth characteristics of deep-sea barophilic bacteria from the Japan Trench. *Fisheries Science (Tokyo), 63*, 228–232.

Yayanos, A. A. (1980). Measurement and instrument needs identified in a case history of deep-sea amphipod research. In F. D. Diemer, F. J. Vernberg, & D. Z. Mirkes (Eds.), *Advanced concepts in ocean measurements for marine biology* (pp. 307–318). Columbia, South Carolina: University of South Carolina Press.

Yayanos, A. A. (1986). Evolutional and ecological implications of the properties of deep-sea barophilic bacteria. *Proceedings of the National Academy of Sciences USA, 83*, 9542–9546.

Yayanos, A. A. (1989). Physiological and biochemical adaptations to low temperatures, high pressures and radiation in the deep sea. In T. Hattori, *et al.* (Eds.), *Proceedings of the 5th international symposium on microbial ecology* (pp. 38–42). Tokyo: Japan Scientific Societies Press.

Yayanos, A. A. (1995). Microbiology to 10,500 meters in the deep sea. *Annual Reviews of Microbiology, 49*, 777–805.

Yayanos, A. A. (1998). Empirical and theoretical aspects of life at high pressures in the deep sea. In K. Horikoshi, & W. D. Grant (Eds.) *Extremeophiles* (pp. 47–92). New York: John Wiley & Sons.

Yayanos, A. A. (2001). Deep-sea piezophilic bacteria. *Methods in Microbiology, 30*, 615–637.

Yayanos, A. A., & DeLong, E. F. (1987). Deep-sea bacterial fitness to environmental temperatures and pressures. In H. W. Jannasch, R. E. Marquis, & A. M. Zimmerman (Eds.) *Current perspectives in high pressure biology* (pp. 17–32). New York: Academic Press.

Yayanos, A. A., & Dietz, A. S. (1982). Thermal inactivation of a deep-sea barophilic bacterium, isolate CNPT-3. *Applied and Environmental Microbiology, 43*, 1481–1489.

Yayanos, A. A., Dietz, A. S., & Van Boxtel, R. (1979). Isolation of a deep-sea barophilic bacterium and some of its growth characteristics. *Science, 205*, 808–810.

ZoBell, C. E. (1952). Bacterial life at the bottom of the Philippine Trench. *Science, 115*, 507–508.

Chapter 27

The Deep Sub-Surface

R.J. Parkes and H. Sass
Cardiff University, Cardiff, UK

Chapter Outline

ABBREVIATIONS

ANME Anaerobic methane-oxidizing
AOM Anaerobic oxidation of methane
DSAG Deep-Sea Archaeal Group
FISH Fluorescence *in situ* hybridization
MBG-D Marine Benthic Group D
MCG Miscellaneous Crenarchaeotic Group
MG-I Marine Group-I
SAGMEG South African Gold Mine Euryarchaeotal Group
TMEG Terrestrial Miscellaneous Euryarchaeotic Group

DEFINING STATEMENT

Marine sediments cover ~70% of the Earth's surface and contain the largest reservoir of organic matter and significant fossil fuels. These sediments have recently been found to be a major microbial habitat to at least 1 km depth. Unique prokaryotic groups are present using ancient organic matter and geosphere energy sources.

INTRODUCTION AND BACKGROUND

Oceans cover ~70% of the area of the Earth and between 5 and 10 billion tons of particulate organic matter are constantly sinking within them, contributing to the formation of marine sediments. Marine sediments are thus very extensive and have an average depth of about 500 m, but can be up to 10 km deep. They contain a huge reservoir of organic carbon (15,000 × 10^{18} g) and other compounds, which affect global chemical distributions (e.g., carbon dioxide, oxygen) and cycles, and hence climate on geological timescales. Methane gas and other hydrocarbons are a significant component of the stored organic carbon, including methane concentrated in hydrate, "methane ice," which forms under the low-temperature and high-pressure conditions characteristic of marine sediments. These hydrates are thought to be a sensitive trigger for climate change and a potential new energy source. The oceans have an average depth of 3800 m and thus an average pressure of 38 MPa (about 380 times higher than atmospheric pressure). At the maximum ocean depth (11,000 m, Marianas Trench) pressure is about

Topics in Ecological and Environmental Microbiology.

110 MPa and pressure increases further with depth into the sediments. This high pressure plus initially low temperature (~2 °C) combined with little surface photosynthetic productivity reaching the sediments (~1%) to provide an energy source, and porosity and nutrients decreasing with depth, resulted in conditions in subsurface sediments being initially considered to be too extreme for life.

This was seemingly confirmed by early researchers in the 1950s who claimed that the limit of the biosphere was reached at 7.47 m depth in marine sediments, as microbes could not be cultured at this or deeper depths. Hence, seemingly, nothing further was happening in these sediments until temperature increases (average 30 °C per km) reached 100–150 °C at kilometer depths, when buried recalcitrant kerogen (preserved macromolecular organic matter) pyrolyzed to produce oil and gas, in some locations. However, subsequent findings that only about 1% of all prokaryotes in the environment can be cultured and that the upper temperature limit for prokaryotes was about 113 °C and suggestions from organic geochemistry (e.g., lipid biomarkers, deep biogenic methane formation) started to question whether subseafloor sediments were really devoid of life.

Comprehensive microbiological investigations since 1986, mainly using the Ocean Drilling Program, have consistently shown the presence of significant prokaryotic populations, globally, in subsurface marine sediments. The presence of these prokaryotes is supported by the detection of microscopic cells, culturable prokaryotes, direct activity measurements (using radiolabeled substrates), presence of high molecular weight prokaryotic DNA, RNA, and lipid biomarkers, isolation of anaerobic prokaryotes with pressure optima matching those *in situ*, and correlation with geochemical changes characteristic of prokaryotic activity. Populations are so large that global estimates suggest they account for between 10% and 30% of all biomass and up to 70% of all prokaryotes on Earth. These findings are consistent with those of significant prokaryotic populations in other deep subsurface environments such as aquifers, consolidated Cretaceous shales, igneous rocks and deep mines, glaciers and ice sheets, lignite and coal formations, and marine and terrestrial oil reservoirs. Hence, there is a globally significant subsurface biosphere on Earth with subseafloor sediments being by far the largest of this type of habitat.

PROKARYOTIC CELL CONCENTRATIONS AND DISTRIBUTIONS IN SUBSEAFLOOR SEDIMENTS

General Depth Trend of Prokaryotic Cells, Growth Rates, and Potential Contamination

Prokaryotic cells both on and off sediment particles are detected after staining with a DNA dye and detecting the bound dye in recognizable cells using UV illumination (epifluorescence microscopy). Generally, prokaryotic cell numbers are highest in near-surface sediments (~10^9 cm^{-3}) and then decrease exponentially with increasing depth (Figure 1). However, at the average oceanic depth of sediment (500 m), the average cell concentration is still very substantial at ~3×10^6 cm^{-3}. The decrease in prokaryotic cells (97%) presumably reflects during burial, the preferential removal of the most degradable organic matter components, and thus the remaining organic matter becomes increasingly recalcitrant, and only able to support a smaller prokaryotic population. Even so, it is remarkable that organic matter tens of millions of years old can still support significant prokaryotic populations. However, this is reinforced by investigations of prokaryotic populations in sapropels, which are discrete layers of high organic matter concentration (up to 30%) that occur in some marine sediments. In Mediterranean sediments prokaryotic cell numbers are significantly stimulated in sampled sapropel layers (Figure 2), the deepest of which is 4.7 Mya, compared to nonsapropel layers. Plotting sapropel age against stimulation in prokaryotic cell numbers to obtain growth rates (expressed as the average time it takes for a cell to divide) results in a remarkable estimated cell division time of ~120,000 years. This extremely low growth rate helps to explain how buried organic matter can continue to provide energy for prokaryotes over geological timescales, but raises new questions about how cells survive between replication and on so little available energy. This situation is totally different from prokaryotes in laboratory culture that can have a division time as fast as every 10 min and slow division times are considered to be in days.

The presence of such discrete zones of high cell concentrations and the preservation of individual sapropel layers also demonstrate that drilling disturbance during sample collection is minimal. This was a concern for microbiological research as prokaryotic cells from shallow layers or from seawater could contaminate the deeper layers. Recently, chemical tracers and fluorescent microspheres (0.5 μm diameter to mimic cells) have been used to directly detect contamination during sampling of deep sediments. This has confirmed that any contamination is usually restricted to the outer surface of the core and thus can be avoided by the normal practice of sampling from the core center. In addition, microscopic approaches have considerable detection limits of about 10^4 cm^{-3}. Hence, even if no cells are detected a population between zero (0) and 10^4 cm^{-3} may actually be present and could represent a considerable prokaryotic community.

Control of Prokaryotic Population by Water Column Depth and Organic Matter Supply but with Stimulation at Deep Geochemical Interfaces and Lithological Layers

Water column depth controls the amount of detrital organic matter reaching and being buried in marine sediments, because the deeper the water column, the longer the residence

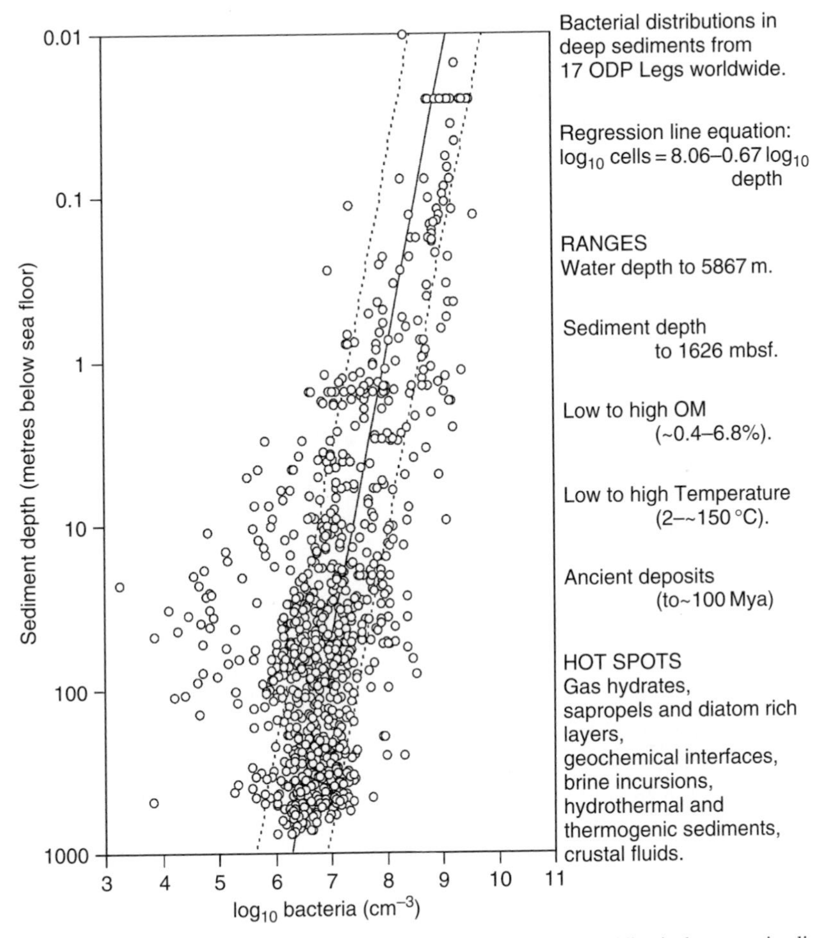

FIGURE 1 Distribution of prokaryotic cells in subseafloor sediments located worldwide. The solid line is the regression line and the dotted lines either side are the 95% prediction limits. ODP, Ocean Drilling Program; OM, organic matter. *Modified from Parkes, R. J., Cragg, B. A, Bale, S. J.,* et al. *(1994). Deep bacterial biosphere in Pacific-Ocean sediments.* Nature, *371, 410–413.*

time and the greater the degradation of sedimenting organic matter. This is reflected in the depth distributions of subsurface prokaryotic cells (Figure 3). These sediments are from the Eastern Pacific (deep water) and Peru Margin sites (shallow water) and show that at the deepest water site (4827 m) the depth distribution of prokaryotic cells is lower than the global average shown in Figure 1 (Figure 3, cell distributions either on or below the 95% prediction limits). As sites become shallower, overall cell concentrations increase and become closer to the global average distributions and at the very shallow site (151 m) cell numbers are either on or above the global average cell distributions (Figure 3). This not only demonstrates that organic matter supply controls overall numbers of subsurface prokaryotic cells, but it also reinforces that organic matter is their main energy supply, and suggests that most cells are active and not dead or dormant. If most cells were inactive or dead, there would be little or no decrease with sediment depth and probably no clear relationship between subsurface cell numbers and organic matter flux or water column depth. The exception to this trend is the deep-water Site 1230 (5086 m), which has cell numbers either on or above the global average cell distributions (Figure 3). This site contains gas hydrates, which have been recognized as deep biosphere hot spots (See "Deep Biosphere Hot Spots"). Clear deviations from the continuing decrease in cell numbers with depth also occur at specific intervals at some of the other sites, Sites 1229 and 1226 (Figure 3, highlighted with a shaded box, prokaryotic activity also increases at these depths, see "Deep Biosphere Hot Spots"). The sharp cell increases at Site 1229 (30 and 90 m) is due to geochemical interfaces (sulfate–methane, bottom interface due to a deep brine incursion), providing a discrete deep energy source and stimulating anaerobic oxidation of methane, AOM). At Site 1226, the three broader zones of stimulation were due to the presence of diatom-rich layers, the deepest 250–320 m and 7–11 Mya, which stimulate prokaryotic populations and activity as described for sapropel layers (Figure 2), but in much deeper and older formations.

Deep Biosphere in Basement Rock and the Deep Ocean Aquifer

In three of the deepest sites in Figure 3 (1231, 1225, 1226) samples were also obtained from the rock (crustal) basement beneath the sediment and this showed that prokaryotic cells

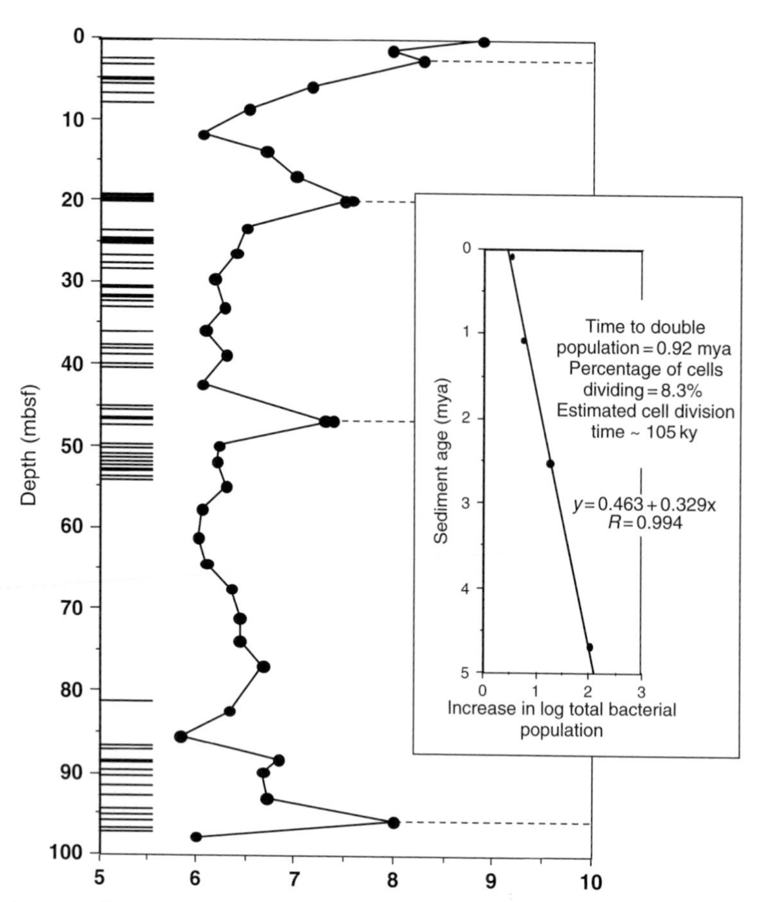

FIGURE 2 Distribution of prokaryotic cells and estimated growth rates in Mediterranean sapropels. Where the high organic matter sapropel layers (short horizontal bars on the depth axis) were sampled (dashed horizontal line), prokaryotic populations increased significantly. Inset, increase in prokaryotic cells in the sapropel layers plotted against sapropel age to calculate growth rates (ky, thousands of years). *Reproduced from Parkes, R.J., Cragg, B. A., & Wellsbury, P. (2000). Recent studies on bacterial populations and processes in subseafloor sediments: A review.* Hydrogeology Journal, *8, 11–28.*

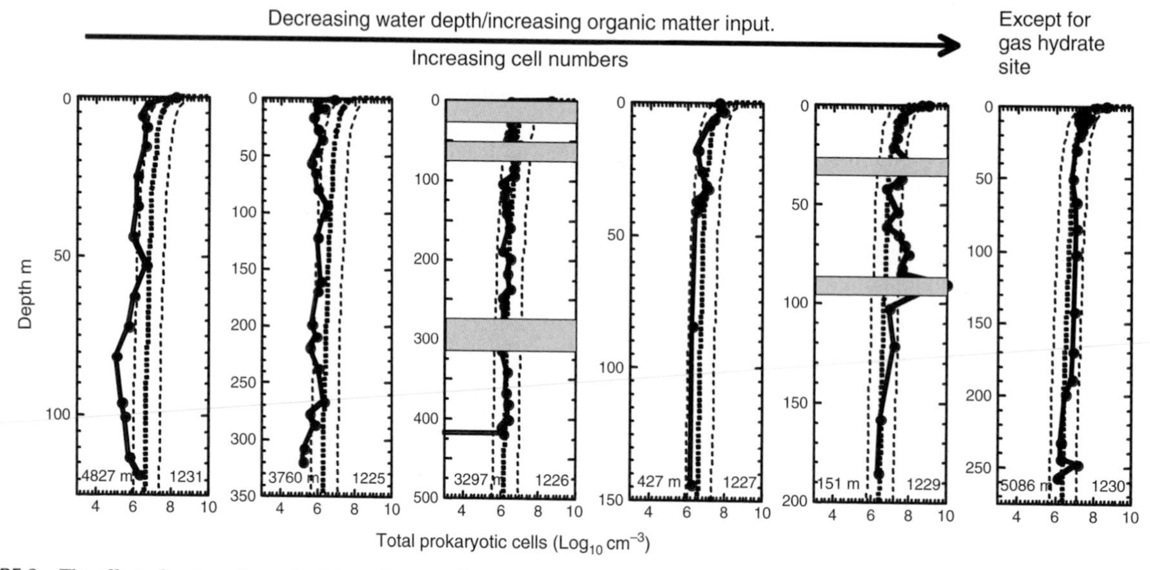

FIGURE 3 The effect of water column depth/organic matter input on the concentration and distribution of prokaryotic cells and subsurface stimulation of populations at geochemical (Site 1229) and lithological interfaces (Site 1226) at East Pacific and Peru Margin Sites. Water column depths on left side of graphs and site number on the right. Shaded boxes are areas of elevated cell numbers at Sites 1226 and 1229.

were present throughout the complete sediment column. This indicates that prokaryotic cells may also be present throughout the complete sediment column at sites with much deeper sediments, as long as conditions are still suitable for prokaryotes. Prokaryotic cells have also been detected in weathered zones of basement rock. Biotic weathering of the rock crust dominates its alteration in the top 250 m, where permeability is high, but steadily decreases down to 500 m depth. The proportion of biotically to abiotically altered basaltic glassy rock does not show any systematic variations with age of the crust. Thus, bioalteration should last as long as abiotic alteration and for as long as water is available for the hydration of the oceanic crust. This water is provided by hydrothermal circulation, which continues for around 70 Mya, and thus, the deep biosphere in basement rock is likely to be at least into crust of this age. Hydrothermal circulation is driven by processes producing new oceanic crust at ocean spreading centers, which results in very high-temperature fluids (up to 400 °C), but even at 115 °C there is biotic weathering and this continues while the crust cools. The crustal fluids themselves contain diverse prokaryotic communities (e.g., 8.5×10^4 cm^{-3} in fluids from a 3.5-Mya-old crust). It is unclear what energy sources are being used in basaltic basement, and despite the huge volume of the upper oceanic crust (10^{18} m^3), globally prokaryotic cell numbers in basement rocks are low and represent much less than 1% of the Earth's biomass. In addition, there is no clear stimulation in prokaryotic cell numbers at the sediment basement interface at the three Pacific Ocean sites where this was sampled (Figure 3, Sites 1231, 1225, 1226). This is consistent with the persistence of nitrate in fluids flowing through deep sediments and basement, because although prokaryotic nitrification (ammonia oxidation) produces this reactive respiratory compound, there are insufficient substrates available to facilitate its subsequent microbial utilization.

Cretaceous Shales, Oil Reservoirs, the Link to Land

More numerous than the discrete sapropel layers are the high organic matter layers of Cretaceous shales. These were deposited some 84–140 Mya during major oceanic anoxic events, which resulted in enhanced global accumulation of sedimentary organic matter. Almost 100 Mya after their deposition, these sediments still provide substrates for continuing prokaryotic activity not only in deep marine sediments (e.g., Equatorial Atlantic, Demerara Rise) but also when uplifted onto the land (see "Deep Biosphere Hot Spots"). Cretaceous shales also have high petroleum generation potential when deeply buried and heated, and intriguingly, many subsurface oil reservoirs have considerable ongoing prokaryotic activity. In some high-temperature (70–110 °C) North Sea oil reservoirs prokaryotic activity is so high that between 3 and 16 kg per day of prokaryotic cells is expelled with production fluids. High prokaryotic activity in oil reservoirs is usually deleterious, resulting in oil degradation and souring (high hydrogen sulfide concentrations).

Metabolic Status of Subseafloor Prokaryotic Cells

Some DNA dyes used to detect prokaryotic cells in sediments do not selectively detect live compared to dormant or recently dead cells (e.g., acridine orange stain); hence, there is debate about whether the huge estimated number of subsurface cells are in fact living. However, the use of techniques that do detect biomolecules characteristic of living cells (e.g., ribosomal RNA using fluorescence *in situ* hybridization, FISH) or provide a quantitative measure of prokaryotic DNA (16S ribosomal RNA gene determination by Q-PCR) suggest that a large fraction of subseafloor prokaryotic cells are alive. This is consistent with the presence of intact phospholipids in deep sediments. These occur in the cell membrane of viable prokaryotes and decompose rapidly after death. So far they have been detected at 845 mbsf (meters below sea floor) and 85 °C. Spores, the survival stage of some bacteria, may contribute to cell numbers as they can be stained with dyes usually used for determination of cell counts like acridine orange and DAPI. However, the actual contribution of endospores to the deep biosphere is unknown and remains a task for future research.

PROKARYOTIC METABOLIC PROCESSES AND ENERGY SOURCES

Metabolism

The metabolism of subseafloor prokaryotes is considerably diverse and reflects the range of prokaryotic activities present in near-surface sediments (Figure 4). Prokaryotic communities usually operate as interacting teams to convert complex sedimented, macromolecular organic matter to carbon dioxide and to obtain energy. A cascade of respiratory compounds (electron acceptors) are preferentially used and become reduced. These are used in order of decreasing energy yield from oxygen to carbon dioxide (Figure 4). Once oxygen is removed the remaining processes are termed anaerobic respiration. In shallow sediments anaerobic respiration seems to occur in the predicted sequence with increasing sediment depth; utilization of nitrate, manganese oxides, iron oxides, sulfate, and carbon dioxide/acetate methanogenesis. In deep sediments, however, reaction rates are much lower, metal oxide minerals with a range of activities are available, and a number of anaerobic reactions occur together (Figure 5). This is influenced by increases in organic matter concentrations and changes in sediment lithology. Anaerobic sulfate reduction and methanogenesis are major prokaryotic processes in subseafloor sediments.

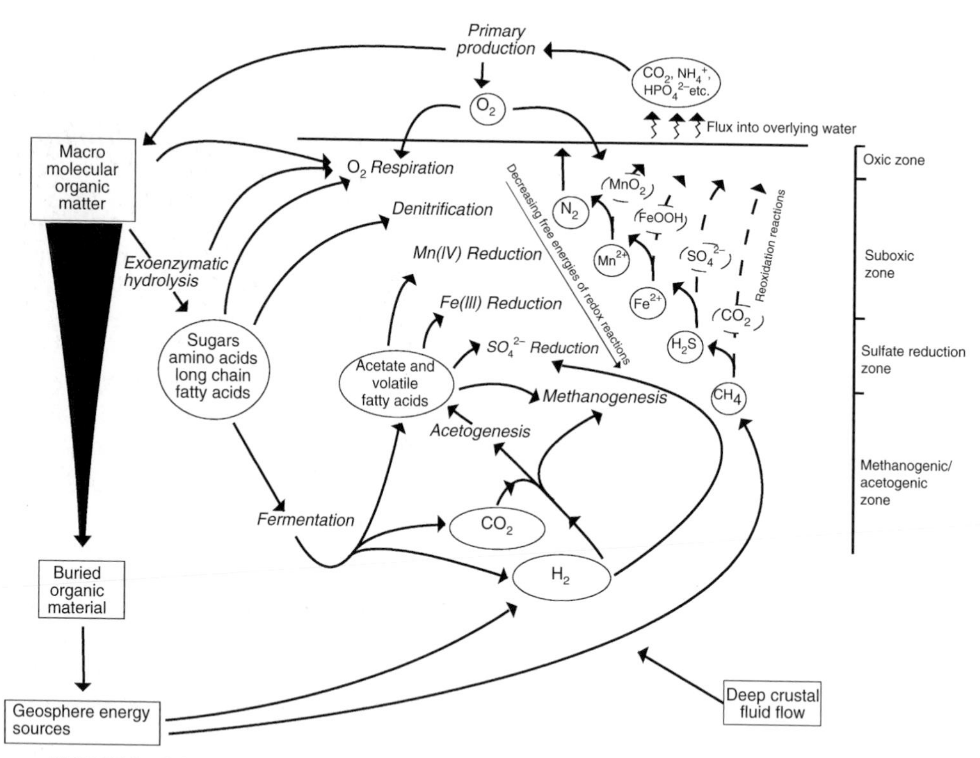

FIGURE 4 Schematic of prokaryotic processes of organic matter degradation in subseafloor sediments.

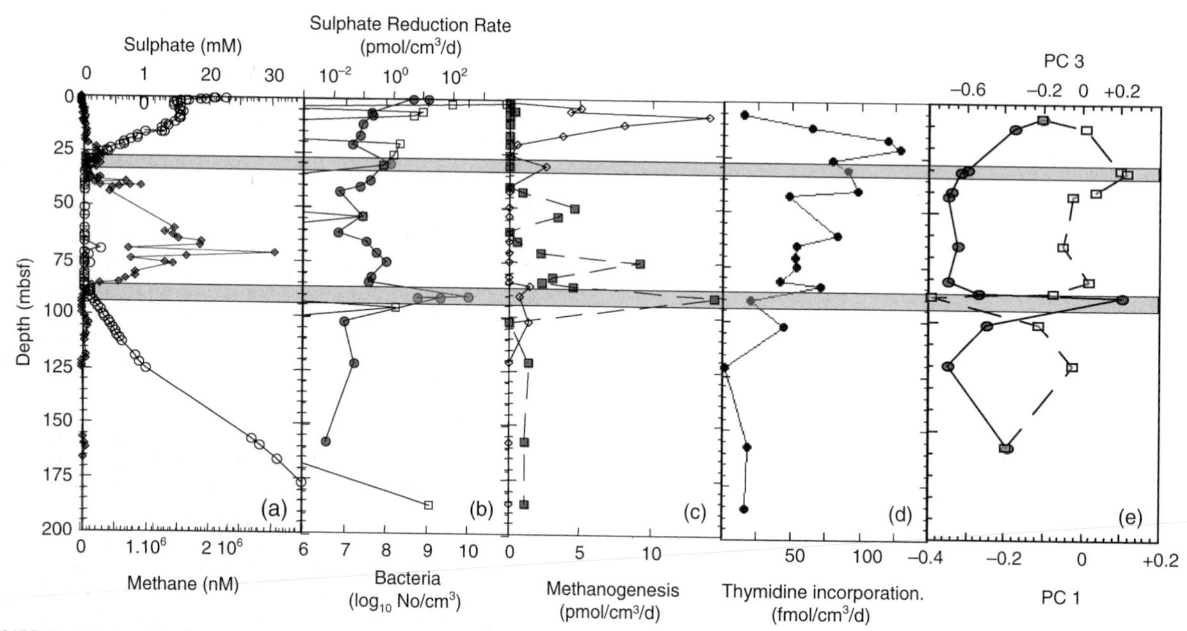

FIGURE 5 Biogeochemical processes and prokaryotic biodiversity profiles in Peru Margin sediments (ODP 1229). (a) geochemistry: Opore water sulfate (mmol l^{-1}), ♦ CH_4 (nmol l^{-1}), (b) □ sulfate reduction rates, • total cell numbers, (c) ♦ methanogenic rates, H_2/CO_2, • acetate, (d) • heterotrophic prokaryotic growth rates (thymidine incorporation), (e) Principal components profile of diversity of *Bacteria* from DGGE analysis of 16s rRNA gene sequences: • Component 1 (56% of variation), □ Component 3 (9% of variation); Component 2 (24% of variation) has a similar profile to Component 1. Shaded boxes highlight elevated prokaryotic processes and sulfate: methane interfaces. *Reproduced from Parkes, R. J., Webster, G., Cragg, B. A., et al. (2005). Deep sub-seafloor prokaryotes stimulated at interfaces over geological time.* Nature, *436, 390–394.*

Energy and Carbon Substrates

Recalcitrant and ancient macromolecular organic matter is a major substrate for subseafloor prokaryotes. As in shallow sediments, this material is initially hydrolyzed and fermented into low molecular weight compounds, particularly acetate and other organic acids, and hydrogen. These compounds are then oxidized by the terminal oxidizing groups using a range of electron acceptors (Figure 4). In deeper layers both acetate and hydrogen can be converted to methane and this produces globally significant amounts of methane gas. Hydrogen is a particularly important substrate as it can diffuse effectively within sediments, doing away with the need for prokaryotes to use scarce energy reserves actively moving toward substrates. Acetate can also be formed from hydrogen and carbon dioxide and this provides a similar amount of energy to that from hydrogen/carbon dioxide methanogenesis. However, it has recently been suggested that acetate itself can be used in the formation of both ethane and propane. Interestingly, these compounds have previously been considered to be of thermogenic origin and their potential prokaryotic origin suggests a wider microbial role in geochemical processes in deep marine sediments than has previously been considered. In fact, biogenic and thermogenic processes overlap as temperatures increase during burial, and at higher temperatures (>100 °C) thermogenic processes can feed the base of the subseafloor biosphere. This includes the anaerobic oxidation of thermogenic methane. When deep sulfate is provided by fluid flow along the basement rock or by deep brine incursion, previously produced biogenic methane can be oxidized deep beneath the sediment surface to provide energy.

Overall, subseafloor sediments are considered a habitat that is under severe energy limitation and may have insufficient energy to even provide conventional amounts of energy for cell maintenance considering the large cell numbers present. However, subseafloor prokaryotes may be well adapted to their habitat where energy supply is on geological timescales. In addition, there may be a greater range of energy sources available than we are currently aware of.

Deep hydrogen can also be produced from inorganic reactions in the sediment, which provides the potential for autotrophic prokaryotic processes. This includes radiolysis of water by the decay of radioactive minerals and, particularly as temperatures increase with sediment depth, weathering of iron silicates can increasingly produce hydrogen. Abiological hydrogen formation and its utilization under anaerobic conditions is extremely important, as it indicates the potential for some subsurface prokaryotic habitats being independent of photosynthetically driven surface life, by using instead geological "dark energy."

DEEP BIOSPHERE BIODIVERSITY

Cultured Diversity

Cultivation efficiencies in subsurface sediments are generally very low, in most cases less than 0.1% of the cells present can be stimulated to grow in microbiological media (Table 1). This low culturability might be explained by the strongly energy-limited nature of deep sediment prokaryotes. Starved cells are often severely harmed (substrate shock) by a sudden exposure to high substrate concentrations as present in most standard microbiological media. This effect may be due to the uncoupling of metabolic reactions. Relatively high viable counts were obtained in more active layers, such as with organic-rich subsurface sediments like sapropels or with deep layers that receive fluids from the underlying ocean crust that supply fresh sulfate and/or growth substrates. In these layers up to 3% of all cells were stimulated to grow. However, this "culturability" is still one order of magnitude lower than in coastal surface sediments when the same media are used.

Isolates obtained from deep subsurface sediments so far belong to the alpha, beta, gamma, delta, and epsilon subgroups of the *Proteobacteria*, *Firmicutes*, *Actinobacteria*, and *Bacteroidetes* (Figure 6). Although these are the same phyla as obtained from marine surface sediments, cultured diversity in subsurface sediments is generally lower. The presence of some of the cultured genera in subsurface sediments, for example, relatives of the alphaproteobacterium *Rhizobium radiobacter*, or the gammaproteobacterial genera *Pseudomonas*, *Halomonas*, and *Acinetobacter*, has also been confirmed by molecular genetic techniques. For example, by quantitative real-time PCR it was shown that *R. radiobacter*-like bacteria can represent up to 5% of the *in situ* microbial community within Eastern Mediterranean sapropels. The presence of some of these cultured bacterial types in deep sediments, however, seems surprising, particularly for bacteria related to the putative soil bacterium *Rhizobium*, but these isolates grew well in marine medium under anoxic conditions, indicating that they may actually thrive in the deep sediments. How these "deep" bacteria differ physiologically from their surface relatives is currently under investigation.

Archaea (except for methanogens) or members of the bacterial candidate (major group of *Bacteria* that currently has no cultured representative and hence is only considered a candidate for taxonomic status) phyla JS1 and OP8 are widespread and abundant in deep sediments, but have not yet been isolated. Representatives of the phylum *Chloroflexi*, also abundant in deep sediments, in contrast, have been successfully enriched but not obtained in pure culture from deep sediments. One reason may be that the supposedly slow-growing *Chloroflexi* are overgrown by opportunistic

TABLE 1 Total and maximum viable cell counts of anaerobes in various deep-sea sediments using complex media containing a variety of different substrates

Deep-Sea Sediment	Total Cell Count	Viable Cell Count	% Cultivated
Japan Sea, ODP Leg 128			
Surface	2.7×10^8 cm^{-3}	5×10^4 cm^{-3}	0.019
0.8 mbsf	7.9×10^7 cm^{-3}	3.1×10^3 cm^{-3}	0.004
500 mbsf	6.8×10^6 cm^{-3}	1.2×10^4 cm^{-3}	0.18
Blake Ridge, ODP Leg 164			
Surface	2.5×10^9 cm^{-3}	2.8×10^6 cm^{-3}	0.11
1 mbsf	4.4×10^7 cm^{-3}	1×10^6 cm^{-3}	2.27
365 mbsf	3.2×10^6 cm^{-3}	1.2×10^5 cm^{-3}	3.75
Woodlark Basin, ODP Leg 180			
Surface	3.3×10^8 ml^{-1}	9.4×10^4 ml^{-1}	0.028
10 mbsf	1.6×10^7 ml^{-1}	100 ml^{-1}	0.001
366 mbsf	4.2×10^6 ml^{-1}	1.3×10^6 ml^{-1}	30.9
Cascadia Margin, ODP Leg 146			
Surface	3.7×10^8 cm^{-3}	6.3×10^4 cm^{-3}	0.017
10 mbsf	1×10^7 cm^{-3}	8.4×10^3 cm^{-3}	0.084
200 mbsf	1×10^7 cm^{-3}	1.6×10^4 cm^{-3}	0.16
Eastern Mediterranean Sea			
Surface	7.5×10^8 cm^{-3}	2.2×10^7 cm^{-3}	2.9
Sapropel S5, 4 mbsf	6.6×10^8 cm^{-3}	2.2×10^7 cm^{-3}	3.3
Juan de Fuca Ridge			
9 mbsf	6.8×10^7 cm^{-3}	40 cm^{-3}	5×10^{-5}
141 mbsf	3.1×10^7 cm^{-3}	1.1×10^6 cm^{-3}	3.5
260 mbsf	2.9×10^7 cm^{-3}	2.1×10^5 cm^{-3}	0.7

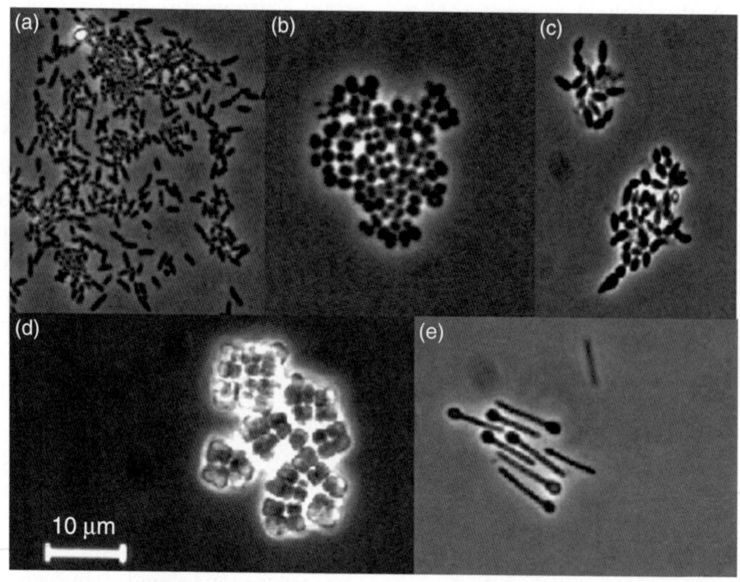

FIGURE 6 Phase contrast micrographs of bacterial isolates from Eastern Mediterranean sapropels (a–c) and carbon-poor hemipelagic sediment layers (d–e). (a) *Halomonas* sp. S6BA, (b) *Alteromonas* sp. S8FS1, (c) *Bacillus* sp. S6BB, (d) *Micrococcus* sp. SM3, (e) *Clostridium* sp. SO1.

fast-growing bacterial types like some *Proteobacteria* or that they fail to form visible colonies. However, as some *Chloroflexi* grow in "typical" anoxic microbiological media, a fermentative metabolism appears likely. The phylogenetic diversity of the yet-to-be-cultured bacterial and archaeal groups (JS1, OP8, MCG, MBG-D, etc., Tables 2 and 3) suggests a wide range of metabolic types. This metabolic diversity may include highly specialized lithotrophic organisms like the newly isolated pelagic Marine Group-I (MG-I) crenarchaeon "*Nitrosopumilus maritimus*." This expected

TABLE 2 Contribution of different bacterial groups to clone libraries in different subsurface and surface sediments

	Percent of all Clones or DGGE bands									
			Proteobacteria							
Site	Chlorofl.	JS1	Alpha	Gamma	Delta	Epsilon	OP8	Actinob.	Others	Total no. of clones or DGGE bands
Peru Margin, ODP Site 1229										
6.7 mbsf	51	2		36			2		9	58
30 mbsf	76	2		1			4		17	86
87 mbsf	34		3	47	1		1		14	92
Nankai Trough, ODP Site 1176										
1 mbsf	22	50	22				6			18
194 mbsf				23	3	53			20	30
Sea of Okhotsk, clay layers										
7.5 mbsf	60	17			17				6	64
58 mbsf	13	73	3		5				6	75
Sea of Okhotsk, ash layer										
46 mbsf	3	3	6	87						150
Mediterranean Sea[a]										
Sapopels S1–S8	100									8
Surface and hemipelagic layers	75				25					4
Wadden Sea, tidal flat sediments[a]										
Surface – 1.8 mbsf	7		5	24	29		2	12	20	41
2–6 mbsf	30	2	2	9	20		4		33	46
Sagami Bay, Japan (1150 *m deep)*										
Sediment surface				24	3	1		71	1	121

Chlorofl., Chloroflexi; Actinob., Actinobacteria; JS1, candidate division JS1; OP8, candidate division OP8.
[a]*DGGE bands.*

metabolic diversity may include some so far unknown metabolic reactions, which makes the design of suitable isolation media challenging.

Gram-positive bacterial isolates of the *Actinobacteria* and *Firmicutes* have frequently been isolated from subsurface sediment and were often dominating. However, these two phyla are rarely detected using molecular genetic methods, suggesting that their cells are present either in very low numbers or in an inactive state. Members of both phyla, *Actinobacteria* (*Micrococcus*) and *Firmicutes* (*Bacillus*), have been isolated from several million years old amber, indicating that they possess extreme longevity.

Possible Contribution of Endospores to the Deep Biosphere

Many *Firmicutes* are able to form spores, resting stages that are inactive, strongly dehydrated, and highly resistant to environmental stresses. Spores do contain enough energy for germination and are specifically adapted to quickly respond to substrate availability and formation of a vegetative cell able to replicate. Therefore, they may be able to easily outgrow other organisms after transfer to microbiological media, explaining their repeated isolation from subsurface sediments. As they do not exhibit active metabolism and do not require energy, it can be expected that they accumulate during sediment burial. The stability and resistance of spores may help to explain the absence of spore-forming bacteria in molecular surveys, as spores possess a very rigid cell wall and it is unclear to what extent they are extracted using standard procedures.

Molecular Diversity of Bacteria and Archaea

Molecular surveys on deep subsurface sediments reveal a microbial community that strongly differs from that known from surface sediments, in contrast to cultivation-based methods that generally detect similar genera. The dominant bacterial groups in the deep biosphere are certain subphyla of the *Chloroflexi*, the *Gammaproteobacteria*, and the candidate division JS1 (Table 2). These can be detected in almost all subsurface sediments and represent the majority of bacterial clones and DGGE bands (DNA sequences of bands after polymerase chain reaction and separation by denaturing gradient gel electrophoresis). *Deltaproteobacteria* are

TABLE 3 Contribution of different archaeal groups to clone libraries in different subsurface and surface sediments

	Percent of all clones or DGGE bands							
		Euryarchaeota			Crenarchaeota			
Site	DSAG	SAGMEG	MBG-D	TMEG	MCG	MG-I	Others	Total No. of clones
Peru Margin, ODP Site 1229								
6.7 mbsf	4	4			92			23
30 mbsf					100			24
87 mbsf		7	4		89			27
Peru Margin, ODP Site 1227								
6–7 mbsf	2	8	5	2	84			127
37–38 mbsf	89	10					1	176
40–46 mbsf					100			109
Nankai Trough, ODP Site 1176								
4.3 mbsf	43		18			15	24	34
Nankai Trough forearc basin								
Above hydrate zone (160–176 mbsf)	86					14		42
Within hydrate zone (209–252 mbsf)						100		35
Below hydrate zone (291–309 mbsf)		71			21	8		49
Mediterranean Sea[a]								
Sapopels S1–S8		3			97			29
Surface, hemipelagic layers	5	16			52	11	16	19
Weddel Sea (2000–3000 m)								
Surface, 0–2 cmbsf						99	1	146

DSAG, Deep-Sea Archaeal Group; SAGMEG, South African Gold Mine Euryarchaeotal Group; MBG-D, Marine Benthic Group D; TMEG, Terrestrial Miscellaneous Euryarchaeotal Group; MCG, Miscellaneous Crenarchaeotal Group; MG-I, Marine Group-I.
[a]*DGGE bands.*

also very widespread but numerically not as numerous. While the detected *Gammaproteobacteria* in most cases are closely related to cultured organisms, the majority of the deltaproteobacterial sequences are only distantly related to known organisms. Less frequently found are sequences which affiliate with the *Alphaproteobacteria*, *Bacteroidetes*, *Planctomycetes*, the candidate division OP8, and the NT-B2 cluster. Other bacterial phyla or clusters have so far only been found in exceptional cases.

Like the *Bacteria*, the archaeal sequences present in deep subsurface sediments are generally not closely related to cultured organisms. The molecular surveys targeting *Archaea* revealed an unexpected high diversity of *Crenarchaeota*, *Euryarchaeota*, and the deep-sea archaeal group (DSAG), which affiliates with none of these two phyla (Table 3). In almost all subsurface sediments studied so far the Miscellaneous Crenarchaeotic Group (MCG) is by far the dominant group. Several other groups are also widespread but numerically less abundant, for example, the crenarchaeotal MG-I and the euryarchaeotal South African Gold Mine Euryarchaeotal Group (SAGMEG), Marine Benthic Group D (MBG-D), and Terrestrial Miscellaneous Euryarchaeotic Group (TMEG). Other archaeal groups (e.g., *Thermococcales*, *Archaeoglobales*, deep-sea hydrothermal vent euryarchaeota) are less frequently found and represent a small percentage of all sequences.

Most of the dominating bacterial and archaeal sequences (e.g., within the *Chloroflexi*, candidate phylum JS1, MCG, SAGMEG) were novel and previously unknown, suggesting that these organisms are specific to subsurface environments. However, more recent investigations have detected many of these groups in shallow subsurface sediments like the younger sapropels or at a few meters depth in coastal sediments (Tables 2 and 3). But even these sediments are at least a few meters deep and already a few hundred to several thousand years old.

Sulfate-Reducing, Methanogenic and Anaerobic Methane-Oxidizing (ANME) Prokaryotes

Sulfate-reducing bacteria and methanogens are the most important terminal oxidizers in anoxic environments. They were therefore expected to be amongst the dominating physiological groups present in subseafloor sediments. However, known sulfate-reducing genera are very rarely detected by molecular genetic approaches, even those using genes specific for sulfate-reducing prokaryotes. This may indicate that the sulfate-reducing population is very small but very active, or alternatively, that sulfate reduction is carried out by so far unknown prokaryotes, for example, within the *Chloroflexi* or by the organisms represented by the unclassified deltaproteobacterial sequences. Although methanogens, similarly, have been found in only some molecular surveys, they are generally more often detected than sulfate-reducing bacteria. The most abundant methanogen sequences affiliate with the genera *Methanosarcina*, *Methanobrevibacter*, and *Methanococcoides*.

Despite the importance of AOM in many gas hydrate sediments and sulfate–methane interfaces (see "Deep Biosphere Hot Spots"), ANME *Archaea* are not amongst the dominating prokaryotes detected. It has to be assumed that AOM in subsurface layers is performed by a small but very active population of ANME prokaryotes or by so far unidentified organisms.

The Archaea–Bacteria Debate

While the phylogenetic composition of bacterial and archaeal communities in the subsurface is becoming well documented, there is uncertainty whether *Archaea* or *Bacteria* are numerically more abundant. Different approaches even on the same sediments (e.g., Peru Margin), have yielded contradictory results, although the vast majority of reports suggest *Bacteria* are dominant in subseafloor sediments. Most of this disagreement may be explained by the use of different methods (clone libraries, FISH, DGGE) and different target molecules (DNA, RNA, lipids). While RNA-targeting methods apparently favor *Archaea*, DNA-targeted methods more often detect *Bacteria*. It is also possible that some approaches may be just measuring dead or dormant cells (e.g., spores), the so-called Paleome. Under some circumstances DNA and some lipid biomarkers show considerable long-term survival. For example, quantifiable amounts of nucleic acids of Chlorobiaceae have been found in 217,000-year-old sapropels of the Mediterranean Sea, representing up to 0.5% of the total community DNA. As the obligately photolithoautotrophic Chlorobiaceae are not able to thrive in the deep-sea sediments, either cells or their DNA must have been persisting since the time of burial. These potential methodological biases and the question about which of the *Domains* is actually dominating is a focus of current deep biosphere research.

Physiological Adaptations

Generally, there is relatively little information available about the physiological adaptations of deep subsurface organisms to their habitat due to the low number of fully characterized isolates. However, a few isolates have been fully investigated and may have characteristic adaptations for deep sediments (Table 4). Some of these isolates have an extraordinary broad temperature range for growth, such as *Desulfovibrio profundus* covering a span of 15–65 °C. This can be seen as a prerequisite to colonize large parts of the deep biosphere, since temperature increases with depth. From the upper sediment layers (<1000 mbsf) predominantly mesophilic and

TABLE 4 Pressure, temperature, and salinity range for growth of some bacterial and archaeal isolates from marine subsurface sediments

	Origin Depth (mbsf)					
Strain	Water	Sediment	Temperature *in situ* (°C)	Pressure Range (MPa)	Temperature Range (°C)	Salinity Range (%)
Shewanella profunda DSM15900[T]	4791	4.2	1–2	0.1–50	4–37	0–6
Rhodoferax ferrireducens DSM 15236[T]	n.a.[a]	5.5	0–8	n.a.	4–30	n.a.
Photobacterium sp. S14	2150	4.5	13–16	n.a.	4–35	1–7.5
Marinilactibacillus piezotolerans DSM 16108[T]	4791	4.2	1–2	0.1–30	4–50	0–12
Desulfovibrio profundus DSM 11384[T]	900	500	16	0.1–40	15–65	0.2–10
Methanoculleus submarinus DSM 15122[T]	950	247	15–16	n.a.	10–50	0.6–9

n.a., data not available.
[a] *Coastal sediments.*

slightly thermophilic bacteria have been isolated. Deeper sediment layers have not yet been analyzed using cultivation-based methods. However, it can be expected that they, like the crustal fluids and basement rock, harbor mainly thermophilic *Bacteria* and Archaea, as suggested by molecular genetic studies.

As pressure increases with water and sediment depth, it is obvious that bacteria in subseafloor sediments have to be pressure-tolerant. As an example, the optimum pressures of sulfate-reducing bacteria (e.g., *D. profundus*) from subsurface sediments of the Japan Sea reflect well the *in situ* pressures as the isolates show no decrease in activity between atmospheric pressure and 20 Mpa and are still active at pressures between 30 and 40 Mpa. However, to date, all samples used for the enrichment and isolation of sulfate-reducing bacteria from subsurface environments have been decompressed during sampling and isolates obtained under atmospheric pressure, thereby potentially selecting for less barotolerant types. Although sampling devices can be recovered slowly to reduce the speed of pressure change, bacteria may still not be able to adapt quickly enough, and barophilic prokaryotes might still be damaged. Recently, coring and sediment handling systems that constantly maintain high pressures have been developed.

Many of the isolated *Bacteria* obtained from subsurface environments are facultative anaerobes. At first sight, this may seem surprising, as subsurface sediments are generally anoxic. However, due to the low metabolic activities and consequently low production of reduced compounds like hydrogen sulfide, the redox potential in many deep-sea and subsurface sediments is not highly reduced. For example, in the Mediterranean Sea nitrate, oxidized manganese, or iron species are present at least as deep as the sapropel S1, which is approximately 12,000 years old. In addition, deep circulation of fluids may provide oxidants to subsurface sediments.

DEEP BIOSPHERE HOT SPOTS

As given in the section titled "Prokaryotic cell concentrations and distributions in subseafloor sediments" describing cell distributions in subseafloor sediments, there are some deep sediments that have elevated cell numbers; often, these also have increased prokaryotic activities, and hence, could be considered deep biosphere "Hot Spots."

Deep Fluid Flow

If deep fluid flow provides substrates and/or electron acceptors into deep sediment layers, there is often a stimulation in prokaryotic activity. For example, in Peru Margin sediments there is a brine incursion that provides sulfate at depth and this stimulates prokaryotic activities, populations, and biodiversity changes in sediments deposited about 0.8 Mya (Figure 5, 90 m). This stimulation also demonstrates that deep sediment prokaryotic communities are active and dynamic, as opposed to being dormant and just surviving burial, as community composition changes at the brine interface. Presumably, AOM activity is stimulated at this interface and this probably also occurs at the upper sulfate–methane interface (Figure 5, 30 m), which is a geochemical interface that is widespread in subseafloor sediments. Similarly, stimulation of deep prokaryotic activity can occur with fluid flow within marine basement rock, diffusing into the sediments above. This can supply sulfate into deep methane-rich sediments, provide nitrate from stimulation of nitrification, seawater-derived dissolved organic matter, or hydrogen, and organic compounds synthesized by abiotic reactions between seawater and basement basalts at temperatures between about 100 and 300 °C.

Gas Hydrates Containing Sediments

Globally, gas hydrates contain huge amounts of methane (about twice the amount of carbon as in other fossil fuels) concentrated in an ice matrix. These hydrates are only stable at low temperatures and high pressures, conditions that are common in subsurface marine sediments. In these sediments there is often a zone of gas hydrate stability (Figure 7, Blake Ridge, Atlantic Ocean), above which they are unstable due to lower pressures and below which higher temperatures cause hydrates to melt. Hence, above and below this hydrate stability zone, methane concentrations can increase and this can stimulate AOM activity (Figure 7). As observed for AOM stimulation at other interfaces (Figure 5), this results in a general stimulation in prokaryotic activity and cell numbers. However, unlike other sites, toward the base of the hydrate stability zone there is a very significant increase in concentrations in the organic acid acetate, from the normal micromolar concentrations to millimolar concentrations (around a 1000 times higher), and there is a considerable increase in rates of acetate methanogenesis (Figure 7). This biogenic methane formation could supply methane to the hydrates above. High acetate concentrations have also been found at other hydrate sites (Hydrate Ridge in the Pacific Ocean and Indian Continental Margin), together with high hydrogen concentrations. A possible explanation for these high concentrations may be the rapid accumulation, burial, and heating of sediments in these regions, stimulating organic matter degradation combined with deep fluid flow. Irrespective of the mechanism, gas hydrate formations do seem to be areas of enhanced subseafloor prokaryotic activity.

Cretaceous Shales, Sapropels, and Other Organic-Rich Layers

As previously described, prokaryotic populations are consistently elevated in high organic matter layers despite these being millions of years old. Prokaryotic activity has also

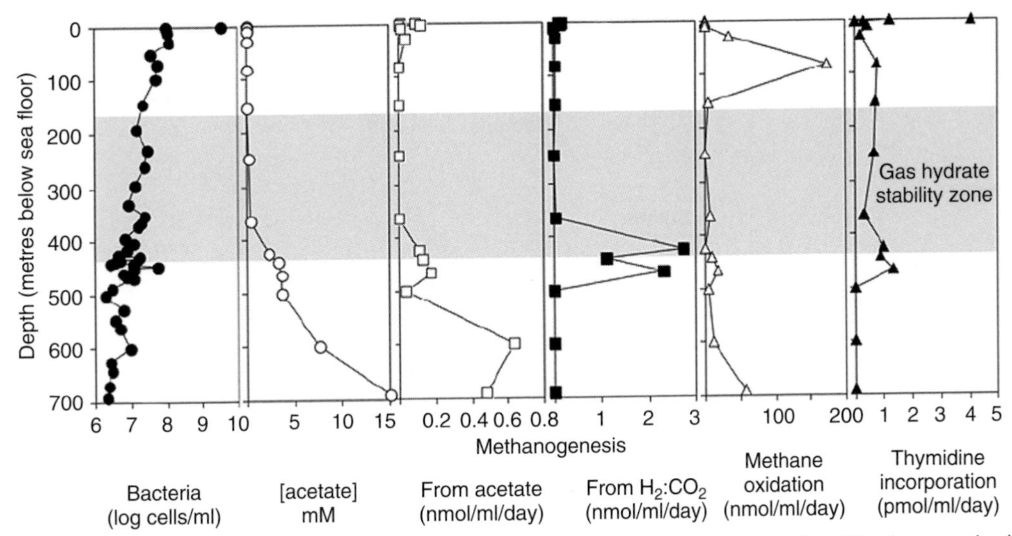

FIGURE 7 Bacterial populations and activities in gas hydrate sediments from Blake Ridge, Atlantic Ocean. Thymidine incorporation into prokaryotic DNA is a measure of heterotrophic growth rates. *Modified from Parkes, R. J., Cragg, B. A., & Wellsbury, P. (2000) Recent studies on bacterial populations and processes in subseafloor sediments: A review.* Hydrogeology Journal, *8, 11–28.*

been shown to be stimulated in these layers (e.g., exoenzyme activity and glucose degradation in 217,000-year-old sapropels; methanogenesis in almost 100 Mya Cretaceous shales) and reinforces that ancient organic matter is being utilized by subseafloor prokaryotes on geological timescales. Even when uplifted on land by tectonic activity, these shales still support prokaryotic activity. But these shales are often highly compacted by the uplifting geological process and porosity and permeability is low. This restricts organic matter degradation to fermentation processes with the organic acid and hydrogen products diffusing slowly to adjacent more permeable sands (Figure 8). Here, oxidation can be completed using electron acceptors supplied to the more permeable layers, and hence, prokaryotic activity is stimulated at the clay shale–sandstone interfaces (e.g., Cerro Negro, New Mexico). Cultures of both sulfate-reducing and acetogenic bacteria have been isolated from these rocks and molecular genetic analysis has demonstrated the presence of a diverse prokaryotic community.

Deep Hot Sediments

As temperatures reach 100–150 °C at kilometer depths, thermogenic breakdown of kerogen occurs to produce oil and gas under some circumstances. However, before these temperatures are reached, kerogen reactivity is incrementally increased, enabling its utilization by prokaryotes (e.g., Nankai Trough, Figure 9). Kinetic modeling predicts increasing reactivity with depth and thermogenic processes are confirmed *in situ* by increases in methane and ethane hydrocarbons with depth. However, prokaryotes are present when this process starts (both cells and intact phospholipids present, Figure 9)

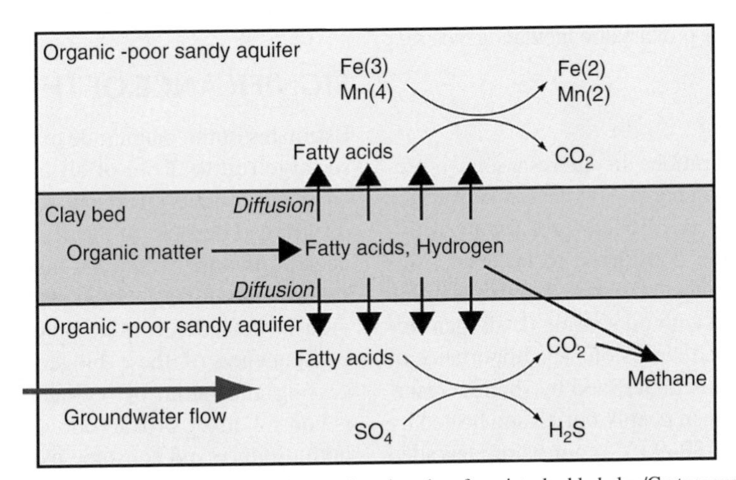

FIGURE 8 Schematic of prokaryotic metabolisms and interactions in subsurface, interbedded clay/Cretaceous shale and sand layers.

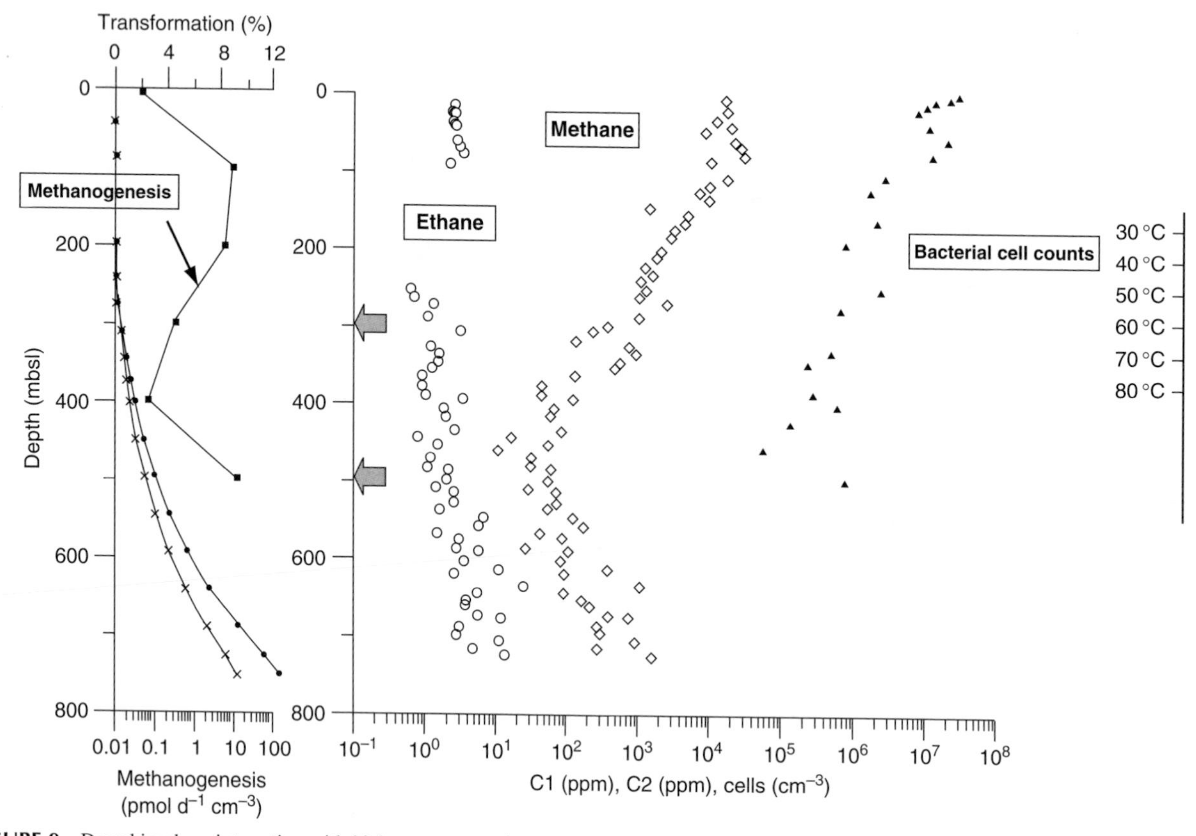

FIGURE 9 Deep biosphere interaction with high-temperature thermogenic processes in Nankai Trough Sediments, Pacific Ocean. Left panel experimentally determined estimates of increasing organic matter reactivity (transformation %) and measured rates of prokaryotic methanogenesis, with meters depth (mbsf). Right panel depth distributions of ethane (circles), methane (diamonds), and total prokaryotic cells (triangles). Temperature range of sediments extreme right. *Reproduced from Horsfield, B., Schenk, H. J., Zink, K.* et al. *(2006). Living microbial ecosystems within the active zone of catagenesis: Implications for feeding the deep biosphere.* Earth and Planetary Science Letters, *246, 55–69.*

and methanogenic activity is stimulated. This demonstrates overlap of biogenic and thermogenic processes in deep, hot sediments and the potential for high-temperature processes to feed the base of the deep biosphere. Prokaryotic substrates that can be produced include organic acids, low molecular weight hydrocarbons, and hydrogen. These can stimulate prokaryotic methanogenesis, which could contribute to methane gas accumulations, potentially at least to 110 °C (maximum current temperature for prokaryotic methanogenesis).

Oil Reservoirs

High hydrocarbon concentrations in oil reservoirs represent a significant potential energy source for subsurface prokaryotes. As previously mentioned, prokaryotic activity in oil reservoirs is often considered to be deleterious due to oil degradation and when associated with anaerobic sulfate reduction, combined with oil souring (hydrogen sulfide formation) and enhanced corrosion. The importance of prokaryotes in this process is underlined by shallow reservoirs that have previously been deeply buried and heated by geological activity (around 80–90 °C), not being degraded in the process, as would be expected, probably because of deep sterilization of hydrocarbon-degrading prokaryotes. However, it has recently been suggested that in many other oil reservoirs hydrocarbon degradation is associated with methanogenesis, which explains the common association of dry gas (methane and other gaseous hydrocarbons) with severely degraded oils worldwide. This also demonstrates a deep biosphere origin for at least some of the methane in subsurface oil reservoirs.

SIGNIFICANCE OF THE DEEP BIOSPHERE

Estimates of the magnitude of the subseafloor biosphere are so large (up to 30%) of all biomass on Earth that it must have a profound effect on shallow global biogeochemical cycles, and because of the deep biosphere's interaction with ancient organic matter and thermogenic processes, the biosphere also affects deeper and longer-term processes previously considered to be purely geological. An important consequence of these biogeochemical cycles is the processing and burial of sedimented organic matter, and this is controlled by prokaryotic activity. This activity can also both produce and consume methane, which is an important greenhouse gas, that potentially could play a major role in future climate change via release from unstable but globally significant gas hydrate deposits. Gas hydrates are one of several subseafloor "hot spot" habitats. Prokaryotic activity probably continues in deep sediments until maximum temperatures are reached (~113 °C, but deep biosphere studies may push this higher). This means that there is overlap between biosphere and geosphere processes, including into the oil window (100–150 °C). Prokaryotes gain additional energy from this interaction and may make a significant contribution to methane gas reserves in the process. Such activity and other enzymatic and biochemical capabilities that enable survival under extreme conditions (e.g., high pressure, temperature extremes, limited energy supply, existence with extremely slow growth, recalcitrant, and ancient organic matter supply) may have biotechnological application, such as recovery of methane from spent oil reservoirs. In addition, the potential response of the deep biosphere to anthropogenic activity, such as sequestration of carbon dioxide, needs to be seriously considered.

Some abiological hydrogen generation "dark energy" reactions provide the possibility of components of the anaerobic deep biosphere to be independent of surface photosynthetically driven life. This has implications for the origin of life, and life on other planets, even those with an inhospitable surface.

FURTHER READING

Coolen, M. J. L., Cypionka, H., Smock, A., Sass, H., & Overmann, J. (2002). Ongoing modification of Mediterranean Pleistocene sapropels mediated by prokaryotes. *Science*, *296*, 2407–2410.

Cowen, J. P., Giovannoni, S. J., Kenig, F. *et al.* (2003). Fluids from aging ocean crust that support microbial life. *Science*, *299*, 120–123.

D'Hondt, S., Jørgensen, B. B., Miller, D. J. *et al.* (2004). Distributions of microbial activities in deep subseafloor sediments. *Science*, *206*, 2216–2221.

Fisk, M. R., Giovannoni, S. J., & Thorseth, I. H. (1998). Alteration of oceanic volcanic glass: Textural evidence of microbial activity. *Science*, *281*, 978–980.

Fry, J. C., Parkes, J. R., Cragg, B. A., Weightman, A. J., & Webster, G. (2008). Prokaryotic biodiversity and activity in the deep subseafloor biosphere. *FEMS Microbiology Ecology in Press*, DOI: 10.1111/j.1574-6941.2008.00566.x.

Hinrichs, K. U., Hayes, J. M., Bach, W. *et al.* (2006). Biological formation of ethane and propane in the deep marine subsurface. *Proceedings of the National Academy of Sciences of the United States of America*, *103*, 14684–14689.

Horsfield, B., Schenk, H. J., Zink, K. *et al.* (2006). Living microbial ecosystems within the active zone of catagenesis: Implications for feeding the deep biosphere. *Earth and Planetary Science Letters*, *246*, 55–69.

Jones, D. M., Head, I. M., Gray, N. D. *et al.* (2008). Crude-oil biodegradation via methanogenesis in subsurface petroleum reservoirs. *Nature*, *451*, 176–180.

Krumholz, L. R., McKinley, J. P., Ulrich, F. A., & Suflita, J. M. (1997). Confined subsurface microbial communities in Cretaceous rock. *Nature*, *386*, 64–66.

Parkes, R. J., Cragg, B. A., Bale, S. J. *et al.* (1994). Deep bacterial biosphere in Pacific Ocean sediments. *Nature*, *371*, 410–413.

Parkes, R. J., Cragg, B. A., & Wellsbury, P. (2000). Recent studies on bacterial populations and processes in subseafloor sediments: A review. *Hydrogeology Journal*, *8*, 11–28.

Parkes, R. J., Webster, G., Cragg, B. A. *et al.* (2005). Deep sub-seafloor prokaryotes stimulated at interfaces over geological time. *Nature*, *436*, 390–394.

Roussel, E. G., Cambon Bonavita, M. A., Querellou, J. *et al.* (2008). Extending the sub-seafloor biosphere. *Science*, *320*, 1046.

Stevens, T. O., & McKinley, J. P. (1995). Lithoautotrophic microbial ecosystems in deep basalt aquifers. *Science*, *270*, 450–454.

Wellsbury, P., Goodman, K., Barth, T., Cragg, B. A., Barnes, S. P., & Parkes, R. J. (1997). Deep marine biosphere fuelled by increasing organic matter availability during burial and heating. *Nature*, *388*, 573–576.

Whitman, W. B., Coleman, D. C., & Wiebe, W. J. (1998). Prokaryotes: The unseen majority. *Proceedings of the National Academy of Sciences of the United States of America*, *95*, 6578–6583.

Chapter 28

Deep-Sea Hydrothermal Vents

Andreas Teske
University of North Carolina at Chapel Hill, Dept. of Marine Sciences, Chapel Hill, NC, USA

Chapter Outline

ABBREVIATIONS

ANME Anaerobic methane-oxidizing
RuBisCO Ribulose-1,5-bisphosphate carboxylase–oxygenase

DEFINING STATEMENT

Hydrothermal vents, hot seafloor springs at mid-ocean ridges, sustain chemosynthetic microbial ecosystems that are independent of photosynthetically produced biomass; these hot, acidic, and chemically toxic habitats harbor a wide, only incompletely recognized spectrum of extremophilic bacteria and archaea that are specifically adapted to the hydrothermal vent environment.

MID-OCEAN RIDGE HYDROTHERMAL VENTS

Hydrothermal vents are hot deep-sea springs at mid-ocean ridge spreading centers, where the extensive basalt plates that form the basement of the seafloor (oceanic crust) are split apart and new ocean crust is constantly generated from melting zones in the Earths upper mantle. They occur globally and follow the mid-ocean ridges and their spreading centers that meander around the ocean surface of Earth, almost as seams around a tennis ball. At slow spreading centers such as the Mid-Atlantic Ridge, the area between the separating oceanic plates just above the geothermally active spreading zone, appears as a valley or trough (the "Rift Valley") framed by the higher ridges of the young basaltic crust on both sides of the spreading center; fast spreading centers such as the East Pacific Rise show often an "Axial High," an upward bulge sandwiched between the separating plates. These areas at the hot, volcanically active center of the mid-ocean ridges are the location of most hydrothermal vents; geologically and chemically distinct types of vents occur off-axis at some distance from the mid-ocean ridge axis. Most hydrothermal vents occur at water depths between 2000 and 3500 m, corresponding to the water depth of most mid-ocean ridges (Kelley et al., 2002).

The fractured and porous basement rock of the mid-ocean ridge entrains cold deep-sea water, which then circulates through the subsurface underneath the spreading center, analogous to water circulating through an aquifer on land. During subsurface passage, seawater is exposed to the geothermal heat of the melting zone underneath the spreading center, the chemistry of the entrained seawater is fundamentally altered as it reacts with the subsurface basalt, and undergoes phase separation under extreme temperature and pressure. The most significant and microbiologically relevant changes include sulfate removal by anhydrite precipitation ($CaSO_4$) and by geothermal reduction of seawater sulfate to hydrogen sulfide; leaching of metals and of additional sulfur from the subsurface basalt; generation of protons and decreasing pH towards 3–4 during water–rock interaction; removal of oxygen due to outgassing; increasing CO_2 concentration as a consequence of magma

Topics in Ecological and Environmental Microbiology.

degassing; elevated hydrogen and methane concentrations. In most cases, H_2S and CO_2 are the dominant component of vent fluids, with concentrations between 3 and >100 mM. The highly altered hydrothermal vent fluids reach temperatures of 300–400 °C, but remain in the liquid state due to the high hydrostatic pressure of the deep-sea (Kelley et al., 2002). The hot, buoyant hydrothermal fluid migrates upwards to the seafloor surface and emerges as a high-temperature vent; alternatively, subsurface seawater mixing can attenuate the temperature and chemistry, resulting in warm vents (Figure 1). High-temperature vent fluids appear as a black or grey cloud, undergoing instant precipitation of dissolved metal sulfides, as the vent fluid mixes into the cold, oxygenated deep-sea water column. The precipitated minerals, generally metal sulfides and anhydrite, accumulate *in situ* and build a friable and porous structure that surrounds the hot fluid flow, channels and focuses it; the result are the hydrothermal chimneys from which the fluid flow emerges (Figure 2). These porous and friable structures can grow to tens of meters in height, depending on local chemistry and the stability of the precipitated minerals. Their walls are only a few cm thick, but separate the channelized flow of hydrothermal vent endmember fluid (300–400 °C) within the chimney from cold, oxygenated deep-sea water surrounding the chimney and cooling its surface (ca. 2 °C). The steep temperature and chemical gradients within the chimney walls are the preferred habitat for anaerobic, hyperthermophilic vent archaea (Figure 3), and provide indeed the source material for enrichment and isolation of the most thermophilic microorganisms on Earth (reviewed in Miroshnichenko and Bonch-Osmolovskaya, 2006).

Mixing of hydrothermal fluids with entrained seawater within the porous subsurface of a hydrothermal vent site produces a wide range of warm vent fluids, which emerge as channelized or diffuse flow (Figure 1). The moderate temperatures and the changes in chemistry favor different microbial populations. Hyperthermophilic archaea originating in the hot subsurface and entrained in the mixed fluids remain detectable, but the dominant populations that are sustained by these mixed vent fluids are sulfur-oxidizing autotrophic bacteria with a mesophilic or moderately thermophilic temperature optimum. These bacteria constitute the dominant primary producers of biomass at hydrothermal vents, and fall into two broad classes: (1) free-living, surface-attached or mat-forming bacteria, and (2) bacterial symbionts of marine invertebrates that contribute to the nutrition of their hosts (Jannasch, 1995).

Hydrothermal vents are subject to sudden geological disturbance in earthquakes and volcanic eruptions that occur frequently at mid-ocean ridges; in addition, the precipitation of minerals, such as metal sulfides, in the subsurface can alter the deep plumbing of a hydrothermal vent system. Thus, individual hydrothermal vent sites have short life spans, in the range of a few years or decades; the microbial communities and the symbiont-dependent vent animals re-colonize new vents quickly as previous vent habitat may disappear at short notice (Kelley et al., 2002). The microbial communities of hydrothermal vents are resilient and have maintained themselves in their volatile and extreme habitat for billions of years.

THE CHEMOSYNTHETIC BASIS OF LIFE AT HYDROTHERMAL VENTS

The hydrothermal vent ecosystem is based on chemolithoautotrophic bacteria and archaea that derive energy from the oxidation of inorganic compounds, mostly sulfide or hydrogen

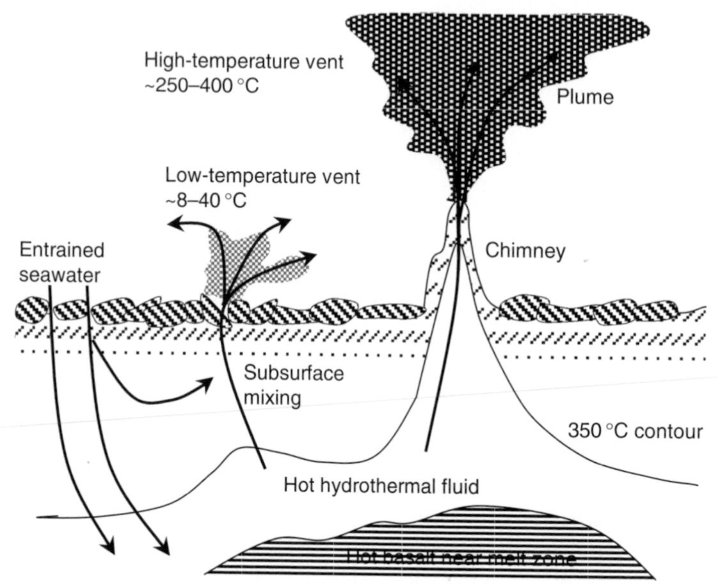

FIGURE 1 Schematic cross-section of hydrothermal circulation at a mid-ocean ridge. Seawater is drawn into the crust, heated to 350–400 °C, chemically altered by interaction with the hot basalt subsurface, and rises buoyantly back to the seafloor. At the seafloor, venting high-temperature endmember fluids precipitate chimneys; subsurface mixing with seawater produces lower temperature vents. *Adapted from McCollom and Shock (1997)*

FIGURE 2 (a). Black Smoker Chimney on the East Pacific Rise (21°N) at 2600 m depth. The orifice of the chimney consists of whitish aragonite ($CaSO_4$); the lower chimney consists the grey-black metal sulfides. (b). A group of black smoker chimneys at East Pacific Rise (21°N). Photos Holger Jannasch, WHOI.

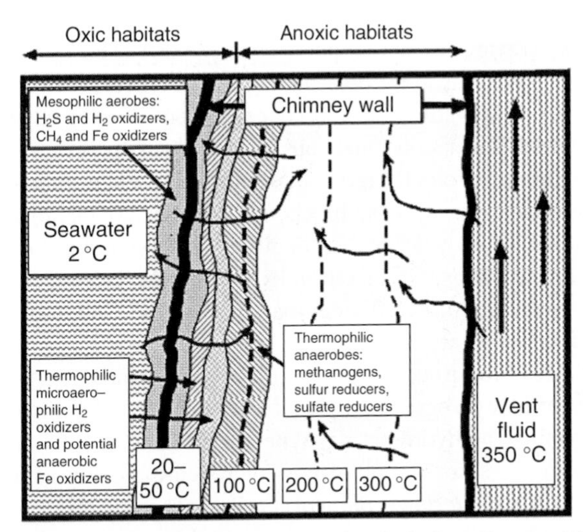

FIGURE 3 Schematic cross-section of a hydrothermal vent chimney wall, showing habitats for thermophilic and mesophilic bacteria and archaea. Anaerobic archaea are found around the 100 °C isothermes and possibly deeper within the chimney wall. Thermophilic, microaerophilic hydrogen-oxidizing bacteria (Aquificales) occur further towards the outer chimney layers where dissolved oxygen penetrates in low concentrations. Mesophilic, aerobic hydrogen-, sulfur- or methane-oxidizing bacteria grow on the chimney surface. *Adapted from McCollom and Shock (1997).*

(lithotrophy), and build up their biomass by assimilation of dissolved inorganic carbon, such as CO_2, CO, HCO_3^-, or CO_3^{2-} (autotrophy). The oxidation reactions of the inorganic electron donors are exergonic and proceed without additional energy input. The energy thus obtained drives proton transport across the cytoplasmic membrane and is conserved as ATP. Also, it drives reverse electron transport and the reduction of physiological hydrogen carriers; these are, together with ATP, essential for autotrophic carbon fixation. The shorthand term "Chemosynthesis" is generally used for this mode of microbial life, in analogy to photosynthesis where the oxidation of an inorganic electron donor requires the input of light energy (Jannasch, 1995).

Depending on the preferred electron acceptor, chemosynthetic metabolism can be aerobic or anaerobic. Oxidation reactions with oxygen or nitrate as terminal electron acceptor give the highest energy yields, which explains the great diversity and abundance of oxygen- or nitrate-dependent chemolithotrophic bacteria at hydrothermal vents (Section "Free-living Chemolithoautrophic Bacteria"). Oxygen-respiring chemolithoautotrophs have also evolved numerous symbiotic associations with marine invertebrates, where the animal host optimizes the supply of electron donor and acceptor in return for nutritional contributions from its interior or exterior symbiont (Section "Symbiotic Chemolithoautotrophic Bacteria"). Chemolithoautotrophic metabolism is also possible with other electron acceptors, such as oxidized metals, oxidized sulfur species, and inorganic carbon in the oxidation state of CO_2, CO and formate. Since chemosynthesis cannot work without oxidized electron acceptors, the hydrothermal vent ecosystem remains geochemically linked to the photosynthetic biosphere near the ocean surface. Only photosynthesis produces oxygen, which in turn is essential for producing other oxidized electron acceptors. For this reason, hydrothermal vent ecosystems are not independent domains of life that could survive and thrive even if some planetary catastrophe extinguishes photosynthetic surface life. The only possible exceptions are chemolithoautotrophic methanogenic archaea, which generate methane from CO_2 and hydrogen and use reactions of the Acetyl-CoA pathway for autotrophic carbon fixation. CO_2 is abundant in hydrothermal fluids and could, together with hydrogen of hydrothermal origin, sustain an autotrophic, autonomous hydrothermal vent biosphere (Jannasch, 1995). For the same reason, CO_2/H_2-autotrophic methanogenesis is a good candidate for a microbial pathway that can sustain autonomous life in the deep subsurface, for example in deep marine sediments or the basaltic ocean crust.

The *in situ* chemical conditions determine which autotrophic pathway or life strategy contributes most to the overall biomass at hydrothermal vents. Where sulfide and oxygen coexist in turbulent mixing of vent water and seawater, symbiotic chemosynthetic bacteria (sulfur-oxidizing, oxygen-respiring chemolithoautotrophs) and their animal hosts make the highest contribution of new organic carbon to hydrothermal vent ecosystems; the symbiotic bacteria sustain not just their own biomass, but their host animals as well. In some cases, symbiont-harboring invertebrates reach unusually large sizes and body mass. The vent clam *Calyptogena magnifica* can reach a length of a foot and weigh over a pound; the vestimentiferan vent worm *Riftia pachyptila* reaches a length of 2 m, two orders of magnitude larger than its closest, nonhydrothermal relatives (Van Dover, 2000). Nonsymbiotic, free-living sulfur-oxidizing bacteria can also grow in an abundance that is not seen anywhere else; for example, the mat-forming, filamentous bacterium *Beggiatoa*, which oxidizes sulfide and elemental sulfur with nitrate as electron acceptor, grows in thick pillows in suitable locations (Section "Unusual Hydrothermal Vents: Loihi, Guaymas, Lost City").

The dominant microbial pathway of autotrophic carbon fixation at hydrothermal vents was assumed to be the Calvin–Benson-Bassham cycle, found in numerous free-living and symbiotic sulfur oxidizers that are phylogenetically related to each other, as members of the Gammaproteobacteria (Jannasch, 1995). However, recently discovered and phylogenetically distinct bacterial populations, especially sulfur- and hydrogen-oxidizing bacteria within the Epsilonproteobacteria, are using the reverse TCA cycle for autotrophic carbon fixation (Campbell et al., 2006); this pathway also extends to symbiotic sulfur oxidizers in hydrothermal vent invertebrates (Markert et al., 2007).

CHEMOLITHOAUTOTROPHIC BACTERIA

Free-living Chemolithoautrophic Bacteria

Most free-living, nonsymbiotic autotrophs in the hydrothermal vent environment belong to three major, physiologically and phylogenegtically distinct groups of sulfur- or hydrogen-oxidizing, aerobic or microaerophilic bacteria.

The first group includes autotrophic bacteria of the genera *Thiomicrospira*, *Beggiatoa*, and *Thiobacillus* that fix carbon with the Calvin–Benson-Bassham cycle; these autotrophs are phylogenetically members of the Gammaproteobacteria. The obligately autotrophic, aerobic, sulfur oxidizing, and mesophilic species of the genus *Thiomicrospira* are found in hydrothermal environments and sulfureta world-wide. The filamentous sulfur oxidizers of the genus *Beggiatoa* form extensive mats on sediments and chimneys; most hydrothermal vent *Beggiatoa* spp. accumulate their electron acceptor nitrate intracellularly; they may be facultative or obligate autotrophs (Jannasch, 1995).

The second group consists of mesophilic or moderately thermophilic sulfur- and hydrogen oxidizers that assimilate carbon through the reverse TCA cycle; these bacteria use oxygen, nitrate, sulfur and sulfite as terminal electron acceptor. Historically, this group was discovered and studied in detail only after the RuBisCO autotrophs; numerous new genera (*Caminibacter*, *Hydrogenimonas*, *Lebetimonas*, *Nautilia*, *Nitratiruptor*, *Sulfurimonas*, *Sulfurovum*) and novel species (candidatus "*Arcobacter sulfidicus*") have been described in recent years almost exclusively from hydrothermal vent habitats. They are consistently members of the Epsilonproteobacteria and are related to epibionts of vent invertebrates, such as the polychaete worm *Alvinella pompeiana* and the vent shrimp *Rimicaris exoculata* (reviewed in Campbell et al., 2006). The epsilonproteobacterial autotrophs generally colonize areas with higher temperatures, higher concentrations of reduced sulfur, and lower concentrations of dissolved oxygen, than the gamma-proteobacterial autotrophs. This environmental preference matches the biochemical characteristics of the reverse TCA cycle; under conditions of oxygen limitation or anaerobiosis, CO_2 fixation by the reverse TCA cycle is energetically more efficient than through the Calvin cycle (Campbell et al., 2006).

The third group consists of thermophilic and hyperthermophilic hydrogen-oxidizing bacteria of the Aquificales phylum. The hydrothermal vent species of the Aquificales are members of the obligately hydrogen-oxidizing genera *Desulfurobacterium*, *Balnearium* and *Thermovibrio*, and of the hydrogen-, sulfur-, sulfite- and thiosulfate-oxidizing genus *Persephonella*. Oxygen, nitrate or elemental sulfur serve as terminal electron acceptors for hydrogen oxidation (Miroshnichenko and Bonch-Osmolovskaya, 2006; Reysenbach and Shock, 2002). As far as known, the Aquificales assimilate carbon through the reverse TCA cycle. The members of the Aquificales grow in temperature ranges from 45 to 95 °C, and thus overlap with the temperature range of the hyperthermophilic archaea. The Aquificales have a wide habitat range, including chimneys and sediments of deep-sea hydrothermal vents (Figure 2), as well as shallow-water marine vents and terrestrial hot springs (reviewed in Reysenbach et al., 2002).

In addition to these major physiological and phylogenetic groups of vent bacteria, other types of chemolithoautotrophic bacteria thrive at hydrothermal vents, such as the recently discovered obligate hydrogen oxidizers *Thermodesulfobacterium hydrogenophilum*, and *Thermodesulfatator indicus*, both sulfate reducers and members of a phylum-level deeply-branching lineage of sulfate-reducing bacteria, *Thermodesulfobacterium* (Miroshnichenko and Bonch-Osmolovskaya, 2006). The potential for discovery of novel chemosynthetic bacteria is certainly not exhausted; molecular surveys demonstrate that only a small portion of the total bacterial diversity in hydrothermal vent sites has been cultured to date (Takai et al., 2006).

Symbiotic Chemolithoautotrophic Bacteria

Symbioses of sulfur-oxidizing, chemolithoautotrophic bacteria with hydrothermal vent animals are characteristic for the vent ecosystem; the host animals can harbor symbionts within their bodies in specialized cells or tissues, or the symbionts live as epibionts on the exterior of the vent animals. Good examples for epibionts are the filamentous Epsilon-proteobacteria that grow on the carapace of the vent shrimp *Rimicaris exoculata*, or on the dorsal bristles of the annelid worm *Alvinella pompeiana* (Campbell et al., 2006). Ongoing metagenomic analyses of the *Alvinella*-associated epibiont community reveal the potential for CO_2 fixation though the reductive TCA cycle (Campbell et al., 2006).

Endosymbiontic sulfur-oxidizing chemolithoautotrophs have induced major modifications in the body plans of their host animals, notably in the relative size of their symbiont-bearing organs and their digestive systems. The digestive systems are greatly reduced or have completely disappeared, while other tissues have been modified to harbor the bacterial symbionts. Symbiotic bivalves provide a good example for this adaptation: Hydrothermal vent bivalves have evolved thickened gill tissues (subfilamental tissues) to accommodate their sulfur-oxidizing intracellular symbionts. As far as detailed studies are available, the sulfur-oxidizing bivalve symbionts assimilate carbon via RuBisCO. Dual symbioses are possible; bivalves of the family Mytilidae that colonize methane-rich seeps also harbor aerobic methane-oxidizing symbionts in addition to sulfur oxidizers (Van Dover, 2000). The two symbionts coexist not just within a single animal but within the same individual host cells; they can adjust their relative dominance in response to the availability of their electron donors, sulfide and methane (Van Dover, 2000).

The most conspicuous hydrothermal vent animal, the large tubeworm *Riftia pachyptila*, is a model system for the study of its unusually versatile symbionts. *Riftia pachyptila* houses its sulfur-oxidizing endosymbionts in a unique, richly vascularized body tissue, the trophosome, where they are simultaneously supplied with oxygen, sulfide and CO_2 (Van Dover, 2000). These sulfur-oxidizing symbionts show an unusual mixture of metabolic strategies; they harbor RuBisCO (ribulose-1,5 bisphosphate carboxylase–oxygenase), indicating carbon fixation via the Calvin cycle; they harbor functioning enzymes of the energy-generating TCA cycle; and they also express the enzymes of the reverse TCA cycle. Thus, the symbionts could switch from RubisCO-autotrophy to heterotrophic oxidation of carbon storage compounds via the TCA cycle to autotrophic growth via the reverse TCA cycle (Markert et al., 2007).

Heterotrophic Vent Bacteria

The chemosynthetic ecosystem at hydrothermal vents also provides a nutritional basis for functionally and phylogenetically diversified heterotrophic microbial communities. Some families and genera of thermophilic, heterotrophic vent bacteria are found consistently at hydrothermal vents and related geothermal habitats. These typical vent heterotrophs include the family Thermaceae (genera *Marinithermus*, *Oceanithermus*, *Vulcanithermus*) and, to a lesser extent, the Thermotogacaea (genera *Thermotoga*, *Fervidobacterium*, *Thermosipho*, *Geotoga* and *Petrotoga*). Both families are obligate organotrophs that require complex substrates (peptides and carbohydrates) or in some cases low molecular weight organic acids. Electron acceptors for members of the Thermaceae are oxygen and nitrate; oxygen is tolerated only in low concentrations. Their microaerophilic growth and wide thermophilic temperature range (total range of different species, 30–80 °C) are consistent with the steep oxygen and temperature gradients of the hydrothermal vent habitat (Miroshnichenko and Bonch-Osmolovskaya, 2006).

In contrast, the Thermotogaceae are anaerobes that grow by fermentation of carbohydrates; these thermophiles are mostly found in terrestrial hot springs, deep oil wells, geothermally heated aquifers, or in shallow marine vents.

Some thermophilic bacteria isolated from hydrothermal vents are capable of lithotrophic as well as autotrophic growth. A particularly interesting example is *Caldanaerobacter subterraneus* (synonymous with *Carboxydobrachium pacificum*), a gram-positive bacterium and member of the Firmicutes phylum that can grow autotrophically by CO oxidation in addition to anaerobic fermentative growth on complex proteinacous substrates, such as peptone or yeast extract, or polymeric or monomeric sugars. Anaerobic fermentation of peptides and sugars is the preferred metabolism of heterotrophic, thermophilic gram-positive bacteria of the genera *Caloanaerobacter*, *Caminicella*, *Tepidibacter* and *Caldanerobacter*, all moderately thermophilic members of the Firmicutes phylum with a growth temperature range of 33–65 °C (Miroshnichenko and Bonch-Osmolovskaya, 2006). Its higher growth temperature range (50–80 °C) distinguishes *Caldanaerobacter subterraneus* from other hydrothermal vent Firmicutes. Interestingly, this bacterium shares the capability for CO oxidation with a hyperthermophilic archaeon, a new strain of the genus *Thermococcus* that grows at temperatures of 45–95 °C (Table 1). Most likely, increased heat tolerance is an adaptation to CO oxidation, since CO is a major gas constituent of hydrothermal vent endmember fluids.

HYPERTHERMOPHILIC ARCHAEA

Soon after the discovery of hydrothermal vents in 1977, hyperthermophilic archaea have been isolated from hydrothermal vent samples. Most archaea are anaerobes that use nitrate, oxidized metals, elemental sulfur, sulfite, sulfate, or CO_2/carbonate as electron acceptors. Table 1 shows the diversity of currently known, metabolically and phylogenetically distinct hydrothermal vent archaea (Miroshnichenko and Bonch-Osmolovskaya, 2006; Reysenbach et al., 2002). The preferred archaeal habitats are the interior matrix of hot chimney walls, hydrothermally heated sediments, or the shallow subsurface of the porous basalt underneath hydrothermal vent areas (Kelley et al., 2002). In these environments, conductive cooling and seawater/vent fluid mixing generate steep chemical and temperature gradients, and provide a suite of electron acceptors for anaerobic metabolism within the growth temperature range of hyperthermophilic archaea, ca. 80–120 °C (Figure 3). Under these conditions, anaerobic reducing reactions are energetically more favorable than aerobic sulfur oxidation, which dominates at mesophilic conditions up to ca. 50 °C (McCollom and Shock, 1997). Thus, thermodynamic constraints in combination with metabolic specialization account for the dominance of archaea in hot vent habitats, for example within chimney walls, and for the dominance of bacteria in warm or moderately hot vent environments, for example the seawater-exposed surface of chimneys.

Chemolithoautotrophic Archaea

Autotrophic CO_2/CO assimilation is widespread among hyperthermophilic vent archaea and can be found in combination with the entire spectrum of anaerobic metabolism: Hydrogen, reduced sulfur species, CO, and oxidized metals serve as electron donors for respiration with nitrate, metals, elemental sulfur, sulfite and sulfate, and methanogenic reduction of CO_2 (Table 1). The current temperature limit for methanogenesis and elemental sulfur reduction is 110 °C, the upper limit for the methanogen *Methanopyrus kandleri* and the sulfur reducer *Pyrodictium occultum*. The upper limit for sulfate reduction appears to be 90 °C, the highest growth temperature for *Archaeoglobus profundus*. Some archaeal metal reducers and nitrate reducers have even higher growth temperatures, as demonstrated by the isolation of the nitrate-reducing and microaerophilic autotrophic archaeon *Pyrolobus fumarii* with a maximal growth temperature 113 °C, and the Iron (Fe-III) reducing archaeon "*Geogemma*" with a maximal growth temperature of 121 °C (Table 1). Thermodynamic stability considerations of essential macromolecules in aqueous solution, such as proteins and DNA, place the upper temperature limit of life near 130 °C; at higher temperatures, cellular macromolecules have half-life times of seconds and cannot be replenished as quickly as they are destroyed.

On the other end of the temperature spectrum, hyperthermophilic vent archaea are also found suspended in the buoyant plumes of hydrothermal fluids that are mixed into seawater and extend hundreds of meters upwards into the cold, oxygenated deep-sea water column. The origin of these archaea is most likely the shallow subsurface underneath the vents and the basalt surface, where subsurface mixing of vent fluids and seawater create a patchwork of niches with suitable environmental conditions. These subsurface reservoirs can be discharged abruptly in major volcanic eruptions; large quantities of hydrothermal fluids and gasses emerge as megaplumes that can reach heights of 1–2 kms and remain detectable for months. After dilution with seawater, these water masses are only minimally warmer than the deepwater background, and are fully oxygenated; yet some vent archaea survive in this environment for several weeks or months, facilitating dispersal (Kelley et al., 2002).

Heterotrophic Archaea

The majority of currently described archaeal hyperthermophilic isolates from hydrothermal vents are heterotrophic (Miroshnichenko and Bonch-Osmolovskaya, 2006; Reysenbach et al., 2002). The mutually related genera *Thermococcus* and *Pyrococcus* are the most frequently isolated representatives; they assimilate and ferment complex organic compounds such as yeast extract, tryptone, peptone, casein, diverse sugars and peptides. Fermentative growth is enhanced by additions of elemental sulfur, or requires sulfur as an essential component in some species; sulfur acts as an auxiliary electron acceptor, and hydrogen sulfide accumulates in high concentrations. Members of the genera *Thermococcus* and *Pyrococcus* grow quickly and robustly on a wide range of liquid, anaerobic laboratory media supplemented with sulfur, and are isolated so frequently from all kinds of hydrothermal vent sample materials, including the subsurface underneath hydrothermal vents, that they are viewed as indicator species of hydrothermal activity (Kelley et al., 2002).

Thermococccus and *Pyrococcus* spp. are particularly interesting as model archaea that can grow as biofilms, often in specific adaptation to chemical or physiological stress. In laboratory simulations and within the porous rock of hydrothermal chimneys, hyperthermophiles grow attached to each other and to their mineral substrate, surrounded by capsular exopolysaccharides. *Thermococcus* and *Pyrococcus* species and strains differ slightly in temperature sensitivity, substrate

TABLE 1 Selected deep-sea hydrothermal vent archaea

Archaeon (genus)	Metabolic Type	Reaction	Temperature Range and Optimum	
Lithoautotrophs				
Methanopyrus kandleri	Methanogenesis	$4H_2 + CO_2 \rightarrow CH_4 + 2H_2O$	84–110 °C	98 °C
Methanocaldococcus jannaschii	Methanogenesis	$4H_2 + CO_2 \rightarrow CH_4 + 2H_2O$	50–86 °C	85 °C
Archaeoglobus profundus[a]	Sulfate reduction	$4H_2 + H_2SO_4 \rightarrow H_2S + 4H_2O$	65–90 °C	82 °C
Archaeoglobus veneficus	Sulfite reduction	$3H_2 + H_2SO_3 \rightarrow H_2S + 3H_2O$	65–85 °C	75–80 °C
Pyrodictium occultum	Sulfur reduction	$H_2 + S^0 \rightarrow H_2S$	82–110 °C	105 °C
Ignicoccus pacificus	Sulfur reduction	$H_2 + S^0 \rightarrow H_2S$	70–98 °C	90 °C
Geoglobus ahangari	Iron reduction	$H_2 + 2Fe^{3+} \rightarrow 2H^+ + 2Fe^{2+}$	65–90 °C	88 °C
Geogemma strain 121	Iron reduction	$H_2 + 2Fe^{3+} \rightarrow 2H^+ + 2Fe^{2+}$	80–121 °C	105–107 °C
Pyrolobus fumarii	Nitrate and oxygen respiration	$4H_2 + NO_3^- + H^+ \rightarrow NH_3 + 3H_2O2H_2$ $+O_2 \rightarrow 2H_2O$ (microaerophilic)	90–113 °C	106 °C
Thermococcus strain AM4	CO oxidation	$CO + H_2O \rightarrow CO_2 + H_2$	45–95 °C	82 °C
Ferroglobus placidus[b]	Fe^{2+}, S^{2-} oxidation with NO_3^-	$S^{2-} + NO_3^- + 2H^+ \rightarrow 2NO_2^- + S^0 + H_2O$ $2Fe^{2+} + NO_3^- + 2H^+ \rightarrow NO_2^- + 2Fe^{3+} + H_2O$	65–95 °C	85 °C
Heterotrophs				
Thermococcus guaymasensis *Pyrococcus abyssi*	reduce S^0 during fermentation of protein-rich substrates, and produce H_2S + org. acids + CO_2; without S^0, H_2 is produced		50–90 °C 67–102 °C	80 °C96 °C
Palaeococcus ferrophilus	reduces Fe^{3+} during fermentation of proteinaceous substrates		60–88 °C	83 °C
Aeropyrum camini	obligately aerobic heterotroph, grows on proteinaceous substrates		70–100 °C	90–95 °C
Pyrodictium abyssi	fermentation to organic acids + CO_2; growth stimulated by H_2		80–110 °C	97 °C
Staphylothermus marinus	reduces S^0 during fermentation of protein-rich substrates and produces H_2S + org. acids + CO_2		65–98 °C	92 °C

Reproduced from Jannasch, H.W. (1995). Microbial interactions with hydrothermal fluids. In S. E. Humphris, R. A. Zierenberg, L. S. Mullineaux, & R. E. Thomson (Eds.). *Seafloor hydrothermal systems: Physical, chemical, biological and geological interactions. Geophysical Monograph* (Vol. 91) (pp. 273–296). Washington, DC: American Geophysical Union; Miroshnichenko, M. L., & Bonch-Osmolovskaya, E. A. (2006). Recent developments in the thermophilic microbiology of deep-sea hydrothermal vents. *Extremophiles, 10*, 85–96; Reysenbach, A. L., G¨tz, D., & Yernool, D. (2002). Microbial diversity of marine and terrestrial thermal springs. In J. Staley, & A. L. Reysenbach (Eds.). *Biodiversity of microbial life* (pp. 345–421). New York: Wiley-Liss.

[a] Obligate H_2 oxidizer, but requires acetate as carbon source for mixotrophic growth.

[b] Isolated from a shallow-water vent.

spectra, and proteolytic repertoire. These physiological differences among *Thermococcus* and *Pyrococcus* isolates, and the numerous variants of their molecular marker genes (16S rRNA genes) in hydrothermal vent samples, indicate that this diversity is a consistently recurring key feature of archaeal vent communities and therefore must have an ecological explanation. Resource partitioning within the chimney matrix is the most likely explanation, analogous to other microbial ecosystems in hot springs and nonthermal habitats (Reysenbach and Shock, 2002).

Uncultured Archaea and Bacteria at Hydrothermal Vents

Molecular surveys based on cloning and sequencing of 16S rRNA and functional genes have revealed an unexpected diversity of uncultured archaeal and bacterial lineages at hydrothermal vents, which remain for now without cultured relatives that could serve as a baseline for physiological inferences or cultivation strategies (Campbell et al., 2006; Reysenbach and Shock, 2002; Teske et al., 2002; see Takai et al., 2006 for a detailed summary and compilation of uncultured vent lineages). Obviously, the evolutionary and physiological diversity of hydrothermal vent microorganisms has been explored only to a limited extent; metagenomic studies and persistent, creative cultivation efforts would enable significant progress. Many deeply-branching archaeal and bacterial lineages at hydrothermal vents cannot be subsumed into the well-defined phylum-level lineages, but represent deeply-branching lineages that emerge from the earliest evolutionary radiation of the bacterial and the archaeal domains. These findings have lent new support to the hypothesis that deep-sea hydrothermal vent microbial ecosystems are among the oldest on Earth, and had sustained microbial life already during its early diversification more than 3.5 billion years ago (Reysenbach and Shock, 2002).

Comparison to Terrestrial Hot Springs

Interestingly, hydrothermal vents and terrestrial hot springs harbor very distinct microbial communities, as a consequence of key differences in the chemical regime. Specific archaeal and bacterial groups show strong biases in environmental preference and can be viewed as signature communities for deep-sea vents and terrestrial hot springs.

The archaeal composition of terrestrial hot springs and deep-sea vents reflects differences in sulfur cycle chemistry and pH. Terrestrial hot springs with reduced sulfur species turn acidic due to proton-releasing sulfur oxidation reactions, once the limited local buffering capacity is exhausted. In contrast, the acidic pH of endmember fluids in marine hydrothermal vents is quickly attenuated by inmixing of slightly basic seawater (McCollom and Shock, 1997), which also buffers protons released by microbial sulfur oxidation reactions. Therefore, terrestrial hot springs harbor a great diversity of acidophilic sulfur-oxidizing archaea (genera *Sulfolobus*, *Acidianus*, *Sulfurococcus*, *Metallosphaera*) that have so far not been found in deep-sea hydrothermal vents (Reysenbach et al., 2001); vice versa, the neutrophilic sulfur reducers of the genera *Thermococcus* and *Pyrococcus* provide examples for almost exclusively marine archaea that are largely missing in terrestrial hot springs. Around twenty recognized *Thermococcus* spp. and a large number of strains have been isolated from marine, deep and shallow hydrothermal vents; only two species have been described from a volcanic lake in New Zealand (Reysenbach et al., 2002).

The second example for strong habitat preference, in favor of terrestrial hot springs and against deep-sea hydrothermal vents, is provided by the Thermotogales, a deeply-branching lineage of thermophilic, heterotrophic bacteria. Some members of the Thermotogales are found exclusively in the terrestrial subsurface (genus *Geotoga*) or in freshwater hot springs (genus *Fervidobacterium*). The ecophysiologically diversified genus *Thermotoga* includes species isolated from shallow-water marine vents, hot springs in salt lakes, and from the geothermally heated terrestrial subsurface. In many cases, the freshwater representatives of *Thermotoga* are more sensitive to high salt concentrations and to high sulfur concentrations. The Thermotogales' preference for sugars and carbohydrates as carbon and energy sources might also link them to terrestrial habitats and hot springs where plant-derived carbon substrates are available (Reysenbach et al., 2002). So far, only *Thermosipho japonicus*, a barophile that grows with thiosulfate as electron acceptor, was isolated from a deep-sea hydrothermal vent (Reysenbach et al., 2002).

The Aquificales are one of the very few groups of hyperthermophiles that occur as a dominant group in deep-sea hydrothermal vents and shallow-water vents, as well as terrestrial hot springs; they are at home in both worlds (Reysenbach et al., 2002). For example, the Aquificales are the dominant hyperthermophilic autotrophic bacteria near-neutral Yellowstone hot springs. Here, hydrogen instead of sulfide is the dominant inorganic electron donor for chemolithoautotrophic metabolism (Spear et al., 2005); atmospheric oxygen is readily available for microaerophilic growth with hydrogen.

At temperatures below 72–74 °C, photoautotrophic bacteria, such as thermophilic cyanobacteria, gain a foothold in terrestrial hot springs and turn them into photoautotrophic microbial ecosystems where the community structure of thermophilic cyanobacteria, and the ecophysiological specialization of specific cyanobacterial community members, reflects the physical and chemical zonation of the microbial habitat (Ward et al., 1998).

UNUSUAL HYDROTHERMAL VENTS: LOIHI, GUAYMAS, LOST CITY

Unusual geological and chemical settings result in several unique hydrothermal vent systems that are very different from the classical mid-ocean ridge, black smoker-type hydrothermal vents. The unique geochemistry of these vents is reflected in their unusual microbial communities. The surprising diversity of geologically, geochemically, and microbiologically different vent sites is illustrated here with three unusual vent sites: The Loihi vents southeast of Hawaii, a vent field dominated by iron-oxidizing microbial communities on a growing seamount that will emerge as the next Hawaiian Island; the Guaymas Basin vents in the Gulf of California, a thickly sedimented spreading center where thermocatalytic degradation of buried organic matter controls the vent fluid chemistry and microbiology; and the Lost City vents located off-axis near the Mid-Atlantic Ridge, where nonvolcanic rock-water chemistry provides the inorganic electron donors for microbial life in huge carbonate chimneys.

The Loihi vents are located at the top of Loihi Seamount on the southeastern end of the Hawaiian Island chain at 900 m depth; the vent fluids are dominated by high concentrations of dissolved reduced iron and CO_2, but contain little or no sulfide; their temperature range is less extreme (10–170 °C) than that at most mid-ocean ridge sites. The iron-rich vent fluids of Loihi emerge into the oxygen minimum layer of the marine water column, which slows down abiogenic Fe-oxidation in oxygenated seawater and favors the development of extensive Fe-oxidizing bacterial communities. The iron-oxidizing microbial communities of Loihi are dominated by mesophilic or psychrophilic, microaerophilic and neutrophilic Fe-II-oxidizing bacteria that grow in thick, rust-colored microbial mats of freshly produced and precipitated amorphous iron oxides (Emerson and Moyer, 2002); the bacteria are members of the different Proteobacterial subdivisions (Figure 4). Phylogenetically and physiologically similar Fe-II-oxidizing bacteria have been isolated also from other vent sites during *in situ* colonization experiments with metal sulfides (reviewed in Takai et al., 2006).

The Guaymas Basin vents are located on a near-shore spreading center in the Gulf of California, between Baja California and the Mexican mainland, and are buried under hundreds of meters of organic-rich sediments that originate from high primary productivity in the upper water column, and from terrestrial runoff. As the vent fluids percolate upwards through the sediment layers, the buried organic material in the sediments is geothermally heated and matures to petroleum compounds, including complex aromatics, organic acids, alkanes and methane. Nitrogen compounds are reduced to ammonia, absent in most vent fluids but abundant at Guaymas. Most of the dissolved metals are precipitated en route to the sediment surface, which reduces the metal toxicity of the fluids. The usually acidic

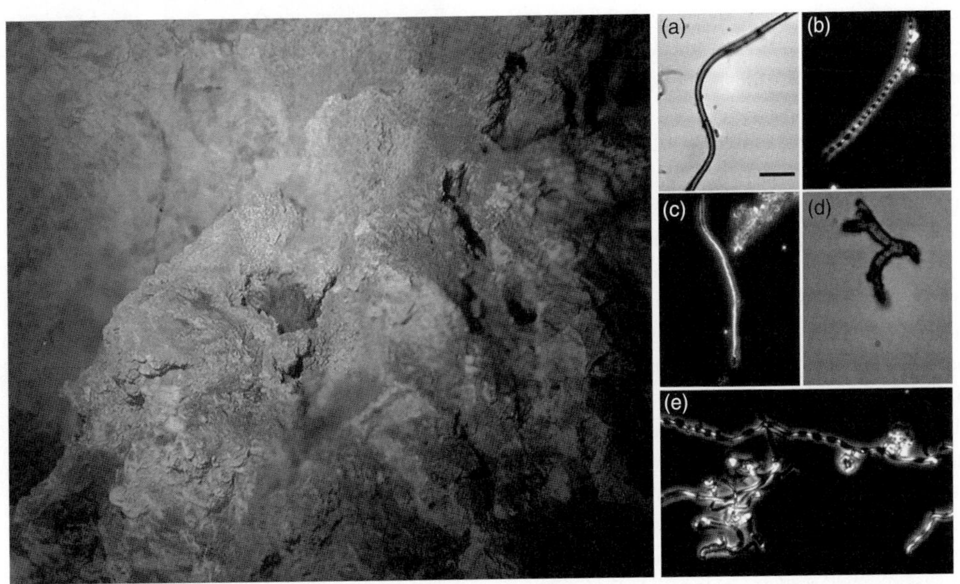

FIGURE 4 Hydrothermal vent orifices at Loihi Seamount surrounded by rust-colored microbial mats that are responsible for depositing iron oxides (left panel). These vents are located at a depth of approximately 1300 m deep in a caldera at near the summit of Loihi. The anoxic vent fluid contains 100s μM Fe(II) and is enriched in CO_2, but has a very low sulfide content; vent fluid temperature at the time of this photograph was approximately 60 °C. The vents support a robust community of Fe-oxidizing bacteria. The right hand panel shows light photomicrographs documenting the morphology of biogenically produced iron oxides typically found in these mat communities. (a) Tubular sheath encrusted with iron oxides; filaments of cells are sometimes visible in these when samples are stained with a fluorescent DNA stain. (b), (c), and (e) show twisted, filamentous stalk-like structures that are formed by a novel member of the Proteobacteria (Emerson and Moyer, 2002), (d) Y-shaped morphology of dense iron-oxides formed by cells that grow at the apical tips of the oxide structures. The scale bar corresponds to 5 μm. Photo credits: Vent, T. Kerby, Hawaiian Undersea Research Laboratory; photomicrographs, D. Emerson. (For interpretation of the references to color in this figure legend, the reader is referred to the Web version of this chapter.)

pH of the vent fluids is buffered into the near-neutral range by buried carbonates. The resulting organic-rich cocktail sustains extensive communities of anaerobic archaea and bacteria in the sediment, including methanogens, sulfate-dependent methane oxidizers, sulfate reducers, and anaerobic heterotrophs (Teske et al., 2002). Sulfide and dissolved inorganic carbon in the vent fluids sustain massive sulfur-oxidizing microbial mats of filamentous *Beggiatoa* spp. at the sediment surface (Jannasch, 1995) (Figure 5). The methane in the vent fluids sustains anaerobic methane-oxidizing communities in the surficial sediments (Teske et al., 2002). Here, methane is oxidized by novel archaea in a consortium with sulfate-reducing bacteria that transfer the methane-derived electrons on sulfate as the electron acceptor. Overall, the Guaymas sediment communities resemble the microbial ecosystems of petroleum and methane seeps more than those of classical, nonsedimented basalt-hosted mid-ocean ridge vent sites (Teske et al., 2002).

The Lost City hydrothermal vent field on the flanks of the mid-Atlantic Ridge owes its existence to deep subsurface mineral–fluid reactions unlike those of mid-ocean spreading centers. The dominant type of rock-fluid interactions are serpentinization reactions in the earth's crust that produce hydrogen and methane; the resulting vent fluids are warm or moderately hot (40–80°C) and contain two orders magnitude less sulfide than hot vent fluids from mid-ocean ridges. The high pH of 9–10, and the high Ca^{2+} content of the vent fluids lead to precipitation of large carbonate mounds, pinnacles and chimneys unlike the previously known metal-sulfide chimneys; with 40–60 m height, the Lost City structures are the largest known hydrothermal chimneys.

The microbial community of the Lost City vents indicates an active microbial methane and sulfur cycle. The archaeal communities of Lost City include members of the Methanosarcinales, uncultured methanogens that form biofilms on surfaces exposed to the hottest vent fluids at Lost City (ca. 80 °C) where they have access to molecular hydrogen. Cooler areas are dominated by anaerobic methane-oxidizing archaea (ANME-1 group) that gain energy by sulfate-dependent oxidation of methane (Brazelton et al., 2006). Thus, the molecular surveys indicate a fully formed anaerobic archaeal methane cycle of hydrogenotrophic methanogens and sulfate-dependent methane oxidizers at the Lost City vents. The bacterial communities include predominantly gamma- and epsilon-Proteobacteria and Firmicutes, in addition to a diverse assemblage of Chloroflexi, Planctomyces, Actinobacteria, Nitrospira, and uncultured subsurface lineages. The gammaproteobacteria at the Lost City vents are closely related to cultured members of genera that aerobically oxidize sulfide (*Thiomicrospira*) and methane (*Methylomonas* and *Methylobacter*). Uncultured epsilonproteobacteria are most likely involved in hydrogen and sulfide oxidation. Several uncultured members of the Firmicutes are most closely related to members of the sulfate-reducing

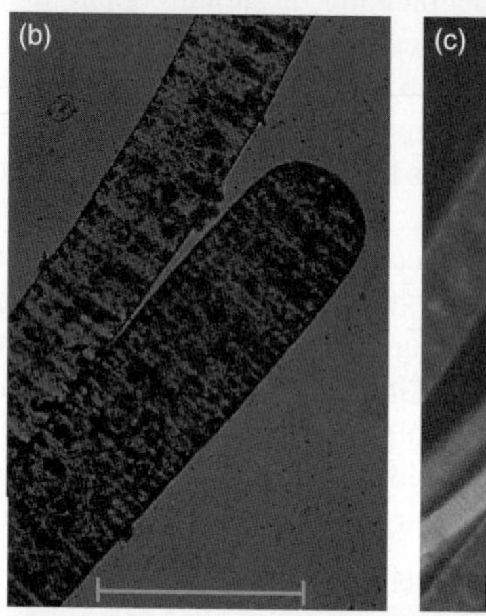

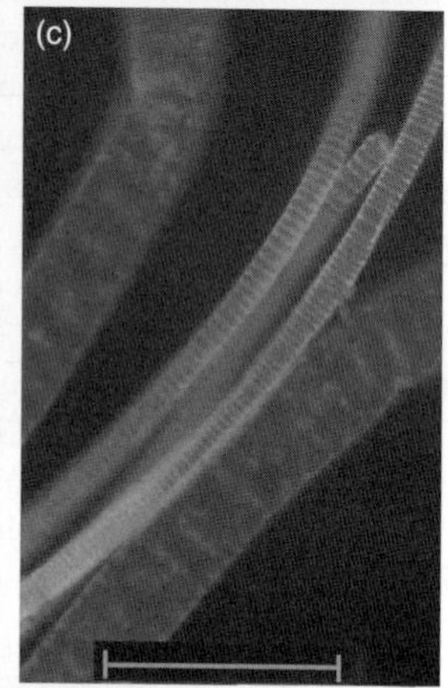

FIGURE 5 Guaymas Basin. (a) White and Orange *Beggiatoa* mats, and a colony of *Riftia pachyptila*, at Guaymas Basin. The diameter of the mats is ca. 1 ft. Photo Courtesy Tim Shank, WHOI. (b) Transmission light Microphotograph of large, white Beggatoa spp., showing the cells within the filaments. Scale bars, 250 μm. Photo, Andreas Teske. (c) Epifluorescence Light microphotograph of white and orange *Beggiatoa* spp. under UV excitation light. The larger, white *Beggiatoa* spp. appears bluish, the smaller orange *Beggiatoa* spp. yellow-orange. Photo, Andreas Teske. (d). Freshly retrieved push cores (2.5 inch diameter) with hydrothermal sediment, covered with mats of orange *Beggiatoa* spp. Photo, Holger W. Jannasch and Douglas Nelson. (For interpretation of the references to color in this figure legend, the reader is referred to the Web version of this chapter.)

genus *Desulfotomaculum* (Brazelton et al., 2006).To summarize, the bacterial communities at Lost City constitute a full sulfur cycle with anaerobic gram-positive sulfate reducers and aerobic or microaerophilic sulfide oxidizers; hydrogen, methane, and sulfide sustain the chemosynthetic microbial communities at the Lost City vents (Brazelton et al., 2006).

The Lost City vent site is the first example of a previously unknown, chemically distinct and unusually long-lived type of hydrothermal vent; the serpentinization reactions in the vent subsurface can yield energy and microbial electron donors over hundreds of thousands of years. Radiocarbon tests of the vent carbonates have shown that the Lost City vents have been active for at least 30,000 years. This long-term habitat stability, and the wide distribution of geological settings that are favorable for serpentinization reactions, have major implications for the evolution of microbial life on Earth and on other planets (Kelley et al., 2002).

SEE ALSO THE FOLLOWING ARTICLES

Archaea (overview); Autotrophic CO_2 Metabolism; Deep Subsurface; Microbial Mats; Methane Cycle; Methanogenesis; Sulfur Cycle; Tree of Life

GLOSSARY

Archaea One of the two prokaryotic domains of life, phylogenetically and also biochemically distinct from the bacteria. The metabolic and physiological properties of archaea from hydrothermal vents reflect adaptations to challenging environmental regimes; vent archaea include anaerobic and microaerophilic thermophiles and hyperthermophiles, methanogens and anaerobic methane oxidizers, heterotrophic and autotrophic archaea.

Black Smoker A common type of hydrothermal vent where extremely hot (ca. 350°C) vent fluid emerges into the water column through channelized flow in a chimney-like structure; the chimneys are formed by precipitation of metal sulfides from the vent fluid and can grow to a height of less than one to more than 10 m, depending on stability of precipitated minerals. The porous chimney matrix is the major natural habitat of hyperthermophilic vent archaea.

Chemosynthesis The use of chemical energy instead of light to drive the assimilation of CO_2 as sole carbon source for biosynthesis of cellular material; chemical energy comes from light-independent oxidation of inorganic electron donors.

Chemolithoautotroph A microorganism that obtains its metabolic energy from oxidation of reduced inorganic compounds, and its carbon for biosynthesis from CO_2 or CO.

Electron Acceptor The oxidant (e.g., O_2, NO_3^-, SO_4^{2}, Fe^{3+}, CO_2) in a biologically mediated redox reaction that accepts electrons from the reductant or electron donor, an organic compound or an inorganic electron donor (H_2S, H_2, reduced metals, CH_4). The combination and concentrations of oxidant and reductant determine the energy yield of a microbial redox reaction.

Electron Donor At hydrothermal vents, inorganic electron donors in the vent fluid (H_2S, H_2, reduced metals, CH_4) are used as the primary electron donors and energy sources for microbial metabolism by chemolithotrophic bacteria and archaea. Organic carbon compounds, mostly from locally produced chemosynthetic biomass, serve as electron donors for heterotrophic bacteria and archaea.

Heterotroph A microorganism that requires organic carbon as energy and carbon source. At hydrothermal vents, heterotrophs degrade biomass and organic substrates derived from chemolithoautotrophic bacteria and archaea; in the surface biosphere, heterotrophs rely on photosynthetic biomass.

Hyperthermophile An extremely thermophilic bacterium or archaeon with a growth optimum temperature above 80 °C.

Mid-Ocean Ridge The geologically active region where two opposite oceanic plates are separating a few cm every year; the split (the spreading center) is characterized by increased volcanic and tectonic activity, and hydrothermal venting. This region is elevated 2–3 km above the deep-sea floor (4–5 km average depth) and appears as an extensive underwater mountain chain or ridge.

FURTHER READING

Brazelton, W. J., Schrenck, M. O., Kelley, D. S., & Baross, J. A. (2006). Methane and sulfur-metabolizing microbial communities dominate the Lost City hydrothermal field ecosystem. *Applied and Environmental Microbiology*, *72*, 6257–6270.

Campbell, B. J., Engel, A. S., Porter, M. L., & Takai, K. (2006). The versatile epsilon-proteobacteria: key players in sulphidic habitat. *Nature Reviews Microbiology*, *4*, 458–468.

Emerson, D., & Moyer, C. L. (2002). Neutrophilic iron-oxidizing bacteria are abundant at the Loihi seamount hydrothermal vents and play a major role in Fe oxide deposition. *Applied and Environmental Microbiology*, *68*, 3085–3093.

Jannasch, H. W. (1995). Microbial Interactions with hydrothermal fluids. In S. E. Humphris, R. A. Zierenberg, L. S. Mullineaux, & R. E. Thomson (Eds.), *Seafloor hydrothermal systems: Physical, chemical, biological and geological interactions Geophysical Monograph* (Vol. 91) (pp. 273–296). American Geophysical Union: Washington, D. C.

Kelley, D. S., Baross, J. A., & Delaney, R. J. (2002). Volcanoes, fluids, and life at mid-ocean ridge spreading centers. *Annual Review of Earth and Planetary Sciences*, *30*, 385–491.

Markert, S., Cordelia, A., Felbeck, H., Becher, D., Sievert, S. M., Hugler, M., et al. (2007). Physiological proteomics of the uncultured endosymbiont of *Riftia pachyptila*. *Science*, *315*, 247–250.

McCollom, T. M., & Shock, E. L. (1997). Geochemical constraints on chemolithoauto-trophic metabolism by microorganisms in seafloor hydrothermal systems. *Geochimica and Cosmochimica Acta*, *61*, 4375–4391.

Miroshnichenko, M. L., & Bonch-Osmolovskaya, E. A. (2006). Recent developments in the thermophilic microbiology of deep-sea hydrothermal vents. *Extremophiles*, *10*, 85–96.

Reysenbach, A. L., & Shock, E. (2002). Merging Genomes with geochemistry in hydrothermal ecosystems. *Science*, *296*, 1077–1082.

Reysenbach, A. L., Götz, D., & Yernool, D. (2002). Microbial diversity of marine and terrestrial thermal springs. In J. Staley, & A. L. Reysenbach (Eds.), *Biodiversity of Microbial Life* pp. 345–421. New York: Wiley-Liss.

Spear, J. R., Walker, J. J., McCollom, T. M., & Pace, N. R. (2005). Hydrogen and bioenergetics in the Yellowstone geothermal ecosystem. *Proceedings of the National Academy of Sciences of the USA*, *102*, 2555–2560.

Takai, K., Nakagawa, S., Reysenbach, A. L., & Hoek, J. (2006). Microbial Ecology of Mid-Ocean Ridges and Back-Arc Basins. In D. M. Christie, C. R. Fisher, S. M. Lee, & S. Givens (Eds.), *Back-Arc Spreading Systems – Geological, biological, chemical and physical interactions Geophysical Monograph*: (Vol. 166) (pp. 185–214). Washington, D. C: American Geophysical Union.

Teske, A., Hinrichs, K. U., Edgcomb, V., de Vera Gomez, A., Kysela, D., Sylva, S. P., et al. (2002). Microbial diversity in hydrothermal sediments in the Guaymas basin: Evidence for anaerobic methanotrophic communities. *Applied and Environmental Microbiology*, *68*, 1994–2007.

Van Dover, C. L. (2000). The ecology of deep-sea hydrothermal vents. Princeton, NJ: Princeton University Press.

Ward, D. M., Ferris, M. J., Nold, S. C., & Bateson, M. M. (1998). A natural view of microbial biodiversity within hot spring cyanobacterial mat communities. *Microbiology and Molecular Biology Reviews*, *62*, 1353–1370.

Chapter 29

Freshwater Habitats

L.G. Leff
Kent State University, Kent, OH, USA

Chapter Outline

ABBREVIATIONS

AOA Ammonia-oxidizing archaea
CPOM Coarse particulate organic matter
DNRA Dissimilatory nitrate reduction to ammonium
DOC Dissolved organic carbon
DOM Dissolved organic matter
FPOM Fine particulate organic matter
RCC River Continuum Concept

DEFINING STATEMENT

Freshwater environments, such as wetlands, lakes, streams, and rivers, are critical components of society and support diverse and complex microbial communities. There is great variation among freshwater habitats, with hydrology, nutrient inputs, water level and movement, the role of allochthonous inputs, and other factors, all influencing the ecology of microorganisms.

OVERVIEW

Freshwater environments are ubiquitous and critical to human success as sources of drinking water, conduits for waste materials, aesthetical and recreational purposes, and so on. The great diversity in form and function of freshwater systems results in significant differences in microbial community structure and processes. Water flow, water depth, light level, nutrient inputs, pH, plant communities, and numerous other factors all influence the microbial ecology of freshwater (aquatic) ecosystems.

The presence of water dictates several key common properties of these environments, as do the interactions with the surrounding terrestrial environment within the watershed. Water provides a means for dispersal of organisms, transport of materials, and structural support, while at the same time limiting light availability as depth increases, altering temperature regimes, and impacting oxygen concentrations. Some freshwater environments are fueled predominantly by allochthonous fixation of C, whereas others are dominated by autochthonous C fixation. At the same time, microbial cells may enter aquatic ecosystems from outside sources and intermingle with autochthonous organisms.

Freshwater habitats are profoundly impacted by anthropogenic disturbance such as alterations in hydrology, land use, disposal of waste materials, contamination with fertilizers, heavy metals, xenobiotic organic compounds, and introduction of exotic species. Global climate change also has great potential impacts on freshwater resources through increases in carbon dioxide, elevated temperatures, an increase in sea level, and alterations in precipitation and snow melt seasonal patterns. Some of these are predicted to impact the hydrology of freshwater systems whereas others may directly impact biota. The combination of these effects is predicted to alter freshwater habitats and thus, there is the potential for the microbial ecology of the systems to also change.

The three main types of freshwater habitats are discussed below and include wetlands, lakes, and lotic ecosystems (streams and rivers). Wetlands are arguably the least studied of these systems relative to the structure of the microbial community, although the functional role of microorganisms in wetlands and other freshwater habitats is well established. For each freshwater habitat type, the nature of the hydrology, geographic location and climate, and origin contribute to the large variations seen in ecosystem properties. The role of allochthonous versus autochthonous sources of organic compounds, nutrients, and microbial cells also varies among ecosystem type along with the contribution of plant detritus to the food web.

FRESHWATER WETLANDS

Wetland Types and Properties

The nature of freshwater wetlands varies based on hydrology, origin of the wetland, plant community composition, and other features; these variations in turn affect the microbial communities of the environment. This complexity also limits the number of generalities that can be drawn about the microbial ecology of these ecosystems.

Wetlands serve as environments that may ameliorate environmental contaminants and manage flooding. At the same time, wetlands are highly threatened environments. For example, the percentage of wetlands in the continental United States is estimated to have declined by about 50% since presettlement times. Wetland loss in Europe as well as in parts of Asia is believed to be much greater. The importance and vulnerability of freshwater wetlands lead to regulations in many countries regarding the use of and destruction of wetlands. This, in turn, contributes to the issues associated with the definition of a wetland.

From a basic point of view, wetlands are shallow aquatic communities dominated by plants in which detritus plays a central role in the food web. From a standpoint of delineation, the definition of wetlands is more complex and has been the subject of great debate. Overall, wetlands, for purposes of this article, can be defined as systems with hydrophytes (wetland-type plants), hydric soils, and, during the growing season, periods of standing surface water or water-saturated soils. The duration and depth of standing water vary greatly among wetlands and, within a wetland, seasonally, and from year to year.

The hydrology, origin, and geographic location of the wetland are major determinants of plant community structure, which in turn influences the microbial communities. Wetlands include those on the fringe of larger bodies of water, such as around the Great Lakes of the United States, prairie potholes, marshes (including freshwater tidal marshes), swamps, riparian zones, and floodplains along rivers, bogs, playas, and billabongs. Each wetland is connected to the surrounding terrestrial environments in a variety of ways and they can also be connected to marine or other freshwater systems. For example, nutrients from a fringing wetland can be transported into lakes linked seasonally by water.

Wetland environments can be stressful and exhibit large seasonal changes in hydrology including periodic drying. This effect, coupled with the development of areas of low oxygen concentration, low redox potential, and the wide array of different types of organic compounds, which in some cases accumulate in the system, defines the microbial ecology of the wetland.

Human use of wetlands is also an important determinant of wetland function and biogeochemistry. Wetlands serve as sources of flood abatement, carbon sequestration, and phytoremediation, and support a rich and unique flora and fauna. Wetlands are also created and used by humans for the production of food, such as rice, fish, and crayfish.

Overall, critical features of wetlands that affect the ecology of microorganisms are as follows:

- Important role of plants and detritus in the food web.
- Critical role of hydrology, which varies greatly among wetlands of different types.
- Seasonal variation in the extent of hydration of soils.
- Profound human impact on wetland extent, nature, and hydrology.
- Creation of new wetlands, perhaps with different ecological properties, to replace wetlands lost through human use or to serve as sites of food production.

Microbial Processes

The importance of microbial processes in wetlands manifests itself in the biogeochemistry of these systems and the central role that detritus decomposition plays in the food web. Dissolved organic matter (DOM) is a major component of the organic matter pool and is made available to higher trophic levels by microorganisms. In some wetlands, often referred to as dystrophic, dissolved organic carbon (DOC) concentrations can be substantial and "stain" the water brown to black because of large amounts of humic substances. These wetlands, such as bogs, are often characterized by unique plant communities, including, in some cases, the occurrence of *Sphagnum*, formation of peat, and low pH.

Inundation with water creates anaerobic conditions in wetland soils, providing the opportunity for alternative metabolic processes. The strong seasonality in the hydrology of many types of wetlands creates circumstances where there is flooding and corresponding chances for anoxia in lower layers of the soil at some times of the year, but not at

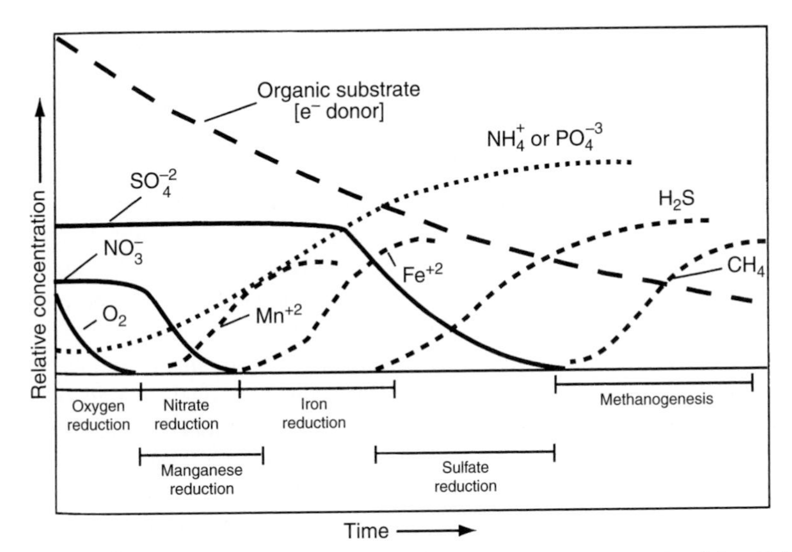

FIGURE 1 Temporal changes in the availability of alternative electron acceptors in wetland soil. *Reproduced from Reddy, K. R, & D'Angelo, E. M. (1994). Soil processes regulating water quality in wetlands. In W. J. Mitsch (Ed.).* Global wetlands: Old world and new *(pp. 309–324). Amsterdam: Elsevier.*

other times (Figure 1). This creates seasonal patterns in key microbial processes, such as methanogenesis and denitrification. In denitrification, a dissimilatory process, microorganisms reduce nitrate (or nitrite) to dinitrogen rather than using oxygen as the electron acceptor. Also, under anaerobic conditions some archaea, known as methanogens, can produce methane when grown on carbon dioxide and hydrogen or when growing on acetate or other simple organic compounds. Lastly, dissimilatory sulfate reduction may occur in anaerobic areas; during this process sulfate is reduced to hydrogen sulfide.

Plants play a seminal role in determination of microbial processes in wetlands, including their impact on soil conditions, the detrital pool, and the oxygen status of the soil. Plant roots lose oxygen (from air-filled tissue in wetland plants called aerenchyma) to the surrounding soil altering the otherwise anaerobic status of the soils. The occurrence of oxygen is variable both spatially and temporally and, thus, the optimal conditions for growth of a particular microorganism may not always be present.

The C cycle of wetlands is characterized by a mixture of anaerobic and aerobic processes occurring at different locations or at different times. Often these processes can occur in spatially close proximity to each other in different portions of the soil. Fermentation is a common process in wetlands and appears to play a critical role in providing suitable organic compounds for other anaerobes. Methanogenesis in wetlands tends to be better studied than fermentation and can result in the release of bubbles of methane gas (referred to as swamp gas). Methane emission from wetlands globally is quite large and is of particular interest because of the role of methane in global climate change and the possibility of management of methane entrance into the atmosphere. Quantification of methane production by wetlands reveals that there is substantial variation among wetland types and geographic locations.

Properties of the N cycle in wetland are highly variable depending on plant community type and hydrology. The sediments of wetlands play a critical role in the N cycle. N can be a limiting nutrient in wetland soil and depending on the N pool, N fixation will be more or less important. For example, the floating fern, *Azolla*, has symbiotic nitrogen-fixing cyanobacterium (*Anabaena*). In wetlands created for rice cultivation the *Azolla–Anabaena* may help rice production. In addition, in some cases, wetland-fringing plants, such as alder (*Alnus*), may host nitrogen fixers that can impact the wetland N cycle. It also appears that nitrogen fixation does occur in association with wetland plants (such as some grasses); however, the organisms responsible for this activity defy cultivation and thus the amount of information on them is still relatively limited.

Denitrification is a potent source of N loss (as N_2) because of the extensive amounts of organic matter and anaerobic zones. The large amounts of N applied to landscapes can work its way into wetlands and perhaps these ecosystems can serve as N sinks in the global N cycle. Denitrification in some wetlands may be limited by low pH, thus loss of N in this fashion, in some wetlands, is more restricted than others. However, generally, denitrification results in a major loss of N from wetlands that exceeds inputs from nitrogen fixation.

Nitrification plays a critical role in the removal of ammonium from the soil, which tends to accumulate because of ammonification. In nitrification, ammonia is oxidized to nitrite, which is in turn oxidized to nitrate; these reactions are performed by ammonia-oxidizing and nitrite-oxidizing bacteria. In addition, ammonia-oxidizing archaea (AOA) have been discovered in marine environments, soil, and wastewater treatment facilities; although the number of studies is limited, AOA have also been detected in freshwater systems. Aerobic zones supportive of nitrification occur in the topmost layer of soil as well as in aerobic patches caused by plants in the rhizosphere. In some wetlands, ammonium is converted into ammonia, which can be lost to the atmosphere, and generally ammonium diffusion rates through wetland soil are a limiting process. In acidic wetlands, nitrification may be slower relative to other wetland types. Bacteria that oxidize ammonium anaerobically (anammox) are present in freshwater wetlands, although their role in the N cycle of these systems is largely unexplored. In addition, the occurrence and importance of dissimilatory nitrate reduction to ammonium (DNRA) has been demonstrated in freshwater wetland sediments.

Sulfur transformations also occur in wetlands, although S is generally not a limiting nutrient and thus it is less frequently studied in freshwater wetlands. One manifestation of these transformations is the distinctive rotten egg smell from the release of hydrogen sulfide when anaerobic wetland sediments are disturbed. Compared with marine wetlands, sulfide emissions from freshwater wetlands are much lower because S concentrations in marine systems greatly exceed those of freshwater systems. Sulfides produced by sulfate reduction can subsequently be oxidized by autotrophs. Sulfides that are not oxidized are toxic to plants and can precipitate with metals.

Unlike N and S, phosphorus lacks alternate oxidation states but is often a limiting nutrient in freshwater ecosystems. Some wetlands, such as bogs, are exceptionally nutrient-limited whereas others receive P input from fertilizer and other sources. Both organic and inorganic P contribute to the P pool and insoluble P complexes that affect bioavailability can be formed. For example, P can complex with iron or calcium or attach to clay or peat. In particular, the sorption of P to clay can form complexes, which may be carried into wetlands. When this material sediments, it helps make P available to the wetland plant community and allows the wetland to retain P.

Microbial Communities

As compared to other freshwater systems, the amount of information on microbial community structure in freshwater wetlands is relatively limited. Although microbe-mediated processes are widely studied, as described above, the underlying communities responsible for a particular function are unknown to varying degrees with more information available on methanogens, methanotrophs, and denitrifying bacteria, for example, relative to other functional groups, such as nitrogen fixers.

Although the role of fungi as decomposers of plant tissue is widely known, there is much less information available on the mycorrhizae associated with wetland plants. However, the potential importance of mycorrhizae in the success of plants may be particularly relevant in wetlands that are nutrient-limited, such as bogs and fens, and can also be important for phytoremediation.

Although the number of investigations on bacterial diversity and community structure in freshwater wetlands is relatively limited, studies have revealed that for the wetlands examined *Proteobacteria* (α-, β-, and γ- as well as the δ-*Proteobacteria*, which are perhaps not as well represented in lakes and streams), Acidobacteria, and Verrucomicrobia are major constituents. Acidobacteria are a widespread but largely understudied group because of limitations in our ability to culture these organisms. Similarly, there are few cultures from the group *Verrucomicrobia* with some of the major subgroups in this taxon having virtually no cultured representatives. *Verrucomicrobium* and other prosthecate bacteria are among the most well-known members of this widespread and understudied group.

Among the archaea, the occurrence of methanogens in wetlands is the most often studied, especially in light of the role of methane as a gas contributing to global climate change. In anoxic wetland soils, methanogenesis is a critical biogeochemical process and varies seasonally; the availability of both suitable substrates and nutrients impacts methanogen function. The nearby occurrence of methanotrophs in oxic areas and their oxidation of methane greatly impact methane release.

The role and diversity of protozoa in wetlands are largely unknown. Generally, the impact of protozoa as bacterial consumers is great in aquatic environments and less so in soil. There is at least some evidence that the protozoan community composition can be very different from that of other freshwater habitats. The number of protozoa and consequently their impact on bacterial numbers fluctuates seasonally in temperate wetlands.

Anthropogenic Disturbance

Human impacts on wetlands are varied and, among these, perhaps the most important is the vast global destruction of wetlands. This leads to the phenomenon of wetland construction whereby new wetlands are created to replace those lost due to conversion to other uses (agriculture, construction, etc.). How well these created wetlands duplicate natural conditions and function as desired is variable and debatable. Hydrological modifications, to reduce flooding for example, are also common and considering the key role of hydrology in wetland microbial ecology undoubtedly greatly impact microbial processes.

Wetlands are also used for various agricultural and horticultural activities, such as peat mining and production of plants, such as rice. Like other freshwater habitats, the impact of humans on water quality can translate into these environments. One manner in which some environmental issues may be addressed is through phytoremediation. Phytoremediation may be accomplished in either natural or constructed wetlands and potentially can remove xenobiotic compounds from water systems. Critical features for the success of phytoremediation in wetland include the redox potential, extent of above- and below-ground plant biomass, and hydrology.

Invasive species represent one area in which the effects on microbial communities in wetlands remain largely unknown. Of particular importance are changes in native plant communities associated with the invasion of exotics, such as reed canary grass (*Phalaris arundinacea*), purple loosestrife (*Lythrum salicaria*), and alligator weed (*Alternanthera philoxeroides*). The role of plant communities in wetlands is a critical defining feature and thus alteration of the plant communities toward potentially nearly monospecific stands of exotic species is of great concern. Native plants are often outcompeted by exotic species of plant, perhaps resulting in a loss of the microflora associated with the native plants (such as mycorrhizae), as well as an alteration in many other factors, such as soil moisture, quality, and nature of the detrital pool. In general, invasive plants are known to alter C, N, and water in soil, such as by altering N availability and fixation, and are different from native species in terms of productivity, growth form, chemistry, and so on. Although the role of these differences in wetlands is less well studied than in terrestrial environments, the potential for a substantial change in wetland microorganisms is clear.

LAKES

Properties and Types

Like wetlands, lakes originate in various ways and the nature of their origin, hydrology, and morphometry impact environmental conditions and biotic functions. Basin morphometry is of particular importance because of the limitations that depth imposes on macrophyte success, the extent of the littoral zone, and the impacts on the relative contributions of allochthonous and autochthonous production. Lake processes are very much dependent on the properties of the surrounding watershed, as in other freshwater ecosystems. Lake age is also highly variable and depends on the lake type. In many ways, human impacts are poorly understood and not always recognized because so many key processes are hidden beneath the water column and thus unseen by the lay person.

One of the most critical determinants of the microbial ecology of a lake is its trophic status, which ranges from eutrophic to oligotrophic (with various modifications of the terms, such as hypereutrophic). The trophic status of the lake profoundly influences the food web, biogeochemistry, and other processes. Oligotrophic lakes are characterized by low nutrient concentrations and low primary production (as reflected in the low abundance of phytoplankton in the lake). Eutrophic lakes, in contrast, are characterized by high-nutrient concentrations and correspondingly high primary production. There is a gradient of trophic status between these two conditions with lakes in the middle of the gradient referred to as mesotrophic. The trophic status is the main determinant of the number, biomass, production, and diversity of microorganisms because of the differences above and their corresponding impacts on oxygen concentration.

Lakes, or lentic ecosystems, are also characterized by their stratification and mixing properties. In temperate zones, many lakes are dimictic with thermal stratification in winter and summer and spring and fall turnover events. In some regions, there are lakes with permanent stratification called meromictic lakes, such as those where salt and freshwater combine, or those that are particularly deep, whereas other lakes never stratify. Amictic lakes are permanently ice covered, such as lakes found in Antarctica. Other lakes are monomictic with stratification once per year or polymictic with mixing and stratification multiple times throughout the annual cycle. The pattern of stratification and mixing depends on latitude and lake properties.

Seasonal stratification in temperate areas occurs because inputs and outputs of heat occur primarily at the surface of the lake. In such temperate zones, the heating of surface waters in summer creates a warm, circulating surface water pool (the epilimnion) with colder, undisturbed water below the thermocline in the hypolimnion (Figure 2). As fall

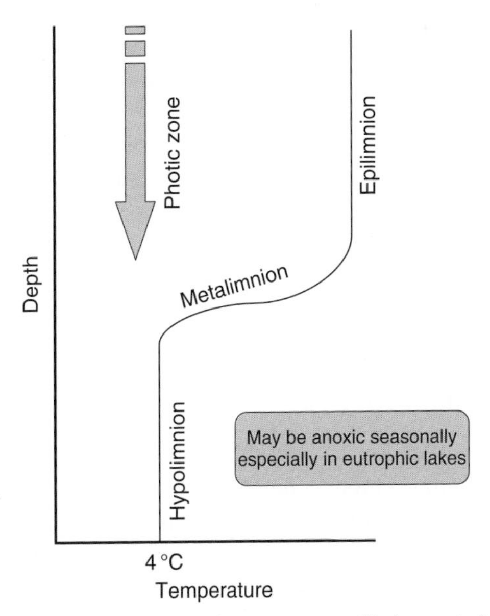

FIGURE 2 The water layers and temperature profile in a typical seasonally stratified lake. The thermocline refers to the plane with the greatest decrease in temperature with depth.

begins, the surface water cools and at some point reaches the same temperature as the underlying water; this results in a loss of density differences and as this surface water becomes cooler and denser it sinks, causing the layers to mix or turn over.

In a dimictic lake that is productive (eutrophic) during times of thermal stratification (Figure 2), the hypolimnion can be depleted of oxygen through biological activity. This happens when this layer is separated from and not mixing with the overlying water; biological activity in this confined layer resulting from decomposition can be great enough to deplete oxygen. This creates microbiologically important gradients with depth that can result in some interesting juxtapositions of biogeochemical processes as described below.

Another main way of categorizing lakes is their vegetation with macrophyte-dominated systems versus algal-dominated systems representing alternative stable states. The dominance of macrophytes versus algae greatly influences the trophic interactions and paths of C transfer. The two alternative states that can occur in shallow eutrophic lakes are (1) clear water with abundant macrophytes and (2) less clear water because of phytoplankton abundance without abundant macrophytes. The clear water state is often associated with desirable fish and invertebrate communities, whereas the turbid water state tends to have lower diversity and undesirable algal blooms. These alternate states represent a shift in the lake community and may result from competition with nutrients, light limitation and turbidity, allelopathy, or other factors. If the food web of a lake (such as by removal of macrophytes) or P concentrations are altered, it is possible to shift the state of a lake.

Overall, important lake properties that influence the microbial ecology of the ecosystem are as follows:

- The trophic status of the lake (as determined by concentrations of N and P) and its point along the oligotrophic:eutrophic gradient.
- The morphometry and size of the lake, which influences the size and relative contribution of the littoral zone and allochthonous inputs.
- The status of macrophytes in the lake; whether it is algal-dominated or macrophyte-dominated.
- The stratification and mixing patterns over the course of a year.

Microbial Processes

Both stratification and trophic status are major determinants of the microbial processes that happen and where they happen. Stratification in a eutrophic lake, in particular, can ultimately lead to large seasonal anoxic zones, which greatly impact microbial processes as well as macrofauna. Development of these so-called "dead zones" is mediated by the microbial consumption of oxygen and significantly impacts lake quality as it may limit fish community success. The nature and extent of stratification as well as nutrient availability creates obvious differences in C and nutrient cycles among different lakes.

C cycle

Taken from a simplistic point of view, lake food webs include both a grazer "food chain" and a microbial "food chain" or microbial loop (Figure 3). The relative amount of C transferred through these two components varies from lake to lake and within a lake (e.g., nearshore vs. offshore). The dichotomy between these two paths is also reflective of the role of detritus in the system with the microbial loop reliant on the extensive dissolved portion of the detrital pool. Although the role of viruses in the microbial loop of lakes is perhaps not as well studied as it is in marine systems, there is ample evidence demonstrating the occurrence of phage-infected bacteria and viruses in lakes. There is less information about the potentially large impact that bacteriophage and algal viruses have on the microbial loop and the contribution of virus activity to the pool of DOM.

The occurrence of anoxic areas in the hypolimnion and benthos creates opportunities for the use of alternative electron acceptors at the appropriate depths within a lake. These processes can affect the C cycle; for example, methane is generated in these anoxic areas by anaerobic archaea and then can be utilized by methanotrophs in the overlying oxic area (the epiliminion). Similarly, fermentation occurs in the anaerobic zones, resulting in a variety of small organic compounds for use by other microbes.

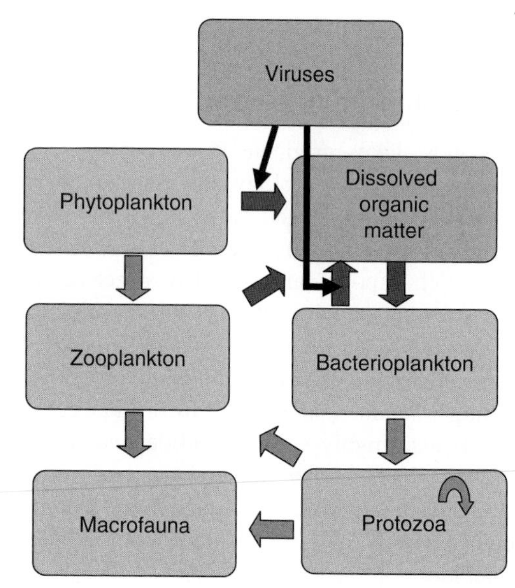

FIGURE 3 Stylized and simplified food web in the water column of a lake illustrating paths of C transfer in grazer and microbial food webs; green arrows indicate grazing whereas brown arrows indicate production and use of detritus in the form of dissolved organic matter (DOM). Note within the protozoa box, multiple trophic transfers may occur. Viral infections may contribute to the DOM pool via lysis of algal or bacterial cells.

On macrophytes, extensive biofilms (often in limnology, such biofilms are referred to as periphyton) can form, which may serve as competitors for nutrients with phytoplankton. These biofilms are structurally complex and feature diatoms interconnected by the extracellular matrix as well as cyanobacteria and heterotrophic bacteria. Some components of this assemblage penetrate into the macrophyte tissue, whereas others are more loosely associated. Within these biofilms, release of extracellular organic compounds by algae supports bacterial growth, and bacteria, in turn, release carbon dioxide and degradative enzymes. The surface area of macrophytes in a lake can be extensive depending on its state, depth, and relative littoral zone size. Thus, the role of these macrophyte-associated biofilms in nutrient acquisition may also be great.

N cycle

The N pool in lakes includes both inorganic and organic components that are utilized and transformed by microorganisms. Cyanobacteria, such as *Anabaena* and *Aphanizomenon*, are important nitrogen fixers, especially, but not always, those that produce heterocysts. For example, members of the Oscillatoriaceae are also common and important in lakes but lack the ability to produce heterocysts. N-fixing cyanobacteria are a valuable surface for attachment of bacteria and the heterocysts may create microzones of low oxygen concentration, creating specialized bacterial niches. There is limited information on the importance of heterotrophs, such as *Azotobacter*, in nitrogen fixation in lakes. This is perhaps a result of the assumption that this process might be limited by the availability of labile organic compounds.

Nitrification in lakes occurs in aerobic zones and thus may be limited in the deeper sediments. In addition, specific dissolved organic compounds (such as tannins) may inhibit nitrification. Denitrification is important in areas with oxygen depletion (although it can also occur aerobically) such as the sediments of littoral zones of lakes and the hypolimnion of stratified eutrophic lakes. The ability to denitrify is widespread among different types of bacteria, including *Pseudomonas*, *Bacillus*, and *Achromobacter*.

Anaerobic ammonia oxidation (anammox) has been demonstrated to occur in lakes; it may represent an additional important mechanism for N loss and has not yet been extensively studied.

P cycle

Each elemental cycle in the lake is intertwined with other cycles; this is illustrated, in particular, for the P cycle. Lake sediments are a major source of P in lentic ecosystems and P is often a limiting nutrient; additions of N and P lead to eutrophication, which greatly impacts biotic processes. Most of the P in a lake is contained in the biomass and in organic compounds and total P concentrations vary considerably from hypereutrophic to ultraoligotrophic lakes. Because P is often a limiting nutrient, much attention has been focused on internal loading from the sediments, the process by which sediment P is released into the water column. Mobilization of P is facilitated by bacteria such as *Pseudomonas* and *Chromobacterium* but chemical processes are dominant. Specifically, complexation with metals plays a major role in the lake P cycle; for example, phosphate can be released from ferric phosphate under anoxic conditions. Formation of iron sulfide associated with sulfate reduction can reduce the complexation of phosphate by iron.

Microbial Communities

The amount of information on lake microbial community structure has increased dramatically over the years. Prokaryotic communities have been studied in a variety of lakes of different sizes and trophic status. These studies have revealed information about the dominant bacterial taxa as well as differences between lake and marine communities.

The water column of microbial community is distinctive in function and also in structure from the benthic community. Depending on the water depth, the benthos may be a viable habitat for phototrophs. Often the benthos may be anaerobic (especially below the sediment surface) even if the overlying water is oxygenated, providing opportunities for methanogens and dentrifyers.

Phototrophic prokaryotes fill a particular series of niches in the lake environment based on their physiological needs and pigments used. Cyanobacteria are abundant and can grow to nuisance levels, especially in eutrophic lakes with abundant P. The gas vesicles produced by cyanobacteria contribute to their buoyancy and ability to maintain position in the water column, which ultimately enhances their success. Many of these cyanobacteria can fix nitrogen, contributing to formation of undesirable blooms, which have negative impacts on water quality. *Microcystis*, which is abundant worldwide, produces microcystins that are of great concern for water quality from a drinking water standpoint and for the success of aquatic animals. The hepatotoxins produced by *Microcystis aeruginosa* have damaging effects on the livers of mammals, and control of this organism can be an important goal of lake management.

In many countries around the world, occurrence of toxin-producing cyanobacteria in lakes is of concern. Cyanotoxins produced include hepatotoxins, neurotoxins, dermatoxins, and endotoxins, which negatively impact water quality. Efforts are made to control the blooms of cyanobacteria by biomanipulation (increasing grazing on the bacteria, top-down control) and by decreasing nutrient availability (bottom-up control). The introduction of exotic species of cyanobacteria that produced toxins, such as *Cylindrospermopsis raciborskii*, can compound the problems associated with these organisms by introducing new types of undesirable algae into lakes.

Beyond the cyanobacteria, green sulfur, purple sulfur, and purple nonsulfur bacteria can occupy niches in lakes. Their niches occur in stratified lakes in which the anoxic hypolimnion has adequate levels of light to support their growth and, for the sulfur bacteria, hydrogen sulfide in the concentrations needed. The occurrence of these ideal conditions creates relatively narrow bands in the water column where these organisms are abundant certain times of the year in temperate dimictic lakes. Factors including oxygen concentrations, temperature, and light penetration are determinants of the vertical stratification of different types of photoautotrophs in lakes. The importance and perhaps diversity of nonoxygenic phototrophs is greater in lakes with permanent stratification; for example, in permanently frozen lakes of Antarctica or in meromictic lakes in other regions.

In the case of bacteria, overall, β-*Proteobacteria* appear to be the dominant group in freshwater, making the community of freshwater systems distinct in some ways from marine systems. The β-*Proteobacteria* include the freshwater ammonia-oxidizing bacteria and are common in high-nutrient conditions and biofilms. In large oligotrophic lakes, there is increasing evidence for the importance of *Actinobacteria*. α-*Proteobacteria* are found in peak numbers, in many cases, under more oligotrophic conditions, and can be correlated with chlorophyll concentration and have a preference for labile organic matter. In addition, the γ-*Proteobacteria* are abundant as are the *Cytophaga–Flavobacterium–Bacteroides* cluster and *Verrucomicrobia*. The *Cytophaga–Flavobacterium* are specialized for utilization of high-molecular weight organic compounds and thus may exhibit spatiotemporal patterns of abundance related to the occurrence of molecules of this type.

The composition of the bacterial community and relative importance of each of these major groups can vary considerably among lakes as related to variations in trophic status, pH, temperature, and nutrient concentrations. Similarly, there are often differences in the relative abundances of major groups between the hypolimnion and the epilimnion in stratified lakes as well as between different portions of lakes (e.g., areas that are shallower and close to riverine inputs might have a different community composition than offshore sites).

Fungi perhaps play a less important role in many lakes than other microorganisms because the role of plant detritus in some lakes is less significant. Aquatic fungi are important in littoral zones and in the degradation of macrophyte tissue. Many fungi that are encountered on plant material may be of allochthonous origin having colonized the tissue prior to submergence. However, the overall contribution of C from littoral zones and macrophytes is highly variable among lakes depending on size and morphometry. In many lakes, pelagic processes and in particular algal–bacterial coupling may be the central component of the microbial food web.

Algae in lakes have been studied for many years and are responsible typically for most of the C fixed in many lakes, especially those that are large and too deep for growth of rooted macrophytes. Algal communities are diverse and different species fill varying ecological roles based on differences in motility, habitat preferences, and palatability to aquatic animals. Algae can be directly consumed by grazers and support bacterial growth via release of exudates (Figure 3). In addition to prokaryotic algae, diatoms, green algae, chrysophytes, and dinoflagellates are common lake flora, whereas some important marine groups (red and brown algae) are not as common in freshwater lakes. The relative balance between the prokaryotic algae (cyanobacteria) and the eukaryotic algae depends on many factors including light and nutrient availability. The stoichiometric ratio between N and P is a particularly critical factor.

Protozoa play an important role in the microbial loop of lakes and are major consumers of bacteria. There is tremendous diversity of freshwater protozoa, including organisms that are mixotrophs (those that are photosynthetic and also consume prey, such as bacteria). There are large spatiotemporal changes in the distribution and abundance of different species, which is facilitated by the ability of some species to form cysts to persist undesirable conditions. There is vast morphological and functional diversity among freshwater protozoa. Overall, from an ecological standpoint, the smaller heterotrophic flagellates may be the most important bacterial grazers.

Anthropogenic Disturbance

Human impacts on lakes include those that are direct and those that are indirect. Indirectly, humans impact hydrology and nutrient loading into lakes whereas anthropogenic disturbance also has direct effects through invasive species. Other anthropogenic disturbances of concern that may impact microbial ecology include changes in lake water level, decreasing pH from acid precipitation, input of gasoline additives from boating, heavy metal pollution (such as mercury), dredging, and so on.

Invasive species in lakes range from fish to invertebrates to plants to zooplankton. Examples include the zebra mussel (*Dreissena polymorpha*), water hyacinth (*Eichhornia crassipes*), and Eurasian watermilfoil (*Myriophyllum spicatum*). There are numerous other examples of invasive species including the widespread practice of stocking of fish into lakes for fishing purposes. The role of exotic species in altering the food web and microbial processes in some cases is largely unknown, whereas in other cases the impacts are at least partially explored. For example, the zebra mussel has invaded areas such as the Great Lakes in North America and greatly alters the environment by clearing the water while filter feeding and depositing materials in sediments through feces and pseudofeces. This increases water clarity and decreases pelagic phytoplankton biomass. This alteration can correspondingly impact the role of the microbial loop in the food web; for example, they may alter the diversity and abundance of protozoa. In addition, these mussels have negative impacts on native bivalves and alter the community of benthic invertebrates.

Lakes are managed to enhance fish productivity, alter the type of vegetation present, reduce undesirable algae, make water bodies more suitable for boating and/or swimming, and improve water clarity. Many lake management strategies are focused on controlling the loading of nutrients into the system. Lake management practices include aeration, biological control, treatment with alum, and treatment with algaecides. Management of undesirable biological features may take top-down or bottom-up approaches. Top-down control includes grazer removal by the addition of planktivorous fish, whereas bottom-up control includes inactivation of P and decreasing external or internal nutrient loading. Species-specific lake management, such as control of a specific undesirable cyanobacterium via cyanophages, is an emerging area.

STREAMS AND RIVERS

Properties of Lotic Ecosystems

Lotic systems (streams and rivers) are characterized by water with a unidirectional flow and are classified based on "size" as represented by stream order. Essentially, as illustrated in Figure 4, the joining of two first-order streams creates a second-order stream, the joining of two second-order streams creates a third order, and so on. The processes at any site along the system are greatly influenced by upstream processes, as described, for example, in the River Continuum Concept (RCC), as well as by processes in the watershed. Water movement imposes a persistent and critical force on the biota, increasing the importance of the benthos of the ecosystem and limiting stratification in contrast to what is typically the case in lakes.

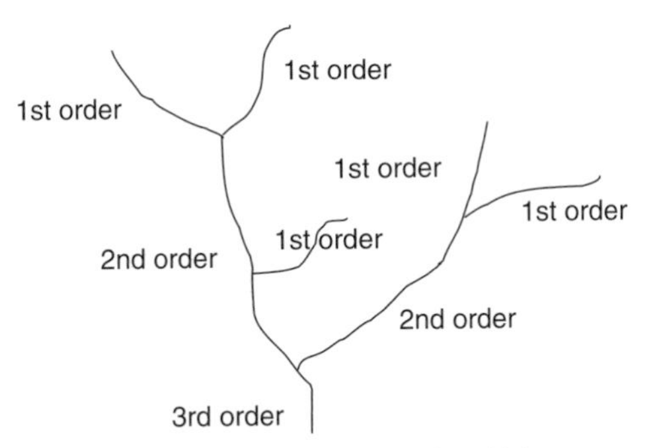

FIGURE 4 Schematic illustrating manner in which stream order is determined.

The RCC illustrates the upstream–downstream connection and the role that riparian vegetation structure can play. Although streams in nature may not follow the RCC model, because of human alterations, like reservoirs, and other factors, the notion of this profound connectivity among parts, longitudinal changes, and gradients of biological processes are conveyed by the model.

In addition to the longitudinal and lateral connections, there are vertical connections with streams and the underlying water in a zone called the hyporheos. The extent of the hyporheos varies based on geomorphology and other features. A unique fauna can develop in these areas, in addition to communities of organisms that live in both the surface and the subsurface. Biofilms formed in the interstitial spaces can be a potent site for microbial activity and create additional niches for microorganisms. Hyporheic processes depend on the redox potential of the surface and underlying waters, and the occurrence of anaerobic hyporheic zones in proximity to aerobic surface substrates can create conditions in a relatively narrow spatial scale for a variety of oxidation and reduction reactions. The occurrence of upwelling and downwelling areas, and whether the hyporheic water is oxic or anoxic, determines the biogeochemistry of the system.

In temperate streams, autumnal leaf inputs are a major source of organic matter for the food web. The quality of this leaf material as a biological resource varies among species based on structural properties and chemical constituents. These leaf materials are conditioned by microorganisms, especially bacteria and aquatic hyphomycetes, and can then be consumed by macroinvertebrates, functionally defined as shredders (Figure 5). The fecal production and fragmentation resulting from the feeding of these invertebrates result in the generation of fine particulate organic matter (FPOM). This FPOM is then available to another functional group of invertebrates, the collectors.

Disturbance via flooding is another major feature of streams that can have varying impacts on the biota depending on the seasonality, magnitude, and duration of the event. Stream life is adapted to the flowing water conditions, but water velocity and depth drastically increase during flooding, causing organisms to be dislodged. In addition, benthic materials can be entrained in the water column and be transported, resulting in scouring of surfaces and loss of biofilm. Flooding also connects the stream proper to its floodplain and the inputs of allochthonous materials jumps (including microorganisms, nutrients, and organic compounds).

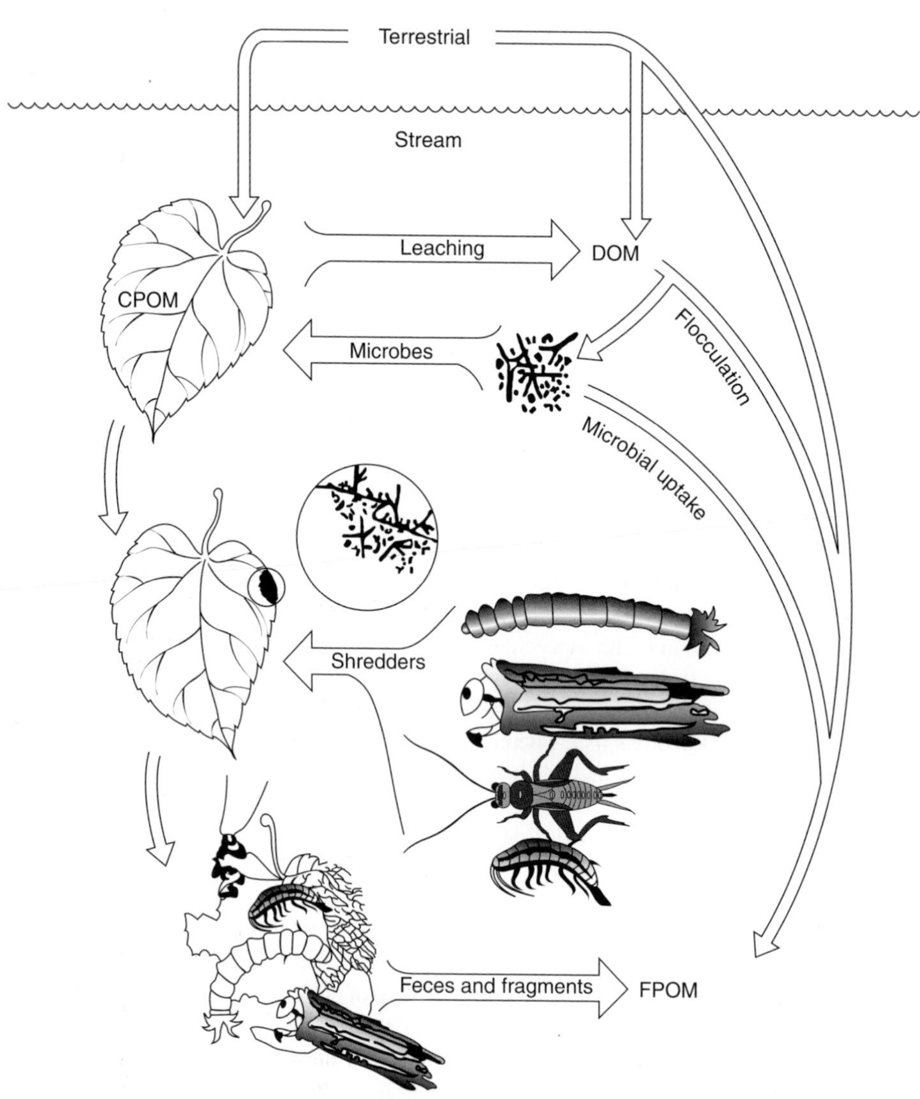

FIGURE 5 Decomposition of leaf material in a stream depicting microbial colonization, consumption by macroinvertebrates shredders, and the generation of fine particulate organic matter (FPOM). *Reproduced from Cummins, K. W. & Klug, M. J. (1979). Feeding ecology of stream invertebrates.* Annual Review of Ecology and Systematics, *10, 147–172.*

Overall, features of lotic ecosystem that are critical to microbial ecology are as follows:

- The unidirectional flow of water.
- The high degree of mixing resulting in a lack of stratification and high aeration.
- Strong interconnection to the surrounding watershed and terrestrial environments.
- Important role of the benthos as a habitat for microorganisms and the site of nutrient uptake.
- The significance of allochthonous inputs of microorganisms, organic compounds, nutrients, and particles.
- Importance to humans as water sources and transporters of waste water.
- Impact of humans on the environment through sewage, industrial pollutants, agricultural runoff, reservoir construction, removal of woody debris, channelization, and so on.

Microbial Processes

Food webs and the C cycle

As in other freshwater environments, DOM is the dominant fraction of the detrital pool. DOM is made available to higher trophic levels through the microbial loop (Figure 6); bacteria utilize the DOM and are in turn consumed by protozoa. Although rates of consumption per individual are much higher for ciliates than flagellates, small flagellates are the predominant protozoan predator of bacteria. Protozoa can then be eaten by meiofauna, such as nematodes and benthic copepods.

Another critical process in the C cycle is the degradation of coarse particulate organic matter (CPOM) such as leaves and wood. Aquatic hyphomycetes play a central role in the degradation of CPOM and can penetrate into

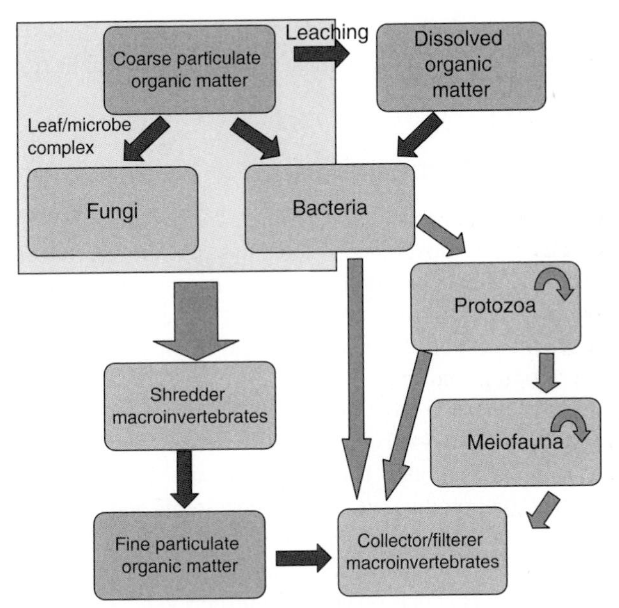

FIGURE 6 Basic illustration of C transfers in a stream where the coarse particulate organic matter (CPOM) is leaves.

the tissue of decomposing leaves, facilitating colonization by bacteria. There is evidence of competition as well as of facilitation among bacteria and fungi on leaves decomposing in streams. As noted above, these conditioned leaves are then palatable to leaf-shredding invertebrates. There are clear patterns of temporal succession within the microbial community of leaves with some fungal and bacterial species serving as early colonizers and others appearing later in the process.

Conditioning of leaf material by microorganisms facilitates consumption by macroinvertebrates that assimilate various fractions of the microbial/leaf assemblage (Figure 6). In addition, leaf-eating invertebrates play host to a variety of microorganisms; for example, the cecum of the crane fly larva, *Tipula*, boasts large numbers of bacteria.

The importance of algal production, and thus autochthonous C sources, varies greatly and light limitation is crucial. Both shading from riparian vegetation and turbidity can limit light and restrict algal production. Many low-order streams with well-developed canopies are considered heterotrophic because the P/R ratios are less than 1. In this case, the food web is dependent on C fixed in the terrestrial environment representing allochthonous sources of organic compounds. In contrast, the RCC predicts that the P/R ratio will reach its maximum at middle orders. The P/R ratio is also expected to be potentially greater than 1 in open canopy streams, especially those with abundant nutrients and suitable substrates for algal attachment.

Algal–bacterial coupling, which is widely observed in marine and lentic ecosystems, can also be evident in streams. Although the absolute contribution of autochthonous organic C to the overall DOC pool may be limited, this fraction generally has higher lability than the allochthonous fraction. The contribution of breakdown products of plant structural materials (lignocellulose) to the allochthonous fraction is high and these products can be highly recalcitrant. The basis for the observed algal–bacterial coupling has two components, namely, growth of bacteria on algal-released organic compounds and use of the algal cells as a physical substrate for attachment.

Although the role of viruses in streams is largely unexplored, they potentially can alter C flow through the microbial loop by attacking bacteria and other organisms. Clearly, in other aquatic environments bacteriophage can be an important component of the microbial loop, and the abundances of viruses observed in lotic ecosystems suggests that this is also the case in flowing waters.

N cycle

One characteristic of nutrient relations in streams is the spiraling manner with which nutrients as well as bacterial cells are transported (Figure 7). The nutrient spiraling concept describes the process by which an atom that is released into the water column travels some distance downstream before being taken up again. The spiraling length varies depending on many stream properties such as retentiveness of the channel and water velocity. Transport of nutrients in this manner is experimentally evaluated by measuring the uptake length of material released into the water column.

In addition to biological processes, geochemical processes impact N dynamics in lotic ecosystems. Reactive sites in the benthos can result in chemical transformation and sorption of nutrients. This impacts nutrient availability to the biota, as well as transport downstream.

Nitrogen, in particular, has been the subject of numerous studies in stream ecology because of its potential to serve as a limiting nutrient, influxes of N from fertilizer, and role in water quality. The quantity of N is highly variable among

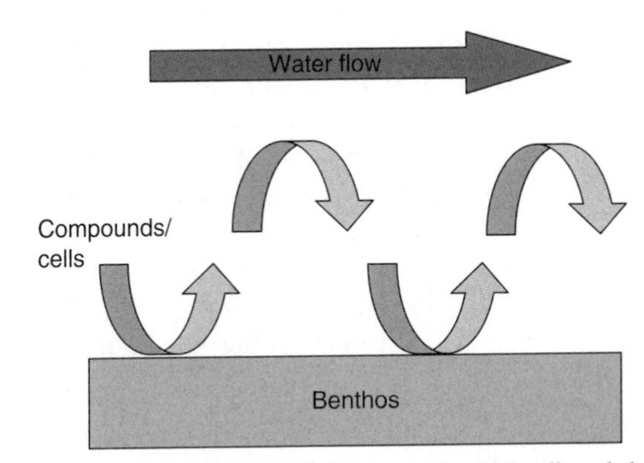

FIGURE 7 Schematic illustrating the manner in which cells and chemical compounds are transported some distance downstream before uptake into the benthos (via attachment, sorption, biological assimilation, etc.). Materials are subsequently released creating a spiral-like pattern.

streams based on influxes from outside sources and biological processes in the riparian zone. Both assimilatory and dissimilatory reactions occur in the N cycle in streams and the role of particular N transformations is variable. Both inorganic and organic N are important N-cycle participants, although most emphasis has been placed on inorganic N.

Sorption of ammonium to sediments can be high and, correspondingly, ammonium concentrations in water of rivers and streams tend to be low. The benthos create circumstances in which ammonification, denitrification, and nitrification can take place within a comparatively narrow fraction of the benthos and there can be steep redox potential gradients (Figure 8). Depending on the redox potential, the hyporheos can be potent sites of denitrification. Denitrification in the soil of the riparian zone can also contribute to the stream N cycle.

N enters streams from both upstream, terrestrial, and groundwater sources where it is subjected to various microbial transformations (Figure 9). Processes are dominated by benthic components and are highly dependent on the oxygen gradient within the sediment.

P cycle

Phosphorous in aquatic ecosystems is often a limiting nutrient and the form of P greatly impacts biological activity. Often soluble reactive phosphate is measured and represents a readily assimilated form; dissolved organic P is also present but is used more slowly. Sorption and desorption of phosphate to sediments (particularly fine particles) act to control the concentration of inorganic P in the water in streams and rivers. In addition, P concentrations can increase greatly after flooding, especially with input from runoff or sewage.

Microbial Communities

Microbial communities in streams are a complex mixture of cells from allochthonous locations and those produced within the stream proper. Distinguishing among these components is quite problematic and the relative contribution of the fractions depends greatly on local events, such as rainfall, flooding, and so on.

Biofilms with complex three-dimensional structures form on stream surfaces including wood, leaves, rocks, and smaller particles. These biofilms are critical habitats for microbial production in lotic ecosystems and consist of complex mixtures of cells of different types embedded in a matrix of extracellular polysaccharide. The matrix is interspersed with channels serving as transport venues and plays host to a mixture of autotrophic and heterotrophic microorganisms. Water velocity is a major determinant of biofilm thickness in streams and physical disruption can "reset" the community by clearing the surface of many of biofilm residents.

The nature and activity of the biofilm may vary depending on substrate type and various terms are applied to biofilms on different surfaces, such as epilithon for biofilms on rocks, epiphyton for biofilms on macrophytes, and epixylon for biofilms on wood. Within the biofilm, various niches occur and there are opportunities for many types of biotic interactions, such as predation and competition. The biofilm creates

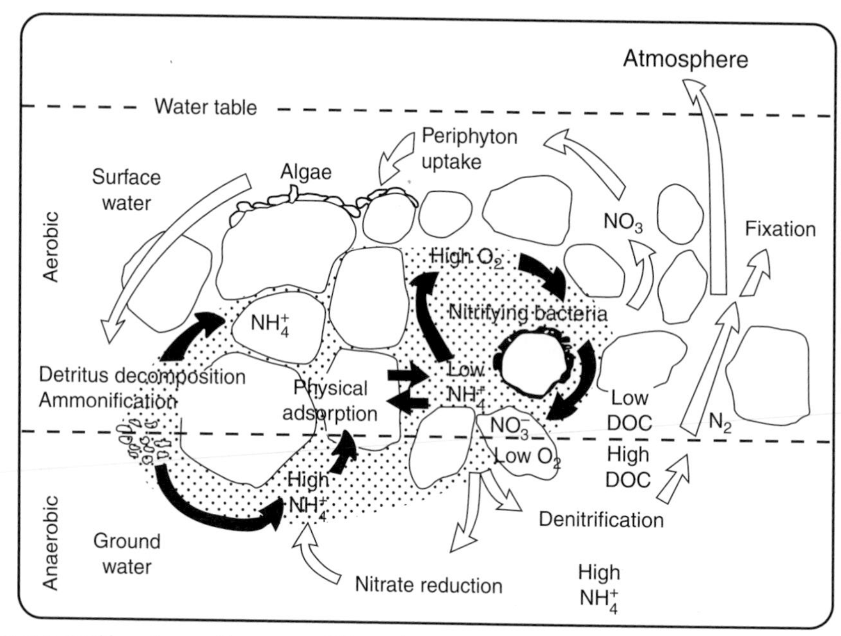

FIGURE 8 Interplay between aerobic and anaerobic zones and N transformations in the benthos of a stream. *Reproduced from Triska, F. J., Jackamn, A. P., Duff, J. H. & Avanzino, R. J. (1994). Ammonium sorption to channel and riparian sediments: A transient storage pool for dissolved inorganic nitrogen.* Biogeochemistry, *26, 67–83.*

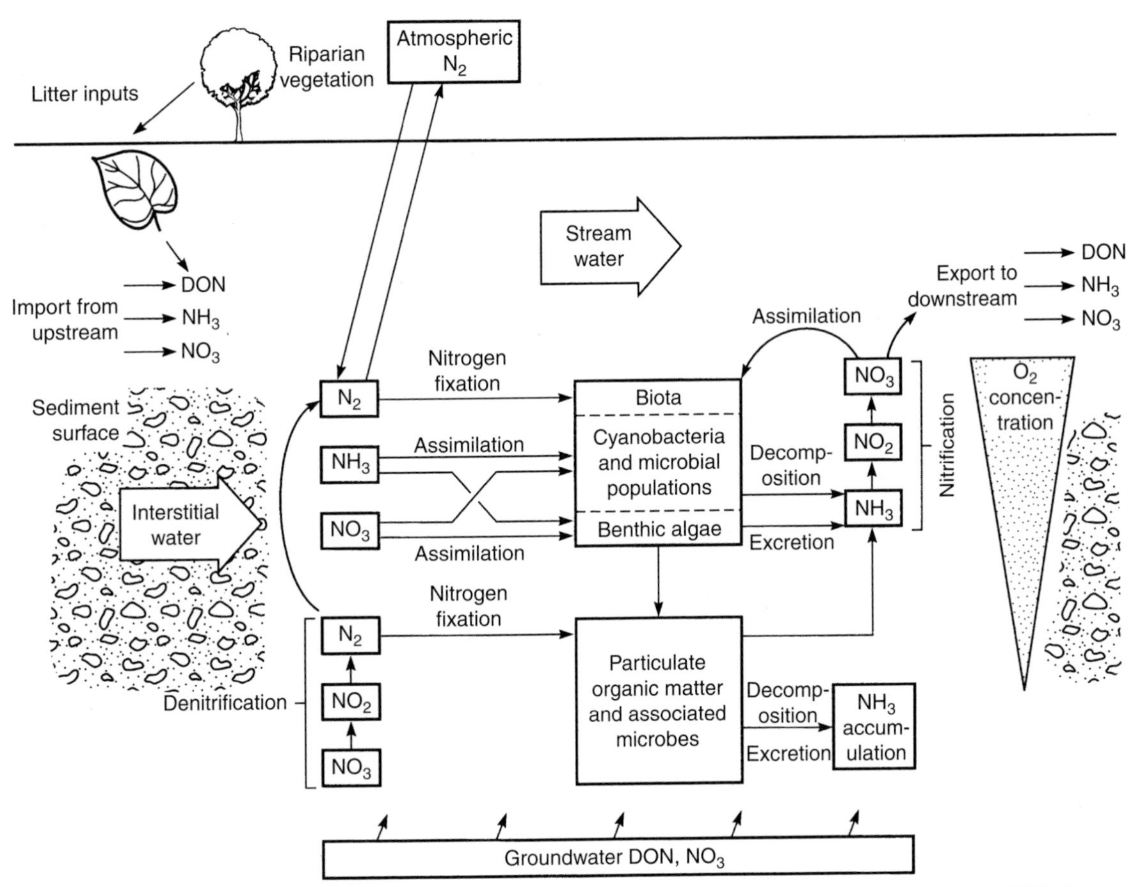

FIGURE 9 Schematic of the nitrogen cycle in a stream; DON, dissolved organic nitrogen. *Reproduced from Allan, J. D. (1995).* Stream ecology: Structure and function of running waters. *London: Chapman and Hall.*

an opportunity for cells to maintain position in the flowing water and to accumulate resources in transport. Of particular importance is the ability of the biofilm to aid in retention of enzymes released from the cells to degrade polymers.

Activity of enzymes in biofilms and other habitats, including phenol oxidase β-D-glucosidase phosphatases, endo-cellulase, cellobiohydrolase, and aminopeptidase, has often been measured to assess the potential for enzymatic degradation. Typically, assays rely on chromogenic or fluorogenic substrates. Although these methods do not allow the researcher to determine the organism that produced the enzyme, they do provide measures of specific activity and degradative potentials.

Like nutrients, microbial cells are also transported along the length of the stream; this process, called information spiraling, builds on the nutrient spiraling concept described above. A variety of processes including the activities of animals, physical disruption, sloughing, and natural dispersal properties of the species influence the release of cells from surfaces. The uptake length of bacterial cells has been measured in streams through experimental addition of bacteria.

Bacterial communities in streams, as in lakes, are numerically dominated by *Proteobacteria*, particularly the β-*Proteobacteria*. Although the diversity of bacteria in streams and rivers is less studied than marine or lake environments, we do know that the *Proteobacteria*, *Actinobacteria*, and *Cytophaga–Flavobacterium* cluster are common in lotic ecosystems. Some bacterial groups may perform specialized roles in nutrient cycling or in degradation of particular organic compounds. For example, the actinomycetes participate in the decomposition of leaf material in streams along with other types of bacteria and the hyphomycetes.

Although there are increasingly good data that describe the bacterial community composition in streams, in which nonculture-based approaches replace the more traditional culture-dependent methods, there are still limitations in our ability to connect structure and function. In streams and rivers (as well as other freshwater habitats), the number of bacterial cells in the community that are active (such as those that are undergoing respiration) is often a small percentage of the total number of cells. This may be a particularly important component in lotic ecosystems because of the inputs of large numbers of allochthonous cells. Although these cells do not contribute necessarily to ecological functions, they are still detected via many of the widely used methods.

The contribution of archaea to lotic ecosystems is largely unknown beyond the role of methanogens. Relative to bacteria and other systems, there is very limited information about the abundance, diversity, and activity of archaea in streams and rivers.

Fungal communities in streams, as well as other freshwater environments, are dominated by aquatic hyphomycetes, which are identified based on the structure of their conidia. Hyphomycetes are not a specific taxonomic group, rather they are a heterogeneous group of fungi. Hyphomycetes are one type of microsporic fungi and produce conidia directly from hyphae or via conidiophores. The group can be further subdivided and, in streams, much focus is on the so-called aquatic hyphomycetes, a polyphyletic group, that play an established role in the decomposition of leaf material. Common genera of hyphomycetes found include *Anguillospora*, *Tetrachaetum*, *Tetracladium*, *Clavatospora*, *Goniopila*, *Helicromyces*, *Lemonniera*, and *Heliscella*.

Algal communities in streams can be light-limited (either by shading of riparian vegetation or by turbidity) and nutrient-limited and their contribution to the C in streams varies based on these limitations. Attachment to surfaces is another critical need as phytoplankton generally develops only in large rivers. Diatoms are both diverse and abundant in streams particularly attached to rocks and cobbles and are generally the most abundant algae in stream biofilms. The distribution of diatom species varies predictably among locations with different water velocities. In addition, the growth form of filamentous green algae (i.e., *Cladophora*) varies with velocity.

Algal stream communities are greatly impacted by light, temperature, herbivory, water velocity, substrate type, and nutrient availability. This latter includes silica, which is essential for diatoms; however, in lotic ecosystems, silica is generally found in adequate supplies. Diatoms, which include genera such as *Achnanthes*, *Gomphonema*, and *Navicula*, can be more successful under low light conditions than other algae; this is a likely contributor to their success in streams. In terms of nutrient limitations, autotrophs in streams are generally thought to be P limited not N (although exceptions to this trend occur).

Protozoa in streams are in many ways less well studied than in lakes. It is known that protozoa play a major role in the microbial loop in streams. Small flagellates are considered to be the most important consumers among the protozoa in streams and can greatly alter bacterial abundance.

Anthropogenic Disturbance

Human impact on streams is profound and has greatly altered the structure and function of streams and rivers globally. These impacts in turn impact the role of the microbial community and essential biogeochemical processes.

Introduction of sewage effluent and fecal waste from human/animal activities into lotic systems is widespread and serves as a source of microorganisms, nutrients, organic compounds, as well as various pollutants used by humans and disposed of in sewage. The input of allochthonous microorganisms in this manner has critical impacts on water quality, human health, and the spread of water-borne diseases. Inputs of organic compounds and nutrients from sewage alter the resource availability to the biota and can, in some cases, result in reduction of oxygen concentrations in the water. Often these inputs are monitored using biological indicators of sewage contamination such as enumerating fecal coliforms. In addition, other fecal-associated organisms and viruses may be monitored to assess different types of risks, like *Cryptosporidium* (from cattle feces) or fecal streptococci (Gram-positive cocci) from human sewage effluent. To reduce the risk of contamination of water by pathogens, treatment of effluent is often carried out, in particular using chlorination. Chlorination does have associated health risks through the production of trichloromethanes from natural-occurring dissolved organic compounds and alternative strategies, and dechlorination can be used to minimize these risks.

Humans also alter the physical structure of streams through channelization for navigation purposes, removal of woody debris, straightening of channels and removing of meanders, dam and reservoir construction, and so on. Such impacts have gone on for many years and thus have had long-term consequences on stream and river function. In particular, water residence time is altered by these activities, which, in turn, alters transport of materials and uptake lengths. For some alterations, such as addition of reservoirs, residence time is greatly increased creating unexpected lentic regions in the middle of otherwise "normal" river continua. At these discontinuities, material in transport can drop out of the water column and pelagic processes can replace benthic processes in relative importance.

Acid mine drainage has a specific impact on streams and has huge impacts on the microbial ecology of the system. The problems start with the mining of coal that is mixed with pyrite. Iron and sulfur oxidation leads to a decrease in pH and favors the growth of specific types of microorganisms, which further this oxygen and maintain the low pH. This greatly lower pH has profound negative impacts on all other forms of stream life.

COMPARISON AMONG FRESHWATER HABITATS

Hydrology, the potential for oxygen depletion, whether organic C is primarily of allochthonous or autochthonous origin, and the role of plant detritus are among the major determinants of variations among freshwater ecosystems in microbial processes. These variations coupled with the relatively limited number of studies on some aspects of ecology of microorganisms in freshwater habitats limits the number of generalities that have emerged.

A critical limitation on our ability to draw generalities about microbial ecology of freshwater habitats is the relatively limited number of studies performed on certain aspects of the topic. This fact, coupled with the limited coverage of the broad range of diversity of ecosystems, suggests that we have yet to reveal many of the underlying tendencies that may unite freshwater habitats. It is not known whether specific locations that have been studied in some detail are representative of the greater array of locations that have never been studied.

Overall, major determinants of the structure and function of the microbial community in freshwater habitats are the occurrence and importance of plant detritus, the nature of the hydrology of the system, and the occurrence of oxic and anoxic zones. The potential for drastic spatial and temporal changes in critical factors, such as redox potential, creates opportunities for a diversity of different microbial niches and interconnectivities. This creates interesting spatiotemporal juxtapositions of complementary microbial processes over comparatively small scales. Each microbe-mediated process can be altered by anthropogenic disturbances, which are all too common in freshwater systems because of the proximity to human populations, strong connections to the surrounding landscape, and vital importance for human success.

RECENT DEVELOPMENTS

The central importance of bacteria, fungi, and other microorganisms to biogeochemical cycles in freshwater systems is clear. There is also tremendous variability among freshwater systems and the diversity of microorganisms in these systems is still relatively understudied. Temporal changes in microbial community composition and diversity can be large and freshwater environments play host to communities of bacteria that are, in many ways, distinct. The number and diversity of freshwater systems that have been investigated continues to grow. Utilization of new methodological approaches is one driver of the expanding knowledge base.

As in other fields of microbial ecology, methodological advances, particularly those that facilitate linkage of community structure and function continue to have an impact on our understanding of freshwater systems. Particular advances have been made in the linkage in structure and function as it relates to key aspects of the nitrogen cycle. In addition, many microbial ecology concepts (such as those related to community assembly) and topics (like metagenomics) that have previously been only occasionally applied to freshwater systems are becoming more established although still dramatically understudied compared to marine systems. Functional genes, such as those involved in nitrogen fixation, denitrification and nitrification, have been studied with increasing frequency both in terms of diversity (e.g., based on terminal fragment length polymorphisms) and quantity (based on quantitative PCR). Some studies of microbial gene expression (such as expression of *nif* genes) in freshwater systems have also been conducted.

Advances in freshwater ecology also have the potential to enhance our understanding of their role as habitats for microorganisms. More specifically the use of embedded wireless sensors has the ability to measure and transmit critical environmental data, like temperature and dissolved oxygen, in real time. Buoy-based sensor platforms placed in freshwater systems (or even mobile platforms) are becoming an increasingly important tool for monitoring of environmental conditions; real time monitoring has become increasingly prevalent in marine systems over the past several years.

Beyond this capacity to measure physical and chemical conditions of the environment, sensor technology can also be used to detect specific microorganisms or their products. For example, there are sensors for fluorometric detection of specific pigments of cyanobacteria, such as phycocyanin, which have application in monitoring of harmful algal blooms as well as those for more general compounds, like chlorophyll. There is also technology now that allows detection of specific bacteria (such as *Escherichia coli*) and may rely on antibody based detection, for example. The next generation of biosensors which is just now being developed for and applied to freshwater systems are nucleic acid-based systems which can be targeted to detect a variety of bacteria and viruses.

Although much focus is placed on microbial ecology in "natural" freshwater systems, be they pristine or impacted to a degree by humans, there is a much needed growing exploration of other systems. Most notable are those in urban areas as well as freshwater systems that are being restored or remediated. Human impacts on freshwater systems are profound and part of the process of restoring natural functions involves creation or alteration of such systems. Such modifications include creation of artificial wetlands as a response to wetland loss or restoration of streams (e.g., morphological modifications). In streams, management strategies may focus on achieving specific biological metrics (as related to fish communities, for example) or physical characteristics of the stream channel. The effects of such management on microbial community function and structure is largely unknown. Generally, study of the structure/function of the microbial community and biogeochemistry in created or restored freshwater systems is still in its infancy.

FURTHER READING

Allan, J. D. (1995). *Stream ecology: Structure and function of running waters*. London: Chapman and Hall.

Bodelier, P. L. E., Frenzel, P., Drake, H. L., *et al.* (2006). Ecological aspects of microbes and microbial communities inhabiting the rhizosphere of wetland plants. In J. T. A. Verhoeven, B. Beltman, R. Booink, & D. F. Whigham (Eds.) *Wetlands and natural resource management*. Berlin: Springer-Verlag.

Findlay, S. (2010). Stream microbial ecology. *Journal of the North American Benthological Society*, *29*, 170–181.

Leff, L. G. (2002). Stream microbiology. G. Bitton (Ed.) *Encyclopedia of environmental microbiology*. New York: Wiley and Sons.

Logue, J. B., Burgmann, H., & Robinson, C. T. (2008). Progress in the ecological genetics and biodiversity of freshwater bacteria. *Bioscience*, *58*, 103–113.

Mitsch, W. J., & Gosselink, J. G. (2000). *Wetlands* (3rd ed.). New York: Wiley and Sons.

Newbold, J. D., Elwood, J. W., O'Neil, R. V., & Van Winkle, W. (1981). Measuring nutrient spiraling in streams. *Canadian Journal of Fisheries and Aquatic Sciences*, *38*, 860–863.

Sigee, D. C. (2005). *Freshwater microbiology*. New York: Wiley and Sons.

Vannote, R. L., Minshall, G. W., Cummins, K. W., Sedell, J. R., & Cushing, C. E. (1980). The river continuum concept. *Canadian Journal of Fisheries and Aquatic Sciences*, *37*, 130–137.

Wetzel, R. G. (2001). *Limnology* (3rd ed.). San Diego: Academic Press.

Zwart, G., Crump, B. C., Kamst-van Agterveld, M. P., Hagen, F., & Han, S. K. (2002). Typical freshwater bacteria: An analysis of available 16S rRNA gene sequences from plankton of lakes and rivers. *Aquatic Microbial Ecology*, *28*, 141–155.

RELEVANT WEBSITES

www.gleon.org
www.lakescientist.com
http://drosophila.biology.kent.edu/IGERT/
http://drosophila.biology.kent.edu/users/lleff/

Chapter 30

Low-Nutrient Environments

J.S. Poindexter
Barnard College, Columbia University, NY, USA

Chapter Outline

ABBREVIATIONS

AODC Acridine orange direct count
Bchl *a* Bacteriochlorophyll *a*
chl *a* Chlorophyll *a*
CFU Colony-forming unit
DAPI 4',6-Diamidino-2-phenylindole
DO Dissolved O_2
DOC Dissolved organic carbon
DOM Dissolved organic matter
NUCC Nucleoid-containing cell
POM Particulate organic matter

DEFINING STATEMENT

Oligotrophic aquatic habitats are defined as low-nutrient-flux environments in which both primary production and microbial densities are distinctly lower than in eutrophic, heavily nourished environments. Progress in design of conditions for laboratory cultivation of oligotrophic bacteria is summarized, and characteristics and identities of some frequently cultivated oligotrophic bacteria are described.

OLIGOTROPHIC AQUATIC HABITATS

The prokaryotic microbes, currently classified as the two domains *Bacteria* and *Archaea*, are the most ubiquitous as well as the oldest forms of life on earth. At present, it is safe to generalize regarding any locale that if it is wet, it is a microbial habitat; that the microbes present will be predominantly or exclusively prokaryotic; and that if the habitat is aerated, those prokaryotic microbes will be predominantly bacteria. Their presence may not be apparent to the casual observer – even to trained ecologists – because the prokaryotic populations of microbial habitats are often so sparse that detection of their presence requires laboratory-assisted techniques. Bacterial populations can be restricted to inconspicuous densities by one or the other of two major causes: either the environmental supply of nutrients is too low to support a dense community, or ambient abiotic conditions are beyond the extremes tolerated by most forms of life. Prokaryotic microbes that thrive in inhospitable, extreme habitats – especially archaea such as the extreme halophiles, thermophiles, and acidophiles – are known to possess unique properties with respect to their membranes, their metabolism, and even their genomes. In contrast, prokaryotic microbes known to occur particularly in communities that are sparse due to nutrient scarcity – oligotrophic microbes – appear to be similar to many other forms of life in their fundamental structural, metabolic, and genetic properties. Because bacteria constitute the great majority of oligotrophic prokaryotic microbes that have been studied, practically all references in this article will be to bacteria, on the understanding that the bacteriological work may or may not eventually prove a helpful guide to the elucidation of the physiology and ecology of oligotrophic archaea.

Parameters of Oligotrophic Waters

With respect to environmental nutrient supply, microbial ecology borrows terms that originated for the classification of freshwater lakes and were promptly extended to marine environments. Extending these terms to soils requires some redefining that will not be done here because this article will emphasize aquatic (freshwater and marine) low-nutrient supply environments. The widely used terms are oligotrophic and eutrophic for the scarce and abundant, respectively, ends of the spectrum of nutrient supply to watery habitats; "mesotrophic" is used for the continuum of intermediate conditions. Generalized ranges for environmental parameters that define these classes are shown in Table 1. (The terms in Table 1 refer to nutritional conditions in the environment. See Table 2 for uses of the terms "oligotrophic" and "copiotrophic" to refer to categories of microbes distinguished by competitiveness in low- and high-nutrient-flux environments, respectively.) The numbers given are averages of values that vary from one site to another within a body of water and may also vary with season of the year.

The most important aspect of lake nutrition is nutrient flux (reflected in C flux, turnover time, and productivity in Table 1), not standing concentrations of nutrients. The concept of a low-nutrient environment should be dynamic: it is the flux, the rate of delivery of utilizable nutrients, that defines a low-nutrient environment, and oligotrophic conditions should be described as low-nutrient flux. A definition based on standing concentrations would not provide a useful or dependable distinction between oligotrophic and eutrophic waters because, except within the bodies of living organisms where constant concentrations and fluxes of soluble nutrients are maintained, microbial habitats generally contain very low concentrations of soluble organic materials. Any lively microbial population will consume available nutrients until the concentration of at least one nutrient (the limiting nutrient) is reduced to a level so low that the organisms cannot detect and transport it for use in metabolism. That nutrient, which may be organic or inorganic, will often also be undetectable by chemical analysis unless large volumes of water are concentrated or extracted for assay.

Nevertheless, it is also evident in Table 1 that two standing concentrations can serve as convenient indicators of oligotrophic/low-nutrient supply conditions: the relative abundance of dissolved O_2 (DO) and the density of demonstrably living bacteria. These two parameters are inversely related to each other because wherever bacteria occur in abundance with adequate organic nutrients to support their respiratory metabolism, oxygen will be consumed to well below its saturation in water at the ambient temperature. With few exceptions, subsaturation levels of DO in natural waters – fresh and marine – are the result of microbial respiration, almost always using organic material as electron donors. Oligotrophy, as a nutritional state, is predictably an aerobic condition.

To distinguish oligotrophic and eutrophic habitats from each other, it is also possible to perform simple organoleptic tests. Oligotrophic waters, whether fresh or marine, do not stink (in human estimation); are optically clear, not cloudy (see Secchi depth, Table 1); and have rocky or sandy, not muddy or mucky, bottoms. Oligotrophic waters are generally at least 15 m deep; on a sunny day appear blue, not green or brown; and harbor a diversity of deep-water animals, including fish, but few if any rooted plants. Microscopical examination of a sample of such water at 100× or higher will reveal very little evidence of life. In contrast, relatively shallow, cloudy, greenish, plant-inhabited, muddy-bottomed waters low in fish diversity are predictably sites supplied with an abundance of nutrients; they are recognizably eutrophic. Microscopical examination of samples of such water at 100× or higher will provide hours of informative observation of the forms and activities of diverse organisms, from bacteria to microfauna and the propagules of plants, animals, algae, and fungi; this (particularly the protozoa) is the entertaining pond water of biology classes.

Nutrient Delivery

There are two general sources of nutrients in aquatic habitats (as in soils): autochthonous (generated within the habitat) and allochthonous (generated elsewhere and transported to the habitat).

As throughout the biosphere, the principal autochthonous source of organic nutrients in aquatic habitats is biotic: photosynthetic CO_2 fixation (primary production) by photoautotrophs – cyanobacteria and algae in deep waters, joined by plants in shallow freshwaters and along the seashore. Because of the central role of primary production in the life of any aquatic habitat, the parameter in Table 1 that best indicates the present level of biotic activity of an aquatic habitat is the concentration of chlorophyll *a* (chl *a*).

Allochthonous nutrients

Major allochthonous sources include the atmosphere, volcanic activity, terrestrial surroundings, and connecting waters. Anthropogenic allochthonous nutrients are delivered as domestic sewage, industrial waste, agricultural runoff, combustion products in the atmosphere, ships, leaching from landfills, and other human discards. The quantity and identity of nutrients from allochthonous sources are strongly influenced by the regional geology, the weather, and the types, density, and activities of the biota. Nutrient delivery by all means and from all sources tends to be discontinuous, and delivery events (surges) vary in frequency and in quantity of nutrients delivered per surge. A nutrient surge will have a detectable effect on the habitat if it significantly relieves limitation by the limiting nutrient, a consequence that is more likely to perturb the habitat when the surge is allochthonous rather than autochthonous.

TABLE 1 Characteristics of aquatic habitats

Habitat	C flux (μg C l^{-1} day^{-1})	Turnover Time (Corg)	Productivity (μg C m^{-2} y^{-1})	Chl a (μg l^{-1})	DOC[a] (μg C l^{-1})	Phosphate (μg P l^{-1})	Total *N* (μg N l^{-1})	Bacteria (CFU ml^{-1})	O_2, Deep Saturation	Turbidity (Secchi, m)	Bottom
Freshwater lakes											
Oligotrophic	10–100	150–1200 h	>150	>3	>0.5	>10	>200	10^2–10^5	>50%	>5	Sandy
Mesotrophic (LW 1971–79[b])				2–7		7–10	380–480				Muddy
Eutrophic(LW 1963–66[b])				24–41		30–35	580–750				Variable
Eutrophic	>100–1600	0.2–168 h	>250	>6	65–950	>20	>500	10^6–10^9	0–10%	>3	Muddy
Marine regions											
Estuarine	3–150	1.5–45		3–40	3–900	1–20					
Coastal	3–150	0.1–2688 h	230–800	2–40	6–450	3–25					
Oceanic	3–150	1 h–6000 y	0.05–50			0.5–100	14–200	10^2–10^6			Sandy
• Surface					0.2–2000						
• Deep					300–600				50–75%		

[a] *DOC is dissolved organic carbon.*
[b] *LW is Lake Washington, Seattle, WA, USA; 1963–1966 was a period of eutrophication, and 1971–1979 were years following recovery of mesotrophic conditions.*

TABLE 2 Correspondence between environmental nutrient flux and probable predominant bacteria

Environment	Nutrient Flux	Bacteria Favored
Oligotrophic	Very low, sporadic, infrequent	Substrate-sensitive oligotrophs
	Very low, frequent or constant	Oligotrophs
Mesotrophic	Low, sporadic, infrequent	Oligotrophs, starvation-resistant copiotrophs
	Low, frequent or constant	Oligotrophs, copiotrophs
Eutrophic	High, sporadic, frequent	Starvation-resistant copiotrophs, spore-formers
	High, constant	Host-independent copiotrophs, parasites

In environments that appear oligotrophic year-round, natural surges are small and infrequent, and it is reasonable to include in the description of oligotrophic waters that blooms of aquatic life are rare, whereas they are characteristic, frequent, and substantial in eutrophic waters. However, the lower the standing concentrations of nutrients and of microbes, the greater the susceptibility of a body of water to disturbance by nutrient surges, and oligotrophic waters are subject to eutrophication. Natural eutrophication ultimately converts a freshwater lake into a marsh, but the conversion may require centuries. Eutrophication is likely to be greatly accelerated when the cause is anthropogenic (cultural eutrophication). However, large bodies of water are resilient; cessation and prevention of large and frequent man-made surges can lead to the restoration of oligotrophy, or at least recovery to mesotrophic condition. This sequence of anthropogenic eutrophication and recovery through protection against nutrient surges has been experienced by freshwater lakes in North America, including both small lakes such as Lake Washington in Seattle (see Table 1) and great lakes such as Eastern Lake Erie and Lake Ontario.

In microbial ecology, it is important to recognize that allochthonous surges are not sterile. Whether the surge arrives from the atmosphere (in rain, snow or dry precipitation, i.e., dust), the embankment soil, a river, or in the form of excreta, shed parts, or decaying carcasses of plants and animals, a bacterial population arrives with it. In most instances, those populations will already be adapted to and feeding on the nutrients with which they travel, and they will continue to multiply in the oligotrophic environment until nutrient concentrations drop to uncapturable levels due to consumption by the introduced microbes, probably joined by the resident microbes. Commensally or competitively, the residents and the immigrants will consume useable nutrients, and the oligotrophs among them will scavenge material that escapes the sensing and transport affinities of copiotrophs present (see text below and Table 2).

Autochthonous nutrients

Primary production, the number one autochthonous source of organic matter, is typically limited in aquatic habitats by light (intensity and duration each day) and by availability of inorganic phosphate; other inorganic nutrients may limit Corg synthesis under some conditions (e.g., Si limitation of diatoms, or nitrate limitation of algae as a consequence of phosphate pollution). Atmospheric nitrogen (N_2) can be fixed into chemical combination autochthonously by both phototrophic and heterotrophic bacteria. Because diazotrophic (N_2-fixing) bacteria are ubiquitous, this process seems to account for photosynthesis being limited more commonly by the availability of P than of N. Most of the exceptional situations are areas of phosphate pollution by humans that makes P abundant relative to N, thereby stimulating N_2 fixation. In the ocean, where the principal diazotrophs are cyanobacteria that use light energy for this process, most of the N_2 fixation is light dependent in time (daytime) and space (photic zone).

There are three other major autochthonous sources of nutrients in aquatic environments. (1) One source is release of soluble substances from aggregates of complex organic substances that are only slowly degraded and solubilized by microbes, and to some extent by animals and protozoa. The slowly utilized substances are regarded as refractory to biodegradation; they may persist for decades in a habitat because of their chemistry, which is often the case with synthetic organic compounds. They may also persist because of their physical state, particularly insolubility or aggregation, and microbial consumption may be limited by the rate of solubilization by enzymes or by weathering. Soluble, dispersed products are considered labile because they are readily consumed by heterotrophic bacteria. (2) Major inorganic autochthonous nutrients commonly cycle seasonally in freshwater. Spring thaw releases useable, soluble forms of nutrients (notably phosphate, calcium salts, and silicates) from the insoluble forms in which they are sequestered during the winter in the sediment. (3) In the ocean, the reaction of seawater with minerals carried in volcanic emissions or exposed as basalt by seabed movement resulting from seismic events releases significant amounts of K, Ca, and Si (if present), which thereby become available to microbes – including the primary producers, cyanobacteria, and algae.

Autochthonous deliveries of nutrients are relatively predictable and steady, lasting throughout certain seasons of the year, yet seasonal swings of the trophic state of aquatic habitats occur. They are most obvious in temperate-zone freshwater lakes, but are also observable in estuaries. Typically, a spring surge of primary production in a lake is initiated by increased duration of daylight, rising atmospheric and water temperature, and – probably most influential – release of inorganic nutrients from winter sequestration. As surface water warms, it sinks and vertical mixing of water and upward redistribution of nutrients from the lake bottom occur. When atmospheric temperature stabilizes during the summer, the lake stratifies, with the surface warm, oxygenated, and illuminated; the bottom relatively cold and anoxic; and a transition layer between them. These layers are the epilimnion, the hypolimnion, and the thermocline, respectively. The springtime changes stimulate primary production that converts CO_2 into particulate organic matter (POM), the cells of phototrophs. Some dissolved organic matter (DOM) arises as exudates from the phototrophs, but most of it appears late in the spring. Its appearance is attributed to activities of protozoa and animals that feed on phototrophs, some of it by sloppy feeding that casts away uningested parts, but most of it as metabolic and digestive excreta of the grazers. In turn, their bodies are nutrients for the next level of feeders in a food chain. At every level, fungi and heterotrophic bacteria absorb DOM, and some of them can solubilize complex organic matter and then absorb the organic nutrients. Ultimately, respiration returns the carbon to the environment as CO_2, completing the production ? mineralization cycle.

Over the season from spring thaw to autumnal decrease in photosynthesis, the lake experiences its major annual surge of organic nutrients, and when primary production is high, the oligotrophic conditions of winter may alternate with meso- or eutrophic conditions in the summer. A second autumnal surge may occur when the biomass accumulated during the spring and summer begins to decline in activity, making it susceptible to disintegration and consumption by heterotrophic bacteria. As surface water (ice) warms, reversing the relative density of the liquid layers, a second period of vertical mixing occurs in dimictic lakes.

Changes that occur according to season, such as fluctuations in temperature, illumination, and sequestering of inorganic nutrients, also accompany deep-ocean currents and disturbances such as storms or floods or exceptional tides, all of which translocate both organic and inorganic nutrients – and, of course, microbes. Consequently, these changes influence local primary production and account for the regular seasonal fluctuations in the chemical and biotic parameters of natural waters.

Transport routes and vehicles

Most allochthonous nutrients are delivered to aquatic habitats in moving water (springs, streams, rivers, tides, sewage effluents, precipitation as rain or snow), although some arrive as dry precipitation (atmospheric dust) or solid masses (falling trees, animal carcasses, discarded shells, sinking ships, plastic trash, etc.). The quantity and identity of the nutrients borne by a river into a lake or estuary depend on natural productivity and population densities in the river itself, on biotic communities along the shore, and on the mineral composition of the local soil and the river bottom. The weather and the contour of the region determine the volume and velocity of the running water, which together determine the rate of delivery of the useable nutrients carried by the water.

Dissolved organic carbon (DOC) is carried by rivers predominantly as small molecules: monomeric sugars and amino acids; small, soluble polymers of these compounds; low-molecular-weight by-products of photosynthesis such as glycolate; and smaller amounts of other nonnitrogenous carboxylic acids. In contrast to compounds such as alcohols, esters, long-chain fatty acids, and hydrocarbons, most of these organic nutrients are not volatile and are not borne by rain. They are readily consumed by heterotrophic microbes during their transport along the river's course so that rivers do not carry substantial amounts of readily metabolizable (labile) organic carbon from lake to lake or from land to the ocean.

Rivers do deliver significant amounts of some readily assimilated inorganic nutrients to lakes and estuaries. Riverwater addition to the seas is a major source of inorganic forms of silicates, sulfate, chloride, sodium, Mg, K, Ca, and bicarbonate. Rain naturally contributes sulfur in various oxidation states, and pollution increases delivery – in rain and/or in rivers – of phosphates, nitrate, ammonium, sulfate, chloride, sodium, and CO_2. Respiration exhales CO_2 into the atmosphere, where the amount of CO_2 was stable for long periods in the past, prior to the Industrial Revolution. Even though atmospheric CO_2 has probably never been the limiting nutrient for primary production, its recent increase (surge) – accelerated by the burning of fossil fuels – is having a significant impact on the ocean. Seawater is mildly alkaline, and its absorption of CO_2 is causing measurable acidification of the open ocean and potential solubilization of the $CaCO_3$ foundation of the shells of marine animals such as corals and of the skeletons of coralline algae.

A river may be as deep as a lake at some points along its course, but mixing – both vertically and horizontally – will be greater and less seasonal than in a lake; lotic (running) waters are therefore less likely to have anoxic regions. However, heavy precipitation that causes a river to overflow its banks, reduction of the river's flow due to withdrawal of large volumes of water for domestic use, or treefalls or dam construction that impedes a river's flow may cause organic matter accumulation and O_2 depletion. Unimpeded rivers are capable of self-cleansing through microbial activities and oxygenation so that assimilable inorganic nutrients added at one point along the river's course are typically consumed within the river and built into biomass. Delivery of nitrate and phosphate to the ocean (or to lakes on the way to the ocean) is low because the salts are assimilated by microorganisms during the river's flow, and most of the N and P that arrives at the seashore is in the form of suspended, relatively refractory POM. When the rivers are polluted, waters downstream from urban areas where human population density is high carry low (>16, atom:atom) N:P ratios, resulting in N limitation in the target estuaries. On the other hand, rivers draining from agricultural regions, particularly where manure is used as fertilizer, carry high levels of ammonium and nitrate. Their N:P ratios are high, and the estuaries into which they flow tend to be P limited. In general, rivers are potential low-nutrient environments and could be sources of bacteria that thrive in such habitats. However, they are subject to more frequent and heavier surges of nutrients than most lakes, and are also the least-studied aquatic habitats with respect to their oligotrophic inhabitants.

Gulfs, bays, bights, marine coastal regions, and estuaries are typically regions of significantly higher nutrient fluxes (per unit volume) than the vast open ocean, and standing concentrations of solutes and of microbes in shore waters fluctuate over wider ranges. Shore waters receive allochthonous nutrients from all directions – rivers flowing into them, clashes of atmospheric temperatures that promote rainfall, and sloshing of the ocean with the tidal cycles. In the near-shore waters, ocean currents colliding with the underwater sides of islands and continents are pushed upward in the phenomenon of upwelling, which delivers nutrients from the ocean bottom and deep currents. Far from shore, vertical translocation of organic nutrients is predominantly downward. Throughout the marine environment, nutrients move due to upwelling, downwelling, currents, waves, and storms. In short, any physical force that moves water moves the nutrients contained in that water, whether dissolved or suspended – and the microbes.

Primary production in most of the ocean is measurable only in the upper 200 m, the illuminated (photic) zone, and the biomass created there is – as in lakes – initially in the form of POM, much of which sinks. As the biomass enters food chains, the more refractory material accumulates in fecal pellets, which also sink. Deep currents translocate nutrients horizontally along the ocean bottom, but typically at very low concentrations. Superficial currents, winds, and storms redistribute nutrients horizontally from shores and from regions of upwelling into and within the photic zone of the open waters. Although nutrient concentrations far from shore are significantly lower than in most freshwater habitats, the volumes are vast (total ocean volume is approximately $1.35 \times 10^{18}\,m^3$ ($1.35 \times 10^{21}\,l$)), and the total amount of nutrients in waters carried up and down and around the oceans is considerable. These features of the open ocean create more persistently and stringently nutrient-limited habitats than the coldest, deepest, most isolated freshwater lakes. If there are any bacteria (or archaea) narrowly specialized for oligotrophic existence, this is where they are likely to be found.

Microbial Population: Definitions of Terms

In contrast to extreme environments, the conditions for classification of habitats based on nutrient supply do not imply upper or lower limits beyond which no known kind of microorganism can survive. Low and high are not absolute values; they are determined by our abilities to detect and measure concentrations of solutes and densities of live microbes. However, they are not entirely arbitrary. Low-nutrient concentrations and low microbial densities can also be regarded as those concentrations and densities that have minimal local environmental effects. Because they are exquisitely small, bacteria do not measurably influence their immediate environments unless they are present and active in densities of at least 10^6 per gram or per milliliter of the environment; on the other hand, the environmental effects of bacterial populations of 10^9 or more per gram or per milliliter may be spectacular, causing rot or disease or unexcelled growth of plants.

A priori, it might be predicted that bacteria that seem to prefer habitats of very low or very high levels of nutrients will not necessarily be restricted to their preference. Nevertheless, explorations of bacterial populations in oligotrophic habitats have revealed the existence of bacteria that are rarely found in eutrophic habitats, and some of them appear to be nutrient sensitive, unable to grow – even to survive – in traditional bacteriological media originally designed for the cultivation of clinically significant bacteria. Many such bacteria can multiply and may grow quite well in highly diluted versions of the same concoctions, strongly implying that it is the concentration not the identity of the nutrients that encourages or inhibits their development. Because these bacteria are naturally distributed in oligotrophic (sparsely nourished) habitats, and because their laboratory cultivation is more likely to succeed in oligotrophic than in rich media, it is quite fitting to call them oligotrophic bacteria. To emphasize the quantitative basis of the important distinction between oligotrophic bacteria and bacteria from eutrophic habitats that are readily cultivated on traditional media, the latter type will be referred to here as copiotrophic (abundantly nourished; see Table 2). The term eutrophic (well nourished; both types grow well when appropriately nourished) will be reserved to refer to nutrient-rich waters. The distinction between these two trophic types of bacteria is based on the greater competitiveness of oligotrophs in/during low-nutrient-flux conditions and of copiotrophs in/during high-nutrient-flux conditions. Because nutrient fluxes are variable in most kinds of natural waters and are especially so in perturbed waters, oligotrophs and copiotrophs can be expected to share habitats, their respective populations rising and falling according to variations in local nutrient flux.

In summary, oligotrophy is the nutritional condition also called low-nutrient flux at the ecosystem level. It tends to be seasonal in freshwater lakes, rivers, estuaries, and coastal regions, and relatively persistent in oceanic regions. The density and diversity of the microbial populations vary from season to season, from year to year, and during versus between nutrient surges. Bacteria that tend to be found dependably during periods between surges, during seasons of oligotrophic conditions, or in waters of persistently low-nutrient flux are oligotrophs.

EXTREME HABITATS

Environmental Limits

Three environmental conditions affect the ability of every kind of microorganism to thrive, that is, to metabolize, grow, and reproduce: the availability of liquid water, heat, and the proportion of water that is dissociated. Each of these conditions can be described quantitatively – as water activity (a_w), temperature (°C), and pH. The vast majority of microbes, like all other life forms, appear to thrive particularly in environments of abundant water, moderate temperature, and roughly neutral pH. Outside these ranges, conditions are incompatible with growth of any known microorganism, and kill most microbes. Nevertheless, diverse *Bacteria* and *Archaea* have evolved mechanisms that enable them to thrive just within the extremes of environmental conditions. It is often the case that some of these organisms are special in ways that impair their competitiveness with more ordinary microbes; they are then labeled obligate and their natural distribution is largely restricted to extreme habitats. Here the concept of extreme environments is examined very briefly to contrast them to low-nutrient environments, which are quite ordinary.

Water activity

Water activity of habitable environments varies from 1.000 in practically pure water, to 0.995 in human blood, to 0.980 in sea water, down to 0.700 in dry cereals and dried fruits. Water activity below 0.980 begins to define the lower extreme of water availability in natural aquatic habitats; the lower limit of a_w known to allow microbial growth in any habitat is near 0.700. When environmental a_w is greater than internal a_w, walled cells are turgid and can be physiologically active. Microbes that can thrive in environments of a_w lower than internal a_w display diverse mechanisms of transport systems, synthesis of compatible solutes, pumping organelles, and others that enable microbes to retain water and thereby survive and/or thrive in dry environments. The efficiency of such systems defines extreme a_w among modern microbes.

Temperature

Temperature is influential both indirectly, through its influence on water, and directly, through its influence on the organic molecules of which living cells are composed. The major influence of environmental temperature on life derives from the dependence of life on water, which must be liquid. Temperature is not the sole determinant of the physical state of water; pressure is also an influence, and depending on the pressure, water can be liquid above the boiling point and below the freezing point of pure water at one atmosphere of pressure. Even so, within the range of temperature where water is liquid, temperature has a strong influence on whether a given kind of organism can survive and/or thrive, and different microbes are suited for different temperature ranges.

Almost all modern living organisms, microbes and others, are best suited for temperatures from slightly above 0 to about 50 °C. Within this range, water is not just liquid; it also associates efficiently with organic molecules such as proteins and nucleic acids as part of the supportive scaffold that maintains macromolecular shape, organization, and function. At higher temperatures, water will begin to leave such associations, and molecular organization will collapse as the molecule denatures. Even more importantly, as temperature rises, the cohesion of water declines, and the ability of water to hold lipid bilayers together in unit membranes weakens; as a membrane's layers dissociate, so does the cell. Different microbes thrive at different temperatures beyond the ordinary range by virtue of diverse molecular mechanisms. These include synthesis of small, relatively heat-stable organic molecules that bind to nucleic acids; covalent rather than hydrophobic connections between membrane layers; relatively high proportions of ionic bonds to stabilize the folding of proteins; proportions of saturated versus unsaturated fatty acids in membrane lipids; and others.

pH

The activities of living organisms are directed almost entirely by proteins and, more remotely, by the nucleic acids that specify the structure of the proteins and that carry messages within cells. Both types of macromolecules are dynamic in shape and with respect to interactions within each molecule and between molecules. There are several ways in which these molecules can interact (attractively or repulsively, binding or repelling, or changing each other) through electrostatic, van der Waals, or hydrophobic forces (actually the action of water on relatively passive molecules); entrapment motions; and probably others we have not yet recognized. Because the charged sites on organic macromolecules are predominantly carboxylate and protonated amino groups, with a smaller and variable proportion of phosphate groups, the relative proportions and actual presence of negative and positive charges on these molecules is strongly influenced by the dissociation equilibrium of water, expressed quantitatively as pH.

The tolerable pH range for almost all forms of life, in their external environment as well as in body fluids such as bloods and plant saps, is roughly pH 5–9. Within this range, some, but not all, carboxyl groups are negatively charged and some, but not all, amino groups are positively charged. Below pH 5, anionic (carboxylate) groups are scarce and below pH 2 are practically unavailable; above pH 9, cationic (protonated amino) groups are scarce and by pH 10 are unavailable. Phosphate groups can provide some anionic sites at pH as low as 2 or even lower, but phosphate groups are not abundant in proteins. Within the ordinary range of intracellular pH, nucleic acid bases bind protons weakly, and they are generally uncharged above pH 4. Modern microbes that can maintain a tolerable internal pH range (i.e., between 5 and 9) in environments of extreme pH conditions exist. Molecular and transport mechanisms are better characterized for microbes that can thrive in extremely low pH environments (acidophiles) than for those that can thrive in high pH environments (alkaliphiles). At both extremes, however, small organic molecules used in cellular metabolism as well as macromolecules become chemically unstable and lose their metabolic and structural functions in cells. Between the extremes, pH conditions allow cells to use proteins as spatially stable electrostatic charges that guide the actions and interactions of large and small molecules in metabolism and in maintenance of the functions of cellular structures.

Oxygen

A fourth environmental feature, the presence and abundance of one specific chemical – molecular oxygen, O_2 – affects every microbe. With very few exceptions, when O_2 appears or disappears, any microbe will enliven, die, stall, or undergo a major switch in its metabolism. Because this chemical can serve as the most electronegative terminal electron acceptor in oxidative metabolism, its respiration provides the greatest efficiency among chemotrophic energy transformations, and usually also the highest rate of organic substrate oxidation to CO_2. As far as is known, any microbe that can use O_2 as a respiratory electron acceptor will do so preferentially, even when alternative substances that it can use for the same purpose are available. O_2-indifferent exceptions are few; they include nonphotosynthetic microbes that can defend themselves against O_2, but do not exploit its electronegativity for chemotrophic growth; when O_2 is present, their energy metabolism will inevitably be less efficient than that of chemotrophs that exploit O_2.

O_2 is, however, a two-edged sword. Unless an organism possesses biochemical protections against the reactivity of O_2, it is likely to experience, spontaneously or with the unintended participation of its own enzymes, oxidative destruction of cellular constituents that leads to cell death. Microbes without such protections are designated obligate anaerobes; as vegetative cells, they do not survive exposure to O_2. In contrast, a high proportion of modern microbes use O_2 exclusively as their terminal electron acceptor; without it, their energy-yielding metabolism halts. These organisms produce enzymes and accumulate reducing agents that protect them from incomplete reduction of O_2 to more reactive, damaging forms such as H_2O_2,O_2^-, and OH when the supply of reducing power overwhelms the supply of O_2.

While absence of O_2 does not necessarily (or even usually) kill an "obligately" aerobic bacterium, it reduces its competitiveness with microbes that have the faculty for switching to O_2-independent metabolism: photosynthesis, fermentation, or use of an alternative electron acceptor such as nitrate or iron. Switching acceptors typically involves a relatively minor change in metabolic capacity, but arrangements to switch to photosynthesis or fermentation are more elaborate, often involving the coordinated gene clusters (operons) of bacterial and archaeal (i.e., prokaryotic) genomes that enable them to make large shifts in energy metabolism swiftly and in an appropriately coordinated manner.

Overall, wherever there are microbes, O_2 will be consumed in energy-transforming or destructive interactions with them, and even in illuminated environments where oxygenic photosynthesis busily produces O_2 during the daytime, O_2-consuming microbes (with a little help from respiring plants and animals) can create and maintain anoxic habitats. The rate and extent to which they do so will depend on the abundance of available reducing power, derived by most microbes in most habitats from DOM.

OLIGOTROPHIC CULTIVATION

The Importance of Cultivation

Since the late nineteenth century, when the technique for inducing a single bacterium to produce at least 20 generations in rapid succession on the surface of an artificial nutritious semisolid substratum was developed in the laboratory of Robert Koch, bacteriologists have turned to this technique first when asking whether, how many, and what kinds of bacteria are present in a given sample. This approach has proved immensely fruitful in the study of living hosts and inanimate materials, revealing to biology the vast diversity and activities of bacteria. The achievement of colony development on agar-gelled media was central to the historical demonstration that microscopic forms of life (germs) can cause disease, and it currently provides bacteriology with the least equivocal basis for the minimal viable count and molecular biology with an essential step in the cloning of genes. Without this technique, we might still agree with Galen that diseases are caused by humors, and that milk sours itself, N_2 is an inert gas, O_2 is essential to life, and plant and animal bodies are discrete biotic units rather than microbial habitats. Even most of our recent molecular techniques provide information that is interpretable only because of previous studies with pure cultures, including (but certainly not limited to) the sizes and numbers of genes, how they are organized within the genome, how their expression is regulated, how they can move from cell to cell, the sequence patterns called genetic codes, and the molecular basis of genetic mutations.

The cultivation of any kind of microbe as genetically homogeneous populations (pure cultures) in liquid or gelled media is a significant and immensely useful accomplishment. It requires success in recognizing that specific kind of microbe, finding its habitat, separating it from the rest of the microbial world, and understanding it well enough to induce it to reproduce in our laboratories supported solely by the concoctions we have devised for its nourishment and under the environmental conditions we have imposed. Because many aspects of a microbe's existence are implied and revealed immediately upon its capture and cultivation, a pure culture is evidence that we have a meaningful grasp of the biotic properties of the organism, and that we may well know enough not just to cultivate it, but to exploit it for human welfare, whether for the management of diseases, for production and protection of our food supply, or for sensible care of the global environment. In short, when a particle can be counted, sometimes specifically stained, and some of its current activities detected, a culture becomes a research tool.

Since the recognition in the early decades of the twentieth century that clear, clean-smelling (i.e., oligotrophic) freshwaters are not free of microbes, and that the blue waters of the open ocean constitute a major bacterial habitat, students of such waters have been attempting cultivation of oligotrophic bacteria. Two large questions have arisen during these studies.

1. Are there bacteria that can survive and thrive only in oligotrophic environments; that is, are there obligate oligotrophs?
2. Why are there almost always many more bacteria-like particles detectable microscopically in environmental samples, particularly from oligotrophic environments, than can be induced to generate pure populations in the laboratory? Are those microscopic particles dead, dormant, or different from known bacteria? (Some surely are archaea.) Are they functioning in the habitat where they are seen, or are they just passing through?

The inferred proportion of uncultivable bacteria has been increased by the introduction of molecular genetic techniques in environmental microbiology that enable the detection of particle-associated genomes, many of which (perhaps one-third or more) appear unlike any genomes known among cultivated microbes. It is not yet clear how or whether these novel genomes function.

The term viable is clearly defined in biology; it means able to reproduce. However, for microscopic organisms being observed by organisms unable to see them, viable becomes an operational term. It is neatly and usefully (and very realistically) defined for bacteria by D. K. Button and coworkers as the "ability of a single cell to attain a population discernible by the observer." Because oligotrophic media generally contain too little organic carbon (not more than 250 mg Corg l^{-1}; see Table 3) to support development of turbidity in liquid culture or macroscopic colonies on agar, discerning that a population has been generated in an oligotrophic culture may require special techniques such as electron microscopy, radioisotope uptake measurements, flow cytometry, chemical assay, or metabolic product indicators. This seems to be especially true for primary cultures, the cultures inoculated with samples of natural populations, which are precisely the cultures that are attempted in order to enumerate viable bacteria in natural samples.

Although the goal of cultivating all the bacteria from oligotrophic habitats is as yet far from realized, decades of work with samples from the open ocean, relatively pristine freshwater lakes, and artificial low-nutrient waters such as tap water and spring water bottled for drinking have generated techniques that have succeeded in providing pure populations of bacteria that predominate and thrive in oligotrophic environments. Successful cultivation has also provided clues regarding their properties and how to improve our ability to capture, cultivate, promote, or inhibit them. Only with pure cultures will it become possible to answer the first question, to learn whether there are bacteria that are intrinsically intolerant of exposure to abundant nutrients (as concentrations or as fluxes), and to elucidate the physiological basis of their hypothesized nutrient sensitivity.

Principles of Oligotrophic Cultivation

Overall, the results of attempts to cultivate the predominant bacteria from oligotrophic waters have implied several general principles. The principles are not significantly different from those that guided the development of cultivation of animal and plant associates, including pathogens, and of fast-growing aquatic and soil bacteria, but they are different in detail. They can be summarized as follows, and example media are shown in Table 3.

Guidelines for primary cultivation of oligotrophic bacteria

1. To encourage the reproduction of oligotrophic bacteria, the explorer should begin by presuming that there will be aerobic chemoorganoheterotrophs able to metabolize in the most energy-efficient manner that is available night and day. They may nevertheless suffer oxidative damage when suddenly provided with an abundance of nutrients and may

TABLE 3 Organic composition of example media tested for enumeration of oligotrophic bacteria

Samples[a]: Coastal, marine, CA, USA						
Medium		2216E	2216E/100			
Substrate added per liter	Peptone	5000 mg	50 mg			
	Yeast extract	1000 mg	10 mg			
Concentration (mg l^{-1})	Total organic	6000	60			
	Organic C	2880	28.8			
	Total N	630	6.3			
CFU ml^{-1}		10–18	50–200			
Samples[b]: Coastal, marine, Australia						
Medium		C′	B′	A′		
Substrate added per liter	Glucose		250 mg	0.5 mg		
	Yeast extract	1000 mg				
	Proteose peptone	1000 mg				
Concentration (mg l^{-1})	Total organic	2000	250	0.5		
	Organic C	960	100	0.2		
	Total N	210	(No added N)	(No added N)		
CFU ml^{-1}		17,000	7000	1500		
Samples[b]: Coastal, marine, Sweden						
Medium		C	B	A		
Substrate added per liter	Peptone	500 mg	50 mg	0.5 mg		
	Yeast extract	250 mg	25 mg	0.25 mg		
	Glucose	250 mg	25 mg	0.25 mg		
	Soluble starch	250 mg	25 mg	0.25 mg		
Concentration (mg l^{-1})	Total organic	1250	125	1.25		
	Organic C	560	56	0.56		
	Total N	78.8	7.88	0.079		
CFU ml^{-1}		200 max	250 max	150 max		
Samples[c]: Oligotrophic lake, Japan						
Medium		LT	LT10-1	LT10-2	LT10-3	LT10-4
Substrate added per liter	Trypticase	5000 mg	500 mg	50 mg	5 mg	0.5 mg
	Yeast extract	500 mg	50 m	5 mg	0.5 mg	0.05 mg
Concentration (mg l^{-1})	Total organic	5500	550	55	5.5	0.55
	Organic C	2640	264	26.4	2.64	0.264
	Total N	580	58	5.8	0.58	0.058
Growth (MPN)		No growth	Low yield	Optimal yield	Yes	Yes
Samples[d]: Drinking water distribution systems, Cincinnati, OH, USA						
Medium		PCA/APHA	R2A			
Substrate added per liter	Tryptone	5000 mg	500 mg			
	Yeast extract	2500 mg	500 mg			
	Casamino acids		500 mg			
	Soluble starch		500 mg			
	Na-pyruvate		300 mg			
	Glucose	1000 mg	500 mg			

TABLE 3 Organic composition of example media tested for enumeration of oligotrophic bacteria—cont'd

Concentration (mg l^{-1})	Total organic	8500	2800	
	Organic C	4000	1224	
	Total N	787.5	157.5	
CFU ml^{-1}		500	4000	
Samples[e]: Drinking water distribution systems, Seattle, WA, USA				
Medium		SMA/APHA	CPS	DP
Substrate added per liter	Peptone	5000 mg	500 mg	100 mg
	Yeast extract	2500 mg		
	Na-caseinate		500 mg	
	Soluble starch		500 mg	
	Glycerol		1 ml	
	Glucose	1000 mg		
Concentration (mg l^{-1})	Total organic	8500	~2500	100
	Organic C	3170	~1080	48
	Total N	788	105	10.5
CFU ml^{-1}		>0.3–3	10–1.7 × 10^5	>3–2.2 × 10^5

[a]*Jannasch and Jones (1959)*. Limnology and Oceanography, 4, *128–139*.
[b]*Kjelleberg, Marshal and Hermansson (1985)*. FEMS Microbiology Ecology, 31, *89–96*.
[c]*Ishida and Kadota (1981)*. Microbial Ecology, *7, 123–130*.
[d]*Reasoner and Geldreich (1985)*. Applied and Environmental Microbiology, 49, *1–7*.
[e]*Maki, LaCroix, Hopkins, and Staley (1986)*. Applied and Environmental Microbiology, 51, *1047–1055*.

need protection by the incorporation of pyruvate or catalase in the medium. In addition, relatively recent research suggests a possible role for light in the energy metabolism of some oligotrophs (see Section "Different?").

2. Media for batch cultures should offer a mixture of organic compounds as nutrients, preferably compounds that provide assimilable carbon and nitrogen and respirable electrons in one transport substrate. Amino acids have generally proved to be preferred over sugars and nonnitrogenous organic acids. In a natural oligotrophic environment, nutrients are likely to arrive as surges of mixtures of compounds, each compound in a small quantity, but the mixture providing a total quantity sufficient to support both energy metabolism and net biosynthesis. The bacteria adapted to this type of nutrient supply will use many components of the mixture and may not be able to depend on a single substrate as sole source of Corg and reducing power.
3. In batch cultures, the preferred total concentration of organic nutrients appears to provide not more than 50 mg C^{-1} l^{-1}; 500 mg C^{-1} l^{-1} may prove inhibitory for many of the bacteria in an oligotrophic sample (see Table 3).
4. In general, incubation temperature should be within 3–10 °C of the temperature of the sampled water, although an incubation temperature of 18–20 °C has proved suitable for bacteria in samples from a wide variety of sources. Lower temperatures require longer incubation times, potentially drying out the cultures or exhausting the patience of the laboratory and resulting in negative results.
5. Because the generation times of isolates of probably oligotrophic bacteria vary from 3 to 4 h up to days, incubation of initial samples, in particular, should not be shorter than 2 weeks. For several reasons, bacteria of interest often require 2–4 weeks to generate macroscopically visible colonies.
   - First, these bacteria tend to grow more slowly than the familiar bacteria of clinical, agricultural, and industrial significance, and so cells pile up more slowly on the agar surface.
   - Second, many of the media suitable for oligotrophic cultivation provide too little material to support the development of conspicuous, macroscopic colonies, especially from small cells not unusual among oligotrophs; a significant number of colonies may be visualized only when the cultures are examined microscopically.
   - Third, the relatively low incubation temperatures that succeed with oligotrophic bacteria support slower growth than routine (i.e., clinical) temperatures such as 35 °C.
   - Fourth, there may be so large a difference between the physical and chemical conditions in the microenvironments at the sampling site and the constant and roomy conditions in the artificial culture conditions that the live bacteria in the sample experience a prolonged lag phase. Once they have become accustomed to the new conditions, the lag phase may not be uncommonly long. This is not, of course, a peculiarity of oligotrophic bacteria.

6. Agar-free methods, for example, extinction–dilution and most probable number protocols, regularly yield higher estimated numbers of viable bacteria in oligotrophic samples than plating for colonies on agar-solidified media. Softer agar (e.g., 0.7–1.0% w/v) can also yield counts that are higher than routine (1.5% agar) media, probably because colony expansion in space is less restricted.
7. Finally, although oligotrophy is presumed in bacterial physiology to refer to a preference for low concentrations or fluxes of organic nutrients, low productivity in oligotrophic waters is more commonly the consequence of inorganic phosphate limitation of primary production by cyanobacteria and algae. Accordingly, attempts to cultivate bacteria that thrive in oligotrophic waters should take notice that phosphate is typically the scarcest nutrient in their natural habitat, and their hypothesized nutrient sensitivity is very likely to include phosphate. Most of the successful attempts at oligotrophic cultivation have employed media with submillimolar phosphate concentrations; when pH buffering is required, organic buffers often serve as tolerable alternatives to phosphate salts.

Possible helper microbes

There is yet another side of cultivation of all kinds of bacteria that are reluctant to reproduce in the laboratory: isolation in pure culture means loss of interaction with other microbes. Bacteria interact through chemical substances, some of which have quite specific and unique structures and are not simply excreted by-products of metabolism from bacteria that lack the metabolic capacity to convert them to CO_2 and H_2O. Many have been identified as signal substances that influence bacterial traits such as virulence or luminescence; others are specific inhibitors of other microorganisms and can even be used in the chemotherapy of infections; still others are enzymes that solubilize organic polymers or mobilize otherwise unavailable material. Such bacterial excretions diffuse into the water and may be essential to the development of populations of microbial neighbors.

Experimental evidence that such influential substances may include stimulatory signals has been obtained by a method in which microcolonies of bacteria develop in a medium containing a soluble polymer, the protein casein, as sole organic nutrient. The chambers allow diffusion of excretions from helper bacteria to bacteria that do not reproduce in the absence of those excretions. This research is just beginning, but has already found that excretions alone – without the helper bacteria – can be sufficient to stimulate activity of the excretion-dependent bacteria. Excretion-independent variants have been generated by cultivation of a dependent clone in the diffusion chambers, and the variant in turn can stimulate division of the dependent clones as well as of itself. While the genetic basis of this variation has not yet been described, it may ultimately prove significant that the putative signals are oligopeptides, the sort of substances that would be appropriate as inducers of proteases that could attack the casein used in this method's primary medium.

It is a common experience of the present author to observe vigorous, morphologically distinctive and readily recognizable, uniform, frequently dividing, swarmer-producing caulobacters (not just particles) that accumulate in primary cultures in surface films on liquid media and as macrocolonies under films of growth of other bacteria on agar media, but then to be unable to detect them upon dilution or streaking-out on the same medium, incubated under the same environmental conditions. It appears that, although caulobacterial clones (such as *Caulobacter* and *Asticcacaulis* from freshwater, and *Maricaulis* from sea water) are readily obtained from oligotrophic samples, some of these bacteria present in the initial sample or produced in the primary mixed culture do not readily subculture as pure populations.

Progress in oligotrophic cultivation

In many studies that have followed some or all of the guidelines summarized above, the cultivability of bacteria from oligotrophic samples has been increased from far less than 0.1% of the microscopically detectable bacteria-like particles (the total counts) to 10–40%, depending on the properties of the microbe and on the insight of the scientist in inferring the organism's needs. This proportion is comparable to the ability of modern analytical chemistry to identify organic substances that occur naturally in aquatic and terrestrial environments. It may be that our limited – although always progressing – success reveals that we underestimate the stressfulness for a 1-μm^3 aquatic organism of being exposed on the surface of an agar gel, where it is unable to dip into a local pool of liquid water trapped in the gel in order to protect itself from the dehydrating pull and oxidizing power of air. It is also far from other microbes whose uninterrupted support or stimulation or protection it requires. It may also reveal that we overestimate its ability to metabolize, grow, and reproduce without rest.

Are There Obligately Oligotrophic Bacteria?

The variability of physical and chemical conditions in all life-supporting environments on earth is readily apparent. Even in the open ocean, which is relatively slow to change, water and the solutes it bears are constantly being translocated. Similarly, the shallower water at the edge of a temperate-zone lake in the winter, its most stringently oligotrophic season, will receive occasional surges of materials leaching from the adjacent exposed soil. With the possible exceptions of the open ocean and high alpine lakes above the plant line, oligotrophic conditions are typically seasonal, not perpetual. It is predictable that the competitiveness and survival of inhabitants of changeable habitats will depend on the flexibility of the organisms themselves.

Among living organisms, bacteria are remarkably flexible. Bacterial populations can change more rapidly than other organisms not only because of their shorter generation times, but also because individual bacteria are capable of swift and coordinated switches in metabolism and behavior within one generation. This kind of change is governed by clusters of genes of related function gathered under the control of a single regulating gene in the genomic regions known as operons that occur widely in prokaryotes, but rarely in organisms with compound genomes enclosed within double-membrane envelopes (eukaryotes). By activating several genes at once, facultative bacteria can switch from living on their own or invading a host; from assimilating ammonium as N source or pulling N into chemical combination from atmospheric N_2; from using light to using chemical energy; from remaining active or entering dormancy; or make any of many other great changes that promote their survival and reproduction. Among the functional molecules known to be included in such regulatory regimes are cell-surface proteins that mediate nutrient transport from the environment to cytoplasm – the sorts of molecules that would enable a bacterium to scavenge sparsely distributed nutrients.

During initial cultivation of bacteria that have been isolated from oligotrophic habitats by oligotrophic cultivation, most of the bacteria grow slowly and only after a long lag; some do not grow at all on media with higher concentrations of the same nutrients, and may not grow when transferred to fresh cultures in the initial medium. However, it is often found that once induced to grow in the laboratory, they become less sensitive to higher organic nutrient concentrations, more tolerant of agar, and exhibit shorter lag times – all without losing their ability to develop in very low oligotrophic concentrations of nutrients. Similarly, copiotrophs from the same original samples that develop in primary cultures only in rich media can often acclimate and eventually be subcultivated in poor media. The bacteria in these derived cultures do not appear to be genetic mutants; the basis for the adaptive behavior of the initially nutrient-sensitive bacteria has not been elucidated.

Such observations imply that restriction of primary growth to media of specific nutrient concentrations may distinguish the physiologic states of bacteria in an oligotrophic sample (their sample state), but it does not necessarily imply an intrinsic sensitivity to or dependence on abundant nutrients. While the distinction between oligotrophic and copiotrophic bacteria as a description of their respective competitiveness in the natural habitat may be valid, it is disappointing to recognize the implication that studies with long-domesticated pure cultures may not help to elucidate why certain kinds of bacteria invariably behave like oligotrophs in primary cultures.

When the interval between periods of relative nutrient abundance is long relative to the periods of abundance (i.e., the frequency of feasts is low, the servings small, and fasting is the usual state of the bacterial population), there are obvious advantages to being able to gather nutrients from very low ambient concentrations and from very small and/or brief surges. It would also be advantageous to have high-affinity, low-specificity transport systems, so that whatever nutrient appeared, it could be efficiently consumed. Physiological studies with bacteria that have been cultivated from oligotrophic habitats have detected transport systems of exceptionally high affinity, and genomic studies of oligotrophic isolates have revealed unusually large numbers of genes tentatively interpreted as structural genes for special transport activities, including TonB systems. A cell form that conferred a high surface-to-volume (supply-to-demand) ratio would benefit nutrient scavenging, and it is not surprising that bacteria cultivated from oligotrophic habitats include relatively high proportions of ultramicrobacteria (very low cell volume), of filamentous cells, and of prosthecate bacteria whose cell-surface protrusions are capable of substrate uptake but engage in little energy-consuming biosynthesis. Caulobacters (Figures 1–3) and other prosthecate bacteria undergo marked changes in the relative length of their prosthecae in response to nutrient availability, growing appendages whose length is inversely related to nutrient flux. In short, sensitive, efficient, and nondiscriminating transport activity that can be increased as nutrient flux declines would seem to be an essential physiologic trait of great value to a bacterium adapted to existence as an oligotroph.

Exceptional nutrient-scavenging capabilities might also account in part for the inability of a significant proportion of bacteria in oligotrophic samples to grow promptly in primary cultures. In many cases, this sensitivity may well be more an aspect of the state of the cells in the sample – their sample state – than an expression of an insurmountable intolerance of nutrient abundance. After a fasting period, as when taken directly from oligotrophic conditions in nature, the surfaces of the bacteria – most importantly their outer membranes – are likely to be well supplied with transport-protein complexes. If this is so, the sudden entry into the cytoplasm of an abundance of nutrients could overwhelm their respiratory metabolism with reducing power that would generate damaging levels of oxygen species such as peroxides, superoxide, or hydroxyl radicals. It is tempting to consider that it could also result in direct structural damage in two ways: by overloading the periplasm with solutes transported so rapidly that the periplasm becomes hypertonic, leading to cell death by dehydration of the protoplast; or by providing an abundance of transporters that serve as entry sites for bacteriophage that would spread through the

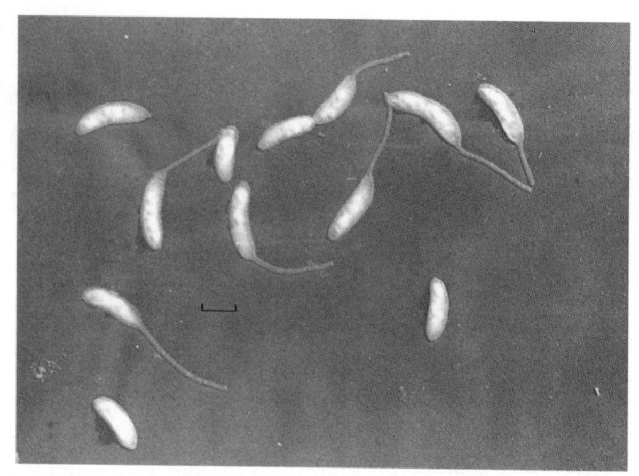

FIGURE 1 Prosthecate bacteria (*Caulobacter crescentus*) growing in phosphate-adequate medium with mean stalk length approximately equal to cell length. EM, Pt-shadowed. Marker = 1 µm.

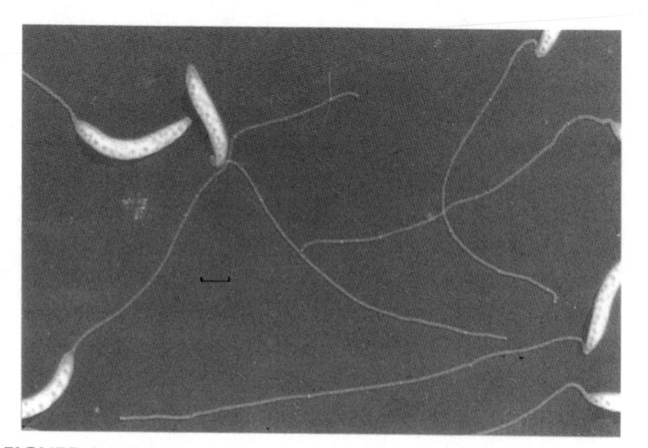

FIGURE 2 Prosthecate bacteria (*Caulobacter crescentus*) growing in phosphate-limited medium with mean stalk length significantly greater than cell length. EM, Pt-shadowed. Marker = 1 µm.

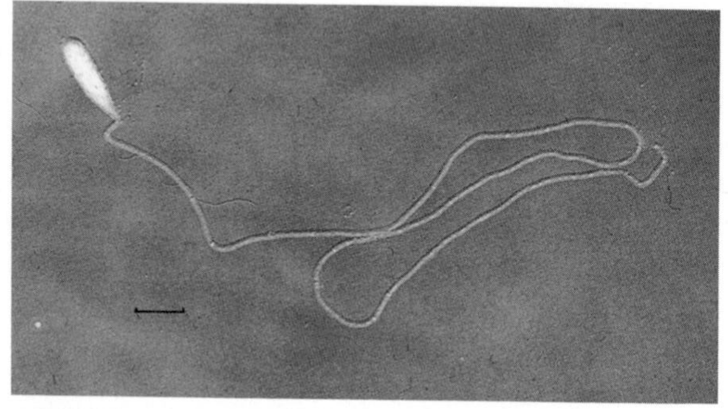

FIGURE 3 Prosthecate bacterium (*Caulobacter* sp.) resident in a laboratory distilled-water reservoir with stalk length 23× cell length. EM, Pt-shadowed. Marker = 1 µm.

population during nutrient abundance sufficient to support bacteriophage propagation. It seems plausible that oligotrophic bacteria may not be dead until they arrive in the laboratory medium, where the sudden abundance of nutrients precipitates inefficient respiration, dehydration, or the lytic cycle and spread of latent bacteriophage.

There are some oligotrophic isolates that appear to be obligate. They develop in oligotrophic media, but they do not acclimate to nutrient concentrations above about 5 mg organic $C l^{-1}$. It is possible that we have focused too narrowly on organic nutrients. In particular, in freshwater lakes in temperate zones, blooms of oligotrophic bacteria begin very shortly after blooms of photosynthetic productivity in the spring. Spring blooms of lacustrine phototrophs, specifically of algae, and among algae specifically of diatoms, is typically a response to the release of inorganic phosphate from a variety of forms of sequestered, insoluble P. Spring-blooming bacteria must compete with phototrophs for that phosphate in order to consume the year's initial, limited supply of organic carbon as it is produced. They will have to be able to take up P from the limited phosphate supply, a property that could cause them to be as sensitive to that inorganic nutrient as they seem to be to organic nutrients. In the open ocean, the photosynthesis-limiting nutrient may be P or N; however, the principal phototrophs of the ocean are cyanobacteria, many of which can fix N_2 from the atmosphere and thereby relieve N limitation. The oligotrophic bacterium's need, once again, is highly competitive uptake of phosphate, a property usually exhibited by cultivated oligotrophs, but not yet explored as possibly having a role in their nutrient sensitivity. It may also prove that even after cultivation is achieved, bacteria isolated from oligotrophic waters may not be able to adapt to high phosphate concentrations or fluxes, whether or not they can adapt to more intense exposure to organic nutrients.

Are Unresponsive Bacteria Dead, Dormant, or Different?

A microscopical count of bacteria-sized particles in any natural sample is invariably greater than the viable (CFU or colony-forming unit) count, as it may be even in laboratory-congenial pure cultures. While there can be no doubt that the colonies develop because bacteria (individuals or clumps) reproduce and their progeny accumulate to visible heaps on or in the agar medium, the nature of the particles is less certain. It is safe to assume that they are indeed bacteria, but they may be dead bacteria that have not yet been devoured and digested by bacteriovores or disintegrated by microbial enzymes or viral infection; they may be dormant in the sense that they are inactive *in situ* and need specific stimulation to resume metabolic and reproductive activities; or they may be different from known bacteria with respect to the cultural conditions and nutrients they require, but investigators have not yet devised laboratory settings suitable for their multiplication.

Dead?

The familiar disparity (2- to 1000-fold or more) between the total (i.e., particle) count and the viable (CFU) count has sometimes been dubbed the "plate count anomaly." (While this article focuses on aquatic habitats, it should be noted that this problem is even greater in soil samples because the density of bacteria-sized nonviable (dead or mineral) particles is far greater in soil than in oligotrophic waters and because there are technical difficulties involved in visualizing and dislodging bacteria from attachment to soil particles. All 100 cells attached to a single soil particle will probably produce, at most, only one colony. Some bacteria may be aggregated in aquatic samples, but the proportions of bacteria in aggregates are generally much lower than in soil samples.)

In the most widely used method for the total count, bacteria-sized particles in a water sample of known volume are collected on membrane filters, fixed chemically, and the membranes stained with the fluorescent dye acridine orange; this is the acridine orange direct count (AODC). Alternatively, the fluorescent dye 4′,6-diamidino-2-phenylindole (DAPI) is used to stain DNA that is visible microscopically within particles and in a mass large and dense enough to bind sufficient DAPI to produce visible fluorescence. A double-staining procedure that exposes the particles to both dyes in series can distinguish AO-staining bacteria that no longer contain nucleoids (ghosts, presumed dead) from bacteria with DAPI-stainable nucleoids (nucleoid-containing cells, NUCCs). Counts of NUCCs are usually about one-half the AODC total counts, reducing the anomaly by a factor of 2. In studies that extend over several months, the ratio that constitutes the AODC:CFU or AODC:NUCCs anomaly varies with season, with the highest ratios usually occurring in the early autumn in lakes and coastal waters. Arithmetically, the ratios rise as AODC increases while CFU and NUCC counts decrease, changes that imply decreased bacteriovore activity in autumn (and winter).

On the whole, studies to date have implied that probably at least 50% of the bacteria-sized, stainable particles in samples from oligotrophic waters are dead – depending on the season of the year.

Dormant?

The possibility that some or most of the bacteria in an oligotrophic sample are in a state of inactivity, but not dead, is not unreasonable, and is implied by studies that reveal that some oligotrophic bacteria are able to acclimate to laboratory conditions. The need to revive bacteria in an environmental sample would imply that something must be supplied in the primary culture in addition to useable, properly balanced nutrients in moderate amounts, or that they may need something other than the substances and conditions they will need for growth and reproduction once they have become acclimated to the culture conditions. Alternatively or additionally, they might need physical stimulation such as brief incubation at a temperature above that of their natural habitat, an observation that has been reported. This could be especially true for bacteria in samples collected during the winter, when viable counts regularly drop in oligotrophic samples. They might also be sensitive to inhibition by particular chemicals that, once acclimated, they will be able to ignore or even to consume as nutrients.

The need to adjust to the cultural conditions and initiate growth and reproduction is familiar to any population biologist who has encountered the lag phase of a typical logistic growth curve. In bacteriology laboratories, for example, when most of the bacteria in a sample of a stored laboratory population have died or become unable to recover swiftly, the offspring of the bacteria that recover relatively promptly will become predominant in the fresh culture, and maintenance of vegetative stock cultures can result in periodic selection of subtypes of the original isolate. This fundamental principle also underlies elective enrichment cultivation to promote multiplication of only certain phenotypes of bacteria from mixed populations or to select specific genetic mutants. However, if the goal of the research is to stimulate the recovery of all the types of viable bacteria in a natural sample, a variety of media and conditions must be used in order to favor each type possibly present.

Differences among bacteria in natural samples with respect to readiness to grow and to grow at different rates has proved particularly frustrating for those who would stimulate the development of populations of oligotrophic bacteria from samples that also contain copiotrophic bacteria. The copiotrophs may be present in an oligotrophic sample in relatively low numbers (depending on the season of the year), but they are likely to respond readily to a fresh supply of nutrients and multiply so rapidly that they will become numerically predominant before the oligotrophs have revived.

One technical solution to the problem of copiotrophic bacteria overwhelming oligotrophs is to mimic the nutrient flux of the natural oligotrophic environment by controlling nutrient flux in the primary culture. Two apparatus are well suited for this purpose: the chemostat and the fed-batch culture. Each method delivers a limiting nutrient more slowly than the population can consume it. In these devices, as in the natural habitat, the ambient concentration of the limiting nutrient is determined by the affinity of the bacterial population for that nutrient. The greatest differences between these perpetual cultures and an oligotrophic habitat is that nutrient supply in the chemostat is steady, not erratic as in nature, and there is no attrition of the accumulating population in a fed-batch culture, from which bacteriovores are intentionally excluded when the purpose is cultivation of oligotrophs for isolation. Nevertheless, this approach discourages rapid growth of copiotrophs present, which both theory and experimental experience predict will have a longer wait between nutrient surges (in this case, delivery of drops of medium) because their transport systems are of lower affinity than those of the oligotrophs. Indeed, if each delivery is very small, dilution of the limiting nutrient from one drop to the total volume of the culture may result in only a brief, low maximum concentration that is not sufficient for the copiotroph to detect. Meanwhile, the oligotrophs may revive and begin to grow and multiply. Once revived, they may be amenable to cultivation in batch cultures or even on agar media.

It appears at present that rather than dormant, some oligotrophic bacteria are in a sample state of arrested metabolic activity from which they emerge sluggishly into an acclimated state in which they respond more readily to fresh nutrients and grow at consistent rates. Why some primary-culture populations fail to propagate upon subcultivation is yet to be resolved experimentally.

Different?

The final prominent explanation for the anomaly proposes that the most dedicated (perhaps obligate) oligotrophic bacteria are unknown kinds of bacteria. They may be genetically and therefore, physiologically quite different from bacteria that have been studied for the past one and a half centuries in pure cultures, but (with the possible exception of substrate transport capacity) it is not clear in what ways they might prove extremely different. Diligent and innovative pursuit of conditions for successful cultivation, summarized above, has significantly decreased the count anomaly without yet yielding any strikingly unique kinds of bacteria. This is in marked contrast to the studies of bacteria and especially of archaea from extreme environments – many of them now successfully cultivated – microbes of such unexpected properties that they are redefining the properties and limits of life on earth and suggesting properties of microbes on other less-inhabitable planets.

Over the past 50 years, approaches to identifying the bacteria that thrive especially in oligotrophic waters have included microscopical, cultivational, and molecular-probe techniques. Although diversity is regarded by many investigators as wide, and indeed almost every type of microbe can be found in such waters, a few bacterial groups – all Gram-negative – appear repeatedly as predominant members of the bacterial populations of oligotrophic waters. Gram-positive oligotrophic bacteria such as *Arthrobacter*, prominent in terrestrial habitats, are rarely encountered in aquatic samples. Gram-negative methylotrophic bacteria such as *Hyphomicrobium* and *Methylocystis* appear in some oligotrophic samples, both aquatic and soil. However, because they can grow at the expense of volatile organic substrates, it is not clear whether their presence depends on soluble organic compounds provided by the water or by the atmosphere. They clearly do not depend on an abundance of vaporized substrates.

Among the cultivated bacteria that appear repeatedly from oligotrophic samples, there appear to be two major classes: bacteria that rise to prominence in incubated samples that are not amended with organic nutrients, and those that multiply rapidly when incubated samples have been enriched by the addition of small amounts of organic substrates. When followed over time in natural waters, the proportion of members of the first group tends to rise during more stringently oligotrophic periods in the habitat such as during the early-summer clear-water period of severe phosphate limitation and low productivity in freshwater lakes. In contrast, members of the second group seem to bloom in response to surges of organic nutrients during the springtime rise in organic material in temperate-zone lakes, the autumnal crash of particulate biomass, and in response to phosphate pollution that stimulates photosynthesis of organic carbon. Members of both groups have been identified by phenotypic characterization and by genomic analysis (almost always by rRNA sequence), allowing both physiological characterization and cultivation-independent preparation of molecular probes for exploration of natural assemblages.

Almost all of the bacterial genera most frequently observed in the first group are assigned to the phylum *Alphaproteobacteria*. The most commonly encountered bacteria are species of *Sphingomonas*, some of which are ultramicrobacteria with cell volumes of 0.006–0.1 μm^3, and species of *Roseobacter* and *Erythrobacter*. Some species of all three of these genera possess bacteriochlorophyll *a* (Bchl *a*) and can grow as aerobic photoorganoheterotrophs. It could be significant that primary cultures inoculated with samples of oligotrophic waters are usually incubated in the dark; illumination of primary cultures might further reduce the count anomaly. Other prominent isolates from oligotrophic waters (which can also be detected and presumptively identified microscopically) are prosthecate bacteria such as *Caulobacter*, *Brevundimonas*, and *Hyphomonas*; spirilla of various genera; a diversity of bacteria of *Microcyclus*-like morphology; and *Cytophaga* and related filamentous types.

Bacteria of phyla *Gammaproteobacteria* and *Bacteroidetes* are prominent members of the second group, accompanied occasionally by *Archaea*. The major genera of *Bacteria* detected (and often cultivated) are *Flavobacterium* and the gammaproteobacteria *Alteromonas/Colwellia*, *Pseudoalteromonas*, *Vibrio*, *Oceanospirillum*, *Pseudomonas*, *Acinetobacter*, and the SAR86 clade. The gene for proteorhodopsin, a light-dependent proton pump that, like Bchl *a* in some alphaproteobacteria, appears to confer the capacity for aerobic photoorganoheterotrophic metabolism; it has been found in *Flavobacterium* and *Pelagibacter* (SAR11 clade) isolates in this group, and in a wide diversity of genomes extracted from natural populations and probed for the gene. Again, while the dependence of such bacteria on the use of light as a source of energy has not yet been evaluated, there is reason to explore the role of light in assisting aerobic chemoheterotrophic bacteria to thrive in oligotrophic waters – and in oligotrophic laboratory cultures. The betaproteobacterial group typified by *Burkholderia* spp. could be sorted into both of these groups, as well as among organisms recognizable as copiotrophs.

There remains a significant proportion (30–50%) of bacteria-sized particles in oligotrophic samples that react with universal probes for bacterial genomes, but do not reproduce under any laboratory conditions offered to date, and whose artificially amplified genes (mainly 16S rRNA genes) are not closely similar to the genes of known bacteria. It is possible that there are, therefore, other kinds of bacteria whose physiology we still infer so inadequately that we cannot cultivate them in the laboratory. They might be obligate oligotrophs – or anaerobes or parasites in transit between hosts.

THE NICHE OF OLIGOTROPHIC BACTERIA IN LOW-NUTRIENT-FLUX AQUATIC HABITATS

It is likely that the most outstanding physiologic feature of oligotrophic bacteria – bacteria that can thrive in oligotrophic habitats, during or between blooms or throughout the year – will prove to be their high-affinity, low-specificity transporters, some of which can be energized even when localized in the outer membrane, and most of which can be repressed by their transport substrates. In an oligotrophic habitat (and therefore in a sample of that habitat), these transport systems may be especially active relative to biosynthetic metabolism. By functioning too well when an abundance of transportable nutrients suddenly appears, they may overburden the cell with solutes, irretrievably perturb metabolism or water regulation, cause extensive oxidative damage, or activate latent viruses. Such a vulnerability would explain why, when cells are transferred from an oligotrophic environment to a relatively rich nutrient environment, they may die, yet when provided with slowly increased amounts of nutrients, they acclimate and become less susceptible to death caused by nutrients suitable for their growth and reproduction.

To date, the only indications of uniqueness reported for oligotrophic bacteria relate to their ability to grow slowly, to consume solutes from concentrations too low for copiotrophs to detect, and to conserve cell structures and reserve materials during periods when nutrient supply is not adequate to support net biosynthesis. As far as is known, they are metabolically efficient, respiring chemoorganoheterotrophs with high-affinity, low-specificity transporters and with a preference for soluble organic compounds that contain reduced nitrogen. Some of them may supplement their respiratory energy metabolism with light energy, but they do not use light energy for CO_2 fixation. Oligotrophic bacteria appear to be specialized for their principal ecologic role as ultimate scavengers that consume organic carbon dispersed to concentrations that are inconspicuous to other microorganisms, recycling that carbon into the biosphere by mineralization and as edible particles. Their work is slow per unit volume of their habitats, but their global effect is as vast as the ocean.

FURTHER READING

Azam, F., Fenchel, T., Field, J. G., Gray, J. S., Meyer-Reil, L. A., & Thingstad, F. (1983). The ecological role of water-column microbes in the sea. *Marine Ecology Progress Series*, *10*, 257–263.

Berner, E. K., & Berner, R. A. (1996). *Global environment: Water, air, and geochemical cycles*. Upper Saddle River, NJ: Prentice Hall.

Button, D. K., Schut, F., Quang, P., Martin, R., & Robertson, B. R. (1993). Viability and isolation of marine bacteria by dilution culture: Theory, procedures, and initial results. *Applied and Environmental Microbiology*, *59*, 881–891.

Edmondson, W. T., & Lehman, J. T. (1981). The effect of changes to the nutrient income on the condition of Lake Washington. *Limnology and Oceanography*, *26*, 1–29.

Eiler, A. (2006). Evidence for the ubiquity of mixotrophic bacteria in the upper ocean: Implications and consequences. *Applied and Environmental Microbiology*, *72*, 7431–7437.

Hirsch, P., Bernhard, M., Cohen, S. S. *et al.* (1979). Life under conditions of low nutrient concentrations. Group report. In M. Shilo (Ed.) *Strategies of microbial life in extreme environments* (pp. 357–372). Berlin: Dahlem Konferenzen.

Kuznetsov, S. I., Dubinina, G. A., & Lapteva, N. S. (1979). Biology of oligotrophic bacteria. *Annual Review of Microbiology*, *33*, 377–387.

Münster, U. (1993). Concentrations and fluxes of organic carbon substrates in the aquatic environment. *Antonie Van Leeuwenhoek*, *63*, 243–274.

Pernthaler, J., Glöckner, F., Unterholzner, S., Alfreider, A., Psenner, R., & Amann, R. (1998). Seasonal community and population dynamics of pelagic bacteria and archaea in a high mountain lake. *Applied and Environmental Microbiology*, *64*, 4299–4306.

Poindexter, J. S. (1981). Oligotrophy: Fast and famine existence. *Advances in Microbial Ecology*, *5*, 63–89.

Pomeroy, L. R. (1984). Significance of microorganisms in carbon and energy flow in marine ecosystems. In M. J. Klug, & C. A. Reddy (Eds.) *Current perspectives in microbial ecology* (pp. 405–411). Washington, DC: American Society for Microbiology.

Schut, F., deVries, E. J., Gottschal, J. C. *et al.* (1993). Isolation of typical marine bacteria by dilution culture: Growth, maintenance, and characteristics of isolates under laboratory conditions. *Applied and Environmental Microbiology*, *59*, 2150–2160.

Schut, F., Prins, R. A., & Gottschal, J. C. (1997). Oligotrophy and pelagic marine bacteria: Facts and fiction. *Aquatic Microbial Ecology*, *12*, 177–202.

Weiss, M., & Simon, M. (1999). Consumption of labile dissolved organic matter by limnetic bacterioplankton: The relative significance of amino acids and carbohydrates. *Aquatic Microbial Ecology*, *17*, 1–12.

Chapter 31

Sediment Habitats

K.H. Nealson and W. Berelson
University of Southern California, Los Angeles, CA, USA

Chapter Outline

ABBREVIATIONS

C_{org} Organic carbon
LMC Layered microbial community
PAR Photosynthetically active radiation
SRB Sulfate-reducing bacteria
SRR Sulfate reduction rate

DEFINING STATEMENT

The sedimentary niche is the largest ecosystem on Earth, harboring ~50% of the biomass on our planet. The processes that occur in these water saturated, physically stabilized niches often lead to the formation of chemical gradients or layers, which can be used to infer modern and/or ancient sedimentary microbial processes.

INTRODUCTION

What is Sediment?

For our purposes, active sedimentary environments are defined as the underlying matrix of aquatic environments in which sedimentary (depositional) processes occur, and in which some form of deposit (sediment) accumulates with time via input of particles from the overlying water. Thus, the entire ocean bottom, excluding rocky outcrops and the active wave-breaking zone, in which sediments seldom permanently accumulate, is a sedimentary environment, as are virtually all lake bottoms. Rivers can accumulate sediments, but depending on the dynamics of the river, these deposits may be short-lived and difficult to predict and/or define. Not all sediments are alike, of course,

Topics in Ecological and Environmental Microbiology.

and their variations are subject to several key factors: (1) the chemical nature of the overlying water; (2) the impact of light (photosynthetically active radiation or PAR) on the sediment surface; (3) the type and amount of organic input to the sediments; and (4) the type and amount of inorganic input to the sediments (Figure 1).

One of the important properties of sediments relates to the fact that the environment is physically stabilized: That is, there is minimal mixing or convection, and the niche is dominated by diffusion. For example, if the oxygen is consumed within a given sediment, then it must be supplied from above by diffusion, and this interplay is extremely important in defining the chemistry and microbiology of the sedimentary niche.

Where are Sediments Found?

Sediments are found at the bottom of virtually every body of water: streams, rivers, lakes, and throughout the ocean. Even in the absence of any obvious sedimentary source, the mean global dust fallout from the atmosphere alone can generate sediment accumulation rates of one or more millimeters per kiloyear. Thus, any body of water on this planet will accumulate sediments, and as material falls through water, it can accumulate a variety of "hitchhikers" including prokaryotic and eukaryotic cells, as well as organic and inorganic materials (metals, minerals, clays, nutrients, etc.).

What are Sediments Composed of?

Dust is the source of ~35% of sediment accumulating in the global ocean: Such dust-dominated sediments are known as red clay deposits, and they are abundant in the deepest ocean and sites furthest removed from continental land masses. In contrast, about 50% of oceanic sediments are termed biogenic (SiO_2 or $CaCO_3$) insofar as the dominant sediment particles are mineral grains formed by either diatoms and/or radiolarians (SiO_2), or coccolithophores and/or foraminifera ($CaCO_3$). These mineral grains, like dust particles, settle through the water column and accumulate an array of "hitchhikers," much like an elevator headed down a skyscraper. Lake sediments and some near-shore oceanic sediments tend to be dominated by particles either carried by rivers or eroded along the shore. These sediments, termed terrigenous, tend to vary considerably both in time and in space, are subjected to mass movement during storm events, and are prone to move offshore and downslope via current dispersion and turbidity flow. When viewed from afar, sediments seem to be slowly and constantly accumulating,

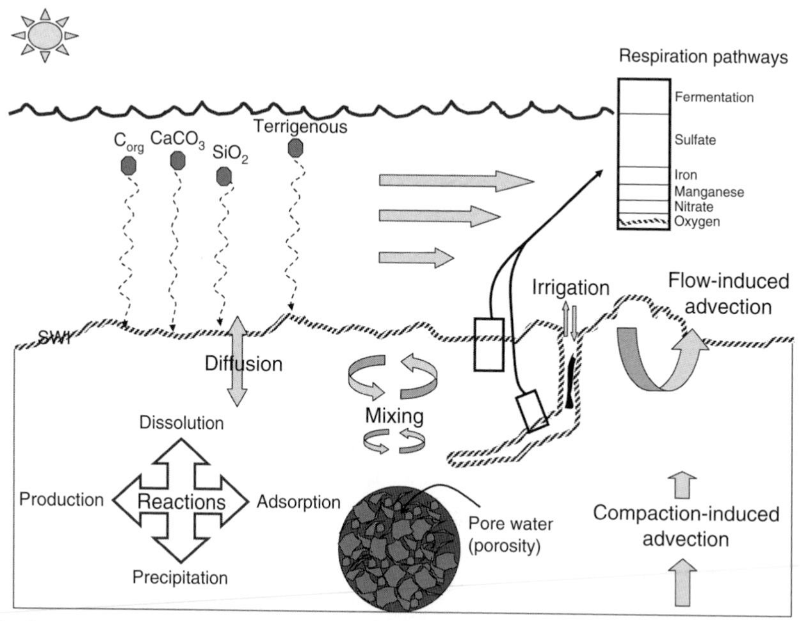

FIGURE 1 Some of the fundamentals of the sedimentary niche. Some of the major processes involved in the formation and growth of sediments are depicted. Delivery of materials is via sedimentation of organic carbon (C_{org}) particles, mineral particles, or continental (terrigenous) runoff. Reactions in the sediments include dissolution and precipitation, biological production, and adsorption onto surfaces, all of which are mediated by the microbial population. Oxygen is delivered to the sedimentary environment via diffusion from the overlying water. Mixing occurs via disturbance by macrofaunal organisms either digging in the sediments, or irrigating the sediments through water-pumping. Pore water, the water in the spaces between particles, is the indicator of the chemistry of the sediments, and as sediments are buried, compaction occurs, which results in upward advection of the water at a rate depending on the rate of sedimentation. Redox gradients (layers) are commonly seen as zones of activity, shown by the boxes in which oxygen respiration and depletion occurs first, followed by nitrate, manganese, iron, sulfate, and CO_2 respiration. If no macrofauna are present, these gradients are horizontal, but irrigation of the macrofaunal burrows can result in complex gradient structures, although the reactants remain the same.

but, in fact, the distribution of sediment types and amounts has varied considerably throughout geologic time, revealing past changes in tectonic activities, climate, and global (especially oceanic) productivity.

THE DIVERSITY OF SEDIMENTS

The Importance of Overlying Water

As opposed to soils, which often alternate between wet and dry states, sediments are constantly wet, and are commonly described by the water that overlays them: marine sediments and lacustrine or freshwater sediments. Superimposed on the marine/lacustrine division is the issue of light. Sediments in shallow water can differ drastically from those in deeper water because of the impact of photosynthesis that can occur in the sediments themselves: Deepwater sediments are nearly always the recipients of photosynthate (photosynthetically produced organic carbon (C_{org})) from above, rather than internal primary production. Since photosynthesis can also be a source of oxygen, the shallow water sediments are prone to being oxic in the daytime, and exhibiting strong day/night differences in terms of redox chemistry, something not seen in deepwater sediments.

Of course, not all lake sediments are overlain by "fresh" water – their chemistries may range widely in pH, salinity, and nutrient chemistry, and in each case, this leads to a different set of conditions in the sediments – conditions that alter the microbial ecology of the sediment basin. Examples of such environments include alkaline lakes such as Walker Lake, NV, alkaline hypersaline lakes such as Mono Lake, CA, and hypersaline lakes such as the Great Salt Lake, UT, and the Salton Sea in the Southern California Desert.

In general, lacustrine sediments receive higher levels of organic input than do marine sediments, and as a result they become anoxic rather quickly (within centimeters or less of the sediment-water interface). In many cases, because lakes are often poorly mixed, the bottom water overlying the sediments can itself be anoxic, leading to sediments lacking macrofauna or other eukaryotes.

Marine Sediments

Marine sediments, with a few notable exceptions, are characterized by a sediment-water interface that includes oxic water overlying sediments that at some level become anoxic. This is because oceanic circulation constantly supplies the deep ocean with oxygen-rich water from the poles, and keeps the deep ocean water often more oxygen-rich than the near-surface waters. Notable exceptions to this "rule" are the Black Sea, which is isolated from oceanic circulation, and is thus anoxic for most of its depth, and marine basins, such as the Cariaco Trench, which is also isolated from deepwater flow and mixing due to the presence of a structural sill that inhibits mixing and results in the water becoming anaerobic far above the sediment-water interface.

As a rule, marine sediments are dominated by the sulfur cycle, due to the fact that seawater usually has ~28 $mmol\,l^{-1}$ sulfate, which, after oxygen is depleted, becomes the major electron acceptor. Thus, marine sediments are notorious for sulfide production and for internal sulfur cycling. Except in cases of very high input of organic matter, sulfate depletion does not occur, and thus, biological methane production in marine sediments is usually not a major process.

Lacustrine Sediments

Lacustrine (i.e., having to do with lakes) sediments vary widely with regard to the oxidation state of the overlying water. In many lakes the waters are well mixed, and very ocean-like in terms of being permanently oxic, while others range from being seasonally stratified (and thus having anoxic bottom waters for some portion of the year) to the so-called meromictic (permanently stratified) lakes, with bottom water that never sees oxygen in abundance. In addition to this rather simple delineation are those sediments beneath lakes that may have chemistries more reminiscent of oceanic environments, such as the Salton Sea, or evaporitic lakes, where salinity can be quite high, and sulfate concentrations similar to those found in the ocean.

In marked contrast to marine sediments, where sulfate reduction dominates, lacustrine sediments are usually dominated by methane production. Sulfate concentration in most lacustrine sediments is in the tens of micromolar range (0.1–1% of the level in seawater), and when it is depleted, carbon dioxide becomes the major electron acceptor, and methane is the major product. Thus, in terms of sediments as microbial habitats, one might regard sulfate concentration as one of the key variables regulating the microbial metabolism and biogeochemical cycling.

SAMPLING AND DESCRIPTION OF SEDIMENTS

Pore Water

Pore water is what it implies: the interstitial water occupying the spaces between the sediment particles (Figure 1). Pore water chemistry is commonly the parameter that is measured to characterize a sedimentary environment and in terms of the microbial ecology of these environments, and is arguably the single most important property of sediments to measure. Pore waters can be obtained by sectioning and centrifuging sediment core samples, by whole-core squeezing, or by allowing diffusion into chambers called "peepers" that are allowed to equilibrate with the sedimentary environment. Sectioning and squeezing are usually performed in an anaerobic glove box to avoid effects of oxygen on sensitive components like sulfide or ferrous iron, but even with this precaution, results can be quite variable. Squeezing sub-cored sediment and expressing the pore waters directly into syringes minimize handling, and often yield much more reproducible results with regard to iron. Peepers, while yielding high quality data, often need to be equilibrated for long periods, not compatible with many field expeditions. Another approach to obtaining pore water chemistry profiles is that of the use of microelectrodes, which directly measure the concentrations (activities) of certain species. This method allows for very high spatial resolution, rapid data acquisition, and high reproducibility. However, the electrodes are solute specific, often very expensive, and can be quite fragile.

Porosity and Tortuosity

Interlinked with the concept of pore water nutrient chemistry are two other key sediment properties: porosity and tortuosity. The porosity is a measure of sediment-water content, defined as the volume of pore space per unit volume of total sediment. Typical surface ocean or lake sediments will have a porosity exceeding 80% (0.80); coarse-grained sediment will have a lower porosity than fine-grained sediment (as clays tend to stand on edge and create large porous volumes in fine-grained sediment); and sediment porosity will decrease with depth due to compaction. Tortuosity describes the path a molecule or any other solute must take in order to move from point A to B within a sediment column. Because sediment grains come in the way, the distance between A and B becomes a tortuous path with a length greater than the straight line between points.

SEDIMENT PROPERTIES

Redox Budgets of Sediments

Sediments arrive at the floor of an aqueous body carrying a load of C_{org} associated with the organic matter. The source of the C_{org} can be productivity within the water or it can be transported in from the continents by erosional processes. Insofar as there is no known sterile body of water on the earth's surface (albeit some ponds subjected to extreme metal loading may be near to being sterile), there will always be this association of adsorbed C_{org} and microbes with sediments.

Sediments can also carry a certain oxidizing potential, most often as particulate iron oxyhydroxides (FeOOH) but also manganese oxides. Terrigenous-dominated sediments have the largest quantity of this solid-phase oxidant, typically 4–7 wt% Fe, whereas Mn might constitute up to 0.1% of sediment mass. While there is no "typical" C_{org} content of sedimenting particles because this parameter will depend on the productivity of the surface ocean or lake, coastal ocean sediments settle to the seafloor with a C_{org} content of 5–10 wt% and many deep-sea sediments settle to the seafloor carrying 3–5 wt% C_{org}. As an example of an electron budget for sediments, assume that particulate material falling to the coastal ocean seafloor is 50% terrigenous and 50% biogenic – the terrigenous material carries 5 wt% Fe as FeOOH and the sedimenting material contains 10 wt% C_{org} (all reasonable values). Such a sediment will be carrying an oxidizing potential (electron acceptors) in deficit of the total reducible load by 2 orders of magnitude! As illustrated here, the reduced load of settling particles is typically much greater than the oxidizing potential of the solid phase, hence the importance of dissolved oxidants and their utilization by heterotrophic prokaryotes in sediment diagenesis.

As indicated by the above mentioned example, an important part of the sedimentary redox budget relates to the dissolved oxidants in the overlying and interstitial spaces. Most sediments are overlain by oxic water, providing ~200–400 $\mu mol\, l^{-1}$ oxygen (depending on salinity and temperature), and minor amounts of nitrate and CO_2. In sediments of high organic content (and high microbial activity), CO_2 and other oxidants are generated *in situ*, further complicating the matter. The major source of oxidizing power in sediments originates from solutes within the overlying water column. Sulfate (SO_4^{2-}) is perhaps the defining redox (and microbiological) difference between marine and freshwater sediments. It is the major oxidant in marine systems, being present in oceanic waters at a concentration of ~28 $mmol\, l^{-1}$ (100× the concentration of oxygen and 4× the number of electrons to be accepted makes sulfate 400 times more potent than oxygen as an oxidant in marine waters), yet often at only ~50 $\mu mol\, l^{-1}$ in freshwater systems. Thus, the chemical reduction of a variety of sulfur species (sulfate, elemental sulfur, thiosulfate) dominates marine sediments, and is comparatively absent in freshwater sediments, which tend to be dominated by methane production as the final redox activity.

Microbial Input to Sediments

Particles arrive at the sediment-water interface carrying a microbial community. Counts of bacteria on sedimenting particles in the ocean yield a wide range in values ($1 \times 10^{5-1} \times 10^{10}$) bacteria per gram of sediment, with the upper surface layers of sediments usually harboring about 10^7–10^9 microbes per gram of sediment. Typically, the bacterial density on sedimenting particles exceeds the bacterial density within sediments. However, the biomass of these attached microbes is much lesser than the biomass of organic matter associated with the particles, with the prokaryotic community contributing typically 3% or less of the total C_{org} burden.

SEDIMENTS AS HABITATS FOR MICROBES

Sediment Properties Relevant to Microbial Ecology

From the point of view of microbial ecology, sedimentary environments are inextricably linked by the facts that they are (1) continuously wet and (2) physically stabilized – that is, the dynamics of the sedimentary environments are defined by diffusion and advection, and not by convection and turbulence more characteristic of the water column. While sediments and soils are often considered to be similar, from a microbiological perspective, they are in fact quite different. Soils are subject to constant changes in wetting and drying, often with extreme dehydration being part of the niche description. In addition, soils lack the constant deposition of new material from above that is so characteristic of sedimentary systems, which are built from above. While diagenetic changes occur within sediments as a result of microbial activity, the primary inputs of material and energy are supplied from above. Soils, on the contrary, are strongly impacted by inputs from within – usually supplied by roots of plants, or decaying plant matter. To some extent, then, sediments can be viewed as much simpler and more predictable environments, once the key variables noted earlier have been characterized (Table 1).

Perhaps one of the most relevant properties of sediments, and one that greatly increases their diversity (and makes them somewhat unpredictable), is that they are composed of particles (clays, minerals, detritus, etc.), and these particles are excellent habitats for microbes and the places where biofilms form. Thus, the biomass in sediments often exhibits a rather nonuniform distribution of microbial biomass, with some particles being covered with thick biofilms, and others being rather devoid of microbes. This nonrandom distribution must surely be of great importance in influencing the way(s) that microbes interact with the geochemistry of the sedimentary niche, and represents one of the great intellectual challenges to understanding the sedimentary niche.

Microbial Abundance in Sediments

Surficial sediments (those near the sediment-water interface) are typically rich in microbes present because of the energy available in the sediments. As mentioned earlier, typical surface sediment numbers have resident

TABLE 1 Comparison of major properties of sediment and soil environments

Sediments	Soils
	Input from within: plant roots (photosynthesis and nutrient transport), decaying cellular material (e.g., roots)
Input primarily from above via sedimentation: clay particles, minerals, and microbial biomass	Some accumulation via wind-blown particles (dust and minerals)
Permanently wet – no airspaces	Alternately wet and dry cycles
Electron acceptors rapidly used up	In dry periods, airspaces are present over all oxic soils
Anoxic zone usually shallow	
Because of anoxia, eukaryotes are rare	Because of airspaces, abundance of eukaryotes (nematodes, protists, and fungi)
Dominated by bacteria and archaea	
Primarily controlled by diffusion	Primarily controlled by mixing (plant roots, fungi, nematodes, small eukaryotes, and water flow during wet periods)
Eukaryotes not common except at surface	
Characterized by chemical layers due to consumption of nutrients and production of products (at rates faster than they can diffuse)	Layers are rare, though occasionally present due to wind-blown events (e.g., clay particles) During wet periods, anoxia is common, and layers can form
Gradients that form tend to be simple, two-dimensional gradients, unless disrupted by bioturbation	Gradients can form around particles, and can be seen as complex 3-D gradients
	Much less predictable

prokaryotes that reach numbers on the order of 10^9 per gram of sediment. One can imagine that such bacteria- and organic-rich environments lead to a rapid depletion of oxygen, and this in fact is a nearly universal feature of sediments. Depending on organic matter input, the so-called oxic–anoxic interface will be present as a result of aerobic activity, rendering the remainder of the subsurface an anoxic environment. Thermodynamic "logic" then prevails, and the other available oxidants are consumed in order, with each oxidant providing a measurable chemical layer as it is consumed.

Generation of Chemical Gradients by Microbial Activities

Dr. Robert Berner of Yale University was the first to describe the succession of chemical reactions that occur when various common oxidants react with C_{org} in sediment. The succession is based on the free energy derived from each reaction; the oxidants are those commonly found in aqueous solutions or associated with sediment solid phases. The succession of reactions that are seen in pore water profiles is the result of microbial metabolism leading to the accumulation of products, or the consumption of reactants at rates greater than they can be removed or supplied by diffusion. The succession is readily seen in pore water chemistry, and it occurs as different oxidants, used in the free energy succession, are sequentially depleted. The free energy paradigm applies to freshwater sediments as well, with the major difference relating to the paucity of sulfate in freshwaters. Hence, sulfate reduction is a minor process in freshwater sediments, where the transition to methane production occurs more rapidly than in seawater.

The cascade of oxidants involves both reactants and products of C_{org} respiration. For example, the oxidation of organic matter with oxygen yields both CO_2 (which can serve as an oxidant in some reactions) and nitrate, as organic NH_3 is oxidized. These product oxidants are also utilized in the cascade sequence of free energy such that even in sediments underlying water containing no nitrate, nitrate can be one of the oxidants utilized by sedimentary heterotrophs.

The succession of geochemical profiles implies that sedimentary bacteria adhere strictly to the "rules" of thermodynamic energetics. The structure of geochemical profiles obtained from pore water analyses generally supports this argument. However, one misguided assumption associated with the layered microbial community (LMC) paradigm is that oxygen consumers "outcompete" nitrate consumers, for example. It has not been demonstrated that competition is involved with the structure of geochemical profiles or bacterial communities. While there is a clear trend in declining bacterial numbers with depth in the sediments, the association of bacterial turnover rates and geochemical zonation has not been explicitly shown.

CHEMICAL PROFILES IN SEDIMENTS (LAYERED MICROBIAL COMMUNITIES)

What is a LMC?

Perhaps the key indicator of an active sedimentary environment is the presence of distinct layers like those described earlier – geochemists call these chemical profiles, while geobiologists prefer to call them LMCs, or layered microbial communities. It is a semantic issue on one hand, and an important intellectual issue on the other. What is measured is pore water chemistry, and the chemical gradients are clearly seen – what is responsible for them is the activity of the resident microbes, feasting on the energy available at these interfaces, and providing the chemical catalysis for their very existence. That is, dead or sterile sediments have chemical profiles that are dictated by diffusion between end-member boundary conditions. Sediments with life often show chemical gradients that overrule diffusion as life will involve consumption or production and modification of the diffusive profiles. The interplay between metabolic activity and the physics of the sedimentary environment leads to the establishment of metabolic layers, and these layers can then be used to establish the processes that led to their establishment. The definition of a metabolic layer does not preclude any other geochemical reaction from occurring within that layer given the right circumstances – it is simply defining the predominant geochemical activity for the given environmental setting.

The chemical and biological consequences of the physical nature of sediments have to do with the solubility and movement of key nutrients through this diffusion-limited environment. Perhaps the key nutrient is oxygen, the solubility of which ranges from about 200–400 $\mu mol\,l^{-1}$ depending on temperature and salinity. Thus, one usually sees oxygen depletion in sedimentary systems, depletion that can occur within a millimeter or less of the sediment-water interface (in eutrophic (highly productive) lakes or marine sediments), to many meters in deep-sea sediments underlying the oligotrophic (minimally productive) ocean gyres. Oxygen limitation, of course, leads to the use of other electron acceptors, including nitrate, metal oxides, sulfate, and finally CO_2. Each of these electron acceptors is used in its thermodynamic "order," and its removal (or the appearance of its reduced product) can be used to characterize the microbial activity of the sediment system, as shown in Figure 2.

The bottom line in terms of defining this environment is that the physical and geochemical conditions set the boundaries for the system; these boundaries define to some degree the chemical and biological consequences on the environment – the processes that will occur can be predicted with some certainty – thus, the biological consequences can be predicted in terms of process. Prediction in terms of species is much less certain: For some of the processes, like

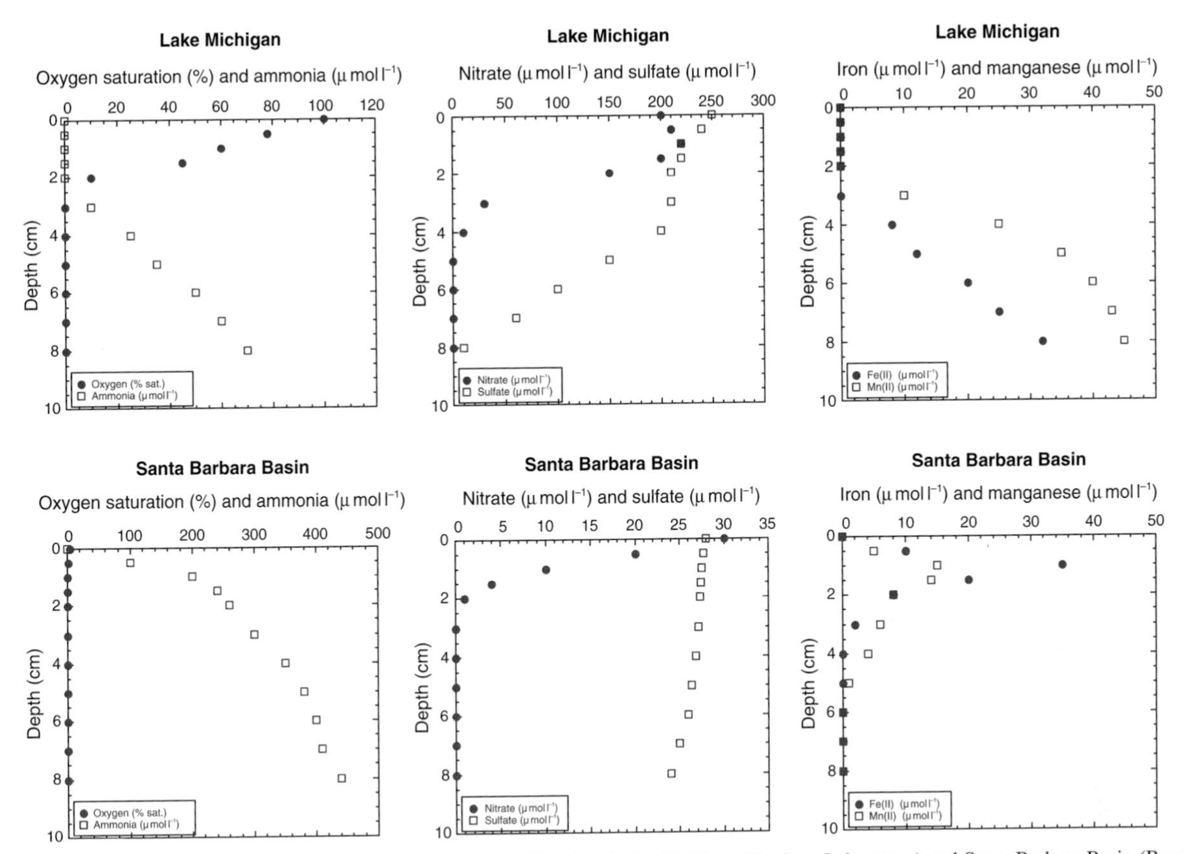

FIGURE 2 Comparison of freshwater and marine pore water profiles from Lake Michigan (Nealson Laboratory) and Santa Barbara Basin (Berelson Laboratory). Of particular note is the lack of oxygen in Santa Barbara Basin sediments, the rapid removal of Fe(II) and Mn(II) in these sediments and the high concentration of ammonia within these marine sediments. The production of ammonia in Lake Michigan sediments is linked to denitrification, sulfate reduction, and metal reduction.

oxygen depletion and nitrate reduction, many taxonomic groups are capable of the catalysis, while for others, like sulfate and CO_2 reduction, the taxa that can be expected are much more well defined.

The Rates of Sediment/Solute Transport Processes

The microbial community associated with sediment solid grains is likely dependent on the movement of solutes to bring various nutrients and energy-yielding chemicals to their habitat. However, it is possible that the sediment grain to which they attach is replete with all the necessary food items. Solid grains including C_{org}, N_{org}, P_{org}, P_{inorg}, and various trace metals are commonly found adsorbed to these and other particles. The oxidants manganese oxide and iron oxide are solids themselves, and are excellent scavengers of trace metals, as well as being found associated with other solid particles. Thus, the sediment grain itself can provide all of life's essential nutrients and the ultimate source of these solid nutrients is the sedimenting particle. As energy-yielding reactions occur, for example, manganese and iron oxide reduction, some of the solid particles (e.g., manganese and iron oxides) can be solubilized, leading to the release of trace metals and other adsorbed nutrients. Such reactions can lead to movement within sediments of layers of metal oxides and the nutrients with which they are associated. Chemical thermodynamic theory predicts that the most easily reduced oxides will be utilized and depleted first, followed by other more crystalline and less reactive oxides. Thus, manganese oxides, which are very easily reduced, will almost always be reduced (solubilized) prior to iron oxides in sedimentary environments, leading to a separation of these two metals in space.

Because sediments represent a record of time, the oldest sediments are deeper within the column than shallower sediments, and because metal oxide depletion occurs close to the sediment-water interface, as particles are buried, the abundance of easily reducible iron or manganese oxides declines. Reduced manganese and iron are soluble and will diffuse away from regions of high concentration. The consequence of this process is that manganese and iron oxides get diagenetically recycled or regenerated when these solutes come in contact with molecular oxygen or nitrate. Alternatively, the reduced manganese and iron may be sequestered as minerals. The reoxidation generally occurs near the sediment-water interface. Thus, a cycle of manganese and iron reduction and oxidation is very common in sediments overlain by water containing oxygen or nitrate.

The reoxidation of reduced metals generates a surplus of oxidants near the sediment-water interface. However, the process of bioturbation, the physical mixing of sediments by the activity of macrofauna, will carry oxidized Fe and Mn particles downward into the deeper sediment column where solute oxidants are absent. Hence, not only are Fe and Mn recycled and renewed when they become oxidized from the soluble reduced form, but they are also injected back into zones that are free of oxidants by sediment mixing. In sediments subjected to bioturbation, there will always be a race between microbes adapting to their particular redox environment before mixing carries them into a new and different redox zone.

Microbial environments also change within sediment habitats because there is a slow transport of pore water due to the compression of sediments as burial progresses. Upon initial settling, sediments are relatively uncompacted – the porosity can approach 0.95 or greater. As sediments become buried, compaction of the grains reduces the porosity. By 10 cm the porosity may be reduced by 10%, by 10 m the porosity is often reduced by 25–40%. This effectively creates an upward flow of pore water, relative to the compacting sediment (the water is incompressible; the grains settle and compact within this medium, giving the mathematical impression of advective "flow" toward the sediment-water interface). This process, termed pore water advection, will juxtapose pore water microbial communities with sediments other than those with which they were originally associated. In shallow coastal regions and some lakes, the rate of sediment accumulation is fast and the upward advection of pore water can be on the order of centimeter per year. However, in most of the ocean this rate is very slow, much lesser than millimeter per year.

Because sediment pore water is generally not subject to turbulent mixing, the most ubiquitous transport mechanism for solutes moving through sediments is molecular diffusion. In the absence of any chemical reaction, diffusion would homogenize the concentration of all solutes throughout the sediment column. And in the vast majority of ocean sediments that are absent of bioirrigation or other mass transport processes, diffusion limits the rate at which solutes can move through pore fluids. An added complication in a pore water–sediment environment is that diffusion cannot take a straight path between high and low concentration. Sediment grains effectively block the most direct diffusive path length; hence, the term tortuosity describes how far from linear the path length has become.

The fact that virtually all sediment pore water solute profiles of the biologically relevant chemicals show curvature, maxima, and minima and approach zero is evidence that there are reactions occurring within the sediments. As diffusion will generate a linear gradient between two fixed boundary conditions, the curvature in a solute profile is easily transformed into an understanding and quantification of reaction rates. The production or consumption of a chemical between two sediment horizons is defined by the flux across each boundary. If more solute is entering the sediment zone (between the two defined horizons) than is leaving the zone, there is net consumption occurring between these horizons. If more solute is leaving a sediment zone than is entering, there is net production occurring within that zone. The rate at which a solute will pass through a sediment horizon (a plane) via diffusion is defined by Fick's first law:

$$J = \frac{\mathrm{d}c}{\mathrm{d}z} D\phi,$$

where J is the flux (moles $\mathrm{cm}^{-2}\,\mathrm{s}^{-1}$), C the concentration in pore water, z the depth in sediment, D the compound's molecular diffusivity (corrected for the tortuosity and considered a physical constant for a given ion, given temperature and salinity), and ϕ the sediment porosity. The difference in flux between two sediment horizons defines the reaction rate. Production or consumption is defined as

$$R = \frac{J_1 - J_2}{z},$$

where R is production or consumption (moles $\mathrm{cm}^{-3}\,\mathrm{s}^{-1}$), J the flux across horizon 1 or 2, and z the distance between horizons 1 and 2.

The ability to translate geochemical gradients into an understanding of fluxes and reaction rates (production and consumption) is extremely important to understanding microbial activity in sediments. A rate measurement made from diffusive flux calculations at two sedimentary horizons will, of course, only be as good as the quality of the gradient data and the assumption that diffusion is the major transport mechanism. Yet these types of data and the diffusive assumption are remarkably robust, and form the basis for much of our thinking about sedimentary dynamics.

It can be informative to compare bulk sediment reaction rate calculations, derived and used by geochemists, to microbial respiration rates as determined by microbiologists. There are numerous estimates of the rates of oxygen consumption in sediments, but to our knowledge, there are no published microbial oxygen respiration rates normalized to a per cell basis. To this end, our own laboratories have made some bacterial oxygen consumption measurements and found that within a factor of 3, microbes during growth phase can consume oxygen at a rate of 50×10^{-10} nmol $\mathrm{CFU}^{-1}\,\mathrm{s}^{-1}$. Given this rate of oxygen consumption per cell and applying a rate of 0.05 nmol $\mathrm{cm}^{-3}\,\mathrm{s}^{-1}$ for bulk sediment oxygen consumption (a high rate indicative of eutrophic coastal environments) suggest that 1×10^7 cells cm^{-3} in growth phase could account for the measured oxygen consumption. This might represent 1–10% of the cells living in a sediment community, or imply that few, if any, of the cells are in the so-called growth phase that we so often study in the laboratory.

There are also many published values for sulfate reduction rates (SRRs) in sediments, generally established using a tracer spike of radiolabeled ^{35}S, applied to recently collected bulk sediments, but in this case, in opposition to the situation with oxygen, there are many published estimates of SRRs per cell, generally measured on pure cultures in the laboratory. The range of SRR per cell is very broad, as are rates measured in the field, but given published values of SRR within bulk sediments, and a range of 3 orders of magnitude in cellular SRRs, it is possible that sulfate-reducing bacteria (SRB) make up a substantial fraction (>10%) of the total sediment inventory.

The two abovementioned examples demonstrate that the connection between cellular activity and bulk sediment activity is not well understood. One outstanding question is how many sedimentary microbes are undergoing exponential growth, and another is whether there are a few superconsumers or if all sedimentary microbes are tuned to respire within a certain range. It is certainly true that sedimentary microbial systems dominated by C_{org} respiration are much more complex than pure cultures examined during optimal growth conditions.

Can Microbial Populations be Predicted from Chemical Layers?

One of the key questions that arises from the study of sedimentary environments is whether it is possible to predict the nature of the microbial populations by the profiles that are present. We propose that it is to some extent possible to predict the general nature of the reactions that must have taken place to account for the observed profiles. However, that is probably as far as it goes. There is a tremendous redundancy in nearly every process – a redundancy that precludes the specification of which phylotypes should be expected to be present. Thus, one has to settle with knowing that oxygen utilizers, denitrifiers, manganese reducers, and so on are present, but other strategies must be employed to identify the organisms actually present and active.

For example, as the geochemists would identify the "zone of manganese reduction" by a profile of manganese as shown in Figure 2, microbiologists would call this the zone of manganese reducer activity. However, there are literally hundreds of different microbial species capable of manganese reduction, thus relegating one to specifying processes rather than species, or even genus, names.

The other part of the predictive process is that most of the processes being studied are in fact cyclical. The oxygen diffusing down into the sediments acts not only as the electron acceptor for heterotrophic metabolism, but as the oxidant of many inorganics, regenerating them for further oxidative activity in the zones of reduction. Thus, near every reductive interface, as described by the geochemists, is an oxidative one, in which the electron acceptors are regenerated, almost always via the microbial activity(ies). As with the reductive activities discussed earlier, these activities can be specified, but the microbes responsible for them can only be designated by metabolic group.

For the microbial ecologist, then, the sedimentary environment remains a great challenge in terms of understanding the diversity and interactions that occur as one proceeds downward through the LMCs.

Complications Introduced by Activities of the Macrofauna

Sediments that are mixed by macrofaunal activities (bioturbation) show complexities in microbial distribution due to the introduction of oxygen (and other oxidants) at depth due to burrowing, pumping of water, and other activities of multicellular macrofauna of a variety of types (Figure 1). Despite these activities, however, which tend to mix the upper surface of the sediments, one sees that bacteria distribute themselves along geochemical gradients and, in fact, are often responsible for the maintenance of these gradients although sometimes pore water chemistry and microbial community structure are uncoupled by this process. Sediments that do not undergo physical mixing provide excellent systems for the calibration of bacterial community activity, gradient formation, and community evolution: Mixed sediments provide sites for understanding the same processes in a much more complex setting of three-dimensional diffusion gradients, perhaps more reminiscent of soil environments.

Intersection of Chemical Gradients

A fascinating feature of sediment communities relates to the places in which chemical gradients of different types intersect with one another, such as can be seen in the sulfate/methane horizon in deep-sea cores. At such interfaces abundant microbes are seen, and the process of anaerobic methane oxidation, which occurs seldom if at all elsewhere, is abundant. The advantage of living near a redox boundary is that the energy-yielding reaction space increases: For example, oxygen utilization in the respiration of C_{org} requires a microbial community capable of breaking C_{org} into oxidizable-sized compounds and it requires that the respiring microbes "find" both the oxidizable compound and the oxygen simultaneously. The presence of oxygen in contact with ammonia, diffusing up from deeper in the sediment column, provides a fixed location where both energy-yielding compounds coexist. In fact, the constancy of geochemical gradients and redox boundaries within the sediment column is evidence that sedimentary microbes adapt to and will aid in the perpetuation of these gradients and boundaries.

Sometimes the net free energy yield of a coupled set of geochemical reactions is sufficient to support a sedimentary microbial community even when some of the "players" are working in a free energy deficit. An example of this is where sulfate is used in the oxidation of methane, a process known to be catalyzed by a consortium of archael methanotrophs and SRB. It has been determined that while the SRB capture significant energy from this coupled reaction, using H_2 as an intermediate, there must be something "in it" for the archaea for their role in this coupled reaction costs them energy.

The further one looks into sedimentary bacterial activities, the more one sees that a simple thermodynamic explanation (i.e., the free energy paradigm) is a rule that is commonly broken. Under eutrophic and low oxygen water columns lie sediments containing little or no oxygen. In these sediments live bacteria that can sequester nitrate from the overlying water column and transport that nitrate within the sediments to zones where sulfide is present in high quantities. Under diffusive conditions, nitrate would not penetrate to the depth of high sulfide, yet these bacteria, for example, *Beggiatoa* and *Thioploca*, are capable of bypassing the diffusive zone and injecting nitrate into sediments rich in sulfide, thereby oxidizing the sulfide to S^0 or SO_4^{2-}. An unsolved enigma is that the greater free energy yield would be obtained if nitrate were to react with Corg compounds. Sedimentary bacteria find ways to utilize unusual combinations of reduced and oxidized compounds, not only those in association with free solutes but also those in the solid phase.

NUTRIENT-POOR SEDIMENTARY ENVIRONMENTS

Recent studies of deep-sea sediments underlying low productivity oceanic zones have revealed that a major part of the deep sea (perhaps 20% or more) consists of sediments that are so organic poor that the numbers of microbes are far less than those usually encountered in near-shore and productive environments. For example, Steve D'Hondt, Bo Joergensen, and their colleagues, in studies of sediments in the South Pacific gyre area, have found surface sediments that typically harbor only 10^3–10^5 bacteria per gram of sediment – orders of magnitude below those found in other more nutrient-rich horizons. Predictably, the chemical gradients typical of more active zones are absent here. Life becomes "cryptic" in terms of using geochemical profiles to define it. The nature of the microbes in such environments is curious – they appear to be alive, but nondividing or extremely slowly dividing, with estimates of doubling times extending to thousands of years.

Subsurface Sedimentary Microbiology

In a situation similar to that seen in nutrient-poor sediments, the deep subsurface of ocean sediments appears to harbor a low-density (10^5 per gram) population that is dividing at very low rates (again, perhaps thousands of years), This so-called starving majority as it has been called by Bo Joergensen constitutes one of the current great mysteries in microbial ecology, and it is tempting to speculate that it may be similar to environments of early Earth, when C_{org} may have been much less abundant, and organisms might have utilized mechanisms for survival not often needed on the nutrient and energy-rich surface of our planet.

Paleontological Properties of Sediments: The Rock Record

As noted earlier, diagenesis is the process of sediment alteration, a process that begins as soon as the sedimentary material arrives, and continues as long as energy is flowing through the system. Superimposed on this are the paleontological consequences of sedimentation. As diagenesis occurs, a number of biosignatures can be deposited as records left to be examined in ancient sediments – records of early metabolism, even in the absence of living biomass, or even cell structures. Biosignatures can consist of molecules indicative of certain taxa and the processes they catalyze (i.e., markers indicative of photosynthetic activity, particularly of oxygenic photosynthesis), as well as stable isotopes of carbon, sulfur, and/or nitrogen that can be fractionated by biological systems. Such molecular fossils are often the only records available in the ancient sediments, and even these are often altered beyond recognition by diagenesis and metamorphosis (cooking and alteration!) of the sediments. However, it is these ancient sediments and their geobiological records that provide hints as to the earliest metabolism(s) on Earth, and even to the first detectable signs of life on our planet.

SEDIMENTS IN SHALLOW WATER AND THE IMPACT OF LIGHT

Almost everything discussed in this article has dealt with the sedimentary environment in the deeper and dark zones of oceans and lakes. Of course, along the continental margin, and in many lakes, sunlight can penetrate the overlying water, adding new complexity to the sedimentary environment. Now the environment becomes even more distinctly layered, often with oxygen-producing cyanobacteria at the surface, and other layers of anoxygenic photosynthetic bacteria below. The oxygen penetration depth can change dramatically between day and night, with oxygen production by the cyanobacteria during the day often pushing the oxic zone to several centimeters, and oxygen consumption by the heterotrophs leading to anoxia, even at the surface at night. This being said, however, the general features of the environment still apply: It is a permanently wet, diffusion-controlled niche that is strongly impacted by physical (porosity, tortuosity, etc.) and chemical (salinity, anion content, etc.) factors.

CONCLUSIONS

In closing, it is appropriate to remind ourselves of the general properties of the sedimentary niche, which is perhaps the largest single niche available to life on the planet.

1. Sediments are not soils – they accumulate biomass by sedimentation from above, and are constantly water saturated.
2. Sediments are physically stabilized, and therefore solute distributions are primarily controlled by diffusion – nutrients must diffuse from above (or sometimes below), and this diffusion controls the overall character of the niche.
3. Sediments can (and often do) have zones of bioturbation, a type of mixing imposed by eukaryotic multicelled organisms that pump water and nutrients into the sediments, as well as physically mixing the environment.
4. Sediments typically become anoxic due to respiration of the microbial population removing the oxygen at a rate faster than it can be supplied by diffusion, thus leading to widespread anoxic sediment distribution.
5. Other oxidants (nitrate, manganese oxides, iron oxides, sulfate, and CO_2) are then used up via respiration, leading to chemical layering of sediments, processes that result in LMC formation.
6. Pore water analyses reveal the existence and nature of the LMCs in sediments, and allow one to specify to some degree the nature of the processes occurring in a given sedimentary habitat.
7. Sediments in the photic zone take on new properties due to the daily cycles of photosynthesis and oxygen consumption, and are further complicated by the input of organic matter directly to the environment by photosynthesis. However, they are still permanently wet, diffusion-controlled systems, subject to the same rules as other sediments.
8. Sediments are in general excellent habitats for prokaryotes, with 10^9 per gram or more cells being common. The sedimentary niche may harbor up to 50% of the biomass on our planet, and interact strongly with the biogeochemical cycles of nearly every major biologically active element.

FURTHER READING

Aller, R. C. (1990). Bioturbation and manganese cycling in hemipelagic sediments. *Philosophical Transactions of the Royal Society of London Series A – Mathematical Physical and Engineering Sciences*, *331*, 51–68.

Berelson, W. M., McManus, J., Coale, K. H. *et al.* (1996). Biogenic matter diagenesis on the sea floor: A comparison between two continental margin transects. *Journal of Marine Research*, *54*, 731–762.

Berner, R. (1980). *Early diagenesis: A theoretical approach*. Princeton, NJ: Princeton University Press.

Brocks, J. J., & Summons, R. E. (2004). Sedimentary hydrocarbons, biomarkers for early life. In H. D. Holland, K. K. Turekian (Eds.) *Treatise on geochemistry* Chichester, UK: John Wiley and Sons.

Burdige, D. J. (2006). *Geochemistry of marine sediments*. Princeton, NJ: Princeton University Press.

Jorgensen, B. B., & D'Hondt, S. (2006). A starving majority deep beneath the seafloor. *Science*, *314*, 932–934.

Madigan, M. T., Martinko, J. M., & Parker, J. (2000). *Brock: Biology of microorganisms* (9th ed.). Prentice Hall: Upper Saddle River, NJ.

Nealson, K. H. (1997). Sediment bacteria: Who's there, what are they doing, and what's new? *Annual Review of Earth and Planetary Sciences*, *25*, 403–434.

Nealson, K. H., & Popa, R. (2005). Metabolic diversity in the microbial world: Relevance to exobiology. *Society for General Microbiology Symposium*, *65*, 1157–1171.

Parkes, R. J., Cragg, B. A., & Wellsbury, P. (2000). Recent studies on bacterial populations and processes in marine sediments: A review. *Hydrogeology Reviews*, *8*, 11–28.

Whitman, B. W., Coleman, D. C., & Wiebe, W. J. (1998). Prokaryotes: The unseen majority. *Proceedings of the National Academy of Science of the United States of America*, *95*, 6578–6583.

Ziebis, W., Forster, S., Huettel, M., & Joergensen, B. B. (1996). Complex burrows of the mud shrimp *Callianassa truncata* and their geochemical impact in the sea bed. *Nature*, *382*, 619–622.

Chapter 32

The Rhizosphere

Frank B. Dazzo, Stephan Gantner

Department of Microbiology and Molecular Genetics, Michigan State University, East Lansing, MI, USA

Chapter Outline

DEFINING STATEMENT

The rhizosphere habitat includes living plant roots and closely associated soil where deposited exudates stimulate microbial activities that significantly influence plant nutrition, health and disease, and therefore are important for terrestrial ecosystems and agriculture. This article highlights studies of rhizosphere microbiology and how various research technologies have enhanced our knowledge of the microbial ecology in this terrestrial habitat.

HISTORICAL DESCRIPTION OF THE RHIZOSPHERE

In 1904, Lorenz Hiltner coined the term rhizosphere, which he defined as the "soil compartment influenced by the root." His thorough investigations showed that microorganisms heavily populate the soil surrounding plant roots, root exudates of crop plants support different microbial communities, and plant roots and their associated microflora can significantly influence each other's activity and development. For instance, some of the microbes whose abundance and metabolic activity are stimulated in the rhizosphere can significantly promote the growth of the plant, whereas other uninvited guests capitalize on root exudates by proliferating in the rhizosphere where they harm the plant. Hiltner's summary of key findings catalyzed the beginning of a century-old era of rhizosphere microbiological research that enjoys an abundance of basic and applied scientific discoveries and will continue far into the future.

GENERAL CHARACTERISTICS OF THE RHIZOSPHERE

Main Differences between the Rhizosphere and Bulk Soil

Most root-free soils are nutritionally oligotrophic, and therefore, microbes there by and large persist on a low calorie diet, with the major limitation in productivity being caused by the low availability of energy-yielding organic nutrients. The rhizosphere includes the root and surrounding soil influenced by living plant roots (Figure 1). Rhizosphere soil contains significantly increased supplies

Topics in Ecological and Environmental Microbiology.

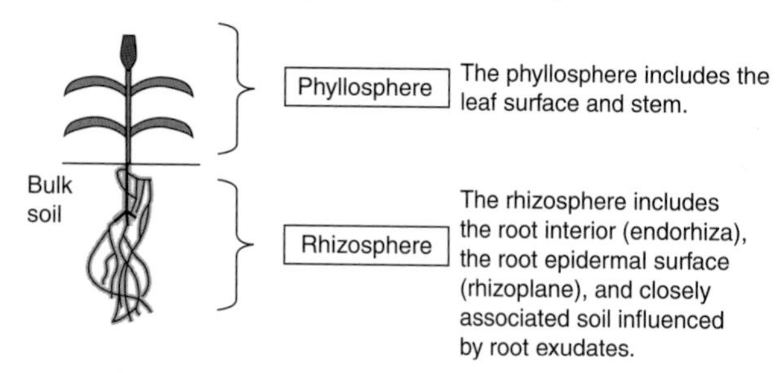

FIGURE 1 Areas of plants colonized by microorganisms.

of organic matter that include readily utilizable organic nutrients. This nutritional enrichment in the rhizosphere results in more intense microbial metabolism and productivity than what is supported in bulk soil, with accompanying changes in respiration, gas exchange, nutrient and moisture availability, and usage.

Importance of the Rhizosphere to Plant Growth

The rhizosphere is immensely important to terrestrial life on Earth. Indeed, all of the physical, chemical, and biological factors or conditions of soil that influence the physiology of plant growth are expressed through the rhizosphere and its associated microbial communities. Thus, the rhizosphere is the component of soil where major, important soil–plant relationships occur. Activities within the rhizosphere impact on the equilibria of microbial immobilization and mineralization of plant nutrients that profoundly influence their cycling in soil, on plant growth and development, and on the susceptibility and resistance of plants to infectious diseases. These interactions manifest as beneficial, neutral, or harmful associations between microorganisms and higher plants. For instance, mycorrhizal fungi colonize roots of 95% of all vascular plants, and their association impacts significantly on plant mineral nutrition and protection against soil-borne root infecting plant pathogens.

Measurements of the Rhizosphere Effect

Early studies on the microbial ecology of the rhizosphere measured the degree of enrichment with respect to cultivated rhizosphere microorganisms and the various environmental factors that influence it. A rhizosphere-to-soil ratio (R:S) was introduced to define the magnitude of the rhizosphere effect. This metric is typically computed as the parameter (e.g., density or activity of microbes) measured in rhizosphere soil divided by the value of the same measurement in nearby nonrhizosphere control soil. Thus, root-associated enrichments would produce R:S values >1, whereas those parameters that are unaffected or suppressed by roots would have R:S values equal to or less than 1, respectively. Typical R:S values based on microbial culturability vary among plant species, soil horizon, and microbial physiological groupings (Table 1). R:S ratios are typically higher in B horizon samples since its nonrhizosphere soil contains much lower organic matter content capable of supporting microbial population growth. High R:S ratios for protozoa reflect the increased abundance of their bacterial prey in this environment, and this higher bacteriovory activity accounts for a significant microbial loop that accelerates mineral nutrient cycling in the rhizosphere, thereby benefiting plant nutrition overall. A high R:S ratio of the denitrifier functional guild reflects this environment's increased abundance of organic matter, lower pO_2 due to the higher biological oxygen demand that accompanies degradation of organic matter and depletion of oxygen by microbial aerobic respiration, and the availability of NO_3^- as the microbes' second choice of alternate electron acceptor used to support ATP-yielding anaerobic respiration.

TABLE 1 Classical analysis data on measurements of the effect of various plant species on the major microbial groups and functional guilds in rhizosphere soil[a]

Microbial Group or Functional Guild	Plant	Rhizosphere Effect (R:S Ratio)[b]
Bacteria	Red Clover	24
Bacteria	Maize	3
Bacteria	Yellow Birch, A horizon	15.1
Bacteria	Yellow Birch, B horizon	57.8
Bacteria	Mangles	120
Protozoa	Mangles	3–23
Filamentous fungi	Wheat	6
Microalgae	Wheat	11
Cellulose-degrading bacteria	Wheat	5
Nitrifying Bacteria	Wheat	1
Ammonifying Bacteria	Wheat	50–125
Denitrifying Bacteria	Wheat	90–1260

[a]*Data are based on viable plate counts and derived from studies by Katznelson (1946).*
[b]*The rhizosphere to soil ratio (R:S) is the metric that defines the magnitude of the rhizosphere effect on its associated microbial community, and is typically computed as the value of the parameter measured (e.g., microbial abundance, diversity or activity) in rhizosphere soil divided by its corresponding value in nearby nonrhizosphere control soil. Thus, root-associated enhancements would produce R:S values > 1, whereas those parameters that are unaffected or suppressed by roots would have R:S values equal to or less than 1, respectively.*

Evidence for a lower *in situ* "*p*O2" in the rhizosphere is supported by manometric measurements of total oxygen uptake in rhizosphere versus nonrhizosphere soil that produce R:S values ranging from 2 to 5, depending on the plant species. This increased indicator of aerobic respiration reflects the combined influence of readily oxidizable nutrient enrichments and the large community of metabolically active, O_2-utilizing rhizosphere microorganisms.

Microbial Diversity in the Rhizosphere

Derived from neighboring soil communities, the rhizosphere habors a vast diversity of prokaryotic and eukaryotic microorganisms, including bacteria, archaea, fungi, protozoa, and green algae. Most studies of rhizosphere microbial ecology focus on the bacterial and fungal (especially mycorrhizal) components, although some new research focuses on the involvement of archaea in soil pathogenic repression and ammonium oxidation, and the association of green algae with desert plants.

Early studies characterized the diversity, morphology, and nutritional groupings of abundant culturable taxa in rhizosphere microbial communities. These studies indicated that the diversity of culturable rhizosphere bacteria was typically dominated by Gram negative, rod-shaped pseudomonads with nutritional requirements satisfied by one or more amino acids or organic acids. Populations of these dominant rhizosphere bacteria increase rapidly and are most competitive when organic nutrient inputs are high, but decline dramatically when they become limiting. They are classified as zymogenous, copiotrophic R strategists. In contrast, the large diversity of culturable bacteria in nonrhizosphere soil are typically dominated by Gram-positive, *Arthrobacter*-like pleomorphic forms with nutritional requirements satisfied by extractable factors in soil organic matter but not by simple sugars, amino acids, or organic acids. These typical nonrhizosphere bacteria are classified as autochthonous, oligotrophic K strategists that exhibit the opposite behavior of successful competition in soil with low mortality in the absence of nutrient enrichments because of their efficient abilities in nutrient uptake, conversion of soil organic matter into microbial biomass, and synthesis/utilization of storage reserve polymers that favor their extended starvation survival in the absence of external nutrient amendments.

Studies using mutant analysis of rhizobacterial pseudomonads indicate that their high competence in ability to colonize tomato rhizospheres involves several physiological systems, including chemotaxis, pili-mediated twitching motility, lipopolysaccharide O-antigen biosynthesis, metabolism of vitamins/nucleotides/organic acids, cell autoaggregation, diverse nutrient uptake and utilization pathways, and protein secretion. Other recent studies using *Pantoea agglomerans* and *Rhizobium leguminosarum* have added exopolysaccharide production, *in situ* biofilm or microcolony formation, and cell surface adhesins to this list of important determinants of bacterial competence in rhizosphere colonization.

Nutrition and Cultivation Strategies of Typical Rhizosphere Microorganisms

The zymogenous nature of typical rhizosphere microorganisms like *Pseudomonas* is also reflected in the significantly higher proportion of the culturable bacteria in the resident community. Interestingly, 40–70% of the rhizosphere community can be isolated and cultured in the laboratory, whereas typically only 0.1–1.0% of the total bacterial community in nonrhizosphere bulk soil is culturable. Thus, the "plate count anomaly" poses much less of a problem when culturing rhizosphere microorganisms than attempting to culture the microbial community in bulk soil, and therefore the rhizosphere represents an accessible reservoir of culturable microorganisms with high phylogenetic and metabolic diversity. This increased culturability for the rhizosphere community predictably reflects a higher proportion of metabolically active microorganisms in the environmental sample, that is, when microorganisms have faster growth rates and less stringent nutritional and physiological growth requirements that can be satisfied by culture media formulations and cultivation conditions in the laboratory.

Root Exudation and Rhizodeposition in the Rhizosphere

Studies by Rovira and colleagues analyzed root exudates of plants grown axenically (without associated microorganisms). These studies identified various low molecular weight organic compounds including numerous sugars, amino acids, vitamins, organic acids, nucleotides, enzymes, and miscellaneous phenolic compounds in root exudates. The diversity and abundance of the excreted plant metabolites varied among different species and ages of plants, and this information helped to solidify the concept that the rhizosphere effect is primarily a consequence of increased microbial metabolic/physiological activity made in response to nutrient enrichments through root exudation of carbon-rich and energy-yielding compounds in the soil environment surrounding roots.

The high metabolic activity and productivity of the microbial community in the rhizosphere is mainly supported by carbon inputs from live roots, which are delivered into that environment in various forms, including water-soluble compounds in root exudates (see above), dead epidermal root cells, root cap cells, and mucilage polymers that slough off to lubricate the root tip during its penetration and geotropic growth through soil. All of the carbon lost from plant roots is termed "rhizodeposition," and, in addition to the above sources, includes respiratory CO_2 and other organic volatiles like ethylene.

Much effort has been expended to measure rhizodeposition and identify factors that influence its magnitude. Values range widely in different experimental systems, reaching as high as 50% of net primary production in some ecosystems. For optimized lab studies, typically, $^{14}CO_2$ is provided to the photosynthetically active aerial plant system, and the amount of ^{14}C-labeled photosynthates released from roots is measured. These studies show that rhizodeposits of amino acids and carbohydrates are higher when roots are grown in a porous solid matrix (rather than in hydroponics), and the proportion of ^{14}C-labeled rhizodeposits is significantly higher (18–25% vs. 4–10%) when plant roots are grown in unsterilized soil (with a rhizosphere microbial community) rather than in sterilized soil. Other studies show that extracellular metabolites of rhizosphere-competent microorganisms like *Pseudomonas* and *Fusarium oxysporum* can significantly increase carbon exudation from plant roots. Important implications of these rhizodeposition studies are that plants release significant amounts of newly fixed photosynthates through their roots without assimilating it into plant biomass, and that rhizosphere microorganisms can significantly increase – even double – the amount of fixed carbon lost from associated plants by root exudation. Such results have profound implications that lead to a provocative question: Who is in charge: the plant or its associated microbes? This rhetorical question generated a lively scientific debate when raised at a recent international congress on the rhizosphere, with approximately equal representation of conference delegates advocating the microbe or the plant component as the main driving force for the association.

Rhizodeposition is not uniformly distributed along plant roots. Various approaches have been used to locate their major sites of nutrient exudation, including ninhydrin staining of filter paper imprints of growing roots to locate concentrated sites of amino acid exudation, localized chemotropism of amino acid-requiring auxotrophs of *Neurospora crassa* fungi on roots, and localization of genetically engineered *lacZ* or Lux reporter strains of rhizobacteria inoculated on plants. All four methods revealed that the major sites of root exudation are just above the meristem in the elongation region of the root and at surrounding sites of lateral root emergence.

The Volume and Spatial Scale of the Rhizosphere

The volume of the rhizosphere that extends from the root cylinder into surrounding soil is a dynamic function of the rate of plant rhizodeposition, the diffusion of nutrients within the soil matrix influenced by their size, solubility in the soil solution, absorption to and mechanical impedance of the soil matrix, and the extent of uptake and utilization of the rhizodeposits by the rhizosphere community. The radial distance of the rhizosphere has been examined by measuring the concentration gradient of carbon rhizodeposits and the density of colony-forming units and individual bacterial cells in soil sampled at various distances from the root surface. These studies indicate that the rhizosphere effect is typically most intense at the root surface (called the rhizoplane) and extends a few millimeters to little more than a centimeter out into the soil. Factors that influence this spatial scale include the species and growth stage of the plant, its root architecture and density, the light intensity and duration of the photoperiod, the texture, depth, fertility, and moisture status of the soil, and the types, abundance, and activities of the microbes, just to name a few. Figure 2 schematically illustrates how root architecture influences the volume of different plant rhizospheres, which is much larger for fibrous and highly branched root systems than for arable and nonfibrous root systems.

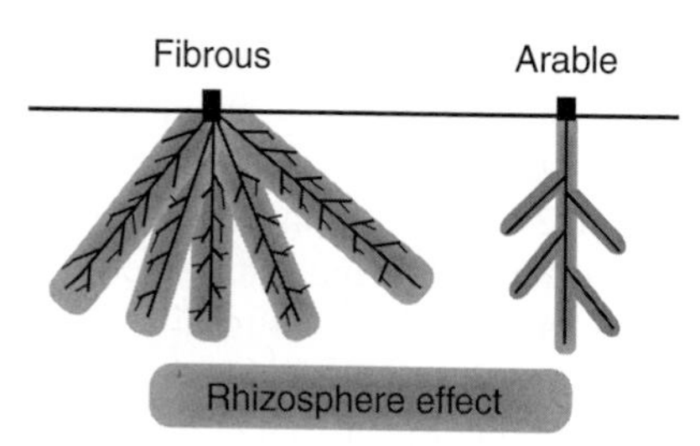

FIGURE 2 Schematic drawing illustrating how the root architecture influences the volume of the rhizosphere.

Rhizospheres of neighboring plants can occupy significant volumes and even overlap when their root densities are sufficiently high. For example, 30–40% of the top 10 cm of the A horizon soil lies within 1 mm of a wheat root at its dough stage in development. Considering the high density of lawn turfgrass, all the soil to depths of their longest roots is rhizosphere.

DIRECT MICROSCOPY OF MICROBIAL COLONIZATION OF ROOTS

Early Microscopy Studies

Starkey, Krasilnikov, Rovira, Bowen, and Foster conducted pioneering studies that utilized microscopy to examine the rhizosphere effect directly. A clever way to examine the rhizosphere effect *in situ* utilized a contact slide modification of the Rossi–Cholodny buried slide technique (Figure 3). Also, roots grown in soil were stained directly and examined by brightfield microscopy, or processed for transmission and scanning electron microscopy to examine the rhizoplane microorganisms.

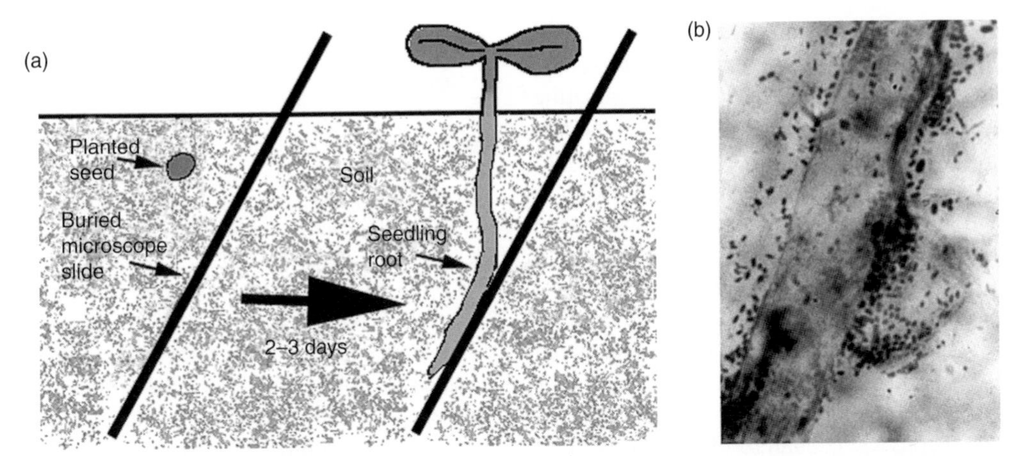

FIGURE 3 (a) Schematic diagram illustrating the experimental design of the seedling contact slide method to examine the rhizosphere effect by direct microscopy. (b) Direct microscopy of the abundant, localized colonization of seedling root tissue by rhizosphere microorganisms on a seedling contact slide. The sample was stained with aniline blue and photographed using brightfield light microscopy.

These studies verified earlier cultivation methods, showing the increased abundance of microbes in soil surrounding roots, their growth into microcolonies, ultrastructural details of the rhizosphere soil fabric, and preliminary descriptions of the microbial distributions on the rhizoplane. Other key findings made by those early microscopy studies were the direct demonstration of a mucilage layer on the root surface that sometimes embeds microorganisms within it, bacteria surrounded by abundant fibrillar capsules whose periphery is covered with particulates of clay envelopes, the morphological diversity of the rhizoplane community, increased abundance of bacteria colonization in grooves between adjacent epidermal cells where exudation is predictably elevated, local sites of eroded epidermal walls allowing microbial penetration, and direct evidence of biological control via the mechanism of hyperparasitism (beneficial microorganisms directly attacking and killing other microbial pathogens of plants). Examination of root cross-sections of field-grown plants typically showed that their root cortex naturally supports large communities of microorganisms without the plant expressing symptoms of disease responses. This benign associated community of microorganisms was initially termed the "endorhizosphere" but later the organisms were called "endophytes" or "internal root colonists."

Figure 4 is a fluorescence micrograph of the microbial community within rhizosphere soil surrounding a white clover seedling root. This image acquired by conventional epifluorescence microscopy is informative in showing the clustered distribution of the morphologically diverse unicellular and filamentous bacteria in close association with particulate autofluorescent organic matter, but it suffers because a major portion of the foreground objects are outside the plane of focus.

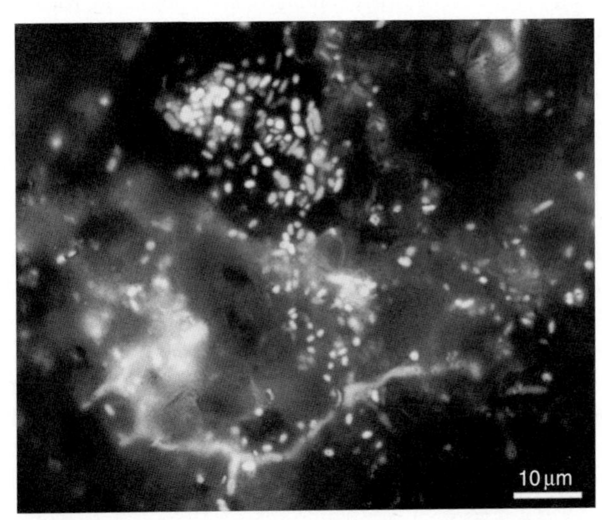

FIGURE 4 Fluorescence micrograph of the microbial community within rhizosphere soil surrounding a white clover seedling root. The sample was stained with the nucleic acid-stain acridine orange, then rinsed with a solution of sodium pyrophosphate, and examined by conventional epifluorescence microscopy. *Reproduced with permission from Prentice Hall.*

Microscopy Studies of the Rhizosphere Using Modern Technologies

Modern technologies provide deeper insights into the understanding of microbial community structures and their interactions in the rhizosphere of plants.

Improvements using CLSM

The most significant advancement in the use of fluorescence microscopy to examine root-associated microbes was the development of confocal laser scanning microscopy (CLSM). The confocal imaging system has an optical design utilizing pinhole apertures at the laser light source and at the detection of the object's image, thus eliminating the stray and out-of-focus light that interferes with the formation of the object's image. This ability to use only signals from the focused plane eliminates a major limitation of the conventional fluorescence microscope when examining plant tissue, soil particles, and organic debris that emits a significant amount of objectionable background autofluorescence and/or absorbs the fluorescent dye itself. Because the light from outside of the plane of focus is not included when the image is formed, the 2-D (x, y) image becomes an accurate optodigital thin section with a thickness that can approach the theoretical 0.2 mm resolution of the light microscope. Also, by digitizing a sequential series of 2-D images while focusing through the specimen in the third (z) dimension (called the "Z series"), an all-inclusive flattened stack or a 3D reconstructed image can be produced, rotated, and quantitatively analyzed. Dazzo and Hartmann independently and simultaneously first reported the successful use of CLSM to visualize the rhizoplane microbial community, in one case, stained with fluorescent acridine orange nucleic acid-staining dye, and the other, stained with homologous fluorescent antibody. Examples are illustrated in Figure 5a and b. By using this approach, one can map the distribution of the rhizoplane microbial community colonized on seedling roots grown in soil. Figure 6 shows a typical result, illustrating that the spatial distribution of microorganisms on the rhizoplane is discontinuous, being almost completely devoid of microbes at the root tip growing in soil, individual cells just above the root tip, some of which eventually develop into microcolonies further up the root. Overall, microbes typically cover no more than ~20% of the rhizoplane area of roots that are actively growing in soil. In some cases, single microbial cells find the rhizoplane environment to be favorable for growth, resulting in localized microcolony biofilms with *in situ* population generation times as short as 2–3 h.

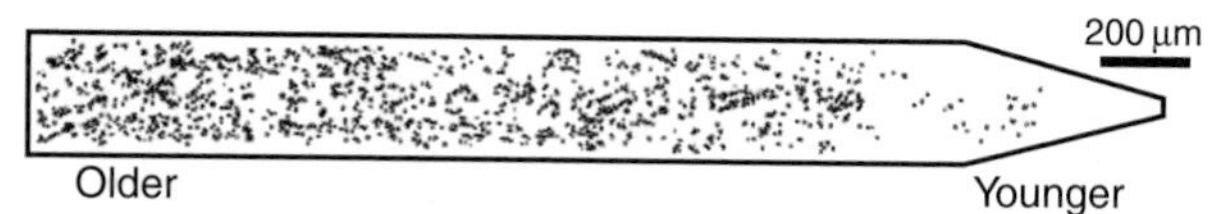

FIGURE 6 Binary montage image depicting the spatial distribution of microorganisms colonized on the rhizoplane of a white clover seedling after 2 days of growth in soil. Note the discontinuous density of colonization that is greater on older regions of the root.

A case study: Rhizobium colonization of rice roots

Rhizobium is the well-known nitrogen-fixing root nodule bacterial symbiont of legume plants. Recent studies have shown that this soil microorganism also develops natural, intimate, and sometimes beneficial endophytic associations with various cereal crops like rice and wheat. This alternate ecological niche of *Rhizobium* can be exploited in sustainable agriculture in the form of efficient biofertilizers that are inoculated on the cereal plant host at planting and can significantly promote its growth, thereby reducing its need for chemical fertilizer applications to achieve maximum grain yield. Figures 7 and 8 illustrate aspects of the colonization and endophytic infection of rice roots by *Rhizobium* using CLSM, scanning and transmission electron microscopies, respectively. These show the

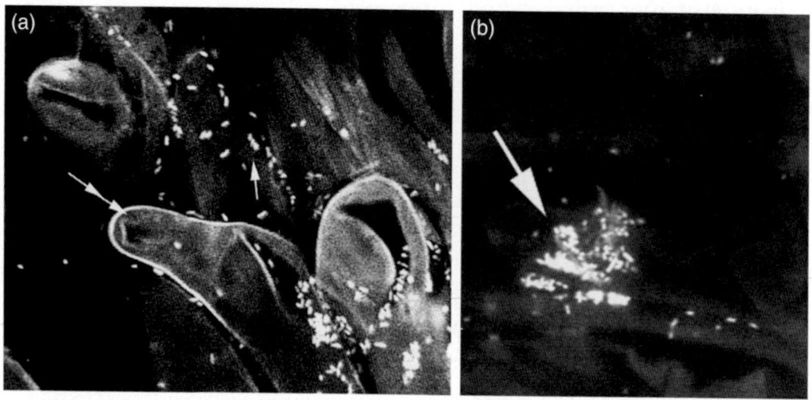

FIGURE 5 Use of confocal laser scanning microscopy to detect rhizoplane microorganisms colonized on roots grown in soil. White clover seedlings were grown in soil for 2 days, then cleared free of rhizosphere soil (Dazzo 2007), then stained with acridine orange, washed and examined by epifluorescence confocal laser scanning microscopy. The image in (A) is historically the first one illustrating the usefulness of confocal microscopy in visualizing the rhizosphere microbial community, and the image in (B) shows direct evidence of bacterial growth and formation of a microcolony biofilm (arrow) localized on the rhizoplane. From Dazzo (2004) and reproduced with permission from Springer-Verlag.

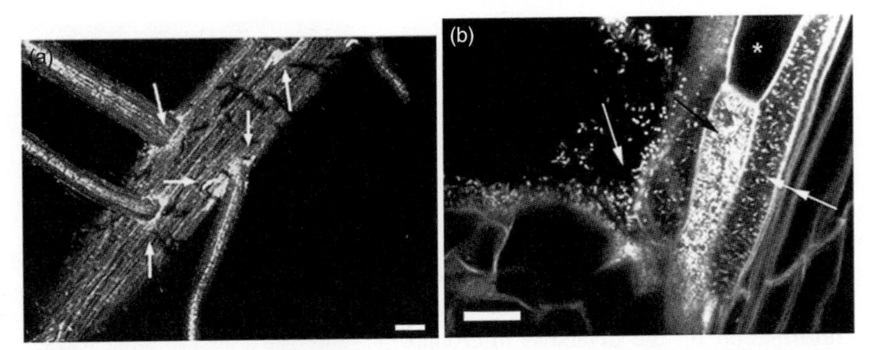

FIGURE 7 Epifluorescence confocal laser scanning microscopy showing colonization of the rhizoplane and internal root structures of rice by rhizobia. (a) Preferential colonization of a dense collar of rhizoplane rhizobia surrounding the base of emerged lateral roots. (b) Endophytic rhizobial colonization of underlying, lysed root cortical cells and the vascular cylinder of rice. Rhizobia are stained with acridine orange. Reproduced with permission from Springer.

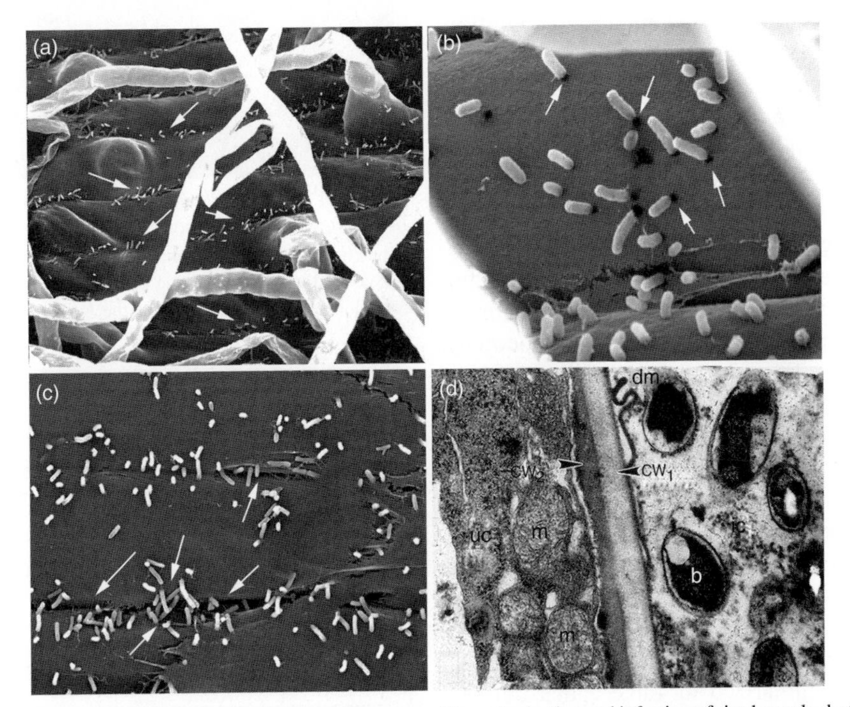

FIGURE 8 Scanning (a, b and c) and transmission (d) electron microscopy of the colonization and infection of rice by endophytic rhizobia. Images illustrate (a) colonization at epidermal root junctions (arrows); (b) localized eroded pits of the epidermal cell wall where rhizobia with cell-bound cellulases have attached (arrows); (c) crack entry mode of infection of the rice root at fissures between epidermal cells (arrows); and (d) endophytic colonization of rhizobia within a dead cortical cell adjacent to an uninfected host cell with intact membranous organelles. Reproduced with permission from CSIRO.

epiphytic colonization of the rhizoplane, especially surrounding lateral root emergence and at junctions between epidermal cells, a portal of crack entry into void spaces created by lateral root emergence and fissures between epidermal cells, endophytic colonization of dead cortical cells within the root interior, and localized pit erosion of root epidermal walls mediated by *Rhizobium* cell-bound cellulases.

Hartmann and colleagues demonstrated the value of CLSM combined with immunofluorescence microscopy (IFM) using fluorescent-labeled antibodies as specific molecular probes to examine colonization of a specific microorganism (*Azospirillum* sp.) on the wheat rhizoplane. In that study, a dual laser system was used to produce the green autofluorescence of the root background upon which the distinctive red immunofluorescent *Azospirillum* cells could be easily seen. The noninvasive optical sectioning ability of the confocal microscope was also used to locate the *Azospirillum* cells within the root mucigel layer.

Use of CMEIAS image analysis software to analyze microbial colonization of rhizoplanes in situ

Computer-assisted digital image analysis can significantly enhance the quantitative analysis of microbial colonization of roots by allowing one to extract, store, retrieve, and electronically transmit numerical information regarding selected and pertinent image features. Detailed morphological analysis of bacterial cells can provide useful information on the diversity, microbial abundance, and spatial distribution of microbial community members. A computer-aided system called CMEIAS (for *C*enter for *M*icrobial *E*cology *I*mage *A*nalysis *S*ystem) is being developed to assist in such assessments. CMEIAS is a semiautomated analysis tool that uses digital image processing and pattern recognition techniques in conjunction with microscopy to gather size and shape measurements of digital images of microorganisms and classify them into their appropriate morphotype, allowing culture-independent quantitative analysis of the morphological diversity and distribution of complex microbial communities. Also, quantitative studies of microbial biogeography are greatly facilitated using CMEIAS computer-assisted microscopy, especially for defining the appropriate *in situ* spatial scale of quorum sensing, niche overlap and competition, and microcolony biofilm colonization at spatial scales directly relevant to individual bacterial cells.

To illustrate applications of CMEIAS for autecological biogeography studies of specific organisms colonized on root surfaces, digital image analysis was performed on the fluorescent antibody-labeled cells of *Rhizobium leguminosarum* bv. trifolii on the white clover rhizoplane shown in Figure 9. In approximately 2 s of computing time, CMEIAS extracted several measurement parameters from each individual, immunofluorescent bacterial cell, producing the data summarized in Table 2 that defined the abundance and pattern of spatial distribution of the bacteria attached to the root surface. Of particular autecological biogeography interest are the image analysis data extracted by CMEIAS needed to perform plot-less distance-based and quadrat-based spatial point pattern analyses, and geostatistical analyses on spatial interactions among the bacteria. These attributes define spatial patterns of distribution, indices of dispersion and mathematical model fitting to predict bacterial colonization behavior *in situ*. The spatial randomness index, computed from the image area, average and standard deviations of the first and second nearest neighbor distances (Table 2), provides quantitative data (values < 1.0) indicating that the bacteria in Figure 9 exhibit a clustered pattern of spatial distribution. This indication of clumped spacing of organisms on the root surface is confirmed quantitatively by five other plot-based spatial distribution indices

FIGURE 9 Epifluorescence micrograph of immunofluorescent cells of *R. leguminosarum* bv. trifolii 0403 colonized on the root surface (especially at junctions between epidermal cells) below the root hair region of a white clover seedling. The bar scale is 20.0 μm. From Dazzo et al. (2007) and reproduced with permission from the American Society for Microbiology.

TABLE 2 CMEIAS image analysis of the abundance and spatial distribution of immunofluorescent bacteria on the white clover root epidermis shown in Figure 9[a]

Parameter Measured	Value Obtained
Image area analyzed	10,434 μm^2
Cell count	136
% Substratum area covered by microbes	2.7%
Total microbial biovolume	259.0 μm^3
Total microbial biomass carbon	51.8 pg
Total microbial biosurface area	1126.6 μm^2
Spatial density	13,034 cells/mm^2
Mean 1st nearest neighbor distance	3.9 ± 2.4 μm
Mean 2nd nearest neighbor distance	5.7 ± 2.9 μm
Spatial randomness index	0.88 (clustered)
Variance: mean ratio	1.796 (clustered)
Morista's dispersion index	1.996 (clustered)
Negative binomial K index	0.950 (clustered)
Lloyd's mean crowding index	1.595 (clustered)
Lloyd's patchiness index	1.995 (clustered)

[a]*Data derived from Dazzo et al. (2007) with permission from the American Society for Microbiology.*

(variance/mean ratio, Morista's dispersion index, negative binomial *K*, Lloyd's mean crowding index, and Lloyd's patchiness index; Table 2) computed from the CMEIAS data. This clumped or aggregated spatial pattern is consistent with the behavior of bacteria that remain *in situ* while actively growing into microcolony biofilms on the root surface. This occurs frequently at fissure junctions between adjacent epidermal cells just below and within the root hair region where localized root exudation is predicted to be high.

This colonization strategy is further indicated by the multimodal (rather than normal) distribution of the first and second nearest neighbor distances (Figure 10a), and by the skewed clustered abundance of short separation distances in the 2-D scatterplot of the first versus second nearest neighbor distance (Figure 10b).

In situ detection and analysis of specific, genetically marked bacterial populations within rhizosphere microbial communities

Several staining techniques can be used with microscopy to detect and measure microbial colonization of the rhizosphere and/or rhizoplane. For instance, immunofluorescence microscopy can be used to track specific inoculant strains of bacteria in the environment if the organisms can be cultured and specific antibodies reactive with them are available (e.g., Figure 9). However, chemical inhibiting factors in soil and nonspecific absorption of the stain by the root can interfere with the IFM approach. Alternative techniques using genetic engineering have been developed to avoid such problems. Genetic markers encoding fluorescent proteins are inserted into their genome, which when produced can be excited by light of specific wavelengths to make the cell autofluoresce. Commonly used genetic markers encode for a green fluorescent protein (Gfp) and red fluorescent protein (Rfp). Such techniques have been applied to study the distribution of bacterial populations colonized on roots. By using different markers and specific fluorescence optics, one can track root colonization by different bacterial populations in the same environment.

Two major concerns arise when applying this autofluorescent genetic marker approach to study bacterial colonization *in situ*. One is whether the metabolic burden of added biosynthesis to constitutively produce the autofluorescent protein influences the colonization behavior of the recombinant strain in ways that differ from its parental wild-type strain. The other is whether the genetic marker is stably maintained within the recombinant bacterial cell during its population growth within the environment. Such markers were originally introduced into the organism using plasmid constructs (extrachromosomal genome sequences) to ensure fast adaptations in their environment. But in this case, the antibiotic whose resistance is conferred by the plasmid must be present within the environment in order to create the selective pressure needed to maintain the plasmid inside the bacteria and its replication in succeeding generations. Such selection pressure may severely disturb the rhizosphere environment and other members of its associated microbial community *in vivo*, and therefore, nowadays these autofluorescent genetic markers are stably inserted into the bacterial chromosome to ensure transfer to each subsequent generation without the need for selective pressure. If such a marker is connected to a specific gene of interest, it can be used to display gene expression activity as a reporter construct as described next.

A case study combining the use of autofluorescent genetic markers in reporter strains and CMEIAS image analysis

Many biocontrol active bacteria modulate their colonization and occupation of ecological niches in natural environments like the rhizosphere using a specific cell-to-cell communication system. This system is based on elaboration of various

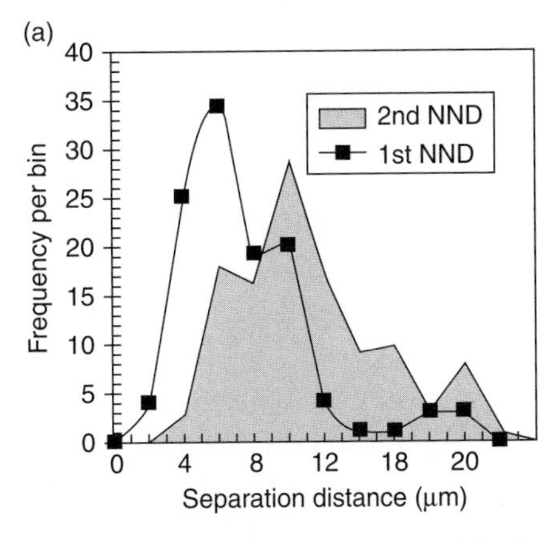

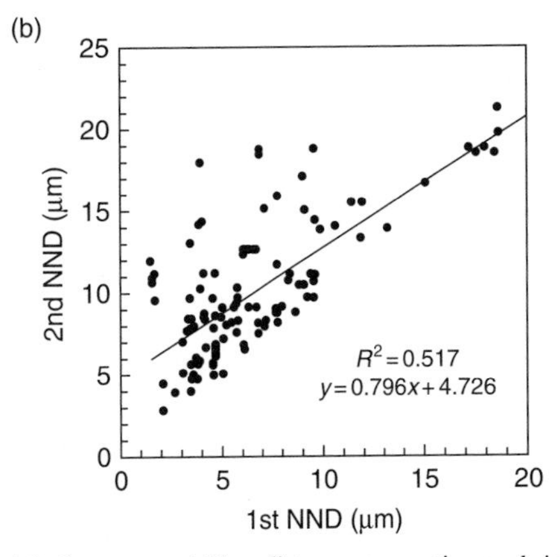

FIGURE 10 CMEIAS plot-less spatial distribution analyses of the 1st and 2nd nearest neighbor distances separating each individual bacterium in Figure 9. (a) Frequency distribution histograms and (b) 2D scatter plots. From (8) and reproduced with permission from Springer-Verlag.

low molecular weight, extracellular signal molecules that allow bacteria to perceive neighboring microbes while colonizing surfaces. The signal molecules are generally oligopeptides for Gram-positive bacteria and *N*-acylhomoserine lactones (AHLs) for Gram-negative bacteria.

The system for Gram-negative bacteria is better understood because most microbial biocontrol agents and plant pathogens are represented in this group. Above threshold concentrations, AHLs activate expression of certain genes to occupy their ecological niches, for example, during biofilm development, production of virulence factors or antibiotics, and persistence and viability on colonized host surfaces. AHL-mediated cross-talk can also be effective across species borders within microbial communities and even between prokaryotes and eukaryotes. Since the bacterial traits activated by AHLs can be beneficial or harmful to hosts, identification of the attributes that govern cell-to-cell communication is very important and has major practical application in the design and implementation of synthetic antimicrobial strategies that expressly target microbial biofilms, plant–microbe interactions, and aquatic environments. It is generally thought that high bacterial population densities are needed to exceed the threshold level of AHL concentrations required to activate genes and their physiological functions. Hence, this specific type of microbial communication has become known as "quorum sensing," functioning primarily as sensors of high population density, thus optimizing the expression of functions that are most beneficial when simultaneously performed by dense populations. This prevailing view is commonly based on measurements of gene expression within populations in stirred liquid suspension or when grown into large biofilms on artificial substrata (e.g., pellicle in unshaken broth cultural media or on plastic surfaces). However, despite its wide appeal, this quorum sensing paradigm has been recently challenged since the methods commonly used to detect it require high populations, and neither the need for group action nor the selective conditions required for its evolution have been demonstrated. Furthermore, unlike planktonic cells, many aspects of microbial physiology are different when cells are colonized on surfaces, because under this condition, the signaling molecules can disperse much faster and further than the attached population itself. For instance, spatial patterns of dispersion and active growth within discontinuous domains are major ecological determinants that govern microbial colonization of host surfaces, especially plants, and during early stages of biofilm development. One could theoretically predict that individual bacteria should benefit significantly if their sensory system could respond efficiently to small, local concentrations of the signal molecule since that would communicate information about competitive and/or cooperative activities by a few neighbors before they reach high population densities in the cell's immediate microenvironment.

Recently developed technologies using fluorescent reporter strain constructions, confocal microscopy, computer-assisted digital image analysis, and geostatistical modeling for single-cell resolution experiments have been used to measure the *in situ* spatial scale and local population requirements of AHL-mediated cell-to-cell communication during bacterial colonization of roots. These studies have shown that this type of cellular communication can be accomplished by individual bacterial cells that are separated from each other and from high population densities by long-range distances. Rhizobacteria were able to conduct cell-to-cell communication on roots with a minimum quorum requirement of two cells (no requirement for high population density) and an *in situ* separation "calling distance" of up to 78 μm (equivalent to two people talking to each other while standing ~130m apart). Also, individual bacteria in small clusters (2–3 cells) communicated with each other even when separated from dense populations by long distances. Thus, during colonization of plant roots, an individual bacterium can produce sufficient AHL signal molecules to communicate with another single bacterial cell neighbor even when those two cells are separated from each other by long distances. Geostatistical modeling of local spatial densities predict that *in situ* AHL-mediated cell-to-cell communication is governed more by the spatial proximity of cells within AHL gradients than by a quorum requirement of high population density.

These new findings of quantitative microscopy challenge the conventional view of a quorum group requirement of closely packed, high cell densities for this type of bacterial communication to occur. Instead, they indicate that AHL-mediated bacterial cell-to-cell communication is more commonplace and effective in nature than hitherto predicted. Stated simply, bacteria can produce, sense, and respond to AHL gradients made by their individual neighbor(s) during colonization of root surfaces, even when separated by long distances or from high population densities. Since these cellular interactions can benefit individual cells (e.g., sensing their competitors or collaborators), they would be evolutionarily selected, thereby explaining why they are rather widespread in the microbial world.

MOLECULAR MICROBIAL ECOLOGY STUDIES OF THE RHIZOSPHERE

The Use of Ribosomal Nucleotide Sequences

The discovery by Carl Woese in 1977 that the ribosomal nucleotide sequence can be used as identification criteria for microorganisms has revolutionized the characterization of microbial communities in environmental studies. Nowadays, many important and even cutting edge techniques use the combination of conservative and variable regions in the ribosomal gene sequence analysis to reveal taxonomical relations between organisms or similarities and differences in microbial community structure. Nucleic acid-based fingerprint techniques based on ribosomal gene sequence identification or detection methods allow one to track microbial community changes in environments without dependence on culturability. Refinements of this approach have made it more sensitive and specific, improving the detection of less-dominant bacteria, and recently, extending its use to include eukaryotic microorganisms within the same community sample. However, since these techniques most often address only partial gene sequences, issues of their accuracy when compared to whole genome sequence analyses such as the Sanger method or the recently described 454 sequence analysis are raised. In general, 16S rRNA gene analysis for prokaryotes and the corresponding 18S rRNA gene analysis for eukaryotes have revealed important taxonomic and environmental relationships of organisms in the rhizosphere. For instance, recent studies have documented the diversity of the portion of the rhizosphere community that is unculturable and the strong impact environmental pressure plays in forming those microbial community structures. Consistent with early findings based on culturing techniques, recent gene-based sequencing studies especially targeting the 16S ribosomal RNA gene of bacteria indicate that the λ-proteobacteria group (of which pseudomonads are representative) embody, on average, ~35% of all bacterial cell clones isolated from rhizosphere samples ($n = 13$ plant species), and their numerical dominance in the rhizosphere of some plant species extends up to ~65% of all bacterial clones analyzed. Many microbial specialists occupying similar or identical ecological niches show closer relations to the rhizosphere community than to other members of their taxonomical orders. Such findings raise the question of the extent to which the taxonomical nomenclature and phylogeny related to the origin and evolution of microbial life can help to explain their real ecological functions *in situ*.

MicroArray Technology

Nanotechnologies have created the ability to differentiate thousands of gene sequences spotted on polymer-coated microchips, allowing for the simultaneous search for gene sequence similarity to environmentally extracted community DNA. Today, several biotechnology companies and research groups provide DNA chips for bacterial community identification.

Functional gene arrays can potentially reveal the diversity of various known metabolic pathways potentially utilized by the rhizosphere microbial communities. However, the ideal approach of using community messenger RNA as the fluorescently labeled sample must await further technologies that overcome the current limitations of instability and therefore reduced accessibility in the rhizosphere.

A Case Study: Use of Functional Gene Arrays and 16S rDNA Analysis to Examine Effects of Climate Change on Rhizosphere Microbial Ecology

The effects of global climate changes are influencing our daily life in an increasing scale. However, many changes in the biosphere remain unnoticed. The rhizosphere community of microbes contains potentially useful indicators of environmental changes, as they are responsive to changes in root exudations of plants, which can be perturbed by environmental changes affecting plant metabolism before variations in growth patterns are discerned. A combination of ribosomal DNA-based techniques and functional gene arrays has recently addressed this environmentally important question. Collaborative studies in a project called the Jasper Ridge Global Climate Change Experiment by scientists at Michigan State University and Stanford University examined the effects of global climate change on grassland plants and their microbial rhizosphere community in a managed natural habitat near Stanford, California. At the experimental site, global climate changes have been simulated for several years by adding combined nitrogen and carbon dioxide. Community analyses using the 16S ribosomal RNA-based fingerprint technique known as the terminal restriction fragment length polymorphism (t-RFLP) assay revealed strong shifts in certain microbial populations in the rhizospheres of treated grassland plant as compared to non-treated controls. These population shifts resulting from the environmental perturbations were grass cultivar and treatment-dependent, and revealed strong connections between the belowground microbial rhizosphere community and the aboveground plant host effects. Functional gene arrays further characterized the potential functional diversity in the microbial rhizosphere community. The results showed strong influences of specific plant species and the ability of microbial communities to adapt in the rhizosphere. These studies demonstrated that different plants selected and defined their own functional rhizosphere communities, and that simulated climate changes have a strong impact on the diversity selected within those microbial communities, which in turn reflects on the physiological conditions of the plants. Indeed, despite their small spatial scale, some rhizosphere processes can impact on large-scale ecosystem responses to various drivers of global change, for example, rising atmospheric CO_2 concentration.

In related studies addressing impacts of global climate change, successful rhizosphere colonizers of *Burkholderia* and *Pseudomonas* spp. were significantly influenced by elevated CO_2 whereas species of *Bacillus* and actinomycetes (representing bacteria more dominant in bulk soil) were not. These findings clearly indicate physiological differences between the lifestyles of rhizosphere-competent bacteria and other microbial groups not adapted to that environment.

A Case Study: Use of Stable Isotope Probing to Analyze Bioremediation Processes within the Rhizosphere

Another modern technique to characterize microbial community activity is the stable isotope probing technology, which can be used to identify and track microorganisms that are actively utilizing certain specialized compounds in the environment of interest. This method utilizes a labeled substrate containing heavy, stable isotopes (e.g., ^{13}C, ^{15}N, ^{34}S) harboring an additional neutron in their atomic structure. Microbial communities can metabolize this labeled substrate qualitatively (not necessarily quantitatively) the same way as unlabeled substrates, thereby incorporating the heavy labeled atoms into their DNA, RNA, or other metabolic products, which are then physically separated from the corresponding unlabeled polymer by density gradient ultracentrifugation. This technique allows one to study specific bacterial groups selected by the substrate they can utilize *in situ*.

Using this approach, one can introduce ^{13}C-labeled xenobiotic compounds into the rhizosphere and identify organisms in that microbial community that are potentially important in degrading them, based on their utilization of the labeled substrate. This is an important tool currently being used to identify organisms involved in bioremediation of environmental pollutants. In terrestrial habitats, bioremediation is enhanced in the rhizosphere due to the increased availability of nutrients that fuel the metabolism of microbial communities responsible for this activity. This molecular tool provides the opportunity to explore more in-depth connections of the microbial communities involved in bioremediation and will likely reveal many more details about rhizosphere microbial ecology in the future.

EFFECTS OF MICROBIAL RHIZOSPHERE COMMUNITIES ON PLANTS

As mentioned earlier, the rhizosphere microbial community can form strong symbioses with plants. The outcomes influence each partner in many ways. Some principles of ecology based on eukaryotic ecosystems also apply to microbial community structures. The ecological interactions between rhizosphere microorganisms and higher plants include commensalisms (one partner benefits), mutualisms (both partners benefit), amensalisms (one partner impedes the success of the other without being affected positively or negatively by the presence of the other), and pathogenic consequences (one partner benefits by harming the other). This section presents brief examples on positive and negative effects of such interactions in the rhizosphere.

Negative (Pathogenic) Effects of Microbes on Plants

Hiltner's research identified the rhizosphere as an important factor affecting plant growth and health in relation to development of plant disease. Both prokaryotic and eukaryotic rhizosphere colonizers can be harmful to their plant host. Some microorganisms successfully compete with plants for nutrients in soil, thereby immobilizing and limiting their availability and causing plant growth inhibition. Other pathogens are necrotrophs that kill the plant hosts and then proliferate while degrading them (e.g., *Erwinia carotovora* degrades the plant cell wall structure to enter carrot roots and causes a root rotting disease). Certain fungi are the major eukaryotic pathogenic microorganisms of plants. For example, the fungus *Pythium* spp. causes a devastating root rot disease of significant agricultural importance. This pathogen invades the roots (particularly of seedlings) of many plant species, eventually killing the plant hosts by sealing off their nutrient and water transport systems for its own use. Such plant pathogens cause enormous economic damage to agricultural productions every year and therefore are of special interests in rhizosphere research. Understanding the mechanisms and characteristics of soil-borne, root-infecting plant pathogens is a major task in plant-associated research to prevent diseases and increase agricultural productivity.

Positive (Beneficial) Effects of Microbes on Plants

Not all organisms interacting with plants are harmful. In fact, many microbial associations can benefit plants, thereby attracting special interest in various fields of microbial ecology, especially in aspects of agricultural sustainability when it involves an important crop plant. The group of bacteria that benefit plant growth while colonizing their roots are called plant growth-promoting rhizobacteria (PGPR). The mechanisms of plant growth promotion used by PGPR are numerous, and can be direct or indirect. One common example is the ability of microorganisms to supply the plant with nutrients that are otherwise limited or unavailable. Examples include nitrogen-fixing bacteria (called diazotrophs) that reduce dinitrogen gas into ammonia and provide this form of fixed nitrogen to supplement the plant's nitrogen nutrition. The root nodule symbiosis between rhizobia and legumes is the classic example that uses this mechanism directly to promote legume plant growth. Another example in this direct benefit category is the mycorrhizal fungus, whose external hyphae extend from the mycorrhizal root out into the soil where they efficiently access nutrients, especially phosphorus, in areas that are normally depleted in these nutrients, and transport them into the plant roots. Examples of the indirect benefit category involving increased bioavalability and enhanced acquisition of mineral nutrients include PGPR like *Azospirillum* or *Rhizobium* that elaborate phytohormones growth regulators, which stimulate elongation and branching of root growth (see Section "A Case Study: *Rhizobium* Colonization of Rice Roots"). The resultant improved root architecture promotes the plant's ability to exploit nutrient reserves in soil much more efficiently than without the association. Other examples using indirect mechanisms in this category include microorganisms that solubilize soil complexes of inorganic and organic phosphates and iron that are otherwise insoluble and hence, unavailable to the plant.

The other major way that rhizosphere microorganisms improve plant health is mediated by their biocontrol against plant diseases. A direct benefit in this category includes the stimulation of plant defense responses triggered by biocontrol active bacteria that induce systemic disease resistance in plants, enabling them to successfully defeat plant pathogenic attacks. This can be achieved by aggressive root colonization by PGPR, outcompeting the niches for plant pathogens. Another direct microbe–plant interaction was revealed in studies using *Serratia liquefaciens* and tomato plants. In that association, the bacteria induce a systemic disease resistance reaction in the plant by a process involving its ability to secrete bacterial cell-to-cell communication signal molecules. In this case, the induced systemic resistance leads to "immunization" of the plant, protecting it against early blight disease. Other microbe–plant interactions are direct biocontrol mechanisms, which act directly on the pathogen itself rather than on the plant host. Examples include hyperparasitism whereby the PGPR aggressively attacks and kills the pathogen thereby reducing its inoculum potential, production of siderophores that restrict iron (F_e) availability and uptake for the pathogen, and production of antimicrobial antibiotics that inhibit the pathogen's growth physiology.

CLOSING REMARKS

The plant rhizosphere harbors a diverse reservoir of culturable microorganisms that can be exploited to benefit mankind. Many rhizosphere microbes benefit crop production, reducing the dependence on chemical fertilizers to achieve high productive yields. Others protect plants from the ravages of pathogens and the diseases they cause. Utilization of these beneficial examples is fully consistent with sustainable agriculture, where the goal of paramount importance is to utilize natural processes that promote the crop's output without irreparably damaging the natural resource base where the crop can be grown.

Among the many recent discoveries in rhizosphere research is the ominous finding that certain human pathogenic microorganisms are more than just opportunistic saprophytes on plants. They can persist and become aggressive, successful inhabitants of the nutrient-enriched rhizosphere and internal plant tissues, and this ecology poses significant implications of potential public health hazards for the people (both producers and consumers) who encounter them. An interesting thought for future exploration is whether the rhizosphere also supports populations of human health-promoting rhizobacteria. We predict it does.

ACKNOWLEDGMENTS

Portions of the authors' work described here were supported by the Michigan State University Center for Microbial Ecology (NSF Grant No. 0221838), Center for Renewable Organic Materials Research Excellence Funds, Kellogg Biological Station Long-Term Ecological Research Program, and the US–Egypt Joint Science & Technology Program Project No. BIO10-001-011.

FURTHER READING

Cardon, Z. G., & Gage, D. J. (2006). Resource exchange in the rhizosphere: molecular tools and the microbial perspective. *Annual Review of Ecology and Evolutionary Systematics*, *37*, 459–488.

Cardon, Z., & Whitbeck, J. (2007). *The rhizosphere: An ecological perspective*. Amsterdam: Elsevier Academic Press.

Dazzo, F. B. (2004). Applications of quantitative microscopy in studies of Plant surface microbiology. In A. Varma, L. Abbott, D. Werner, & R. Hampp (Eds.), *plant surface microbiology* (pp. 503–550). Germany: Springer-Verlag.

Dazzo, F. B. (2007). Visualization of the rhizoplane microflora by computer assisted microscopy and spatial analysis by CMEIAS image analysis. In R. Findley, & J. Luster, (Eds.) *Handbook of methods in rhizosphere research*. COST 631, Section 4.1 Microbial Growth and Visualization of Bacteria and Fungi. WSL Biomensdorf Switzerland. http://cme.msu.edu and http://www.rhizo.at/default.asp?id=574&lid=2.

Dazzo, F. B., Schmid, M., & Hartmann, A. (2007). Immunofluorescence microscopy and fluorescence *in situ* hybridization combined with CMEIAS and other image analysis tools for soil- and plant-associated microbial autecology. In C. Hurst, R. Crawford, J. Garland, D. Lipson, A. Mills, & L. Stetzenbach (Eds.), *Manual of environmental microbiology*. (3rd ed.) (pp. 595–792). Washington, DC: American Society for Microbiology Press.

Dazzo, F. B., & Yanni, Y. G. (2006). The natural *Rhizobium*-cereal crop association as an example of plant-bacteria interaction. In N. Uphoff, A. Ball, E. Fernandes, H. Herren, O. Husson, & M. Laing, et al. (Eds.) *Biological approaches to sustainable soil systems* (pp. 109–127). Boca Raton: CRC Taylor & Francis.

Drigo, B., Van Veen, J., & Kowalchuk, G. (2009). Specific rhizosphere bacterial and fungal groups respond differently to elevated atmospheric CO_2. *International Society for Microbial Ecology Journal*, *3*, 1204–1217.

Foster, R., Rovira, A., & Cock, T. (1983). *Ultrastructure of the root–soil interface*. St. Paul, MN: American Phytopathological Society.

Gantner, S., Schmid, M., Dürr, C., Schuhegger, R., Steidle, A., Hutzler, P., *et al.* (2006). *In situ* quantitation of the spatial scale of calling distances and population density-independent N-acylhomoserine lactone-mediated communication by rhizobacteria colonized on plant roots. *Federation of European Microbiological Sciences Microbiology Ecology*, *56*, 188–194.

Hartmann, A. (2004). Historical perspective of Prof. Hiltner's contributions to rhizosphere research. In A. Hartmann, M. Schmid, W. Wenzel, & P. Hinsinger (Eds.), *Rhizosphere 2004: Proceedings of the 1st International Congress on the Rhizosphere* (pp. 1–4). Munich, Germany: GSF-National Research Center for Environmental Health.

Katznelson, H. (1946). The rhizosphere effect of mangles on certain soil microorganisms. *Soil Science*, *62*, 343–354.

Kent, A., & Triplett, E. (2002). Microbial communities and their interactions in soil and rhizosphere ecosystems. *Annual Review of Microbiology*, *56*, 211–236.

Lugtenberg, B., Dekkers, L., & Blumebang, G. (2001). Molecular determinants of rhizosphere colonization by Pseudomonas. *Annual Review of Phytopathology*, *39*, 461–490.

Muresu, R., Maddau, G., Delogu, G., Cappuccinelli, P., & Squartini, A. (2010). Bacteria colonizing root nodules of wild legumes exhibit virulence-associated properties of mannalian pathogens. *Antonie van Leeuwenhoek*, *97*, 143–153.

Woese, C., & Fox, G. (1977). Phylogenetic structure of the prokaryotic domain: the primary kingdoms. *Proceedings of the National Academy of Sciences USA*, *74*, 5088–5090.

Part IV

Biogeochemical Cycles and their Consequences

Chapter 33

The Nitrogen Cycle

P. Cabello, M.D. Roldàn, F. Castillo and C. Moreno-Viviàn
Universidad de Còrdoba, Còrdoba, Spain

Chapter Outline

ABBREVIATIONS

Amo Ammonia monooxygenase
Amt Ammonium transporters
AOA Ammonia-oxidizing archaea
AOB Ammonia-oxidizing bacteria
ATase Adenylyltransferase
DNRA Dissimilatory nitrate reduction to ammonium
DraG Dinitrogenase reductase-activating glycohydrolase
DraT Dinitrogenase reductase ADP-ribosyltransferase
FDP Flavodiiron proteins
GDH Glutamate dehydrogenase
GOGAT Glutamine: 2-oxoglutarate amidotransferase
GS Glutamine synthetase
Hao Hydroxylamine oxidoreductase
Hcp Hybrid cluster protein
Hzh Hydrazine hydrolase
Hzo Hydrazine oxidoreductase
MFS Major facilitator superfamily
Mo-bis-MGD Mo-*bis*-molybdopterin guanine dinucleotide
Nap Periplasmic nitrate reductases
Nar Respiratory membrane-bound nitrate reductases
Nas Assimilatory nitrate reductases
Nir Nitrite reductase
NOB Nitrite-oxidizing bacteria
Nor Nitric oxide reductase
Nos Nitrous oxide reductase
Nrf Pentaheme-nitrite reductase
Nxr Nitrite oxidoreductase
UTase/UR Uridylyltransferase/uridylyl-removing enzyme

DEFINING STATEMENT

This article summarizes the current knowledge in microbiology and biochemistry, and regulation of the main processes of the nitrogen cycle driven by prokaryotic organisms. Ecological and environmental aspects and evolution of the nitrogen cycle are also discussed.

INTRODUCTION

Nitrogen is an essential element for life and is present in proteins, nucleic acids, and many biomolecules and, on average, represents about 6% dry weight of organisms. Therefore, nitrogen availability limits microbial and plant growth. Nitrogen is distributed between three major pools: atmosphere, soils/groundwater, and biomass. The complex nitrogen exchange among these three pools constitutes the so-called nitrogen cycle. In the ecosystems, nitrogen may be present as both organic and inorganic compounds. Inorganic nitrogen is found in several oxidation states, from −3 in the most reduced form (ammonium) to +5 in the most oxidized compound (nitrate). However, in organic compounds, nitrogen is usually found in the fully reduced state, as amino, amido, or imino groups.

THE BIOLOGICAL NITROGEN CYCLE

The biological nitrogen cycle consists of diverse processes that interconvert the different nitrogenous compounds in the biosphere (Figure 1). Prokaryotic organisms play a predominant role in all these reactions. However, specialized fungi also perform some nitrogen cycle processes; algae and higher plants assimilate nitrate; and glutamine synthetase (GS), the key enzyme in ammonium assimilation, is ubiquitously found in living organisms. In addition to biotic processes, the global biogeochemical nitrogen cycle also includes abiotic reactions such as the production of nitrogen oxides by combustion and the atmospheric conversion of dinitrogen (N_2) to nitric compounds by lightning and photochemical reactions.

The biological nitrogen cycle includes both oxidative and reductive reactions, catalyzed by enzymes with different redox cofactors (Table 1), which serve assimilatory, dissimilatory, or respiratory purposes. In general, a process is assimilatory if it generates molecules that are further incorporated in the cell material. Ammonium assimilation, N_2 fixation, and assimilatory nitrate reduction are the main assimilatory processes in the nitrogen cycle. Ammonium is the only inorganic nitrogenous compound that is incorporated in carbon skeletons. Glutamine, the first organic nitrogenous compound formed, is used as the main nitrogen donor for the synthesis of cellular nitrogenous compounds. Therefore, ammonium is the preferred nitrogen source for microbial growth. Although other inorganic nitrogenous compounds, such as N_2 or nitrate, are used as a nitrogen

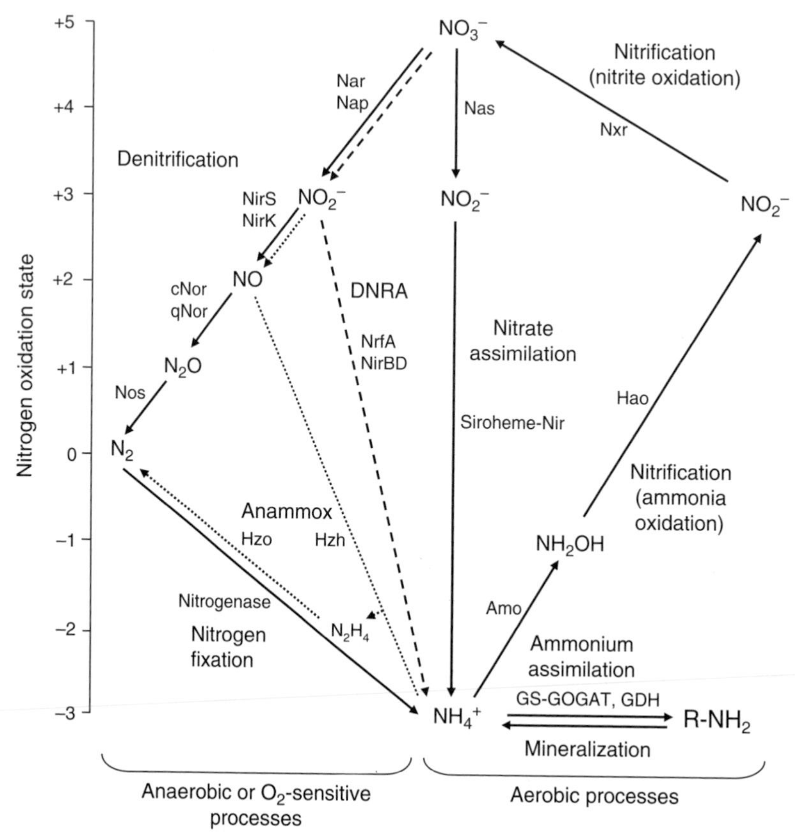

FIGURE 1 The biological nitrogen cycle. The different nitrogen compounds are arranged according to their oxidation states, and the main oxidative or reductive pathways are indicated by arrows. Anaerobic ammonia oxidation (anammox) and dissimilatory nitrate reduction to ammonia (DNRA) are indicated by dotted or dashed arrows, respectively. In the left are drawn the oxygen-sensitive reactions and anaerobic processes and in the right are shown the aerobic processes.

TABLE 1 Structure and cofactors of the main prokaryotic enzymes involved in the processes of the nitrogen cycle

Processes and Enzymes	Enzyme Structure	Molecular Mass (subunit, kDa)	Cofactors	Subcellular Location	Relevant Organisms
Ammonium assimilation					
Glutamine synthetase					
GSI	Dodecamer	50	Mg^{2+} or Mn^{2+}	Cytoplasm	Heterotrophic bacteria, cyanobacteria, phototrophic bacteria, archaea
GSII	Octamer	45–50	Mg^{2+} or Mn^{2+}	Cytoplasm	*Rhizobium*, *Frankia*, *Streptomyces*
GSIII	Hexamer	80	Mg^{2+} or Mn^{2+}	Cytoplasm	*Bacteroides*, *Butyrivibrio*, cyanobacteria
Glutamate synthase					
NADPH-GOGAT	Heterooctamer ($\alpha_4\beta_4$)	150 (α)50 (β)	FMN, [3Fe–4S]FAD, 2 × [4Fe–4S]	Cytoplasm	Most bacteria. Archaea have truncated forms (α- or β-like subunits)
Ferredoxin-GOGAT	Monomer	170	FMN, [3Fe–4S]	Cytoplasm	Cyanobacteria
Glutamate dehydrogenase					
GDH1, GDH2	Hexamer	45–55	NAD or NADP	Cytoplasm	Most bacteria and archaea
Nitrogen fixation					
Mo nitrogenase	MoFe protein (NifDK): heterotetramer $\alpha_2\beta_2$	MoFe protein:55–60	MoFe protein: 2 × FeMoCo,	Cytoplasm<c rspan = “2”>*Klebsiella*, *Azotobacter*, *Clostridium*, *Desulfovibrio*, *Rhizobium*, *Frankia*, anoxygenic phototrophs (*Rhodobacter*), cyanobacteria, methanogenic archaea	
	Fe protein (NifH): dimer	Fe-protein: 64	2 × P clusters Fe protein: [4Fe–4S]		
V nitrogenase	VFe protein (VnfDKG): $\alpha_2\beta_2\delta_2$	VFe protein: 55–60 (α, β); 15 (δ)	VFe protein: 2 × FeVCo, 2 × P clusters	Cytoplasm	*Azotobacter chroococcum*, *Azotobacter vinelandii*
	Fe protein (VnfH): dimer	Fe protein: 64	Fe protein: [4Fe–4S]		
Fe nitrogenase	FeFe protein (AnfDKG): $\alpha_2\beta_2\delta_2$	FeFe protein: 55–60 (α, β); 15 (δ)	FeFe protein: 2 × FeFeCo, 2 × P clusters	Cytoplasm	*Rhodobacter capsulatus*, *A. vinelandii*
	Fe protein (AnfH): dimer	Fe protein: 64	Fe protein: [4Fe–4S]		

(Continued)

TABLE 1 Structure and cofactors of the main prokaryotic enzymes involved in the processes of the nitrogen cycle—cont'd

Processes and Enzymes	Enzyme Structure	Molecular Mass (subunit, kDa)	Cofactors	Subcellular Location	Relevant Organisms
Nitrate assimilation					
Nitrate reductase					
NADH-Nas	Heterodimer	90–95 (NasA); 45 (NasC)	NasA: [4Fe–4S], MGD NasC: FAD	Cytoplasm	*Klebsiella*, *Bacillus subtilis*, *R. capsulatus*
Ferredoxin-Nas	Monomer	80–105	[4Fe–4S], MGD	Cytoplasm	Cyanobacteria, *A. vinelandii*, *Haloferax mediterraneii* (105 and 50 kDa dimer)
Nitrite reductase					
NADH–Nir	Monomer/heterodimer	87 (NasB) 12 (NirD)	NasB: FAD, [4Fe–4S], siroheme	Cytoplasm	*Klebsiella*, *Bacillus subtilis*, *R. capsulatus*
Ferredoxin-Nir	Monomer	55–66	[4Fe–4S], siroheme	Cytoplasm	Cyanobacteria, *Haloferax mediterraneii*
Denitrification					
Nitrate reductase					
Nar	Heterotrimer (αβγ)	140 (α or NarG) 60 (β or NarH) 25 (γ or NarI)	α: [4Fe–4S], MGD β: [3Fe–4S], 3 × [4Fe–4S] γ: 2 × heme *b*	Membrane-bound (active site facing cytoplasm)	*Escherichia coli*, *Paracoccus denitrificans*, *Pseudomonas*. In archaea, the active site faces outside cytoplasmic membrane
Nap	Heterodimer/monomer	90 (NapA) 15 (NapB)	NapA: [4Fe–4S], MGD NapB: 2 × heme *c*	Periplasm	*E. coli*, *Rhodobacter sphaeroides*, *Bradyrhizobium*, *Paracoccus*, *Desulfovibrio* (70 kDa NapA)
Nitrite reductase					
cd_1-Nir (NirS)	Dimer	60	Heme *c*, heme d_1	Periplasm	*Pseudomonas*, *Paracoccus*, *Kuenenia*
Cu-Nir (NirK)	Trimer	40	Cu_I, Cu_{II}	Periplasm	*Rhodobacter sphaeroides*, *Alcaligenes*, *Achromobacter*
Nitric oxide reductase					
cNor	Heterodimer	53 (NorB) 17 (NorC)	NorB: 2 × heme *b*, Fe_B NorC: heme *c*	Cytoplasmic membrane	*Pseudomonas stutzeri*, *P. denitrificans*, *Bradyrhizobium*
qNor	Monomer	80	2 × heme *b*, Fe_B	Cytoplasmic membrane	*Ralstonia eutropha*, *Pyrobaculum aerophilum*, *Neisseria*

Nitrous oxide reductase					
Nos	Dimer	65	Cu_A, Cu_Z	Periplasm	*Pseudomonas stutzeri*, *Paracoccus*
Nitrification					
Ammonia monooxygenase	Heterotrimer?	27 (AmoA) 38 (AmoB) 32 (AmoC)	Cu, Fe?	Cytoplasmic membrane	*Nitrosomonas* and AOB. In *P. denitrificans* is a dimer (38 and 46 kDa)
Hydroxylamine oxidoreductase	Trimer	63	8 × heme *c* (24 heme *c* in total)	Periplasm	*Nitrosomonas* and AOB. In *Paracoccus* is a 20-kDa monomer with nonheme iron
Nitrite oxidoreductase	Heterodimer	115 (NorA) 65 (NorB)	Mo (MGD), FeS	Cytoplasmic membrane	*Nitrobacter* and NOB, anammox bacteria (*Kuenenia*)
Anammox					
Hydrazine hydrolase	Trimer	61	8 × heme *c* (~26 heme *c* in total)	Anammoxosome	*Kuenenia*, *Brocardia*, *Anammoxoglobus*, *Scalindua*
Hydrazine oxidoreductase	Dimer	62	8 × heme *c* (16 heme *c* in total)	Anammoxosome	*Kuenenia*, *Brocardia*, *Anammoxoglobus*, *Scalindua*
DNRA (ammonification)					
Nitrite reductase Nrf	Heterodimer NrfAH/ heterotetramer NrfABCD	60 (catalytic NrfA)	5 × heme *c* (NfrA)	Membrane-bound (active site facing periplasm)	*E. coli*, *Desulfovibrio*, *Wolinella*
Nitrite reductase NirBD	Heterodimer	90 (NirB)	NirB: FAD, [4Fe–4 S], siroheme [2Fe–2 S] 12 (NirD)	Cytoplasm	*E. coli*

source, their reduction to ammonium is required before their incorporation into carbon skeletons. N_2 constitutes about 78% of the atmosphere, but only a few prokaryotic organisms can use this stable molecule as a nitrogen source. Nitrogen fixation involves the reductive breakage of the triple bond of N_2 to generate NH_4^+, a process carried out only by some free-living or symbiotic prokaryotes. Nitrate assimilation takes place through two sequential reductive reactions catalyzed by the assimilatory nitrate and nitrite reductases (Nirs) ($NO_3^- \rightarrow NO_2^- \rightarrow NH_4^+$). This process is performed by some archaea and bacteria and also by fungi, algae, and plants. Degradation of organic matter by soil decomposer organisms is also a source of ammonium in many environments. Decomposition of organic matter to inorganic compounds by saprophytic bacteria and fungi is globally known as "mineralization," but the term "ammonification" is used to describe ammonium release from organic compounds. Ammonium ions are dissolved in soil water, retained in soils, and immobilized or absorbed and assimilated by organisms, but in alkaline media, a significant amount is volatilized to atmosphere as gaseous ammonia. It is worth noting that the term ammonification is also used for other nitrogen cycle processes, such as dissimilatory reduction of nitrate to ammonium that serves for energy conservation (respiration) or detoxification and electron sink. The concepts "dissimilatory" and "respiratory" are commonly used equivalently in the literature. However, *sensu stricto*, a dissimilatory process is a nonassimilatory reaction or route that is not directly coupled to generation of protonmotive force and serves other purposes such as redox balancing or detoxification, whereas a redox process should be referred to as respiratory when electron transfer is coupled to protonmotive force generation (ATP synthesis). Some prokaryotes use respiratory or dissimilatory processes involving inorganic nitrogenous compounds, such as nitrate respiration and denitrification, anaerobic ammonia oxidation (anammox), and nitrification. Denitrification is the sequential respiration of nitrate, nitrite, nitric oxide, and nitrous oxide ($NO_3^- \rightarrow NO_2^- \rightarrow NO \rightarrow N_2O \rightarrow N_2$), catalyzed by the corresponding reductases. Most denitrifiers are facultative bacteria that grow in anoxic environments using these compounds as electron acceptors to generate an electrochemical gradient, but there are also aerobic denitrifiers. Nitrification involves the two-step oxidation of ammonia to nitrite ($NH_4^+ \rightarrow NH_2OH \rightarrow NO_2^-$) and of nitrite to nitrate ($NO_2^- \rightarrow NO_3^-$), predominantly performed by ammonia-oxidizing bacteria (AOB) and nitrite-oxidizing bacteria (NOB), respectively, that use energy released in these reactions to support CO_2 fixation and growth. However, a number of archaea, heterotrophic bacteria, and fungi are also capable of nitrification. Anammox is an anaerobic process that generates N_2 by coupling ammonium oxidation to nitrite reduction ($NH_4^+ \rightarrow NO_2^- \rightarrow N_2$), performed by some planctomycetes. Whereas assimilatory pathways make nitrogen available for organisms, the combined action of nitrification, denitrification, and anammox causes losses of fixed nitrogen from the environment.

Some nitrogenous compounds, including xenobiotic nitroaromatic chemicals and natural compounds such as ammonium, nitrite, NO, hydroxylamine, hydrazine, cyanate, and cyanide, are toxic for living organisms. Toxicity is generally due to the high reactivity of these molecules, which inactivate or damage proteins (particularly metalloenzymes) and other biomolecules. Therefore, organisms have different detoxification pathways to transform these compounds. In some cases, detoxification reaction products are used as nitrogen sources. Thus, reductive detoxification of nitrite or hydroxylamine may generate ammonium, and degradation of nitroaromatic compounds may release either ammonium or nitrite, which is further assimilated. Cyanide and its derivatives are also used as a nitrogen source by some bacteria and fungi. In the past few decades, human activities have drastically altered the natural balance of the nitrogen cycle. Thus, huge amounts of nitrogenous compounds enter annually into the ecosystems by anthropogenic processes such as production of ammonia fertilizers by the Haber–Bosch industrial process, release of nitrogen oxides by combustion of fossil fuels, cultivation of N_2-fixing crops, and production of toxic cyanides and nitroaromatics in metallurgical, explosives, dyes, pharmacological, and chemical industries.

AMMONIUM ASSIMILATION

Ammonium assimilation requires transport of this ion inside the cells and its further incorporation into carbon skeletons. In alkaline environments, ammonia may enter the cells by diffusion through the membranes, but ammonium transporters (Amt) are usually present in both prokaryotic and eukaryotic organisms. Inside the cells, ammonium is incorporated in carbon skeletons mainly through the sequential action of GS and glutamate synthase (glutamine: 2-oxoglutarate amidotransferase, GOGAT). GS is a ubiquitous enzyme that catalyzes the ATP-dependent amidation of glutamate to generate glutamine. GOGAT catalyzes the reductive transfer of the amide group of glutamine to 2-oxoglutarate, yielding two molecules of glutamate. Therefore, the global process of the GS–GOGAT pathway (Figure 2) results in the incorporation of ammonium into 2-oxoglutarate to produce a molecule of glutamate with consumption of ATP (GS activity) and reducing power (GOGAT activity). An alternative pathway for ammonium assimilation is its direct incorporation into 2-oxoglutarate catalyzed by the glutamate dehydrogenase (GDH). In general, this reaction plays a minor role in ammonium assimilation because, although it does not consume ATP, it is less effective at low ammonium concentrations. Once glutamine and glutamate are formed by the GS–GOGAT pathway or the alternative GDH reaction, other

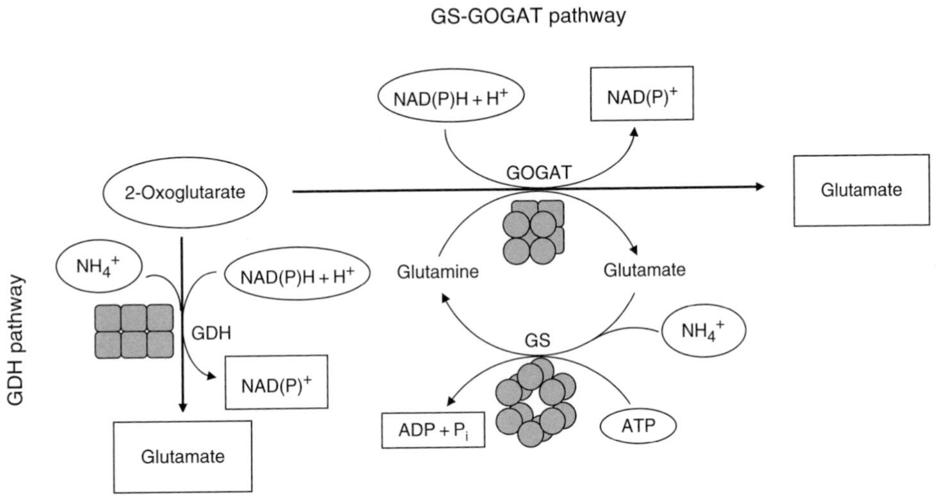

FIGURE 2 Ammonium assimilation pathways. The main glutamine synthetase (GS) and glutamate synthase (glutamine: 2-oxoglutarate amidotransferase, GOGAT) route and the alternative glutamate dehydrogenase (GDH) pathway are shown. Substrates and final products are placed within ovals and rectangles, respectively, and enzymes are drawn according to their subunit structures. GDH also catalyzes the oxidative deamination of glutamate to 2-oxoglutarate.

nitrogenous compounds are synthesized by a set of secondary transfers. 2-Oxoglutarate, the main carbon skeleton for ammonium assimilation, represents the connection between carbon and nitrogen metabolism. In bacteria and archaea, 2-oxoglutarate is also the main molecule used as signal of nitrogen limitation, whereas glutamine, the fully aminated product of ammonium assimilation, is often a signal of nitrogen sufficiency. PII proteins (GlnB and GlnK) are ubiquitously distributed among prokaryotes as 2-oxoglutarate sensors and, in some bacteria, are covalently modified by enzymes that respond to glutamine.

Ammonium Transport

Amt are integral membrane proteins of 45–50 kDa with 10–12 membrane-spanning helices. Amt permeases seem to mediate an active ammonium uniport that depends on protonmotive force, although it has been also proposed that Amt proteins facilitate the bidirectional diffusion of ammonia. These permeases also transport an ammonium methyl analog, and, in fact, most transport studies have been undertaken using [^{14}C] methylammonium. As methylammonium transport is inhibited by ammonium, it is believed that Amt proteins play a significant role at low ammonium concentrations, a normal condition in natural environments. Putative *amt* genes are present in archaea, bacteria, yeast, and plants. In most prokaryotic genomes, with exception of cyanobacteria, the *amt* genes are linked to the regulatory *glnK* gene that codes for a PII protein. In *Escherichia coli*, *amtB* and *glnK* gene products interact to regulate transport activity in response to nitrogen availability. In *Synechocystis* sp. PCC 6803, three *amt* genes have been identified; *amt1* is the main ammonium permease whereas *amt2* and *amt3* contribute very little to the uptake activity.

The GS–GOGAT Pathway

GS catalyzes the ATP-dependent synthesis of glutamine from glutamate and ammonium in the presence of divalent cations (Mg^{2+} or Mn^{2+}). Three different types of GS are known (Table 1). GSI is a dodecameric protein with identical 50 kDa subunits (*glnA* gene product) arranged in two hexagonal rings with a cylindrical aqueous channel. GSI is found in prokaryotes, including archaea, and low-GC Gram-positive and most Gram-negative bacteria. Some bacterial species such as *Bacteroides fragilis*, *Butyrivibrio fibrisolvens*, and the cyanobacterium *Pseudanabaena* sp. PCC 6903 contain GSIII, a hexameric enzyme with 80 kDa identical subunits encoded by the *glnN* gene. Other cyanobacteria, such as *Synechocystis* and *Synechococcus* sp. PCC 7942, contain both GSI and GSIII, although GSIII is only observed under nitrogen deficiency. These two types of GS are different in their amino acid sequence, but they share five conserved domains that include residues involved in the catalytic site and divalent cation binding. Finally, GSII is an octameric enzyme with identical 45–50 kDa subunits commonly found in eukaryotic organisms and some Gram-negative or Gram-positive bacteria such as *Rhizobium*, *Frankia*, and *Streptomyces*.

GOGAT is an iron–sulfur flavoprotein that synthesizes two molecules of glutamate from glutamine and 2-oxoglutarate, in a two-electron reductive reaction. There are three different types of GOGAT depending on the electron donor. The NADPH-dependent enzyme is commonly found in heterotrophic bacteria. It is composed of two different subunits arranged in $\alpha_4\beta_4$ structure. The large α subunit (150 kDa) is encoded by the *gltB* gene and contains flavin mononucleotide (FMN) and one

[3Fe–4 S] center, whereas the small β-subunit (50 kDa) is encoded by the *gltA* gene and contains flavin adenine dinucleotide (FAD). Two additional [4Fe–4 S] centers are located in the interface of the subunits in the holoenzyme. Archaeal genomes contain putative genes encoding "truncated" GOGAT of about 55 kDa, which are similar to *gltB* or *gltA* genes and could represent ancestral forms of the large or small subunits of bacterial NADPH-GOGAT, respectively. Ferredoxin- and NADH-dependent GOGAT are present in cyanobacteria. Ferredoxin-GOGAT is a 170-kDa monomer (*glsF* gene product) with FMN and one [3Fe–4 S] center. Therefore, ferredoxin-GOGAT is similar both in size and cofactors to the large subunit of NADPH-GOGAT. The ferredoxin-GOGAT of *Plectonema boryanum* has a relevant role in ammonium assimilation at high light intensity. In addition to the ferredoxin-GOGAT, *Synechocystis* and *P. boryanum* also contain an NADH-GOGAT with two subunits of 160 and 60 kDa encoded by *gltB* and *gltD*, respectively, which are homologous to the large and small subunits of NADPH-GOGAT. An NADH-GOGAT is also present in some plants, such as *Medicago sativa*, but in this case the enzyme is a 200-kDa monomer that appears to derive from the fusion of both bacterial NADPH-GOGAT subunits because it is similar to the large subunit in its N-terminal domain and to the small subunit in its C-terminal domain.

Alternative Pathways for Ammonium Assimilation

GDH catalyzes both the oxidative deamination of glutamate and the reductive amination of 2-oxoglutarate. However, GDH is usually a minor pathway for ammonium assimilation which plays a role only at high ammonium concentrations because the enzyme has an apparent K_m value for ammonium between 1 and 5 mmol l^{-1}. The GDH of prokaryotes and lower eukaryotes usually react with a particular coenzyme, NAD^+ or $NADP^+$, whereas the enzyme of higher eukaryotes shows dual coenzyme specificity. There are four distinct types of GDH. GDH1 and GDH2 are small hexameric proteins composed of identical subunits of 45–55 kDa, which use either NAD^+ or $NADP^+$ and may be involved in ammonium assimilation. GDH1 is found in bacteria, fungi, and algae, and GDH2 is present in all three life domains. GDH3, mainly found in eukaryotes, contains identical 115 kDa subunits and has a catabolic role in glutamate deamination. Finally, GDH4 is a 180-kDa, NAD^+-specific enzyme present only in bacteria. Sequence analyses of putative *gdh* genes present in the genomes of bacteria, archaea, and lower eukaryotes suggest that lateral gene transfer has played an important role in the evolution of GDH families. In some bacteria, alanine dehydrogenase catalyzes the reductive amination of pyruvate to alanine as another alternative minor pathway for ammonium assimilation.

Regulation of Ammonium Assimilation

Ammonium assimilation through GS–GOGAT is tightly regulated. GS, the primary ammonium-assimilating enzyme, is the target of a complex regulation at both transcriptional and posttranslational levels (Figure 3). Although there are differences in the regulatory mechanisms depending on the organism, in general, nitrogen sufficiency promotes negative regulation of GS activity and *glnA* gene transcription, whereas maximal gene expression and GS activity are reached under nitrogen starvation.

In enterobacteria, the activity of dodecameric GSI is regulated by accumulative allosteric inhibition and by covalent modification (adenylylation/deadenylylation) of the enzyme monomers with participation of three proteins (ATase, GlnB, and UTase/UR) and two effectors (2-oxoglutarate and glutamine). Under nitrogen excess (low 2-oxoglutarate and high glutamine), GS is adenylylated and shows low activity, whereas under nitrogen starvation (high 2-oxoglutarate and low glutamine) GS is deadenylylated and fully active. A bifunctional adenylyltransferase (ATase) uses ATP to adenylylate a tyrosine residue of each GS subunit and also removes this AMP moiety by phosphorylytic cleavage. The GS adenylylation state is controlled by the trimeric PII protein (GlnB), a 2-oxoglutarate sensor that may be in two forms, uridylylated or deuridylylated. A bifunctional UTase/UR enzyme catalyzes both the transference of UMP from UTP to a tyrosine residue of each GlnB subunit and the hydrolytic removal of UMP groups. GlnB binds to 2-oxoglutarate and interacts with a variety of target proteins to promote the nitrogen response. The UTase/UR enzyme senses glutamine as signal of nitrogen excess for modulating GlnB activity. Unmodified GlnB stimulates GS adenylylation (inactivation), whereas uridylylated GlnB promotes GS deadenylylation (activation). Expression of the *glnA* gene is controlled by the general nitrogen system Ntr, a two-component regulatory system. At low nitrogen, the NtrB sensor protein phosphorylates an aspartic residue of the NtrC regulator, which results in activation. Phosphorylated NtrC activates transcription of *glnA* and other nitrogen-regulated genes by binding to target sites in the corresponding promoters, which are recognized by the specific RNA polymerase σ^{54} factor (RpoN). At high nitrogen concentrations, deuridylylated GlnB stimulates dephosphorylation of NtrC by NtrB, and this determines a low transcription of *glnA* and nitrogen-regulated genes.

GlnK is a GlnB paralog that is present in most bacteria and archaea. The *glnK* gene is usually clustered together with the *amt* gene and both gene products interact to regulate ammonium transport. The formation of an AmtB–GlnK membrane-bound complex is controlled by covalent modification of GlnK. Thus, excess ammonium causes deuridylylation of GlnK that binds AmtB permease to block ammonium transport, and, conversely, nitrogen deficiency

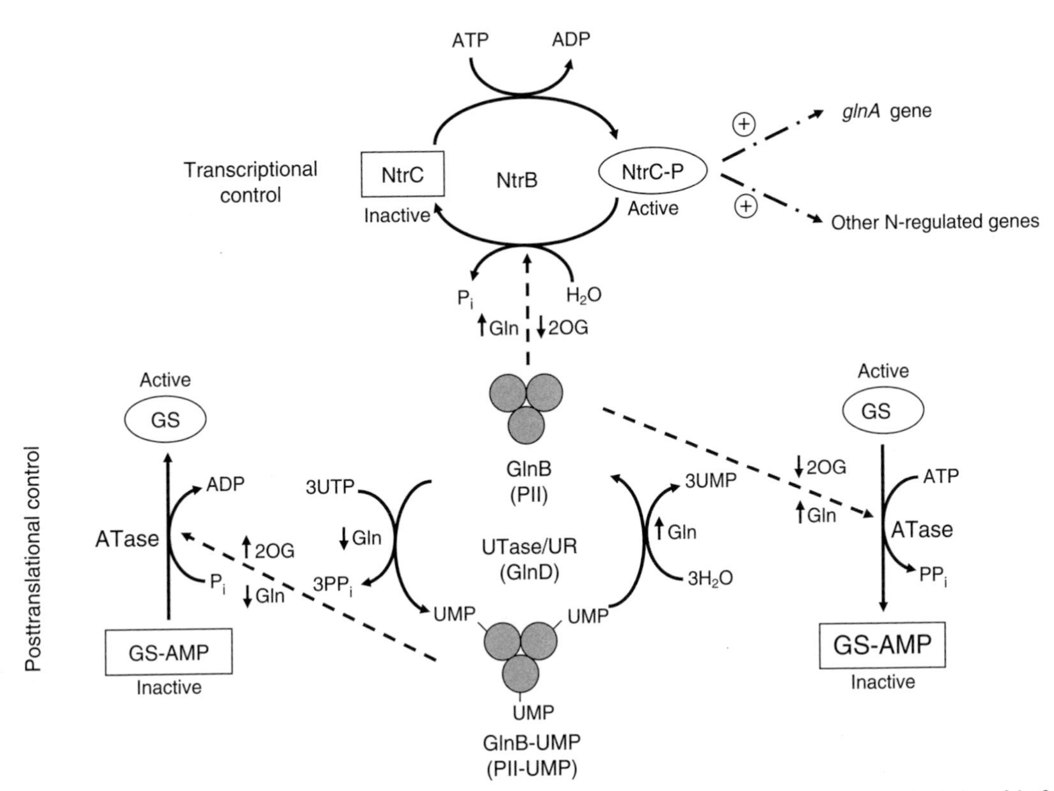

FIGURE 3 Control of glutamine synthetase (GS) activity and *glnA* gene expression in enterobacteria. Uridylylation/deuridylylation of the 2-oxoglutarate sensor GlnB (PII protein) regulates the adenylylation state of the GS subunits (posttranslational control) and the activity of NtrB/NtrC bicomponent system that controls expression of *glnA* and other nitrogen-regulated genes (transcriptional control). Nitrogen deficiency (high 2-oxoglutarate and low glutamine) leads to uridylylation of PII and deadenylylation of GS (activation), whereas nitrogen sufficiency (low 2-oxoglutarate and high glutamine) causes PII deuridylylation, which leads to GS adenylylation (inactivation) and NtrC dephosphorylation by NtrB to decrease *glnA* gene expression.

provokes GlnK uridylylation and dissociation of AmtB–GlnK to facilitate ammonium transport.

This regulation model for ammonium assimilation is widely distributed among bacteria, although with specific differences among organisms. Thus, some bacteria lack the Ntr system. In cyanobacteria, the main nitrogen regulator is the NtcA transcription factor belonging to the cAMP receptor protein family. NtcA activates some genes, such as *amt1* (ammonium permease) and *glnA* (GS), but represses two genes (*gifAB*) that code for small polypeptides of 7 and 17 kDa (IF7 and IF17), respectively, that inactivate GS under nitrogen excess. Therefore, GS activity is not covalently modified by adenylylation but controlled by the inactivating factors IF7 and IF17. In addition, cyanobacteria have only one PII gene, which is not associated to *amt* genes, and this PII protein is modified by phosphorylation, rather than by uridylylation, depending on 2-oxoglutarate concentration irrespective of glutamine levels. Both PII and NtcA proteins directly bind to 2-oxoglutarate, and their signaling and transcriptional activities seem to be mutually dependent.

Archaea may use a different mechanism to control gene expression in response to the C–N balance. The NrpR protein of *Methanococcus maripaludis* is a tetrameric (60 kDa subunit) DNA-binding protein that acts as repressor of *glnA* (GS) and *nif* (nitrogen fixation) genes. 2-Oxoglutarate decreases binding of NrpR to the corresponding operator sequences, thus activating gene expression. PII proteins are present in some archaea since *glnK*-like genes are found linked to *amtB* genes.

NITROGEN FIXATION

Biological nitrogen fixation is an essential process in the nitrogen cycle and provides a major source of available (fixed) nitrogen for organisms. Thus, it is estimated that about 2.2×10^{11} kg nitrogen is biologically fixed annually (1.4×10^{11} kg nitrogen per year in terrestrial ecosystems and 0.8×10^{11} kg nitrogen per year in marine environments). Nitrogenase is the enzyme complex that catalyzes the formation of two molecules of NH_4^+ by reduction of N_2. Proton reduction to H_2 also occurs during the catalysis, and nitrogenase can also reduce several substrates with double or triple bonds, such as acetylene or cyanide. N_2 fixation is very expensive in biological energy equivalents since it requires large amounts of ATP and reducing power. The reaction stoichiometry may be represented as follows:

$$N_2 + 8e^- + 10H^+ + 16MgATP \rightarrow 2NH_4^+ + H_2 + 16MgADP + 16P_i$$

Nitrogenase is present only in a variety of prokaryotic organisms (diazotrophs) that fix N_2 free-living, in symbioses, or in association with plants. Free-living organisms may grow diazotrophically under anaerobic, microaerobic, or aerobic conditions. In this case, as nitrogenase is oxygen-labile, different mechanisms are used to protect the enzyme against oxygen. Anaerobic diazotrophs include clostridia, sulfate-reducing bacteria (*Desulfovibrio*), anoxygenic phototrophic bacteria (*Rhodobacter*), and methanogenic euryarchaeota (*Methanococcus*). Facultative anaerobes (*Klebsiella*) fix N_2 only anaerobically. Obligate aerobes such as *Azotobacter* respire at very high rates to maintain low internal oxygen concentrations and have additional protective mechanisms that prevent nitrogenase damage. Unicellular cyanobacteria show circadian rhythm, fixing N_2 during the night and performing oxygenic photosynthesis during the day. Filamentous cyanobacteria fix N_2 inside specialized cells, called heterocysts, that contain a thick envelop that makes O_2 diffusion difficult and that lack the normal photosynthetic apparatus, avoiding O_2 generation. Symbiotic diazotrophs carry out N_2 fixation inside plant root nodules, which are specialized structures that allow bacterial growth under microaerobic conditions. The most representative is the *Rhizobium*–legume symbiosis, although some actinomycetes (*Frankia*) establish symbioses with nonleguminous trees (alder) and certain cyanobacteria with lichens, pteridophytes, and gymnosperms. Also, some diazotrophs (*Herbaspirillum*, *Azospirillum*, and *Azotobacter*) form associations with plant grasses.

The Nitrogenase Enzymes

Four types of nitrogenases are known; three of them are similar and closely related (Mo, V, and Fe nitrogenases), and one is a different enzyme only found in *Streptomyces thermoautotrophicus*. With this exception, all diazotrophic organisms have Mo nitrogenase either exclusively (*Klebsiella* and *Rhizobium*) or together with other alternative form. Thus, *Azotobacter chroococcum* possesses Mo and V nitrogenases, *Rhodobacter capsulatus* has Mo and Fe nitrogenases, and *Azotobacter vinelandii* contains the three enzymes (Table 1). These three related nitrogenases have probably arisen from a common ancestor and consist of two separate metalloproteins, known as component 1 or dinitrogenase (MoFe, VFe, or FeFe protein) and component 2 or dinitrogenase reductase (Fe protein). More than 20 *nif* genes are involved in Mo nitrogenase components and cofactors biosynthesis, electron transfer, and regulation, and homologous *vnf* and *anf* genes are required for alternative V and Fe nitrogenases, respectively.

In the Mo nitrogenase, the MoFe protein is a 230-kDa $\alpha_2\beta_2$ tetramer, encoded by the *nifD* and *nifK* genes, that contains two pairs of two unusual and unique metallocenters, the FeMo cofactor and the P cluster. Each FeMo cofactor is composed of $Mo_1Fe_7S_9$ and homocitrate, since it consists of two partial cubanes [4Fe–3 S] and [Mo–3Fe–3 S] bridged by three sulfides and with homocitrate coordinated to the Mo atom. Each P cluster is composed of Fe_8S_7 and consists of a [4Fe–4 S] center that shares one S atom with a [4Fe–3 S] center. Therefore, the tetrameric MoFe protein contains a total of 2 Mo, 30 Fe, and 32 S atoms. The Fe protein is a 64-kDa homodimer, encoded by the *nifH* gene, with a single [4Fe–4 S] center that bridges the two subunits. An MgATP-binding site is also found in each subunit.

The MoFe protein contains the catalytic site for N_2 reduction, and the Fe protein is the electron donor to the MoFe protein. In addition, Fe protein is also required for biosynthesis and maturation of the MoFe protein. MoFe protein and Fe protein form a complex during catalysis. Sequence of catalytic events includes two connected cycles: the Fe protein cycle describes complex formation, electron transfer from Fe protein to MoFe protein, MgATP hydrolysis, and reduction of oxidized Fe protein with the physiological electron donor (ferredoxin or flavodoxin); and the MoFe protein cycle includes the partial reactions within the MoFe protein for substrate reduction.

Alternative nitrogenases contain either vanadium or iron instead of molybdenum. In these enzymes, the VFe or FeFe proteins contain additional 15 kDa δ-subunits, resulting in an $\alpha_2\beta_2\delta_2$ structure, and also contain P clusters and the corresponding cofactors in which V or Fe replaces the Mo atom. The unique nitrogenase of *S. thermoautotrophicus* has also two components: an αβγ heterotrimeric MoFeS protein and a Mn-containing superoxide oxidoreductase that replaces the Fe protein and oxidizes superoxide anion to oxygen for electron transfer to MoFeS protein. The electron donor is a Mo-containing CO dehydrogenase that couples CO oxidation to O_2 reduction, forming the superoxide anion.

Regulation of Nitrogen Fixation

N_2 fixation is regulated at both transcriptional and post-translational levels in response to ammonium and oxygen, although the effect of these molecules and the regulatory mechanisms are characteristic of each diazotrophic organism (Figure 4). Expression of *nif* genes in free-living diazotrophs is repressed by ammonium, but nitrogenase of symbiotic bacteria is less sensitive to ammonium since this product is exported to the host cells. In *Klebsiella*, *nif* gene expression is controlled by the *nif*-specific regulatory genes *nifLA*, which, in turn, are controlled by the general Ntr system. NifA is a σ^{54}-dependent transcriptional activator required for the expression of all *nif* promoters. NifL is a flavoprotein that binds and inactivates NifA in response to oxygen or fixed nitrogen. The *nifLA* genes are not present in all diazotrophs and there are differences in regulation among organisms. In cyanobacteria, NtcA is involved in heterocysts differentiation, and, during adaptation to

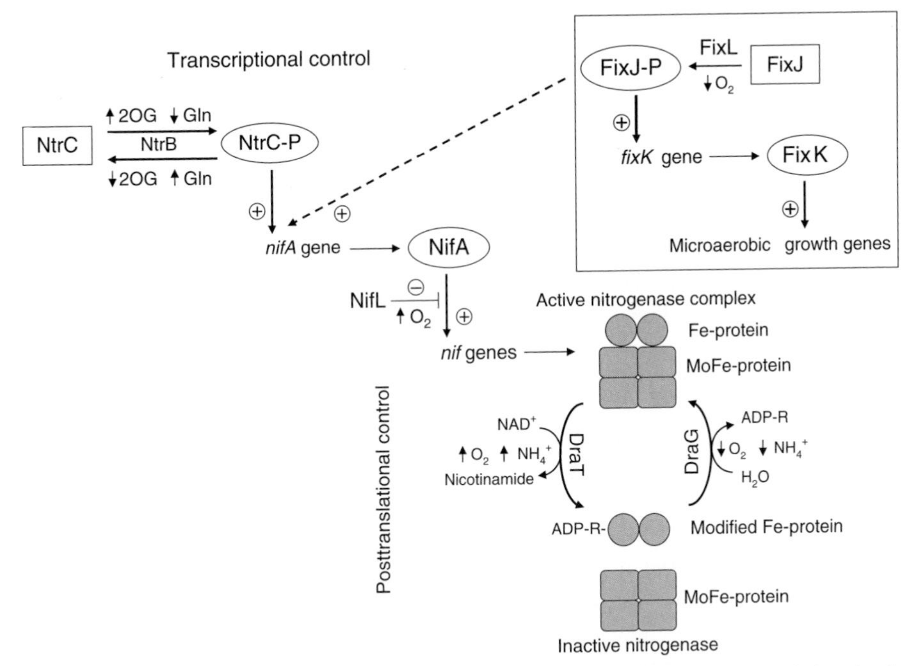

FIGURE 4 Levels of regulation of nitrogen fixation. Transcriptional control operates by a regulatory cascade with three levels: activation of NtrC by NtrB in response to low nitrogen, induction of *nifA* gene expression by phosphorylated NtrC, and induction of *nif* genes by NifA. Oxygen regulates *nif* gene expression by affecting NifA activator either directly or through the NifL protein, and, in some bacteria (*Sinorhizobium* and *Bradyrhizobium*), additional bicomponent systems such as FixL/FixJ also control expression of *nifA* and/or microaerobic growth genes (box). In *Rhodobacter* and *Rhodospirillum*, nitrogenase activity is regulated by a switch-on/switch-off mechanism (posttranslational control) in response to oxygen and ammonium in which the DraT/DraG bicomponent system participates. See text for details.

diazotrophic growth, the PII protein is phosphorylated in vegetative cells but unphosphorylated in heterocysts. In euryarchaeota, *nif* gene expression is regulated by a repressor (NrpR) rather than by an activator. NrpR binds to operator sequences preventing *nif* gene transcription in presence of ammonium.

In anoxygenic phototrophic bacteria, the posttranslational regulation of nitrogenase activity is known as switch-off/switch-on as it allows a rapid and reversible response to O_2 and NH_4^+. This regulatory mechanism is mediated by the dinitrogenase reductase ADP-ribosyltransferase (DraT)/dinitrogenase reductase-activating glycohydrolase (DraG) system and involves reversible ADP-ribosylation of one Fe protein subunit at a specific arginine residue. In response to ammonium, DraT is activated and catalyzes the transference of ADP-ribose from NAD^+ to this arginine residue, which prevents the formation of the complex with MoFe protein leading to enzyme inactivation. At low ammonium, DraG becomes active to remove the ADP-ribose of the Fe protein subunit and nitrogenase activity is recovered. The DraT/DraG system is also regulated at posttranslational level by PII protein. In the euryarchaeota *M. maripaludis*, nitrogenase activity is also regulated by ammonium, but this control does not involve covalent modification of NifH and requires two PII-like proteins ($NifI_1$ and $NifI_2$), which act as 2-oxoglutarate sensors. These proteins bind reversibly to the enzyme and inhibit nitrogenase activity, but 2-oxoglutarate prevents this inhibition.

Plant Symbioses and Associative Nitrogen Fixation

The most important N_2-fixing symbiosis is established between leguminous plants and rhizobia, a group of Gram-negative soil bacteria. Establishment of symbiosis requires infection of legume roots by the bacteria and formation of a specialized organ, the nodule, where the N_2-fixing process occurs. Nodulation is a host-specific interaction in which each rhizobial strain infects a defined plant host. Thus, *Rhizobium leguminosarum* biovar *phaseoli*, *Sinorhizobium meliloti*, *Bradyrhizobium japonicum*, and *Mesorhizobium loti* establish specific symbioses with bean (*Phaseolus*), alfalfa (*Medicago*), soybean (*Glycine*), or bird's-foot trefoil (*Lotus*), respectively. The symbiotic process occurs with a close coordination between plant and microorganism based on exchange of diffusible signal molecules. Plants secrete flavonoids (flavones or isoflavones), which in the rhizosphere act as specific signals recognized by the compatible bacterial strain. Bacteria respond by attaching to root hairs and inducing nodulating (*nod*) genes. The *nodD* gene codes for a transcriptional regulator, which, in presence of the specific flavonoid, induces expression of *nod* genes encoding enzymes that produce a specific lipo-chitin that is excreted and acts as nodulation signal recognized by the plant host. Lipo-chitin is a β1 → 4-linked polymer of *N*-acetyl-D-glucosamine with a fatty acid replacing the acetyl group at the molecule end. This Nod signal activates a sequence of events in the root hair allowing infection. During this process, curling of the root hair is observed and bacteria enter the plant through an intracellular tubular structure known as "infection thread." Cortical cells are activated to induce cell divisions forming the nodule primordium, and the infection thread continues its growth into the cortex and ramifies into the nodule primordium. Then, bacteria are released into the cells and enclosed in a vacuole-like compartment called "symbiosome" in which they differentiate into bacteroids to carry out N_2 fixation.

There are two types of nodules depending on the plant host. Determinate nodules are initiated from cortical cell divisions and grow by cell expansion, resulting in a globular structure with peripheral vascular tissue. Indeterminate nodules appear as modified lateral roots, have persistent apical meristem with lateral vascular tissue, and show clear zones corresponding to nodule development stages, which are difficult to distinguish in determinate nodules. Temperate climate legumes (alfalfa) usually form indeterminate nodules, whereas tropical legumes (soybean) commonly form determinate nodules. During nodulation, some enzymes for nitrogen assimilation (GS) and other processes are expressed in the nodules. These proteins are called "nodulins" and depending on their expression time may be early nodulins, which function in infection and nodule development, or late nodulins, which function in N_2 fixation and nodule maintenance. Nodules maintain low oxygen concentrations to avoid nitrogenase inhibition. Leghemoglobin, a protein similar to hemoglobin, plays an important role in this process binding oxygen and controlling its release to the bacteroids, which possess a high-affinity respiration system that uses oxygen transferred directly from leghemoglobin. In rhizobia, two bicomponent regulatory systems (FixL–FixJ and RegR–RegS) may be involved in the control of *nifA* gene expression by oxygen. In addition, NifA itself senses oxygen to activate *nif* gene expression at low oxygen concentration (Figure 4).

Sucrose synthesized photosynthetically by the plant is converted into dicarboxylic acids, which are used by the bacteroids to generate ATP and reducing power for N_2 fixation. Ammonia formed in this process diffuses out of the bacterium into the symbiosome space. After protonation, ammonium is transported through an ion channel into host cell cytoplasm, where it is assimilated by the GS–GOGAT pathway. The final step in symbiotic N_2 fixation is the export of fixed nitrogen from the nodules to the rest of the plant through the xylem. In temperate climate legumes, fixed nitrogen is translocated as amides, mainly asparagine, whereas in tropical legumes it is exported predominantly as ureides such as allantoin and allantoic acid.

Other N_2-fixing symbioses between prokaryotes and nonlegume plants also exist. *Frankia*, a Gram-positive soil bacterium, forms actinorhizal nodules with dicotyledoneous plants (trees or woody shrubs). Also, filamentous cyanobacteria (*Anabaena* and *Nostoc*) establish symbiosis with lichens, bryophytes, pteridophytes, and one angiosperm (*Gunera*). In all these cases, the microsymbionts fix N_2 using carbon sources supplied by the plant, and the resulting fixed nitrogen is exported to the host plant. In addition, some gramineous crops and grasses can also obtain part of their nitrogen demand by association with N_2-fixing bacteria. In general, diazotrophs associated with plant roots belong to α-, β- and γ-proteobacteria, such as *Azospirillum*, *Derxia*, *Azoarcus*, *Herbaspirillum*, *Burkholderia*, *Azotobacter*, *Klebsiella*, and *Serratia*. Some of them are able to colonize intercellularly the plant tissues without damaging the host. These diazotrophs are known as endophitic (*Azoarcus* and *Herbaspirillum*). Facultative endophitic species (*Azospirillum*) may enter the root tissues but usually are found in the root surface.

NITRATE ASSIMILATION

Nitrate assimilation is a main process of the nitrogen cycle that mobilizes more than 2×10^{13} kg nitrogen per year. Nitrate is used as nitrogen source for growth by higher plants, fungi, algae, archaea, and bacteria. This process requires uptake of nitrate by a transport system and its further reduction to ammonium, via nitrite, by assimilatory nitrate and Nirs (Figure 5). Ammonium formed in this process is incorporated into carbon skeletons mainly by the GS–GOGAT pathway. Nitrate assimilation has been studied in heterotrophic bacteria (*Klebsiella*, *Bacillus*, and *Azotobacter*), oxygenic phototrophs (cyanobacteria), and anoxygenic phototrophic bacteria (*Rhodobacter*). In cyanobacteria, both nitrate and Nirs use photosynthetically reduced ferredoxin or flavodoxin as physiological electron donor, but the enzymes of *R. capsulatus* and most heterotrophic bacteria are NADH-dependent. In general, nitrate assimilation is induced by nitrate and repressed by ammonium through different control systems. Nitrate assimilation genes are usually clustered, but gene cluster organization varies depending on the bacteria.

The Nitrate Uptake Systems

Nitrate may be incorporated by an ATP-dependent ABC-type transporter or by a membrane potential-energized permease of the major facilitator superfamily (MFS). ABC-type transporters of *Klebsiella* and *R. capsulatus* include a periplasmic substrate-binding protein, a homodimeric hydrophobic protein with six transmembrane helices, and a homodimeric cytoplasmic ATPase, which are encoded by the *nasFED* genes, respectively. In cyanobacteria, the ABC-type transporters are encoded by four genes (*nrtABCD*) and have some structural

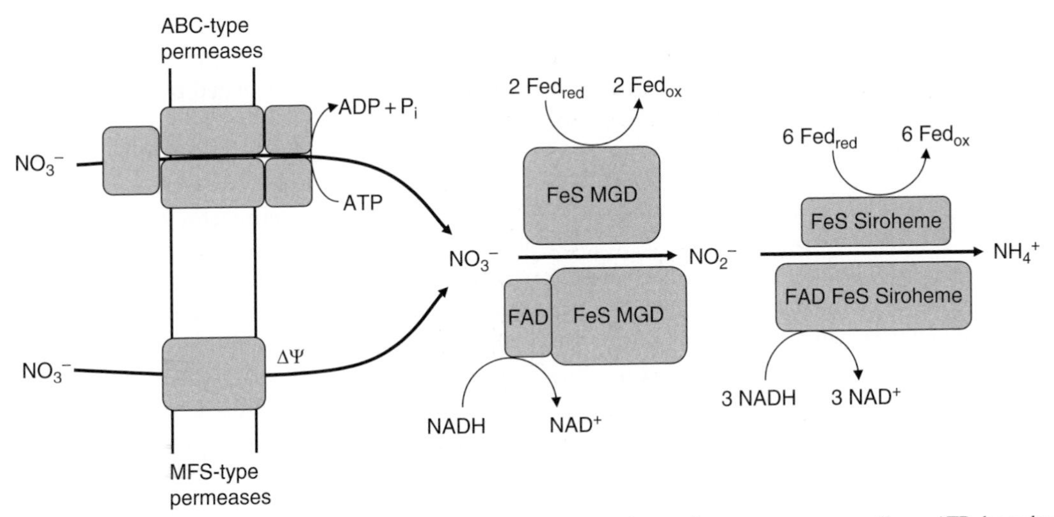

FIGURE 5 Bacterial assimilatory nitrate reducing systems. Nitrate assimilation requires a nitrate transporter, usually an ATP-dependent, ABC-type transporter, although an MFS-type permease energized by membrane potential ($\Delta\Psi$) operates in some bacteria. NADH or reduced ferredoxin (or flavodoxin) may be used as electron donors for nitrate and nitrite reductases. Generally, ferredoxin-dependent nitrate reductases are monomers with [4Fe–4S] and Mo-*bis*-molybdopterin guanine dinucleotide (Mo-*bis*-MGD) cofactor, whereas NADH enzymes have an additional subunit with FAD. Ferredoxin-nitrite reductases contain [4Fe–4S] and siroheme, and NADH-dependent enzymes also contain FAD. See text for details.

differences. The periplasmic substrate-binding protein NrtA has a lipid moiety for membrane attachment, the transmembrane protein NrtB has only five helices, and the ATPase is a heterodimer of NrtC and NrtD proteins. NrtC has an additional C-terminal domain homologous to NrtA and may have a regulatory function. MFS permeases are present in *Bacillus subtilis* and *Paracoccus denitrificans* (NasA) and in two cyanobacterial strains (NrtP). Like other MFS permeases, NasA and NrtP proteins have 12 transmembrane segments, but its fueling mechanism has not been determined. However, as these proteins are homologous to NarK permeases, it is assumed that they are H^+ / NO_3^- symporters.

Assimilatory Nitrate Reductases (Nas)

Nas are molybdoenzymes that catalyze the two-electron reduction of nitrate to nitrite. Prokaryotic and eukaryotic Nas share no significant sequence similarity. The eukaryotic enzymes are cytosolic homodimeric proteins that use NAD(P)H as reductant. Each 100kDa monomer has three functional domains: the N-terminal domain binds a molybdopterin or Mo cofactor (6-alkyl pterin derivative with a phosphorylated C4 chain and two thiol groups binding the Mo atom), the central domain has a *b*-type heme, and the C-terminal domain binds NAD(P)H and FAD. On the other hand, bacterial Nas are cytoplasmic proteins structurally and functionally different from the dissimilatory periplasmic nitrate reductases (Nap) and the respiratory membrane-bound nitrate reductases (Nar) that participate in denitrification. These three types of bacterial nitrate reductases bind a Mo-*bis*-molybdopterin guanine dinucleotide (Mo-*bis*-MGD) cofactor and at least one [4Fe–4S] cluster.

There are two types of bacterial Nas: the ferredoxin- or flavodoxin-dependent enzyme found in cyanobacteria, *Azotobacter*, and the archaeon *Haloferax mediterraneii*; and the NADH-dependent enzyme present in heterotrophic bacteria and *R. capsulatus*.

The cyanobacterial enzyme is an 80kDa monomer, encoded by *narB*. Electrons from ferredoxin or flavodoxin are transferred to the [4Fe–4S] cluster and the Mo-*bis*-MGD cofactor, the nitrate reduction site. In *A. vinelandii*, the enzyme is a 105-kDa monomeric protein with [4Fe–4S] and Mo-*bis*-MGD that uses flavodoxin as physiological reductant. The haloarchaeon *H. mediterraneii* has a dimeric nitrate reductase (105 and 50kDa subunits) that also uses reduced ferredoxin as electron donor. None of these ferredoxin (flavodoxin)-dependent enzymes contains flavin as prosthetic group.

NADH-dependent Nas are present in *Klebsiella*, *Bacillus*, and *Rhodobacter*. In *Klebsiella*, the enzyme is composed of a 92-kDa catalytic subunit and a small electron-transfer subunit (43kDa). The large subunit (*nasA* gene product) binds Mo-*bis*-MGD, one [4Fe–4S] cluster in the N-terminal end, and one additional [2Fe–2S] center in a C-terminal extension, which is absent in other bacterial Nas. The small subunit (*nasC* gene product) is an FAD-containing NADH oxidoreductase (diaphorase) that mediates electron transfer from NADH to the catalytic subunit NasA. Electron flow between redox cofactors follows the sequence: NADH → FAD → [2Fe – 2S] → [4Fe – 4S] → Mo-*bis*-MGD → NO_3^-. *B. subtilis* also has a dimeric NADH-dependent enzyme. However, the large catalytic subunit lacks the [2Fe–2S] cluster, and the small electron-transfer subunit contains two putative [2Fe–2S] centers in addition to FAD. Finally, in *R. capsulatus*, the assimilatory NADH-dependent nitrate reductase is a heterodimer composed of a 45-kDa FAD-containing diaphorase and a 95-kDa catalytic subunit with Mo-*bis*-MGD and one N-terminal [4Fe–4S] cluster, without additional [2Fe–2S] centers.

Assimilatory Nir

Assimilatory Nir are cytoplasmic enzymes that catalyze the reduction of nitrite to ammonium. Although NO and NH_2OH are catalytic intermediates, they are not released as free products and the process occurs as a single six-electron reaction. Ferredoxin-dependent Nir are found in oxygenic photosynthetic organisms, either eukaryotes (algae and plants) or prokaryotes (cyanobacteria), and in the haloarchaeon *H. mediterraneii*, whereas NADH-dependent Nir are present in fungi, heterotrophic bacteria, and the anoxygenic phototroph *R. capsulatus*. Both types of Nir contain siroheme as prosthetic group (special heme with eight carboxylic acid-containing peripheral side chains, also found in sulfite reductases) and one [4Fe–4S] center. These assimilatory Nir are completely different in structure and function in comparison to the respiratory pentaheme Nir pentaheme-nitrite reductase (Nrf) and the cytochrome cd_1 and copper Nir of denitrifying organisms.

Cyanobacterial Nir are 55 kDa monomers (encoded by the *nirA* gene) that receive electrons from photosynthetically reduced ferredoxin or flavodoxin. Then, the [4Fe–4S] center and the siroheme group mediate electron transfer to nitrite. However, the enzyme from *P. boryanum* contains a C-terminal domain similar to plant-type [2Fe–2S] ferredoxin which is probably involved in the electron flow to the [4Fe–4S] cluster. Curiously, the haloarchaeon *H. mediterraneii* has also a 66-kDa monomeric ferredoxin–Nir with siroheme and the [4Fe–4S] center.

NADH–Nir of heterotrophic bacteria contain FAD in addition to siroheme and the [4Fe–4S] cluster. Both FAD- and NADH-binding sites are located in an N-terminal extension that is not present in the ferredoxin-dependent enzymes. Two additional [2Fe–2S] centers are also present in the Nir of *Klebsiella* and *Bacillus*. However, whereas the enzyme of *Klebsiella* is a monomer encoded by the *nasB* gene, the Nir from *B. subtilis* is composed of two subunits (*nasDE* gene products). *R. capsulatus* is the only phototrophic organism described so far with an NADH–Nir. This enzyme, which also contains FAD, [4Fe–4S], and siroheme, is a heterodimer composed of a 87-kDa catalytic subunit (NasB) and a 12-kDa subunit homologous to the *E. coli* NirD protein.

Regulation of Nitrate Assimilation

Although the regulatory mechanisms that control bacterial nitrate assimilation depend on the organism, this process is usually regulated by nitrate induction (pathway-specific control) and ammonium repression (general nitrogen control). In *Klebsiella*, ammonium control of the nitrate assimilation (*nas*) genes is mediated by the Ntr system, whereas nitrate induction is mediated by a transcription antitermination protein (NasR). In response to nitrogen limitation, the Ntr system activates *nas* gene expression, but without nitrate transcription terminates shortly downstream of a leader sequence. However, in presence of nitrate, NasR protein binds to this *nas* mRNA leader sequence and acts as a transcription antiterminator to allow *nas* gene expression. In *A. vinelandii*, the general nitrogen control is also mediated by the Ntr system, but nitrate induction requires a two-component system composed of a nitrate sensor (NasS), which is similar to the NrtA/NasF substrate-binding components of nitrate transporters, and a response regulator (NasT) similar to NasR. Without nitrate, the NasS sensor inhibits the transcription antitermination effect of NasT, but in presence of nitrate, NasS does not block NasT, and binding of NasT to the mRNA leader sequence allows *nas* gene transcription. Similar NasS/NasT proteins are also present in *R. capsulatus*, but a programmed translational frameshift −1 in the *nasS* transcript could allow the synthesis of NasS only under nitrogen deficiency.

In *B. subtilis*, which lacks the Ntr system, nitrate assimilation genes are under the control of the general nitrogen regulator TnrA. In cyanobacteria, global nitrogen control is exerted by the NtcA transcription factor, which is influenced by the 2-oxoglutarate sensor PII. In some cyanobacteria, there is also a pathway-specific positive control by nitrate/nitrite in which NtcB, a protein of the LysR family, and other regulators such as CnaT and NirB participate. However, the molecular mechanisms through which these proteins affect gene expression or enzyme activity remain to be elucidated.

DENITRIFICATION

Denitrification is a process in which nitrate, nitrite, and the gaseous nitric and nitrous oxides serve as alternative terminal acceptors for electron transport phosphorylation, forming N_2 as final product. This process is sequentially catalyzed by nitrate reductase (Nar and Nap), Nir, nitric oxide reductase (Nor), and nitrous oxide reductase (Nos) (Table 1). Nitrate reductase receives electrons directly from the quinol pool, but for nitrite and nitrogen oxide reduction the H^+-pumping cytochrome *bc1* complex mediates the electron flow to cytochrome *c* or small copper proteins (azurin and pseudoazurin), which are the enzyme electron donors (Figure 6). As denitrification allows organisms an oxygen-independent respiration, it is usually carried out under anaerobic conditions. Denitrifying organisms belong to proteobacteria, halophilic and hyperthermophilic archaea, and even certain fungi, but the best studied denitrifiers are some *Pseudomonas* and *Paracoccus* strains. Some bacteria,

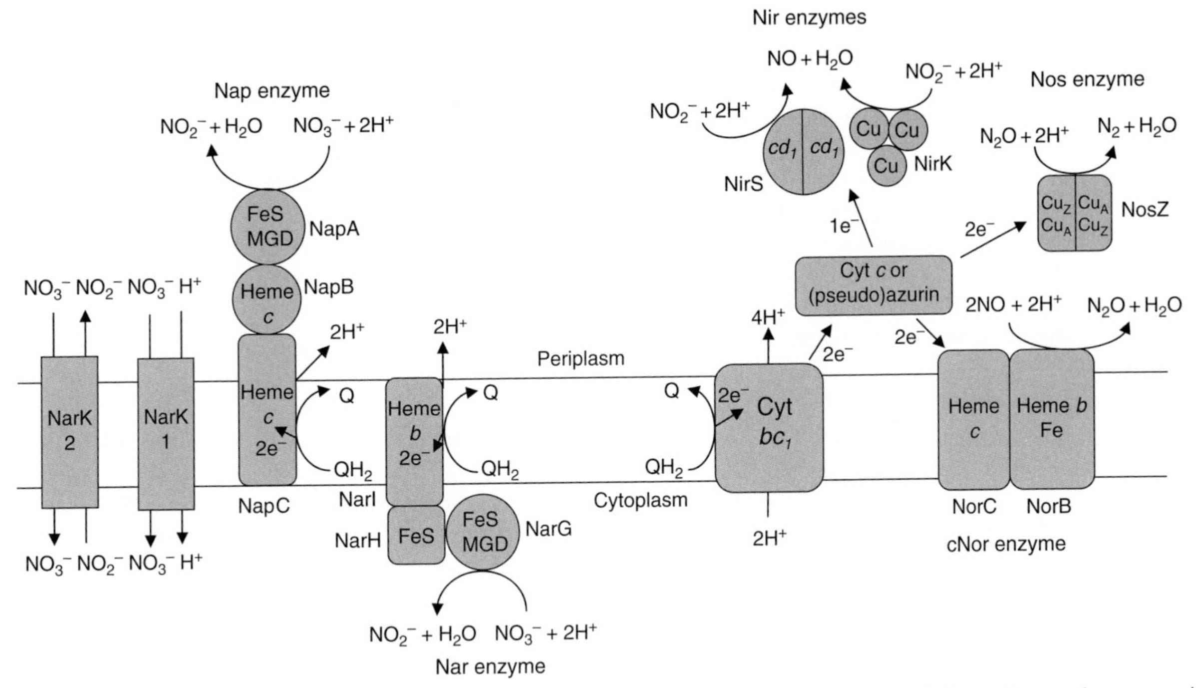

FIGURE 6 The denitrification pathway. Membrane-bound nitrate reductase (Nar), periplasmic nitrate reductase (Nap), cytochrome cd_1 or copper nitrite reductases (Nir), nitric oxide reductase (Nor), and nitrous oxide reductase (Nos) are represented according to their location, subunit structure, and cofactors. Note that not all components of the denitrification pathway are present in the same organism. Nitrate transporters and proton-pumping cytochrome bc_1 complex are also shown. QH2/Q represents the membrane quinol/quinone pool. See text for details.

such as *E. coli*, are able to respire nitrate and/or nitrite, but do not develop a complete denitrification and are not considered true denitrifiers.

Nitrate Reductases on Denitrification (Nar and Nap)

Two types of nitrate reductases may be involved in denitrification: the respiratory membrane-bound enzyme (Nar), which faces the active site to the cytoplasm in bacteria but to the periplasm in archaea, and the periplasmic Nap. The catalytic subunit of both enzymes has a Mo-*bis*-MGD cofactor and one [4Fe–4S] cluster, such as bacterial Nas.

Membrane-bound Nar use the quinol pool as the physiological electron donor and generate a protonmotive force by a redox loop mechanism. In this process, two protons are consumed in the cytoplasm and two protons are released into the periplasm, which generates an electrochemical gradient that can be coupled to ATP synthesis. The Nar complex is composed of three subunits: NarG or α-subunit (140 kDa), the catalytic subunit with the Mo-*bis*-MGD cofactor and one [4Fe–4S] cluster; NarH or β-subunit (60 kDa), the electron-transfer subunit with one [3Fe–4S] and three [4Fe–4S] centers; and NarI or γ-subunit (25 kDa), the quinol-oxidizing membrane subunit with two *b*-heme groups, a low-potential group located at the periplasmic site and a high-potential group facing the cytoplasm. NarG and NarH have a cytoplasmic location and contact with NarI by the C-terminal end of NarH. Two electrons are transferred from reduced quinones in the membrane to nitrate via low-potential and high-potential heme groups of NarI, [3Fe–4S] and [4Fe–4S] clusters of NarH, and [4Fe–4S] center and Mo-*bis*-MGD of NarG. These structural characteristics make Nar electrogenic since it contributes to the generation of a proton electrochemical gradient across membrane (redox loop). Crystal structure has been solved for the Nar enzyme of *E. coli*, which respires nitrate but is not a true denitrifier.

Since the active site of NarG faces the cytoplasm, nitrate must be incorporated inside the cells, usually by transport proteins of the NarK family with 12 membrane-spanning helices. This type of nitrate/nitrite permease is also present in *E. coli* and in other nondenitrifiers. In *Paracoccus*, the *narK* gene codes for a transmembrane protein with 24 helices organized in two domains, NarK1 and NarK2. NarK1 functions as a NO_3^-/H^+ symporter to initiate respiration, whereas NarK2 is a NO_3^-/NO_2^- antiporter that maintains the nitrate respiration process.

Respiratory nitrate reductases in archaea are not well characterized, but may exist as two types of Nar systems: an nNar similar to the bacterial Nar with the active site in the cytoplasm and a pNar (potential positive environment), which is exclusively found in archaea and faces the active site on the outside of the membrane. The pNar proteins have twin arginine motifs for translocation across cytoplasmic membrane by the Tat pathway. A quinone cycle mechanism for energy conservation may be coupled to this pNar system through a putative diheme protein (NarC) present in some archaeal species.

Periplasmic Nap were initially described in phototrophic and denitrifying bacteria, but they are widespread among Gram-negative bacteria. Nap enzymes are usually composed of a 90-kDa catalytic subunit (NapA) with Mo-*bis*-MGD and one [4Fe–4 S] centre, and a 15-kDa biheme cytochrome *c* (NapB) that reduces NapA. NapB receives electrons from a 25-kDa quinol-oxidizing, membrane-anchored tetraheme *c* protein (NapC). Crystal structures of some Nap enzymes have been solved. Although this system is linked to membrane quinol oxidation, it does not generate protonmotive force. There are some bacterial strains without NapB or NapC proteins and, depending on the organism, the Nap system includes additional proteins. Thus, NapF protein participates in assembling the [4Fe–4 S] center into the catalytic NapA subunit, and NapD is involved in NapA maturation.

Since oxygen inhibits nitrate transport to the cytoplasm, some denitrifiers can perform aerobic denitrification coupling the Nap enzyme to nitrite and nitrogen oxide reductases. Besides their role in denitrification, Nap systems also have other functions depending on the organism. In *Rhodobacter sphaeroides* and *P. denitrificans*, Nap is a dissimilatory enzyme used to dissipate excess reducing power (redox balancing), which is required for an optimal growth under phototrophic conditions or on highly reduced carbon sources. In addition, *E. coli* and *Haemophilus influenzae* use the Nap system for nitrate scavenging in environments with low nitrate concentration.

Nir on Denitrification

Nir involved in denitrification are periplasmic-located enzymes that catalyze the one-electron reduction of NO_2^- to NO. Therefore, at difference of other bacterial Nir, these respiratory enzymes generate NO instead of NH_4^+. Two types of Nir are found in denitrifying organisms: the cytochrome cd_1-Nir, a heme protein encoded by the *nirS* gene, and the Cu-Nir encoded by the *nirK* gene. Both types of enzymes are never found in the same bacterium, so that denitrifiers with the cd_1-Nir lack the Cu-Nir.

Cytochrome cd_1-Nir are homodimeric proteins, each 60 kDa subunit with a covalently bound *c*-type and a noncovalently bound d_1-type heme groups organized in two separate domains; the d_1-heme-binding domain has α-helices, whereas the *c*-heme domain has an eight-bladed β-propeller structure. The catalysis occurs at the d_1-heme, whereas the *c*-heme is the electron entry site. Catalysis starts with substrate binding to the d_1-heme iron, followed by a nitrosyl intermediate (d_1 – Fe^2 + NO^+) to the end product NO. The cd_1-Nir also catalyzes the reduction of NH_2OH to NH_3.

The Cu-Nir are trimeric enzymes in which each monomer (40 kDa) contains two different copper centers – type I Cu and type II Cu. Type I site is the redox active center from which electrons are transferred to type II site, where substrate binds. The catalytic mechanism is similar to that described for cd_1-Nir, including a nitrosyl intermediate (Cu^+NO^+) formed at the type II site. Cu-Nir also produce N_2O or NH_3 but at a very low rate.

Nor and Nos

Nor are membrane proteins that produce N_2O from NO. Three types of Nor have been described: cNor, qNor, and qCu_ANor. The cNor from Gram-negative denitrifying bacteria are the best studied enzymes and consist of cytochrome *bc* complexes. cNor use *c*-type cytochromes, azurin, or pseudoazurin as electron donors and have two subunits, the small heme *c*-containing subunit NorC (17 kDa) and the large *b*-heme subunit NorB (53 kDa). The low spin *c*-heme of NorC is located to the periplasmic side of this subunit and receives electrons from the electron donor. NorB is a hydrophobic protein with 12 transmembrane helices and contains six conserved histidine residues for cofactor binding, coordinating heme *b* (*bis*-His), heme *b3* (*mono*-His), and nonheme iron Fe_B (*tri*-His). The heme *b* (*bis*-His) transfers electrons to the dinuclear center composed of the heme *b3* (*mono*-His) and the nonheme iron Fe_B (*tri*-His), where NO binds and N_2O is released.

The qNor enzymes have been isolated from *Ralstonia eutropha* and the archaeon *Pyrobaculum aerophilum* and are also present in some nondenitrifying pathogenic bacteria where they play a role in NO detoxification. qNor use ubiquinol or menaquinol as electron donors and contain *b*-heme, but lack heme *c*. They are composed of an 80 kDa single subunit similar to the dinuclear center of cNor, where heme-*b* and a nonheme Fe_B cofactors are present in a 2:1 ratio. The Gram-positive bacterium *Bacillus azotoformans* has a third type of Nor, qCu_ANor, that uses both menaquinol and cytochrome *c551* as electron donor. The enzyme contains a large subunit similar to NorB, with one Fe_B and two *b*-heme groups, and a small subunit without heme *c* but with a Cu_A site.

The reduction of N_2O to N_2 catalyzed by the periplasmic Nos is the last step in denitrification. Nos enzymes are homodimers with two multinuclear Cu centers – Cu_Z and Cu_A. Each subunit (65 kDa) has two domains; the N-terminal domain binds the catalytic Cu_Z site and the C-terminal domain carries the electron transfer center Cu_A, which also contributes to enzyme stabilization. The Cu_A center has a cupredoxin fold with a Cu_2S_2 rhomb, and spectroscopic analysis reveals that there is also a sulfide group in the Cu_Z site. Depending on the organism, Nos is exported to the periplasm as holoprotein by the Tat translocase (*Pseudomonas stutzeri*) or as unfolded protein via the Sec pathway (*Wolinella succinogenes*). In *P. stutzeri*, Nos is encoded by the *nosZ* gene, but other *nos* gene products are required for copper and sulfur incorporation and electron donation. However, these components are not present in all denitrifiers. Some archaea show Nos activity but, although accessory *nos* genes are present, *nosZ* is not found. Therefore, a different type of Nos must be present in these organisms.

Regulation of Nitrate Respiration and Denitrification

Nitrate respiration and denitrification are controlled by multiple regulatory systems that show differences among organisms but basically respond to O_2, nitrate/nitrite, and NO (Figure 7). Some of these regulatory mechanisms belong to the so-called two-component regulatory systems. FixL and Fnr proteins are oxygen sensors. FixL contains an O_2-sensitive heme group at the N-terminal domain that controls autophosphorylation and P-group transfer to FixJ, the response regulator. Phosphorylated FixJ induces the expression of $FixK_2$, which in turn induces transcription of genes required for microaerobic growth. Fnr is a transcription activator with a [4Fe–4S] cluster that is sensitive to O_2 and NO. Fnr activates transcription of many genes in response to anaerobiosis, but it can also act as a repressor. The Anr protein is an O_2-sensitive ortholog of Fnr present in some denitrifiers. The ArcA/ArcB two-component system also participates in the anaerobic respiratory control. When O_2 is depleted, kinase activity of ArcB is activated, ArcA is phosphorylated, and genes required for aerobic growth are repressed.

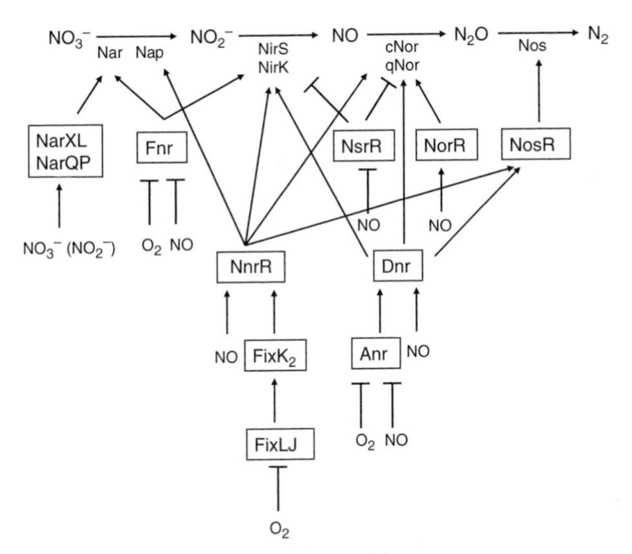

FIGURE 7 Regulation of denitrification. The main positive or negative regulators of the denitrification genes and their activating or inactivating effector molecules are shown. Positive effects are represented by arrows and negative effects are indicated by perpendicular bars. Note that not all regulatory proteins are present in the same organism.

NarXL, NarQP, and NarR are well-known nitrate/nitrite response systems. NarX and NarQ are nitrate and nitrite sensors that phosphorylate the response regulators NarL and NarP, respectively. NarR belongs to the Fnr family, but lacks the cysteine motif for binding the [4Fe–4S] cluster and probably responds to nitrate/nitrite rather than to O_2. In addition, different nitrogen oxide sensors have been described. NnrR and Dnr are NO sensors that belong to the Fnr family and lack the [4Fe–4S] center. These proteins regulate respiratory nitrite and NO reduction. NorR is a σ^{54}-transcriptional activator with nonheme iron that senses NO and shows ATPase activity, which is stimulated in the presence of NO to allow transcription activation of Nor and NO detoxification genes. NirI protein is a NorR homolog with a specific role in *nir* gene transcription in response to nitrogen oxides and oxygen limitation. The NosR regulator has a role in NO respiration and is required for *nos* gene expression. Finally, the NsrR protein is a nitrite and NO-sensitive regulator that controls expression of genes encoding respiratory Nir and Nor and detoxification proteins, such as flavohemoglobin and hybrid cluster protein (Hcp).

NITRIFICATION

The whole nitrification process involves the aerobic oxidation of ammonium to nitrate. The classical process, which involves the sequential oxidation of ammonium to nitrite, via hydroxylamine, and of nitrite to nitrate, is carried out by two different groups of chemolithoautotrophs known as AOB and NOB, respectively, represented by *Nitrosomonas* and *Nitrobacter* species (Figure 8). However, ammonia oxidation is also found in some planctomycetes, heterotrophic bacteria, and fungi. Several soil and marine crenarchaeota are also capable of chemolithoautotrophic ammonia oxidation. Nitrite-oxidizing process has been less studied and it is restricted to only four proteobacteria genera. Under normal conditions, ammonium oxidation to nitrite is the limiting step and nitrite is oxidized to nitrate more rapidly.

Nitrosomonas and *Nitrobacter* are nitrifiers that use aerobic oxidation of ammonium or nitrite as the sole source of energy and reductant for bacterial growth, fixing CO_2 by the Calvin cycle. Electrons from either ammonium or nitrite flow through the respiratory chain to oxygen and these processes are coupled to protonmotive force generation, allowing ATP synthesis. However, oxidation of ammonium or nitrite does not provide enough reducing power to form the necessary NAD(P)H and, therefore, these bacteria perform the so-called reverse electron transport. Thus, the electron flow is mainly directed to ATP synthesis, but protonmotive force is also used to drive reverse electron transfer to $NAD(P)^+$.

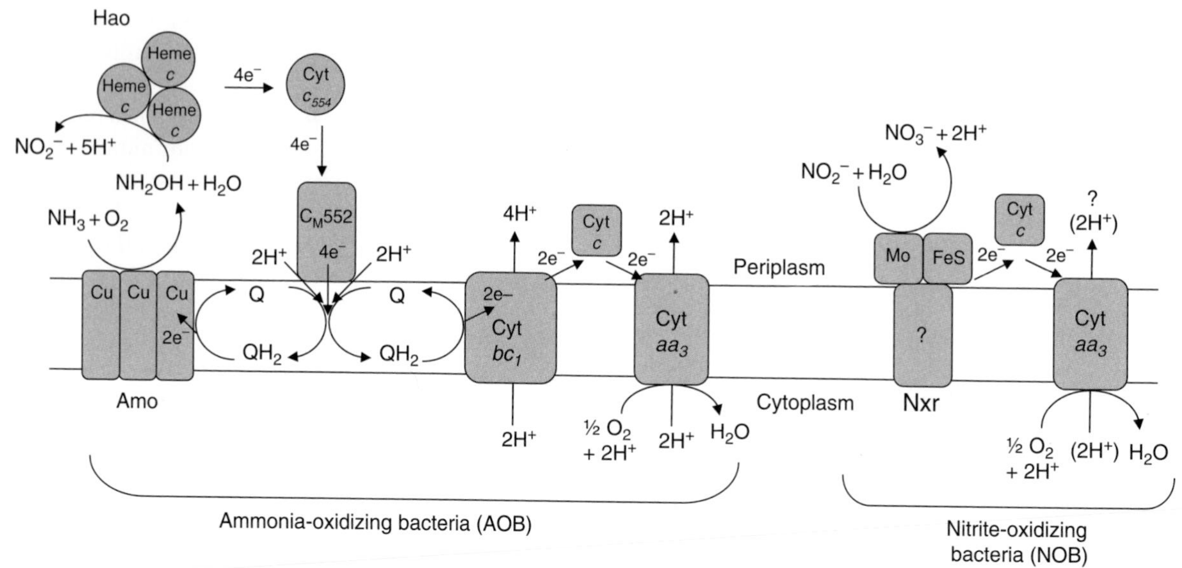

FIGURE 8 The nitrification pathway in chemolithoautotrophic bacteria. Left, ammonia oxidation by ammonia-oxidizing bacteria (AOB); right, nitrite oxidation by nitrite-oxidizing bacteria (NOB). Location, putative subunit composition, and cofactors of the enzymes ammonia monooxygenase (Amo), hydroxylamine oxidoreductase (Hao), and nitrite oxidoreductase (Nxr) are indicated. The proton-pumping cytochrome bc_1 and cytochrome aa_3 complexes are also shown. QH_2/Q represents the membrane quinol/quinone pool. The location of the Nxr active site (periplasm or cytoplasm) is still controversial and the H^+-pumping activity of cytochrome aa_3 in NOB is also under discussion.

Ammonia oxidation is a complex process that requires several enzymes and proteins. In *Nitrosomonas europaea*, ammonia is first oxidized to hydroxylamine by ammonia monooxygenase (Amo), a multimeric transmembrane copper protein that receives two electrons directly from the quinol pool. This reaction also requires O_2. Amo is homologous to methane monooxygenase, a particulate copper enzyme of methane-oxidizing bacteria. These bacteria are also able to oxidize ammonia as a result of the broad substrate specificity of methane monooxygenase. In the second step, a trimeric multiheme *c*-type hydroxylamine oxidoreductase (Hao) converts hydroxylamine into nitrite in the periplasm with production of four electrons that are transferred through a soluble cytochrome c_{554} to the membrane-anchored protein c_M552, a *c*-type cytochrome belonging to the NapC family that donates the electrons to the quinol pool. Ubiquinol is also oxidized by the cytochrome bc_1 complex, which transfers two electrons to an aa_3-type oxidase and finally to oxygen with protonmotive force generation. In addition, ubiquinol is also oxidized by a third route based on the reverse electron transfer to an NADH-ubiquinone oxidoreductase to generate NAD(P)H. In all AOB, the *amoCAB* genes encoding Amo are organized into a cluster, whereas another gene cluster *hao-orf2-cycAB* codes for the Hao enzyme and two cytochromes. Both *amo* and *hao* genes are induced by ammonium. The *hao* gene has probably evolved from the *nrfA* gene, which codes for the pentaheme cytochrome *c* Nir, involved in respiratory nitrite ammonification. Ammonia-oxidizing archaea (AOA) contain *amo*-like genes but lack other genes required for bacterial ammonia oxidation, suggesting that AOA use a different mechanism for this process.

Nitrite oxidation is performed by *Nitrobacter* and other three genera of NOB. *Nitrobacter* contains an aa_3-type cytochrome oxidase, but it is not clear whether H^+-pumping is coupled to the nitrite-oxidizing enzyme. The aa_3-type oxidase has a Cu_A center that receives electrons from soluble or membrane *c*-type cytochromes. The nitrite-oxidizing enzyme is known as "nitrite oxidoreductase" (Nxr). The enzyme contains at least molybdenum and iron–sulfur centers (cytochromes a_1 and c_1 have been also proposed), but the molecular mechanism of nitrite oxidation is an open question. In the *Nitrobacter* genome, there are two genes (*norAB*) that code for the putative 115 and 65 kDa subunits of Nxr, which are related to the respiratory nitrate reductase *narGH* genes. It is believed that Nxr is anchored to an unknown transmembrane protein, but it is unknown whether nitrite is oxidized in the cytoplasmic side of the membrane, coupled to the proton-pumping aa_3-type cytochrome oxidase, or the Mo-active site of Nxr is located in the periplasm and electrons flow to the aa_3-type oxidase without H^+-pumping activity. Genome analysis reveals the presence of other putative quinol oxidases (*bo*- and *bd*-type oxidases), but their role in nitrite oxidation is not yet demonstrated.

Nitrite is an intermediate of both nitrification and denitrification, so that ammonium oxidation to nitrite and further nitrite denitrification (nitrifier denitrification) may occur (Figure 9). *Nitrosomonas* also contain Cu-Nir and Nor, encoded by *nirK* and *norCB* genes, respectively.

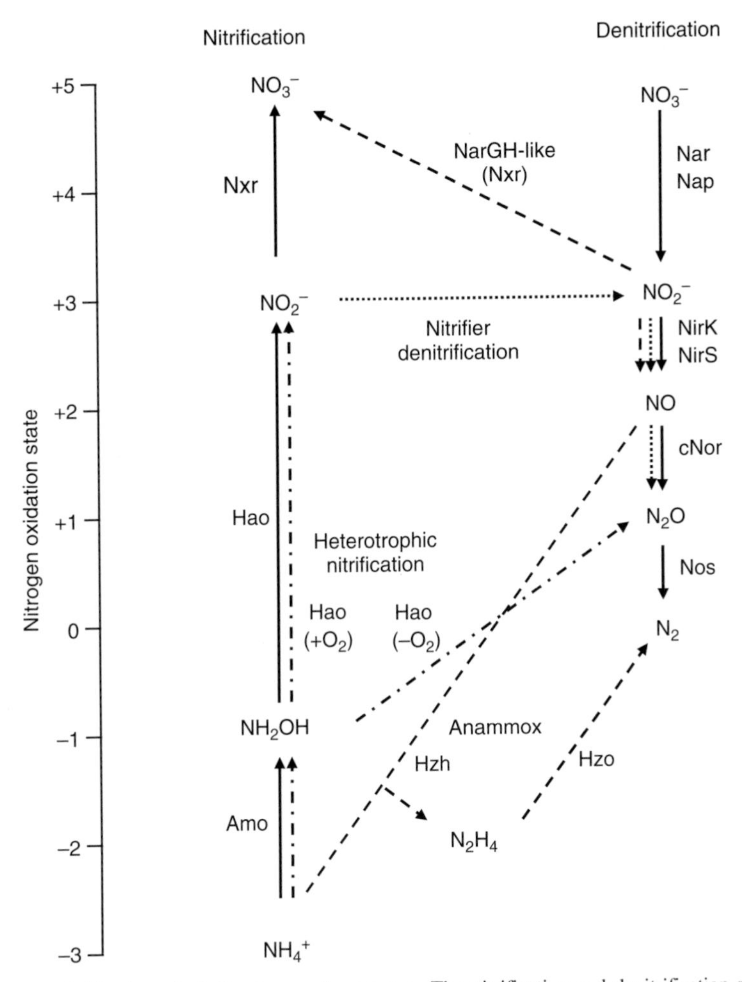

FIGURE 9 Relationships between nitrification, denitrification, and anammox. The nitrification and denitrification pathways are indicated by normal arrows. The anammox process is represented by dashed arrows. In nitrifier denitrification (dotted arrows), nitrite formed by ammonia oxidation is converted into nitrogen oxides by NirK and cNor enzymes. In heterotrophic nitrification (dashed–dotted arrows), hydroxylamine formed by ammonia monooxygenase (Amo) is converted into nitrous oxide (anaerobically) or nitrite (aerobically) by the Hao enzyme. See text for details.

Denitrifying enzymes can be used for detoxification of nitrite and NO, but also to generate protonmotive force at low oxygen. Nitrification is not only restricted to chemolithoautotrophic organisms but also to heterotrophic bacteria such as *P. denitrificans*. However, little is known about heterotrophic nitrification. It seems that in anoxia, a 20-kDa Hao with nonheme iron catalyzes a two-electron oxidation of hydroxylamine and generates nitroxyl radicals that dimerize to form N_2O, instead of nitrite. In aerobiosis, nitrite can be formed by oxidation of nitroxyl with oxygen. In chemolithoautotrophic hydroxylamine oxidation, four electrons are transferred to the cytochrome c_{554}: two electrons are passed into the quinol pool and used by Amo, and the other two electrons are used to generate protonmotive force or to form NAD(P)H by reverse electron transfer. In heterotrophic nitrification, as the Hao product is nitroxyl and nitrite generation involves reaction with O_2, only two electrons are produced and this does not allow autotrophic growth. In *P. denitrificans*, it has been proposed that these two electrons are transferred via cytochrome c_{500} or pseudoazurin to the denitrifying reductases without protonmotive force generation. In this case, nitrification seems to have a role in redox balancing rather than in energy conservation.

ANAMMOX

Anammox involves the anaerobic oxidation of ammonium to N_2 using nitrite as electron acceptor (Figure 9). Therefore, this process couples the oxidation of ammonium to the reduction of nitrite and includes features of both nitrification (ammonium oxidation) and denitrification (nitrite reduction and N_2 production) in a single process that is performed by bacteria of the Planctomycetales order (*Brocardia*, *Kuenenia*, *Anammoxoglobus*, and *Scalindua*). Anammox bacteria are chemolithoautotrophs that use the exergonic reaction $NH_4^+ + NO_2^- \rightarrow N_2 + 2H_2O$ as the primary source of energy and reductant for autotrophic growth, fixing carbon by the acetyl-CoA pathway. Hydrazine (N_2H_4)

is a free intermediate in the process. Anammox was discovered in 1995 in a wastewater treatment plant in Delft (NL), but now it is clear that it is the main nitrogen cycle process that contributes up to 50% to the removal of fixed nitrogen in the oceans. However, anammox bacteria grow very slowly, with only about 15 generations per year. Anammox bacteria have unique lipids (ladderane), which constitute the membranes of a special compartment termed "anammoxosome" that contains the anammox enzymes. Ladderane lipids may confer impermeability and rigidity to the membrane, so that the function of anammoxosome is the internalization of toxic hydrazine to limit its losses. Environmental genomics has allowed the assembling of the complete genome of *Kuenenia stuttgartiensis*. Genomic data suggest that a *cd1*–Nir (NirS) reduces nitrite to NO. Hydrazine hydrolase (Hzh) is a Hao-like multiheme *c* trimeric protein that uses NO as oxidant for ammonium oxidation to form hydrazine, which is finally converted into N_2 by the hydrazine-oxidizing enzyme (Hzo), a homodimer with eight heme *c* in each 62kDa subunit. The four electrons released in the Hzo reaction are passed through the quinol pool and the proton-pumping cytochrome bc_1 complex to small cytochrome *c* molecules. This generates a protonmotive force across anammoxosome membrane, allowing ATP synthesis. Cytochrome *c* donates the three electrons required for the Hzh reaction and the single electron consumed in the NirS reaction (Figure 10). In addition, anaerobic oxidation of nitrite to nitrate by a NarGH-like Nxr similar to Nxr from aerobic NOB (Figure 9) provides the electrons required for carbon fixation by the acetyl-CoA pathway.

OTHER PROCESSES OF THE NITROGEN CYCLE

Nitrogen Mineralization

Organic matter is present in soils and some water environments. Nitrogen mineralization is the process by which inorganic nitrogen is obtained by decomposition of dead organisms and degradation of organic nitrogenous compounds. As this process releases ammonium, it is also known as "ammonification," although this term is also used for other dissimilatory processes. Glutamate deamination by GDH and hydrolysis of urea by urease are important ammonification reactions. Also, extracellular proteolytic enzymes produced by saprophytic bacteria and fungi are of great importance in nitrogen turnover in the ecosystems since they have high substrate affinity and are responsible for protein degradation to small peptides and amino acids that are finally a source of ammonium in soils. Main fungal extracellular proteases belong to cysteine and aspartic proteases groups, whereas bacterial extracellular proteases are usually serine proteases and alkaline and neutral metalloproteases. Activity of these enzymes shows seasonal variations, with higher values in spring and autumn. Proteolytic activity increases after application of fertilizers and is also affected by soil composition, texture, and profile. Thus, the activity is high in soils with an elevated proportion of clay and humic colloids and decreases with increasing depth due to a decline in nutrient contents.

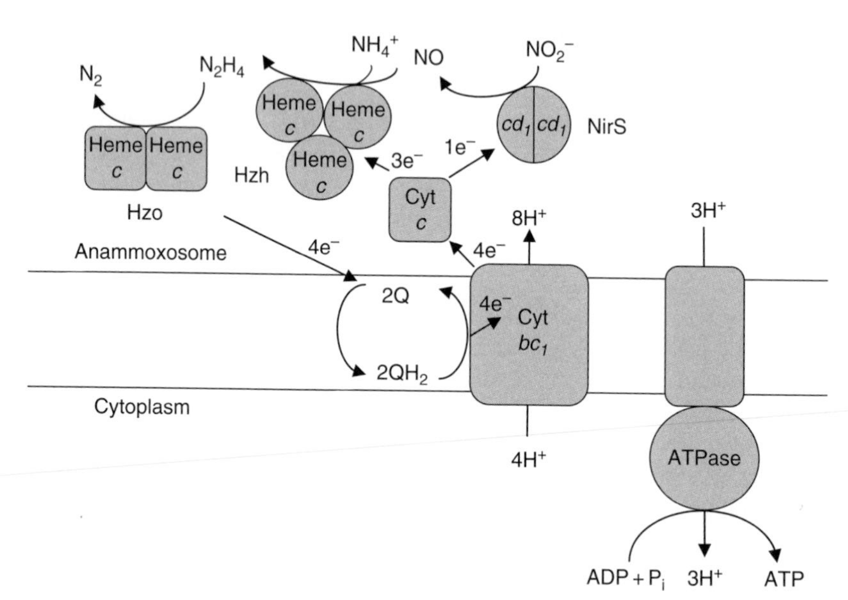

FIGURE 10 The anammox process in planctomycetes. The enzymes nitrite reductase (NirS), hydrazine hydrolase (Hzh), and hydrazine oxidoreductase (Hzo) are represented with their putative subunit structures and cofactors. The proton-pumping cytochrome bc_1 and the ATP synthase (ATPase) complexes are also shown in the anammoxosome membrane. QH_2/Q represents the quinol/quinone pool.

Dissimilatory Reduction of Nitrate to Ammonium (Nitrate Ammonification)

The dissimilatory nitrate reduction to ammonium (DNRA) or nitrate ammonification is an anaerobic process in which nitrate reduction to nitrite is followed by the six-electron reduction of nitrite to ammonium. This strictly anaerobic process occurs in reductant-rich environments such as anaerobic marine sediments, sulfide-thermal vents, and gastrointestinal tracts of animals. There are two types of DNRA: a periplasmic energy-conserving (respiratory) nitrate ammonification, which uses the pentaheme cytochrome *c* Nir Nrf to catalyze electron transfer from formate or H_2 to nitrite; and a cytoplasmic dissimilatory system that acts as electron sink and detoxifies nitrite formed in nitrate respiration, which uses the siroheme NADH-dependent Nir to reduce nitrite (Table 1). Therefore, DNRA is catalyzed by different enzymes (Nap–Nrf and Nar–Nir), occurs in different subcellular compartment (periplasm and cytoplasm), and plays different functions (ATP generation, dissipation of excess reducing power, and nitrite detoxification), depending on the organism. Although ammonium formed in the anaerobic DNRA process may be potentially assimilated, this is not its primary role. Therefore, regulation of this process is completely different from assimilatory nitrate reduction but similar to denitrification since it is repressed by oxygen and unaffected by ammonium.

In respiratory ammonification, nitrate reduction is usually performed by the Nap and nitrite reduction to ammonium is catalyzed by the Nrf using nonfermentable substrates (formate or H2) as reductants. This periplasmic system allows nitrate ammonification without nitrate/nitrite transport across the membrane and generates protonmotive force by coupling nitrite reduction to formate dehydrogenase or hydrogenase. Nrf is found in proteobacteria of the groups γ (*E. coli*), δ (*Desulfovibrio*), and ε (*Wolinella*), but not in α- and β-proteobacteria, which are mainly aerobic or facultative bacteria able to denitrify. In δ- and ε-proteobacteria, the pentaheme catalytic subunit NrfA (60kDa) is located at the periplasmic side, anchored to the membrane by the NrfH protein, a 22-kDa tetraheme *c* cytochrome that oxidizes menaquinol and mediates electron transfer to NrfA. In γ-proteobacteria, NrfH protein is replaced by the *nrfBCD* gene products. NrfB is a pentaheme cytochrome *c* that reduces the NrfA catalytic subunit, and NrfCD are integral membrane proteins with quinol oxidase activity. It has been proposed that NrfA plays a role in nitrosative stress defense since NO is substrate of this enzyme, which also produces N_2O under certain conditions. Denitrification and nitrate ammonification allow ATP generation under anaerobic conditions, but these processes do not occur in the same bacterium. Energy yield per reduced nitrate is slightly higher for respiratory nitrate ammonification than for denitrification, and, probably for this reason, bacteria living in environments with nitrate limitation prefer nitrate ammonification over denitrification.

Some bacteria use nitrite as an electron sink to consume excess of reductant in anaerobic environments, allowing regeneration of oxidized redox coenzymes (NAD^+) to sustain an optimal growth. This nitrite reduction to ammonium also serves to detoxify nitrite formed in nitrate respiration in nondenitrifying facultative anaerobes, such as *E. coli*. The Nir involved in this process, which in contrast to NrfA should be considered dissimilatory rather than respiratory, is the siroheme-containing Nir protein. The *E. coli* enzyme has a large catalytic subunit (NirB) and a small subunit (NirD) that is similar to the C-termini of monomeric Nir of *Klebsiella* (NasB) and fungi. It should be emphasized that *E. coli* NirBD is not a proper assimilatory enzyme, although it uses NADH as reductant and shows the biochemical characteristics of assimilatory siroheme-containing Nir. In contrast to assimilatory enzymes, the *E. coli* ammonifying Nir is only induced under anoxia and is not repressed by ammonium. Although nitrite ammonification by Nir is not coupled to energy conservation, facultative bacteria obtain ATP in anaerobiosis by nitrate respiration through membrane-bound Nar. In addition, when fermentative bacteria reduce nitrite with the Nir enzyme, this compound is not only detoxified but also used as an electron sink replacing intermediates of fermentation that would be reduced in the absence of nitrite. Thus, as acetate rather than ethanol is the final product of the C metabolism, additional ATP is generated by substrate-level phosphorylation.

Detoxification of Nitrogenous Compounds

Many nitrogenous compounds are toxic for organisms. Thus, ammonium is an uncoupler of phosphorylation; NO induces the formation of highly reactive species, such as peroxynitrite, that generates the so-called nitrosative stress; nitrite, hydrazine, and hydroxylamine are potent mutagens; cyanide is an inhibitor of cytochrome oxidase and other metalloproteins; and xenobiotic nitroaromatic compounds have cytotoxic, mutagenic and, xenoestrogenic effects. Therefore, there are a number of enzymes that play a role in detoxification of these molecules, including some bacterial enzymes of nitrogen cycle processes. Thus, Nor, Nrf, and siroheme-Nir detoxify NO. Other enzymes are also involved in NO detoxification, such as bacterial flavohemoglobins, which oxidize NO to nitrate in aerobiosis, and flavodiiron proteins (FDP), which are widespread among prokaryotes and reduce NO to N_2O under microaerobic conditions. *E. coli* has a type of FDP (flavorubredoxin) that contains a nonheme diiron site FMN and a rubredoxin-like domain with a Fe–S center. Electrons for NO reduction are transferred to this flavorubredoxin by an FAD-containing reductase that uses NADH. The monoheme cytochrome P460 is also a detoxification protein, which oxidizes NO and hydroxylamine in AOB and methane-oxidizing bacteria. In addition, the Hcp (or prismane) is a protein with an unusual hybrid [4Fe–2S–2O] cluster that plays a role in detoxification of NO and hydroxylamine in bacteria and archaea. The Hcp protein of *R. capsulatus* reduces hydroxylamine to ammonium, which is further assimilated by the GS–GOGAT pathway. In this bacterium, the *hcp* gene is clustered together with the *nas* genes required for nitrate assimilation. Hydroxylamine is also detoxified by its reduction to ammonium catalyzed by cd_1-Nir and NrfA, or by its oxidation to nitrite catalyzed by Hao. As NrfA and Hao are homologous proteins, it is possible to speculate that they evolved from a primitive hydroxylamine detoxification system. Curiously, in *E. coli*, Hcp and Nrf are regulated by the NO-sensitive repressor NsrR, which also regulates expression of NirK in *Nitrosomonas*. Therefore, NsrR may be a general regulator of proteins involved in defense against nitrosative stress.

Some bacteria are resistant to cyanide and use this toxic compound as nitrogen source. Thus, the alkaliphilic bacterium *Pseudomonas pseudoalcaligenes* CECT5344 grows with cyanide, cyanate, and nitriles, and its response to cyanide includes induction of alternative oxidases for cyanide-insensitive respiration, putative enzymes for cyanide and cyanate degradation (nitrilase and cyanase), proteins involved in siderophores biosynthesis for iron uptake, and regulatory proteins (GlnK) that sense nitrogen starvation.

Nitroaromatic compounds are xenobiotic chemicals produced by human activities that may be degraded by bacterial oxidative or reductive pathways. Nitroreductases are enzymes widespread among prokaryotes that reduce the nitro groups of these compounds to the corresponding hydroxylamino or amino derivatives, but other enzymes are also involved in the different degradative pathways. In general, nitroaromatic compound degradation releases either nitrite or ammonium, which may be used as a nitrogen source for bacterial growth.

ECOLOGICAL AND ENVIRONMENTAL ASPECTS OF THE NITROGEN CYCLE

The nitrogen cycle reactions have important environmental, agronomic, and health implications and may be also used for technical applications. Thus, nitrification, denitrification, and anammox are important processes for removal of nitrogenous compounds in the treatment of industrial or domestic wastewater.

Nitrogen is usually an essential nutrient that limits plant growth. Therefore, to produce plants of quality with high yield and sufficient protein content, it is necessary to fertilize agricultural soils with mineral fertilizers (synthesized by the Haber–Bosch process, which needs high energy consumption), or manure, a mixture of both inorganic and organic nitrogenous compounds that are mineralized by soil microorganisms. The *Rhizobium*–legume symbiosis has great agricultural and ecological importance as it provides a source of fixed nitrogen for plant growth without the need for fertilizers. Also in rice paddies, the symbiotic N_2 fixation based on *Anabaena azollae*, together with rice rhizosphere-attached diazotrophic strains of *Azotobacter* and *Azospirillum*, allows rice growth without fertilizer addition.

In fertilized croplands, a certain amount of ammonium is oxidized to nitrite and nitrate, which leach easily from soils into ground and surface waters. Thus, excessive use of fertilizers leads to accumulation of nitrate and nitrite in groundwater causing environmental and health problems. The main environmental threat is the eutrophication of aquatic ecosystems by excessive growth of bacteria and algae. Consumption of drinking water with high nitrate content may cause methaemoglobinaemia and gastric cancer. Denitrification has a positive effect because it decreases nitrate leaching to groundwater, but has also negative effects since it causes important losses of nitrogen from soils and is a source of the greenhouse gas N_2O. In agricultural soils, losses by denitrification may represent up to 30% of applied fertilizer. Nitrogen oxides formed by incomplete denitrification and nitrification have a great impact on the atmosphere chemistry. N_2O is a stable gas (over 100 years of lifetime) that shows 320-fold more greenhouse effect than CO_2, contributing to global warming. Also, in the stratosphere, NO and N_2O participate in reactions that destroy the ozone layer, which protects life against UV radiation. Atmospheric NO is also oxidized to NO_2, which after hydration generates HNO_3, causing acid rainfalls. As N_2O is not harmful for organisms, the absence of Nos activity in some denitrifiers does not represent a problem for the bacteria but has serious environmental consequences. Nitrifier denitrification, mainly by marine AOB, also contributes to N_2O emissions. If denitrification is the main pathway for nitrogen losses in terrestrial ecosystems, anammox may account for up to 67% of N_2 production in marine environments. Nitrite availability likely controls the abundance of anammox bacteria, which are usually linked to the activity of other bacterial groups that produce nitrite by nitrification or denitrification.

Interactions among organisms affect significantly the nitrogen cycle processes. N_2 fixation in natural ecosystems is greatly affected by associative bacteria that potentially benefit the plants. More than 80% of higher plants are colonized by arbuscular mycorrhizal fungi, which interact with diazotrophic bacteria and assimilate nitrate and ammonium that are finally transferred to the plant. Another special interaction among organisms occurs in the gut of earthworms, an anoxic habitat for denitrifying soil bacteria. Understanding of the role of microbial consortia in nitrogen cycle was limited by the fact that only 1% of the bacteria present in an ecosystem may be cultured by traditional microbiology techniques. However, modern molecular tools and environmental genomics (metagenomic) are changing our knowledge of these communities in natural habitats and their effects on the nitrogen cycle. Metagenomic studies have revealed several networks of bacterial populations catalyzing different steps of the nitrogen cycle in natural environments. Thus, although no known methanotroph is able to denitrify, various microorganism consortia may use methane as carbon source to carry out denitrification aerobically or anaerobically. Aerobic methane oxidation coupled to denitrification is performed by aerobic methanotrophs that release soluble organic compounds (methanol) which are used as electron donor by denitrifiers. A bacterium–archaeon consortium also couples denitrification to methane oxidation under anaerobiosis, although this microbial consortium grows very slowly and its potential contribution to the global nitrogen cycle is unknown.

The nitrogen cycle is profoundly affected by anthropogenic impacts, such as the high input of nitrogen to the biosphere through application of synthetic fertilizers, use of N_2-fixing crops in agriculture, release of nitrogen oxides by combustions, and industrial production of different nitrogenous compounds that pollute worldwide regions. This input of human-produced nitrogen has greatly stimulated nitrifier denitrification, with concomitant increases in N_2O emissions. Mitigation of imbalances in nitrogen cycle will be required to reduce microbial production of greenhouse gases and global warming.

EVOLUTION OF THE NITROGEN CYCLE

In the course of evolution, the nitrogen cycle has suffered drastic changes. At beginning of life and provided a strict anoxic scenario, the relatively abundant ammonium was probably used in an assimilation/mineralization cycle by protocellular organisms. To replenish ammonium pool, a vast N_2 reservoir coming from the primitive protosolar nebula was chemically reduced to ammonium at hydrothermal vents by FeS/H_2S of abiotic origin. At the same time, N_2 was also oxidized by atmospheric lightning to nitric nitrogen, which was used as ammonium source through nitrite respiration by Nrf and/or nitrite and NO detoxification by incomplete denitrification (ancient Nar, Nir, and Nor). To detoxify hydroxylamine, a highly toxic, dead-end byproduct of nitrite reduction to ammonium, hydroxylamine reductase (Hcp), could be evolved, as well as anammox through Hzh coupled to NirS and Hzo, which recycles part of the fixed nitrogen back to N_2. Appearance of assimilatory Nas and Nir was also subject to strong evolutionary pressure to replenish the ammonium pool. Nitrate reductases are distributed in three clades: the assimilatory eukaryotic cytosolic enzyme and the prokaryotic enzymes Nas/Nap and Nar. Although the eukaryotic enzyme seems to have a monophyletic origin, Nap and Nar were probably acquired by horizontal gene transfer, Nap being a subclade of Nas. Eukaryotic assimilatory Nir is a chloroplastic siroheme enzyme, whereas bacterial Nir may contain different cofactors and are either NO-producing periplasmic enzymes (NirS and NirK) or NH_4^+-producing membrane-bound (NrfA) or cytosolic (NirB) enzymes.

Another step could be added to the biological nitrogen cycle when the nitrogenase systems began to catalyze a six-electron transfer to N_2 to yield ammonium, thus contributing to replenish ammonium reservoirs. This system could be evolved from a primitive Fe nitrogenase to the actual MoFe or VFe enzymes. The process probably included duplication of ancient gene pairs to yield the present *nifDK* and *nifEN* operons. Although the wide distribution of *nif* genes among bacteria and archaea suggests an early appearance of N_2 fixation, this could have been due to horizontal gene transfer. The complexity and energetic cost of the process and the lack of *nif* genes in eukaryotic organelles strongly support a late emergence of biological N_2 fixation as a key step of the nitrogen cycle.

Two billions years ago, evolving of oxygenic photosynthesis dramatically changed the whole earth's ecosystem for a sharp transition from anoxic to oxic environments. Oxygen appearance made possible bioavailability of copper, sequestered as sulfidic minerals, and evolution of high-yielding aerobic respiration based on heme–copper oxidases. Appearance of the Cu enzymes NirK and Nos allowed complete denitrification and closed the nitrogen cycle by producing N_2. Explosion of aerobic life forms reinforced the early route of nitrate assimilation to ammonium, and the increasing O_2 pressure made possible the emergence of aerobic nitrification through Amo and Hao, thus adding the last step to the extant nitrogen cycle.

In conclusion, evolution of the biological nitrogen cycle could be envisaged from a noncyclic route including an incomplete denitrification step that was supplied with nitrogen from enormous reservoirs of N_2 and nitrate and that evolved to produce ammonium to satisfy nitrogen requirements of protocells. Recycling of ammonium was first accomplished by ammonium assimilation/mineralization, nitrite respiration, and nitric nitrogen assimilation, and then by N_2 fixation, whereas N_2 was mainly recycled through anammox. Some processes also evolve for detoxifying purposes, such as hydroxylamine or NO-handling enzymes. Transition from anoxic to oxic environment and copper bioavailability bring about closing of the nitrogen cycle by complete N_2-producing denitrification and oxygen-dependent nitrification.

Actually, over half of the nitrogen fixed that enters into the ecosystems has anthropogenic origin. Human activities are modifying very rapidly the amounts of nitrogenous compounds at the level of all nitrogen reservoirs and hence dynamics of these compounds and composition of bacterial and archaeal populations in terrestrial, freshwater, and marine environments. As the composition of these complex ecosystems and their dynamics are not known exactly, it is not possible to make accurate predictions about evolution of the nitrogen cycle in the long time, but it is necessary to include human being as one of the most important factors affecting the nitrogen equilibrium in the biosphere at a global scale in the next future.

RECENT DEVELOPMENTS

Modeling is an emerging methodology in systems biology research. In a recent study, a model analysis for nitrogen assimilation by *E. coli* suggests that the ammonium transport and the GS activity are coupled to maximize ammonium assimilation flux, avoiding possible negative ammonia diffusion and allowing a quick response to environmental changes. Also, the overall regulatory mechanisms of *E. coli* nitrogen metabolism have been clarified by knocking out different nitrogen assimilation genes and by studying the effect of carbon and nitrogen limitations on transcriptional mRNA levels and enzyme activities under aerobic continuous cultures.

In the last year, significant advances have been made in the field of nitrogen fixation. At the molecular level, some relevant studies have contributed to elucidate the dual function of the NifEN proteins in nitrogenase cofactor assembly and catalysis. The biosynthesis of the FeMo cofactor is an intricate and composite process in which the NifEN protein complex, which is homologous to the NifDK nitrogenase MoFe protein, plays a key role acting as scaffold for the maturation of the FeMo cofactor precursor before its delivery to the MoFe protein. Very recently, a homocitrate-free, all-iron form of the FeMo cofactor associated to the NifEN scaffold has been identified that may serve as a precursor for the nitrogenase cofactor. Mo, V and Fe can be also incorporated into this NifEN precursor complex, suggesting that this form may be used as precursor for all nitrogenase cofactors. In addition, the NifEN complex shows catalytic activity with acetylene and azide, although it is unable to reduce N_2 or to evolve H_2. This result provides important implications for the study of the catalytic mechanism of the nitrogenase and for enzyme evolution. Homocitrate, the organic component of the nitrogenase FeMo cofactor, is generated by a homocitrate synthase encoded by the *nifV* gene. However, it is known that most rhizobial species that fix nitrogen in symbiotic association with legumes lack this NifV protein. Recently, it has been reported that the FEN1 gene of *Lotus japonicum* encodes a homocitrate synthase which overcomes the absence of NifV in *Mesorhizobium loti*. Therefore, the host plant supplies the bacteroids with the homocitrate required for the synthesis of a functional nitrogenase enzyme, thus providing a new molecular basis for the essential complementary partnership between legumes and rhizobia in symbiotic nitrogen fixation.

Recent evidence also suggests that epigenetic modification of rhizobial DNA is required for an efficient plant–microbe symbiosis since the initial events of nodulation are affected by the activity of the CcrM DNA methyltransferase. Transport of branched-chain amino acids is also essential for effective nitrogen fixation in the nodules, and it seems that legumes may control development and persistence of bacteroids by regulating the supply of these amino acids to the microbes. The RNA-binding protein Hfq is a chaperone involved in stress resistance and pathogenicity in many bacteria. It has been demonstrated that the Hfq protein also plays a key role in the establishment of the symbiosis between *Sinorhizobium meliloti* and *Medicago sativa* since the Hfq defective mutants form small, ineffective nodules that are unable to fix nitrogen. In *R. leguminosarum*, the Hfq protein seems to participate in the posttranscriptional regulation of *nifA* gene expression, and it has been reported that the Hfq-dependent RNaseE cleaves the *nifA* mRNA upstream of the translational start codon. On the other hand, the irreversible terminal differentiation of rhizobia during symbiosis is caused by some plant factors which are transported into the bacteria. These plant factors have been recently identified as nodule-specific, cysteine-rich peptides. The ability of legumes to exchange carbon and nitrogen compounds with the nitrogen-fixing rhizobia suggests that legumes could have competitive advantages under elevated CO_2 atmospheres. Also, CO_2 fixation in the nodule via phosphoenolpyruvate carboxylase may allow effective nitrogen fixation since it provides oxaloacetate for nitrogen assimilation. Recent data reveal that nitrogen fixation in alfalfa increases when the nodules are exposed to high CO_2, thus supporting the hypothesis that elevated CO_2 concentrations improve symbiotic nitrogen fixation.

In *Bradyrhizobium japonicum*, the regulatory protein NifA is affected by the oxygen-response regulatory cascade FixLJ-FixK (see Figure 4). Recently, it has been reported that NifA is also involved in the expression of denitrification genes in this bacterium and, therefore, it is required for maximal growth under denitrifying conditions. It seems that the NifA protein is needed for the full synthesis of different denitrification enzymes including periplasmic nitrate reductase (NapABC), nitrite reductase (NirK), and nitric oxide reductase (NorC). Other relevant advances on denitrification are the recent report of the high-resolution crystal structure of the copper-containing nitrite reductase (NirK) with its cognate cytochrome *c* electron donor and the demonstration that the catalytic mechanism of the NirK nitrite reductase from *Alcaligenes xylosoxidans* involves proton-coupled electron transfer.

The terminal step of DNRA is carried out by a novel octaheme cytochrome *c* Nir in the haloalkaliphilic bacterium *Thioalkalivibrio nitratireducens*. The high-resolution crystal structure of this protein has shown that the enzyme forms a stable hexamer containing a total of 48 heme groups. Each subunit consists of an N-terminal domain with three heme groups and a catalytic C-terminal domain with the remaining five heme groups, which is structurally and catalytically similar to the pentaheme NrfA Nir. Concerning the anammox process, a recent proteomic study has allowed the identification of the Hzh, the hydrazine-oxidizing enzyme (Hzo), and the membrane-bound ATP synthase, the three key proteins involved in this process, thus supporting the anammox reactions proposed in Figure 10. Immunogold labeling has also revealed that the Hzh is located in the anammoxosomes.

There is no doubt that modern molecular tools including proteomic and genomic approaches applied to the study of bacterial populations in natural habitats will contribute to the knowledge of the complex interactions between the organisms and processes affecting the nitrogen cycle in different environments. A large number of works have been carried out in the last year to study signaling pathways and regulatory networks; phylogenetic relationships; the diversity, abundance, and distribution of bacterial populations and their responses to environmental changes, genome variations, and gene transfer among organisms; and the contribution of the different processes of the nitrogen cycle to nitrogen losses and nitrogen oxides emissions in different ecosystems and biogeochemical niches. Finally, significant efforts have been made to improve the biological removal of nitrate and ammonia from industrial and agricultural wastewaters, and in the development of processes in bioreactors based on nitrification, denitrification, and anammox for nitrogen removal.

FURTHER READING

Arp, D. J., Chain, P. S. G., & Klotz, M. G. (2007). The impact of genome analyses on our understanding of ammonia-oxidizing bacteria. *Annual Review of Microbiology*, *61*, 503–528.

Bothe, H., Ferguson, S. J., & Newton, W. E. (Eds.) (2007). *Biology of the nitrogen cycle*. Amsterdam: Elsevier.

Cabello, P., Roldàn, M. D., & Moreno-Viviàn, C. (2004). Nitrate reduction and the nitrogen cycle in archaea. *Microbiology*, *150*, 3527–3546.

Flores, E., Frías, J. E., Rubio, L. M., & Herrero, A. (2005). Photosynthetic nitrate assimilation in cyanobacteria. *Photosynthesis Research*, *83*, 117–133.

Jetten, M. S. M., Wagner, M., Fuerst, J., van Loosdrecht, M., Kuenen, G., & Strous, M. (2001). Microbiology and application of the anaerobic ammonium oxidation ('anammox') process. *Current Opinion in Biotechnology*, *12*, 283–288.

Klotz, M. G., & Stein, L. Y. (2008). Nitrifier genomics and evolution of the nitrogen cycle. *FEMS Microbiology Letters*, *278*, 146–156.

Leigh, J. A., & Dodsworth, J. A. (2007). Nitrogen regulation in bacteria and archaea. *Annual Review of Microbiology*, *61*, 349–377.

Moreno-Viviàn, C., Cabello, P., Martínez-Luque, M., Blasco, R., & Castillo, F. (1999). Prokaryotic nitrate reduction: Molecular properties and functional distinction among bacterial nitrate reductases. *Journal of Bacteriology*, *181*, 6573–6584.

Moreno-Viviàn, C., & Ferguson, S. J. (1998). Definition and distinction between assimilatory, dissimilatory and respiratory pathways. *Molecular Microbiology*, *29*, 664–666.

Muro-Pastor, M. I., Reyes, J. C., & Florencio, F. J. (2005). Ammonium assimilation in cyanobacteria. *Photosynthesis Research*, *83*, 135–150.

Richardson, D. J., & Watmough, N. J. (1999). Inorganic nitrogen metabolism in bacteria. *Current Opinion in Chemical Biology*, *3*, 207–219.

Rubio, L. M., & Ludden, P. W. (2005). Maturation of nitrogenase: A biochemical puzzle. *Journal of Bacteriology*, *187*, 405–414.

Simon, J. (2002). Enzymology and bioenergetics of respiratory nitrite ammonification. *FEMS Microbiology Reviews*, *26*, 285–309.

Spiro, S. (2007). Regulators of bacterial responses to nitric oxide. *FEMS Microbiology Reviews*, *31*, 193–211.

Zumft, W. G. (1997). Cell biology and molecular basis of denitrification. *Microbiology and Molecular Biology Reviews*, *61*, 533–616.

Chapter 34

The Phosphorus Cycle

K.R.M. Mackey[1] and A. Paytan[2]

[1]*Stanford University, Stanford, CA, USA*

[2]*University of California Santa Cruz, Santa Cruz, CA, USA*

Chapter Outline

ABBREVIATIONS

ELF Enzyme-labeled fluorescence
PhoA Alkaline phosphomonoesterase
Pit Phosphate transport system
PNP *Para*-nitrophenyl phosphate
Pst Phosphate-specific transport
SRP Soluble reactive phosphorus

DEFINING STATEMENT

This article includes a discussion of the global phosphorus cycle, including sources, sinks, and transport pathways of phosphorus in the environment, microbially mediated transformations of phosphorus and their genetic regulation, and microbial responses to anthropogenic changes to the phosphorus cycle.

INTRODUCTION

Phosphorus is an essential nutrient for all living organisms, and the phosphorus cycle is an important link between earth's living and nonliving entities. The availability of phosphorus strongly influences primary production, the process by which photosynthetic organisms fix inorganic carbon into cellular biomass. Therefore, knowledge of the phosphorus cycle is critically important for understanding the global carbon budget and hence how biogeochemical cycles impact and are influenced by global climate.

Global Significance of the Phosphorus Cycle

Natural assemblages of microbes have a critical role in the phosphorus cycle because they forge a link between phosphorus reservoirs in the living and nonliving environment. Microbes facilitate the weathering, mineralization, and solubilization of nonbioavailable phosphorus sources, making orthophosphate available to microbial and plant communities and hence to higher trophic levels within the food web. However, microbes also contribute to immobilization of phosphorus, a process that diminishes bioavailable phosphorus by converting soluble reactive forms into insoluble forms. The mechanisms of microbial involvement in these processes vary from passive (e.g., resulting from microbial metabolic byproducts) to highly regulated active contributions (e.g., the regulation of gene expression in response to environmental cues).

Microbially mediated processes also link the phosphorus cycle to the carbon cycle and hence to global climate. During photosynthesis, photoautotrophs incorporate phosphorus and carbon at predictable ratios, with approximately 106 carbon atoms assimilated for every one phosphorus atom for marine photoautotrophs and terrestrial vegetation.

The phosphorus cycle therefore plays an important role in regulating primary productivity, the process in which radiant energy is used by primary producers to form organic substances as food for consumers, as in photosynthesis.

Because phosphorus is required for the synthesis of numerous biological compounds, its availability in the environment can limit the productivity of producers when other nutrients are available in excess. This fact carries important implications for the global carbon budget because the rate of incorporation (fixation) of carbon dioxide into photosynthetic biomass can be directly controlled by the availability of phosphorus; if phosphorus is not available, carbon dioxide fixation is halted. The intersection of the phosphorus and carbon cycles is of particular significance for global climate, which is affected by atmospheric carbon dioxide levels. Hence, through the growth of producers, the phosphorus cycle contributes to the regulation of global climate.

In aquatic environments, photosynthetic microbes (i.e., phytoplankton) may comprise a substantial portion of the photosynthetic biomass and hence primary production. Marine and freshwater phytoplankton are characterized by extensive biodiversity and, as a group, inhabit an incredible number of different niches within aquatic environments. Phytoplankton, like other microbes, have strategies that enable them to adapt to changes in the amounts and forms of phosphorus available within their environment. For example, the production of phosphatase enzymes, which hydrolyze organic P compounds to inorganic phosphate, helps to mediate the mineralization of organic phosphorus compounds in surface waters and allows cells to adapt when phosphate levels are low.

Recent estimates suggest that phytoplankton are responsible for as much as half of global carbon fixation, thereby contributing significantly to the regulation of earth's climate. Phytoplankton productivity is strongly influenced by nutrient availability, and which nutrient ultimately limits production depends on (1) the relative abundance of nutrients and (2) the nutritional requirements of the phytoplankton. With some exceptions, production in most lakes is limited by the availability of phosphorus, whereas limitation of primary production in the ocean has traditionally been attributed to nitrogen, another nutrient required by cells in large quantities. However, phosphorus differs from nitrogen in that the major source of phosphorus to aquatic environments is the weathering of minerals on land that are subsequently introduced into water bodies by fluvial and aeolian sources. In contrast, microbially mediated nitrogen fixation, in which bioavailable forms of nitrogen are generated from nitrogen gas in the atmosphere, is a major pathway by which phytoplankton gain access to nitrogen (in addition to continental weathering). Because there is no phosphorus input process analogous to nitrogen fixation, marine productivity over geological timescales is considered to be a function of the supply rate of phosphorus from continental weathering and the rate at which phosphorus is recycled in the ocean. Accordingly, the phosphorus cycle influences phytoplankton ecology, productivity, and carbon cycling in both marine and freshwater ecosystems.

Characterizing and Measuring Environmental Phosphorus Pools

Occurrence of elemental phosphorus in the environment is rare, given that it reacts readily with oxygen and combusts when exposed to oxygen; thus phosphorus is typically found bound to oxygen in nature. Phosphorus has the chemical ability to transfer a 3s- or p-orbital electron to a d orbital, permitting a relatively large number of potential configurations of electrons around the nucleus of the atom. This renders the structures of phosphorous-containing molecules quite variable and relatively reactive. These properties are likely responsible for the ubiquity and versatility of phosphorus-containing compounds in biological systems.

Because phosphorus exists in many different physical and chemical states in the environment, specific definitions are needed to clarify different parts of the phosphorus pool. Chemical names reflect the chemical composition of the phosphorus substance in question, whereas other classifications are based on methodological aspects of how the substance is measured. For example, "orthophosphate" is a chemical term that refers specifically to a phosphorus atom bound to four oxygen atoms – forming the orthophosphate molecule (PO_4^{3-}, also referred to as phosphate), whereas the term "soluble reactive phosphorus" (SRP) is a methodological term referring to everything that gets measured when an orthophosphate assay is performed (such as the ascorbic acid method, described below). The majority of measured SRP comprises orthophosphate and other related derivatives (e.g., $H_2PO_4^-$, HPO_4^{2-} depending on pH) but other forms of phosphorus may also be included because of experimental inaccuracy. Therefore, although SRP tends to closely reflect the amount of orthophosphate in a sample, the values may not be identical.

The most common assay for measuring SRP is the ascorbic acid method, which is approved by the U.S. Environmental Protection Agency for monitoring phosphate in environmental samples. In this method, ascorbic acid and ammonium molybdate react with SRP in the sample, forming a blue compound that can be observed visually or determined spectrophotometrically. Assays for measuring total phosphorus are also based on the ascorbic acid method, but begin with a step to transform all of the phosphorus in the sample to orthophosphate (typically through digestion by heating the sample in the presence of acid). After digestion, the sample is analyzed by the ascorbic acid method. A filtration step is typically not included in either of these methods, and accordingly in such cases all size fractions are measured.

When phosphorus is measured in water samples, distinguishing forms that are part of particulate matter from those that are in solution is often useful. Phosphorus is therefore classified as "soluble" or "insoluble," a distinction based on the method used to measure the sample. Soluble phosphorus includes all forms of phosphorus that are distributed in solution and that pass through a filter with a given pore size (typically 0.45 μm), whereas the insoluble, or particulate, fraction is the amount retained on the filter. Measurements of soluble phosphorus include both the "dissolved" and "colloidal" fractions. Dissolved phosphorus includes all forms that have entered a solute to form a homogeneous solution. (For example, orthophosphate that is not bound to a cation is considered dissolved because it is associated with water molecules homogeneously in solution rather than being held within a salt crystal.) By contrast, colloidal forms include any tiny particles that are distributed evenly throughout the solution but are not dissolved in solution. Soluble phosphorus is commonly reported because colloidal phosphorus particles are very small, and differentiating between colloidal and dissolved phosphorus is methodologically difficult.

When living organisms assimilate phosphorus into their cells, the resulting phosphorus-containing compounds are collectively called "organic phosphorus." It must be stressed that this definition of the biologically associated phosphorus as "organic phosphorus" is not identical to the chemical definition of organic compounds (e.g., containing carbon). For example, some intracellular biologically synthesized compounds such as polyphosphate may not contain C bonding. The term "organic phosphorus" encompasses molecules within living cells as well as molecules that are liberated into the environment after the decay of an organism. The major source of terrestrial organic phosphorus is plant material, which is released as vegetation undergoes decay; however, microbial and animal sources also contribute significantly. In the marine environment, organic phosphorus comes from a variety of sources (e.g., plankton, fish excrement, advection from land), the relative contributions of which differ widely depending on location in the ocean. In contrast to organic forms, inorganic phosphorus compounds are not always directly of biogenic origin. Rather, they also encompass phosphorus derived from the weathering of phosphate-containing minerals, including dissolved and particulate orthophosphate.

Phosphorus Sources, Sinks, and Transport Pathways

The phosphorus cycle encompasses numerous living and nonliving environmental reservoirs and various transport pathways. In tracing the movement of phosphorus in the environment, the interplay between physical and biological processes becomes apparent. In addition to acting as reservoirs of phosphorus in the environment (as discussed in this section), microbes contribute to the transformation of phosphorus within other reservoirs such as in soil or aquatic environments (see Section "Microbially Mediated Processes").

Within the earth's crust, the abundance of phosphorus is 0.10–0.12% (on a weight basis), with the majority of phosphorus existing as inorganic phosphate minerals and phosphorus-containing organic compounds. A phosphate mineral is any mineral in which phosphate anion groups form tetrahedral complexes in association with cations, although arsenate (AsO_4^{3-}) and vanadanate (VO_4^{3-}) may also be substituted in the crystalline structure. Apatite is the most abundant group of phosphate minerals, comprising hydroxyapatite, fluorapatite, and chlorapatite (Table 1). These three forms of apatite share nearly identical crystalline structures, but differ in their relative proportions of hydroxide, fluoride, and chloride, each being named for the anion that is most abundant in the mineral. Phosphate minerals generally form in the environment in magmatic processes or through precipitation from solution (which may be microbially mediated), and the chemical composition of the minerals depends on the ion or ions present in solution at the time of precipitation. For this reason, it is not uncommon for natural deposits of phosphate minerals to be heterogeneous,

TABLE 1 Phosphate minerals and their chemical compositions

Apatite	$Ca_5(PO_4)_3(F, Cl, OH)$
Hydroxylapatite	$Ca_5(PO_4)_3OH$
Fluorapatite	$Ca_5(PO_4)_3F$
Chlorapatite	$Ca_5(PO_4)_3Cl$
Frankolite	$Ca_{10-a-b}Na_aMg_b(PO_4)_{6-x}(CO_3)_{x-y-z}(CO_3F)_y(SO_4)_zF_2$
Lazulite	$(Mg, Fe)Al_2(PO_4)_2(OH)_2$
Monazite	$(Ce, La, Y, Th)PO_4$
Pyromorphite	$Pb_5(PO_4)_3Cl$
Strengite	$FePO_4{\cdot}2H_2O$
Triphylite	$Li(Fe, Mn)PO_4$
Turquoise	$CuAl_6(PO_4)_4(OH)_8{\cdot}5H_2O$
Variscite	$AlPO_4{\cdot}2H_2O$
Vauxite	$FeAl_2(PO_4)_2(OH)_2{\cdot}6H_2O$
Vivianite	$Fe_3(PO_4)_2{\cdot}8H_2O$
Wavellite	$Al_3(PO_4)_2(OH)_3{\cdot}5H_2O$

Apatite is the general term for the three minerals hydroxylapatite, fluorapatite, and chlorapatite rather than composed of one homogeneous type of phosphate mineral. These natural deposits of phosphate minerals are collectively called "phosphorites" to reflect variations in their chemical compositions.

Soils and lake sediments are another terrestrial reservoir of phosphorus, comprising primarily inorganic phosphorus from weathered phosphate minerals, along with organic phosphorus from the decomposition, excretion, and lysis of biota (Figure 1). The behavior of phosphorus in soils largely depends on the particular characteristics of each soil, and, besides microbial activity, factors such as temperature, pH, and the degree of oxygenation all influence phosphorus mobility. In soils, inorganic phosphorus is typically associated with Al, Ca, or Fe, and each compound has unique solubility characteristics that determine the availability of phosphate to plants. The mobility and bioavailability of phosphate in soils are limited primarily by adsorption (the physical adherence or bonding of phosphate ions onto the surfaces of other molecules) and the rate of microbially mediated mineralization of organic forms of phosphorus. Mineralization is discussed in detail in Section "Microbially Mediated Processes."

Marine sediments also represent an important phosphorus reservoir, but, because the physical and chemical factors affecting marine sediment differ considerably from those on land, processes controlling phosphorus dynamics in marine sediments are somewhat different from that of soils. In marine sediment, phosphate can be present in insoluble inorganic phosphate minerals (such as phosphorites), which are relatively immobile. Phosphate can also be sorbed onto iron or manganese oxyhydroxides. The sorbed phosphate can regain mobility in response to changes in the redox potential at the sediment–water interface and thus is considered more mobile. As in terrestrial sediments, phosphorus in marine detrital organic matter can also become remobilized as decomposition progresses through microbially mediated processes.

Biota (i.e., microbes, plants, and animals) serve as another reservoir of phosphorus in the environment, as they assimilate phosphorus within their cellular biomass. Biota can contribute significantly to environmental phosphorus levels; for example, microbial communities contribute 0.5–7.5% of total phosphorus in grassland and pasture topsoil, and up to 26% in indigenous forests. Microbes are also responsible for generating the myriad

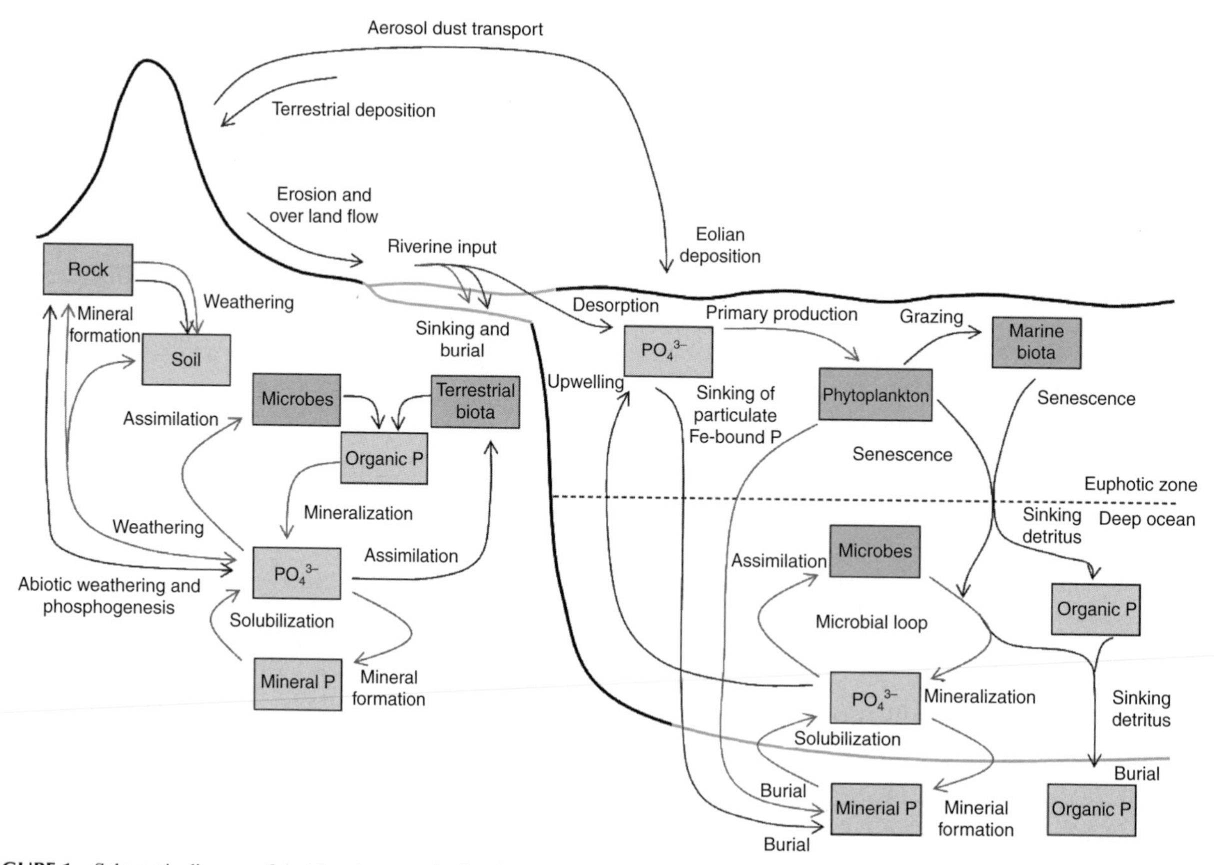

FIGURE 1 Schematic diagram of the phosphorus cycle showing phosphorus reservoirs (living in green boxes; nonliving in gray boxes), physical transport pathways (blue arrows), and microbially mediated transformations (green arrows). (For interpretation of the references to color in this figure legend, the reader is referred to the Web version of this chapter.)

of organic phosphorus compounds found throughout the environment. In particular, microbes and primary producers play an important role in providing nutrition, including phosphorus, to higher trophic levels by making it biologically available (bioavailable). Phosphorus assimilation is a microbially mediated process, which is discussed in Section "Transitory Immobilization."

Phosphorus is transported within the environment through various mass-transfer pathways. For example, rivers are important in the phosphorus cycle as both reservoirs and transport pathways. Phosphorus that has weathered from minerals and has leached or eroded from soils enters rivers through a variety of vectors, including dissolved and particulate forms in water from overland flow and in groundwater, and particulates brought by wind. Approximately 95% of phosphorus in rivers is particulate, and approximately 40% of that is bound within organic compounds. Rivers influence the distribution of phosphorus in soils and lakes by contributing or removing phosphorus, and riverine input is the single largest source of phosphorus to the oceans.

A number of outcomes are possible for phosphorus entering the ocean. Much of the riverine phosphorus flux is trapped in near-shore areas of the ocean, such as continental margins and estuaries, through immediate sedimentation and biological assimilation. The remaining phosphorus enters the dynamic surface ocean, also called the "euphotic zone," in which nearly all bioavailable phosphorus is sequestered within biota through primary production. Upon death of the organisms, a fraction of the biologically sequestered phosphorus sinks below the euphotic zone and most of it is regenerated into bioavailable forms such as orthophosphate by heterotrophic organisms. This recycling is part of the so-called microbial loop. Physical processes such as upwelling and deep convective mixing draw the deep water, which, in most parts of the ocean, is nutrient-rich compared to the surface waters in the euphotic zone, where up to 95% of it is reused in primary production. The remainder is removed from the ocean reservoir through particulate sedimentation, mineral formation (which may be microbially mediated), and scavenging by iron and manganese oxyhydroxides, all of which deposit phosphorus as a component of ocean sediment.

The phosphorus cycle differs from the cycles of other biologically important elements, such as carbon, nitrogen, and sulfur, in that it lacks a significant gaseous component; nearly all phosphorus in the environment resides either in solid or in aqueous forms. The one exception to this rule is the volatile compound phosphine (PH3, also called phosphane), a colorless, poisonous gas formed in the environment from the breakdown of alkali metal or alkali earth metal phosphides with water. This process is poorly characterized and likely comprises various multistage chemical reactions. Microbially mediated phosphine production can be a major source of the gas in engineered systems (e.g., sewage treatment facilities and constructed wastewater treatment wetlands) where organic phosphorus is abundant and reducing conditions are common, suggesting that microbes could also play a role in phosphine formation in natural systems (although the direct enzymatic production of phosphine has not yet been identified). Although phosphorus can exist as phosphine, the gas does not persist in the environment owing to rapid autoxidation, precluding significant accumulation of phosphine in the atmosphere. Phosphine is therefore a minor component of the environmental phosphorus pool.

The absence of a significant gaseous phase does not eliminate the atmosphere as an important reservoir in the phosphorus cycle. When weathering and erosion of soils generate inorganic and organic particulate phosphorus, wind transports some of the particles from their source to a new location. These particles can include mineral dust, pollen and plant debris, insect fragments, and organic phosphorus bound to larger particles. This distribution of terrestrial particulate phosphorus, termed "eolian deposition," plays an important role in delivering nutrients to the oceans. In oligotrophic ocean waters where nutrient levels are naturally low, such as in the open ocean gyres where riverine inputs do not extend and significant upwelling does not occur, eolian deposition may comprise a large portion of the nutrient flux that is available for primary production. The eolian phosphorus flux to the oceans is approximately $1 \times 10^{12}\,g\,year^{-1}$, of which approximately half is organic and the other half is inorganic. The solubility, and therefore bioavailability, of the phosphorus in eolian particulate matter differs significantly depending on its source; however, estimates suggest that approximately 15–50% is typically soluble.

MICROBIALLY MEDIATED PROCESSES

Weathering

Rock material exposed to the atmosphere breaks down, or weathers, as a result of numerous environmental processes. Weathering processes are classified into two categories. In mechanical weathering, physical processes (including thermal expansion, pressure release, hydraulic action, salt crystal formation, freeze–thaw, and frost wedge events) cause deterioration of rock material without changing the chemical composition of the parent material. In contrast, chemical weathering causes deterioration by altering the chemical structure of the minerals that the rock is made of. Chemical weathering processes include dissolution, hydrolysis, hydration, and oxidation–reduction (redox) reactions. Biological organisms can contribute to mechanical weathering by altering the microenvironments at the surface of the parent material (e.g., by increasing local humidity or by forming biofilms on surfaces); however, most biological weathering processes are classified as chemical weathering because they chemically alter the composition of the parent rock material directly or indirectly. These biological weathering processes are also referred to as "solubilization."

Solubilization

Inorganic phosphorus can occur in nature in soluble and insoluble forms. The solubility of the most abundant form of inorganic phosphorus, orthophosphate, is determined by the ambient pH and the cation to which it is bound as a mineral (e.g., Ca^{2+}, Mg^{2+}, Fe^{2+}, Fe^{3+}, and Al^{3+}). Microbially mediated phosphorus solubilization plays an important role in the conversion of insoluble phosphorus minerals into soluble forms of phosphorus. Solubilization directly benefits the microbes that perform it by providing the bioavailable phosphorus needed for growth. Similarly, the process benefits other organisms (including other cells, fungi, and higher plants) that are able to utilize the surplus of solubilized phosphorus.

Production of organic and inorganic acids is the primary mechanism of microbial phosphorus solubilization. In this process, biogenic acid interacts with phosphorus minerals to form mono- and dibasic phosphates, thereby bringing phosphorus into solution. Chemoautotrophic bacteria (e.g., nitrifying bacteria and *Thiobacillus* spp.) generate nitric and sulfuric acids by oxidizing ammonium and sulfur, respectively, and these acids are able to liberate soluble phosphorus from apatite, the most abundant phosphorus mineral. Production of organic acids occurs in numerous microbial taxa, and contributes to the solubilization of phosphorus minerals.

In addition to acid production, microbially mediated redox reactions contribute to phosphorus solubilization through the reduction of iron oxyhydroxides and associated ferric phosphate (strengite). In this process, dissimilatory iron reduction of ferric phosphates liberates soluble ferrous iron as well as orthophosphate associated with it. This occurs under reducing conditions, such as in flooded, anoxic soils and in some benthic aquatic environments. In another redox process, hydrogen sulfide (H_2S) produced by sulfur-reducing bacteria reduces the ferric iron in iron phosphate ($FePO_4$) to ferrous iron. In this reaction, iron sulfide and elemental sulfur are precipitated, and orthophosphate is generated.

Microbes also produce chelating compounds that contribute to phosphorus mineral solubilization. Chelation is the reversible binding (complexation) of a ligand to a metal ion. Chelators increase the solubility of insoluble phosphate mineral salts by complexing the metal cations, thereby making dissolution of the salt more energetically favorable. Examples of common chelators produced by microbes include citrate, oxalate, lactate, and 2-ketogluconate.

Mineralization

Plant and animal detritus comprises a large reservoir of organic phosphorus in the soil environment. However, because organically bound phosphorus sources are generally unable to cross cell membranes, most of the organic phosphorus from the detrital pool is not directly available to many living organisms. In order to become bioavailable, phosphorus bound to organic material must first be mineralized to phosphate. Mineralization is the process in which organically bound phosphorus is converted to inorganic phosphate, and is accomplished through the activity of a suite of microbial enzymes. Because this process makes available nutrients that would otherwise be sequestered in nonreactive forms, mineralization provides a vital link between the detrital pool and living organisms. It is estimated that approximately 70–80% of soil microbes are able to participate in phosphorus mineralization.

In general, mineralization is optimal and more phosphorus is liberated in uncultivated soils than in soils undergoing extensive cultivation, and a higher proportion of the organic phosphorus pool is mineralized in uncultivated soils. Further, mineralization rates tend to be higher in soils where inorganic phosphates are actively taken up and sequestered in plants and where microbial grazers are present, as would be expected in mature, uncultivated soils with fully developed, autochthonous microbial communities and trophic structures. As in many enzyme-catalyzed systems, mineralization is encouraged by higher levels of available substrate; however, high levels of inorganic phosphate (the product) do not impede the reaction, and mineralization will occur even if an abundance of phosphate is present. Other ambient conditions favoring phosphorus mineralization include warm soil temperatures and near-neutral pH values, which are conditions that also favor mineralization of other elements. Accordingly, phosphorus mineralization rates tend to reflect rates of ammonification and carbon mineralization in soils, and together these microbially mediated mineralization processes yield a C:N:P ratio that is similar to the ratio of these elements in humus (i.e., the organic soil fraction consisting of decomposed vegetable or animal matter.)

Enzymes involved in mineralization comprise a diverse group of proteins, called "phosphatases," with a broad range of substrates and substrate affinities, and varying conditions for optimal activity. In addition, phosphatases can either be constitutively expressed by an organism, or the expression can be upregulated under conditions of low phosphate (and in some cases, low carbon). Enzyme synthesis allows microbial cells to access organic phosphorus during periods of phosphate limitation, thereby avoiding the growth limitation and physical stress associated with nutrient deprivation. Phosphatases can be classified on the basis of the type of carbon–phosphorus bond they cleave, but any given phosphatase enzyme may catalyze reactions for numerous organic phosphorus compounds. In other words, a phosphatase enzyme has specific substrate requirements for a class of compounds, but lacks specificity in selecting substrates from within that class. The most common categories of microbial phosphatases contributing to phosphorus mineralization include phosphomonoesterases, phosphodiesterases, nucleases, and nucleotidases, as well as phytases.

Phosphomonoesterases catalyze reactions with phosphomonoesters, which are compounds in which one phosphate group is covalently bound to one carbon atom. The reaction involves the hydrolysis of the phosphorus–carbon bond, generating a free phosphate molecule and an alcohol as products. One example of a phosphomonoester is glycerol phosphate, a source of phosphate and carbon for some microbes. Phosphomonoesterases are further classified as "acid" or "alkaline" on the basis of their optimal pH ranges for maximum catalytic activity. Probably as a result of their ubiquity and importance in phosphorus mineralization, phosphomonoesterases tend to be referred to simply as phosphatases in the scientific literature, rather than by their full name, and context is often necessary to determine which group is being referenced. Therefore, a phosphomonoesterase with a pH optima near 8 might be referred to as an "alkaline phosphatase" rather than an "alkaline phosphomonoesterase" (PhoA) in the scientific literature.

A similar class of enzymes, phosphodiesterases, attack diester bonds in which a phosphate group is bonded to two separate carbon atoms, such as in phospholipids and nucleic acids. For example, the sugar–phosphate backbone in DNA comprises phosphodiester bonds. With water added across the phosphorus–carbon bond, cleaving of a diester proceeds similar to the monoester reaction, yielding phosphate and an alcohol. Once a diester has undergone hydrolysis, the resulting alcohol phosphomonoester must undergo another hydrolysis step, catalyzed by a phosphomonoesterase, for phosphorus mineralization to be complete.

Nucleic acids represent an important source of organic nutrients and are released from a cell upon lysis. Their rapid degradation and relatively low concentrations in the environment suggest an important role for nucleic acids as microbial nutrient sources. Many heterotrophic microbes are able to use nucleic acids as their only source of phosphorus, nitrogen, and carbon, and numerous others can use nucleic acids to supplement their nutritional requirements. The mineralization of phosphorus from nucleic acids proceeds in a two-step process involving two different enzymes. In the first step, depolymerizing nuclease enzymes such as DNase for DNA and RNase for RNA cleave the nucleic acid molecules into their constituent monomer nucleotides. Complete phosphorus mineralization of the resulting fragments proceeds via the activity of nucleotidase enzymes, which yield a phosphate group and a nucleoside molecule after hydrolysis.

Phytins, which are complex organic molecules containing up to six phosphate groups, are mineralized by a class of enzymes called "phytases." Phytases catalyze hydrolysis of the phosphate ester bonds that attach phosphate groups to the inositol ring, yielding reactive phosphate and a series of lower phosphoric esters. The location on the ring of the first hydrolysis reaction catalyzed by a phytase determines its classification; 3-phytases initiate hydrolysis at the phosphate ester bond of the ring's third carbon atom, whereas 6-phytases initiate at the sixth carbon atom. Hydrolysis by 6-phytases always leads to complete dephosphorylation of the inositol ring, whereas the 3-phytases may lead to incomplete dephosphorylation. Phytases are produced broadly by microbes, plants, and animals. In general, plants produce 6-phytases and microbes produce 3-phytases; however, 6-phytase activity has been observed in *Escherichia coli*. Microbial phytase activity is optimal over a broader range of pH values as compared to plant phytases, with pH optima spanning from 2 to 6 (plant phytases are optimal near pH 5). In addition to being affected by ambient pH, hydrolysis by phytases is also influenced by the degree of complexation of the phytin substrate with metal cations.

Because many phosphatase enzymes are synthesized in response to low environmental phosphate levels (i.e., when the cells experience phosphate limitation), phosphatase activity has been used extensively as a metric for determining the nutrient status of microbial communities. The activity of phosphatase enzymes has been measured in many terrestrial, limnic, and marine environments using a variety of methods, and numerous laboratory studies have also been conducted. In the environment, the bulk phosphatase activity of an entire microbial community is commonly measured by incubating soil, sediment, or water samples with a phosphate-bound substrate in which the hydrolytic product undergoes a color change that can be observed visually or spectrophotometrically, such as *para*-nitrophenyl phosphate (PNP) (Figure 2a and b), phenolphthalein phosphate, glycerophosphate, and 5-bromo-4-chloro-3-indolyl-phosphate (Figure 2c). In addition, measurements can be made fluorometrically for the substrates 3-*O*-methylfluoresein phosphate and 4-methylumbellyferyl phosphate. Radiometric analyses can similarly be made using ^{32}P-labeled glycerol phosphate (or an equivalent molecule). Chemical analysis of the hydrolytic products of glycerol phosphate and other bioenergetically important molecules has also been used to estimate phosphatase activity in bulk populations.

A major drawback to measuring bulk phosphatase activity is that it provides limited information, if any, about which members of the microbial community experience phosphate limitation at the time of sampling. This can be addressed to some extent by size-fractionating cells (as on a filter) before incubation with the substrate, thereby allowing phosphatase activity to be assigned to gross taxonomic classes of organisms. However, size fractionation introduces other obstacles for data interpretation, and results must be interpreted with care. For example, activity in different size fractions can be skewed by groups of bacteria that coalesce to form larger particles, although the individual cells are small and would otherwise be grouped with smaller size fractions. Studies with mixed populations of bacteria and green algae showed

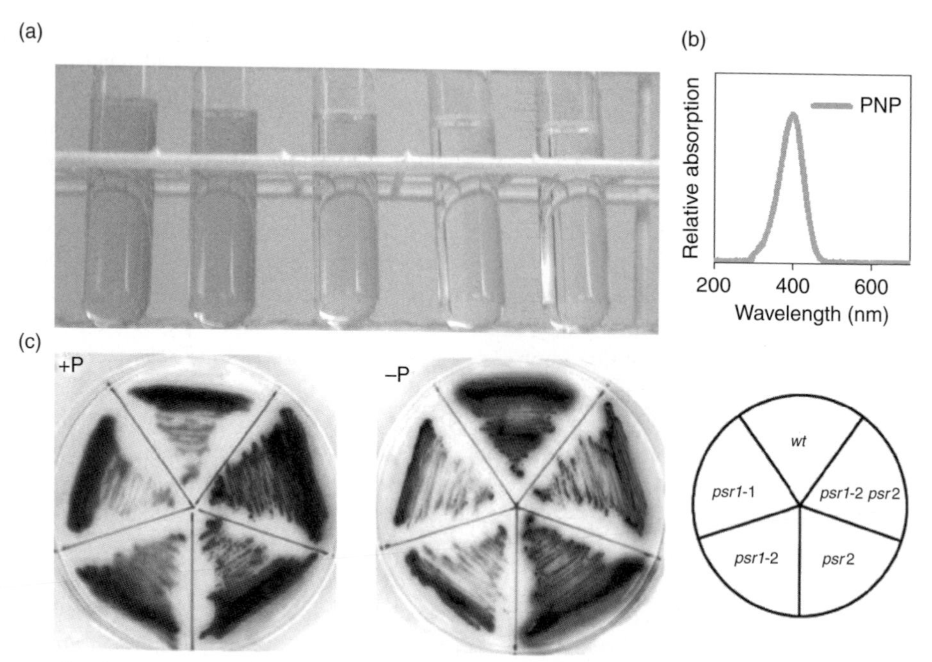

FIGURE 2 (a) The *para*-nitrophenyl phosphate (PNP) assay for alkaline phosphatase activity produces a yellow color in the presence of the enzyme. (b) The PNP assay quantifies alkaline phosphatase activity based on the absorption of light at 380 nm. (c) The 5-bromo-4-chloro-3-indolyl-phosphate assay with *Chlamydomonas* algae under phosphate-replete (left panel) and phosphate-limited (middle panel) conditions shows phosphatase activity using the blue coloration formed around the cells when they express the enzyme. Key (right panel) identifies mutants used in the study (wt is wild type). The phosphatase of the wild-type cells is induced in phosphate-free medium. *Reproduced from Shimogawara, K., Wykoff, D. D., Usuda, H., & Grossman, A. R. (1999).* Chlamydomonas reinhardtii *mutants abnormal in their responses to phosphorus deprivation.* Plant Physiology, 120, *685–693, with permission from the American Society of Plant Biologists.* (For interpretation of the references to color in this figure legend, the reader is referred to the Web version of this chapter.)

that 44% of the measured phosphatase activity was attributable to aggregated groups of cells. Moreover, in the marine environment, substantial phosphatase activity has been shown to persist for 3–6 weeks at 50% of initial levels in water samples filtered to remove particles. These observations suggest that phosphatases free in solution or bound to soluble organic material can contribute a significant amount of phosphatase activity, potentially leading to overestimates of phosphatase activity in the small cell size fraction. A difficulty common to both bulk and size-fractionated samples is that phosphatases can persist for long periods without being bound to a living cell. It is not uncommon for microbial cells to retain phosphatase activity for months or years after being dried or preserved, indicating that cell viability is not critical for maintaining phosphatase enzymes over these time periods, and dead cells may contribute to the overall phosphatase activity in a sample.

Several methods have been developed to overcome the limitations of bulk measurements by directly labeling cells when phosphatases are present. These methods allow phosphatase activity to be attributed to individual cells or taxa, allowing greater resolution of the phosphate status of organisms within a mixed community. Direct cell staining with azo dyes or precipitation of lead phosphate at the site of enzyme-mediated phosphate release has been used together with light microscopy to visualize phosphatase activity on individual cells. Similarly, enzyme-labeled fluorescence (ELF) labels individual cells with a fluorescent precipitate (ELF-97) after hydrolysis of the nonfluorescent substrate molecule (2-(5′-chloro-2′phosphoryloxyphenyl)-6-chloro-4-(3H)-quinazolinone) at the site of the enzyme (Figures 2c and 3).

Immobilization

Immobilization refers to the process by which labile phosphorus is sequestered and removed from the environmental reservoir of reactive phosphorus for a period of time. Immobilization processes can generally be grouped into two categories. The first category, transitory immobilization or cellular assimilation, includes all processes that sequester phosphorus within living microbial cells and is rapidly reversible upon cell death. The second category, mineral formation, encompasses processes that generate phosphorus-containing minerals.

Transitory immobilization

Transitory immobilization, or assimilation, is an important mechanism of phosphorus sequestration in soil and freshwater environments. Within cells, phosphorus is incorporated in numerous essential biological molecules and is required

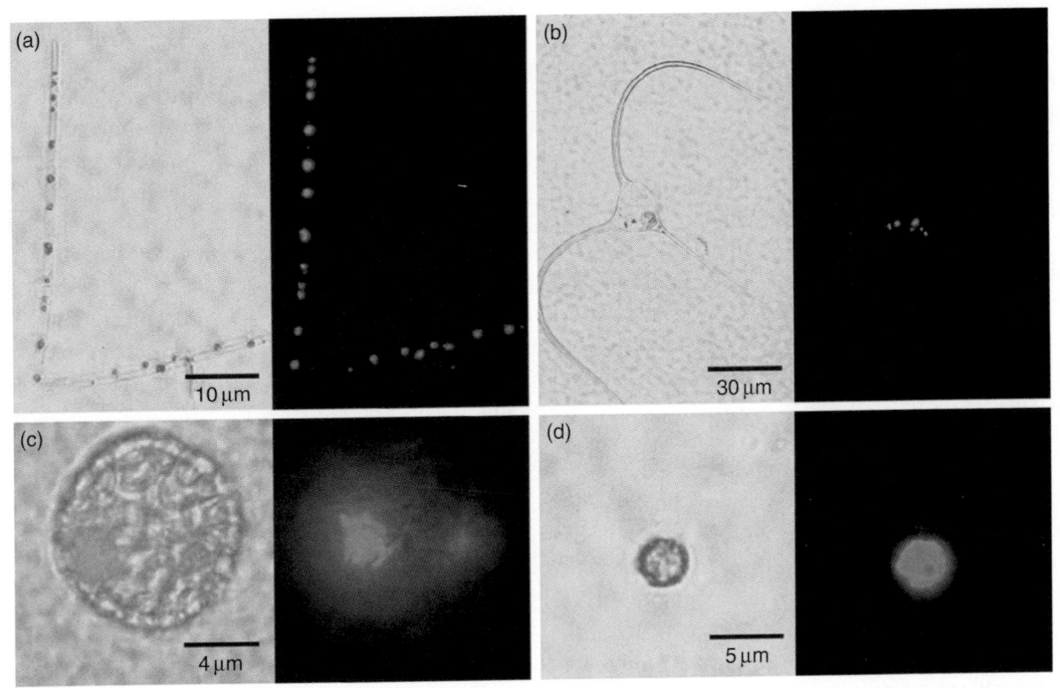

FIGURE 3 Micrographs of ELF-97-labeled phytoplankton from the euphotic zone in the Gulf of Aqaba (a) *Trichodesmium* sp., (b) *Ceratium* sp., (c) coccolithophore, and (d) *Cyanothece* sp. For each pair, the left panel is a view under visible light and the right panel under UV illumination. ELF-97-labeled areas appear as bright areas under UV illumination and show the location of phosphatase enzymes on the cells. *Reproduced from Mackey, K. R. M., Labiosa, R. G., Calhoun, M., Street, J. H., Post, A. F., & Paytan, A. (2007). Phosphorus availability, phytoplankton community dynamics, and taxon-specific phosphorus status in the Gulf of Aqaba, Red Sea.* Limnology and Oceanography, 52, *873–885, with permission from the American Society of Limnology and Oceanography, Inc.*

in larger quantities than many other elements. However, unlike other biologically important nutrients such as nitrogen and sulfur that must first undergo reduction before being incorporated into the cell, phosphorus remains oxidized before and after assimilation. Because mineralization of cellular material occurs rapidly after cell death, cellular assimilation of phosphorus into biological macromolecules leads to relatively short-term phosphorus retention in living cells where the duration is related to the characteristics of the microbial community in question. Within the cellular reservoir, phosphorus is present as different compounds and serves various functions.

Phospholipids are lipids in which a phosphate group has replaced one of the fatty acid groups. Lipids are generally hydrophobic; however, phospholipids are amphipathic, meaning that each molecule has a hyrdrophilic portion and a hydrophobic portion. The phosphate group is responsible for giving phospholipids their partially hydrophilic character, hence imparting a wide range of biochemical properties. In the cell, phospholipids are important in the formation of biological membranes and in some signal transduction pathways.

Nucleotides are biological compounds consisting of a pentose sugar, a purine or pyrimidine base, and one or more phosphate groups. Nucleotides are the structural subunits (monomers) of RNA and DNA, and alternating bonds between the sugar and phosphate groups form the backbones of these nucleic acids. Specifically, the phosphate groups form phosphodiester bonds between the third and fifth carbon atoms of adjacent sugar rings, thereby imparting directionality to the molecule. In addition, nucleotides are present as several major cofactors in the cell (e.g., flavin adenine dinucleotide, nicotinamide adenine dinucleotide phosphate) that have important functions in cell signaling and metabolism.

Adenosine triphosphate (ATP) is a nucleotide that performs multiple functions of considerable importance to the cell. The primary function of ATP is to aid in intracellular energy transfer by storing energy that is generated during photosynthesis and respiration so that it can be used in other cellular processes that require energy (e.g., cell division and biosynthetic reactions). In the cell, phosphate can be assimilated in ATP through substrate-level phosphorylation, oxidative phosphorylation, as well as photophosphorylation, and via the adenylate kinase reaction. ATP is also active in signal transduction pathways, where it can be used as a substrate for kinase enzymes in reactions that transfer phosphate groups to proteins and lipids, forming phosphoproteins and phospholipids, respectively. Phosphorylation is an important and ubiquitous form of signal transduction in many organisms.

Phytic acid (also called inositol hexaphosphate, IP6, phytate, and myoinositol 1,2,3,4,5,6-hexakis dihydrogen phosphate) is an organic, phosphorylated, cyclic, sugar alcohol produced by plants and found in high concentrations in seeds, and may also be produced by microbes. In its fully phosphorylated form, six phosphate groups attach to the inositol ring; however, various isomers of the less highly substituted inositol phosphate molecules also exist. Phytic acid is a highly reactive compound that forms stable complexes with a variety of mineral cations (e.g., Zn^{2+}, Fe^{2+}, Mn^{2+}, Fe^{3+}, Ca^{2+}, Mg^{2+}), as well as proteins and starches within mature seeds and when present in the environment. The complexed form of phytic acid, also known as "phytin," can be a persistent phosphorus reservoir in the environment because it is less accessible to degradation by hydrolytic enzymes when complexed.

In addition to biological molecules such as nucleic acids and phospholipids, which by weight consist primarily of carbon along with smaller portions of phosphorus, the intracellular accumulation of polyphosphate granules in microbial cells is an important type of transitory phosphorus immobilization. Polyphosphates are chains of 3–1000 phosphate residues connected by anhydride bonds that form by the activity of polyphosphate kinase enzymes, and therefore represent a highly concentrated reservoir of phosphorus in the cell. The energy stored in the anhydride bonds can be used by the microbial cell during periods of starvation. Polyphosphates may also serve as ligands for metal cations such as calcium, aluminum, and manganese in some microbes; however, the extent and the elemental stoichiometry of intracellular polyphosphate cation complexation vary among organisms.

Phosphorus immobilized within cells is considered transitory because it is able to rapidly reenter the reactive phosphate pool after cell death through microbially mediated mineralization processes. The immobilization of phosphorus via cellular assimilation is therefore relatively brief (i.e., within the lifetime of a cell) compared to other processes within the phosphorus cycle, some of which occur over geological timescales.

Phosphate mineral formation

Phosphate mineral formation represents another phosphorus sink that is influenced by microbial activity; however, it encompasses processes other than cellular assimilation and short-term storage of phosphorus as biological molecules. In mineral formation (also called "phosphogenesis"), phosphate anions react with cations in the environment to form insoluble precipitates. Sediments that comprise significant amounts of phosphorus-containing minerals are called "phosphorites," and may contain apatite, francolite, and a number of other phosphorus minerals (Table 1). Mineral formation is generally an important mechanism of phosphorus sequestration in marine environments where the high seawater calcium levels facilitate microbially mediated formation of insoluble phosphorites. The sequestration of phosphorus by this process retains (immobilizes) phosphorus for longer periods than does transitory immobilization.

For an insoluble mineral to form, the concentration of the ions forming the mineral must be high enough such that supersaturation is reached and equilibrium of the precipitation reaction is shifted toward the product. In addition, the physical and chemical characteristics of the environment must be conducive to precipitation of that mineral based on its solubility characteristics, and factors such as pH, redox state, and concentrations of co-occurring ions all influence mineral precipitation. The mineral must also be stable in the environment for mineral formation to constitute a long-term sink for phosphorus.

Mineral formation occurs both authigenically and diagenetically in the environment. Authigenic mineral formation is the formation of insoluble precipitates (minerals) *in situ* rather than by having been transported or deposited in a location through secondary processes. In contrast, diagenetic mineral formation is the alteration of existing minerals by chemical changes occurring after the initial deposition of a mineral (i.e., during or after burial and lithification). The primary type of diagenetic phosphorus mineral formation in marine environments is the substitution of phosphate into calcium carbonate minerals such as calcite or aragonite. Microbes contribute to this process by mineralizing organic phosphorus to reactive phosphate, which then substitutes diagenetically into calcium carbonate.

In authigenic mineral formation, microbes may generate reactive phosphate from the mineralization of organic phosphorus sources, and the resulting localized high phosphate concentrations favor precipitation of phosphate minerals. In soils, microbes convert organic phosphorus into phosphate, increasing its concentration in the soil and promoting formation of stable minerals such as apatite. In productive areas of the ocean, microbial mineralization of detrital matter at the sediment–water interface generates reactive phosphate, some of which reacts with seawater calcium to form phosphorite. (Reactive phosphate that does not contribute to mineral formation is available for biological assimilation by benthic microbes, or may be reintroduced to the euphotic zone by diffusion and upwelling for use by phytoplankton.)

The accumulation of phosphorus in microbial cells during transitory immobilization also contributes to mineral formation by increasing the pool of phosphate that could react with cations to form minerals. This is particularly important in anoxic areas of the ocean and soils where reactive phosphorus levels are low. Under oxic conditions where reactive phosphate is more abundant, luxury uptake and storage of phosphate as polyphosphate molecules occur in some microbes (e.g., *Pseudomonas* spp., *Actinobacter* spp.), as discussed above. Cells use the energy stored in polyphosphates to activate an alternative organic electron acceptor when conditions shift toward anoxia, freeing substantial levels of reactive orthophosphate in the process. The sequestration and release of phosphate by the cell under oxic and anoxic conditions, respectively, represent a mechanism by which microbes contribute to mineral formation because it generates locally elevated reactive phosphate concentrations in the vicinity of the cells that are high enough to induce precipitation of minerals. Accumulation of phosphate within cells may also lead to phosphorus immobilization via mineral formation if phosphorus minerals are generated and stored within the cell. Intracellular formation of mineral apatite is an example of phosphorus immobilization that has been observed in some microbes (e.g., *E. coli*, *Bacterionema matruchotii*) after incubation with calcium phosphate at a slightly basic pH. This process occurs in living and dead cells, suggesting that the locally elevated reactive phosphate concentrations within the cell help initiate apatite formation. Similarly, the formation of carbonate fluorapatite after cell death has been observed in Gram-negative rods, possibly pseudomonads, in coastal marine sediment, and is believed to be an important phosphorite formation process in locations where sedimentation rates are low.

GENETIC REGULATION OF MICROBIALLY MEDIATED PROCESSES

Microbially mediated processes, including those involved in the phosphorus cycle, are the outcome of numerous biological pathways occurring in concert across diverse microbial communities. Even cursory observations of natural microbial communities demonstrate that, while microbially mediated processes influence and change the environment, the environment likewise shapes the activity of microbes, in many cases by providing feedback that either inhibits or enhances the processes. (An example of this type of feedback is the synthesis of phosphomonoesterase enzymes, many of which are only present during periods of orthophosphate deprivation but not when orthophosphate is abundant in the environment.) Similarly, microbially mediated processes can also be controlled indirectly by secondary factors (other than phosphorus) that influence growth and metabolism, such as the availability of oxidized nitrogen or sulfur in some chemoautotrophic microbes.

These processes, which are manifest in the environment as the combined outcome of activities from a diverse microbial community, are in fact a result of genetic mediation within single cells. The regulation of genes in response to environmental stimuli determines how a cell will respond to its environment, including if and how it will contribute to microbially mediated processes in the phosphorus cycle. To understand gene regulation in greater detail, highly sensitive genetic and molecular methods have been developed. Under laboratory conditions, these methods have elucidated pathways important in the immobilization (i.e., phosphorus assimilation into cell biomass) and mineralization (i.e., phosphatase production) of phosphorus, as well as countless other pathways and processes.

In *E. coli*, two major phosphate assimilation pathways have been identified. The phosphate transport system (Pit), which comprises a hydrogen phosphate symport powered by protonmotive force, is expressed constitutively and provides the cell with sufficient phosphorus for growth when phosphate concentrations in the media are not limiting. When media phosphate concentrations decrease below a threshold concentration, the high-affinity phosphate-specific transport (Pst) system becomes engaged. This system has a 100-fold greater affinity for phosphate than Pit, enabling the cell to acquire phosphate from a limited reservoir. Uptake of phosphate through Pst is an ATP-dependent process (e.g., requires energy input from the cell).

Pst is activated as part of the Pho regulon, a group of operons that is expressed when phosphate levels are low. Activation of the Pho regulon is initiated through phosphorylation of the PhoB cytoplasmic protein, which, in its phosphorylated state, is a transcriptional activator of the operons within the Pho regulon. In addition to Pst, the Pho regulon also includes genes encoding PhoA, outer-membrane porin proteins that facilitate diffusion of phosphate into the periplasm (PhoE), and proteins for the uptake and processing of glycerol-3-phosphate (*ugp* operon) and phosphonates (*phn* operon).

Metabolism of glycerol-3-phosphate is an interesting strategy for heterotrophic microbes, such as *E. coli*, because it is a potential source of both phosphate and carbon for the cell. However, when grown under phosphate-deplete conditions and expressing *ugp* genes, cells are only able to use glycerol-3-phosphate as a phosphate source, not as a carbon source. For cells to grow with glycerol-3-phosphate as the only phosphate source, another carbon source must also be provided. The *ugp* system is less efficient when internal cell phosphate levels are high, and is no longer expressed if external phosphate levels increase above a threshold level. Another system that is not part of the Pho regulon, the *glp* transport system, is regulated by external and internal glycerol-3-phosphate levels rather than by phosphate concentrations. Unlike in the *ugp* system, glycerol-3-phosphate acquired by the *glp* system is able to serve as the sole source of carbon and phosphate for the cell. Both *ugp* and *glp* systems facilitate the direct cellular uptake of glycerol-3-phosphate; however, each is regulated by different internal and external cues (i.e., phosphate or glycerol-3-phosphate levels), and has a different nutritional strategy (i.e., supplying phosphate alone vs. phosphate and carbon together).

These two systems are an example of how microbes, by developing multiple interrelated pathways, are able to contribute to microbially mediated processes in the phosphorus cycle under a range of environmental and physiological conditions. Experimental evidence shows that the phosphate assimilation pathways in other heterotrophic bacteria are similar to that in *E. coli*, and many microbes are known to have portions of the Pho regulon. In particular, the *phoA* gene and homologs have been identified in numerous microbial taxa, and although the primary function of the protein remains the same, factors that influence its expression and activity vary from organism to organism. The diversity of organisms and environmental conditions in which this gene exists allow microbially mediated mineralization of phosphorus to occur in nearly every environment where microbes are found. For example, photosynthetic cyanobacteria in the genus *Synechococcus*, which populate freshwater environments, coastal waters, and vast areas of the open ocean, have the *phoA* gene along with many of the other genes encoded in the Pho regulon, highlighting the global ubiquity of microbially mediated processes in the phosphorus cycle.

ANTHROPOGENIC ALTERATION OF THE P CYCLE: EUTROPHICATION IN AQUATIC ECOSYSTEMS

As discussed above, microbes have an important role in nearly every aspect of the phosphorus cycle, and their activities help control the relative rates at which phosphorus is mobilized and immobilized within the environment. However, humans also influence the phosphorus cycle and alter the structure of microbial communities, with devastating ecological consequences.

Postindustrial human activities, including deforestation, phosphorus mining, and agricultural practices, affect the phosphorus cycle by increasing the mobility of phosphorus in the environment and causing it to accumulate in soils and aquatic environments. Several factors contribute to the mobilization of phosphorus by these activities. Deforestation and mining expose phosphate (and other) minerals in rock and soil to the atmosphere, leading to increased rates of weathering and erosion. Agricultural soils are also highly susceptible to erosion, making the localized elevation of phosphorus levels from application of fertilizers a particularly large source of the anthropogenic phosphorus flux. As a result of these practices, recent estimates suggest that the net storage of phosphorus in terrestrial and freshwater habitats has increased 75% over preindustrial levels, and the total reactive phosphorus flux to the ocean is twofold higher than prehuman levels. Consequently, eutrophication (the excessive growth of phytoplankton in response to overenrichment of a growth-limiting nutrient) has become a widespread problem in lakes and estuaries throughout the world, carrying serious environmental, economic, esthetic, and human health consequences. Eutrophication has been observed in many ecosystems, including freshwater lakes such as Lake Erie, large estuaries such as the Chesapeake Bay, and coastal areas such as the hypoxic "dead zone" of the Gulf of Mexico.

Organic fertilizers (e.g., poultry litter, manure) are typically applied to crops on the basis of the rate of crop nitrogen uptake, resulting in the overapplication of phosphorous and its rapid accumulation in soils. Elevated soil phosphorus levels increase the amount of phosphorus in runoff and ultimately lead to the accumulation of phosphorus in lakes and estuaries. When phosphorus from agriculture application is washed into water bodies where phosphorus limits production, substantial changes in the microbial community occur. Reversal of phosphorus limitation leads to the rapid growth of bloom-forming phytoplankton, some of which are toxic or nuisance species (such as *Pfiesteria* sp.) that are harmful to aquatic organisms and humans. As the bloom exhausts the supply of phosphorus, the phytoplankton senesce, sink to the bottom of the water body, and are decomposed by the heterotrophic microbial community. At depth, where light levels are low, photosynthetic phytoplankton are not able to balance the metabolic oxygen demands of the heterotrophs, and anoxia occurs in the bottom waters. Anoxia damages the benthic environment, leading to fish kills and harming benthic invertebrate communities. Loss of submerged aquatic vegetation, coral reef death, human shellfish poisoning, and a reduction in biodiversity are among the possible outcomes caused by microbial responses to the anthropogenic introduction of excess phosphorus to sensitive aquatic ecosystems.

CONCLUSION

Microbially mediated processes in the phosphorus cycle forge a critical link between the geosphere and biosphere by assimilating phosphorus within biological molecules and contributing to chemical transformations of phosphorus in the environment. In addition to acting as living reservoirs of phosphorus, microbes also contribute to the transformation of phosphorus within other nonliving reservoirs, such as rock, soils, rivers, lakes, and oceans. Microbially mediated phosphorus transformation includes processes that increase the bioavailability of phosphorus in the environment, such as weathering, solubilization, and mineralization, as well as those that decrease its bioavailability, such as assimilation and mineral formation. These large-scale environmental processes are the outcome of numerous biological pathways occurring in concert across diverse microbial communities. Genetic diversity and finely tuned regulation of gene expression allow microbes to adapt to harsh environments, and to contribute to the phosphorus cycle under numerous and diverse environmental conditions. Human alteration of the natural phosphorus cycle causes unintended consequences in microbial communities, and serious environmental, economic, esthetic, and human health problems are caused by microbial responses to the anthropogenic introduction of excess phosphorus to sensitive aquatic ecosystems.

RECENT DEVELOPMENTS

Microbially mediated processes in the phosphorus (P) cycle link the living and nonliving environment by storing P in and releasing P from organic matter, as well as by contributing to chemical transformations of P in the environment. Microorganisms contribute to nearly all pathways in the P cycle, including weathering, solubilization, mineralization, transitory immobilization, and mineral formation. In recent years, a number of advances have been made in our understanding of the role of microorganisms in the P cycle. New methods for characterizing P compounds in the environment have been developed, and these have led to a greater appreciation for the role of microorganisms in the remineralization of inorganic phosphate (Pi) from P-containing compounds such as polyphosphates and phosphonates. In addition, several novel adaptations have been identified in microorganisms that allow them to cope with Pi starvation and make them more competitive in environments where Pi is scarce.

Until recently, characterization of dissolved and particulate P pools has been accomplished using nuclear magnetic resonance (NMR) techniques. These analyses indicate that environmental P reservoirs comprise a wide array of P compounds, including P monoesters, diesters, polyphosphates, and phosphonates. However, NMR-based analyses are somewhat limited in that, while they can identify different types of P bonds (e.g., ester bonds, anhydride bonds), they are unable to identify specific types of compounds or minerals.

Two new methods have been developed to characterize environmental P pools, and have been used to shed light on novel roles of microorganisms in the P cycle. X-ray spectromicroscopy can be used to identify particulate organic, mineral, and polymeric P compounds at very high resolution (<1 μm), and has been used to characterize P compounds in particulate material ranging from cells to sediments. Coupled reverse osmosis and electrodialysis (RO/ED) can be used to isolate and concentrate dissolved organic P compounds, and has been used to help characterize small molecular weight organic molecules.

One application of these new techniques has revealed an important role of polyphosphates in the authigenic formation of apatite minerals (chiefly calcium phosphate) in marine sediments. Polyphosphates are linear polymers of phosphate molecules linked by phosphoanhydride bonds. In *E. coli*, polyphosphate genesis is catalyzed by the enzyme polyphosphate kinase 1 (PPK1) through the reversible polymerization of ATP, while polyphosphate degradation is modulated by the exopolyphosphatase enzyme (PPX) during Pi starvation. The genes encoding both of these enzymes are part of the Pho regulon, which controls the cell's Pi starvation response, and have been identified in numerous prokaryotes including the freshwater cyanobacteria *Synechocystis*.

In the ocean, certain eukaryotic phytoplankton also store P in polyphosphate molecules associated with calcium in their cytoplasm. Polyphosphate accounts for ~20–40% of the total cellular phosphorus in the diatoms *Skeletonema* and *Thallasiosira*, and these molecules remain protectively encased within the cells' silica frustules during sinking and once arriving at the sediment. Microbial decomposition of diatom frustules allows calcium and polyphosphate molecules to be released to the sediment concurrently and in a molecular configuration conducive to apatite formation. Apatite formation from these diatom-derived polyphosphate granules appears to occur rapidly, initiating in sediment less than 3 years old, possibly because the mineral "scaffolding" of polyphosphate molecules facilitates nucleation of apatite. Fine-grained authigenic apatite comprises 9–13% of total sedimentary phosphorus in open ocean sediments. The packaging and transport of P in polyphosphates by diatoms therefore represents a major role of marine microbes in P mineral formation.

Another group of P compounds that has recently received more attention is the phosphonates. Phosphonates are organic molecules containing one or more covalent bonds between carbon and P that are resistant to chemical hydrolysis, thermal degradation, and photolysis, though they can be broken by enzymatic activity. The genes involved in the phosphonate uptake (PhnCDE) and degradation via a C–P lyase (PhnFGHIJKLMNOP) are part of the Pho regulon in *E. coli*, and are therefore upregulated under conditions of Pi starvation.

In the environment, many phosphonates derive from anthropogenic sources, including the herbicide glyphosate (*N*-(phosphonomethyl)glycine, trade name Roundup), flame retardants, plasticizers, and pharmaceuticals. Phosphonates also occur naturally in many different microbial taxa where they function as glycolipids, glycoproteins, antibiotics, and phosphonolipids; however, to date only prokaryotes appear to have the ability to break C–P bonds. In freshwater environments, the use of phosphonates as a P source by prokaryotes is well documented, and the resistance of at least six ubiquitous genera of cyanobacteria to glyphosate has been observed, suggesting an important role for prokaryotes in the mineralization of Pi from otherwise recalcitrant phosphonate compounds. In addition, phosphonate utilization by freshwater *Synechecoccus* isolated from the hot springs of Yellowstone National Park demonstrates that extremophiles may also meet their P requirements by using phosphonates. In the marine environment, where phosphonates comprise up to 25% of the high molecular weight Dissolved Organic Phosphorus pool, phosphonates have been identified as an important P source for nitrogen fixing *Trichodesmium* cells, as well as in the picocyanobacteria *Synechococcus* and *Prochlorococcus*. Interestingly, while the expression of phosphonate utilization genes is induced under Pi starvation in the vast majority of microorganisms studied to date, their expression appears to be constitutive.

In liberating Pi from highly chemically stable phosphonate compounds, microorganisms play an important role in the P cycle by contributing to P remineralization. However, microbially mediated breakage of the C–P bond has recently been shown to have major implications for global climate via the production of the greenhouse gas methane in the ocean. In surface waters, methane is supersaturated with respect to the atmosphere, suggesting a local source must exist. Until recently, only anaerobic processes were known to produce methane. Because the surface ocean is generally aerobic, these processes were not likely to be the source. However, recent observations in the open ocean where Pi is very scarce indicate that methane is generated under aerobic conditions when microorganisms utilize methylphosphonate, the simplest phosphonate compound. Methane was generated by mixed assemblages of marine microorganisms amended with methylphosphonate, glucose, and nitrate, but only when Pi was scarce, thus in keeping with phosphonate utilization genes being under control of the Pho regulon. The liberation of methane from methylphosphonate by marine microorganisms explains the supersaturation of methane in surface ocean waters, a characteristic that makes the ocean a net source of methane to the atmosphere. Phosphonate degradation by microorganisms is therefore an important step in the P cycle, but also has important implications for C cycling and the regulation of global climate.

Utilization of organic P sources such as phosphonates and storage of Pi as polyphosphate within the cell are two microbial adaptations that affect the P cycle by allowing microbes to survive in environments where Pi is scarce. (The induction of high-affinity P transporters and phosphatase enzymes under Pi stress is well documented for numerous microorganisms, and is discussed in detail in the main text.) However, other adaptations have recently been identified that have increased our understanding of the role of microbes in the P cycle. For instance, marine phytoplankton has been shown to substitute sulfur- and nitrogen-containing lipids for phospholipids in their membranes under conditions of Pi limitation. This adaptation allows cells to minimize their P requirements, presumably by directing the limited P that is available to other vital functions within the cell.

FURTHER READING

Adams, M. M., Gomez-Garcia, M. R., Grossman, A. R., & Bhaya, D. (2008). Phosphorus deprivation responses and phosphonate utilization in a thermophilic *Synechococcus* sp. from microbial mats. *J. Bacteriology*, *190*, 8171–8184.

Alexander, M. (1967). Microbial transformations of phosphorus. In M. Alexander (eds.), *Introduction to soil microbiology* (2nd ed.), New York, NY: John Wiley and Sons.

Brandes, J. A., Ingall, E., & Paterson, D. (2007). Characterization of minerals and organic phosphorus species in marine sediments using soft X-ray fluorescence spectromicroscopy. *Marine Chemistry*, *103*, 250–265.

Compton, J., Mallinson, C. R., Glenn, D. *et al.* (2000). Variations in the global phosphorus cycle. In: C. R. Glennal, *et al.* (Eds.) *Marine authigenesis: From global to microbial*. Tulsa: SEPM (Society for Sedimentary Geology). Special Publication 66.

Diaz, J. *et al.* (2008). Marine polyphosphate: A key player in geologic phosphorus sequestration. *Science*, *320*, 652–655.

Dyhrman, S. T., Ammerman, J. W., & Van Mooy, B. A. S. (2007). Microbes and the marine phosphorus cycle. *Oceanography*, *20*, 110–116.

Forlani, G., Pavan, M., Gramek, M., Kafarski, P., & Lipok, J. (2008). Biochemical bases for a widespread tolerance of cyanobacteria to the phosphonate herbicids glyphosate. *Plant Cell Physiol*, *49*, 443–456.

Ehrlich, H. L. (1999). Microbes as geologic agents: Their role in mineral formation. *Geomicrobiology Journal*, *16*, 135–153.

Ehrlich, H. L. (2002). *Geomicrobial interactions with phosphorus. Geomicrobiology* (4th ed.). New York, NY: CRC Press, Inc.

Follmi, K. B. (1996). The phosphorus cycle, phosphogenesis and marine phosphate-rich deposits. *Earth-Science Reviews*, *40*, 55–124.

Heath, R. T. (2005). Microbial turnover of organic phosphorus in aquatic systems. In B. L. Turner, E. Frossard, & D. S. Baldwin (Eds.) *Organic Phosphorus in the Environment* Cambridge, MA: CABI Publishing.

Ilikchyan, I. N., *et al.* (2009). Detection and expression of the phosphonate transporter gene phnD in marine and freshwater picocyanobacteria. *Environmental Microbiology*, *11*, 1314–1324.

Karl, D. *et al.* (2008). Aerobic production of methane in the seas. *Nature Geosciences*, *1*, 473–478.

Oberson, A., & Joner, E. J. (2005). Microbial turnover of phosphorus in soil. In B. L. Turner, E. Frossard, & D. S. Baldwin (Eds.), *Organic phosphorus in the environment*. Cambridge, MA: CABI Publishing.

Paytan, A., & McLaughlin, K. (2007). The oceanic phosphorus cycle. *Chemical Reviews*, *107*, 563–576.

Ruttenberg, K. C. (2003). The global phosphorus cycle. In H. S. William (Eds.) *Treatise on Geochemistry* (Vol. 8) (pp. 585–643). Elsevier H. D. Holland, & K. K. Turekian (Executive eds.) p. 682.

Sharpley, A. N., Daniel, T., Sims, T. *et al.* (2003). *Agricultural phosphorus and eutrophication* (2nd ed.). U.S. Department of Agriculture, Agricultural Research Service, ARS. p. 149.

Stewart, J. W. B., & Tiessen, H. (1987). Dynamics of soil organic phosphorus. *Biogeochemistry*, *4*, 41–60.

Sundby, B., Gobeil, C., Silverberg, N. *et al.* (1992). The phosphorus cycle in coastal marine sediments. *Limnology and Oceanography*, *37*, 1129–1145.

USEPA. (1983). *Methods for chemical analysis of water and wastes* (2nd ed.). Washington, DC: Environmental Protection Agency.

USEPA. (1996). *Environmental indicators of water quality in the United States*. Washington, DC: Environmental Protection Agency.

Vetter, T. A. *et al.* (2007). Combining reverse osmosis and electrodyalysis for more complete recovery of dissolved organic matter from seawater. *Separation Purification Technology*, *56*, 383–387.

Whitton, B. A., Al-Shehri, A. M., Ellwood, N. T. W. *et al.* (2005). Ecological aspects of phosphatase activity in cyanobacteria, eukaryotic algae and bryophytes. In B. L. Turner, E. Frossard, D. S. Baldwin (Eds.) *Organic Phosphorus in the Environment*. Cambridge, MA: CABI Publishing.

Chapter 35

The Sulfur Cycle

P. Lens
Wageningen University, Wageningen, The Netherlands

Chapter Outline

ABBREVIATIONS

COS Carbonyl sulfide
CS_2 Carbonyl disulfide
DMS Dimethyl sulfide
NMR Nuclear magnetic resonance
SRB Sulfate-reducing bacteria

DEFINING STATEMENT

The sulfur cycle plays an important role in global planetary geochemical cycles, in which microorganisms play a key role.The sulfur bioconversions are carried out by specialized, mainly Eubacterial genera. Unbalanced sulfur cycling in ecosystemscan have devastating effects, whereas environmental biotechnology engineers sulfur cycle activities for benign processes.

PLANETARY SULFUR FLUXES

Sulfur is the 14th most abundant element in the earth's crust. Sulfur is present in several of the large environmental compartments on earth (Table 1). The lithospheric compartment is the largest and contains roughly 95% of this element. The second largest compartment is the hydrosphere, with the earth's oceans containing approximately 5% of the total sulfur. Sulfate is the second most abundant anion in seawater. The other compartments in which sulfur is found together comprise <0.001% of the remaining sulfur.

Within the framework of global biogeochemical cycling (Figure 1), sulfur is transformed with respect to its oxidation state, formation of organic and inorganic compounds, and its physical (gas, liquid, or solid; soluble or insoluble) status. In oxidizing conditions, the most stable sulfur species is sulfate. In reducing environments, sulfide is formed. Sulfide and some organic compounds containing reduced sulfur, for example, dimethyl sulfide (DMS), carbonyl sulfide (COS), and carbonyl disulfide (CS_2), are volatile and can escape into the atmosphere. From there, sulfur compounds are redeposited into the litho-, hydro-, and pedosphere, either directly or after conversion to sulfate, via the gaseous intermediate SO_2.

Sulfide has high chemical reactivity with some metal cations, leading to the formation of poorly soluble metal sulfides. This results in an accumulation of solid-state reduced sulfur stocks within the global planetary sulfur cycling, due to the formation of insoluble metal sulfides in most anaerobic environments, such as marshes, wetlands, freshwater, and sea sediments, all over the globe. The formation of solid-state reduced sulfur has continued from the early history of planetary biogeochemical cycling. Accumulation of sulfur in anaerobic, highly organic rich deposits of biomass resulted in the contamination of coal by metal sulfides as well as organic bond sulfur compounds (sulfur content ranging between 0.05% and 15.0%). Similarly, mineral oils and petroleum can contain substantial amounts of sulfur compounds (0.025–5%). Under certain conditions, large – commercially exploitable – quantities of elemental sulfur

Topics in Ecological and Environmental Microbiology.

TABLE 1 Amount of sulfur present in various components of the earth

Component	Amount of Sulfur (kg)
Atmosphere	4.8×10^{9}
Lithosphere	2.4×10^{19}
Hydrosphere	
Sea	1.3×10^{18}
Freshwater	3.0×10^{12}
Marine organisms	2.4×10^{11}
Pedosphere	
Soil	2.6×10^{14}
Soil organic matter	1.0×10^{13}
Biosphere	8.0×10^{12}

can accumulate in petroleum reservoirs (see Section "Sulfur Cycling Within the Ecosystems"). Another stock of accumulated solid-state reduced sulfur comprises the sulfidic ores, for example, pyrite (FeS_2), covellite (CuS), chalcopyrite ($CuFeS_2$), galena (PbS), and sphalerite (ZnS).

Increasing anthropogenic extraction of sulfur-containing compounds from the lithosphere considerably perturbs the global sulfur cycle. The most important anthropogenic flux is emission into the atmosphere (113 Tg S/year, i.e., 113×10^{12} g S/year). In the continental part of the sulfur cycle, this flux is comparable only with weathering (114.1 Tg S/year) and river runoff to the world oceans (108.9 Tg S/year). The second major anthropogenic flux of sulfur is the pollution of rivers and a subsequent runoff of the river waters (104 Tg S/year). The anthropogenic sulfur inputs on the globe result in an acceleration of sulfur cycling. This is manifested in elevated levels of sulfate in runoff waters, buildup of sulfide in anaerobic environments, and, after exposure of reduced sulfur stocks to air, acidification of the environment and leaching of toxic metals (see Section "Environmental Consequences and Technological Applications"). At the same time, the increased anthropogenic emissions to the atmosphere bring about other adverse environmental effects, for example, acid rains. The negative consequences of acid rains are well known and include extensive damage to forests and wildlife, as well as detrimental effects on buildings, constructions, and artifacts such as art works and statues.

THE MICROBIAL SULFUR CYCLE

The behavior of sulfur compounds in the environment is highly influenced by the activity of living organisms, particularly microbes. In Figure 2, the stocks of sulfur with

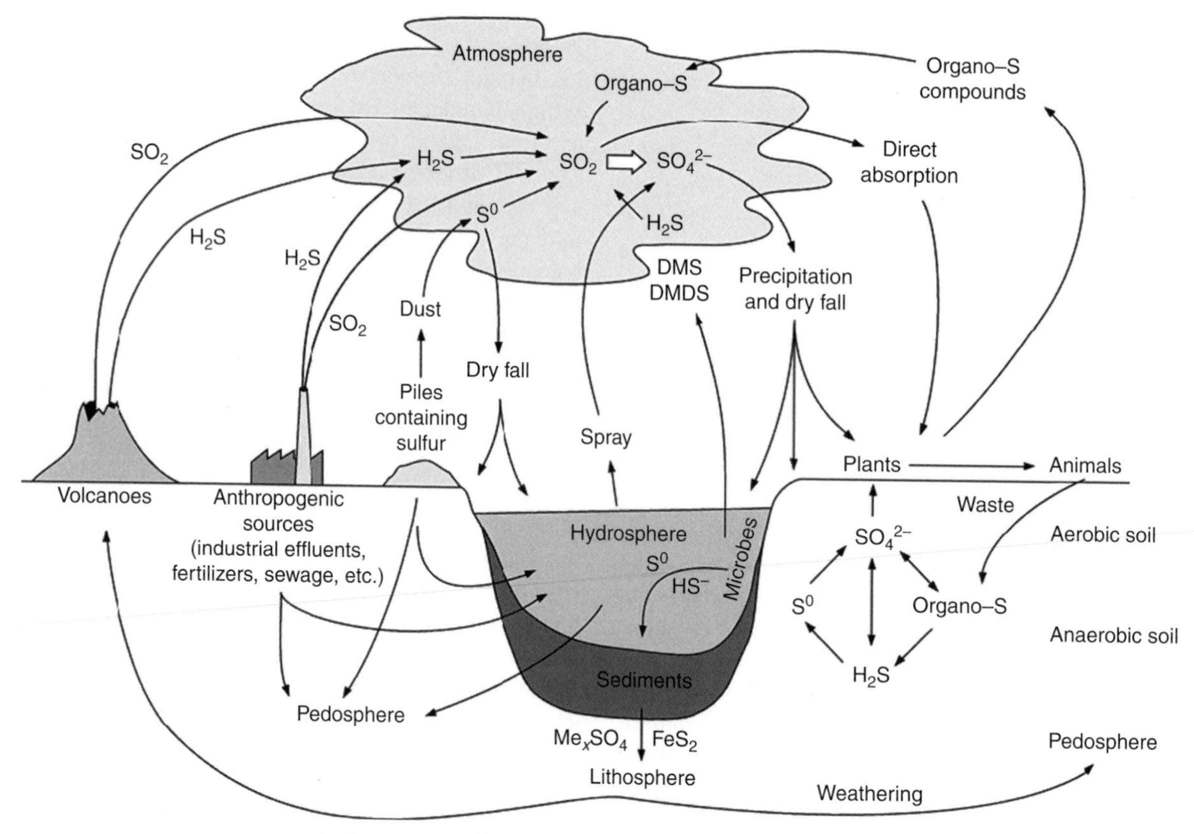

FIGURE 1 Simplified version of the overall sulfur cycle in nature.

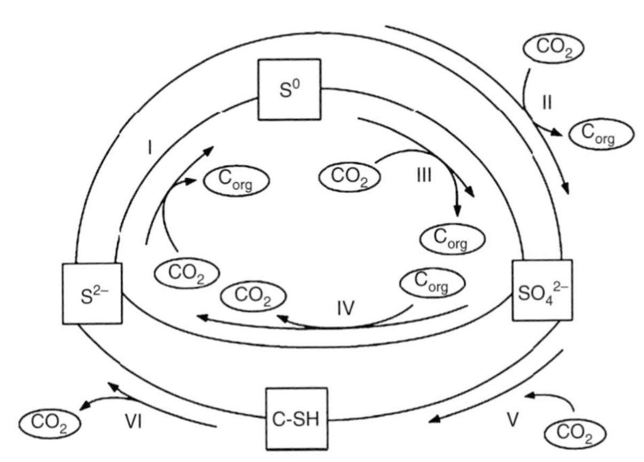

FIGURE 2 Schematic representation of the different pathways occurring in the microbial sulfur cycle. *Reproduced from Tichy, R., Lens, P., Grotenhuis, J. T. C., & Bos, P. (1998). Solid-state reduced sulfur compounds: Environmental aspects and bioremediation.* Critical Reviews in Environmental Science and Technology, 28, *1–40.*

different oxidation states (marked by squares) are given: S^{2-}, the sulfidic form; S^0, elemental sulfur; SO_4^{2-}, sulfate; C–SH, the stock of organic sulfur compounds. Shaded arrows indicate the tropical status of microbes in each process, distinguishing autotrophic (using inorganic CO_2) and heterotrophic (using organic carbon compounds, Corg).

In the last few decades, the microbial ecology of bacteria having a role in the sulfur cycle has received considerable attention from different scientific fields, for example, microbial mats and sediments, wastewater treatment biofilms, and corrosion. This involved various advanced analytical techniques, for example, quantification of reaction products at the micrometer scale using microelectrodes for oxygen, sulfide, pH, and glucose; determination of metabolic and transport processes using ^1H, ^{13}C, and ^{31}P nuclear magnetic resonance (NMR) techniques; and studies of population dynamics using 16S rRNA-based detection methods.

Microbial Sulfate Reduction

Sulfate salts are the major stock of mobile sulfur compounds. They are mostly highly soluble in water, and considerable amounts can be transported in the environment. In the microbial sulfur cycle, sulfate is converted into sulfide by sulfate-reducing bacteria (SRB) via dissimilatory sulfate reduction (pathway I). This process of bacterial respiration occurs under strictly anaerobic conditions and uses sulfate as a terminal electron acceptor. Electron donors are usually organic compounds, eventually, hydrogen:

$$8H_2 + 2SO_4^{2-} \rightarrow H_2S + HS^- + 5H_2O + 2OH^- \quad (1)$$

SRB consist of the traditional sulfate-reducing genera *Desulfovibrio* and *Desulfotomaculum*, in addition to the morphologically and physiologically different genera *Desulfobacter, Desulfobulbus, Desulfococcus, Desulfonema,* and *Desulfosarcina*. In the presence of sulfate, SRB are able to use several intermediates of the anaerobic mineralization process. Besides the direct methanogenic substrates molecular hydrogen (H_2), formate, acetate, and methanol, they can also use propionate, butyrate, higher and branched fatty acids, lactate, ethanol, and higher alcohols, fumarate, succinate, malate, and aromatic compounds. In the sulfidogenic breakdown of volatile fatty acids, two oxidation patterns can be distinguished. Some SRB are able to completely oxidize volatile fatty acids to CO_2 and sulfide as end products. Other SRB lack the tricarboxylic acid cycle and carry out an incomplete oxidation of volatile fatty acids with acetate and sulfide as end products.

In addition to the reduction of sulfate, reduction of sulfite and thiosulfate is also very common among SRB. *Desulfovibrio* strains have been reported to be able to reduce di-, tri-, and tetrathionate. A unique ability of some SRB, for example, *Desulfovibrio dismutans* and *Desulfobacter curvatus*, is the dismutation of sulfite or thiosulfate:

$$4SO_3^{2-} + H^+ \rightarrow 3SO_4^{2-} + HS^-, \quad (2)$$

$$S_2O_3^{2-} + H_2O \rightarrow SO_4^{2-} + HS^- + H^+. \quad (3)$$

Some SRB were found to be able to respire oxygen, despite being classified as strictly anaerobic bacteria. Thus far, however, aerobic growth of pure cultures of SRB has not been demonstrated. The ability of SRB to carry out sulfate reduction under aerobic conditions nevertheless remains intriguing and could be of significance for microscale sulfur cycles (see Section "Sulfur Cycling Within the Ecosystems").

In the absence of an electron acceptor, SRB are able to grow through a fermentative or acetogenic reaction. Pyruvate, lactate, and ethanol are easily fermented by many SRB. An interesting feature of SRB is their ability to perform acetogenic oxidation in syntrophy with hydrogenotrophic methanogenic bacteria, as described for cocultures of hydrogenotrophic methanogenic bacteria with *Desulfovibrio* sp. using lactate and ethanol or with *Desulfobulbus*-like bacteria using propionate. In the presence of sulfate, however, these bacteria behave as true SRB and metabolize propionate as an electron donor for the reduction of sulfate.

Microbial Sulfur Oxidation

Different bacteria can oxidize various reduced sulfur compounds, for example, sulfide, elemental sulfur, or thiosulfate. Oxidation of sulfide into elemental sulfur (pathway II) is performed by autotrophic bacteria. Equation (4) gives the stoichiometry of the chemoautotrophic process, which proceeds aerobically or microaerobically. In addition, photoautotrophic sulfide oxidation can also occur under anaerobic conditions. Photosynthetic sulfur bacteria are capable of photoreducing CO_2 while oxidizing H_2S to S^0 (Equation 5), in a striking analogy to the photosynthesis of eukaryotes (Equation 6).

$$2H_2S + O_2 \rightarrow 2S^0 + 2H_2O, \quad (4)$$

$$2H_2S + CO_2 + h\nu \rightarrow 2S^0 + [CH_2O] + H_2O, \quad (5)$$

$$2H_2O + CO_2 + h\nu \rightarrow 2O_2 + [CH_2O]. \quad (6)$$

Sulfide can also be completely oxidized to sulfate. Equation (7) gives a formula for the chemoautotrophic process, although photoautotrophic oxidation of sulfide to sulfate can also occur.

$$H_2S + 2O_2 \rightarrow SO_4^{2-} + 2H^+ \quad (7)$$

This oxidation reaction, catalyzed by, for example, *Thiobacillus*, involves a series of intermediates, including sulfide, elemental sulfur, thiosulfate, tetrathionate, and sulfate:

$$SH^- \rightarrow S^0 \rightarrow S_2O_3^{2-} \rightarrow S_4O_6^{2-} \rightarrow SO_4^{2-}. \quad (8)$$

Eventually, oxidation of sulfide may proceed in oxygen-free conditions, using nitrate as an electron acceptor (Equation 9). Oxidation of sulfide in anoxic conditions is mostly carried out by bacteria from the genus *Thiobacillus*, such as *T. albertis* and *T. neapolitanus*.

$$\begin{aligned}&0.422H_2S + 0.422HS^- + NO_3^- + 0.437CO_2\\&+0.0865HCO_3^- + 0.0865NH_4^- \rightarrow 1.114SO_4^{2-}\\&+0.5N_2 + 0.0842C_4H_7O_2N(\text{biomass}) + 1.228H^+\end{aligned} \quad (9)$$

Elemental sulfur can be oxidized to sulfate via chemoautotrophic or photoautotrophic microorganisms (Figure 2, pathway IV). The stoichiometry of the chemoautotrophic process is given in Equation (10).

$$2S^0 + 3O_2 + 2H_2O \rightarrow 2SO_4^{2-} + 4H^+. \quad (10)$$

The biological oxidation of reduced sulfur is mediated by a diverse range of bacterial species (Table 2). They can be divided into two main groups: the aerobic and microaerobic

TABLE 2 Sulfide- and sulfur-oxidizing bacteria

Genus or group	Habitat	Comments
Chemotrophs		
Thiobacillus	Soil, water, marine	Mostly lithotrophs, one Fe(II) oxidizer, some thermophiles,deposit S^0 outside the cell
Sulfobacillus	Mine tips	Lithotroph, sporer, Fe(II) oxidizer, thermophilic
Thiomicrospira	Marine	Lithotroph
Beggiatoa	Water, soil, marine	
Thiothrix	Water, soil, marine	Gliding bacteria, deposit S^0 inside the cell, difficult to isolate
Thioploca	Water, soil	
Achromatium	Water	
Thiobacterium	Water	
Macromonas	Water	Deposit S^0 inside the cell, not grown in pure cultures
Thiovulum	Water	
Thiospira	Water	
Sulfolobus	Geothermal springs	Archaebacterium, lithotroph, Fe(II) oxidizer, thermophilic
Phototrophs		
Chlorobiaceae (Green S bacteria)	Water, marine	Lithotrophs, deposit S^0 outside the cell
Chromatiaceae (Purple S bacteria)	Water, marine	Mixotrophs, deposit S^0 inside the cell, except for *Ectothiorhodospira* spp., which deposit S^0 outside the cell
Chloroflexaceae	Geothermal springs	Mixotrophs, deposit S^0 outside the cell, thermophiles
Oscillatoria (Blue-green algae)	Water	S^0 deposited outside the cell under anoxic conditions

chemotrophic sulfur oxidizers (sometimes called the "colorless sulfur bacteria") and the anaerobic phototrophic sulfur oxidizers (sometimes called the "purple and green sulfur bacteria").

The most common microorganisms associated with sulfide oxidation are *Thiobacillus* spp. These non-spore-forming bacteria belong to the colorless sulfur bacteria. Thiobacilli are Gram-negative rods about 0.3 μm in diameter and 1–3 μm in length. Most thiobacilli species are motile by polar flagella. All thiobacilli species grow aerobically, although anaerobic growth (see Equation 9) has been observed for some species as well. Elemental sulfur accumulates on the cell surface, in contrast to filamentous colorless sulfur bacteria, for example, *Thiothrix* sp., *Beggiatoa* sp., and *Thioploca* sp., which accumulate elemental sulfur intracellularly. *Thiothrix* spp. are commonly found in flowing sulfidic water containing oxygen in both marine and freshwater environments such as outlets of sulfidic springs and wastewater treatment plants. They differentiate ecologically from other filamentous colorless sulfur bacteria in that *Thiothrix* sp. prefers hard substrates to which it attaches with a holdfast, whereas *Beggiatoa* sp. and *Thioploca* sp. do not attach and prefer soft bottom sediments. The diameter of the filaments is variable in all the three genera and is used to define species.

The phototrophic bacteria comprise the Chlorobiaceae (green sulfur bacteria), the Chromatiaceae (purple sulfur bacteria), and the filamentous thermophilic flexibacteria, exemplified by *Chloroflexus aurantiacus*. Phototrophic bacteria contribute substantially to both the sulfur cycle and the primary productivity of shallow aquatic environments, but are less significant in the mid-ocean where they are limited by light availability.

In addition to the specialized bacteria listed in Table 2, a considerable number of common bacteria (e.g., *Bacillus* spp., *Pseudomonas* spp., and *Arthrobacter* spp.) and fungi (e.g., *Aspergillus* spp.) have also been shown to oxidize significant amounts of reduced sulfur compounds when grown in pure cultures. Even though such transformations are incidental, these organisms are present in large numbers in soils and in the aquatic environment compared to *Thiobacillus* spp., suggesting that nonlithotrophic sulfur oxidation may be quantitatively important.

Thiobacilli, as the main representatives of the acidophiles, play a key role in the degradation of sulfidic materials. In addition to the oxidation of reduced sulfur compounds as their energy source (see Equations 7 and 10), they can also gain energy from the conversion of ferrous (Fe^{2+}) into ferric (Fe^{3+}) iron. The best known acidophile is *Thiobacillus ferrooxidans*, which combines the ability to oxidize both sulfur compounds and ferrous iron. The bacterium is able to oxidize sulfidic minerals such as pyrite at low pH values (pH 1–4) according to the following reaction (Equation 11):

$$4FeS_2 + 15O_2 + 2H_2O \rightarrow 2Fe_2(SO_4)_3 + 2H_2SO_4 \quad (11)$$

T. ferrooxidans has been isolated from acid mine drainage water in the late 1940s, together with *Thiobacillus thiooxidans*. The latter species can oxidize only sulfur and reduced sulfur compounds and lack the ferrous iron oxidizing capacity. In contrast to *T. ferrooxidans*, *T. thiooxidans* cannot attack sulfidic minerals on its own, but it can contribute to their solubilization in a syntrophic relation with ferrous iron oxidizers such as *Leptospirillum ferrooxidans*. The latter microorganism can convert only ferrous iron and has no sulfur-oxidizing capacity.

The group of acidophiles also consists of facultative autotrophs or even obligate heterotrophs. A representative of this subgroup is *Thiobacillus acidophilus*, a facultative autotrophic sulfur compound oxidizer, which can also grow on sugars. Apart from solubilizing sulfidic minerals, facultative autotrophic acidophiles are also crucial in removing low molecular weight organic acids, which are toxic to the obligate autotrophs even at very low concentrations (5–10 mg l^{-1}). These toxic effects are due to the uptake of organic compounds with carboxylic groups, which are undissociated in the outer medium with low pH but are dissociated in bacterial cytoplasm with circum-neutral pH.

Transformations of Organic Sulfur Compounds

The formation and the degradation of organic sulfur compounds (C–SH) are not solely microbial processes, but numerous other organisms also participate. Particularly, the formation of organic sulfur (Figure 2, pathway V) is accomplished by all photosynthesizing organisms, such as algae or green plants. DMS is the most common product of oceanic green algae sulfur conversions. Green plants and many microorganisms assimilate sulfate as their sole sulfur source. Therefore, they reduce sulfate via a reductive process, namely, assimilatory sulfate reduction, in which the formed sulfide is incorporated into organic matter via a condensation reaction with serine derivates to generate the amino acid cysteine.

Conversion of organic sulfur into sulfide occurs during the decomposition of organic matter (Figure 2, pathway VI). Considerable environmental risks are encountered in these processes, especially regarding the volatilization of organic sulfur compounds and associated odor pollution.

SULFUR CYCLING WITHIN THE ECOSYSTEMS

The sulfur cycle, together with the carbon and nitrogen cycle, is distinguished from most other mineral cycles (e.g., P, Fe, Si) by exhibiting transformations from gaseous to ionic (aqueous or solid) forms. Each of the transformations (see Section "The Microbial Sulfur Cycle") relies on appropriate cellular and ambient oxygen (or redox) conditions. Thus, cycling of these elements depends on the presence of oxygen (or redox) gradients. These gradients vary in size from the smallest, which can be generated over distances of only a few micrometers – for example, biofilms and sediments – to larger (a few millimeter) zones such as in soil crumbs and microbial mats. In exceptional circumstances, these gradients may extend over several tens of meters, for example, geothermal springs, stable stratified lakes, or sewer outfalls. Of particular interest for the sulfur cycle are the steep gradients present in microenvironmental or microzonal conditions, as they allow interactions between cells that normally cannot coexist, that is, anaerobic sulfate reducers and aerobic sulfide oxidizers.

In some cases, very neat cyclic reactions are possible. The clearest example is the sulfur cycle as illustrated in Figure 3. Habitats with a complete sulfur cycle are known as "sulfureta." Both groups of bacteria may be found close to the border between the aerobic and the anaerobic habitats, and neither organism can grow in the other's space. Yet, they are totally dependent on sulfur compounds that diffuse between them. Such cycles are common in marine or estuarine sediments and are also evident in stratified water bodies and fixed-film wastewater treatment systems (Figure 4).

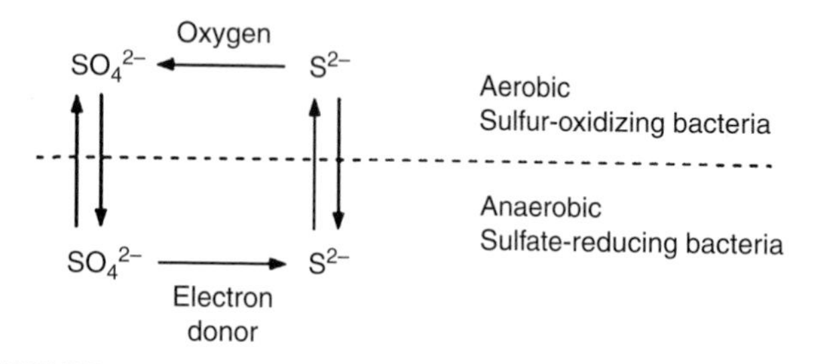

FIGURE 3 Schematic representation of the cyclic reactions prevailing in a microbial sulfur cycle between sulfate-reducing and sulfide-oxidizing bacteria.

Figure 4 depicts possible blocks of the cyclic activity, for example, sulfide precipitating with heavy metals (e.g., FeS) or the accumulation of elemental sulfur in sulfide-oxidizing bacteria. The latter also prevails in natural, light-exposed environments, as illustrated in Figure 5, which depicts some scenarios for the biogenic deposition of elemental sulfur in a lake and in the vicinity of petroleum reservoirs. Formation of the colloid/solid elemental sulfur does not necessarily imply a complete blocking of the sulfur cycling activity. The sulfur can be reduced to sulfide by, for example, *Desulfovibrio desulfuricans*, thus allowing the sulfur cycle to proceed (Figure 6).

ENVIRONMENTAL CONSEQUENCES AND TECHNOLOGICAL APPLICATIONS

Disrupture of sulfur cycling can also have serious environmental consequences. An overview of the major environmental effects when the sulfur cycle is unbalanced is given here. This imbalance can be a result of both natural and anthropogenic processes. A number of technological applications that utilize bacteria of the sulfur cycle are also presented.

Environmental Consequences

Microbial transformations of solid-state reduced sulfur compounds, which are used or affected by anthropogenic activities, represent major environmental risks. Materials such as fossil fuels, ores, anaerobic sediments, or solid

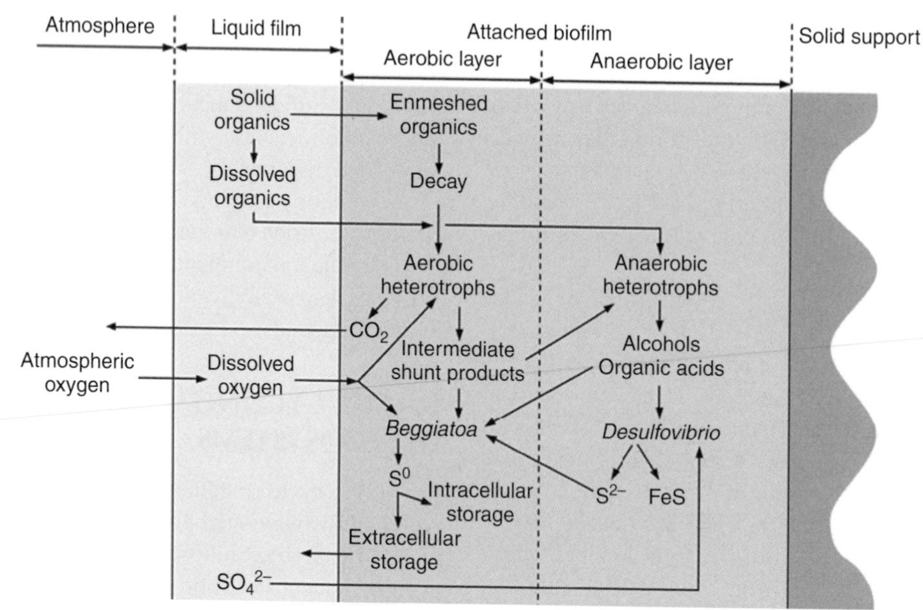

FIGURE 4 Reactions involved in a sulfur cycle in a biofilm of a rotating biological contactor type wastewater treatment system. *Reproduced from Alleman, J. E., Veil, J. A., & Canaday, J. T. (1982). Scanning electron microscope evaluation of rotating biological biofilm.* Water Research, 16, *543–550.*

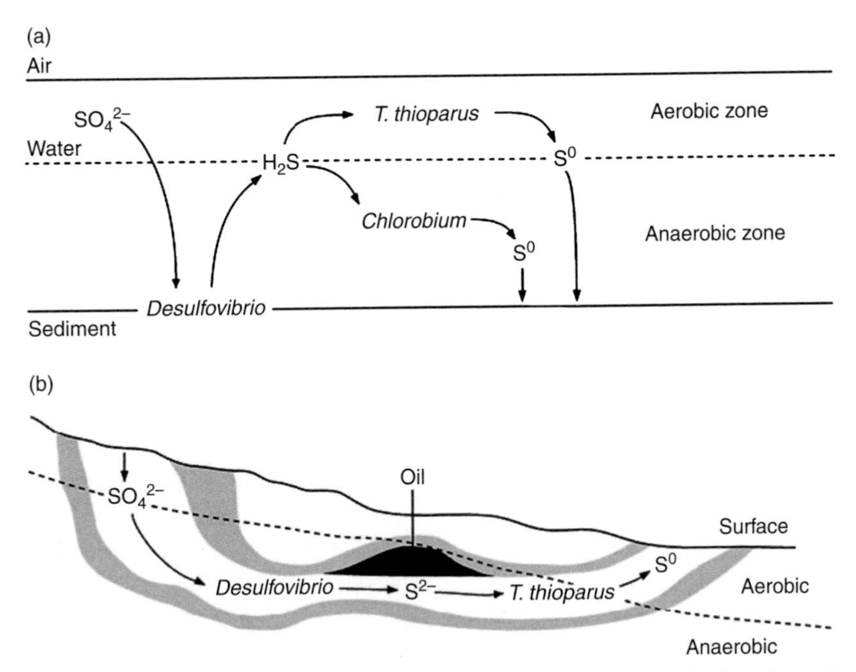

FIGURE 5 The biological deposition of sulfur (a) in a lake and (b) in a geological stratum mediated by the disruption of the sulfur cycling activities.

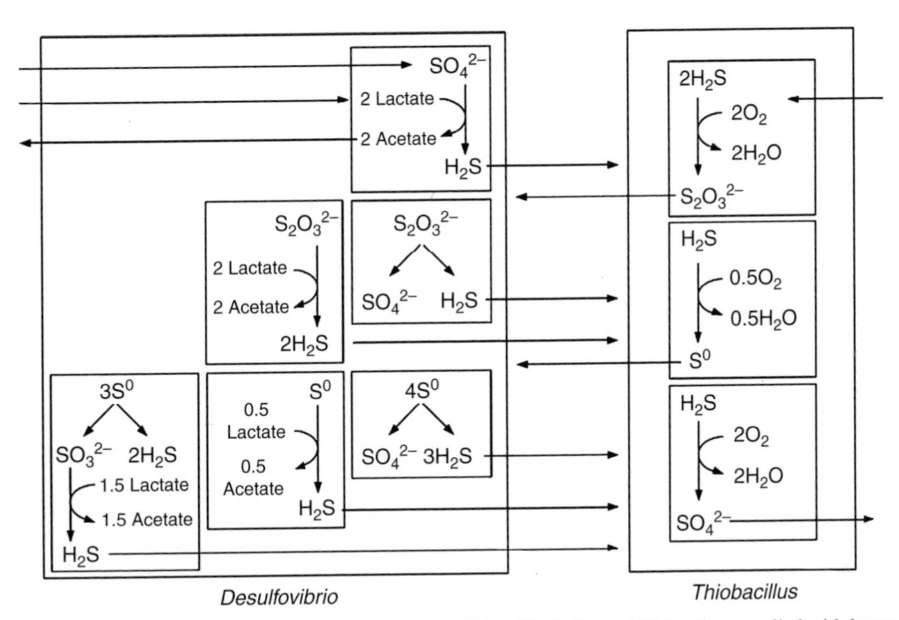

FIGURE 6 Possible pathways of sulfur transformations in mixed cultures of *Desulfovibrio* and *Thiobacillus* supplied with lactate, sulfate, and oxygen. *Reproduced from Van den Ende, F. P., Meier, J., & Van Gemerden, H. (1997). Syntrophic growth of sulfate-reducing bacteria and colorless sulfur bacteria during oxygen limitation.* FEMS Microbiology Ecology, 23, *65–80.*

waste may undergo oxidative changes, resulting in solubilization of sulfur from the solid phase. Thus, large amounts of sulfuric acid are formed (see Equation 11), which are transported off-site in the so-called acid mine drainage. The acid inhibits plant growth and aquatic life. Its major environmental consequence is, however, the solubilization of cationic heavy metals.

In tropical regions of the world, periodic flooding and draining of maritime estuarine soils leads to the accumulation of reduced sulfur and its subsequent oxidation. This results in the production of sulfuric acid and the appearance of acid sulfate soils, sometimes called "cat clays." In many cases, a proper treatment (neutralization) of these sites is required, for example, before allowing farming.

Sulfate-rich wastewater is generated by many industrial processes that use sulfuric acid or sulfate-rich feed stocks (e.g., fermentation or seafood processing industry). Also, the use of reduced sulfur compounds in industrial processes, that is, sulfide (tanneries, Kraft pulping), sulfite (sulfite pulping), thiosulfate (fixing of photographs), or dithionite (pulp bleaching), contaminates wastewater with sulfurous compounds.

Pollution by humans has greatly increased the sulfur dioxide levels in the atmosphere. When dissolved in water in the clouds, it acidifies rainwater. Acidic rainfall has been implicated in the deterioration of the growth of forests in North America and Europe. Acid rain also has a detrimental effect on buildings, constructions, and other artifacts such as ancient artworks, causing etching and destroying their original beauty.

Many of the volatile, both organic and inorganic, sulfur compounds are extremely odorous. Farmyard feedlots and improperly managed rendering and composting facilities are examples of places where offensive sulfur gases are generated.

Technological Applications

Hydrometallurgical processes play an important role in the extraction of metals from certain low-grade ores, for example, <1% Cu by weight. Hydrometallurgy consists of the dissolution of metals from minerals, usually by constant percolation of a leaching solution through beds of ore. The process is most effective in ores that contain substantial amounts of pyrite and is based on the rapid oxidation of S^{2-} and Fe^{2+} promoted by acidophiles. In the literature, two mechanisms by which acidophiles attack the insoluble metal sulfides are proposed: indirect leaching and direct leaching. In the indirect mechanism, the ferric iron acts as a chemical oxidizer of the sulfidic minerals. The ferric iron is a product of bacterial oxidation. In this mechanism, the biological sulfur-oxidizing capacity of the acidophiles is of no relevance. In the direct mechanism, sulfidic mineral solubilization involves both the biological ferrous iron and sulfur compound oxidation. Pyrite oxidation by *T. ferrooxidans* was shown to be mediated by the biological oxidation of both ferrous iron and sulfide.

Microbial leaching (bioleaching) is also proposed for the decontamination of polluted solid waste, soils, or sediments. Successful demonstration of the bioleaching of toxic metals from an anaerobically digested sewage sludge was demonstrated on a technological scale. Experimental attempts to use bioleaching for heavy metal removal from freshwater sediments are reported as well. The process may

TABLE 3 Environmental biotechnological applications using processes of the microbial sulfur cycle

Application	Sulfur Conversion Utilized	Waste Stream
Wastewater treatment		
Sulfate removal	Sulfate and/or sulfite reduction + partial sulfide oxidation to S^0	Industrial wastewaters, acid mine drainage, and spent sulfuric acid
Sulfide removal	Partial sulfide oxidation to S^0	Industrial wastewaters
Heavy metal removal	Sulfate reduction	Extensive treatment in wetlands or anaerobic ponds High rate reactors for process water, acid mine drainage, and groundwater
Microaerobic treatment	Internal sulfur cycle in the biofilm	Domestic sewage
Off-gas treatment		
Biofiltration of gases	Oxidation of sulfide and organosulfur compounds	Biogas, malodorous gases from composting and farming
Treatment of scrubbing waters	Sulfate and/or sulfite reduction + partial sulfide oxidation to S^0	Scrubbing waters of SO_2-rich gases
Solid waste treatment		
Bioleaching of metals	Sulfide oxidation	Sewage sludge, compost
Desulfurization	Oxidation	Rubber
Gypsum processing	Sulfate reduction	
Treatment of soils and sediments		
Bioleaching of metals	Sulfide oxidation	Dedged sediments and spoils
Degradation xenobiotics	Sulfate reduction	Polychlorinated biphenyl (PCB) contaminated soil slurries

use the autoacidification potential of the sediment, driven by the presence of reduced sulfur and ferrous compounds. If the latter is insufficient relative to the sediment's buffering capacity, additional sulfur can be added to achieve satisfactory extraction yields.

Reduced sulfur compounds are commonly added to alkaline soils so that the produced sulfuric acid decreases the soil pH to a level acceptable for plant growth. Moreover, sulfur compounds can be used as fertilizer to improve crop production in those areas of the world where sulfur is the limiting nutrient. This is useful, for example, in Western European countries, where atmospheric deposition of sulfur over farmlands has decreased as a result of the stringent control of sulfur dioxide emissions.

A better insight into sulfur transformations has enabled the development of a whole spectrum of new biotechnological applications for the bioremediation of polluted waters, gases, soils, and solid wastes (Table 3). These technologies rely both on bacterial sulfate reduction and on sulfide/sulfur oxidation. They allow the removal of sulfur and organic compounds as well as heavy metals and nitrogen. Until recently, biological treatment of sulfur-polluted waste streams was rather unpopular because of the production of hydrogen sulfide under anaerobic conditions. Gaseous and dissolved sulfides cause physical–chemical (corrosion, odor, increased effluent chemical oxygen demand), or biological (toxicity) constraints, which may lead to process failure. However, anaerobic treatment of sulfate-rich wastewater can be applied successfully provided a proper treatment strategy is selected, which depends on the aim of the treatment: (1) removal of sulfur compounds, (2) removal of organic matter, or (3) removal of both.

Microbial sulfur transformations can provide a unique tool to control pollution by both sulfur compounds and heavy metals. In some cases, extensive techniques using natural processes are applied, for example, the use of wetlands or anaerobic ponds to treat voluminous aqueous streams such as acid mine drainage or surface runoff waters. These systems require low maintenance and can sustain their function for prolonged time intervals. Subsequent regeneration of these systems using bioleaching and treatment of spent extraction liquor by SRB may be a logical step. In general, technological applications using various processes of the microbial sulfur cycle may provide many beneficial effects in the future.

FURTHER READING

Alena, L., & Kusnierova, M. (2005). Bioremediation of acid mine drainage contaminated by SRB. *Hydrometallurgy, 77*, 97–102.

Alleman, J. E., Veil, J. A., & Canaday, J. T. (1982). Scanning electron microscope evaluation of rotating biological biofilm. *Water Research, 16*, 543–550.

Begheijn, L. Th., van Breemen, N., & Velthorst, E. J. (1978). Analysis of sulfur compounds in acid sulfate soils and other marine soils. *Communications in Soil Science and Plant Analysis, 9*, 873–882.

Clark, M. E., Batty, J. D., van Buuren, C. B., Dew, D. W., & Eamon, M. A. (2006). Biotechnology in minerals processing: Technological breakthroughs creating value. *Hydrometallurgy, 83*, 3–9.

Dopson, M., Baker-Austin, C., Koppineedi, P. R., & Bond, P. L. (2003). Growth in sulfidic mineral environments: Metal resistance mechanisms in acidophilic micro-organisms. *Microbiology, 149*, 1959–1970.

Evangelou, V. P., & Zhang, Y. L. (1995). Pyrite oxidation mechanisms and acid mine drainage prevention. *Critical Reviews in Environmental Science and Technology, 25*, 141–199.

Johnson, D. B., & Hallberg, K. B. (2005). Acid mine drainage: Remediation options. *The Science of the Total Environment, 338*, 3–14.

Kaksonen, A. H., Dopson, M., Karnachuk, O. V., Tuovinen, O. H., & Puhakka, J. A. (2008). Biological iron oxidation and sulfate reduction in the treatment of acid mine drainage at low temperatures. In R. Margesin, F. Schinner, J. C. Marx, & C. Gerday (Eds.) *Psychrophiles: From biodiversity to biotechnology* (pp. 429–454). Berlin: Springer Verlag.

Kelly, D. P., Shergill, J. K., Lu, W. P., & Wood, A. P. (1997). Oxidative metabolism of inorganic sulfur compounds by bacteria. *Antonie Van Leeuwenhoek, 71*, 95–107.

Lens, P. N. L., Meulepas, R. J. W., Sampaio, R., Vallero, M., & Esposito, G. (2007). Bioprocess engineering of sulfate reduction for environmental technology. In C. Dahl, & C. G. Friedrich (Eds.) *Microbial sulfur metabolism* (pp. 285–295). Berlin: Heidelberg/Springer.

Lens, P. N. L., Vallero, M., & Esposito, G. (2006). Bioprocess engineering of sulphate reduction for environmental technology. In L. L. Barton, & W. A. Hamilton (Eds.) *Sulphate-reducing bacteria: Environmental and engineered systems* (pp. 324–336). Cambridge, UK: Cambridge University Press.

Smet, E., Lens, P., & van Langenhove, H. (1998). Treatment of waste gases contaminated with odorous sulfur compounds. *Critical Reviews in Environmental Science and Technology, 28*, 89–116.

Tichy, R., Lens, P., Grotenhuis, J. T. C., & Bos, P. (1998). Solid-state reduced sulfur compounds: Environmental aspects and bioremediation. *Critical Reviews in Environmental Science and Technology, 28*, 1–40.

Van den Ende, F. P., Meier, J., & Van Gemerden, H. (1997). Syntrophic growth of sulfate-reducing bacteria and colorless sulfur bacteria during oxygen limitation. *FEMS Microbiology Ecology, 23*, 65–80.

Widdel, F. (1988). Microbiology and ecology of sulfate- and sulfur reducing bacteria. In A. J. B. Zehnder (Ed). *Biology of anaerobic microorganisms* (pp. 469–586). New York, USA: John Wiley & Sons.

Chapter 36

Cycling of Some Redox-Active Metals

C. Rensing[1] and B.P. Rosen[2]

[1]*University of Arizona, Tucson, AZ, USA*

[2]*Wayne State University, School of Medicine, Detroit, MI, USA*

Chapter Outline

ABBREVIATIONS

AQP Aquaporin
CCA Chromated copper arsenate
c-Cyts c-Type cytochromes
DMA Dimethylarsenate
DMS Dimethyl sulfide
DMSO Dimethyl sulfoxide
EXAFS Extended X-ray absorption fine structure spectroscopy
FMN Flavin mononucleotide
Grx Glutaredoxin
GSH Glutathione
LMW PTPase Low molecular weight protein tyrosine phosphatases
MCO Multicopper oxidase
MMA Monomethylarsenate
MRP Multidrug resistance-associated protein
NADPH Nicotinamide adenine dinucleotide phosphate
N_2OR N_2O reductase
PHM Peptidylglycine-alphahydroxylating monooxygenase
PQQ Pyrroloquinoline
RNAP RNA polymerase
SOD Superoxide dismutase
Tat Twin-arginine translocation
TMAO Trimethylarsine oxide
Tmp Trimethyl purine methylase
Trx Thioredoxin
UNICEF United Nations Children's Fund
WHO World Health Organization

DEFINING STATEMENT

Microbial enzymes are significantly involved in transforming metal(loid)s in the environment. Recent advances in identification and characterization of As-, Mn-, Cu-, and Se-transforming enzymes and remaining challenges are described.

INTRODUCTION

Life may have first originated in deep oceanic hydrothermal vents that were rich in metals such as arsenic, lead, copper, manganese, and zinc. Microorganisms not only had to deal with the toxic effects of metal(loid)s but were also actively involved in transforming metal(loid)s, thereby shaping the environment. This ancient environmental challenge was a driving force for the evolution of mechanisms for metal ion homeostasis and detoxification. For example, copper is an essential nutrient that serves as a cofactor for numerous enzymes and yet is cytotoxic because of its role in generating reactive oxygen species and other reactive compounds. Organisms in all kingdoms of life have evolved elaborate copper homeostatic mechanisms to take advantage of copper chemistry while preventing unwanted side reactions. The same principle holds true for other essential metals. The four metal(loid)s discussed here are all toxic in excess but can perform biologically important roles, for example, as terminal electron acceptors. Arsenic (As) is an abundant, ubiquitous, and dynamic trace element that is involved in a variety of microbial metabolic processes including several detoxification and energy conservation pathways. The environmental processes contributing to arsenic's fate and transport have been studied intensively during the last decade, in part due to global water quality crises arising from As-contaminated drinking water. For example, arsenic in the water supply in southern and western Bangladesh and the adjacent regions of India has triggered a health catastrophe (http://bicn.com). Manganese oxides are often produced by bacteria and fungi and affect many biological processes including carbon fixation, photosynthesis, and scavenging of reactive oxygen species. In addition, manganese minerals are often utilized as a terminal electron acceptor. Finally, selenocysteine incorporated into proteins as the 21st amino acid makes selenium a required trace element in many organisms. However, the higher valence states of Se(VI) and Se(IV) are toxic at elevated concentrations, and microorganisms are involved in transformations such as reduction and methylation, making the products less toxic.

$^{33}As_{75}$

The Arsenic Biogeocycle

An overview of the arsenic biogeocycle is shown in Figure 1. Although arsenic is relatively rare in the earth's crust, ranking 20th in abundance, it is widespread and found in high concentration in association with other minerals such as arsenides, sulfides, and sulfosalts with other metals. The free metal is uncommon, and arsenic occurs primarily in three oxidation states: arsenates (As(V)), arsenites (As(III)), and arsenides (As(–III)). Volcanic activity and hydrothermal sources are major contributors to human exposure to arsenic. For example, the Yellowstone Caldera, the largest volcanic system in North America, discharges massive amounts of arsenic from geysers and fumaroles. Some of

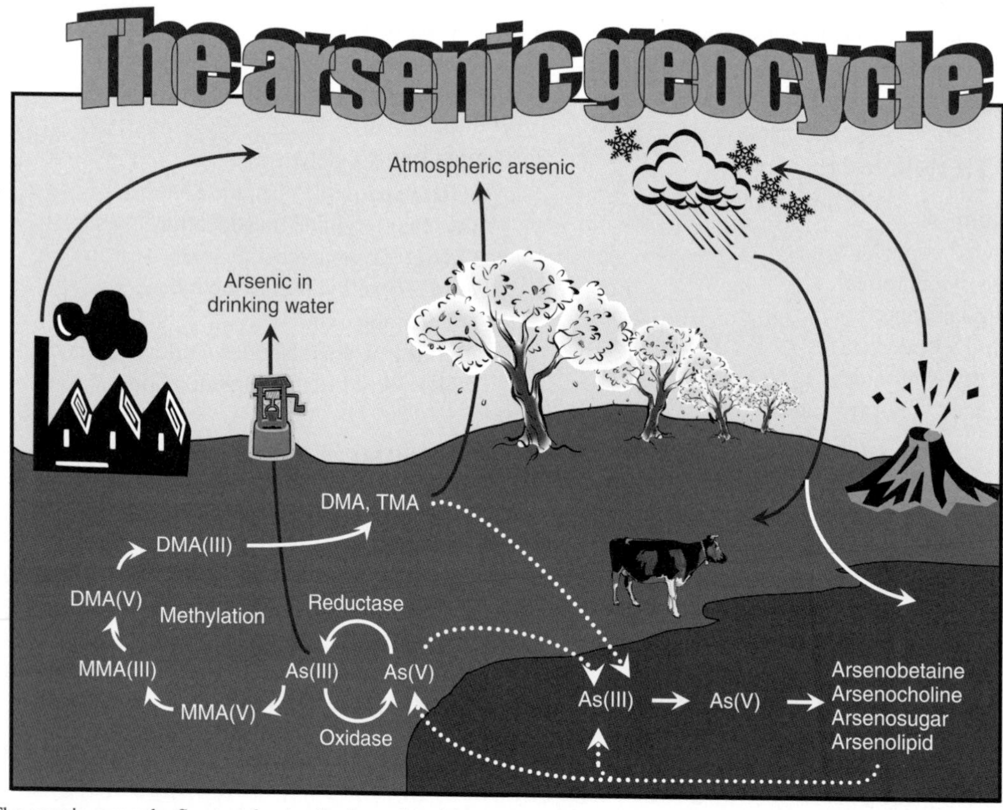

FIGURE 1 The arsenic geocycle. See text for details. *Reproduced from Bhattacharjee, H., & Rosen, B. P. (2007). Microbial arsenic metabolism. In D. H. Nies, & S. Silver (Eds.)* How bacteria handle heavy metals *(pp. 371–406). Heidelberg: Springer.*

the arsenic is in the form of insoluble sulfides, but soluble arsenic can be 4 mmol l^{-1}, one of the highest concentrations of soluble arsenic in the world. There is abundant microbial life in these streams and pools in spite of these extraordinary amounts of arsenic. Arsenic begins as the reduced form, arsenite, but is rapidly oxidized biologically by microorganisms with arsenite oxidases. Further from the source, the arsenite can be reduced by microorganisms with respiratory arsenate reductases. Some of the arsenic is volatilized, and organisms that methylate the arsenic to trimethylarsine have been identified. Mining, copper smelting, and burning of coal also introduce arsenic into the environment. Arsenopyrite (FeAsS), the most common mineral form of arsenic, is often associated with other elements, including gold and copper. This can complicate bioleaching of ores using organisms such as thiobacilli because the sulfuric acid produced by the bacteria to dissolve the gold also dissolves the arsenic, which is toxic to the organism. Arsenic derived from an ancient Roman copper mine is also a major environmental contaminant in the Rio Tinto in southwestern Spain. Despite the millimolar concentrations of arsenic in the river, organisms proliferate, including highly arsenic resistance fungi.

Anthropogenic sources of arsenic include herbicides, insecticides, rodenticides, wood preservatives, animal feeds, paints, dyes, and semiconductors. Some contain inorganic arsenic such as chromated copper arsenate (CCA), which has been used for many decades to treat wood against attack by fungi and insects. If the wood is not sealed, the arsenic can find its way into human water and food supply. Both inorganic and organic arsenicals are used for agriculture and animal husbandry. During the last century, arsenic acid (H_3AsO_4), sold as Desiccant L-10 by Atochem/Elf Aquitaine, was euphemistically called "harvest aid for cotton" because it was used to defoliate cotton to allow planting of the next cotton crop. While it is no longer used agriculturally, the inorganic arsenic remains in fields throughout the southern United States. That land is now used for planting rice, and grocery store rice from those states constitutes the largest non-seafood source of arsenic in the American diet. The sodium and calcium salts of monomethylarsenate (MMA) and dimethylarsenate (DMA) are currently widely used as herbicides and pesticides. The sole active ingredient in Weed-B-Gone Crabgrass Killer, which is sold in neighborhood lawn and garden stores, is calcium MMA. DMA and MMA are also widely used as fungicides on golf courses in Florida, and the resulting arsenic enters the water supply of Florida municipalities. MMA is also degraded to arsenate and arsenite, presumably by microorganisms. During the Vietnam War, the United States sprayed DMA (in the form of the "rainbow herbicide" Agent Blue) on rice paddies and other crops in Vietnam to eliminate the food supply. Some bacteria can use DMA as a source of carbon, with demethylation to inorganic arsenic. Organic arsenicals are also used as growth enhancers and feed supplements in animal husbandry. For example, Roxarsone (4-hydroxy-3-nitrophenylarsonic acid) is fed to half a billion chickens each year in Maryland alone to control coccidial intestinal parasites and is excreted unaltered. The manure is used as fertilizer for crops, and, during composting, the roxarsone is degraded anaerobically, at least in part by bacteria such as clostridia, into inorganic arsenic, which contaminates water supplies. The use of organic arsenicals as feed supplements also has the potential to spread antibiotic resistance.

Arsenic contamination of human water and food supplies is a worldwide problem, but the worst situation is in West Bengal, India, and Bangladesh, where tens of millions of inhabitants are exposed to dangerous levels of arsenic. While this arsenic is from natural sources and is not anthropogenic, it is an unintended consequence of human intervention for humanitarian reasons. To help prevent the spread of infectious diseases such as cholera and malaria, the United Nations Children's Fund (UNICEF) recommended and financed the construction of tube wells for water for drinking and irrigation in lieu of surface waters that carried infectious organisms. While this practice reduced the incidence of infections, the well water has dissolved arsenic concentrations that exceed the World Health Organization's (WHO) maximum allowable level of 10 ppb by 10- to 100-fold, leading to skyrocketing rates of cancer and other arsenic-related disorders. Where the arsenic in well water comes from is still somewhat controversial. Most of the arsenic is immobilized in bedrock, so how it entered the water supply was unclear. Recent data suggest that the irrigation of crops allows organic material to percolate downward, providing anaerobic organisms with carbon and reducing potential to reduce and solubilize arsenic-containing minerals. Indeed, the location of the immobilized arsenic may actually be in a higher stratum than originally thought, such that the arsenic actually enters the wells from above.

What is becoming clear, however, is that human agriculture coupled with microbial activity is related to this environmental catastrophe. This emphasizes the substantial role that microbes play in cycling arsenic between oxidized and reduced forms and between inorganic and organic species. Arsenate is taken up by microbes, primarily via phosphate transport systems. Intracellular arsenate is reduced to arsenite, which is then extruded out of the cell, through either channels or active efflux systems. Bacteria can also use arsenate as a terminal electron acceptor in anaerobic respiration, generating arsenite as the product. Their action in arsenate-rich sediments can lead to arsenic contamination of ground water. Arsenite-oxidizing microbes utilize the reducing power from As(III) oxidation to gain energy for cell growth. Soil bacteria can also methylate arsenite to the gas trimethylarsine, returning soil arsenic to the atmosphere. Marine microorganisms convert inorganic arsenicals to various water- or lipid-soluble organic arsenicals, including di- and trimethylated arsenic derivatives, arsenocholine, arsenobetaine, arsenosugars, and arsenolipids. Metabolism of arsenobetaine by marine microbes completes the arsenic cycle in marine ecosystems.

Arsenic Uptake Systems

Inorganic arsenic has two biologically important oxidation states, pentavalent (As(V)) and trivalent (As(III)). Since arsenic is only a toxic element and has no nutritional or metabolic role, both trivalent and pentavalent inorganic arsenic are taken into cells adventitiously by uptake systems for other compounds. In solution, pentavalent inorganic arsenic is arsenate (H_3AsO_4), an analog of phosphate, and bacteria take up arsenate via phosphate transporters. In *Escherichia coli*, both phosphate transporters, Pit and Pst, take up arsenate, with the Pit system being the major system.

Solid trivalent arsenic (As_2O_3 or arsenic trioxide) dissolves to form the undissociated acid, $As(OH)_3$, at neutral pH. With a p*Ka* of 9.2, it would be present in significant amounts as the anion arsenite only in alkaline environments (which could be important for the arsenic biology of alkalophiles). Even so, inorganic As(III) is usually called "arsenite" in the literature, so this term will be used interchangeably with arsenic trioxide. Two pathways for uptake of trivalent metalloids As(III) and Sb(III) have been identified in prokaryotes and eukaryotes. The *E. coli* glycerol facilitator GlpF was the first identified uptake system for As(III) (and Sb(III)). Bacterial GlpF was also the first member of the aquaporin (AQP) superfamily to be identified, even before the identification of the human water channel Aqp1, which led to the award of the Nobel Prize in Chemistry to Peter Agre in 2003. The superfamily has two branches, the classical AQPs, which are water channels with small channel openings and aquaglyceroporins with channels large enough for molecules as big as glycerol (Figure 2). Most bacteria have GlpF homologs for glycerol uptake that also allow inadvertent arsenite entry, rendering them sensitive to arsenite. Eukaryotic microbes similarly have AQPs that conduct uptake of As(III): for example, Fps1p, the yeast homolog of GlpF, conducts arsenite uptake in *Saccharomyces cerevisiae*. Another eukaryotic microbe, the human pathogen *Leishmania major*, takes up As(III) via aquaglyceroporin LmAQP1. It also takes up the related metalloid Sb(III) by LmAQP1, which is relevant to the treatment of leishmaniasis. The pentavalent antimonial drug Pentostam is reduced, at least in part by macrophage to Sb(III), the active form of the drug. In the phagolysosome of the infected macrophage, the *Leishmania amastigote* takes up the activated drug by LmAQP1.

Surprisingly, some organisms have *ars* operons encoding an aquaglyceroporin homolog termed AqpS. Obviously, this arsenic-inducible AQP is related to arsenic resistance and did not evolve for glycerol uptake. This presents a quandary: How does a channel that can conduct solutes only down a concentration gradient confer resistance when homologs actually make cells sensitive? The answer appears

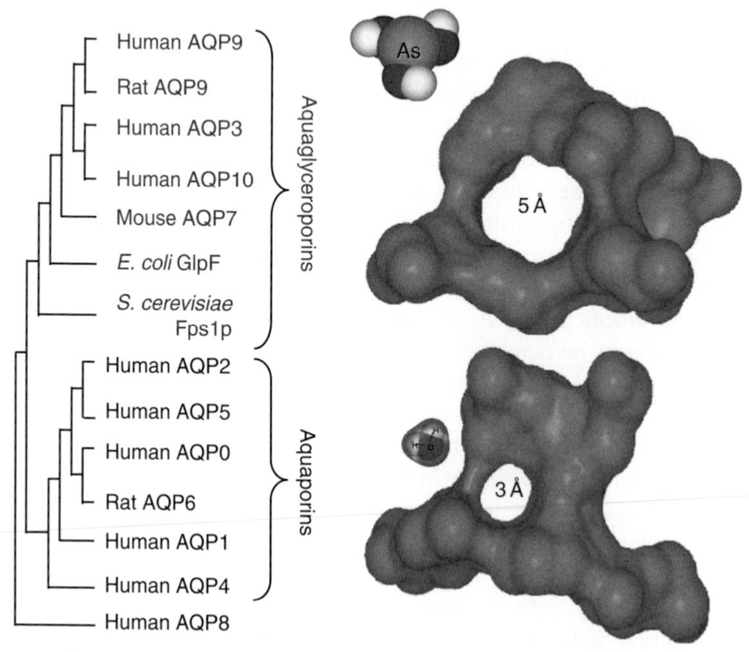

FIGURE 2 Members of the aquaporin superfamily conduct $As(OH)_3$. Left: The aquaporin superfamily is very large, and only selected members are shown. The lower branch (left) includes classical aquaporins that conduct only water. The upper branches are aquaglyceroporins from bacteria, yeast, and mammals. Of those, *Escherichia coli* GlpF, *Saccharomyces cerevisiae* Fps1p, and mammalian AQP7 and AQP9. Right: Space-filling models of $As(OH)_3$ (top) and water (bottom) next to cross-sections of the X-ray crystal structure of AQP1 (bottom) and GlpF (top) at the narrowest point of the channel (diameter indicated).

to be that AqpS does, indeed, render the cells sensitive to arsenite but confers arsenate resistance in conjunction with the ArsC arsenate reductase, about which more is discussed below. As arsenate enters the cell, it is reduced to arsenite. Lacking an arsenite efflux system, the cell substitutes the AqpS, which allows the internally generated arsenite to flow down its concentration gradient out of the cell.

Another group of *S. cerevisiae* permeases that adventitiously facilitates arsenite uptake is the Hxt glucose transporter permease family. *S. cerevisiae* has 18 hexose transporters, Hxt1p–Hxt17p, Gal2p, and two glucose sensors, Snf3p and Rgt2p, and a number of them transport arsenite. While no Hxt takes up As(III) as rapidly as Fps1p, in the aggregate about 75% of the arsenite gets into yeast by Hxts and about 25% by Fps1p when glucose is absent from the medium. In the presence of glucose, which competitively inhibits arsenite uptake by Hxts, most of the arsenite gets in by Fps1p. It is not known whether bacterial glucose permeases or other sugar transporters also catalyze arsenite uptake. A *glpF* deletion of *E. coli* still takes up about 20% of the arsenite compared to the wild type, suggesting the presence of one or more as-yet-unidentified arsenite uptake systems.

Glucose (Hxt) permeases in *Saccharomyces cerevisiae* had been shown to adventitiously facilitate the uptake of arsenite. This observation was recently extended to mammals, where the rat Hxt homolog, glucose transporter isoform 1 or rGLUT1, was shown to facilitate uptake of the trivalent arsenicals arsenite as $As(OH)_3$ and methylarsenite as $CH_3As(OH)_2$. GLUT1 is expressed in the neonatal heart and the epithelial cells that form the blood–brain barrier, and may be the major pathway for arsenic uptake into heart and brain, where the metalloid causes cardiotoxicity and neurotoxicity. The translocation properties of GLUT1 were compared for trivalent arsenicals and glucose. Substitution of Ser^{66}, Arg^{126} and Thr^{310}, residues critical for glucose uptake, led to decreased uptake of glucose but increased uptake of $CH_3As(OH)_2$. The K_m for the uptake of $CH_3As(OH)_2$ of three clinically identified mutants, namely, S66F, R126K, and T310I, were decreased 4–10 fold compared to native GLUT1. The osmotic water permeability coefficient (P_f) of GLUT1 and the three clinical isolates increased in parallel with the rate of $CH_3As(OH)_2$ uptake. GLUT1 inhibitors Hg(II), cytochalasin B, and forskolin reduced uptake of glucose but not $CH_3As(OH)_2$. The conclusion of this study is that that trivalent arsenicals and water use a common translocation pathway in GLUT1 that is different from that of glucose transport.

Arsenic Respiration

Respiratory arsenate reductases

As discussed above, arsenate reduction plays an important role in arsenic geochemistry and may lead to arsenic contamination of drinking water supplies. Under anaerobic growth, the reduction of arsenate as a terminal electron acceptor by respiratory enzymes generates energy, with the production of arsenite. For example, *Chrysiogenes arsenatis* uses acetate as the electron donor for arsenate respiration by an arsenate reductase enzyme ArrAB. The *C. arsenatis* arsenate reductase is located in the periplasm and is a dimer composed of the 87-kDa ArrA and the 29-kDa ArrB subunits. ArrAB couples to the respiratory chain and provides energy for oxidative phosphorylation. ArrA is a Mo/Fe protein homologous to dimethyl sulfoxide (DMSO) reductases. ArrB probably has an [4Fe–4S] cluster that transfers electrons to the Mo cofactor of ArrA. ArrA from *Shewanella* sp. strain ANA-3 has a twin-arginine translocation (Tat) motif for translocation to the periplasm. ArrAB from *Bacillus selenitireducens* strain MLS10 is a heterodimer of 150kDa composed of ArrA (110kDa) and ArrB (34kDa), again with a putative Tat signal in the gene for ArrA. In contrast to the energy-generating ArrAB arsenate reductases, cytosolic ArsC arsenate reductases are involved in arsenic detoxification (see below), and some bacteria such as *Shewanella* sp. strain ANA-3 have both. The *arrAB* genes are expressed anaerobically at nanomolar arsenate or arsenite. Only when environmental arsenic builds up to toxic levels is the *ars* resistance operon expressed both aerobically and anaerobically.

Arsenite oxidases

Arsenite oxidases are found in both heterotrophic and chemoautotrophic bacteria. Arsenite oxidases have been cloned from β-proteobacteria, the heterotrophic *Hydrogenophaga* sp. strain NT-14, and the chemolithoautotrophic bacterium, NT-26. The latter two use the reducing power from As(III) oxidation for growth, suggesting a bioenergetic function for arsenite oxidases. The arsenite oxidase genes of *Agrobacterium tumefaciens* are regulated by a two-component sensor kinase system, composed of the *aoxS* (sensor) and *aoxR* (response regulator) genes. AoxR does not contain an identifiable As(III) binding site, so what is actually sensed by AoxR is not known. The best characterized arsenite oxidase enzyme is from the soil bacterium *Alcaligenes faecalis*. The *A. faecalis* arsenite oxidase is a 100-kDa dimer comprising a large and a small subunit. The large 825-residue catalytic subunit is structurally related to DMSO reductases and is homologous to the ArrA subunit of the respiratory arsenate reductases. The small subunit (134) is homologous to Rieske proteins. The oxidase is bound to the periplasmic surface of the inner membrane by an N-terminal transmembrane helix of the small subunit.

Arsenic Detoxification

Regulation of ars operons

Expression of *ars* genes is controlled by ArsR regulatory proteins, which are members of the ArsR/SmtB family of metal(loid)-responsive repressors. The first identified members were the 117-residue As(III)/Sb(III)-responsive ArsR repressor of the *arsRDABC* operon of plasmid R773 and the 122-residue Zn(II)-responsive SmtB repressor. Over 200 homologs have been identified in Gram-positive and Gram-negative bacteria and in archaea. These include proteins that respond to As(III)/Sb(III) (ArsR), Pb(II)/Cd(II)/Zn(II) (CadC), Cd(II)/Pb(II) (CmtR), Zn(II) (SmtB and ZiaR), and Co(II)/Ni(II) (NmtR). The well-characterized members (and by extrapolation, probably all) are homodimers that bind the operator/promoter DNA in the absence of inducing metal ion, repressing transcription. Upon binding metal, they dissociate from the DNA, resulting in derepression.

Given the diversity of the ArsR/SmtB family, how can As(III)-responsive repressors be recognized? It is reasonable to assume that a homolog within or adjacent to an *ars* operon or to arsenic resistance genes is an ArsR repressor. Even though the arsenic-responsive repressors are all homologs, comparison of several sequences shows that As(III)-binding sites arose independently at spatially distinct locations in their structures. The best characterized is the R773 ArsR. Each ArsR subunit has a metal-binding site formed by Cys32, Cys34, and Cys37. The distance between As(III) and each of the three cysteine thiolates is 2.25 Å, as determined by extended X-ray absorption fine structure spectroscopy (EXAFS). From crystallography of small-molecule As(III)-thiol compounds, the sulfur-to-sulfur distances can be predicted to be 3.5 Å. A structural model of the R773 ArsR aporepressor (Figure 3a) was constructed on the crystal structure of the homologous CadC repressor (Figure 3c). The As(III) (inducer)-binding site is located in the first (α3) helix of the helix–loop–helix DNA-binding domain. In this model, the sulfur atoms of the three cysteines are linearly arrayed along the α3 helix, with more than 10 Å from Cys32 to Cys37, which is considerably more than the 2.25 Å from the EXAFS data. To bring the cysteine thiolates that close to each other, binding of As(III) must induce a large conformational change that breaks the helix (Figure 3b), resulting in dissociation of ArsR from the operator/promoter site and transcription of the resistance genes.

Another ArsR that has a completely different As(III)-binding site is found in an arsenic resistance operon from

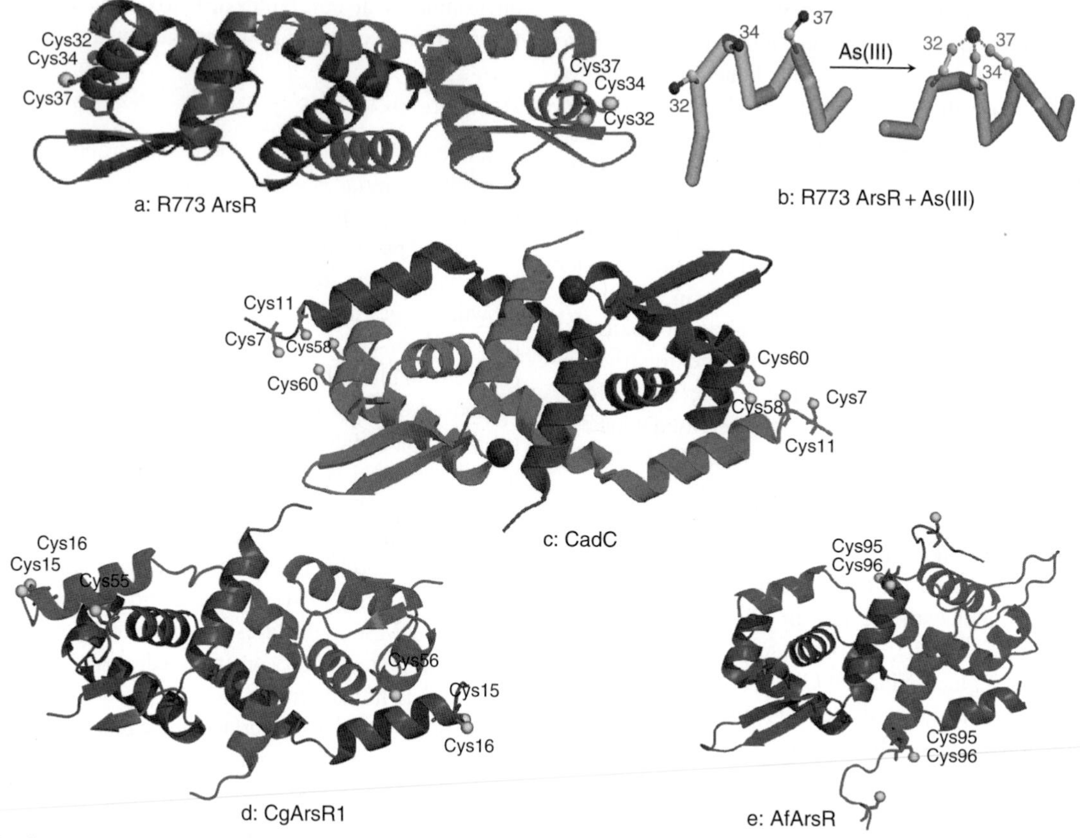

FIGURE 3 Structural models of three ArsR As(III)-responsive repressors. Shown are structures of (a) R773 ArsR, (d) *Acidithiobacillus ferrooxidans* ArsR, and (e) *Corynebacterium glutamicum* ArsR1, all of which were modeled on the 1.9-Å CadC crystal structure (e). The CadC dimer is shown as a ribbon diagram with secondary structural units N-α1-α2-α3-β1-α4-α5-β2-β3-α6-C. One As(III)- or Cd(II)-binding site in each repressor is circled. In the R773 and *A. ferrooxidans* ArsRs, the intrasubunit As(III)-binding sites are formed by three cysteine residues in each subunit. In the *C. glutamicum* ArsR1 and CadC repressors, the intersubunit metal-binding sites are composed of one or two residues from each monomer. The cysteine residues that form the As(III)- or Cd(II)-binding sites are represented as sticks with the sulfurs as yellow balls. (b) On the left is the R773 ArsR α3 helix of the DNA-binding site from (a). On the right is a model of the α3 helix after binding As(III), as deduced from EXAFS data. This suggests that derepression by As(III) binding is the result of distortion of the α3 helix, resulting in ArsR dissociation from the operator/promoter DNA.

Acidithiobacillus ferrooxidans. AfArsR lacks the R773 $C^{32}VC^{34}DLC^{37}$ sequence in the DNA-binding site. Instead, it has three cysteine residues, Cys95, Cys96, and Cys102, that are not present in the R73 ArsR. EXAFS results show that these three cysteine residues form a three-coordinate As(III)-binding site. DNA-binding studies indicate that binding of As(III) to these cysteine residues produces derepression. From a homology model built on the CadC structure, the As(III)-binding sites in AfArsR are located at the ends of antiparallel C-terminal helices in each monomer that form a dimerization domain (Figure 3d). Thus the As(III)-S_3 binding sites in AfArsR and R773 ArsR are located in spatially distinct positions in their three-dimensional structures, which implies that their mechanism of derepression by As(III) differs.

A third ArsR with yet another and completely different As(III)-binding site is encoded by an *ars* operon from *Corynebacterium glutamicum* ATCC 13032. EXAFS results show that CgArsR1 binds As(III) in a three-coordinate sulfur environment to Cys15, Cys16, and Cys55. These three cysteine residues do not correspond to As(III) ligands in either of the other ArsRs, nor do they correspond to metal-binding residues in CadC or SmtB. However, from homology modeling, CgArsR1 has the N-terminal extension characteristic of CadC repressors (Figure 3e). In CadC, the Cd(II)-binding site is composed of two cysteine residues from the N terminus of one subunit and two cysteine residues from the first (α4) helix of the DNA-binding site of the other subunit. This suggests that the CgArsR1 inducer-binding site is formed by binding As(III) between Cys15 and Cys15 of one subunit and Cys55 from the other subunit. Even though it resembles the intrasubunit Cd(II)-binding site of CadC, the location of the three cysteines in CgArsR1 is different from the four cysteines CadC, indicating that the two metal-binding sites are not derived from a common ancestor. Moreover, the As(III)-binding sites of R773 ArsR, AfArsR, and CgArsR1 likewise did not evolve from a common ancestral binding site but appear to be the result of three independent and relatively recent evolutionary events, building on the same backbone repressor protein.

Secondary arsenite efflux carrier proteins: ArsB and Acr3

Even though the most common extracellular form of arsenic under aerobic conditions is arsenate, in the highly reducing environment found inside cells, arsenite, which is the more toxic form, predominates. Thus, most detoxification mechanisms have arisen to cope with arsenite, and, most commonly, these are efflux systems. In bacteria, at least two different families of arsenite permeases have evolved, ArsB and Acr3. The first identified arsenite efflux protein, ArsB, is encoded by the R773 *arsRDABC* operon. ArsB is widespread in bacteria and archaea. It has 12 membrane-spanning segments, similar to many carrier proteins, and functions as an $As(OH)_3/H^+$ antiporter coupled to the protonmotive force. ArsB can also associate with the ArsA ATPase to form a pump that confers a higher level of arsenite resistance than ArsB alone. Thus, the Ars efflux system exhibits an unusual and unique dual mode of energy coupling that depends on its subunit composition.

Members of the Acr3 family are similarly widespread in bacteria, archaea, and fungi. Many Acr3s have been annotated as ArsBs, which confuses the literature – Acr3 and ArsB show no significant sequence similarity. The first Acr3s to be identified were a bacterial one encoded by the *Bacillus subtilis ars* operon located in the skin (*sigK* intervening) element and a fungal Acr3p encoded by the *S. cerevisiae acr123* gene cluster.

Members of the Acr3 family of arsenite permeases confer resistance to trivalent arsenic by extrusion from cells, with members in every phylogenetic domain. Recently, bacterial Acr3 homologs from *Alkaliphilus metalliredigens* and *Corynebacterium glutamicum* were characterized. The *acr3* genes from both organisms were cloned and expressed in *E. coli*. Modification of a single cysteine residue that is conserved in all analyzed Acr3 homologs resulted in loss of transport activity, indicating that it plays a role in Acr3 function. The results of treatment with thiol reagents suggested that the conserved cysteine is located in a hydrophobic region of the permease. A scanning cysteine accessibility method was used to show that Acr3 has 10 transmembrane segments, and the conserved cysteine would be predicted to be in the fourth transmembrane segment.

Arsenite efflux pumps: ArsAB ATPases and multidrug resistance-associated proteins (MRP)

ArsAB ATPases

Cells of *E. coli* expressing the chromosomally encoded ArsB are resistant to moderate levels of arsenite, while cells expressing the R773 *arsRDABC* operon produce the ArsAB ATP-coupled efflux pump and are resistant to much higher levels of arsenite. The different levels of resistance are due to the fact that ArsB is a secondary transporter that uses the protonmotive force, while the ArsAB ATPase uses ATP to drive active transport of As(III) against much higher gradients than ArsB alone. The ArsA ATPase has been extensively studied at the biochemical and structural levels (Figure 4).

MRPs

ABC transporters are ubiquitous in nature. The MRP subfamily catalyzes ATP-coupled extrusion of drug conjugates and other solutes. *S. cerevisiae* Ycf1p (yeast cadmium factor) is a close homolog of the human MRP1 and catalyzes the vacuolar sequestration glutathione (GSH) conjugates of Cd(II) or As(III). Thus far, bacterial MRP homologs that confer arsenic resistance have not been identified.

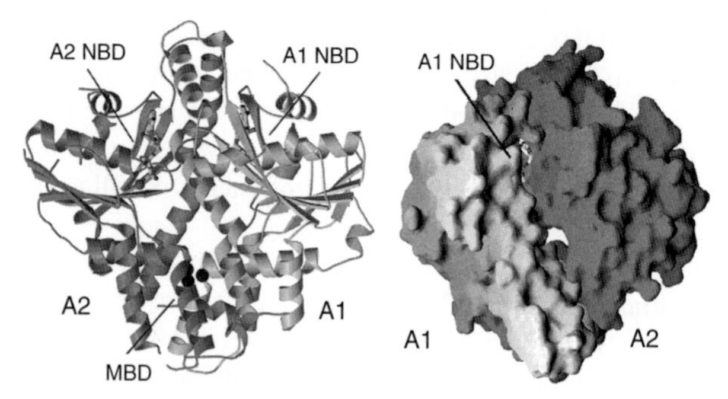

FIGURE 4 Structure of R773 ArsA ATPase. Left: The overall structure of ArsA is shown as a ribbon diagram. Mg^{2+}ADP is bound to each of the two NBDs in the A1 and A2 halves of ArsA, while three Sb(III) is bound at the single MBD. Right: A view of the molecular surface of ArsA showing the relative positions of the A1 and A2 halves and details of ADP bound in the NBD1.

ArsD: an As(iii) chaperone

Metallochaperones are proteins found in all kingdoms that buffer cytosolic metals and deliver them to protein targets such as metalloenzymes and extrusion pumps. For example, the *S. cerevisiae* chaperone Atx1 transfers copper to Ccc2p, a trans-Golgi Cu(I)–ATPase that is required for incorporation of copper into the multicopper oxidase (MCO) Fet3p. Until recently, no arsenic chaperones had been identified. However, those *ars* operons that contain both *arsD* and *arsA*, such as the *arsRDABC* operon of plasmid R773, always have the two genes adjacent to each other, suggesting that they might have interrelated functions. Recently, the product of the *arsD* gene, a 120-residue polypeptide that is a functional homodimer, was shown to be a chaperone that delivers As(III) to the ArsAB As(III)-translocating ATPase. The presence of ArsD, in addition to the ArsAB pump, gives cells a clear growth advantage over cells with only the ArsAB pump (such as in an *arsD* deletion). Expression of ArsD increases the ability of the ArsAB pump to extrude As(III). ArsD and ArsA physically interact with the transfer of As(III) from the chaperone to the pump. In doing so, the affinity of ArsA for As(III) is increased several orders of magnitude. In the absence of ArsD, the ArsAB pump gives resistance to concentrations of arsenite found only in the highly contaminated parts of the world, but ArsD makes the ArsAB pump more effective at much lower concentrations of arsenic such as those found more widely environmentally.

ArsD is a metallochaperone that delivers trivalent metalloids (As(III) or Sb(III)) to the ArsA ATPase, the catalytic subunit of the ArsAB pump encoded by the *arsRDABC* operon of the *E. coli* plasmid R773. Interaction with ArsD increases the affinity of ArsA for As(III), conferring resistance to environmental concentrations of arsenic. Previous genetic analysis suggested that ArsD residues Cys12, Cys13, and Cys18 are involved in As(III) transfer to ArsA. A recent study characterized the biochemical properties of ArsD (9). X-ray absorption spectroscopy was used to show that As(III) is coordinated with three sulfur atoms, consistent with the three cysteine residues forming the As(III) binding site. Two single tryptophan derivatives of ArsD exhibited quenching of intrinsic protein fluorescence upon binding of As(III) or Sb(III), which allowed estimation of the rates of binding and affinities for metalloid. Substitution of Cys12, Cys13, or Cys18 decreased the affinity for As(III) more than 10-fold. Reduced GSH greatly increased the rate of binding As(III) to ArsD but did not affect binding of As(III) to ArsA, indicating that *in vivo* cytosolic As(III) might be initially bound to GSH, transferred to ArsD, and then to the ArsAB, which pumps the metalloid out of the cell.

Arsenate reductases: ArsCS and Acr2s

In contrast to the periplasmic respiratory arsenate reductases, cytosolic arsenate reductases are usually (but not always) encoded by *ars* operons or in fungi in arsenic gene clusters, and confer resistance to arsenate. There are three independent and unrelated families of arsenate reductases. One family of arsenate reductase that includes the chromosomal *E. coli* and plasmid R773 ArsCs uses glutaredoxin (Grx) and GSH as reductants. The crystal structure of the enzyme has been determined with and without bound substrates and products (Figure 5a). R773 ArsC is related to Spx of *B. subtilis*, a transcriptional repressor that interacts with the C-terminal domain of RNA polymerase (RNAP) α-subunit and is essential for growth under disulfide stress. The second family includes the *S. aureus* plasmid pI258 ArsC and the *B. subtilis* chromosomal ArsC, both of which use thioredoxin (Trx) as reductant. (Again, like ArsB, unrelated arsenic resistance proteins have unfortunately been given the same name.) The structure of this

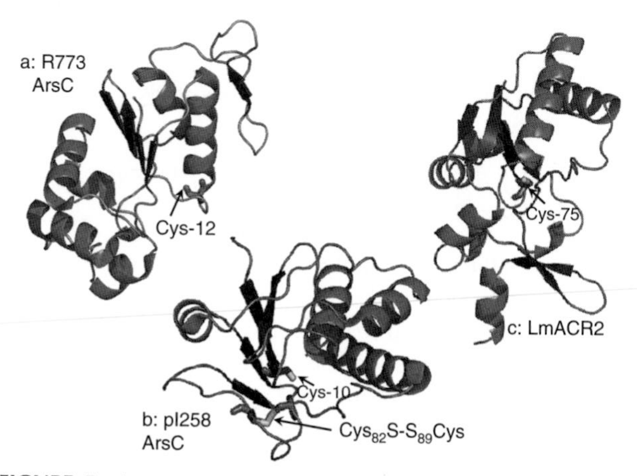

FIGURE 5 Arsenate reductase structures. The structure of R773 and pI258 ArsC are shown identifying their secondary structural elements. The catalytic cysteines at the active sites of either enzymes complexed with As(V) are also indicated. The *Leishmania* LmACR2 has a similar active site as the human Cdc25a.

ArsC has been determined in various conformations that describe the catalytic cycle (Figure 5b). This bacterial family of arsenate reductases is related to low molecular weight protein tyrosine phosphatases (LMW PTPase). The third family of arsenate reductases is found primarily in eukaryotic microorganisms and includes *S. cerevisiae* Acr2p and LmACR2 from the protozoan *L. major*. The structure of the *Leishmania* enzyme has been solved recently (Figure 5c). These eukaryotic arsenate reductases are related to the catalytic domain of the Cdc25 cell-cycle protein tyrosine phosphatase. From a comparison of the three structures, which show no similarity to each other, it is clear that these arsenate reductases arose by convergent evolution.

Arsenite methylases: ArsM

Members of every kingdom have the ability to methylate arsenic. The series of reactions, called the "Challenger Pathway," alternate reduction of pentavalent to trivalent arsenicals with oxidative methylations (Figure 6). The pentavalent species are monomethylarsenate (MMA(V)), dimethylarsenate (DMA(V)), and trimethylarsine oxide (TMAO(V)) and the trivalent are MMA(III), DMA(III), and TMA(III). Humans and some other mammals methylate inorganic arsenic and excrete methylated species such as DMA(V) and, to a lesser extent, monomethylated MMA(V) in the urine, leading to the proposal that methylation is a detoxification process. However, the trivalent intermediates in the pathway, MAs(III), DMAs(III), and TMAs(III), are considerably more toxic than inorganic arsenate or arsenite, suggesting that methylation may activate inorganic arsenic to more toxic metabolites. Whether arsenic methylation is a detoxification processes in mammals is an unresolved question.

In contrast, it is clear that bacterial methylation detoxifies arsenic. Bacteria and fungi produce volatile and toxic arsines, but, until recently, the biochemical basis and physiological role of arsenic methylation in microorganisms were not known. There are a large number of genes homologous of the human arsenic methylase in bacteria and archaea. Those that are controlled by an *arsR* gene have been termed *arsM* and their protein product as ArsM (arsenite S-adenosylmethyltransferase). The *arsM* gene (accession number NP_948900.1) from the soil bacterium *Rhodopseudomonas palustris* for the 283-residue ArsM (29,656 Da) was cloned and expressed in an arsenic hypersensitive strain of *E. coli*. Expression of *arsM* conferred As(III) resistance in *E. coli* in the absence of any other *ars* genes, demonstrating that methylation is sufficient to detoxify arsenic. ArsM converted As(III) into DMA(V), TMAO(V), and TMA(III). While TMA(III) is more toxic than arsenite, it is a gas and volatilizes as it is being formed so that it does not accumulate in the cells or media. Since homologs of ArsM are widespread in every kingdom, microbial-mediated transformation may have a significant impact on the global arsenic cycle.

Bacteria have been shown to detoxify arsenic by methylation using an As(III)-S-adenosylmethionine methyltransferase enzyme, ArsM. Mammals also methylate arsenic,

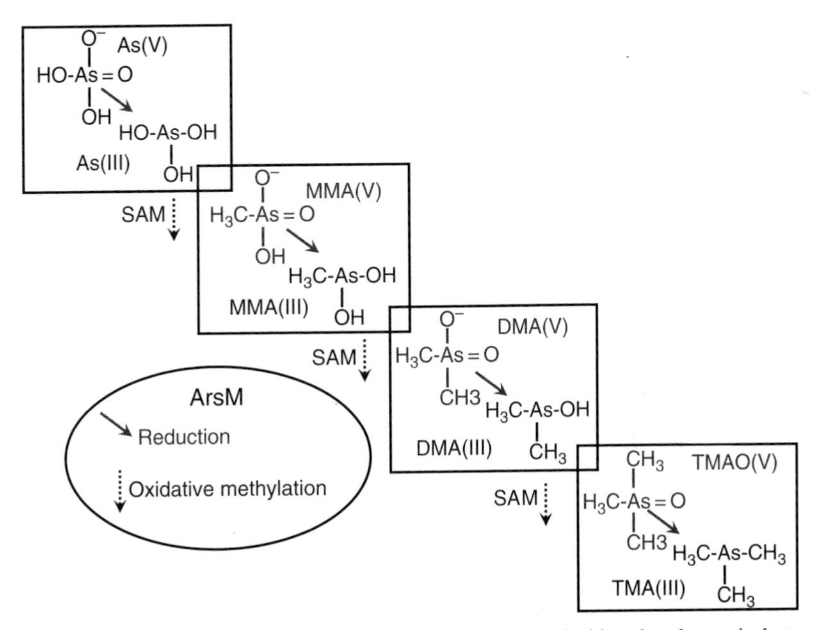

FIGURE 6 The Challenger Pathway of arsenic methylation. In each step a pentavalent arsenical is reduced to a trivalent arsenical (red), which is then oxidatively methylated with *S*-adenosylmethionine to form the pentavalent form (blue). The overall scheme involves four reductive steps and three methylations to form the gas TMA(III) from inorganic As(V). *Reproduced from Bhattacharjee, H., & Rosen, B. P. (2007). Microbial arsenic metabolism. In D. H. Nies, & S. Silver (Eds.)* How bacteria handle heavy metals *(pp. 371–406). Heidelberg: Springer.* (For interpretation of the references to color in this figure legend, the reader is referred to the Web version of this chapter.)

but it is not clear whether that is a detoxifying mechanism. Geothermal environments are known for their elevated arsenic content and thus provide an excellent setting in which to study microbial redox transformations of arsenic. Recently, an extremophilic eukaryotic alga of the order *Cyanidiales* was shown to detoxify arsenic by conversion to the gas trimethylarsine (TMAs(III)) at elevated temperatures. *Cyanidioschyzon* sp. isolate 5508 oxidized arsenite [As(III)] to arsenate [As(V)], reduced As(V) to As(III), and methylated As(III) to form trimethylarsine oxide (TMAO) and dimethylarsenate [DMAs(V)]. Two arsenic methyltransferase genes, CmarsM7 and CmarsM8, were cloned from this organism and demonstrated to confer resistance to As(III) in an arsenite hypersensitive strain of *E. coli*. The two recombinant CmArsMs were purified and shown to transform As(III) into monomethylarsenite (MMAs(III)), DMAs(V), TMAO, and (TMAs(III)) gas at 60–70 °C. These studies illustrate the importance of eukaryotic microorganisms to the biogeochemical cycling of arsenic in geothermal systems, and offer a molecular explanation for how algae tolerate arsenic in their environment.

An NADPH–FMN reductase: ArsH

ArsH is widely distributed in bacteria and found sparsely in fungi, plants, and archaea. From its sequence, it shows conserved domains related to the NADPH (nicotinamide adenine dinucleotide phosphate) dependent flavin mononucleotide (FMN) reductase class of proteins. ArsH appears to confer resistance to arsenicals, although this is controversial, and the mechanism is not clear. ArsH from *Yersinia enterocolitica* or from the IncH12 plasmid R478 appears to confer resistance to both arsenite and arsenate. However, *A. ferrooxidans* and *Synechocystis* ArsHs do not seem to participate in arsenic resistance. The chromosomal *ars* operon of the legume symbiont *Sinorhizobium meliloti* Rm1021 displays a cluster of four genes: *arsR*, *aqpS*, *arsC*, and *arsH*. Inactivation of the *S. meliloti arsH* gene results in increased As(III) sensitivity. Overexpression of ArsH in Δ*ars* strain of *S. meliloti* confers resistance to trivalent arsenicals, showing that ArsH does not need the other *ars* gene products for resistance, indicating that it has a novel mechanism that does not require AqpS and ArsC activities. The protein has been purified and shown to catalyze azo dye reduction and H_2O_2 formation. Its crystal structure has been solved, showing it to be a tetramer with an FMN-binding site in each subunit (Figure 7).

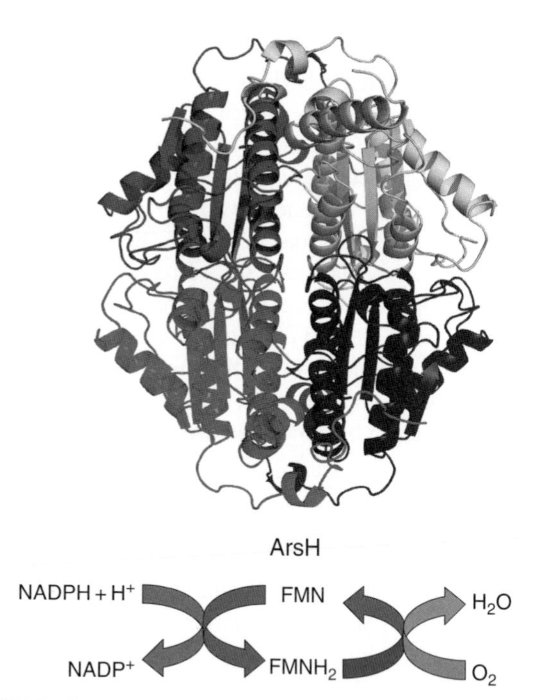

FIGURE 7 Structure of the ArsH NADPH–FMN oxidoreductase. On the top is shown the 1.8-Å crystal structure of the ArsH tetramer. The monomers A, B, C, and D are colored red, green, blue, and yellow, respectively. At the bottom is shown the NADPH–FMN oxidoreductase reaction that ArsH catalyzes with the generation of H_2O_2. (For interpretation of the references to color in this figure legend, the reader is referred to the Web version of this chapter.)

^{29}CU$_{63}$

Introduction

Copper speciation and availability in the environment is critical for a number of biogeochemical cycles exemplified here by nitrogen. Key steps in nitrogen cycling are redox reactions carried out by metalloenzymes of denitrifiers, nitrifiers, and nitrogen fixers. It has been shown that denitrifying bacteria are affected by copper limitation because copper is required in nitrous oxide reductase and some nitrite reductases (*nirK*). This is of importance since nitrous oxide (N_2O) is a major greenhouse gas contributing to global warming. For example, the respiratory decomposition of organic matter in marine and freshwater environments leads to an almost complete exhaustion of available oxygen, stimulating the use of alternative terminal electron acceptors. Under oxygen-limiting conditions, denitrifying bacteria can thus reduce nitrate (NO_3^-), in a stepwise fashion, to nitrite (NO_2^-), nitric oxide (NO), nitrous oxide (N_2O), and dinitrogen (N_2). Most significantly, denitrification is the largest sink of oceanic fixed nitrogen and is estimated to account for up to 67% of total global denitrification. It is therefore not surprising that many genomes of representative recently sequenced marine microorganisms contain genes involved in denitrifications. Genes for nitrous oxide reduction have been found in diverse marine bacteria including *Roseobacter denitrificans*, *Marinobacter aquaeolei*, *Silicibacter pomeroyi*, and *Stappia aggregata*, to name a few. In addition, agricultural land contributes significantly to the net increase in atmospheric N_2O through denitrification and nitrification. Several rhizobia such as *Bradyrhizobium japonicum* USDA110 and *S. meliloti* are capable of denitrification including the last biochemical step mediated by N_2O

reductase (N_2OR). Recently, it has been shown that soybean roots nodulated with *B. japonicum* carrying the *nos* genes are able to remove very low concentrations of N_2O. However, several groups have reported that cultivation of legume crops often enhances N_2O emissions from fields of alfalfa, soybean, white clover, and Bengal gram. We believe this apparent paradox can in part be explained by lack of copper bioavailability in these soils and therefore the inability to assemble a catalytically active N_2OR enzyme.

Microorganisms can actively reduce and oxidize copper. Oxidation of copper has not been coupled to energy generation but, in all cases studied, is a copper resistance mechanism. Likewise, reduction of oxidized copper minerals, while thermodynamically feasible, as terminal electron acceptor has not been observed. In contrast, reduction of cupric ion to cuprous ion as a prerequisite for copper uptake is widespread.

Oxidation of Cuprous Copper by MCOs

MCOs couple the one-electron oxidation of substrate(s) to full reduction of molecular oxygen to water by employing a functional unit formed by three types of copper-binding sites with different spectroscopic and functional properties. Type 1 blue copper (T1) is the primary electron acceptor from the substrate, while a trinuclear cluster formed by type 2 copper and binuclear type 3 copper (T2/T3) is the oxygen-binding and oxygen-reduction site. Prominent bacterial enzymes involved in metal transformations include CueO involved in copper resistance by oxidizing Cu(I) to Cu(II) and CumA from *Pseudomonas putida* MnB1 and GB-1 responsible for manganese oxidation of Mn(II) to Mn(III)/(IV).

CueO, a bacterial MCO, protects periplasmic enzymes from copper-induced damage and possesses laccase-like activity. CueO has antioxidant activity by inhibiting the Cu(I)-induced Fenton reaction due to removal of the primary reactants to drive the Fenton reaction (Equation 1) and hydroperoxide-dependent lipid peroxidation (Equation 2).

$$Cu(I) + H_2O_2 \rightarrow Cu(II) + OH + OH^-, \quad (1)$$

$$Cu(I) + LOOH \rightarrow Cu(II) + LO + OH^- \quad (2)$$

Cu(I) oxidation could be shown not only for CueO but also for Fet3 from yeast. In addition, the oxidation of catecholate siderophores such as enterobactin by CueO was shown to protect cells by preventing reduction of Cu(II) to Cu(I) and subsequent oxidative damage. The activity of CueO is regulated by copper. Deletion of an extra α-helical region near the type 1 copper site converts CueO into a laccase with better access for bulkier substrates. As described later, CumA and other Mn-oxidizing MCOs might also oxidize siderophores.

CueO is a 53-kDa periplasmic protein that confers copper tolerance in *E. coli* under aerobic conditions. CueO is capable of oxidizing a wide variety of substrates, including 2,6-dimethylphenol, enterobactin (a catecholate siderophore found in *E. coli*), and Fe(II). Intriguingly, very little oxidase activity occurs in the absence of additional copper in solution. Cu(I) is an excellent CueO substrate, with higher activity than that described for the homologs Fet3p and ceruloplasmin, suggesting this is the *in vivo* target of the protein.

The crystal structure of CueO was obtained at 1.4 Å resolution, which is by far the highest resolution achieved for this structural family and sufficient for defining the structures of trapped reaction intermediates (Figure 8). The CueO structure resembles those of laccase, ascorbate oxidase and ceruloplasmin, although the oligomeric state and addition of extra protein modules differ among these proteins. The CueO fold consists of three azurin-like copper-binding domains connected by linker peptides. The trinuclear copper center (TNC, T2 + T3 coppers) lies between domains 1 and 3, while the T1 copper occupies the azurin-like position in domain 3.

Two aspects of CueO differ from other MCOs. First, CueO requires a fifth copper adjacent to the T1 site for activity (labeled rCu in Figure 8). This copper atom is labile and so copper concentration regulates activity. Our working hypothesis is that the fifth copper-binding site is for binding Cu(I) as substrate and Cu(II) for oxidation of other substrates such as enterobactin. Second, CueO displays a methionine-rich helix that lies over the T1 site and may function by sequestering additional Cu(I) atoms. Methionine-rich regions are found in numerous proteins involved in copper homeostasis, leading to the suggestion that such regions are involved in copper ligation; however, the functional role of such sites has not been resolved for any of these proteins. There is preliminary data demonstrating that at least two additional Cu(I) atoms can bind to the methionine-rich site.

Identification of reaction intermediates in MCOs has been hampered by low resolution and insufficient spectroscopic analyses of crystalline complexes. Two MCO structures have been published, containing unusual electron density at or near the trinuclear copper centers. A structure of laccase at 2.4 Å resolution from *Melanocarpus albomyces* was interpreted as having a dioxygen molecule bridging the T3 copper atoms in the trinuclear center. A structure of the endospore coat laccase from *B. subtilis* at 2.45 Å resolution contained unexplained electron density between two histidine ligands of the T3 copper atoms, but not bound to the copper atoms, which was interpreted as an oxygen molecule. Although intriguing, the resolutions of both structures are modest and spectroscopic analyses of the complexes were not undertaken, making identification of the unexplained electron density difficult. Better resolution has been achieved in other non-MCO copper proteins. Side-on Cu–NO coordination was reported for nitrite reductase in a 1.3-Å structure with corresponding electron paramagnetic resonance (EPR) data supporting this interpretation. End-on dioxygen binding

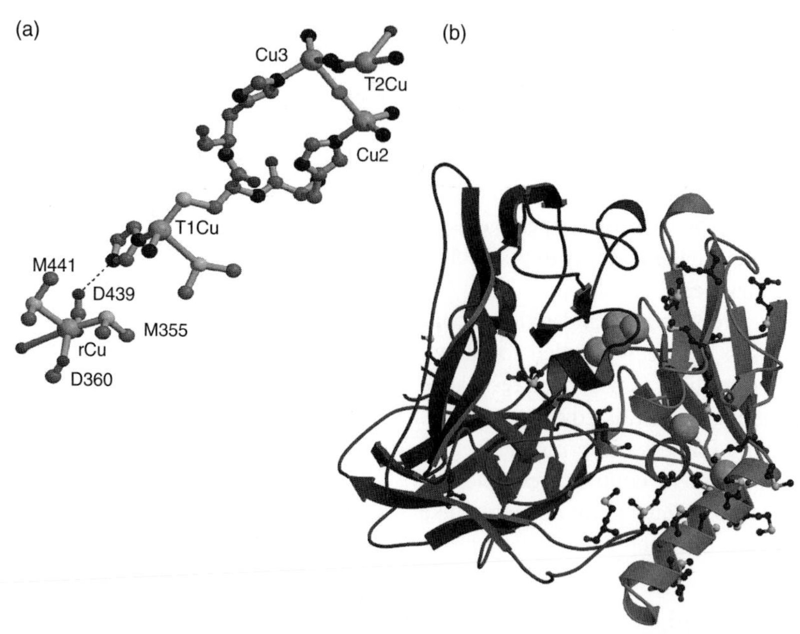

FIGURE 8 CueO structure. (a) Five-copper electron transfer path. (b) Overall fold with copper atoms is shown as blue spheres and methionines as ball and stick. The methionine-rich helix is in the lower right corner.

to copper has also been reported for the enzyme peptidylgly cine-alphahydroxylating monooxygenase (PHM); however, even at 1.85 Å resolution this interpretation is not unambiguous and spectroscopic data were not reported.

The possible physiological roles of Cu(I) oxidizing MCOs such as the well-studied CueO from *E. coli* and related enzymes have been broadened. CueO from *Salmonella enterica serovar typhimurium* has recently been implicated in playing a role in virulence. It had also been observed that a combination of copper and silver had an additive antimicrobial effect on many bacteria. Results from the labs of Montfort and of ours suggest that this is due to silver inhibition of CueO even at very low silver concentrations (personal communication). This could explain the existence of additional periplasmic resistence sytems in many microorganisms. There were also more detailed biochemical studies of CueO establishing CueO as a model MCO with significant unique characteristics.

Reduction and Acquisition of Cupric Copper

Copper minerals containing cupric copper could potentially function as terminal electron acceptor and be reduced to Cu(I) or Cu(0). However, this has not been reported yet. In contrast, Cu(II) has to be reduced in order to be taken up as Cu(I). Yeast is probably the best studied microorganism regarding copper uptake. Copper uptake in *Saccharomyces cerevisiae* is mediated by separate high- and low-affinity systems. High-affinity copper uptake requires plasma membrane reductases to reduce Cu(II) to Cu(I). This reduction is mediated by the same plasma membrane reductases, Fre1 and Fre2, that are involved in iron uptake. Once reduced to Cu(I), the ion is taken up by separate high-affinity transporter proteins encoded by the *CTR1* and *CTR3* genes. High-affinity copper uptake is energy-dependent and specific for Cu(I) over other metals. A third, lower affinity system for copper uptake has also been detected in yeast, but neither the biochemical properties nor the gene(s) responsible for this activity is known.

$^{25}MN_{55}$

Introduction

In a variety of environments, manganese is cycled between the main oxidation states, namely, Mn(II), Mn(IV), and also Mn(III). These states are capable of influencing chemical gradients in environments such as the coastal ocean. Mn oxides have been found to be strong environmental oxidants able to influence the speciation of other redox-sensitive metals and metalloids such as Cu and As. Manganese can perform these roles because bacterial processes enable rapid cycling. It is therefore not surprising that much work has been done to understand the molecular mechanisms of manganese reduction and oxidation. Redox-active proteins involved in electron transfer to and from Mn(hydr)oxides are localized to the outer membrane in Gram-negative bacteria and include c-type cytochromes (c-Cyts) and MCOs. In addition, Mn-oxidizing activity has also been found on the spore surface of Gram-positive bacteria. In this article, we intend to understand the mechanisms and enzymes involved in these transformations. These processes are still not well understood. Other aspects such as mineralization have been extensively covered in recent reviews.

Oxidation of Mn(ii) to Mn(iii, iv)

The capability to oxidize Mn(II) to Mn(III, IV) has been detected in bacteria and fungi. In bacteria, an MCO is almost always an essential part of Mn(II) oxidation. *Bacillus* SG-1, *Pedomicrobium* ACM3067, *Leptothrix discophora*, and *P. putida* all have MCO genes, which when disrupted the organisms no longer oxidize Mn(II). However, as described in more detail below, the MCO often is part of a complex and the activity of the bacterial MCO alone is not sufficient for Mn(II) oxidation. In the systems that have been well studied using genetics, Mn(II) oxidation seems to require an MCO, but only in the case of *Bacillus* SG-1 has the connection between the MCO gene and the enzyme activity been shown. Here, the MCO MnxG is thought to oxidize Mn(II) → Mn(III) → Mn(IV) in sequential one-electron steps. In other microbes such as *P. putida* MnB1 and GB-1, it might work a bit differently. The MCO CumA is also required for Mn oxidation to take place but other proteins are also involved. In addition, Mn(II) is oxidized to Mn(III) spontaneously only in the presence of siderophores and probably stoichiometrically 1:1 Mn(II):siderophore. In the presence of sufficient Fe and therefore the absence of siderophores, Mn(II) is rapidly oxidized to Mn(IV), probably via a Mn(III) intermediate. At limiting but not depleted Fe, the reaction slows since the Mn(III):PVD complex can accumulate. Since Mn oxides also form at this time, the Mn(III):PVD complex could be oxidized by the MnOx or by another enzyme that acts either on the Mn(III) or by oxidizing the PVD and releasing Mn(III), which would be disproportionate to Mn(II) + Mn(IV). MCOs have been shown to oxidize catecholate siderophores such as enterobactin. Pyoverdin is a mixed catecholate and hydroxamate siderophore, so this reaction is possible. It is not known whether PVD or some other molecule is also acting as a carrier. Briefly, it appears that the reason behind this might be that the reaction occurs as a sequence of two enzymatically mediated one-electron transfer reactions.

In *Erythrobacter* SD-21, the enzyme oxidizes Mn(II) → Mn(III). At this point, it is not known whether the organism also produces a siderophore that can oxidize Mn(II). However, a recent report suggests in *E.* SD-21, a copper-dependent MCO did not stimulate Mn(II) oxidation activity but rather by pyrroloquinoline (PQQ). Even more surprising is that PQQ could rescue a non-Mn(II)-oxidizing *P. putida* MnB1 insertional mutant of the anthranilate synthase gene. How these proteins and molecules act together to oxidize Mn(II) is not known, but one could envision an electron pathway from the respiratory chain in cytoplasmic membrane proteins such as MCOs located at the outer membrane.

In fungi, MCOs such as laccases often have Mn(II) to Mn(III) oxidation activity. In addition, fungi can also oxidize Mn(II) to Mn(III) by heme-containing Mn peroxidase. However, a recent paper describes the complete oxidation of Mn(II) to Mn(IV).

Mn-oxidizing activity and the genes responsible are widespread among different bacterial groups, indicating these genes must have an important function. At this point this is unknown, but rather it appears that this metabolic activity seems to be accidental since there have been no reports of Mn-dependent induction of any of the genes thought to be responsible for Mn(II) oxidation. However, this "accidental" activity has significant environmental impact. For example, the Mn oxides present in desert varnish are largely catalyzed by bacterial activity.

A careful reexamination revealed that the MCO CumA is not responsible for Mn oxidation in *Pseudomonas putida* GB-1 and probably also other Gram-negative bacteria. Although a two-component system was shown to be required for Mn-ox in the same organisms, the actual mechanism of Mn oxidation in bacteria remains elusive. This might be due to the fact that it is entirely adventitious.

Reduction of Mn(iv)

In anaerobic environments, insoluble minerals such as Mn(IV) can be used as a terminal electron acceptor by a number of bacteria and archaea. Dissimilatory reduction of Mn(III, IV) usually requires multiheme c-Cyts often located on the outer membrane facing the extracellular space. However, the molecular mechanisms are completely unknown. Aspects about the molecular details of this process have been extensively studied in *Shewanella* and *Geobacter* species. Since Mn(IV) is insoluble, the actual process is a mineral–microbe interaction. There are a number of possibilities on how electrons can be transferred from the microbe to the mineral (Figure 9).

- One or more proteins located at the cell surface directly interact with the mineral and transfers electrons, usually originating from the electron transport chain, to the substrate. This has been observed with outer membrane multiheme c-Cyts such as OmcA from *S. oneidensis*.
- A small soluble electron shuttle such as phenazine picks up electrons from the cell and delivers them to the substrate. The electrons could be transferred at the inner or outer membrane.
- Electrically conductive pili (geopili) could directly transfer electrons from the cells to the substrate.
- Substrates are bound by a ligand such as a chelator and brought to the cell for reduction.

It is clear that at this point many of the proposed mechanisms have not been proven. For example, a MCO OmpB was shown to be involved in anaerobic iron (and probably Mn) reduction in *Geobacter sulfurreducens*. What role this enzyme plays is unclear since MCOs usually require oxygen to oxidize their substrates.

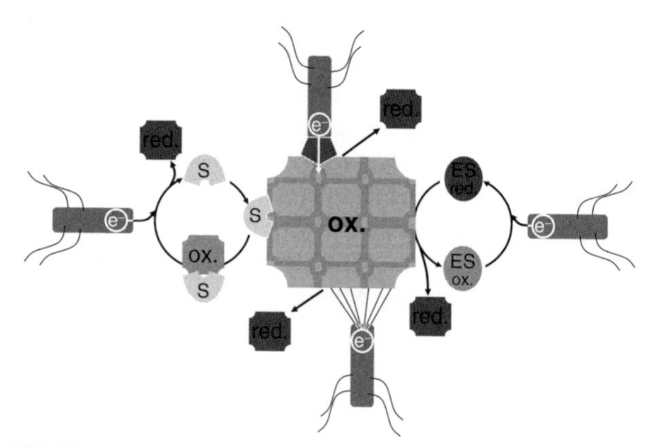

FIGURE 9 Four potential strategies for extracellular Mn respiration. The substrate is shown as a solid mineral in the center (e.g., a Mn (hydr)oxide), and its reduced product (e.g., Mn(II)) is shown in red. These strategies can apply to other substrates, and the flow of electrons can also be reversed. From the top clockwise:

- A protein that resides on the cell surface (blue) directly interacts with the extracellular substrate to transfer electrons from the cell to the substrate.
- An electron shuttle interacts with the substrate outside the cell; the substrate remains at a distance from the cell, with the shuttle catalyzing transfer of electrons between the cell and the substrate. The shuttle, in principle, could be reduced on the outside or inside of the cell.
- A cellular appendage such as an electrically conductive pilus forms a bridge between the cell and the substrate, catalyzing electron transfer.
- A chelator binds the substrate and transports the complex to the cell for reduction (either on the outside or on the inside); in the case of an electron shuttle (closed green oval, oxidized state; open green oval, reduced state). (For interpretation of the references to color in this figure legend, the reader is referred to the Web version of this chapter.)

$^{34}SE_{80}$

Introduction

Microbial transformations including oxidation, reduction, methylation, and demethylation reactions affect the solubility, toxicity, sorption, volatility, and specific gravity of metals and metalloids. Bacteria are responsible for many transformations of selenium. Some of the genes and enzymes involved have been identified and characterized. However, at this point it is still not clear whether these reactions represent active responses to the presence of selenium or are more or less accidental. For example, methylated species are volatile and the presence of genes encoding selenium methylases makes cells more resistant. If these genes could be induced by the presence of Se, they would constitute an active resistance mechanism. However, this is not the case. Other selenium transformations have not been studied in great detail and the genes involved in selenium oxidation and demethylation have not been identified. In contrast, much has been learned about selenium incorporation into proteins via selenocysteine. Better tools to analyze genomic databases for the presence of selenoproteins have yielded an ever-growing diversity, most of them being redox proteins containing catalytic selenocysteines.

Reduction of Selenate

Microorganisms are largely responsible for the reduction of soluble selenate (Se(VI), SeO_4^{2-}) and selenite (Se(IV), SeO_3^{2-}) to the insoluble mineral form Se(0) in most environments. The *serABDC* operon was shown to encode the genes responsible for dissimilatory selenium respiration in the anaerobic bacterium *Thauera selenatis*.

The *serABDC* operon is closely related to operons involved in chlorate reduction, dimethyl sulfide (DMS) oxidation to DMSO, and possibly nitrate reduction. The homology *serABDC* to *chrABDC* is quite high and both SerA and ChrA are capable of chlorate and selenate reduction albeit with differing substrate affinities. In addition, they are probably both flanked by a transposase or other elements indicating mobility. No regulatory elements have been identified.

The catalytic subunit SerA is predicted to be a periplasmic molybdenum-containing enzyme belonging to the DMSO reductase family. SerB is a Fe–S protein containing four cysteine residues that might bind 4[Fe–S] clusters. SerD is of unknown function but might be a molecular chaperone aiding protein maturation. Finally, SerC is thought to encode a b-type cytochrome located in the periplasm in analogy to DdhC from *Rhodovulum sulfidophilum*.

Methylation of Selenite (and Selenocysteine) to Dimethyl Selenide and Dimethyl Diselenide

Dimethyl selenide and dimethyl diselenide are the prevalent volatiles of bacterial methylation of selenium. The enzymes involved can all methylate selenite and selenocysteine to dimethyl selenide and dimethyl diselenide, but some such as the trimethyl purine methylase (Tmp) from *Pseudomonas syringae* can also methylate selenate. Two very different selenium methyltransferases could be isolated from a freshwater *Pseudomonas* strain. While the first is a bacterial thiopurine methyltransferase, the other, MmtA, is a homolog of calichaemicin methyltransferase with homologs in many bacterial species.

ACKNOWLEDGMENT

This work was supported by United States Public Health Service Grants AI43428, GM52216, and GM55425 (to BPR) and United States Public Health Service Grant GM079192 (to CR).

FURTHER READING

Achard, M. E., Tree, J. J., Holden, J. A., Simpfendorfer, K. R., Wijburg, O. L., Strugnell, R. A. *et al.* The multi-copper-ion oxidase CueO of *Salmonella enterica serovar Typhimurium* is required for systemic virulence. *Infection and Immunity*, *78*, 2312–2319.

Bhattacharjee, H., & Rosen, B. P. (2007). Arsenic metabolism in prokaryotic and eukaryotic microbes. In D. H. Nies, & S. Simon (Eds.) *Molecular microbiology of heavy metals* (Vol. 6). (pp. 371–406). Heidelberg/New York: Springer-Verlag.

Djoko, K. Y., Chong, L. X., Wedd, A. G., & Xiao, Z. (2010). Reaction mechanisms of the multicopper oxidase CueO from Escherichia coli support its functional role as a cuprous oxidase. *Journal of the American Chemical Society*, *132*, 2005–2015.

Fu, H. L., Meng, Y., Ordonez, E., Villadangos, A. F., Bhattacharjee, H., Gill, J. A. *et al.* (2009). Properties of arsenite efflux permeases (Acr3) from *Alkaliphilus metalliredigens* and *Corynebacterium glutamicum*. *Journal of Biological Chemistry*, *284*, 19887–19895.

Geszvain, K., & Tebo, B. M. Identification of a two-component regulatory pathway essential for Mn(II) oxidation in Pseudomonas putida GB-1. *Applied and Environmental Microbiology, 76*, 1224–1231.

Gralnick, J. A., & Newman, D. K. (2007). Extracellular respiration. *Molecular Microbiology*, *65*, 1–11.

Jiang, X., McDermott, J. R., Abdul Ajees, A., Rosen, B. P., & Liu, Z. (2010). Trivalent arsenicals and glucose use different translocation pathways in mammalian GLUT1. *Metallomics*, *2*, 211–219.

Kataoka, K., Sugiyama, R., Hirota, S., Inoue, M., Urata, K., Minagawa, Y. *et al.* (2009). Four-electron reduction of dioxygen by a multicopper oxidase, CueO, and roles of Asp112 and Glu506 located adjacent to the trinuclear copper center. *Journal of Biological Chemistry*, *284*, 14405–14413.

Kim, E. H., Rensing, C., & McEvoy, M. M. Chaperone-mediated copper handling in the periplasm. *Natural Product Reports*, *27*, 711–719.

Messens, J., & Silver, S. (2006). Arsenate reduction: Thiol cascade chemistry with convergent evolution. *Journal of Molecular Biology*, *362*, 1–17.

Mukhopadhyay, R., & Rosen, B. P. (2002). Arsenate reductases in prokaryotes and eukaryotes. *Environmental Health Perspectives*, *110*, (supplement 5), 745–748.

Mukhopadhyay, R., Rosen, B. P., Phung, L. T., & Silver, S. (2002). Microbial arsenic: From geocycles to genes and enzymes. *FEMS Microbiology Reviews*, *26*, 311–325.

Qin, J., Lehr, C. R., Yuan, C., Le, X. C., McDermott, T. R., & Rosen, B. P. (2009). Biotransformation of arsenic by a Yellowstone thermoacidophilic eukaryotic alga. *Proceedings of the National Academy of Sciences USA*, *106*, 5213–5217.

Rensing, C., & Grass, G. (2003). *Escherichia coli* mechanisms of copper homeostasis in a changing environment. *FEMS Microbiology Reviews*, *27*, 197–213.

Shi, L., Squier, T. C., Zachara, J. M., & Fredrickson, J. K. (2007). Respiration of metal (hydr)oxides by Shewanella and Geobacter: A key role for multiheme c-type cytochromes. *Molecular Microbiology*, *65*, 12–20.

Stolz, J. F., Basu, P., Santini, J. M., & Oremland, R. S. (2006). Arsenic and selenium in microbial metabolism. *Annual Review of Microbiology*, *60*, 107–130.

Tebo, B. M., Bargar, J. R., Clement, B. G. *et al.* (2004). Biogenic manganese oxides: Properties and mechanisms of formation. *Annual Review of Earth and Planetary Sciences*, *32*, 287–328.

Tebo, B. M., Johnson, H. A., McCarthy, J. K., & Templeton, A. S. (2005). Geomicrobiology of manganese(II) oxidation. *Trends in Microbiology*, *13*, 421–428.

Tottey, S., Harvie, D. R., & Robinson, N. J. (2005). Understanding how cells allocate metals using metal sensors and metallochaperones. *Accounts of Chemical Research*, *38*, 775–783.

Yang, J., Rawat, S., Stemmler, T. L., & Rosen, B. P. Arsenic binding and transfer by the ArsD As(III) metallochaperone. *Biochemistry*, *49*, 3658–3666.

Zumft, W. G., & Kroneck, P. M. (2007). Respiratory transformation of nitrous oxide (N_2O) to dinitrogen by bacteria and archaea. *Advances in Microbial Physiology*, *52*, 107–227.

RELEVANT WEBSITE

http://bicn.com. – Bangladesh International Community News.

Chapter 37

Heavy Metal Pollutants: Environmental and Biotechnological Aspects

G.M. Gadd
University of Dundee, Dundee, Scotland, UK

Chapter Outline

ABBREVATIONS

DMSe Dimethyl selenide
EPS Extracellular polymeric substances
SRB Sulfate-reducing bacteria
γ-Glu-Cys γ-Glutamyl-cysteinyl

DEFINING STATEMENT

This article describes the roles of microorganisms in the transformation of heavy metals, as well as metalloids, organometals, and radionuclides, between soluble and insoluble phases, and the environmental and biotechnological importance of these transformation processes in biogeochemical cycles and in new biotechnologies for the treatment of metal, metalloid, and radionuclide pollution.

INTRODUCTION

Heavy metals comprise an ill-defined group of more than 60 metallic elements, of density higher than 5g cm^{-3}, with diverse physical, chemical, and biological properties, but generally having the ability to exert toxic effects toward microorganisms. Many metals are essential for microbial growth and metabolism at low concentrations (e.g., Cu, Fe, Zn, Co, and Mn); yet they are toxic in excess amounts, and both essential and nonessential metal ions may be accumulated by the microbial cells by physicochemical and biological mechanisms. Thus, "toxic metals" and "potentially toxic metals" are useful general terms. In this article, the term "heavy metal" is used in a broad sense, and the discussion will include actinides, metal radionuclides, and organometal(loid) compounds. All these substances have a common potential for microbial toxicity and bioaccumulation, and are of environmental significance as pollutants or because of introduction as biocides and other substances.

ENVIRONMENTAL ASPECTS OF HEAVY METAL POLLUTION

Heavy Metals in the Environment

Although elevated levels of toxic heavy metals can occur in natural locations (e.g., volcanic soils and hot springs), average environmental abundances are generally low, with most of that immobilized in sediments and ores being biologically unavailable. However, anthropogenic activities have disrupted natural biogeochemical cycles, and there is increased atmospheric release as well as deposition into

Topics in Ecological and Environmental Microbiology.

aquatic and terrestrial environments. The major sources of pollution include fossil fuel combustion, mineral mining and processing, nuclear and other industrial effluents and sludges, brewery and distillery wastes, biocides, and preservatives including organometallic compounds. In fact, almost every industrial activity can lead to altered mobilization and distribution of heavy metals in the environment. Because of the fundamental microbial involvement in biogeochemical processes, as well as in plant and animal productivity and symbioses, toxic metal pollution can have significant short- and long-term effects and ultimately affect higher organisms, including humans, for example, by accumulation and transfer through food chains.

Effects of Heavy Metals on Microbial Populations

Metals exhibit a range of toxicities toward microorganisms, depending on physicochemical and biotic factors; while toxic effects can arise from natural activities, toxic effects on microbial communities are more commonly associated with anthropogenic contamination or redistribution of toxic metals. This can arise from aerial and aquatic sources, as well as agricultural practices, industrial activity, and domestic and industrial wastes. In some cases, microbial activity can result in remobilization of metals from other wastes and transfer into aquatic systems. It is commonly accepted that toxic metals (and their chemical derivatives and related substances) can have significant effects on microbial populations, and almost every index of microbial activity can be affected.

For toxicity to occur, heavy metals must directly interact with microbial cells and/or indirectly affect growth and metabolism by interfering with, for example, nutrient uptake, or by altering the physicochemical environment of the cell. A variety of nonspecific and specific mechanisms (e.g., biosorption and transport, respectively) determine the entry of mobile metal species into cells, and, if toxic thresholds are exceeded, cell death will result unless mechanisms for detoxification are possessed. The plethora of intracellular metal-binding ligands ensures that many toxic interactions are possible. Thus, practically every index of microbial activity can be adversely affected by toxic metal concentrations, including primary productivity, methanogenesis, nitrogen fixation, respiration, motility, biogeochemical cycling of C, N, S, P, and other elements, organic matter decomposition, and enzyme synthesis and activity in soils, sediments, and waters.

Despite potential toxicity, many microorganisms still survive, grow, and flourish in apparently metal-polluted locations, and a variety of mechanisms, both active and incidental, contribute to resistance and tolerance. However, general conclusions about heavy metal effects on natural populations are difficult to make because of the complexity of metal speciation, toxicity in the environment, and the morphological and physiological diversity encountered in microorganisms. Furthermore, environmental perturbations associated with industrial metal pollution (e.g., extremes of pH, salinity, and nutrient limitations) may also have adverse effects on microbial communities. Nevertheless, it is commonly assumed that microbes are able to respond to metal contamination and maintain metabolic activity through changes in microbial community structure and selection for resistance. Resistance and tolerance are arbitrarily defined, frequently interchangeable terms, and often based on whether particular strains and isolates can grow in the presence of selected heavy metal concentrations in laboratory media. It is probably more appropriate to use "resistance" to describe a direct mechanism resulting from heavy metal exposure, for example, bacterial reduction of Hg^{2+} to Hg^0 or metallothionein synthesis by yeasts. "Tolerance" may be a result of intrinsic biochemical and structural properties of the host, such as possession of impermeable cell walls, extracellular slime layers or polysaccharide, metabolite excretion, as well as environmental modification of toxicity. However, distinctions are difficult in many cases because several direct and indirect mechanisms, both physicochemical and biological, can contribute to microbial survival. Thus, although heavy metal pollution can qualitatively and quantitatively affect microbial populations in the environment, it may be difficult to distinguish metal effects from those of environmental components, environmental influence on metal toxicity, and the nature of microbial resistance/tolerance mechanisms involved. Although some gross generalizations are possible regarding toxic metal influence on microbial communities, individual cases are likely to be site-specific and potentially complex.

Environmental Modification of Heavy Metal Toxicity

The physicochemical characteristics of a given environment determine metal speciation and, therefore, chemical and biological properties of heavy metals. Concentration and speciation of metals in solution are governed by many processes, including inorganic and organic complexation, oxidation–reduction reactions, precipitation–dissolution, and adsorption–desorption, some of these being mediated by microbial activities. Because major mechanisms of metal toxicity are a consequence of strong coordinating properties, a reduction in bioavailability may reduce toxicity and enhance microbial survival. Such parameters as pH, temperature, aeration, soluble and particulate organic matter, clay minerals, and salinity can influence heavy metal speciation, mobility, and toxicity. Metals can exist in solution as free cations (e.g., Cu^{2+}, Cd^{2+}, and Zn^{2+}), as soluble complexes with inorganic or organic ligands (e.g., $ZnCl^+$, $CdCl_3^-$, and metal citrates), or in association with colloidal material. Common inorganic ligands that can complex metals include SO_4^{2-}, Cl^-, OH^-, PO_4^{3-}, NO_3^- and CO_3^{2-}. Organic complexing agents include humic and fulvic acids, aromatic and aliphatic compounds, and carboxylic acids. Acidic conditions may increase metal availability, although H^+ may successfully compete with and reduce or prevent binding and transport. Environmental pH also affects metal complexation with organic components and inorganic anions (e.g., Cl^-). With increasing pH, there may be formation of hydroxides, oxides, and carbonates of varying solubility and toxicity. Some hydroxylated species may associate more efficiently with microbial cells than the corresponding metal cations. The oxidation state of several metals also determines solubility, for example, Cr(VI) being soluble and toxic and Cr(III) being immobile and less toxic. Such reductive transformations may be mediated by microbes, with accompanying consequences for survival. Colloidal materials of significance in affecting metal bioavailability and transport include iron and manganese oxides, clay minerals, and organic matter. Metals can precipitate as solid phases, for example, $CdCO_3$, $Pb(OH)_2$, ZnS, CuS, as well as mixed compounds. Toxic metals may also substitute for other metals in indigenous minerals: for example, Cd may substitute for Ca in $CaCO_3$. In addition, toxic metals may sorb onto preexisting minerals. A reduction in toxicity in the presence of elevated concentrations of anions such as Cl^-, CO_3^{2-}, S^{2-}, and PO_4^{3-} is frequently observed. Mono-, di-, and multivalent cations may affect heavy metal toxicity by competing with binding and transport sites. Other synthetic and naturally produced soluble and particulate organic substances, including microbial metabolites, may influence toxicity by binding and complexation. Anionic contaminants, such as arsenic, selenium, and chromium, and oxyanions, such as AsO_4^{3-}, AsO^{2-}, SeO_4^{2-}, SeO_3^{2-}, and CrO_4^{2-}, can sorb to positive charges on insoluble organic matter and iron, manganese, aluminum oxides, and carbonates. The removal of heavy metal species by intact living and dead microbial biomass through physicochemical and/or biochemical interactions may also be significant in some locations. In more general terms, microbial growth and activity is influenced by environmental parameters, including the availability of organic and inorganic nutrients, and this can clearly affect the responses to potentially toxic metals.

Mechanisms of Microbial Heavy Metal Detoxification

Extracellular metal complexation, precipitation, and crystallization can result in detoxification. Polysaccharides, organic acids, pigments, proteins, and other metabolites can remove metal ions from solution and/or convert them into less toxic species. Iron-chelating siderophores may chelate other metals and radionuclides and possibly reduce their toxic effects. The production of H_2S by microorganisms, for example, by *Desulfovibrio* sp., results in the formation of insoluble metal sulfides and also disproportionation of organometallics to volatile products as well as insoluble sulfides: for example,

$$2CH_3Hg^+ + H_2S \rightarrow (CH_3)_2Hg + HgS$$

$$2(CH_3)_3Pb^+ + H_2S \rightarrow (CH_3)_4Pb + (CH_3)_2PbS$$

Many other examples of metal crystallization and precipitation are known, and mediated by processes dependent and independent of metabolism. Some of these are of great importance in biogeochemical cycles and involved, for example, in microfossil formation, iron and manganese deposition, silver and uranium mineralization, and formation of stable calcareous minerals.

Decreased accumulation, sometimes as a result of efflux, and impermeability may be important survival mechanisms. Impermeability may be a consequence of cell wall and/or membrane composition, lack of transport mechanism, or increased turgor pressure. Bacterial plasmids have resistance genes to many toxic metals and metalloids, for example, Ag^+, AsO_2^-, AsO_4^{3-}, Cd^{2+}, Co^{2+}, CrO_4^{2-}, Cu^{2+}, Hg^{2+}, Ni^{2+}, Sb^{3+}, TeO_3^{2-}, Tl^+, and Zn^{2+}. Related systems are frequently located on bacterial chromosomes, for example, Hg^{2+} resistance in *Bacillus*, Cd^{2+} efflux in *Bacillus*, and arsenic efflux in *Escherichia coli*. Copper tolerance genes are generally genome-located. General conclusions in bacterial metal resistance include the following: (1) plasmid-determined resistances are highly specific; (2) resistance systems have been found on plasmids in all bacterial groups tested; and (3) resistance mechanisms generally involve efflux from the cells or enzymatic detoxification. However, other less specific interactions, for example, sorption, may contribute to the overall response. Many bacterial metal resistance mechanisms, for example, Cd, Cu, and As, depend on efflux. Efflux pumps, determined by plasmid and chromosomal systems, are either ATPases or chemiosmotic systems, with mechanisms often showing similarity in different types of bacteria. Cd^{2+} resistance may involve (1) an efflux ATPase in Gram-positive bacteria, (2) cation–H^+ antiport in Gram-negative bacteria, and (3) intracellular metallothionein in cyanobacteria. Arsenic-resistant Gram-negative bacteria have an arsenite efflux ATPase and an arsenate reductase (which reduces arsenate [As(V)] to arsenite [(As(III)], which comprise the underlying biochemical mechanism. A Cd^{2+} efflux ATPase is widely found in Gram-positive bacteria, including species of *Bacillus*. Systems for Hg^{2+} resistance occur on plasmids from Gram-positive and Gram-negative bacteria with component genes being involved in the transport of Hg^{2+} to the detoxifying enzyme mercuric reductase, which reduces Hg^{2+} to elemental Hg^0. The enzyme organomercurial lyase can break the C–Hg bond in organomercurials. The large plasmids of *Alcaligenes eutrophus* have several toxic metal resistance determinants, for example, three for Hg^{2+}, one for Cr^{6+}, and two for divalent cations, *czc* (Cd^{2+}, Zn^{2+}, and Co^{2+} resistance) and *cnr* (Co^{2+} and Ni^{2+} resistance). *Czc* functions as a chemiosmotic divalent cation/H^+ antiporter. In *Enterococcus hirae* (previously *Streptococcus faecalis*), copper resistance is determined by two genes, *copA* and *copB*, which determine uptake and efflux P-type ATPases, respectively. Plasmid-determined Cu^{2+} resistance has been described in *Pseudomonas* sp., *Xanthomonas* sp., and *E. coli*. Chromosomal genes also affect Cu^{2+} transport and resistance by determining uptake, efflux, and intracellular Cu^{2+} binding. Bacterial arsenic resistance is plasmid-mediated in Gram-positive bacteria, and several mechanisms of plasmid-mediated tellurite resistance have been suggested, including reduction, reduced uptake, and enhanced efflux, although, as with Ag^+, resistance does not appear to depend on reduction to the elemental form (Te^0).

As with bacteria, intracellular metal concentrations in fungi may be regulated by transport, including efflux mechanisms. Such mechanisms are involved in normal metal homeostasis but also have a role in the detoxification of potentially toxic metals. Reduced heavy metal uptake has been observed in many tolerant microbes, including bacteria, algae, and fungi, although this is dependent on environmental factors, including pH and ion competition. However, some resistant strains may accumulate more metal than sensitive parental strains because of more efficient internal detoxification. Inside the cells, metal ions may be detoxified by chemical components, which include metal-binding proteins, or compartmentalized into specific organelles. Metal-sequestering organic and inorganic molecules, for example, polyphosphate, have been implicated in several microbial groups, whereas metal-binding peptides and proteins, including metallothioneins and γ-Glu-Cys peptides (phytochelatins, cadystins), have been detected in all microbial groups examined. Metallothioneins are small cysteine-rich polypeptides that can bind essential metals (e.g., Cu and Zn), in addition to nonessential metals (e.g., Cd). Metal-binding γ-glutamyl-cysteinyl peptides are short peptides of the general formula $(\gamma\text{-Glu-Cys})_n$-Gly. Peptides of n = 2–7 are most common, and these are important detoxification mechanisms in algae as well as several fungi and yeasts. In *Schizosaccharomyces pombe*, the value of n ranges from 2 to 5, whereas in *Saccharomyces cerevisiae*, only an n_2 isopeptide has been observed. Although $(\gamma EC)_nG$ can be induced by a wide variety of metal ions, including Ag, Au, Hg, Ni, Pb, Sn, and Zn, metal binding has only been shown for a few, primarily Cd and Cu. For Cd, two types of complexes exist in *S. pombe* and *Candida glabrata*. A low molecular weight complex consists of $(\gamma EC)_nG$ and Cd, whereas a high molecular weight complex also contains acid-labile sulfide. The $(\gamma EC)_nG\text{–Cd–}S^{2-}$ complex has greater stability and higher Cd-binding capacity than a low molecular weight complex, and consists of a CdS crystallite core and an outer layer of $(\gamma EC)_nG$ peptides. The higher binding capacity of sulfide-containing complex confers tolerance to Cd. In *S. pombe*, evidence has also been presented for vacuolar localization of $(\gamma EC)_nG\text{–Cd–}S^{2-}$ complexes. The main function of *S. cerevisiae* metallothionein (yeast MT) is cellular copper homeostasis. However, induction and synthesis of MT as well as amplification of MT genes leads to enhanced copper resistance in *S. cerevisiae*. The fungal vacuole also has an important role in the regulation of cytosolic metal ion concentrations and the detoxification of potentially toxic metal ions. Metals preferentially sequestered by the vacuole include Mn^{2+}, Fe^{2+}, Zn^{2+}, Co^{2+}, Ca^{2+}, Sr^{2+}, Ni^{2+}, and the monovalent cations K^+, Li^+, and Cs^+. The absence of a vacuole or a functional vacuolar H^+-ATPase in *S. cerevisiae* is associated with increased sensitivity and largely decreased capacity of the cells to accumulate Zn, Mn, Co, and Ni, the metals known to be mainly detoxified in the vacuole.

Chemical transformations of metal and metalloid species by microorganisms may also constitute detoxification mechanisms, for example, bacterial Hg^{2+} reduction to Hg^0. However, plasmid-determined chromate resistance appears unconnected with chromate [Cr(VI)] reduction to Cr(III), the resistance depending on reduced CrO_4^{2-} uptake. Similarly, plasmid-mediated Ag^+ resistance appears not to involve Ag^+ reduction to Ag^0. In addition to these, other examples of reduction are carried out by bacteria, algae, and fungi (e.g., Au^{3+} to Au^0). Methylated metal and metalloid species may be volatile and lost from a given environment, for example, dimethyl selenide (DMSe). Methylation of Hg^{2+}, by direct and indirect microbial action, can result in the formation of CH_3Hg^+ and $(CH_3)_2Hg$. Arsenic methylation can be mediated by many organisms with compounds having the general structure $(CH_3)_nAsH_{3-n}$, and mono-, di-, and trimethylarsine (n = 1–3, respectively) being major volatile compounds. The reduction of arsenic oxyanions by reductase enzymes is also frequent and a determinant of As resistance. However, there appears no involvement of such reductases in biomethylation.

Organometallic compounds may be detoxified by sequential removal of alkyl or aryl groups. Organomercurials can be degraded by organomercurial lyase, whereas organotin detoxification involves sequential removal of organic groups from the tin atom:

$$R_4Sn \rightarrow R_3SnX \rightarrow R_2SnX_2 \rightarrow RSnX_3 \rightarrow SnX_4.$$

It should be stressed that abiotic mechanisms of metal methylation and organometal(loid) degradation also contribute to their transformation and redistribution in aquatic, terrestrial, and aerial environments. The relative importance of biotic and abiotic mechanisms is often difficult to establish.

BIOTECHNOLOGICAL ASPECTS OF HEAVY METAL POLLUTION

Microbial Processes for Metal Removal and Recovery

Certain microbial processes can solubilize metals, thereby increasing their bioavailability and potential toxicity, whereas others immobilize them and thus reduce their bioavailability (Figure 1). The relative balance between mobilization and immobilization varies depending on the organisms and their environment. As well as being an integral component of biogeochemical cycles for metals, these processes may be exploited for the treatment of contaminated solid and liquid wastes. Metal mobilization can be achieved by autotrophic and heterotrophic leaching, chelation by microbial metabolites and siderophores, and methylation, which can result in volatilization. Similarly, immobilization can result from sorption to cell components or exopolymers, transport and intracellular sequestration, or precipitation as insoluble organic and inorganic compounds, for example, oxalates, sulfides, or phosphates. In addition, microbiologically mediated reduction of higher valency species may effect either mobilization, for example, Mn(IV) to Mn(II), or immobilization, for example, Cr(VI) to Cr(III), and U(VI) to U(IV). In the context of bioremediation, solubilization of metal contaminants provides a means for removal of metals from solid matrices such as soils, sediments, and industrial wastes. Alternatively, immobilization processes enable metals to be transformed *in situ* and in bioreactors into insoluble, chemically inert forms. Biotechnological development of microbial systems may provide an alternative or adjunct to conventional physicochemical treatment methods for contaminated effluents and wastewaters. Growing evidence suggests that some biomass-related processes are economically competitive with existing treatments in mining and metallurgy.

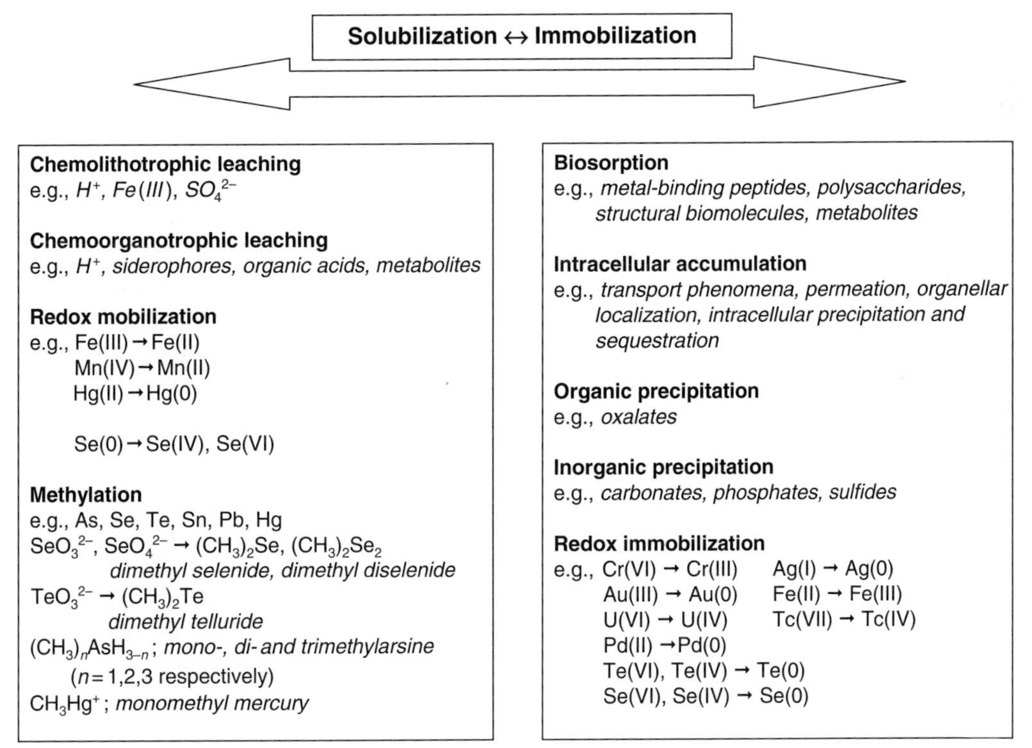

FIGURE 1 Diagram depicting the major mechanisms of microbial metal transformation between soluble and insoluble metal species. The relative balance between such processes will depend on the environment and associated physicochemical conditions, the microorganism(s) involved, as well as relationships with plants, animals, and anthropogenic activities. Chemical equilibrium between soluble and insoluble phases is influenced by abiotic components, including dead biota and their decomposition products, as well as other physicochemical components of the environmental matrix, for example, pH, water, inorganic and organic ions, molecules, compounds, colloids, and particulates. Solubilization can occur by chemolithotrophic (autotrophic) and chemoorganotrophic (heterotrophic) leaching; siderophores and other complexing agents; redox reactions; methylation and demethylation; and biodegradation of organoradionuclide complexes. Immobilization can occur by biosorption to cell walls, exopolymers, other structural components, and derived/excreted products; precipitation can be a result of metabolite release, for example, sulfide, oxalate, or reduction, transport, accumulation, intracellular deposition, localization and sequestration, and adsorption and entrapment of colloids and particulates. The overall scheme is also affected by reciprocal interactions between biotic and abiotic components of the ecosystem, such as abiotic influence on microbial diversity, numbers and metabolic activity, ingestion of particulates and colloids (including bacteria) by phagotrophs, biotic modification of physicochemical parameters including redox potential, pH, O_2, CO_2, other gases and metabolites, temperature, and nutrient depletion.

Metal Solubilization

Chemolithotrophic (autotrophic) leaching

Most chemolithotrophic metal leaching is carried out by chemolithotrophic, acidophilic bacteria, which obtain energy from oxidation of Fe(II) or reduced sulfur compounds and solubilize metals because of the resulting production of Fe(III) and H_2SO_4. The microorganisms involved include sulfur-oxidizing bacteria (e.g., *Acidithiobacillus thiooxidans*), iron- and sulfur-oxidizing bacteria (e.g., *Acidithiobacillus ferrooxidans*), and iron-oxidizing bacteria (e.g., *Leptospirillum ferrooxidans*). As a result of sulfur and iron oxidation, metal sulfides are solubilized, concomitant with the decrease in pH of their immediate environment, thereby resulting in solubilization of other metal compounds, including metals sorbed to soil and mineral constituents. Chemolithotrophic leaching of metal sulfides is well established for industrial-scale biomining processes, but it has also been used to solubilize metals from sewage sludge as well as remediate other metal-contaminated solid materials, including soil and red mud, the main waste product of Al extraction from bauxite. One two-stage soil treatment process used a mixture of sulfur-oxidizing bacteria to acidify metal-contaminated soil before treatment of the metal-loaded leachate with sulfate-reducing bacteria (SRB).

Chemoorganotrophic (heterotrophic) leaching

Many chemoorganotrophic (heterotrophic) fungi (and bacteria) can leach metals from industrial wastes, low-grade ores, and metal-bearing minerals. This occurs as a result of proton efflux, siderophores (for Fe(III)), and organic acids, for example, citric and oxalic. Organic acids provide a source of protons and a metal-complexing anion, for example, citrate, oxalate, with complexation being dependent on the metal/anion concentrations, pH, and metal complex stability constants. Organisms such as *Aspergillus niger* and *Penicillium simplicissimum* have been used to leach Zn, Cu, Ni, and Co from a variety of solid materials, including industrial filter dust, copper converter slag, lateritic ores, red mud, manganiferous minerals, and municipal waste fly ash. Citrate and oxalate can form stable complexes with a large number of metals. Many metal citrates are highly mobile and not readily degraded. Oxalic acid can act as a leaching agent for metals that form soluble oxalate complexes, including Al and Fe.

Siderophores are highly specific Fe(III) ligands (formation constant often $>10^{30}$) that are excreted by microorganisms to aid iron assimilation. Such assimilation may be improved by attachment to solid Fe oxides in soil. Although primarily produced as a means of obtaining iron, siderophores are also able to bind other metals such as magnesium, manganese, chromium (III), gallium (III), and radionuclides such as plutonium (IV).

Metal Immobilization

Biosorption

Biosorption (defined here as the microbial uptake of organic and inorganic metal, metalloid, and radionuclide species, both soluble and insoluble, by physicochemical mechanisms) may be influenced by metabolic activity (in living cells), and may also provide nucleation sites for the formation of stable minerals, including phosphates, sulfides, and oxides. Crystallization of elemental gold and silver may occur as a result of reduction, whereas the formation of hydrolysis products can enhance precipitation of U and Th. All biological macromolecules have an affinity for metal species, with cell walls and associated materials being of the greatest significance in biosorption (Table 1). Moreover, mobile cationic species can be accumulated by cells via transport systems of varying affinity and specificity, and internally bound, transformed, precipitated, localized within organelles, or translocated to specific structures depending on the metal concerned and the organism.

Biosorption by Cell Walls and Associated Components

In bacteria, peptidoglycan carboxyl groups are the main cationic binding sites in Gram-positive species, with phosphate groups contributing significantly in Gram-negative species. Chitin, phenolic polymers, and melanin are important structural components of fungal walls, and these are also effective biosorbents for metals and radionuclides. Fungi can be efficient sorbents of metal ions over a wide range of pH values, and although they may take up less metal per unit dry weight than clay minerals (the most important metal-sorbing component in soil), they are more efficient sorbents per unit surface area. Variations in the chemical behavior of metal species as well as the composition of microbial cell walls and extracellular materials can result in wide differences in biosorptive capacities (Table 1). Extracellular polymeric substances (EPS), a mixture of polysaccharides, mucopolysaccharides, and proteins, can bind significant amounts of potentially toxic metals and entrap precipitated metal sulfides and oxides. One process uses floating mats of cyanobacteria, the metal-binding process being due to large polysaccharides (>200,000 Da).

The ability of surface-associated macromolecules to effect the immobilization of aqueous metal(loid) species may be of great importance, particularly where organisms grow as surface-attached biofilms, enmeshed in a matrix of EPS. The biofilm mode of growth is now widely accepted to be the predominant form in which natural bacterial populations occur, and it appears that natural mixed-species biofilms can act as sinks for precipitated minerals, including

TABLE 1 Examples of microbial metal and actinide accumulation to industrially significant levels

Microorganism	Metal	Accumulation (% Dry Weight)
Bacteria		
Streptomyces sp.	Uranium	2–14
Streptomyces viridochromogenes	Uranium	30
Thiobacillus ferrooxidans	Silver	25
Bacillus cereus	Cadmium	4–9
Zoogloea sp.	Cobalt	25
	Copper	34
	Nickel	13
Citrobacter sp.[a]	Lead	34–40
	Cadmium	170
	Uranium	900
Pseudomonas aeruginosa	Uranium	15
Mixed culture	Silver	32
Cyanobacteria		
Anabaena cylindrica	Cadmium	0.25
Anacystis nidulans	Nickel	1
Spirulina platensis	Gold	0.52
Plectonema boryanum	Zirconium	0.16
Nostoc sp.	Cadmium	1
Algae		
Chlorella vulgaris	Gold	10
	Lead	8.5
Chlorella regularis	Uranium	15
	Zinc	2.8
	Manganese	0.8
Scenedesmus sp.	Molybdenum	2.3
Euglena sp.	Aluminum	1.5
Sargassum natans[b]	Gold	25
	Lead	8
	Silver	7
	Uranium	4.5
	Copper	2.5
	Zinc	2
	Cobalt	6
	Cadmium	8.3
Ascophyllum nodosum[b]	Gold	4
	Cobalt	15
	Cadmium	10
Fungi		
Phoma sp.	Silver	2
Penicillium sp.	Uranium	8–17
Rhizopus arrhizus	Lead	0.6
	Cadmium	3
	Lead	10
	Uranium	20
	Thorium	19
	Silver	5
	Mercury	6
Aspergillus niger	Thorium	19
	Uranium	22
	Gold	6–18
	Zinc	1–10
	Silver	10
Saccharomyces cerevisiae	Thorium	12
	Uranium	10–15
	Cadmium	7
	Copper	1–3
Ganoderma lucidum	Copper	1
Mucor miehei	Zinc	3.4
	Uranium	18

Data derived from a number of sources and presented without reference to important experimental conditions, for example, metal and biomass concentration, pH, and whether freely suspended, living, dead or immobilized, or the mechanism of accumulation. In most cases, highest uptake levels are due to general biosorptive mechanisms
[a] *Phosphatase-mediated metal removal.*
[b] *Macroalgae (seaweeds).*

potentially toxic metals, in aqueous environments. The biofilm EPS matrix can act as a direct adsorbent of dissolved metal ions, with the ionic state and charge density of EPS components determining the ionic binding and electrostatic immobilization properties. Bacterial EPS are dominated by polysaccharides, but secreted polymers also include proteins, nucleic acids, peptidoglycan, lipids, and phospholipids. This heterogeneous matrix generally has a net negative charge, with polyanionic moieties acting as an ion-exchange matrix for metal cations. Well-characterized examples include the propensity of uronic acid-containing polysaccharides to bind with carboxyl groups and thus bind metals, whereas neutral carbohydrates can bind metals by the formation of weak electrostatic bonds around the hydroxyl groups. Cross-linking of extracellular polysaccharides by metal ions themselves may alter the mechanical and chemical properties of EPS.

The biofilm growth mode appears to further enhance metal removal in various ways. Biosorption and bioprecipitation can be interrelated phenomena, such that ionic concentration by sorption at low-energy cellular or EPS surface sites within biofilms can initiate mineral formation and immobilization within biofilms. Mineral precipitates formed in the bulk solution may also be physically entrapped or chemically adsorbed by the biofilm EPS matrix.

Biosorption by Free and Immobilized Biomass

Both freely suspended and immobilized biomass from bacterial, cyanobacterial, algal, and fungal species have received attention, with immobilized systems appearing to possess several advantages over "free" biomass, including higher mechanical strength and easier biomass/liquid separation. Living or dead biomass of all groups has been immobilized by encapsulation or cross-linking using supports, which include agar, cellulose, alginates, cross-linked ethyl acrylate-ethylene glycol dimethylacrylate, polyacrylamide, silica gel, and cross-linking reagents such as toluene diisocyanate and glutaraldehyde. Immobilized living biomass has mainly taken the form of bacterial biofilms (see Section "Biosorption by Cell Walls and Associated Components") on inert supports, and has been used in a variety of configurations, including rotating biological contactors, fixed-bed reactors, trickle filters, fluidized beds, and airlift bioreactors.

Metal Desorption

Biotechnological exploitation of biosorption may depend on the ease of biosorbent regeneration for metal recovery. Metabolism-independent processes are frequently reversible by nondestructive methods, and hence can be considered analogous to conventional ion exchange. Most work has concentrated on nondestructive desorption, which should be efficient, cheap, and result in minimal damage to the biosorbent. Dilute mineral acids (~0.1 M) can be effective for metal removal, although more concentrated acids or lengthy exposure times may result in biomass damage. It may be possible to apply selective desorption of metalloid species from a loaded biosorbent using an appropriate elution scheme. For example, metal cations (e.g., Cu^{2+}, Cr^{3+}, Ni^{2+}, Pb^{2+}, Zn^{2+}, Cd^{2+}, and Co^{2+}) were released from algal biomass using eluant at pH 2, whereas at higher pH values anionic metal species (e.g., SeO_4^{2-}, CrO_4^{2-}, and MoO_4^{2-}) were removed. Au^{3+}, Ag^{+}, and Hg^{2+}, however, remained strongly bound at pH 2, and these were removed by the addition of ligands that formed stable complexes with these metal ions. Carbonates and/or bicarbonates are efficient desorption agents with the potential for cheap, nondestructive metal recovery. Operating pH values for bicarbonates cause little damage to the biomass, which may retain at least 90% of the original uptake capacity.

Metal-binding proteins, polysaccharides, and other biomolecules

A diverse range of specific and nonspecific metal-binding compounds are produced by microorganisms. Nonspecific metal-binding compounds are metabolites or by-products of microbial metabolism and range from simple organic acids and alcohols to macromolecules, such as polysaccharides and humic and fulvic acids. Specific metal-binding compounds may be produced in response to external levels of metals. Siderophores are low molecular weight Fe(III) coordination compounds (500–1000 Da) excreted under iron-limiting conditions by iron-dependent microorganisms. Although specific to Fe(III), siderophores can also complex Pu(IV), Ga(III), Cr(III), scandium (Sc), indium (In), nickel, uranium, and thorium. Specific low molecular weight (6000–10,000 Da) metal-binding metallothioneins are produced by animals, plants, and microorganisms in response to the presence of toxic metals. Metal-binding γ-Glu-Cys peptides (phytochelatins and cadystins) contain glutamic acid and cysteine at the N-terminal position, and have been identified in plants, algae, and several microorganisms. The metal-binding abilities of siderophores, metallothioneins, phytochelatins, and other similar molecules may have the potential for bioremediation of waters containing low metal concentrations, although few examples have been rigorously tested.

Transport and accumulation

Microbial metal transport systems are of varying specificity, and essential and nonessential metal(loid) species may be accumulated. The rates of uptake can depend on the physiological state of cells as well as the nature of the environment or growth medium. Integral to the transport of metal ions into cells are transmembrane electrochemical gradients, for example of H^{+}, resulting from the operation of enzymatic pumps (ATPases) that transform the chemical energy of ATP into this form of biological energy. ATPases are also involved in ion efflux in a variety of organisms and organellar ion compartmentation in eukaryotes via operation across vacuolar membranes. Metals may also enter (and leave) cells via pores or channels. With toxic heavy metals, permeabilization of cell membranes can result in exposure of intracellular metal-binding sites and increase passive accumulation. Intracellular uptake may result in death of sensitive organisms, unless a means of detoxification is possessed or induced (Figure 2). Other mechanisms of

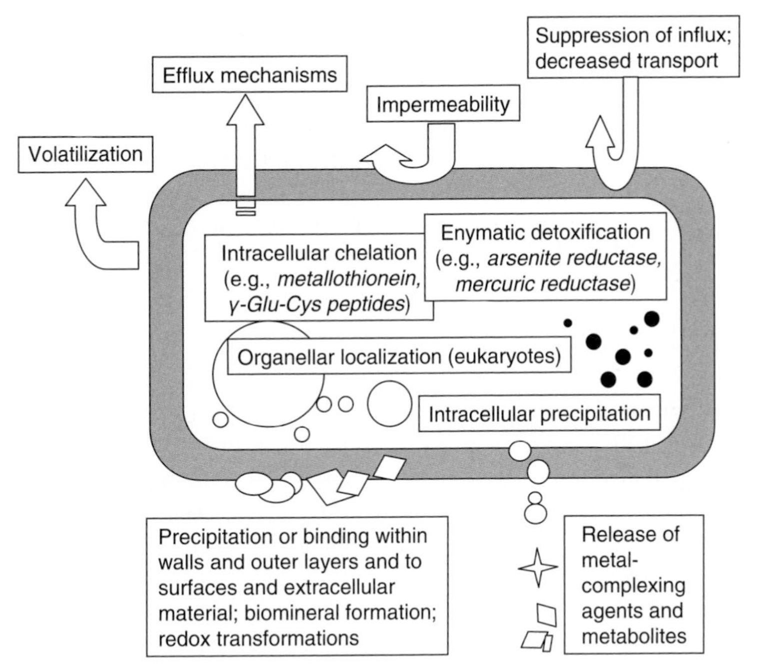

FIGURE 2 Mechanisms involved in the detoxification of metals including mechanisms that restrict entry into the cell and intracellular detoxification or organellar compartmentation, the latter occurring in some eukaryotes, for example, fungi. Operation of a number of mechanisms is possible depending on the organism and the cellular environment, some dependent and/or independent of metabolism. A variety of mechanisms may be involved in transport phenomenon, contributing to decreased uptake and/or efflux. A variety of specific or nonspecific mechanisms may also affect redox transformations, intracellular chelation, and intracellular precipitation.

microbial metal accumulation include iron-binding siderophores and cotransport of metals with organic substrates.

Metal precipitation

Precipitation by Redox Processes: Metal-Reducing Bacteria and Iron Oxidizers

A diverse range of microorganisms can use oxidized metallic species, for example, Fe(III), Cr(VI), or Mn(IV), as terminal electron acceptors. Many use more than one metal or anion, such as nitrate or sulfate. Fe(III) and Mn(IV) appear to be the most commonly used metals as terminal electron acceptors in the biosphere. However, since the solubility of both Fe and Mn is increased by reduction, other metals have been targeted in waste treatment, for example, molybdenum(VI) and Cr(VI). The reduction of, for example, Cr(VI) to Cr(III), by organisms including *Enterobacter cloacae* and *E. coli*, may facilitate removal by biosorption or (bio)precipitation. One potential application of dissimilatory biological metal reduction is uranium precipitation by reduction of soluble U(VI) compounds to U(IV) compounds, such as the hydroxide or carbonate, which have low solubility at neutral pH. Strains of *Shewanella* (*Alteromonas*) *putrefaciens* and *Desulfovibrio* sp. can produce a very pure precipitate of U(IV) carbonate. Such bacterial uranium reduction can also be combined with chemical extraction methods. The solubility of some other radionuclides, for example, Ra and Pu, may be increased by reduction, which may favor removal from, for example, contaminated soil.

Bacterial Fe oxidation is ubiquitous in environments with sufficient Fe^{2+} and conditions to support bacterial growth such as drainage waters and tailing piles in mined areas, pyritic and hydric soils (bogs and sediments), drain pipes and irrigation ditches, and plant rhizospheres. Iron oxidizers commonly found in acidic soil environments are acidophilic chemolithotrophs, such as *A. ferrooxidans*, significant for their role in generating acid mine drainage. Fungi, too, oxidize metals in their environment. Desert varnish is an oxidized metal layer (patina) of few millimeter thickness found on rocks and in soils of arid and semiarid regions and is believed to be of fungal and bacterial origin.

Sulfate-Reducing Bacteria

SRB are strictly anaerobic heterotrophic bacteria found in environments where carbon substrates and sulfate are available. These utilize an energy metabolism in which the oxidation of organic compounds or hydrogen is coupled to the reduction of sulfate as the terminal electron acceptor, producing sulfide which often reach significant concentrations in sediments or bioreactors:

$$SO_4^{2-} + 9H^+ + 8e^- \rightarrow HS^- + 4H_2O.$$

Sulfur in S(VI) oxidation state is stoichiometrically reduced to S(–II) and, under circumneutral conditions in which SRB are generally encountered, the main product is bisulfide (HS^-), with a small proportion of volatile H_2S. Bisulfide is a highly reactive species, with the propensity to bind metal cations in solution-forming metal sulfide solids. This is the main mechanism whereby SRB remove toxic metals from solution, for example,

$$M^{2+} + SO_4^{2-} + 2CH_3CH_2OH \rightarrow 2CH_3COOH + 2H_2O + MS\downarrow$$

$$M^{2+} + SO_4^{2-} + 2CH_3CHOHCOOH \rightarrow 2CH_3COOH + 2CO_2 + 2H_2O + MS\downarrow$$

The solubility products of most heavy metal sulfides are very low, in the range of 4.65×10^{-14} (Mn) to 6.44×10^{-53} (Hg), so that even a moderate output of sulfide can remove metals to safer levels permitted in the environment. SRB can create extremely reducing conditions, which can chemically reduce metals such as uranium(VI). In addition, sulfate reduction partially eliminates acidity from the system, which can result in further precipitation of metals, for example, Cu, Al, as hydroxides as well as increasing the efficiency of sulfide precipitation.

Secondary metal(loid) removal by adsorption on SRB-produced metal sulfides deposited within, or immobilized by, the biofilm matrix can also contribute to overall removal. SRB-generated metal sulfides can adsorb a range of cations and anions. Fe(II) sulfides can bind a range of metals, allowing the level of metals in solution to be reduced from original concentrations in the order of milligrams per liter to micrograms per liter.

Processes Utilizing Metal Sulfide Precipitation

Sulfate reduction can provide both *in situ* and *ex situ* metal removal from acid mine drainage and, together with other mechanisms such as biosorption, contributes to the removal of metals and acidity in artificial and natural wetlands. Large-scale bioreactors have also been developed using bacterial sulfate reduction for treating metal-contaminated waters. The best known commercial application involving metal sulfide precipitation is the THIOPAQ technology, developed and marketed by Paques Bio Systems B.V., Balk, the Netherlands (http://www.paques.nl/), and first applied in 1992 for the treatment of contaminated groundwater at the Budelco zinc refinery in the Netherlands. The basic THIOPAQ system consisted of a two-stage biological process in series: anaerobic sulfate reduction to sulfide followed by aerobic sulfide oxidation to elemental sulfur. Since the solubilities of most metal sulfides are much lower than those of their hydroxides, an advantage of the THIOPAQ system is that considerably lower effluent metal concentrations can be achieved than in the neutralization processes, which immobilize metals by hydroxide precipitation. In addition, the metal sulfide precipitate formed may be reprocessed in a smelter or a refinery.

Electron donors suitable for small-scale THIOPAQ installations were ethanol, various fatty acids, and organic waste streams. For large-scale applications, where more than 2.5 tons of hydrogen sulfide is produced per day, hydrogen gas is preferentially used as a reductant. Hydrogen gas can be produced onsite by cracking methanol or by steam-reforming natural gas or liquefied petroleum gas. The main reaction that occurs in a reactor operated with H_2 is:

$$H_2SO_4 + 4H_2 \rightarrow H_2S + 4H_2O.$$

Hydrogen sulfide in the reactor gas (3–15%, v/v) can be employed for metal precipitation. Compared to the addition of a NaHS or Na_2S solution, an advantage with the use of H_2S is that sodium is not introduced into the system. It is also possible that careful control of the pH and the redox potential of the process liquid may allow selective recovery of metals. Thus, sulfide precipitation makes it possible to separate copper from zinc, arsenic from copper, iron from nickel, and so on, in multiple reaction stages at different pH values. Alternatively, metals may also be precipitated as sulfides inside the anaerobic bioreactor.

At the Budelco zinc refinery in Budel-Dorplein, the Netherlands, a THIOPAQ system capable of processing approximately $300\,m^3\,h^{-1}$ of polluted groundwater has been in operation since 1992. Its products, a metal sulfide sludge (mainly ZnS) and sulfur slurry, are fed back to the roasters in the refinery. The capacity of the installation was increased to $400\,m^3\,h^{-1}$ (in 1998), with the feed also including a mixture of groundwater and process water. The most recent THIOPAQ installation (called Budelco II, in operation since 1999) treats several bleed streams and process water: sulfate reduction occurs in a 500-m^3 bioreactor where hydrogen is used as the electron donor.

Another process, integrating bacterial sulfate reduction with bioleaching by sulfur-oxidizing bacteria, was developed to remove contaminating toxic metals from soils. In this process, sulfur- and iron-oxidizing bacteria are employed to release metals from soils by the breakdown of sulfide minerals and production of sulfuric acid, which liberates acid-labile forms such as hydroxides, carbonates, or sorbed metals. Metals are liberated in the form of an acid sulfate solution, which enables both the large proportion of acidity and almost the entirety of metals to be removed by bacterial sulfate reduction. Precipitation efficiency is further increased by the addition of flocculating agents.

In confined systems, SRB can also bring about significant increases in bulk pH, which can enhance sulfide precipitation and lead to precipitation of hydroxides and carbonates of transition metals. When an organic substrate acts as the electron donor, bicarbonate is also generated:

$$2CH_3CHOHCOO^- + SO_4^{2-} \rightarrow 2CH_3COO^- + 2HCO_3^- + HS^- + H^+.$$

This has useful implications for the use of SRB in the remediation of acidic metal-processing waters and mine wastes, particularly where they are active within suspended or surface-attached mesophilic biofilms. SRB are reported to contribute significantly to metal removal in constructed wetlands as well as in alkalization of acidic mine wastes. Both sulfide generation and pH-related precipitation appear to be important. However, some studies have questioned the contribution of SRB in these broad-scale systems, arguing that Fe(III)-reducing bacteria make greater contribution where carbon is limiting, in terms of both metal removal and ameliorating low pH.

There is also evidence that extremely reducing conditions that develop during sulfate reduction can lead to chemical conversion of oxyanions into cationic species, which can be more easily precipitated or biosorbed. The indirect chemical reduction of Cr(VI), as soluble chromate (CrO_4^{2-}), to much less soluble Cr(III) cationic species by sulfide and/or Fe^{2+} in SRB culture appears to be at least partially responsible for the removal of chromate from solution by SRB.

Phosphatase-Mediated Metal Precipitation

In this process, metal or radionuclide accumulation by bacterial (*Citrobacter* sp.) biomass is mediated by a phosphatase enzyme induced during metal-free growth, which liberates inorganic phosphate from a supplied organic phosphate donor molecule, for example, glycerol 2-phosphate. Metal/radionuclide cations are then precipitated as phosphates on the biomass to high levels. In addition, metal precipitation by secreted phosphate generated from polyphosphate hydrolysis has also been suggested as a mechanism to remove metals and actinides from aqueous waste streams.

High-Gradient Magnetic Separation

Metal ion removal from solution has been achieved using bacteria rendered susceptible to magnetic fields. "Nonmagnetic" bacteria can be made magnetic by the precipitation of metal phosphates (aerobic) or sulfides (anaerobic) on their surfaces, as described previously. For those organisms producing iron sulfide, it has been found that this compound is not only magnetic but also an effective adsorbent for metallic elements.

Metal, Metalloid, and Organometal Transformations

Microorganisms can transform certain metal, metalloid, and organometallic species by oxidation, reduction, methylation, or dealkylation. Biomethylated derivatives are often volatile and may be eliminated from a system by evaporation. The two major metalloid transformation processes described are reduction of metalloid oxyanions to elemental forms and methylation.

Microbial reduction and oxidation of metalloid oxyanions

The reduction of selenate (Se(VI)) and selenite (Se(IV)) to elemental selenium can be catalyzed by numerous microbes, which can result in a red precipitate deposited around the cells and colonies. Some bacteria use SeO_4^{2-} as a terminal e^- acceptor in dissimilatory reduction as well as reduce and incorporate Se into organic components, for example, selenoproteins (assimilatory reduction). Selenate (SeO_4^{2-}) and selenite (SeO_3^{2-}) can be reduced to Se^0, with SeO_3^{2-} reduction appearing more ubiquitous than SeO_4^{2-} reduction. However, only SeO_4^{2-} can support bacterial growth under anaerobic conditions: SeO_4^{2-} reduction to Se^0 is a major sink for Se oxyanions in anoxic sediments. Anaerobic SRB such as *Desulfovibrio desulfuricans* can reduce selenate/selenite to Se^0, but neither oxyanion could be used for respiratory growth. Reduction to Se^0 can be considered a detoxification mechanism. Reduction of TeO_3^{2-} to Te^0 is also a means of detoxification found in bacteria and fungi, with Te^0 being deposited in or around the cells, resulting in black colonies. The opposite process of Se^0 oxidation can occur in soils and sediments. It is possible that Se^0 oxidation is similar to S oxidation, and may be mediated by heterotrophs and autotrophs. In aerobic soil slurries, Se^{4+} is the main product with lower amounts of Se^{6+} being produced; heterotrophic and autotrophic thiobacilli were believed to be the active organisms.

The extreme reducing conditions that SRB creates can also result in indirect chemical reduction of metal(loid) and radionuclide species. This appears to be the case for U(VI), which is reduced chemically to U(IV) under highly reducing sulfidic conditions. The ability of SRB to enzymatically reduce uranium is now established, and there has been some debate as to the relative contribution of chemical and enzymatic uranium reduction, with chemical reduction rates appearing relatively low. Nevertheless, while enzymatic reduction is effective in nonmetabolizing cells, some studies appear to support a role for chemical reduction in the presence of growing cells.

Technetium (which, as ^{99}Tc, is a long half-life product of the nuclear fuel cycle) is present in many environments as Tc(VII) in the form of highly mobile pertechnetate ion (TcO_4^-). Chemical and enzymatic reductive precipitation of Tc has been demonstrated in SRB, in that chemical precipitation appears to be more efficient for uranium, with sulfide as a reductant. Under sulfidogenic conditions, chemical precipitation is operated in preference to enzymatic reduction, and *D. desulfuricans* was able to precipitate Tc extracellularly, probably as sulfide.

In contrast to the reductive precipitation of metal(loid) described above, arsenic reduction frequently increases the solubility of this toxic element. Microbial dissimilatory reduction of As(V) to As(III) has been identified as an important route for increased As toxicity in the environment. This capacity appeared, for some time, to be phylogenetically and metabolically separate from dissimilatory sulfate reduction, but a *Desulfotomaculum* strain has the capability to simultaneously reduce arsenic and sulfate and to stimulate the precipitation of As(III) sulfide. The capacity of SRB to reduce and solubilize As and for soluble As(III) to precipitate with sulfide has further potential for bioremediation.

Methylation of metalloids

Microbial methylation of metalloids to yield volatile derivatives, for example, DMSe, dimethyl telluride, or trimethylarsine, can be effected by a variety of bacteria, algae, and fungi. Bacteria and fungi are the most important Se methylaters in soil, with the most frequently produced volatile being DMSe. Selenium methylation appears to involve the transfer of methyl groups such as carbonium (CH_3^+) ions via the S-adenosyl methionine system. Arsenic compounds such as arsenate (As(V), AsO_4^{3-}), arsenite (As(III), AsO_2^-), and methylarsonic acid ($CH_3H_2AsO_3$) can be methylated to volatile dimethylarsine($(CH_3)_2HAs$) or trimethylarsine ($(CH_3)_3As$). Environmental factors that affect microbial activity can markedly affect Se methylation, for example, pH, temperature, organic amendments, and Se speciation; however, the addition of organic amendments can stimulate methylation. The opposite process of demethylation can also occur in soil and water systems. Anaerobic demethylation may be mediated by methylotrophic bacteria.

Microbial metalloid transformations and bioremediation

In situ immobilization of SeO_4^{2-} by reduction to Se^0 has been achieved in Se-contaminated sediments. Microbial methylation of selenium, resulting in volatilization, has also been used for *in situ* bioremediation of selenium-containing land and water at Kesterson Reservoir in the United States. Selenium volatilization from soil was enhanced by optimizing soil moisture, particle size, and mixing, while in waters it was stimulated by the growth phase, salinity, pH, and selenium concentration. Se-contaminated agricultural drainage water was evaporated to dryness until the sediment selenium concentration approached 100 mg Se kg^{-1} dry weight. Conditions such as carbon source, moisture, temperature, and aeration were then optimized for selenium volatilization, and the process continued until selenium levels in sediments declined to acceptable levels. Some potential for *ex situ* treatment of selenium-contaminated water has also been demonstrated.

Mercury and organometals

Key microbial transformations of inorganic Hg^{2+} include reduction and methylation. The mechanism of bacterial Hg^{2+} resistance is enzymic reduction of Hg^{2+} to nontoxic volatile Hg^0 by mercuric reductase. Hg^{2+} may also arise from the action of organomercurial lyase on organomercurials. Since Hg^0 is volatile, this could provide one means of mercury removal. Methylation of inorganic Hg^{2+} leads to the formation of more toxic volatile derivatives; the bioremediation potential of this process (as for other metals and metalloids, besides selenium, capable of being methylated, e.g., As, Sn, and Pb) has not been explored in detail. In addition to organomercurials, other organometals may be degraded by microorganisms. Organoarsenicals can be demethylated by bacteria, while organotin degradation involves sequential removal of organic groups from the tin atom. In theory, such mechanisms and interaction with bioremediation possibilities described previously may provide a means of detoxification.

Concluding Remarks

Microorganisms play important roles in the environmental fate of toxic metals, metalloids, and radionuclides, with physicochemical and biological mechanisms effecting transformations between soluble and insoluble phases. Such mechanisms are important components of natural biogeochemical cycles for metals and associated elements, for example, sulfur and phosphorus, with some processes being of potential application to the treatment of contaminated materials. The removal of such pollutants from contaminated solutions by living or dead microbial biomass and derived or excreted products may provide a means for element recovery and environmental protection. Although the biotechnological potential of some of these processes has only been explored in the laboratory or on a pilot scale, some mechanisms, notably bioleaching, biosorption, and precipitation, have been employed at a commercial scale. Of these, chemolithotrophic leaching is an established major process in mineral extraction but has also been applied to the treatment of contaminated land. There have been several attempts to commercialize biosorption using microbial biomass but success has been short-lived, primarily due to competition with commercially produced ion-exchange media. Bioprecipitation of metals as sulfides has achieved large-scale application, and this holds out promise for further commercial development. Exploitation of other microbiological processes will undoubtedly depend on a number of scientific, economic, and political factors.

RECENT DEVELOPMENTS

The fate of heavy metals in the environment, microbial responses, and applications in bioremediation are now often viewed within the context of geomicrobiology. Geomicrobiology can simply be defined as the roles of microbes in geological processes, and all kinds of microbes, including prokaryotes and eukaryotes, and their symbiotic associations with each other and "higher organisms" can contribute actively to geological phenomena. Most of these processes, such as mineral dissolution or formation, biogeochemical cycles, bioweathering, and soil formation, involve metal transformations and/or redistribution in the environment. Apart from being important in natural biosphere processes, metal–microbe interactions can have beneficial or detrimental consequences in a human context. Bioremediation refers to the application of biological systems to the cleanup of organic and inorganic pollution, with bacteria and fungi being the most important organisms in this context for reclamation, immobilization, or detoxification of metals. Some metal-containing biominerals or metallic elements deposited by microbes may also have catalytic and other properties in nanoparticle, crystalline, or colloidal forms, and these are relevant to the development of biomaterials for structural, technological, environmental, and antimicrobial purposes. In contrast, metal (and mineral) transformations by microbes may result in degradation and spoilage of natural and synthetic materials, rock and mineral-based building materials, acid mine drainage and associated metal pollution, biocorrosion of metals, alloys, and related substances, and adverse effects on radionuclide speciation, mobility, and containment. Most microbial survival mechanisms depend on some change in metal speciation leading to decreased or increased mobility. These include redox transformations, the production of metal-binding peptides and proteins (e.g. metallothioneins, phytochelatins), organic and inorganic precipitation, active transport, efflux, and intracellular compartmentalization, while cell walls and other structural components have significant metal-binding abilities. Other microbial properties lead to metal solubilization from organic and inorganic sources. It is now appreciated that these processes are central to metal biogeochemistry and emphasize the link between microbial responses and geochemical cycles for metals. Such metal–mineral–microbe interactions are especially important in the so-called terrestrial "critical zone", defined as "the heterogeneous, near-surface environment in which complex interactions involving rock, soil, water, air, and living organisms regulate the natural habitat and determine the availability of life sustaining resources".

Organic matter decomposition is one of the most important microbial activities in the biosphere, and the ability of microbes, mainly bacteria and fungi, to utilize a wide spectrum of organic compounds is well known. Degradation of such substances results in redistribution of component elements between organisms and environmental compartments. The vast majority of elements in plant, animal, and microbial biomass (>95%) comprise carbon, hydrogen, oxygen, nitrogen, phosphorus and sulfur; in addition to these, several other elements are typically found in living organisms most with essential biochemical and structural functions, for example K, Ca, Mg, B, Cl, Fe, Mn, Zn, Cu, Mo, Ni, Co, Se, Na, and Si. However, all 90 or so naturally occurring elements may be found in plants, animals, and microbes, including Au, As, Hg, Pb, Cd and U. Some of these elements will be taken up as contaminants in food and from the environment. It should therefore be stressed that all decomposing, degradative, and pathogenic microbial activities are linked to cycling of these constituent elements, most of which are metals and some of which may be radionuclides accumulated from anthropogenic sources. This emphasizes the global involvement of microbes in almost all elemental cycles, including those for metals.

Many microbial metal and mineral transformations have the potential for the treatment of environmental pollution, and some processes are in commercial operation. However, many processes are still at the laboratory scale and have still to be tested in a rigorous applied and/or commercial context. Microbial activities in anaerobic subsurface environments also offer possibilities for metal and radionuclide bioremediation. Metal(loid)s that form insoluble precipitates when reduced include Se(0), Cr(III), Tc(IV), and U(IV). Microbial reduction of U(VI) to U(IV) has been proposed as a bioremediation strategy for uranium-contaminated groundwaters, as reduction of U(VI) under anaerobic conditions produces U(IV), which precipitates as the insoluble mineral uraninite. Biogenic uraninite is an important nanoscale biogeological material and crucial to the viability of microbial bioremediation strategies for subsurface uranium contamination by stimulated uranium reduction because it is orders of magnitude less soluble than most other U species. In addition to bioremediation, microbe metal/mineral transformations have applications in other areas of biotechnology and bioprocessing, including biosensors, biocatalysis, electricity generation, and nanotechnology. Metal micro/nanoparticles, with appropriate chemical modification, have applications as new metal composites or structured materials for a variety of applications. The use of metal-accumulating microbes for the production of nanoparticles, as well as their assembly, may allow control over size, morphology, composition, and crystallographic orientation of the particles. The potential of such biomimetic materials appears great and is relevant to production of new advanced materials, with applications in metal and radionuclide bioremediation, antimicrobial treatments (e.g., nanosilver), solar energy and electrical battery applications, and microelectronics. Bacterial reduction of Pd(II) to Pd(0) (Bio-Pd) can be carried out by several anaerobic bacteria, and the application of bio-Pd nanoparticle catalysts for dehalogenation, reduction and (de)hydrogenation reactions for the treatment of contamination has been demonstrated. Applications of microbially produced nanosilver particles include their use as antimicrobial agents, catalysts in chemical synthesis, biosensors, and electrodes. Microbial interactions with and transformations of metals are a vital part of natural biosphere processes and can also have beneficial or detrimental consequences for human society. Our understanding of this important area of microbiology and exploitation in bioremediation and other areas of biotechnology will clearly require a multidisciplinary approach.

FURTHER READING

Bargar, J. R., Bernier-Latmani, R., Glammar, D. E., & Tebo, B. M. (2008). Biogenic uraninite nanoparticles and their importance for uranium remediation. *Elements*, *4*, 407–412.

Brantley, S. L., Goldhaber, M. B., & Ragnarsdottir, K. V. (2007). Crossing disciplines and scales to understand the critical zone. *Elements*, *3*, 307–314.

Burford, E. P., Fomina, M., & Gadd, G. M. (2003). Fungal involvement in bioweathering and biotransformation of rocks and minerals. *Mineralogical Magazine*, *67*, 1127–1155.

Chasteen, T. G., & Bentley, R. (2003). Biomethylation of selenium and tellurium: Microorganisms and plants. *Chemical Reviews*, *103*, 1–26.

Ehrlich, H. L., & Newman, D. K. (2009). *Geomicrobiology* (5th ed.). Boca Raton, FL: CRC Press/Taylor and Francis Group.

Gadd, G. M. (1993). Interactions of fungi with toxic metals. *New Phytologist*, *124*, 25–60.

Gadd, G. M. (1993). Microbial formation and transformation of organometallic and organometalloid compounds. *FEMS Microbiology Reviews*, *11*, 297–316.

Gadd, G. M. (2000). Bioremedial potential of microbial mechanisms of metal mobilization and immobilization. *Current Opinion in Biotechnology*, *11*, 271–279.

Gadd, G. M. (2000). Microbial interactions with tributyltin compounds: detoxification, accumulation, and environmental fate. *Science of the Total Environment*, *258*, 119–127.

Gadd, G. M. (2001). Accumulation and transformation of metals by microorganisms. In H. J. Rehm, G. Reed, A. Puhler, & P. Stadler (Eds.) *Biotechnology, a multi-volume comprehensive treatise vol. 10: Special Processes* (pp. 225–264). Weinheim, Germany: Wiley-VCH Verlag GmbH.

Gadd, G. M. (Ed.), (2001). *Fungi in bioremediation*. Cambridge: Cambridge University Press.

Gadd, G. M. (2004). Microbial influence on metal mobility and application for bioremediation. *Geoderma*, *122*, 109–119.

Gadd, G. M. (2007). Geomycology: biogeochemical transformations of rocks, minerals, metals and radionuclides by fungi, bioweathering and bioremediation. *Mycological Research*, *111*, 3–49.

Gadd, G. M., & Sayer, G. M. (2000). Fungal transformations of metals and metalloids. In D. R. Lovley (Ed.) *Environmental Microbe-Metal Interactions* (pp. 237–256). Washington, DC: American Society for Microbiology.

Gadd, G. M., & White, C. (1993). Microbial treatment of metal pollution – a working biotechnology. *Trends in Biotechnology*, *11*, 353–359.

Gadd, G. M. (2009). Biosorption: critical review of scientific rationale, environmental importance and significance for pollution treatment. *Journal of Chemical Technology and Biotechnology*, *84*, 13–28.

Gadd, G. M. (2010). Metals, minerals and microbes: geomicrobiology and bioremediation. *Microbiology*, *156*, 609–643.

Hennebel, T., Gusseme, B. D., & Verstraete, W. (2009). Biogenic metals in advanced water treatment. *Trends in Biotechnology*, *27*, 90–98.

Karlson, U., & Frankenberger, W. T. (1993). Biological alkylation of selenium and tellurium. In H. Sigel, & A. Sigel (Eds.) *Metal ions in biological systems* (pp. 185–227). New York: Marcel Dekker.

Konhauser, K. (2007). *Introduction to geomicrobiology*. Oxford: Blackwell.

Lloyd, J. R., Lovley, D. R., & Macaskie, L. E. (2004). Biotechnological applications of metal-reducing microorganisms. *Advances in Applied Microbiology*, *53*, 85–128.

Lloyd, J. R., & Renshaw, J. C. (2005). Bioremediation of radioactive waste: radionuclide-microbe interactions in laboratory and field-scale studies. *Current Opinion in Biotechnology*, *16*, 254–260.

Lloyd, J. R., Pearce, C. I., Coker, V. S., Pattrick, R.A.D.P., van der Laan, G., Cutting, R. *et al.* (2008). Biomineralization: linking the fossil record to the production of high value functional materials. *Geobiology*, *6*, 285–297.

Macaskie, L. E. (1991). The application of biotechnology to the treatment of wastes produced by the nuclear fuel cycle – biodegradation and bioaccumulation as a means of treating radionuclide-containing streams. *Critical Reviews in Biotechnology*, *11*, 41–112.

Morley, G. F., & Gadd, G. M. (1995). Sorption of toxic metals by fungi and clay minerals. *Mycological Research*, *99*, 1429–1438.

Nies, D. H. (2003). Efflux-mediated heavy metal resistance in prokaryotes. *FEMS Microbiology Reviews*, *27*, 313–339.

Sparks, D. L. (2005). Toxic metals in the environment: the role of surfaces. *Elements*, *1*, 193–197.

Stolz, J. F., & Oremland, R. S. (1999). Bacterial respiration of arsenic and selenium. *FEMS Microbiology Reviews*, *23*, 615–627.

White, C., Sharman, A. K., & Gadd, G. M. (1998). An integrated microbial process for the bioremediation of soil contaminated with toxic metals. *Nature Biotechnology*, *16*, 572–575.

White, C., Wilkinson, S. C., & Gadd, G. M. (1995). The role of microorganisms in biosorption of toxic metals and radionuclides. *International Biodeterioration & Biodegradation*, *35*, 17–40.

RELEVANT WEBSITE

http://www.paques.nl/. – PAQUES Proven Technology based on natural solutions.

Part V

Biotechnological Topics

Chapter 38

Biodeterioration - Including Cultural Heritage

G. Ranalli[1], E. Zanardini[2] and C. Sorlini[3]

[1]*University of Molise, Campobasso, Italy*

[2]*University of Insubria, Como, Italy*

[3]*University of Milan, Milan, Italy*

Chapter Outline

ABBREVIATIONS

DGGE Denaturing gradient gel electrophoresis
EPS Extracellular polymeric substances
MCF Microcolonial fungi
MIC Microbially influenced corrosion
SRB Sulfate-reducing bacteria

DEFINING STATEMENT

This article gives an overview of the biodeterioration of cultural heritage caused by biodeteriogen agents and their mechanisms and processes. Basic information on the material composition of objects of art is provided to address the problem of biodeterioration. Finally, it stresses and highlights the relevance of preservation and conservation practices.

BIODETERIORATION

Biodeterioration is a phenomenon that affects all materials, including those used in buildings, metals, monuments, pigments, frescoes, and so on. Stone deterioration is a natural process resulting from exposure to atmospheric agents and is referred to as stone/rock disintegration, with debris (gravel, sand, and clay) being formed that, through transportation and sedimentation processes, can then constitute new rock.

Outdoor artwork in particular suffers deterioration processes, especially in urban and industrialized areas where atmospheric pollutants and climatic conditions act together. Indeed, rain and pollution act directly on stone surfaces, altering their chemical composition, while wind and thermal differences tend to lead to an increase in specific types of stone surfaces, resulting in the formation of microfissures and microfractures. This can give rise to opportune conditions for the development and growth of microbes. Biodeterioration is defined as any form of irreversible alteration, implying a modification in the properties of materials due to metabolic activity and the growth of organisms called biodeteriogens. Atmospheric pollutants play a determining role in deterioration processes, and industrial development over the past century has led to an increase in the number of pollutants and caused accelerated deterioration in artwork exposed to the open air.

Apart from corrosive substances present as atmospheric contaminants, other compounds can also accumulate on surfaces and favor the colonization of specific microbial populations able to utilize such compounds for growth (Table 1).

Deterioration is a phenomenon that occurs in materials of every type, including those used in buildings, metals, stones of monuments, fresco pigments, and so on. The deterioration of stone on the geological temporal scale is part of a natural cycle that leads to the disintegration of rock exposed to atmospheric agents, forming detritus material (gravel, sand, and clay) that, through processes of transportation, sedimentation, and diagenesis, can constitute new rock. Artwork out in the open air in highly industrialized areas is especially subjected to deterioration as there is direct interaction with atmospheric polluting agents as well as climatic conditions. Rain and pollutants act directly on the stone surfaces, altering the chemical composition of the material. Wind, extreme temperatures, and aggressive pollutants present in the atmosphere provoke over time a deterioration that leads to an increase in the specific surfaces of marble, with the phenomenon of decohesion and the formation of microfissures. After the surface has deteriorated, conditions become favorable for microbial attack and the development of living organisms (Figure 1).

TABLE 1 Activities of main microbial deteriogens capable to use organic and inorganic substrates related to several matrices of cultural heritage

Microbial deteriogens	Activities	Substrates	Matrices	Genera
Proteolytic bacteria (ammonifying)	Hydrolyze, peptidase	Casein, egg yolk, collagen	Parchment, leather, wool, silk, frescoes	*Pseudomonas, Sarcina, Bacteroides*
Cellulolytic bacteria	Cellulase	Cellulose, tannins, resins, shellac, waxes, etc.	Textile fibers, (cotton, linen, jute, hemp), paper	*Cytophaga, Sporocytophaga, Sorangium, Vibrio, Cellvibrio, Cellfalcicula*
Amylolytic	Amylase	Starch	Frescoes	*Bacillus, Clostridium*
Lipolytic bacteria	Lipase	Fats, walnut and linseed oils, waxes	Wood, frescoes	*Bacillus, Alcaligenes, Staphylococcus, Clostridium*
Nitrifying bacteria	Corrosion by NO_2^- and NO_3^- yield	NH_3	Stone and metal artworks, varnish	*Nitrosomonas, Nitrobacter*
Denitrifying bacteria	Denitrification (anaerobiosis)	Several organic substrates	Stone artworks	*Pseudomonas, Bacillus, Vibrio*
Sulfate-oxidizing bacteria	Oxidations	S compounds	Stone artworks	*Thiobacillus, Thiosphaera*
Sulfate-reducing bacteria	Corrosion by H_2S (anaerobiosis)	S compounds	Stone and metal artworks	*Desulfovibrio, Desulfobacter, Desulfococcus*
Heterotrophic bacteria	Red/ox; weak corrosion	Inorganic compounds (air pollutants); cell lyses of organotrophs	Several materials	*Idrogenomonas, Thiobacillus*
Actinomycetes	Hydrolytic; Mechanical action	Cellulose, hemicellulose, etc.	Textile fibers (cotton, linen, jute, hemp)	*Streptomyces, Nocardia Cellulomonas*
Yeasts	Fermentative	Carbohydrates	Frescoes, textile fibers	*Candida, Lypomyces Cryptococcus, Torulopsis*
Fungi	Several (hydrolyze, proteinase, lipase); Mechanical action by hyphae penetration	Starch, cellulose, hemicellulose, lignin	Wood and stone artworks, frescoes, mortar, plaster, varnish, paper	*Geotrichum, Pullularia, Cladosporium, Tricoderma, Aspergillus Alternaria, Penicillium*
Algae	Photosynthetic	Inorganic compounds	Stone artworks, frescoes, mortar, varnish, metals	*Chlorococcales* order *Scenedesmus Chlorella*

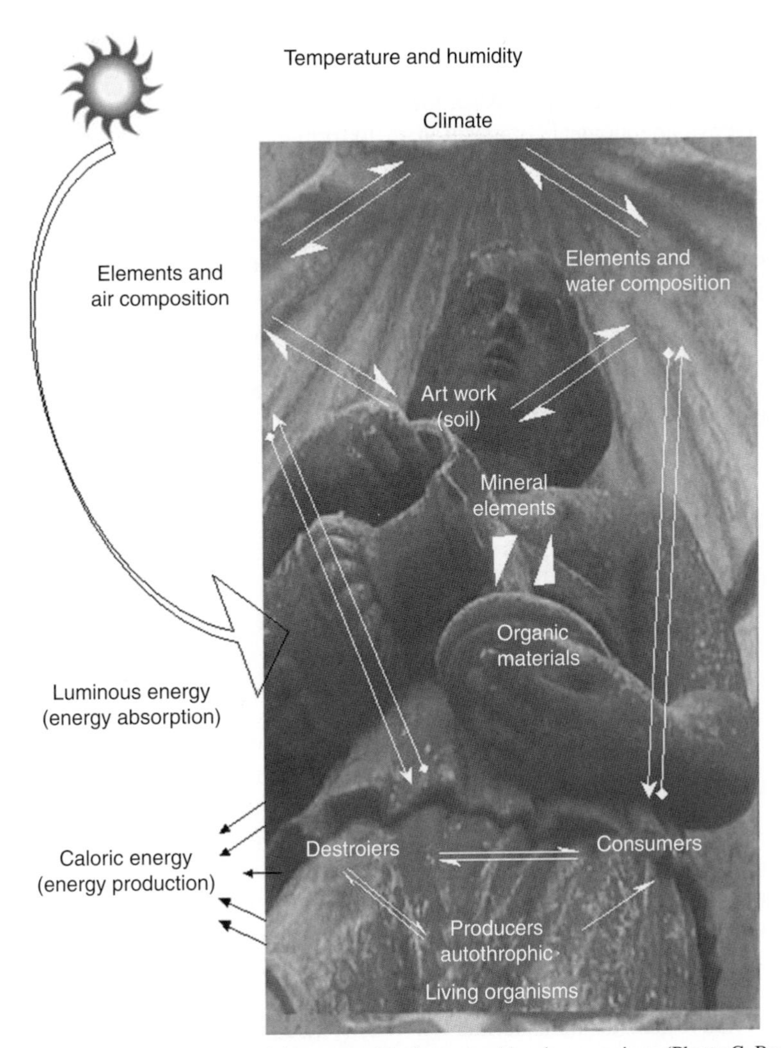

FIGURE 1 Ecological succession on the artwork by autotrophic–heterotrophic microorganisms (Photo: G. Ranalli, G. Lustrato).

Biological agents such as bacteria, algae, fungi, lichens, mosses, and plants can colonize different substrates (paper, wood, stone, frescoes, paintings, etc.) and damage both surfaces and deep layers. Indeed, microflora able to grow on objects of art have a very complex ecosystem, and their development varies depending on the environmental conditions and the chemico-physical properties of the substrate. Also, climatic conditions play an important role. Actinomycetes and mycetes growing on a stone surface produce acids and substances, and these react with the substrate, causing its disintegration and allowing penetration of the hyphae into internal layers. Thus, such processes represent both chemical and mechanical deterioration. Furthermore, organisms often aggregate and produce biofilm and biological patina.

Laboratory experiments have demonstrated that biofilm formation on surfaces occurs in two steps: organism adhesion on a surface followed by organism aggregation. The first step involves organisms and materials and is connected to the properties of the material (e.g., hydrophilic or hydrophobic areas) including Van der Waals and Debye forces.

The phenomenology of the microbial attack and the type of alteration depend on both the environmental conditions and the characteristics and state of the stone surface. Indeed, biodeterioration is a secondary deterioration process that occurs in substrates that have already been altered by chemical and physical damages.

Abiotic Factors

The main abiotic factors that affect the growth of biodeteriogens are humidity, light, temperature, the type of substrate, ventilation, and air pollution. Humidity strongly favors the growth of algae and fungi. Light, both natural and artificial, is necessary for photosynthetic microorgansisms (algae and cyanobacteria), lichens, mosses, and plants. Temperatures higher than 30 °C together with high relative humidity accelerate all biological processes, while low temperatures (4 °C) delay them. The substrate composition of artwork plays an important role in the growth of organisms. A material that is mainly composed of organic substances favors the growth of heterotrophs, particularly mycetes, while materials consisting mainly of inorganic substances, such as stone, favor the growth of chemolithotrophs and phototrophs. Also, the properties of materials, such as mineralogical composition, porosity, and structural characteristics, along with the conservation state, can affect microbial colonization. This tendency, defined as the bioreceptivity of a surface, includes all the properties and characteristics of a material that contribute to, and favor, microbial growth. Ventilation is very important in the colonization process to reduce humidity, prevent condensate formation on the surface, avoid stagnancy of polluted air, and prevent atmospheric particulate and microorganism deposition on surfaces. The organic chemical pollutants in the atmosphere serve as an important carbon source for organisms. There is a strong correlation between aerobiology and biodeterioration. In fact, air represents the most important medium for the transportation of microorganisms (vegetative cells and spores) that can reach the surface of artistic manufacts and give rise to colonization. Other factors that can affect this process are the type of microorganism, the chemical composition of the substrate, and climatic conditions.

Mechanisms of Biodeterioration

Deterioration is the term used to define the modification of a material that results in a worsening in the material's characteristics from the point of view of conservation. On the other hand, the term "alteration" indicates a modification of a material, but it does not necessarily imply a worsening in its characteristics from the conservation point of view. Indeed, the aggression of living organisms on materials can occur in two different ways that, however, act in synergy toward the disintegration of the material.

Chemical aggression

Chemical aggression occurs through the decomposition of a substrate and its transformation due to the direct action of intermediate and final products of microbial metabolism. Such deterioration, exclusive to stonework, is defined as biocorrosion. The metabolites mainly involved in this deterioration mechanism are organic and inorganic acids and chelating substances. The acids produced by the organisms react with the material of the stone, generally the insoluble salts, dissolving them and exposing them to be washed away by rainwater. All living aerobic organisms produce, through respiration, CO_2 as a final metabolism product, and when this reacts with water on limestone material, it is transformed to carbonic acid (H_2CO_3). Although weak, this acid still dissolves the relatively insoluble calcium and magnesium carbonates to form their bicarbonates that are very soluble and easily removed by heavy rain. The sulfuric acid produced biologically by the sulfur-oxidizing bacteria, particularly the *Thiobacillus* genera, and the nitric acid produced by the nitrifying *Nitrobacter* and *Nitrosomonas* bacteria act as powerful corrosives, reacting with calcium ions to form the soluble and easily washed away calcium sulfate (chalk) and calcium nitrate. All acidic substances taking part in chemical attack processes act through an acidolysis mechanism. The chelating substances, some of which are organic acids, dissolve the trivalent metallic cations of the stone and thus provoke its disintegration. On materials containing organic substances (frescoes, tapestries, and paintings), there are biochemical processes due to the activity of exoenzymes from heterotrophic colonizers, mostly fungi, that use the organic substances present (casein, solvents, organic acids, and pigments of natural origin) as growth substrates. Indeed, bacteria developing on wall paintings can provoke variations in the color of paintings not only through pigment production but also through the excretion of metabolic products that provoke a loss of color in parts of the painting on the wall.

Aggression via mechanics

Stones exposed to the open air suffer accelerated deterioration of a mechanical type – washing by the rain, thermal excursions in the form of freezing and thawing cycles, and the action of the wind. Mechanical aggression by living organisms occurs through the growth of microorganisms or parts thereof (hyphae of fungi, rizine of licheni, and superior plant roots). Such action is manifested in the form of fissures and fractures of the attacked substrate, but there is no variation in the chemical composition of the material. This mechanical aggression becomes much more important when the colonizing organisms are not limited to the development on the surface of the material (epilytic organisms) but actively penetrate the stone (endolytic organisms), provoking disintegration of the material. Toward the end, there is a physical/mechanical attack by the filamentous microorganisms, which, taking advantage of the material's porosity, penetrate and cause its disintegration. Some microorganisms cause damage due to variation in the water mass volume in the tissues; as such tissues are subject to dehydration and rehydration. A particular type of degradation, associated indirectly with biodeterioration, can be seen in restoration interventions. Indeed, damage is often due to operators who are not fully aware of the many pitfalls of dealing manually with plants anchored to walls or the cleaning of surfaces covered by biological patina.

BIODETERIOGENS

Biodeteriogens are organisms that participate in the biodeterioration process of materials, including historical and artistic manufacts. Such organisms are autotrophs and heterotrophs and colonize different types of inorganic (stonework, metals, and glass) and organic materials (paper, wood, parchment, leather, textile, etc.). Until the last decade, it was always considered that the first colonizers of stone were autotrophic organisms (both photoautotrophs and chemolithotrophs) able to utilize CO_2 as the carbon source, while the colonization by heterotrophs on the surface occurred only subsequently, with these organisms growing on the organic compounds released by the cellular lyses of autotrophs.

It was recently demonstrated that microbial symbiotic association can occur on stone surfaces. Under certain environmental conditions (i.e., in polluted areas), specific and adapted heterotrophic microflora have been evidenced in cometabolic processes associated with the degradation of organic atmospheric pollutants accumulated on exposed surfaces. In these cases, pollutants acting directly in the deterioration process and causing chemical damage on the superficial layers can also act indirectly as a nutritional source for microorganisms, favoring the selection of a specific heterotrophic microflora. Indeed, it has been demonstrated that some of these microorganisms growing on stonework exposed to air pollution are able to degrade aliphatic and aromatic hydrocarbons, including polycyclic compounds, as the carbon source for their growth. Furthermore, the metabolic potentiality of the colonizing microflora has been confirmed by the presence of functional genes that encode enzymes involved in the degradation of aromatic compounds, such as monoxygenases and dioxygenases. Thus, on the basis of this metabolic capability, heterotrophic microflora can also be considered the first colonizer of pollutant-exposed stonework. Note also that the microorganisms can grow using the very compounds applied to the stonework as a protective and consolidating agent (Figure 2).

Finally, some microorganisms that grow on art objects can also exhibit particular metabolic conditions, lying dormant or in a noncultivable state under oligotrophic conditions for many years. The principal groups of microorganisms that are biodeteriogenic with regard to cultural properties are listed below. They are classified according to autotrophic or heterotrophic nutritional characteristics.

Bacteria (Eubacteria and Archaea)

Bacteria are unicellular microorganisms, principally characterized by the prokaryotic organization of their cells, that show a wide range of metabolic functions and capabilities. As per the phylogenetic tree of life, two prokaryotic domains, Bacteria and Archaea (unicellular organisms that have a nucleus but no membrane) are present. The main differences in the cellular structure of bacterial and archaeal microorganisms are seen in the composition of the cell wall: peptidoglycan in most Eubacteria and pseudomurein, polysaccharides, proteins, or glycoproteins in the Archaea, while

FIGURE 2 Scanning electron micrograph of biofilm on altered marble surface (Photo: G. Ranalli).

the RNA polymerase is made up of a greater number of subunits for the bacteria, between 8 and 12, as compared to the 4 found in Eubacteria. Given the different cell wall composition, the Eubacteria are subdivided into Gram-positive and Gram-negative. The groups most frequently associated with artwork deterioration are considered here.

Chemolithotrophic bacteria

Sulfur-oxidizing bacteria are often considered one of the most dangerous groups for the conservation of stonework (chemolithotrophic) as they produce sulfuric acid, an inorganic acid that has a strong degrading action through the oxidation of hydrogen sulfide, elemental sulfur, and thiosulfates. Indeed, the most dangerous for artifacts are the thiobacilli (*Thiobacillus* and *Thiomicrospira*) as they directly oxidize sulfides to sulfates and their ecological characteristics favor a greater production of sulfuric acid.

Nitrifying bacteria include ammonia-oxidizing bacteria (*Nitrosomonas* and *Nitrosococcus*) that oxidize ammonia to nitrous acid and nitrite-oxidizing bacteria that oxidize nitrous acid to nitric acid (*Nitrobacter* and *Nitrococcus*). The metabolic process is called nitrification. Nitrifying bacteria play an important role in stone weathering because of nitric acid production.

Hydrogen bacteria use molecular hydrogen as a source of energy, converting CO_2 to organic carbon. Their importance lies in the fact that they remain active under limited environmental conditions and even in the temporary absence of the primary energy source. Almost all are facultative chemolithotropic organisms, having the capacity to use organic compounds as the energy source (chemoorganotrophs). Hydrogen bacteria are active in both the presence and the absence of organic substances, as long as the hydrogen concentration is high. Both Gram-positive (*Bacillus*) and Gram-negative organisms are found, the most common and well known being the *Pseudomonas*, *Paracoccus*, and *Alcaligenes*.

Ferrobacteria are microorganisms that obtain their energy from the aerobic oxidation of iron: from ferrous (Fe^{2+}) to ferric (Fe^{3+}). Ferrobacteria are found on stone containing pyrites (ferric sulfide), frescos and wall paintings, and so on, where reduced iron compounds were employed, and on iron artifacts. In acid environments, *Thiobacillus ferrooxidans* is the most frequently found, growing autotrophically by utilizing both ferrous ions and compounds reduced from sulfur as electron donors.

Heterotrophic bacteria

A great number of heterotrophic bacteria (Eubacteria and Archaea) are able to colonize artwork. Such bacteria are responsible for various types of enzymatic activity that can directly affect colonized materials, especially organic material. Heterotrophic bacteria have also been found frequently on inorganic materials such as stone monuments and mural paintings, and the great majority of them have been Gram-positive bacteria belonging to the genera *Bacillus*, *Brevibacillus*, *Micrococcus*, *Kocuria*, *Clostridium*, *Frankia*, *Geodermatophilus*, *Blastococcus*, *Staphylococcus*, *Streptomyces* and related genera (such as *Nocardia*, *Rhodococcus*, *Streptoverticillium*, and *Micromonospora*), and Gram-negative bacteria (*Pseudomonas* and *Acinetobacter*). Recently, halophilous Archaeal bacteria such as *Halomonas* and *Halococcus* were found on frescos. Heterotrophic bacteria can be distinguished by the nature of their enzymatic activity:

Proteolytic and Ammonifying Bacteria

These bacteria hydrolyze proteinaceous substances to peptides and the peptides to amino acids, producing extracellular hydrolytic enzymes (i.e., protease and peptidase). The amino acids are then broken down with the liberation of ammonia. The most frequently identified species belong to the genera *Pseudomonas*, *Sarcina*, *Bacteroides*, and *Streptomyces*.

Cellulolytic Bacteria

These bacteria can be distinguished as primary cellulolytic bacteria and as facultative cellulolytic bacteria. The primary cellulolytic bacteria specialize in cellulose degradation to such an extent that they are unable to grow if this compound is not present (*Cytophaga*, *Sperocytophaga*, and *Sporangium*), while the facultative cellulolytic bacteria are able to also utilize other organic compounds (*Vibrio*, *Cellvibrio*, and *Cellfalcicula*). Cellulolytic bacteria are quite common and have the capacity to break down lignin and other wood components such as resins, gums, dye, tannic acid, waxes, and fats.

Amylolytic Bacteria

They are essential for starch breakdown. Among them certain species of *Bacillus* and *Clostridium* are able to break down starches, and the speed of the breakdown will vary according to the chemical composition of the different molecules constituting the starch, particularly the percentage of amylopectin present. Amylopectin is a branched compound and is the component that is the slowest to be broken down by bacteria.

Lipolytic Bacteria

Relatively few genera are able to break down lipids, resulting in the production of lipase (a specific esterase that hydrolyzes the ester bonds between glycerol and the fatty acids), but among these particular mention must be made about *Bacillus*, *Alcaligenes*, *Staphylococcus*, and *Clostridium*. These microorganisms are able to degrade artifacts in which fatty substances are present as natural components (e.g., in wood) or else introduced during artifact manufacture (e.g., pastels).

Denitrifying Bacteria

These are anaerobic bacteria that are relatively easily found on the surface of artifacts, provided organic substances are present. In fact, they are facultative anaerobes and metabolize the substrate by means of aerobic respiration in the presence of oxygen; however, to reduce nitrates they require anaerobic conditions. They can also utilize other electron acceptors such as ferrous ion (Fe^{3+}). The biodeteriogenic activity of denitrifying bacteria is, in all respects, similar to that of other heterotrophic bacteria and is always bound to a specific enzymatic activity (amylase, protease, and esterase) and the production of catabolites such as organic acids, regardless of whether metabolism occurs in aerobic or anaerobic conditions (*Bacillus denitrificans*, *Pseudomonas stutzeri*, *Achromobacter severina*, etc.).

Fungi

Eumycota

They are eukaryotic organisms that are heterotrophic and are characterized by the presence of a rigid cell wall made principally of chitin; they belong to the kingdom of Fungi or Mycota. Fungi play an important role in the degradation of cultural heritage as they act either directly on organic materials that are utilized nutritionally or indirectly on materials of an inorganic nature that, although not metabolized directly, can nevertheless have organic fractions able to support fungal growth. In addition, fungi adapt easily to varying environmental conditions, and this means that environments for the conservation of artifacts of a historical and artistic interest, for example, hypogean environments, churches, museums, libraries, and archives, can provide ideal habitats for their development. In fungi with a pigmented mycelium (e.g., *Dematiaceae*), melanin is present. This confers the fungi with a resistance to numerous chemical and physical agents such as reducing substances, UV radiation, gamma and X-rays, and attack by enzymes. This resistance can be explained by the chemical nature of melanin, a quinone and hydroquinone combination that limits the damage caused by the formation of free radicals. However, the presence of melanin in the cell wall also seems to play a fundamental role in biodeterioration processes affecting stone in that it favors the penetration of the fungus into the substrate.

Many fungi such as *Cladosporium* have wide metabolic diversity and are able to flourish even in conditions of oligotrophy (scarcity of nutrient presence). Thus, they have often been identified as the fungi responsible for fresco degradation, the particles on the fresco surface representing the sole organic source. Degradation caused by microfungi, which in the case of artifacts of an organic origin often coincides with the degradation of the artwork itself, is due to the production of intracellular and extracellular enzymes specific to the individual substances to be consumed. Furthermore, the development of mycelium on a material, namely on the artwork, can also give rise to physical and mechanical damage caused by hyphae penetration.

Meristematic Fungi

Among the fungi, meristematic fungi are of particular interest in stonework biodeterioration processes. Black fungi of meristematic growth (or microcolonial fungi, MCF) are a broad and heterogenous group of microorganisms, grouped together because of their ability to grow meristematically and to produce a melanin pigment, generally olive-black in color.

Both the production of melanin and meristematic growth endow the cell with resistance to extreme environmental conditions such as strong sunlight and water and nutrient scarcity, conditions frequently encountered by artwork in stone that is left out in the open in Mediterranean regions. Black fungi with meristematic growth form small black colonies (60–100 μm in diameter) on rocks and grow very slowly (1–2 months) both under natural conditions and in the usual culture media. Moreover, they are extremely versatile and are able to grow on the most varied substrates, as well as in oligotrophic conditions. Such fungi are resistant to environmental stress such as high temperatures (>100 °C), as well as to prolonged periods of UV radiation exposure and osmotic stress (up to 40%). It is also important to consider the biodeterioration actions of organisms like insects, rodents, and birds on art objects made using different materials.

Until now, microorganisms on cultural heritage have generally been considered to be connected to actions of alteration. However, it was recently demonstrated that some microorganisms can be used in biocleaning and biorestoration procedures to remove harmful compounds from artistic objects. For example, the sulfate-reducing bacteria (SRB), *Desulfovibrio vulgaris*, has been used to remove a black crust containing sulfates and specific heterotrophic bacteria, *P. stutzeri*, to remove glue from frescoes. In addition, such bacteria can also be employed in calcite bioprecipitation for the conservation of monumental stone and for stone reinforcement.

Phototrophic microorganisms

This group includes algae and cyanobacteria and there is remarkable diversity in the organization of cells and vegetative structures. Algae are eukaryotic cells, exhibiting a full range of pigments (chlorophylls, carotenoids, and phycofibilins), highly specific reserve materials, and wall components that can be used as discriminating elements during placement within systematic classification. Cyanobacteria are prokaryotic organisms linked to the kingdom of Eubacteria, and are able to carry out photometabolism similar to that of higher plants. They colonize archaeological remains, stonework, frescoes, wall paintings, and mosaics, producing chromatic alterations, patinas, and biofilms of different colors (green, gray, orange, brown, and violet-red). Their deteriorative action is seen not only in aesthetical damage but also in chemical damage, because of substrate corrosion. Cyanobacteria can be divided into epilithic, growing on the surface of the stone, chasmoendolithic, in the fissures and cavities of the stone but in contact with the surface, and endolithic, inside the stone itself. The genera of algae most frequently referred to in the biodeterioration of stone belong to the division *Chlorophyta*. The matrix may form a structured sheath called a sheath, capsule, or calyx, or else simply be an amorphous mucilage. It constitutes a fundamental characteristic for the colonization and the survival of the organisms: it permits adhesion to the substrate, functioning as a water reserve as it slowly swells with water and then gradually releases it, allowing the organism to overcome periods of adversity, inhibits predation, and, most importantly, it is the cement that holds together the microbial cells, thus giving rise to the formation of the biofilm. Investigations carried out on stone monuments have shown that most of the cyanobacteria present are forms endowed with a gelatinous matrix. The chemical nature of such a matrix, generally made up of negatively charged polysaccharides due to the presence of acid groups, can also contribute to the deterioration of the stone through complex-forming processes and leaching of the constituent elements; it also represents a source of organic material that supports the growth of the nutritionally more demanding heterotrophic microorganisms.

Lichens

The most frequent organisms found on stone monuments in an outdoor environment are the lichens. They tend to enrich environments lacking in biodiversity, for example, archeological sites and urban centers, but can also be agents of chromatic alteration and surface biodeterioration. Lichens, small symbiotic ecosystems formed by a fungus (mycobiont), depend nutritionally on photosynthetic partners (photobionts). In addition to their biodeteriogenic action, lichens can also play a bioprotective role: in fact, in the presence of high atmospheric pollution levels, they provide an effective barrier against pollutants that attack the mineral components of a substrate.

Bryophytes

Bryophytes are photosynthetic eukaryotic organisms that can play a significant role in the biodeterioration of materials, especially stonework, as they penetrate the substrate by means of rhizoidal structures that cause physical and mechanical damage. Like algae and lichens, many moss species can be considered pioneer organisms as they can colonize exposed rocky surfaces, and in a few cases are able to survive high temperature levels and radiation from the sun as well as long drought periods. Because of their capacity to survive in a latent form, and return to normal metabolic activity with the restoration of favorable conditions of humidity, Bryophytes are also considered poikilohydric organisms.

Vascular plants

They are an important consideration in biodeterioration processes. Indeed, their roots form a complex system that, during growth, can exert considerable pressure on a substrate. In the event of the substrate being ancient walls, architectural structures, monuments, or hypogeum chambers, root growth can result in severe physico-mechanical damage.

MATERIALS RELATED TO CULTURAL HERITAGE

Considering the most accepted and first proposed definition of biodeterioration by Hueck (1965) as "any undesirable change in the properties of a material caused by the vital activities of organisms", it is important to illustrate the different kinds of materials that can be attacked.

Materials are any form of matter, with the exception of living organisms, that are used by humankind, and all materials have an intrinsic value and thus there is an important economic dimension to biodeterioration. Our cultural heritage is made up of almost all types of materials, both natural and artificial, organic (that may be natural, semisynthetic, or synthetic compounds) and inorganic, and include stone, metals, textiles, paper, wood, and so on; in more recent times, photographic documents and magnetic and optical carriers have been introduced and widely established all over the world.

The biodeterioration of organic materials is an essential process in the environment that helps in recycling complex organic matter and is an integral component of life. This process, however, also destroys historical structures, resulting in the loss of valuable cultural properties stored in libraries, archives, and repositories. Many enzymes break the bonds of polymers; some polyesters are degraded by lipases and esterases, proteins by proteases, starch by amylases, and cellulose and cellulose derivatives by cellulases. The organisms that pose major concerns to memorable heritage objects are microorganisms (bacteria, fungi, yeasts, and algae) and insects.

The biodeterioration of inorganic materials used in the works of art is essentially a process that involves monumental rocks and archaeological remains, glasses, and metals. These cultural heritage objects are influenced by environmental parameters, which can modify their structure and composition by biological mechanisms. As already cited, the main types of damage derived from the metabolic activity of organisms are associated with physical, chemical, and aesthetical mechanisms, while the intensity of damage is strictly related to the type and dimension of the organism involved, the kind of material and state of its conservation, environmental conditions, microclimatic exposure, and the level and types of air pollutants (Figure 3). Information on materials related to cultural heritage, separated on the basis of the inorganic, organic, and composite materials, are reported.

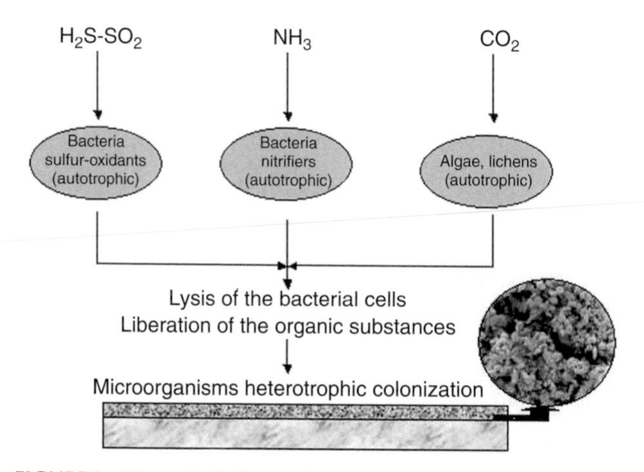

FIGURE 3 Theoretical scheme of ecosystemic view of a cultural heritage (Photo: G. Ranalli, G. Lustrato).

Inorganic Materials

Historical stone objects (sculptures, buildings, and rocks) and frescoes

Deterioration of building stone begins from the moment it is quarried as a result of natural weathering processes. Other factors, sometimes acting synergistically, including a progressive dissolution–crystallization of soluble salts, pollution, and biological colonization, can accelerate natural deterioration of the mineral matrix. Decay is a consequence of the weathering action of several physical, chemical, and biological factors. In the case of calcareous stones, the materials, due to calcite leaching, increase their porosity in time and decrease their mechanical characteristics. All building walls contain soluble salts, dispersed within the porous materials or locally concentrated. These salts are solubilized and migrate with the water in and out of the stone. The drying out of the solution at the exposed surface results in the formation of efflorescences. Moreover, climate plays an important part in influencing the activity of organisms on stone in monuments and other objects of cultural value. The presence of living organisms on stone increases its susceptibility to damage through their water-binding capacity. Then, the mineralogy, porosity, surface roughness, and capacity to collect water and organic materials control its bioreceptivity and tendency to biodeterioration. However, environmental factors such as temperature, light intensity, pH, and relative humidity affect the number and type of colonizing species and hence the progress of colonization. Different hygroscopic salts (carbonates, chlorides, nitrates, sulfates, etc.) can be found on the surface of decayed monuments, which have extremely saline environments. Halotolerant bacteria and Archaea can grow in many different niches including deteriorated mural paintings; the detection and identification of an Archaeal community by DNA extraction, fragment amplification with specific primers using denaturing gradient gel electrophoresis (DGGE) of PCR-amplified DNA encoding 16S rRNA, confirm important ecological implications of Archaea on monuments because of the considerable production of efflorescences. Water is very important in causing the mechanical degradation of stone, assessed mainly by porosity and bulk density, while its circulation through stone is controlled by its porosity size, distribution, specific surface, and capacity. Rainfall intensity and amount greatly influence erosion rates of marble and limestone, with a minor effect on sandstone. The presence and amount of water control biological activity and growth; moreover, hydrodynamic forces bring microorganisms (i.e., bacteria) close to substrates and their sorption onto solid surface induces the formation of extracellular polymeric substances (EPS), biofilms, and patina.

Patina includes many environmentally induced surface changes such as oxalate film, lacquer, crust, deposit, karst, rock varnish, microstromatolite, efflorescence, carbonate, gypsum, iron, manganese, oxalate and silica skin, and others. The formation of patina is a very complex process. Since it takes a very long period of time to develop, the patina itself is often seen as an object of historical interest. The main process can be summarized as an exchange of matter and energy between two open systems: the solid matrix and the surrounding environment (indoor or outdoor atmosphere, water soil, an intermediate cover of plaster or paint). The formation of patina can come to a standstill under environmental conditions. However, in cases in which patina formation becomes a historic evolution of the surface area of a monument, the real danger is in the addition of new layers on historical deposition on top of the original material. These deposits (crusts, skins, sinters, gypsum crust, etc.) may represent an increasingly dangerous hazard for the original material they cover; in fact, the mass increase may lead to fissures, exfoliation, and loss of the original surface material. Among the additional factors that accelerate deterioration of stone, the effects of anthropogenic sources of pollution on building stone are still to be definitively characterized. An example of particular interest is the formation and constituents of black crusts (Figure 4).

It has long been acknowledged that burning of fossil fuels has led to an increase in the concentrations of acid gases in the atmosphere; of these, perhaps the most important is sulfur dioxide, which forms sulfurous acid when dissolved in water (acidic rain). Sulfurous acid is oxidized to sulfuric acid, which in turn reacts with calcium carbonate to form calcium sulfate, the mineralized form of which is known as gypsum. The formation of gypsum then leads to the creation of cavities below the surface as a result of migration of calcium ions to the surface. Thus, if soluble gypsum is washed away, it takes with it some of the stone itself, initially causing loss of surface details but eventually leading to loss of structural integrity. Particulate matter from the atmosphere can combine with gypsum to leave unsightly black crusts

FIGURE 4 Black crust patinas on marble surface lunetta (Milan Cathedral, Italy). (Photo: G. Ranalli, E. Zanardini).

on the surface of the stone. Carbonaceous particles were thought to be the most significant element in black crusts, but they also contain a complex mixture of aliphatic and aromatic carboxylic acids and polycyclic aromatic hydrocarbons. Finally, stones in urban areas act as "passive repositories for any kind of gaseous and particulate air pollutant present in the surrounding atmosphere."

Glass

Glass is a ceramic material (solid, inorganic, and nonmetallic) obtained by vitrification (fusion and cooling without crystallization) of primitive substrate (amorphous silicate, about 70%) and by adding several agents for fusion, stabilizing, decoloring, coloring, and so on. The exact origin of glass is still unknown, but there exist objects dated 1700 BC from Mesopotamia. The biodeterioration of glass is related to several factors, including both the glass composition and the environmental conditions on exposure. Although preliminary case studies of glass biodeterioration date back to the early nineteenth century, the phenomenon is understood only now. The presence of fungi, bacteria, lichens, and algae as biodeteriogen agents poses a serious problem for the glass surfaces found inside damp ancient structures. Free silicon bonds can react with organic acids produced by specialized microorganisms, which are capable of growing on the glass surface and finding nutrients for their metabolism (acidic attack), on glass surfaces. A hazard is the alkaline attack leading to the rapid destruction of the Na reticulum of the glass, accompanied by silicon release. This process is accelerated by temperature and exists even when glass is exposed to air, but not for a long time. Glass may be severely damaged by biodeterioration because of initial fungal attack that leads to etching, increase in opacity, and the presence of black spots on the surface. Among the bacteria capable of colonizing objects in glass, the iron bacteria (*Spaerotilus* and *Gallinella*) and the sulfur-oxidizing bacteria play an important role in the metabolism of sulfur and manganese, respectively. In the marine environment, algae and bacteria can play an important role inducing phenology of more alterations in glass as surface erosion, microfractures, and pitting. In other cases, isolated bacteria related to the genus *Flexibacter*, exhibit their capability to grow on the surface of glass and adapt to an oligotrophic environment and to develop themselves inside a thin layer of biofilm produced by other microorganisms.

Metals

Deterioration of metal materials appears as a complex matter from the physical, chemical, and biological points of view. Their specific mineralogical, structural, mechanical, and electrical properties are further influenced by the fact that metal materials utilized in the artistic fields are represented by alloys from the fusion of a single metal with almost another element (e.g., ancient bronze, made of copper and tin). All metal materials, except gold, are chemically reactive in the presence of compounds present in air, soil (O_2, CO_2, SO_2, salts, etc.), and, electrochemically, in water and moisture. Chemical corrosion of metals occurs when the structural elements undergo a chemical change, transitioning from the ground state to an ionized state. The phenomenon consists of two equivalent reactions, for example, for a ferrous metal in an aqueous oxygenated environment:

$$2Fe \rightarrow 2Fe^{2+} + 4e \text{ (anodic reaction, oxidation)}$$

$$O_2 + 2H_2O + 4e \rightarrow 4OH^-$$

(cathodic reaction, reduction).

In the absence of oxygen, either hydrogen ions or water is reduced at the cathode:

$$4H^+ + 4e \rightarrow 2H_2,$$

$$2H_2O + 4e \rightarrow H_2 + 2OH^-.$$

Microbial activities can influence the above reactions (biocorrosion), but the basic mechanism is still electrochemical. The microbially influenced corrosion (MIC), estimated as 20% of all metal corrosion, is generally associated with the microbial colonization of the metal surfaces and the formation of nonhomogeneous biofilms; the resulting surface appears as a pitting corrosion (biopitting) where the microbial metabolites, CO_2, S, H_2S, NH^{4+}, and acids (citric, oxalic, succinic, and fumaric), are the responsible agents. Aerobic and anaerobic bacteria are the major microbial groups involved in metal biocorrosion.

Under aerobic conditions, the sulfur-oxidizing bacteria related to genus *Thiobacillus* can oxidize elemental sulfur to obtain energy, solubilize metals producing sulfuric acid, which is highly corrosive. Metal biocorrosion by *T. ferrooxidans* and *Thiobacillus thiooxidans*, *Metallogenium* spp., *Sulfolobus* spp. is a serious problem, for example, in acid mine waters where, in presence of Fe^{3+} ions, the pH value drops to between 2 and 3, causing ferric sulfate deposition and finally severe corrosion in concrete and pipes.

Under anaerobic conditions and in a reduced environment with redox potential higher than −100 mV, the SRB related to genera *Desulfovibrio* and *Desulfotomaculum* produce H_2S, which is highly corrosive, inducing precipitation of metal sulfur salt. In the presence of iron ions, the overall reaction, where FeS and $3Fe(OH)^{2-}$ are final corrosion products, is as follows:

$$4Fe + SO_4^{2-} + 4H_2O \rightarrow FeS + 3Fe(OH)^{2-} + 2OH^-.$$

Organic Materials

Paper

Its invention is usually credited to a Chinese court official, Tsai Lun, in about AD 105. In the eighth century, paper was produced in the Middle East by the Arabs. Papermaking techniques were slightly changed by the Arabs, who introduced linen into the raw material and the use of paper mills. The art of papermaking gradually spread and in the twelfth century paper became the most common writing material in the West as the Europeans learnt the art from the Arabs. The principal compound in paper is cellulose, which is made of vegetable fibers. Until the nineteenth century, pure cellulose was mainly obtained from rags generally made of linen, hemp, and cotton. The handmade process consisted of converting the cellulose into pulp. After dispersion in water, the pulp was drained through a mold, dried, and pressed to form sheets. As the demand for paper grew, wood became the main raw material from the nineteenth century. Fibers were then obtained either by mechanical or by chemical processes. So, the composition of paper changed from almost pure cellulose to cellulose and significant amounts of hemicellulose and lignin. However, the production of a paper sheet also needs other compounds. For example, in order to reduce the spread of ink, the paper surface needs to be filled, the most used sizing materials being starch, gelatin, resin, and various synthetic polymers. Fillers, such as clays or chalk, are added to fill the pores and make the paper opaque. Other products used during papermaking include dyes, pigments, and fluorescent whitening agents.

The scientific and conservation literature claim that paper is an easy source of organic carbon for microorganisms. Paper biodegradation is well known. A major concern related to fungal colonization is pigment that may stain the substrate. Some pigments are soluble in certain solvents and can be removed through the application of a specific solvent. In addition, some permanent discoloration may also be caused by weak acids produced by fungi. Interestingly, there is one type of degradation named foxing that is still not fully understood. Some foxing stains are claimed to have a biotic origin and others an abiotic one. Foxing is a deterioration of paper occurring as brownish, reddish stains that could be the result of the activity of microorganisms or the oxidation and/or formation of heavy metal deposits. After many years of research and discussion, it is believed that the majority of foxing stains are due to microorganisms.

Wood

The most serious damage to wooden objects in indoor environments (museums) is caused by insects; they utilize wood as a nutrient source and for depositing eggs. Many insects (Coleoptera and Lepidoptera orders) use as substrate cellulose, as they have microorganisms in their gut that supply the cellulase enzyme. In an outdoor environment, when the moisture content of the wood is above 20%, it becomes susceptible to microbial attack by microfungi and bacteria. The predominant fungal infections, by exoenzyme production and subsequent destruction of natural polymers, are white rot (*Fomes* sp., and *Pleurotus* sp.), brown rot (*Poria* spp., *Merulis* sp., and *Coniophora* sp.), and soft rot (*Chaetomium* sp., *Alternaria* sp., and *Humicola* sp.). Because both bacteria (*Pseudomonas* sp. and *Achromobacter* sp.) and Actinomycetes (*Micromonospora* sp.) require a higher water content, their role is important in outdoor and marine environments.

Papyrus

The papyrus plant belongs to a widely distributed family of plants found in wet environments in tropical and subtropical areas. Papyrus was the material used by ancient and classical civilizations to write their texts. The raw material of papyrus comes from the pith of *Cyperus papyrus*, the largest member of the family. The papyrus pith is composed mainly of cellulose (54–68%) and lignin (32–24%); the proportions are principally based on age, manufacturing process, and environmental effects. It is worth noting that papyrus contains more lignin and less cellulose than paper and also contains uronic acid. The papyrus-cementing material is a polymer whose monomers are galactose, arabinose, and rhamnose.

Papyrus samples from different museums in Cairo were isolated and degrading microorganisms were identified. Some were tested for their cellulolytic activity, and it was found that the cellulose was attacked by all *Chaetomium* fungi that attack cellulose as well as by *Emericellopsis minima* and *Botryodiplodia theobromae*. In particular, *Chaetomium globosum*, *Chaetomium ochraceum*, and *Chaetomium elatum* along with *E. minima* intensely depolymerized cellulose with values ranging from 76% to 100%. The majority of microorganisms are able to hydrolyze the sugars DL-arabinose and L-rhamnose. Papyrus may be completely decomposed by fungi (genus *Chaetomium*) or Actinomycetes; the latter are especially dangerous when the conditions are unsuitable for other microorganisms, as in the desert. *Penicillium* is very often detected on paper but high temperature does not favor its growth. The order of the utilization of inorganic nitrogen sources is ammonium nitrate, ammonium phosphate, ammonium sulfate, and sodium nitrate. Ammonium nitrate represents a good source of nitrogen as nitrogen is present as ammonium and nitrate ions. The suitable temperature for fungal growth, under slightly acidic pH and relative humidity of 95%, is 2430 °C, but it is 42 °C for only seven species.

Parchment

Perhaps, parchment was invented in the second century BC in Pergamum, or probably known 100 years before; parchment replaced papyrus almost completely, from the third or fourth century AD. Parchment had many advantages, especially over papyrus. It was stronger and more durable than fragile papyrus and the raw materials for making it, animal skins, were available everywhere and not limited to one geographic location, namely Egypt. Parchment could be written on both sides (*recto* and *verso*).

Collagen is the major component of most connective tissues in animals and therefore is the bulk material for parchment. The main distinction between parchment and leather is that parchment preparation involves drying the wet skin while under tension. In contrast, the properties of leather are mainly due to the addition of tanning agents. As for the other supports, the susceptibility of parchment to microbial deterioration is dependent on the raw material, methods of production, and conditions of preservation. The most active microbial agents that cause damage to parchment belong to the genera *Streptomyces*, *Cladosporium*, *Scopulariopsis*, *Fusarium*, *Sporendonema*, *Ophiostoma*, *Aspergillus*, *Mucor*, *Penicillium*, *Nocardia*, *Alternaria*, *Trichoderma*, *Botryotrichum*, *Micrococcus*, *Bacillus*, and *Serratia*. The biodeterioration of these materials (parchment and leather) was mainly caused by fungal attack. In contrast, the increase in the pH during the deacidification treatment, generally necessary for parchment with a pH value lower than 5, by using ammonia greatly increased the microbiological deterioration suffered by parchment.

Biodeterioration of these materials (parchment and leather) was mainly due to fungal contamination, and less due to bacterial contamination. However, bacteria with proteolytic activity and in particular collagenolytic activity were sought and most of the isolated strains were able to grow on media containing collagen or parchment as the sole source of carbon and nitrogen. SEM observations demonstrated that the bacteria colonized subsurface layers along the collagen fibers. Among several biocides, Preventol R-80 and Catamin AB proved to be the more active biocides. In addition, mixed treatment using gamma radiation after a 3% pretreatment with Catamin AB is suggested.

Composite Materials

Paintings

Paintings, including easel and mural objects, contain a wide range of organic and inorganic compounds, many of which are biodegradable by a large variety of microflora. Additives such as glue, emulsifiers, and thickeners are used to prepare the support, to facilitate the application of paint layers, or to increase the final aesthetic object. In easel, the organic components of the paintings are sugars, gums, polysaccharides, proteins, oils, waxes, and undefined mixtures such as egg yolk. In mural paintings, pigments are mixed in oil and water, in the presence of casein or milk, in order to guarantee better adhesion on the plaster. This implies that microorganisms can grow inside different ecological niches on several substrates. The biological attack on these materials occurs in confined environments only under poor conservation conditions such as high humidity level, soil contact, poor ventilation, and rare maintenance operations. Moreover, the differences in the materials used and the chemical nature of the substrate may influence different evolution of microbial taxa colonizing works of art. Several studies on microbial succession in mural paintings show that bacteria can be considered among the first colonizers of frescoes, able, in some cases, to oxidize inorganic elements like lead present in pigments, and leading to the formation of lead oxides that cause damage by changing color (brown-black spot production). Moreover, the most frequent bacterial species present are members of the genus *Alcaligenes*, *Bacillus*, *Flavobacterium*, and *Pseudomonas*. Fungi normally are considered secondary colonizers. However, when frescoes in hypogean sites (tombs and grottes) are excavated, the predominant species isolated are members of the order Actinomycetales (genera *Streptomyces* and *Noocardia*); then, in very short time (months), photosynthetic bacteria, fungi, and algae become associated with the first colonizers, conferring typical colored crusts (green to black stain); these phenomena are quite evident after exposure to air, on illumination, and when hypogean rooms are open to visitors. Many are the mechanisms of aggression and microbial successions on paintings. Generally, microorganisms colonizing paintings and frescoes may be effected by environmental conditions and by the presence and concentration of atmospheric pollutants, especially sulfur dioxide. The sulfuric acid present in air containing moisture dissolves calcium carbonate in works of art (fresco, plaster, and stone) and leads to the precipitation of dihydrous calcium sulfate (gypsum). Gypsum deposition, by the formation of white crystal aggregates, is usually responsible for efflorescence, and for colonization by sulfur-oxidizing bacteria that cause direct and indirect mechanical damage such as the detachment of portions of painted layers. This is true if we consider that the death and lysis of these bacteria provide the organic substrate necessary for heterotrophic bacteria and fungi.

Photography

Photography is a method that produces a visible image by the interaction of light with silver halide salts contained in a binder applied to a substrate. Photographic materials include still and moving images and microfilms; many other different methods have been introduced and used since 1831. However, from a microbiological point of view, it is very interesting to know about the materials that photographic materials are made of. Photographic supports can be made of metals, for example,, a copper plate covered with a fine silver layer as in the daguerreotype or a black varnished metallic plate as in the ferrotype; glass as for the collodium plate; paper, typical of calotype and albumin print; and, finally, semisynthetic and synthetic polymers, including cellulose nitrate, cellulose acetate, and polyethylene terephthalate. The light-sensitive materials are silver halides: silver chloride, silver bromide, and silver iodide. Other important photographic materials are binders such as collodium, which contains cellulose nitrate, albumin, and gelatin, sensitizers, antifoggants, and hardenings. Gelatin is the most satisfactory binder in terms of cost, chemical processing, and physical and optical features. Fungal attack is the main form of biodeterioration that occurs in photographic materials. The presence of substances of animal origin such as albumin and gelatin represent an important source of nutrition for microorganisms, in particular fungi that cause, for example, an increase in gelatin solubility in water. Thus, water or water solutions should not be used to clean photographic materials containing gelatin when there is a fungal attack. The use of cotton soaked in adequate film cleaner is suggested. In addition, the contact between fingers and photographic documents should be avoided as salts in the fingerprints absorb moisture and create favorable conditions for fungal growth. The Krieger collection includes thousands of glass plate negatives made in the years 1880–1926: several glass negatives showed fungal contamination, and among the isolated fungi, *Aspergillus nidulans*, *Aspergillus versicolor*, *Cladosporium cladosporioides*, *Penicillium cyclopium*, and *Penicillium janthinellum* were the most frequent.

Photographic prints were protected against microflora by immersing the prints for 3 min in 0.5% water solution of sodium salt of 4-chloro-3-cresol. For both black-and-white and color films and prints, fungicidal treatment by immersion in 1% solution of zinc fluosilicate yields positive results. Use of activated film coating reduces scratches, fungal growth, and color dye fading, where humidity control of storage environment is not feasible. On microfilm objects based on cellulose acetate and polyester, the susceptibility to biodeterioration and subsequent isolation and identification of biodeteriogens showed the presence of *Aspergillus niger*. Different species and strains of fungi contaminate microfilms. One way to prevent fungal and insect attack is either to hermetically seal the photographs in laminated envelopes or to laminate/encapsulate the photograph in plastic. However, the best method to prevent fungal deterioration of photographic material is to keep it in a cool, dry, and dark place.

Magnetic tape, optical disk

The magnetic tape is made of a thin magnetic layer that comprises magnetic particles, such as iron oxide and chromium dioxide suspended in a lubricant, and a polymer binder (generally a polyester- or polyether polyurethane-based system). The most commonly used tape substrate is polyethylene terephthalate. Among several sources of damage (heat, light, grit, moisture, and atmospheric pollutants), microbial growth is considered a main concern for the preservation of magnetic media. The most frequently isolated fungi belong to the genera *Alternaria*, *Aspergillus*, *Chaetomium*, *Penicillium*, and *Stemphylium*, and these have been detected on floppy disks in tropical countries and also in temperate countries under humid conditions. Many of these microorganisms are widespread fungi that are able to grow on a variety of products. For the prevention of fungal growth on videotape, it is recommended that the interior of the tape container and the edges of the tape itself be checked for black, brown, and greenish stains that might indicate the presence of fungi. Moldy tapes should not be used until the microorganisms have been eradicated. Recent experiences suggest that the material should be placed in a desiccating container and after equilibrium is reached the surface should be vacuumed. After this treatment, the film can be cleaned with a chlorinated solvent cleaning machine. However, the use of a UV source, except for some black-and-white films, is not recommended as UV damages dyes.

Compact disk (CD-ROM) and digital versatile disks (DVDs) introduced into the market are based on an improved polymer, made of aromatic polycarbonates, more durable than aliphatic. CD producers generally do not care about biological deterioration of CDs. However, a fungus of the *Geotrichum* type was able to grow on a common CD under tropical conditions (about 30 °C and UR 90%). Biodeterioration occurred as bioturbation traces. A fractal structure and the destruction of the pits, which in turn means the loss of the information stored in the disk, were observed. It is worth noting that *Geotrichum* is a very common genus commonly found in food products, water, and animals, and it has also been isolated from air. Thus, fungal growth is not related to a particular geographical area but to the environmental conditions that permitted the growth.

METHODS OF CONTROL AND PREVENTION

The microbial contamination of cultural heritage is still an underestimated concern. One reason may be the difficulty in establishing a definitive causative relation between the biological agents and the damage to materials, especially in those cases where there is damage but not much biological growth.

The prevention of biodeterioration of a cultural heritage includes all the operations needed to avoid the growth of microorganisms and a consequent biological attack by biodeteriogen microorganisms on the materials composing the heritage. These so-called indirect methods do not act directly on the microorganisms but hinder or slow down biological growth, modifying where possible the environmental conditions so that they become unsuitable for biological growth. Because of the close environmental relationship and dependence between the biological species and the environment, the most efficient and easily realized methods for impeding undesirable growth act on the causal factors, particularly humidity, temperature, ventilation, light, that can inhibit or condition their presence. These factors cannot always be controlled, whereas it is possible to control the conditions of confined environments (museums, libraries, and churches). It is decidedly more difficult to control an external environment (monuments and archeological areas). In fact, whereas some parameters (light, humidity, and temperature) can theoretically be modified, when nutritional factors present in the artwork support the biological growth they cannot be changed without damaging the nature of the object itself. On the contrary, factors encouraging biodeterioration (dirt, dust, and deposits of varying kind) can be removed. It is not necessary to act on all the parameters that restrict biological growth and it is sufficient to lower or to raise just one of these until there is no increase. "Direct" methods of biodeterioration control provide for direct action on the organisms, and this can vary.

Biochemical

Such methods use compounds of biological origin: antibiotics (active at very low doses, but there is loss of effectiveness over time), enzymes (used especially for cleaning) and pheromones (used to induce males to leave the structures where they nest) are used for infestations in the deep layers of the material where biocides cannot penetrate.

Biological

These are methods that use antagonist or parasitic species to limit the growth of other animal or vegetable species. In the case of flora, phytophage insects can be used, while for animals, above all avifauna, the choice lies with predator species of eggs and the young to reduce the population density of the other species.

Physical

These methods foresee the use of X-rays with biocidal action or noxious gamma radiation and UV rays. Indeed, UV rays have low penetration and are thus not effective for alterations involving the interior of objects. The most problematic aspect of UV treatment is bound to its ability to alter, from the chemical point of view, the treated materials (pigments, cellulose, and proteic materials) and to the difficulty of transporting the irradiating source *in situ*. Note that gamma rays have a greater biocidal activity than UV, but their application is limited to treating organic materials (paper, parchment, and wood), and is aimed more at resolving microflora disinfection problems than those connected to the presence of insects.

Mechanical

These methods foresee the removal of biodeteriogen organisms by the frequent use of manual instruments like scalpels, scrapers, spatulas, and so on. Nevertheless, they have the disadvantage of failing to guarantee results over the long term, in that it is very difficult to completely remove the vegetative or reproductive structures of the species present, at least without seriously damaging the surface of the substrate. In fact, on dealing with a fungal mycelium, a lichen thallus, or even with roots rather than a plant, it is rare for the colonization to be only superficial, so to avoid damage to the surface of the stone mechanical removal ends up being only partial. However, mechanical methods do have the advantage of adding nothing that can cause further degradation. Generally speaking, these methods can be used with chemical methods and can thus be very useful.

Chemical

Chemical methods foresee the use of synthetic chemical substances like pesticides and disinfectants, more generally called biocides. The requirements of products to be used in the conservation of artwork are high efficiency against biodeteriogens, absence of interference with the constituent materials, low toxicity for human health, and low pollution risk for the environment. Biocides must be employed with caution as they can react with the substrate and cause macroscopic effects like spots, yellowing and bleaching, and increased brilliance or opacity. In fact, the recent application of more accurate investigative methodologies showed how, in reality, many of these products produce negative collateral effects, though these are not always obvious. Furthermore, biocide substances can interact with other products used in previous restoration.

For protecting monumental stones, it is practically impossible to control weather variables, while for rock sites the construction of protective structures, preventing water runoff of surfaces, can effectively reduce biological growth.

In order to reduce biological risk for the organic objects maintained in confined environments, the control of temperature and relative humidity is frequently cited as the first step in environmental control especially in museums, archives, and libraries. Poor storage conditions impede their preservation. High relative humidity, for example, above 60%, along with high temperature encourage the growth of microorganisms. Damage to historical records due to bacteria and fungi can be reduced by limiting access of visitors to storage areas and monitoring the collections environment. However, it is worth noting that not only improper storage conditions but also disasters caused by water, such as broken pipes, leaking roofs, blocked drains and fire extinguishing, favor microbial growth, if prompt action is not taken. In indoor environments, another important goal should be the monitoring of the microflora of the air throughout the year. This can then be used as a standard reference to determine changes that had happened in a building or a room. It is worth noting that a close similarity has been found between the aeromicroflora and the library and archival materials biodeteriogens. Another important issue is maintenance, such as dust removal; dust and dirt can contain spores and bacteria and provide the nutrients required for microbial growth.

No less important is the protection of workers from risks related to exposure to biological agents at work. In the past few decades, growing concerns about the use of chemical compounds have resulted in an in-depth evaluation of the effects of pesticides on human health and safety. Any process should be safe for the operator and the environment and no chemical residues should remain in the treated materials. The awareness of pesticides hazardous to human health and the environment has led to a search for alternatives. Among the main nonchemical treatment alternatives used for collections including freezing, gamma rays, microwaves, and modified atmospheres, only high-energy radiation can kill microflora. However, high-energy radiation is generally recognized as a harmful process for library and archival materials. Possible future alternatives can include biological control to stop the growth of harmful microorganisms and bioremediation/biorecovery of cultural heritage.

RECENT DEVELOPMENTS

Recent findings suggest and confirm that microorganisms also contribute to the degradation of all materials and buildings by utilizing pollutants deposited from the atmosphere as primary substrates. The production of large quantities of exopolymeric substances (EPS), biofilm formation, and the increase in microbial community population colonizing artifacts cause deterioration and the corrosion of materials through chemical and mechanical processes.

In the last decade, the introduction of molecular-based detection techniques to study the roles of microorganisms in biodeterioration of different kinds of materials has permitted more information on microbial diversity in deteriorating artworks to be gathered and has increased the range of microorganisms that can be detected.

In particular, biofilm formation, biocorrosion caused by organic and inorganic acids, redox processes and physical penetration by microbial communities have all been highlighted and discussed in many cases of biodeterioration of artistic and historic works.

Further biodeterioration studies, as well as further research into the preservation and restoration of historic stonework must be considered in order to guarantee the most appropriate, long-term conservation of our cultural heritage. Most of the methods are based on the use of preventive and remedial approaches.

New nanotechnology tools and nanomaterials are examined and proposed as intricate and ingenious solutions for the cleaning, consolidation, and conservation of cultural heritage.

A few studies of the biorestoration of deteriorated historic stonework based on the use of sulfate and nitrate-reducing bacteria, hydrocarbon-degrading bacteria, biocalcifying bacteria, and on the hydrolytic activity of lipases, proteases and carbohydrases have been reported. Traditional diagnostic, molecular methods combined with recently-developed biocleaning techniques (e.g. bioremediation with *D. vulgaris*) have been successfully applied to Demetra and Cronos limestone sculptures. Black crusts, which are principally gypsum, and biological discoloration demonstrated that microbial communities were complex and included heterothopic bacteria, microfungi and algae. The authors reported that, after three bioapplications, the black crusts were no longer detectable.

The use of fluorescent, internal spacer-PCR was chosen from the various techniques available for the cluster analysis of bacteria isolated from both air and from deteriorated fresco surfaces.

DNA and RNA–based studies on colonization present in palaeolithic paintings in caves, indicate differences between the microorganisms present and those metabolically active in the process.

Gorbushina and Broughton specifically reviewed the "subaerial biofilm" (SAB) formation and development on solid mineral surfaces exposed to the atmosphere and its interaction with stone materials. A strict complex association of fungi, algae, cyanobacteria, and heterotrophic bacteria including specialized actinobacteria (e.g., *Geodermatophilus*) and rock-inhabiting fungi including the black (melanized) MCF is present in the subaerial biofilm. This ecosystem favors the retention of water, protection of the cells from variations in environmental conditions and from solar radiation and atmospheric aerosols, gases, and particles accumulated in the biofilm can be used as a source of nutrients. The role of fungi in the deterioration of cultural heritage has also been discussed and confirmed in a recent review. Moreover, an integrated, diagnostic approach combining biological, microclimatic, and geophysical tools for the evaluation of the state of conservation in a crypt was proposed by Cataldo et al. 2009.

Greater importance is now given to a multidisciplinary approach to the study of the biodeterioration processes in Cultural Heritage. Several papers report the contributions on microscopic observations, culture-dependent and culture-independent methods in order to better identify and describe the microbial communities inhabiting surface materials, to predict the potential risk and to plan an efficient control of the growth and expansion of these colonizations. The roles of applied microbiology and biotechnology in the preservation of artwork (stone, frescoes and other materials) is gradually being acknowledged and better linked to, and integrated with, physical and chemical-based methods.

In general, conservation and restoration in Cultural Heritage use traditional materials and techniques. However, taking into consideration that weathering, environmental conditions including climate change, use in public or private spaces or in exhibitions and museums and the associated transportation, as well as unpredictable events can all cause damage and or degradation of both movable and immovable cultural heritage artwork. More specifically, research in the cultural heritage field needs to be increased (goals elucidated in 7FP-ENV 2007–2013). Therefore, the compatibility between old and new materials and techniques (including bio- and nanotechnology) may offer new solutions that are innovative, more effective and longer-lasting. The development and application of compatible materials with suitable properties (e.g., consolidants, coatings, substitutes, etc.) and/or techniques (e.g., biocleaning methods, reinforcement, etc.) in the future could assure the conservation of original objects and/or improve the physical state of damaged objects. Future research should also assess the long term behavior of materials with regard to durability, and/or the performance of the chosen techniques; moreover, materials with novel properties such as self-healing or self-cleaning, or materials with an improved resistance to degradation mechanisms should be considered.

Finally, multidisciplinary research contributing to the conservation and safeguarding of Cultural Heritage and responding to the challenges from the changes in our natural environment, as well as from man-made activities, needs to be augmented.

ACKNOWLEDGMENT

The authors thank MIUR-Prin 2007 for financial assistance.

FURTHER READING

Allsopp, D., Seal, K., & Gaylarde, C. (Eds.) (2003). *Introduction to biodeterioration*. (2nd ed.)Cambridge: Cambridge University Press.

Caneva, G., Nugari, M. P., Pinna, D., & Salvadori, O. (Eds.) (1996). *Il Controllo del Degrado Biologico – I Biocidi nel Restauro dei Materiali Lapidei*. Fiesole: Nardini.

Caneva, G., Nugari, M. P., & Salvadori, O. (Eds.) (2005). *La Biologia Vegetale per i Beni Culturali*. Firenze: Nardini Publisher.

Achal, V., Mukherjee, A., Basu, P. C., & Sudhakara Reddy, M. (2009). Strain improvement of *Sporosarcina pasteurii* for enhanced urease and calcite production. *Journal Industrial Microbiology Biotechnology*, *36*, 981–988.

Cappitelli, F., Principi, P., & Sorlini, C. (2006). Biodeterioration of modern materials in contemporary collections: Can biotechnology help? *Trends in Biotechnology*, *24*, 350–354.

Cappitelli, F., Shashoua, Y., & Vassallo, E. (Eds.) (2006). *Macromolecoles in cultural heritage Macromolecoles Symposia* (Vol. 238) . (pp. 1–104).

Cappitelli, F., & Sorlini, C. (2005). From papyrus to compact disk: The microbial deterioration of documentary heritage. *Critical Reviews in Microbiology*, *31*, 1–10.

Cappitelli, F., & Sorlini, C. (2008). Microorganisms attack synthetic polymers in items representing our cultural heritage. *Applied and Environmental Microbiology*, *74*, 564–569.

Cappitelli, F., Toniolo, L., Sansonetti, A. *et al.* (2007). Advantages of using microbial technology over traditional chemical technology in the removal of black crust from stone surface of historical monuments. *Applied and Environmental Microbiology*, *73*, 5671–5675.

Cappitelli, F., Zanardini, E., Ranalli, G., Mello, E., Daffonchio, D., & Sorlini, C. (2005). Improved methodology for bioremoval of black crusts in historical stone artworks by use of sulphate-reducing bacteria. *Applied and Environmental Microbiology*, *72*, 3733–3737.

Castanier, S., Le Metayer-Levrel, G., & Perthuisot, J. P. (1999). Ca-carbonates precipitation and limestone genesis – The microbiologist point of view. *Sedimentary Geology*, *126*, 9–23.

Ciferri, O., Tiano, P., & Mastromei, G. (Eds.) (2000). *Of microbes and art: The role of microbial communities in the degradation and protection of cultural heritage*. New York: Kluwer Academic/Plenum Publishers.

Cataldo, R., Leucci, G., Siviero, S., Paliotti, R., & Angelici, P. (2009). Diagnostic of the conservation state of the crypt of the Abbey of Montecorona: Biological, microclimatic and geophysical evaluations. *Journal of Geophysics and Engineering*, *6*(3), 205–220.

Diakumaku, E., Gorbushina, A. A., Krumbein, W. E., Panina, L., & Soukharjevski, S. (1995). Black fungi in marble and limestones – An aesthetical, chemical and physical problem for the conservation of monuments. *The Science of the Total Environment*, *167*, 295–304.

Dornieden, T., Gorbushina, A. A., & Krumbein, W. E. (2000). Biodecay of cultural heritage as a space/time – Related ecological situation – An evaluation of a series of study. *International Biodeterioration &. Biodegradation*, *46*, 261–270.

Gaylarde, C., & Morton, L. H. G. (2002). Biodegradation of mineral materials. In G. Bitton (Ed.), *Encyclopedia of environmental microbiology* (pp. 516–527). New York: Wiley.

Gonzales, J. M., & Saiz-Jimenez, C. (2005). Application of molecular nucleic acid-based techniques for the study of microbial communities in monuments and artworks. *International Microbiology*, *8*, 189–194.

Gorbushina, A. A., & Broughton, W. J. (2009). Microbiology of the atmosphere-rock interface: How biological interactions and physical stresses modulate a sophisticated microbial ecosystem. *Annual Review of Microbiology*, *63*, 431–450.

Griffin, P. S., Indictor, N., & Koestler, R. J. (1991). The biodeterioration of stone: A review of deterioration mechanisms, conservation case histories, and treatment. *International Biodeterioration*, *28*, 187–207.

Gurtner, C., Heyrman, J., Piñar, G., Lubitz, W., Swings, J., & Rölleke, S. (2000). Comparative analyses of the bacterial diversity on two different biodeteriorated wall painting by DGGE and 16s rRNA sequence analysis. *International Biodeterioration & Biodegradation*, *46*, 229–239.

Jroundi, F., Fernandez-Vivas, A., Rodriguez-Navarro, C., Bedmar, E. J., & Gonzalez-Munoz, M. T. (2010). Bioconservation of deteriorated monumental calcarenite stone and identification of bacteria with carbonatogenic activity. *Environmental Microbiology*, *60*(1), 39–54.

Michaelsen, A., Pinar, G., Montanari, M., & Pinzari, F. (2009). Biodeterioration and restoration of a 16th-century book using combination of conventional and molecular techniques: A case study. *International Biodeterioration and Biodegradation*, *63*(2), 161–168.

Pangallo, D., Chovanova, K., Drahovska, H., De Leo, F., & Urzì, C. (2009). Application of fluorescence internal transcribed spacer-PCR (f-ITS) for cluster analysis of bacteria isolated from air and deteriorated fresco surfaces. *International Biodeterioration and Biodegradation*, *63*(7), 868–872.

Piñar, G., Ramos, C., Rölleke, S. *et al.* (2001). Detection of indigenous *Halobacillus* populations in damaged ancient wall paintings and building materials: Molecular monitoring and cultivation. *Applied and Environmental Microbiology*, *67*, 4891–4895.

Saiz-Jemenez, C. (Ed.) (2003). *Molecular biology & cultural heritage*. Lisse, The Netherlands: Swets & Zeitlinger B.V.

Portillo, M. C., Gonzalez, J. M., & Saiz-Jimenez, C. (2008). Metabolically active microbial communities of yellow and grey colonizations on the walls of Altamira Cave, Spain. *Journal of Applied Microbiology*, *104*, 681–691.

Portillo, M. C., Saiz-Jimenez, C., & Gonzalez, J. M. (2009). Molecular characterization of total and metabolically active bacterial communities of "white colonizations" in the Altamira Cave, Spain. *Research in Microbiology*, *160*, 41–47.

Polo, A., Cappitelli, F., Brusetti, L., Principi, P., Villa, F., Giacomucci, L. *et al.* (2010). Feasibility of removing surface deposits on stone using biological and chemical remediation methods. *Microbial Researches*, *60*(1), 1–14.

Ranalli, G., Zanardini, E., & Sorlini, C. (2009). Biodeterioration – Including Cultural Heritage. In M. Schaecter (Ed.), *Encyclopedia of microbiology* (pp. 191–205). Oxford: Elsevier.

Ruffolo, S. A., La Russa, M. F., Malagodi, M., Oliviero Rossi, C., Palermo, A. M., & Crisci, G. M. (2010). ZnO and $ZnTiO_3$ nanopowered for antimicrobial stone coating. *Applied Physics A: Materials Science and Processing*, 1–6.

Scheerer, S., Ortega-Morales, O., & Gaylarde, C. (2009). Microbial deterioration of stone monuments – An updated overview. *Advanced in Applied Microbiology*, *66*, 97–139 (Chapter 5).

Saiz-Jimenez, C., & Laiz, L. (2000). Occurrence of halotolerant/halophilic bacterial communities in deteriorated monuments. *International Biodeterioration & Biodegradation*, *46*, 319–326.

Sand, W. (1997). Microbial mechanisms of deterioration of inorganic substrates – A general mechanistic overview. *International Biodeterioration & Biodegradation*, *2–4*, 187–190.

Sterflinger, K. (2010). Fungi: Their role in the deterioration of cultural heritage. *Fungal Biology Reviews*, *24*(1–2), 47–55.

Warscheid, T., & Braams, J. (2000). Biodeterioration of stone: A review. *International Biodeterioration & Biodegradation*, *46*, 343–368.

Webster, A., & May, E. (2006). Bioremediation of weathered-building stone surface. *Trends Biotechnology*, *24*, 255–260.

RELEVANT WEBSITE

http://www.irnase.csic.es/. – Instituto De Recursos Naturales y Agrobiología de Sevilla, Sevilla, Spain..

Chapter 39

Biosensors

G.D. Griffin[1], D.N. Stratis-Cullum[2]
[1]Fitzpatrick Institute for Photonics, Duke University, Durham, NC, USA
[2]US Army Research Laboratory, Adelphi, MD, USA

Chapter Outline

ABBREVIATIONS

BOD Biological oxygen demand
BWAs Biowarfare agents
ECL Electrochemiluminescence
ELISA Enzyme-linked immunoabsorbent assay
FETs Field effect transistors
FPW Flexural-plate wave
gfp Green fluorescent protein
GMOs Genetically modified organisms
LAPS Light addressable potentiometric sensor
LB Langmuir–Blodgett
LNAs Locked nucleic acids
LPS Lipopolysaccharide
MIP Molecularly imprinted polymers
NASBA Nucleic acid sequence-based amplification
PCR Polymerase chain reaction
PNA Peptide nucleic acid
QCM Quartz crystal microbalance
SAW Surface acoustic wave
scFv Single-chain variable fragments
SERS Surface enhanced Raman spectroscopy
SPR Surface plasmon resonance

DEFINING STATEMENT

Biosensors are integrated analytical devices which use a recognition element, generally a biomolecule, to bind an analyte, and some transduction mechanism to detect this binding event. Details regarding recognition and transduction, as well as applications of such devices to food safety, environmental sensing, and biological warfare agent detection are discussed.

INTRODUCTION/LIMITATION OF SCOPE

The subject of biosensors is voluminous, with thousands of journal articles (English language only) treating the subject in the last 10 years. Even taking into account that some of these articles use a rather loose definition of biosensors (e.g., not integrated systems), the literature is too extensive to provide an in-depth review for this article. The focus in this article will be on discussion of biosensors of relevance to microbiology and then on a subset of all potential microbiological applications. The overwhelming majority of the current biosensors market is made up of glucose monitors that are used by diabetics for monitoring blood sugar. Since this application has been covered so extensively in numerous reviews, it is omitted here. Also excluded are discussions of biosensors for the wine industry (ethanol sensing) and applications of biosensors to petroleum microbiology.

Since understanding of the biorecognition and transduction processes is critical to understanding biosensors, extensive discussion of these topics is included. The various types of molecules ,which may serve as biorecognition elements are reviewed, and the transduction systems based on the four major categories (i.e., electrochemical, thermal, mass, and optical) are discussed. This article also focuses on the applications of biosensors to detection of pathogenic organisms or organism toxins in the areas of food safety and biowarfare/terrorism, since these applications have become of great concern due to events in the recent past. A brief overview of biosensors for environmental monitoring is also provided, although in many such applications, the biosensor is in fact a bioreporter organism, and not integrated into a sensor device. There is also a discussion of future directions in biosensor development, in which topics such as integrated systems (lab-on-a-chip/biochip devices), and nanotechnology applications to biosensors are briefly mentioned. This article is divided into four main sections: (1) Biosensor Overview, (2) Biorecognition Elements, (3) Transduction Mechanisms, and (4) Applications.

BIOSENSOR OVERVIEW

Definition

The IUPAC definition of a biosensor is that a biosensor is a self-contained integrated device that is capable of providing specific quantitative or semiquantitative analytical information using a biological recognition element (biochemical receptor) that is retained in direct spatial contact with a transduction element. See Figure 1 for a schematic of how a biosensor works. A common working definition is that biosensors are analytical devices that use biological macromolecules to recognize an analyte and subsequently activate a signal that is detected with a transducer. However, this common working definition does not suggest an integrated device in which the transducing element is in spatial contact with the biorecognition element. Some authorities subdivide biosensors into affinity biosensors and catalytic biosensors, based on the activity of the biorecognition element. Thus, affinity biosensors have as their fundamental property the recognition (binding) of the analyte by the biorecognition element (e.g., antibody–antigen). Catalytic biosensors have as their biorecognition element proteins (or microorganisms) that not only bind the analyte but also catalyze a reaction involving the analyte to produce a product (e.g., glucose biosensors).

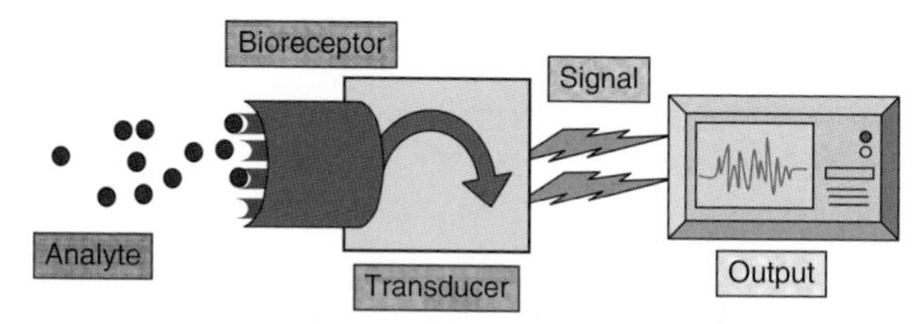

FIGURE 1 Schematic diagram of a biosensor.

Ideal Biosensor Characteristics

Table 1 lists the ideal characteristics of a biosensor. In many cases, biosensors are single-use, as the definitions of biosensors do not, in general, explicitly require that the biosensor be regenerable, although for many applications, this is desirable. The ideal biosensor is frequently described as being robust, selective, reproducible, and sensitive with a large dynamic range. This ideal is seldom attained. Bioanalytical systems, such as immunoassays like ELISA (enzyme-linked immunoabsorbent assay), which may well use the same elements (i.e., bioreceptors and transduction devices), but require additional processing steps, such as reagent addition and washings, and usually involve reading the completed assay on some separate piece of instrumentation, are to be distinguished from biosensors. Thus, the very large opus of experimental literature describing the use of immunoassays, nucleic acid-based assays, and so on for detection of various analytes does not necessarily describe biosensors, *per se*, although the techniques described in such assays may well have application in the development of actual biosensors.

Assay Formats

In general, a variety of assays can be employed with the many different biorecognition and transduction elements described below. There is not sufficient space to provide anything more than a very general overview of the assay formats involved in biosensing. Briefly, the assay formats can be either direct or indirect. In the direct methods, analyte binding to the biorecognition element is detected directly, by such techniques as mass changes, changes in refractive index or impedance, pH, and so on (Figure 2). For indirect techniques, an additional reaction (other than the biorecognition binding event) has to occur in order to detect the binding of analyte and bioreceptor. In a competitive indirect strategy, the analyte and a competitor molecule (labeled in some fashion) compete for limited numbers of biorecognition binding sites (Figure 3). In the noncompetitive indirect assay, a second biorecognition element with a label is added

TABLE 1 Ideal biosensor characteristics

Characteristic	Description
Analysis time	A fast analysis time is desired, with "real-time" responses to target analytes ideally desired
Sensitivity	A sensitive analysis allows for detection of low concentrations of target analytes and leads to low false-negative analyses
Specificity	A specific analysis allows for discrimination between target analytes and other closely related species and minimizes false-positive analyses
Reproducibility	The analysis should be highly reproducible, to provide for a reliable analysis that is easy to calibrate
Accuracy	The biosensor device should be highly accurate, meaning false-positives and false-negatives are minimized
Robustness	The biosensor must be insensitive to environmental conditions (temperature, pH, electronic interferences, etc.)
Unit and operational costs	A lower unit cost and operational costs for reagents, and so on will allow for more wide spread implementation of the biosensor systems
Size and weight	The ability to miniaturize a biosensor is desired, particularly for integration into process monitoring applications as well as for field portable applications
Regeneration	The ability to regenerate the binding surface, allowing multiple measurements by the same sensor element is preferable, although single-use platforms are sometimes sufficient
Multianalyte detection	A biosensor that can detect multiple analytes simultaneously is highly desired for efficient cost, time, and size and weight utilization
User interface	Ideally, fully automated systems are desired, or at the minimum require little-to-know operator skills for routine analysis

to the analyte sample, so that when the analyte binds to the immobilized biorecognition element, the second biorecognition element also binds to the analyte (Figure 4). In the case of labels for the indirect type of assay, the label can be optical-, electrochemical-, or mass-related. In the case of direct assays and noncompetitive indirect assays, the signal is directly proportional to the analyte concentration, while in the case of competitive indirect assays, the signal is inversely proportional to analyte concentration. Frequently

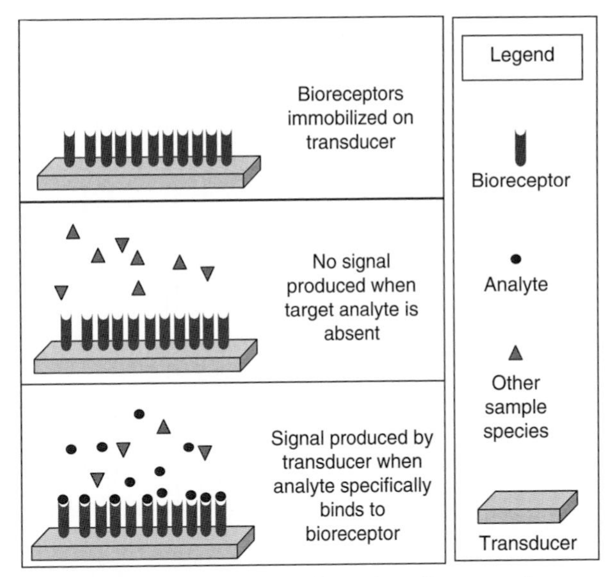

FIGURE 2 Schematic diagram illustrating a direct assay format.

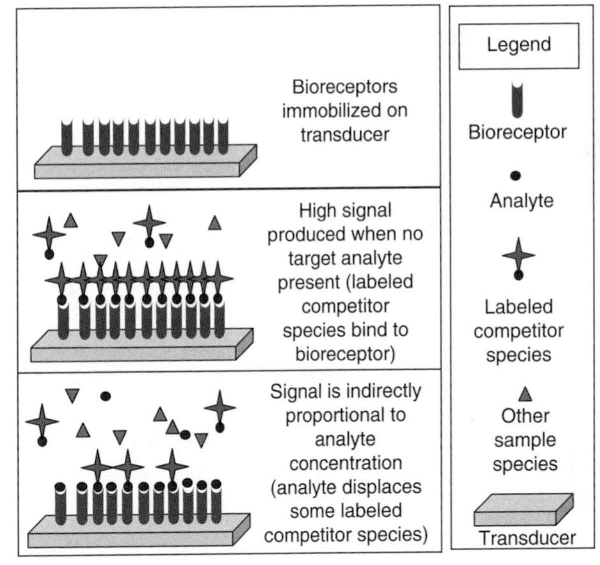

FIGURE 3 Schematic diagram illustrating an indirect assay with competitive binding.

also, biosensor assay formats involve some form of amplification, the rationale for this being increased sensitivity of the assay. Either the analyte molecules themselves may be amplified (e.g., polymerase chain reaction (PCR) of DNA sequences) or the biorecognition event may be amplified (e.g., ELISA immunoassays).

BIORECOGNITION ELEMENTS OF BIOSENSORS

The biorecognition element of biosensors has classically been some natural biomolecule, antibodies being the most common example. However, biorecognition elements also

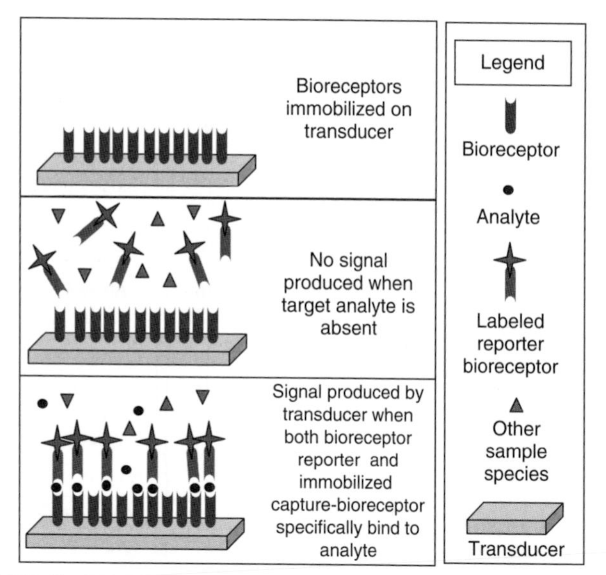

FIGURE 4 Schematic diagram illustrating an indirect assay with noncompetitive binding.

include bioreceptors such as enzymes, nucleic acids, and receptors that reside in the membranes of cells, and even larger biological units, such as whole cells. The biorecognition element plays a crucial role to the overall biosensor performance, imparting selectivity for a particular analyte. If the definition of biorecognition elements only includes natural biomolecules (up to the collection of biomolecules comprising a whole cell), such artificial recognition elements as imprinted polymers or peptide nucleic acids (PNAs) would be excluded. This rigorous definition seems too restrictive, however, and in this article, the authors include biosensors using such biorecognition elements as aptamers, PNAs, liposomes, and molecularly imprinted polymers (MIP). A brief description of the various types of biorecognition elements will now be given.

Antibodies

Antibodies are proteins produced by the immune system that have as their unique feature antigen recognition sites that bind antigens by noncovalent interactions in an exquisitely specific fashion and often with relatively high affinity. The antigen-binding portions of the molecule are the V_H (heavy chain variable region) and the V_L (light chain variable region) regions, which both fold to provide a sort of "lock and key" fit for the specific antigen. Included in antibody categories would be polyclonal and monoclonal antibodies. Antibodies are probably the most commonly used of all biorecognition elements for sensor applications. Frequently, sensor design makes use of antibody "sandwiches," where, for example, capture antibodies immobilize the analyte, while other labeled detector antibodies bind to the analyte and thus provide a sensing signal. It is also common to use some sort of amplification technique (e.g., enzymes-ELISA assays) to provide a more robust signal of the recognition process. Ingenious combinations of nucleic acid and immunological techniques (e.g., immuno-PCR and immuno-rolling circle amplification) can also be used for amplification.

Antibody fragments/engineered antibody fragments

For biosensing applications, the molecular recognition site(s) of the antibody (the antigen combining sites) are of overwhelming importance, while the function of the Fc region, of importance physiologically, may not be required for sensing applications. Also, antibodies are often immobilized on solid substrates for sensing applications, and here the orientation of the antibodies is as critical as the antigen combining site should ideally be oriented in a spatially accessible manner. Thus, smaller antibody fragments may have certain advantages, such as more defined points of immobilization through thiol groups liberated after cleavage of the whole immunoglubulin molecule, and also these fragments could be more densely packed upon a sensor surface, thus increasing the epitope density over that found with intact immunoglobulin and so perhaps enhancing sensitivity. Enzymatic cleavage of immunoglobulin molecules to produce fragments such as Fab (antigen-binding fragment) is one way to produce these smaller recognition elements. Using the techniques of phage display or ribosome display for antibody gene cloning and expression, antigen-binding fragments of antibody molecules called single-chain variable fragments (scFv) can be obtained. These fragments feature the V_H and V_L domains joined by a flexible polypeptide linker, which can be engineered to have desirable properties for immobilization (e.g., a thiol group of cysteine). Some studies have found the scFv antibody fragments to be superior to either the Fab fragment or the whole immunoglobulin molecule for sensor applications.

Enzymes

Enzymes are catalytic proteins that have an active site that shares some of the features of the antigen-binding site of antibodies, that is, exquisite specificity for certain molecular structures, referred to as their substrates. Again, genetically engineered enzymes with altered properties can be used. Enzymes are not frequently used as biorecognition elements, *per se*, but are usually a component of a multiple molecular biorecognition package, where the enzyme is included to provide amplification (such as in ELISA immunoassays or coupling to an electrochemical detection transduction system (e.g., horse radish peroxidase and urease for electrochemical detection schemes)). An example where an enzyme functions as a direct biorecognition element would be the use of cholinesterase enzymes to detect certain organic phosphate pesticides/chemical warfare agents that are cholinesterase inhibitors.

Proteins/Peptides with Specific Binding Properties

Phage display

Using techniques developed in recent years, it is now possible to produce and evolve peptides/proteins with strong binding affinity for specific proteins. These techniques include phage display and ribosome display. It is outside the scope of this article to provide an in-depth review of these procedures. In each technique, iterative rounds of affinity purification enrich for proteins/peptides with desired binding properties for a specific ligand. The unique feature of phage display is that the production of 10^7–10^8 short peptides on the surface of filamentous bacteriophage can be achieved through the fusion of an epitope library of DNA sequences with the gene(s) coding for one of several coat proteins (the minor coat protein, phage gene III, was used initially). An epitope library of such cloning can display 10^7 or greater peptides that can be panned for peptides that bind specifically to an analyte of interest. Proteins made from the phage display technique have been used to detect bacteria, bacterial spores, toxins, and viruses.

Nucleic Acids

Included here are oligo- or polynucleotides such as DNA and RNA. DNA is used more frequently due to its inherent greater stability under a variety of conditions. The biorecognition process consists of noncovalent interactions between bases of complementary nucleic acid strands by Watson–Crick base pairing rules and is manifested by hybridization of two single strands having such complementary base sequences. Exquisite specificity can be obtained due to the cumulative specific interaction of many bases to their complementary units on the other polynucleotide chain. Nucleic acid sensors are frequently coupled to schemes that utilize amplification of diagnostic nucleic acid sequences (e.g., PCR, nucleic acid sequence-based amplification (NASBA), and rolling circle amplification) with the potential for extreme sensitivity, since a large number of copies of an initially low copy number sequence can be obtained by amplification techniques.

Aptamers

Aptamers can also be included as nucleic acid biorecognition elements, since they are made up of single strands of nucleic acid (DNA or RNA), but here the biorecognition is not via base pairing but by folding to form a unique structure. In contrast to base pairing as the biorecognition feature, the aptamer folds in such a manner that it forms a highly specific three-dimensional structure that recognizes a specific analyte molecule, somewhat similar to the nature of the antibody–antigen interaction. Aptamer selectivity and binding affinity can be similar to the specificity and affinities of antibodies. An advantage of aptamers over antibodies is that the aptamer structure is inherently more robust than the antibody quaternary structure, so that aptamers can be subjected to numerous rounds of denaturation and renaturation, thus providing for easy regeneration of the sensor as well as an overall more robust biosensor. Antibodies are fundamentally products of animal immune systems, with all the limitations that implies. Aptamers are produced by chemical synthesis after selection rather than via immune system cells, and so can be readily modified to enhance stability and affinity. Also, aptamers can be selected against certain targets that can cause the immune system to fail to respond, such as bacterial toxins or prions. Because of all these advantages, aptamers are being used increasingly as biosensor elements. A recent development has been designing RNA sensors, which combine aptamers with ribozymes (catalytic RNAs) to produce RNA molecules (these structures have been called aptazymes) whose catalytic activity depends in some manner upon binding of a specific ligand, the analyte. The binding of the analyte can either stabilize or destabilize the catalytic domain and so adjust the catalytic activity. Thus, detection of the analyte species is signaled by changes in the enzyme's activity.

Peptide nucleic acids

PNA structures are not strictly natural biorecognition molecules, since they are a hybrid of a series of *N*-(2-aminoethyl)-glycine units making up the backbone structure (instead of the sugar–phosphate backbone of natural nucleic acids) with the standard bases bound to the peptide backbone. All intramolecular distances and configurations of the bases are similar to those of natural DNA. Thus, hybridization to DNA or RNA molecules is readily achieved. Since the PNA backbone is uncharged, PNA–DNA duplexes are more thermally stable than DNA–DNA duplexes, and so single-base mismatches have a more destabilizing effect in PNA–DNA hybrids. The PNA structure is stable over a wide temperature and pH range and resistant to nuclease and protease activity and so has obvious applications in the biosensor area. Also, PNAs can be used for the detection of double-stranded DNA directly, eliminating the requirement for thermal denaturation, due to the ability of PNAs to form higher order complexes (three- and four-stranded) with DNA.

Also to be mentioned are another type of nucleotide analogue called locked nucleic acids (LNAs). These nucleic acid analogue have modified RNA nucleosides where a ribonucleoside has a methylene linkage between the 2′-oxygen and the 4′-carbon. These LNAs have the usual hybridization properties expected of nucleic acids with other desirable properties such as highly stable base pairing, high thermal stability, compatibility with most enzymes, and predictable melting behavior.

Molecular beacons

This is a subset of nucleic acid biorecognition molecules, generally synthetic oligonucleotides that are designed to have a hairpin (stem-loop) structure. The loop contains a nucleotide sequence complementary to the analyte sequence to be detected, while the stem structure has only a small number of complementary bases, forming a short double-stranded portion. Fluorophore and quencher molecules are held in close proximity on the two ends of the double-stranded stem, when the molecular beacon is in the closed position. Upon binding of the analyte sequence to the loop portion, the stem opens, the quenching effect on the fluorophore is relieved and the resulting fluorescence signals the presence of the analyte sequence. Aptamers can also serve as molecular beacon biosensors, with this important distinction, that they can be used to detect nonnucleic acid analytes, such as proteins or small organic molecules. PNAs can also be used to form molecular beacon-type structures, and these can be stem-less, in contrast to the usual stem-loop structure found in natural DNA molecular beacons. This stemless structure has advantages (less sensitive to ionic strength, quenching not affected by DNA-binding proteins) as compared to the initially designed DNA beacons.

Cell Surface Receptors/Glycoproteins, Glycolipids/Glycans

Many of the molecular recognition events having to do with cell–cell interaction, pathogen attack upon cells, and so on take place in the glycocalyx layer of the cell membrane. The glycocalyx layer consists primarily of oligosaccharide headgroups of glycoproteins and glycolipids. The lipid or protein portion of these molecules is embedded in the membrane, while the hydrophilic oligosaccharide chain is extended into the outer environment. As the molecular biology of pathogen and toxin interaction with the cell surface is elucidated at the molecular level, these receptor molecules are receiving increasing attention as biorecognition elements. One of the major difficulties heretofore with exploitation of these receptors is the complexity of the cell membrane, making mimicry of this structure a daunting task. Simpler models of cell membranes with defined chemical compositions include liposomes (vesicles) and Langmuir–Blodgett (LB) monolayers. Recent advances have formulated biorecognition molecules into these synthetic lipid structures. In some cases, pathogens or toxins produced by pathogens target specific cell surface receptor sites. These same cell-based receptors can then become biosensor recognition elements. Thus, biorecognition elements seeing use in sensors include gangliosides, glycosphingolipids that are localized upon certain cell types, as well as carbohydrate/oligosaccharide structures found on the cell surface. For example, ganglioside GT1b has been used as a recognition element for botulinum toxin, while *N*-acetylneuraminic acid and *N*-acetylgalactosamine bound cholera and tetanus toxin, respectively, in a semispecific manner. Uropathogenic *Escherichia coli* strains attach to uroepithelial cells by adhesion to the glycolipid globotetraosylceramide(globo side), and so sensors for these *E. coli* strains can be formulated using this globoside. Human gastric mucin has been used in a biosensor format to study the interaction of *Helicobacter pylori* with the extracellular matrix components. Lectins, a broad class of proteins/glycoproteins that bind with high affinity to specific carbohydrate moieties, have also been incorporated into biosensing devices.

Often, oligosaccharides form primary molecular components and markers on pathogen surfaces, and thus may form the basis for diagnostic sensors for specific bacteria or groups of bacteria. For example, the lipopolysaccharide (LPS) molecules (commonly known as bacterial endotoxin), which are found on the cell surface of Gram-negative bacteria, can serve as targets for biorecognition by lectins or other proteins. See Sections "Liposomes" and "Biomimetic Materials" for further discussion on this topic.

Whole Cells

In this article, whole cells refer exclusively to bacterial cells. In some ways, whole cells do not fit the strict definition of a biosensor biorecognition element, since they are most often used to function as "bioreporters." That is, the cells are genetically altered to synthesize a certain marker that produces a detectable signal of some sort (luminescence and fluorescence) when they encounter specific compounds/environmental conditions. Such whole cell sensors have been applied to environmental analysis and monitoring. A brief description of such applications is included later in this article. There are, however, some examples of whole cells being used as biosensors, particularly microbial cells. Often in these instances, the whole cellular machinery functions in a sense as a biorecognition/processing system for the analyte, as, for example, using bacterial cells induced to constitutively produce an amino acid-metabolizing enzyme to detect that particular amino acid. Another innovative approach using whole cells is to genetically engineer cells of the immune system to respond to specific pathogen binding events by producing a detectable signal from some genetically inserted reporter construct.

Bacteriophage

Bacteriophage, by virtue of the fact that they recognize in a specific manner their host bacteria, can become biorecognition elements. The bacteriophage themselves may be used, with appropriate labels (e.g., fluorescent dyes) to detect the binding event, or the phage may be selected for binding to a specific analyte via fusion proteins on the phage coat (i.e., phage display). This latter paradigm has even been extended to expression of the biorecognition peptide as an N-terminal add-on to the major coat protein of the phage. In this case, there is a display of thousands of copies of the recognition protein on the phage surface, thus potentially greatly increasing the sensitivity of the assay by increasing the density of the epitopes displayed. Thus, the phage themselves become both biorecognition (by virtue of peptides displayed on their surface) elements and amplification (multiple copies of fluorescent dye per phage) elements. This type of biosensor could be considered a whole cell biosensor (bacteriophage) that has a bioreceptor on its surface.

Biomimetic Materials

Liposomes

Liposomes are synthetic lipid vesicles, mimicking in some respects cell membranes, although not nearly as complex. Liposomes are frequently used as components of biorecognition systems, but usually the liposomal membrane structure is modified with a standard biorecognition element such as an antibody or oligonucleotide, and these molecules provide the specific recognition. In such cases, the liposome may have encapsulated fluorescent dyes, enzymes, electrochemically active species, and so on, which, upon biorecognition, may be released, and thus the liposome itself functions as an amplification system. More direct involvement of liposomes as sensing elements *per se* are also occasionally encountered. For example, insertion of the ganglioside GM1 into the liposomal membrane has been used to develop a sensor for cholera toxin, since the GM1 has specific affinity for cholera toxin. In a similar vein, liposomes containing cholesterol in the membrane could function as sensors for the toxin listeriolysin.

Molecularly imprinted polymers

A recent development that may, in the future, provide artificial macromolecular bioreceptors is the technique of molecular imprinting. In this procedure, a target molecule (i.e., the analyte), acting as a molecular template, is used as a pattern to direct the assembly of specific binding elements (i.e., monomeric units) that are subsequently polymerized with a cross-linker. The interaction of the monomeric units with the template molecule may be either through noncovalent (the more usual) or covalent bonding. Following polymerization, the target molecule is removed from the polymeric matrix (by suitable solvent or chemical cleavage) leaving a cavity within the polymeric matrix, which is a molecular "negative" image of the target molecule. The monomeric subunits that interact with the target molecule in a noncovalent molecular imprinting scheme do so by interactions such as hydrogen bonding, van der Waal's forces, and hydrophobic interactions, just as antibody–antigen interactions take place. The subsequent interaction with the cross-linker freezes the binding groups in a specific orientation. Ideally, once the target molecule is removed from the polymer, a binding site now exists which is complementary in size and chemical functionality to the original target molecule. In a number of studies, MIPs have been shown to have binding characteristics similar to those of antibodies and other bioreceptors. A potentially important advantage of MIPs is enhanced stability compared to natural biomolecules. The use of MIPs for the analysis of small molecules is becoming established, but detection of larger biomolecules (e.g., proteins and protein toxins) is more problematic. Here, the formation of the template cavity to conform to the three-dimensional image of the total protein may be difficult (e.g., individual antibodies to proteins respond to one antigenic determinant or epitope). Similarly, use of MIPs for sensing whole microorganisms can be challenging.

Biomimetic chromic membranes

A specific biomimetic system that has been used for convenient colorimetric sensing of bacterial toxins, bacteria or viruses is the use of cell membrane-mimicking materials ("smart" materials) into which are inserted cell surface receptor residues. These systems use polydiacetylenic membranes (LB or vesicle bilayers) containing either cell surface gangliosides, sialic acid, or carbohydrate residues to detect the presence of toxins such as cholera toxin or influenza virus, the binding event being signaled by a color change (bichromic shift). The specificity of the sensor resides in the ganglioside or sialic acid residues in the polymeric assembly. Ganglioside GM1 is the primary target of cholera toxin on the intestinal cell surface, while gangliosides GT1b are located at the neuromuscular junction and form the target for botulinum toxin. In the case of influenza virus, the sialic acid residues on glycolipids form targets for the viral surface hemagglutinin of the influenza virus, by which pathway the virus is endocytosed into the cell. When these receptor molecules are incorporated into the polydiacetylenic structure, binding of the virus or bacterial toxins apparently induces a distortion of the conjugation plane of the polymer, leading to the color change.

TRANSDUCTION MECHANISMS

As described above, the transducer is the portion of the biosensor responsible for converting the biorecognition event into a measurable signal. As shown in Figure 5, it is possible

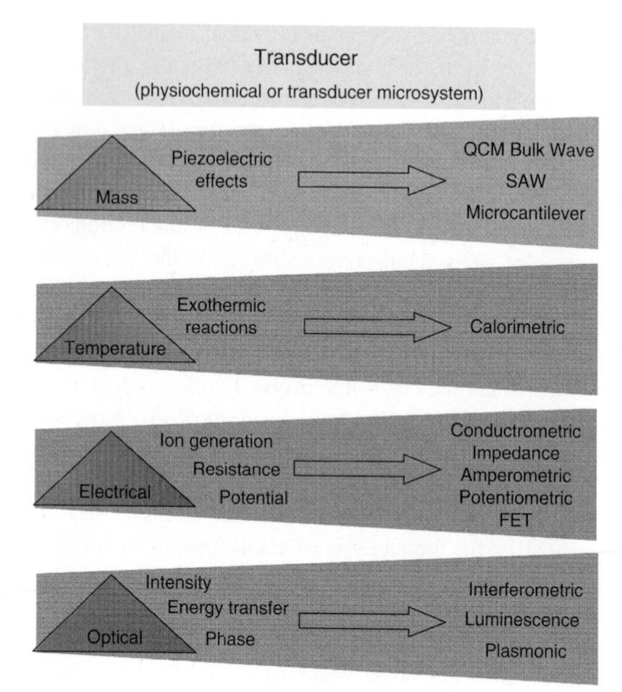

FIGURE 5 Schematic diagram of typical transduction formats employed in biosensors.

to exploit a change in a number of physical and chemical properties (mass, temperature, electrical properties, and optical properties) to allow for different transduction formats. The four basic transduction types: electrochemical, piezoelectric, calorimetric, and optical are reviewed here, with emphasis on the advantages and disadvantages to each, and recent advances toward idealizing biosensor performance.

Electrochemical

Electrochemical transduction is one of the most popular transduction formats employed in biosensing applications. One of the main advantages of biosensors which employ electrochemical transduction is the ability to operate in turbid media and often in complex matrices. Another distinct advantage of electrochemical transduction is that the detection components are inexpensive and can be readily miniaturized into portable, low cost devices. In general, electrochemical-based sensing can be divided into three main categories, potentiometric, amperometric, and impedance. Potentiometric sensors typically rely on a change in potential caused by the production of an electroactive species that is measured by an ion selective electrode. For a biosensor system, this change in electroactive species concentration is usually brought about by an enzyme. In an amperometric sensor system, a change in current is directly measured. Electrochemical sensors based on impedance, most commonly utilize impedance spectroscopy since controlled AC electrical stimulus over a range of frequencies is used to detect variations in the sensor surface properties (i.e., charge transfer and capacitance at the interface layer). In this way, the resistance to flow of an alternating current is measured as voltage/current. For example, metabolic changes (e.g., growth and metabolism) have been shown to correspond to an increase or decrease in impedance. Some of the many variations of potentiometric, amperometric, and impedance biosensors that provide for improved biosensor performance include field effect transistors (FETs) and electrochemiluminescence (ECL).

Field effect transistors

Many researchers have recently looked to FETs as a means to miniaturize potentiometric sensors, while providing increased sensitivity due to minimal circuitry. A particularly promising FET advance includes the light addressable potentiometric sensor, or LAPS, which consists of an n-type silicon semiconductor-based sensor and an insulating layer that maintains contact with the biological solution. Detection of potential changes occurring at this silicon interface are monitored as differences in charge of distribution between the insulator surface and the FET. To accomplish this, alternating photocurrents generated by a light-emitting diode are utilized, so that changes in potential can be transduced into voltage per time differentials. Successful application of the LAPS has been demonstrated with commercially available systems, for the detection of biological threats of interest to defense applications, as well as foodborne pathogen threats of concern to food safety applications.

Electrochemiluminescence

ECL combines the advantages of chemiluminescence (high sensitivity and low background) with electrochemical transduction. ECL utilizes a controlled voltage potential at the surface of an electrode to power a luminescent redox reaction. The redox reaction most commonly employs a ruthenium (II) trisbipyridal chelate coupled with a tripropyl amine, although recent studies have demonstrated success with other fluorophore species.

Mass

Biosensors that detect the change in mass due to target and biorecognition element interactions predominately rely on piezoelectric transduction. Piezoelectric transduction relies on an electrical charge produced by mechanical stress, which is correlated to a biorecognition binding event causing a change in the mass on the piezoelectric device. The main advantage to the piezoelectric transduction (i.e., mass sensor) approach includes the ability to perform label-free measurements of the binding events, including real-time analysis of binding kinetics.

Bulk wave

The most commonly employed transducer is the quartz crystal microbalance (QCM), which relies on a bulk wave effect, illustrated in Figure 6. A QCM device consists of a quartz disk that is plated with electrodes. Upon introduction of an oscillating electric field, an acoustic wave propagates across the device. The change in mass associated with bioreceptor–target interactions causes a decrease in the oscillation frequency that is directly proportional to the amount of target. This transduction format can be coupled to a wide variety of bioreceptors (e.g., antibody, aptamer, and imprinted polymer), provided that the mass change is large enough to produce a measurable change in signal. Not surprisingly, QCM transduction is not capable of small molecule detection directly, and usually requires some sort of signal amplification to be employed.

Surface acoustic wave (SAW)

Changes in the overall mass of the biomolecular system due to association of the bioreceptor with the target analyte can be measured using alternative piezoelectric transducer devices that offer some advantages over bulk wave sensing. For example, SAW devices exhibit increased sensitivity compared to bulk wave devices and transmit along a single crystal face, where the electrodes are located on the same side of the crystal and the transducer acts as both a transmitter and a receiver. SAW devices can directly sense changes in mass due to binding interactions between the immobilized bioreceptor and target analytes and exhibit increased sensitivity compared to bulk wave devices. However, the acoustic wave is significantly dampened in biological solutions, limiting its utility for biosensing applications. Some improvements using dual channel devices, and special coated electrode systems allowing for noncontact SAW devices, which can function in biological solution interfaces have been produced. However, reliable biosensor application incorporating these devices is still under pursuit, as improvements in sensitivity are still required for specific microbial analyses.

Micro- and nanomechanical devices

Micro- and nanomechanical cantilevers can be manufactured through silicon fabrication techniques developed in the electronics industry. Under ambient temperature conditions, the cantilever devices naturally oscillate, and this resonant frequency can be monitored (Figure 7). Through modifications with bioreceptors, target interactions can be monitored as changes in mass, and hence the resonant frequencies, upon binding. Obvious advantages of this form of transduction include low cost and mass production. However, practical applications to biosensing are limited due to oscillation dampening in liquid solutions, compared to air- and vacuum-packed cantilever systems. To address this limitation, there have been recent advances toward incorporating hollow channels inside the microcantilever in order to immobilize the bioreceptors internal to the device. In other words, high resonant efficiencies obtained through vacuum packaging can be maintained, while specific mass changes can be monitored by flowing the sample through the inside of the device. Further signal enhancements can be obtained through use of nanoparticle and magnetic bead amplification to cause larger frequency shifts, and overall a more sensitive analysis.

Flexural-plate wave

A flexural-plate wave (FPW) transducer contains a thin membrane that can propagate a SAW. The FPW is in contact with a thin film of liquid that vibrates with the membrane and hence the change in mass of a vibrating element can be

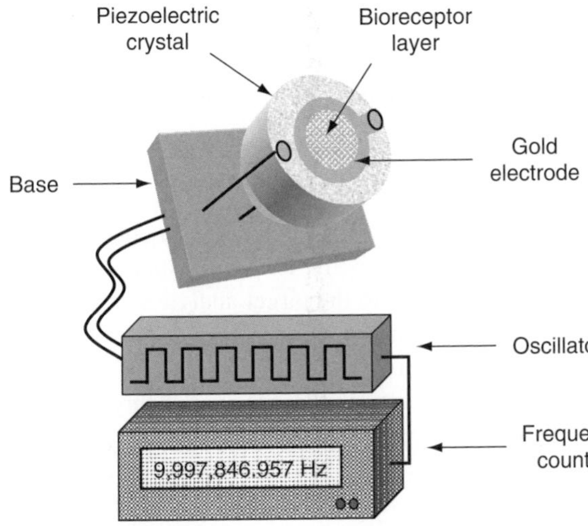

FIGURE 6 Schematic diagram of a biosensor based on piezoelectric/acoustic wave transduction.

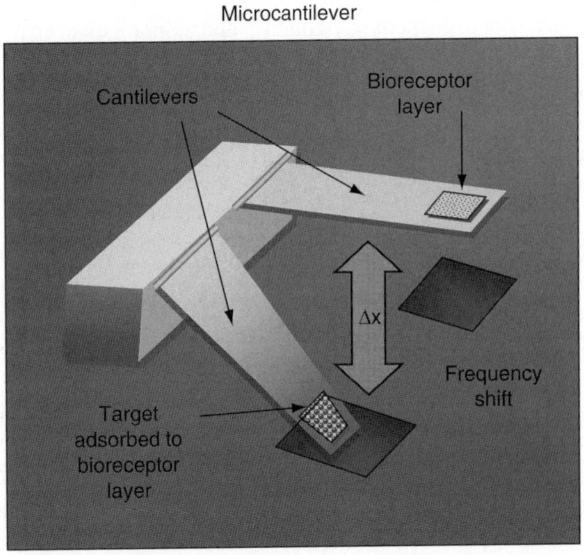

FIGURE 7 Schematic diagram of a microcantilever-based biosensor.

detected to indicate biological interactions. This technique is not very sensitive, but similar to SAW devices can provide for real-time analyses.

Calorimetric

Calorimetric sensors utilize thermistor probes to monitor changes in temperature due to exothermic chemical reactions. Many biological reactions are exothermic (e.g., enzyme reactions), and hence calorimetric detection allows for a near universal transduction format. One key disadvantage of this approach is that environmental temperature fluctuations must be shielded from the sensor system. Traditionally calorimetric biosensors have been large and bulky, although advances in silicon microfabrication technologies and microfluidics have allowed for miniaturization and improved performance.

Optical

Due to a number of advantages, optical transduction is one of the most widely used biosensor transduction formats. For example, optical transduction can be very rapid where the limiting factor for the speed of detection is often a diffusion-limited process of the biomolecular recognition event, rather than the optical transducer. Another advantage of optical transduction is the interferences that can hinder electrochemical transduction measurements such as voltage surges, harmonic induction, corrosion of electrode elements, and radio frequency interferences are not present. Some of the disadvantages of using optical transduction formats include: detection challenges when analyzing turbid samples, and the cost associated with detection system components. A wide variety of optical transduction formats have been employed, where changes of the interaction of light with the biomolecular system is used to produce a measurable signal. These changes can be based on differences in refractive index, production of chemiluminescent reaction products, fluorescence emission, fluorescence quenching, radiative and nonradiative energy transfer, temporal changes in optical emission properties, scattering techniques as well as other optical effects. These effects can be monitored using a variety of optical platforms including total internal reflectance and evanescent wave technologies, interferometric, resonant cavities, biochip devices, and so on. The following paragraphs review the most common as well as most popular emerging optical transduction formats: fluorescence, interferometric, chemiluminescence, surface plasmon resonance (SPR), and surface enhanced Raman spectroscopy (SERS).

Fluorescence

Fluorescence is the most popular form of optical transduction due to the high sensitivity that is fundamental to this type of optical process. Another advantage of fluorescence-based methods is that they generally do not have the interference issues that SPR and other refractive index-based methods possess. However, in most cases, the intrinsic sample fluorescence is not sufficient for analysis, and a fluorogenic reporter is used to label an affinity reagent to create a bioreporter. By monitoring the intensity of the fluorogenic reporter, it is possible to determine the presence and concentration of the target analytes, as illustrated in the bioassay techniques section described previously. It is also possible to monitor shifts in the wavelength of the fluorophore reporter, as well as energy transfer phenomena, and time dependence of the fluorescence emission, all of which can be related to binding interactions depending on the assay employed. The distinguishing features between fluorescence-based biosensors, besides the above mentioned properties that can be monitored, include the optical detection format used. For example, it is possible to utilize fiber optic probes, often termed optrodes, to immobilize bioreceptors at the tip of the fiber, and use total internal reflectance properties of the fiber to transmit excited and emitted light. Total internal reflectance can also be employed in an evanescent wave format, where a residual amount of (evanescent) light that escapes at the reflectance point is used to excite immobilized bioreceptors that are in close proximity to the surface, rather than in bulk solution. This format allows for controlled excitation, and can allow for minimal fluorescence background. However, a key disadvantage is the lack of evanescent excitation power, and sometimes poor coupling of the emission when using similar collection geometries. Fluorescence detection can be used with a wide variety of detection formats. For example, it is routinely coupled with flow cytometry and microfluidic platforms or imaging array systems such as biochips, which utilize spatial patterning of biological recognition elements to match fluorescence location to target species.

Interferometry

Interferometers can measure biomolecular interactions that take place at a surface within an evanescent field that causes a refractive index change and a corresponding change in the phase of the propagating light. To accomplish this, two channels are used, where one channel serves as a continuous reference that does not experience refractive index changes due to the target analyte's presence. By combining light from the reference and sample beams, an interference pattern is created, which can be used to determine the presence of the associated target species. Although this approach has been in existence for quite some time, it is primarily a laboratory technique since it is not a very robust biosensor technology, and suffers from significant false-positive results.

Chemiluminescence

Chemiluminescence is one type of optical sensor technology, which relies on a series of chemical reactions, usually employing oxidation of a luminol reactant, to produce a reaction product that gives off characteristic light. The intensity of this light is correlated to a sample target of interest, by coupling some of the reaction products to biorecognition elements, to serve as chemiluminescent reporters. The advantage of chemiluminescent transduction includes very low optical background. Theoretically, the background in the absence of analyte for a chemiluminescent biosensor should be very low or non-existent, leading to a very sensitive (high signal-to-background) measurement but practically, nonspecific interactions commonly occur, resulting in some low to very low level of chemiluminescence even in the absence of analyte. The reaction usually employs signal amplification and bioassays with multiple washing and labeling steps, which leads to a somewhat time-consuming analysis.

Surface plasmon resonance

SPR is a phenomenon that can exist at the interface between a metal and a dielectric material such as air or water. A surface plasmon (commonly excited by optical radiation) of the electrons at the surface of the metal results in an associated surface bound and evanescent electromagnetic wave of optical frequency. This evanescent wave has a maximal intensity at the interface and decays exponentially with distance away from the interface. In SPR-based biosensing, changes at the interface, that is, biological recognition element and analyte binding, causes changes in the local refractive index, which in turn causes changes in the resonance excitation of the surface plasmon.

SPR is a form of reflectance spectroscopy, where change in a SPR that occurs during optical illumination of a metal surface is harnessed to detect biomolecular interactions. A schematic diagram illustrating a basic SPR chip platform is illustrated in Figure 8. The SPR chip consists of a prism with a thin gold film upon which the bioreceptors are immobilized. Light is totally internally reflected from the prism face coated with metal and the changes in reflectivity measured. Surface plasmons are excited at a specific incident angle and result in a massive reduction in reflectivity at that angle. Changes in the refractive index at the interface results in a change of the optimal angle required for excitation.

Any change in the optical properties of the medium in contact with the immobilized layer will cause changes in the characteristics of the surface Plasmon wave. Specifically, these changes can be measured as changes in wavelength, intensity, phase, or angle of incidence. SPR techniques are widely popular in contemporary biosensor development because it is a surface-sensitive technique that can measure real-time interactions between unlabeled species.

Although SPR is a simple technique with a number of advantages, it is not the most sensitive. However, one variation of SPR includes a resonant mirror format, which utilizes a series of polarizing filters to block light that is internally reflected. At a particular incident resonant angle, the light is diverted through a spacer layer that has a low refractive index, into a higher refractive index guide so that the signal peak appears on a dark background. Despite these advances, one of the primary limitations of SPR-based biosensors is that anything which alters the refractive index at the sensing surface will interfere with the analysis including nonhomogenous (complex) sample matrices, and nonspecific binding interactions.

Surface enhanced Raman spectroscopy

Another type of plasmonic spectroscopy that is gaining popularity in biosensing applications is SERS. SERS is an enhanced form of Raman spectroscopy, where a

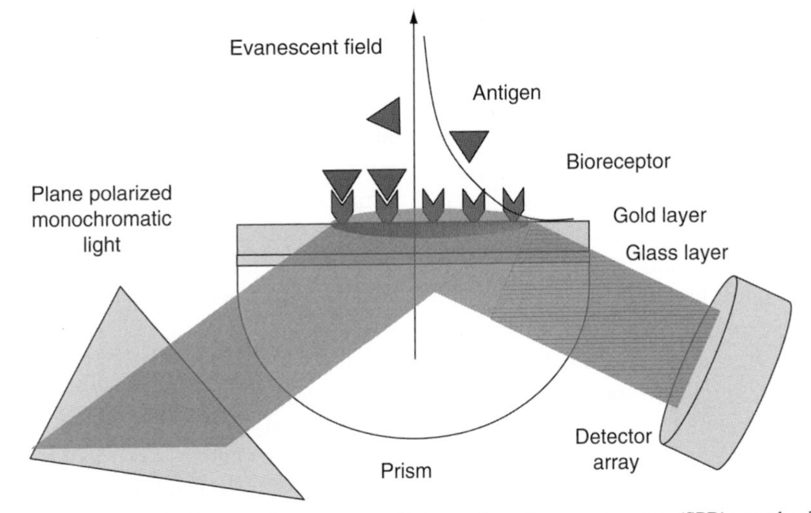

FIGURE 8 Schematic diagram of a biosensor utilizing surface plasmon resonance (SPR) transduction.

nano-roughened metal surface is used to enhance the scattered Raman signal. The SERS enhancement is thought to be the result of a combination of intense localized fields arising from SPR in metallic nanostructures and chemical effects.

Advantages of this approach include the ability to obtain a spectroscopic fingerprint, similar to infrared absorption spectroscopy, while being relatively interference-free from water. Also, since SERS is an enhanced technique, it can be used to see very low concentrations of even biological materials, as evidenced by recent interest in a variety of immuno-SERS methods under development. The difficulties often encountered with SERS biosensors, primarily stem from signal reproducibility challenges, which are directly tied to the reproducibility of nanostructured SERS substrate fabrication. Of the transduction methods covered in this article, SERS is one of the least established for biosensor development but shows much promise for continued development.

APPLICATIONS OF BIOSENSORS

Bacterial Biosensors for Environmental Monitoring

Whole cell bioreporters which respond to certain pollutants/toxicological conditions are beginning to find applications in environmental sensing. Applications range from detection of contaminants, measurement of pollutant toxicity, monitoring of genetically engineered bacteria released into the environment, and even uses for detection of stresses such as ionizing radiation and oxidative damage. In general, such whole cell sensors should not be called biosensors unless the cells are integrated somehow into a stand-alone sensing/transduction/readout system. Some whole cell sensors use genetically engineered bacteria into which a *lux* gene construct (coding for an active luciferase) under the control of an inducible promoter is introduced. The general sensing strategy is to put the luciferase construct under control of a promoter that recognizes the analyte of interest, although in some cases, the bioreporter organism can be engineered to respond to many analytes (broad specificity) by using a heat shock promoter fused to the lux cassette.

For example, the seminal work in this arena describe a naphthalene biosensor organism (genetically engineered *Pseudomonas fluorescens*) that detects the bioavailability of naphthalene (serving as a surrogate for polyaromatic hydrocarbons) and salicylate in contaminated soil by induction of the lux reporter, and subsequent bioluminescence as a result of expression of the reporter gene. The naphthalene biosensor described above is an example of a catabolic bioreporter. Catabolic bioreporters produce luminescence only in the presence of a specific contaminant, since the *lux* genes are linked to a degradation pathway promoter. Metabolic bioreporters, on the other hand, have reporter genes downstream of a strong constitutive promoter, and thus ,luminescence is directly affected by toxic agents that interfere with metabolism. These concepts have been subsequently used to develop a variety of bioreporter organisms. Other investigators have used genetically engineered bacteria in which the green fluorescent protein (gfp), (or derivatives thereof), from the jellyfish *Aequorea Victoria* functions as a bioreporter in the same manner as the *lux* gene. Some arguments have been advanced that the expression of bioluminescence is a more direct bioreporter than gfp expression, since the expression of gfp is only detected upon excitation with the correct wavelength of light. However, in practical terms, bioluminescence must be quantified by some sort of photodetector/luminometer, so the difference between that and a fluorescence excitation/emission detection system seems minimal. Beta-galactosidase enzymes have also been used as biomarkers. Others have used naturally luminescent bacteria (i.e., *Vibrio fischeri*) as bioreporter organisms, since toxic compounds can disturb metabolic processes, thus causing a reduction in their bioluminescence. Microbial biosensors have been developed to assay biological oxygen demand (BOD), a value related to total content of organic materials in wastewater. These sensors often measure BOD by evaluating the rate at which microbial cells deplete oxygen. It is not possible to provide a comprehensive list of whole cell sensor applications in this article. Very seldom are these bioreporter organisms integrated into a complete sensing package where a microculture environment, integrated luminometer or flurometer, and light-tight enclosure can produce a true biosensor device.

Food Biosensing

Some of the factors of interest in regard to analysis of foods include monitoring nutritional parameters, food additives, contaminants, microbial contamination, shelf life assessment and factors such as smell and odor. Many of the aforementioned factors have to do with quality of the food, while others reflect directly on food safety. There are obvious overlaps between food quality and food safety. The assessment of food quality can be subjective, as factors such as appearance, taste, smell, and texture may enter into the overall evaluation. Sensor technologies that measure specific parameters such as sugar content and composition, total titratable acidity (e.g., for fruit), specific chemicals such as glucose, sucrose, lactate, alcohol, glutamate, and ascorbic acid provide more objective means of evaluating food quality and freshness or spoilage. Also, quantitative detection of contaminants of food such as growth hormones fed to animals, antibiotics, and pesticides is important. Finally, there is the issue of pathogen and/or pathogen toxin contamination of food. This last is of great concern because of the potential for outbreaks of foodborne illnesses, such as are seen periodically. Fatalities can result from such exposures, and recent outbreaks have generated much media interest.

There is a clear role for biosensors in many areas relating to food quality and food safety. The space limitations of this article will not permit consideration of all aspects, and so the authors have chosen to focus on pathogen detection in food products. Recent reviews of statistics regarding pathogen testing (all types of tests including biosensors) in the food industry prove interesting in regard to future biosensor development. For example, for 1999, *c.* 16% of all microbial tests in the industry were for specific pathogens, *c.* 16% were for yeast and mold, *c.* 31% were for coliform and *E. coli*, and *c.* 37% were for total viable organisms. The same statistical survey found that microbial testing in each of the food sectors was as follows: 36% for the processed food sector, 10% for fruits and vegetables, 22% for meat, and 32% for dairy. The US food industry performed *c.* 144.3 million microbiological tests in 1999. Unfortunately, heavy reliance is still placed upon conventional culturing techniques, so there are significant time gaps between sampling for microbial contamination and detection of microbial contamination, and these techniques are often labor-intensive and require expertise in interpreting results. Other investigators also point out the importance of pathogen detection in the food industry, that is, that the food industry accounts for *c.* 38% of total research in the field of pathogen detection. The other major areas of research interest for biosensors are water and environment quality control (16%) and clinical applications (18%). In terms of research articles describing biosensors applied to pathogen detection, a review article, "Pathogen detection: A perspective of traditional methods and biosensors," summarizes a number of interesting points. Thus, of research articles concerning pathogen detection, techniques for detection of Salmonellae species are the most abundant (33%), followed by *E. coli* (27%), with *Listeria* (14%) and *Campylobacter* (11%) species, the other major pathogens. *Legionella* accounts for 7% of research articles discussing detection techniques with all other bacterial species accounting for 8%. This relative abundance of research articles published with regard to detection schemes clearly indicates the emphasis upon pathogen detection for food safety. This same review also discusses some relevant statistics regarding detection technologies for pathogens, gleaned from a review of the relevant literature over the past 20 years. The most popular methods by far relate to culture/colony counting with PCR following closely behind. ELISA assays are in third place in terms of abundance, with biosensors following in fourth place. In the biosensor category, the most used techniques are optical (35%), followed by electrochemical (32%), piezoelectric (16%), and all other categories (16%).

The tried and true methods of microbiology involve concentration of microbes from food samples. Plating, culture, growth in selective media, and colony counting are still the gold standard with regard to identification and quantification. Nevertheless, as pointed out above, these techniques are time- and labor-intensive and do not provide answers in a timely manner. Even PCR often requires a time investment of several hours (at minimum) before analysis is complete, and also relies on technical expertise. There is clearly a need for rapid, low cost techniques that provide automated or semi-automated pathogen analysis. Even better would be systems which integrate all aspects of the analysis from sampling to final quantitative result. Biosensors seem well-suited to fill at least some of this niche in the food safety testing market. Yet, a main problem facing all such attempts to move pathogen testing from classical microbiological procedures to biosensor-based analysis is the issue of sensitivity. Recent literature indicates that the infectious dose of *Salmonella* or *E. coli* 0157:H7 is ten organisms, while the existing coliform standard for *E. coli* in water is 4 cells/100 ml. Culture methods can and do attain this sensitivity. From a review of recent biosensor literature the present authors carried out, it is clear that this level of sensitivity is not achievable by most of the biosensor research devices currently proposed. It seems clear that biosensors will only see their full potential realized with regard to pathogen detection for food safety, when specificity and sensitivity can compare to established methods and such biosensors can also be cost-competitive (or cost-saving) with current techniques. It should also be pointed out that testing for genetically modified organisms (GMOs) will probably become a more frequent analysis with regard to food and some authorities expect this area to see the fastest growth of any testing market in the food industry. Biosensor research for GMO detection seems to be mainly focused on DNA-based detection technologies at present.

Biodefense Biosensing Applications

Events of recent years have indicated the need for sensors of pathogens/toxins which could be used by military enemies or for terrorist purposes. Much research effort is currently focused on analytical strategies to detect these agents. A major difficulty with biological attacks is actually determining whether the attack has occurred, in a timely manner, to enable early response, and minimize casualties. The ideal sensor for biological agents, which could be used for attack (hereafter called biowarfare agents (BWAs)) would provide highly sensitive and selective identification of threat organisms in virtually real time. Also, since the potential release of BWAs could be on the battlefield or in urban settings, the surrounding environment/atmospheric milieu may be highly complex (smokes, dust, and particulate matter), and the analytical technique must be able to detect organisms of interest without interference from background material. To date, these requirements have provided a daunting analytical challenge.

Attempts to meet this challenge have employed molecular techniques that can identify chemical markers of BWAs. In general, detection of BWAs can follow one of two paths.

In one, positive identification of a BWA must be obtained in a few hours, the so-called detect-to-treat option. Such detection would give medical personnel the means to successfully treat individuals that have been exposed to BWAs. The time frame of several hours makes the analytical task easier than is the case with the other detection scenario. In this latter case, a detect-to-warn sensor must be able to provide a warning within a few minutes that a BWA release has occurred. If such sensors could be developed, then perhaps therapeutic treatment of all members of an exposed population might be an option. The variety of BWA that potentially could be used, that is, bacteria (vegetative or spores, Gram-negative or Gram-positive), viruses, and toxins, adds difficulty to the sensing task. Specificity and sensitivity are both critical for useful BWA sensors. False positives are unacceptable due to the consequences attendant upon an assumed BWA incident. False negatives are also unacceptable for obvious reasons. It is true that nucleic acid-based biosensing technologies have been demonstrated to have the sensitivity to detect one organism or one spore. Also, in general, nucleic acid-based sensing systems are more sensitive than antibody-based detection systems. Such analyses, however, depend on amplification of the target nucleic acid, a process that takes time (although intensive efforts have been and are being made to shorten this step). Further, nucleic acids from nonrelevant sources may interfere in PCR amplifications, and nucleic acid-based detection will not work for toxins. Detection of BWA with extreme sensitivity is important since some of the organisms manifest infectivity at extremely low levels. Nucleic acid-based detection also offers the best chance for detection of novel, engineered organisms or so-called stealth organisms, where virulence genes are inserted into an otherwise innocuous microbe. In these cases, the nucleic acid sequence of the virulence elements still provides a diagnostic signature. Integration of all steps (lysing cells/spores to prepare nucleic acids, amplification of target sequences, and detection of the same) required for nucleic acid sensing is crucial for a true biosensor, and so far, reducing the entire assay time to a few minutes has been unreachable. Thus, there has been interest in nucleic acid-based techniques that provide gene-based specificity, but do not require amplification steps to attain detection sensitivity to the required levels. Such exquisitely sensitive detection almost precludes techniques like SPR and recommends procedures like fluorescence detection.

This difficulty has led others to suggest that detectors based on surface features (e.g., immunogenic molecular species) of the threat organisms may be a more tractable route for development of detect-to-protect devices. Such devices must have a rapid response, must not give false positives, must have high sensitivity and the ability to detect the target in aerosol samples, when the natural environment for most detecting biomolecules is aqueous. The major factors for successful immunosensing of BWA are the efficiency of the antigen–antibody complex formation and the ability to sensitively detect these complexes. Nonspecific binding can be a problem for such techniques, particularly as sensitivity is pushed. There is also an urgent need for multiplexed detection in BWA sensors, since the list of potential BWAs, while not overly long, still numbers more than a handful.

CONCLUDING REMARKS

The above discussion of biosensors shows that this is an emerging field that has not by any means reached its full potential. During preparation of this article, the authors reviewed many research papers with very ingenious ideas for biosensors. Unfortunately, the gap between biosensor prototypes that work in the laboratory and commercial biosensors which can see routine use in real-world environments is often still large. The many varied biorecognition elements, transducers, and detection strategies lead to a wealth of potential applications. The field of biosensor research/applications is still probably in the stage where it is not clear what biorecognition elements/transducers/detection schemes will be most productive, and indeed, this may well continue to evolve in different directions as techniques like MIP become more established. Also, the large number of potential applications may well demand a quite varied armamentarium of biosensor techniques. In general, robustness of the biorecognition element continues to be a concern for many applications. The use of transducers/detection mechanisms that involve sophisticated/expensive instrumentation also will preclude such devices from many routine applications. Simpler is not necessarily always better, but dipstick sensors for certain applications are attractive.

A few comments regarding two other areas which the authors of this article believe will have increasing importance in the field of biosensor applications in the future will close out this article. Array biosensors, biochips, labs-on-a-chip, that is, devices that permit multianalyte detection (multiplex analysis) and, ideally, that can integrate all processing steps into a microanalytical system, will certainly see increased interest in the future. Such integrated devices have a number of advantages, such as ability to assay large numbers of samples in minutes, savings of expensive reagents, and reduction in time for individual assay steps due to decreased assay volumes. Unfortunately, there are a number of obstacles still to be overcome with regard to these systems. Among these are such things as further development of microfluidics systems, sample preparation modules, sensitive detection modules, and robust assay methodologies. Another area receiving much recent interest is in the area of nanotechnology, more specifically the use of nanomaterials such as nanoparticles in biosensor development. Nanoparticles do not serve as biorecognition elements *per se*, but can be modified by attachment of appropriate molecular species (e.g., antibodies and DNA fragments) to become nanosized carriers of biorecognition elements, which can also serve valuable transduction functions (i.e., magnetic properties and fluorescence). The unique properties of nanoparticles, that is, colloidal gold and quantum dots, also recommend them for novel sensing applications. A number of investigators have mentioned the potential use of nanoparticles, particularly quantum dots, as molecular bar codes. Metallic nanoshells and semiconductor quantum dots are highly promising markers for use in high throughput screening and barcoding, by virtue of spectral multiplexing. However, much work remains to be done before the full promise of these materials is applied to practical and useful biosensors.

RECENT DEVELOPMENTS

In this section, we wish to highlight several areas of biosensor research and development which, in our opinion, either already have demonstrated, or have the potential to demonstrate novel and powerful sensing capabilities. We will briefly discuss each of the three technologies we wish to highlight in the following text. The three areas which we see as having unusual potential for application to a variety of sensing needs are: (1) virus-based biosensors; (2) genetically encoded biosensors based on fluorescent proteins; and (3) nanomaterials, and their applications as components of biosensors.

VIRUS-BASED BIOSENSORS

Viruses continue to intrigue researchers because of their varied structures and range of capabilities, all contained in packages of such small size. As structural entities, viruses can be thought of as highly structured nanosystems with a well-defined three-dimensional structure which can occur in a wide variety of shapes and sizes. Of particular importance for sensing applications, viruses are characterized by multiple identical repeating units of coat proteins (thus providing numerous identical sites for scaffolding of various ligands and tags), water solubility, stability in aqueous buffers and perhaps most importantly, particles of a single type of virus are essentially identical and thus mono-disperse in size and shape. Because of the repeating nature of the coat proteins, attachment of modifying molecules at precisely defined positions is possible. Thus, fluorescent dyes can be bound at defined positions to make a "super-bright" viral particle in which the spacing between dye molecules is such as to mostly eliminate fluorescence quenching, while multivalent display of ligands on the viral surface can improve binding affinity for cellular targets. By using site-directed mutagenesis it is also possible to express new reactive residues on the viral coat proteins. It is also possible to modify the viral coat protein to display not only fluorescent tags, but also targeting ligands, such as antibodies. The use of phage display to genetically engineer the capsid proteins of filamentous bacteriophages to produce recognition peptides of almost unlimited specificity for any target analyte has already been mentioned in this article and will not be discussed further here. Bacteriophages (afterward shortened to phage) are not the only viruses of interest as sensing moieties, however, as a variety of plant viruses such as cowpea mosaic virus, tobacco mosaic virus, and turnip yellow mosaic virus have also been employed. It is also possible to use the interior compartment of viruses, once the nucleic acid is removed to entrap entities such as nanoparticles, polymers, enzymes, fluorophore tags, etc.

Only in recent times has the potential usefulness of viruses as sensing entities been exploited. Not surprisingly, phage-based sensors have been applied in a variety of ways: (1) as target-recognizing probes (think antibody substitute); (2) peptides isolated from phage display techniques used directly as probes; (3) lytic phages used directly as sensors by means of lysing their target bacteria and detection of components of lysed contents; and (4) phages conjugated to other nanomaterials to form a composite sensor. Phages have been immobilized on the surface of a QCM and used for the detection of pathogenic bacteria. Phages displaying specific peptides on their major coat protein have also proven to be effective replacements for antibodies in ELISA-type assays. In a similar manner, phages have also been used as sensing elements in immunofluorescence assays, in SPR detection and in electrochemical sensors. As already mentioned, removing the interior contents of the virus and using the interior cavity for encapsulation of various reagents is a novel modification which is just now being explored. Such constructs are referred to as virus-like particles (VLP). A natural property of intact virus is binding between viral coat proteins and specific receptors on cell surfaces. Thus, VLP can retain this binding specificity, while carrying labeling reagents on the interior surface. For example, canine parvovirus, which has a natural affinity for transferrin receptors, could be modified into VLP containing a fluorescent label. These labeled VLP became a biosensor for certain tumor cells which express transferrin on their surface.

GENETICALLY ENCODED BIOSENSORS USING ENGINEERED FLUORESCENT PROTEINS

This remarkable group of naturally fluorescent proteins, occurring in the biological realm in certain sea creatures, have proved to be powerful tools for elucidating information regarding complex biological interactions. Perhaps, some measure of their significance to biological research can be gleaned from the fact that the 2008 Nobel Prize in Chemistry was awarded to three investigators (Shimomura, Chalfie and Tsien) who played pivotal roles in isolation, study, and application of these proteins to biological systems. Isolation of the first gfp from the *Aequorea victoria* jellyfish was followed by eventual sequencing of the gene coding for the fluorescent protein, and the discovery that the fluorescent protein (FP) could be expressed in virtually any cell, not just in species of origin. Clever site-directed mutagenesis has produced variants of GFP, which have enhanced fluorescent properties relative to the native protein, and which have different excitation/emission maxima, for example CFP (cyan) and YFP (yellow) being among the most useful. Further studies of various Anthozoa species (reef corals) have permitted the color palette of FP's to be greatly extended into the orange and red regions of the visible spectrum. These studies have spawned a wide variety of modified fluorescent proteins with enhanced fluorescence emissions and with fluorescence properties spanning essentially the whole visible spectrum. Recently, expansion of the FP spectrum into the near IR has also been reported, using a bacteriophytochrome from a microorganism.

To understand the applications of these FP's to biological science, a few words need to be said about the FP structure. The tripeptide responsible for the fluorescence of FP's is in a central alpha helix, which is enclosed in a 11-stranded beta barrel structure. The size of the FP varies between 20 and 30 kDa. Mutations within the chromophore site, as already alluded to, can result in changes in fluorescence properties. Once the gene for the GFP was identified and sequenced, it was found possible to place it under control of promoters and express the protein in, for instance, animal cells. This application, however, was by no means the most significant use of FP's. It was found possible to fuse the FP gene to other protein genes, whilst still retaining the fluorescent properties of the FP. This immediately led to tremendously diverse applications, as the FP could be fused to protein scaffolds, which could act as molecular recognition elements or subcellular localization signals. Thus, the FP's can become biosensors, reporting on molecular interactions with heretofore unachievable spatial and temporal resolution.

A powerful advantage of these fluorescent proteins (FP's), as compared to sensor molecules such as Quantum dots or fluorescent probes, which by definition are exogenous to the cell, and must be introduced by various techniques, is that FP's are introduced into cells as a genetic construct, and therefore are manufactured directly by the cellular machinery. Thus, these genetically encoded fluorescent probes, by definition, are noninvasive. A further advantage of these FPs is that they can be directed to certain subcellular compartments by incorporation of signal sequences ,which control subcellular localization, while exogenous fluorophores may be constrained to only certain subcellular compartments. Fusion constructs of PF genes and other protein genes are easy to construct by standard molecular biology techniques. The potential to resolve molecular events on the time scale of 100s of milliseconds exists using these FP probes. Further, the use of FP's coupled with FRET-based techniques (see below) allows one to capture the intricacies of spatial interactions on a distance scale which has been difficult or impossible to achieve by other techniques. Also, using FP's allows one to gain information about biorecognition processes in the natural habitat of the proteins of interest, thus providing information about processes occurring within the complex environment of living cells, as opposed to the test tube.

Applications of FP's span an enormous spectrum of investigations. Here, we can only provide a generalized overview with emphasis on common thematic elements. Fusion chimeras of FP's and a molecular recognition element (MRE) can either involve an endogenous or exogenous MRE, with the changes in FP properties being the sensor endpoint. On the other hand, FRET-based sensors, by analogy may involve either intramolecular or intermolecular FRET interactions occurring as MRE's on the same protein chain or separate protein chains interact, with alterations in FRET properties as the endpoint. Also, the FP chromophore may be fragmented into two parts, with the biorecognition event reconstituting the intact chromophore. Some further explanation of these possibilities with examples will now be provided.

Perhaps the simplest application of an FP/single polypeptide chain fusion is in studying colocalization of separate proteins to subcellular compartments. Here, the use of two FP's with quite different fluorescent properties, each fused to one of a pair of proteins under study, permits the imaging of the chimeras in various subcellular compartments and evidence either for or against colocalization. In terms of FP's with endogenous MRE's, a variety of biosensors have been developed where changes in pH or ion concentration, etc., can produce changes in fluorescence properties of the FP. In these cases, the FP itself serves as the sensor, that is, the analyte directly influences the FP's fluorescence properties. Fusion to other polypeptides may or may not be necessary for its intended use. Due to the chromophore's environment within the protein, GFP demonstrates alterations in fluorescence properties as pH changes. Thus, with FP fused to organelle-specific targeting peptides, a subcellular pH sensor has been constructed, providing information about pH of various cellular compartments. FP constructs have been made to serve as halide sensors, in which fluorescent properties of the FP change upon halide interaction, although here the additional requirement of engineering an ion channel into the chromophore part of the FP structure is needed to make the sensor effective. Also, interestingly, a redox sensor has been constructed in which site-specific mutagenesis has produced two cysteines in close enough apposition to the chromophore element of the FP so as to alter the fluorescent properties depending on the redox state of the cysteines.

In the case of exogenous MRE's fused to FP's in which the total construct is still one polypeptide chain, an ingenious technique is used to generate the sensing structure (these constructs are sometimes referred to as circularly permuted GFP's). Here, the original amino and carboxyl terminal amino acids of the FP are connected with a flexible peptide linker, while in some other location within the FP structure, a peptide bond is broken and an exogenous MRE is inserted at this point. The FP structure is not so perturbed as to lose all fluorescence properties, but when analyte binds to the MRE, inducing a conformational change, this change is relayed to the chromophore, resulting in altered fluorescence properties. Thus, for example, a Ca sensor was devised where calmodulin was inserted in the fusion construct in such a location, that upon Ca binding, a large change in fluorescence emission was seen. A sensor giving information about relative ATP:ADP concentrations was developed by inserting an adenylate-binding protein into the YFP structure. Shifts in excitation spectra resulted when different ratios of the two ligands bound to the adenylate-binding protein. Other sensors in this class include hydrogen peroxide and Zn sensors.

We now discuss the FRET-based biosensors, which employ various FP pairs. These sensors provide potential for incredible spatial resolution of molecular interactions. This is by virtue of the fact that FRET is an energy transfer process, which is critically dependent upon both distance and orientation between donor and acceptor groups. In fact, effective distances where FRET interaction occurs are usually considered to be 10 nm or less. This is well below the standard fluorescence imaging resolution of 100s of nm. Thus, FRET provides an insight into molecular interactions which is hardly to be obtained by other means. In intramolecular FRET configurations of FP sensors, two different FP's (most often CFP and YFP) are fused in tandem to a polypeptide sequence which is an MRE. Upon binding of the analyte to the MRE, there is a change of structure of the peptide chain, which results in an alteration of the distance between the donor and acceptor fluorophores of the FRET pair. This in turn causes a change in the efficiency of FRET, which is the signal change that is detected. For example, if two FP's are within FRET interaction distance and separated by a peptide sequence which is sensitive to a particular proteolytic enzyme, presence of that enzyme will result in cleavage at the peptide site, thus separating the two FP's and abolishing FRET interaction, a result easily distinguished by loss of acceptor emission. Intramolecular FRET is also a powerful tool for studying the action of kinases, enzymes which phosphorylate various sites in proteins and have critical roles in regulatory cascades. In this sensor format, a peptide substrate for the kinase of interest and the binding domain for phorphorylated substrate are both fused to two FP's which form a FRET pair. When the kinase encounters this substrate, and phosphorylation occurs, the binding domain binds to the substrate, there is a rather large change in conformation, and the FRET interaction between the FP's is altered, which is detected by spectroscopy. This same general paradigm has been applied to other types of sensors. Bacterial periplasmic binding proteins, for instance, can have binding pockets specific for certain sugars. When the specific sugar binds, the more open configuration of the protein becomes more condensed, and FP's on either side of the binding pocket can come into FRET apposition. Calcium sensors called cameleons have been developed in which calmodulin, and M13 peptide are fused to CFP and YFP. In the absence of Ca^{++}, the fusion protein assumes a dumbbell shape and distances between the FP's are too great for FRET, but upon Ca^{++} binding, configuration changes to a compact shape and FRET is now detectable.

Analogous to the above FRET analysis system, it is possible to use intermolecular FRET-based assays, in which the two FP's are fused to two distinct polypeptide chains. Here again, if protein/protein interaction occurs upon presence of the analyte, FRET interaction can occur. These assays become somewhat more complicated because two separate fusion constructs must be co-transfected. A biosensor for cAMP has been constructed using this principle, in which two FP's of a FRET pair were fused respectively to the regulatory RII and catalytic subunits of protein kinase A, the main effector of cAMP. Protein kinase A consists of a holotetramer composed of two regulatory and two catalytic subunits. Activation of protein kinase A by cAMP causes dissociation of the catalytic from the regulatory subunits. In the FP – labeled intact protein kinase A, FRET interaction occurs. Upon binding of cAMP, dissociation of the two subunits is induced, resulting in loss of FRET energy exchange. The same type of intermolecular FRET can be used to examine other protein-protein interactions.

Yet another type of sensor strategy is available for FP-based sensors. This is referred to as bimolecular fluorescence complementation. The principle here is that a biorecognition event (analyte binding) brings together two fragments of a split FP into close enough apposition so as to effectively constitute an active chromophore (i.e. by refolding into the beta barrel structure). Another way to conceptualize this is to think of a MRE fused to one FP fragment and an analyte protein fused to another FP fragment. Upon interaction of analyte portion with MRE, a "reconstituted" active chromophore is generated. This type of sensing strategy has been used to study the interaction between three protein subunits that constitute the influenza A polymerase complex, and to examine oligomerization between adenosine A_{2A} and dopamine D_2 receptors.

It is not possible in this brief review to give more than a taste of what can be done with these very versatile FP reporters. The subject should not be left, however, without mentioning some potential difficulties associated with use of FP fusion constructs, as well as very recent developments.

Unfortunately, because the FP's themselves are quite large proteins, the size of fusion constructs can cause deterioration of the fused construct. Also, the cell may mislocate the fusion protein, again probably because of the hybrid structure of the construct. The FRET-based applications, while potentially very powerful, are very dependent not only on distance but *orientation* of the adjoining fluorophores, and orientation may be critically changed in complex protein constructs. Also, FRET, while seemingly conceptually straightforward, is, in practice, somewhat more complex, and ratios of fluorescence changes between donor and acceptor molecules must commonly be examined, when attempting to assess conformational changes as monitored by FRET.

In spite of these caveats, the use of genetically encoded fluorescent biosenors bids fair to provide us with ever greater insights. One can envision a complete repertoire of these biosensors, each with specific properties that will allow interrogation of the status of the event under examination. Furthermore, other more advanced spectroscopic techniques like fluorescence lifetime imaging and fluorescence recovery after photobleaching, will permit more detailed analysis of these various FP fusion hybrids as they interact in sensor configurations. Finally, recent developments herald the applications of so-called photoactivatable or photoswitchable FP, which have been termed optical highlighters, in which illumination of the FP at certain wavelengths can produce photochemical reactions that change the properties of the chromophore. These optical highlighters have enabled imaging of fast protein dynamics and subdiffusion limit imaging.

NANOMATERIALS AND APPLICATIONS TO BIOSENSORS

The recent development of nanomaterials science has produced an incredibly rich array of novel materials which are likely to find multiple applications, one among them being as components of biosensors. Various inorganic nanomaterials have been produced and characterized, including nanoparticles, nanocrystals (quantum dots), nanorods, nanowires, and carbon nanotubes (both single walled and multi-walled). This list is not by any means all-inclusive, as more complex geometric shapes and various hybrid structures are being continuously developed. The field is so vast that it is not possible to cover every recent development in a short review. Rather, in this article, we hope to focus on general principles which are being utilized in numerous applications of these nanomaterials to biosensing, and to provide a few examples illustrating such applications.

What then are the potential advantages which nanomaterials bring to biosensing applications? Due to their extremely small size, these materials have unique physical properties which recommend them as scaffolds for biosensing. Included among these properties must be items such as electrical conductivity, unique optical properties, unique structural characteristics (i.e. very high surface area to volume ratio), and certain biological properties predicated upon their nano-size. Indeed, the nanometer-size scale of these materials allows relatively facile entry into cells, providing potential for intracellular sensing, although the innocuousness of these materials within the cell is still somewhat of an open question. The large surface area of these materials allows derivitization with many biologically active ligands, and the resultant high density of such bioreceptors may increase affinity toward target molecules by a multivalent binding effect, while also shortening the response time for sensing. In the ideal case, nanomaterial-based sensors would selectively bind to the intended target, without interfering with the underlying biological process, and produce a strongly amplified signal which could be detected with commercially available detection technology.

How are nanomaterials utilized for biosensing applications? This is a complex question to answer, as innovative paradigm shifts are being constantly developed, but some general principles can be elucidated. First, nanomaterials may be used as scaffolds/platforms upon which to immobilze various biomolecules which are the actual biosensing element. Thus, biomolecules such as antibodies, aptamers, nucleic acids, carbohydrates, proteins, lipids, and subunits of the preceeding, may all be immobilized (absorbed, covalently linked, etc.) onto nanomaterials. Second, nanomaterials may bring unique capabilities to transducer output (i.e. signal which is detected) compared to conventional transducer mechanisms. For example, carbon nanotubes (CNT) have unique electrical conducting activities, making them highly useful for electrical sensing applications such as FETs, electrochemical sensors, etc. Also, gold nanoparticles show unique optical changes depending upon degree of aggregation, thus forming the basis for colorimetric sensing applications. Gold nanoparticles provide ready-made surfaces for SPR sensing, while gold nanorods can serve as interdigitating electrodes for electrical sensing applications. Nanoparticles can also serve as intermediaries to increase sensitivity of detection in such sensing devices as QCM's and microcantilevers. Here the nanoparticles, appropriately derivitized so as to bind to a specific analyte, add further mass to the microbalance or cantilever, thus hopefully increasing the sensitivity without sacrificing selectivity.

Third, nanomaterials, by virtue of their size, are potentially very attractive intracellular biosensing probes. Entry into the cell can be a problem, but not necessarily always is. For instance, cells may endocytose nanomaterials, particularly if they are decorated with biomolecules or bio-like-molecules. Once inside, the nanomaterials can provide unique advantages for sensing applications, for example, the outstanding fluorescence intensity of the quantum dots. Further, biomaterials (nucleic acids for example) show increased stability and protection from intracellular degradation when coupled to nanomaterials. Some studies have indicated that the nanomaterials have little or no toxicity upon intracellular introduction, but this issue still needs further investigation. Coating the nanoparticle with an inert or biocompatible matrix has been investigated as an approach to mitigate any toxicity issues. Finally, nanomaterials may serve as molecular barcodes, that is, as identification elements which can serve to uniquely tag specific analytes. As such, they can provide multiplexing capabilities for bioassays. We now provide some examples, by no means comprehensive, of how nanomaterials have been utilized in biosensing applications.

Because of electron or energy transfer, single walled CNT's can serve as very effective quenchers for a variety of fluorophores. Thus, single walled CNT's have been substituted for organic molecules used as quenchers in molecular beacons, and have better quenching efficiency, low background and high signal-to-noise ratio. Other analytical arrangements in which single or double-stranded nucleic acids are linked to single walled CNT's have also been explored as platforms for optical sensing. Aptamer-based molecular beacons for protein sensing can also be linked to single walled CNT's. Single walled CNT's can also be functionalized with several different probes for multiplexed sensing. Single walled CNT's have also been used as transducing elements with a thrombin-specific aptamer, but in this case the CNT was employed as a FET. CNT's have also been used as electrochemical sensors with applications to a wide variety of analytes.The features of CNT's which make them most suitable for these applications include high electrical conductivity, rapid electrode kinetics, chemical stability and mechanical strength. Also, a biosensor based upon antibodies immobilized on CNT's for detection of specific antigens has been described in which electrochemical sensing is the detection endpoint.

CNT's are not the only nanomaterials being employed for biosensor use with electrical detection. For example, ingenious detection schemes for nucleic acids have been developed wherein gold microelectrodes were fabricated on silicon wafers, capture DNA strands were immobilized in the electrode gaps, and target DNA was captured within the electrode gaps. Then gold nanoparticles with an oligonucleotide complementary to the unhybridized end of the target DNA were added, hybridization took place, and gold nanoparticles were now immobilized between the electrodes. Subsequent coating with a silver layer increased electrical conductivity and sensitivity of the assay and lower limits of detection in the FM region were attained.

Optical nanoparticle sensors for intracellular detection of a variety of analytes, called felicitously PEBBLES, have been developed by Kopelman's group. Most of these sensors are not strictly biosenors by the definition of this article since they do not involve a biomolecule as a recognition element. Nevertheless, the principles involved lend themselves to broader applications. The nanomaterials involved in these sensors include quantum dots, polymeric nanoparticles (i.e. organic molecule polymer or silica-based) as well as metallic nanoparticles. The sensing element is generally an organic fluorophore, or in the case of quantum dots, the nanoparticle itself. A unique aspect of these sensor nanosystems is the inclusion of a reference fluorescent dye, along with the detecting fluorophore, which allows some measure of quantification. Another research group has developed a maltose biosensor using a maltose-binding protein as a receptor attached to a quantum dot. Maltose binding was detected by FRET changes, as a quencher molecule was displaced from the monosaccharide binding site.

The development of semiconductor nanocrystals, called quantum dots, has had great impact upon the applications of fluorescence sensing for biosenors. The inherent weaknesses of organic molecule fluorophores can be largely overcome by these fluorescent nanoparticles. Quantum dots (CdSe, CdTe, CdS) have broad excitation spectra and also feature very narrow emission bandwidths, which can be tuned to various wavelenghts depending on the size of the nanoparticle. Among the advantages of quantum dots are photostability, high quantum yield, and the narrow emission profiles, thus permitting multiple analyte detection in a given sample. Modification of the surfaces of quantum dots is usually necessary to produce sites for biomolecules to bind, as quantum dots, in general, are hydrophobic after synthesis.

Noble metal (Ag,Au) nanoparticles have also proven their utility for optics-based methods of detection. It is possible to distinguish well-separated from aggregated Au nanoparticles on the basis of optical properties (fundamentally based on SPR). Thus, by attaching complementary DNA oligomers to two separate batches of nanoparticles, the hybridization event could be detected by a change in color of the solution. Interestingly, the melting point profile for the hybridized DNA oligomers attached to nanoparticles showed very sharp transitions, apparently due to the nanoparticle attachments. Thus, due to the sharp transitions in melting temperature, it was possible to adjust the sensitivity of this assay so that single base pair mismatches could be detected. To detect specific nucleic acid sequences (target sequences), an assay format was developed in which capture oligonucleotides, complementary to part of the target sequence ,were immobilized to a solid substrate, and gold nanoparticles derivitized with oligonucleotides complementary to another portion of the target formed a sandwich, binding the target sequence between them. To increase sensitivity, a silver coating (so-called silver enhancement) was deposited upon the gold nanoparticles and the resulting surface was scanned in a flatbed scanner. This assay was found to be 100-fold more sensitive than a assay based upon conventional fluorophore techniques.

Another optical technique which has been adopted to nanomaterial-based sensors is SERS. In this case, metallic nanoparticles provide the basis for strong SERS enhancement of the normally weak Raman signal, thus providing the possibility of sensitive detection. Organic molecules, which are efficient Raman scatterers can be used in an analogous manner to fluorescent dyes to provide SERS-sensitive labels for biosensing strategies. Thus, protein and nucleic acid analysis strategies have been developed using SERS sensing and metallic nanoparticles. Molecular sentinels, analogous to molecular beacons have also been developed, wherein the detector oligonucleotide has a metallic nanoparticle on one end of the hairpin loop and a SERS-active dye on the other end. In this case, hybridization with the target DNA results in decrease or loss of SERS signal.

Investigators have developed striped metallic micron-length nanorod particles, termed nanobarcodes. Identification of the stripe pattern can be made by means of a microscope. Labeling these nanorods with various bioreceptors allows one to identify and quantify a number of biomolecule targets simultaneously. Mirkin's group have developed a technique, called nanoparticle-based bio-bar codes for ultrasensitive detection of proteins. Here two types of particles are used, one a magnetic particle, which has a capture antibody attached to it, and another a gold nanoparticle with an antibody against another epitope of the antigen of interest attached to it. The antigen (protein) of interest is first captured by the magnetic particle, and subsequently the gold nanoparticle with second antibody also attaches to the antigen. The key to this assay is that the gold nanoparticle has also attached to it double-stranded (complementary) oligonucleotides, in which only one of the oligonucleotide strands is bound to the nanoparticle. Following separation by a magnetic field of the particle/particle complex from reactants and a dehybridization step, the released oligonucleotide becomes a "bio-bar code" specific for the protein of interest and also the means by which that particular protein is detected. Since many oligonucleotides can be bound to a single nanoparticle, there is an enormous potential for amplification. The released oligonucleotide strand can be readily detected by a variety of techniques such as the silver enhancement assay discussed above or PCR. The sensitivity of this method for detection of a clinically relevant protein (PSA) was found to be approximately 6 orders of magnitude better than standard immunological (ELISA) techniques.

The list of applications in which nanomaterials are playing a crucial role could go on and on. This short summary will, perhaps, provide a flavor of the scope of the potential uses of nanomaterials. As more nanomaterials are developed, with more unique properties, the list of biosensor applications can only grow.

FURTHER READING

Alocilja, E. C., & Radke, S. M. (2003). Market analysis of biosensors for food safety. *Biosensors and Bioelectronics*, *18*, 841–846.

Arora, K., Chand, S., & Malhotra, B. D. (2006). Recent developments in bio-molecular electronics techniques for food pathogens. *Analytica Chimica Acta*, *568*, 259–274.

Asefa, T., Duncan, C. T., & Sharma, K. K. (2009). Recent advances in nanostructured chemsosensors and biosensors. *The Analyst*, *134*, 1980–1990.

Cullum, B. M. (2007). *Nanoscale optical biosensors and biochips for cellular diagnostics*. Smart Biosensor Technology (pp. 109–133). Boca Raton: CRC Press.

D'Orazio, P. D. (2003). Biosensors in clinical chemistry. *Clinica Chimica Acta International Journal of Clinical Chemistry*, *334*, 41–69.

Frommer, W. B., Davidson, M. W., & Campbell, R. E. (2009). Genetically encoded biosensors based on engineered fluorescent proteins. *Chemical Society Reviews*, *38*, 2833–2841.

Gooding, J. J. (2006). Biosensor technology for detecting biological warfare agents: Recent progress and future trends. *Analytica Chimica Acta*, *559*, 137–151.

Iqbal, S. S., Mayo, M. W., Bruno, J. G., Bronk, B. V., Batt, C. A., & Chambers, J. P. (2000). A review of molecular recognition technologies for detection of biological threat agents. *Biosensors and Bioelectronics*, *15*, 549–578.

Lazcka, O., Del Campo, F. J., & Munoz, F. X. (2007). Pathogen detection: A perspective of traditional methods and biosensors. *Biosensors and Bioelectronics*, *22*, 1205–1217.

Lee, Y. E., & Kopelman, R. (2009). Optical nanoparticle snesors for quantitative intracellular sensing. *Wiley Interdisciplinary Reviews, Nanomedicine and Nanobiotechnology*, *1*, 98–110.

Leonard, P., Hearty, S., Brennan, J., *et al.* (2003). Advances in biosensors for detection of pathogens in food and water. *Enzyme and Microbial Technology*, *32*, 3–13.

Lim, D. V., Simpson, J. M., Kearns, E. A., & Kramer, M. F. (2005). Current and developing technologies for monitoring agents of bioterrorism and biowarfare. *Clinical Microbiology Reviews*, *18*, 583–607.

Mao, C., Liu, A., & Cao, B. (2009). Virus-based chemical and biological sensing. *Angewandte Chemie Int*, *48*, 6790–6810.

Nakamura, H., & Karube, I. (2003). Current research activity in biosensors. *Analytical and Bioanalytical Chemistry*, *377*, 446–468.

Pejcic, B., De Marco, R., & Parkinson, G. (2006). The role of biosensors in the detection of emerging infectious diseases. *The Analyst*, *131*, 1079–1090.

Rodriguez-Mozaz, S., Lopez de Alda, M. J., & Barcelo, D. (2006). Biosensors as useful tools for environmental analysis and monitoring. *Analytical and Bioanalytical Chemistry*, *386*, 1025–1041.

Sapsford, K. E., Shubin, Y. S., Delehanty, J. B., *et al.* (2004). Fluorescence-based array biosensors for detection of biohazards. *Journal of Applied Microbiology*, *96*, 47–58.

Thaxton, C. S., Georganopoulou, D. G., & Mirkin, C. A. (2006). Gold nanoparticle probes for the detection of nucleic acid targets. *Clinica Chimica Acta*, *363*, 120–126.

Velasco-Garcia, M. N., & Mottram, T. (2003). Biosensor technology addressing agricultural problems. *Biosystems Engineering*, *84*, 1–12.

Chapter 40

Gnotiobiotic and Axenic Animals

A.J. Macpherson, M.B. Geuking, J. Kirundi, S. Collins and K.D. McCoy
McMaster University Medical Centre, Hamilton, ON, Canada

Chapter Outline

ABBREVATIONS

HEPA High efficiency particulate air

DEFINING STATEMENT

Axenic and gnotobiotic animal husbandry is an essential and powerful technique to investigate microbial–host interactions under defined conditions. This article describes the methodology and protocols used in a successful axenic/gnotobiotic research facility. The maintenance of germ-free status in these facilities requires several layers of sterility barriers and scrupulous microbiological testing.

INTRODUCTION

An important area of modern microbiology is to understand the way that consortia of microbes adapt to different habitats. A particularly important example of this, which is highly relevant to us as higher mammals, is the enormous number of commensal microbes that are present in our lower intestines (ileum and colon). Whereas the microbial densities found in soils, or other geological or marine habitats, typically reach densities of up to 10^8 microbes per gram, the mammalian lower intestine is home to 10^{12} microbes per gram of luminal contents. These microbes therefore outnumber our own cells by an order of magnitude, and we constitute the best of good culture media.

The motivation for maintaining animals either entirely germ free (axenic) or with a limited defined microbiota (gnotobiotic) is to be able to carry out precise experiments to look at the effect of these bacteria on the mammalian host when a microbiota is present or absent. This includes the way that the structure and function of the intestine is shaped in the presence of commensal bacteria, including epithelial cell gene expression and mucosal immunity. There are also widespread effects on distant body systems. We have recently reviewed these diverse effects of colonization by an intestinal microbiota (and the presence of microbes on other body surfaces) on many body systems. The experiments are typically carried out by taking germ-free animals and either deliberately introducing selected microbes by gavage or inoculation or by causing the animals to acquire a commensal microbiota by housing them in the same cage as animals that are already colonized. These experiments can be made more powerful and informative with a strain combination approach: where colonization of a wild-type strain can be compared with a strain on the same genetic background but with a targeted deletion of a selected gene function. For example, secretion of antibodies in the intestine can be abrogated by deletion of the joining segments of the immunoglobulin heavy chain locus, so the function of immunoglobulins in the adaptation to commensal microbes during the colonization can be compared by contrasting the effects in this strain with wild-type mice that are capable of inducing normal antibody responses. Because very large numbers of gene-targeted mice are available, experiments *in vivo* to understand the way in which the microbiota achieves a mutualistic relationship with the host are possible.

These experiments on host microbial mutualism are predicated on the husbandry of the axenic and gnotobiotic animals. This has a history of nearly 100 years. Early attempts to maintain rodents germ free were stimulated by the question of whether mammalian life was possible in the absence of commensal microorganisms. These experiments were carried out after cesarean sections and hand rearing of the neonates in an aseptic environment. Initially, animals could only be maintained short term, but later it became possible to rear rodents as axenic adults, when they could be interbred to maintain the germ-free colony without the laborious stage of aseptic hand rearing. This is the way in which germ-free rodents are normally maintained today, and the technology and methods involved is described in this article. Some experiments are still carried out with piglets that have been delivered by aseptic cesarean section and reared in sterile conditions, but these methods will not be described further.

In the following sections, we have used general descriptions of the equipment and manipulations required for germ-free or axenic husbandry: specific operating protocols are referenced on our website (given below).

FLEXIBLE FILM ISOLATORS

General Design of Isolators

Germ-free rodents are normally maintained in flexible film isolators. These are essentially large plastic bubbles with integrated gloves for manipulating animals inside. They are ventilated with sterile air passed through a high efficiency particulate air (HEPA) filter to generate a positive pressure inside the isolator of 5–10 mm H_2O. At one end, the isolator contains a port that is mounted through the flexible film: this has coverings or doors inside and outside the isolator. Air leaves the isolator through an exhaust tube, which is mounted with a second HEPA filter to avoid contaminated air entering the isolator if the positive pressure should temporarily fail.

The principle is to keep the inside of the isolator sterile at all times. Before the isolator is brought into service it is checked for leaks, and the inside is thoroughly sterilized before any materials or animals are introduced. Only sterilized food, water, and bedding are brought in, using methods described in more detail below.

It is possible to keep colonies of germ-free rodents almost indefinitely, provided the husbandry is carried out without mistakes and there is no equipment failure. Unfortunately, the footprint of the isolators and the space required for connecting the transport drums means that a large floor area is required for a relatively small number of animals in the vivarium. The space available can be used most efficiently with a two-tier design of isolators, although this requires special systems to hoist the drums used to supply the isolators, and the animal technologists must work on special platforms to care for the animals in the upper tier.

Initial Setup of Isolators

Cleaning and preparation

Before bringing an isolator into service it has to be thoroughly cleaned to remove any old bedding material or waste and washed inside and out with soapy water. Once the isolator is clean the gloves are normally renewed, and the ports are sealed in order to carry out a leak test.

Leak testing

This is done by putting a bung into the exhaust port and running the ventilator motor for a few minutes to inflate the isolator over pressure. Once this has reached 20 mm H_2O (15.2 mbar) the air inflow port is also blocked with a bung (and the ventilator motor turned off) and the isolator is allowed to stand to monitor the speed of fall of pressure. We accept a pressure fall of less than 6 mm H_2O (4.56 mbar) in the first hour and less than 9 mm H_2O (6.84 mbar) after the second hour as constituting a satisfactory leak test. When the isolator fails the leak test the source of the leak must be found and repaired, and it is normally convenient to employ soapy water for this purpose. In cases of serious leaks where repairs might prove unreliable in service, it is normally sensible to change the canopy (the protocol for doing this depends on the exact design of the isolator and should be obtained from the manufacturer).

Internal sterilization

Once the leak test is satisfactory, the next stage is to sterilize the interior of the isolator. This is done with 3% (v/v) peracetic acid sprayed under high pressure from a stainless steel gun, as the acid will rapidly corrode normal steel or other metals present in spray guns sold for cellulose-based paints. Peracetic acid should be made up fresh for each use. We normally use approximately 350 ml to generate the mist for sterilizing a 2.7-cm^3 isolator prior to service and this is handled with precautions to prevent contact with skin or inhalation of fumes. Eye protection and a respirator should be used. The pressure hose is passed through the port of the isolator and loosely sealed with laboratory cling film to limit fumes from the peracetic acid mist entering the surrounding area of the animal unit. The mist is sprayed all around the inside of the isolator, and briefly into the ventilation port and the exhaust port (excess peracetic acid in the ports may damage the HEPA filters). The gloves are manipulated so that the internal surfaces are fully exposed, and any cage racks and the internal port door are also misted on both sides. Once misting is complete, the ventilation port and the exhaust port are sealed with bungs and the spray equipment is transferred to the gap between the internal and external port doors, and the internal port door is closed. Closing the internal door seals the isolator and maintains internal sterility (and contains the peracetic mist) allowing the spray equipment to be removed from the gap by opening the external port door. The isolator is left sealed for at least 18 h for the peracetic acid to work; after this the bungs are removed from the ventilation and exhaust ports and the ventilator motor is started to bring the isolator to 5–10 mm H_2O positive pressure. At this stage the empty isolator can be prepared for the import of supplies and animals.

Importation of Supplies

Connecting a supply drum

The principle of importation of sterile supplies relies on the use of large sealable drums that are used to autoclave materials, food, water, and bedding. These drums are constructed with a perforated frame, which is covered by a mesh filter, which allows penetration of steam when the drum is in the autoclave (Figure 1a). One end is sealed with Mylar film: this seal can be broken to access the supplies inside.

To achieve a sterile connection between the autoclaved drum and the isolator, a plastic sleeve is used (Figure 1b). This may be fitted with an integral glove to simplify transfer of supplies into the isolator. Once the drum has been brought into position, the external door of the isolator is removed (leaving the internal door sealed until the inside of the sleeve has been sterilized). The sleeve is taped or clamped over the outside ring of the port and over the end of the drum.

To sterilize the sleeve, it is sprayed with 2% (v/v) peracetic acid using the same high pressure stainless steel misting system described for preparation of the isolators in the previous section, with appropriate precautions to avoid exposure of personnel to the peracetic acid or its fumes. There is a special plastic port to spray in the peracetic acid mist, but before doing this, the port and the bung that will seal it after spraying are wiped with neat 2% peracetic acid using a swab mounted in a stainless steel holder (Figure 1c). If there are uneven surfaces on the port connection, these may also be wiped with peracetic acid before connecting the sleeve. The mist is then sprayed in through the port, being careful

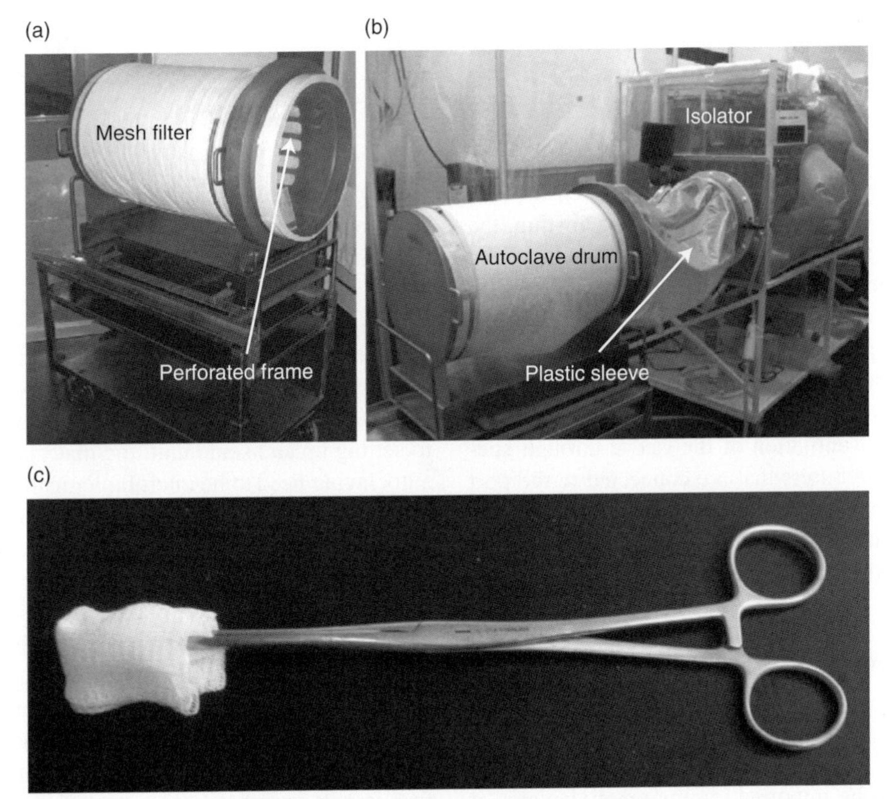

FIGURE 1 Sealable drums are used to autoclave materials, food, water, and bedding to import into germ-free isolators. (a) A sealable drum constructed with a perforated frame and covered with a white mesh filter is shown. (b) A sealable drum is shown connected to an isolator with a plastic sleeve, which can be sprayed with peracetic acid mist to sterilize it. (c) A swab mounted in a stainless steel holder is used to wipe the sleeve pot and the bung with 2% peracetic acid.

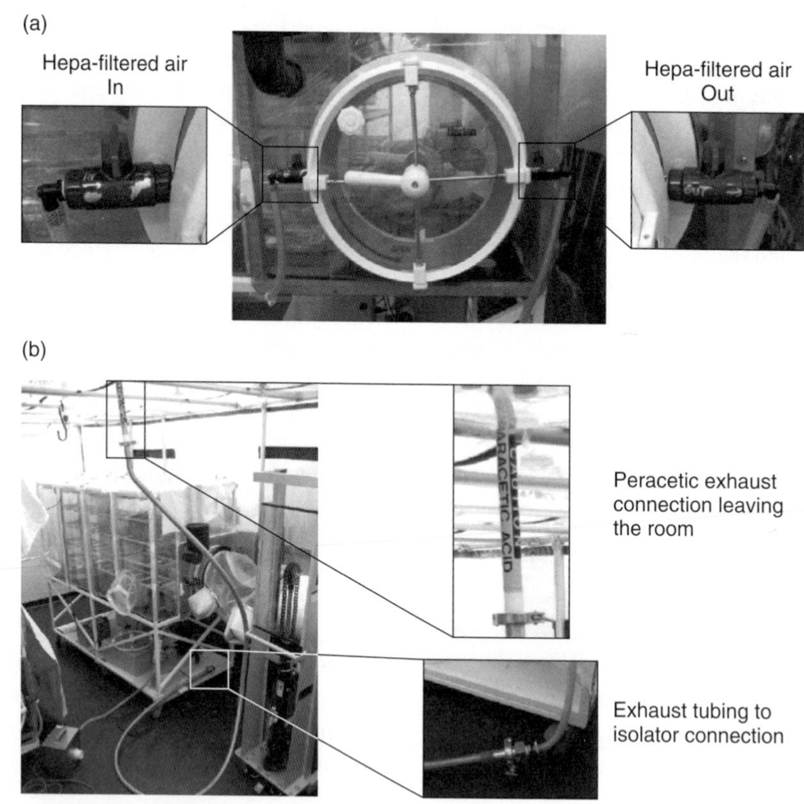

FIGURE 2 After sterilization with peracetic acid the plastic sleeve is vented through specially installed HEPA filters in order to avoid exposure of animals to residual peracetic acid fumes. (a) The outer quadralock door of an isolator is shown before attachment of the sleeve. The "In" and "Out" valves are shown on either side of the port door. (b) In order to vent the sleeve, air is blown into the sleeve through a HEPA filter via the "In" valve and is collected through the "Out" vent. The air exhausting from the sleeve is again HEPA filtered to prevent contamination by outside air in the case of accidental negative pressure built up, and collected in a special exhaust hose that is vented externally.

to move the integral glove during spraying to ensure full coverage of the internal sleeve surfaces. After misting, the bung is placed into the spray port and held in position with plastic tape. We allow the peracetic acid 60 min to sterilize the sleeve before proceeding further.

After the development time, the supplies may be directly imported by opening the internal port door and breaking the Mylar seal on the drum. Many units, including ours, will provide a period of ventilation of the sleeve through specially installed HEPA filters that are connected to the port door (Figure 2), in order to avoid exposure of animals inside the isolator to residual peracetic acid fumes from the sleeve. The sleeve is vented with HEPA-filtered air for 15 min.

Imports to Bring an Empty Sterile Isolator into Service

Order of importation

Before animals can be imported into an empty isolator, it must be charged with the necessary supplies for husbandry. Usually, the order of importation (each with a separate drum) is as follows:

1. Cages, bedding, and instruments
2. Food
3. Water
4. Animals.

We normally autoclave each of these categories on different settings (see website), although the specifics of our autoclave settings must only be considered as a guide, and prior to setting up an axenic unit, the materials in the drum after autoclaving need to be microbiologically controlled to optimize the operating protocols. Microbiological controls are also needed for each of the imports: on each occasion 4–6 × 100 ml Duran bottles are imported with the materials from the drum and are used to take samples. The protocols for processing these are given on our website.

We normally load bedding into thick stockings, tied at each end, for autoclaving. Cages are autoclaved clean in sections, and assembled with grids, water sippers, and bedding once they have been transferred into the isolator. The first import includes some simple autoclavable husbandry equipment including dustpan and brush, long forceps for picking up mice, cage labels, wiping cloths, blunt-ended scissors, empty cloth bags to bag food as it is imported and empty paper bags that can be used for waste bedding, and other materials that need to be re-exported in the future.

Food is autoclaved in a special drum with perforated metal trays. We have not found irradiated "sterile" food to be sufficiently sterile for reliable axenic use, and it is important to spread the food chow in single layers over the trays. Perforated waxed paper can be spread on the trays before the chow, to avoid caramelized material contaminating the metal grids of the trays, making them hard to clean properly. We use Harlan Teklad S-2335 Mouse Breeder Diet #7004 and Harlan Teklad Irradiated LM-485 Mouse/Rat Diet #7012 with the following autoclave protocol (see website). Every drum that is used for import of food must contain the 135 C indicator strip, a mortar and pestle, a series of paper bags, and six 100 ml Duran bottles and matching caps. Once the sleeve has been sterilized and the isolator is open, the food is gathered up into bags and stored in the isolator. Two of the sample bottles simply have their caps screwed on as negative controls for the microbiological cultures and the remaining four bottles are each loaded with two pellets chosen at random from the imported bags and crushed with the pestle and mortar. These sample bottles are re-exported before the internal isolator door is closed and taken to the laboratory for microbiological quality control.

We autoclave water in polycarbonate bottles, with the caps loosely screwed on. The bottles stand in a rat cage, which has been drilled with perforations in order to allow steam to circulate freely through the drum. Inevitably there is additional condensation when autoclaving water, and some units use drums with an unperforated base to catch the condensed water and to avoid soaking the sterile filter. We use a standard drum, but the base contains a water trap of silver foil (Figure 3).

Finally, the animals themselves must be imported. This is done from another isolator, using a drum that allows use of a removable cap when animals are imported into the drum. The drum is initially sealed with a Mylar membrane with the removable cap and ring inside the drum. The sealed drum containing the cap is then autoclaved. The drum has a special distance piece over-ring that allows the transfer sleeve to be attached, leaving room for the cap to be mounted when the drum is attached to the isolator (Figure 4a and b). The procedure is to mount the presterilized over-ring drum to the isolator with a transfer sleeve and spray with 2% peracetic acid as described in the previous section. After 60 min, the sleeve is ventilated, the Mylar is ruptured open, the drum cap is retrieved from inside the drum and placed inside the sleeve, and the internal isolator port door is removed. The required animal cages are exported from the isolator into the drum, and the internal isolator door is replaced and the drum cap is mounted. The sleeve is now taken down and washed, and the whole process of connecting the drum to the import isolator is repeated. Once the sleeve is sterilized and vented, the internal port door is opened, the resealable cap removed from the drum, and the animal cages are transferred. There is a special modification of this method that is used to transfer animals between different axenic units. Germ-free animals are shipped in special shipping containers that have plastic sleeves at both ends that can be directly mounted to the isolator (Figure 4c).

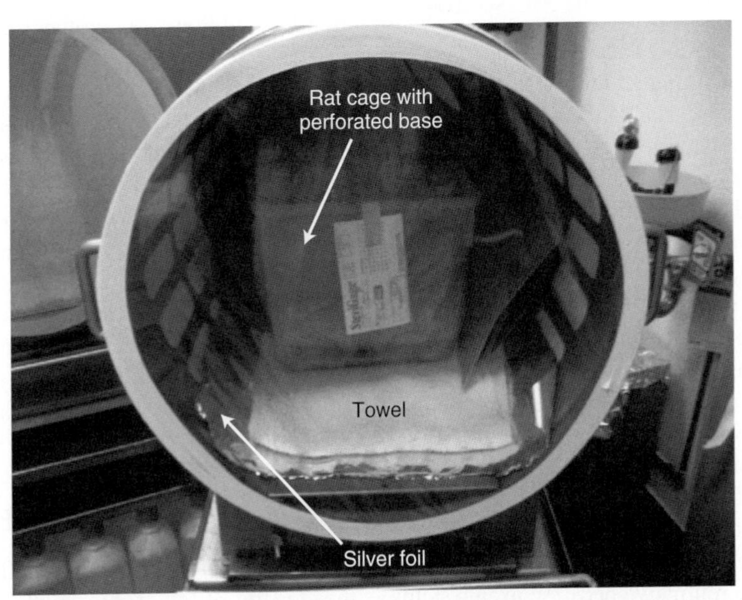

FIGURE 3 Sealable drums that are used for autoclaving water require a water trap. These drums are fitted with a layer of silver foil and a towel in order to soak up any excess condensation to avoid soaking the sterile mesh filter. The water bottles are placed in a large rat cage, which has had perforations drilled into the base to allow steam to freely circulate.

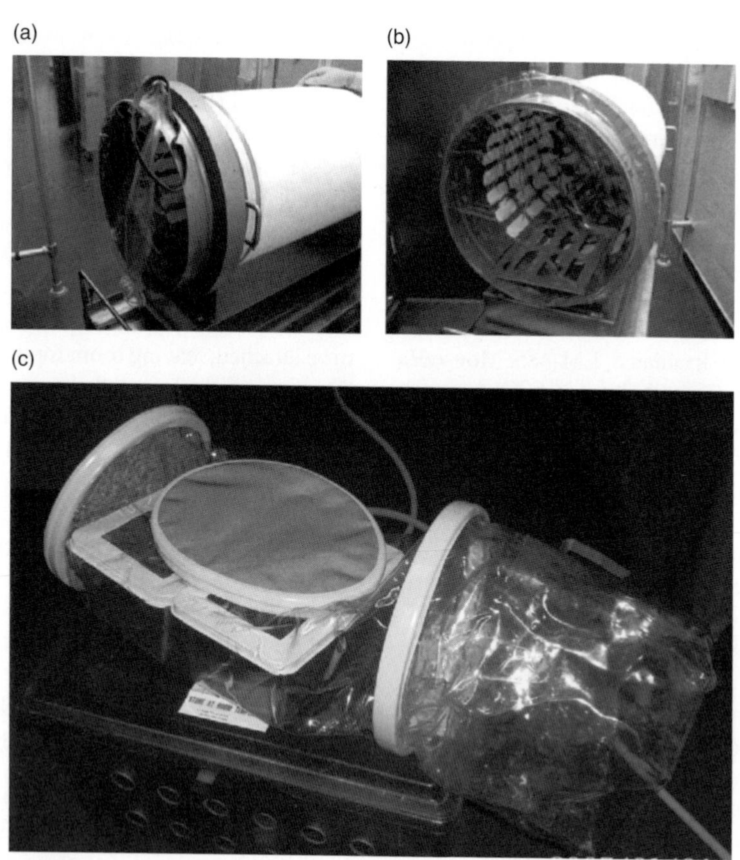

FIGURE 4 Importation of animals requires a sealable drum with a special over-ring to allow a removable cap and a sleeve to be attached. (a) The drum is shown with the removable plastic cap partially connected. Note the over-ring on the drum that allows space for the sleeve to be attached. (b) The drum is shown with the removable cap mounted. (c) This is a specialized transport container for germ-free animals. It has a sleeve on both ends to allow direct connection to an isolator for the export and import of mice.

Husbandry Once Animals are in an Isolator

Having set an isolator in service, rodent husbandry is reasonably standard, except working within an isolator will be more time-consuming than in a standard unit. It is extremely important to keep each isolator very tidy, with supplies set out of the way of the manipulation gloves, and cages in a rack. Each cage is tagged for identification. There are daily spot checks at which the health status of animals is controlled, and food and water are topped up. Dust on the floor of the isolator is cleared on a daily basis using the dustpan and brush. There is a weekly cage change, moving the mice between cages using long forceps to catch the tail. Small quantities of waste materials can be exported by spraying the gap between the internal and external doors of the isolator port. The isolator has to be recharged with supplies of different sorts using the same protocols as for the initial setup described in "Importation of Supplies," and fecal pellets are taken from some of the cages to check sterility on every import. We also routinely check the axenic status of the animals by exporting two sentinels every month and opening the cecum to sample and culture its contents under aseptic conditions (see protocol on website). Sentinels are also exported to verify that there are no pathogens in the standard schedule used in specific pathogen-free units.

A further consideration in standard axenic or gnotobiotic husbandry is to back up the different mouse strains, so that breeding stocks are available in at least two isolators. This means that in the unlikely event of contamination, one can clear the affected isolator, clean and prepare it from scratch as given in "Initial Setup of Isolators and in Importation of Supplies" and restock from the backup isolator. In practice, this is only carried out for the most important strains, and units that are capable of germ-free axenic embryo transfer will choose to rederive some strains if necessary for space reasons.

MICROBIOLOGICAL CHECKS

The microbiological protocols that are used to control the sterility of the axenic isolators are given on our website. Generally, at every connection between a supply drum and an isolator, fecal samples should be taken for microbiological controls, and every month at least two sentinel animals are taken out of the isolator for cecal cultures and staining for unculturable microbes. Unfortunately, the only way to recover sterility in a contaminated isolator is to eliminate the animals inside, and to restock with germ-free animals from one of the other isolators.

EXPERIMENTATION WITH GERM-FREE OR GNOTOBIOTIC ANIMALS

In the following section, we assume that all procedures will have received written approval by the appropriate institutional and national regulatory authorities before work is commenced. This includes the specifics of the extra manipulations required because of the special microbiological status of the animals.

We generally carry out two sorts of experimental protocols. The first, in which unmanipulated animals are studied using *ex vivo* methods, is relatively simple, because animals are merely exported from the isolator and euthanized with an ethically approved method within an hour. In this time, one can assume that the animals are effectively sterile. This can be verified by carrying out the dissection under aseptic conditions in a laminar flow hood after spraying the fur with 70% ethanol.

To carry out *in vivo* experimentation is more complicated. Clearly the risk of contamination is higher under these conditions, so the first step is to export the required number of animals into a smaller mobile "surgical" isolator (Figure 5a) using the techniques described in "Importation of Supplies." The surgical isolator must be set up in a standard way beforehand, and supplies either imported, as given in "Importation of Supplies," or directly imported from the stocks in the supply isolator at the same time as the animals. Where space (and height) permit, it is also possible to directly link a surgical isolator with a breeding isolator via a transfer sleeve, and carry out the imports directly.

The animals are allowed to settle in the surgical isolator, and it is moved to an area where a special laminar flow hood is available, with a port similar to those on the isolators themselves (Figure 5b). This means that the surgical isolator can be connected to the hood and cages of animals transferred in and out of the work surface in the hood for manipulations under aseptic conditions. This includes surgery where the hood is configured with operating microscopes (see Figure 5b). For simple manipulations, such as intraperitoneal injections or gavage, materials can be sprayed into the surgical isolator through the two doors, or transferred via the lamina flow hood port without needing to move the animal cages into the hood itself.

AXENIC REDERIVATION

One of the advantages of using germ-free mice to address the mechanisms of mutualism between the host and the microbiota in the intestine and on other body surfaces is that many different genetically defined strains are available to investigate the role of different host molecules through strain combination experiments *in vivo*. For this, each of the relevant strains needs to be rederived germ-free. There are potentially three ways of doing this. (1) Caesarian section of a colonized animal with aseptic hand rearing. (2) Caesarian

(a)

(b)

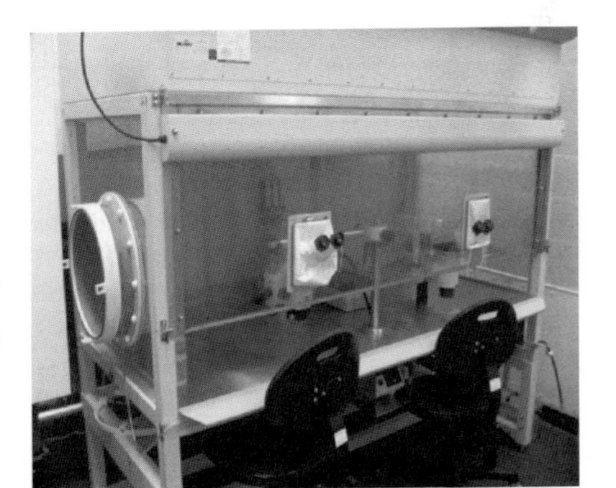

FIGURE 5 Small surgical isolators and specially designed laminar flow hoods are used for *in vivo* experimentation with germ-free animals. (a) A small mobile "surgical" isolator is shown. This isolator contains a port door on both the right and the left side and two sets of gloves on the front and back for greater manipulation. The port on either side can be connected to a laminar flow hood with a transfer sleeve. (b) A specially designed laminar flow hood is depicted with a quadralock door on the left side where surgical isolator can be connected. Animals can then be transferred in and out of the hood where sterile manipulations can be performed. This laminar flow hood is configured with two microscopes and has a water-jacketed base, which can be kept at 37 °C.

section of a colonized animal with fostering of the pups by a germ-free lactating dam. (3) Axenic embryo transfer. Of these three methods the first one is never used, because germ-free stocks of a strain suitable for the second or third way can be readily imported into a new unit. Many units use Caesarian section with cross-fostering, although this is highly unreliable because the stage of the pregnancy at which the Caesarian section is carried out is critical to the survival of the pups, and getting the foster mother to accept them can be demanding. Axenic embryo transfer requires more technology and skill, but is much more reliable and easier to control, so this is the method used exclusively in our unit.

Embryo transfer is routinely carried out using SPF (colonized) pseudopregnant female recipients, and the protocols have been extremely well described so they will not be repeated here. In this section, the outline of the procedure and the modifications that are necessary to carry it out germ-free is presented.

The principle is that colonized animals of the required strain are used to produce fertilized embryos for transfer into a germ-free pseudopregnant recipient during aseptic surgery. In order to increase the number of embryos available, young (4–6 weeks old) females are superovulated under SPF or conventional husbandry conditions and mated with proven stud males of that strain (Figure 6a). We prefer to set approximately ten pairs of animals to be sure of sufficient embryos, and freeze the unused embryos. The female mice are checked for vaginal plugs early on the morning of the day before the embryo transfer. Thirty-six hours after mating, the donor female mice are euthanized and the oviducts are dissected (Figure 6b) and placed in M2 medium. These are flushed under a dissecting microscope and the two cell embryos (Figure 6c) are collected and washed at least five times in antibiotic-containing M2 medium before being transferred into a sterile drop of medium in a sterile Petri dish in a laminar flow hood using a new sterile transfer capillary. This Petri dish is aseptically transferred into a larger sterile Petri dish and transported to the axenic unit, where it is held in a 5% CO_2 cell culture incubator before transfer.

The recipient females are prepared in a large germ-free isolator dedicated to the purpose, also stocked with about 25 germ-free males that have been sterilized by vasectomy under aseptic conditions. Records are kept of the plugging efficiency of the males so that inefficient studs can be replaced. Females are selected for their stage in estrus from the external vaginal appearance and these are mated with the vasectomized stud males on a 1:1 ratio, and daily checks are made for vaginal plugs. The timing of the setup

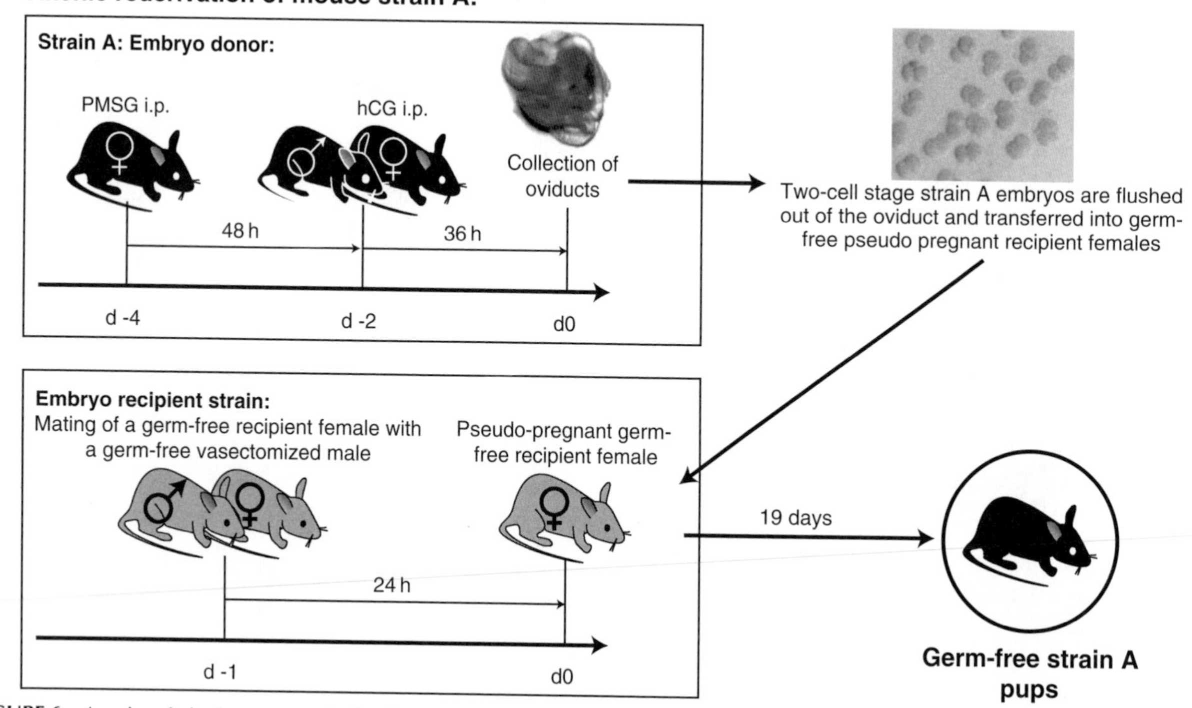

FIGURE 6 Axenic rederivation uses standard embryo transfer protocols but requires germ-free pseudopregnant recipient females and embryo transfers must be performed aseptically. The upper panel indicates the superovulation protocol and time line for collection of two-cell embryos from the donor strain; the oviduct and the two-cell embryos are shown. The lower panel illustrates the protocol and time-line for generating the pseudopregnant germ-free recipient. This part of the protocol is done in a germ-free isolator dedicated for this purpose. On the morning of the transfer, the plugged females are transferred into a small surgical isolator, which is then connected to a laminar flow hood for implantation of the embryos.

and protocol inside and outside the axenic unit is shown in Figure 6a. Suitable pseudopregnant females are transferred into a prepared surgical isolator on the morning of the embryo transfer, and this is linked by a transfer sleeve to a laminar flow hood equipped with two microscopes, one for microsurgery and the other for embryo handling. The hood is scrupulously prepared and sterilized prior to the procedure.

The embryo transfer protocol is carried out under standard operating conditions, but with strict aseptic precautions. We exclusively use instruments, transfer capillaries, anesthetic pipework and an anesthetic base plate that has been sterilized by autoclaving or baking. The operator scrubs up and dons a sterile gown and gloves using a closed technique as usual for aseptic surgery. We usually transfer 8–12 two cell embryos into the oviduct on each side, depending on the donor embryo strain. After transfer, the females are housed singly or in pairs, but either kept in the surgical isolator until the pups are born or transferred to a nursery isolator. Once the pups are weaned the recipient females are euthanized and the cecal contents are checked for sterility before the offspring are moved into a breeding isolator.

GNOTOBIOTIC HUSBANDRY

The principle of keeping gnotobiotic animals with a limited defined bacterial flora, especially after monocolonization with a single organism, is similar to that of axenic animals using isolators. Gnotobiotic animals with a slightly more diverse flora (such as the Schaedler flora of eight defined organisms) can also be kept in individually ventilated cages, provided there is scrupulous attention to general hygiene within the gnotobiotic unit. These individually vented cages also have a supply of sterile filtered HEPA air and can be of a design where the water bottle can actually be contained within the cage lid itself. Cages are normally changed in a HEPA filter lamina flow hood to ensure sterility. Any germ-free strain can be easily colonized with an intestinal flora, either by deliberate inoculation with the defined organism in pure culture or by cross-contamination through cohousing the germ-free source animals with other mice of the required microbiological status. The methods for inoculating germ-free strains with pure cultures of Schaedler organisms are given on the Taconic farms website (http://www.taconic.com.).

CONCLUSIONS

In this article, we have outlined the methods for maintaining and carrying out experiments in germ-free or gnotobiotic animals, and provided references to detailed protocols. In most animal units experiments are carried out under specific pathogen-free conditions, where the intestinal microbiota is assumed to be of the modified Schaedler type, or is undefined. It is uncommon to control colonies for aerobic or anaerobic intestinal commensal bacteria, and those that breeders supply as modified Schaedler may not be very stable after transfer, particularly if there is considerable human traffic into the animal rooms, or the hygiene precautions in handling are not scrupulous. The use of animals of carefully defined microbiological status is essential for studies of host microbial mutualism, and highly desirable to eliminate variability in other animal models, especially of infectious organisms.

FURTHER READING

Jackson, I. J., & Abbott, C. M. (2000). *Mouse genetics and transgenics: A practical approach* (1st ed.). Oxford: Oxford University Press.

Macpherson, A. J., Geuking, M. B., & McCoy, K. D. (2005). Immune responses that adapt the intestinal mucosa to commensal intestinal bacteria. *Immunology*, *115*, 153–162.

Macpherson, A. J., & Uhr, T. (2004). Compartmentalization of the mucosal immune responses to commensal intestinal bacteria. *Annals of the New York Academy of Sciences*, *1029*, 36–43.

Nagy, A., Gertsenstein, M., Vintersten, K., & Behringer, R. (2003). *Manipulating the mouse embryo: A laboratory manual* (3rd ed.). New York: Cold Spring Harbor Laboratory Press.

Schaedler, R. W., Dubos, R., & Costello, R. (1965). The development of the bacterial flora in the gastrointestinal tract of mice. *Journal of Experimental Medicine*, *122*, 59–67.

Smith, K., McCoy, K. D., & Macpherson, A. J. (2007). Use of axenic animals in studying the adaptation of mammals to their commensal intestinal microbiota. *Seminars in Immunology*, *19*, 59–69.

Whitman, W. B., Coleman, D. C., & Wiebe, W. J. (1998). Prokaryotes: The unseen majority. *Proceedings of the National Academy of Sciences*, *95*, 6578–6583.

RELEVANT WEBSITE

http://www.fhs.mcmaster.ca. – MUGSI.

http://www.taconic.com. – Taconic.

Chapter 41

Microbial Corrosion

J.-D. Gu
The University of Hong Kong, Hong Kong, PR China

Chapter Outline

ABBREVIATIONS

SRBs Sulfate-reducing bacteria
MIC Microbial influenced corrosion
GNP Gross national product
EIS Electrochemical impedance spectroscopy
PMS Phenyazine methosulfate

DEFINING STATEMENT

Microorganisms are known for their role in the corrosion of a wide range of metallic materials, but the mechanisms, including depolarization of metals, biomineral formation, complexation through exopolymeric materials, H_2 embrittlement, and electron shuttling, are complicated. The best-known model is the involvement of sulfate-reducing bacteria (SRBs), but the understanding about the specific enzymes and their involvement is still not complete. Other proposed mechanisms need substantive experiments to confirm the specific biological basis of those processes. Prevention of microbial influence corrosion (MIC) is a challenge and is discussed in this article.

HISTORY AND SIGNIFICANCE

Corrosion has severe economic consequences and this fact was recognized at a very early stage of research development. It has been estimated that ~4% of the gross national product (GNP) is lost due to corrosion and 70% of the corrosion in gas transmission pipelines is caused by microorganisms. The American refinery industry loses $1.4 billion a year from microbial corrosion. A wide variety of microorganisms are capable of corrosion and degradation and the causative microorganisms include both aerobic and anaerobic bacteria. The SRBs have been the chosen organisms in a large number of studies on biocorrosion mechanism investigations. In addition to SRBs, exopolymer (slime)- and acid-producing bacteria are also recognized for their active participation in corrosion processes through mechanisms by which metal ions are complexed with functional groups of the not-well-defined exopolysaccharides, resulting in release of metallic species in the solution. Similarly, fungi were observed to be involved in the corrosion of aluminum and its alloys by a process in which organic acids of fungal origin attack the material matrices through the acids produced. Fungi are also shown to be the causative organisms in degradation of concrete and polymeric materials that are widely used for structure as well as protective purposes against metal corrosion.

The importance of microorganisms in the corrosion of metallic materials was first reported by Dutch scientist von Wolzogen Kuhr in 1922 when anaerobic SRBs were postulated to contribute to iron corrosion through removal of hydrogen accumulated at the cathodic site on surfaces. Since then, microbial-influenced corrosion (MIC) has been reported for a wide range of industrial materials, including those in oil fields, offshore, pipelines, pulp and paper industries, armaments, nuclear and fossil fuel power plants, chemical manufacturing facilities, wastewater treatment, drinking water distribution system, food industries, and facilities with membrane

Topics in Ecological and Environmental Microbiology.

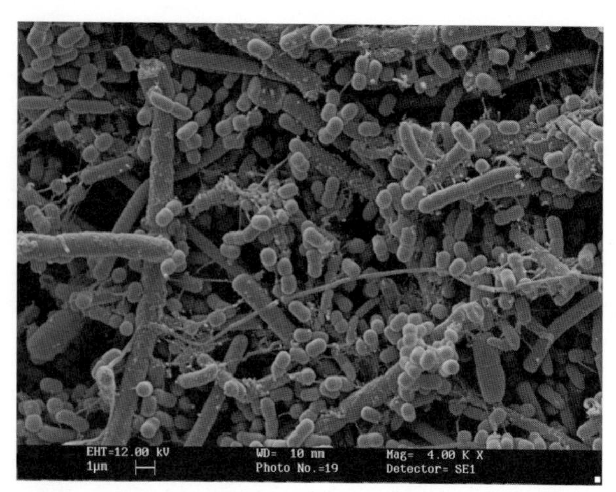

FIGURE 1 A scanning electron micrograph showing a biofilm community developed on surfaces of carrier materials treating organic wastes. The sample was dehydrated and critical point-dried before being coated with palladium and gold.

applications. The terminology of microbiological corrosion has frequently been used interchangeably with microbiological fouling, but the two are not synonymous. MIC is still not clearly defined on a scientific basis as the biochemical mechanisms involving bacteria are not well defined, and ambiguity and misuse are common in the literature.

A wide range of microbial processes can cause corrosion of metals. In general, the presence of inorganic deposits and differential concentrations of oxygen and chloride are important parameters determining the corrosion mechanisms and the extent of corrosion. The presence of microflora and fauna on surfaces of materials alters the local environment, providing appropriate conditions for initiation of metal dissolution. Some of the processes involved are discussed below. Microbial involvement in corrosion of metals is a result of adhesion and subsequent metabolic activity on surfaces. Microorganisms form complex communities on surfaces of materials, and develop complex heterogeneous biofilms under natural conditions (Figure 1). Such microbial associations on surfaces are also responsible for degradation of the underlying materials. Specific aerobic microorganisms obtain electrons from metal oxidation and, at the same time, reduce CO_2 for synthesis and growth. Bacteria growing on these surfaces alter the microenvironments, resulting in acidity and oxygen depletion, and changing the diffusivity of metabolites and nutrients. During growth, the anaerobic SRBs produce H_2S, which reacts and corrodes metals. Corrosion is typically mediated by the hydrogenase activity of the SRBs, particularly the genus *Desulfovibrio*.

MICROBIAL BIOFILMS AND CORROSION

Natural microorganisms are mostly associated with surfaces in the environment. This is especially true for those in aquatic and terrestrial environments where nutrients are in either oligotrophic or eutrophied conditions. Microorganisms adhere to both nonliving and living (tissue) surfaces under submerged or moist conditions, in industrial environments exposed to moisture, and enclosed conditions. Physical surfaces under nutrient-poor conditions accumulate and concentrate nutrients, which may chemotactically attract certain microorganisms. Microbial growth on such surfaces may result in well-defined colonies which may further utilize the nutrients available more efficiently from surfaces. Adhesion of microorganisms onto surfaces of metals alters the electrochemical characteristics of the material by establishing microelectrochemical concentration cells, in which electron flow can be initiated from anode to cathode. The resultant microbial biofilm can lead to cathodic depolarization due to oxygen depletion near the microbial colonies because of microbial activity and diffusion limitation of O_2 in aqueous phase coupling with an increasingly localized acidity around the microbial colonies. The structure of a biofilm community on any surface is highly heterogeneous in composition and in structure as well as over time and space. The community composition reflects changes in the local environment, nutritional conditions, and selective pressure under the specific environments. Bacterial attachment to surfaces can have several effects: (1) enhance the initiation of corrosion, (2) recruit invertebrates for settlement, and (3) passivate metallic surfaces. An understanding of bacterial adhesion processes and characteristics of biofilms is essential for better understanding of initiation and control of corrosion.

Biofilms can affect corrosion by means of any or a combination of the following factors: (1) direct effects on cathodic or anodic processes, (2) changes in surface film resistivity by microbial metabolites and exopolymeric materials, (3) generation of microenvironments promoting corrosion, including low oxygenation conditions and acidic microenvironments, (4) establishment of concentration cells facilitating electrochemical reaction to take place, and (5) microbial products in the forms of H_2S and other organic acids that promote the aggressive corrosion of underlying metals. Because microorganisms are ubiquitous and biofilms are present in various environments, their influence on materials covers a wide range of temperature, humidity, salinity, acidity, alkalinity, and barometric conditions. In some cases biofilms may also cause ennoblement of metal rather than corrosion, but the underlying biological mechanisms are still not well understood. Obviously, there is an apparent lack of information to reveal the basic biology involved.

The microflora forms patches on the material, inducing the formation of differential concentration cells on metals. After adhesion by electrostatic attraction or random collision, the organisms divide and form colonies, which deplete oxygen and release hydrogen ions. Microorganisms have a tendency to select specific sites for initial attachment and roughness of material surfaces is fundamentally important in the establishment of biofilms. Such physical factors effects can be identified from investigation by using atomic force microscopy, but the lack of a biological basis for such observations is a big problem for further substantiation of the results obtained. For example, surface treatment can modify the metal surfaces to be smoother at the microscopic level and attachment of bacteria on these surfaces is positively correlated to surface roughness. Such information is of practical application as it allows modification of metal surfaces so that minimum attachment of bacteria can be achieved because of the choice of materials. In food processing and pharmaceutical industries, this approach can lead to significant savings in cleaning and disinfection. The adhesive process may induce the expression of genes responsible for polysaccharide synthesis. Exopolymeric materials are synthesized, yielding a complex biofilm community. Under oligotrophic conditions, microorganisms synthesize large quantities of exopolysaccharides that serve as protectants from desiccation and energy reserves. When nutrients are further depleted, the cells can recycle these polymers as a source of carbon and energy.

The existence of biofilms on submerged surfaces has gained wide acceptance; knowledge on biofilm formation on a whole range of material surfaces has advanced significantly in recent years. The initial attachment of bacterial cells on any surface and mechanisms governing the processes are still not resolved; lack of fundamental understanding of adhesion processes prevents us from formulating effective preventive strategies for the control of bacterial biofilms. Biomolecules have been implicated in these processes. Unfortunately, large quantities of biocides and antibiotics are used in industrial and medical areas. Because of the challenge, selective microorganisms may develop mechanisms for resisting toxic chemicals. Selective chemicals are capable of repelling marine bacteria under experimental conditions. Strong repulsion was observed with α-amino-*n*-butyric acid, *N,N*-dimethylphenylene diamine, hydroquinone, and acrylamide, and to a lesser extent with *n*-amyl acetate, benzene, benzoic acid, 2,4-dichlorophenoxy acetic acid, indole, 3-methylindole (skatole), tannic acid, *N,N,N′,N′*-tetramethylethylene diamine, thioacetic acid, phenythiourea, and thiosalicylic acid. The threshold concentration for effective repulsion of bacteria from surfaces was between 0.1 and 1.0 mmol l^{-1}. It is interesting that tannin and tannic acid have been tested for antifouling activity; the difficult problem is to retain these potentially effective chemicals in a formulated coating for slow release over an extended period of time.

Active movement of bacteria away from a particular chemical substance is called negative chemotaxis. Any chemical agent causing this phenomenon can prevent adhesion and then the growth of the targeted microorganisms at an early stage. Toxic chemicals such as chloroform, toluene, ethanol benzene, $CuSO_4$, and $Pb(NO_3)_2$ have been found to induce negative chemotaxis in marine microorganisms. Obviously, selective chemicals can also serve as cues to attract certain bacteria; nutrients like acetate and glucose are effective attractants.

AEROBIC CORROSION PROCESSES

Molecular oxygen (O_2) serves as an electron acceptor for microorganisms to achieve maximal energy for growth under aerobic conditions. When molecular oxygen becomes limited, bacteria have an alternative strategy for utilizing other electron acceptors including NO_3^- and Fe^{2+} and Mn^{2+}. Since microorganisms living under natural conditions tend to adhere to surfaces, some of the cells within a biofilm face shortage in oxygen availability due to diffusion limitation in an aquatic environment. Adhesion of bacteria on surface is by means of long-range and short-range forces operating between the bacterial cells and the surfaces from a physics point of view. At a distance, attractive forces dominate and affect the closeness between a bacterium and the surface. When moving to a critical distance near to a surface, the repulsion forces become dominant and keep bacteria away from the surface.

During aerobic corrosion, the area of a metal beneath these colonies acts as an anode, while the area further away from the colonies, where oxygen concentrations are relatively higher, serves as a cathodic site (Figure 2). Electrons flow from anode to cathode and the corrosion process is initiated. Electrolytes affect the distance between the anode and cathode, being shorter at low and longer at high salt concentrations. An electrochemical potential is eventually developed across the two sites and corrosion reactions take place, resulting in the dissolution of metal. Dissociated metal ions form ferrous hydroxides, ferric hydroxide, and a series of Fe-containing minerals in the solution phase, depending on the biological species present and the chemical conditions. It should be noted that oxidation, reduction, and electron flow must all occur to consume electrons produced for corrosion to proceed. However, the electrochemical reactions never proceed at theoretical rates because the rate of oxygen supply to cathodes and removal of products from the anodes limit the overall reaction. In addition, impurities and contaminants of the metal matrices also stimulate corrosion by initiating the formation of differential cells and accelerated electrochemical reactions.

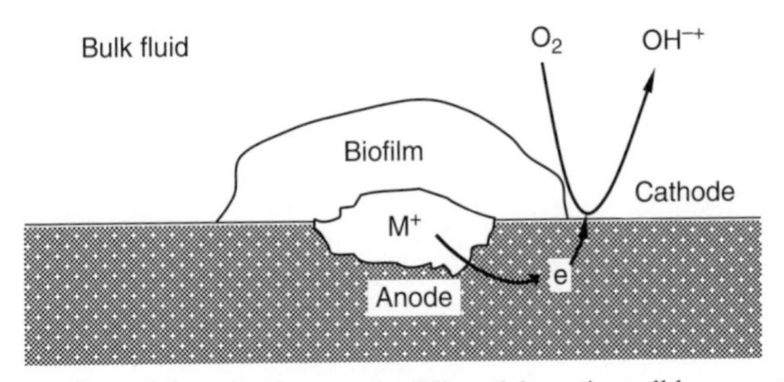

FIGURE 2 Schematic diagram of a differential aeration cell by oxygen depletion under aerobic conditions on a metal surface.

When aerobic corrosion occurs, corrosion products usually form a structure consisting of three layers called tubercles (Figure 3). The inner green layer is mostly ferrous hydroxide ($Fe(OH)_2$). The outer one consists of orange ferric hydroxide ($Fe(OH)_3$). In between these two, magnetite (Fe_3O_4) forms a black layer. The most aggressive form of corrosion is tuberculation caused by the formation of differential oxygen concentration cells on metal surfaces. The overall reactions are expressed as follows:

$$Fe^0 \rightarrow Fe^{2+} + 2e^- \text{ (anode)}, \quad (1)$$

$$O_2 + 2H_2O + 4e^- \rightarrow 4OH^- \text{ (cathode)}, \quad (2)$$

$$2Fe^{2+} + \frac{1}{2}O_2 + 5H_2O \rightarrow 2Fe(OH)_3 + 4H^+ \text{ (tubercle)}. \quad (3)$$

Initial oxidation of Fe of mild steel at near-neutral pH is driven by dissolved O_2. Subsequent oxidation of Fe^{2+} to Fe^{3+} is an energy-producing process carried out by a few specialized species. The amount of free energy from this reaction is small, approximately −31 kJ. Large quantities of Fe^{2+} are oxidized to support microbial growth. This is especially true in intertidal environment where alternation of oxic and anaerobic conditions is facilitated by tidal activity. Because the Fe^{2+} oxidative reaction is rapid under natural conditions, microorganisms compete with chemical processes for Fe^{2+}. As a result, biological involvement under aerobic conditions may be underestimated.

A number of aerobic microorganisms play an important role in corrosion, including the sulfur bacteria, iron- and manganese-depositing and slime-producing bacteria, fungi, and algae. At neutral pH, Fe^{2+} is not stable in the presence of O_2 and is rapidly oxidized to the insoluble Fe^{3+} state. In fully aerated freshwater at pH 7, the half-life of Fe^{2+} oxidation is less than 15 min. Because of this, the only neutral pH environments where Fe^{2+} is present are interfaces between anoxic and oxic conditions. Improved techniques allowed the isolation of new Fe^{2+}-oxidizing bacteria under microaerophilic conditions at neutral pH. Ferric oxides may be enzymatically deposited by *Gallionella ferruginea* and nonenzymatically by *Leptothrix* sp., *Siderocapsa*, *Naumanniella*, *Ochrobium*, *Siderococcus*, *Pedomicrobium*, *Herpetosyphon*, *Seliberia*, *Toxothrix*, *Acinetobacter*, and *Archangium*. Questions remain as to the extent of microbial involvement in specific processes of corrosion involving iron oxidation.

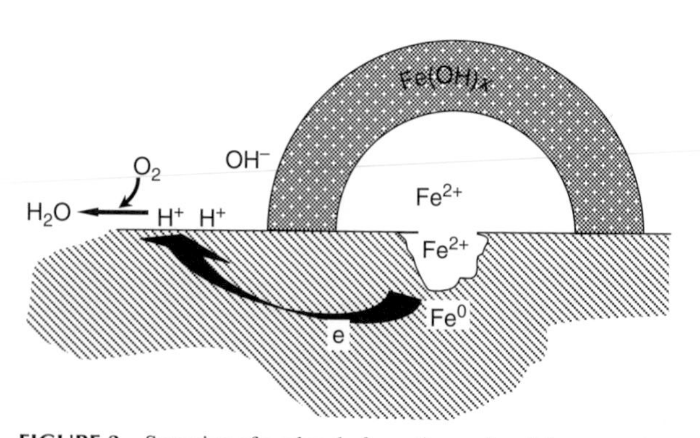

FIGURE 3 Scanning of a tubercle formation under pitting corrosion.

Other microorganisms in the genus of *Thiobacillus* are also responsible for oxidative corrosion. Because they oxidize sulfur compounds to sulfuric acid metabolically, the acid around the cells may attack alloys. *Thiobacillus* is the most common and *Thiobacillus ferrooxidans* oxidizes Fe^{2+} to Fe^{3+}, but the product limits growth of the organisms. SO_4^{2-} is required by the Fe-oxidizing system in T. ferrooxidans at the surface of the bacteria. The electrons removed from Fe^{2+} are passed to periplasmic cytochrome *c*. The reduced cytochrome *c* binds to the outer plasmic membrane of the cell, allowing transport of electrons across the membrane to cytochrome oxidase located in the inner membrane. Recent findings also emphasize the importance of bacterial physiology and water chemistry as the most important determinants of microbial corrosion through generation of mutants defect of biofilm formation and measurement of electrochemical impedance spectroscopy (EIS).

Most microorganisms accumulate Fe^{3+} on their outer surface by reacting with acidic polymeric materials. Such mechanisms have important implications not only for corrosion of metals, but also for the accumulation of metals in natural habitats. *Aquaspirillum magnetotacticum* is capable of taking up complexed Fe^{3+} and transforms it into magnetite (Fe_3O_4) by reduction and partial oxidation. The magnetite crystals are single-domain magnets and they play an important role in bacterial orientation to the two magnetic poles of the earth in natural environments. However, magnetite can also be formed extracellularly by some nonmagnetotactic bacteria. The role of these bacteria has been articulated in carbon cycling in a natural environment but their role in metal corrosion is still unknown. Several studies conducted recently have indicated that such microorganisms may be involved in corrosion through electron shuttling molecules in the environment, but such mechanism must be confirmed to realize their importance in the corrosion of metals.

Manganese deposition by microorganisms also affects the corrosion behavior of alloys. Growth of *Leptothrix discophora* resulted in ennoblement of stainless steel by elevating the open circuit potential to +375 mV. Further examination of the deposits on surfaces of coupons using X-ray photoelectron spectroscopy (XPS) confirmed that the product was MnO_2. The MnO_2 can also be reduced to Mn^{2+} by accepting two electrons from metal dissolution, and the intermediate product is MnOOH.

ANAEROBIC CORROSION PROCESSES

In submerged or polluted environments, oxygen is limited and surfaces are often covered with anaerobic microorganisms and their exopolymeric layers. Within this gelatinous matrix of a highly heterogeneous biofilm, there may be oxic and anoxic zones, permitting aerobic and anaerobic processes to take place simultaneously depending on the hydrodynamic conditions. In the absence of oxygen, anaerobic bacteria, including methogens, SRBs, and acetogens, may actively participate in corrosion processes. Emphasis has traditionally been on SRBs and on hydrogenases, in which the hydrogen on metal surfaces is consumed by microbial metabolism. Severe damage by SRBs can be found in oilfield drilling steel, materials exposed to deep wells, buried pipelines, and immersed structural materials. However, the involvement of SRBs and their hydrogenases in corrosion of mild steel is still controversial.

SRBs are among the most intensely investigated groups of microorganisms as regards their role in biological corrosion. Under anaerobic conditions, oxygen is not available to accept electrons produced. Instead, SO_4^{2-} or other compounds are used as electron acceptors under such conditions. Each type of electron acceptor is unique in the pathway of metabolism. When corrosion begins, the following reactions take place:

$$4Fe^0 \rightarrow 4Fe^{2+} + 8e^- \text{(anodic reaction)}, \quad (4)$$

$$8H_2O \rightarrow 8H^+ + 8OH^- \text{(water dissociation)}, \quad (5)$$

$$8H^+ + 8e^- \rightarrow 8H\text{(adsorbed)}\text{(cathodic reaction)}, \quad (6)$$

$$SO_4^{2-} + 8H \rightarrow S^{2-} + 4H_2O\text{(bacterial consumption)}, \quad (7)$$

$$Fe^{2+} + S^{2-} \rightarrow FeS\downarrow \text{(corrosion products)}, \quad (8)$$

$$4Fe + SO_4^{2-} + 4H_2O \rightarrow 3Fe(OH)_2\downarrow + FeS\downarrow + 2OH^- \quad (9)$$

von Wolzogen Kuhr and van der Vlugt suggested that the above set of reactions is caused by SRBs. This electrochemical generalization has been accepted and is still prevalent. During corrosion, the redox potential of the bacterial growth medium is −52 mV. After inoculation of a corrosion testing cell with SRBs, the overall internal resistance decreases from the initial value of 15 ohms to approximately 1 ohm, while the sterile cell actually shows an increase in resistance. Several phases of change in the electrical potential of steel can be observed after inoculation with SRBs. Before inoculation, the value is determined by the concentrations of hydrogen ions in the medium. A film of hydrogen forms on surfaces of Fe and steel, inducing polarization. Immediately after inoculation, SRBs begin growth and depolarization occurs, resulting in a drop of 50 mV in the anodic direction. The SRBs by means of their hydrogenase system remove the adsorbed hydrogen, depolarizing the system. The overall process was described as depolarization, based on the theory that these bacteria remove hydrogen that accumulates on the surface of iron. The electron removal as a result of hydrogen utilization results in cathodic depolarization and forces more iron to be dissolved at the anode. A typical polarization curve is shown in Figure 4 where both current density and polarization potential of the metal specimen are shown.

Since direct removal of hydrogen from the surface is equivalent to lowering the activation energy for hydrogen removal by providing a depolarization reaction, the enzyme hydrogenase in many species of Desulfovibrio spp. is believed to be involved in this specific depolarization process. Under anaerobic conditions, particularly in the presence of SRB, SO_4^{2-} in the aqueous phase can be reduced to S_2^- microbiologically. The biogenically produced S_2^- reacts with Fe^{2+} to form precipitate of FeS on metal surfaces. Controversy surrounding the mechanisms of corrosion includes more complex mechanisms involving both sulfide and phosphide and processes related to hydrogenase activity. The addition of chemically prepared Fe_2S and fumarate as electron acceptors also depolarizes. However, higher rates are always observed in the presence of SRBs. As a result of the electrochemical reactions, the cathode always tends to be alkaline with an excess of OH^-. These hydroxyl groups also react with ferrous irons to form precipitates of hydroxy iron. Precipitated iron sulfites are frequently transformed into minerals, such as mackinawite, greigite, pyrrhotite, marcasite, and pyrite. Biogenic iron sulfides are identical to those produced by purely inorganic processes under identical conditions. Biogenic minerals can be useful microbiological signature markers.

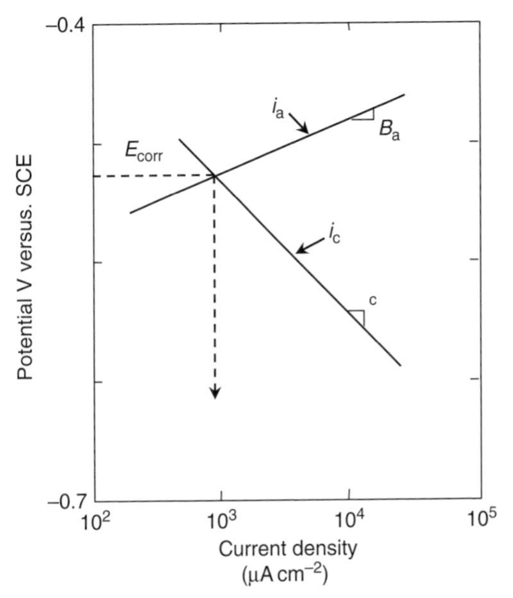

FIGURE 4 A polarization curve showing the relationship between corrosion potential and current density.

SRBs are divided into two physiological groups: one group utilizes lactate, pyruvate, or ethanol as carbon and energy sources, reducing sulfate to sulfide. Examples are *Desulfovibrio*, *Desulfomonas*, *Desulfotomaculum*, and *Desulfobulbus*. The other group oxidizes fatty acids, particularly acetate, reducing sulfate to sulfide. This group includes *Desulfobacter*, *Desulfococcus*, *Desulfosarcina*, and *Desulfonema*. Some species of *Desulfovibrio* lack hydrogenase. For example, *Desulfovibrio desulfuricans* is hydrogenase negative and *Desulfovibrio salexigens* is positive. The rate of corrosion by these bacteria is correlated with their hydrogenase activity. Hydrogenase-negative SRBs were completely inactive in corrosion. Apparently, hydrogenase-positive organisms utilize cathodic hydrogen depolarizing the cathodic reaction, which controls the kinetics. In contrast to this theory, ferrous sulfide (FeS) is suggested to be the primary catalyst for initiation of corrosion.

It is evident that corrosion by SRBs is mediated by the hydrogenase enzyme system. Available information suggests that a protective film of atomic hydrogen is formed on the surface of metals and this neutralizes any further electrochemical reaction, for example, initial exposure of metal to soil and water. However, SRBs disrupt this delicate balance by utilizing this hydrogen as an energy source, resulting in further oxidation of iron or metals to maintain the initial electrochemical steady state. In this process, biochemically, an iron hydrogenase takes an electron from the steel surface and the enzyme is thought to be located on the exterior surface of the outer membrane in order to utilize the surface-adsorbed hydrogen on metals. Hydrogenases have been found only in the periplasm, cytoplasmic membrane, and cytoplasm of the SRBs. *Desulfovibrio vulgaris* contains three hydrogenases, (Fe), (NiFe), and (NiFeSe), in various quantities and in specific locations of the cells. Approximately 95% of the hydrogenase activity is found to be associated with periplasmic (Fe) hydrogenases and the remaining with cytoplasmic membrane-bound (NiFe) hydrogenase and a very small fraction by the (NiFeSe) cytoplasmic hydrogenase. To explain this, Fe^{2+} regulation may play an important role in controlling the synthesis of key proteins associated with outmembrane and periplasm for hydrogen generation. The depolarization theory suggests that bacteria, for example, SRBs, utilize the protective hydrogen layer on metal surfaces through the hydrogenase. This requires the hydrogenase to be localized on the outer membrane's exterior surface to utilize the hydrogen. Since only a very small fraction, that is, 1%, of periplasmic (Fe) hydrogenase activity is associated with the outmembrane, high molecular weight cytochrome in the outmembrane may be the initial electron transporter for hydrogen to be passed on to (Fe) hydrogenase, which is located next to the cytochrome but inward.

Other microorganisms should be noted for their potential role in anaerobic corrosion. They include methanogens, acetogens, thermophilic bacteria, and obligate proton reducers. Among them, only methanogens have been implicated for their role in corrosion of metals. As all these anaerobic microorganisms have defined metabolic capability with regard to the use of hydrogen, it would not be a big surprise to confirm their role in biocorrosion of metals. More research work is needed to elucidate the role of these organisms in corrosion.

ALTERNATING AEROBIC AND ANAEROBIC CONDITIONS

Corrosion under natural conditions may not be constantly anoxic or anaerobic over time, and most of the time alternation of aerobic and anaerobic conditions may take place. In this case, corrosion can be accelerated greatly when transformation of element sulfur is considered as sulfate is reduced to H_2S by SRBs under anaerobic conditions, but H_2S is oxidized to elemental SO and sulfate and other S species, which generates acidity. Acid attack is a significant process contributing to the corrosion of metals and also inorganic materials including stone and concrete. Since S can be recycled further, such process can be carried out continuously, resulting in dissolution of metals and significant deterioration of inorganic materials. Since constant oxic or anoxic conditions are rare in natural or industrial environments, alternating of the two conditions, depending on oxygen gradient and diffusivity in a specific environment and microbial activity, is common. Microbial corrosion under such conditions is quite complex, involving two generally different groups of microorganisms; an interface that serves as a transition boundary of the two conditions can also be identified. Such system is complicated for scientific investigation and a recognized and generally agreed model system should be established for such research to gather information on the relative contribution of each group of microorganisms. Resultant corrosion rates are often higher than those observed under either continuous oxic or anoxic conditions. Microbial activity can reduce the oxygen level at the oxic–anoxic interface, facilitating anaerobic metabolism when nutrients are available. The corrosion products resulting from anaerobic processes, such as FeS, FeS_2, and SO, can be oxidized when free oxygen is available.

Stainless steel is more resistant to corrosion than mild steel because of the formation of a passivation film on the metal surfaces. Pitting corrosion of stainless steel is often initiated, in particular, in areas of welding and crevices as commonly observed in applications. Corrosion of stainless steel AISI 304 and AISI 316 is often associated with marine environments. During oxidation of reduced sulfur compounds more corrosive sulfides are produced under anoxic conditions through microbial physiological processes, causing cathodic reactions. The corrosion rate increases as the reduced and oxidized FeS concentrations increase. Corrosion rate of iron by H_2S is accelerated through formation of FeS, which decreases hydrogen concentration at the cathode site of the metal surface. In such a system, metal serves as an anode. During corrosion of iron and steel by SRB, a thin layer of ~1 μm thickness is always observed as an adherent layer of tarnish and has been confirmed to be mackinawite (tetragonal FeS_{1-x}). This layer can loosen when it grows. When Fe^{2+} concentration is low, the mackinawite changes to greigite. This transformation of minerals is only observed when microorganisms are involved. If Fe^{2+} concentration is high, mackinawite is accompanied by green rust, which is a ferrosoferric oxyhydroxide. It should be noted that mackinawite under condition of SRB can be further transformed to greigite (Fe_3O_4), smythite (Fe_9S_{11}), and finally to pyrrhotite (FeS_{1+x}). Formation of these minerals also accelerates the equilibrium established between soluble fraction and metals.

CORROSION BY MICROBIAL EXOPOLYMERS

Bacteria are known to produce copious quantities of exopolymers, which are important because of their pathogenecity and survival in natural and artificial environments. Such organic materials, in addition to their ecosystem function in recruitment of larvae in aquatic environment and formation of mutualistic or pathogenic relationship with plants and animals, appear to be implicated in their role in corrosion process. These exopolymers consisting mostly of polysaccharides and proteins are acidic and contain reactive functional groups that may bind metal ions tightly. These exopolymers play an important role in facilitating the adhesion of bacterial cells onto surfaces during their initial natural development. These materials have been found to be involved in severe corrosion of copper pipes and water supplies in buildings and hospitals after water stagnation for a period of time. Using surface analysis employing XPS, the organic fraction of the chemicals influences the electrochemical characteristics of metals through functionality-rich materials to complex metal ions from the surface of metal matrix, releasing them into aqueous solution in soluble and complexed forms. As a result, corrosion is initiated. Proteins in bacterial exopolymeric materials use their disulfide-rich bonds to induce corrosion very effectively through a chemical reaction. It is a fact that current understanding about the chemical constituents in the exopolymeric materials of any bacteria is still very rudimentary as the analysis has been on several different categories of chemicals, for example, polysaccharides, proteins, uronic acid, 2-keto-3-deoxyoctanoic acid, and deoxyribonucleic acid. Only when the specific chemical structure and also their quantity are known is it possible to know the detailed mechanisms these chemicals play in the corrosion of metals.

Purified exopolymeric materials of *Deleya* (*Pseudomonas*) *marina* were used in an investigation. Proteins in the exopolymeric materials are responsible for reduction of molybdate (MoO_4^{2-}) under deaerated conditions through comparison with the same exopolymeric materials where proteins have been removed. The reduction process is thought to be due to the presence of proteins containing disulfide moieties. The resultant Mo^{5+} species can be reoxidized when molecular oxygen is available in the ambient environment. Such reduction process can also be observed on the surface of the MoO_4^{2-}-treated austenitic type 304 SS and formation of MoO_2 was detected. Though exposure to the exopolymer results in surface depletion of Fe and enrichment of Cr, rendering an increased hydration of Cr, corrosion resistance of the SS is not compromised by the exopolymer attachment on the surface. Since the duration of the experiment was short and the chemicals were purified, such factors must be taken into account when considering long-term exposure and life-span assessment. On the other hand, aluminum is very susceptible to corrosion by fungi and results have shown that fungal hyphae formed imprints into the aluminum alloys through acids produced during fungal metabolism. It is clear here that microorganisms may play at least two roles in inducing corrosion, one by the exopolymeric materials synthesized and the other by direct attack through the acidic molecules released.

In addition to their genetic basis, production of bacterial exopolymers is affected by several environmental factors. Polysaccharide production by Enterobacter aerogenes is stimulated by the presence of Mg^{2+}, K^+, and Ca^{2+} ions in the culture media. Toxic metal ions, for example, Cr^{6+}, can also enhance the synthesis of polysaccharides and other exopolymeric materials in selective bacteria as a response to the chemical in the culture medium. In particular, synthesis of the exopolymeric materials is positively correlated with Cr^{6+} concentration in *Vogesella indigofera*. It is clear that specific chemical molecules responsible for corrosion need to be identified and the corrosion mechanisms by these selective chemical molecule can be postulated and confirmed to further advance our understanding of the science of corrosion and then the controlling mechanisms against corrosion in various applications.

MICROBIAL HYDROGEN EMBRITTLEMENT

Biologically produced hydrogen (H_2) has been more widely known in energy-related research and development in recent years. This process of release of hydrogen ion or molecular hydrogen by microorganisms is a common phenomenon associated with many different groups of microorganisms.

However, the role of bacteria in embrittlement of metallic materials is not fully understood and the limited number of publications only postulates the potential importance of such biological process on material integrity. During the growth of bacteria, fermentation processes covert complex organic compounds to simpler organic acids and molecular hydrogen (H_2). Such hydrogen, when generated and released on the surface of materials, can be absorbed into material matrix, causing polarization. Some bacteria, particularly the methanogens, sulfidogens, and acetogens, are also capable of hydrogen utilization.

Initially, a possible role for microbial hydrogen in hydrogen embrittlement of metal was proposed. Permeation of microbial hydrogen into metal was measured using a modified Devanathan cell. In a mixed microbial community commonly found in nature, hydrogen production and consumption occur simultaneously. In the complex natural system, competition for hydrogen between microbial species determines the ability of hydrogen to permeate into metal matrices, causing crack initiation. In a pure culture of *Clostridium acetobutylicum* ATCC 824, further confirmation that hydrogen produced by bacteria can be absorbed by palladium foil was demonstrated through monitoring of current density during bacterial growth. Microbial hydrogen involved in material failure may be explained by two distinctively different hypotheses: pressure and surface energy changes. The kinetic nature of hydrogen embrittlement of cathodically charged mild steel is determined by the competition between diffusion and plasticity. The strength level and the susceptibility of the alloy are positively correlated. However, microstructures were also proposed to be the more critical determinant of material susceptibility. Hydrogen permeation may increase the mobility of screw dislocations, but not the mobility of edge dislocations.

CORROSION BY OTHER MICROBIAL METABOLITES

The microbial world is complex and organisms can be less than a micrometer in size or visible to the naked eye. Bacteria have a complex metabolic network and they produce a diverse group of chemicals as their degradation intermediates. Some of the chemicals can accumulate and may be corrosive depending on the metabolic process. Fermentative microorganisms thriving under reduced availability of oxygen and in the presence of sufficient carbon source can produce simple organic acids, formic, acetic, butyric, and so on. In addition to these physiological and biochemical capabilities, microorganisms are also reported to secrete extracellular electron shuttling molecules, which may have a significant role in the corrosion of metals.

Fungi have been shown to cause deterioration of a wide array of polymeric materials, including electronic polyimides, packaging cellulose acetate, epoxy resins, protective coatings, and concrete. Degradation of protective coating needs to be considered because the underlying metals are dependent on the protective properties of the polymer for corrosion control. Fungi also produce highly corrosive metabolites including a wide range of organic acids. These acids have been shown to corrode fuel tanks. They survive very well at the water–fuel interface, metabolizing the fuel hydrocarbon as carbon and energy sources. They are also capable of generating corrosive oxidants, including hydrogen peroxide.

ELECTRON SHUTTLING IN CORROSION

Electron transfer is the most fundamental and basic electrochemical process in the corrosion of metallic materials regardless of the participation of microorganisms. Such electron transfer process can be facilitated by natural molecules, for example, humic substances, commonly found in the environment, both aquatic and terrestrial. More recently, results showed that bacterial quinone, an effective electron transferring molecule, can be excreted by the bacterium *Shewanella putrefaciens*, suggesting that extracellular process of electron transport may initiate and participate actively in the corrosion of metals. It would be interesting to further investigate the extent of such microbial electron transport process in corrosion and the mechanisms involved using candidate electron transfer chemical molecule.

Direct contact between bacterial cells and the metal oxide is required for the energy conservation process and such a system may provide insight for further research in understanding the mechanisms involved in metal corrosion. For example, enzymatic reduction of Cr(VI) has been observed in a number of bacteria and the reduction of Cr(VI) can be achieved by either soluble enzyme systems or the membrane-bound system. Membrane-associated chromate reductase activity was first observed in *Enterobacter cloacae* HO1 where the insoluble form of reduced chromate precipitates was detected on the cell surfaces. In the presence of ascorbate reduced phenazine methosulfate (PMS) as electron donor, active chromate reduction has been shown in membrane vesicles of *E. cloacae* HO1. Membrane-associated constitutive enzyme that mediated the transfer of electrons from NADH to chromate was later elucidated. In the case of *Sh. putrefaciens* MR-1 chromate reductase activity is associated with the cytoplasmic membrane of anaerobically grown cells. Formate and NADH served as electron donors for the reductase. No activity can be observed when NADPH or L-lactate is provided as the electron donor. However, in *Pseudomonas putida*, unlike in *Sh. putrefaciens*, NADPH serves as an electron donor. The presence of soluble chromate reductase was possible in *Escherichia coli*. Cr(VI) reduction in another Gram-negative bacteria, *Pseudomonas* species CRB5, was found to be mediated by a soluble enzyme in cytoplasm. In addition to

Gram-negative bacteria, soluble chromate reductases have also been observed in Gram-positive strains. NADH was the preferred electron donor for the reduction of chromate by the soluble enzyme in *Bacillus coagulans*.

Enzymes capable of Cr(VI) reduction are often referred to as "chromate reductases" in the literature. Several bacterial Cr(VI) reductases, some conferring resistance to chromate, have been characterized. These enzymes commonly show a NADH:flavin oxidoreductase activity and can use Cr(VI) as electron acceptor. The ability to reduce chromate may be a secondary function for Cr(VI) reductases, which have a primary role other than Cr(VI) reduction. This is likely true as the gene sequences available from GenBank indicate that the *P. putida* chromate reductase is a quinone reductase (AF375641) that has a bound flavin and *Pyrus ambigua* ChrR is a flavin reductase (D83142). The nitroreductases NfsA/NfsB from *Vibrio harveyi* possess a nitrofurazone nitroreductase as primary activity and a Cr(VI) reductase activity as a secondary function. Similarly, ferric reductase FerB from *Paracoccus denitrificans* uses both Fe(III)-nitrilotriacetate and Cr(VI) as substrates. These secondary functions may be related to the bacterial enzymatic adaptation as a result of the relatively recent increase of Cr(VI) content in the environment due to anthropogenic activities.

YieF is a flavoprotein containing the FMN cofactor and the enzyme is able to reduce chromate *in vitro*. YieF is shown to possess quinone reductase activity, which appears to guard against oxidative stress by preventing redox cycling of quinones, which would otherwise generate ROS, and by maintaining a pool of reduced quinone in the cell that is able to quench ROS directly. The quinone reductase activity of YieF is likely the primary biological role of this enzyme. ChrR of *P. putida* is the currently best-studied Cr(VI) reductase. During Cr(VI) reduction, ChrR shows a quinone reductase activity that generates a flavin semiquinone. Cr(VI) is reduced to Cr(V) by ChrR, previously reduced by NADH; Cr(V) is next converted to Cr(III) by diverse biomolecules generating reactive oxygen species (ROS). ROS may be eliminated by alternative mechanisms (i.e., catalases or peroxidases) or by the additional function of ChrR. ChrR in a reduced status may reduce quinines (such as vitamin K or coenzyme Q), which may then detoxify previously formed ROS. The soluble bacterial quinone reductase has the ability to reduce a variety of quinone substrates. The above model may provide a unique system for verification of electron shuttling in corrosion of metals.

PREVENTION AND CONTROL

In engineering systems and industries, it is a common practice to introduce biocides to eradicate the initiation and development of microbial biofilm on the surfaces of a wide range of materials. Chlorination and glutaldehyde are also used routinely as a preventive measure in cooling water and oil drilling. Such treatment practices give rise to secondary halogenated by-products in water, resulting in environmentally unacceptable residues of chemicals. In addition, planktonic and biofilm bacteria are very different in terms of their biochemistry and physiology for biofilm bacteria are more resistant to toxic chemicals and environmental shock than the planktonic ones. Similarly, the biocidal effect is compromised when a biofilm has already been formed on surfaces of materials. However, this has not been taken into account when testing the effectiveness of biocides in very different industries, and consequently biocides are not as effective as predicted.

Biofilms are advantageous because they limit diffusion, prevent penetration of the disinfectant, and facilitate more effective exchange of genetic materials between members. Selective organic biocides, used to prevent bacterial growth in industrial systems for an extended period of time, may enrich a population capable of biocide resistance by genetic modification in bacteria, for example, by recruitment of extracellular DNA. However, a new generation of environmentally acceptable biocides is becoming available and their application and effectiveness have to be tested over time. It is probable that some of these biocides will be capable of either preventing biofilm formation or killing the microorganisms in already formed biofilms. Based on the knowledge gained about biofilms, it is unrealistic to expect a magic biocide for eradicating microbial biofilms over a long period of time.

Corrosion protective coatings also have wide applications because of the development of metallic materials and susceptibility to both environmental and microbiological corrosion. Polymeric coatings are designed to prevent contact of the underlying materials with corrosive media and microorganisms. However, microbial degradation of coatings may accelerate and severely damage the underlying metals. A typical example is the corrosion of underground storage tanks. Natural bacterial populations were found to readily form microbial biofilms on surfaces of coating materials, including epoxy and polyamide primers and aliphatic polyurethanes. Surprisingly, the addition of the biocide diiodomethyl-*p*-tolylsulfone in polyurethane coatings did not inhibit bacterial attachment or growth of bacteria effectively due to development of biofilm and bacterial resistance.

Metals are protected from corrosion by a number of mechanisms. Among them, polyurethaner coating is commonly used. Polyurethane-degrading microorganisms include *Fusarium solani*, *Curvularia senegalensis*, *Aureobasidium pullulans*, and *Cladosporidium* sp. through esterase activity detected with *C. senegalensis*. A number of bacteria, including four strains of *Acinetobacter calcoaceticus*, *Arthrobacter globiformis*, *Pseudomonas aeruginosa*, *Pseudomonas cepacia*, *P. putida*, and two other *Pseudomonas*-like species, are believed to be capable of degrading polyurethane. In addition, *Pseudomonas chlororaphis* encodes a lipase responsible for degradation. Using EIS, both primers and aliphatic polyurethane coatings can be monitored for their response to biodegradation by bacteria and fungi. Results indicated that primers are more susceptible to degradation than polyurethane. The degradation process has similar mechanisms as polyimides. Aliphatic polyurethane-degrading bacteria have been isolated and one of them is *Rhodococcus globerulus* P1 base on a 16S rRNA sequence.

Microbial growth and propagation on material surfaces can be controlled by physical and chemical manipulations of the material and the artificial environments. Surface engineering, so that attachment by and susceptibility to microorganisms and then to the fouling organisms is greatly reduced, can be used to prevent biodeterioration. As a control measure, lowering humidity is a very effective means to slow down the growth of microorganisms on surfaces in an enclosed environment and prevention against potential contamination will prolong the life time of the objects.

Basic measures for controlling biodeterioration should focus on the surface especially the initial population of organisms. Without a better understanding of what is on the surface, subsequent protection measures cannot be target specific. Recent development in molecular techniques involving DNA-based information allows a better examination of any surface due to the shortcomings associated with traditional microbiological techniques. By coupling the understanding of surface microbial ecology using molecular techniques and the controlling measures, better results can be achieved. By modification of the microbial community, oxygenation as a means of alleviating the propagation of SRBs can be achieved under anoxic conditions. At the same time, biocides can be effective in controlling biofilms and subsequent deterioration of materials to some extent. Other attempts at community modification include precipitation of microbial-produced H_2S by ferrous chloride ($FeCl_2$), and displacement of *Thiobacillus* sp. by heterotrophic bacteria. All of these efforts have met with limited success.

Biocides are commonly applied in repairing, cleaning, and maintenance of artworks. Chlorine, iodine, and other organic biocidal compounds are used widely and routinely in controlling biofilms that cause corrosion and deterioration of a wide range of materials in industries and for conservation of art. These chemicals have been shown to be ineffective in killing biofilm bacteria. In addition to their environmental unacceptability, most of the time because of toxicity, biocides induce the development of biofilms that are highly resistant to the levels of chlorine normally utilized to prevent biocorrosion. Organic biocides, used to prevent bacterial growth in industrial systems, may selectively enrich populations of microorganisms capable of biocide resistance. No solution to these problems is currently available and alternative biocides have been screened from natural products. Current research by materials scientists focuses on the prevention of adhesion of corrosive microorganisms to surfaces through surface treatments and modification.

Since bacteria are capable of forming biofilms on surfaces of materials, future tests should focus on the dynamics of biofilm and quantification rather than descriptively show biofilms of scanning electron micrographs. In particular, tests for assaying efficacy of biocides should be conducted based on biofilm condition than for liquid culture efficacy. This major discrepancy has not been resolved fully. Because biofilm bacteria are more resistant to antibiotics and biocides, tests based on planktonic cells are not truly representative of their actual conditions on surfaces of materials. New initiative is needed for innovative methodology to assess biocidal effects using surface-oriented assays.

CONCLUSIONS

Under both aerobic and anaerobic conditions, microorganisms can directly or indirectly contribute to the corrosion of metals. Under anaerobic conditions bacteria corrode metals by cathodic depolarization and the formation of FeS, or by consuming hydrogen produced by polarization. Other corrosion mechanisms, including microbial hydrogen embrittlement, complexation of metals from matrices by microbial exopolymeric materials, or extracellular electron respiration, have also been implicated in recent years. It is clear that we have only begun to understand the complexity of interactions between microflora and metals that lead to corrosion, but little about the specific mechanisms involved. Molecular biology-based investigation coupled with electrochemistry and chemical biology should permit a better understanding of the fundamental biochemical processes and the role of microorganisms in corrosion of metals. Equipped with such knowledge, it would be feasible to formulate strategies for controlling biofilms and MIC in industrial and other environments. Use of biocides will not solve the MIC problems, but spreading of antibiotic-resistant microorganisms is a much more serious public health issue to be comprehensively addressed by the professionals and also the general public.

ACKNOWLEDGMENTS

Financial support for this research was partially provided by Innovative Technology Fund. I would like to thank Zhenye Zhao for the drawings and Wensheng Lan for the discussion on metal reductase.

FURTHER READING

Bond, D. R., Holmes, D., Tender, L. M., & Lovley, D. R. (2002). Electrode-reducing microorganisms that harvest energy from marine sediments. *Science*, *295*, 483–485.

Daniel, L., Belay, N., Rajagopal, B. S., & Weimer, P. J. (1987). Bacterial methanogenesis and growth from CO_2 with elemental iron as the sole source of electrons. *Science*, *237*, 509–511.

Dexter, S. C. (1993). Role of microfouling organisms in marine corrosion. *Biofouling*, *7*, 97–127.

Dubiel, M., Hsu, C. H., Chien, C. C., Mansfeld, F., & Newman, D. K. (2002). Microbial iron respiration can protect steel from corrosion. *Applied and Environmental Microbiology*, *68*, 1440–1445.

Ford, T., & Mitchell, R. (1990). The ecology of microbial corrosion. *Advances in Microbial Ecology*, *11*, 231–262.

Ford, T., Sacco, E., Black, J., Kelley, T., Goodacre, R., Berkeley, R. C., *et al.* (1991). Characterization of exopolymers of aquatic bacteria by pyrolysis-mass spectrometry. *Applied and Environmental Microbiology*, *57*, 1595–1601.

Ford, T. E., Searson, P. C., Harris, T., & Mitchell, R. (1990). Investigation of microbiologically produced hydrogen permeation through palladium. *Journal of the Electrochemical Society*, *137*, 1175–1179.

Gu, J. D., Ford, T. E., & Mitchell, R. (2000). Microbial corrosion of metals. In W. Revie (Ed.) *The Uhlig corrosion handbook*. (2nd ed.) (pp. 915–927). New York: John Wiley & Sons.

Gu, J. D., Ford, T. E., & Mitchell, R. (2000). Microbial degradation of materials: General processes. In W. Revie (Ed.) *The Uhlig corrosion handbook*. (2nd ed.) (pp. 349–365). New York: John Wiley & Sons.

Hamilton, W. A. (1985). Sulfate-reducing bacteria and anaerobic corrosion. *Annual Review of Microbiology*, *39*, 195–217.

Little, B. J., Wagner, P. A., Characklis, W. G., & Lee, W. (1990). Microbial corrosion. In W. G. Characklis, & K. C. Marshall (Eds.) *Biofilms* New York: John Wiley & Sons, Inc.

Lovley, D. R., Coates, J. D., Blunt-Harris, E. L., Phillips, E. J. P., & Woodward, J. C. (1996). Humic substances as electron acceptors for microbial respiration. *Nature*, *382*, 445–448.

Walch, M. (1992). Corrosion, microbial. In J. Lederberg (Ed.) *Encyclopedia of microbiology* (pp. 585–591). San Diego, CA: Academic Press.

Chapter 42

Petroleum Microbiology

O.P. Ward[1], A. Singh[1], J.D. Van Hamme[2] and G. Voordouw[3]

[1]*University of Waterloo, Waterloo, ON, Canada*

[2]*Thompson Rivers University, Kamloops, BC, Canada*

[3]*University of Calgary, Calgary, AB, Canada*

Chapter Outline

ABBREVIATIONS

AMO Anaerobic methane oxidation
BP British Petroleum
BT Benzothiophene
BTEX Benzene, toluene, ethylbenzene, xylenes
DBT Dibenzothiophene
DNB Denitrifying bacteria
HDB Hydrocarbon degrading bacteria
HDS Hydrodesulfurization
hNRB Heterotrophic nitrate-reducing bacteria
IB Iron bacteria
MEOR Microbial enhanced oil recovery
MPB Methane producing bacteria
NDO Naphthalene dioxygenase
NR-SOB Nitrate-reducing, sulfide-oxidizing bacterium
OWC Oil–water contact
PAHs Polycyclic aromatic hydrocarbons
SB Sulfur bacteria
SRB Sulfate-reducing bacteria
VFA Volatile fatty acids
VOC Volatile organic compound

DEFINING STATEMENT

Environmental bioremediation technologies may be applied to clean-up marine oil spills as well as to remediate contaminated sludges, soils, water and air. Oil production and processing strategies include microbial enhanced oil recovery (MEOR), de-emulsification, promotion of methanogenic oil degradation and control of reservoir souring, desulfurization, denitrogenation, demetallation, and biosynthesis of novel compounds. Physiological aspects of petroleum microbiology include phenomena of taxis, hydrocarbon accession, transport and efflux, and pathways for aerobic and anaerobic degradation of saturated and aromatic hydrocarbons.

Topics in Ecological and Environmental Microbiology.

INTRODUCTION

Petroleum comprises a very complex combination of hydrocarbons and other organic constituents including some organometallic components. Many of these compounds are metabolizable by microorganisms, and indeed some are products of microbial metabolism. Understanding the ways in which microorganisms attack, degrade, and utilize petroleum components for growth was a strong early focus of petroleum microbiology, with a view to addressing oil spills and other environmental problems. Investigation of the kinds of microbial transformations and the associated microbial species, which reside in the subsurface and oil reservoirs has represented a more recent research thrust. As a result of developments in molecular techniques, we are gaining new insights into hydrocarbon catabolism and the many novel catalytic mechanisms that have been elucidated. The molecular, cellular, and physiological responses of microbes to hydrocarbon-containing environments are also being intensively investigated. While applied petroleum microbiology research continues to be directed to development of improved and more reliable biodegradation and bioremediation technology, there is a growing interest in the application of microbiological techniques to enhance oil recovery and in the processing and refining of petroleum products (Table 1). The field of microbiology plays an important role in petroleum industry (Figure 1).

TABLE 1 Applied petroleum microbiology and biotechnology

Application	Process	Key microorganism/biocatalyst	Role of microbes
Oil recovery	Microbial-enhanced oil recovery	*Acinetobacter calcoaceticus* *Arthobacter paraffineus* *Bacillus licheniformis* *Clostridium acetobutylicum* *Leuconostoc mesenteroides* *Xanthomonas campestris* *Zymomonas mobilis*	Biosurfactants and chemicals produced by microbes help in oil dissolution, viscosity reduction, selective biomass plugging, permeability increase, oil swelling, and pressure increase
	Microbial de-emulsification	*A. calcoaceticus* *Bacillus subtilis* *Corynebacterium petrophilum* *Nocardia amarae* *Pseudomonas aeruginosa* *Rhodococcus globerulus*	De-emulsification of oil emulsions, oil solubilization, viscosity reduction, wetting
	Other oil recovery applications	*A. calcoaceticus* RAG-1	Crude oil recovery from tank bottoms using biosurfactants
Biorefining and bioprocessing	Biodesulfurization	*Agrobacterium* MC501 *Arthrobacter* sp. *Corynebacterium* sp. SY1*Gordona* CYKS1 *Nocardia* sp. *Rhodococcus erythropolis* H2 *Rhodococcus* sp. IGTS8	Biotransformation of organic sulfur compounds, selective removal of sulfur from crude oil or refined petroleum products
	Biodenitrogenation	*Comamonas acidovorans* *Nocardioides* sp. *P. aeruginosa* *Pseudomonas ayucida* *Rhodococcus* sp.	Biotransformation of organic nitrogen compounds in crude oil, nitrogen removal
	Biodemetallation	*Bacillus megaterium* *Caldariomyces fumago* *Escherichia coli*	Enzymatic removal of Ni and V from crude oil using microbial enzymes chloroperoxidase, cytochrome C reductase, and heme oxygenase
	Bioprocessing	General microbial enzymes Cytochrome p450-dependent Monooxygenases Dioxygenases Lipoxygenases and peroxidases	Biotransformation of petroleum compounds to produce fine chemicals

TABLE 1 Applied petroleum microbiology and biotechnology—cont'd

Application	Process	Key microorganism/biocatalyst	Role of microbes
Biodegradation and bioremediation	Bioremediation	*Acinetobacter* spp. *Pseudomonas* spp. *Rhodococcus* spp. *Sphingomonas* spp. Some fungi	Emulsification through adherence to hydrocarbons; dispersion; foaming agent; detergent; soil flushing
	Monitoring of contaminated sites	*E. coli* DH5α *Pseudomonas fluorescens* HK44 *Pseudomonas putida* RB1401	Bacterial biosensors in monitoring of contaminant and bioremediation progress
	Biofiltration of VOC	*Acinetobacter* spp. *Pseudomonas* spp. *Rhodococcus* spp. *Sphingomonas* spp.	Biodegradation of volatile hydrocarbons
	Biological removal of H_2S and SO_x	*Thioalcalobacteria* spp. *Thiobacillus* spp. *Thiocalovibrio* spp. *Thiomicrospira* spp.	Biotransformation of H_2S to elemental sulfur and sulfate

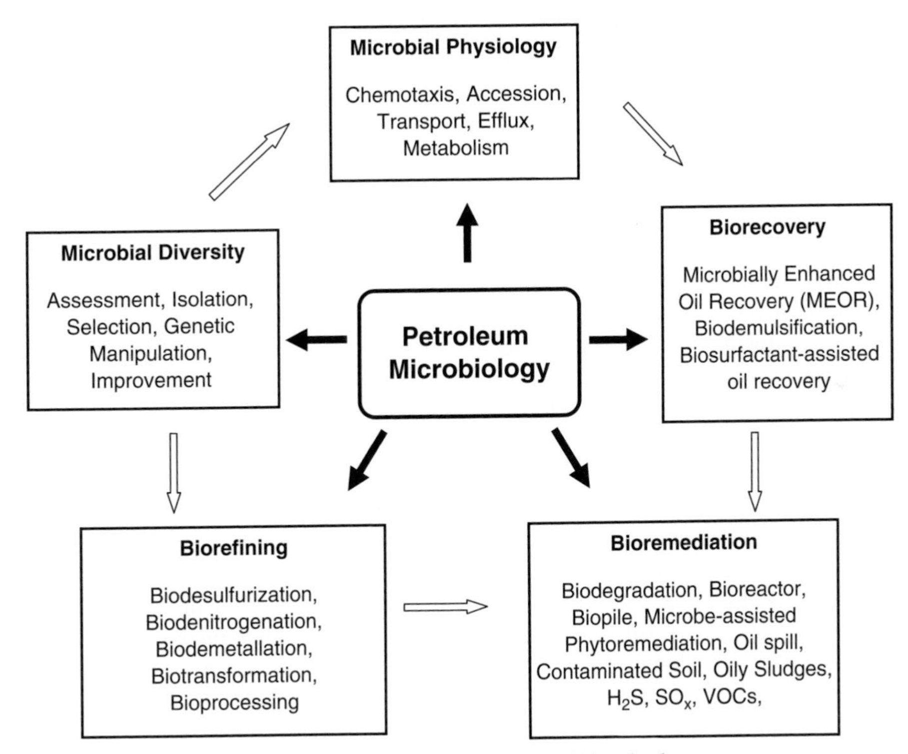

FIGURE 1 Petroleum microbiology and biotechnology.

BIODEGRADATION AND BIOREMEDIATION

During the various operations of drilling, production, processing, transport and storage of petroleum, accidental oil spills occur and large volumes of petroleum-contaminated waste materials are generated as oily sludges. Petroleum hydrocarbon oils and molecules exhibit a tendency to bind strongly to solid surfaces, including soils, and this attachment often increases the challenges of remediation of these materials. The lighter and more volatile hydrocarbon components often volatilize into the atmosphere, and since they are often more toxic, they reduce air quality and impact human and animal health. In turn, degradation of some higher molecular weight petroleum compounds is mediated by cometabolism, which relies on the microbes using the low molecular weight molecules as a source of carbon for both energy production and growth. Thus, volatilization of these smaller molecules reduces the capacity of microorganisms to degrade some of the higher molecular weight contaminants.

Bioremediation technology is effective in the treatment of oil pollutants since the majority of molecules in petroleum hydrocarbons are biodegradable. Because petroleum components are so widely distributed in the environment, oil-degrading microorganisms are ubiquitous. The following sequence of petroleum components represents the order of decreasing biodegradability: *n*-alkanes > branched-chain alkanes > branched alkenes > low molecular weight *n*-alkyl aromatics > monoaromatics > cyclic alkanes > PAHs > asphaltenes.

Microbial Bioremediation of Marine Oil Spills

The first high profile example of bioremediation of marine oil occurred along the shoreline following the Exxon Valdez spill off the coast of Alaska in March 1989. Oleophilic nutrient formulations for stimulation of microbial growth were successfully used in strategies to bioremediate the oils. Supplementation with bacterial oil degrading inoculants showed no advantage over treatment with fertilizers alone. After a 2-year period, most of the oil contaminating the shoreline was remediated largely due to microbial biodegradation. Further monitoring has indicated that after 19 years there is still oil that can be bioremediated on the beaches.

The environmental, scientific, and technical challenges resulting from marine oil spills recently captured the World's attention when an explosion of the Deepwater Horizon drilling rig in the Gulf of Mexico (April 2010) produced the largest ever marine oil spill (http://en.wikipedia.org/wiki/Deepwater_Horizon_oil_spill; http://en.wikipedia.org/wiki/List_of_oil_spills). In its September 2010 issue, Thompson Reuters ScienceWatch addressed "Oil Spills" as its special topic, provided lists of relevant papers addressing this area, many of which dealt with petroleum contamination, oil biodegradation and remediation methods (http://sciencewatch.com/ana/st/oil-spills/). ScienceWatch noted that in 2010 alone there have been substantial oil spills in China, India, and Egypt and three others in the United States in addition to the Deepwater Horizon spill. An interactive world map identifies locations of the largest global oil spills (http://sciencewatch.com/ana/st/oil-spills/map/index.html). http://news.sciencemag.org/oilspill/ contains a large collection of articles on the gulf oil spill, one of five subsets of which is relevant to petroleum microbiology as it focuses on the fate of the oil. The following are additional links relevant to this topic: http://en.wikipedia.org/wiki/Oil_spill; http://www.theoildrum.com/node/6724.

Between April and July 15, 2010, 4.28 million barrels of light Louisiana crude oil spilled into the Gulf of Mexico following an explosion at the British Petroleum's (BP's) Deepwater Horizon drilling rig. This spill was approximately 10 times larger than the Exxon Valdez spill off the coast of Alaska. It was estimated that about 35% of this oil would evaporate into the atmosphere with as much as 50–60% remaining on or below the surface of the sea. Scientists reported the presence of large undersea oil plumes, the development of which was likely to be promoted by the large quantities of dispersants used to emulsify the light crude oil. Microbial biodegradation has long been recognized as one of the main mechanisms for removal of oil from aquatic environments and rates of microbial biodegradation of Louisiana light crude oil were known to be faster than biodegradation rates for heavier crudes from locations such as Alaska. Laboratory studies indicated a half-life of 12–17 days for Louisiana light crude in seawater. Higher temperatures in the Gulf of Mexico might have been expected to further accelerate biodegradation of the spilled Lousiana light crude, over biodegradation rates observed in the Alaska cleanup, although temperatures in deep water locations are only about 5°C. However, conditions were quite different in that the principal oil biodegradation in the Gulf of Mexico was under the sea surface rather than on the shoreline, such that relative oxygen levels may also have been biodegradation rate determinants.

Several teams of scientists deployed ships to the Gulf to investigate microbiological and other aspects of the oil plumes from the spill, and early scientific reports on the rates and extents of biodegradation of the sub-surface oil plumes were conflicting. The research findings received high profile media coverage. Scientific American (August 2010) described the "boom times for oil-eating microbes in the deep waters of the Gulf of Mexico." The New York Times (August 4, 2010) wrote that "among the stars of the gulf cleanup is an oil-hungry bacterium.... Alcanivorax." The Wall Street Journal (25 Aug 2010) spoke of microbes "making quick work of a vast oily plume from the Deepwater Horizon disaster." Scientists from the Woods Hole Oceanographic Institute provided evidence of a 35 km deep oil plume that does not appear to be degrading at a fast rate, which may threaten marine life deep in the ocean. In contrast, scientists from the Lawrence Berkeley National Laboratory identified a new 20 micron long oil-degrading bacterium from deep water locations in the spill area which represented about 95% of the bacterial community present. It was related to the Oceanospirillales family, which includes known hydrocarbon degraders such as *Oleispirea antarctica, Oceaniserpentilla haliotis,* and *Thalassoliticus oleivorans.* It was presumed that natural populations of these organisms have developed in the Gulf area as a result of natural oil seepage. Evidence of oil degradation caused by these plume-associated bacteria was supported by the fact that oxygen saturation rates were lower within the plume than outside the plume. Estimated high alkane biodegradation rates (half life ~1.2–6.1 d) resulted in concentrations of oil in the plume that were so low after the active degradation period that oxygen was not being depleted.

Chemical dispersants are widely used to facilitate breakup and biodegradation of oil spills in the sea, typically at dispersant:oil dose rates ratios of 1:10–1:50. It is estimated that a volume of about 7600 m^3 dispersant was used in the Gulf of Mexico oil spill. The actual dispersants used were Corexit 9500 and Corexit 9527 (Nalco) which contained components such as: 1,2-propanediol; 2-butoxy-ethanol; butanedioic acid, 2-sulfo-, 1,4-*bis*(2-ethylhexyl) ester, sodium salt (1:1); various sorbitan-(9Z)-9-octadecenoates and derivatives; 2-butoxy-1-methylethoxy-2-propanol and hydrotreated light petroleum distillates. The dispersants promote formation of smaller oil droplets thereby increasing surface area and hydrocarbon bioavailability. Biodegradation of Corexit-9500-dispersed oil has been shown to be much more rapid than nondispersed oil.

There were concerns that these dispersants are toxic to marine life, including corals and to the food chain. Recent reports from the EPA (http://www.epa.gov/bpspill/dispersants-qanda.html) indicated that the 8 dispersants used on the BP spill were generally less toxic than the oil itself. Corexit 9500 and 9627 were neither the most effective nor the least toxic dispersants among products approved by the EPA (12 other products ranked more favourably, and indeed Corexit 9500 and 9627 are banned for use on oil spills in the UK) but were used in the Gulf based on availability at the time of the spill. The dispersants generally get biodegraded within a few weeks or months.

Bioremediation of Contaminated Soils and Sludges

Biological methods for treatment of oil-contaminated soils and sludges include landfarming, biopiling/composting, bioventing, and use of bioreactors. Biodegradation rate of oil depends on a variety of factors including the nature of the contaminated medium, type of soil, nature and concentration of different structural classes of hydrocarbons present, bioavailability of the substrate, and the properties of the biological system involved. Bioremediation of petroleum hydrocarbon-contaminated sites typically involves a process called landfarming, in which contaminated soil is augmented with nutrients and sometimes microorganisms, to enhance microbial growth and metabolic processes for biodegradation of the contaminants. Field bioremediation experiments with inorganic nitrogen and phosphate fertilizers, various organic nitrogen fertilizers, biosurfactants, and bulking agents were often found to be successful. Especially in treatment of petroleum hydrocarbon oil spills in soil bioremediation, use of added microbial cultures (bioaugmentation) has generally not improved rates or extents of contaminant degradation over and above what was attained through nutrient enrichment (biostimulation) alone.

The biodegradation rate of a contaminant in soil depends on its bioavailability to the metabolizing organisms, which is influenced by factors such as desorption, diffusion, and dissolution. The often observed very tight binding of hydrophobic petroleum hydrocarbon molecules to soil particles may render them inaccessible to degrading microbes and their low water solubility may limit the ability of bacteria present in aqueous phases to access and participate in degradation of these substrates. Through production of cell-associated or extracellular biosurfactants, hydrocarbon-degrading microbes increase the apparent aqueous solubility of hydrophobic contaminants, and thereby enhance removal of petroleum hydrocarbons from soil or solid surfaces. Similarly, chemical surfactants may be exploited to emulsify or pseudosolubilize poorly water-soluble petroleum hydrocarbon compounds, with typical surfactant aqueous concentrations of 1–2% and 0.1–0.2% being required to wash contaminants from soil and to simply solubilize contaminants in an aqueous solution, respectively. However, when the effects of surfactants on biodegradation of petroleum hydrocarbon contaminants have been investigated, both stimulatory and inhibitory effects of surfactants on degradation have been reported, and hence there is a need to develop a much greater understanding of the various physical, chemical, and biological factors that are at play in these situations.

Treatment time in landfarming methods of petroleum waste treatment is prolonged due to lack of control of parameters affecting microbial activity (temperature, pH, moisture, aeration, and mixing). Accelerated bioremediation of petroleum waste can be achieved in advanced bioreactors where greater control over pH, temperature, moisture, mixing, bioavailability of nutrients, energy source (substrate), and oxygen promotes optimal microbial growth and activity. These bioreactors can be used to treat soils or sludges with solid concentrations of 1–50% w/v with improved engineering approaches to aeration and mixing, which can be exploited to disintegrate solid aggregates and desorb and disperse more water-insoluble hydrocarbons, thereby promoting greater microbial accession and degradation. Bioreactor-based petroleum sludge degradation processes also allow for greater retention of volatile organic hydrocarbons so that they are biodegraded rather than simply released into the atmosphere, and surfactants may be exploited to enhance their retention. As indicated above, retention and degradation of these low molecular weight hydrocarbons support microbial growth and cometabolic degradation of some high molecular weight hydrocarbons. Typical degradation rates of petroleum hydrocarbons, which may be achieved in landfarms, simple aerated bioreactor-based

Untreated oily sludge **Biotreated oily sludge**

FIGURE 2 Petroleum oily sludge treated to nonhazardous level in a slurry bioreactor.

systems, and more optimized bioreactors, are 0.01–0.0%, 0.1–0.3%, and 1% per day, respectively. Typical samples of untreated oily sludge and microbial-treated oily sludge in a slurry bioreactor system are shown in Figure 2.

Microbe-Assisted Phytoremediation

Some plants and their rhizospheric microorganisms participate in petroleum hydrocarbon remediation through a process known as phytoremediation. Examples of microbes that may participate in these processes include *Agrobacterium tumefaciens* and *A. rhizogenes*, *Rhizobium* spp., *Enterobacter cloaceae*, *Pseudomonas*, and mycorrhizal fungi. Plants function in phytoremediation in two ways, the major one being facilitation of favorable conditions for microbial degradation, specifically by plant root colonizing microbes. The second mechanism involves the plant root itself, which may provide a simple and inexpensive means of accessing contaminants existing in subsurface soils and water. In the rhizosphere, the microbial population is supported by plant nutrients in the form of decaying biomass and root exudates. Plant root exudates also stimulate cometabolic transformations, leading to degradation of some organic contaminants.

Plants facilitate soil phytoremediation of contaminants using the following strategies: phytostabilization, phytovolatilization, phytoextraction, phytostimulation, and phytotransformation. In phytostabilization, plants essentially stabilize the contaminated soil environment by providing a physical cover of vegetation over the moderately to heavily contaminated site, and hence reduce or prevent wind and water erosion and therefore limit contaminant mobility. Consequently, these plants must possess good tolerance to the contaminating metals or organics and preferably cause their immobilization. Phytostabilization typically relies on local plant communities occurring at contaminated sites. In the case of phytovolatilization, the plant root system removes the contaminant from the soil and releases it in its original form or after modification to the atmosphere by volatilization. This process can hardly be described as environmental remediation as the contaminant is simply transferred from a soil medium to the atmosphere. In the process of phytotransformation, the plants mediate the transformation of the contaminants to less toxic forms.

Arguably, the most effective phytoremediation strategy is phytoextraction. This involves cultivation of selected plants that can absorb the contaminants and concentrate them in their above-ground tissues. At the end of the growth period, the plant material containing the contaminants is harvested and dried or incinerated, and the resulting contaminant-enriched material is disposed of appropriately or further processed. The energy gained from burning the biomass may contribute to process economic viability, provided the contaminants are not released into the atmosphere during burning. For phytoextraction to be worthwhile, the dry biomass or the ash derived from above-ground tissues of a crop should contain substantially higher concentrations of the contaminant than the polluted soil. Phytostimulation processes involve the biodegradation of contaminants in the soil by the microbes found in the rhizosphere. Root exudates from the plant support the growth and activity of soil bacteria.

Plant root exudates can supply carbon and nitrogen sources for microbial growth thereby substantially raising the population counts of rhizospheric bacteria relative to the general microbial population in the surrounding soil, and enzymes may be produced that degrade organic contaminants. Phytoremediation is a suitable method for treatment of contaminated soils but is not suitable for high volume oily sludges.

Biodegradation of H_2S and SO_X Compounds

Various petrochemical gas and liquid waste streams produce high quantities of H_2S and SO_X that need to be properly treated prior to their disposal to the environment. Microbiological treatment processes have been commercialized that purify these waste streams and convert these by-products to elemental sulfur. The Thiopaq process.

$$HS^- + \frac{1}{2}O_2 \rightarrow S^0 + H_2O$$

describes a sulfur removal process for the production of elemental sulfur from H_2S-containing gas streams by sulfur-oxidizing bacteria. In this process, gas streams are first washed with an aqueous liquid to dissolve the sulfur components into an aqueous phase

$$H_2S + OH^- \rightarrow HS^- + H_2O$$

At neutral pH in the presence of a suitable electron acceptor, the sulfides are converted to elemental sulfur

$$HS^- + \frac{1}{2}O_2 \rightarrow S^0 + OH^-$$

mediated by sulfide oxidizing *Thiobacilli*, *Thiocalovibrio*, or *Thioalcalobacteria*, bacteria that accumulate the elemental sulfur in the extracellular medium. A separator system facilitates recovery of the sulfur. This method removes sulfide *ex situ*. *In situ* sulfide removal by nitrate injection was already discussed in "Control of Reservoir Souring."

Biodegradation of Volatile Organic Compound (VOC) Vapors

The major VOCs in petroleum are the Benzene, toluene, ethylbenzene, xylenes (BTEX) compounds, which are well-known hazardous environmental priority pollutants, and a variety of physical separation or destruction methods are used to control emission of these contaminants into the atmosphere. However, public pressures to improve air quality are leading to increased environmental regulation of release of these contaminants, and the high costs of VOC treatment are encouraging the development of alternative and/or more competitive treatment technologies, including development of improved biological methods, especially through air biofiltration processes. Biological oxidation of VOCs in air biofilters is mediated by bacteria and fungi, immobilized on a solid support medium as biofilms, and placed in biofilter reactors. The VOCs, contained in a humid gas phase, are passed through the high-surface porous solid support phase containing microbial films configured so that there is little resistance to gas flow. The contaminants are sorbed into the medium-containing biofilms where contaminant degradation occurs. The contaminant and/or medium organic and inorganic compounds supply the microbial population present in the biofilter with constituents to support microbial growth and metabolism. Generally, retention time for the gas stream in the biofilter is of the order of 30–90 s while achieving >90% removal. Biofilter microbial activity needs to be able to operate at gas flow rates of around 1–2 l gas per liter biofilter capacity per minute and degrade around 1–2 kg VOCs per 1000 l of biofilter capacity per day (0.1–0.2% per day).

Concluding Remarks

Petroleum hydrocarbon bioremediation processes are considered to be advantageous in terms of their relatively low cost, process flexibility, benign nature, and on-site utility. However, the processes are often inadequately designed or engineered and/or are often operated by personnel lacking proper technical training. As a result, processes have often failed to achieve the required low contaminant concentration criteria, which has limited general confidence in bioremediation. As a result, the technology is only applied to address a small portion of the US$ 10 billion annual remediation market.

MICROORGANISMS IN OIL RECOVERY

Microbial Enhanced Oil Recovery

Major factors limiting oil recovery from oil reservoirs are high crude oil viscosity resulting in poor mobility and uneven permeability, causing injection of water to enhance recovery to continuously travel the same path of least resistance. The concept of MEOR was first proposed nearly 80 years ago but received only limited attention until the early 1980s in laboratory-based studies, and greater attention in the early 1990s for field applications. Some indigenous or injected microorganisms can produce fermentation products, which can assist in the process of oil recovery from reservoirs, and this has been exploited in MEOR processes. Microorganisms participating in MEOR also act by producing polymeric products such as biosurfactants and polysaccharides, or gases such as carbon dioxide, methane, and hydrogen, or by plugging highly permeable watered-out regions of oil reservoirs with bacterial cells and biopolymers. Microbial application is followed by reservoir repressurization, interfacial tension (IFT)/oil viscosity reduction, and selective plugging of the most permeable zones to move additional oil to producing wells.

Biosurfactants that have a proven capacity to emulsify oil and remove oil films from rocks may have application in MEOR processes. Microbial surfactants can facilitate enhanced petroleum recovery in oil reservoirs by increasing the apparent solubility of petroleum components through reduction of the IFTs of oil and water in oil wells and oilfield emulsions. To be successfully applied in MEOR, the microbes need to have physiological characteristics that allow them to thrive in the environmental conditions of the reservoir in terms of salinity, pH, temperature, and pressure. In general, only bacteria are useful in MEOR applications. Other microorganisms lack the physiological characteristics necessary for growth and survival in the often high salt (1.3–2.5%), high temperature (70–90 °C), and high pressure (2000–2500 psi) reservoir environment. Some microorganisms have been discovered that can grow at temperatures up to 121 °C. Viable microorganisms have also been shown to be present in an oil-bearing formation at a temperature of 118–124 °C. Application of *Pseudomonas aeruginosa*, *Xanthomonas campestris*, *Bacillus licheniformis*, and *Desulfovibrio desulfuricans* along with nutrients and biosurfactants to injection wells can increase oil recovery by 30–200%. Promising MEOR tests in laboratory columns and reservoirs have been implemented using biosurfactants produced by thermo- and halotolerant species of *B. licheniformis* JF-2 and *Bacillus subtilis* with resulting enhanced oil recoveries. Application of suspensions of organisms to reservoir strata from a variety of taxanomic groups including *Bacillus*, *Desulfovibrio*, *Clostridium*, *Micrococcus*, *Pseudomonas*, *Arthrobacter*, *Peptococcus*, and *Microbacterium* has been suggested. According to the National Institute of Petroleum & Energy

Research, 27% of oil reservoirs and 40% of oil-producing carbonate reservoirs may represent candidate target sites for MEOR in the United States. By way of example, it has been suggested that a single-well stimulation treatment might have potential to increase the rate of production from 0.2 to 0.4 tons of oil per day for a period of 2–6 months without additional treatments.

Laboratory studies at Shengli Oilfield in China have indicated the existence of indigenous microorganisms such as hydrocarbon degrading bacteria (HDB), denitrifying bacteria (DNB), methane producing bacteria (MPB), and detrimental microorganisms such as sulfate-reducing bacteria (SRB), iron bacteria (IB) and sulfur bacteria (SB) in crude oil and formation water. Indigenous bacteria beneficial to MEOR can be selectively stimulated while the detrimental bacteria can be restricted to some extent by a selective nutrient system of corn syrup and nitrate injection. The corn syrups contained enough nutrition for the beneficial microbes, including sources of carbon, nitrogen, phosphorus and trace elements, while the nitrate inhibited the growth of the detrimental bacteria through electron acceptor transferred from oxygen atom to nitrogen atom. The injection of H_2O_2 or water–air mixture with nitrogen and phosphorus salts resulted in an increase in the number of aerobic and anaerobic organotrophic bacteria, and reduction in the rates of sulfate reduction and methangonesis in formation water.

The North Blowhorn Creek Unit field in Lamar, Alabama had 20 injector wells and 32 producer wells. The MEOR process used there involved the addition of KNO_3 and NaH_2PO_4 to the water flood. In 2001, the DOE reported that the project had added reserves of 400,000–600,000 bbl, extending the economic life of the field by 5–11 years. While injection of nutrients stopped in January 2002, the field is still producing in 2009, even though it was scheduled to be abandoned in 1998.

Advanced mathematical models have been developed to simulate MEOR process. The models included multidimensional multi-component transport in a two-phase oil-water system and included separated terms to account for the dispersion, convection, injection, growth and death of microbes. Both wettability alteration of reservoir rock from oil wet to water wet and reduction in IFT on relative permeability and capillary pressure curves were included in advanced MEOR simulation models.

The tendency for surfactants to adsorb on the surface of reservoir rock by rock–oil–brine interactions is considered one of the major drawbacks of chemical or biosurfactant use in MEOR and is a phenomenon that likely dictates the economic feasibility of the chemical flooding processes. Factors that affect surfactant adsorption include rock surface charges and fluid interfaces where positively charged cationic surfactants attract to the negatively charged surface, and negatively charged anionic surfactants attract to positively charged surfaces.

Thus, potential for application of biosurfactants in MEOR processes remains uncertain. The level of *in situ* biosurfactant production is hard to predict because it is a metabolic side stream. In contrast to mainstream fermentation products such as alcohols or organic acids, the production of which can be accurately predicted, the production of biosurfactant may vary substantially. Also, there is a need to design MEOR processes suitable for each reservoir condition. The particular MEOR process to be applied may cause deleterious effects to the injection wellbore and lead to disastrous plugging of the injection well with a major reduction in oil delivery. Formation of slime, fouling of internal pipe surfaces, and general cleaning problems in the oil- or fuel-handling hardware are all possible pitfalls leading to the development of unforeseen microbial problems following attempts to implement MEOR. An area for further research will be the development of a universal additive mixture suitable for extreme reservoir conditions consisting of a combination of suitable microbial strains, nutrients, biosurfactants, and buffering agents in appropriate proportions.

Microbial De-Emulsification

Stable oil emulsions, which are produced at various steps during oil production and processing operations, are a major problem in the oil industry. Stability of these emulsions is affected by viscosity, droplet size, phase–volume ratio, temperature, pH, age of emulsion, type of emulsifying agent present, presence or absence of particulate matter, density difference, and agitation. Water and sediments in oil can corrode tanks and pipelines, and the maximum allowable limit of a basic sediment and water content is 0.5–2.0% in crude oil for transportation through pipelines. Traditional de-emulsification methods include capital-intensive methods such as centrifugation and treatments with heat, electricity, and chemicals containing soap, fatty acids, and long-chain alcohols.

Microbial de-emulsification processes can potentially be used under nonextreme conditions to treat emulsions at the wellhead, thus saving on transport and high capital equipment costs. Species of *Nocardia*, *Corynebacterium*, *Rhodococcus*, *Bacillus*, *Micrococcus*, *Torulopsis*, *Acinetobacter*, *Alteromonas*, *Aeromonas*, *Alcaligenes*, *Brevibacillus*, *Dietzia*, *Ochrobactrum*, *Pusillimonas*, *Sphingopyxis*, *Achromobacter* and some mixed bacterial cultures are known to possess de-emulsification properties. Microbes exploit their hydrophobic cell surfaces, or the dual hydrophobic/hydrophilic nature of biosurfactants produced, to disrupt emulsifiers and other emulsion-stabilizing agents present at the oil–water interface. Some biologically produced polymers, especially polysaccharides, glycolipids, glycoproteins, lipopetides, phospholipids, and rhamnolipids, exhibit de-emulsification properties.

Among various screening methods, including surface tension, oil-spreading and blood-plate hemolysis tests, surface tension measurement was proposed and successfully implemented for the screening of demulsifying bacteria and their application in demulsification tests in laboratory or oilfield emulsions. A preliminary screening standard of 50 mN/m surface tension of culture was proposed for potential demulsifying bacteria. Based on demulsification performance, twenty biodemulsifier-producing strains were isolated from various environmental sources.

Demulsifying bacteria utilize hydrophobic substrates as carbon source for the production of biodemulsifiers, including crude oil, diesel, tetradecane, hexadecane, kerosene, and paraffin. Alternative substrates such as waste streams from agriculture and food industries have also been tested including lactic whey, potato process effluent, straw, cashew apple juice, and waste oils. Elevating the incubation temperature generally accelerates the microbial de-emulsification process by reducing the viscosity of the oil phase, increasing the density difference between the phases, weakening the stabilizing interfacial film, and causing an increased rate of droplet collision leading to coalescence. However, inconsistencies are experienced in the performance of different bio-demulsifiers due to variability in the properties of crude oil emulsions. In order to realize commercial applications, further research on microbial de-emulsification processes with oilfield emulsions needs to be aimed at development of more reliable and universally effective systems.

Other Applications of Biosurfactants

Because microbial biosurfactants enhance the apparent solubility and reduce the surface tension of crude oil, they can be used as bioemulsifiers in oil recovery from petroleum tank bottom sludges and in facilitating transport of heavy crude oil through pipelines. The Kuwait Oil Company has used biosurfactants for crude oil storage tank cleanup with up to 90% oil recovery. *Acinetobacter calcoaceticus* RAG-1 produces an excellent polyanionic heteropolysaccharide bioemulsifier, Emulsan, which consists of *N*-acetyl-D-galactosamine, *N*-acetyl galactosamine uronic acid, and an amino sugar linked covalently with fatty acid side chains of α- and β-hydroxydodecanoic acid. Application of Emulsan can reduce the viscosity of Boscon heavy crude oil from 200,000 to 100 cP, for easier pumping of heavy oil in commercial pipelines. Microbial enhanced separation of oil with 97% removal efficiency from petroleum refinery sludge by two biosurfactant producing strains of *Bacillus* has also been demonstrated. The bacteria could affect the separation of oil so as to form a floating scum within 48 h. Rhamnolipid biosurfactants have been used to remove oil from used oil sorbents. As compared to chemical surfactants, biosurfactants are generally perceived as being more environmentally friendly because of their biocompatibility and biodegradability. Physicochemical processes for oil recovery are generally capital intensive. Disposal of the chemical emulsifier/de-emulsifier in the aqueous phase and removal of the emulsifier/de-emulsifier from the oil phase are major problems in oil recovery methods.

Methanogenic Oil Degradation

Much of the World's oil appears to have been biodegraded *in situ* over geological time. This biodegradation appears to result largely from the action of methanogenic consortia, consisting of (1) syntrophs, (2) CO_2-reducing methanogens, and (3) acetotrophic methanogens. The syntrophs attack the oil with water, producing CO_2, H_2, and acetate. The acetotrophic methanogens convert acetate to CO_2 and methane, and the CO_2-reducing methanogens convert H_2 and CO_2 to methane. The overall result is a water-driven cracking of hydrocarbons in the subsurface, for example, for hexadecane the overall reaction is

$$4C_{16}H_{34} + 30H_2O \rightarrow 49CH_4 + 15CO_2$$

The discovery of these methanogenic conversions was important because it explains how biodegradation of oil can proceed continuously in the subsurface over geological time in the absence of significant concentrations of electron acceptors (oxygen, nitrate, Fe(III), or sulfate). Water is plentiful in the subsurface. An oil reservoir typically consists of a layer in which oil is trapped in the pores of sedimentary rock, underlayered by water. The methanogenic degradation takes place in the zone of oil–water contact (OWC). Because the syntrophs initiating the oil hydrolysis prefer the lower molecular weight hydrocarbons (alkanes and aromatics), this reaction removes the light end of the spectrum of compounds that make up the oil creating an increasingly viscous, heavy oil. The produced methane and CO_2 move upward until they reach an impermeable rock layer. Hence, the oil in a reservoir held in porous rock may have a gas cap on top, a water layer at the bottom, and may have changing physical properties as a function of depth. In other words, the oil becomes heavier from top to bottom in the direction of the OWC. The endpoint of this methanogenic biodegradation may be the very heavy, highly viscous, tar-like material that is presently found in the Alberta oil sands near Fort McMurray. By comparing compositions of light oil, heavy oil, and the tar sands, and assuming that these derived from each other by methanogenic transformation, one can deduce that these transformations can convert half the oil to gas, which escapes into the atmosphere if the reservoir is not overlain with a sufficiently impermeable cap rock. In the case of the Alberta oil sands, this amounts to 10^{12} barrels of oil having been transformed and largely lost over geological time.

Methanogenic consortia can also be active above ground, for example, in oil storage tanks. The presence of water in such tanks should be avoided. If present, methane and CO_2 may build up forcing the tank operator to periodically release gas, which may lead to odor complaints from neighboring municipalities. This indicates that water-driven conversion of oil into methane can take place in a short time. Hence, yet another route to use microorganisms in oil recovery may be to stimulate subsurface conversion of residual oil, which cannot economically be recovered, into methane. Recent studies have indicated that the alkanes in oil are the main component converted into methane and that this process could cover 10% of current use of natural gas in the US.

Control of Reservoir Souring

Another area in which miroorganisms affect oil recovery is the phenomenon of reservoir souring, which is becoming increasingly well understood along with how reservoir-wide treatment technologies, especially nitrate injection, can be used to remedy this problem. Souring, the often gradual increase in H_2S concentration in oil reservoirs, occurs in many reservoirs that are subject to water injection to pressurize the reservoir and produce the oil. In most offshore oil production operations seawater is injected, which has a high sulfate concentration (~28 mmol L^{-1}, 2700 ppm). Mixing of cold, sulfate-containing seawater and hot hydrocarbon-containing formation water creates ideal conditions for the proliferation of SRB. Depending on the reservoir temperature, which in turn depends on reservoir depth, this proliferation may occur close to the wellbore or throughout the reservoir. Souring is not limited to fields injected with seawater. Freshwater contains a much lower concentration of sulfate (typically 10–100 ppm). However, conversion of such concentrations to sulfide together with inputs of geological sulfide, or conversion of organic sulfur to H_2S, may similarly lead to unacceptably high sulfide levels. These represent a variety of risks to production, including increased corrosion. Depending on sulfide concentration, regular steel piping may have to be replaced with more expensive sour service steel, providing a strong incentive for operators to try to keep the souring problem under control.

A very interesting biotechnology-based method to control souring, which has gained significant popularity in the last 10 years, has been the injection of nitrate. One of the first field tests in this regard was the injection of nitrate in the Coleville field in Saskatchewan in Canada. Injection of 300 ppm nitrate over 5 weeks through a limited number of water injection wells reduced the sulfide concentration by 40–100%. Since then, nitrate injection has been successfully applied in the remediation of souring in fields in the North Sea and elsewhere. The technology is considered so successful that in some cases, nitrate is now being applied prophylactically, that is, in situations where souring has not yet occurred, with the clear objective of preventing the process from the start.

North Sea fields typically have a high bottom-hole temperature (60–80 °C). Coleville is unusual in having a much lower bottom-hole temperature (30 °C). Several years of nitrate injection in the low temperature Enermark field, which like Coleville field is located in western Canada, has indicated that sulfide levels do indeed drop by 70% in the first 5 weeks. However, this is followed by recovery due to the ability of SRB to grow back deeper in the reservoir. Injection of high nitrate pulses appeared to be able to retarget these deeper SRB-inhabited zones. In Coleville, nitrate injection was shown to strongly boost the population of the nitrate-reducing, sulfide-oxidizing bacterium (NR-SOB) *Thiomicrospira* sp. strain CVO, which as the name implies oxidizes sulfide with nitrate to sulfur and sulfate, as well as nitrite and nitrogen. The products depend on the nitrate to sulfide ratio. If this ratio is high (excess nitrate), sulfate and nitrite are formed, whereas when this ratio is low (excess sulfide) sulfur and nitrogen are formed. In addition to NR-SOB, oil fields contain heterotrophic nitrate-reducing bacteria (hNRB). These use water-soluble oil organics to reduce nitrate to nitrite and nitrogen or ammonia. These bacteria are believed to contribute to souring control by competitive exclusion, that is, by using the same oil organic electron donors as used by the SRB, preventing SRB activity. hNRB may also contribute to lowering the sulfide concentrations, because the nitrite that they form can chemically react with sulfide, forming a variety of oxidized sulfur compounds and ammonia. Because thermophilic NR-SOB have never been isolated, this mechanism may be a main contributor to souring control in higher temperature oil fields (resident temperature 60–80 °C), as found in the North Sea and elsewhere. An unresolved question, which is relevant for the nitrate dose that should be injected, is which water-soluble oil organics need to be counted as hNRB and SRB substrates. Often, only volatile fatty acids (VFA; including acetate, propionate, and butyrate) are included. However, other low molecular weight organics, for example, toluene, benzene, and xylenes, are good hNRB and SRB substrates and may have to be considered as well. Hence, considerable research still needs to be done to make nitrate injection a successful souring remediation technology in every possible situation. In the process, we will learn a lot on the role microorganisms can play in oil recovery, which may boost production through newly designed MEOR processes.

OIL BIOREFINING AND BIOPROCESSING

Crude oil contains about 0.05–5% sulfur and 0.5–2.1% nitrogen; indeed sulfur is usually the third most abundant element in crude oil, but in heavier oils it can be present at concentrations up to 14%. Most of the sulfur in crude oil is contained in condensed thiophenes. Thus, the removal of sulfur from crude oil requires costly and extreme conditions, including hydrodesulfurization (HDS). Nitrogenous compounds in crude oil consist of pyrroles, indoles, and carbazole, which is a potent inhibitor of HDS. These toxic and mutagenic carbazole compounds also denature petroleum cracking catalysts and have deleterious environmental impacts by contributing to formation of air-polluting NO_x. High temperature- and high pressure-requiring hydrotreatment processes are used to remove nitrogenous compounds from petroleum. Thus, the momentum pushing the search for alternative biological methods for removal of sulfur and nitrogen from crude oil is the high cost of current physicochemical treatments. The major barrier to using a bioprocess to remove nitrogen or sulfur from crude oil appears to be the necessary involvement of water in such systems and hence the need to form a water-in-oil two-phase system. Major developments in this area are discussed below.

Microbial Desulfurization and Denitrogenation

A variety of microbial strains including *Rhodococcus*, *Nocardia*, *Agrobacterium*, *Mycobacterium*, *Gordona*, *Klebsiella*, *Xanthomonas*, and *Paenibacillus* are capable of selective desulfurization of organic sulfur. When certain species such as *Rhodococcus erythropolis* are cultured aerobically, they exhibit the ability to desulfurize compounds such as dibenzothiophene (DBT) without degrading the carbon ring structure. These strains can use the DBT-released sulfur as sole source of sulfur for growth. The sequence of catabolism of DBT by *Rhodococcus* is mediated by two monooxygenases and a desulfinase, and results in successive production of dibenzothiophene-5-oxide (DBTO), dibenzene-5,5-dioxide ($DBTO_2$), 2-(2-hydroxybiphenyl)-benzenesulfinate (HPBS), and 2-hydroxybiphenyl (HBP) with associated release of inorganic sulfur in the form of sulfite.

Desulfurization genes have been manipulated by directed evolution and gene shuffling approaches to broaden substrate specificity and improved biocatalysts have been engineered. Deletions in desulfurizing bacteria of enzymes in the biodesulfurization pathway such as dibenzothiophene sulfone monooxygenase (DszA) or hydroxyphenyl benzene sulfinase (DszB) have created opportunities to use these microbial cell biocatalysts for production of potentially valuable sulfur-containing metabolic intermediates as products. Molecular manipulations, involving use of a *Rhodococcus–E. coli* shuttle vector, were used to construct the recombinant strain, *Rhodococcus* sp. T09, which could desulfurize alkylated DBT and benzothiophene (BT) and could use both DBT and BT as the sole sulfur source. Resting cells of these strains could also desulfurize alkylated DBT in oil–water, two-phase systems.

In-situ coupling of adsorptive desulfurization and biodesulfurization is a novel desulfurization technology for petroleum oil. It has the merits of high-selectivity of biodesulfurization and high-rate of adsorptive desulfurization. It is carried out by assembling nano-adsorbents onto surfaces of microbial cells. In-situ post reaction cell separation was facilitated when bacterial cells (*Rhodococcus erythropolis*) for biodesulfurization were decorated with superparamagnetic Fe_3O_4 nanoparticles (45–50 nm size). Scanning electron microscopy (SEM) showed that the magnetic nanoparticles substantially coated the surfaces of the bacteria. Compared to non decorated cells, the coated cells had 56% higher desulfurizing activity due to permeabilization of the bacterial membrane, facilitating the entry and exit of reactant and product. The one-step magnetically immobilized cells exhibited good catalytic activity and repeated-batch desulfurization operational stability. Gamma-Al_2O_3 nano-particles modified using gum arabic to avoid agglomeration in aqueous solutions is another example of a nano-adsorbant used in this application. Immobilization of bacterial cells and biomodification of inorganic supports by the adsorption process can increase the bioavailability of sulfur substrates for bacterial cells, improving biodesulfurization activity.

A novel adsorption–bioregeneration system has been developed recently by combining adsorption and biodesulfurization processes for the deep desulfurization of diesel. The sequence of adsorption capacity of DBT was activated carbon (AC) > NiY > AgY > alumina > 13X. For hydrotreated diesel, mesoporous aluminosilicates (MAS) showed high adsorption capacity, while MCM-41 and NaY showed low adsorption capacity. The bioregeneration process with *P. delafieldii* R-8 cells improved DBT desorption from adsorbents. The desorption of DBT from adsorbents by bioregeneration followed the sequence: 13X > alumina > AgY > NiY > AC. Ag-MAS can be completely regenerated in the *in situ* adsorption–bioregeneration system.

Biodesulfurizations carried out in two-phase aqueous-alkane solvent systems exhibited increased sulfur removal rates as compared with aqueous systems. In treating crude oil, it is necessary to apply intensive high energy mixing and/or addition of a surfactant to create a two-phase microbial biodesulfurization system with high interfacial areas, and after desulfurization, a de-emulsification step is required. The key technoeconomic challenge to the viability of biodesulfurization processes is to establish cost-effective means of implementing the two-phase bioreactor system and de-emulsification steps as well as the product recovery step. It has been found that use of multiple-stage airlift reactors can reduce mixing costs, and centrifugation approaches facilitate de-emulsification, desulfurized oil recovery, and cell recycling. Other identified goals relate to improving microbial cell reaction kinetics and to achieving continuous growth and regeneration of the biocatalyst in the same system, rather than in a separate reactor. Extent of biodesulfurization varies dramatically with the nature of the oil feedstock, and especially with feedstock physical properties and the extent to which the feedstock has been refined. Biodesulfurization extents in the ranges of 20–60%,

20–60%, 30–70%, 40–90%, 65–70%, and 75–90% were observed for crude oil, light gas oil, middle distillates, diesel, hydrotreated diesel, and cracked stocks, respectively.

While the 1990 Clean Air Act Amendment set sulfur content of diesel fuel at a maximum of 500 ppm, lower sulfur standards are enforced in some jurisdictions and future values for diesel fuel are expected to be in the region of 30 ppm. HDS technologies cannot achieve the future required 30 ppm levels and, the de-emulsification extent ranges above suggest current microbial desulfurization technology is not cost-effective for heavy or middle distillates of crude oil. However, a combination of biodesulfurization and HDS technology has the potential to achieve the future required 30 ppm level.

A single-stage reactor was adopted to investigate desulfurization of DBT at high cell densities, in which the growth of bacterial culture (*Rhodococcus erythropolis* IGTS8), induction of desulfurizing enzymes, and desulfurization reactions were carried out in a single step. This system concept has potential to simplify and lower the operating cost of the bio-desulfurization process. A micro-channel reactor continuous system with smaller inner diameters has shown high biodesulfurization rates in the oil/water by more than nine-fold that in a batch (control) reaction.

Bacteria exhibit some general similarities in the pathways where oxygenases play an important role in the initial attack in the transformation of nitrogen compounds. Some species of *Alcaligenes*, *Bacillus*, *Beijerinckia*, *Burkholderia*, *Comamonas*, *Mycobacterium*, *Pseudomonas*, *Serratia*, and *Xanthomonas* can utilize indole, pyridine, quinoline, and carbazole compounds. Pyrrole and indole are easily degraded, but carbazole is relatively resistant to microbial attack. The genes responsible for carbazole degradation by *Pseudomonas* sp. have been identified and cloned. Gene manipulations have created recombinant strains able to transform a wide range of nitrogen-containing aromatic compounds including carbazole, *N*-methylcarbazole, *N*-ethylcarbazole, dibenzofuran, DBT, dibenzo-*p*-dioxin, fluorene, naphthalene, phenanthrene, anthracene, and fluoranthene.

Klebsiella sp. LSSE-H2, isolated from dye-contaminated soil based on its ability to metabolize carbazole as a sole source of carbon and nitrogen, can efficiently degrade carbazole from either aqueous and biphasic aqueous–organic media, displaying a high denitrogenation activity and a high level of solvent tolerance. A co-culture of *Klebsiella* sp. LSSE-H2 and *Pseudomonas delafieldii* R-8 strains degraded approximately 92% of carbazole (10 mmol/L) and 94% of DBT (3 mmol/L) from model diesel in 12 h. Biofilm-immobilized cells of *Burkholderia* sp. IMP5GC in a packed reactor successfully operated in semi-continuous mode and efficiently degraded carbazole present in a mixture of gas oil and light cycle oil.

Effective biodesulfurization and biodenitrogenation require removal of sulfur and nitrogen through specific enzymatic attack of the C–S and C–N bonds, respectively, but without C–C bond attack, thereby preserving the fuel value of the residual products. Critical to biorefining process development will be the design of cost-effective, two-phase bioreactor systems with subsequent oil–water separation and product recovery. From a practical implementation perspective, denitrogenation and desulfurization processes need to be integrated and in practice may have to be combined with physical HDS treatments.

Microbial Demetallation

Asphaltenes are high molecular weight compounds containing aromatic and aliphatic constituents, heteroatoms (S, O, and N), and heavy metals (Ni and V). Although microorganisms have been shown to associate with bitumens and asphaltenes, only some fractions are susceptible to enzymatic or microbial attack.

Some hemeproteins such as chloroperoxidase, cytochrome C peroxidase, cytochrome C reductase, and lignin peroxidase from *Caldariomyces fumago*, *Bacillus megaterium*, *Catharanthus roseus*, and *E. coli* can perform biocatalytic modifications of the asphaltene fraction and removal of nickel and vanadium from petroporphyrins and asphaltenes. The enzymatic treatment with chloroperoxidase can remove up to 93% of Ni from nickel octaethylporphine, 53% of V from vanadyl octaethylporphine, and 20% of the total Ni and V from the asphaltene fraction of heavy oil. Metal-containing fossil fuels, which can be treated, include petroleum, distillate fractions, coal-derived liquid shale, bitumens, Gilsonite, tars, and synthetic fuels derived from them. While cytochrome C reductase and chloroperoxidase enzymes have potential applications in metal removal from petroleum, further investigations on the biochemical mechanisms and bioprocessing aspects are required for development of a commercially feasible biodemetallation process.

Biosynthesis of Novel Compounds

Opportunities exist to exploit the unique regio- and stereospecificity properties of enzymes and their capacities to catalyze reactions in nonaqueous media in systems for biotransformation of petroleum compounds into wholly novel, high value chemicals. These approaches may be extended through applications of molecular methods to expand substrate specificity, to enhance enzyme stability in nonaqueous media, and to manipulate reaction rates. Through advances in X-ray crystallography and protein engineering techniques, more powerful biocatalysts may be created with applications for specific transformations or for upgrading petroleum fractions or pure hydrocarbon compounds. Enantiospecific conversions of petrochemical substrates and their derivatives can be achieved by stereoselective, biocatalytic hydroxylation reactions using cytochrome P450-dependent monooxygenases, dioxygenases, lipoxygenases, and peroxidases. A range of attractive diol precursors for chemical synthesis can be produced by naphthalene dioxygenase (NDO), which also catalyzes a variety of other reactions including monohydroxylation, desaturation, O- and N-dealkylation, and sulfoxidation.

PHYSIOLOGICAL MECHANISMS OF ACCESSION AND EFFLUX

Hydrocarbons as Substrates

Most aerobically and anaerobically growing microorganisms require a reduced electron donor and a terminal electron acceptor linked by an electron transport chain in order to extract fuel from substrates for maintenance and growth (some grow fermentatively). Hydrocarbons, natural compounds found in petroleum reservoirs, as well as produced by living organisms, such as microorganisms, insects, and plants, are highly reduced, energy-rich molecules common in the environment – thanks to both anthropogenic and nonanthropogenic release. This structurally diverse class of molecules includes saturated and unsaturated linear, branched and cyclic alkanes, mono- and polycyclic aromatics, and nitrogen- and sulfur-containing heterocyclics that are used by eukaryotic and prokaryotic microorganisms for energy. In addition to energy-processing mechanisms, microorganisms also possess pathways to funnel catabolic products into central metabolic pathways. Given the structural similarities between some hydrocarbons and cellular macromolecules, for example straight-chain alkanes and fatty acids, microorganisms may incorporate hydrocarbons directly into membranes and vesicles without shuttling through central pathways and typical anabolic steps.

Historical culture-based studies, and more recent molecular-based inquiries, have clearly illustrated microbial hydrocarbon metabolism in bacteria, cyanobacteria, yeast, fungi, and algae in a diversity of habitats. A number of obligate and near-obligate hydrocarbonoclastic microorganisms such as *Alcanivorax* spp., *Cytoclasticus* spp., *Oleiophilus* spp., *Oleispira* spp., *Planomicrobium* spp., and *Thalassolituus* spp. have recently been isolated, and molecular ecology studies have found these organisms to be ubiquitous in marine environments throughout the world. As is often the case, these organisms may be present below detectable levels prior to hydrocarbon exposure, but they achieve near dominance for periods during which preferred substrates are available.

With low to intermediate water solubilities, for example, from 0.000282 to 1800 mg/L for tetradecane and benzene, respectively, compared to hexoses with solubilities of 100 g L or more, microorganisms have evolved mechanisms to exploit hydrocarbons as nutrients and to minimize toxic solvent effects on membrane structure and function. Mechanistically, the stages leading up to hydrocarbon catabolism may include sensing and taxis, substrate accession and uptake, and, if toxicity is an issue, efflux to maintain tolerable levels within the cell.

An excellent foundation of laboratory and field studies describing these steps is now being strengthened through genome sequencing of important petroleum-associated microorganisms and communities (Table 2) and identification of genetic elements involved in each behavior.

TABLE 2 A sample of plasmid, complete genome, metagenomic DNA and RNA sequence libraries, and peptidomes available for important petroleum hydrocarbon-related microorganisms

Organism	Properties	Accession no.
Alcanivorax borkumensis SK2	Aerobic saturated hydrocarbon metabolism; marine environments; biosurfactant and biofilm formation	AM286690
Acinetobacter venetianus plasmid pAV1 and pAV2	Diesel fuel biodegradation	NC_010309–NC_010310
Arthrobacter sp. FB24	Xylene metabolism	CP000454–CP000457
Aromateoleum aromaticum EbN1	Aromatic hydrocarbon metabolism; denitrifying	NC006513; NC006823–NC006824
Burkholderia vietmaniensis G4	Benzene and toluene metabolism; facultative	CP000614–CP000616 (chromosomes); CP000617–CP000621 (plasmids)
Dechloromonas aromatica RCB	Anaerobic benzene metabolism	CP000089
Desulfatibacillum alkenivorans AK-01	Anaerobic alkane-degrading; sulfate reducing	CP001322
Desulfococcus oleovorans Hxd3	Anaerobic alkane metabolism	CP000859
Desulfovibrio desulfuricans subsp. desulfuricans str. G20 & ATCC 27774	Anaerobic hydrogen sulfide production; oil well corrosion	CP001358 CP000112
Desulfovibrio salexigens DSM 2638		CP001649

(Continued)

TABLE 2 A sample of plasmid, complete genome, metagenomic DNA and RNA sequence libraries, and peptidomes available for important petroleum hydrocarbon-related microorganisms—cont'd

Organism	Properties	Accession no.
Desulfovibrio vulgaris DP4		CP000527–CP000528
Desulfovibrio vulgaris str. "Miyazaki F"		CP001197
Desulfovibrio vulgaris Hildenborough		AE017285.1–AE017286.1
Geobacter metallireducens GS15	Participate in anaerobic aromatic hydrocarbon metabolism	CP000148–CP000149
Geobacter sulfurreducens KN400		CP002031
Geobacter sulfurreducens PCA		AE017180
Geobacter metallireducens GS-15		CP000148–CP000149
Marinobacter aquaeoli VT8 also known as M. hydrocarbonoclasticus	Aerobic hexadecane, pristane, crude oil metabolism; moderate halophile; iron oxidation	CP000514–CP000516
Methylobium petrophilum	MTBE metabolism to CO_2; benzene, toluene, ethylbenzene metabolism	CP000555–CP000556
Mycobacterium gilvum PYR-GCK	Pyrene and other PAHs	CP000656–CP000659
Mycobacterium sp. KMS	PAH metabolism	CP000518–CP000520
Mycobacterium sp. MCS		CP000384–CP000385
Mycobacterium vanbaalenii strain PYR-1	High molecular weight PAH metabolism (pyrene)	CP000511
Novosphingobium aromaticivorans	Aerobic aromatic hydrocarbon metabolism	CP000248, CP000676, CP000677
Pelobacter propionicus DSM 2379	Anaerobic unsaturated hydrocarbon metabolism	CP000482–CP000484
Petrotoga mobilis SJ95	Oil well community member (high temperature and salinity)	CP000879
Polaromonas naphthalenivorans CJ2	Aerobic naphthalene metabolism; >20 °C	CP000529–CP000537
Pseudomonas putida F1	Benzene, toluene, ethylbenzene metabolism	CP000712
Pseudomonas putida HS1 TOL plasmid pDK1	Toluene biodegradation	AB434906
P. putida KT2440	Toluene metabolism	AE015451
P. putida	Naphthalene degradation plasmid	AF491307
NCIB 9816–4		
Rhodococcus sp. RHA1	Aerobic alkyl benzene metabolism (best known for PCB degradation)	CP000431.1–CP000434.1
Rhodopseudomonas palustris	Anaerobic aromatic hydrocarbon metabolism; facultative phototroph	CP000250
Thauera sp. MZ1T	Anaerobic aromatic hydrocarbon metabolism; facultative	CP001281–CP001282
Tolumonas auensis DSM 9817	Toluene biosynthesis	CP001616
Yarrowia lipolytica CLIB122	Alkane metabolizing yeast	CR382127–CR382132 (6 chromosomes); AJ307410 (mitochondrion)
Yarrowia lipolytica W29		AJ307410 (mitochondrion sequence)
Metagenomic and Proteomic Libraries:		
Anaerobic sulfate reducing enrichment culture N47	Anaerobic naphthalene and methylnaphthalene metabolism	GU080116–GU080137; GU080090–GU080115
"Candidatus Methylomirabilis oxyfera" dominated culture "Twente"	Anaerobic methane-oxidizing, bacterium that produces N_2 and O_2 by a novel mechanism during nitrate reduction without nitrous oxide reductase. Methane is oxidized under anaerobic conditions using O_2.	SRR023516.1, SRR022749.2 Short read archives; FP565575 Assembled genome; GSE18535 RNA sequences in gene expression omnibus; PSE127 Peptidome
"Candidatus Methylomirabilis oxyfera" dominated culture "Ooij"		SRR022748.2 Short Read Archive; PSE128 Peptidome

For example, the genome of *Alcanivorax borkumensis* includes genes for alkane metabolism, biosurfactant production and biofilm formation, uptake of a variety of inorganic nutrients, and efflux.

Taxis

As a phenomenon, microbial taxis are movement resulting from direct or indirect response to external environmental conditions. For example, chemotaxis is the response to specific chemical attractants and repellents, which relies on two-component signaling pathways consisting of transmembrane chemoreceptor proteins and signal kinases linked to response regulators that switch flagellar motor rotational direction. Well understood for model organisms in response to water-soluble substrates, signal cascades include a variety of other regulators that allow for a short-term memory function to ensure that microbial movement generally tends to move toward more favorable locations. Other signals such as light, dark, environmental redox potential, and cellular energy levels can also direct microbial taxis. There are examples of microoganisms that exhibit chemotactic responses to a number of hydrocarbons including benzene, toluene, ethylbenzene, and naphthalene. In unmixed or poorly mixed experimental systems, hydrocarbon chemotaxis has been shown to increase substrate accession and biodegradation. Putative chemoreceptors, some linked to biodegradation genes, have been proposed for other hydrocarbon substrates in both aerobic and anaerobic microorganisms based on whole genome studies. For example, *Pseudomonas putida* strain NCIB 9816–4 harbors the 83 kb pDTG1 plasmid that carries genes for naphthalene catabolism (*nah* genes) as well as the gene for NahY, which is homologous to methyl-accepting chemotaxis proteins. Mutants lacking *nahY* are able to metabolize naphthalene but lose chemotactic abilities.

Accession

Once a microorganism is in the vicinity of dissolved- or nonaqueous-phase hydrocarbon, the next stage prior to metabolism or efflux is accession. Distinct from uptake, that is, transport across the cell membrane, accession includes any process by which a hydrocarbonoclastic microorganism reduces mass transfer problems associated with poorly soluble substrates. To this end, microorganisms may produce biosurfactants to pseudosolubilize hydrocarbons in the aqueous phase, or increase cell-surface hydrophobicity in order to adhere directly to hydrocarbon droplets. These accession modes are illustrated in Figure 3, which shows the growth of a hydrophobic *Rhodococcus* sp. and a biosurfactant-producing *Pseudomonas* sp. on Bow River crude oil, in pure and co-cultures. The associated micrographs reveal that the *Rhodococcus* sp. adheres to the oil explaining why the culture aqueous phase is free of cells. In the mixed culture, it is interesting to note that the emulsion formed is more stable, and that the hydrophilic *Pseudomonas* cells remain in the aqueous phase but are closely associated with the Rhodococci that are still adhered to oil droplet surfaces.

On the whole, biosurfactants have a range of physiological functions including roles in cell signaling, biofilm formation, cellular differentiation, motility, and, of particular interest here, increasing substrate bioavailability. Analogous to their chemical counterparts, biosurfactants may increase the apparent solubility of hydrocarbons by decreasing interfacial tensions between oil–water phases and via micellization. These effects are achieved due to the amphipathic nature of these cell-associated or extracellular compounds,

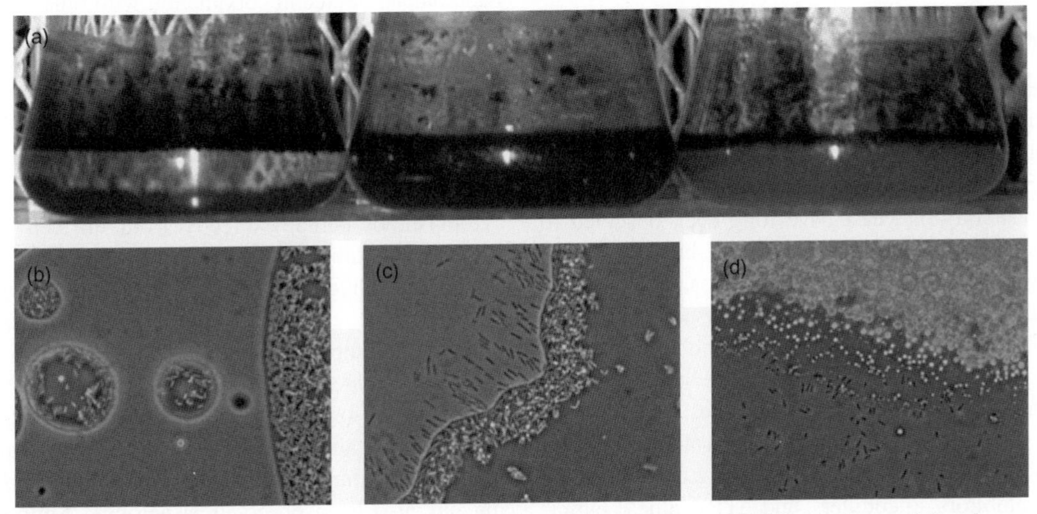

FIGURE 3 Photographs of 48h crude oil cultures inoculated with (left to right): *Rhodoccus* sp. F9-D79, coculture and *Pseudomonas aeruginosa*. Corresponding phase contrast micrographs are shown below each flask. *(Micrographs reproduced from Van Hamme, J. D., & Ward, O. P. (2001). Physical and metabolic interactions of Pseudomonas sp. strain JA5-B45 and Rhodococcus sp. strain F9-D79 during growth on crude oil and effect of a chemical surfactant on them.* Applied and Environmental Microbiology, 67, *4874–4879. Copyright © American Society for Microbiology, DOI: 10.1128/AEM.67.10.4874–4879.2001).*

which may be composed of a combination of fatty acids, peptides, and carbohydrates. A dissolved hydrocarbon, with or without low concentrations of biosurfactant present, may be directly accessible to a microorganism in the aqueous phase and, once in close contact, the hydrocarbon may partition into the interior of the cell membrane via passive uptake. In some cases, initial oxidation steps are carried out by membrane-associated enzymes whereby the substrate is oxygenated and passed into the cytoplasm for further metabolism, the AlkB monooxygenase being a well-characterized example as described below. As the biosurfactant concentration increases above the critical micellization concentration, a microorganism with a hydrophilic cell surface may interact with hydrophilic micellar head groups and access nonpolar hydrocarbons partitioned in the micellar core. It is possible for the biosurfactant micelle to recombine with the cell membrane at this point and deliver hydrocarbon into the cell. If the micelle concentration becomes too high, then mass transfer may be limited by substrate dilution in the micellar phase.

If a microorganism has expressed a hydrophobic cell surface in order to adhere directly to the oil–water interface, then a biosurfactant may disrupt cell–hydrocarbon contact, especially in mixed communities. There are many examples of microorganisms that become highly hydrophobic when growing on hydrocarbons. For example, some Rhodococci adhere so tightly to oil–water interfaces that centrifugation will not remove them. Rhodococci are well known to produce complex cell walls that include nonpolar materials such as mycolic acids and have been shown to directly incorporate unaltered and partially oxidized alkanes into the cell wall. Finally, a microorganism with a hydrophobic cell surface may shed outer biosurfactant-type cell wall material in order to make the cell surface more hydrophilic and detach from a used oil droplet.

Transport in the Cell

With respect to uptake, that is, passage of substrate across the cell membrane, it was historically believed that microorganisms rely solely on passive hydrocarbon partitioning into the cell rather than by facilitated diffusion or active uptake. However, bacteria and yeast have long been observed to store unaltered hydrocarbons in membrane-bound vesicles, which may indicate a more active process, especially in Gram-negative bacteria where the lipopolysaccharide layer creates a barrier to the passage of hydrophobic molecules. In addition, some metabolic flux rates have been calculated to be greater than passive partitioning rates, and selective uptake and storage of specific alkanes from alkane mixtures has been observed. Experiments employing electron transport chain inhibitors, uncouplers, and ATP synthase inhibitors indicate good evidence for active uptake of naphthalene and phenanthrene.

Genome studies have revealed putative hydrocarbon transporters, mostly related to the FadL long-chain fatty acid transporter in *Escherichia coli*, in a number of microorganisms. Evidence for facilitated diffusion of toluene by TbuX and TodX, *m*-xylene by XylN, and for *n*-hexadecane, phenanthrene, and DBT is available, although no transport proteins have been isolated to date. The FadL transporter has been crystallized and structural studies indicate that this ß-barrel protein contains a hatch that opens spontaneously due to conformational changes upon substrate binding. StyE is a facilitated uptake membrane protein for styrene related to FadL, and other putative permeases have been proposed for related compounds such as benzoate, phthalate, and γ-hexachlorocyclohexane. For the latter, the putative transporter coded by *linKLMN* appears to consist of a permease, ATPase, periplasmic protein, and lipoprotein.

It has been proposed that AlkL is involved in alkane uptake in *Pseudomonas* and that *pbhD* in *Sphingomonas paucimobilis* var. EPA505 may code for a fluoranthene metabolite transporter. Of course, the transport of other limiting nutrients such as nitrogen and phosphorus is also important in oil-impacted environments. The genome of the obligate hydrocarbonoclast *A. borkumensis* carries approximately 50 permeases, many of which are presumed to be associated with hydrocarbon uptake.

Proteome comparisons of the fully sequenced *Marinobacter hydrocarbonoclasticus* SP17 grown on acetate in suspended cultures versus hexadecane-grown biofilms revealed the upregulation of type VI secretion systems, normally seen in virulence and symbiotic interactions, as well as a protein (MARHY0478) that was found to have significant homology to TodX, TbuX and FadL. This is a good potential candidate as an alkane uptake transporter. Analysis of cDNA libraries and generation of mutants suggest that OmpW in *Pseudomonas fluorescens* is involved in naphthalene uptake in conjunction with other binding proteins and permeases.

Efflux

In contrast to uptake, a number of efflux pumps have been described that are able to expel hydrocarbons from the cell cytoplasm or perhaps directly from the cell membrane. Related to antibiotic efflux pumps, these pumps can often efflux both hydrocarbons and antibiotics; hydrocarbon efflux pumps are constructed of an outer membrane protein linked to a pump protein via a periplasmic channel protein. Pumps for toluene (e.g., *srpABC, ttgABC, ttg2ABC, ttgK and ttgGHI*), some of which also pump styrene, *m*-xylene, ethylbenzene, and propylbenzene, have been characterized. Phenanthrene, anthracene, and fluoranthene (*emhB*) pumping has also been observed in aerobes. Antibiotic efflux pumps have been identified in medically important anaerobes, and putative efflux genes in anaerobic hydrocarbon degrading microorganisms have been discovered through genome sequencing.

For a nonhydrocarbon-metabolizing microorganism, being able to efflux membrane-disrupting hydrocarbons is essential. For hydrocarbonoclasts, efflux pumps are important for maintaining homeostatic levels of metabolizable substrate, and may also serve to excrete nonmetabolizable hydrocarbons from mixtures.

METABOLISM

Aerobic Alkane Metabolism

As a catabolic group, microorganisms possess metabolic tools to exploit a diverse range of hydrocarbons as sources of energy, carbon, nitrogen, and sulfur. Microorganisms have been isolated that are able to metabolize C1 to C40 alkanes, cycloalkanes, BTEX, and related compounds, two- to six-ring polyaromatic hydrocarbons, and N- and S-containing heterocyclics such as carbazole and DBT. Generally, hydrocarbon catabolism begins with mono- and dioxygenase reactions involving molecular oxygen under aerobic conditions, and via addition of other groups such as hydroxyl from water, fumarate, or carboxyl groups under anaerobic conditions.

Aerobic hydrocarbon metabolism by bacteria and fungi has been studied for almost a century. Due to interest in biodegradation and production of single-cell protein and thanks to recent advances in high-throughput genomics, a clearer image of catabolic diversity is emerging. For alkanes, plasmid-borne or chromosomal gene products have been described and a well-characterized system is the *alk* operon carried on the OCT plasmid in *P. putida* GPo1. This system utilizes AlkL, which may be an outer membrane transport protein, followed by conversion of the alkane into a terminal alcohol by an inner membrane-associated monooxygenase (AlkB). A dehydrogenase (AlkJ) on the inner leaflet of the inner membrane converts the alcohol to an aldehyde prior to conversion to an acid by an aldehyde dehydrogenase (AlkH). This is a logical pathway as it converges with normal fatty acid metabolism via ß-oxidation following the addition of an acetyl-CoA moiety by AlkK. Important in this system are soluble rubredoxins (AlkF and AlkG) and a rubredoxin reductase (AlkT), which serve as electrons shuttles.

Similar systems using different cofactors and under different regulatory control have been described in other organisms, and it appears that more than one, and often many, broad substrate-specificity monooxygenases are commonly found in a single organism. Evidence to date indicates that alkane hydroxylases can be classified based on the carbon chain length of substrates used. Soluble and particulate methane monooxygenases act on C1–C8 and C1–C5 alkanes and alkenes, respectively. AlkB and other membrane-bound hydroxylases monooxygenate the C5–C16 alkanes, cyclic alkanes, and alkane-substituted aromatics among other substrates. The P450 oxygenases of eukaryotes and prokaryotes attack C10–C16 and C5–C16 alkanes, respectively, while dioxygenases have been characterized for C10–C30 alkanes. Recently, genes from *Acinetobacter* sp. strain DSM 17874 were shown to code for AlmA, which appears to be a flavin-binding monooxygenase that acts on C32 alkanes.

At this time, the substrate specificity of these hydroxylases is not understood, but as more crystal structures are obtained trends will appear. In addition, given the industrial applications of these enzymes, more information should emerge on other pathways such as the Finnerty pathway that relies on a dioxygenase to produce *n*-alkyl hydroperoxides, and the alkane desaturase-mediated pathway for subterminal alkane attack.

Aerobic Aromatic Metabolism

When considering aromatic compounds from BTEX to polycyclic aromatic hydrocarbons (PAHs), an intermediate often seen in both prokaryotic and eukaryotic metabolic pathways is a catechol formed by the addition of molecular oxygen across a double bond. In many prokaryotes, catechol is formed from a dioxygenase-generated *cis*-dihydrodiol prior to either *ortho*- or *meta*-cleavage producing a muconic acid or semialdehyde, respectively. Alternatively, eukaryotes such as nonligninolytic fungi and certain prokaryotes will use cytochrome P450 monooxygenases to form an arene oxide, which is converted to a *trans*-dihydrodiol prior to catechol formation. While nonligninolytic fungi produce a variety of oxidized metabolites, which are generally less toxic than the starting material, prokaryotes will shunt oxidized products into central metabolic pathways via common metabolites such as acetaldehyde, pyruvic acid, succinic acid, and acetyl-CoA. PAHs, having up to five rings, may be degraded, although solubility is a challenge and in some cases cometabolism is required.

Ligninolytic fungi produce extracellular peroxidases such as magnesium and lignin peroxidase as well as laccase that oxidize PAHs via nonspecific radical reactions. These microorganisms are important where substrate transport across cell membranes is limited. The ligninolytic fungi can mineralize PAHs with the help of cytochrome P450 monooxygenases and epoxide hydrolases.

Anaerobic Hydrocarbon Metabolism

Given the widespread distribution of anaerobic habitats, and the fact that hydrocarbons are often found in low oxygen environments, it is important to understand the role played by anaerobic hydrocarbon metabolic processes. Further, anaerobes that are active in petroleum environments are of interest due to their impacts on oil recovery and processing. A diverse set of microorganisms including *Azoarcus* spp., *Dechloromonas* spp., *Pseudomonas* spp., *Thauera* spp., *Vibrio* spp., *Geobacter* spp., *Desulfobacula* spp., and *Desulfobacterium* spp. have been isolated from anaerobic hydrocarbon-metabolizing communities. While generally slower than aerobic hydrocarbon metabolism, the list of target substrates is growing rapidly as interest in this area increases. Linear and branched alkanes from C6 to C20 have been shown to be metabolized by pure cultures along with cycloalkanes, BTEX, naphthalene, alkylnaphthalenes, phenanthrene, and tetralin. Nitrate reducers, Fe(III) reducers, sulfate reducers, and anoxygenic photoheterotrophs have all been shown to use hydrocarbons as electron donors. Other terminal electron acceptors that can be used include humic acids, fumarate, and manganese. Methanogens acting in reverse and in partnership with sulfate reducers can catalyze sulfate-dependent, anaerobic methane oxidation (AMO).

Mechanistically, the initial reactions in anaerobic hydrocarbon metabolic pathways studied to date all begin with the addition of an activating substrate other than molecular oxygen. These reactions include hydroxylations using water with molybdenum-based dehydrogenases, formation of hydrocarbon succinates via fumarate addition by glycyl-radical enzymes, methylation using unknown methyl donors apparently generated from CO_2 reduced in the acetyl-CoA pathway, and addition of methanogenic coenzymes M and B to methane during AMO.

Ethylbenzene dehydrogenase has been purified and crystallized, and the metabolic pathway from this enzyme leads to acetyl-CoA and benzoyl-CoA for entry into anaerobic central metabolic pathways. Addition of fumarate to aromatic hydrocarbons, *n*-hexane, and cycloalkanes via enzymes such as benzylsuccinate synthase appears to lead to succinyl-CoA and benzoyl-CoA as has been shown for toluene and 2-methylnaphthalene.

The molecular diversity of genes responsible for these reactions is being revealed through construction and analysis of genomic and metagenomic libraries. For example, metagenomic analysis of the anaerobic hydrocarbon-degrading N47 enrichment culture lead to the description of the *b*eta-oxidation of *n*aphthyl-2-methyl-*s*uccinate operon (*bnsAB-CDEFGH*) which converts naphthyl-2-methyl-succinate to 2-naphthoyl-CoA and contains genes that are relatives of those involved in anaerobic toluene degradation. Other genes in the culture include the recently discovered *nmsABC* operon for fumarate addition to the methyl group of 2-methylnaphthalene, the 2-naphtholy-CoA reductase (*ncr*) genes, and genes for aromatic ring cleavage. Proteomic studies led to the discovery of the first genes described for anaerobic carboxylation of benzene (Abc proteins), opening up the field for a better understanding of how naphthalene is activated under anaerobic conditions given that evidence for methylation as a first step has not yet been found. For short-chain alkanes, it is significant that anaerobic *n*-propane biodegradation has been shown to proceed via fumarate addition in terrestrial environments.

A major shift in our understanding of how hydrocarbons may be metabolized in anaerobic environments came about through a study that used metagenomic analysis of community DNA and expressed RNA, in conjunction with community proteomics and stable isotope probing, to revitalize an old hypothesis that anaerobic denitrification can produce O_2 as a by-product during the conversion of nitrate to N_2 gas. A reconstruction of the genome of the dominant un-cultured organism in the consortium has shown that denitrification proceeds without the production of N_2O, and that the oxygen produced as an end product can be used to metabolize methane via a traditional aerobic methane oxidation pathway. It is possible that this form of oxygen production may predate oxygenic photosynthesis and that this type of activity may be utilized for the anaerobic metabolism of other hydrocarbons.

Many interesting questions remain in this area as illustrated by the recent sequencing of the *Dechloromonas aromatica* genome, which lacks genes for known anaerobic hydrocarbon metabolism enzymes but is able to convert benzene to phenol using nitrate as the terminal electron acceptor.

FURTHER READING

Abu Laban, N., Selesi, D., Rattei, T., Tischler, P., & Meckenstock, R. U. (2010). Identification of enzymes involved in anaerobic benzene degradation by a strictly anaerobic iron-reducing enrichment culture. *Environmental Microbiology*, *12*, 2783–2796.

Ansari, F., Grigoriev, P., Libor, S., Tothill, I. E., & Ramsden, J. J. (2009). DBT Degradation enhancement by decorating *Rhodococcus erythropolis* IGST8 with magnetic Fe_3O_4 nanoparticles. *Biotechnology and Bioengineering*, *102*, 1505–1512.

Bamforth, S. M., & Singleton, I. (2005). Bioremediation of polycyclic aromatic hydrocarbons: Current knowledge and future directions. *Journal of Chemical Technology and Biotechnology*, *80*, 723–736.

Biello, D. (2010). The job of cleaning up after the Gulf oil spill will fall to the microbes. *Scientific American*, *303*, 14–17.

Bao, M., Kong, X., Jiang, G., Wang, X., & Li, X. (2009). Laboratory study on activating indigenous microorganisms to enhance oil recovery in Shengli Oilfield. *Journal of Petroleum Science and Engineering*, *66*, 42–46.

Brown, L. R. (2010). Microbial enhanced oil recovery (MEOR). *Current Opinion in Microbiology*, *13*, 316–320.

Dinamarca, M. A., Ibacache-Quiroga, C., Baeza, P., Galvez, S., Villarroel, M., Olivero, P.,*et al.* (2010). Biodesulfurization of gas oil using inorganic supports biomodified with metabolically active cells immobilized by adsorption. *Bioresource Technology*, *101*, 2375–2378.

Ettwig, K., Butler, M., Le Paslier, D., Pelletier, E., Mangenot, S., Kuypers, M., *et al.* (2010). Nitrite-driven anaerobic methane oxidation by oxygenic bacteria. *Nature*, *464*, 543–548.

Fuller, C., Bonner, J., Page, C., Ernest, A., McDonald, T., & McDonald, S. (2004). Comparative toxicity of oil, dispersant, and oil plus dispersant to several marine species. *Environmental Toxicology and Chemistry, 23*, 2941–2949.

Gieg, L. M., Duncan, K. E., & Suflita, J. M. (2008). Bioenergy production via microbial conversion of residual oil to natural gas. *Applied and Environmental Microbiology, 74*, 3022–3029.

Greene, E. A., Hubert, C., Nemati, M., Jenneman, G. E., & Voordouw, G. (2003). Nitrite reductase activity of sulfate-reducing bacteria prevents their inhibition by nitrate-reducing, sulfide-oxidizing bacteria. *Environmental Microbiology, 5*, 607–617.

Head, I. M., Jones, D. M., & Larter, S. (2003). Biological activity in the deep subsurface and the origin of heavy oil. *Nature, 426*, 344–352.

Heider, J. (2007). Adding handles to unhandy substrates: Anaerobic hydrocarbon activation mechanisms. *Current Opinion in Chemical Biology, 11*, 188–194.

Huang, X. F., Guan, W., Liu, J., Lu, L. J., Xu, J. C., & Zhou, Q. (2010). Characterization and phylogenetic analysis of biodemulsifier-producing bacteria. *Bioresorce Technology, 101*, 317–323.

Kaster, K. M., & Voordouw, G. (2006). Effect of nitrite on a thermophilic, methanogenic consortium from an oil storage tank. *Applied Microbiology and Biotechnology, 72*, 1308–1315.

Savage, K., Krumholz, L., Gieg, L., Parisi, V., Suflita, J., Allen, J., *et al.* (2010). Biodegradation of low-molecular-weight alkanes under mesophilic, sulfate-reducing conditions: metabolic intermediates and community patterns. *FEMS Microbiology Ecology, 72*, 485–495.

Selesi, D., Jehmlich, N., von Bergen, M., Schmidt, F., Rattei, T., Tischler, P., *et al.* (2010). Combined genomic and proteomic approaches identify gene clusters involved in anaerobic 2-methylnaphthalene degradation in the sulfate-reducing enrichment culture N47. *Journal of Bacteriology, 192*, 295–306.

Schneiker, S., dos Santos, A. P., Bartels, D., Bekel, T., Brecht, M., Buhrmester, J., *et al.* (2006). Genome sequence of the ubiquitous hydrocarbon-degrading marine bacterium *Alcanivorax borkumensis*. *Nature Biotechnology, 24*, 997–1004.

Shafir, S., Van Rijn, J., & Rinkevich, B. (2007). Short and long term toxicity of crude oil and oil dispersants to two representative coral species. *Environmental Toxicology and Chemistry, 41*, 5571–5574.

Singh, A., & Ward, O. P. (Eds.). *Applied bioremediation and phytoremediation*. Soil Biology Series (Vol. 1), (2004). Berlin: Springer-Verlag.

Singh, A., & Ward, O. P. (Eds.). *Biodegradation and bioremediation*. Soil Biology Series (Vol. 2), (2004). Berlin: Springer-Verlag.

Stroud, J. L., Paton, G. I., & Semple, K. T. (2007). Microbe-aliphatic hydrocarbon interactions in soil: Implications for biodegradation and bioremediation. *Journal of Applied Microbiology, 102*, 1239–1253.

Sunde, E., & Torsvik, T. (2005). Microbial control of hydrogen sulfide production in oil reservoirs. In B. Ollivier, & M. Magot (Eds.). *Petroleum Microbiology* (pp. 201–213). Washington, DC: ASM Press.

Telang, A. J., Ebert, S., Foght, J. M., Westlake, D. W. S., Jenneman, G. E., Gevertz, D., *et al.* (1997). The effect of nitrate injection on the microbial community in an oil field as monitored by reverse sample genome probing. *Applied and Environmental Microbiology, 63*, 1785–1793.

Throne-Holst, M., Wentzel, A., Ellingsen, T. E., Kotlar, H. K., & Zotchev, S. B. (2007). Identification of novel genes involved in long-chain n-alkane degradation by *Acinetobacter* sp. strain DSM 17874. *Applied and Environmental Microbiology, 73*, 3327–3332.

van Beilen, J. B., & Funhoff, E. G. (2007). Alkane hydroxylases involved in microbial alkane degradation. *Applied Microbiology and Biotechnology, 74*, 13–21.

van den Berg, B. (2005). The FadL family: Unusual transporters for unusual substrates. *Current Opinion in Structural Biology, 15*, 401–407.

Van Hamme, J. D., Singh, A., & Ward, O. P. (2003). Recent advances in petroleum microbiology. *Microbiology and Molecular Biology Reviews, 67*, 503–549.

Van Hamme, J. D., Singh, A., & Ward, O. P. (2006). Physiological aspects – Part 1. *Biotechnology Advances, 24*, 604–620.

Van Hamme, J. D., & Ward, O. P. (2001). Physical and metabolic interactions of Pseudomonas sp. strain JA5-B45 and Rhodococcus sp. strain F9-D79 during growth on crude oil and effect of a chemical surfactant on them. *Applied and Environmental Microbiology, 67*, 4874–4879.

Vaysse, P., Prat, L., Mangenot, S., Cruveiller, S., Goulas, P., & Grimaud, R. (2009). Proteomic analysis of *Marinobacter hydrocarbonoclasticus* SP17 biofilm formation at the alkane-water interface reveals novel proteins and cellular processes involved in hexadecane assimilation. *Research in Microbiology, 160*, 829–837.

Venosa, A. D., Campo, P., & Suidan, M. T. (2010). Biodegradability of Lingering Crude Oil 19 Years after the Exxon Valdez Oil Spill. *Environmental Science and Technology, 44*, 7613–7621.

Venosa, A. D., & Holder, E. L. (2007). Biodegradability of dispersed crude oil at two different temperatures. *Marine Pollution Bulletin, 54*, 545–553.

Voordouw, G., Grigoryan, A. A., & Lambo, A. (2009). Sulfide remediation by pulsed injection of nitrate into a low temperature Canadian heavy oil reservoir. *Environmental Science and Technology, 43*, 9512–9518.

Youssef, N., Simpson, D. R., & Duncan, K. E. (2007). In sit*u* biosurfactant production by *Bacillus* strains injected into a limestone petroleum reservoir. *Applied and Environmental Microbiology, 73*, 1239–1247.

Zengler, K., Richnow, H. H., Rossello-Mora, R., Michaelis, W., & Widdel, F. (1999). Methane formation from long-chain alkanes by anaerobic microorganisms. *Nature, 401*, 266–269.

Chapter 43

Metal Extraction and Biomining

C.A. Jerez

University of Chile and ICDB Millennium Institute, Santiago, Chile

Chapter Outline

ABBREVIATIONS

2D-PAGE Two-dimensional polyacrylamide gel electrophoresis
AMD Acid mine drainage
DGGE Denaturing gradient gel electrophoresis
ESI-MS Electron spray ionization MS
FISH Fluorescence *in situ* hybridization
FT-ICR MS Fourier transform ion cyclotron resonance mass spectrometer
HPLC High performance liquid chromatography
MS Mass spectrometry
QS Quorum sensing
SDO Sulfur dioxygenase
SOR Sulfite oxidoreductase
TEM Transmission electron microscopy

DEFINING STATEMENT

Microorganisms interact with heavy metals, transforming them by uptake, bioaccumulation, bioprecipitation, bioreduction, biooxidation, and other mechanisms. Some of these activities result in the solubilization or extraction of metals, which are successfully used in industrial biomining processess to recover valuable metals or in biorremediation to remove toxic metals from contaminated soils.

MICROBIAL TRANSFORMATIONS OF METALS

Microorganisms interact with metals by several mechanisms, most of which are shown in Figure 1. All bacteria require several metals that are essential for their functioning and the uptake of most of them is metabolism or energy dependent. For this bioaccumulation, they possess specific or general energy-dependent metal transporters to directly incorporate them or through chelation by means of organic compounds such as siderophores in the case of iron. Some of these transporters are used to bioaccumulate metals inside the cell. This can be done by sequestering the metal by proteins rich in cysteine or histidines or they can be chelated by inorganic polyphosphates (polyP), which are long chains of phosphate molecules joined through phosphodiester bonds and are highly negatively charged at neutral pH. Microorganisms such as *Acinetobacter* spp. bioprecipitate metals in the form of metal phosphates via hydrolysis of stored polyP depending upon alternating aerobic (polyP synthesis) and anaerobic (polyP hydrolysis and phosphate release) periods.

The metabolism-independent sorption of heavy metals or biosorption refers to the binding of metal ions by whole biomass (living or dead). It can take place by adsorption in which metals accumulate at the surface of the biomass and by absorption or a rather uniform penetration of the metal ions from a solution to another phase. Biomolecules present

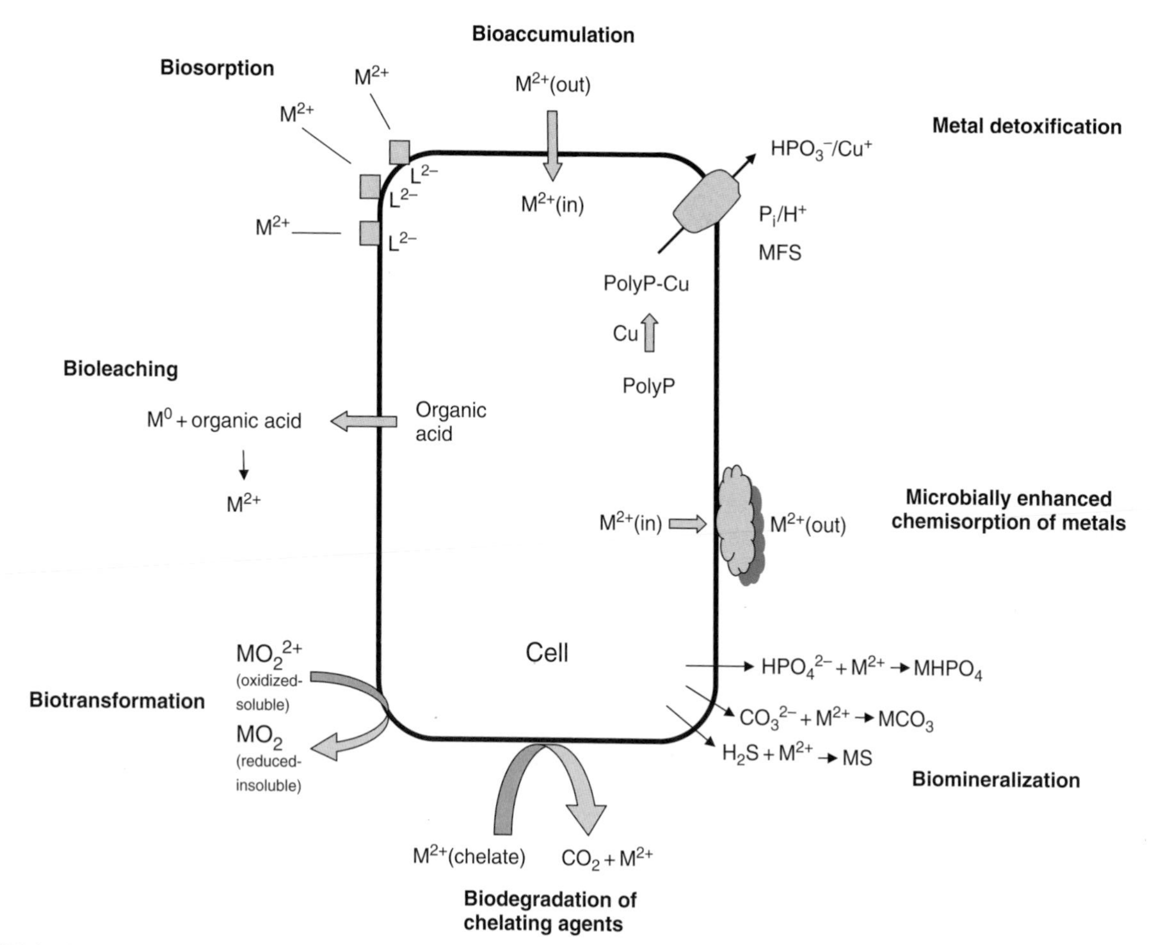

FIGURE 1 A general scheme showing the most common bacteria–metals interactions. *Modified from Lloyd, J. R., Anderson, R. T., & Macaskie, L. E. (2005). Bioremediation of metals and radionuclides. In R. M. Atlas, & J. C. Philp (Eds.)* Bioremediation: Applied microbial solutions for real-world environmental cleanup *(pp. 293–317). Washington, DC: ASM Press.*

in the biomass contain several chemical groups that act as ligands (L^{2-} in Figure 1) for the biosorption of the metal ions. The most common are the amino, carboxyl, phosphate, and sulfhydryl groups present in proteins, nucleic acids, and polysaccharides.

Microorganisms can also catalyze several biotransformations. They transform most toxic metals into less soluble or less volatile forms. One example is the reduction of metals such as Cr(VI) to Cr(III), U(VI) to U(IV), Te (VI) to Te(0), and many others that results in the precipitation of the metal under physiological conditions. Another very well-studied biotransformation of a toxic metal is the bioreduction of Hg(II) to the relatively nontoxic volatile elemental Hg(0).

Microorganisms also precipitate metals in the form of carbonates, hydroxides, or insoluble sulfides, and phosphates. This constitutes a biomineralization, and due to the very low solubility products of these compounds formed, most of the soluble metals would be removed by precipitation in their liquid medium (Figure 1).

MICROORGANISMS THAT SOLUBILIZE METALS

A basic prerequisite for life is the existence of a chemical redox gradient to convert energy. This allows energy flow and it is used in the biochemical reactions of the cells. Most cells use organic compounds like sugars as fuels (electron donors). Their oxidation in the presence of oxygen or respiration provides carbon to the cell, generating as end products CO_2 and H_2O and energy in the form of electrons. This energy is stored mainly in the form of ATP to be used in cell metabolism. A group of special microorganisms (bacteria and archaea) known as chemolithoautotrophs are capable of using minerals as fuels. Their oxidation generates electrons to obtain ATP and the carbon is obtained by fixing CO_2 from the air. During this aerobic mineral oxidation metals are solubilized or bioleached.

Anaerobic respiration can also solubilize metals. In this case, the mineral acts as an electron acceptor and the metal is solubilized under reducing conditions. Several microorganisms are capable of reducing heavy metals and couple this reduction with the oxidation of energy sources such as hydrogen or organic compounds (formate, lactate, amino acids, and others). Characteristic examples are the reduction of Fe and Mn, which account for a good part of organic carbon turnover in several environments. Several of the dissimilatory metal-reducing bacteria not only reduce Fe and Mn but also other metals such as the toxic Cr(VI) and U(VI) and therefore, microorganisms like *Shewanella oneidensis* could also be used for bioremediation.

The reactions just mentioned are part of normal biogeochemical processes in nature. They are performed under neutral conditions or as the majority of microorganisms do, at very high acidic values (pH 1–3 usually). This article will concentrate mainly on reviewing metals mobilization by acidophilic microorganisms.

Some of the general oxidation reactions that acidophiles are able to catalyze are given below.

$$\text{Ferrous iron}: 2Fe^{2+} + 0.5O_2 + 2H^+ \rightarrow 2Fe^{3+} + H_2O$$

$$\text{General metal sulfide}: MS + 2O_2 \rightarrow M^{2+} + \rightarrow SO_4^{2-}$$

$$\text{Pyrite}: 4FeS_2 + 15O_2 + 2H_2O \rightarrow 2Fe_2(SO_4)_3 + 2H_2SO_4$$

$$\text{Chalcopyrite}: CuFeS_2 + 4Fe^{3+} \rightarrow 5Fe^{2+} + Cu^{2+} + 2S^0$$

$$\text{Sulfur}: S^0 + 1\tfrac{1}{2}O_2 + H_2O \rightarrow H_2SO_4$$

These reactions not only solubilize the metals present in the minerals but also generate sulfuric acid. The acidophilic microorganisms, therefore, are able to stand not only low pH values but also very high metal concentrations. They possess heavy metal resistance or detoxification mechanisms. Neutrophilic bacteria have metal resistance mechanisms involving either an active efflux or a detoxification of metal ions by different transformations (Figure 1). In the case of copper, for example, these include intracellular complexation, decreased accumulation, extracellular complexation, or sequestration in the periplasm. Neutrophilic bacteria are able to grow in a range of copper concentration between 1 and 8 $mmol\,l^{-1}$ depending on the species. However, acidophiles such as *Acidithiobacillus ferrooxidans* or archaeons such as *Sulfolobus metallicus* can resist concentrations of copper up to 800 and 200 $mmol\,l^{-1}$, respectively. Most likely, they possess additional mechanisms to those present in neutrophiles, allowing them to have such dramatic metal resistance. It has been proposed that in microorganisms possessing large amounts of polyP, such as some chemolithoautotrophic acidophilic bacteria and archaea, these polymers may be actively involved in the elimination of toxic heavy metals such as Cu. This detoxification would take place through the enzymatic hydrolysis of polyP that would generate free phosphate that would bind the excess of cytoplasmic metal to form a metal–phosphate complex that is transported outside the cell through phosphate transporters (Figure 1). These properties make these microorganisms very appropriate for biomining and also for the bioremedation or removal of heavy metals from polluted places (see below).

MICROBIAL EXTRACTION OF METALS FROM ORES AND BIOMINING

Microbial solubilization of metals is widely and successfully used in industrial processes called bioleaching of ores or biomining, to extract metals such as copper, gold, uranium, and others. This process is done by using chemolithoautotrophic microorganisms. There is a great variety of microorganisms capable of growth in situations that simulate biomining commercial operations and many different species of microorganisms are living at acid mine drainage (AMD) sites. The most studied leaching bacteria are from the genus *Acidithiobacillus. A. ferrooxidans* (Figure 2b) and *Acidithiobacillus thiooxidans* are acidophilic mesophiles and together with the moderate thermophile *Acidithiobacillus caldus*, they belong to the Gram-negative γ-proteobacteria. Figure 2c shows *A. ferrooxidans* cells growing in the form of a biofilm on the surface of an elemental sulfur particle, most likely as a monolayer. When this biofilm is removed from the solid particle by using a detergent (Figure 2d), only those cells attached more strongly to the sulfur particle remain. They are seen as attacking the sulfur surface through a "pitting" in which some cells are still tightly bound to the cavities and others have been released, leaving empty cavities. A similar attack has been observed on the surface of other minerals such as pyrite.

Members of the genus *Leptospirillum* are other important biomining bacteria that belong to a new bacterial division. Some Gram-positive bioleaching bacteria belonging to the genera *Acidimicrobium*, *Ferromicrobium*, and *Sulfobacillus* have also been described. Biomining using extremely thermophilic archaeons capable of oxidizing sulfur and iron (II) has been known for many years, and the archaeons are mainly from the genera *Sulfolobus* (Figure 2b), *Acidianus*, *Metallosphaera*, and *Sulfurisphaera.* Recently, some mesophilic iron (II)-oxidizing archaeons belonging to the Thermoplasmatales have been isolated and described – *Ferroplasma acidiphilium* and *Ferroplasma acidarmanus*. In fact, a consortium of different microorganisms participates in the oxidative reactions, resulting in the extraction of dissolved metal values from ores.

Industrial biomining operations are of several kinds depending on the ore type and its geographical location, the metal content, and the specific minerals present (metal oxides, metal sulfides of different kinds). One of the most used setups for the recovery of gold or copper is the irrigation type of processes. These involve the percolation

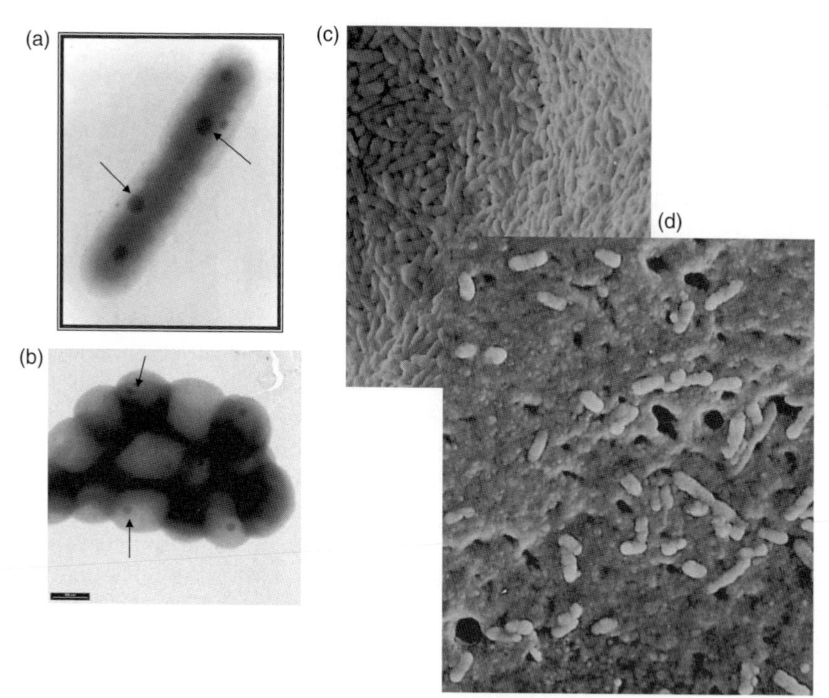

FIGURE 2 Some examples of acidophilic microorganisms that participate in metal extraction through biomining. (a) *Acidithiobacillus ferrooxidans* cells. (b) A group of *Sulfolobus metallicus* cells. (c) A biofilm, possibly a monolayer of *A. ferrooxidans* cells growing on the surface of an elemental sulfur prill. (d) Most of the biofilm of *A. ferrooxidans* seen in (c) was removed from the solid particle by using a detergent and vigorous shaking of the sample. The remaining cells seen are those more strongly adhered to the particle. Cells in (a) and (b) were observed by transmission electron microscopy (TEM) of unstained preparations and those in (c) and (d) by scanning electron microscopy. The arrows in each case point to electron dense polyphosphate granules that may help these microorganisms in their extremely high metal tolerance.

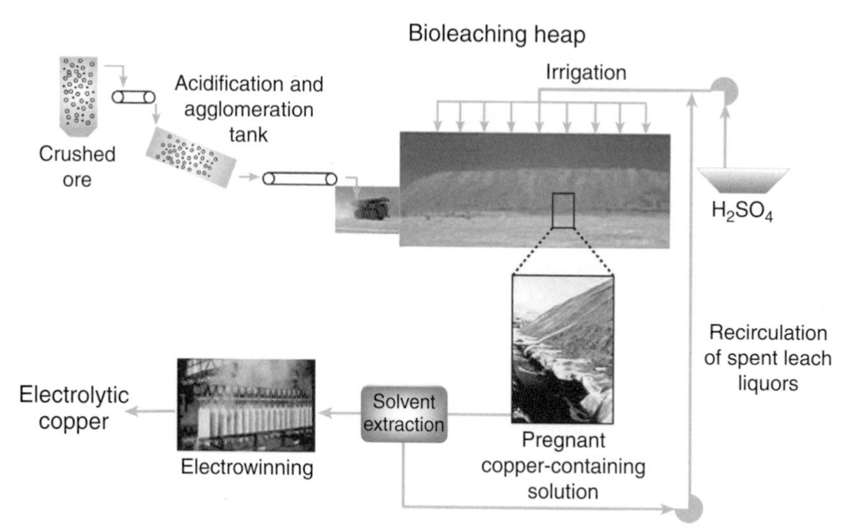

FIGURE 3 A scheme showing the construction of a heap bioleaching process to obtain copper in a large scale.

of leaching solutions through the crushed ore that can be contained in a column, a heap or a dump. In Figure 3, we can see a scheme in which the crushed ore to bioleach is transported to an agglomeration tank or drum where it is acidified. This process is the key one since the bigger ore particles are surrounded by the very fine particles that stick to them thus preventing all the particles, especially, the fine material sediment to the bottom of the heap. In this way, irrigation and aeration of the heap takes place from the top to the bottom, allowing a much more homogeneous growth of the microorganisms and therefore a better metal solubilization. The heap can be 6–10 m tall and 100 or more meters long and wide and is constructed over irrigation pads lined with high-density polyethylene to avoid losses of the pregnant copper-containing solution (Figure 3). This solution containing copper sulfate generated by the microbial solubilization of the insoluble copper sulfides present in the ore is subjected to solvent extraction to have a highly concentrated copper sulfate solution from which the metal is recovered in an electrowinning plant to generate electrolytic copper of high purity (Figure 3). Since most mining operations are located in areas where water is scarce, the spent leach liquors or raffinates are recirculated to the heap for further irrigation. This has to be controlled because the liquor is being enriched in salts that can select for those microorganisms able to stand the high salt and are not necessarily the most fit for the biooxidation reactions.

Bioleaching bacteria can also be used for gold recovery. Gold is usually found in nature associated with minerals containing arsenic and pyrites (arsenopyrites). During gold bioleaching, the iron- and sulfur-oxidizing microorganisms attack and solubilize the arsenopyrite, releasing the trapped gold particles. Following this release, the gold is complexed with cyanide according to standard gold-mining procedures. Instead of using big leaching heaps or dumps as in the case of bioleaching of copper ores, gold bioleaching is usually done by using highly aerated stirred tank bioreactors connected in series. Since these reactors are expensive to build, they are used with high-grade ores or with mineral concentrates. The advantage of tank reactors over heaps and dumps, which are "open bioreactors," is that in the tanks conditions can be controlled, thus having a much faster and efficient metal extraction process.

Currently, there are operations using both mesophilic and thermophilic microorganisms. Biomining has distinctive advantages over the traditional mining procedures. For example, it does not require the high amounts of energy used during roasting and smelting and does not generate harmful gases such as sulfur dioxide. Nevertheless, AMD can be generated, which if not properly controlled, pollutes the environment with acid and metals. Biomining is also of great advantage since not only discarded low-grade ores from standard mining procedures can be leached in an economically feasible way but some high-grade ores also can likewise be leached. In countries like Chile, which is actually the first world copper producer, many mining operations process from 10,000 to 40,000 tons of ore per day and produce between 10,000 and 200,000 tons of copper per year by using heap or dump bioleaching of minerals such as oxides, chalcocite, covellite, chalcopyrite, and others. Similar situations take place in the United States, Australia, and other countries. The most successful ones have been those processing copper oxides and secondary copper sulfides. However, chalcopyrite is the most abundant copper sulfide in the world. Since it is the most difficult to be solubilized by microorganisms, there is actually great interest in developing processes mainly using thermophilic biomining microorganisms.

ACIDIOPHILIC MICROORGANISM–MINERAL INTERACTION

The microorganisms used in biomining belong to those known as extremophiles since they live in extremely acidic conditions (pH 1–3.0) and in the presence of very high toxic heavy metals concentrations. Considerable effort has been spent in the past few years to understand the biochemistry of iron and sulfur compounds oxidation, bacteria–mineral interactions, and the several adaptive responses that allow the microorganisms to survive in a bioleaching environment. All of these are considered key phenomena for understanding the process of biomining.

How do the bacterial cells recognize the site of attachment in the ores from which they will extract the metals? Do they have a specific way to detect or sense where the oxidizable substrate is present in the rocks? Ample evidence has shown that attachment of the bacterial cells to metal sulfides does not take place randomly. For example, *A. ferrooxidans* cells preferentially adhere to sites with visible surface scratches and are seen to form pitting at the surface of minerals as already seen in Figure 2.

Bacteria such as *A. ferrooxidans* and *Leptospirillum ferrooxidans* have been shown to possess a chemosensory system that allows them to have chemotaxis, that is, the capacity to detect gradients of oxidizable substrates being extracted from the ores such as Fe (II)/Fe (III) ions, thiosulfate, and others (Figure 4). The response is the positive chemotactic attraction to the sites that will constitute specific favorable mineral attachment sites.

It is known that most leaching bacteria grow attached to the surface of the solid substrates such as elemental sulfur and metal sulfides. This attachment is predominantly mediated by extracellular polymeric substances (EPS) surrounding the cells and whose composition is adjusted according to the growth substrate (Figure 4). Thus, planktonic or free-swimming cells grown with soluble substrates such as iron (II) sulfate produce almost no EPS.

Biofilms are organized layers of bacteria associated to a solid surface by means of the matrix of EPS. The environment within this community favors intercellular interactions between bacteria. Thus, in the presence of oxygen or aerobiosis, *A. ferrooxidans* can respire this element and oxidize Fe (II). Some of these microorganisms are also able to perform Fe (III) respiration under anaerobiosis, regenerating Fe (II) that can be used by those microbial cells closer to the oxygen within the biofilm structure. Other types of bacteria forming part of the microbial biofilm consortium will generate compounds useful to other members of the community and they can themselves benefit from the metabolic byproducts of other microbes present in the biofilm.

Several microorganisms are known to monitor their own population density through processes collectively described as quorum sensing (QS), and in several cases there is strong

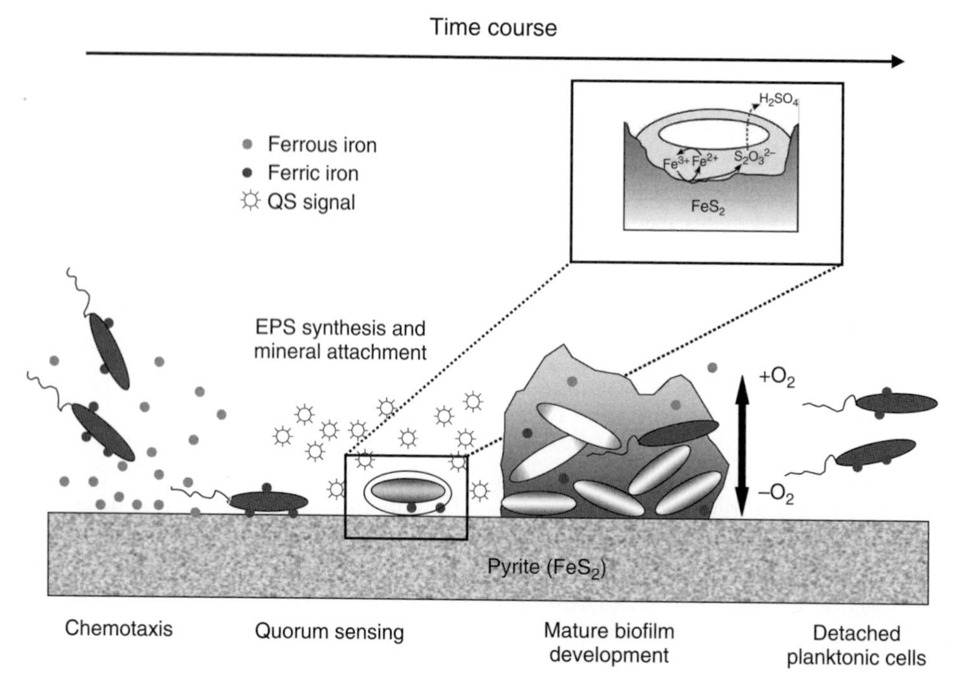

FIGURE 4 A diagrammatic representation showing the main steps of the bacteria–mineral interaction. Details of the attack of *Acidithiobacillus ferrooxidans* on the mineral pyrite are shown in the amplified inset. The cell is embedded in an extracellular polymeric substances (EPS) biofilm in which the indirect mechanism generates ferrous iron and thiosulfate, which is finally oxidized to sulfuric acid. Inset *Reproduced from Rohwerder, T., Gehrke, T., Kinzler, K., & Sand, W. (2003). Bioleaching review part A: Progress in bioleaching: Fundamentals and mechanisms of bacterial metal sulfide oxidation.* Applied Microbiology and Biotechnology, 63, *239–248.* (For interpretation of the references to color in this figure legend, the reader is referred to the Web version of this chapter.)

evidence indicating that biofilm formation is affected by QS. In most cases, specific genes within the bacterium are switched on at a defined population density (defined bacterial quorum) and the result obtained is the activation of functions under the control of a quorum sensor. In almost all cases, the capacity to detect a bacterium quorum depends on the release of a signal molecule from the microorganism that accumulates in proportion to the cell number (Figure 4). Thus, QS represents a multicellular action in bacteria, where each cell communicates with each other to coordinate their behavior. Very recently, it has been demonstrated that *A. ferrooxidans* not only contains quorum sensor signal molecules but may induce the expression of genes related to QS and EPS production when grown attached to a solid substrate. Thus, modulation of the attachment of the microorganisms to ores through interferences of their QS responses can be envisaged as a new way to control metal extraction by these microorganisms.

A. ferrooxidans is also able to develop biofilm structures when growing in solid substrates such as elemental sulfur or metal sulfides and presents morphological modifications during the cellular adhesion process. For example, new proteins related to sulfur metabolism appear in the surface of *A. ferrooxidans* when grown in sulfur. This is clearly seen in Figure 5, in which a primary antibody specific against a protein related to sulfur metabolism is bound to the surface

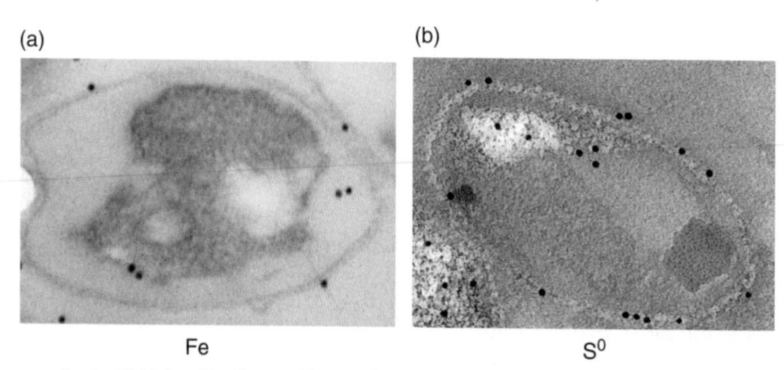

FIGURE 5 Example of changes on the *Acidithiobacillus ferrooxidans* surface depending on growth conditions. It is known that *A. ferrooxidans* produces a protein in high amounts when the bacterium grows on elemental sulfur and not in ferrous iron (Fe). To see the location of that protein, an immunological system was used in which a gold particle (black dots) indicates the presence of the protein in the cells grown in ferrous iron (Fe) (a) or elemental sulfur (b).

of *A. ferrooxidans* cells grown in sulfur but is almost absent in ferrous iron-grown cells. The primary antibody bound to the protein was recognized by using a secondary antibody labeled with gold particles and specific to recognize the primary antibody bound to the sulfur-induced protein. The presence of the protein changing its expression is seen as the black dots of gold by transmission electron microscopy (TEM).

MECHANISMS INVOLVED IN METAL SOLUBILIZATION BY ACIDOPHILES

Traditionally, it has been proposed that microorganisms oxidize metal sulfides by either a direct or an indirect mechanism. In the first one, the bacteria have to attach to the mineral and the electrons would be directly extracted by an enzymatic reaction of the microorganism. However, there is no evidence how the bacteria break the metal–sulfide bond of a given mineral. In the indirect mechanism, the role of bacteria would be to oxidize ferrous ions present in the solution to ferric ions. Ferric ions are strong oxidants, therefore will oxidize chemically the sulfide metals, being an indirect mode of attack. Actually, many researchers consider the existence of a contact mechanism, since the bacteria attaches to the surface of the mineral carrying Fe (III) bound to its exopolysaccharides, and when the microorganism forms a biofilm being embedded in its EPS, this metal would chemically attack the metal sulfide generating, in the case of pyrite, ferrous iron that is reoxidized to Fe (III) and thiosulfate that can be further oxidized to sulfuric acid (Figure 4). The mechanism still is indirect (known also as indirect contact mechanism). This close contact of the bacterium with the mineral makes more efficient and specific the sulfide oxidation.

The insoluble metal sulfides are oxidized to soluble metal sulfates by the chemical action of ferric iron, the main role of the microorganisms being the reoxidation of the generated ferrous iron to obtain additional ferric iron (Figure 4).

As already mentioned, *A. ferrooxidans* is a chemolithoauthotrophic bacterium that obtains its energy from the oxidation of ferrous iron, elemental sulfur, or partially oxidized sulfur compounds and it has been considered a model biomining microorganism. The reactions involved in ferrous iron oxidation by *A. ferrooxidans* have been studied in detail, however, the electron pathway from ferrous iron to oxygen has not been entirely established. The terminal electron acceptor is assumed to be a cytochrome oxidase anchored to the cytoplasmic membrane. The transfer of electrons would occur through several periplasmic carriers, including at least the blue copper protein rusticyanin and cytochrome c552. Recently, a high molecular weight c-type cytochrome, Cyc2, has been suggested to be the prime candidate for the initial electron acceptor in the respiratory pathway between ferrous iron and oxygen (Figure 6). This pathway would be Cyc2 → rusticyanin → Cyc1(c552) → aa3

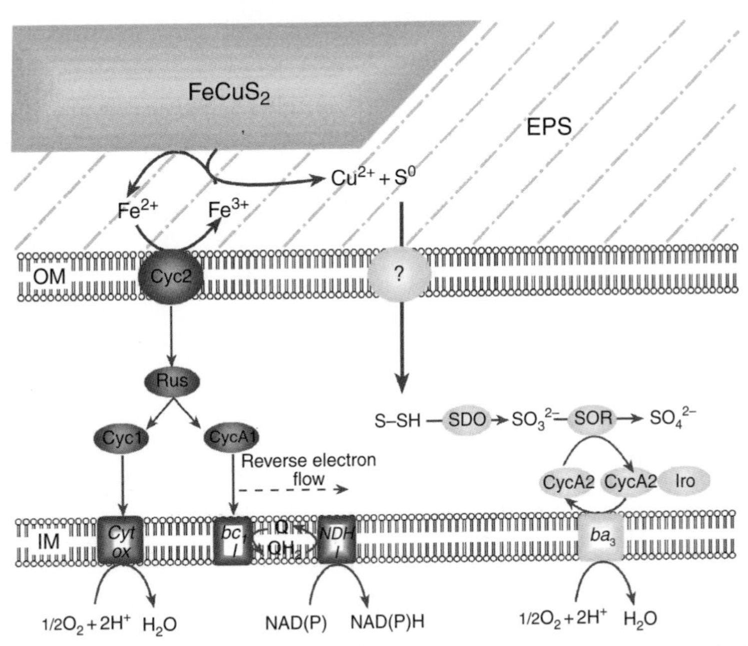

FIGURE 6 Simplified model showing the main protein components (in red) present in the periphery of the cell and involved in the transfer of electrons from ferrous iron to oxygen and those oxidizing elemental sulfur (in yellow) and transferring the electrons to oxygen. This model represents the indirect mechanism of mineral attack by the ferric iron bound to the EPS of the *Acidithiobacillus ferrooxidans* cells. As an example, chalcopyrite is shown from which ferrous iron, copper (II), and elemental sulfur are generated upon its oxidation. SDO, sulfur dioxygenase; SOR, sulfite oxidoreductase. *Adapted from Rawlings, D. E. (2005). Characteristics and adaptability of iron- and sulfur-oxidizing microorganisms used for the recovery of metals from minerals and their concentrates.* Microbial Cell Factories, 4, *13*. (For interpretation of the references to color in this figure legend, the reader is referred to the Web version of this chapter.)

cytochrome oxidase. In addition, there is an apparent redundancy of electron transfer pathways via bc(1) complexes and terminal oxidases in *A. ferrooxidans*.

As already mentioned, thiosulfate has been postulated as a key compound in the oxidation of the sulfur moiety of pyrite (Figure 4). Iron (III) ions are exclusively the oxidizing agents for the dissolution. Thiosulfate would be consequently degraded in a cyclic process to sulfate, with elemental sulfur being a side product. This explains why only Fe(II) ion-oxidizing bacteria are capable of oxidizing these metal sulfides. All reactions comprising this oxidation have been shown to occur chemically. However, sulfur compound-oxidizing enzymes such as the tetrathionate hydrolase of *A. ferrooxidans*, *A. thiooxidans*, or *T. acidophilus* may also be involved in the process.

The oxidation of some metal sulfides such as chalcopyrite generates elemental sulfur as a side product instead of thiosulfate (Figure 6). The aerobic oxidation of elemental sulfur by *A. ferrooxidans* and other microorganisms is carried out by a sulfur dioxygenase (SDO) and a sulfite oxidoreductase (SOR) (Figure 6).

It is important to remark here that the ultimate oxidizing agent for iron (II) and reduced inorganic sulfur compounds is oxygen, since often the transport of dissolved oxygen is the rate-limiting step in commercial bioleaching operations.

BIOMINING IN THE POSTGENOMIC ERA

Systems microbiology, which is part of systems biology, is a new way to approach research in biological systems. By this approach, it may be possible to explore the new properties of microorganisms that arise from the interplay of genes, proteins, other macromolecules, small molecules, and the environment. This is particularly possible today due to the large numbers of genomic sequences that are becoming increasingly available. However, additional genomic sequences of the different biomining microorganisms will be required to define the molecular adaptations to their environment and the interactions between the members of the community.

The use of genomics, metagenomics, proteomics and metaproteomics (Figure 7), together with metabolomics to study the global regulatory responses that the biomining community uses to adapt to their changing environment is just beginning to emerge. These powerful OMICS approaches will have a key role in understanding the molecular mechanisms by which the microorganisms attack and

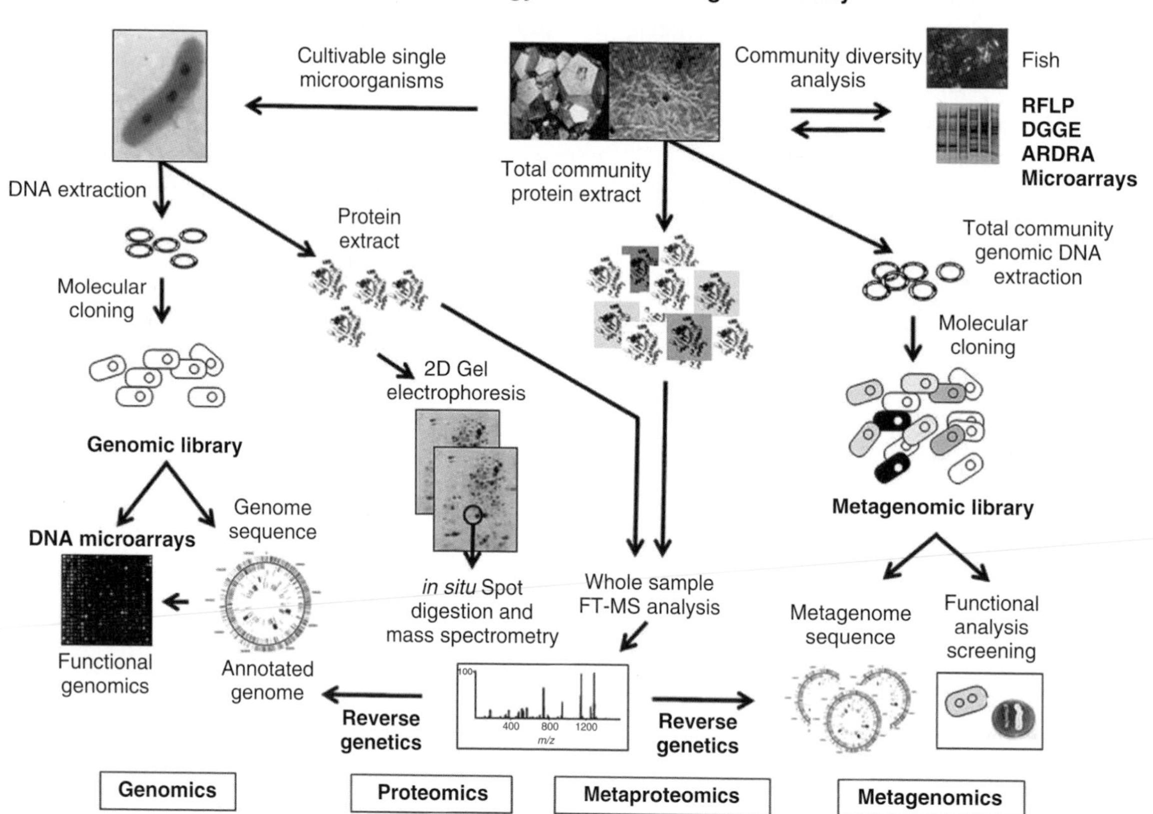

FIGURE 7 Systems microbiology and the use of OMICS approaches to study microbial communities as applied to biomining microbes. *Modified from Valenzuela, L., Chi, A., Beard, S.,* et al. *(2006). Genomics, metagenomics and proteomics in biomining microorganisms.* Biotechnology Advances, 24, *197–211.*

solubilize ores. Furthermore, they offer the possibility of discovering exciting new findings that will allow analyzing the community as a microbial system, determining the extent to which each of the individual participants contributes to the process, and how they evolve in time to keep the conglomerate healthy. This, taken together with the physicochemical, geological, and mineralogical aspects of the process, will allow improving the efficiency of this important biotechnology.

Biomining Community Diversity Analysis

Of extreme importance is not only to know the microorganisms present in a bioleaching operation, but to be able to monitor their behavior during the process, to determine the predominant species and the way they evolve in time with the changing environment as the metals are solubilized. In recent years, several molecular methods have been developed for other microorganisms and these have been successfully applied to many biomining operations. The most common techniques employed to explore the bacterial diversity use 16S rRNA and rDNA profiles and are culture-independent methods. Among these methods are included fluorescence *in situ* hybridization (FISH), denaturing gradient gel electrophoresis (DGGE), real time PCR, DNA microarrays, and others (Figure 7).

Genomics

A. ferrooxidans was the first biomining microorganism whose genome was sequenced by TIGR. This information has been very useful to do genome-wide searches for candidate genes for important metabolic pathways and several important physiological functions, which can now be addressed. Furthermore, predictions for the functions of many new genes can be done. The main focus of research has been the energy metabolism that is directly responsible for bioleaching. Genes involved in phosphate, sulfur, and iron metabolism, QS, those potentially involved in several other functions such as metal resistance and amino acid biosynthesis pathways, and those involved in the formation of EPS precursors have been studied. Having the genomic sequence of a given biomining microorganism is very important, since it is then possible to formulate hypothesis about the regulation of the expression of most of these genes under different environmental conditions. Metabolic reconstruction and modeling provides an important preliminary step in understanding the unusual physiology of this extremophile especially given the severe difficulties involved in its genetic manipulation and biochemical analysis. However, all these bioinformatic predictions will have to be demonstrated experimentally by using functional genomics, proteomics, and other approaches. The existence of paralog genes that show sequence similarities but may have different functions in the same microorganism becomes obvious only when the genome sequence is available.

So far, there are few genomic sequences of acidophilic microorganisms usually found in biomining or AMD places. These are for the bacteria *A. ferrooxidans*, *Leptospirillum* group II, and *Acidiphilium cryptum*. It has also been mentioned that although not yet publicly available, the genome sequences of *A. thiooxidans* and *A. caldus* have been determined. Amongst the archaea, *Metallosphaera sedula* and *F. acidarmanus* genome sequences are available. *Sulfolobus acidocaldarius* and *Sulfolobus tokodaii* are also known, although they have not been reported as having a role in bioleaching. It is therefore not difficult to predict in the very near future the generation of new DNA microarrays to monitor not only all the microorganisms present in samples from industrial operations (Figure 7) but also specific genes such as those indicating the nutritional or stress state of the microorganisms or those involved in iron or sulfur compound oxidation whose products will be predominant during different stages of active bioleaching.

Functional genomics

One of the most used techniques to study differential genome expression at the level of mRNA synthesis, transcriptomics, or the functional analysis of new and characterized genes is the use of DNA microarrays. However, a prerequisite to apply microarrays is to know the genomic sequence of the organism to be analyzed.

The use of microarrays based on the entire genome of *A. ferrooxidans* and other microorganisms will enable a nearly complete view of gene expression of the members of the microbial community under several biomining conditions, helping to monitor their physiological state and adjustment made during the bioleaching process. A first preliminary pilot DNA macroarray formed with 70 different genes has been used to study the relative variations in mRNA abundance of some genes related with sulfur metabolism in *A. ferrooxidans* grown in different oxidizable substrates. A genome-wide microarray transcript profiling analysis (approximately 3000 genes of the *A. ferrooxidans* ATCC 23270 strain) has also been performed. The genes preferentially transcribed in ferrous iron growth conditions or in sulfur conditions were studied. The results obtained supported and extended models of iron and sulfur oxidation (Figure 6) and supported the possible presence of alternate electron pathways and that the oxidation of these two kinds of oxidizable substrates may be coordinately regulated. By using the same approach, the expression of the genes involved in carbon metabolism of *A. ferrooxidans* has been studied in response to different oxidizable substrates.

As already mentioned, some mining companies are currently interested in doing transcriptomic analysis of their newly isolated microorganisms with improved capacities to leach copper since they already have obtained their genomic sequences.

A very interesting alternative approach can be used to analyze gene function in environmental isolates without knowing the sequence of the microorganism of interest. A random genomic library from the isolated microorganism can be printed on a microarray. Gene expression by using total RNA extracted from the microorganism grown under different conditions can be determined. With this approach, it is possible to select and sequence only those clones bearing the genes that showed an altered expression pattern. Shotgun DNA microarrays are very powerful tools to study gene expression with environmental microorganisms whose genome sequence is still unknown.

In the near future, the use of microarrays based on the entire genomes of biomining microorganisms will allow having a nearly complete view of gene expression of the members of the microbial community under several biomining conditions, helping to monitor their physiological state and adjustment made during the bioleaching process.

Metagenomics

Metagenomics is the culture-independent genomic analysis of microbial communities. In conventional shotgun sequencing of microbial isolates, all shotgun fragments are derived from clones of the same genome. To analyze the genomes of an environmental microbial community (Figure 7), the ideal situation is to have a low diversity environment. Such systems were found when analyzing the microbial communities inhabiting a site of extreme AMD production, in which few types of organisms were present. Still, variation within each species might complicate assembly of the DNA fragments. Nevertheless, random shotgun sequencing of DNA from this natural acidophilic biofilm was used. It was possible to reconstruct the near-complete genomes of *Leptospirillum* group II and *Ferroplasma* type II and partially recover three other genomes. The extremely acidic conditions of the biofilm (pH about 0.5) and relatively restricted energy source combine to select for the small number of species found.

The analysis of the gene complement for each organism revealed the metabolic pathways for carbon and nitrogen fixation and energy generation. For example, genes for biosynthesis of isoprenoid-based lipids and for a variety of proton efflux systems have been identified, providing insights into survival strategies in the extreme acidic environment. Clearly, the metagenomic approach for the study of microbial communities is a real advancement to fully understand how complex microbial communities function and how their component members interact within their niches. A full understanding of the biomining community also will require the use of all these current molecular approaches.

Proteomics

Proteomics provide direct information of the dynamic protein expression in tissue or whole cells, giving us a global analysis. Together with the significant accomplishments of genomics and bioinformatics, systematic analysis of all expressed cellular components has become a reality in the post genomic era, and attempts to grasp a comprehensive picture of biology have become possible.

One important aspect of proteomics is to characterize proteins differentially expressed by dissimilar cell types or cells imposed to different environmental conditions. Two-dimensional polyacrylamide gel electrophoresis (2D-PAGE) in combination with mass spectrometry (MS) is currently the most widely used technology for comparative bacterial proteomics analysis (Figure 7). The high reproducibility of 2D-PAGE is particularly valuable for multiple sample comparisons. In addition, it directly correlates the changes observed at the peptide level to individual protein isoforms.

2D-PAGE separates a complex mixture of proteins such as those present in a bacterial cell extract based on the isoelectric point of the proteins in the first dimension (isoelectrofocusing gel). These proteins or groups of proteins are further resolved in a second dimension (SDS-PAGE), in which the proteins are all negatively charged and are separated based on their molecular masses (Figure 7). Actual 2D-PAGE procedures can resolve around 1000 protein spots in a single run. One of the limitations of 2D-PAGE is that only the most abundant proteins in the cell can be detected. Therefore, to increase the resolution of the method, it is also possible to analyze a subproteomic cell fraction instead of the total cell proteins as shown for the periplasmic proteins from *A. ferrooxidans*.

The high reproducibility of 2D-PAGE is particularly valuable for multiple sample comparisons. In addition, it directly correlates the changes observed at the peptide level to individual protein isoforms. This is a typical "reverse genetics" approach in which (after isolating from a 2D-PAGE) an individual protein is differentially expressed in a condition of interest, and the amino acid sequence of a peptide from the protein is obtained to identify its possible homologue in databases. With this information, its coding gene and genomic context can be searched using the genome DNA sequence (Figure 7). Depending on these results, a suggested function could be hypothesized. It will be of great importance to demonstrate the expression of putative genes related to EPS synthesis in *A. ferrooxidans* cells grown on different metal sulfides and to find out if these genes are involved in cell attachment and biofilm formation. In this regard, *A. ferrooxidans* is known to form biofilms on solid substrates (Figure 2). In an AMD biofilm analyzed by the proteomic approach, it was not known which microorganisms are responsible for the production of the polymer embedding the community. However, the presence of numerous glycosyltransferases and polysaccharide export proteins in the predominant bacterial species also suggests a role in biofilm formation.

Several studies have used 2D-PAGE to study changes in protein expression of *A. ferrooxidans* under different growth conditions. Proteins induced under heat shock, pH stress, phosphate limitation, or the presence of copper have been reported. A set of proteins that changed their levels of synthesis during growth of *A. ferrooxidans* in metal sulfides, thiosulfate, elemental sulfur, and ferrous iron has been characterized by using 2D-PAGE.

During growth of *A. ferrooxidans* in metal sulfides containing iron, such as pyrite and chalcopyrite, proteins upregulated both in ferrous iron and in sulfur compounds were synthesized, indicating that the two energy-generating pathways are simultaneously induced depending on the kind and concentration of the available oxidizable substrates.

In the past decade, an increasing number of sequenced genomes provided good options for high-throughput functional analysis of proteomes. Proteomic studies are well advanced for diverse bacteria, such as the model bacterium *Escherichia coli* and others. Nevertheless, there is still a lack of data for identification of proteins from organisms with unannotated or unsequenced genomes, which makes large-scale microbacterial proteomics analysis a challenge. With the development of highly sensitive and accurate computational gene-finding methods, new microbial genomes could be explored and scientific knowledge of them could be maximized.

Traditional 2D gel electrophoresis coupled with MS is time consuming as a result of the nature of spot-by-spot analysis and it is biased against low abundance proteins, integral membrane proteins, and proteins with extreme pI or molecular weight (MW). Alternatively, solution-based approaches offer unbiased measurement of relative protein expression regardless of their abundance, subcellular localization, or physicochemical parameters (Figure 7). This methodology, however, results in extremely complex samples. For instance, of the 4191 predicted genes in the complete genome of *E. coli*, 2800 of them are believed to be expressed at any one time. Additional complexity is introduced upon enzymatic digestion, which generates multiple peptide species for each protein. To obtain comprehensive protein expression information from the samples, a chromatographic separation step prior to MS protein analysis is often necessary. High performance liquid chromatography (HPLC) coupled with online electron spray ionization MS (ESI-MS/MS) has been proved to be a valid approach for analyzing protein expression in complex samples. Alternatively, Fourier transform ion cyclotron resonance mass spectrometer (FT-ICR MS) is well suited for the differential analysis of protein expression due to the high mass accuracy and high resolution as well as its inherent wide dynamic range. Peptide charge state can be readily derived with the accurate isotopic peak distribution information provided by FT MS experiment, and coeluting species with the same nominal *m/z* ratio can be resolved. The differentially expressed *m/z* values identified can be assigned to the peptide sequences and, subsequently, differentially expressed proteins can be identified (Figure 7).

In the periplasm of *A. ferrooxidans*, 216 proteins were identified, several of them changing their levels of synthesis when the bacterium was grown in thiosulfate, elemental sulfur, or ferrous iron media. Thirty four percent of them corresponded to unknown proteins. Forty one proteins were exclusively present in sulfur-grown and 14 in thiosulfate-grown cells. The putative genes coding for all the proteins were localized in the available genomic sequence of *A. ferrooxidans* ATCC 23270. The genomic context around several of these genes suggests their involvement in sulfur metabolism and possibly in sulfur oxidation and formation of Fe–S clusters.

Metaproteomics

Recently, the term "metaproteomics" was proposed for the large-scale characterization of the entire protein complement of environmental microbiota at a given point in time. High-throughput MS has been used in a metaproteomic approach to study the community proteomics in a natural AMD microbial biofilm. Two thousand and thirty-three proteins from the five most abundant species in the biofilm were detected, including 48% of the predicted proteins from the dominant biofilm organism *Leptospirillum* group II. It was also possible to determine that one abundant novel protein was a cytochrome central to iron oxidation and AMD formation in the natural biofilm. This novel approach together with functional metagenomics can offer an integrated study of a microbial community to establish the role each of the participant plays and how they change under different conditions.

The goal of functional proteomics is to correlate the identification and analysis of distinct proteins with the function of genes or other proteins. With the discovery of a variety of modular protein domains that have specific binding partners, it has become clear that most proteins occur in protein complexes and that the understanding of a function of a protein within the cell requires the identification of its interacting partners. In the case of a biomining bacterium such as *A. ferrooxidans*, the identification of protein complexes involved in oxidative reactions is of high priority. What complexes are formed by cytochrome-like proteins such as rusticyanin (Rus in Figure 6) with other proteins in the periplasm? Do some periplasmic proteins form a complex involved in sulfur compound oxidation in the periplasm? What other complexes of oxidative reactions are present in this microorganism? Proteomics may answer these and many other questions that will help to understand better the biomining process.

The OMICS procedures briefly analyzed and summarized in Figure 7 should be used in close conjunction with the known physiological functions of the microorganisms being studied. It should be possible to better control the activity of the bacteria and archaea by giving them the appropriate nutritional and physicochemical conditions, and by interfering with some of these microbiological functions in order to enhance their action (for metal extraction) in the case of biomining or to inhibit their capacities to control AMD. As already mentioned (Figure 4), some of these key physiological behaviors are chemotaxis, QS, and biofilm formation. Bacteria such as *A. thiooxidans* and *L. ferrooxidans* clearly possess chemotactic systems and are attracted by a concentration gradient of thiosulfate or ferrous iron such as the one generated on the surface of pyrite (Figure 4). This sensing ability is very important for the specific bacterial adherence at the places where the substrates to be solubilized through oxidation are present.

The development of efficient transformation or conjugation systems to introduce DNA to generate mutations affecting a gene of interest by using gene knockout systems is almost entirely lacking for biomining microorganisms. These tools will be essential not only to perform functional genomics and have experimental demonstrations for the suggested gene functions based on bioinformatics analysis of the postgenomic data, but also to eventually improve some physiological bacterial capabilities. At the same time, the new OMICS methods are greatly helping to monitor more precisely the biomining consortia, and it is expected that providing the right physiological conditions to the community together with proper chemical or physical manipulations of the bacterial environment will further improve bioleaching rates in biomining operations.

ENVIRONMENTAL EFFECTS OF METALS SOLUBILIZATION AND BIOREMEDIATION

The acidophilic microorganisms mentioned, mobilize metals and generate AMD, causing serious environmental problems. AMD should be remediated or abated. Often there is a sealing of the contaminated sites or the location of barriers to contain the acidic fluids. Many approaches use prevention techniques to avoid further spillage of acidic effluents in the contaminated area. It can be controlled by chemical treatments such as the use of calcium oxide that neutralizes the acid pH. It is also possible to inhibit the acidophilic microorganisms responsible for the acid generation. This can be done by using certain organic acids, sodium benzoate, sodium lauryl sulfate, or quaternary ammonium compounds that will affect the growth of bacteria such as *A. ferrooxidans*.

Bioremediation or removal of the toxic metals from these contaminated soils can be achieved by a very interesting combination of two opposite biological activities: that of sulfur-oxidizing bacteria with the one of sulfate-reducing microorganisms. In a first step, the sulfur-oxidizing bacteria generate sulfuric acid which bioleaches or solubilizes the metals in the solid phase of the soil. The leachate metals are then precipitated in a second step by using a bioreactor in which the hydrogen sulfide generated by the sulfate-reducing bacteria under neutral and anaerobic conditions forms insoluble metal sulfides, which are eliminated (Figure 8). Metal contaminants such as Cu, Cd, Ni, and others can be efficiently leached from contaminated soils. The effluents obtained from such a process are clean enough of the metals that they can be reused in the environment.

Bioleaching microorganisms such as *A. ferrooxidans* can also have other uses to help avoiding metal contaminations in modern societies. For example, this bacterium has been successfully used to recover metals such as cadmium from spent batteries. By using bioreactors, *A. ferrooxidans* is grown attached on elemental sulfur. The bacteria generate sulfuric acid through the oxidation of sulfur that is then used for the indirect dissolution of spent nickel–cadmium batteries recovering after 93 days 90–100% of cadmium, nickel, and iron. Bioleaching of spent lithium ion secondary batteries, containing lithium and cobalt, has also been explored. These approaches are not only economically valuable but may be an effective method which could be considered the first step to recycle spent and discarded batteries preventing one of the many problems of environmental pollution.

CONCLUSIONS

Microorganisms (bacteria and archaea) require several metals that are essential for their life. However, when the metal concentrations reach toxic levels, they also possess metal resistance mechanisms that involve active efflux or a detoxification of metal ions by different transformations. They can transform most toxic metals to less soluble or less volatile forms by intracellular complexation, decreased accumulation, extracellular complexation, or sequestration in the periplasm. Some of these activities result in the solubilization or extraction of metals, which are successfully used in environmental biotechnological applications.

The aerobic mineral oxidation or the anaerobic respiration of different microorganisms can result in the solubilization or extraction of metals. All these bacteria–metal interactions are part of the normal biogeochemical processes in nature.

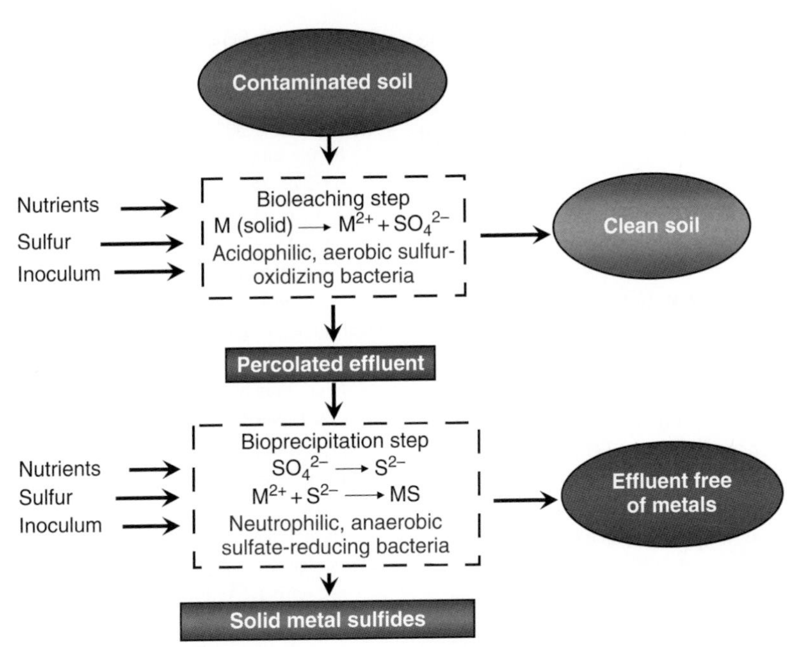

FIGURE 8 A schematic diagram illustrating a process for the bioremediation of soils polluted with metals.

The microbial solubilization of metals in acid environments is successfully used in industrial processes called bioleaching of ores or biomining, to extract metals such as copper, gold, uranium and others. On the contrary, the acidophilic microorganisms mobilize metals and generate AMD, causing serious environmental problems. However, bioremediation or removal of the toxic metals from contaminated soils can be achieved by using the specific properties of microorganisms interacting with metals.

Current approaches to study microorganisms consider the microorganism or the community as a whole, integrating fundamental biological knowledge with genomics, proteomics, metabolomics, and other data to obtain a global picture of how a microbial cell or a community functions. This new knowledge will help not only in understanding microbiological phenomena but it will also be useful to improve applied microbial biotechnologies.

RECENT DEVELOPMENTS

Very recently, some interesting new findings have been reported in the metal extraction area. It is expected that this new knowledge will allow further improvements in the biomining capacities of microbes for industrial applications.

Regarding bioleaching genomics, several genome sequencing projects of microorganisms involved in oxidation/reduction of iron and/or reduced inorganic sulfur compounds have been completed and several others are ongoing or available as draft sequences. These advances are very important since new whole-cell models of functions encoded in these genomes are beginning to emerge. Furthermore, knowing the genomes of several of the biomining microorganisms allows comparative genomics to help in the identification of common mechanisms of adaptation to their environment and also the discovery of new mechanisms characteristic of acidophilic bacteria and archaea. A system to transform *Acidithiobacillus caldus* has recently been reported by several research groups. However, an efficient system to introduce DNA into almost all known biomining microorganisms is still missing, although strongly needed for the generation of knockout mutations to perform functional genomics.

In addition, high-throughput proteomic studies of these biomining microorganisms have allowed the elaboration of a detailed model for the location and putative function of many of the periplasmic proteins, including those involved in the oxidative reactions that these bacteria use in dissolving minerals. On the other hand, strain-resolved community proteomics has revealed that acidophilic bacteria have recombining genomes. This interesting finding suggests that exchange of large blocks of gene variants is crucial for adaptation to specific ecological niches within the very acidic, metal rich environment of bioleaching microorganisms.

Some new insights into the mechanisms of copper homeostasis in biomining bacteria such as *Acidithiobacillus ferrooxidans* have been described. The current view is that biomining extremophiles respond to the extremely high copper concentrations in their habitat by using simultaneously all or most of the following key elements: (a) a wide repertoire of copper-resistance determinants, (b) duplication of some of these copper-resistance determinants, (c) existence of novel copper chaperones, (d) an inorganic polyphosphate-based copper-resistance system, and (e) an oxidative stress defense system. Further knowledge of the response to copper and other metals in each member of the biomining community will be of great relevance, since this information could be used to select the more fit microorganisms of the bioleaching community to attain more efficient industrial biomining operations.

FURTHER READING

Auernik, K. S., Maezato, Y., Blum, P. H., & Kelly, R. M. (2008). The genome sequence of the metal-mobilizing, extremely thermoacidophilic archaeon *Metallosphaera sedula* provides insights into bioleaching-associated metabolism. *Applied and Environmental Microbiology*, *74*, 682–692.

Baker-Austin, C., & Dopson, M. (2007). Life in acid: pH homeostasis in acidophiles. *Trends in Microbiology*, *15*, 165–171.

Chi, A., Valenzuela, L., Beard, S., Mackey, A. J., Shabanowitz, J., Hunt, D. F., *et al.* (2007). Periplasmic proteins of the extremophile *Acidithiobacillus ferrooxidans*: a high throughput proteomic analysis. *Molecular and Cellular Proteomics*, *6*, 2007, 2239–2251.

Farah, C., Vera, M., Morin, D., Haras, D., Jerez, C. A., & Guillani, N. (2005). Evidence of a functional quorum sensing type AI-1 system in the extremophilic bacterium *Acidithiobacillus ferrooxidans*. *Applied and Environmental Microbiology*, *71*, 7033–7040.

Jerez, C. A. (2009). Biomining microorganisms: molecular aspects and applications in biotechnology and bioremediation. In A. Singh, R. C. Kuhad, & O. P. Ward (Eds.) *Advances in applied bioremediation* (pp. 239–256). The Netherlands: Springer.

Jerez, C. A. (2007). Proteomics and metaproteomics applied to biomining microorganisms. In E. Donati, & W. Sand (Eds.) *Microbial processing of metal sulfides* (pp. 241–251). The Netherlands: Springer.

Lloyd, J. R., Anderson, R. T., & Macaskie, L. E. (2005). Bioremediation of metals and radionuclides. In R. M. Atlas, & J. C. Philp (Eds.) *Bioremediation: Applied microbial solutions for real-world environmental cleanup* (pp. 293–317). Washington, DC: ASM Press.

Lo, I., Denef, V. J., VerBerkmoes, N. C., Shah, M. B., Goltsman, D., DiBartolo, G., *et al.* (2007). Strain-resolved community proteomics reveals recombining genomes of acidophilic bacteria. *Nature*, *446*, 537–541.

Lovley, D. K. (2000). Fe(III) and Mn(IV) reduction. In D. K. Derek (Ed.) *Environmental microbe–metal interactions* (pp. 3–30). Washington, DC: ASM Press.

Navarro, C. A., Orellana, L. H., Mauriaca, C., & Jerez, C. A. (2009). Transcriptional and functional studies of *Acidithiobacillus ferrooxidans* genes related to survival in the presence of copper. *Applied and Environmental Microbiology*, *75*, 6102–6109.

Nealson, K. H., Belz, A., & McKee, B. (2002). Breathing metals as a way of life: Geobiology in action. *Antonie van Leeuwenhoek*, *81*, 215–222.

Olson, G. J., Brierley, J. A., & Brierley, C. L. (2003). Bioleaching review part B: Progress in bioleaching: Applications of microbial processes by the mineral industries. *Applied Microbiology and Biotechnology*, *63*, 249–257.

Parro, V., & Moreno-Paz, M. (2003). Gene function analysis in environmental isolates: The nif regulon of the strict iron oxidizing bacterium *Leptospirillum ferrooxidans*. *Procceedings of the National Academy of Sciences of the United States of America*, *100*, 7883–7888.

Quatrini, R., Appia-Ayme, C., Denis, Y., *et al.* (2006). Insights into the iron and sulfur energetic metabolism of *Acidithiobacillus ferrooxidans* by microarray transcriptome profiling. *Hydrometallurgy*, *83*, 263–272.

Ram, R. J., VerBerkmoes, N. C., Thelen, M. P., *et al.* (2005). Community proteomics of a natural microbial biofilm. *Science*, *308*, 1915–1920.

Rawlings, D. E. (2005). Characteristics and adaptability of iron-and sulfur-oxidizing microorganisms used for the recovery of metals from minerals and their concentrates. *Microbial Cell Factories*, *4*, 13.

Rohwerder, T., Gehrke, T., Kinzler, K., & Sand, W. (2003). Bioleaching review part A: Progress in bioleaching: Fundamentals and mechanisms of bacterial metal sulfide oxidation. *Applied Microbiology and Biotechnology*, *63*, 239–248.

Siezen, R. J., & Wilson, G. (2009). Bioleaching genomics. *Microbial Biotechnology*, *2*, 297–303.

Tyson, G. W., Chapman, J., Hugenholtz, P., *et al.* (2004). Community structure and metabolism through reconstruction of microbial genomes from the environment. *Nature*, *488*, 37–43.

Valdes, J., Pedroso, I., Quatrini, R., & Holmes, D. S. (2008). Comparative genome analysis of *Acidithiobacillus ferrooxidans*, *A. thiooxidans* and *A. caldus*: insights into their metabolism and ecophysiology. *Hydrometallurgy*, *94*, 180–184.

Valenzuela, L., Chi, A., Beard, S., *et al.* (2006). Genomics, metagenomics and proteomics in biomining microorganisms. *Biotechnology Advances*, *24*, 197–211.

van Zyl, L. J., van Munster, J. M., & Rawlings, D. E. (2008). Construction of arsB and tetH mutants of the sulfur-oxidizing bacterium *Acidithiobacillus caldus* by marker exchange. *Applied and Environmental Microbiology*, *74*, 5686–5694.

Watling, H. R. (2006). The bioleaching of disulphide minerals with emphasis on copper sulphides – A review. *Hydrometallurgy*, *84*, 81–108.

RELEVANT WEBSITE

http://cmr.jcvi.org. – Comprehensive Microbial Resource.

Chapter 44

Industrial Water Treatment

A.R. Bielefeldt
University of Colorado – Boulder, Boulder, CO, USA

Chapter Outline

ABBREVIATIONS

AF Anaerobic filters
AMW Apparent molecular weights
AS Activated sludge
ASB Aerated stabilization basins
BAFs Biological aerated filters
BOD Biological oxygen demand
COD Chemical oxygen demand
CSTR Completely stirred tank reactor
DGGE Denaturing gradient gel electrophoresis
EGSB Expanded granular sludge bed
EPA Environmental Protection Agency
FAME Fatty acid methyl ester
FB Fluidized bed
FC Fecal coliforms
FISH Fluorescence *in situ* hybridization
FWS Free water surface
GSS Gas–solids separators
HRT Hydraulic residence time
IC Internal circulation
IFFAS Integrated fixed film activated sludge
JLRs Jet loop reactors
MBBR Moving-bed biofilm reactor
MBR Membrane bioreactors
MF Microfiltration
NOB Nitrite-oxidizing bacteria
PCR Polymerase chain reaction
RBC Rotating biological contactor
SALR Surface area loading rates
SBBR Sequencing batch biofilm reactor
SBR Sequencing batch reactor
SRT Solids residence time
SSF Subsurface flow
TOC Total organic carbon
TRI Toxics Release Inventory
UASB Upflow anaerobic sludge blanket
UF Ultrafiltration
WET Whole effluent toxicity

DEFINING STATEMENT

An overview of microbiological methods to treat industrial wastewaters is provided. The range of characteristics of common industrial wastewaters is provided. Commonly used aerobic, anaerobic, coupled, and wetland treatment processes are described. Case studies from full-scale systems are highlighted, along with key emerging results from laboratory and pilot-scale studies.

Topics in Ecological and Environmental Microbiology.

BIOLOGICAL TREATMENT OF INDUSTRIAL WASTEWATER

Industrial wastewaters are generated from a wide variety of sources and have a broad diversity of chemical properties and constituents. In the United States, the Federal Environmental Protection Agency (EPA) has developed wastewater discharge standards for over 60 industries (40 CFR Chapter 1, Subchapter N Effluent Guidelines and Standards; http://www.epa.gov/.). There are also pretreatment standards that regulate the quality of wastewater discharged from industries into public sewers under the National Pretreatment Program (EPA-833-B-98-002). To achieve the wastewater characteristics required by the standards, more industries are treating their wastewater on-site by a variety of processes. Most of these processes are aimed at reducing the concentrations of total organics, total solids, and/or specific chemicals down to acceptable concentrations. Many treatment processes can be quite expensive, and biological processes are considered to be cost-effective alternatives when the primary contaminants of concern are biodegradable organics or nitrogen compounds. Treatment designs are continually evolving to provide greater treatment efficiency. Microbial methods that are used to treat industrial wastewaters must be more robust than traditional municipal wastewater treatment processes due to the greater variability in flow rates and chemical characteristics, and the potential presence of toxic or inhibitory compounds. More innovative approaches to achieve treatment objectives include coupling anaerobic and aerobic processes, and coupling biological and chemical treatment processes. Molecular methods are beginning to be used to characterize the microorganisms present in the wastewater and present in treatment reactors.

CHARACTERISTICS OF INDUSTRIAL WASTEWATER

Unlike municipal wastewater, the wastewater generated from different industrial activities has widely varying composition. Even within a single type of industry, specific processes and chemicals used to produce similar products can differ resulting in significant changes in wastewater characteristics over time. For example, the wastewater stream at a brewery is actually made up of a combination of wastes from different processes, including waste from malting, brewing, fermenting, yeast drying, conditioning, and packaging. Similarly, textile mill effluent may be a combination of bleaching, fiber dyeing, yarn dyeing, and rinsing wastewater streams. A sample of reported wastewater composition is provided in Table 1.

Industrial wastewaters may be caustic, with extreme pH of <2 or >12, which may be due to different cleaning agents used. Without buffering to more neutral conditions, these extreme pH values would be inhibitory to microorganisms. Some industrial wastewaters contain extremely high concentrations of organic compounds (measured as biological oxygen demand (BOD) or chemical oxygen demand (COD)); more than ten times the concentrations in municipal wastewater. High salt concentrations are also present in some types of industrial wastewaters. Nutrient limitations

TABLE 1 Examples of typical industrial wastewater characteristics

Wastewater Type	Average pH Range	Suspended Solids ($mg\,l^{-1}$)	BOD_5 ($mg\,l^{-1}$)	COD ($mg\,l^{-1}$)	TKN ($mg\,Nl^{-1}$)	Total P ($mg\,l^{-1}$)	Salt ($g\,l^{-1}$)
Brewery	3.3–7.6	500–3000	1400–2000	815–12,500	14–171	16–124	
Dairy milk-cheese plants	5.2–11.3	350–1082	709–10,000	189–20,000	14–450	37–78	0.5
Dairy parlour	2–11	100–300	166–477	470–820	25–45	17–21	0.05–0.7
Dying	8.2–12	56–70	140–840	70–3,200	27–42	5–7	
Food pickling	2.6–3	40–110	7000–8000	20,000–22,000	4–6	22–25	30–150
Metal working fluids	9		1500–11,400	5300–40,000	160–440	28–77	
Pulp and paper	6.6–10	21–1120	77–1150	100–3500	1–3	1–3	~0.05
Tannery	8–11	2070–4320	1000–7200	3500–13,500	250–1000	4–107	6–40
Textile mills	4.5–10.1	20–210	700–1650	1900–100,000	14–72	1–18	0.5–0.9
Winery	3.9–5.5	170–1400	210–8000	320–27,200	21–64	16–66	0.1–1
Municipal	6–8	100–350	110–400	250–1000	20–85	4–15	>0.5

are a potential problem for biological treatment of some industrial waste streams, including nitrogen (N) limitation in brewery wastewater, phosphorus (P) limitation in pulp and paper wastes including Kraft pulp and cellulose production wastewaters, and N and P limitation in winery and pickling wastewaters. In other cases, a significant excess of nutrients may be present, such as in the case of pectin factory wastewaters with 1280–2990 mg l $^{-1}$ nitrate-N and COD/N ratios of 7 to 12:1.

In addition to bulk chemical constituents, many industrial wastewaters also contain specific organic and inorganic compounds that are toxic. For example, tannery wastewaters may contain 50–100 mg l $^{-1}$ chromium. The US EPA tracks the industrial use, treatment, and release of more than 600 chemicals via the Toxics Release Inventory (TRI). These chemicals tend to fall into a few main classes: metals or inorganics, aromatic hydrocarbons, or chlorinated hydrocarbons. Microbial-based treatment is only a viable option for organic compounds and nutrients (N, P). Metals may change redox state during biological treatment, but are not destroyed and will be present in the effluent wastewater or associated with the sludge. Therefore, physical and chemical processes are generally used to treat metals in the wastewater. As such, further discussion will be limited to organic compounds. The top chemicals on the TRI list, in terms of quantity of treated on- or off-site (including air, water, and solids) are summarized in Table 2. Many of the most prevalent TRI compounds are relatively nontoxic. However, the uncontrolled release of these substances into sanitary sewers or the environment can disrupt treatment or ecology. Therefore, pretreatment or permits regulate the quantities of specific chemicals allowable in the waste discharged from an industrial facility, in addition to the more commonly regulated bulk parameters (COD, TSS, and total N, e.g.). Many industries have modified their processes to use more environmentally compatible (less toxic and more degradable) compounds. In addition to regulation of the concentration of specific compounds in industrial wastewater effluents, the aggregate toxicity of the wastewater may be regulated. This whole effluent toxicity (WET) is generally tested by short-term and chronic tests on aquatic life such as water fleas, shrimp, and fish.

TABLE 2 Industrially produced compounds on the US toxics release inventory

Chemical	TRI Rank[a]	Class[b]
Methanol	2	VOC
Ammonia	3	Inorganic
Ethylene	5	Organic
Toluene	7	VOC, AHC
Nitrate compounds	8	Inorganic
n-Hexane	14	VOC
Xylenes	16	VOC, AHC
Benzene	20	VOC, AHC

[a]*2004 TRI; rank by total weight treated on- or off-site in air, water, and/or solids.*
[b]*VOC, volatile organic compound; AHC, aromatic hydrocarbon.*

Regardless of the specific compounds and sources, biological treatment processes typically offer the lowest cost alternative for degrading organic compounds in the wastewater. Biological processes can also be used to remove nitrogen and phosphorus. Nitrogen can be removed from the water via the following processes: organic-N is mineralized by aerobic heterotrophs to ammonia; ammonia-N is aerobically biotransformed to nitrate-N; and under anaerobic conditions, heterotrophic bacteria degrade nitrate-N to N_2, which equilibrates out of the water into the atmosphere. Phosphorus can be removed from the wastewater via accumulation in particular types of bacteria. The processes used to biologically treat industrial wastewaters are often similar to those that are widely used for municipal wastewater. However, the composition range of industrial wastewaters often requires a somewhat more engineered approach to the wastewater treatment. Many industrial wastewater treatment facilities include pre- and post treatment by various physical and chemical methods.

In addition to biodegradation of waste components, a variety of microorganisms may be present in the wastewater itself. Pathogenic species are of concern due to regulatory limits. The removal of pathogens during wastewater treatment is usually tracked using "indicator organisms" such as total coliform or fecal coliform bacteria, which are generally associated with the gut of mammals. Other microorganisms in the wastewater are of interest because they can contribute to or compete against the waste-degrading bacteria. For example, winery and dairy wastewater typically contain significant quantities of yeast and bacteria, which have been found represented as significant populations in the biological treatment units. In addition, there is an emerging concern that microorganisms in the wastewater may carry antibiotic-resistant genes. In particular, antibiotic-resistant bacteria have been found associated with animal wastes, and as such could be present in water used to washout dairy stalls, and so on. Although the animals are fed different antibiotics than are used for humans, the genes associated with resistance to animal antibiotics may also impart resistance to the antibiotics used for humans.

The use of molecular methods to characterize bacterial populations has been gaining popularity, although these methods have as yet been only sparingly applied to characterize biological treatment processes for industrial wastewaters. These approaches may provide tools to better understand the microbiology of a treatment system, particularly under upset conditions. Examples of molecular methods include those targeting bacterial cell membrane components and genetic material. Fatty acid methyl ester (FAME) is a method that extracts and quantifies the fatty acid components of cellular membranes, generating a profile or "fingerprint" of the microbial community. Denaturing gradient gel electrophoresis (DGGE) creates a community fingerprint based on the stage at which DNA melts in the presence of a denaturing agent into fragments; the fragments migrate through an electrophoresis gel to different extents creating different banding patterns. By comparing band location and intensity, variability in the structure of a microbial community can be determined. Further information can be gathered by removing the DNA from specific bands of interest and sequencing it to identify specific bacteria. These methods have been used to characterize microbial diversity in full-scale industrial wastewater treatment plants, and varying levels of species richness and diversity have been found to be associated with different processes.

Genetic methods can target microbial DNA or RNA with probes to identify specific species or groups of bacteria. The so-called full-cycle 16S rRNA approach has the potential to identify previously uncharacterized microbial populations in samples from natural and engineered environments. First, genomic DNA is extracted from all members of the microbial community, followed by 16S rDNA gene amplification by the polymerase chain reaction (PCR), cloning, and finally, sequencing of 16S rDNA gene fragments. These sequences are then compared with a database to identify the bacteria and/or their closest relatives. This method is therefore able to determine the diversity of the members of the microbial community in a treatment system. Using the sequence information available from the clone library, 16S rRNA-targeted oligonucleotide hybridization probes can be developed. These probes can be used for fluorescence *in situ* hybridization (FISH). Examples of studies that have used molecular methods to characterize the microbiology at full-scale industrial wastewater treatment facilities are provided in Table 3.

AEROBIC

In general, aerobic biodegradation processes of organic compounds are faster than anaerobic. Aerobic treatment processes require the supply of oxygen, and result in oxidation of organics to carbon dioxide and biomass. The biomass or "sludge" is a secondary by-product that may require further treatment. The oxygen supply can be expensive; it is usually supplied by blowers and aeration equipment, involving an energy input. This chapter covers typical aerobic treatment reactors used for industrial wastewater treatment, and highlights some specific examples of wastewater treatment by each of the most common reactor types: activated sludge (AS), sequencing batch reactor (SBR), rotating biological contactor (RBC), moving-bed biofilm reactor (MBBR), and submerged membrane bioreactors (MBR). In order for processes using suspended bacterial growth (AS, SBR) to work optimally, biomass must be "settleable" in the form of flocs so that high biomass concentrations (and long solids retention times) can be maintained in the system. Alternatively, a membrane can be used to retain biomass in the system. Biomass can also grow attached to an inert media, such as in an RBC or biotrickling filter. The wastewater to be treated flows over the biofilms on the media, and leaves the reactor. These reactors, therefore, have the ability to retain slow growing bacteria. However, too much biomass will eventually clog the media. Newer treatment systems are hybrids of attached and suspended growth, such as MBBR.

Activated Sludge

AS processes are among the most common biological methods used to treat domestic wastewater in the United States. The process is also frequently used to treat various

TABLE 3 Examples of molecular methods used to characterize microbiology at industrial wastewater treatment plants

Wastewater Type (Treatment)	Methods Used	Reference
Dairy food processing; activated sludge	Full-cycle 16S rRNA	Oerther et al. (2002)
Dairy; wetland	DGGE, PCR	Ibekwe et al. (2003)
Fibers, plastics, chemical manufacturing; activated sludge	Real-time PCR Competitive PCR	Robinson et al. (2003) and Dionisi et al. (2002)
Pharmaceutical; activated sludge	DGGE, rDNA sequence analysis	LaPara et al. (2000)
Rendering; activated sludge with nitrification and denitrication (AS-N)	Full-cycle 16S rRNA DGGE	Juretschko et al. (2002) and Kreuzinger et al. (2003)
Tannery and fish processing; AS-N	FISH	Moussa et al. (2006)
Slaughterhouse; anaerobic SBR and anaerobic static filter	FAME	Mach and Ellis (2002)

types of industrial wastewaters. Conventional AS has an aeration basin where oxygen is supplied and the contents of the reactor are mixed by this aeration process, followed by clarifiers where biomass solids are removed by gravity settling. The settled biomass is generally returned to the beginning of the aeration basin as "return activated sludge." This results in an average biomass retention time (so-called solids residence time (SRT)), which is significantly longer than the hydraulic residence time (HRT). Because the treatment performance is dictated by maintaining high biomass concentrations in the aeration basin due to biomass recycle, it is important that the biomass forms aggregates termed "floc" that have good settling properties. When the biomass does not settle well, reactor upsets occur and treatment performance usually decreases. Molecular methods are being used to determine which bacteria might be linked with poor biomass settling.

Another important operational consideration is the amount of sludge wasted from the system; this stream of concentrated biosolids is a waste that must be treated further (such as per the Part 503 Biosolids regulations in the United States). Frequently, these solids are treated and used as soil amendments. However, the sorption of toxic compounds, particularly metals, to the biomass can limit the safe uses of these biosolids.

The aerobic bacteria in AS systems biodegrade organic compounds, with a high degree of mineralization to carbon dioxide typically expected. The amount of oxygen that must be supplied increases with increased organic loading. When the total concentration of carbonaceous chemicals in the wastewater is measured as BOD or COD, the approximate amount of oxygen that must be supplied to the system is known. Oxygen supply involves the bulk of the operational costs in these systems.

Nitrogen and phosphorus must be present at sufficient quantities in wastewater relative to organic loading to allow heterotrophic bacteria to assimilate the carbon and grow in an uninhibited manner. Many industrial wastewaters are deficient in N and P relative to carbon, requiring the addition of supplemental nutrients to meet cellular demand. Examples of nutrient-deficient industrial wastewaters can be found in Table 1; nitrogen will generally be deficient if the COD/TKN ratio is greater than 35:1, and phosphorus limiting when COD/P ratio is greater than 300:1. Thus, the brewery, food pickling, textile, and winery wastewaters are commonly N limited, while the metal working fluids, food pickling wastewater, and winery wastewaters are often P limited.

While most AS processes are aerated via fine bubble aeration, jet loop systems are an alternative method for supplying oxygen and mixing. Jet loop reactors (JLRs) are vertical systems that claim to achieve higher oxygen transfer at lower energy than conventional aeration. JLRs have been studied at a pilot scale for treatment of winery wastewater, olive oil wastewater, and brewery wastewater. The systems have high velocity of liquid exiting the venturi nozzle, which could exert selective pressure on the microbiology of the system via a combination of physical shear stress and temperature effects. Specifically, filamentous fungi and bacteria have not been reported in JLRs. However, these systems do not appear widely applied in full-scale industrial wastewater treatment.

In addition to removing carbon compounds efficiently via biodegradation, AS processes have also been used to remove nutrients including nitrogen and phosphorus via sequential processes. Aerobic autotrophic bacteria can nitrify ammonia to nitrite (so-called ammonia-oxidizing bacteria (AOB); typically β-Proteobacteria) to nitrate (so-called nitrite-oxidizing bacteria (NOB); such as *Nitrospira* spp.). Then under anoxic conditions, heterotrophic bacteria can denitrify nitrate to nitrogen gas (N_2). Sequential anoxic – aerobic processes with a "single sludge" circulated/recirculated through the reactors in series are generally used for N-removal from wastewaters, combining the benefits of dentrification in the presence of sufficient organic carbon with nitrification under lower carbon loading. Some biological phosphorus removal is also possible by starting the single sludge system with an anaerobic selector; the sequential anaerobic/aerobic processes encourage intracellular accumulation of P-deposits within the bacteria, such that biosolids waste from the system removes soluble P from the wastewater.

Due to the more complex microbiology that is needed to achieve these multiple goals in wastewater treatment (organic removal by aerobic heterotrophs, nitrification by aerobic autotrophs, and denitrification by anoxic heterotrophs), molecular methods have been used to characterize the microbial populations present in these treatment systems. These methods can be used to help determine the source of treatment problems such as incomplete nitrification due to toxicity or extended anaerobic periods that inhibit nitrifiers (see examples in Table 3).

Aerated Stabilization Basins(ASB)

ASB are a potentially cost-effective alternative to AS treatment of industrial wastewater. ASB systems have typically lower capital, energy, and operation and maintenance costs than comparable AS systems. In particular, pulp and paper industries appear to be selecting ASBs for wastewater treatment over AS systems. Pulp and paper industry wastewater contains variable-quantity contributions from four to five major process areas with significantly different wastewater chemistries, a wide range of apparent molecular weights (AMW) for organic wastewater components, and a brown color due to high-AMW dissolved lignin. Because AS systems do not handle shock loads efficiently, the installation of flow equalization, neutralization, and sedimentation units prior to the AS unit is common. Mechanically ASB represent about half of the lagoon systems and have the best performance of the lagoon types. There are two classes of mechanically ASB: completely mixed and partially mixed. The large majority of ASBs in the pulp and paper industry are partially mixed. Mixers in the systems also achieve aeration. The mixers used include jet aerators, surface aerators, static aerators, and submerged fine bubble diffusers. In the ASB process, the HRT and SRT are the same at ~5–15 days. For pulp and paper wastewater treatment, supplemental nutrient addition is needed, but at lower rates than in AS systems. Properly operated ASB systems can reach comparable effluent BOD and nutrient concentrations to AS systems. ASB systems are more forgiving in responding to influent quality and flow variations and are simpler to operate. Settled secondary solids in an ASB system reduce in volume through hydrolysis to typically 10% concentrations. At periods up to 20 years the accumulated solids must be dredged, dewatered, and disposed. ASBs, because of their longer hydraulic retention times, tend to have their effluent wastewater temperature approach ambient air conditions. This is desirable during summer conditions, but can be a treatment problem in cold climates. Loss of an insulating foam cover is a greater problem for ASB systems. ASB systems typically require supplemental cooling equipment to attenuate the high influent temperatures of the wastewater. A naturally aerated "polishing" pond sometimes follows ASBs.

Sequencing Batch Reactors

In SBRs, the treatment and settling of biomass are done within a single tank, unlike AS reactors. This may result in a smaller footprint overall than the AS process. The technology is becoming more popular in Germany, with 51 SBR plants for industrial wastewater treatment in operation; however, SBRs yet represent less than 1% of the German wastewater treatment plants. A review of 29 full-scale SBR plants treating industrial wastewater in Turkey was presented by Artan et al. (2003). SBRs are used for food processing, pharmaceutical, textile, and other wastewaters treating 12–4800 $m^3\ d^{-1}$ of flow, inlet COD 600–12,000 $mg\ l^{-1}$, cycle times of 6–168 h, HRT 12–200 h, and recycle ratios (carry-over volume to new feed volume) of 0.2–4.8. These plants achieved effluent COD of 50–2000 $mg\ l^{-1}$, representing 77–98.5% removal. The advantage of the SBR systems was largely for high strength and variable quality wastewaters. SBRs are generally more automated in their operation than AS systems. SBR effluents generally contain less SS than AS systems. SBRs are somewhat less common at large wastewater plants due to the fairly high area required for the multiple reactors needed to treat continuous inflow. The operation of the SBR can be varied within a treatment cycle to enable different metabolic processes (such as anaerobic or aerobic) to predominate. This makes it particularly suited to simultaneous carbon and nitrogen removal.

Tannery wastewater is an example of a strong wastewater with high carbon and TKN. High chloride, sulfide, and chromium in the wastewater are potentially inhibitory to heterotrophic and nitrifying bacteria. Lab and pilot-scale studies have found comparable SBR performance to the full-scale AS systems on-site. These systems were operated with supplemental P addition, cycle times of 8–24 h, HRT 0.8–3 days, and SRT 15–17 days. Full-scale SBR systems have been used to treat brewery, winery, and pulp mill wastewaters and tanker truck wash water. These systems had 12–24-h cycles and achieved >93% COD removal.

Specific industrial chemicals have been shown to upset biological treatment in SBR systems. This is a disadvantage compared to complete mixed AS systems, which quickly dilute toxic concentrations of inlet spikes, versus the batch treatment where inlet concentrations are only attenuated by the carry-over biomass from the previous treatment cycle. Nitrification and low-temperature conditions seem particularly sensitive to these problems.

Rotating Biological Contactor

RBC treatment uses partially submerged rotating disks on which biomass can grow. As the disks rotate, the attached bacteria are sequentially exposed to the air, where they acquire oxygen and wastewater. Additional bacteria are also suspended in the wastewater treatment reactor. Energy is consumed by the large motor required to turn the biodisks, which can range in size from 3.5 to 5 m in diameter. Rotational speeds are slow, at about 1 rpm. RBCs are fairly uncommon in industrial wastewater treatment, and have also largely fallen out of use in municipal wastewater treatment. A search of peer-reviewed literature found only five citations to full-scale RBC and industrial wastewater treatment. All except one citation were RBCs treating combined dairy wastewater with municipal wastewater, and all citations were before 1997. Supplemental aeration in the first and second stages that were often overloaded significantly improved the RBC treatment performance and were more robust against changing organic loading rates due to variable inlet wastewater quality. Winery wastewater during grape harvest has been treated in pilot-scale RBC reactors. Yeast species that are naturally associated with grapes (*Saccharomyces cerevisiae*, *Candida intermedia*, and so on) were found to be codominant in the attached biofilms with bacteria, indicating the potential importance of inlet microorganisms on the treatment.

Hybrid Suspended and Attached Growth Systems

Hybrid systems combine some characteristics of both suspended growth and attached growth biotreatment systems. There are three main types of hybrid treatment systems used for industrial wastewater treatment: the MBBR, integrated fixed film activated sludge (IFFAS), and biological aerated filters (BAFs). In each case, a few key companies have patented versions of the technology and market specific varieties. In general, all of the hybrid systems contained a solid media inside the traditional AS basin. This media is retained in the reactor and provides a support matrix on which attached biofilms form. Therefore, higher organic loadings can be treated in the same overall reactor volume due to the higher overall biomass in the reactor. This allows a smaller footprint for the wastewater treatment facility. In addition, the process is less reliant on the performance of the secondary clarifier and upsets in biomass sludge settling, because the attached biomass is retained regardless. In addition, specialized subpopulations of microorganisms can be retained in the fixed films rather than washed out of the reactor, and therefore, better able to respond to intermittent loading of various compounds.

Moving-bed biofilm reactor

Since the early 1990s, MBBRs have been in development and use, beginning in Norway and Europe, and more recently, expanding to applications in the United States. There are more than 400 large full-scale operations in 22 different countries. For industrial wastewater treatment, MBBR systems have been used to treat food industry, dairy, flavor production, slaughterhouse, refinery, and pulp and paper wastewaters. Buoyant plastic media is aerated and suspended in a tank. Polyethylene Kaldnes media (2–50 mm long × 9–64 mm diameter) is one example, and has been the most widely reported for use at full-scale industrial treatment facilities. At typical carrier loading, this results in a specific area of biofilm growth of about 350 $m^2\,m^{-3}$ reactor volume. A sieve at the reactor outlet (with approx. 5–7 mm openings) is used to retain the media in the reactor. Alternatively, the LINPOR system uses foam cubes (~1.5 cm) as packing media. Circulation in the tanks by aeration or submerged mixers maintains movement of the media, and this agitation dislodges excess biomass from the carrier media and prevents clogging. Waste sludge sloughed from the media is collected in secondary clarifiers, similar to a trickling filter. This sludge is not recirculated. Oxic zones can remove BOD and ammonia, and anoxic zones can remove nitrate. Phosphorus removal will not occur in the bioreactor.

One advantage of these systems is that the media can be added into existing AS basins to increase the biomass in the system. Municipal wastewater treatment systems have increased the total biomass in retrofitted AS basins by as much as a factor of 2. Most of the biomass in an MBBR (>90%) is attached to the media rather than suspended in the liquid. The MBBR system is also less prone to process upsets from poorly settling biomass than AS systems, since much of the total biomass in the system is attached on the retained media rather than being returned to the reactor as underflow from a secondary clarifier. LINPOR reports full-scale use of their technology for primarily municipal wastewaters, with one paper industry installation (9000 $m^3\,d^{-1}$, inlet COD = 400 $mg\,l^{-1}$, and effluent >70 $mg\,l^{-1}$) and a coke oven installation (125,000 $m^3\,d^{-1}$, COD = 280 $mg\,l^{-1}$ and N = 50 $mg\,l^{-1}$).

In some cases, MBBRs have been used as roughing filters ahead of downstream treatment processes, including conventional AS. In this configuration, significantly higher loadings can be applied since high removal efficiency of carbon and/or nitrogen is not required. For example, surface area loading rates (SALR) of 20–53 g COD $m^{-2}\,d^{-1}$ and 10–25 g BOD $m^{-2}\,d^{-1}$ have been reported (~2–5 kg COD $m^{-3}\,d^{-1}$). Given the observed removal efficiency, this translates to removal rates of 17–42 g COD $m^{-2}\,d^{-1}$ and 8–18 g BOD $m^{-2}\,d^{-1}$. These values are expected to vary widely based on the amount of soluble biodegradable organic matter in the total COD. Nitrogen removal up to 0.83 g NH_3-N $m^{-2}\,d^{-1}$ has been reported.

Another variation on MBBR technology is to add the plastic media into a SBR. This so-called sequencing batch biofilm reactor (SBBR) combines the advantages of an SBR with the biomass retention properties of attached biofilms. Most of the published studies on SBBRs are from laboratory-scale experiments with various synthetic wastewaters. A few pilot-scale studies have been reported with industrial wastewater, with reported removal rates of 0.3- to 9.1-kg COD $m^{-3}\,d^{-1}$.

Integrated fixed film activated sludge

IFFAS is a hybrid process in which buoyant media is aerated and suspended along with mixed liquor in AS basins. A variety of medium types can be used. Rope-like Ringlace was developed in Japan in the 1970s and has been used in more than 400 locations worldwide. Biofilms develop on the medium, increasing the overall biomass concentration in the AS basins. Frequently, this medium is added to existing AS basins to retrofit the plant to treat greater flow rates and/or achieve better removal of BOD and nitrification. Incorporation of the medium into an AS basin in sequence with traditional anaerobic and anoxic stages can achieve denitrification and biological phosphorus removal. Minimal peer-reviewed studies on IFFAS have been published. Most of the applications appear to be associated with municipal wastewater treatment.

Biological aerated filters

In a BAF, aerated wastewater flows through medium which acts as a filter. Most large systems are for municipal wastewater treatment, although there are at least 18 full-scale systems treating pulp and paper mill effluent and a number of smaller plants in Japan treating various industrial wastewaters. These systems have also been used for pretreatment of industrial wastewaters from food production, carpet dying, and pharmaceuticals to reduce high organic loads, with reported surface loadings of 15–30 g COD m^{-2} d^{-1} and volumetric loadings of ~4.5 kg COD m^{-3} d^{-1}. Higher organic loadings can be applied compared to conventional AS systems. These systems have also been reported to be more robust to changes in influent wastewater characteristics. No secondary clarifier is required. Phosphorus removal cannot be directly incorporated into the bioreactor. Two companies market the majority of the BAF processes.

Kruger/ John Meunier, Inc.'s Biostyr uses plastic beads called BioStyrene (3.6 mm diameter; specific biofilm surface area ~276 $m^2 m^{-3}$). Reported typical nitrification and denitrification rates in municipal wastewater treatment are about 0.9–1.2 kg N m^{-3} d^{-1}, achieving <5–20 mg l^{-1} effluent N. Tertiary denitrification rates are higher at 3 kg N m^{-3} media d^{-1} achieving <2 mg l^{-1} effluent N. In highly loaded facilities with ~0.5 d SRT, effluent BOD is <20 mg l^{-1}. Gravity backwashing about once a day is important due to headloss accumulation. This type of system has been used to treat dairy wastewater at a loading rate of up to 12 kg COD m^{-3} d^{-1}. The Biostyr system was also used to retrofit an AS plant to achieve nitrification at water temperatures as low as 7 °C. The wastewater is first treated in AS basins, such that the BAF unit received ~44 mg l^{-1} BOD, 58 mg l^{-1} SS, and 32 mg l^{-1} NH_4-N loadings.

Infilco's BIOFOR BAF uses expanded clay media with a mean diameter of 2.7 mm. In lab- and pilot-scale tests it has been determined that the typical packed-bed macroporosity is ~0.4, with 1333 m^2 of carrier interfacial surface area per m^3 of reactor volume. The media bed depth is typically 3 m. In municipal wastewater treatment applications, the system can achieve effluent inorganic N less than 2 mg l^{-1} and 4–5 mg l^{-1} total N. Backwashing normally occurs once every 24–48 h and can be controlled based on elapsed time and/ or accumulated headloss across the filter bed. BIOFOR systems are in operation at over 100 locations worldwide, treating from 380 to 416,000 m^{-3} d^{-1} each. This type of media was used in a pilot study to treat citrus industry wastewater with inlet COD = 2400–5800 mg l^{-1}, BOD_5 = 1580 = 3900 mg l^{-1}, and pH 6–7.2. To reliably meet the effluent standard of 1100 mg l^{-1} effluent COD, volumetric load less than 72 kg COD m^{-3} d^{-1} and hydraulic load less than 0.98 m h^{-1} was recommended. To achieve less than 600 mg l^{-1} effluent COD, a maximum load of 20 kg COD m^{-3} d^{-1} and hydraulic load of 0.36 m h^{-1} was recommended. Cocurrent flow was more effective than countercurrent flow.

Another alternative media used is 2–3 mm diameter sand. This process has been used at a full-scale treatment facility treating up to 654 m^3 d^{-1} of ammonia liquor from a coke plant. The six aerobic upflow filters are each 2.9 m wide, 6.5 m long, and 5.5 m high, with a total of 42.5 m^3 min^{-1} of air added. Removal efficiencies are greater than 99, 78, and 99.9% for thiocyanate, ammonium, and phenols, respectively.

Membrane Bioreactors

MBRs are rapidly gaining popularity for industrial wastewater treatment. In North America, the first full-scale industrial treatment MBRs was implemented in 1991. In Europe, more than 100 full-scale MBR plants for industrial wastewater treatment are in operation, compared with only 39 systems in North America. MBR have the advantage of retaining biomass in the treatment reactor via the filtration of the effluent wastewater through the membrane. Most commonly the membranes are in the ultrafiltration (UF) range, with pore sizes 0.1–0.4 μm. In some cases, microfiltration (MF) membranes are used. Originally, the systems were operated with external membranes following the AS basin, replacing conventional secondary clarifiers. Biomass was returned to the reactor using a recirculation loop. In the newer configuration, the membranes are submerged directly in the bioreactor. The biomass in these systems is both suspended in the liquid and attached to the surfaces of the membranes. The systems therefore combine the aeration basin and clarifier common to AS systems in a single basin. Very high biomass concentrations can be retained in the system, with 12–15 g l^{-1} typical and as high as 30 g l^{-1}, without worrying about upsets from poorly settling flocs. Very long SRT is usually maintained in the system, and overall sludge wasting is generally less. In North America, zenon is the most popular manufacturer of submerged hollow-fiber membranes for industrial applications.

The primary disadvantage of the MBR system is the higher overall operating cost due to energy consumption, largely for pulling water through the membranes and the frequent cleaning cycles to minimize membrane fouling. Cornel and Krause (2006) noted that 0.4–4 kWh m^{-3} treated wastewater is typically needed for pump energy. For aerobic treatment systems, oxygen supply is also required. This can be provided via conventional bubble aeration or using submerged membranes for aeration. The energy needed for oxygen supply is typically on the order of 0.3 kWh m^{-3} for municipal wastewater with higher amounts needed for wastes with greater COD.

A key concern with MBRs is avoiding membrane fouling. Fouling reduces the flux through the membranes and increases the energy consumption. To avoid problems, the influents wastewater is generally screened first, often through a sieve of ~0.5 mm. Common approaches to minimize fouling with submerged hollow fiber membranes include: *in situ* backwash with permeate every few minutes on an automatic schedule; chemically enhanced backwash on an automatic schedule approximately every day; maintenance cleaning with acids on a weekly basis, and infrequent (1–2*x*/year) *ex situ* intensive cleaning.

Most of the industrial wastewater treatment applications for MBRs in Europe are used for pharmaceutical, chemical, and food industry wastewaters, treating 50–1840 $m^3 d^{-1}$. In North America, MBRs are used primarily for food and beverage processing and chemical plant wastewater, treating 19–18,925 $m^3 d^{-1}$. Case studies reported include a pharmaceutical/chemical industry wastewater treatment facility with three AS tanks followed by four zenon hollow fiber immersed external membrane units with 15,840 m^2 membrane surface. Inlet average COD and ammonia-N were 3500 $mg\,l^{-1}$ and 100 $mg\,l^{-1}$, respectively; effluent contained 300 and >1 $mg\,l^{-1}$, respectively. The averaged flux was 12.5 $l\,m^{-2}h^{-1}$ and MLSS was 12,000 $mg\,l^{-1}$. Another example is the full-scale MBR plant treating pharmaceutical wastewater in California. The operating liquid volume is approximately 147,000 gallons and the system is designed to treat 5 gallons per minute of wastewater at a feed total organic carbon (TOC) concentration of 14,000 $mg\,l^{-1}$. The system contains four Model 500 C Zeeweed modules (pore size, 0.03–0.04 μm) with a total membrane area of 186 m^2. The system is designed to operate at 10,000 $mg\,l^{-1}$ biomass as TSS in the reactor. To avoid fouling that significantly increases the vacuum required to draw the liquid across the membranes, the system automatically backpulses permeate plus 3–5 $mg\,l^{-1}$ sodium hypochlorite (bleach) for 60 s every 20 min. Once per day, a more vigorous cleaning procedure of maximum 20 min with ~200 $mg\,l^{-1}$ hypochlorite is conducted.

In a case study treating tannery wastewater, Reemtsma et al. (2002) found that the MBR with UF membranes did not improve the removal of poorly biodegradable polar organic compounds, such as naphthalene disulfonate, and benzothiazoles, compared to conventional AS treatment. Other laboratory studies have explored the suitability of MBR technology for chemical production, food production, paper mill, pharmaceutical, soap production, and various textile wastewaters with varying degrees of success.

Irreversible fouling of membrane surfaces is a significant problem. It is hypothesized that different microbial communities may be associated with irreversible clogging, and that different operating conditions can be used to select for microorganisms less associated with irreversible clogging. This is an active area of research.

ANAEROBIC

Anaerobic processes degrade organics in the absence of oxygen. A significant advantage is that the high energy required for aeration of aerobic treatment systems is not needed; a potential savings of ~1 $kWh\,kg^{-1}$ COD removed. Another notable feature is the potential production of methane, which can be captured and used as an energy source, at a theoretical maximum energy yield of 3.8 $kWh\,kg^{-1}$ COD removed. The biomass yield during organic degradation is also lower than aerobic processes, resulting in as much as 90% less by-product sludge. These systems can be used to encourage water reuse in factories. These factors indicate that anaerobic treatment may have significant advantages over aerobic processes and overall may be more sustainable for the environment.

Anaerobic treatment may be better suited to some waste types given the organic and nutrient concentrations present. The C/N/P ratio in typical winery wastewater is unbalanced for aerobic treatment and requires N and P addition; however, the 800:5:1 C/N/P ratio (BOD_5/N/P 333:3.3:1) is suitable for anaerobic treatment. The seasonal nature of the wastewater production at small wineries poses an additional challenge for anaerobic systems because they generally have longer start-up times due to lower cell yield. It has also been reported that the presence of a high concentration of sodium (such as 4600–14,000 $mg\,l^{-1}$ and 200–250 $mmol\,l^{-1}$) is inhibitory for anaerobic biological treatment.

VanLier conducted a review of anaerobic technologies and implemented full-scale industrial wastewater treatment. From 1998 to 2004, of 519 systems, the most popular were upflow anaerobic sludge blanket (UASB), internal circulation (IC), expanded granular sludge bed (EGSB), lagoon, hybrid, completely stirred tank reactor (CSTR), and anaerobic filters (AF) or fluidized-bed (FB) technologies at 36, 34%, 17%, 4%, 2%, 2%, 1%, and 1%, respectively. Note that this represented a shift to more granular sludge based and super high rate based systems (FB + EBSG + IC = 52%) compared to the 1383 anaerobic treatment technologies implemented from 1981 to1998 (super high rate = 8%). High rate anaerobic treatment was commonly applied to food processing wastewater from industries such as sugar, potato, distilleries, wineries, fruit juice, starch, beer, and soft drinks. Anaerobic biotreatment applications in paper mills and chemical wastewater are also increasing. In another survey paper, Austermann-Haun et al. (1999) noted that of the 125 full-scale methane reactors treating industrial wastewater in Germany; 43 were sludge blanket reactors, 33 fixed-film reactors, 11 CSTRs, and the rest were various hybrid systems. Examples of anaerobic treatment of industrial wastewaters in various anaerobic reactors are provided in Table 4. Note that AF and FB systems are attached growth while the remainder are suspended growth, with hybrid systems containing a combination of attached and suspended biomass.

Upflow Anaerobic Sludge Blanket

The most widely applied form of anaerobic wastewater treatment is the UASB. In these systems, the biomass tends to form granular agglomerates that have very good settling properties. The systems tend to work best with high strength wastewaters and at higher temperatures. The conventional UASB process can accommodate a fairly high organic

TABLE 4 Examples of anaerobic treatment of industrial wastewaters

WW type	Reactor	Scale (Volume)	HRT (days)	OLR (g COD l^{-1} d^{-1})	Supplements?	% COD removal	Notes	Reference
Brewery	UASB	Full 500 m^3	1	6	N, P, soda ash for pH	40–75	50% TS removal	Parawira et al. (2005)
Brewery	UASB	Pilot	3.5	7		95	0.30 l methane yield g^{-1} COD	Oktem and Tufekci (2006)
Brewery	EGSB-AF	Lab; 3.4 l	0.75	4.5		93–95	0.28 l CH_4 g^{-1} COD d^{-1}	Connaughton et al. (2006)
Dairy	AFFBSBRUASB	Lab; 2–14 l Lab; 0.6–3 l Lab; 3.5 l Lab; 1–8 l	1–60.3–1.33.20.2–2	3–91–3.86.259–24		94–9880–929678–98		Omil et al. (2003)
Dairy	UASB	Full; 4000 m^3	8	0.55		63		Omil et al. (2003)
Dairy parlor	UASB	Lab; 12 l	3.5	0.2		51–73	10 °C lower COD removal; followed by septic tank	Luostarinen and Rintala (2005)
Distillery	UASB	Full scale 150 m^3 300 m^3	~2	15		>90	grain distillation caused floatings cum layer problem	Laubscher et al. (2001)
Food processing	MBR	Lab; 0.4 m^3	2–4	4.5		80–85	UF membranes; gas yield 0.14 l g^{-1}COD	He et al. (2005)
Pharmaceutical	CSTR	Lab; 5 l	0.5	13		13	44% acidification	Oktem et al. (2006)
Potato waste	UASB	Lab; 0.84 l	3.3	6.1		>90	0.23 l CH_4 g^{-1} COD	Parawira et al. (2006)
Textile sizing	UASB	Lab; 2.5 l	0.5	4.8		85		Spanjers (2006)
Winery	UASB	Lab; 2.6 l	1	7–16		60–85		Kalyuzhnyi et al. (2000)
Winery	UASB	Full;2@138 m^3	3	~7	N, P		Prestorage tank with some acidification; postaerobictmt w/two basins	Moeltta (2005)
Winery	AF	Full scale 4 m^3	1.2	4.6–11		88–98 overall	Pretreat in acidogenic reactor, follow with aerobic tmt	Moeltta (2005)
Winery	SBR	Lab; 5 l	2.2	8.6	N, P, NaOH	>98 sol		Ruiz et al. (2002)
Synthetic high salt ww	AF	Lab	2	5–6.2		>80	35 g l^{-1} NaCl, 35 C; higher salt or OLR failed	Riffat and Krongthamchat (2002)

loading rate, in the range of 5–15 kg COD m^{-3} d^{-1} for readily biodegradable wastewaters. UASB systems have been used to treat wastewater from alcohol distilleries, fruit juice, meat packing, paper mills, potato processing, slaughterhouses, and sugar processing plants. Problems sometimes noted with UASB systems include difficulties when treating wastewater containing high TSS content, foaming when treating high amounts of fats, lipids, and proteins; clogging of the sludge bed when treating fatty effluents; and biomass washout.

UAB reactors can be operated under a variety of temperature regimes ranging from psychrophilic (<20 °C), to mesophilic (35 °C), to thermophilic (55 °C). Mesophilic is the most common. External heating is generally needed to operate at mesophilic or thermophilic temperatures; this may be supplied by using the methane gas produced in the digestion. The high temperature of effluent water from the pulp and paper industry (50–60 °C) may be particularly suited to thermophilic anaerobic digestion, although at present only laboratory studies on these systems are known. Different bacterial species will generally survive at mesophilic versus thermophililc temperatures.

To improve the process performance of UASB reactors, modifications including expanded granular sludge bed (EGSB) and internal circulation (IC) have been made. EGSB is designed for a faster upflow velocity of water through the sludge bed (4–10 m h^{-1}) than traditional UASB reactors (0.5–2 m h^{-1}). This increased flux partially fluidizes the granular sludge bed, resulting in improved contact between the wastewater and the sludge. The IC process has two UASB compartments on top of each other. In the first, bottom stage, most of the biogas is produced. This gas is trapped in an internal gas hood, rises through a riser pipe to a gas liquid separator, and then liquid flows down a downcomer pipe to the first state. In this manner, the biomass production drives IC of the wastewater. The second top stage functions as biomass retention and polishing. The number of full-scale plants using these enhanced processes has been increasing year by year. In these EGSB and IC systems, higher organic loading rates (15–45 kg m^{-3} d^{-1}), tall reactors 12–20 m high, and high upflow velocities of 8–30 m h^{-1} are common. At higher organic loading rates, the higher intensity of biogas production tends to cause sludge washout from the reactor. This problem can be addressed using a compartmentalized UASB reactor, which is equipped with multiple numbers of gas–solids separators (GSS).

Anaerobic Sequencing Batch Reactor

Anaerobic sequencing batch reactors were first developed in the early 1990s to treat landfill leachate and high strength organic wastewater. Like aerobic SBRs, there is a treatment cycle of four discrete steps: fill, react, settle, and withdrawal. These cycles should be as frequent as possible, without idle time. The reactors are amenable to automatic process control, enabling operational flexibility. These systems have been used to treat piggery and slaughterhouse wastewater, winery wastewater, and landfill leachate.

Anaerobic MBR

Most of the literature citations of anaerobic MBRs are for domestic wastewater treatment. Cake formation is noted as a major cause of flux decline. This may result in irreversible clogging and flow membrane flux per headloss. One contributor to this problem may be that the selective pressure to maintain microbial granules is gone so that the biomass quickly degenerates into loose floc even if seeded with granular sludge material. Scaling due to high concentrations of carbonates may also be a problem. Lab-scale studies treating starch wastewater found that a membrane significantly improved the performance of the acid fermentation reactor but had little impact on the efficiency of the methane fermentation tank; other lab-scale studies have been conducted with brewery, fermentation, food, and pulp and paper wastewaters.

COUPLED ANAEROBIC/AEROBIC TREATMENT PROCESSES

By combining anaerobic and aerobic treatment processes, the overall treatment of industrial wastewater may consume less energy, more fully mineralize complex organic contaminants, and enable optimum nutrient removal. Typically, the anaerobic stage is first. This achieves organics removal without requiring energy to supply oxygen and a lower sludge yield. Useful by-products such as methane gas can be generated. The step can generally remove 80–90% of the organics. The second aerobic stage can polish the wastewater to meet stringent discharge limits for organics, and may also remove nitrogen and phosphorus. The specific reactor types selected for the anaerobic and aerobic stages can vary. Some examples are provided in Table 5. The UASB-cyclic AS systems for brewery wastewater treatment have been reported to reduce sludge production by a factor of 5 and lower energy consumption by a factor of 3 compared to existing two-stage aerobic treatment systems, with 11 full-scale systems reported.

At full-scale plants in Germany, treating wastewater from beet sugar factories, potato processing, breweries, and potato and wheat starch factories, the wastewater is anaerobically pretreated and then followed by an aerobic stage with nitrification and denitrification. In these cases, there is generally a by-pass of some fraction of the wastewater around the methane reactor to supply enough organic carbon for the denitrification stage. The authors recommend that 50–70% of the AS tank be used for denitrification.

Some recalcitrant organic compounds can only be mineralized by anaerobic and aerobic biodegradation processes in series. For example, azo dye decolorization can occur via

TABLE 5 Examples of sequential anaerobic:aerobic treatment of industrial wastewater

Wastewater Source	Scale; Q & OLR	Anaerobic Stage	Aerobic Stage	Reference
Brewery	Full; 24–417 $m^3 h^{-1}$ 2200–32,280 kg COD d^{-1}	UASB ~7 h HRT 9 kg COD $m^{-3} d^{-1}$	Cyclic AS ~20 hHRT 0.4 kg BOD $m^{-3} d^{-1}$	Gerards et al. (2005)
Dye containing Remazol Black 5	Lab; 5–25 kg COD $m^{-3} d^{-1}$	UASB 3–7 kg COD removed $m^{-3} d^{-1}$; 92–87% color removal	CSTR SRT 1.7 to 11 days resulted in 28–90% COD removal	Sponza *and* Isik (2002)
Milk quality control lab	Lab; 10.5 $g l^{-1}$ COD;5–6 kg COD $m^{-3} d^{-1}$	Anaer filter PVC packing, upflow;2 d HRT; 90% COD rem	SBR, 24 h cycle effluent<200 $mg l^{-1}$ COD, <10 $mg l^{-1}$ N	Omil et al. (2003)
Kraft mill	Full 5000 $m^3 h^{-1}$ 19,000 kg BOD d^{-1}	Anaerobic basin; 2 km^2; 3–1.2 m deep	Two partial mixed aerobic ponds	Starke (2001)
Pharmaceutical plant	Pilot; 9.7–20 $g l^{-1}$ COD	Baffled1.25–2.5 d HRT 4–10 kg COD removed $m^{-3} d^{-1}$	Biofilm airlift suspension 5–12.5 hHRT 3–8 mg COD removed $m^{-3} d^{-1}$	Zhou et al. (2006)
Slaughter-house	Pilot 0.14–0.5 $m^3 h^{-1}$	UASB 2–7 h HRT	RBC 5.3 g sBOD $m^{-2} d^{-1}$	Torkian et al. (2003)

anaerobic reduction or cleavage of the azo bond followed by mineralization of aromatics and amines under aerobic conditions. Therefore, these sequential systems have been studied for treatment of dye water, such as sequential UASB/AS systems.

NATURAL TREATMENT: WETLANDS

In addition to engineered treatment of industrial wastewaters in tanks and constructed basins, there are various examples of industrial treatment in lagoons and constructed wetlands. These processes are most often used in agricultural areas where large areas of land are available, rather than urban industrial settings. There are two predominant types of constructed wetlands: free water surface (FWS) and subsurface flow (SSF) wetlands. In the FWS wetland, also called surface flow wetlands, there is freestanding water over the soil bedding of the wetland. In contrast, the SSF wetland has a greater depth of granular media into which the plants root, and no water flows above the surface of the media. Typical depths of granular media are 1–1.5 m in SSF wetlands, although the actual water depth flowing through the bed may be much less (as low as 0.15 m reported). In both types of wetlands, good engineering design dictates that the wetland is constructed on a low permeability liner. The liner can be natural material (such as clay) or more commonly a geomembrane material such as high-density polyethylene. The liner prevents wastewater that has yet to be fully treated from leaching out of the bottom of the wetland and contaminating the underlying groundwater.

Both systems achieve removal of organics and nutrients by a combination of plants and associated microorganisms. The types of plants in the wetland are critical to the type of treatment efficiency achieved. The plants evapotranspirate water, uptake nitrogen and phosphorus, and can impact the redox conditions. Common plants in constructed treatment of wetlands include bulrush (Typha), grasses, and reeds. Excess plant growth over time may require harvesting. This has been shown to be true in water hyacinth ponds.

The redox conditions in a wetland range from aerobic to anaerobic. Oxygen is available near the FWS due to dissolution from the bulk atmosphere and near plant roots that excrete oxygen. Anaerobic conditions occur at depth and in the lower soil layers. Pathogens are also attenuated through the treatment process via filtration and natural die-off under environmental conditions and bacterial competition. Removal of fecal coliforms (FC) typically range from one to three orders of magnitude.

Unless a large amount of land is available to allow long residence times, these wetland treatment processes are used for polishing and nutrient removal after the bulk of organics and solids have been removed. Typical natural processes used for wastewater treatment before wetlands include solids settling and various types of lagoons (anaerobic, facultative, and aerobic). Solids removal is particularly important in SSF wetlands, where high solids loading may otherwise clog the porous media. Therefore, presettling is a common pretreatment. Even still, the inlet zone of SSF wetlands often contains coarse gravel to minimize clogging problems. The final effluent from wetlands is often reused for irrigation, agricultural "washing," or other uses. Newman et al. (2000) cite the presence of ten constructed wetlands for agricultural wastewater treatment in North America. Examples of wetland treatment systems are provided in Table 6. Note that residence times are normally determined by nonreactive tracer studies.

TABLE 6 Examples of wetland treatment of industrial wastewaters

Waste-water Source	Location (min.water temp.)	HRT (d)	Pre-treatment	Wetland type; total area (m^2); depth, m	Plants	OLR (g/m2/d)	% Treatment efficiency	Reference
Dairy	Arizona	4.5	Solid separation, anaer. pond, aerobic pond	SSF 5000; 1.5	Cattail, bulrush, reeds	12 BOD5	0 BOD5, 30 TSS, 23 TKN	Karpiscak et al. (1999)
Dairy	CA	7	Facultative pond	SSF 1200; 1	Reeds, bulrush	15 BOD	50 BOD, 40 COD, 87 SS, 26 TKN, 35 PO4-P	Ibekwe et al. (2003)
Dairy	Canada; 1.6 °C	45–52	None	FWS 100; 1–0.15	Cattails, grasses	450 BOD	62–99 BOD5, TSS, TP, NH4-N	Smith et al. (2006)
Dairy	Lab in India	4	None	FWS not reported; 0.1	Water hyacinth	6–20 BOD	86 BOD, COD 32 TSS; 58 TDS, 67 TN	Trivedy and Pattanshetty (2002)
Milk-house	CT 2.5 °C	>12 (ave 41)	None	FWS 400; actual ave. 0.3	Cattail or reed or bulrush	15 BOD5	76 BOD5, 90 TSS,28 TKN, 45 TP, 98 FC	Newman and Clausen (1997), Newman et al. (2000)
Batik Dye	Tanzania	1.2	None	FWS 3 @ 1.7 ea; 0.6	Cattail or cocoyam	11 COD	70 COD, 56 sulfate, 75 color	Mbuligwe (2005)

Poor treatment efficiency in colder winter months outside the plant growing season may be a problem. However, some treatment occurs even at air temperatures as low at −8 °C, water temperatures as low as 1.6 °C, and with ice on the top of the wetlands. It has been reported that more TSS, TP, TKN, ammonia, and nitrate/nitrite are removed during the plant growing season than during plant senescence. In fact, effluent ammonia-N may be higher than inlet wastewater during the winter due to plant senescence. Owing to nitrifying bacteria, it is expected that N removal will be more temperature sensitive than carbon removal. Settling and increased storage may account for much of the contaminant removal during winter months.

COUPLED BIOLOGICAL WITH CHEMICAL OR PHYSICAL TREATMENT

The coupling of biological treatment with physical or chemical treatment may improve the overall treatment of the wastewater. One of the most common approaches is to couple chemical oxidation with biological treatment. The most commonly used oxidant is ozone, with a few examples of Fenton's reagent and hydrogen peroxide use. Examples are summarized in Table 7. Ozone added into water causes both direct oxidation of organics with ozone and indirect oxidiation with the hydroxyl radicals that ozone forms. In Fenton's oxidation, iron serves as a catalyst for hydrogen peroxide solution to form hydroxyl radicals. These radicals can oxidize most organic substances. Future ideas under exploration are ways to use solar collectors to power photocatalysis by TiO_2 or photo-Fenton oxidation. These types of systems would be more environmentally sustainable than traditional oxidant technologies.

Biological treatment processes may also be combined with absorption on activated carbon, to remove recalcitrant organic compounds. The bulk liquid concentration of toxic wastewater components is decreased by adsorption so that inhibitory effects on the bacteria are minimized. Volatile organics may be kept available to the bacteria instead of being stripped. Over time, if the recalcitrant organics are intermittently loaded, biodegradation may regenerate the sorptive capacity of the activated carbon. Kolb and Wilderer (1997) explored this type of system in a lab-scale SBR system treating benzene and 2-chlorophenol or trichloroethene plus phenol wastewater. They found that it was more efficient to sequentially load the activated carbon and then biologically regenerate it, rather than trying to simultaneously sorb and biodegrade the compounds in a single step. Another potential combination is steam stripping followed by biological treatment. This process has been used to treat Kraft Mill wastewater.

CONCLUSION

Microbiological-based wastewater treatment processes can degrade many of the organic compounds in industrial wastewaters, mineralizing them to carbon dioxide and/or methane. Specific reactor configurations are better suited to some types of wastewaters, although in general all of the main types of industrial wastes have been treated under both aerobic and anaerobic conditions with various reactor configurations for each. Therefore, the optimum waste treatment reactor will depend on site-specific constraints including land availability, operator skill and time, energy, and sludge disposal, just to name a few. Although typical reactor sizing and bulk organic loading rates are available, due to the variability of wastewater characteristics and potential presence of specific target compounds that may or may not be recalcitrant or inhibitory to bacteria, bench scale or pilot scale testing are often conducted prior to sizing the full-scale treatment facility. In general, these smaller scale results seem to be predictive of full-scale performance. If problems arise during full-scale treatment, molecular methods to characterize the microbiology may provide a good tool along with traditional physical (suspended solids) and chemical analyses (pH, nutrients, and organics) to diagnose the source of the problem. Owing to the vast breadth of information briefly overviewed in this article, readers are encouraged to read further on the specific technologies.

RECENT DEVELOPMENTS

Basic literature searches were conducted on peer-reviewed literature, with 1738 articles on industrial wastewater found from 2007 to 2010 (1124 on industrial wastewater treatment), compared to 4978 over all years of the index. The specific kinds of industrial wastewater treated by various methods are summarized in Table 8. Many of the wastewater treatment studies use non biological methods. Of the biological treatment approaches, aerobic treatment such as AS was more widely explored than anaerobic treatment for industrial wastewater in general. However, for wastewaters with very high concentrations of organics, such as from brewery and dairy sources, anaerobic treatment was more common. The rising costs of energy should lead to expanded interest in anaerobic methods, as the energy expenses associated with aeration are avoided. Novel reactors such as MBR are being researched, as well as natural systems such as wetlands. The ability to extract energy from industrial wastewater is also being actively studied, particularly with brewery wastewater to power microbial fuel cells.

The peer reviewed literature from 2007 to 2010 contained minimal references to full scale industrial wastewater treatment. Examples of full scale studies include four studies

TABLE 7 Examples of Coupled Oxidation and Biological Treatment Systems

Wastewater Source	Scale (Wastewater Characteristics)	Oxidation Treatment	Biological Treatment	Reference
Dairy and pharmaceutical	Full COD 3000–11,000 $mg\,l^{-1}$; 54 $m^3\,h^{-1}$	Ozone post biotreatment 2 g $O_3\,g^{-1}$ COD; 68 kg $O_3\,d^{-1}$, 10 min contact time	Two SBRs, 1.2 MG ea, jet-aeration 77 kg urea d^{-1}added for NSRT 10–20 d; 8 h cycle	Helmig et al. (2002)
Formaldehyde-urea adhesives factory	Bench & full COD 460–3900 $mg\,l^{-1}$, formaldehyde 220–4000 $mg\,l^{-1}$,TKN 110–805 $mg\,l^{-1}$	Post ozonation recycled to aerobic chamber 60 min HRT partially degraded>8000 g/mol^{-1} organics	Single-sludge anoxic-aerobic activated sludge	Helmig et al. (2002)
Olive processing	Lab phenolic substances up to 500 $mg\,l^{-1}$	Pre wet air oxidation w/ H_2O_2 6 h at443–483 K and 3–7 MPa; converted46–99% COD	AS 2x increase in kinetics with pre-oxidation	Rivas et al. (2001)
Optoelectronic	Lab; contains dimethylsulfoxide (DMSO)	Ozone gas-induced reactor 240 min, complete DMSO removal	SBR achieved effluent limit of <100 $mg\,l^{-1}$ COD	Lin and Chang (2006)
Pharmaceutical	Lab and full COD 1000 to 7000 $mg\,l^{-1}$	Fenton's oxidation; H_2O_2/Fe^{+2}ratio 170; 20 l H_2O_2 + 0.33 kg$FeSO_4$ per m^3 ww; 45–50% COD removal	Two SBR in series overall 98%COD removal	Tekin et al. (2002)
Plastics manufacturing	Full; 5000 $m^3\,d^{-1}$	Ozonation at 6 mgO_3/h post AS reactor	Upflow BAF (six 3.8 m dia ×4.7 m high BAFs with sand media)	Chen et al. (2000)
Tannery	Lab; COD ~3500 $mg\,l^{-1}$, NH_4-N 250 $mg\,l^{-1}$	Ozone18 or 50% of SBBR volume moved, ozonated 1 h at 520 mgO_3/h, returned to the SBBR	SBBR with Kaldnes media 8 h cycles, 1 d HRTOLR 3 kg COD $m^{-3}\,d^{-1}$, NLR 0.19 kg N $m^{-3}\,d^{-1}$, 31–44 g TSS/l filter	Di Iaconi et al. (2004)
Textile wastewater	Full scale AS Lab scale ozone	Ozone pre or post 40 mg O_3/min for 15 min 85–99% color, 14–19% COD removed	AS 7 d SRT and ~20 h HRT	Orhon et al. (2002)
3-methyl-pyridine	Lab	Pre ozone bubble column produced formate, acetate, NH_3	SBBR	Carini et al. (2001)

TABLE 8 Number of peer reviewed publications on different types of industrial wastewater treatment methods from 2007 to 2010[a]

Wastewater Source	Number of Articles that Included the Following Key Words:						
	Treatment	Anaerobic	Activated Sludge	Aerobic	Wetland	Membrane/MBR	Fuel Cell
Industrial	1124	154	253	112	42	200/68	6
Brewery	35	28	10	4	1	5/0	7
Dairy	137	62	29	35	30	12/5	0
Dying	535	53	67	42	11	65/7	2
Pulp	138	18	27	13	1	10/2	1
Paper	1239	198	278	109	63	203/84	14
Tannery	93	7	24	8	8	21/5	0
Textile	397	56	74	44	9	74/18	2
Winery	35	13	18	11	5	1/1	0

[a] *Based on topic searches performed in ISI Web of Knowledge, April 2010, using the keywords "wastewater" and the source and treatment keywords noted above.*

on dairy wastewater treatment in wetlands, with clogging of sub-surface flow systems of particular concern. The publications on full scale treatment of tannery wastewater included a SBR, MBR, and two AS models.

Interest in MBRs has grown over the past 3 years relative to other technologies. Membrane costs have dropped rapidly over the past 10 years, making MBR technologies cost-competitive with other biological wastewater treatment methods. References were found to extensive use of immersed MBRs in Europe and Japan for industrial wastewater treatment. For example, approximately 240 full scale industrial treatment plants use Kubota membranes alone, with 45% of the industrial users in the food and beverage industry (Ferre et al. 2009). At 23 wineries, treatment capacity ranged from 14 to 200 m^3 per day. General Electric-Zenon Environmental are the other main supplier of membranes used for MBR systems. These MBR systems may be either aerobic or anaerobic. Immersed membranes can often be retrofitted into existing AS basins, with a footprint savings due to elimination of the standard clarification step after AS and the higher biomass concentration in the MBR compared to AS. Anaerobic MBR (AnMBR) offer an opportunity for energy recovery from the generated methane. Full-scale AnMBR applications in Japan include a 1st stage AnMBR followed by an aerobic MBR to treat wastewater from the Shochu Distillery (20–60 ton/d) and Awamori Distillery in Okinawa (15 ton/d). In addition, AnMBR have been used to treat wastewater from a wide variety of food processing plants, such as ~0.13 MGD of wastewater from a salad dressing/BBQ sauce manufacturer and wastes from potato, milk, bean, and confectionary processing (Page 2008).

Microbial fuel cells (MFC) powered by industrial wastewater is an area of active study, although no full-scale MFC systems were found. The basic system relies on anaerobic conditions in the anode chamber, which produce electrons that transfer via an external circuit to the cathode chamber. The chambers can be separated by an ion selective membrane, the MFC may be a single chamber with the cathode exposed to the air, or alternative configurations can be used. Brewery wastewater has been commonly studied, and a 1 m^3 pilot-scale MFC reactor at the Fosters Brewing Company in Queensland, Australia, was operated from 2007 to 2009. Other industrial wastewater types studied include chocolate industry wastewater, meat processing wastewater, and starch processing wastewater (Pant et al. 2010). MFC technology is expected to become a viable industrial wastewater treatment method in the future.

FURTHER READING

Anreottola, G., Foladori, P., & Ziglio, G. (2009). Biological treatment of winery wastewater: An overview. *Water Science & Technology*, *60*(5), 1117–1125.

Artan, N., Yagci, N. O., Artan, S. R., & Orhon, D. (2003). Design of sequencing batch reactors for biological nitrogen removal from high strength wastewaters. *Journal of Environmental Science and Health, Part A*, *38*(10), 2125–2134.

Austermann-Haun, U., Meyer, H., Seyfried, C. F., & Rosenwinkel, K. H. (1999). Full scale experiences with anaerobic/aerobic treatment plants in the food and beverage industry. *Water Science and Technology*, *40*(1), 305–312.

Bryant, C. W., & Barkley, W. A. (2001). Comparative virtues of ASB and AS systems. Proceedings of the Water Environment Federation 74th Annual Conference & Exposition on Water Quality and Wastewater Treatment 2001, Atlanta: Water Environment Federation.

Carini, D., von Gunten, U., Dunn, I. J., & Morbidelli, M. (2001). Ozonation as pretreatment step for the biological batch degradation of industrial wastewater containing 3-methyl-pyridine. *Ozone-Science &. Engineering*, *23*(3), 189–198.

Cervantes, F. J., Pavlostathis, S. G., & van Haandel, A. C., (Eds.) (2006). *Advanced biological treatment processes for industrial wastewaters: Principles and applications*. London: IWA Publishing.

Chan, Y. J., Chong, M. F., Law, C. L., & Hassell, D. G. (2009). A review on anaerobic-aerobic treatment of industrial and municipal wastewater. *Chemical Engineering Journal*, *155*(1–2), 1–18.

Chen, J. J., McCarty, D., Slack, D., & Rundle, H. (2000). Full scale case studies of a simplified aerated filter (BAF) for organics and nitrogen removal. *Water Science and Technology*, *41*(4–5), 1–4.

Connaughton, S., Collins, G., & O'Flaherty, V. (2006). Psychrophilic and mesophilic anaerobic digestion of brewery effluent: A comparative study. *Water Research*, *40*(13), 2503–2510.

Cornel, P., & Krause, S. (2006). Membrane bioreactors in industrial wastewater treatment – European experiences, examples and trends. *Water Science and Technology*, *53*(3), 37–44.

Di Iaconi, C., Bonemazzi, F., Lopez, A., & Ramadori, R. (2004). Integration of chemical and biological oxidation in a SBBR for tannery wastewater treatment. *Water Science and Technology*, *50*(10), 107–114.

Dionisi, H. M., Layton, A. C., Harms, G., Gregory, I. R., Robinson, K. G., & Sayler, G. S. (2002). Quantification of *Nitrosomonas oligotropha*-like ammonia-oxidizing bacteria and *Nitrospira* spp. from full-scale wastewater treatment plants by competitive PCR. *Applied and Environmental Microbiology*, *68*(1), 245–253.

Ferre, V., Trepin, A., Gimenez, T., & Lluch, S. (2009). Design and performance of full-scale iMBR plants treating winery wastewater effluents in Italy and Spain. In *5th International Specialized Conference on Sustainable Viticulture: Winery Waste and Ecologic Impacts Management*. Trento and Verona, Italy; April. http://www.kubota-mbr.com/resources/Winery%20MBR%20plants%20in%20Europe.pdf.

Gerards, R., Gils, W., & Vriens, L. (2005). Upgrading of existing aerobic plants with the LUCAS (R) anaerobic system based on full-scale experiences. *Water Science and Technology*, *4*(52), 39–46.

He, Y., Xy, P., Li, C., & Zhang, B. (2005). High-concentration food wastewater treatment by an anaerobic membrane bioreactor. *Water Research*, *39*, 4110–4118.

Helmig, E., Malik, J., Bradford, M., & Jun, L. G. (2002). Advanced treatment and reuse of high strength industrial wastewater. In *Proceedings of the Water Environment Federation 75th Annual Conference & Exposition on Water Quality and Wastewater Treatment*. Chicago: Water Environment Federation.

Ibekwe, A. M., Grieve, C. M., & Lyon, S. R. (2003). Characterization of microbial communities and composition in constructed dairy wetland wastewater effluent. *Applied and Environmental Microbiology*, *69*(9), 5060–5069.

Juretschko, S., Loy, A., Lehner, A., & Wagner, M. (2002). The microbial community composition of a nitrifying-denitrifying activated sludge from an industrial sewage treatment plant analyzed by the full-cycle rRNA approach. *Systematic and Applied Microbiology, 25*(1), 84–99.

Kalyuzhnyi, S. V., Gladchenko, M. A., Sklyar, V. I., Kurakova, O. V., & Shcherbakov, S. S. (2000). The UASB treatment of winery wastewater under submesophilic and psychrophilic conditions. *Environmental Technology, 21*(8), 919–925.

Kantardjieff, A., & Jones, J. P. (1997). Practical experiences with aerobic biofilters in thermomechanical pulping, sulfite and fine paper mills in Canada. *Water Science and Technology, 35*(2–3), 227–234.

Karpiscak, M. M., Freitas, R. J., Gerba, C. P., Sanchez, L. R., & Shamir, E. (1999). Management of dairy waste in the Sonoran desert using constructed wetland technology. *Water Science and Technology, 40*(3), 57–65.

Kolb, F.R. and Wilderer, P.A., 1997. Activated carbon sequencing batch biofilm reactor to treat industrial wastewater. *Water Science and Technology, 35*(1), 169-176.

Kreuzinger, N., Farnleitner, A., Wandl, G., Hornek, R., & Mach, R. (2003). Molecular biological methods (DGGE) as a tool to investigate nitrification inhibition in wastewater treatment. *Water Science and Technology, 47*(11), 165–172.

LaPara, T. M., Nakatsu, C. H., Pantea, L., & Alleman, J. E. (2000). Phylogenetic analysis of bacterial communities in mesophilic and thermophilic bioreactors treating pharmaceutical wastewater. *Applied and Environmental Microbiology, 66*(9), 3951–3959.

Laubscher, A. C. J., Wentzel, M. C., Le Roux, J. M. W., & Ekama, G. A. (2001). Treatment of grain distillation wastewaters in an upflow anaerobic sludge bed (UASB) system. *Water SA, 27*(4), 433–444.

Lin, S. H., & Chang, C. S. (2006). Treatment of optoelectronic industrial wastewater containing various refractory organic compounds by ozonation and biological method. *Journal of the Chinese Institute of Chemical Engineers, 37*(5), 527–533.

Luostarinen, S. A., & Rintala, J. A. (2005). Anaerobic on-site treatment of black water and dairy parlour wastewater in UASB-septic tanks at low temperatures. *Water Research, 39*(2–3), 436–448.

Mace, S., & Mata-Alvarez, J. (2002). Utilization of SBR technology for wastewater treatment: An overview. *Industrial and Engineering Chemistry Research, 41*(23), 5539–5553.

Mach, K. F., & Ellis, T. G. (2002). Fatty acid methyl ester (FAME) analysis of high rate anaerobic wastewater treatment systems. In *Proceedings of the Water Environment Federation 75th Annual Conference & Exposition on Water Quality and Wastewater Treatment*. Chicago: Water Environment Federation.

Mbuligwe, S. E. (2005). Comparative treatment of dye-rich wastewater in engineered wetland systems (EWBs) vegetated with different plants. *Water Research, 39*(1), 271–280.

Moeltta, R. (2005). Winery and distillery wastewater treatment by anaerobic digestion. *Water Science and Technology, 51*(1), 137–144.

Moussa, M. S., Fuentes, O. G., Lubberding, H. J. *et al.* (2006). Nitrification activities in full-scale treatment plants with varying salt loads. *Environmental Technology, 27*(6), 635–643.

Newman, J. M., & Clausen, J. C. (1997). Seasonal effectiveness of a constructed wetland for processing milkhouse wastewater. *Wetlands, 17*(3), 375–382.

Newman, J. M., Clausen, J. C., & Neafsey, J. A. (2000). Seasonal performance of a wetland constructed to process dairy milkhouse wastewater in Connecticut. *Ecological Engineering, 14*(1–2), 181–198.

Oerther, D. B., Stroot, P. G., & Butler, R. *et al.* (2002). Impact of influent microorganisms upon poor solids separation in the quiescent zone of an industrial wastewater treatment system. In *Proceedings of the Water Environment Federation 75th Annual Conference & Exposition on Water Quality and Wastewater Treatment*. Chicago: Water Environment Federation.

Oktem, Y., & Tufekci, N. (2006). Treatment of brewery wastewater by pilot scale upflow anaerobic sludge blanket reactor in mesophilic temperature. *Journal of Scientific and Industrial Research, 65*(3), 248–251.

Oktem, Y. A., Ince, O., Donnelly, T., Sallis, P., & Ince, B. K. (2006). Determination of optimum operating conditions of an acidification reactor treating a chemical synthesis-based pharmaceutical wastewater. *Process Biochemistry, 41*(11), 2258–2263.

Omil, F., Garrido, J. M., Arrojo, B., & Mendez, R. (2003). Anaerobic filter reactor performance for the treatment of complex dairy wastewater at industrial scale. *Water Research, 37*(17), 4099–4108.

Orhon, D., Dulkadiroglu, H., Dogruel, S., Kabdasli, I., Sozen, S., & Babuna, F. G. (2002). Ozonation application in activated sludge systems for a textile mill effluent. *Water Science and Technology, 45*(12), 305–313.

Page, I. C. (2008). Energy production from waste using the anaerobic membrane bioreactor (AnMBR) process. ADI Systems Inc. Energy from Biomass and Waste Conference. Pittsburgh, PA. Oct. http://www.freesen.de/ebw/2008/conf/adi_page.pdf.

Pant, D., Bogaert, G. V., Diels, L., & Vanbroekhoven, K. (2010). A review of the substrates used in microbial fuel cells (MFCs) for sustainable energy production. *Bioresource Technology, 101*(6), 1533–1543 doi: 10.1016/j.biortech.2009.10.017

Parawira, W., Kudita, I., Nyandoroh, M. G., & Zvauya, R. (2005). A study of industrial anaerobic treatment of opaque beer brewery wastewater in a tropical climate using a full-scale UASB reactor seeded with activated sludge. *Process Biochemistry, 40*(2), 593–599.

Parawira, W., Murto, M., Zvauya, R., & Mattiasson, B. (2006). Comparative performance of a UASB reactor and an anaerobic packed-bed reactor when treating potato waste leachate. *Renewable Energy, 31*(6), 893–903.

Reemtsma, T., Zywicki, B., Stueber, M., Kloepfer, A., & Jekel, M. (2002). Removal of sulfur-organic polar micropollutants in a membrane bioreactor treating industrial wastewater. *Environmental Science & Technology, 36*(5), 1102–1106.

Riffat, R., & Krongthamchat, K. (2002). Anaerobic treatment of high saline wastewater using Halophilic methanogens in anaerobic filters (AF). *Proceedings of the Water Environment Federation 75th Annual Conference & Exposition on Water Quality and Wastewater Treatment*. Chicago: Water Environment Federation.

Rivas, F. J., Beltran, F. J., Gimeno, O., & Alvarez, P. (2001). Chemical-biological treatment of table olive manufacturing wastewater. *Journal of Environmental Engineering, 127*(7), 611–619.

Robinson, K. G., Dionisi, H. M., Harms, G., Layton, A. C., Gregory, I. R., & Sayler, G. S. (2003). Molecular assessment of ammonia- and nitrite-oxidizing bacteria in full-scale activated sludge wastewater treatment plants. *Water Science and Technology, 48*(8), 119–126.

Ruiz, C., Torrijos, M., Sousbie, P., Martinez, J. L., Moletta, R., & Delgenes, J. P. (2002). Treatment of winery wastewater by an anaerobic sequencing batch reactor. *Water Science and Technology, 45*(10), 219–224.

Rusten, B., Eikebrokk, B., Ulgenes, Y., & Lygren, E. (2006). Design and operations of the Kaldnes moving bed biofilm reactors. *Aquacultural Engineering, 34*(3), 322–331.

Schlegel, S., & Teichgr¨ber, B. (2000). Operational results and experience with submerged fixed-film reactors in the pretreatment of industrial effluents. *Water Science and Technology, 41*(4–5), 453–459.

Smith, E., Gordon, R., Madani, A., & Stratton, G. (2006). Year-round treatment of dairy wastewater by constructed wetlands in Atlantic Canada. *Wetlands*, *26*(2), 349–357.

Spanjers, H. (2006). *IDS water online. Anaerobic treatment of industrial wastewater*. White Paper. http://www.idswater.com.

Sponza, D. T., & Isik, M. (2002). Ultilmate azo dye degradation in anaerobic/aerobic sequential processes. *Water Science and Technology*, *45*(12), 271–278.

Starke, T. M. (2001). Treatment system upgrade at a Kraft mill. In *Proceedings of the Water Environment Federation 74th Annual Conference & Exposition on Water Quality and Wastewater Treatment*. Atlanta: Water Environment Federation.

Stephenson, T., Cornel, P., & Rogalla, F. (2004). Biological aerated filters (BAF) in Europe: 21 years of full scale experience. In *Proceedings of the Water Environment Federation 77th Annual Conference & Exposition on Water Quality and Wastewater Treatment*. New Orleans: Water Environment Federation.

Teichgraber, B., Schreff, D., Ekkerlein, C., & Wilderer, P. A. (2002). SBR technology in Germany – An overview. *Water Science and Technology*, *43*(3), 323–330.

Tekin, H., Gulkaya, I., & Bilkay, O. *et al*. (2002). Treatment of pharmaceutical wastewaters by Fenton's oxidation followed by activated sludge: A case study. In *Proceedings of the Water Environment Federation 75th Annual Conference & Exposition on Water Quality and Wastewater Treatment*. Chicago: Water Environment Federation.

Torkian, A., Alinejad, K., & Hashemian, S. J. (2003). Posttreatment of upflow anaerobic sludge blanket-treated industrial wastewater by a rotating biological contactor. *Water Environment Research*, *75*(3), 232–237.

Trivedy, R. K., & Pattanshetty, S. M. (2002). Treatment of dairy waste by using water hyacinth. *Water Science and Technology*, *45*(12), 329–334.

van Lier, J. B. (2006). Anaerobic industrial wastewater treatment; perspectives for closing water and resource cycles. In *Proceedings of the ACHEMA 2006 Congress* 7. Paper 1713.

Yang, W. B., Cicek, N., & Ilg, J. (2006). State-of-the-art of membrane bioreactors: Worldwide research and commercial applications in North America. *Journal of Membrane Science*, *270*(1–2), 201–211.

Zhou, P., Su, C. Y., Li, B. W., & Qian, Y. (2006). Treatment of high-strength pharmaceutical wastewater and removal of antibiotics in anaerobic and aerobic biological treatment processes. *Journal of Environmental Engineering*, *132*(1), 129–136.

RELEVANT WEBSITE

http://www.achema.de/. – ACHEMA (2009).

http://www.idswater.com/. – idswater, The Information Resource for the Water Industry.

http://www.epa.gov/. – United States Environmental Protection Agency.

Chapter 45

Wastewater Treatment of Non-Infectious Hazards

T.K. Graczyk[1], T.E.A. Chalew[1], Y. Mashinski[1] and F.E. Lucy[2]

[1]*Johns Hopkins Bloomberg School of Public Health, Baltimore, MD, USA*

[2]*Institute of Technology, Sligo, Ireland Environmental Services Ireland, Co. Leitrim, Ireland*

Chapter Outline

DEFINING STATEMENT

The origin of chemical constituents in wastewater depends on the composition of water users, the quality and type of wastewater systems, and the urban storm water runoff. Disinfection by-products (DBPs), endocrine disruptors (EDCs), and pharmaceutically active chemicals (PhACs), 1,4-dioxane, perchlorate (ClO_4^-), methyl tertiary-butyl ether (MTBE), and engineered nanoparticles, represent chemical constituents of public health and environmental concern.

WASTEWATER IN PUBLIC WATER SUPPLIES

Increasing demand for water, driven by both population growth and urban expansion, forces water agencies to seriously evaluate new water reuse systems and technologies for alternative water resource options. Current and future demand for water cannot be fully achieved by traditional hydrological cycle-dependent water resources; therefore purified wastewater has to be reclaimed and recycled. Reclaimed water is a resource available at the doorstep of the urban environmental world, where water is needed most and often priced the highest. Effective ways of water recycling are already being implemented in many parts of the world, with a high quality of reclaimed water essential for public health. The quantity and quality of municipal wastewater derived from a variety of sources such as households, schools, hospitals, offices, and commercial and industrial facilities varies among communities depending on the composition of water users; the quality and type of wastewater collection systems; and the urban storm water runoff (in the case of combined sewer systems). Thus, untreated wastewater potentially contains a variety of chemical constituents hazardous to human health and the environment. In highly industrialized countries, reliable wastewater treatment and public health-related water reuse policy and regulations dictate the feasibility and public acceptance of water reuse (Table 1). Public perception and understanding of

Topics in Ecological and Environmental Microbiology.

TABLE 1 Comparison of United States Environmental Protection Agency (USEPA) drinking water standards with reclaimed water quality parameter values in San Diego, CA, USA; Tampa, FL, USA; and Denver, CO, USA

Constituent[a]		USEPA drinking water standards	Reclaimed water[b]		
			San Diego	Tampa	Denver
Physical					
	TOC		0.27	1.88	0.2
	TDS	500	42	461	18
	Turbidity (NTU)		0.27	0.05	0.06
Nutrients					
	Ammonia-N		0.8	0.03	5
	Nitrate-N		0.6	0	0.1
	Phosphate-P		0.1	0	0.02
	Sulfate	250	0.1	0	1
	Chloride	250	15	0	19
	TKN		0.9	0.34	5
Metals					
	Arsenic	0.05	<0.0005	0[c]	ND[d]
	Cadmium	0.005	<0.0002	0[c]	ND
	Chromium	0.1	<0.001	0[c]	ND
	Copper	1.0	0.011	0[c]	0.009
	Lead	[e]	0.007	0[c]	ND
	Manganese	0.05	0.008	0[c]	ND
	Mercury	0.002	<0.0002	0[c]	ND
	Nickel	0.1	0.0007	0.005	ND
	Selenium	0.05	<0.001	0[c]	ND
	Silver	0.05	<0.001	0[c]	ND
	Zinc	5.0	0.0023	0.008	0.006
	Boron		0.29	0	0.2
	Calcium		<2.0		1.0
	Iron	0.3[f]	0.37	0.028	0.02
	Magnesium		>3.0	0	0.1
	Sodium		11.9	126	4.8

[a]*NTU, nephelometric turbidity units; TDS, total dissolved solids; TKN, total Kjeldahl nitrogen; TOC, total organic carbon. All reported values with the exception of turbidity are expressed in mg l^{-1}.*
[b]*San Diego physical and nutrient concentration values are arithmetic means. Any nondetected observations were assumed to be present at the corresponding detection limit. Metal concentration values are geometric means determined through probit analysis. Tampa values are arithmetic means of detected values. Denver values are geometric means of detected values.*
[c]*Not detected in seven samples.*
[d]*Not detected in more than 50% of samples.*
[e]*Lead is regulated according to a treatment standard.*
[f]*Noncorrosive limit for iron.*

potable water reuse varies and depends on a number of factors including awareness, general education, self-education about water issue, and access to multimedia and mass media. The role of the latter in shaping public opinion and perception should not be underestimated since the media do not usually objectively present reasons for potable water recycling and in fact frequently portray return flow as "treated sewage" rather than water; reinforce the understanding of the public that "sewage is a sewage until we lose track of it, and then it becomes water"; and support public aversion to reclaimed water. There are two types of potable water recycling: direct potable reuse and indirect unplanned and planned potable reuse.

Direct Potable Reuse

Direct potable reuse is the most extreme case of water recycling and is at present used only in water-critical situations. There are four issues related to direct reuse: public perception; health risk concerns; technological capacities;

and cost considerations. The technology is technically and societally challenging as the effluent of a wastewater treatment plant (WWTP) is routed directly to the drinking water treatment plant; thus, reclaimed water is piped directly to the water supply system. Because of the relatively closed-loop cycle involved, it is also called "toilet-to-tap" or "pipe-to-pipe." The wastewater in this system requires extensive advanced physical and chemical treatment prior to reintroduction in the drinking water treatment plant, and therefore is usually discharged to source water reservoirs. Direct potable reuse is associated with negative public perception, due to existing public health and hygiene concerns about using wastewater for drinking purposes. Any system that loops back a large quantity of wastewater, without advanced physical and chemical treatment for drinking water production, carries the risk of concentrating chemical pollutants over time, including those that are not on the monitoring list. Although "toilet-to-tap" and "pipe-to-pipe" systems do work theoretically and although some pilot plants are presently being monitored, they are generally an option of the last resort. There are some system designs, which include "time lapse" in the "closed-loop architecture," that may involve an open aerated tank or alternatively a pond with no water release from the tank or pond except to the drinking water plants. Currently, in the United States, there is no imperative for the use of reclaimed water for direct potable reuse; however, it is inevitable that potable reuse will occur. There are four study cases for direct reuse: emergency potable reuse in Chanute, KS, USA; direct potable reuse in Windhoek, Namibia; direct potable reuse demonstration project in Denver, CO, USA; and Orange County Ground Water Replenish System, FL, USA.

Indirect Potable Reuse

Indirect potable reuse is a traditional system in which raw water is abstracted, treated, distributed to drinking water customers, and after use, the wastewater is collected, treated, and discharged to a water reservoir, which is usually the same one from which source water was abstracted. In this system, the reclaimed water is used to augment potable water supplies by mixing it with natural water (Table 1). Treated wastewater is discharged downstream of the raw water abstraction.

Unplanned indirect reuse

Unplanned indirect reuse generally occurs at the community level, when the treated wastewater discharged from an upstream community is abstracted for drinking water production purposes by another community at downstream locations, and source water quantity and quality is not controlled by the user. The abstracted fraction of wastewater volume for drinking water production can be high, sometimes reaching close to 50%.

Planned indirect reuse

In planned reuse systems, treated wastewater is intentionally used to augment source water supplies. The treated wastewater is discharged at upstream water abstraction locations to mix with raw water, prior to diversion to drinking water production. Reclaimed water is used directly or indirectly without losing control of quantity and quality. Planned indirect potable reuse involves multiple safety measures and barriers to remove contaminants such as wastewater treatment, dilution, mixing, and natural cleansing in a wastewater-receiving reservoir, efficacious drinking water treatment, and effective source and finished drinking water monitoring.

CHEMICAL CONSTITUENTS IN UNTREATED WASTEWATER

Chemical constituents are classified as inorganic and organic aggregate and individual constituents. Inorganic chemical constituents include dissolved constituents such as nutrients, nonmetallic constituents, metals, and gases. Aggregate organic constituents consist of a number of individual components that cannot be distinguished separately. Untreated wastewater contains known and unknown organic and inorganic constituents that are present naturally in the water supply source, present in treated drinking water, added by the water users, added from storm water in combined collection systems, formed in the collection system as a result of abiotic and biotic reactions, and added in the wastewater collection system in order to control odor or corrosion.

Constituents in Natural Water

Natural water contains both organic and inorganic constituents. Inorganic constituents originate from the dissolution of rock and minerals with enhanced concentrations due to the evaporation process. The main inorganic cations include calcium (Ca^{2+}), magnesium (Mg^{2+}), potassium (K^+), and sodium (Na^+), with the corresponding anions, bicarbonate (HCO_3^-), sulfate (SO_4^{2-}), and chloride (Cl^-). Concentrations of these constituents vary depending on both the geological characteristics and the land-use patterns. Most waters also contain natural organic matter (NOM) with varied concentration depending on the water origin, for example, reservoir surface water versus aquifer groundwater. NOM is typically composed of humic materials from plants and algae, and high molecular weight aliphatic and aromatic hydrocarbons. Most of the NOM represents nuisance constituents that cause aesthetic taste, color, and odor-related concerns. However, a greater concern exists as some high molecular weight aliphatic and aromatic hydrocarbons may pose adverse health effects. Humic materials serve as precursors in the formation of trihalomethanes (THMs) and other organohalogen oxidation by-products during water disinfection process.

Constituents in Drinking Water

Residual inorganic chemicals in drinking water, such as known or suspected carcinogens, arsenic, lead, and cadmium, represent health concerns to varying degrees. Several inorganic chemicals (i.e., aluminum, chromium, copper, manganese, molybdenum, nickel, selenium, zinc, and sodium) are essential to human nutrition at a low level, but may, however, have adverse health effects at high doses. Additional constituents can originate from piping and plumbing materials such as lead, copper, zinc, and asbestos. Other organic constituents include acrylamide and epichlorohydrin, that is, components of coagulants; polynuclear aromatic hydrocarbons (PAHs) originating from pipe lining and joint adhesives; and DBPs. The aggregated residual organic compounds in drinking water are of little human health concern.

Constituents Added Through Domestic, Commercial, and Industrial Use

Organic constituents are typically composed of combinations of carbon, hydrogen, and oxygen together with nitrogen, phosphorus, and sulfur. The organic matter in wastewater typically consists of proteins (40–60%), carbohydrates (25–50%), and oils and fat (8–12%). Urea decomposes rapidly and therefore is only found in fresh wastewater. Inorganic components include anions, cations, metals, and nonmetallic constituents (Table 2). In addition to proteins, carbohydrates, oils and fats, and urea, fresh wastewater contains a large number of different synthetic organic chemicals (SOCs) originating from commercial and industrial use, of which a vast majority remain unidentified.

Constituents Added from Storm Water in Combined Collection Systems and via Infiltration

The constituents derived from combined sewer system include oils, PAHs, grease, tars, rubber, salts, suspended solids, and metals from roadway runoff; pesticides, herbicides, and fertilizers; animal feces; and decayed humic materials. Predominant infiltration-derived constituents include salts from marine and brackish water intrusion when the collection system is aged or not watertight and dissolved organic substances (DOSs) that significantly interfere with the disinfection process.

TABLE 2 Typical increase of inorganic constituents in fresh wastewater related to domestic water use

Constituent		Increment range, mg l^{-1} [a]
Anions		
	Bicarbonate (HCO_3)	50–100
	Carbonate (CO_3)	0–10
	Chloride (Cl)	20–50
	Sulfate (SO_4)	15–30
Cations		
	Calcium (Ca)	6–16
	Magnesium (Mg)	4–10
	Potassium (K)	7–15
	Sodium (Na)	40–70[b]
Other Constituents		
	Aluminum (Al)	0.1–0.2
	Boron (B)	0.1–0.2
	Fluoride (F)	0.2–0.4
	Manganese (Mn)	0.2–0.4
	Silica (SiO_2)	2–10
Total alkalinity (as $CaCO_3$)		60–120
Total dissolved solids (TDS)		150–380

[a] *Based on 460 l/capita × day (or 120 gal/capita × day), which represents a medium strength wastewater.*
[b] *Excluding the addition from domestic water softeners.*

Constituents Formed in the Collection System as a Result of Abiotic and Biotic Reactions

When the water travel time in the collection system is greater than 6 h, a number of biotic and abiotic reactions can occur in wastewater. The most common reaction is the formation of hydrogen sulfide that frequently occurs due to microbial breakdown of organic matter, under anoxic conditions.

Constituents Added to Wastewater in Collection System for Odor or Corrosion Control

In some cases, pure chemicals are added to wastewater to prevent formation of odor and to mitigate corrosion of the collection system. Pure oxygen is added to suppress anoxic and anaerobic reactions, which frequently lead to the formation of hydrogen sulfide.

IMPACT OF CHEMICAL CONSTITUENTS REMAINING AFTER WASTEWATER TREATMENT

Wastewater Treatment Levels

Primary treatment is used to physically remove floating and settleable materials. Because of removal of these materials, there is an appreciable reduction in biochemical oxygen demand (BOD), total suspended solids (TSSs), total organic carbon (TOC), and some metals associated with TSSs. Biological and chemical processes used in the secondary treatment remove most of the organic matter, and hence the BOD and TSS. The principal application of tertiary treatment is for the removal of residual TSSs and nutrients remaining after secondary sedimentation, typically by cloth or media filtration, disinfection, or engineered wetland polishing. Advanced wastewater treatment (AWWT) is used to remove residual suspended solids and other constituents that are not reduced significantly by conventional secondary treatment. Reverse osmosis (RO) and microfiltration (MF) may be used to remove emerging constituents of concern (Table 3).

Disinfection By-Products

The impact of the constituents that remain after various treatment processes can be of profound importance to the long-term protection of public health and the environment. Currently, it is possible to measure the concentration of emerging constituents listed in Table 3 in the range of 10^{-9}–10^{-12} g l^{-1}, but their exact health and environmental impact remains unknown. The chemical oxidation processes such as chlorination used for the disinfection of wastewater effluents during tertiary treatment produce DBPs. Most DBPs are dissolved organohalogens derived from the oxidative breakdown of organic substances in wastewater. The DBPs are grouped into THMs, haloacetonitriles, haloketones, haloacetic acids, chlorophenols, aldehydes, trichloronitromethane, chloral hydrate, iodoacetic acid, and cyanogen chloride. Trichloromethane and haloacetic acids are the most common DBPs and are often present at higher concentration than the other less frequently detected DBPs. Another DBP found in wastewater and reclaimed water is *N*-nitrosdimethylamine (NDMA), a potent carcinogen (Table 3). Chlorine disinfection is the most common method for wastewater disinfection worldwide. The extend of DBP formation by chlorination depends on pH, temperature, reaction time, free and combined chlorine concentrations, ammonia concentration, DBP precursor concentration, and precursor types. Organic matter that is highly aromatic, and contains chloride reactive sites such as phenol, 2,4-pentanedione, organic nitrogen, and *meta*-dihydroxybenzene is a precursor of DBPs.

Human health impact

The human health impact of DBPs is derived from the ingestion of chlorinated water; however, because of crop irrigation, consumers can be indirectly exposed to DBPs through the food chain as well. Fortunately, most of the DBPs volatilize in the ambient environment and are readily biodegradable via chemical and environmental biological reactions. Following discharge to the environment, natural processes and systems will assimilate the compounds and the reclaimed water will soon have the chemical signature of the environment. The degree of treatment and environmental factors will control the rate and extent of assimilation of DBPs. However, anthropogenic constituents, such as DBPs, can be found in reclaimed water in systems with a short retention time in the environment, such as water flowing into streams with a low environmental reaction time. Formation of DBPs in reclaimed water is of greater concern when indirect or direct potable reuse is considered.

Use of Surrogate Parameters

Because reclaimed water can contain hundreds of compounds that can be traced to human origin, rigorous analytical methods are needed to characterize even a portion of these chemicals quantitatively. To overcome limitations associated with expensive sampling and time-consuming laboratory work, several surrogate parameters have been developed to assess the chemical makeup of treated effluents

TABLE 3 Reported range of removal of selected emerging constituents from influent wastewater

	Removal range (%)[a]		
Constituent	Secondary treatment	Microfiltration (MF)	Reverse osmosis (RO)
N-nitrosdimethylamine (NDMA)	50–75	50–75	50–75
17B-Estradiol			50–100
Alkyphenols ethoxylates (APEOs)	40–80	40–80	40–80

[a]*Significant variations have been observed in the concentrations of these constituents in the influent wastewater.*

and reclaimed water. The primary surrogates for aggregate organic trace constituents in water include assimilable organic carbon (AOC), biodegradable dissolved organic carbon (BDOC), TOC, and BOD (Table 1).

Emerging Chemical Constituents in Wastewater

The term "emerging pollutants" is used for chemicals that have been identified in water only recently and are under regulative consideration. These constituents are inherent to municipal wastewater, and knowledge of their occurrence in reclaimed water is limited. When reclaimed water is discharged to aquatic environments, these chemical constituents are indivertibly released, potentially resulting in an adverse impact on the environment and on human health. The fate and transport in surface water and in groundwater and the risk associated with unintentional transfer of these chemicals to humans are virtually unknown.

EDCs and PhACs

Certain synthetic and natural compounds that mimic, block, stimulate, or inhibit natural hormones in the endocrine system of animals and humans are called EDCs and are linked to adverse effects in humans and wildlife. EDCs have a wide variety of origin including pharmaceuticals, personal care products, household chemicals, pesticides and herbicides, industrial chemicals, DBPs, naturally occurring hormones, and metals. Chemicals classified as pharmaceutically active (PhAC) are synthesized for medical purposes such as antibiotics, anti-inflammatory drugs, X-ray contrast media, and antidepressants. Some PhACs, such as contraceptives and steroids, are also EDCs. PhAC and EDCs have been discovered in various types of surface and groundwater and some of them have been linked to ecological impacts at trace concentrations. A substantial number of studies have shown that conventional drinking and WWTP cannot completely remove many EDCs and PhACs. Future research needs include more detailed fate and transport data, standardized analytical methodology, removal kinetics, predictive models, and determination of the toxicological relevance of trace levels of EDCs and PhACs in water.

N-nitrosdimethylamine

NDMA (Table 3) is a member of a group of extremely potent carcinogens, the *N*-nitrosamines. Until recently NDMA contamination was of concern in food, consumer products, and air. However, drinking water contamination resulting from reactions occurring during chlorination and/or via direct industrial contamination is currently being detected. UV radiation can effectively remove NDMA; however, there is considerable interest in developing less expensive alternative treatments. These technologies include approaches for removing organic nitrogen-containing NDMA precursors to chlorination, use of sunlight photolysis, and *in situ* bioremediation to remove NDMA and its precursors. Effluents from conventional treatment and AWWT can contain relatively high concentrations of NDMA. NDMA is usually present in untreated municipal wastewater prior to chlorination, with concentrations as high as 105,000 ng l^{-1} being reported. Facilities with advanced treatment capacities typically use MF-RO and/or UV treatment. This treatment train has been shown to be effective in removing NDMA itself and NDMA precursors.

1,4-Dioxane

1,4-Dioxane is classified as a probable human carcinogen and has been reported as a water contaminant common in industrial and commercial products due to its high aqueous solubility and resistance to biodegradation. 1,4-Dioxane is used as a stabilizer of chlorinated solvents and is formed as a by-product during the manufacturing of polyester and various polyethoxylated compounds. At present, advanced oxidation processes (AOPs) are the only proven technology for removal of 1,4-dioxane from wastewater. Chemical and energy costs of AOPs, however, may be substantial, and thus, their use is not widespread. UV is also commonly used as part of AOP; however, 1,4-dioxane is a relatively weak absorber of UV light and it is degraded poorly via direct photolysis. UV light, in combination with titanium oxide (TiO_2) catalyst, has also been demonstrated to effectively degrade 1,4-dioxane.

Perchlorate

Perchlorate, ClO_4^- is a highly oxidized (+7) chlorine oxyanion manufactured for use as the oxidizer in solid propellants for rockets, missiles, explosives, and pyrotechnics. Perchlorate release into the environment has occurred primarily in association with the manufacture and usage of solid rocket propellants. When released into water, perchlorate can spread over large distances because it is highly soluble and adsorbs poorly to soil. Two proven techniques to remove perchlorate from drinking water include anaerobic biological reactors and ion exchange. Perchlorate contamination of the environment affects agricultural and natural plants. Perchlorate is toxic to some plants, and when absorbed by them is not degraded, thus accumulating in plants, and can be released back to the environment when they decompose. Perchlorate accumulation in plants represents a significant human health risk. Accumulation of perchlorate has been found in plants irrigated with contaminated water and when fed to cattle resulted in high concentration of perchlorate in dairy products.

MTBE and other oxygenates

Currently, MTBE is the most widely used oxygenate in gasoline, followed by ethanol. Accidental gasoline releases from underground storage tank and pipelines are the most significant point sources of oxygenates in groundwater. Because of their polar characteristics, oxygenates migrate through aquifers with minimal retardation, raising concerns nationwide about the risk of reaching drinking water sources.

Toxic Potential of Engineered Nanoparticles and Nanomaterials in Wastewater

The industrial production and the commercial and personal use of nanoparticles and nanomaterials have raised concerns about human exposure in the absence of technologies for the removal of these compounds from wastewater (all technologies are in the developmental or experimental phases); data on public health safety of nanoparticles produced for industrial (including in the water industry) and personal use are also disturbing. Contamination of source waters is based on the assumption that nanoparticles are highly mobile in porous media due to their size, although some data demonstrated that retention of smaller nanoparticles in porous media is enhanced by their high ionic strength and the presence of small quantities of divalent ions. A wide variety of *in vitro* and *in vivo* animal studies postulated negative effects of engineered nanoparticles present in water on human health. Results of *in vivo* studies demonstrate various toxicological effects of water-suspended nanoparticles such as enhanced inflammation, tumorgenesis, and oxidative stress. Multiple studies demonstrated toxicological effects of nanoparticles (i.e., silver nanoparticles, quantum dots, carbon buckey balls (C60), and titanium dioxide (TiO_2) nanoparticles) on aquatic organisms, including fish and invertebrates. Engineered nanoparticles and nanomaterials may have adverse effects on WWTP bacteria, since these compounds are being used because of their antibacterial properties. More research is needed to substantiate the toxicological effects of engineered nanoparticles and nanomaterials present in treated wastewater on humans

RECENT DEVELOPMENTS

Currently, there are no accepted methods for detection of NPs in wastewater or surface waters due to low concentrations of NPs, similarity to natural colloids, and difficulty in conducting *in situ* measurements (MacCormack and Goss, 2008; Nowack, 2009). Researchers have estimated the environmental concentrations of nanoparticles to be up to 100 μg/L based on the quantity of NPs produced, usage of consumer products containing NPs, percentage NPs leached from those products, population size, and estimated removal of NPs during wastewater treatment (Blaser *et al.*, 2008; Luoma, 2008; Mueller and Nowack, 2008). Concentrations in surface waters are expected to increase over time with greater usage and disposal of NP-containing products (Boxall *et al.*, 2007; Klaine *et al.*, 2008; MacCormack and Goss, 2008).

Leaching of NPs from products during usage is being investigated in order to estimate the concentration of NPs in the domestic waste stream and surface waters. During a simulated laundering of NP-containing socks with ultrapure water, these socks released ionic and nano-silver particles (Benn and Westerhoff, 2008). Benchtop experiments were conducted to simulate NP removal during wastewater treatment. Using estimated influent silver NP loads and WWTP removal of silver NPs, Benn *et al.* concluded that the WWTP removal was sufficient, that released effluent would not violate EPA's water quality standards for silver (Benn and Westerhoff, 2008). Silver NPs were found to partition into the biosolids. Dissolved silver from dissolution of NPs are expected to be removed by WWTP during precipitation or biomass adsorption. Other NP-containing products, such as food storage containers, cleaners, toothpaste, and clothing have NPs imbedded within them or on the surface, which are likely to leach from products into the domestic waste stream. In the future, influent concentrations of NPs may impact surface waters or usage of biosolids.

The degree of NP removal from WWTPs is related to NP concentrations and characteristics, such as composition, surface funtionalization, zeta potential, aggregation, and size (MacCormack and Goss, 2008 Moore, 2006; Oberdorster *et al.*, 2005). Water characteristics, such as organic matter and pH, will also affect NP aggregation and removal from the water (Christian *et al.*, 2008; Farré *et al.*, 2009; French *et al.*, 2009). TiO_2 NPs have a high affinity to biosolids, with an estimated removal during WWTPs between 70% and 85% (Kiser *et al.*, 2009). Cerium oxide NP removal from a model WWTP ranged from 92% to 95% (Limbach *et al.*, 2008). The affect of functionalization of silica dioxide NPs during WWTP removal was investigated and it was determined that unfunctionalized SiO_2 NPs were not significantly removed during sedimentation and flocculation within relevant time scales for WWTP but tween coated NPs were less stable and were effectively removed during sedimentation (Jarvie *et al.*, 2009).

Since many NPs are antibacterial, it is unknown whether their presence in the wastewater stream will impact the bacteria within the WWTP or the environment. The toxicity of silver NPs to *Pseudomonas putida* biofilms was affected by the presence of organic matter (Fabrega *et al.*, 2009). Silver NPs alone yielded a reduction in bacterial biofilm biomass but silver NPs with Suwannee River humic acid did not cause a change in biofilm biomass or cell viability compared to the control (Fabrega *et al.*, 2009). Interaction of bacteria from WWTP effluent samples with single walled carbon nanotubes and aqueous C60 NPs yielded resulted in cell death (Kang *et al.*, 2009).

Due to their novel properties, NPs will be increasingly used in consumer products, especially those that wash down the drain. With increased usage of these products, NPs within WWTP influent and effluent will increase, since there is not 100% removal. This will yield increased NPs in the environment.

FURTHER READING

Asano, T., Burrton, F. L., Leverenz, H. L., Tsuchihashi, R., & Tchobanoglous, G. (2007). *Water reuse: Issues, technologies, and applications*. New York: Metcalf & Eddy, AECOM.

Benn, T. M., & Westerhoff, P. (2008). Nanoparticle silver released into water from commercially available sock fabrics. *Environmental Science & Technology*, *42*, 4133–4139.

Blaser, S. A., Scheringer, M., Maclead, M., & Hungerbuhler, K. (2008). Estimation of cumulative aquatic exposure and risk due to silver: Contribution of nano-functionalized plastics and textiles. *Science of The Total Environment*, *390*, 396–406.

Bouwer, H., Fox, P., & Westerhoff, P. (1998). Irrigating with treated effluents – How does this practice affect underlying groundwater. *Water Environment & Technology*, *10*, 115–118.

Boxall, A. B. A., Tiede, K., & Chaudhry, Q. (2007). Engineered nanomaterials in soils and water: How do they behave and could they pose a risk to human health? *Nanomedicine*, *2*, 919–927.

Christian, P., Von der Kammer, F., Baalousha, M., & Hofmann, T. (2008). Nanoparticles: Structure, properties, preparation and behaviour in environmental media. *Ecotoxicology*, *17*, 326–343.

Deeb, R. A., Chu, K. H., Shih, T., et al. (2003). MTBE and other oxygenates: Environmental sources, analysis, occurrence, and treatment. *Environmental Engineering Science*, *20*, 443–447.

Fabrega, J., Renshaw, J. C., & Lead, J. R. (2009). Interactions of silver nanoparticles with pseudomonas putida biofilms. *Environmental Science & Technology*, *43*, 9004–9009.

Farré, M., Gajda-Schrantz, K., Kantiani, L., & Barceló, D. (2009). Ecotoxicity and analysis of nanomaterials in the aquatic environment. *Analytical and Bioanalytical Chemistry*, *393*, 81–95.

French, R. A., Jacobson, A. R., Kim, B., Isley, S. L., Penn, R. L., & Baveye, P. C. (2009). Influence of ionic strength, pH, and cation valence on aggregation kinetics of titanium dioxide nanoparticles. *Environmental Science & Technology*, *43*, 1354–1359.

Jarvie, H. P., Al-Obaidi, H., King, S. M., Bowes, M. J., Lawrence, M. J., Drake, A. F., *et al.* (2009). Fate of silica nanoparticles in simulated primary wastewater treatment. *Environmental Science & Technology*, *43*, 8622–8628.

Kang, S., Mauter, M. S., & Elimelech, M. (2009). Microbial cytotoxicity of carbon-based nanomaterials. Implications for River Water and Wastewater Effluent. *Environmental Science & Technology*, *43*, 2648–2653.

Kiser, M. A., Westerhoff, P., Benn, T., Wang, Y., Pérez-Rivera, J., & Hristovski, K. (2009). Titanium nanomaterial removal and release from wastewater treatment plants. *Environmental Science & Technology*, *43*, 6757–6763.

Klaine, S. J., Alvarez, P. J. J., Batley, G. E., Fernandes, T. F., Handy, R. D., Lyon, D. Y., *et al.* (2008). Nanomaterials in the environment: Behavior, fate, bioavailability, and effects. *Environmental Toxicology and Chemistry*, *27*, 1825–1851.

Kolpin, D. W., Skopec, M., Meyer, M. T., Furlong, E. T., & Zaugg, S. D. (2004). Urban contribution of pharmaceuticals and other organic wastewater contaminants to streams during differing flow conditions. *The Science of the Total Environment*, *328*, 119–130.

Krasner, S. W., McGuire, M. J., Jackangelo, J. G., *et al.* (1989). The occurrence of disinfection byproducts in U.S. drinking water. *Journal of American Water Works Association*, *81*, 41–53.

Lahnsteiner, J., & Lempert, G. (2007). Water management in Winhoek, Namibia. *Water Science and Technology*, *55*, 441–448.

Lauer, W. C., Rogers, S. E., LaChance, A. M., & Nealy, M. K. (1991). Process selection for potable reuse health effect studies. *Journal of American Water Works Association*, *83*, 52–63.

Limbach, L. K., Bereiter, R., Mueller, E., Krebs, R., Gaelli, R., & Stark, W. J. (2008). Removal of oxide nanoparticles in a model wastewater treatment plant: Influence of agglomeration and surfactants on clearning efficiency. *Environmental Science & Technology*, *42*, 5828–5833.

Lin, A., Plumlee, M. H., & Reinhard, M. (2006). Natural attenuation of pharmaceuticals and alkylphenol polyethoxylate metabolites during river transport: Photochemical and biological transformation. *Environmental Toxicology and Chemistry*, *25*, 1458–1464.

Lovern, S. B., & Klaper, R. D. (2006). *Daphia magna* mortality when exposed to titanium dioxide and fullerene (C60) nanoparticles. *Environmental Toxicology and Chemistry*, *25*, 1132–1137.

Luoma, S. N. (2008). *Silver nanotechnologies and the environment: Old problems or new challenges?*. Project on Emerging Nanotechnologies, Washington DC.72 p).

MacCormack, T. J., & Goss, G. G. (2008). Identifying and predicting biological risks associated with manufactured nanoparticles in aquatic ecosystems. *Journal of Industrial Ecology*, *12*, 286–296.

Moore, M. N. (2006). Do nanoparticles present ecotoxicological risks for the health of the aquatic environment? *Environment International*, *32*, 967–976.

Mueller, N. C., & Nowack, B. (2008). Exposure modeling of engineered nanoparticles in the environment. *Environmental Science & Technology*, *42*, 4447–4453.

Mitch, W. A., Sharp, J. O., Trussell, R. R., *et al.* (2003). *N*-nitrosodimethylamine (NDMA) as a drinking water contaminant, a review. *Environmental Engineering Science*, *20*, 389–404.

Nowack, B. (2009). The behavior and effects of nanoparticles in the environment. *Environmental Pollution*, *157*, 1063–1064.

Oberdorster, G., Oberdorster, E., & Oberdorster, J. (2005). Nanotoxicology: An emerging discipline evolving from studies of ultrafine particles. *Environ Health Persp*, *113*, 823–839.

Oberdorster, G., Sharp, Z., Atudorei, V., *et al.* (2004). Translocation of inhaled ultrafine particles to the brain. *Inhalation Toxicology*, *16*, 437–445.

Rebhun, M., Heller-Grossman, L., & Manka, J. (1997). Formation of disinfection byproducts during chlorination of secondary effluents and renovated water. *Water Environmental Research*, *69*, 1154–1162.

Snyder, S. A., Westerhoff, P., Yoon, Y., & Sedlak, D. K. (2003). Pharmaceuticals, personal care products, and endocrine disruptors in water: Implications for water industry. *Environmental Engineering Science*, *20*, 449–469.

Stevens, A. A., Moore, L. A., & Miltner, R. S. (1989). Formation and control of non-trihalomethane disinfection byproducts. *Journal of American Water Works Association*, *81*, 54–60.

Tchobanoglous, G., Burton, F. L., & Stensel, H. D. (2003). *Wastewater engineering: Treatment and reuse* (4th ed.). New York: McGraw-Hill.

Theron, J., Walker, J. A., & Cloete, T. E. (2008). Nanotechnology and water treatment: Applications and emerging opportunities. *Critical Reviews in Microbiology*, *34*, 43–69.

Waller, K., Swan, S. H., DeLorenze, G., & Hopkins, B. (1998). Trihalomethane in drinking water and spontaneous abortion. *Epidemiology*, *9*, 134–140.

Wiesner, M. R., Lowry, G. V., Alvarez, P., Dionysiou, D., & Biswas, P. (2006). Assessing the risk of manufacturing nanomaterials. *Environmental Science & Technology*, *40*, 4336–4345.

Zenker, M. J., Borden, R. C., & Barlaz, M. A. (2004). Biodegradation of 1,4, dioxane using trickling filter. *Journal of Environmental Engineering*, *130*, 926–931.

Zhu, S., Oberdorster, E., & Haasch, M. L. (2006). Toxicity of an engineered nanoparticle (fullerene, C(60)) in two aquatic species, *Daphnia* and fatheaded minnow. *Marine Environmental Research*, *62*, 85–89.

Chapter 46

Drinking Water

P.S. Berger[1], R.M. Clark[1], D.J. Reasoner[1], E.W. Rice[2], J.W. Santo Domingo[2]
[1]US Environmental Protection Agency (Retired) Cincinnati, OH, USA
[2]US Environmental Protection Agency Cincinnati, OH, USA

Chapter Outline

ABBREVIATIONS

AOC Assimilable organic carbon
CCL Contaminant Candidate List
CFU Colony-forming units
DE Diatomaceous earth
DOC Dissolved organic carbon
ED Electrodialysis
EPA Environmental Protection Agency
EU European Union
FR Federal Register
GAC Granular activated carbon
GWR Groundwater Rule
HACCP Hazard Analysis and Critical Control Point
HPC Heterotrophic plate count
MCL Maximum contaminant level
MF Microfiltration
MST Microbial source tracking
NF Nanofiltration
NTU Nephelometric turbidity units
POE Point-of-entry
POU Point-of-use
PVC Polyvinyl chloride
RO Reverse osmosis
SWTR Surface water treatment rule
TCR Total coliform rule
UF Ultrafiltration
WHO World health organization

DEFINING STATEMENT

The primary objective of drinking water microbiology is to prevent waterborne disease. This objective can be achieved through proper water treatment and control practices and by monitoring their effectiveness. This article addresses characteristics of the microbial flora of drinking water, water treatment practices in use, and monitoring and control strategies.

INTRODUCTION

Safe drinking water is one of the oldest public health concerns known. Ancient civilizations practiced water treatment, as evidenced by Egyptian inscriptions and Sanskrit writings. Sand filtration was in use in some cities as early as the sixteenth century. Chlorination of drinking water, so widely used today, was not introduced until the first decade of the twentieth century. These two treatment practices dramatically decreased the incidence of waterborne disease, although waterborne disease is still a problem in the United States and elsewhere.

As with many other countries, the United States takes a multibarrier approach in preventing pathogen entry into drinking water. This means that a drinking water system is required to use several barriers to counter waterborne disease, throughout the entire process from source water intake to the tap. Systems using surface water may control human activities on the watershed supplying the system. In groundwater systems natural barriers such as soil and bedrock function in preventing passage of pathogens into well water. As part of this approach, a system should use the cleanest source water available, provide appropriate water treatment for the water source used, and minimize pathogen entry into the distribution system via cross-connections, pipe breaks, and other distribution system deficiencies. It may even include watershed control. In addition, the multibarrier approach includes a good management plan, minimum operator standards and training, a state laboratory accreditation program, periodic on-site sanitary surveys, strong plumbing system codes, and a host of other quality assurance programs that collectively reduce the potential for drinking water contamination. An integral part of this approach is for a system to understand the potential sources of contamination in the watershed, including human activity and livestock practices.

This article discusses how water is treated and monitored to insure its microbiological safety for human consumption. The primary focus is on water treatment practices in the United States, although those of other countries will be mentioned. The article also includes a brief description of the drinking water regulations for microbiology in the United States.

INDICATOR ORGANISMS IN MICROBIAL MONITORING

Ideally, specific detection of the various waterborne pathogens would be the most direct approach in determining their presence and the need for control. Unfortunately, in actual practice, this is impractical for several reasons. The variety of potential waterborne pathogens (including various species of bacteria, viruses, and protozoa) makes a search for all – especially on a routine basis – extremely difficult, time-consuming, and expensive. Moreover, the efficiency of current techniques for recovering and detecting known waterborne pathogens in drinking water is often very low, primarily due to the low levels of these organisms in drinking water. In addition, detection does not always translate into risks as some strains of the same species are more pathogenic than others, and current detection methods cannot easily discriminate between pathogenic and nonpathogenic subpopulations.

Although culture techniques for isolating many bacterial pathogens are available, some are nonselective, thus allowing nontarget organisms to proliferate in numbers that overgrow the pathogen. Viral pathogens are fastidious in their growth requirements and grow only in special tissue cultures that are expensive and often difficult to maintain. Some viruses cannot yet be cultivated in the laboratory. Currently, recovery and assay methods for pathogenic protozoa are inefficient. Extended delays can be involved in carrying out specific identification procedures for pathogens, and only certain laboratories with specially trained technologists may have the expertise to carry out some of these procedures. Moreover, if pathogens were present in drinking water, their concentrations would usually be sufficiently small to require analysis of large-volume samples. Finally, extended delays would usually be involved in carrying out identification procedures, under the extremely doubtful assumption that laboratories involved in water testing would have such resources.

Because of the problems associated with trying to detect specific enteric pathogens, indicator organisms are used as surrogates. Such indicator organisms are used to assess the microbiological quality of drinking water. The ideal drinking water indicator has the following attributes

- It is suitable for all types of drinking water.
- It is present in sewage and polluted waters at much higher densities than fecal pathogens.
- Its survival time in water is at least that of waterborne pathogens.
- It is at least as resistant to disinfection as the waterborne pathogens.
- It is easily detected by simple, inexpensive, laboratory tests in the shortest time with accurate results.
- It is stable and nonpathogenic.
- It is generally not present in waters contaminated by mammalian feces.

The indicators generally employed worldwide are total coliforms, fecal coliforms, and/or *Escherichia coli*. Total coliforms constitute a group of closely related bacteria in the family *Enterobacteriaceae* that are usually not pathogenic and are widespread in ambient water. They are often, but not necessarily, associated with sewage. Total coliforms are not defined in precise taxonomic terms, but by their ability to produce acid and gas from lactose on selective culture media designed to recover enteric bacteria. These media vary by country. This bacteria group include most species of the genera *Enterobacter*, *Klebsiella*, *Citrobacter*, and *Escherichia*, although some species of *Serratia*, and other genera may also be included. Total coliforms are used to determine the effectiveness of water treatment processes in removing enteric microorganisms to monitor the integrity of the underground pipe network called the distribution system, and as a screen for fresh fecal contamination. Treatment and other water system practices that provide coliform-free water should also reduce pathogens to minimal levels. A major shortcoming of using total coliforms as an indicator is that they are only marginally adequate for predicting the potential presence of some pathogenic protozoa and viruses, since total coliforms may be more susceptible to disinfection than are these other organisms.

Fecal coliforms are a subset of the total coliform group, primarily including *E. coli* and a few thermotolerant strains of *Klebsiella*. Fecal coliforms and *E. coli* are more suitable indicators of fresh fecal contamination than are total coliforms. This is important as most waterborne pathogens are associated with fecal contamination. However, total coliforms are a more suitable indicator than fecal coliforms or *E. coli* for determining the vulnerability of a system to fecal contamination, especially in the absence of fecally contaminated samples at the times and locations of sample collection. Total coliforms are usually present at much higher concentrations in source waters than these other indicators, and they are relatively more resistant to chlorine and other environmental stresses. Fecal coliform or *E. coli* monitoring may be preferable in countries where monitoring for total coliforms is impractical due to consistently high densities of these organisms in the drinking water. Most countries that have formal drinking water rules, including the United States, monitor for total coliforms. Many of these countries, including the United States, also require fecal coliform or *E. coli* monitoring.

In addition to the indicators just described, other drinking water indicators also have been used or suggested, usually as a supplement to the initial indicators. These supplemental indicators include fecal streptococci, enterococci, coliphage, *Clostridium perfringens*, *Bacteroides*, heterotrophic bacteria (as measured such as the heterotrophic plate count (HPC)), and other organisms. Several chemicals have also been mentioned, for example, fecal sterols, caffeine, assimilable organic carbon (AOC), and disinfectant residuals. For monitoring ambient freshwaters, the US Environmental Protection Agency (EPA) suggests the use of enterococci or *E. coli* as indicators, while enterococci are recommended for marine waters.

EPA DRINKING WATER REGULATIONS

General

The EPA publishes enforceable regulations under the Safe Drinking Water Act, which was passed by the US Congress in 1974 and subsequently amended. Regulations under this Act apply to all public water systems, that is, to those systems that regularly serve 25 or more people at least 60 days out of the year, or that have 15 or more service connections. There are about 158,000 public water systems, of which about 14,000 are surface water systems and the rest are groundwater systems. About 9000 systems serve 3300 people or more, and these 9000 systems serve a total of approximately 258 million people.

The EPA has published three regulations to protect the public against waterborne pathogens: the Total Coliform Rule, the Surface Water Treatment Rule, and the Groundwater Rule(GWR). These three rules are summarized here. The Code of Federal Regulations (40 CFR 141 and 142) provides detailed requirements about the three existing rules. The *Federal Register* (FR) also provides detailed requirements, along with the rationale for these requirements. (The general website for the Code of Federal Regulations and the *FR* is http://www.gpoaccess.gov.).

Total Coliform Rule (TCR)

The TCR (54 FR 27544; 29 June 1989) sets a maximum contaminant level (MCL) for total coliforms as follows. For systems that collect 40 or more samples per month, no more than 5.0% may have any coliforms; for systems that collect fewer than 40 samples per month, no more than one sample may be total coliform-positive. If a system exceeds the MCL for a month, it must notify the public using mandatory language developed by the EPA. The required monitoring frequency for a system depends on the size of the population served and whether the system operates full time or not. This frequency ranges from 480 samples per month for the largest systems to once annually for certain of the smallest systems. The regulation also requires all systems to have a written plan identifying where samples are to be collected.

If a system has a total coliform-positive sample, it must take three (for small systems, four) repeat samples within 24h of being notified of the positive sample. In addition, systems that collect fewer than five samples per month must, with some exceptions, collect at least five routine samples the next month of operation. Both routine and repeat samples count toward calculating compliance with the MCL.

If any sample is total coliform-positive, the system must also test the positive culture for the presence of either fecal coliforms or *E. coli*. Any positive fecal coliform or *E. coli* test must be reported to the state. If two consecutive samples at a site are total coliform-positive and one is also fecal coliform- or *E. coli*-positive, the system is in violation of the MCL and must notify the public using more urgent mandatory language than used for the presence of total coliforms alone.

The TCR also requires each system that collects fewer than five samples per month to have the system inspected every 5 years (10 years for certain types of systems using only protected and disinfected groundwater). This on-site inspection (referred to as a sanitary survey) must be performed by the state or by an agent approved by the state. The regulation also defines when a state may invalidate a total coliform-positive sample, and when a laboratory must invalidate a total coliform-negative sample. It also specifies which analytical methods are approved for compliance samples. The sample volume examined must be 100 ml, regardless of the method used.

Surface Water Treatment Requirements

The Surface Water Treatment Rule (SWTR) (54 FR 27486; 29 June 1989) was designed to protect all public water systems that use surface water, or groundwater under the direct influence of surface water, against waterborne pathogens, specifically including *Giardia lamblia*, viruses, and *Legionella*. The SWTR requires systems to use sufficient water treatment (removal and/or inactivation) to reduce *G. lamblia* cysts by at least 99.9% (3 logs) and enteric viruses by at least 99.99% (4 logs) of their source water densities. The SWTR specifies criteria for meeting these standards, and criteria for avoiding filtration. More specifically, the SWTR requires all systems covered by the rule to disinfect. It also requires all such systems to filter their water, unless (1) the system has an effective watershed control program; (2) it complies with the TCR and MCL for trihalomethanes; (3) it uses a good-quality source water, as defined as having no more than 20 fecal coliforms per 100 ml or no more than 100 total coliforms per 100 ml, and a turbidity (opaqueness) no greater than five nephelometric turbidity units (NTU), and (4) it meets stringent disinfection conditions.

The regulation and associated EPA guidance assist the system in meeting the specified pathogen reduction levels by identifying pertinent *CT* values for disinfection inactivation (*C*, disinfectant concentration in mg l^{-1}; *T*, time of disinfectant contact with the water in min). *CT* values are provided for *Giardia* and enteric viruses by disinfectant type (e.g., chlorine, chloramines, ozone, or chlorine dioxide), water pH, and temperature. The regulation also requires a system using surface water to have at least a 0.2 mg l^{-1} disinfectant residual continually entering the distribution system and at least a detectable disinfectant residual in at least 95% of the sample sites throughout the distribution system. The rule specifies the monitoring frequency and locations for determining these disinfection residuals and (for unfiltered systems) testing source water quality.

The SWTR has been strengthened by three subsequent rules, the Interim Enhanced SWTR (63 FR 69478, 16 December 1998), Long-Term 1 Enhanced SWTR (67 FR 1812; 14 January 2002), and Long-Term 2 Enhanced SWTR (LT2ESWTR) (71 FR 6135; 30 January 2006). These three rules collectively tighten the turbidity standards, require systems to monitor the turbidity levels leaving each individual filter, require periodic on-site sanitary surveys by the state, and require the system either to cover all finished drinking water storage facilities (e.g., reservoirs) or treat the water in those facilities, as well as add other new requirements. The primary purpose of these new requirements is to control *Cryptosporidium*. Another purpose is to ensure that utilities do not degrade pathogen control while complying with other regulations that control for disinfection by-products (formed by the reaction of disinfectants with organic material in the source water). Under these three rules, a system must provide sufficient water treatment to reduce *Cryptosporidium* by at least 99% (2 logs). However, because of a concern that systems drawing water from a poor-quality source might be exposing the public to a greater pathogen risk than is reasonable, the LT2ESWTR requires systems to provide additional *Cryptosporium* treatment if required source water monitoring indicates the average density of *Cryptosporidum* oocysts is 0.075 l^{-1} or greater (for unfiltered systems, >0.01 l^{-1}). The actual required minimum level of additional treatment depends upon how much greater than 0.075 l^{-1} the average *Cryptosporidum* density is. In this manner, in theory, the consumer would have approximately the same risk of infection, regardless of the system used.

Groundwater Rule

The GWR (71 FR 65574; 8 November 2006) applies to all public water systems that use groundwater (about 142,000 systems). However, systems that use groundwater that are under the direct influence of surface water must instead comply with the SWTR. Under the GWR, the State will determine whether a system is fecally contaminated or is vulnerable to fecal contamination. This determination may be made by a variety of means, including direct monitoring of the source water (usually the well), periodic on-site sanitary surveys by a trained inspector, and an examination of the site's hydrology. A threatened site is required to take corrective action such as providing an alternate water source, eliminating the contamination source, correcting all significant deficiencies found during a sanitary survey, and/or providing treatment (removal, inactivation, or both) that reliably achieves at least a 4-log (99.99%) virus reduction. Under the rule, a system may use *E. coli*, enterococci, or coliphage for source water monitoring; the rule approves specific analytical methods for each of the three. The rule specifies when, where, and how often a system must monitor; the frequency of required on-site sanitary surveys, minimum disinfectant requirements; and other provisions.

Rules and Guidelines of Other Countries

Several multicountry organizations have strong drinking water directives or guidance. These include the World Health Organization (WHO) and the European Union (EU). WHO's *Guidelines for Drinking-water Quality*, third edition

(2006), contains a comprehensive strategy for ensuring safe drinking water, from the water source though treatment and distribution to the consumer. The guidelines do not set specific numerical goals for removal/inactivation of pathogens or indicators of water quality. Rather WHO takes the position that no single approach to drinking water standards is appropriate, that such standards will vary among countries according to local needs and capacities. Guidelines are provided for a large variety of drinking water systems and circumstances, including large metropolitan systems, small community piped systems, nonpiped systems in communities and in individual dwellings, large buildings, food production, emergencies and disasters, and travelers and commercial transport. The document describes the approaches used in deriving the guidelines, including guideline values, and includes fact sheets on significant microbial and chemical hazards.

A key component of the WHO guidelines is the Water Safety Plan, which has three key elements: the development of an individual water safety plan that includes an assessment of risk from source to tap; the identification of control measures and, for each measure, appropriate monitoring that, collectively, will control the identified risks in a timely manner; and the development of management plans describing actions to be taken during normal operation or incident conditions, as well as supporting programs (http://www.who.int.).

EU directive 98/83/EC (3 November 1998) specifies that no *E. coli*, enterococci, or coliform bacteria should be detected in 100 ml of drinking water. It also states that no abnormal change in colony counts (22 °C) should occur. The directive also specifies that if water originates from or is influenced by surface water, it must also monitor *C. perfringens* (including spores). If the density of *C. perfringens* exceeds 0/100 ml, the utility must investigate to ensure that water has not been contaminated by pathogens. In contrast to US standards, the monitoring frequency for bacterial indicators is based upon the volume of water produced or distributed each day, rather than by the number of people served by the utility (http://europa.eu.).

Among individual countries with an effective drinking water program are Canada (http://www.hc-sc.gc.ca.), Australia (http://www.waterquality.crc.org.au.), and New Zealand (http://www.mfe.govt.nz.).

One strategy that is gaining increased attention worldwide (including the United States) for controlling pathogens in drinking water is known as the Hazard Analysis and Critical Control Point (HACCP) system. This quality assurance program, which originated in the food and beverage industry in the 1960s, involves identifying critical points in the system or operation whose failure would result in greater public health risk than other points, and directing resources to these critical locations and processes.

SOURCES, TREATMENT, AND DISTRIBUTION

Sources of Water Supply

Drinking water sources can be divided into two categories: surface water and groundwater; treatment requirements vary with the source. Microbiologically, groundwaters tend to be of better quality than surface waters and, consequently, require less intensive treatment. Most groundwater supplies used for drinking water are pumped wells (dug, drilled, or driven), artesian wells, or springs.

Surface water sources include streams, rivers, ponds, lakes, and manmade impoundments and reservoirs. These sources represent precipitation runoff that is not lost to the atmosphere by evaporation and does not enter the ground via infiltration and percolation. Surface waters become contaminated by human pathogens through direct and indirect inputs of municipal sewage and other sources of human and animal excreta. Removal or inactivation of these microorganisms to provide safe drinking water involves a multi-barrier concept, using a variety of treatment processes, in addition to measures aimed at controlling or reducing source water pollution (source protection).

Bacterial concentrations in groundwater, as measured by the HPC, are usually below 100 colony-forming units (CFU) ml^{-1}; coliforms are normally absent. Bacterial concentrations in surface waters depend on the degree of contamination with human and animal fecal material. Pristine and relatively uncontaminated surface waters commonly contain bacterial concentrations of 10^1–10^4 CFU ml^{-1}, whereas contaminated surface waters may contain more than 10^6 CFU ml^{-1}. Coliform bacteria concentrations in surface waters range from >1/100 to >10^6/100 ml, and fecal coliforms range from >1/100 to >10^5/100 ml. Raw water containing more than 2000 fecal coliforms/100 ml should not be used as a drinking water source if at all possible.

Water Treatment

The major water treatment processes for improving the microbiological quality of surface waters for human consumption include coagulation and flocculation, sedimentation, filtration, and disinfection. Most groundwaters are only disinfected, if they are treated at all. Figure 1 is a schematic of the unit processes in a conventional drinking water treatment system used for surface waters. Several additional treatment processes are widely used, for example, softening, fluoridation, and iron removal, but these are not addressed in this article.

Another process called 'bank filtration' is used in Germany and the Netherlands to provide water for drinking water supply purposes. Bank filtration wells are installed adjacent to a river or other source of water. The wells capture groundwater from the landward side and surface water that

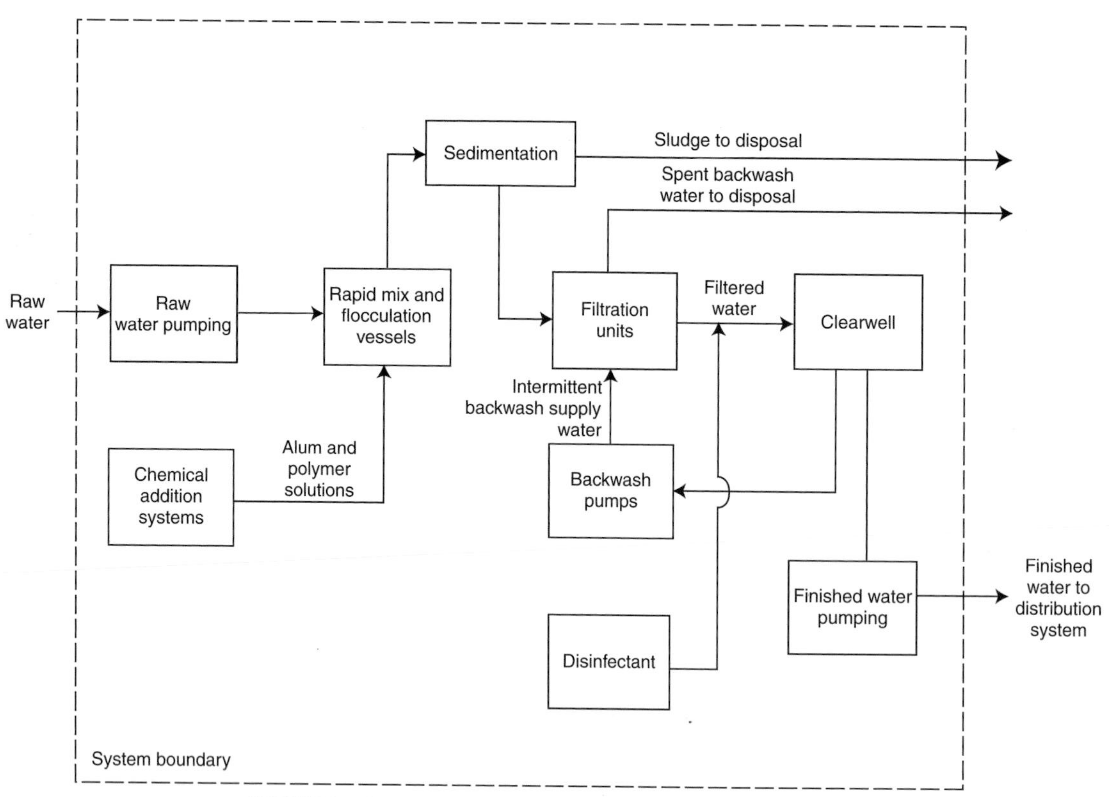

FIGURE 1 Conventional filtration system for drinking water treatment.

is induced by well pumping to infiltrate through the subsurface and into the well. The produced water is processed through other treatment steps for drinking water. The use of 'bank filtration' in the United States has seen limited application and will not be discussed further.

Coagulation, flocculation, and sedimentation

In the first step, raw water is pumped into a rapid mix unit, where chemicals are added to destabilize particles in the water electrostatically (i.e., make them 'sticky'). This step is referred to as coagulation. Then water enters a flocculation basin, which is often a series of chambers with slow moving paddles. During flocculation, the destabilized particles are brought into contact with each other so that aggregation into larger particles, or flocs, can occur. Microorganisms become trapped by, or attached to, the flocs. The optimal coagulation practice varies with water pH and temperature, raw water turbidity, and type of coagulant(s) used. Commonly used coagulants include aluminum sulfate (alum), calcium oxide (lime), ferrous sulfate, and ferric chloride. Often, coagulant aids such as activated silica, bentonite clay, and polyelectrolytes (synthetic polymers of varying charge) are also used to reduce the concentration of primary coagulants employed.

After flocculation, water enters the sedimentation basin(s), where flocs are given time to settle by gravity. Some systems omit this step and feed the flocculated water directly to filters (direct filtration). If the source waters are highly turbid, systems may have an additional sedimentation step before coagulation (presedimentation). Sludge in the sedimentation basins is removed and discharged to a municipal sewer, lagooned, or dewatered and hauled to a landfill. The sedimentation basins are generally equipped with sludge removal mechanisms; if not, they usually have a sloping bottom so that most of the sludge flows out with the water when the basin is drained for cleaning.

Filtration

The next treatment step is filtration, which removes suspended and colloidal material that has not settled. The filtration unit consists of steel or concrete vessels containing granular materials such as graded sand, anthracite, and/or gravel. Three types of filters are commonly used: rapid granular filters, slow sand filters, and diatomaceous earth (DE) filters.

Rapid Granular Filters

The most commonly used filter in the United States is the rapid granular filter. These filters may consist of silica sand, anthracite coal, and/or other materials. Filters are usually set up to provide gradation in filter media, with the coarsest particles on top and the smallest on the bottom. In a rapid granular filter, the rate at which water is applied is at least 2 gal min^{-1} per square foot of surface area.

Rapid granular filters gradually accumulate a large amount of particles that impede water flow; thus, the filters must be periodically cleaned. In this process, water flow through the filter is reversed in a process termed 'backwashing.' Backwashing expands the filter media, thereby releasing trapped particles into the water. Jets of water, air injection, or mechanical agitation at the surface may improve the process. The filter backwash water should be disposed of in an environmentally acceptable manner.

Slow Sand Filters

Slow sand filters are similar to the rapid granular filters, except that water flows through the filter at a much slower rate, 0.05–0.15 gal min^{-1} per square foot of surface area, and pretreatment (coagulation, flocculation, and sedimentation) is often omitted. Removal occurs primarily in the upper portion of the sand, by straining, sedimentation, adsorption, and chemical and microbiological action. As contaminants are removed, a layer of deposited material called a 'schmutzdecke' forms. Schmutzdecke contains a large number of bacteria that break down organic material in the water. When the schmutzdecke clogs the filter, it must be removed by scraping the top layer of sand to improve water flow. Normally, slow sand filters are used by systems with relatively clean source water (turbidity of 10 NTU or less and no undesirable chemical contaminants). They require far less maintenance than rapid granular filters, but must be much larger in surface area to produce a given volume of drinking water.

Diatomaceous Earth Filtration

In DE filtration, solids are removed by passing water through a thin filter consisting of a layer of DE supported on a rigid base (septum). DE is composed of crushed siliceous shells of diatoms (microscopic algae). To maintain adequate water flow, additional DE called body feed is continually added to the raw water during operation. Like slow sand filters, systems that use DE are usually small with relatively clean source water, and often pretreatment is omitted.

Granular Activated Carbon (GAC) Filters

GAC filters are sometimes used in conjunction with rapid granular filters to control taste and odor problems. In Europe, many systems treat the water with ozone (ozonation) before applying it to the GAC filters. Ozonation produces more easily utilized organic carbon which promotes the growth of heterotrophic bacteria in the filter. The high density of bacteria that grow on the surface of the filter media significantly reduces the level of organic substances in the water as it passes through the filter.

Membrane Processes

Membrane technology has become increasingly popular in potable water treatment. It has been used for desalinization, removal of dissolved inorganic chemicals, water softening, and removal of solids. The simplest way to describe membrane technology is that water is forced through a porous membrane under pressure while suspended solids, larger molecules, or ions are held back or rejected by the membrane.

Membrane processes may be pressure- or electricity-driven. The pressure-driven processes are microfiltration (MF), ultrafiltration, nanofiltration, and reverse osmosis. In MF water is forced through a porous membrane with a pore size of 0.45 μm, which is relatively large compared with other membrane processes. It is probably best used for small particles such as bacteria or protozoa. Ultrafiltration (UF) uses a membrane with a pore size less than 0.1 μm, which removes colloids and other high-molecular-weight materials. It is effective for removal of most organic compounds. MF or UF will not remove dissolved organic and/or inorganic species without advanced pretreatment to transform the dissolved species into particulate form. Nanofiltration (NF) rejects even smaller molecules than UF, and has been used for water softening and removal of total dissolved solids and viruses. Reverse osmosis (RO) has the highest rejection rate of all of the membrane processes. It has been used primarily for desalinization of seawater. The electricity driven processes are electrodialysis (ED) and ED reversal. Electrodialysis transfers ions though a membrane as a result of direct electric current, and ED reversal is similar to ED except that the polarity of the direct current process is periodically reversed. These types of systems are used primarily for the treatment of brackish water. Of these processes, NF seems to have the greatest potential for increased use in water treatment.

Disinfection

Disinfection is the primary means for inactivating pathogenic microorganisms in water. For systems using groundwater, disinfection is generally the only treatment practiced. Several disinfectants are available, including chlorine, chloramines (chlorine combined with ammonia or organic amines), ozone, chlorine dioxide, and ultraviolet light. All are oxidizing agents.

Chlorination is, by far, the most common disinfection technique practiced in the United States. The dose of chlorine usually applied is sufficiently high to meet chlorine demand (i.e., the tendency of organic substances and ammonia to react with chlorine) and still leave a sufficient concentration (chlorine residual) to inactivate microorganisms throughout the distribution system. A major shortcoming is that chlorine combines with organic substances in the water to produce chlorinated by-products, some of which are toxic (e.g., trihalomethanes such as chloroform). Unlike chlorine, chloramines do not result in any significant trihalomethane formation, but their microbial inactivation rates are substantially less than that of chlorine. Chlorine dioxide also does not generate significant by-product formation, but its intermediates (chlorite and chlorate) are toxic.

Like chlorine, chloramines and chlorine dioxide are useful because they provide disinfectant residuals in the water throughout the distribution system.

Ozone is the most effective disinfectant generally available, and is widely used in water treatment in Europe. Fewer toxic by-products have been identified for ozone than for chlorine, but ozone is more expensive and does not leave a significant residual. There is increased interest in the United States in using ozone as a predisinfectant, followed by chlorine or chloramine. This process would allow a system to control the formation of toxic by-products, yet maintain a disinfectant residual in the distribution system.

Mostly used by smaller systems and individual domestic systems (especially those using groundwater), ultraviolet light is presently being considered by some larger drinking water treatment facilities. The optimum wavelength for biocidal effectiveness is 254 nm. Dosage is expressed as the product of radiation intensity (ΦW) and the time (s) per unit area (cm^2). Ultraviolet light does not produce disinfectant by-products in water and leaves no disinfectant residual. Its effectiveness is reduced by high turbidity, air bubbles, some dissolved chemicals that block light penetration, the lack of reliable methods or meters to measure dosage, and equipment maintenance and reliability considerations.

It has been demonstrated that UV irradiation is very effective against (oo)cysts of *Cryptosporidium* and *Giardia*. These two organisms are pathogenic organisms of major importance for the safety of drinking water. Results of inactivation studies have shown that UV is effective against all waterborne pathogens; however viruses, specifically adenoviruses, and bacterial spores show some resistance to UV inactivation.

Microbiology of Treatment Processes

The removal of microorganisms by the treatment processes through the filtration step is variable due to a variety of factors, including fluctuating source water quality, coagulant dose, pH, water temperature, depth of filter, filter medium particle size, and filtration rate. Drinking water treatment processes and the approximate percentage removal/inactivation of microorganisms achieved by each of the processes are shown in Table 1. Removal percentages represent only removal percentages for that stage, not the cumulative percentage. It is difficult to establish the cumulative removal percentages from one step to the next because of considerable variation in the data reported from different studies. These differences reflect different source waters and qualities, different treatment processes, and different operating conditions. Thus, direct comparison of study results is difficult, but sufficient studies have been conducted to permit general characterization of the effectiveness of water treatment processes for removing microorganisms.

TABLE 1 Microorganism removal efficiency by water removal (%)[a]

Unit process		Bacteria	Viruses	Protozoa	Helminths
Storage reservoirs		80–90	80–90		
Aeration					
Pretreatment[b]		90–99	90–99	>90	>90
Hardness reduction	High lime	90–99.9	99–99.9		
	Low lime	90–99	90–99		
Slow sand filtration	Without pretreatment[b]	35–99.5	10–99.9	59–94	
	With pretreatment[b]	90–99.9	90–99.9	59–99.98	
Rapid granular filtration	Without pretreatment[b]	0–90	0–50	0–90	
	With pretreatment[c]	90–99	90–99	90–99	
	With pretreatment[b]	90–99.9	90–99	90–99.9	
Diatomaceous earth filtration[d]		90–99.9	99–99.96	99–99.999	
Activated carbon			10–99		
Microfiltration		>99.999	>99	>99.9999	

[a]*Removal can also expressed using log units. 1 log = 90% removal, 2 log = 99% removal, 3 log = 99.9% removal, and so on.*
[b]*Pretreatment includes coagulation, flocculation, and sedimentation.*
[c]*Pretreatment, except no sedimentation.*
[d]*With pretreatment and precoating of filter.*
Source: *Reprinted in part from Amirtharajah, A. (1986). Variance analyses and criteria for treatment regulations.* Journal of the American Water Works Association, *78, 34–49.*

Slow sand filters

The efficiency of microorganism removal by slow sand filters is influenced by several factors, including the particle size of the filter medium and the extent to which the scum layer (schmutzdecke) has developed. A filter with a smaller particle size is more efficient in microbe removal, but results in shortened filter runs, that is, more frequent filter cleaning is necessary to maintain a suitable water flow rate. In addition, new or newly cleaned slow sand filters require some conditioning (ripening) before they are effective in removing microorganisms. The ripening period allows a biologically active schmutzdecke to build up on the particles and in the filter bed. This layer then assists in the filtration of other particles and colloids from the water. The schmutzdecke is important in microbe removal, particularly for bacteria and viruses. Virus removal by sterile sand is negligible, and removal by clean nonsterile sand is variable but poor.

Removal of coliform bacteria by slow sand filters ranges from about 83% with new filter sand to nearly 100% for sand with an established schmutzdecke. Removal of poliovirus type 1 by slow sand filters ranges from 22% to 96% for clean sand to >99.9% for sand with an established biological population. Removal of *Giardia* cysts (protozoan) by slow sand filtration ranges from 59% to 99.98%.

Cold water temperatures generally decrease microorganism removals by slow sand filtration. At water temperatures >5 °C, removal of heterotrophic bacteria (measured by HPC) and coliforms decreases by about 2% and 2–10%, respectively, compared with removals at temperatures greater than 15 °C. Virus removals decrease by about 0.5% and protozoan cyst removals decline by 0–6.2%. Increased water flow through the filter (filtration rate) also results in decreased microorganism removals.

Rapid granular filters

Systems that use rapid granular filtration must first pretreat the raw source water (coagulation, flocculation, and sedimentation) for effective microbial removal, unless the water is clear (turbidity >10 NTU). Effective pretreatment should reduce the turbidity of muddy surface water to a level well below 1 NTU. Effective pretreatment and filtration collectively should reduce the turbidity to 0.1 NTU or less.

Microorganism removals achieved by pretreatment and rapid granular filtration are high. Bacterial removals (heterotrophic bacteria and coliforms) range from 86% to 98.8%, and virus removals range from 90% to >99.99%. The coagulation and filtration processes generally can achieve removals of protozoan cysts ranging from 83% to 99.99%. Efficient removal of *Giardia* and other protozoan cysts by filtration is dependent on an adequate coagulant dose. A change in alum dose from 5 to 10 mg l^{-1} can increase the removal of *Giardia* cysts from 96% to >99%.

Factors that adversely affect removal efficiency include interruption of chemical feed (coagulants and polymers), poor filter efficiency at the beginning of a filter run, sudden increases in water flow, and turbidity breakthrough that can occur with higher filter head loss (resistance to water passage through the filter) at the end of a filter run. Any of these factors can seriously degrade the microbiological quality of the filtered water. In addition, some source waters are so highly polluted that full treatment (pretreatment, filtration, and disinfection) may not achieve a suitable level of microbiological reduction.

Diatomaceous earth

Bacterial removals by DE filtration are affected by the grade of DE used. When fine DE is used that yields a median pore size of 1.5 μm in the filter cake, bacterial removals of nearly 100% can be achieved. Lower percentage removals occur when coarser DE is used that provides increased median pore size. However, by chemical conditioning of coarser grades of DE, good bacterial removals can be obtained.

DE filtration can satisfactorily remove viruses from water, but to be most effective, the raw water must be pretreated with a coagulant aid (polymer) or the DE filter cake must be chemically conditioned to enhance virus attachment to the filter material during filtration. Overall, DE filtration is most effective for removal of microorganisms in the size range of *Giardia* or *Entamoeba histolytica* cysts. *Giardia* cyst removals of 99% or greater can be achieved by proper DE filter operation. The use of DE to remove smaller organisms (bacteria and viruses) can be improved by coating it with aluminum hydroxide precipitate. This coating gives the DE a positive surface electrical charge. Bacteria and viruses with a negative electrical charge are then removed by surface attachment.

Membranes

Membrane processes can be effective at removing various microbial contaminants. For example, MF and UF are generally capable of removing 6–7 logs of protozoan cysts and greater than 5.5 logs of bacteria. Lower removal rates are achieved for viruses: less than 2.5 logs removal and greater than 3 logs removal for MF membranes and UF systems, respectively.

Disinfection

General Considerations

The final treatment process for drinking water is chemical or physical disinfection intended to inactivate pathogens that enter the distribution system. The effectiveness of disinfection is a function of the types of organisms to be inactivated, the quality of the water, the type and concentration of the disinfectant, the exposure or contact time, and the temperature of the water.

TABLE 2 *CT* value ranges for inactivation of microorganism by disinfectants

Microorganism	Free Chlorine (pH 6–7)	Preformed Chloramine (pH 8–9)	Chlorine Dioxide (pH 6–7)	Ozone (pH 6–7)
E. coli	0.034–0.05	95–180	0.4–0.75	0.02
Poliovirus-1	1.1–2.5	768–3740	0.2–6.7	0.1–0.2
Rotavirus	0.01–0.05	3806–6476	0.2–2.1	0.006–0.06
Phage f2	0.08–0.18	ND	ND	ND
G. lamblia cysts	47–150	2200[a]	26[a]	0.5–0.6
G. muris cysts	30–630	1400	7.2–18.5	1.8–2.0
Cryptosporidium parvum	7200[b]	7200[c]	78[c]	5–10[b]

ND, No data.
All *CT* values are for 99% inactivation at 5 °C except where noted.
[a]*Values for 99.9% inactivation at pH 6–9.*
[b]*99% inactivation at pH 7 and 25 °C.*
[c]*90% inactivation at pH 7 and 25 °C.*

As stated previously, *CT* values are used to identify the level of inactivation provided by a given disinfectant for an organism under a specific environmental condition. These values are useful for comparing biocidal efficiency. Table 2 provides *CT* values for several microorganisms. Most of the available *CT* data for microorganisms of health concern were developed from laboratory studies that might not be indicative of actual *CT* values achieved in operating water treatment plants.

Water temperature can affect disinfection rates (and thus *CT* values). Microorganism inactivation rates decrease as water temperature decreases. Water pH can also affect disinfection rates. In most water systems, the pH is kept in the range of 7–9. Water pH, for example, determines the proportions of the most important chlorine species, hypochlorous acid (HOCl) and hypochlorite (OCl^-). Lower pH values (pH 6–7) result in the formation of HOCl, which is favorable for rapid inactivation, whereas higher pH values (pH 8–10) result in formation of OCl^-, which results in slower inactivation. For chlorine dioxide (ClO_2), which does not dissociate, inactivation is more rapid at higher pH values (pH 9) than at lower pH values (pH 7). Ozone disinfection efficacy does not appear to be affected by pH.

An important factor in bacterial inactivation is the phenomenon of cell injury. Disinfection and other environmental stresses may cause nonlethal physiological injury to waterborne bacteria. This phenomenon causes problems for monitoring water quality and calculating *CT* values, because injured bacteria may not grow on selective media normally used to detect or enumerate the bacteria. Thus, the actual number of viable cells may be underestimated. In some cases, injured pathogens remain infective. Problems with detecting injured cells can be mitigated by the use of media and procedures that remain selective, yet permit the injured cells to repair metabolic damage.

Microorganism Inactivation

Table 2 shows the effect of disinfectant concentration, contact time, pH, and temperature on microbial inactivation.

Chlorine. Table 2 shows that enteric viruses (represented by poliovirus type 1) are more resistant to inactivation by chlorine than are bacteria (represented by *E. coli*), and protozoan cysts are nearly two orders of magnitude more resistant than the enteric viruses.

Chloramines. Comparison of chloramines with chlorine for disinfection of microorganisms (Table 2) shows that, in general, for all types of microorganisms, *CT* values for chloramines are higher than *CT* values for free chlorine species.

Chlorine dioxide. Chlorine dioxide *CT* values in Table 2 show that at pH 7.0, ClO_2 is not as strong a bactericide and virucide as HOCl. However, as the pH is increased, the efficiency of ClO_2 for inactivation of viruses increases. CT data for protozoan cyst inactivation is not available.

Ozone. Overall, comparison of *CT* values for ozone with those for chlorine and ClO_2 indicates that ozone is a much more effective biocide than the other disinfectants. *E. coli* is about 10-fold (1 $\log_{10}$) more sensitive to ozone (Table 2) than poliovirus type 1. *Giardia muris* cysts are about 10-fold more resistant to ozone than poliovirus type 1. Since ozone is a powerful oxidant, it reacts rapidly with both microorganisms and organic solutes and is very useful as a primary disinfectant.

The order of microbial disinfectant efficiency is O_3 > ClO_2 > HOCl > OCl^- > NH_2Cl > $NHCl_2$ > RNHCl (organic chloramines). However, for technical reasons, practical handling considerations, cost and effectiveness, the frequency of use of disinfectants by utilities in the United States is generally chlorine >> chloramines > O_3 > ClO_2.

Ultraviolet Light. The first reliable application for using ultraviolet (UV) light for disinfecting drinking water occurred in Switzerland and Austria in 1955. However, use of UV light was limited by the low cost of chorine, early operational problems, and the absence of a disinfectant residual that would continue to disinfect the drinking water in the distribution system. In addition, organisms that are sublethally injured by UV light exposure may, under appropriate conditions, be able to repair the damage (i.e., photoreactivation or dark repair). Nevertheless, as of 2010, more than 9,000 water systems in the US use UV light.

Recently, it was found that two important protozoan pathogens in drinking water, *Cryptosporidium* oocysts and *Giardia* cysts (but not *Acanthameba* cysts) were susceptible to UV inactivation, prompting a growing interest in the use of UV disinfection by water systems. The primary reasons for this interest in the US is that water systems are required to comply with *Cryptosporidium* and *Giardia* standards as well as meet standards that limit the levels of chlorine and toxic disinfection byproducts in drinking water that form when chlorine disinfectant interacts with natural substances in the source water. Among the largest systems in the US that have or are constructing UV facilities are those supplying the cities of Seattle, New York City, and Cincinnati.

Enteric bacteria are most sensitive to UV inactivation, while enteric viruses and bacterial sporeformers the most resistant. Spores of *Bacillus anthracis* are especially resistant, with about a 40 mJ/cm^2 dose required for a 99% (2-log) inactivation. For comparison, 99.99% (4-log) inactivation was achieved for a number of nonsporeforming bacterial pathogens with a UV dose of less than 12 mJ/cm^2. A 99% inactivation for viruses, *Giardia*, and *Cryptosporidum* requires a UV dose of 100, 5.2, and 5.8 mJ/cm^2, respectively.

Distribution Systems

Description

Water transmission and distribution systems are needed to deliver water to the consumers. In 2006, the US National Academy of Sciences published a report, "Drinking Water Distribution Systems: Assessing and Reducing Risks" that recommended the following measures to control waterborne pathogens entering the distribution system.

1. Storage facilities should be inspected on a regular basis.
2. Better sanitary practices are needed during installation, repair, replacement, and rehabilitation of distribution system infrastructure.
3. Water residence times in pipes, storage facilities, and premise plumbing should be minimized.
4. Positive water pressure should be maintained.
5. Distribution system monitoring and modeling are critical to maintaining hydraulic integrity.
6. Microbial growth and biofilm development in distribution systems should be minimized.
7. Residual disinfectant choices should be balanced to meet the overall goal of protecting public health.
8. Standards for materials used in distribution systems should be updated to address their impact on water quality, and research is needed to develop new materials that will have minimal impacts.
9. Although it is difficult and costly to perform, condition assessment of buried infrastructure should be a top priority for utilities.
10. Cross-connection control should be in place for all water utilities.
11. Where feasible, surge protection devices should be installed.
12. Prior to distribution, the quality of treated water should be adjusted to minimize deterioration of water quality.

Distribution systems represent the major investment of a municipal water works and consist of large mains that carry water from the source or treatment plant, service lines that carry water from the mains to the buildings or properties being served, and storage reservoirs that provide water storage to meet demand fluctuations, for firefighting use, and to stabilize water pressure. The branch and loop (or grid) are the two basic configurations for most water distribution systems.

The layout of a branch system is similar to that of a tree branch, with smaller pipes branching off from larger pipes throughout the area served. This system, or a derivative of it, is normally used to supply rural areas where water demand is relatively low and long distances must be covered. Disadvantages of this configuration are the possibility that a large number of customers will be without service should a main break occur, and the potential water quality problems in parts of the system resulting from the presence of stagnant water. System flushing should be accomplished at regular intervals to reduce the possibility of water quality problems.

The loop configuration currently is the most widely used distribution system design. Good design practices for smaller systems call for feeder mains to form a loop approximately 1 mi (1600 m) in radius around the center of the town with additional feeder loops according to the particular layout and geography of the area to be served. The area inside and immediately surrounding the feeder loops should be gridded with connecting water mains on every street.

The most commonly used pipes for water mains are ductile iron, pre-stressed concrete cylinders, polyvinyl chloride (PVC), reinforced plastic, steel, and asbestos cement.

Microbiology

Microbiologically, water distribution systems are interesting bacterial ecosystems that present a real challenge to the water utilities in terms of maintaining good-quality water with low bacterial densities. The construction characteristics, operation, and maintenance of a water distribution system provide ample opportunities for microbial recontamination of the treated water during distribution. Pipe joints, valves, elbows, tees, and other fittings as well as the vast amount of pipe surface provide both changing water movement and stagnant areas where bacteria can attach and colonize. Water distribution systems are susceptible to cross-connections that may allow entry of pathogens into the system. A cross-connection is any direct connection between the drinking water distribution system and any nonpotable fluid or substance.

Biofilms in Water Distribution Systems

Bacteria found in water distribution systems can be classified into indigenous (autochthonous) and exogenous (allochthonous) populations. The indigenous organisms are well-adapted biofilm-forming bacteria that represent a stable ecosystem that is difficult to eradicate. The exogenous bacteria are contaminants that are transported into the system by a variety of mechanisms.

The development of a permanent biofilm in the distribution system occurs because the bacteria find physical and chemical conditions conducive to colonization and growth at the solid surface/water interface. These conditions include an ample supply of nutrients (AOC) for growth, a relatively stable temperature, and some degree of protection from exposure to harmful chemicals such as the disinfectant(s) used to treat the water.

When an adequate disinfectant residual is maintained in the water throughout a distribution system, growth of bacteria is usually well controlled and the density of bacteria in the bulk water traveling through the pipes will remain low – in the range of >10 to several hundred CFU ml^{-1} (HPC). The density of bacteria in the biofilm, however, may be several orders of magnitude higher, ranging upward of 1×10^5 CFU/cm^2 (HPC). The disinfectant residual concentration needed to control the growth of the bacteria varies from one water system to another. The choice of the medium and method used to determine the bacterial density may result in low or high counts of heterotrophic bacteria. In general, rich culture media and incubation at 35 °C will yield lower counts than dilute nutrient media and incubation at 20–28 °C. Factors that appear to be critical are pH, temperature, dissolved organic carbon (DOC) concentration, AOC concentration, and type of disinfectant used. The disinfectant residual will also reduce or suppress extensive biofilm growth if the residual is maintained throughout the system.

Disinfectant residuals should be at least 0.2 mg l^{-1} for chlorine and 0.4 mg l^{-1} for monochloramine (NH_2Cl). Higher concentrations of disinfectant residual may be applied and maintained, but if the water contains high levels of DOC, it may be difficult to maintain an adequate disinfectant residual to control bacterial growth and still have water that is aesthetically acceptable to the consumers.

Bacterial concentrations in distribution water vary from >1 CFU ml^{-1} in the water leaving the treatment plant to as high as 10^5–10^6 CFU ml^{-1} in water from slow flow or stagnant areas of the distribution system. The concentrations of bacteria in the water and on the pipe surfaces vary spatially and temporally in the distribution system. Bacterial densities in the pipe wall biofilm and in sediments may reach 10^7 CFU/cm^2. The biofilm contributes bacteria to the flowing water through shear loss (erosion) and by migration of actively motile bacterial cells into the water.

Some of the bacteria commonly found in drinking water, sediments, and biofilms are listed in Table 3. Many of these bacteria are found in both the water and the biofilm, indicating the influence of the biofilm on the bacterial quality of distribution system water. Indeed, biofilms are where the most significant bacterial growth occurs in the distribution

TABLE 3 Microorganisms found in treated distribution water, sediment, and distribution system biofilm

Microorganism	Distribution Water	Sediment	Biofilm
Pseudomonas spp.	X	X	X
Alcaligenes spp.	X		X
Acinetobacter spp.	X		X
Moraxella spp.	X	X	X
Arthrobacter spp.	X	X	X
Corynebacterium spp.			X
Bacillus spp.	X	X	X
Enterobacter spp.	X	X	X
Micrococcus spp.	X		
Flavobacterium spp.	X	X	X
Klebsiella spp.	X	X	X
Mycobacterium spp.	X	X	X
Iron/sulfur bacteria	X	X	X
Nitrifying bacteria	X	X	X
Yeasts/fungi	X	X	X
Invertebrates/protozoa	X	X	X
Enteric viruses	X	X	X

system. Moreover, such multispecies biofilms form relatively protected microniches where pathogens can seek refuge. Bacteria in drinking water are detected and enumerated using traditional culture media and methods (coliforms, fecal coliforms, HPC, and so on). The recent development of culture-independent molecular methods (DNA-based) has begun to reveal the great diversity of bacteria in water, biofilms, and sediments. Genes of many bacteria that cannot yet be cultured have been detected using the DNA-based methods.

Iron and sulfur bacteria are 'nuisance' organisms that cause taste and odor problems and are often associated with groundwater sources. Nitrifying bacteria and fungi cause problems in chloraminated drinking water systems, including depletion of total chlorine residual, conversion of reduced nitrogen compounds such as ammonia to nitrite and nitrate, and high levels of heterotrophic bacteria. Invertebrates and protozoa are nuisance organisms associated with aesthetic complaints about water quality. Invertebrates and protozoa may also harbor pathogenic and nonpathogenic bacteria, either internally or attached to their surfaces, and this association provides the bacteria with some degree of protection from the inactivation by disinfectant residuals.

Information about the development and control of biofilms in drinking water is sparse. In addition, there is meager data on other issues related to biofilms, including the role of iron and sulfur bacteria in biofilm development, the role of biofilms in the corrosion of pipe material, the role of biofilms in nitrification for systems that use chloramines as a secondary disinfectant, and the effect of added corrosion inhibitors on bacterial populations in biofilms.

The presence of chlorine or chloramine retards the development and affects the spatial distribution of biofilm. The type of disinfectant residual provided selects for bacteria that are more tolerant to the specific disinfectant used. The biofilm environment provides protection for the cells by diffusional resistance and neutralization of the disinfectant. Therefore, biofilm organisms are less inhibited by the disinfectant residual than are planktonic cells. Differential effectiveness of chlorine and monochloramine for control of biofilm growth has been shown. Monochloramine, because it is less reactive and therefore more persistent, apparently can penetrate the biofilm and is more effective than chlorine in controlling biofilm growth.

In some cases, coliforms that gain entry into the distribution system may attach to pipes or pipe sediments and proliferate, thus becoming a biofilm constituent. The intermittent, sporadic, or persistent sloughing of coliform bacteria from biofilms into the water of the distribution system may cause systems to repeatedly violate standards for total coliforms. This problem is most frequently associated with utilities that use a surface water supply where the water temperature is 15 °C or greater. Public safety dictates that all coliforms detected by testing be regarded as representing system vulnerability, unless strong evidence suggests otherwise.

The utility should review its treatment operations and increase monitoring of the quality of water entering the distribution system to be sure that inadequate or failed treatment is not responsible for the total coliform occurrences and that *E. coli* is not present in the water. The water utility should also review the operation and management of the water distribution system to ensure that no *E. coli* are present, a disinfectant residual is present, and an adequate cross connection control program is in effect. Finally, the water utility should insure that large volumes of water held in storage in reservoirs, standpipes, or above ground tanks are not the source of the total coliform problem.

During warm-water periods, it is more difficult to maintain a disinfectant residual in the water in all parts of the distribution system. With increasing temperature, the disinfectant reacts more rapidly with dissolved organic chemicals in the water and in the biofilm, and bacteria grow faster. Increased reaction rates and bacterial growth are often compounded by a lack of knowledge of system hydraulics and of the actual water movement throughout the system. In many cases, there are large areas where minimal movement of the water occurs. Many water utilities may create microbial or chemical contaminant problems by their failure to understand how their systems work; better system management can reduce those problems.

Biofilms in Household and Building Plumbing Systems

Biofilms develop in household and building plumbing systems and can degrade water quality more rapidly and to a greater extent than in the distribution mains. In such systems, pipe materials, long residence times, stagnation/low-flow conditions, residual disinfectant concentration, and water temperature may differ considerably from those found in the main distribution piping. Increased disinfectant consumption in household and building networks may be related to the smaller diameters of service line piping, which results in a greater surface-to-volume ratio than in the pipes of the main distribution system. Depletion of the disinfectant residual due to water stagnation, organic deposits on the pipe surfaces and increased water temperature result in increased biofilm bacterial growth.

In general, there is little evidence that high levels of heterotrophic bacteria in drinking water from plumbing systems have adverse health effects. However, for specific organisms such as *Legionella pneumophila* and nontuberculous *Mycobacterium* spp. that have been found as part of the biofilm in such systems, there is evidence of adverse health effects. Plumbing system components such as hot water storage tanks and shower heads may permit amplification of the bacteria to significant levels. For example, outbreaks of Legionnaire's disease in healthcare facilities have been related to amplification of *L. pneumophila* in hot-water storage tanks and shower heads.

In the case of nontuberculous mycobacteria, some evidence suggests that the type of residual disinfectant used in the drinking water treatment process may select for colonization of biofilms by these organisms, and species found in treated drinking water have been linked to human infections in immunocompromised patients.

ALTERNATIVE WATER SOURCES

In areas where both surface and groundwater sources may not be suitable as potable source water, other sources of drinking water may be necessary. Bottled water is one alternative source, but is very expensive and is therefore not viewed as a long-term solution.

Bottled Water

The bacterial quality of noncarbonated bottled water varies both among brands and from lot to lot within a brand. Chemical quality varies as well. Bottled water in the United States is usually treated by R O, ozone, and/or ultraviolet light. Bacterial densities in noncarbonated bottled water, determined by plate count methods, vary from 0 to >1.0 $\times 10^5$ CFU ml^{-1}. Fresh noncarbonated bottled waters often have low bacterial densities, but during storage prior to sale the bacteria multiply, often by several orders of magnitude. Carbonated (sparkling) bottled waters generally have low or no detectable bacteria because of the low pH. In most bottled waters examined, coliform bacteria have been absent. Bottled water quality is regulated by the US Food and Drug Administration, but must comply with EPA drinking water regulations to the extent it is possible to do so.

Emergency Drinking Water

Treatment of water for drinking use in an emergency generally involves either boiling the water or chemically disinfecting the water using liquid chlorine laundry bleach, tincture of iodine, or iodine or chlorine tablets. Water to be disinfected should be strained through clean cloth to remove particles and floating matter.

Boiling is the most effective means of inactivating pathogens. For heat disinfection, the strained water should be boiled vigorously for at least 1 min and allowed to cool before use. For chemical disinfection, add liquid chlorine bleach (4–6% chlorine) at a rate of two drops (clean clear water) or four drops (cloudy water) per quart, mix thoroughly, and allow to stand for 30 min. A slight chlorine odor should be detectable in the water; if not, repeat the dosage and let stand for an additional 15 min before using. If tincture of iodine is used, add five drops (clean clear water) or ten drops (cloudy water) of 2% tincture of iodine solution per quart of water. Allow the water to stand for 30 min before use. For chlorine or iodine tablets (obtained from drug or sporting goods stores), follow the instructions on the package. Keep all purified water in clean closed containers. Chlorine and iodine may not inactivate the protozoan pathogens, especially *Cryptosporidium*.

Point-of-Use/Point-of-Entry (POU/POE) Treatment

POU/POE devices (or filters) are small package units designed to treat water at a designated tap in a household or to treat all of the water entering the household. These devices are becoming increasingly sophisticated and are growing in popularity. POU devices are generally placed under a sink and POE devices might be located in a basement or a garage. Currently, POU/POE treatment is used to reduce the levels of a wide variety of contaminants in drinking water, including organic material, turbidity, fluoride, iron, radium, chlorine, arsenic, nitrate, ammonia, and pathogens (including cysts/oocysts). Taste, odor, and color can also be improved with POU/POE treatment.

The types of technology that may be used in POU/POE treatment devices include adsorption, ion exchange, R O filtration, chemical oxidation, distillation, aeration, and disinfection such as chlorination, iodination, ozonation, and ultraviolet light. Activated carbon filtration, a type of adsorption device, is the most widely used POU system for home treatment of water. The performance of an individual unit depends on a combination of factors, such as unit design, type and amount of activated carbon, and contact time of the water with the carbon. Most units use GAC in their design, although other types of carbon such as pressed block, briquettes of powdered carbon, and powdered carbon are available.

Bacterial growth in activated carbon units may be a problem. The organic material adsorbed onto the carbon provides nutrients for bacterial growth. Stagnation periods between operations allow time for bacteria to grow and high densities will develop in newly installed filters within 2 weeks. To help minimize this problem, some units contain GAC impregnated with silver or some other bacteriostatic agent. Silver, although useful against some bacteria (e.g., coliforms and related enteric bacteria) as a bacteriostatic agent, does not effectively inhibit the growth of all bacteria in GAC, POU, or POE filters. Bacterial levels in water leaving GAC filters, with or without silver, range from 10^2 to 10^5 CFU ml^{-1}, but it may take longer for bacterial levels in a GAC filter with silver to reach the higher concentrations. When a GAC filter is used as the key treatment technique and the incoming water is not microbiologically safe, water from the unit should be disinfected either chemically or by UV irradiation to reduce bacterial levels in the water and to inactivate opportunistic pathogens that may colonize the carbon particles. Silver resistant nontuberculous mycobacteria have been shown to colonize point of use water filters containing silver, suggesting a potential risk to immunocompromised persons who consume water from such devices.

Some carbon block units do not contain silver but claim bactericidal effects, probably due to straining through small carbon pore size (0.5 μm). Some POU/POE systems use a disinfectant such as ozone or UV irradiation to control bacteria.

Household Water Treatment in Developing Countries

Developing countries often lack the resources for the types of water treatment systems used in developed countries. The WHO, US Centers for Disease Control and Prevention, and other organizations are developing and testing simple and inexpensive alternatives that households in such countries can use to control waterborne disease. Among the more promising alternative disinfection methods being tested and used are boiling, solar disinfection using clear plastic containers (combined UV radiation and thermal effects), solar disinfection using dark, or partially dark, opaque containers (thermal effects alone), UV lamps, chlorine tablets, and liquid solutions of sodium hypochlorite.

Sometimes, a filtration process is needed, given that high organic matter levels in the water affect chlorine efficiency and high water turbidity levels affect UV treatment efficiency. Simple filtration processes used by households in the third world include adsorption by clay, activated carbon, charcoal, porous ceramic filters, fibers, cloth, paper, and crushed vegetative matter such as seeds. To minimize post-treatment contamination, the WHO recommends household water storage in narrow-mouth containers with a spigot or spout.

The candidate treatment alternatives mentioned above are by no means exhaustive. The alternative selected depends upon such issues as raw water quality, treatment effectiveness and reliability, and availability of resources such as electricity (e.g., for UV lamps), wood or other fuel (for boiling water), and hours and days of sunlight per year (solar heating). It also depends upon broader issues such as affordability, technical simplicity, education level, cultural acceptance of candidate alternatives, and the prospect for continuing the use of candidate alternatives over time. More information on household treatment systems in developing countries is available at http://www.who.int. and http://www.cdc.gov.

A meta-analysis conducted on 28 separate randomized controlled trials of household water treatments found that ceramic filters were effective in preventing disease, whereas chlorination, coagulation-chlorination, and solar disinfection had little if any benefit Hunter, P. (2009).

OTHER ISSUES ASSOCIATED WITH WATER TREATMENT

Drinking Water Microbial Diversity

For nearly 100 years, information about the microbial diversity in the distribution system has relied on culture-based techniques. This means that, until recently, characterization of the bacterial flora has been based solely upon bacteria that can form colonies on agar media under the selected conditions of incubation. However, only a small fraction of the total bacterial population can grow under such conditions, and this has led to a biased picture of the microbial diversity. A similar situation holds for the viruses.

Recent developments in molecular methods have deepened our understanding of the microbial community dynamics in water distribution systems. For example, studies using methods targeting phospholipid fatty acids and RNA have demonstrated that different disinfection treatments (i.e., chlorination vs. chloramination) modify the characteristics of the active and dynamic microbial community in drinking water biofilms. Unlike DNA, microbial phospholipid fatty acids and RNA are thought to degrade fairly rapidly in environmental settings and thus their detection can be used as evidence for the presence of metabolically active cells and hence disinfection efficiency in drinking water.

DNA-based methods have also been used to detect microbes in drinking water, often aided by polymerase chain reaction amplification. For example, such methods have revealed previously unidentified bacteria closely related to *Legionella* and *Mycobacterium* spp. in drinking water (Figure 2). While species of the latter genera have been associated with waterborne illness, their detection does not necessarily imply a health risk or a deficiency in water treatment, as most species within these genera are not pathogenic to humans. Nucleic-acid-based methods have also detected fastidious organisms in drinking water such as nitrifying bacteria, whose presence may degrade drinking water quality and thus may reflect inadequate water treatment practices.

Other biotechnological advances such as next generation sequencing, microarrays, bioinformatics, biocomputing, and "omics" (e.g., genomics, metagenomics, metatranscriptomics, metabolomics, and proteomics) are providing researchers with new tools to help describe the composition of water microbial communities for both natural and engineered systems in a more comprehensive and economical fashion. This information will lead to the development of more robust detection methods and a better understanding of how microorganisms interact with environmental factors within water distribution systems.

Source Water Protection

Preventing the pollution of waterbodies that serve as sources of drinking water (e.g., groundwater, lakes, and rivers) is becoming an important challenge to both developed and

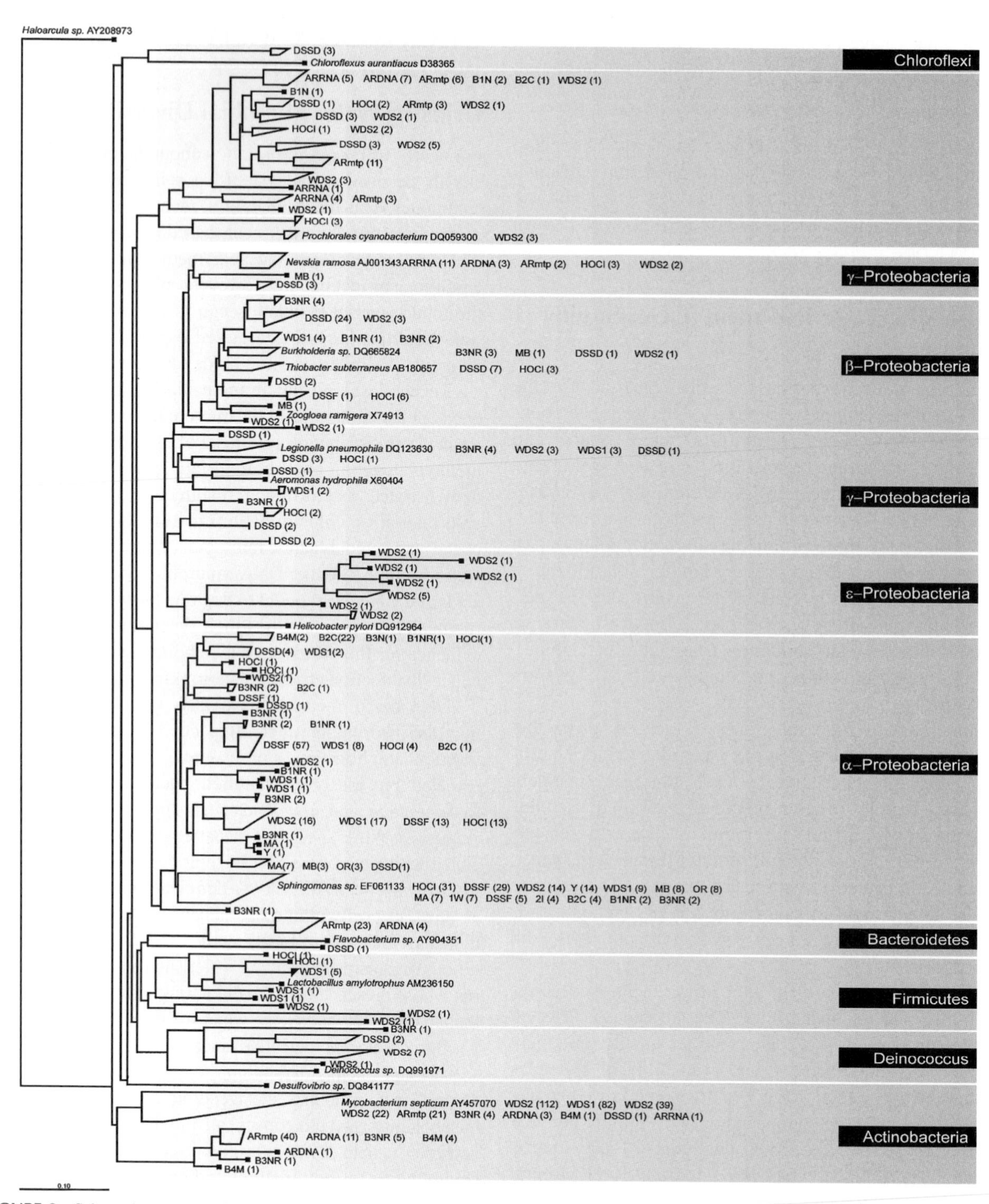

FIGURE 2 Schematic representation of the distribution of 16S rRNA gene sequences derived from drinking water biofilm and planktonic samples. HOCl, tap water; DSSF, distribution system simulator feed water; DSSD, distribution system simulator discharge water; WDS2, distribution system – surface water source; WDS1, distribution system – groundwater source; OR/Y/MA/MB, distribution system simulator R2A isolates; ARRNA, model biofilm using 16S rRNA; ARDNA, model biofilm using 16S rRNA gene; ARmtp, model biofilm DNA; B1N, model biofilm no disinfectant; B2C, model biofilm chlorinated; B3N, model biofilm no disinfectant; B4M, model biofilm monochloramine.

developing countries. Source water pollution has a direct impact on treatment plants, particularly as the level of pollution can have an effect on the removal and disinfection efficiency of enteric pathogens and their eventual intrusion into the distribution system. In the United States, under the Clean Water Act, each state sets water quality standards that include limits on the concentration of specific pollutants that can be present in an ambient water body (usually a lake or a river) and this limit depends on its designated use (e.g., swimming, boating, fishing, drinking water source, and shellfish harvesting). To limit pathogens, normally a fecal indicator such as *E. coli* or enterococci is used. Then discharges (e.g., a fecal indicator) into that water body are set at levels that meet that water quality standard. One goal of this requirement is to limit the level of fecal matter that reaches a water system intake. This limit facilitates treatment and reduces the health risk in the event of a treatment failure. In spite of these standards, a significant portion of US surface waters exceed regulatory limits. Several microbial source tracking (MST) methods have recently been developed to identify the primary sources of pollution and to establish adequate pollution control practices. The use of MST data could lead to measures that decrease the level of enteric pathogens entering water bodies used by drinking water systems.

Homeland Security

Concerns regarding the vulnerability of municipal water supplies to biological attacks have prompted renewed interest in treatment issues, particularly as they relate to the survival of select agents in potable water (e.g., *Bacillus anthracis*, *Francisella tularensis*, and smallpox virus). Data on the inactivation of bacterial bioterrorism agents by chlorination are available. The detection of bioterrorism agents in drinking water requires methods with low detection limits as these agents need to be detected at very low densities, given that they have low infection doses. As a result, future early warning detection systems will rely on the development of efficient concentration methods as well as rapid molecular procedures with low detection limits (e.g., polymerase chain reaction assays).

CONCLUSIONS

Overall, the treatment technology available today is capable of providing microbiologically safe drinking water to all consumers. In practice, however, many systems are still vulnerable to waterborne disease. Reasons include (1) inability to fund the installation and use of adequate technology, (2) ignorance of the need for adequate technology, (3) inadequate or improper monitoring to insure that barriers against waterborne disease remain intact, (4) lack of real-time monitoring data, and (5) lack of adequate data on the effectiveness of various treatment practices against certain pathogens. In addition, with increased population pressure, source water quality may decline over time, especially in areas where water is scarce; thus the technology used becomes inadequate. These problems are common on a global basis. The challenge globally is to highlight the imperative of safe drinking water and to find solutions to these problems.

FURTHER READING

American Public Health Association. (2005). *Standard methods for the examination of water and wastewater* (21st ed.). Washington, DC.

Amirtharajah, A. (1986). Variance analyses and criteria for treatment regulations. *Journal American Water Works Associaction, 78*, 34–49.

Berry, D., Xi, C., & Raskin, L. (2006). Microbial ecology of drinking water distribution systems. *Current Opinion in Biotechnology, 17*(3), 297–302.

Clark, R. M., & Tippen, D. L. (1990). Water supply. In R. Corbitt (Ed.) *Standard handbook of environmental engineering* (pp. 5.1–5.225). New York: McGraw-Hill.

Geldreich, E. E. (1996). *Microbial quality of water supply in distribution systems*. New York: Lewis Publishers.

Hijnen, W. A. M., Beerendonk, E. F., & Medema, G. J. (2006). Inactivation credit of UV radiation for viruses, bacteria and protozoan (oo)cysts in water: A review. *Water Research, 40*, 3–22.

Hrudey, S. E., & Hrudey, E. J. (2004). *Safe drinking water lessons from recent outbreaks in affluent nations*. London, UK: IWA Pub.

Hunter, P. (2009). Household water treatment in developing countries: Comparing different intervention types using meta-regression. *Environmental Science & Technology, 43(23)*, 8991–8997.

McFeters, G. A. (1989). *Drinking water microbiology*. New York: Springer-Verlag.

National Research Council. (2006). *Drinking water distribution systems: assessing and reducing risks*. Washington, DC: The National Academies Press.

O'Melia, C. R. (1985). Coagulation. In R. L. Sanks (Ed.) *Water treatment plant design* (pp. 65–81). Ann Arbor, MI: Ann Arbor Science Publishers.

Rose, L. J., Rice, E. W., & Jensen, B. (2005). Chlorine inactivation of bacterial bioterrorism agents. *Applied and Environmental Microbiology, 71*, 566–568.

USEPA (US Environmental Protection Agency). (1991). *Manual of small public water supply systems*. Washington, DC: EPA 570/9–91–003.

WHO (World Health Organization) (2006). *Guidelines for drinking-water quality* (3rd ed.). Geneva, Switzerland: World Health Organization.

RELEVANT WEBSITE

http://www.gpoaccess.gov. – A Service of the U.S. Government Printing Office.

http://www.cdc.gov. – Centers for Disease Control and Prevention.

http://europa.eu. – Europa.

http://www.hc-sc.gc.ca – Health Canada Santé Canada.

http://www.mfe.govt.nz. – Ministry for the Environment New Zealand.

http://www.epa.gov/safewater, the http://www.epa.safewater.gov website of the US Environmental Protection Agency's Office of Ground Water and Drinking Water.

http://www.wqra.com.au/. – Water Quality Research Australia Limited.

http://www.who.int/en. – World Health Organization.

Chapter 47

Spacecraft Associated Microbes

J. Bruckner[1], S. Osman[1], K. Venkateswaran[1] and C. Conley[2]
[1]*California Institute of Technology, Pasadena, CA, USA*
[2]*National Aeronautics and Space Administration, Washington, DC, USA*

Chapter Outline

ABBREVIATIONS

ATLO Assembly, Test, and Launch Operations
CFU Colony-forming units
COSPAR Committee on Space Research
DHMR Dry heat microbial reduction
HEPA High efficiency particle air
JPL Jet Propulsion Laboratory
JSC Johnson Space Center
KSC Kennedy Space Center
LAL Limulus amebocyte lysate
LMA-MTF Lockheed Martin Assembly Facility
LPS Lipopolysaccharide
MER Mars Exploration Rovers
MPN Most probably number
MRO Mars Reconnaissance Orbiter
NAC NASA Advisory Council
NPR NASA Procedural Requirements
NRC National Research Council
PCR Polymerase chain reaction
PP Planetary protection
RFLP Restriction fragment length polymorphism
SAF Spacecraft assembly facility
SSB Space Studies Board
ULPA Ultra low particle air

DEFINING STATEMENT

Preventing forward contamination of extraterrestrial environments with terrestrial microbes is essential in achieving one of NASA's strategic goals, "the advancement of scientific knowledge of the origin and history of the solar system, and the potential for life elsewhere".

PLANETARY PROTECTION (PP) POLICIES AND PRACTICES

PP covers policies and practices that (1) protect solar system bodies from contamination by terrestrial biological material (forward contamination), thus preserving future opportunities for scientific investigation and (2) protect the Earth from harmful contamination by materials returned from outer space via robotic or human missions (backward contamination). International policy and requirements for PP are promulgated by the Committee on Space Research (COSPAR), which oversees implementation of guidelines laid out in Article IX of the 1967 Outer Space Treaty. NASA has engaged in PP activities since the early exploration of the Moon (then called Planetary Quarantine) with the Apollo program implementing the first quarantine procedures for returned samples and astronauts.

Topics in Ecological and Environmental Microbiology.
Published by Elsevier Inc.

By COSPAR and NASA policy, individual missions are assigned a PP Category ranging from I to V dictated by a combination of the planetary target and the individual mission's goals. Mission Categories and requirements are detailed in the NASA Procedural Requirements (NPR) document NPR 8020.12C and, briefly, are as follows:

- Category I: any mission to a body considered not of direct interest for understanding chemical evolution or the origins of life; beyond initial categorization these missions have no additional PP requirements.
- Category II: any mission to a body of significant interest relative to chemical evolution that has only remote chances of forward contamination; PP implementation requirements of documentation only.
- Category III: any mission planning to fly by or orbit a body of significant interest relative to chemical evolution and the origins of life or for which contamination would jeopardize future exploration. Requirements imposed on Category III missions include significant documentation, as well as cleanliness and/or orbital lifetime restrictions.
- Category IV: any mission including landers and probes to the surfaces of planetary bodies of significant interest to chemical evolution and the origins of life or for which contamination would jeopardize future exploration. Requirements for Category IV missions include thorough documentation as well as cleanliness requirements designed to minimize biological contamination of the target body.
- Category V: all missions with a sample return component are assigned Category V for the return leg, with the outbound leg assigned the appropriate category for that mission and target combination. Category V missions are further categorized as:
  - *Unrestricted Earth Return* – includes samples from locations not of biological concern, in which case documentation is the only requirement.
  - *Restricted Earth Return* – includes samples from planetary bodies of biological concern. Samples assigned "restricted Earth return" are automatically considered hazardous to Earth until demonstrated otherwise by appropriate testing. For these samples the highest possible containment that will protect the Earth from the sample and vice versa is mandated. This requires the construction of a facility operating at better than Biosafety Level 4 (the most stringent containment available for diseases like the smallpox or Ebola viruses).

Missions assigned Category III or IV are required to meet specific constraints on spacecraft cleanliness before launch and during mission operations. However, the specific requirements a mission must meet to minimize forward biological contamination depend on the target planetary body and are determined based on advice from the Space Studies Board (SSB) of the National Research Council (NRC) and the Planetary Protection Subcommittee of the NASA Advisory Council (NAC). For example, missions to Europa must fulfill the requirement of maintaining a less than 10^{-4} probability of contaminating the Europan ocean over the lifetime of the spacecraft, both during and after the active mission. Other icy moons in the outer Solar System are currently protected using a similar probabilistic approach.

In the case of Mars, limits on bioburden are based on requirements first imposed on the Viking missions. Appendix B of NPR 8020.12C contains individual specification sheets that provide detailed requirements, including absolute limits on the number of heat-resistant "spores" that would be allowed on flyby, orbiting, or lander missions. In the case of certain Category III missions (spacecraft intended to flyby or orbit the planet only), bioburden limits may be relaxed provided there is a 95% probability that the minimum orbital lifetime around Mars will be at least 50 years from mission launch. Category IV landed missions are assigned one of the three subcategories (IVa, IVb, and IVc) that impose differing cleanliness requirements depending on the location of the landing site and the specific objectives of the mission.

In general, the surface of Mars is considered too cold and dry to allow Earth life to propagate. There are however some areas on the surface or within the subsurface where conditions might allow terrestrial microorganisms to reproduce – these areas are considered "special regions" and spacecraft or subsystems accessing these areas must adhere to strict bioburden limits. Category IVa is assigned to missions that will land on a "non-special" region in the surface of Mars. Category IVb is assigned to any lander mission to Mars that carries *in situ* life detection instrumentation. Category IVc is assigned to spacecraft or lander subsystems that are intended to access a "special region" through either vertical or horizontal mobility. Spacecraft or subsystems that are assigned either Category IVb or Category IVc have the strictest level of bioburden limits and may require that certain subsystems (or even the spacecraft itself) not have more than 30 spores on free surfaces. This level of cleanliness is obtained by processes such as aggressive cleaning, protection from recontamination, and aseptic assembly, and then performing a sterilization step, such as dry heat microbial reduction (DHMR), which has been demonstrated to reduce the initial bioload by four orders of magnitude.

Although the traditional culture-based assay has been shown to be a valid proxy for microbial abundance, NASA recognizes that it can offer only a limited assessment of the microbial communities present. To mitigate this limitation in future missions, the SSB put forth several recommendations in a document entitled "Preventing the Forward Contamination of Mars." NRC recommendation #5 states that NASA should "plan for the effective implementation of PP requirements at the earliest stages of Mars mission and instrument design, and should provide engineers with a selection of effective, certified bioburden reduction tools". Given the limitations of culture-based analyses (including the NASA standard assay), it was clear that molecular-based methods must be implemented to comprehensively evaluate the microbial burden of spacecraft surfaces. In fact, NRC recommendation #5 suggested that "NASA should require the routine collection of phylogenetic data to a statistically appropriate level to ensure that the microbial diversity of Assembly, Test, and Launch Operations (ATLO) environments and of all NASA spacecraft to be sent to Mars is reliably assessed. NASA should also require the systematic archiving of environmental samples for ATLO environments and for all spacecraft to be sent to Mars".

To address these recommendations, NASA has implemented numerous strategies to be used in parallel with the culture-based NASA standard assay. Given that spacecraft assembly environments are highly selective for microorganisms that tolerate desiccation, chemical oxidizing agents, and UV irradiation, several culture-based methods have been utilized to enumerate and isolate not only aerobic spore formers, but also microbes capable of growth under low temperature, high salt, anaerobic, and space radiation conditions. These remarkably hardy species may persist despite sterilization procedures and therefore have the potential to remain viable on spacecraft hardware after launch. The use of culture-based methods will continue to be an important component of PP activities, as the study of microbial isolates not only expands knowledge regarding the occurrence and physiology of these extremophilic organisms but also shapes bioburden reduction policies to better target problematic, hardy microbes (thus addressing NRC recommendation #3).

Molecular detection methods have revealed the presence of microbes on spacecraft and in associated environments that are (so far) not cultivable, suggesting that cultivation approaches estimate only 1–10% of the actual microbial diversity. To address the inherent (and demonstrated) limits of culture-based assays, NASA is exploring the use of molecular-based evaluations of microbial burden on and around spacecraft and spacecraft payloads at various stages of assembly. In addition to the NASA-certified limulus amebocyte lysate (LAL) and ATP quantification assays, techniques utilizing the polymerase chain reaction (PCR) have been used to quantify (Q-PCR) and qualify (molecular cloning of amplified bacterial and archaeal 16S rDNA) the microorganisms present. In addition to the greatly reduced selective bias, these methods have proved to be more sensitive than culture-based assays and therefore well suited for analyses from low biomass environments. Data obtained by using these techniques provide a better understanding of spacecraft-associated microbial communities and aids in the evaluation and development of more effective sterilization protocols, thereby addressing both the previously mentioned NRC recommendations.

The characterization of spacecraft subsystems and their associated environments will additionally assist in achieving NASA's strategic subgoal 3C, namely, the advancement of scientific knowledge of the origin and history of the solar system and the potential for life elsewhere. The NASA Planetary Protection Research program focuses on (1) evaluating the capability of terrestrial microbes to grow in other planetary environments, (2) developing techniques to assess biological contamination on spacecraft, and (3) improving bioburden reduction techniques both (2) and (3) include validation of new standards and techniques. The approaches described in this article facilitate NASA's strategic objective of progressing in identifying and investigating past or present habitable environments on Mars and other worlds and determining if there is or ever has been life elsewhere in the solar system.

METHODOLOGIES INVOLVED IN MICROBIAL DIVERSITY STUDIES

Characteristics of Spacecraft Assembly Clean Rooms

The reduction of biocontamination and maintenance of a clean environment is of particular importance to the assembly of spacecraft hardware. The search for extraterrestrial life relies heavily on the authenticity of cells and/or biomarkers detected in recovered samples. Contamination of these samples with terrestrial organic matter (e.g., forward contamination) would inherently confound the interpretation of such biosignatures. Clean rooms are therefore of immense value in the fabrication and assembly of spacecraft components. Clean room certification is based on the maximum number of particles greater than 0.5 μm per cubic foot of air per minute (ISO 14644–1; www.iest.org.). The low density of aerosolized particulates within these rooms reduces both inorganic and biological contamination on and within assembled hardware. Current PP protocols require that spacecrafts are constructed and assembled under stringent quality-controlled clean room conditions. Clean room classification requirements are mission-specific.

In addition to air quality, several environmental parameters of spacecraft-associated clean rooms, such as humidity, temperature, and air circulation, are also controlled. Spacecraft assembly facilities are maintained with daily cleaning regiments consisting of the replacing tacky mats, wiping surfaces and support hardware fixtures, and vacuuming or mopping floors using clean room-certified sanitizing agents (disinfectants, alcohol, or ultrapure water). Prior to entering the clean room, personnel take actions to minimize the influx of particulate matter. Although the specific entry procedures vary depending on the clean room certification level and the presence or absence of mission hardware, precautions generally include the donning of clean room-certified garments. Cosmetics, fragrances, and hair gels are prohibited in clean rooms certified to be 5K or less. The air of spacecraft assembly facilities is maintained using high efficiency particle air (HEPA) or ultra low particle air (ULPA) filters. NASA quality assurance engineers perform periodic audits to ensure that certified-facility cleanliness levels conform to the requirements delineated.

Sampling and Sample Processing Technology

Surface samples from spacecraft assembly facilities are obtained using a number of materials. Polyester and cotton swabs are generally used to collect sample from small, defined areas while larger polyester wipes are used to collect sample from areas as large as a square meter. The use of an alternative material, macrofoam, is currently being evaluated. Regardless of the collection material, sample-containing swabs and wipes are individually placed in sterile phosphate-buffered saline to minimize the loss of biologically viable cells. Biological molecules and viable cells are released from sampling materials by using a combination of vortexing and sonication. Sample-containing buffer can then be aliquoted appropriately and used for culture- or molecular-based biological assays.

A number of molecular-based assays utilize DNA as biomarker for assessing microbial burden. Nucleic acid extractions are therefore important and are performed using various methods. Manual extractions using organic solvents (i.e., chloroform and phenol) are often utilized, but they are labor-intensive and result in the generation of hazardous chemical waste. Commercially available extraction kits relying on mechanical or chemical disruption of cells to release nucleic acids typically generate less hazardous waste, but may provide lower extraction efficiency. The use of automated nucleic acid extraction instruments is ongoing and may provide a suitable alternative to manual or kit-based extractions. These technologies (e.g., AutoLyser A-20; IGene, Menlo Park, CA) are generally stand-alone, fully automated, field deployable, and capable of performing nucleic acid extractions without additional manual steps (aside from initial loading). Disposable sample-handling cartridges minimize contact with samples and corrosive agents.

Culture-Based Methodologies

NASA standard assay

PP approaches for current missions are largely based on heritage derived from the significant body of data collected during studies supporting the Viking missions in the early 1970s. Scientists of this era determined that the risk of terrestrial microorganisms surviving the journey to Mars and propagating when they arrived would be constrained by their hardiness with respect to the environments they encountered. For this reason, NASA selected aerobic, heterotrophic mesophilic spore-forming organisms as a proxy for the overall bioburden present on the spacecraft. Briefly, the NASA standard assay is performed as follows: Microbes are extracted from swab or wipe samples into a collection buffer via sonication and vortexing. Aliquots of sample collection buffer are initially heat shocked (80 °C, 15 min) to eliminate vegetative cells. The heat shocked fluid is then pour-plated with TSA media and incubated at 32 °C. Colony-forming units (CFU) are enumerated at 24, 48, and 72 h intervals. Counts of cultivable spores are then compared against a maximum limit that is derived from a combination of mission parameters, the extraterrestrial target, and a calculation of risk that a transferred organism will result in "harmful contamination."

Enumeration of total cultivable bacteria

Heterotrophic plate count assays may also be used to enumerate and assess microbial burden. Aliquots of surface samples are spread-plated onto R2A media and incubated at 25 °C; CFUs are enumerated after 2 and 7 days. Although the heterotrophic plate counts give a better assessment of overall microbial diversity than the NASA standard assay, there remains a significant selective bias toward microbes capable of growth under these conditions. A variety of modified cultivation assays have therefore been employed as a means of selecting for physiologically diverse microbial populations. Modification of media and/or incubation conditions has also been used to screen for extremotolerant and extremophilic bacteria, including thermophiles, psychrophiles, halophiles, and alkaliphiles.

Identification of microbial isolates

Multiple methods are utilized for the identification of microbes isolated from spacecraft and their associated assembly environments. A variety of phenotypic tests that can classify prokaryotes based on metabolic activity, growth in various media conditions, and morphological characteristics are available. Unfortunately, these tests lack the ability to resolve differences among very closely related bacterial species. An alternative to phenotypic tests is the use of DNA sequencing and molecular taxonomy to identify and classify bacterial isolates. Although these are molecular-based methods, they would not be possible without the initial cultivation of isolates. Utilizing the conserved nature of the 16S small subunit rRNA gene, the relatedness of two prokaryotes can be determined based on the extent to which their DNA sequences differ. DNA extracted from isolates is used as a template for the PCR and the entire 16S rRNA gene (~1400 base pairs) is amplified and subsequently sequenced. These sequences are compared against public sequence databases by using sequence comparison software, and the relative phylogenies of isolates determined.

Culture-Independent Microbial Burden Estimation

ATP-based bioluminescence assay

Using ATP-based microbial detection, it is possible to obtain a measure of all viable microorganisms present in an environmental sample based on the presence of ATP, a form of chemical energy common to all life. Commercially available kits using a combination of enzymes and detergents make it possible to distinguish the ATP present in viable cells from the extracellular ATP background. This technology is useful for a wide variety of applications, as the dynamic range of the assay is from 10 to 10^5 cells. In comparison to cultivation-based methods, which require days to provide an estimation of bioburden, this assay is rapid and can be completed in as little as 60 min.

Estimation of endotoxin-producing bacterial population

When microorganisms invade arthropods the immune system may respond by initiating a highly specific enzyme cascade in some of its blood cells (amebocytes). This cascade is initiated by the presence of lipopolysaccharide (LPS) found in the outer membrane of Gram-negative bacteria and β-glucan in yeasts, and results in the formation of a gelclot that destroys the invading microbes. The LPS-based microbial detection assay exploits this principle as it occurs in amoebocytes of the horseshoe crab (*Limulus polyphemus*) coupled with a chromogenic substrate. Sensitive spectrographic analyses measure the amount of endotoxin present in a sample; this measurement can then be used to estimate microbial diversity. Despite the molecular sensitivity of this method, it is only a measure of Gram-negative bacteria and yeast – Gram-positive bacteria, archaea, and fungi are not accounted for.

Taqman quantitative PCR

Quantitative PCR is a real-time analytical technique that provides rapid and accurate quantification of target DNA sequences in a sample. In Q-PCR, a short fragment of the rRNA gene (~100 base pairs) is amplified. The amount of DNA amplified is measured as a function of fluorescence from a target-specific probe released during the amplification process. Fluorescence intensity for each sample is compared to a set of known standards to estimate the initial number of gene copies present. One advantage of the Q-PCR assay for microbial enumeration is its specificity. Primers used to generate DNA fragments during the Q-PCR can be designed with broad range of specificity, allowing the user to specifically target bacterial strains or universally target broad groups of microbes. Unfortunately, due to the nature of the assay, Q-PCR is unable to distinguish whether the amplified signal represents DNA from viable cells or the extracellular DNA fragments of dead cells.

Molecular microbial community analysis

The use of molecular methods to determine the phylogeny of cultured isolates has been discussed previously (see Section "Identification of Microbial Isolates"). Molecular community analyses use this same basic methodology to determine the phylogeny of microbes present in a sample "without" the selective bias of culture-based techniques. Instead, molecular cloning methods are utilized to isolate individual strands of amplified DNA, each strand theoretically representing an individual organism present in the microbial community. Specific fragments of DNA amplified from individual clones can be subjected to phylogenetic analyses just as DNA amplified from individual isolates. Because the process does not require cultivation, a more thorough and complete biodiversity study can be performed.

As with all techniques, molecular community analyses have their disadvantages. There is an inherent susceptibility to bias toward organisms that disproportionately dominate a particular sample or organisms with DNA sequences that are preferentially amplified. As with other molecular techniques, this analysis cannot distinguish between live and dead cells. Additionally, prepping and sequencing large clone libraries can be cost prohibitive, although this can be partially mitigated by using methods of screening to minimize the number of redundant species identified. Restriction fragment length polymorphism (RFLP) analysis, for example, utilizes restriction endonucleases to cut amplified DNA strands at specific nucleotide iterations. The restriction products from individual clones are analyzed and the subsequent patterns compared. Using this methodology, only representative clones are sequenced and identified. This approach has successfully been used in a number of spacecraft-associated biodiversity studies to maximize the number of unique 16S rRNA genes observed in each sample.

BIOBURDEN AND BIODIVERSITY OF SPACECRAFT AND ASSOCIATED SURFACES

As discussed previously, PP approaches for current missions are largely dictated by the significant body of data collected during studies supporting the Viking missions of the early 1970s. Scientists of this era utilized aerobic, heterotrophic mesophilic spore-forming bacteria as benchmark organisms against which spacecraft cleanliness was evaluated. As such, sterilization procedures were designed to reduce the number of cultivable spores below a predetermined maximum limit. This mission-specific limit was derived from a combination of factors including the mission objectives, the target planet, and an evaluation of the risk that any microbe transferred would have resulted in "harmful contamination" of the planet.

Culture-based analyses are effective as a proxy to assess microbial contamination; however, they do not provide a complete picture of microbial burden as it is widely accepted that ≤1% of the microorganisms present in an environment cannot be cultured. Molecular detection methods have revealed a much larger microbial diversity on spacecraft hardware and in spacecraft assembly facilities, confirming that cultivation approaches were estimating only a fraction of the actual microbial diversity. Additionally, the rapid turnaround associated with many molecular assays (particularly ATP- and LPS-based) results in contamination evaluation significantly quicker than that can be seen with culture-based enumeration. This is not to say that cultivation methods do not yield important contributions; the study of microbes from spacecraft and spacecraft-associated environments has revealed numerous novel organisms. Clearly, both traditional culture-based and modern molecular methods need to be used in tandem to best assess spacecraft-associated microbial bioburden.

Microbial Community Analysis of Spacecraft Assembly Facilities

The microbial communities of spacecraft and their associated assembly facilities have been the focus of many studies. Research has been directed at community structure and the unique properties of individual isolates obtained from these environments. The data obtained from these studies are presented in the following sections, with similar studies grouped accordingly.

Bacterial communities associated with clean room dust particles collected using witness plates

Using both traditional culture-based methods and modern molecular techniques, bacterial communities associated with clean room dust particles were evaluated. Witness plates composed of spacecraft materials were allowed to collect dust particles within the Jet Propulsion Laboratory (JPL)–spacecraft assembly facility (SAF) for periods of approximately 7–9 months. Total cultivable aerobic heterotrophs and heat-tolerant spore formers were enumerated as described previously. The data indicated that witness plates coated with spacecraft-qualified paints attracted more dust particles than uncoated plates and that there was a positive correlation between the amount of dust particles deposited and the total number of cultivatable bacteria. Although phylogenetic analyses of isolates suggested the predominance of Gram-positive, spore-forming *Bacillus* species (*B. licheniformis*, *B. pumilus*, *B. cereus*, and *B. circulans*), subsequent molecular cloning and 16S rDNA sequence analyses revealed an equal representation of both Gram-positive and Gram-negative bacteria.

Characterization of cultivable bacteria from NASA clean room facilities

As mentioned previously, the traditional culture-based methods utilized to enumerate microbes in spacecraft assembly facilities have yielded numerous isolates that belong to the genus *Bacillus*. To further evaluate the microbial communities present within various NASA facilities, cultivation approaches were employed to select for extremophilic/extreme-tolerant microbes able to survive the nutrient-limited conditions of clean rooms. Additionally, a variety of environmental stresses were employed in an attempt to isolate hardy organisms that potentially represent a more significant forward contamination threat (especially considering their general proximity to spacefaring objects and resistance to sterilization procedures). When samples collected from NASA clean rooms were challenged with UVC irradiation, 5% hydrogen peroxide, heat shock, pH extremes (pH 3.0 and 11.0), temperature extremes (4–65 °C), and hypersalinity (25% NaCl) prior to and/or during cultivation, numerous varieties of bacteria were cultured. Isolated members of the Bacillaceae family were found to be more physiologically diverse than those reported in previous studies, and included thermophiles (*Geobacillus*), obligate anaerobes (*Paenibacillus*), and halotolerant, alkalophilic species (*Oceanobacillus* and *Exiguobacterium*). Non-spore-forming microbes (α- and β-proteobacteria as well as actinobacteria) exhibiting tolerance to the selected stresses were also encountered. Molecular methods utilized in parallel with the cultivation studies suggested that the samples contained bioburdens ranging from below detection limit (~10 cells) to 10^6 cells m^{-2}, although only a percentage of the total community was able to grow on the media employed.

A number of the microbes isolated from spacecraft or the clean rooms in which they are assembled or readied for launch have been newly discovered species. Environmental sampling of the Mars Odyssey Orbiter and associated clean rooms at the Kennedy Space Center (KSC), for example, yielded two previously uncharacterized species of *Bacillus*. A variety of genotypic tests including multiple gene phylogenetic analysis and DNA–DNA hybridization clearly distinguished these isolates from their nearest evolutionary neighbors. The name *Bacillus safensis* was proposed for one of the novel species, indicating its isolation from a SAF. The other species was isolated from the space craft surface itself and hence was named *Bacillus odysseyi*. Both species possess characteristics relevant to PP, including resistances to ultraviolet radiation and hydrogen peroxide (typical spacecraft sterilants). The JPL-SAF served as an assembly site for several robotic missions (e.g., Viking, Mars Pathfinder, Mars Exploratory Rovers) and has likewise been a location of discovery for new species of bacteria. One facultatively anaerobic species, *Bacillus nealsonii*, showed high resistance to ultraviolet and gamma radiation as well as H_2O_2 and desiccation. Two novel species of spore-forming bacteria, *Paenibacillus pasadenensis* and *Paenibacillus barengoltzii*, were additionally discovered in the JPL-SAF.

Anaerobic bacteria in spacecraft assembly facilities

The presence of anaerobic spore-forming bacteria has been documented in previous NASA studies by the isolation of anaerobic microbes. Of particular interest was the isolation of three obligate anaerobic strains identified as *Paenibacillus wynnii* and two strains most closely related to *B. licheniformis*. The isolation of strains of strictly anaerobic *B. licheniformis*, thought to be facultative anaerobes, emphasizes the lack of knowledge on this class of bacteria. Furthermore, the presence of anaerobic isolates further demonstrates the importance of developing expanded cultivation methods beyond the standard assay to assess the unique microbial populations present in these clean room facilities.

Fungal community structures

The traditional culture assays employed by NASA are biased to the detection of endospore-forming bacteria as a primary PP concern. The extreme hardiness of these microbes allows them to tolerate inhospitable conditions for long periods of time and makes them particularly good candidates for surviving the journey to planetary bodies that may support life. Yet, while fungal species also produce protective structures (spores, conidia, or cysts) as part of their normal life cycle or as a response to environmental stress, few reports have examined their survival under simulated space conditions. No published report has studied fungal distribution in assembly facilities, even though these eukaryotic microorganisms may have serious impact on a variety of space missions. While there has been a significant increase in the understanding of bacterial populations, much work remains in determining the distribution of eukaryotic microorganisms present in these locations and their potential for contaminating spacecraft. Fungal species in particular are a risk for inadvertent colonization of extraterrestrial environments. In addition to their capacity for producing protective structures, they also possess unique mechanisms of DNA repair and damage tolerance that can allow them to thrive under extreme radiation conditions. The hardiness of fungi and their ability to withstand extreme environments signifies them as potential spacecraft contaminants putatively possessing the ability to survive outside Earth's biosphere. These eukaryotic microbes are therefore important organisms with respect to PP concerns and their abundance, community structure, and resistance capabilities warrant further study.

Geographic variability of bacterial communities

Based on the variety of microbes isolated from different spacecraft and spacecraft assembly facilities, molecular bacterial diversity studies were undertaken at three geographically distinct spacecraft-associated clean rooms (facilities were located at Jet Propulsion Laboratory (JPL), Johnson Space Center (JSC), and KSC). The focus of these studies was to investigate if bacterial communities were influenced by the surrounding geographically specific environments or if the conditions of the clean rooms themselves selected the observed diversity. A total of nine clone libraries representing different surfaces within the spacecraft facilities and three clone libraries from the surrounding air were created and ~1200 clones were analyzed. Despite the highly desiccated, nutrient-bare conditions within these clean rooms, a broad diversity of bacteria was detected, covering all the main bacterial phyla. Furthermore, the bacterial communities were significantly different from each other, revealing only a small subset of microorganisms common to all locations (i.e., *Sphingomonas*, *Staphylococcus*). Samples collected from JSC assembly room surfaces showed the greatest diversity of bacteria, particularly within the α-, γ-proteobacteria and actinobacteria. The bacterial community structure of KSC assembly surfaces revealed a high presence of proteobacterial groups, while the surface samples collected from the JPL-SAF showed a predominance of *Firmicutes*. Clearly, geographically distinct clean rooms contain distinct bacterial communities although no evidence directly suggests that the environment outside the clean room was responsible for the diversity within. It was, however, interesting to note that the NASA facility located in the most arid environment (JPL is located in southern California) contained the majority of microbes predicted to have desiccation resistance mechanisms (the *Firmicutes*).

Archaeal communities within NASA facilities

Despite the wealth of information regarding bacterial contaminants in NASA clean room facilities, relatively little is known about archaeal communities. Archaea comprise the third domain of life and are considered by some to represent a form of terrestrial life capable of surviving in the Martian atmosphere. In fact, diverse lithoautotrophic archaeal communities have been discovered in the Earth's permafrost regions thriving in very low temperature, anaerobic conditions. It has also been suggested that the presence of organisms performing methanogenesis (a unique archaeal process on Earth) could help to explain the calculated methane imbalance in Mars' atmosphere. To investigate the possibility of archaeal communities within NASA clean rooms, numerous samples were collected and screened for the presence of archaeal 16S rRNA gene sequences; many of these samples tested positive for archaeal contaminants. Sequence analyses of clone libraries suggested the presence of microbes most closely related to other uncultivated archaea within the eury- and chrenarcheaota lineages. Attempts to cultivate archaeal members using state-of-the art technologies were not successful.

Resistance Characteristics of Bacterial Species Isolated from Spacecraft-Associated Environments under Several Simulated Conditions

The presence of viable microbes within facilities and on surfaces subjected to sterilization procedures has demonstrated the resistance capabilities of numerous organisms within spacecraft-associated microbial communities. Investigations on the resistance capabilities of these organisms are important to NASA not only from the academic standpoint of increased understanding but also for the development of more stringent bioburden reduction techniques.

UV_{254} resistance

One of the more prevalent microbes isolated from surfaces (including spacecraft) within the clean assembly rooms was the spore-forming *B. pumilus*. Due to its ubiquitous nature and recurrent isolation, *B. pumilus* strains isolated from several NASA clean rooms were utilized in a study to assess the efficacy of various spacecraft sterilants, including treatment with biocidal UV light. Purified spores of several *B. pumilus* isolates were subjected to 254-nm UV and their UV resistance was compared to spores of a standard *Bacillus subtilis* biodosimetry strain. Spores of six of the ten isolates were significantly more resistant to UV than the biodosimetry strain, and one of the isolates, *B. pumilus* SAFR-032, exhibited the highest degree of spore UV resistance observed by any *Bacillus* spp. encountered to date. Results of this study support revision of UV-sterilization requirements for spacecraft components.

H_2O_2 resistance

During microbial diversity studies of the NASA clean rooms, numerous H_2O_2-resistant *B. pumilus* were isolated from various surfaces. This was a concern, as H_2O_2 is utilized as a sterilant for spacecraft hardware due to its compatibility with various modern-day spacecraft materials, electronics, and components. To assess the degrees of resistance, multiple *B. pumilus* isolates (both vegetative cells and spores) were exposed to 5% liquid H_2O_2 for 60 min. Although the results indicated that the extent of resistance was strain-specific, in all cases spores proved hardier than their respective vegetative cells. Additionally, some of the *B. pumilus* strains tested were shown to have 2–3 times more resistance than a standard *B. subtilis* dosimetry strain. It is interesting to note that the type strain of *B. pumilus*, ATCC 7061, was shown to be sensitive to H_2O_2. The H_2O_2-resistance mechanism employed by these isolates is not currently understood, although it is hoped that ongoing genomic studies will provide some insight.

Gamma radiation resistance

During missions, microbial communities on the surfaces of spacecraft and landers should be subjected to intense UV irradiation; microbes within these vehicles, however, would be shielded from UV-mediated damage. Gamma irradiation, in contrast, would penetrate most areas and therefore potentially affect any internal microbial stowaways. To assess the potential for gamma-radiation resistance, surface samples were collected from internal circuit boards of X-2000 (avionics) spacecraft while within the SAF. These samples were irradiated with 1 Mrad gamma radiation for 5.5 h (50 rads s^{-1} dose rate). Culture-based studies of the pre- and postirradiated samples demonstrated the effectiveness of gamma radiation at inhibiting the growth of most microbes. Molecular characterization of the cultivable microbes (e.g., 16S rDNA sequence analyses) indicated that while the initial samples were predominantly populated with *Bacillus*, *Exiguobacter*, and *Staphylococcus* species, only *B. megaterium* and *Aureobasidium pullulans* survived 1 Mrad gamma radiation. Interestingly, molecular characterization using DNA extraction and 16S rDNA PCR directly from the collected samples demonstrated little change in microbial diversity before and after irradiation. This suggests that while gamma irradiation can inhibit the growth of most of the cultivable microbes, insufficient damage is applied to microbial DNA to prevent PCR amplification.

UV Resistance Under Simulated Mars Conditions

The discovery of Martian environments with the potential to support life highlights the importance of PP. As such, several studies have been undertaken to evaluate the potential risk of forward contamination with terrestrial microbes. Microbial species isolated from the surfaces of several spacecrafts, including Mars Odyssey, X-2000 (avionics), and the International Space Station, along with their respective assembly facilities, were exposed to simulated Martian UV irradiation. The effects of UVA (315–400 nm), UVA-B (280–400 nm), and the full UV spectrum (200–400 nm) on microbial survival were studied at UV intensities expected to strike several Martian surfaces. Of the 43 *Bacillus* spore lines screened, 19 isolates showed resistance to UVC irradiation (200–280 nm) after exposure to 1000 J m^{-2} at 254 nm. As with previous studies, spores of *Bacillus* strains isolated from spacecraft-associated surfaces were more resistant than standard dosimetric strains. *B. pumilus* SAFR-032, for example, was shown to be six times as resistant as *B. subtilis* 168. It was additionally observed that 35-fold longer exposure times were required for UVA-B irradiation to reduce viable spores to the same degree as the full UV spectrum, suggesting that UVC is the primary biocidal bandwidth.

To further evaluate the resistance of spores to Martian UV irradiation, seven *Bacillus* species were exposed to simulations of Martian normal UV fluence rates equivalent to clear-sky conditions of equatorial Mars. To approximate microbial contamination on spacecraft surfaces, bacterial endospores were deposited as thin monolayers on aluminum coupons and exposed to simulated Martian UV irradiation for various durations of time (from 0 to 180 min). Spore survival was assayed using culture techniques coupled with most probably number (MPN) methodologies. The results indicated that although *B. pumilus* SAFR-032 again exhibited remarkable UV resistance compared with other strains, no strain tested was immune to UV-mediated damage. It was concluded that all strains would be inactivated on Sun-exposed surfaces within a few tens of minutes to a few hours on sol 1 under clear-sky conditions at equatorial Mars. It was however noted that the presence of UV-resistant microbes was still a concern due to non-Sun-exposed surfaces and the potential shielding effect created by dust deposited during landing operations.

Microbial Community Structures of Several Missions

Mars Odyssey orbiter mission

The Mars Odyssey orbiter was designed to make global observations of Mars and to aid in the search for water and environments that may be capable of supporting life. Microbial characterization of the spacecraft's surface and the KSC Spacecraft Assembly and Encapsulation Facility II (SAEF-II) utilized both culture-based and molecular methods. The most dominant cultivable microbes were species of *Bacillus*, but comamonads, microbacteria, and Actinomycetales were also represented. A number of bacterial isolates were resistant to ionizing and UV radiation, as well as hydrogen peroxide and desiccation. Sequences arising in clone libraries were fairly consistent between the spacecraft and facility, confirming that the environment directly surrounding a spacecraft strongly correlates with the microbial diversity observed on that craft's surface. This study furthered NASA's understanding of the microbial community structure, diversity, and survival capabilities of microbes in an encapsulation facility physically associated with co-located spacecraft.

Mars Exploration Rovers (MER) lander mission

Originally planned as a 3-month mission, the *MER* Spirit and Opportunity have far surpassed expectations, transmitting data for more than 3 years. The primary objective of MER was to discover clues indicating the potential for past and/or present water on the surface of Mars. To estimate the microbial burden and identify potential contaminants, air samples of the MER assembly facility at KSC were analyzed utilizing a variety of diagnostic tools including culturing, cloning, and sequencing of 16S rRNA genes, DNA microarray analysis, and ATP assays to assess viable microorganisms. In addition to the cultivation of *Agrobacterium*, *Burkholderia*, and *Bacillus* species, molecular cloning suggested the presence of oligotrophs, symbionts, and γ-proteobacteria members.

Genesis curation laboratory clean room

The *Genesis* discovery mission was the first sample return mission to be performed beyond the lunar orbit. Its objective was the collection of solar wind samples that could be compared with known compositions of the planets to help elucidate the origins of the solar system. Because the goal of the *Genesis* mission was particulate collection, meticulous cleaning and assembling of spacecraft components took place in very low particulate Class 10 clean rooms. Microbial burden of samples collected from these clean rooms was assessed via traditional culture-based and molecular biomarker-targeted methods (LPS assay, ATP assay, and Q-PCR). The Cleaning Laboratory and Assembly Laboratory had extremely low bioburdens with most (but not all) samples below the detection limits of the assays employed. Spikes in microbial incidence did occur on plenums, light fixtures, and in the sub-floors. Total cultivable heterotroph counts were congruent with observed spikes in ATP and endotoxin. Ultimately, the data suggested that the GCL class 10 clean room cleaning procedures coupled with air filtration maintained a very low bioburden, thus decreasing the likelihood of spacecraft contamination. Nevertheless, the detection of viable bacteria in these clean rooms despite stringent cleaning and air-handling protocols exemplifies the potential for microbial survival.

Mars Reconnaissance Orbiter (MRO) mission

Launched in August 2005, the main goals of the MRO were to provide data that could determine the extent to which water persisted on the surface of Mars and to help locate potential landing sites for future lander missions. To carry out these objectives the orbiter underwent several months of aerobraking, using the Martian atmosphere to slow the craft and ease it into a circular orbit low enough for scientific observations. This process and the resulting operational altitude required compliance with more stringent PP cleanliness requirements than typical Mars orbiters. Prior to launch, samples from the spacecraft and two of its assembly facilities were made available for cultivation and molecular assays of bioburden.

Traditional cultivation revealed that spore-forming bacteria common to both the spacecraft and its assembly facilities included *B. pumilus*, *B. licheniformis*, and *B. megterium* (Figure 1), which have been repeatedly isolated in previous studies of spacecraft assembly facilities. Although spore formers were present in all sample sets, they constituted only 16% of the total culturable isolates (Figure 2) within the MRO spacecraft assembly facilities (~10^2 CFU m^{-2}). Intracellular ATP values indicated a viable cell count that was higher than the culturable count (~10^4 CFU m^{-2}) while measures of total ATP and 16S rRNA copy numbers suggested an even higher degree of bioburden (~10^7 CFU m^{-2}). However, the recognition that inviable organisms may still contain both

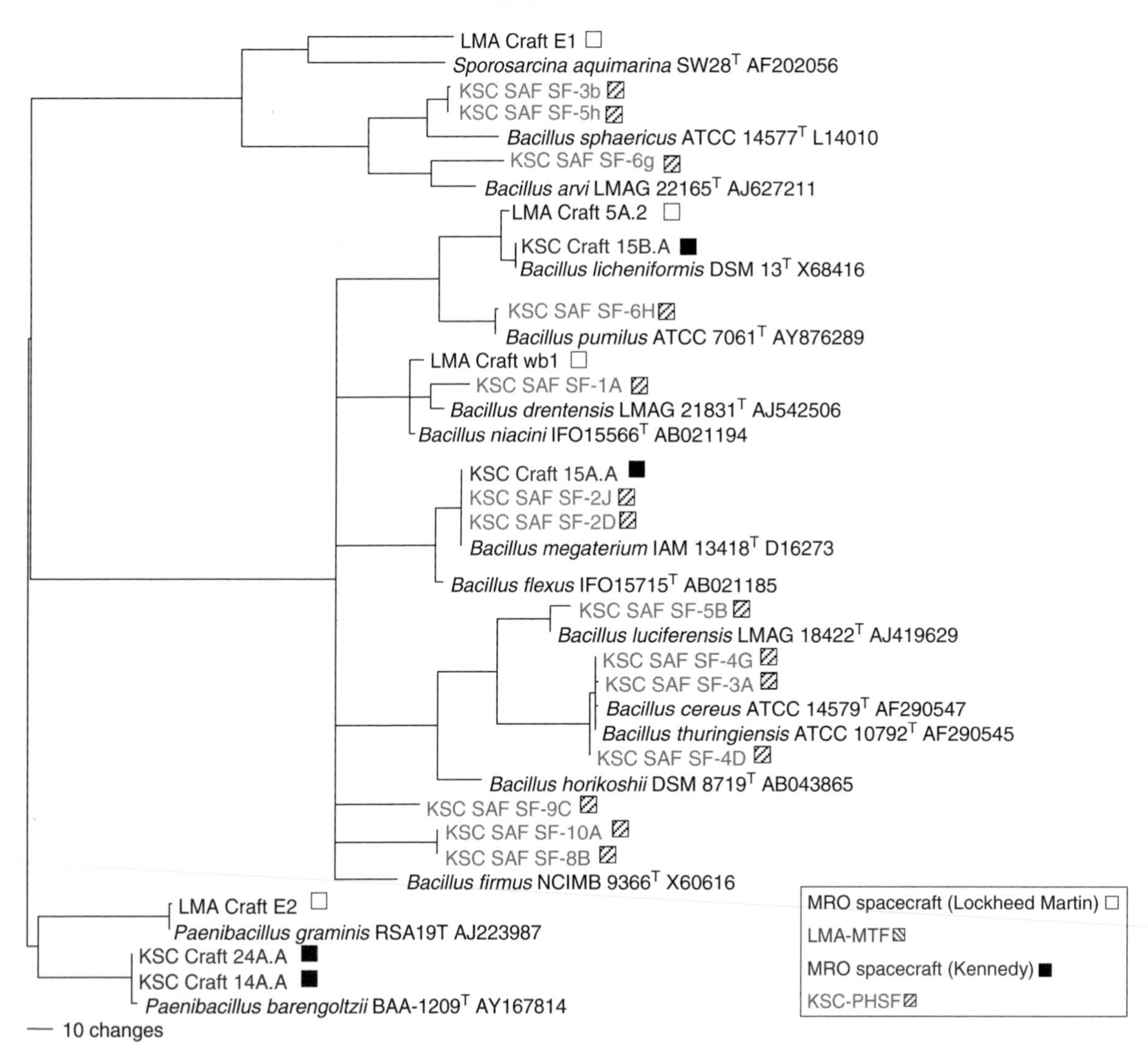

FIGURE 1 The NASA standard assay yielded a number of *Bacillus* species commonly found in spacecraft assembly facilities. A *Paenibacillus* spore-forming bacteria previously discovered and described at the Jet Propulsion Laboratory was also discovered on the *Mars Reconnaissance Orbiter* (MRO) spacecraft.

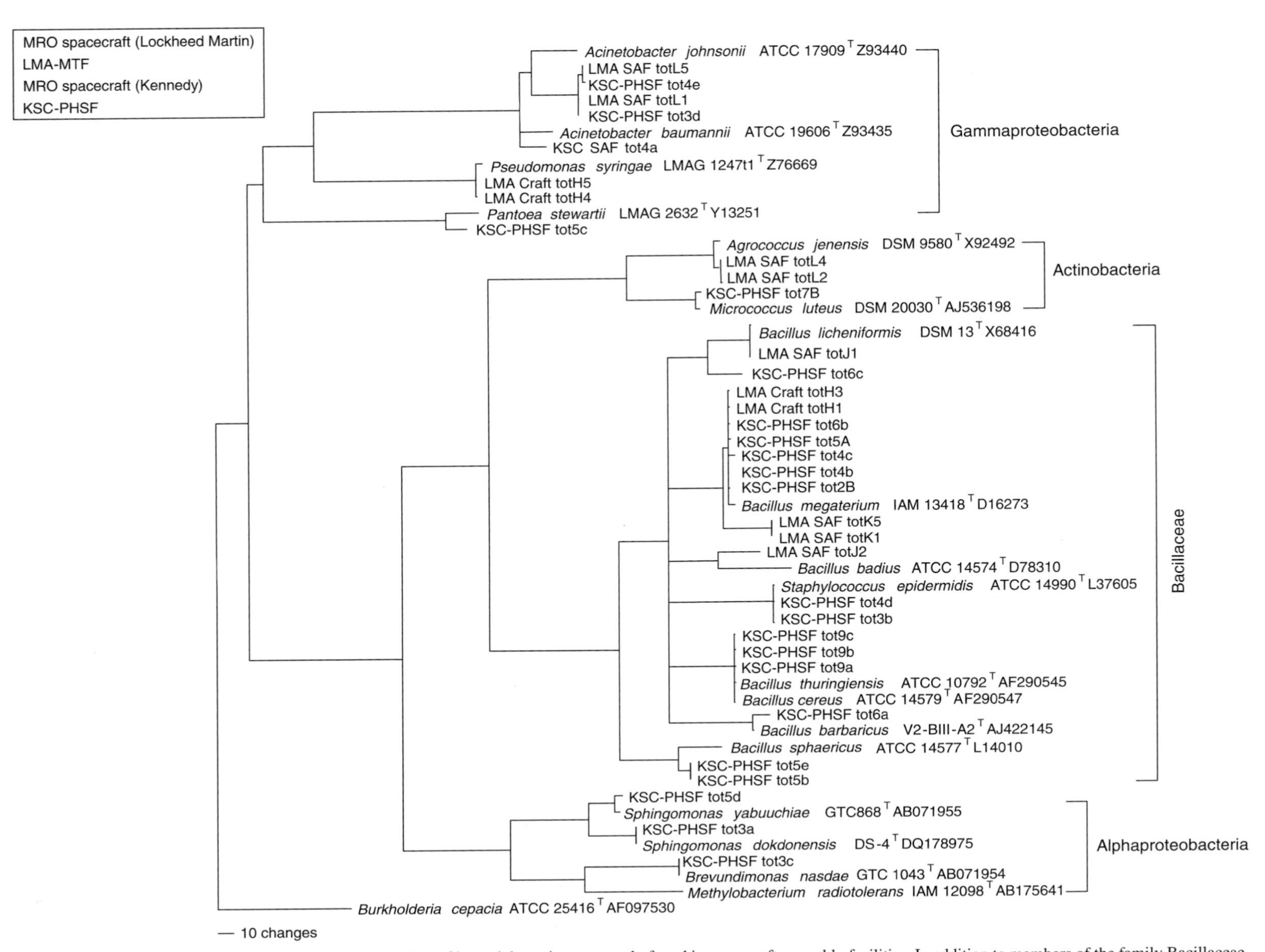

FIGURE 2 Total cultivable bacterial assay yielded a number of bacterial species commonly found in spacecraft assembly facilities. In addition to members of the family Bacillaceae, several isolates belonging to actinobacteria and proteobacteria were also isolated on the spacecraft and associated environments.

ATP and DNA suggests that this latter value is likely to be an overestimate of the viable microbes present.

The DNA-based biodiversity study of the MRO and its assembly facilities revealed a broad spectrum of microbes that varied depending on the geographic location of the spacecraft. The vast majority of clones sequenced from the MRO at the Lockheed Martin Assembly Facility (LMA-MTF, 89%) and from the LMA-MTF itself (99%) were microbes associated with human skin (Figure 3). The majority of clones from the MRO spacecraft at KSC were predominantly β- and γ-proteobacteria (40% and 51%, respectively) while clones from the KSC Payload Hazard Spacecraft Facility had an even distribution of clones among the bacterial classifications. The absence of Gram-positive, particularly of the genus *Bacillus*, from the MRO spacecraft while present at LMA or KSC further illustrated the low concentration of *Bacillus* with respect to other bacterial genera.

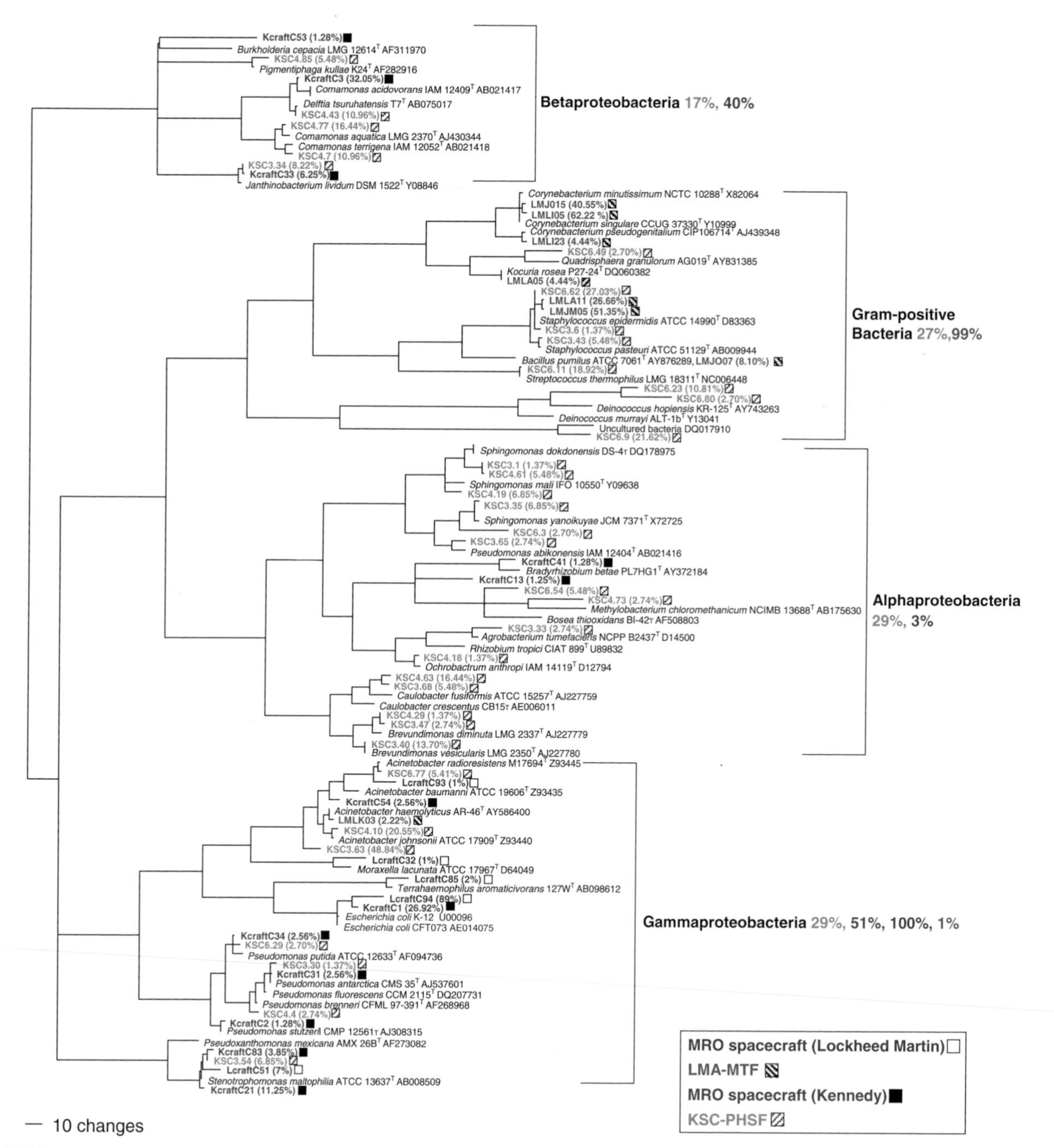

FIGURE 3 Molecular phylogenetic analysis using cloning yielded a much broader bacterial biodiversity than the NASA standard assay. The clone libraries from each of the *Mars Reconnaissance Orbiter* (MRO) assembly facilities and the two samplings of the MRO itself showed an uneven distribution among bacterial groups. The percent abundance of clones in each sample set is indicated.

Significance of Whole Genome Analysis of "Hardy" Bacteria

Currently, the process of completing the whole genome annotation of the UV radiation-resistant bacteria, *B. pumilus* SAFR-032, is being undertaken. A comparison of this genome with other known *Bacillus* genomes will enable the use of whole genome or proteome DNA microarrays and/or PCR-based amplification of targeted genes corresponding to predicted biomarkers to examine various sterilization regimes. Such protein or DNA-based approaches will facilitate the development of appropriate countermeasures and help to set PP standards. The value of studies on the *B. pumilis* SAFR-032 genome is further illustrated by the beneficial impact on a number of diverse applied research areas:

- From a PP perspective it is essential to minimize the microbial contamination of spacecraft and their associated hardware as discussed previously.
- From an astrobiology perspective it is known that microorganisms can live under unusual/extreme conditions such as low and high pH, low and high temperatures, high salinity, and/or high hydropressure. By understanding the mechanism(s) of spore resistance to H_2O_2 and UV, we will enhance our understanding of the possibility of terrestrial microbes to survive the Mars environment.
- From a biodefense perspective it is important to understand the mechanisms of resistance associated with *B. pumilis* SAFR-032 in the context of lateral transfer between species. In a worst-case scenario, the transfer of resistance against environmental stresses could be utilized to augment spore agents such as *Bacillus anthracis* to persist in the presence of H_2O_2 and UV, both of which are currently being used as sterilants in bioterrorism prevention and remediation efforts. An understanding of the resistance capabilities and mechanisms is therefore important.
- From a medical and environmental health perspective, repeated isolation of H_2O_2- and UV-resistant strains of *B. pumilus* in clean room environments is problematic because persistence represents a potential threat to health. Likewise, clean rooms such as medical operation theaters, semiconductor assembly, and spacecraft assembly facilities might exert selective pressures on indigenous microbes whereby certain strains or species could survive in the low-nutrient, controlled humidity, and clean conditions and later proliferate. Our results indicate that some strains of *B. pumilus* spores are more resistant to H_2O_2 than others. These observations have implications not only in the preparation of spacecraft for life-detection missions but also in the medical and pharmaceutical fields, where H_2O_2 is routinely used in sterilizing sensitive equipment.

CONCLUDING REMARKS

The primary missions of NASA's PP program are the prevention of forward and backward biological contamination, with the goals of protecting the Earth, and preserving other planetary bodies for scientific investigation including the search for life. Spacecraft hardware fabricated for missions to Mars, Europa, or Enceladus are of particular concern, as these solar system bodies present the greatest likelihood of sustaining terrestrial life and affording it the ability to (1) colonize and proliferate and/or (2) complicate subsequent searches for extraterrestrial life forms. By isolating and identifying organisms with the potential to survive Martian conditions and by elucidating the microbial community structures from which such organisms arise, the PP program facilitates NASA's science objective 3C.3 of progressing in identifying and investigating past or present habitable environments on Mars and other worlds, and determining if there is or ever has been life elsewhere in the solar system. Furthermore, PP program goals assist in meeting NRC recommendations such as the "development or adaptation of modern molecular analytical methods to rapidly detect, classify, and/or enumerate the widest possible spectrum of Earth microbes carried by spacecraft before, during, and after assembly and launch processing". NASA's strategic science objectives and commitment to meet NRC recommendations ensure the integrity of life science missions and prevent the contamination of "pristine" extraterrestrial environments.

FURTHER READING

Bruckner, J. C., & Venkateswaran, K. (2007). Overview of methodologies to sample and assess microbial burden in low biomass environments. *Japanese Journal of Food Microbiology*, *24*, 61–70.

Colwell, R. R. (2000). Viable but nonculturable bacteria: A survival strategy. *Journal of Infection and Chemotherapy*, *6*, 121–125.

Crawford, R. L. (2005). Microbial diversity and its relationship to planetary protection. *Applied and Environmental Microbiology*, *71*, 4163–4168.

La Duc, M. T., Dekas, A. E., Osman, S., Moissl, C., Newcombe, D., & Venkateswaran, K. (2007). Isolation and characterization of bacteria capable of tolerating the extreme conditions of clean-room environments. *Applied and Environmental Microbiology*, *73*, 2600–2611.

La Duc, M. T., Kern, R., & Venkateswaran, K. (2004). Microbial monitoring of spacecraft and associated environments. *Microbial Ecology*, *47*, 150–158.

La Duc, M. T., Nicholson, W., Kern, R., & Venkateswaran, K. (2003). Microbial characterization of the Mars Odyssey spacecraft and its encapsulation facility. *Environmental Microbiology*, *5*, 977–985.

Moissl, C., Bruckner, J., & Venkateswaran, K. (2008). Archaeal community analysis of spacecraft assembly facilities. *The ISME Journal*, *2*, 115–119.

Moissl, C., La Duc, M. T., Osman, S., Dekas, A. E., & Venkateswaran, K. (2007). Molecular bacterial community analysis of clean rooms where spacecraft are assembled. *FEMS Microbiology Ecology*, *61*, 509–521.

NASA (1967). Harmful contamination: Moon treaty United Nations Treaty on Principles Governing the Activities of States in the Exploration and Use of Outer Space, Including the Moon and other Celestial Bodies Article IX, UN Doc. A/RES/222/(XXI), TIAS No. 6347.

NASA. (1980). *Standard procedures for the microbiological examination of space hardware*. NHB 5340.1B, Rev. B. Washington, DC: National Aeronautics and Space Administration.

NASA. (2005). *Planetary protection provisions for robotic extraterrestrial missions*. Washington, DC: National Aeronautics and Space Administration. NPR 8020.12C, April 2005.

Newcombe, D. A., Schuerger, A. C., Benardini, J. N., Dickinson, D., Tanner, R., & Venkateswaran, K. (2005). Survival of spacecraft-associated microorganisms under simulated martian UV irradiation. *Applied and Environmental Microbiology*, *71*, 8147–8156.

NRC. (2002). Signs of life. National Research Council of the National Academies. In N. R. Council (Ed.) (pp. 1–52). Washington, D.C.: The National Academic Press.

NRC. (2003). New frontiers in the solor system: An integrated exploration strategy. National Research Council. In M. Belton (Ed.) Washington, DC: NASA. The National Academies press.

NRC. (2005). *Sensor systems for biological agent attacks: Protecting buildings and military bases*. Washington, DC: National Academies Press.

NRC. (2000). *Preventing the forward contamination of Europa, task group on the forward contamination of Europa. Task group on the forward contamination of Europa, space studies board,* (pp. 1–41). National Research Council. Washington, DC: National Academies Press.

NRC. (2006). *Preventing the forward contamination of mars. Committee on preventing the forward contamination of mars:*. National Research Council of the National Academies.Washington, DC: National Academies Press.

Schuerger, A. C., Richards, J. T., Newcombe, D. A., & Venkateswaran, K. (2006). Rapid inactivation of seven *Bacillus* spp. under simulated Mars UV irradiation. *Icarus*, *181*, 52–62.

UN-Treaty (1967). Harmful contamination: Moon treaty United Nations Treaty on Principles Governing the Activities of States in the Exploration and Use of Outer Space, Including the Moon and other Celestial Bodies Article IX, UN Doc. A/RES/222/(XXI), TIAS No. 6347.

RELEVANT WEBSITE

http://www.iest.org. – Institute of Environmental Sciences and Technology.

http://www.planetaryprotection.nasa.gov. – National Aeronautics and Space Administration.

Chapter 48

Biological Warfare

J.A. Poupard[1] and L.A. Miller[2]

[1]Pharma Institute of Philadelphia, Inc., Philadelphia, PA, USA

[2]GlaxoSmithKline Collegeville, PA, USA

Chapter Outline

ABBREVIATIONS

ACDA U.S. Arms Control and Disarmament Agency
BARDA Biomedical Advanced Research and Development Authority
BDRP Biological Defense Research Program
BW Biological warfare
CBRN Chemical, biological, radiological, and nuclear agents
CDC Centers for Disease Control and Prevention
DARPA Defense Advanced Research Project Agency
DoD Department of Defense
NATO North Atlantic Treaty Organization
TW Toxin weapon
USAMRIID U.S. Army Medical Research Institute of Infectious Diseases
WMD Weapon of mass destruction

DEFINING STATEMENT

The most general concept of biological warfare (BW) involves the use of any biological agent as a weapon directed against humans, animals, or crops with the intent to kill, injure, or create a sense of havoc against a target population. This agent could be in the form of a viable organism or a metabolic product of the organism, such as a toxin. This article will focus on the use of viable biological agents because many of the concepts relating to the use of toxins are more associated with chemical warfare. The use of viable organisms or viruses involves complex issues that relate to containment. Once such agents are released, even in relatively small numbers, the area of release has the potential to enlarge to a wider population due to the ability of the viable agent to proliferate while spreading from one susceptible host to another.

INTRODUCTION

During the last decade of the twentieth century and early years of the twenty-first century, several events marked significant alterations in the concept of biological warfare (BW). These events include the end of the Cold War, the open threat by Iraq of using BW agents in the First Gulf War, the events of 9/11 followed by dissemination of *Bacillus anthracis* spores through the U.S. Mail system, and the war with Iraq over possible weapon of mass destruction (WMD). These events lead to the full realization that, in addition to BW between nations, the developed world is quite susceptible to attack by radical terrorists employing BW agents. This was a major expansion in the concept of BW and transformed the subject, once limited to the realm of political

Topics in Ecological and Environmental Microbiology.

and military policy makers, to a subject that must be considered by a wide range of urban disaster planners, public health officials, and the general public. BW is a complex subject that is difficult to understand without a basic knowledge of a long and convoluted history. BW can be traced to ancient times and have evolved into more sophisticated forms with the maturation of the science of bacteriology and microbiology. It is important to understand the history of the subject because one often has preconceived notions of BW that are not based on facts or involve concepts more related to chemical rather than biological warfare. Many of the contemporary issues relating to BW deal with third-world conflicts, terrorist groups, or nonconventional warfare. An understanding of these issues becomes important because many of the long-standing international treaties and conventions on BW were formulated in an atmosphere of either international conflict or during the Cold War period of international relations. Many of the classic issues have undergone significant alteration by more recent events. The issue of BW is intimately bound to such concepts as offensive versus defensive research or to the need for secrecy and national security. It is obvious that BW will continue to be a subject that will demand the attention of contemporary and future students of microbiology as well as a wide range of policy and scientific specialists.

HISTORICAL REVIEW

300–1925 BC

Many early civilizations employed a crude method of warfare that could be considered BW as early as 300 BC, when the Greeks polluted the wells and drinking water supplies of their enemies with the corpses of animals. Later, the Romans and Persians used the same tactics. All armies and centers of civilization need palatable water to function, and it is clear that well pollution was an effective and calculated method for gaining advantage in warfare. In 1155, at a battle in Tortona, Italy, Barbarossa broadened the scope of BW, using the bodies of dead soldiers and animals to pollute wells. Evidence indicates that well poisoning was a common tactic throughout the classical, medieval, and Renaissance periods. In more modern times, this method has been employed as late as 1863 during the U.S. Civil War by General Johnson, who used the bodies of sheep and pigs to pollute drinking water at Vicksburg.

The wide use of catapults and siege machines in medieval warfare introduced a new technology for delivering biological entities. In 1422 at the siege of Carolstein, catapults were used to project diseased bodies over walled fortifications, creating fear and confusion among the people under siege. The use of catapults as weapons was well established by the medieval period, and projecting diseased bodies over walls was an effective strategy employed by besieging armies. The siege of a well-fortified position could last for months or years, and it was necessary for those outside the walls to use whatever means available to cause disease and chaos within the fortification. This technique became commonplace, and numerous classical tapestries and works of art depict diseased bodies or the heads of captured soldiers being catapulted over fortified structures.

In 1763, the history of BW took a significant turn from the crude use of diseased corpses to the introduction of a specific disease, smallpox, as a weapon in the North American Indian wars. It was common knowledge at the time that the Native American population was particularly susceptible to smallpox, and the disease may have been used as a weapon in earlier conflicts between European settlers and Native Americans. In the spring of 1763, Sir Jeffrey Amherst, the British Commander-in-Chief in North America, believed the western frontier, which ran from Pennsylvania to Detroit, was secure, but the situation deteriorated rapidly over the next several months. The Indians in western Pennsylvania were becoming particularly aggressive in the area around Fort Pitt, near what is now Pittsburgh. It became apparent that unless the situation was resolved, western Pennsylvania would be deserted and Fort Pitt isolated. On 23 June 1763, Colonel Henry Bouquet, the ranking officer for the Pennsylvania frontier, wrote to Amherst, describing the difficulties Captain Ecuyer was having holding the besieged Fort Pitt. These difficulties included an outbreak of smallpox among Ecuyer's troops. In his reply to Bouquet, Amherst suggested that smallpox be sent among the Indians to reduce their numbers. This well-documented suggestion is significant because it clearly implies the intentional use of smallpox as a weapon. Bouquet responded to Amherst's suggestion stating that he would use blankets to spread the disease.

Evidence indicates that Amherst and Bouquet were not alone in their plan to use BW against the Indians. While they were deciding on a plan of action, Captain Ecuyer reported in his journal that he had given two blankets and a handkerchief from the garrison smallpox hospital to hostile chiefs with the hope that it would spread the disease. It appears that Ecuyer was acting on his own and did not need persuasion to use whatever means necessary to preserve the Pennsylvania frontier. Evidence also shows that the French used smallpox as a weapon in their conflicts with the native population.

Smallpox also played a role in the American Revolutionary War, but the tactics were defensive rather than offensive: British troops were inoculated against smallpox, but the rebelling American colonists were not. This protection from disease gave the British an advantage for several years, until Washington ordered inoculation against smallpox for all American troops.

It is clear that by the eighteenth century BW had become disease-oriented, even though the causative agents and mechanisms for preventing the spread of diseases were largely unknown. The development of the science of bacteriology in the nineteenth and early twentieth centuries considerably expanded the scope of potential BW agents. In 1915, Germany was accused of using cholera in Italy and plague in St. Petersburg. Evidence shows that Germany used glanders and anthrax to infect horses and cattle, respectively, in Bucharest in 1916 and employed similar tactics to infect 4500 mules in Mesopotamia the following year. Germany issued official denials of these accusations. Although there apparently was no large-scale battlefield use of BW in World War I, numerous allegations of German use of BW were made in the years following the war. Britain accused Germany of dropping plague bombs, and the French claimed the Germans had dropped disease-laden toys and candy in Romania. Germany denied the accusations.

Although chemical warfare was far more important than BW in World War I, the general awareness of the potential of biological weapons led the delegates to the Geneva Convention to include BW agents in the 1925 Protocol for the Prohibition of the Use in War of Asphyxiating, Poisonous or Other Gases, and of Bacteriological Methods of Warfare. The significance of the treaty will be discussed later (see Section "International Treaties").

1925–1990

The tense political atmosphere of the period following the 1925 Geneva Protocol and the lack of provisions to deter biological weapons research had the effect of undermining the treaty. The Soviet Union opened a BW research facility north of the Caspian Sea in 1929; the United Kingdom and Japan initiated BW research programs in 1934. The Japanese program was particularly ambitious and included experiments on human subjects prior to and during World War II.

Two factors were significant in mobilizing governments to initiate BW research programs: (1) a continuing flow of accusations regarding BW and (2) the commitment of resources for BW research by several national adversaries, thus creating a feeling of insecurity among governments. The presence of BW research laboratories in nations that were traditional or potential adversaries reinforced this insecurity. Thus, despite the Geneva Protocol, it was believed that it was politically unwise for governments to ignore the threat of BW, and the result was the use of increasingly sophisticated biological weapons.

In 1941, the United States and Canada joined other nations and formed national programs of BW research and development. Camp Detrick (now Fort Detrick) became operational as the center for U.S. BW research in 1943, and in 1947 President Truman withdrew the Geneva Protocol from Senate consideration, citing current issues such as the lack of verification mechanisms that invalidated the underlying principles of the treaty. However, there was no widespread use of BW in a battlefield setting during World War II. BW research, however, continued at an intense pace during and after the war. By the end of the decade, the United States, the United Kingdom, and Canada were conducting collaborative experiments involving the release of microorganisms from ships in the Caribbean. In 1950, the U.S. Navy conducted open-air experiments in Norfolk, Virginia, and the U.S. Army conducted a series of airborne microbial dispersals over San Francisco using *Bacillus globigii*, *Serratia marcescens*, and inert particles.

Not surprisingly, the intense pace of BW research led to new accusations of BW use, most notably by China and North Korea against the United States during the Korean War. In 1956, the United States changed its policy of "defensive use only" to include possible deployment of biological weapons in situations other than retaliation. During the 1960s, all branches of the U.S. military had active BW research programs, and additional open-air dissemination experiments with stimulants were conducted in the New York City subway system. By 1969, however, the U.S. military concluded that BW had little tactical value in battlefield situations, and since it was felt that in an age of nuclear weapons dominated the strategic equation, the United States would be unlikely to need or use BW. Thus, President Nixon announced that the United States would unilaterally renounce BW and eliminate stockpiles of biological weapons. This decision marked a turning point in the history of BW: Once the U.S. government made it clear that it did not consider biological weapons a critical weapon system, the door was opened for negotiation of a strong international treaty against BW.

Once military strategists had discounted the value of BW, an attitude of openness and compromise on BW issues took hold, leading to the 1972 Convention on the Prohibition of the Development, Production, and Stockpiling of Bacteriological (Biological) and Toxin Weapons (TWs) and on Their Destruction (see Section "International Treaties"). The parties to the 1972 Convention agreed to destroy or convert to peaceful use all organisms, toxins, equipment, and delivery systems. Following the signing of the 1972 treaty, the U.S. government generated much publicity about its compliance activities, inviting journalists to witness destruction of biological weapons stockpiles.

The problem of treaty verification, however, beleaguered the 1972 Convention. Press reports accusing the Soviet Union of violating the treaty appeared as early as 1975. When an outbreak of anthrax was reported in Sverdlovsk, Soviet Union, in 1979, the United States claimed it was caused by an incident at a nearby Soviet biological defense laboratory that had released anthrax spores into the surrounding community. The Soviet government denied this allegation, claiming the outbreak was caused by contaminated black market meat.

BW continued to be discussed in the public media throughout the 1980s. In 1981, reports describing the American "cover-up" of Japanese BW experiments on prisoners of war began to surface in the public and scientific literature. In 1982, *The Wall Street Journal* published a series of articles on Soviet genetic engineering programs that raised many questions about the scope of Soviet BW activities. The environmental effects of testing biological agents at Dugway Proving grounds in Utah received considerable press attention in 1988, leading to a debate over the need for such a facility.

The 1980s also were characterized by debate over larger issues relating to BW. A public debate in 1986 considered the possible role of biological weapons in terrorism. Scientific and professional societies, which had avoided the issues for many years, began considering both specific issues, such as Department of Defense (DoD) support for biological research and more general issues, such as adopting ethical codes or guidelines for their members.

1990 and Contemporary Developments

The last decade of the twentieth century and the early years of the twenty-first century saw a remarkable transition in the concepts relating to BW. With the fall of the Soviet Union and the rise of the United States as the only remaining superpower, many policies that were in place in a dual superpower setting became altered. Although the concept of BW as an alternative offensive weapon among developing and nonnuclear States remained a serious concern, the new emphasis was on the threat of these weapons being used by bioterrorist groups. The events of 11 September 2001, followed by a U.S. anthrax scare, dramatically shifted awareness to the need for improved Public Health protection for the general population, the initiation, and placement of early and rapid detection systems for potential biological agents and increased border protection. The concept of BW agents in the hands of terrorists was not new, however, and the foundation for shift in emphasis from nations to independent terrorist groups had begun in the 1990s.

Prior to the 1990s, most U.S. defensive research and BW-related policies were directed to counter potential BW use by the Soviet Union. As the wall of Soviet secrecy eroded during the 1990s, the extent of the Soviet BW program became apparent. There was international concern that a large number of unemployed BW researchers could find work as advisors for developing countries that viewed BW as a rational defense strategy, especially those countries without nuclear capability or those without restrictive laws against radical terrorist groups. Also, the open threat by the Iraqi military to use BW agents raised serious concerns and changed attitudes about BW. The plans for Operation Desert Storm included provisions for protective equipment and prophylactic administration of antibiotics or vaccines to protect against potential biological weapons. Many of the critics of the U.S. Biological Defense Research Program (BDRP) were now asking why the country was not better prepared to protect its troops against biological attack. BW was not used during the First Gulf War, but the threat of its use provided several significant lessons. Although there was considerable concern that genetic engineering would produce new, specialized biological weapons, most experts predicted that "classical" BW agents, such as anthrax and botulism, would pose the most serious threats to combat troops in Operation Desert Storm. Efforts by the United Nations after the war to initiate inspection programs demonstrated the difficulty of verifying the presence of production facilities for BW agents; these difficulties highlight the need for verification protocols for the BW Convention. Verification and treaty compliance are prominent contemporary BW issues. Following the Gulf War, the actual extent of the intense Iraqi BW research programs was understood based on information from defectors. The actual programs were huge compared to predicted estimates of the Iraqi BW program as well as the postwar "verification" programs that essentially uncovered very little. This vast discrepancy demonstrated the inadequacy of verification procedures.

The World Trade Center attack and the anthrax incidents in 2001 were the most significant contemporary developments contributing to the realization that urban centers, public facilities, and the general population are all vulnerable to attack by terrorists employing BW agents. Local and national governments realized the extent of the vulnerability and began taking extensive measures to formulate policies to address potential bioterror attacks. In the United States, it was felt that the various agencies involved in protecting the population against the threat of bioterrorism had to be strengthened. The U.S. Patriot Act of October 2001 and the formation of the Department of Homeland Security are just some of the steps taken to address the needs of responding to and limiting the threat of a BW or bioterrorist attack in the United States. Canada and several European countries have created new or greatly enhanced existing government agencies, such as the European Centre for Disease Control, to protect the population against bioterrorism and other major potential disasters. Much work remains to be accomplished in this area.

Additionally, in the United States, a number of legislative acts and directives that were designed to increase the nation's biodefense capabilities and responsiveness were approved. Increased resources have been made available through the Department of Health and Human Services in support of medical countermeasure preparedness. On 21 July 2004, President George W. Bush signed the Project BioShield Act of 2004 (Project Bioshield). The purpose of this bill was to accelerate the research, development, acquisition, and availability of safe and effective countermeasures against Chemical, Biological, Radioactive, and

Nuclear Threats by providing the funding to purchase countermeasures. Under this bill, the HHS pursued acquisition of countermeasures for anthrax, smallpox, botulinum toxins, and radiological/nuclear agents. Project Bioshield created a $5.6 billion special reserve fund for use over 10 years (FY04-FY13) to acquire medical countermeasures. New legislation, the Pandemic and All-Hazards Preparedness Act was approved in December 2006. This act provided for the establishment of the Biomedical Advanced Research and Development Authority (BARDA) within HHS. BARDA is intended to address the inadequacy of Project Bioshield which, while providing funds for procurement of the actual end product, did not allow for funding of the actual research and development. BARDA should have addressed the so-called valley of death where there is no funding available to support the expensive later stage development. BARDA is intended to provide direct funding of medical countermeasure advanced research and development (Federal Register/ Vol. 72, No. 53/Tuesday, 20 March 2007, p. 13109). Most recently an accent has been placed on awarding contracts for producing medical countermeasures to protect against natural and manmade biological threats. One set of BARDA grants awarded $55 million for initial phase and up to $100 million over 3 years for projects that may lead to faster processes for making vaccines, rapid high-throughput surveillance and molecular diagnostics of microbial pathogens. (MICROBE, volume 5, No.11, page 466, 2010)

INTERNATIONAL TREATIES

The 1925 Geneva Protocol

The 1925 Geneva Protocol was the first international treaty to place restrictions on BW. The Geneva Protocol followed a series of international agreements that were designed to prohibit the use in war of weapons that inflict or prolong unnecessary suffering of combatants or civilians. The St. Petersburg Declaration of 1868 and the International Declarations concerning the Laws and Customs of War, which was signed in Brussels in 1874, condemned the use of weapons that caused useless suffering. Two major international conferences were held at The Hague in 1899 and 1907. These conferences resulted in declarations regarding the humanitarian conduct of war. The conference regulations forbid nations from using poison, treacherously wounding enemies, or using munitions that would cause unnecessary suffering. The so-called Hague Conventions also prohibited the use of projectiles to diffuse asphyxiating or deleterious gases. The Hague Conventions still provide much of the definitive law of war as it exists today.

The Hague Conventions did not specifically mention BW, due in part to the lack of scientific understanding of the cause of infectious diseases at that time. The Conventions have, however, been cited as an initial source of the customary international laws that prohibit unnecessary suffering of combatants and civilians in war. While biological weapons have been defended as humanitarian weapons, on the grounds that many biological weapons are incapacitating but not lethal, there are also biological weapons that cause a slow and painful death. It can be argued, therefore, that the Hague Conventions helped to set the tone of international agreements on laws of war that led to the 1925 Geneva Protocol.

The 1925 Geneva Protocol, formally called the Prohibition of the Use in War of Asphyxiating, Poisonous or Other Gases, and of Bacteriological Methods of Warfare, was opened for signature on 17 June 1925 in Geneva, Switzerland. More than 100 nations have signed and ratified the protocol, including all members of the Warsaw Pact and North Atlantic Treaty Organization (NATO). The 1925 Geneva Protocol was initially designed to prevent the use of chemical weapons in war; however, the protocol was extended to include a prohibition on the use of Bacteriological Methods of Warfare. The Geneva Protocol distinguishes between parties and nonparties by explicitly stating that the terms of the treaty apply only to confrontations in which all combatants are parties and when a given situation constitutes a "war." Additionally, a number of nations ratified the Geneva Protocol with the reservation that they would use biological weapons in retaliation against a biological weapons attack. This resulted in the recognition of the Geneva Protocol as a "no first use" treaty.

The 1972 BW Convention

International agreements governing BW have been strengthened by the 1972 BW Convention, which is officially called the 1972 Convention on the Prohibition of the Development, Production and Stockpiling of Bacteriological (Biological) and TW and on Their Destruction. The Convention was signed simultaneously in 1972 in Washington, London, and Moscow and entered into force in 1975. The preamble to the 1972 BW Convention states the determination of the state's parties to the treaty to progress toward general and complete disarmament, including the prohibition and elimination of all types of WMD. This statement places the Convention in the wider setting of international goals of complete disarmament. The 1972 BW Convention is also seen as a first step toward chemical weapons disarmament.

The 1972 BW Convention explicitly builds on the Geneva Protocol by reaffirming the prohibition of the use of BW in war. The preamble, although not legally binding, asserts that the goal of the Convention is to completely exclude the possibility of biological agents and toxins being used as weapons and states that such use would be repugnant to the conscience of humankind. The authors of the 1972 Convention, therefore, invoked societal attitudes as justification for the existence of the treaty.

In 1972 BW Convention evolved, in part, from a process of constant reevaluation of the Geneva Protocol. From 1954 to the present, the United Nations has periodically considered the prohibition of chemical and biological weapons. The Eighteen-Nation Conference of the Committee on Disarmament, which in 1978 became the Forty-Nation Committee on Disarmament, began talks in 1968 to ban chemical weapons. At this time, chemical, toxin, and biological weapons were being considered together, in an attempt to develop a comprehensive disarmament agreement. However, difficulties in reaching agreements on chemical warfare led to a series of separate negotiations that covered only BW and TWs. The negotiations resulted in the drafting of the 1972 BW Convention.

The 1972 BW Convention consists of a preamble, followed by 15 articles. Article I forms the basic treaty obligation. Parties agree never, under any circumstances, to develop, produce, stockpile, or otherwise acquire or retain the following:

1. Microbial or other biological agents, or toxins irrespective of their origin or method of production, of types and in quantities that have no justification for prophylactic, protective, or other peaceful purposes.
2. Weapons, equipment, or means of delivery designed to use such agents or toxins for hostile purposes or in armed conflict.

Article II requires each party to destroy, or divert to peaceful purposes, all agents, toxins, equipment, and delivery systems that are prohibited in Article I and are under the jurisdiction or control of the party. It also forbids nations from transferring, directly or indirectly, materials specified in Article I and prohibits nations from encouraging, assisting, or inducing any state, group of states, or international organizations from manufacturing or acquiring the material listed in Article I. There is no specific mention of subnational groups, such as terrorist organizations, in the treaty.

Article IV requires each party to the Convention to take any measures to ensure compliance with the terms of the treaty. Article IV has been interpreted by some states as the formulation of civil legislation or regulations to assure adherence to the Convention. This civil legislation could regulate activities by individuals, government agencies, universities, or corporate groups.

Articles V–VII specify procedures for pursuing allegations of noncompliance with the 1972 BW Convention. The United Nations plays an integral part in all of the procedures for investigating allegations of noncompliance. According to Article VI, parties may lodge a complaint with the Security Council of the United Nations if a breach of the treaty is suspected. All parties must cooperate with investigations that may be initiated by the Security Council. Article VII requires all parties to provide assistance or support to any party that the Security Council determines has been exposed to danger as a result of violation of the Convention. Articles VII–IX are general statements for obligations of the parties signing the protocol. Article X gives the parties the right to participate in the fullest possible exchange of equipment, materials, and scientific or technological information of the use of bacteriological (biological) agents and toxins for peaceful purposes. Article XI allows parties to propose amendments to the Convention. The amendments only apply to those states that accept them and enter into force after a majority of the states parties to the Convention have agreed to accept and be governed by the amendment.

Article XII requires that a conference be held 5 years after the entry into force of the BW Convention. Article XIV states that the 1972 BW Convention is of unlimited duration. A state party to the treaty is given the right to withdraw from the treaty if it decides that extraordinary events, related to the subject matter of the Convention, have jeopardized the supreme interests of the country. This article also opens the Convention to all nations for signature. Nations that did not sign the Conventions before its entry into force may accede to it at any time.

Review Conferences

The 1972 Convention contained a stipulation that a conference be held in Geneva 5 years after the terms of the Convention entered into force. The purpose of the conference was to review the operation of the Convention and to assure that the purposes of the Convention were being realized. The review was to take into account any new scientific and technological developments that were relevant to the Convention. The first review conference was held in Geneva in 1980. Several points contained in the original Convention were clarified at this conference. There was general agreement that these conferences would serve a definite function in solving contemporary problems, such as issues of verification and compliance that need clarification based on changing events. While limited in scope, these conferences made some progress in keeping the 1972 Convention relevant to the needs of a changing world situation.

Select U.S. Laws and Acts

A number of U.S. laws have been enacted since 1989 that impact on BW:

Biological Weapons and Anti-Terrorist Act (1989): Established as a federal crime, the development, manufacture, transfer, or possession of any biological agent, toxin, or delivery system for use as a weapon.

Chemical and Biological Weapons Control Act (1991): Places sanctions on companies that knowingly export goods or technologies relating to biological weapons to designated prohibited nations.

The Defense Against WMD Act (1996): Designed to enhance federal, state, and local emergency response capabilities to deal with terrorist incidents.
Antiterrorism and Effective Death Penalty Act (1996): Established as a criminal act, any threat or attempt to develop BW or DNA technology to create new pathogens or make more virulent forms of existing organisms.
National Laboratory Response Network (1999): A joint effort by the Centers for Disease Control and Prevention (CDC) and U.S. Army Medical Research Institute of Infectious Diseases (USAMRIID) to establish a network of public health laboratories throughout the United States, and was reinforced and expanded following the anthrax letter incident of 2001.
The USA Patriot Act (2001): Greatly expanded the ability of law enforcement agencies for fighting terrorism in the United States and abroad. New crime categories, such as domestic terrorism, were established. Under this act laws on immigration, banking, money laundering, and foreign intelligence have been amended.
Public Health Security and Bioterrorism Preparedness and Response Act of 2002: Encouraged technological solutions to prevent the threat of bioterrorism and the stockpiling of vaccines and supplies.
Homeland Security Act (2003): Attempt to consolidate 25 agencies and tens of thousands government employees into a new department to prevent terrorist attacks, reduce vulnerability, and minimize damage and recover from attacks that may occur.
Project BioShield Act of 2004 (Project Bioshield): Provides funds to purchase countermeasures against Chemical, Biological, Radioactive, and Nuclear Threats.
Pandemic and All-Hazards Preparedness Act of 2006: Included provisions for the establishment of the BARDA within HHS. Funding provided through BARDA will support medical countermeasure advanced research and development.
CDC 42 CFR Part 1003 Possession, use, and Transfer of Select Agents and Toxins: Established a rule regarding possession, use, and transfer of select agents and toxins that pose a significant risk to public health and safety. Infectious agents have been added to this list at several intervals over time. Some of the organisms regulated by CDC are listed in Table 1.

TABLE 1 Representative organisms regulated by CDC (Center for Disease Control)

Bacteria	Viruses
Bacillus anthracis	Arenaviruses
Brucella species	Crimean-Congo hemorrhagic fever virus
Burkholderia mallei, *B. pseudomallei*	Eastern equine encephalitis virus
Clostridium botulinum, *B. perfringens*	Ebola virus
Escherichia coli 0157:H7	Equine morbillivirus
Francisella tularensis	Hantavirus
Salmonella species	Lassa fever virus
Shigella species	Marburg virus
Vibrio cholerae	Monkeypox virus Nipah virus
Yersinia pestis	Rift Valley fever virus South American hemorrhagic fever viruses
Coccidioides immitis	Tick-borne encephalitis complex viruses
Rickettsiae and Chlamydia	Variola (smallpox) major virus
Chlamydia psittaci	Venezuelan equine encephalitis virus
Coxiella burnetii	Western equine encephalitis virus
Rickettsia prowazekii	Yellow fever virus
Rickettsia rickettsii	
Select Genetic Elements, Recombinant Nucleic Acids, and Recombinant Organisms	

CURRENT RESEARCH PROGRAMS

The U.S. BDRP today is headquartered at the USAMRIID at Fort Detrick, Maryland. USAMRIID is an organization of the U.S. Army Medical Research and Materiel Command. In accordance with official U.S. policy, the BDRP is solely defensive in nature, with the goal of providing methods of detection for, and protective countermeasures against, biological agents that could be used as weapons against U.S. forces or civilians by hostile states or individuals. USAMRIID plays a key role as the only laboratory in the U.S. DoD equipped for the safe study of highly hazardous infectious agents that require maximum containment at biosafety level (BSL)-4.

Current U.S. policy stems from the 1969 declaration made by President Nixon that confined the U.S. BW program to research on biological defense such as immunization and measures of controlling and preventing the spread of disease. Henry Kissinger further clarified the U.S. BW policy in 1970 by stating that the U.S. biological program will be confined to research and development for defensive purposes only. This did not preclude research into those offensive aspects of biological agents necessary to determine what defensive measures were required. The BDRP expanded significantly in the 1980s, in an apparent response to alleged treaty violations and perceived offensive BW capabilities in the Soviet Union. These perceptions were espoused primarily by representatives of the Reagan Administration and the Department of State. At congressional hearings in May 1988, the U.S. government reported that at least 10 nations, including the Soviet Union, Libya, Iran, Cuba, Southern Yemen, Syria, and North Korea, were developing biological weapons. Critics of the U.S. program refuted the need for program expansion.

The BDRP is focused in three sites, the USAMRIID at Fort Detrick, Maryland; Aberdeen Proving Ground in Maryland; and the Dugway Proving Ground in Utah. USAMRIID is designated as the lead laboratory in medical defense against BW threats. Research conducted at the USAMRIID focuses on medical defense such as the development of vaccines and treatments for both natural diseases and potential BW agents. Work on the rapid detection of microorganisms and the diagnosis of infectious diseases are also conducted. The primary mission at the Aberdeen Proving Ground is nonmedical defense against BW threats including detection research, such as the development of sensors and chemiluminescent instruments to detect and identify bacteria and viruses, and development of methods for material and equipment decontamination. The U.S. Army Dugway Proving Ground is a DoD major range and test facility responsible for development, test, evaluation, and operation of chemical warfare equipment, obscurants and smoke munitions, and biological defense equipment. Its principal mission with respect to the BDRP is to perform developmental and operational testing for biological defense material, including the development and testing of sensors, equipment, and clothing needed for defense against a BW attack.

The BDRP focuses on five main areas:

1. Development of vaccines.
2. Development of protective clothing and decontamination methods.
3. Analysis of the mode of action of toxins and the development of antidotes.
4. Development of broad-spectrum antiviral drugs for detecting and diagnosing BW agents and toxins.
5. Utilization of genetic engineering methods to study and prepare defenses against BW and toxins.

The BDRP has often been a center of controversy in the United States. One BDRP facility, the Dugway Proving Ground, was the target of a lawsuit that resulted in the preparation of the environmental impact statement for the facility. A proposal for a high-level containment laboratory (designated P-4) was ultimately changed to new plans for a lower-level (P-3) facility.

The use of genetic engineering techniques in BDRP facilities has also been a focus of controversy. The BDRP position is that genetic engineering will be utilized if deemed necessary. The DoD stated that testing of aerosols of pathogens derived from recombinant DNA methodology is not precluded if a need should arise in the interest of national defense.

Many secondary sites have received and continued to obtain contracts for biological defense research. Once specific program requires a special note. Defense Advanced Research Project Agency (DARPA) is a Pentagon program that invests significantly in pathogen research through grants to qualified institutions. This project initially focused on engineering and electronics (computer) projects; however, starting in 1995 biology became a key focus, and several BW-defensive research grants are now in operation at many academic and private institutions.

An important issue in biological defense has been the convergence in the late 1990s and the first decade of this century of the increased need for new biological countermeasures at the same time that private pharmaceutical company research and development in infectious diseases has diminished. Numerous factors have been outlined by a variety of organizations, including the Infectious Disease Society of America, that provide reasons for the decreasing activity of the private sector in infectious diseases research and development. A major component of this is the low return on investment of infectious disease pharmaceutical agents compared to agents in other therapy areas such as oncology and diabetes. The concern about diminished research and development in infectious diseases has been met with a number of initiatives by the U.S. government and other groups to try to stimulate companies to invest in infectious disease countermeasures. Most of these attempts have had little effect on large pharmaceutical companies, but a degree of success in stimulating research in the smaller biotechnology industry.

The two major programs in the United States have been (1) the 2004 Project Bioshield and (2) the 2006 Pandemic and All-Hazards Preparedness Act that provided for the establishment of the BARDA within HHS. These actions have the goal of obtaining medical countermeasures for both potential bioterror agents and also naturally occurring infectious disease outbreaks, either by providing funds for purchasing the countermeasures or by providing financial support for research and development activities (see Section "1990 and Contemporary Developments").

Very little is written in the unclassified literature on BW research conducted in countries other than the United States. Great Britain has maintained the Microbiological Research Establishment at Porton Down; however, military research is highly classified in Great Britain and details regarding the research conducted at Porton are unavailable.

During the 1970s and 1980s, a great deal of U.S. BW policy was based on the assumption of Soviet offensive BW capabilities. Most U.S. accounts of Soviet BW activities were unconfirmed accusations or claims about treaty violations. The Soviet Union was a party to both the 1925 Geneva Protocol and the 1972 BW Convention. According to Pentagon sources, the Soviet Union operated at least seven top-security BW centers. These centers were reported to be under strict military control. While the former Soviet Union proclaimed that their BW program was purely defensive, the United States has consistently asserted that the Soviet Union was conducting offensive BW research.

CONTEMPORARY ISSUES

Genetic Engineering

There has been considerable controversy over the potential for genetically engineered organisms to serve as effective BW agents. Recombinant DNA technology has been cited as a method for creating novel, pathogenic, microorganisms. Theoretically, organisms could be developed that would possess predictable characteristics, including antibiotic resistance, altered modes of transmission, or altered pathogenic and immunogenic capabilities. This potential for genetic engineering to significantly affect the military usefulness of BW has been contested. It has been suggested that because a large number of genes must work together to endow an organism with pathogenic characteristics, the alteration of a few genes with recombinant DNA technology is unlikely to yield a novel pathogen that is significantly more effective or usable than conventional BW agents.

The question of predictability of the behavior of genetically engineered organisms was addressed at an American Society for Microbiology symposium held in June 1985. Some symposium participants believed that the use of recombinant DNAs increases predictability because the genetic change can be precisely characterized. Other participants, however, felt that the use of recombinant DNA decreases predictability, because it widens the potential range of DNA sources. Other evidence supports the view that genetically engineered organisms do not offer substantial military advantage over conventional BW. Studies have shown that genetically engineered organisms do not survive well in the environment. This fact has been cited as evidence that these organisms would not make effective BW agents.

Despite the contentions that genetic engineering does not enhance the military usefulness of BW, a significant number of arguments support the contrary. At the 1986 Review Conference of the BW Convention, it was noted that genetic engineering advances since the Convention entered into force may have made biological weapons a more attractive military option.

Several authors have contended that the question of the potential of genetic engineering to enhance the military usefulness of BW is rhetorical, because the 1972 BW Convention prohibits development of such organisms despite their origin or method of production. Nations participating in both the 1980 and 1986 review conferences of the BW Convention accepted the view that the treaty prohibitions apply to genetically engineered BW agents. An amendment to the treaty, specifically mentioning genetically engineered organisms, was deemed to be unnecessary. Additionally, the United States, Great Britain, and the Soviet Union concluded in a 1980 briefing paper that the 1972 BW Convention fully covered all BW agents that could result from genetic manipulation.

While the utility of genetic engineering for enhancing the military usefulness of BW agents has been questioned, the role of genetic engineering for strengthening defensive measures against BW has been clear. Genetic engineering has the potential for improving defenses against BW in two ways: (1) vaccine production and (2) sensitive identification and detection systems. The issues of the new technologies in defensive research have been evident in the U.S. BW program. Since 1982, U.S. Army scientists have used genetic engineering to study and prepare defenses against BW agents. Military research utilizing recombinant DNA and hybridoma technology include the development of vaccines against a variety of bacteria and viruses, methods of rapid detection and identification of BW agents, and basic research on protein structure and gene control. By improving defenses against BW, it is possible that genetic engineering may potentially reduce the risk of using BW.

The primary effect of BW on the government regulations on genetic engineering is the tendency toward more stringent control of the technologies. The fear of genetically engineered BW agents has prompted proposals for government regulation of BW research utilizing genetic engineering research. The DoD has released a statement indicating that all government research was in compliance with the 1972 BW Convention. The government has also prepared an environmental impact statement of research conducted at Fort Detrick.

Government regulations on genetic engineering also affect BW research through limitations on exports of biotechnology information, research products, and equipment. In addition to controls of exports due to competitive concerns of biotechnology companies, a substantial amount of information and equipment related to genetic engineering is prohibited from being exported outside the United States. The Commerce Department maintains a "militarily critical technology" list, which serves as an overall guide to restricted exports. Included on the list are containment and decontamination equipment for large production facilities, high-capacity biological reactors, separators, extractors, dryers, and nozzles capable of disseminating biological agents in a fine mist.

Genetic engineering has altered the concept of BW. A current, comprehensive discussion of BW would include both naturally occurring and potential genetically engineered agents. Many current defenses against BW are developed with genetic engineering techniques. Government regulations on biotechnology have limited BW research, while fears of virulent genetically engineered BW agents have strengthened public support for stronger regulations. Future policies related to BW will need to be addressed in light of their altered status. Recent concerns have been raised relating to synthetic biology and its potential to produce wholly new infectious threat agents. A 2010 report of the National Research Council of the National Academy of Sciences concluded that despite concerns about synthetic science, for the foreseeable future, threats from this new area of research are more likely to come from assembling or modifying known agents. (MICROBE, volume 5, No. 10, page 418, 2010.)

Mathematical Epidemiology Models

While genetic engineering may potentially alter characteristics of BW agents, mathematical models of epidemiology may provide military planners with techniques for predicting the spread of a released BW agent. One of the hindrances that has prevented BW from being utilized or even seriously considered by military leaders has been the inability to predict the spread of a BW agent once it has been released into the environment. Without the capability to predict the spread of the released organisms, military planners would risk the accidental exposure of their own troops and civilians to their own weapons. The development of advanced epidemiology models may provide the necessary mechanisms for predicting the spread of organisms that would substantially decrease the deterrent factor of unpredictability.

Low-Level Conflict

Another important factor that has affected the current status of BW is the increase in low-level conflict or the spectrum of violent action below the level of small-scale conventional war, including terrorism and guerrilla warfare. In the 1980s, the low-intensity conflict doctrine, which was espoused by the Reagan administration, was a plan for U.S. aid to anti-Communist forces throughout the world as a way of confronting the Soviet Union without using U.S. combat troops. The significant changes in the world since the inception of the low-intensity conflict doctrine have only increased the probability of increasing numbers of small conflicts. Although no evidence indicated that the United States would consider violating the 1972 BW Convention and support biological warfare, the overall increase in low-level conflicts in the future may help create an environment conducive to the use of BW.

While BW may not be assessed as effective weapons in a full-scale conventional war, limited use of BW agents may be perceived as advantageous in a small-scale conflict. While strong deterrents exist for nuclear weapons, including unavailability and, most formidably, the threat of uncontrolled worldwide "nuclear winter," BW may be perceived as less dangerous. Additionally, the participants of low-level conflicts may not possess the finances for nuclear or conventional weapons. BW agents, such as chemical weapons, are relatively inexpensive compared to other weapon systems and may be seen as an attractive alternative to the participants and leaders of low-level conflicts. Low-level conflict, therefore, increases the potential number of forums for the use of BW.

Terrorism

The most significant factor that has altered the concept of BW is the global threat of terrorism. Although there have been isolated incidents to date of the use of biological pathogens by terrorist groups, these attempts have only had minimum impact. However, the possibility of a major incident by bioterrorists is predicted as inevitable by many experts.

The relationship of terrorism and BW can be divided into two possible events. The first involves terrorist acts against laboratories conducting BW-related research. The level of security at Fort Detrick is high, the possibility of a terrorist attack has been anticipated, and contingency plans have been made. Complicating the problem of providing security against terrorist attack in the United States is the fact, that, while most BW research projects are conducted with the BW research program of the DoD, an increasing number of projects are supported by the government that are conducted outside of the military establishment. These outside laboratories could be potential targets.

The second type of terrorist event related to BW is the potential use of BW by terrorists against urban areas or major public facilities. Biological weapons are relatively inexpensive and easy to develop and produce compared to conventional, nuclear, or chemical weapons. BW agents can be concealed and easily transported across borders or within countries. Additionally, terrorists are not hampered by a fear of an uncontrolled spread of the BW agent into innocent civilian populations. To the contrary, innocent civilians are often the intended targets of terrorist activity, and the greater chance for spread of the BW is considered to be a positive characteristic for a bioterrorist. The use of these agents to attack agricultural operations or food supplies, although not directly targeting humans, can have an enormous economic impact. All aspects of BW are receiving needed attention as the potential threat of their use by terrorists is further realized by policy makers. Environmental monitoring in public places along with new technologies for rapid identification have been greatly stimulated by the threat of bioterrorism and will continue to receive much attention in the foreseeable future.

Offensive Versus Defensive BW Research

The distinctions between "offensive" and "defensive" BW research have been an issue since 1969, when the United States unilaterally pledged to conduct only defensive research. The stated purpose of the U.S. BDRP is to maintain and promote national defense from BW threats. Although neither the Geneva Convention nor the 1972 Convention prohibits any type of research, the only research that nations have admitted to conducting is defensive. The problem is whether or not the two types of research can be differentiated by any observable elements.

Although production of large quantities of a virulent organism and testing of delivery systems has been cited as distinguishing characteristics of an offensive program, a substantial amount of research leading up to these activities including isolating an organism, then using animal models to determine pathogenicity, could be conducted in the name of defense.

Vaccine research is usually considered "defensive," whereas increasing the virulence of a pathogen and producing large quantities is deemed "offensive." However, a critical component of a strategic plan to use biological weapons would be the production of vaccines to protect the antagonist's own personnel (unless self-annihilation was also a goal). This means that the intent of a vaccine program could be offensive BW use. Furthermore, research that increases the virulence of an organism is not necessarily part of an offensive strategy because one can argue that virulence needs to be studied to develop adequate defense.

The key element distinguishing offensive from defensive research is intent. If the intent of the researcher or the goals of the research program is the capability to develop and produce BW, then the research is offensive BW research. If the intent is to have the capability to develop and produce defenses against BW use, then the research is defensive BW research. While it is true that nations may have policies of open disclosures (i.e., no secret research), "intent" is not observable.

Although the terms "offensive BW research" and "defensive BW research" may have some use in describing intent, it is more a philosophical than a practical distinction, one that is based on trust rather than fact.

Secrecy in Biological Warfare-Related Research

Neither the Geneva Protocol nor the 1972 BW Convention prohibits any type of research, secret, or nonsecret. While the BDRP does not conduct secret or classified research, it is possible that secret BW research is being conducted in the United States outside of the structure of the BDRP. The classified nature of the resource material for this work makes it impossible to effectively determine if secret research is being conducted in the United States or any other nation.

It is not, however, unreasonable to assume that other nations conduct significant secret BW research. Therefore, regardless of the facts, one cannot deny the perception that such research exists in a variety of countries and that this perception will exist for the foreseeable future.

Secrecy has been cited as a cause of decreased quality of BW research. If secret research, whether offensive or defensive, is being conducted in the United States or other nations, it is unclear if the process of secrecy affects the quality of the research. If the secret research process consists of a core of highly trained creative and motivated individuals sharing information, the quality of the research may not suffer significantly. It must be stated, however, that secrecy by its very nature will limit input from a variety of diverse observers.

Secrecy may increase the potential for violations of the 1972 BW Convention; however, violations would probably occur regardless of the secrecy of the research. Secrecy in research can certainly lead to infractions against arbitrary rules established by individuals outside of the research group. The secret nature of the research may lure a researcher into forbidden areas. Additionally, those outside of the research group, such as policy makers, may push for prohibited activities if the sense of secrecy prevails. Secrecy also tends to bind those within the secret arena together and tends to enhance their perception of themselves as being above the law and knowing what is "right." As in the case of Oliver North and the Iran–Contra affair, those within the group may believe fervently that the rules must be broken for a justified purpose and a mechanism of secrecy allows violations to occur without penalty.

The distrust between nations exacerbates the perceived need for secret research. The animosity between the United States and the Soviet Union during the 1980s fueled the beliefs that secret research leading to violations of the 1972 BW Convention was being conducted in the Soviet Union. As the belligerence of the 1980s fades into the New World Order, the questions will not focus on the Soviet Union as much as on the Middle East and third-world countries. There are factions in the United States that believe strongly that other countries are conducting secret research that will lead to violations of the Convention. There is also a tendency to believe that the secrecy in one's own country will not lead to treaty violations, while the same secret measures in an enemy nation will result in activities forbidden by international law.

The importance of the concept of secrecy in BW research is related to the perception of secrecy and arms control agreements. Regardless of the degree of secrecy in research, if an enemy believes that a nation is pursuing secret research, arms control measures are jeopardized. The reduction of secrecy has been suggested as a tool to decrease the potential for BW treaty violations. A trend toward reducing secrecy in BW research was exemplified by the 1986

Review Conference of the 1972 BW Convention, which resulted in agreements to exchange more information and to publish more of the results of BW research.

Whether or not these measures have any effect on strengthening the 1972 BW Convention remains to be seen.

Organizations and individuals have urged a renunciation by scientists of all secret research and all security controls over microbiological, toxicological, and pharmacological research. This action has been suggested as a means of strengthening the 1972 BW Convention. The belief that microbiologists should avoid secret research is based on the assumption that (1) secret research is of poor quality due to lack of peer review and (2) secrecy perpetrates treaty violations.

While it may be reasonable to expect microbiologists to avoid secret research, it is not realistic. Secrecy is practiced in almost every type of research including academic, military, and especially industrial. Furthermore, there will always be those, within the military and intelligence structures, who believe that at least some degree of secrecy is required for national security.

Secrecy in BW research is a complex issue. The degree to which it exists is unclear. Individuals are generally opposed to secrecy in BW research although other examples of secrecy in different types of research exist and are generally accepted. The effect of the secrecy on the quality of research, the need for the secrecy, and the choice of microbiologists to participate in secret BW research remain unanswered questions.

Problems Relating to Verification

One of the major weaknesses of the 1972 BW Convention has been the lack of verification protocols. Problems with effectively monitoring compliance include the ease of developing BW agents in laboratories designed for other purposes, and the futility of inspecting all technical facilities of all nations. Measures that have been implemented with the goal of monitoring compliance have included (1) open inspections, (2) intelligence gathering, (3) monitor research, (4) use of sampling stations to detect the presence of biological agents, and (5) international cooperation. The progress achieved with the Chemical Weapons Convention has renewed interest in strengthening mechanisms for verification of compliance with the 1972 BW Convention. While this renewed interest in verification of compliance with the emergence of the Commonwealth of Independent States from the old Soviet Union has brought optimism to the verification issue, the reticence of countries such as Iran and North Korea to cooperate with United Nations inspection teams is a reminder of the complexities of international agreements. The examples herein are typical of the many issues attached to the concept of BW.

FURTHER READING

Atlas, R. M. (1988). Biological weapons pose challenging for microbiology community. *ASM News*, *64*, 383–389.

Buckingham, W. A. Jr. (Ed.) (1984). *Defense planning for the 1990s*. Washington, DC: National Defense University Press.

Cole, L. (1996). The specter of biological weapons. *Scientific American*, *6*, 60–65.

Frisna, M. E. (1990). The offensive-defensive distinction in military biological research. *The Hastings Center Report*, *20*(3), 19–22.

Gravett, C. (1990). *Medieval siege warfare*. London: Osprey Publishing Ltd.

Guillemin, J. (1999). *Anthrax*. London: University of California Press.

Guillemin, J. (2005). *Biological weapons*. New York: Columbia University Press.

Harris, R., & Paxman, J. (1982). *A higher form of killing*. New York: Hill and Wang.

Livingstone, N. C. (1984). Fighting terrorism and "dirty little wars". In W. A. Buckingham Jr. (Ed.). *Defense planning for the 1990s* (pp. 165–196). Washington, DC: National Defense University Press.

Livingstone, N. C., Douglass, J. Jr. (1984). *CBW: The poor man's atomic bomb*. Medford, MA: Institute of Foreign Policy Analysis, Tufts University.

Meselson, M., Guillemin, J., Hugh-Jones, M. *et al.* (1994). The Sverdlovsk anthrax outbreak of 1979. *Science*, *266*, 1202–1208.

Milewski, E. (1985). Discussion on a proposal to form a RAC working group on biological weapons. *Recombinant DNA Technical Bulletin*, *8*(4), 173–175.

Miller, L. A. (1987). The Use of Philosophical Analysis and Delphi Survey to Clarify Subject Matter for a Future Curriculum for Microbiologists on the Topic of Biological Weapons. Thesis, University of Pennsylvania, Philadelphia. University Micro-films International, Ann Arbor, MI. 8714902.

Murphy, S., Hay, A., & Rose, S. (1984). *No fire, no thunder*. New York: Monthly Review Press.

Poupard, J. A., Miller, L. A., & Granshaw, L. (1989). The use of smallpox as a biological weapon in the French and Indian War of 1763. *ASM News*, *55*, 122–124.

Smith, R. J. (1984). The dark side of biotechnology. *Science*, *224*, 1215–1216.

Stockholm International Peach Research Institute. (1973). In *The problem of chemical and biological warfar* (Vol. II). New York: Humanities Press.

Taubes, G. (1995). The defense initiative of the 1990's. *Science*, *267*, 1096–1100.

Wright, S. (1985). The military and the new biology. *The Bulletin of the Atomic Scientists*, *42*(5), 73.

Wright, S., & Sinsheimer, R. L. (1983). Recombinant DNA and biological warfare. *The Bulletin of the Atomic Scientists*, *39*(9), 20–26.

Zilinskas, R. (Ed.) (1992). *The microbiologist and biological defense research: Ethics, Politics, and International Security*. New York: The New York Academy of Sciences.

Index

Note: Page numbers followed by *f* indicate figures and *t* indicate tables.

A

C

D

E

F

G

H

N

O

P

Q

R

S

T

U

V

W

X

Y

Z